Reactive Oxygen Species in Biological Systems: An Interdisciplinary Approach

Reactive Oxygen Species in Biological Systems: An Interdisciplinary Approach

Daniel L. Gilbert
National Institutes of Health
Bethesda, Maryland

and

Carol A. Colton
Georgetown University Medical School
Washington, D.C.

Springer Science+Business Media, LLC

Library of Congress Cataloging in Publication Data

Reactive oxygen species in biological systems: an interdisciplinary approach / Daniel L. Gilbert and Carol A. Colton.

p. cm.

Includes bibliographical references and index.

ISBN 978-1-4757-8190-8 ISBN 978-0-306-46806-3 (eBook)

DOI 10.1007/978-0-306-46806-3

1. Active oxygen in the body. 2. Active oxygen—Pathophysiology. I. Gilbert, Daniel L. II. Colton, Carol A.

QP535.O1R43 1999 98-49453

572'.53—dc21 CIP

ISBN 978-1-4757-8190-8

Originally published by Kluwer Academic / Plenum Publishers, New York in 1999

Softcover reprint of the hardcover 1st edition 1999

10 9 8 7 6 5 4 3 2 1

A C.I.P. record for this book is available from the Library of Congress.

Contributors

R. John Aitken MRC Reproductive Biology Unit, Centre for Reproductive Biology, Edinburgh EH3 9EW, Scotland

Bismark Amoah-Apraku Division of Nephrology and Hypertension, Georgetown University Medical Center, Washington, D.C. 20007

Patrik Andrée Department of Biochemistry, Arrhenius Laboratories for Natural Sciences, Stockholm University, S-106 91 Stockholm, and Clinical Research Center, NOVUM, Karolinska Institute, S-141 86 Huddinge, Sweden

Saimar Arif International Antioxidant Research Centre, UMDS–Guy's Hospital, London SE1 9RT, England

Bernard M. Babior Division of Biochemistry, Department of Molecular and Experimental Medicine, The Scripps Research Institute, La Jolla, California 92037

C. Jacyn Baker U.S. Department of Agriculture, Agricultural Research Service, Molecular Plant Pathology Laboratory, Beltsville, Maryland 20705

Eduard Berenshtein Department of Cellular Biochemistry, Hebrew University–Hadassah Schools of Medicine and Dental Medicine, Jerusalem 91120, Israel

Barbara S. Berlett Laboratory of Biochemistry, National Heart, Lung, and Blood Institute, National Institutes of Health, Bethesda, Maryland 20892-0342

Alberto Boveris Department of Physical Chemistry, School of Pharmacy and Biochemistry, University of Buenos Aires, Buenos Aires, Argentina

Robert H. Brown, Jr. Day Neuromuscular Research Laboratory, Massachusetts General Hospital, Charlestown, Massachusetts 02129

D. Allan Butterfield Department of Chemistry, Center of Membrane Sciences, and Sanders-Brown Center on Aging, University of Kentucky, Lexington, Kentucky 40506-0055

Enrique Cadenas Department of Molecular Pharmacology and Toxicology, School of Pharmacy, University of Southern California, Los Angeles, California 90033

Mordechai Chevion Department of Cellular Biochemistry, Hebrew University–Hadassah Schools of Medicine and Dental Medicine, Jerusalem 91120, Israel

Gerald Cohen Department of Neurology and Center for Neurobiology, Mount Sinai School of Medicine, New York, New York 10029

Carol A. Colton Department of Physiology and Biophysics, Georgetown University Medical School, Washington, D.C. 20007

Dana R. Crawford Department of Biochemistry and Molecular Biology, The Albany Medical College, Albany, New York 12208

Gustav Dallner Department of Biochemistry, Arrhenius Laboratories for Natural Sciences, Stockholm University, S-106 91 Stockholm, and Clinical Research Center, NOVUM, Karolinska Institute, S-141 86 Huddinge, Sweden

Bruce Demple Division of Toxicology, Department of Cancer Cell Biology, Harvard School of Public Health, Boston, Massachusetts 02115

Lars Ernster Department of Biochemistry, Arrhenius Laboratories for Natural Sciences, Stockholm University, S-106 91 Stockholm, Sweden

Martin Feelisch Wolfson Institute for Biomedical Research, London W1P 9LN, England

Jon Fukuto Department of Molecular Pharmacology, University of California, Los Angeles, California 90269

Daniel L. Gilbert Unit on Reactive Oxygen Species, BNP, NINDS, National Institutes of Health, Bethesda, Maryland 20892-4156

Cecilia Giulivi Department of Molecular Pharmacology and Toxicology, School of Pharmacy, University of Southern California, Los Angeles, California 90033; *present address*: Department of Chemistry, University of Minnesota, Duluth, Minnesota 55812

Beatriz González-Flecha Physiology Program, Department of Environmental Health, Harvard School of Public Health, Boston, Massachusetts 02115

Matthew B. Grisham Department of Physiology, Louisiana State University Medical Center, Shreveport, Louisiana 71130

John M. C. Gutteridge Oxygen Chemistry Laboratory, Critical Care Unit, Royal Brompton Hospital, London SW3 6NP, England

Nicolas J. Guzman Division of Nephrology and Hypertension, Georgetown University Medical Center, Washington, D.C. 20007

Stephen M. Hahn Department of Radiation Oncology, University of Pennsylvania, Philadelphia, Pennsylvania 19104

Barry Halliwell Biochemistry Department, National University of Singapore, Singapore 119260

Jay W. Heinecke Departments of Medicine and of Molecular Biology and Pharmacology, Washington University, St. Louis, Missouri 63110

Robert E. Huie Physical and Chemical Properties Division, National Institute of Standards and Technology, Gaithersburg, Maryland 20899

Murali C. Krishna Radiation Biology Branch, National Cancer Institute, National Institutes of Health, Bethesda, Maryland 20892

Sasha Madronich Atmospheric Chemistry Division, National Center for Atmospheric Research, Boulder, Colorado 80307

Dianne M. Meacher Department of Community and Environmental Medicine, University of California, Irvine, Irvine, California 92697-1825

Daniel B. Menzel Department of Community and Environmental Medicine, University of California, Irvine, Irvine, California 92697-1825

James B. Mitchell Radiation Biology Branch, National Cancer Institute, National Institutes of Health, Bethesda, Maryland 20892

P. Neta Physical and Chemical Properties Division, National Institute of Standards and Technology, Gaithersburg, Maryland 20899

Harry S. Nick Department of Biochemistry and Molecular Biology, University of Florida, Gainesville, Florida 32610

Elizabeth W. Orlandi Department of Microbiology, University of Maryland, College Park, Maryland 20742-5815

Dale A. Parks Department of Anesthesiology, University of Alabama at Birmingham, Birmingham, Alabama 35233

Burkhard Poeggeler Department of Pathology, University of South Alabama, Mobile, Alabama 36617

Catherine Rice-Evans International Antioxidant Research Centre, UMDS–Guy's Hospital, London SE1 9RT, England

Angelo Russo Radiation Biology Branch, National Cancer Institute, National Institutes of Health, Bethesda, Maryland 20892

Amram Samuni Molecular Biology, School of Medicine, Hebrew University, Jerusalem 91010, Israel

Henry B. Skinner Department of Anesthesiology, University of Alabama at Birmingham, Birmingham, Alabama 35233

Kelly A. Skinner Department of Anesthesiology, University of Alabama at Birmingham, Birmingham, Alabama 35233

Earl R. Stadtman Laboratory of Biochemistry, National Heart, Lung, and Blood Institute, National Institutes of Health, Bethesda, Maryland 20892-0342

Sidhartha Tan Department of Anesthesiology, University of Alabama at Birmingham, Birmingham, Alabama 35233

John F. Valentine Department of Medicine, University of Florida, and Gainesville VA Medical Center, Gainesville, Florida, 32610

Yoram Vodovotz Tumor Biology Section, Radiation Biology Branch, National Cancer Institute, Bethesda, Maryland 20892

David A. Wink Tumor Biology Section, Radiation Biology Branch, National Cancer Institute, Bethesda, Maryland 20892

Ben-Zhan Zhu Department of Cellular Biochemistry, Hebrew University–Hadassah Schools of Medicine and Dental Medicine, Jerusalem 91120, Israel

Preface

This volume with its many internationally recognized contributors shows that the importance of reactive oxygen species (ROS) in biology has finally become recognized. Back in 1954 when free radicals were theorized to occur in living organisms, most scientists did not take this theory seriously. Even in 1981, when D. L. Gilbert edited the book *Oxygen and Living Processes: An Interdisciplinary Approach*, there were a significant number of scientists who did not take this theory seriously. In 1999, the situation has changed; even the lay public knows about the importance of antioxidants. In 1981 there was no mention of molecular biology; that has changed in the present volume. The field of reactive oxygen species has come a long way, baby!

In this volume it is suggested that the unpaired electron in the free radicals be designated by a superscript dot to the right of the atom which contains the unpaired electron. There are two exceptions. The first exception occurs when there is a charge on the atom containing the unpaired electron, then the superscript dot appears on the left. The second exception occurs when there are two or more atoms in the free radical, the superscript dot appears to the left of the atom containing the unpaired electron. In conclusion, the superscript dot is always associated with the atom containing the unpaired electron. Thus, the following are acceptable: $\cdot O_2^-$ for the superoxide radical anion, $RO_2\cdot$ for the peroxyl radical, $HO_2\cdot$ for the hydroperoxyl radical, $\cdot OH$ for the hydroxyl radical, $RO\cdot$ for the alkoxyl radical, $\cdot NO$ for nitric oxide, and $\cdot NO_2$ for nitrogen dioxide.

This book is divided into 8 parts. Part I, the Introduction, provides the background in Chapters 1 and 2 for the other chapters by covering the history and chemistry of ROS.

Part II is concerned with General Biochemistry and Molecular Biology; six chapters are in this part. Chapter 3 is about the steady state production of ROS by mitochondria. This is followed by Chapter 4 on the transition metals, so important in redox chemistry. Chapters 5 and 6 cover the molecular biology or genetics of the activation of antioxidant enzymes. Chapter 7 discusses how inflammation regulates manganese superoxide dismutase. The final chapter in this part, Chapter 8, discusses antioxidant proteins and signaling by oxygen radicals.

Part III on Nitrogen Reactive Species covers the importance of nitrogen radicals. C. C. Chieuh, D. L. Gilbert, and C. A. Colton edited a volume on this subject, entitled *The*

Neurobiology of NO· and ·OH in 1994. Chapter 9 discusses the biological enzyme, nitric oxide synthase, which produces nitric oxide, a free radical. Next Chapter 10 discusses the beneficial and deleterious interactions of nitric oxide in biological organisms. The concluding chapter in this part, Chapter 11, deals with the protective role of nitroxides against oxidative stress.

Part IV presents Environmental Pro- and Antioxidants. Chapter 12 discusses the importance of stratospheric ozone, which filters out damaging ultraviolet radiation. The next chapter, Chapter 13, is on the damaging influences of ozone and nitrogen dioxide, a free radical, acting directly on biological organisms. This concludes with Chapter 14 on antioxidants in nutrition.

Part V contains three chapters on Internal Pro- and Antioxidants. Chapter 15 points out how xanthine oxidase can contribute to a prooxidant condition in the biological organism, and also how it generates antioxidants. The next chapter, Chapter 16, discusses how the hormone, melatonin, can act as an antioxidant; the final chapter, Chapter 17, in this part points out that ubiquinol is a very effective lipid-soluble antioxidant.

Part VI is concerned with ROS in Specific Tissues. Chapter 18 is on plant tissue. Chapter 19 takes up the production of ROS by phagocytes. Chapters 20 and 21 point out that both spermatozoa and fertilized ova produce ROS. Chapters 22 and 23 are concerned with nervous tissue. Brain chemiluminescence is the topic in Chapter 22 and Chapter 23 takes up the role of ROS in neuronal function.

The title of Part VII is Pathological States and Aging. Chapter 24 is on Parkinson's disease, which is followed by Chapter 25 on Alzheimer's disease. The third chapter, Chapter 26, is on amyotrophic lateral sclerosis or Lou Gehrig's disease. The final chapter, Chapter 27, is on the oxidation of proteins in aging.

Part VIII, the Conclusion in Chapter 28 gives an overall summary. It also very briefly includes some other topics that have not been covered, including the role of ROS in programmed cell death, cataracts, hypoxia, peroxisomes, arachidonic acid, and reperfusion injury.

Daniel L. Gilbert
Carol A. Colton

Acknowledgments

The editors would like to thank the reviewers, especially Dr. Douglas R. Spitz and Dr. David Wink.

Contents

Part I. Introduction

Chapter 1

From the Breath of Life to Reactive Oxygen Species

Daniel L. Gilbert

Chapter 2

Chemistry of Reactive Oxygen Species

Robert E. Huie and P. Neta

Part II. General Biochemistry and Molecular Biology

Chapter 3

The Steady-State Concentrations of Oxygen Radicals in Mitochondria

Cecilia Giulivi, Alberto Boveris, and Enrique Cadenas

Chapter 4

The Role of Transition Metal Ions in Free Radical-Mediated Damage

Mordechai Chevion, Eduard Berenshtein, and Ben-Zhan Zhu

Chapter 7

Inflammatory Regulation of Manganese Superoxide Dismutase

John F. Valentine and Harry S. Nick

Chapter 8

Antioxidant Protection and Oxygen Radical Signaling

John M. C. Gutteridge and Barry Halliwell

Part III. Nitrogen Reactive Species

Chapter 9

Nitric Oxide Synthase

Nicolas J. Guzman and Bismark Amoah-Apraku

Chapter 10

The Chemical Biology of Nitric Oxide

David A. Wink, Martin Feelisch, Yoram Vodovotz, Jon Fukuto, and Matthew B. Grisham

Chapter 11

Nitroxides as Protectors against Oxidative Stress

James B. Mitchell, Murali C. Krishna, Amram Samuni, Angelo Russo, and Stephen M. Hahn

Chapter 14

Dietary Antioxidants and Nutrition

Catherine Rice-Evans and Saimar Arif

Part V. Internal Pro- and Antioxidants

Chapter 15

Xanthine Oxidase in Biology and Medicine

Dale A. Parks, Kelly A. Skinner, Sidhartha Tan, and Henry B. Skinner

Chapter 20

Production and Effects of Reactive Oxygen Species by Spermatozoa

R. John Aitken

Chapter 21

Respiratory Burst Oxidase of Fertilization: Peroxidative Mechanisms in Sea Urchin Eggs and Human Phagocytes

Jay W. Heinecke

Chapter 22

Brain Chemiluminescence as an Indicator of Oxidative Stress

Alberto Boveris and Enrique Cadenas

Chapter 23

Reactive Oxygen Species and Neuronal Function

Carol A. Colton and Daniel L. Gilbert

Part VII. Pathological States and Aging

Chapter 24

Oxidative Stress and Parkinson's Disease

Gerald Cohen

Chapter 25

Alzheimer's β-Amyloid Peptide and Free Radical Oxidative Stress

D. Allan Butterfield

Part VIII. Conclusion

Chapter 28

An Overview of Reactive Oxygen Species

Daniel L. Gilbert and Carol A. Colton

Part I

Introduction

Chapter 1

From the Breath of Life to Reactive Oxygen Species

Daniel L. Gilbert

Oxygen was not an element like any other, but, to use his [Schönbein's] own expression, it was the king of elements, the Jupiter of the scientific Olympus. He [Schönbein] spoke of oxygen as his hero, which he regarded as omnipotent.
Florkin (1975)

1. INTRODUCTION

It has long been known that air was essential for human life. Many cultures have referred to this as the *breath of life* (Gandevia, 1970a; Gilbert, 1981). The oldest reference to this concept is found in the Sumerian creation myth of about 3000 BC, which states: "For the sake of the good things in their pure sheepfolds Man was given breath." A statue of Prince Gudéa, dated at about 2200 BC, is inscribed with the following phrase: "generously endowed with the breath of life." The Indian Hindu *Rigveda*, which dates from 1500 to 1000 BC, makes reference to "Giver of vital breath." The Chinese also have a breath of life concept, represented by *chhi*. The Egyptian book *The Physician's Secret: Knowledge of the Heart's Movement and Knowledge of the Heart*, written about 1600 to 1550 BC, states that "the breath of life enters into the right ear, and the breath of death enters into the left ear." About 1350 BC, Akhenaton, the heretic Egyptian pharaoh who worshipped the sun, had in his Hymn to the Sun: "Who givest breath to animate every one that he maketh." In the Old Testament, Genesis, chapter 2, verse 7, states that "God Yahweh formed man from clods in the soil and blew into his nostrils the breath of life. Thus man became a living being." More recently, Gandevia (1970a) has written that "Queensland

Daniel L. Gilbert Unit on Reactive Oxygen Species, BNP, NINDS, National Institutes of Health, Bethesda, Maryland 20892-4156.

Reactive Oxygen Species in Biological Systems, edited by Gilbert and Colton. Kluwer Academic / Plenum Publishers, New York, 1999.

[Australia] (Cape Bedford) aborigines relate the spirit wau-wau to the breath; it never leaves the body during life."

Parmenides, who was born about 515 BC, believed that fire within living organisms was necessary for life. Fire was thus related to the biosphere. Empedocles, who lived about 490–430 BC, believed the elements to be earth, water, air, and fire. These four elements were accepted by the ancient Greeks (Gandevia, 1970b; Gilbert, 1981). Other cultures had similar conceptions about the elements. Until space flight permitted us to go to the moon, our environment had been the crust of the Earth. It is natural that the ancients recognized these components of the Earth's crust as elements. The recognized components of the planet's crust are the lithosphere or earth (as defined by the ancients), hydrosphere or water, atmosphere or air, and biosphere. The ancients also believed that fire was related to living organisms, i.e., the biosphere. Aristotle, who lived from 384 to 322 BC, wrote: "Hence, of necessity, life must be coincident with the maintenance of heat, and what we call death is its destruction." The Romans in the first century BC were aware of the vital flame or flamma vitalis. Leonardo da Vinci (1452–1519) realized that air that cannot support combustion is also air that cannot support life.

FIGURE 1. Robert Boyle, 1627–1691. He performed many experiments on the effect of changing gas pressures on animals. From the History of Medicine Division, National Library of Medicine, Bethesda, Maryland.

Empirically, it was known that some component of the air was necessary for human life. Today, we know that oxygen is that component. The only places on Earth that have significantly less than 158 torr (21.07 kPa) of oxygen are in the mountainous regions of the world. The first reference to the difficulty of reaching the high mountain passes was by Too Kin in China sometime between 37 and 32 BC (Gilbert, 1983a). It was not until 1590 that the deleterious effects of altitude on humans were recorded firsthand by Acosta as he climbed the high-altitude pass by the twin peaks of Pariacaca in Peru (Gilbert, 1983b, 1991; Bonavia *et al.*, 1984, 1985); Acosta's description of mountain sickness influenced Robert Boyle (Figure 1) to investigate the effects of pressure on animals. Boyle and his followers in England were interested in the different types of gases (Frank, 1980) and their effects on humans. In fact, Boyle in 1673 had heated lead oxide until oxygen was released. He had also noted that a partial vacuum caused both a flame to be extinguished and an animal to die. Thus, there was something in the air that supported both life and combustion. In the same decade of the seventeenth century, Hooke had shown that respiration required some atmospheric constituent; shortly thereafter, Mayow, referring to respiration, wrote: "an aërial something essential to life, whatever it may be, passes into the mass of the blood." Hooke theorized there was a substance mixed in air that is like the substance in potassium nitrate; Mayow theorized that the substance in air was a nitro-aërial spirit. Gunpowder was already known at this time and is a mixture of potassium nitrate (nitre) and sulfur with carbon.

2. PHLOGISTON THEORY

In continental Europe, the idea was that some fire material was contained in the earth. The Greeks and Chinese thought that sulfur was this fire material. Alchemists were trying to convert earth into noble metals. Paracelsus in the sixteenth century named sulfur the *combustible principle*. Both Hooke and Mayow thought that combustible materials contained sulfur. Becher in 1667 (Figure 2) regarded combustion to be a loss of fatty combustible earth, which corresponded to Paracelsus's sulfur. In 1697, Stahl, a highly respected German scientist (Figure 3), called this fire material *phlogiston*. After the wooden house in Figure 4 is burned, very little solid material remains, indicating that something is lost. Indeed, something is lost and that is a combination of carbon dioxide and water; this fact was not recognized at that time. The small amount of solid material surviving the fire contained charcoal; the charcoal could subsequently be burned indicating that it was a source of phlogiston. Combustion was thought to be just a release of phlogiston from a combustible material. Heated metal oxides were reduced in the presence of charcoal to their respective metals:

$$MO_2 + C \rightarrow M + CO_2 \qquad (1)$$

where M represents the metal. According to this theory, the phlogiston was released from the charcoal: The liberated phlogiston combined with the metal oxide to form the reduced metal (Conant, 1957).

FIGURE 2. Johann Joachim Becher, 1635–1682. He thought that combustion was a loss of fatty combustible earth. From the History of Medicine Division, National Library of Medicine, Bethesda, Maryland.

FIGURE 3. Georg Ernst Stahl, 1660–1734. He called Becher's fire material *phlogiston* and was a strong proponent of the phlogiston theory. From the History of Medicine Division, National Library of Medicine, Bethesda, Maryland.

FIGURE 4. Effect of fire. When a wooden house is burned, the only visible object remaining is ash composed of charcoal.Almost all of the solid structure is lost. The phlogiston theory attributed this loss to phlogiston.The small part that remained was charcoal, which was thought to contain a large amount of phlogiston.

3. OXIDATION THEORY

Lavoisier (Figure 5) first became interested in phlogiston and combustion after he learned about effervescence from Buffon's translation of Stephen Hales's (Figure 6) *Vegetable Staticks* (Guerlac, 1961, 1975). Stephen Hales (1677–1761) published *Vegetable Staticks* in 1727 in England. He published an appendix for the second edition, appearing in 1733 in his *Haemastaticks.* However, Buffon inserted this appendix in his French translation of *Vegetable Staticks* in 1735. Hales, in this appendix, discussed the topic of effervescence, which had a tremendous influence on Lavoisier. Hales wrote about gases being released from solids (effervescence). It was commonly accepted at this time that the four elements were earth, air, fire, and water, but here was release of air from earth. In England, scientists were concentrating on the different kinds of air, whereas in the rest of Europe, scientists were concentrating on the different kinds of earth. When Lavoisier (Holmes, 1985, 1994) learned about this effervescence, he was already trained in geology. What was being released from this effervescence? His first experiments were on heating phosphorus and sulfur in air to see if there was any increase in weight caused by the heating. These experiments led to the origin of a revolutionary idea in chemistry, namely, the theory of oxidation. Even though there were many reports on the gain in weight when metals were heated in air to produce oxides, or *calxes* as they were called in the eighteenth century, this fact was still controversial at the time when Lavoisier did his experiments. Later, he heated tin and lead in air and found that these metals also gained in weight when heated. Because he could not read English, he was unaware when he started these investigations of the experiments of Joseph Black (Figure 7) in Scotland on carbon dioxide, which was called *fixed air* in 1756. Black found that calcium hydroxide precipitates carbon dioxide as the white calcium carbonate, and so this reaction became a test for the gas, carbon dioxide. Black discovered that both combustion and respiration produce carbon dioxide; he also noted that carbon dioxide killed birds rapidly. Lavoisier had his wife learn English, so that he could keep current with the English writings of Joseph Black, Joseph Priestley (Figure 8), and Henry Cavendish. In 1766, Cavendish, a student of Black, isolated a new gas, *inflammable air,* which we now know as hydrogen

FIGURE 5. Antoine-Laurent Lavoisier, 1743–1794, and Marie Anne Pierrette Paulze Lavoisier, his wife, 1758–1836. A black-and-white print of a color painting by Jacques Louis David (1748–1825) painted in 1788. David was a political enemy of Lavoisier and a friend of Robespierre. According to Poirier (1993), David charged an equivalent of 1.4 million 1993 francs (U.S. $240,000). Poirer (1993) poses the question of whether David portrayed Antoine Lavoisier as a subordinate of his wife. The description in the Museum states that his wife's portfolio of her drawings is on the armchair on the left and that Lavoisier withdrew this painting from the Salon in 1789. From the Metropolitan Museum of Art, New York. The Metropolitan Museum of Art, Purchase, Mr. and Mrs. Charles Wrightman Gift, in honor of Everett Fahy, 1977. (1977.10)

FIGURE 6. Stephen Hales, 1677–1761. He noted effervescence; this phenomenon influenced Antoine Lavoisier to begin his thoughts about oxidation. From the History of Medicine Division, National Library of Medicine, Bethesda, Maryland.

(Partington, 1962). Priestley found both carbon dioxide and hydrogen to be lethal (Holmes, 1985).

In the meantime, the Swedish pharmacist Carl Wilhelm Scheele (Figure 9) discovered that oxygen was released when he heated silver carbonate in the presence of alkali to absorb the carbon dioxide. In 1774, Scheele wrote to Lavoisier about this method of obtaining oxygen. Priestley realized that respiration finally made the air unfit for further respiration and combustion. He reasoned there had to be some way for the air to be regenerated to its original state, otherwise animal life on Earth would cease to exist. In 1772, Priestley demonstrated that plants were the source of this regeneration. In 1779, Ingenhousz found that plants exposed to sunlight released oxygen. Also, Priestley devised the nitric oxide test for determining the percentage of oxygen in the atmosphere. The nitric oxide reacts with oxygen rapidly (Gilbert, 1981, 1994), and by measuring gas volume changes, Priestley could determine the purity of air in 1772. Thus, this reaction became the quantitative test for oxygen. Qualitative tests were observing how long a flame and an animal could survive in oxygen. Priestley obtained oxygen by heating red mercuric

FIGURE 7. Joseph Black, 1728–1799. He is noted for his studies on the detection of carbon dioxide. From the History of Medicine Division, National Library of Medicine, Bethesda, Maryland.

oxide and told Lavoisier about it at a dinner in Paris in 1774. Both Scheele and Priestley called oxygen *dephlogisticated air*.

Lavoisier obtained oxygen by heating mercuric oxide and then he reversed the reaction by heating the products, mercury and oxygen, to form the red mercuric oxide. This oxide releases oxygen when heated to over 360°C; the oxide is formed when mercury is heated at about 350°C. Mercury boils at 357°C (Sanderson, 1967). The reaction of both decomposing the oxide at high temperatures and forming it at lower high temperatures is

FIGURE 8. Joseph Priestley, 1733–1804, the independent discoverer of oxygen. From the Library of Congress, Washington, D.C.

FIGURE 9. Carl Wilhelm Scheele, 1742–1786, the other independent discoverer of oxygen. From the Library of Congress, Washington, D.C.

$$2\,HgO \leftrightarrow 2\,Hg + O_2 \tag{2}$$

Lavoisier also noted that the mercuric oxide did not require as much heat in the presence of charcoal to reduce the oxide to mercury. He analyzed the liberated gas by using the tests for oxygen and carbon dioxide as developed by Priestley and Black, respectively (Conant, 1957). In the presence of charcoal, he found that the liberated gas was carbon dioxide, as in Reaction (1).

Later, he and his collaborators combined hydrogen with oxygen to produce water. He also heated water in the presence of a metal; the metal was used to pick up the liberated oxygen to produce the metal oxide, as follows:

$$2\,H_2O + M \rightarrow MO_2 + 2\,H_2 \tag{3}$$

Lavoisier now had proved that the combustion reaction was an oxidation process and had replaced Stahl's phlogiston theory with his oxidation theory involving oxygen (Lavoisier, 1783). Mme. Marie Lavoisier, dressed as a priestess, burned the works of Stahl in Paris (Partington, 1962), a disgraceful act (Mahaffy, 1995; Gilbert, 1996b). On the other hand, German supporters of phlogiston burned her husband, Antoine Lavoisier, in effigy in Berlin (Partington, 1962).

However, Cavendish and others believed that hydrogen was phlogiston (Partington, 1962). Today, we recognize that oxidation can also occur by dehydrogenation reactions. If phlogiston is hydrogen, then dephlogistonation reactions become dehydrogenation reactions. Indeed, Cavendish's theory had some merit; the biggest obstacle to it was the fact that metals gained weight when they were oxidized.

4. OXYGEN POISONING

The discoverers of oxygen, Priestley and Scheele, and the discoverer of the oxidation process, Lavoisier, also reported on oxygen toxicity (Gilbert, 1981) (Figure 10). All three of them realized the dual role of oxygen (dephlogisticated air): On the one hand, life requires oxygen, and on the other, life is destroyed by oxygen. Scheele noted that peas would not grow in oxygen; Priestley found that both plants and mice died in oxygen. Priestley (1775) wrote: "for, as a candle burns out much faster in dephlogisticated [oxygenated] than in common air, so we might, as may be said, *live out too fast*, and the animal powers be too soon exhausted in this pure kind of air. A moralist, at least, may say, that the air which nature has provided for us is as good as we deserve." Lavoisier (1782–1783) killed guinea pigs exposed to oxygen and concluded: "Healthy air is therefore composed of a good proportion between vital air [oxygen] and atmospheric moffete [nitrogen], . . . when there is an excess of vital air [oxygen], the animal only undergoes a severe illness; when it is lacking, death is almost instantaneous." Later, when Lavoisier did some experiments with his young protégé, Seguin, on respiration using guinea pigs, they reported that no toxic effects of oxygen were observed after several days in pure oxygen (Seguin and Lavoisier, 1789). More recently, experiments on survival times in oxygen of guinea pigs showed that they probably could not survive for several days in oxygen (Clark and Lambertsen, 1971; Frank *et al.*, 1978; Pocidalo *et al.*, 1983; Sosenko and Frank, 1987). We believe that an explanation for this inconsistency is that Seguin and Lavoisier (1789) could have used an oxygen atmosphere less than 95%.

5. THE DISCOVERERS OF OXYGEN AND OF OXIDATION

Scheele was a brilliant chemist, employed as a pharmacist. During his 5-year employment in Uppsala, Sweden, he met and collaborated with Torbern Bergman, a well-known chemist and naturalist. Scheele left Uppsala in 1775 to operate a pharmacy in Köping after its owner died. His only ambition was to pursue scientific truth; however,

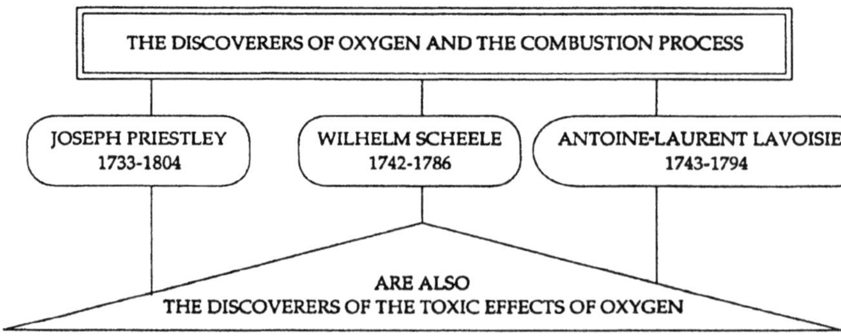

FIGURE 10. Discoverers of oxygen and combustion. Joseph Priestley, the codiscoverer of oxygen; Carl Wilhelm Scheele, the independent codiscoverer of oxygen; and Antoine-Laurent Lavoisier, the discoverer of the mechanism of oxidation, were the first ones to report on the toxic effects of oxygen.

FIGURE 11. Political caricature of Joseph Priestley: "Doctor Phlogiston, The Priestley Politician or the Political Priest!" Note the evil smile. He holds two burning papers, "Political Sermon" and "Essay on Government." His pockets contain papers entitled "Revolution Toasts," "Essays on Matlin—Spirit," and "Gunpowder." He steps on papers including one on "Bible explained away" (Attic Miscellany, 1791). July 1, 1791. From the Library of Congress, Washington, D.C.

FIGURE 12. Political caricature of Joseph Priestley by Gregor Gillray in 1791. At the bottom is written "A BIRMINGHAM TOAST, as given on the 14th of July, by the ———Revolution Society." July 21, 1791. For simplicity, I have shown only the left-hand portion here: The right-hand legend has been transferred to the bottom of this figure. A toast is given by Priestley: "The———[King's] Head, here!" Sheridan bends forward, filling his glass from a pitcher of sherry, saying, "Damn my Eyes! but I'll pledge you that Toast tho Hell gapes for me." There is also Home Tooke, saying, "I have not drank so glorious a Toast since I was Parson of Brentford, & kept it up with Balf & McQuirk!" The Constitutional Society of Birmingham met on July 14, 1791, to commemorate the second anniversary of the storming of the Bastille. This date marks the beginning of the French Revolution. Priestley was warned not to attend this meeting, but even though he did not attend, his home was attacked by a mob because of his liberal views (Gillray, 1791). From the Library of Congress, Washington, D.C.

his genius was recognized and he was elected to the Royal Academy of Science of Sweden in 1777. It has been reported that he conducted 15,000 to 20,000 experiments (Frängsmyr, 1986). He was a bachelor, but on his deathbed, he married the widow of his predecessor so as to allow her to retain the pharmacy. He suffered from rheumatism and gout, and died at age 43 (Partington, 1962).

Both Lavoisier and Priestley were subjected to political persecution. Although Lavoisier revolutionized chemistry and was considered an outstanding scientist, he was also a hated tax collector. He was independently wealthy and he helped the French government by serving as the head of a commission to increase gunpowder production. He served as director of the Académie des Sciences and was a member of the commission responsible for introducing the metric system for the new French government. He was guillotined in 1794 during the "Reign of Terror" of the French Revolution while at the height of his career (Partington, 1962).

Priestley was a liberal clergyman who supported the ideals of the American and French revolutions (Gibbs, 1965). Because of his political beliefs, many unfavorable caricatures of Priestley appeared (Roberts, 1989) (Figures 11 and 12). His greatest patron was Shelburne, who later as prime minister negotiated the treaty with the American colonies. Priestley's home was burned on July 14, 1791. Thomas Paine, George Washington, and Priestley were made citizens of France (Gibbs, 1965). Because of Priestley's support of the French Revolution, he was forced to flee England. However, he was against the violence of the Revolution. France declared war against England on February 1, 1793, and Priestley sailed for the United States on April 8, 1794 (Gibbs, 1965); he was in the middle of the Atlantic Ocean on his way to his new home in the United States when Lavoisier was executed on May 8, 1794.

6. ACID PRODUCER

Oxygen refers to "acid producer," because Lavoisier thought that oxygen is present in all acids. Most languages use a root meaning "acid" for the word for the element oxygen; the exceptions are the Danish, Polish, and Chinese languages. The Chinese use the word meaning "nourishing gas" for the element oxygen whereas the Danish and Polish languages use words meaning "fire" and "smother" (Gilbert, 1981). All of the acids that Lavoisier tested contained oxygen except for hydrochloric acid; but he thought that his method for detecting oxygen in hydrochloric acid was not good enough. It was not until 1810 when Sir Humphry Davy proved that hydrochloric acid did not contain oxygen that Lavoisier's theory of acids was discarded. However, today we know that hydrogen is present in all acids. Lavoisier was mistaken in this belief, but oxidation does occur more easily both on a thermodynamic basis (Figure 13) as well as on a kinetic basis in acid media (Funahashi *et al.*, 1994).

7. OXYGEN THERAPY

Because oxygen was shown to be indispensable both for aerobic living organisms and for combustion, Thomas Beddoes, James Watt, Robert John Thornton, and others believed that excess oxygen could be used as a therapeutic agent to cure many diverse

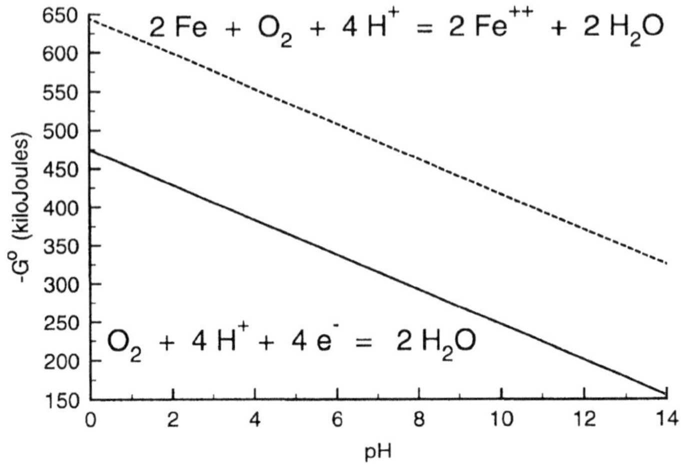

FIGURE 13. Redox reactions as a function of pH. The standard thermodynamic potential for these two reactions is plotted as a function of pH. Note that as the pH is decreased, the thermodynamic potential is increased for both of these reduction–oxidation reactions. These reactions, one being the oxidation of metallic iron to ferrous iron and the other being the reduction of oxygen to water, are just examples. At pH 0, the standard thermodynamic potential is 644.2 kJ for the iron reaction and is 474.4 kJ for the oxygen reaction. The slope for both lines is –22.81 kJ/pH. Generally, this statement is true for reactions in which the hydrogen ion participates in the chemical reactions.

diseases (Gilbert, 1981). Most physicians soon realized that this extra oxygen had no effect. Beddoes in 1797 wrote, "I have now no chemical theory of any one disease . . . I started conjectures to be compared with facts; and now I think all these conjectures are shewn to be erroneous by facts." Nonetheless, Thornton was the best known advocate for oxygen therapy in the early part of the nineteenth century. In fact, he was called "vital air [oxygen] and gas mad." Mr. Morton, an actor, mentions "one DR. OXYGEN, who gives his patient, by mistake, instead of a certificate of Cures, the bills of 'Mortality'!" (Gilbert, 1981). Samuel Parkes in his early nineteenth-century book, *A Chymical Catechism*, romanticizes about the role of oxygen as a cure in a poem (Mahaffy, 1995), the idea being that the more plentiful oxygen is, the better off are living organisms. In the 1980s, the musical group The Sweet sang the following lyrics in their song about oxygen: "Love is like oxygen, You get too much, You get too high, Not enough and you're going to die."

8. OZONE

Schönbein in 1840 discovered ozone; ozone is trioxygen, a molecule containing three oxygen atoms, whereas ordinary oxygen is dioxygen. Both forms are gases under ordinary temperature and pressure. However, ozone is much more reactive than dioxygen. Schönbein believed that dioxygen had to be activated into ozone in biological oxidation processes (Florkin, 1975).

FIGURE 14. Louis Pasteur, 1822–1895, the discoverer of unicellular organisms that cannot exist in the presence of oxygen. From the Library of Congress, Washington, D.C.

9. LOUIS PASTEUR

Louis Pasteur (Figure 14) in 1861, while studying fermentation, wrote: "Not only do these infusoria [type of microorganisms] live without air, but the air kills them." He found the microorganisms to survive very well when carbon dioxide was introduced, but when air was substituted, they died within 1 or 2 hours. He wrote: "This is, I think, the first example known of . . . animals who live without free oxygen gas" (Pasteur, 1861). Thus, Pasteur discovered the existence of organisms so sensitive to oxygen toxicity that the presence of any oxygen kills them. He proposed the name *anaérobies* for these organisms, and *aérobies* for those unicellular organisms that can multiply in air (Pasteur, 1863). Pasteur was an outstanding scientist who believed that chance favors only a prepared mind. His first major contributions were in the fields of crystallography and optical activity. He became interested in applied research during his research on fermentation (Geison, 1974). He discovered that microorganisms were the cause of the type of fermentations he was studying, disproving the theory of spontaneous generation. Finally, he made major contributions in the treatment and protection against infectious diseases by killing pathogenic organisms using heat; this process is called *pasteurization.* He also developed vaccines for chicken or fowl cholera, anthrax, swine fever, and rabies. He was the first director of the Institut Pasteur (1888) and remained as director until his death in 1895. He was glorified for his brilliant contributions during his life, and this worldwide glorification still exists today. A political conservative, he supported Emperor Louis Napoleon during France's Second Empire. After Louis Napoleon abdicated in 1870, Pasteur ran for the Senate and was humiliated with a crushing defeat. Pasteur ordered his family not to show his laboratory notebooks to anyone. However, these notebooks, which were recently released, reveal some unethical aspects of his behavior (Geison, 1995).

10. PAUL BERT

The first one to prove that the effects of altering the barometric pressure are related to the effects of changing the oxygen pressure was Paul Bert (Figure 15) (Bert, 1878). He

FIGURE 15. Paul Bert, 1833–1886, who discovered that oxygen was toxic to all living organisms. Bert was also the first to note the deleterious effects of increased oxygen on the central nervous system. From the History of Medicine Division, National Library of Medicine, Bethesda, Maryland.

stated that "the diminution of barometric pressure acts upon living beings only by lowering the oxygen tension in the air they breathe," and "we have given abundant proof that they are the consequence not of the barometric pressure as a physico-mechanical agent, but of the increase in the tension of the ambient oxygen." In addition, he demonstrated that *oxygen can kill animals rapidly with convulsive symptoms.* The effect of oxygen producing disorders has been termed the *Paul Bert effect* (Bean, 1945). He demonstrated that these convulsions were not related to a direct action on the muscles, but mediated by the motor nerves. He proved this by cutting only the sciatic nerve of one leg and observing that this procedure "prevented all convulsive movement, fibrillary or generalized, from appearing in the corresponding muscles." He next demonstrated that the anesthetic chloroform prevented these convulsions, and concluded that the convulsions "indicate that the toxic action produces its effect on the nervous centers." He summarized his findings on high pressures by stating that "all living beings . . . perish more or less rapidly in air that is sufficiently compressed. . . . For the higher animals, death is preceded by tonic and clonic convulsions of extreme violence." Bert was the first to demonstrate that the nervous system is especially sensitive to oxygen toxicity.

Like Priestley and Lavoisier, Paul Bert (1833–1886) was also active politically (Fulton, 1943; Kellogg, 1978). Following his father, Bert also became a lawyer in 1857, but found that he was more interested in studying natural sciences and proceeded to obtain his M.D. degree in 1863. His thesis was on tissue transplantation, which he continued to do in Claude Bernard's laboratory. He received the degree of Doctor of Natural Sciences in 1865 and won the experimental physiology prize from the Académie des Sciences for his transplantation experiments. He married a Scottish woman and settled in Auxerre

(Kellogg, 1978). He became active in politics in Auxerre when the Second Empire under Emperor Napoleon III was defeated in the Franco-Prussian War of 1870. He was elected to the National Assembly and in 1881 became Minister of Education. His views were predominately liberal. He became the governor-general of French Indochina in 1886 and died that same year in Hanoi, Vietnam. A museum to honor him was inaugurated in Auxerre in 1977 (Kellogg, 1978).

11. LORRAINE SMITH

Lorraine Smith in 1899 observed that increased oxygen pressure causes lung inflammation, a condition first noted by Lavoisier (1782–1783). Smith noted that "the toxic effects described by Bert occur at a tension which is much higher than that required to produce the inflammatory effect on the lungs." The lung effect is now known as the *Lorraine Smith effect* (Bean, 1945).

Why are both the brain and the lungs noted for their sensitivity to these adverse oxygen effects? The lungs' sensitivity to oxygen reflects the fact that the lungs are exposed to one of the highest oxygen pressures in the body. The brain's sensitivity to oxygen is large because of the organ's high oxygen consumption and decreased antioxidant defense.

12. PRELUDE TO THE FREE RADICAL THEORY OF OXYGEN POISONING

Guyton de Morveau, in an oral paper on May 2, 1787, was the first chemist to use the term *radical* to mean a chemical entity that forms an acid when combined with oxygen. This paper presented part of the new chemical nomenclature advanced by de Morveau, Lavoisier, Berthollet, and Fourcroy (Partington, 1962). Lavoisier used de Morveau's definition in his widely read book, *Elements of Chemistry* (Lavoisier, 1789). The term *radical* was used until 1810 when Davy demonstrated that Lavoisier's acid theory was incorrect (Ihde, 1967). In 1815, Gay-Lussac discovered cyanogen (C_2N_2), which was believed to be the *free radical,* CN. Dumas, Liebig, Piria, Berzelius, Bunsen, and other chemists (Pryor, 1966, 1968; Ihde, 1967) theorized that they had produced free radicals. Other chemists, including some of the proponents of the free radicals, were confused and did not accept free radicals as entities in the 1830s. In the late 1840s, both Kolbe and Frankland thought that they had produced free radicals. In the second half of the nineteenth century, gas densities were used to obtain molecular weights; this technique showed that the so-called free radicals were really dimers. However, V. and C. Meyer had shown in 1879 that the chlorine atom did exist; in the following year, V. Meyer obtained the iodine atom; and with H. Züblin he obtained the bromine atom (Mellor, 1927). By the end of the nineteenth century, chemists no longer believed in free radicals (Ihde, 1967). However, Moses Gomberg actually produced the first free radical, which he reported at the turn of the century, and which was the triphenylmethyl (trityl) free radical (Gomberg, 1900). Even then, it took about another decade before the theory of free radicals was generally accepted. Gomberg (1914, 1925) pointed out that the dimer did exist, but that it was in equilibrium with the free radical. The current definition of a free radical is any atom or group of atoms that has an unpaired electron in an outer orbital. Filled electron orbitals contain two electrons with opposite spins, so that the net spin is zero. Usually a very strong tendency exists for unpaired electrons to pair together in an orbital so as to eliminate the electron spin. Thus, the hydrogen atom as well as the halogen atoms, i.e., chlorine, iodine, and bromine, are free radicals. Generally, free radicals are extremely reactive, so they exist in the free state for only a very short period of time. During the 1920s, they were demonstrated to be responsible for chain reactions. In the 1930s, they were shown to cause polymerizations (Flory, 1937; Pryor, 1966) as well as depolymerizations (Walling, 1957). Haber and Weiss (1934) demonstrated that hydrogen peroxide is catalytically decomposed by iron salts and proposed a free radical chain mechanism in which the hydroxyl radical, ·OH, plays an important role. The Haber–Weiss reactions are probably very important in biological mechanisms (Cohen, 1977; Fridovich, 1981). Haber and Weiss were unaware that 40 years earlier, Fenton (1894) had noted that catalytic concentrations of ferrous sulfate and hydrogen peroxide oxidize tartaric acid. Farmer *et al.* (1943) proposed that lipid peroxidation proceeded by free radical intermediates. Michaelis postulated that oxygen would be reduced by univalent oxidation states; these reactive oxygen intermediates in the presence of hydrogen would be as follows (Michaelis, 1946):

$$O \rightarrow HO_2\cdot \rightarrow H_2O_2 \rightarrow H_2O + \cdot OH \rightarrow H_2O \tag{4}$$

At neutral pH, the acid $HO_2\cdot$ dissociates into H^+ and $\cdot O_2^-$, i.e., the superoxide radical anion. The other free radical generated in this cascade is ·OH, i.e., the neutral hydroxyl radical. Thus, as oxygen is reduced to water, there are two free radicals generated as well as hydrogen peroxide and these three species are the reactive oxygen intermediates. The overall reaction is

$$2\,H_2 + O_2 \rightarrow 2\,H_2O \tag{5}$$

Respiration is the process of reducing oxygen in the body to water:

$$1/6\,(CH_2O)_6 + O_2 \rightarrow CO_2 + H_2O \tag{6}$$

The thermodynamic potential of Reaction (5) is almost the same as Reaction (6). From an energetic point of view, hydrogen is stored in the sugar, i.e., $(CH_2O)_6$, and the carbon dioxide acts just as a sponge absorbing the hydrogen. Photosynthesis is the reverse process of reaction (6); the splitting of the water molecule occurs before carbon becomes involved. Cellular respiration processes involve a hydrogen transport system to cytochrome *c*, where the hydrogen combines with oxygen (Gilbert, 1960).

The discovery of radioactivity made by Becquerel in 1892 opened up other means of producing free radicals. Ionizing radiation produced by radioactive materials ionized the water into

$$H_2O \rightarrow e^- + H_2O^+ \tag{7}$$

$$e^- + H_2O \rightarrow H_2O^- \tag{8}$$

The resulting charged water molecules are unstable and give rise to free radicals as follows:

$$H_2O^+ \rightarrow H^+ + \cdot OH \tag{9}$$

$$H_2O^- \rightarrow OH^- + \cdot H \tag{10}$$

The result is that water is dissociated into hydrogen atoms and hydroxyl radicals. The first workers to note that irradiated water gives rise to these two free radical entities were Debierne in 1914 (Spinks and Woods, 1990) and Risse in 1929 (Dale, 1954). However, Fricke during the 1930s wrote about radiation causing water to be activated, and it was this *activated water* that had an effect on solutes dissolved in the water (Dale, 1954; von Sonntag, 1987).

There was much interest in ionizing radiation during World War II and the remainder of the 1940s. Irradiated gases are simpler to analyze than irradiated liquids and solutions, because the free radicals and other excited species can diffuse more freely in the gaseous state than in the liquid state (Spinks and Woods, 1990). Weiss (1944) pointed out that *activated water* induced by radiation was really the same as the free radicals of the hydrogen atom and the hydroxyl free radical. It was known that living organisms contain about 60 to 70% water; therefore, a good part of the biological effects of ionizing radiation were thought to be related to the free radicals produced in water. It was also known that decreasing the oxygen decreased the biological damage caused by ionizing radiation. This *oxygen effect* was reasoned to reflect the following process:

$$\cdot H + O_2 \rightarrow HO_2\cdot \tag{11}$$

It was only recognized after the 1950s that the pK of the hydroperoxyl radical, $HO_2\cdot$, was only 4.7, so that at a pH of 7, the $HO_2\cdot$ would exist as the superoxide radical anion, $\cdot O_2^-$ (Spinks and Woods, 1990).

13. ORIGIN OF THE FREE RADICAL THEORY OF OXYGEN POISONING

Such was the state of knowledge regarding free radicals when Wallace O. Fenn suggested to Rebeca Gerschman (Figure 16) that she look for a possible decrease in the adrenal gland ascorbic acid, as an index of adrenal cortex activation, in rats exposed to high oxygen stress. Gerschman confirmed that high oxygen did produce such a decrease, as did occur for other stresses (Gerschman and Fenn, 1954). However, her oral report at the September 1952 American Physiological Society meeting in New Orleans revealed that adrenalectomy protected against oxygen poisoning (Gerschman and Fenn, 1952). What sets apart oxygen stress from other stresses? She knew that the metabolic rate is increased when the adrenal cortex is activated; she read that an increased metabolic rate was a factor in decreasing the resistance to high oxygen (Campbell, 1938). Could it be that adrenalectomy produced a decreased metabolic rate, which would be protective against oxygen poisoning? Gerschman was scheduled to deliver a regular seminar in late 1952 or 1953 in the Department of Physiology at the University of Rochester on why the adrenals were protective against all kinds of stress except that of high oxygen. Being an

FIGURE 16. Rebeca Gerschman, the originator of the theory that free radicals were responsible for the toxic effects of oxygen. She died on April 4, 1986, in Buenos Aires, Argentina.

avid reader and possessing an excellent memory, she remembered an old paper reporting that oxygen and X-irradiation produced similar histological changes in the testes (Ozorio de Almeida, 1934). What was the fundamental basis for this similarity? She read that the biological effects were related to free radicals, and according to the univalent theory of Michaelis (Michaelis, 1946), free radicals were intermediates in oxidation processes. Hence, an increase in metabolism, or oxidation processes, should result in a greater production of free radicals. She was supposed to present her data on the protective action of adrenalectomy, which she barely mentioned. Instead, she presented some preliminary data showing that these effects were similar to the biological effects of ionizing radiation: She theorized that free radicals were responsible for both of these effects (Gerschman, 1981; Gilbert, 1996c).

Previously, I learned about the biological effects of radiation in experiments at the State University of Iowa for my M.S. degree using ^{32}P as an index of blood flow in dystrophic muscle (Gilbert *et al.*, 1950). I performed the radioactivity measurements in the Radiation Research Laboratory of Dr. Titus C. Evans. Being in this laboratory was a great introduction into radiation biology. He was the first managing editor of *Radiation Research,* the official publication of the Radiation Research Society (Barr, 1972; Evans, 1972). Evans was the managing editor for 50 volumes of the journal from 1954 to 1972; no other editor has held this position longer than Evans.

After graduation, I decided to study muscle electrolytes under Dr. Wallace O. Fenn at the University of Rochester. An outstanding physiologist, Fenn received many honors in his life (Rahn, 1979). I was unaware that the federal government had chosen the University of Rochester to investigate the biological effects of ionizing radiation during World War II. When I arrived there as a graduate student in 1950, the university in conjunction with the Atomic Energy Commission set up a program to train health physicists. Because I had previous experience with radioisotopes, I decided to take advantage of this program and almost had taken enough courses to qualify as a health physicist prior to Rebeca Gerschman's seminar. I believe I was the only one present at the seminar who knew about free radicals. The theory that free radicals are responsible for the deleterious actions of both X-irradiation and oxygen toxicity excited me so much that I actively became involved in the research, although my Ph.D. thesis was on the calcium pump in muscle (Gilbert and Fenn, 1957). I knew about the so-called *oxygen effect* in radiation biology. However, I had no idea that high concentrations of oxygen were poisonous; on the contrary, I felt that more oxygen was always helpful to living organisms. As a result of her stimulating and provocative seminar, I became her principal research collaborator. Another indication suggesting a common mechanism of action for both oxygen poisoning and irradiation was the observation that some thiol enzymes could be inactivated by high oxygen pressures (Stadie *et al.*, 1944; Haugaard, 1946) and by irradiation (Barron *et al.*, 1949). Gerschman persuaded Dr. Henry A. Blair, director of the Department of Radiation Biology, to see if rats could be subjected to high doses of X-irradiation and high oxygen pressures. Blair was very encouraging to Rebeca and even had the experiment conducted in his department. The results showed a small, but statistically highly significant, synergism between the effects of X-irradiation and high oxygen. Rats, given antioxidants and then exposed to high oxygen, survived longer than their respective controls.

Immediately after her seminar, we began having stimulating discussions about oxygen toxicity; these discussions often lasted into the early morning hours. As we read and discussed many published papers, our views about free radicals were greatly strengthened. When our manuscript was submitted to *Science*, we were thrilled to read the referee's report, which stated: "We are here lifted to a higher plane of observation in which the similarity of the two effects is established, first by citation of the literature, secondly by the submission of new data showing cumulative effects." In his letter of acceptance dated January 28, 1954, Duane Roller, the editor, wrote: "I am very glad to say that the paper by Gerschman *et al.* has not only been accepted for publication in *Science*, but also will be scheduled as a sub-lead article." The article was published in the May 7, 1954, issue (Gerschman *et al.*, 1954). It was exciting to see it published in *Science* as well as to read the comments from the journal's editor. We stated that "free-radical formation is also expected in normal oxidative metabolism." On the other hand, it was also a discouraging time for us as many came by to voice their strong objections. Essentially, their view was that a significant concentration of free radicals was not possible, because they are very reactive and exist for very limited spans of time. However, we argued they were intermediates in metabolism despite their predicted low concentrations. Further, we argued that as the free radicals are removed rapidly because of their high reactivity, the precursors of free radicals would rapidly react with each other and regenerate more free radicals. Moreover, in a steady state, short life spans of free radicals imply high production rates of them. The free radicals could produce deleterious effects, such as chain reactions. As already mentioned, at the beginning of the twentieth century, Gomberg (1914, 1924) received similar criticisms for his ideas about the existence of free radicals in chemistry, until he isolated a stable free radical. Fridovich (1975) wrote: "Similarities between the lethality of oxygen and of ionizing radiation led, in 1954, to the theory that the undisciplined reactivities of free radicals were the root cause of oxygen toxicity [Gerschman *et al.*, 1954]. This was a remarkable prescient theory considering the paucity of information concerning the generation and scavenging of specific free radicals in biological systems available at that time. The developments of recent years have provided a firm foundation for a reasonable discussion of the basis of oxygen toxicity."

Furthermore, this *Science* publication contained experiments showing that antioxidants given to mice subjected to high oxygen tensions increased their survival times. Such experiments performed at lower oxygen tensions are called *aging experiments*. However, Gerschman did not mention the term *aging* until 1959 (Gerschman, 1959). The thought occurred to Harman in November 1954 that free radicals could be involved in aging (Harman, 1992); he did show that antioxidants increased the survival times of animals at normal oxygen tensions (Harman, 1957).

Following this publication in *Science*, Gerschman and I continued to collaborate, which resulted in several publications. Commoner headed a group that actually detected free radicals by electron spin resonance in dry biological materials in a paper submitted on July 12, 1954 (Commoner *et al.*, 1954). In 1957, this same group (Commoner *et al.*, 1957) found that living organisms produce free radicals. However, McCord and Fridovich (1969) recognized that the enzyme superoxide dismutase (McCord and Fridovich, 1977) served a protective function by getting rid of the superoxide anion radical; now the free radical theory became a serious theory. Antioxidant mechanisms are required by aerobic organisms for their very existence. *Reactive oxygen species* (ROS) is a broader expression than *free radicals* or *oxidizing free radicals*. ROS includes the superoxide radical anion ($\cdot O_2^-$), hydrogen peroxide (H_2O_2), the hydroxyl radical ($\cdot OH$), lipid peroxides, the peroxyl radicals ($RO_2\cdot$), the alkoxy radicals ($RO\cdot$), the radicals of nitric oxide ($\cdot NO$) and nitrogen dioxide ($\cdot NO_2$), ozone (O_3), and possibly singlet oxygen, either in its low-energy form ($^1\Delta O_2$) or in its high-energy form ($^1\Sigma O_2$).

Although my Ph.D. thesis was on muscle calcium, I was constantly thinking about a possible connection between muscle calcium and free radicals. I (Gilbert, 1955) postulated that "vitamin E [a known antioxidant] deficiency may . . . be analogous to a chronic form of oxygen poisoning," which could possibly block the calcium pump, and produce an increase in muscle calcium. Fenn, my Ph.D. thesis adviser, not only allowed me to collaborate with Gerschman, but also encouraged this collaboration. I (1996c) wrote: "[Rebeca Gerschman] never received the recognition that she richly deserved in her lifetime."

14. ANTIOXIDANT DEFENSES AND THE ROLE OF REACTIVE OXYGEN SPECIES IN NORMAL PHYSIOLOGICAL PROCESSES

De Saussure in 1820 noted that walnut oil became rancid while absorbing air over a period of 8 months (Halliwell and Gutteridge, 1989). Antioxidants have been used to delay the rancidity that occurs when oils are exposed to the air. It would appear that a greater antioxidant defense would result in a better ability to cope with an oxidative stress. However, under some circumstances, the situation appears to be more complicated. In 1926, Moureu and Dufraisse (1926) wrote about anti- and prooxidants and pointed out that antioxidants in one system can become prooxidants in another system. Others have also pointed out this phenomenon (Mattill, 1947; Gilbert, 1963). Farmer and others at the

British Rubber Producers Association during the 1940s studied the lipid peroxidation reactions responsible for this rancidity (Halliwell and Gutteridge, 1989). The peroxidation reactions are free radical chain reactions. These reactions occur *in vivo* as well as *in vitro.*

When I first became involved in this field, I naively thought oxygen was needed only for energetics and for synthesis in the cell. There was also the realization that oxygen will of necessity have a destructive influence because of its potential. I did not realize that this could be used against unwanted foreign invaders and cellular debris. It was noted in 1933 that phagocytic cells have an increased respiration (Ado, 1933; Baldridge and Gerard, 1933). Almost three decades later, it was observed that this respiratory burst or oxygen burst resulted in a release of hydrogen peroxide (Iyer *et al.*, 1961). Klebanoff (1967) noted that leukocytes, which are phagocytic cells, kill bacteria by this mechanism. Forty years after the first observation of an oxidative burst, Babior *et al.* (1973) reported that the superoxide anion radical was released from leukocytes. Tissue-resident macrophages are phagocytic cells and exist in many areas of the body. Microglia, the macrophages in the brain, can produce superoxide radical anions (Giulian and Baker, 1986; Colton and Gilbert, 1987).

There are other examples where both the predator and the prey use and defend against ROS (Gilbert, 1996a).

15. REDOX CONTROL

It is now realized that some ROS have nondestructive actions. For example, in fertilization, ROS are produced by spermatozoa for two purposes: (1) maturation of spermatozoa in the female reproductive tract and (2) membrane fusion with the oocyte (see Chapter 20). The oocyte also produces ROS that cause cross-reactions within the first few minutes after fertilization, resulting in a hard fertilization membrane. This phenomenon is accompanied by a transient increase in the oocyte respiration (Warburg, 1908) and is similar to the oxygen burst observed in phagocytic cells (see Chapter 21).

It is known that sulfhydryl groups are sensitive to ROS (Stadie *et al.*, 1944; Haugaard, 1946). The *N*-methyl-D-aspartate receptor contains sulfhydryl groups (Janaky *et al.*, 1993) and is under redox control; reducing conditions activate the receptor and oxidizing conditions inhibit the receptor in a reversible manner. Activation of the inducible transcription factor, nuclear factor κB, occurs when there is a mild oxidative stress and this activation is inhibited by antioxidants (Schreck and Baeuerle, 1991; Hayashi *et al.*, 1993). Schreck and Baeuerle (1991) proposed that ROS can act as second messengers in low concentrations. Some gene families are activated by an increased oxygen stress and respond by increasing antioxidant defenses (see Chapters 5, 6, and 8). However, the hydroxyl radical is so reactive that it reacts with most everything. The result is that it principally causes damage.

16. SUMMARY

Certainly the idea that air is necessary for life was understood by humans living in prehistoric times. Many diverse cultures refer to the breath of life; the earliest written document on this appears 5000 years ago. Primitive humans recognized that their environment was composed of air, water, and earth. Fire was dynamic and was capable of destroying objects as well as providing heat for cooking and comfort. It was only natural that diverse primitive cultures considered elements to be air, water, earth, and fire. Fire was certainly the most mysterious, not being tangible like the others. Fire was associated with life by the Greeks and was later expressed as the vital flame (*flamma vitalis*) by the Romans.

In the Middle Ages, alchemists were performing experiments attempting to convert the earth into noble metals. Some of the alchemists unwittingly produced oxygen. English scientists in the seventeenth century began investigating air, in part inspired by an association between high altitude and mountain sickness. Robert Boyle investigated the effect of changing air pressure on living organisms and found that when the atmospheric pressure was greatly reduced, the animals could no longer survive. Something present in the air was necessary for life. Robert Hooke and Mayow thought this something to be related to nitrate. Why? Because it was known that potassium nitrate was a major ingredient of gunpowder. In other European countries, scientists became interested in the chemistry and physics of solids, probably because of the interest in alchemy. The Greeks and Chinese believed that sulfur was the combustible substance, possibly in part reflecting their knowledge that sulfur compounds are released in some volcanoes. Thus, a combustible substance was sulfur. Both Becker and later Stahl believed that some fatty sulfureous earth was lost during fire; Stahl named it *phlogiston*. It was a reasonable theory at that time, considering that after a fire much is lost.

In the eighteenth century, both Joseph Priestley in England and Carl Scheele in Sweden discovered oxygen. Antoine Lavoisier, trained as a geologist, read about gases being given off in solids, the phenomenon called *effervescence* by Stephen Hales in England. Lavoisier became interested in gases and proved that an oxidized metal is heavier than the metal alone. This proof as well as other considerations led to the downfall of the phlogiston theory and its replacement by our present combustion theory of oxidation. Priestley, Scheele, and Lavoisier soon realized that oxygen can be toxic. Nonetheless, Beddoes and others at the end of the eighteenth century thought that administering an increased oxygen tension could cure many diverse diseases. They soon learned that an excess of oxygen did not cure diseases.

Louis Pasteur in 1861 noted that some unicellular organisms cannot exist in the presence of oxygen. In the next decade, Paul Bert showed that an increased oxygen pressure was toxic, but that this toxicity was not related to the pressure per se. Specifically, Bert demonstrated that an increased oxygen pressure caused nervous system pathology as evidenced by convulsions. Lorraine Smith rediscovered the damaging effect that excess oxygen has on lung pathology.

The modem era of oxygen toxicity studies was begun by Rebeca Gerschman, who theorized that X-irradiation damage and oxygen toxicity were mediated by oxidizing free radicals. Gerschman *et al.* published a paper in 1954 in *Science* giving support to this theory, both by experiment and by literature arguments. However, the scientific community generally did not accept this free radical theory of oxygen toxicity. The biggest objection to the theory was that free radicals, being so reactive, could not exist *in vivo* in significant concentrations. Gerschman and I argued that even if the free radical concentrations were small, their production could be important in producing deleterious biological effects. Commoner *et al.* used electron spin resonance to detect free radicals *in vivo* in 1957. It was not until McCord and Fridovich's discovery in 1969 of the function of superoxide dismutase that there was much thought that free radicals can exist *in vivo*.

Antioxidants were produced by aerobic organisms to serve defensive purposes. ROS, besides having detrimental effects, can in some circumstances be protective. Babior *et al.* in 1973 showed that leukocytes can produce the superoxide radical anion as a means of killing invading bacteria. Clearly, the immune system has used ROS to protect organisms against bacterial attack. Some ROS can act as second messengers. ROS have come a long way in the past 45 years!

ACKNOWLEDGMENTS. I wish to thank Dr. Claire Gilbert for her French translations and for her assistance in editing this manuscript. I also wish to thank Raymond L. Gilbert for showing me The Sweet's lyrics about oxygen.

17. REFERENCES

Ado, A. D., 1933, Über den Verlauf der oxydativen und glykolytischen Prozesse in den Leukocyten des entzundeten Gewebes wahred der Phagocytose, *Z. Gesamte Exp. Med.* **87**:473–480.

Attic Miscellany, 1791, 7887 DOCTOR PHLOGISTON, THE PRIESTLEY POLITICIAN OR THE POLITICAL PRIEST, in *Catalogue of Political and Personal Satires Preserved in the Department of Prints and Drawings in the British Museum. Vol. VI. 1784–1792, printed 1947* (M. D. George, ed.), p. 806, Department of Prints and Drawings, British Museum, London.

Babior, B. M., Kipnes, R. S., and Curnutte, J. T., 1973, Biological defense mechanisms. The production by leukocytes of superoxide, a potential bactericidal agent, *J. Clin. Invest.* **52**:741–744.

Baldridge, C. W., and Gerard, R. W., 1933, The extra respiration of phagocytosis, *Am. J. Physiol.* **103**:235–236.

Barr, N. F., 1972, Titus C. Evans. Managing Editor 1954–1972, *Radiat. Res.* **50**:iii.

Barron, E. S. G., Dickman, S., Muntz, J. A., and Singer, T. P., 1949, Studies on the mechanism of action of ionizing radiations. I. Inhibition of enzymes by x-rays, *J. Gen. Physiol.* **32**:537–552.

Bean, J. W., 1945, Effects of oxygen at increased pressure, *Physiol. Rev.* **25**:1–147.

Bert, P., 1878, *Barometric Pressure. Researches in Experimental Physiology* (M. A. Hitchcock and F. A. Hitchcock, Trans.), College Book Co., Columbus, Ohio, 1943.

Bonavia, D., Leon-Velarde, F., Monge, C. C., Sanchez-Grinan, M. I., and Whittembury, J., 1984, Tras las huellas de Acosta 300 anos despues. Consideraciones sobre su descripcion del "Mal de altura," *Historica* **8**(1):1–31.

Bonavia, D., Leon-Velarde, F., Monge, C. C., Sanchez-Grinan, M. I., and Whittembury, J., 1985, Acute mountain sickness: Critical appraisal of the Pariacaca story and on-site study, Respir. Physiol. **62**:125–134.

Campbell, J. A., 1938, Effects of oxygen pressure as influenced by external temperature, hormones and drugs, *J. Physiol. (London)* **92**:29P–30P.

Clark, J. M., and Lambertsen, C. J., 1971, Pulmonary oxygen toxicity: A review, *Pharmacol. Rev.* **23**:37–133.

Cohen, G., 1977, In defense of Haber–Weiss, in *Superoxide and Superoxide Dismutases* (A. M. Michelson, J. M. McCord, and I. Fridovich, eds.), pp. 317–321, Academic Press, New York.

Colton, C. A., and Gilbert, D. L., 1987, Production of superoxide anions by CNS macrophage, the microglia, *FEBS Lett.* **223:** 284–288.

Commoner, B., Townsend, J., and Pake, G. E., 1954, Free radicals in biological materials, *Nature* **174**:689–691.

Commoner, B., Heise, J. J., Lippincott, B. B., Norberg, R. E., Passonneau, J. V., and Townsend, J., 1957, Biological activity of free radicals, *Science* **126**:57–63.

Conant, J. B., 1957, Case 2. The overthrow of the phlogiston theory. The chemical revolution of 1775–1789, in *Harvard Case Histories in Experimental Science,* Volume 1 (J. B. Conant and L. K. Nash, eds.), pp. 67–115, Harvard University Press, Cambridge, MA.

Dale, W. M., 1954, Basic radiation chemistry, in *Radiation Biology. Volume 1: High Energy Radiation* (A. Hollaender, ed.), pp. 255–281, McGraw–Hill, New York.

Evans, T. C., 1972, Editorial. Fifty volumes of *Radiation Research*, *Radiat. Res.* **50:**v–xvi.
Farmer, E. H., Koch, H. P., and Sutton, D. A., 1943, The course of autoxidation reactions in polyisoprenes and allied compounds. Part VII. Rearrangement of double bonds during autoxidation, *J. Chem. Soc.* **1943:**541–547.
Fenton, H. J. H., 1894, Oxidation of tartaric acid in presence of iron, *J. Chem. Soc.* **65:**899–910.
Florkin, M., 1975, *A History of Biochemistry. Part III. History of the Identification of the Sources of Free Energy in Organisms*, in *Comprehensive Biochemistry*, Volume 31 (M. Florkin and E. H. Stotz, eds.), Elsevier, Amsterdam.
Flory, P. J., 1937, The mechanism of vinyl polymerizations, *J. Am. Chem. Soc.* **59:**241–253.
Frängsmyr, T., 1986, Carl Wilhelm Scheele (1742–1786), *Chem. Scr.* **26:**507–511.
Frank, L., Bucher, J. R., and Roberts, R. J., 1978, Oxygen toxicity in neonatal and adult animals of various species, *J. Appl. Physiol.* **45:**699–704.
Frank, R. G. J., 1980, *Harvey and the Oxford Physiologists. Scientific Ideas and Social Interaction,* University of California Press, Berkeley.
Fridovich, I., 1975, Oxygen: Boon and bane, *Am. Sci.* **63:**54–59.
Fridovich, I., 1981, Superoxide radical and superoxide dismutases, in *Oxygen and Living Processes: An Interdisciplinary Approach* (D. L. Gilbert, ed.), pp. 250–272, Springer-Verlag, Berlin.
Fulton, J. F., 1943, Foreword, in *Barometric Pressure. Researches in Experimental Physiology*, pp. v–ix, College Book Co., Columbus, Ohio.
Funahashi, T., Floyd, R. A., and Carney, J. M., 1994, Age effect on brain pH during ischemia/reperfusion and pH influence on peroxidation, *Neurobiol. Aging* **15:**161–167.
Gandevia, B., 1970a, The breath of life: An essay on the earliest history of respiration. Part I, *Austr. J. Physiother.* **16:**5–11.
Gandevia, B., 1970b, The breath of life: An essay on the earliest history of respiration. Part II, *Austr. J. Physiother.* **16:**57–69.
Geison, G. L., 1974, Louis Pasteur, in *Dictionary of Scientific Biography*, Volume X (C. C. Gillespie, ed.), pp. 350–416, Scribner's, New York.
Geison, G. L., 1995, *The Private Science of Louis Pasteur*, Princeton, Princeton University Press, Princeton, NJ.
Gerschman, R., 1959, Oxygen effects in biological systems, Symp. Spec. Lect., XXI Int. Congr. Physiol. Soc., pp. 222–226.
Gerschman, R., 1981, Historical introduction to the "free radical theory" of oxygen toxicity, in *Oxygen and Living Processes: An Interdisciplinary Approach* (D. L. Gilbert, ed.), pp. 44–46, Springer-Verlag, Berlin.
Gerschman, R., and Fenn, W. O., 1952, Ascorbic acid content of the adrenal in oxygen poisoning, *Am. J. Physiol.* **171:**726.
Gerschman, R., and Fenn, W. O., 1954, Ascorbic acid content of the adrenal in oxygen poisoning, *Am. J. Physiol.* **176:**6–8.
Gerschman, R., Gilbert, D. L., Nye, S.W., Dwyer, P., and Fenn, W. O., 1954, Oxygen poisoning and x-irradiation: A mechanism in common, *Science* **119:**623–626.
Gibbs, F. W., 1965, *Joseph Priestley. Adventurer in Science and Champion of Truth*, Thomas Nelson and Sons, Camden, NJ.
Gilbert, D. L., 1955, The permeability of isolated frog skeletal muscle to calcium, Ph.D. thesis, University of Rochester, Rochester, NY.
Gilbert, D. L., 1960, Speculation on the relationship between organic and atmospheric evolution, *Perspect. Biol. Med.* **4:**58–71.
Gilbert, D. L., 1963, The role of pro-oxidants and antioxidants in oxygen toxicity, *Radiat. Res. Suppl.* **3:**44–53.
Gilbert, D. L., 1981, Perspective on the history of oxygen and life, in *Oxygen and Living Processes: An Interdisciplinary Approach* (D. L. Gilbert, ed.), pp. 1–43, Springer-Verlag, Berlin.
Gilbert, D. L., 1983a, The first documented report of mountain sickness: The China or Headache Mountain story, *Respir. Physiol.* **52:**315–326.
Gilbert, D. L., 1983b, The first documented description of mountain sickness: The Andean or Pariacaca story, *Respir. Physiol.* **52:**327–347.
Gilbert, D. L., 1991, The Pariacaca or Tullujuto story: Political realism? *Respir. Physiol.* **86:**147–157.
Gilbert, D. L., 1994, Keeping reactive oxygen species (ROS) in their proper place, *Ann. N.Y. Acad. Sci.* **738:**1–7.

Gilbert, D. L., 1996a, Evolutionary aspects of atmospheric oxygen and organisms, in *Handbook of Physiology, Section 4. Adaptation to the Environment. Volume II* (M. J. Fregly and C. M. Blatteis, eds.), pp. 1059–1094, Oxford University Press, London.

Gilbert, D. L., 1996b, The eradication of the phlogiston theory by book burning, *FASEB J.* **10:**A31 (Abstract 177).

Gilbert, D. L., 1996c, Rebeca Gerschman: A personal remembrance, *Free Radical Res. Biol.* **21:**1–4.

Gilbert, D. L., and Fenn, W. O., 1957, Calcium equilibrium in muscle, *J. Gen. Physiol.* **40:**393–408.

Gilbert, D. L., Janney, C. D., and Hines, H. M., 1950, Circulatory transfer of P^{32} to skeletal muscles under various experimental conditions, *Am. J. Physiol.* **163:**575–579.

Gillray, G., 1791, 7894 A BIRMINGHAM TOAST, AS GIVEN ON THE 14TH OF JULY, BY THE——— REVOLUTION SOCIETY, in *Catalogue of Political and Personal Satires Preserved in the Department of Prints and Drawings in the British Museum. Vol. VI. 1784–1792, printed 1947* (M. D. George, ed.), p. 82, Department of Prints and Drawings, British Museum, London.

Giulian, D., and Baker, T. J., 1986, Characterization of ameboid microglia isolated from developing mammalian brain, *J. Neurosci.* **6:**2163–2178.

Gomberg, M., 1900, An instance of trivalent carbon: Triphenylmethyl, *J. Am. Chem. Soc.* **22:**757–771.

Gomberg, M., 1914, The existence of free radicals, *J. Am. Chem. Soc.* **36:**1144–1170.

Gomberg, M., 1924, Organic radicals, *Chem. Rev.* **1:**91–141.

Guerlac, H., 1961, *Lavoisier—The Crucial Year. The Background and Origin of His First Experiments on Combustion in 1772*, Cornell University Press, Ithaca, NY.

Guerlac, H., 1975, *Antoine-Laurent Lavoisier. Chemist and Revolutionary*, Scribner's, New York.

Haber, F., and Weiss, J., 1934, The catalytic decomposition of hydrogen peroxide by iron salts, *Proc. R. Soc. London Ser. A* **147:**332–351.

Hales, S., 1727, *Vegetable Staticks: Or, An Account of Some Statical Experiments on the Sap in Vegetables: Being an Essay Towards a Natural History of Vegetation. Also, a Specimen of an Attempt to Analyse the Air, By a Great Variety of Chymio-Statical Experiments; Which Were Read at Several Meetings Before the Royal Society*. London: W. and J. Innys, T. Woodward.

Hales, S., 1733, *Statical Essays: Containing Haemastaticks; or an Account of Some Hydraulick and Hydrostatical Experiments Made on the Blood and Blood-Vessels of Animals. Also an Account of Some Experiments on Stones in the Kidneys and Bladder; With an Enquiry Into the Nature of Those Anomalous Concretions. To Which is Added, an Appendix, Containing Observations and Experiments Relating to Several Subjects in the First Volume. The Greatest Part of Which Were Read at Several Meetings Before the Royal Society. With an Index to Both Volumes. Vol. II.* London: W. Innys and R. Manby, T. Woodward.

Hales, S., 1735, *La Statique des Vegetaux, et L'Analyse de L'Air. Experiences Nouvelles Lûes à la Societé Royales de Londres* (d. Buffon, Trans.), Chez DEBURE l'aîne, à l'entrée du Qauay des Augustins, du côté du Pont Saint Michel, à Saint Paul, Paris.

Halliwell, B., and Gutteridge, J. M. C., 1989, *Free Radicals in Biology and Medicine*, 2nd ed., Oxford University Press, London.

Harman, D., 1957, Prolongation of the normal life span by radiation protection chemicals, *J. Gerontol.* **12:**257–263.

Harman, D., 1992, Free radical theory of aging: History, in *Free Radicals and Aging* (I. Emerit and B. Chance, eds.), pp. 1–10, Birkhauser Verlag, Basel.

Haugaard, N., 1946, Oxygen poisoning XI. The relation between inactivation of enzymes by oxygen and essential sulfhydryl groups, *J. Biol. Chem.* **164:**265–270.

Hayashi, T., Ueno, Y., and Okamoto, T., 1993, Oxidoreductive regulation of nuclear factor κB. Involvement of a cellular reducing catalyst thioredoxin, *J. Biol. Chem.* **268**:11380–11388.

Holmes, F. L., 1985, *Lavoisier and the Chemistry of Life. An Exploration of Scientific Creativity*, University of Wisconsin Press, Madison.

Holmes, F. L., 1994, Antoine Lavoisier. The conservation of matter, *Chem. Eng. News* **72**(Sept. 12, No. 37):38–45.

Ihde, A. J., 1967, The history of free radicals and Moses Gomberg's contributions, *Pure Appl. Chem.* **15:**1–13.

Iyer, G. Y. N., Islain, M. F., and Quastel, J. H., 1961, Biochemical aspects of phagocytosis, *Nature* **192:**535–541.

Janaky, R., Varga, V., Saranssri, P., and Oja, S. S., 1993, Glutathione modulates the N-methyl-D-aspartate receptor-activated calcium influx into cultured rat cerebellar granule cells, *Neurosci. Lett.* **156:**153–157.

Kellogg, R. H., 1978, La Pression Barométrique: Paul Bert's hypoxia theory and its critics, *Respir. Physiol.* **34**:3–28.

Klebanoff, S. J., 1967, A peroxidase-medicated anti-microbial system in leukocytes, *J. Clin. Invest.* **46**:1078.

Lavoisier, 1782–1783, Sur les altérations qui arrivent à l'air dans plusieurs circonstances où se trouvent les hommes réunis en société, *Hist. Soc. Méd.* **5** (Read in 1785):569–582; *Mémoires de Médecine*, **5** (Read in 1785).

Lavoisier, 1783, Réflexions sur le phlogistique, pour servir de suite [développement] à la théorie de la Combustion et [&] de la Calcination, Publiée en 1777, in: *Le Ministre de l'Instruction Publique. Oeuvres de Lavoisier. Vol. II. Mémoires de Chimie et Physique.* Imprimerie Impériale, Paris, 1862, pp. 623–655. (Histoire de l'Académie Royale des Sciences avec Les Mémoires de Mathématique & de Physique, pp. 505–538).

Lavoisier, A. L., 1789, *Elements of Chemistry, in a New Systematic Order, Containing All the Modern Discoveries* (D. McKie, ed.; R. Kerr, Trans.), Dover, New York, 1965.

Mahaffy, P. G., 1995, Breathing life into chemists. Resuscitating chemistry with insights from 19th century textbooks, *J. Chem. Educat.* **72**:767–773.

Mattill, H. A., 1947, Antioxidants, *Annu. Rev. Biochem.* **16**:177–192.

McCord, J. M., and Fridovich, I., 1969, Superoxide dismutase. An enzymic function for erythrocuprein (hemocuprein), *J. Biol. Chem.* **244**:6049–6055.

McCord, J. M., and Fridovich, I., 1977, Superoxide dismutases: A history, in *Superoxide and Superoxide Dismutases* (A. M. Michelson, J. M. McCord, and I. Fridovich, eds.), pp. 1–10, Academic Press, New York.

Mellor, J. W., 1927, Part 10. The physical properties of chlorine, bromine, and iodine, in *The Halogens: A Comprehensive Treatise on Inorganic and Theoretical Chemistry,* Volume II, pp. 46–70, Longmans, Green, London.

Michaelis, L., 1946, Fundamentals of oxidation and respiration, *Am. Sci.* **34**:573–596.

Moureu, C., and Dufraisse, C., 1926, Catalysis and auto-oxidation. Anti-oxygenic and pro-oxygenic activity, *Chem. Rev.* **3**:113–162.

Ozorio de Almeida, A., 1934, Recherches sur l'action toxique des hautes pressions d'oxygène, *C.R. Soc. Biol.* **116**:1225–1227.

Partington, J. R., 1962, *A History of Chemistry*, Volume 3, St. Martin's Press, New York.

Pasteur, L., 1861, Animalcules infusoires vivant sans gas oxygène libre et déterminant des fermentations, *C.R. Acad. Sci.* **52**:344–347.

Pasteur, L., 1863, Recherches sur la putréfaction, *C R. Acad. Sci.* **56**:1189–1194.

Pocidalo, J., Braun-Pascaud, M., Blayo, M., and Azoulay-Dupuis, E., 1983, Respiratory effects of normobaric oxygen toxicity in awake guinea-pig, *Comp. Biochem. Physiol.* **74B**:831–836.

Poirier, J., 1993, *Antoine Laurent de Lavoisier. 1743–1794,* Paris: Pygmalion, Gérard Watelet.

Priestley, J., 1775, Experiments and observations on different kinds of air, Volume II, Sections III–V, in *Foundations of Anesthesiology*, Volume 1 (A. Faulconer and T. C. Keys, eds.), pp. 39–70, Thomas, Springfield, IL.

Pryor, W. A., 1966, *Free Radicals*, McGraw–Hill, New York.

Pryor, W. A., 1968, Organic free radicals, *Chem. Eng. News* **46**(Jan. 15):70–89.

Rahn, H., 1979, Wallace Osgood Fenn. August 27, 1893–September 20, 1971 in *Biographical Memoirs*, Volume 50 (National Academy of Sciences, ed.), pp. 140–173, National Academy of Sciences, Washington, DC.

Roberts, R. M., 1989, *Serendipity. Accidental Discoveries in Science*, p. 30, Wiley, New York.

Sanderson, R. T., 1967, *Inorganic Chemistry*, Reinhold, New York.

Schreck, R., and Baeuerle, P. A., 1991, A role for oxygen radicals as second messengers, *Trends Cell Biol.* **1**:39–42.

Seguin and Lavoisier, 1789 (read Nov. 17, 1791, published 1791), Premier mémoire sur la respiration des animaux [Also in *Mémoires sur la Respiration et la Transpiration des Animaux*, Paris, 1920, Gauthier-Villars, pp. 31–51, 185]. *Mém. Acad. Sci.* (in *Hist. Acad. Sci.* [A. Lavoisier, ed.], pp. 566–584).

Smith, J. L., 1899, The pathological effects due to increase of oxygen tension in the air breathed, *J. Physiol. (London)* **24**:19–35.

Sosenko, I. R. S., and Frank, L., 1987, Guinea pig lung development: Antioxidant enzymes and premature survival in high O_2, *Am. J. Physiol.* **252**:R693–R698.

Spinks, J. W. T., and Woods, R. J., 1990, *An Introduction to Radiation Chemistry*, 3rd ed., Wiley, New York.

Stadie, W. C., Riggs, B. C., and Haugaard, N., 1944, Oxygen poisoning, *Am. J. Med. Sci.* **207**:84–114.
von Sonntag, C., 1987, *The Chemical Basis of Radiation Biology*, Taylor & Francis, London.
Walling, C., 1957, *Free Radicals in Solution*, Wiley, New York.
Warburg, O., 1908, Beobachtungen über die Oxydationsprozesse im Seeigelei, *Z. Physiol. Chem.* **57**:1–16.
Weiss, J., 1944, Radiochemistry of aqueous solutions, *Nature* **153**:748–750.

Chapter 2

Chemistry of Reactive Oxygen Species

Robert E. Huie and P. Neta

1. INTRODUCTION

Aerobic organisms are constantly subjected to a variety of reactive entities derived from molecular oxygen, often collectively referred to as *reactive oxygen species* (ROS). These include the species derived from the reduction of molecular oxygen: superoxide/hydroperoxyl radicals ($\cdot O_2^-/HO_2\cdot$), hydrogen peroxide (H_2O_2), and the hydroxyl radical ($\cdot OH$); the species derived from the reaction of carbon-centered radicals with molecular oxygen: peroxyl radicals ($RO_2\cdot$), alkoxyl radicals ($RO\cdot$), and organic hydroperoxides (ROOH); and other oxidants that can result in free radical formation such as hypochlorous acid (HOCl), peroxynitrite ($ONOO^-$), and singlet oxygen ($O_2{}^1\Delta_g$). Sometimes, ozone (O_3), nitric oxide ($\cdot NO$), and nitrogen dioxide ($\cdot NO_2$) are also included because they are important exogenous radical sources and the latter two may be produced endogenously. There appear to be numerous possible sources of these reactive oxygen species, many of which are discussed elsewhere in this volume. Possibly the most prolific source of reactive oxygen species, particularly $\cdot O_2^-$ and H_2O_2, is leakage from the mitochondrial electron-transport chain (Halliwell and Gutteridge, 1989). Superoxide is also formed in various autoxidation reactions and by various enzymes, such as peroxidases, cytochrome P450, and xanthine oxidase (Halliwell and Cross, 1994). It is produced by phagocytes through the respiratory burst oxidase to initiate a defense mechanism (Babior, 1994; Allen, 1994).

A ubiquitous source of reactive radicals is ionizing radiation, which arises from both terrestrial and cosmic sources. The interaction of ionizing radiation with water leads to the production of several reactive species

---

Robert E. Huie and P. Neta Physical and Chemical Properties Division, National Institute of Standards and Technology, Gaithersburg, Maryland 20899.

Reactive Oxygen Species in Biological Systems, edited by Gilbert and Colton. Kluwer Academic / Plenum Publishers, New York, 1999.

$$H_2O \rightarrow \rightarrow \rightarrow e^-_{aq}, \cdot OH, \cdot H, H_3O^+, H_2, H_2O_2$$

with the following yields: 0.28 e^-_{aq} + 0.06 $\cdot H$ + 0.28 $\cdot OH$ + 0.045 H_2 + 0.08 H_2O_2 + 0.28 H_3O^+ in units of micromoles per joule, where the energy unit refers to the absorbed dose (Spinks and Woods, 1990). Reactions of these initial reactive species lead to other free radicals.

The physiological effects of reactive oxygen species are discussed in detail elsewhere in this volume. In this chapter, we will concentrate on some of the basic chemistry of these species—particularly the reactions involving free radicals. A large number of rate constants have been compiled for free radicals and other transient species, including $HO_2\cdot/\cdot O_2^-$ (Bielski *et al.*, 1985); hydrated electrons, hydrogen atoms, and hydroxyl radicals (Buxton *et al.*, 1988); several inorganic radicals (Neta *et al.*, 1988); organic peroxyl radicals (Neta *et al.*, 1990); transient metal ions and metal complexes (Buxton *et al.*, 1995); and aliphatic carbon-centered radicals (Neta *et al.*, 1996). In this review, an attempt will be made to generalize and point out reactivity patterns, rather than just list rate constants. To this end, reactions of the radicals with nonphysiological molecules may be mentioned, but the bulk of the discussion will be on the types of molecules found in the physiological milieu.

Free radicals, in general, react with each other and with molecules by several possible mechanisms. Radicals may combine to form a dimeric product or may disproportionate to form a reduced and an oxidized product, e.g.,

$$\cdot OH + \cdot OH \rightarrow H_2O_2$$

$$HO_2\cdot + HO_2\cdot \rightarrow H_2O_2 + O_2$$

The latter may occur by the transfer of an electron or of a hydrogen atom. Cross-reactions between different free radicals also can take place. Those involving relatively stable radicals such as NO are increasingly being recognized as important, e.g.,

$$\cdot NO + \cdot O_2^- \rightarrow ONOO^-$$

Reactions between an oxidizing and a reducing radical can be important, e.g.,

$$\cdot O_2^- + \cdot CO_3^- \rightarrow O_2 + CO_3^{2-}$$

Radicals react with molecules by hydrogen abstraction, addition, or electron transfer.

Some radicals in the singlet electronic state, the best known being the oxygen atom, O^1D, can also react by insertion into C–H bonds, but this reaction mode does not appear to be likely for physiologically important radicals. Most reactions of radicals with molecules lead to formation of product radicals, which will react further to yield stable products. These various reactions proceed with wide variations in the rate constants and have different dependences on temperature, solvent, and other system parameters. Despite the wide variations in reactivities, general patterns of behavior have been elucidated for various mechanisms and a large number of rate constants have been reported so as to permit a reasonable prediction of the behavior of many radical reactions.

Hydrogen-abstraction reactions are a specific case of atom transfer reactions. In these reactions, the bond between an atom and the rest of the molecule is broken and a new bond is formed between that atom and the radical:

$$M{-}X + \cdot R \rightarrow \cdot M + R{-}X$$

The rate constants of such reactions are related to their energetics, i.e., the difference between the bond dissociation energies for R–X and M–X. Other considerations, such as spin multiplicity, can have dramatic effects on the rate constants, but the bond-energy relationship holds well within a class. Bond energies have been reported for various C–H and O–H bonds, which can help in predicting the likelihood or the direction of a reaction. Such reactions do not involve strongly polar intermediates and, generally, have little solvent effect. For example, the reactivity of ·OH radicals toward many saturated compounds is the same in water and in the gas phase. Therefore, hydrogen-abstraction reactions may be expected to have similar rate constants in the aqueous phase and in the organic phase.

Addition reactions of a free radical include not only the addition of the radical to an unsaturated center in an organic compound, but also substitution into the coordination sphere of a transition metal ion or the formation of charge-transfer complexes. Of greatest interest in physiological chemistry is the addition of a radical to a carbon–carbon double bond, such as those found in fatty acids. These reactions often are in competition with hydrogen-abstraction reactions because both paths, in certain molecules, have similar rate constants (Alfassi *et al.*, 1997). In some cases, addition to the double bond can be followed by the abstraction of a nearby hydrogen atom. The addition of an oxidizing radical to an unsaturated bond tends to be electrophilic and is enhanced by electron donating substituents on that bond. Four alkyl groups situated on a double bond may increase its reactivity toward addition by several orders of magnitude (Alfassi *et al.*, 1993a). Solvent effects have been observed for addition reactions, but they are not large.

Electron transfer reactions have been studied widely and their behavior is generally predictable on the basis of the Marcus theory (Marcus and Sutin, 1985). The equations that have been developed relate the rate of a reaction with its driving force and with the molecular and solvent reorganization associated with that reaction. The driving force is the difference between the reduction potentials of the two species involved; this value may be known or is measurable independent of the system under study. Reorganization energies are less readily available and are often determined by comparison with other systems. For many systems, however, rate constants are predictable, to a first approximation, from the driving force and by comparison with known reactions of similar species. Electron-transfer reactions clearly involve polar and ionic species and thus are more strongly affected by solvent polarity than the other types of reactions.

The above discussion was meant to stress that certain dependences and trends are known for various types of reactions, which may be helpful for estimating the behavior of reactants in a complex system. More accurate predictions can be made from experimental measurements on specific reactions in isolated systems. One should take into account, however, that a rate constant for a reaction measured in a homogeneous solution is not necessarily equal to that in the biological system and that the effect of the medium may be dramatically different for different reactions. Specific examples will be discussed below.

2. FENTON CHEMISTRY

The centennial of the Fenton reaction was celebrated in 1993. Although the first brief report on the oxidation of tartaric acid by a mixture of ferrous iron and hydrogen peroxide appeared in 1876, the isolation of the reaction product was not reported until 1893 (Koppenol, 1993). This centennial elicited a number of reviews on this subject (Koppenol, 1993; Goldstein *et al.*, 1993; Stohs and Bagchi, 1995; Wardman and Candeias, 1996). Basic Fenton chemistry involves the oxidation of an organic substance by a mixture of Fe(II) and H_2O_2. In the initial study, ClO^- was used also as an oxidant and in subsequent investigations many different transition-metal ions and their complexes have been employed. The mechanism of this system is still a subject of active investigation (Bakac and Espenson, 1983; Koppenol, 1985; Moffett and Zika, 1987; Eberhardt and Colina, 1988; Masarwa *et al.*, 1988; Johnson *et al.*, 1988; Sawyer *et al.*, 1993, 1995; Wink *et al.*, 1994b; Kawanishi *et al.*, 1994; Shi *et al.*, 1994a; Gunther *et al.*, 1995; Pogozelski *et al.*, 1995; Buettner and Jurkiewicz, 1996; Held *et al.*, 1996; Biaglow *et al.*, 1996). The two basic competing mechanisms were proposed in the early 1930s (Koppenol, 1993) the first of which has the release of a free hydroxyl radical into the solution,

$$\mathrm{Fe(II)} + H_2O_2 \rightarrow \mathrm{Fe(III)} + OH^- + \cdot OH$$

and the second of which involves the formation of the iron(IV)-oxo ion,

$$\mathrm{Fe(II)} + H_2O_2 \rightarrow \mathrm{Fe^{IV}O} + H_2O$$

Hydrogen peroxide reacts preferentially with transition-metal ions by one-electron transfer (Bakac and Espenson, 1983). The rate constants for reactions with octahedral aquo ions correlate with their reduction potentials, with the exception of V^{2+}, which only slowly undergoes substitution into the coordination sphere. The reactions of substitutionally inert complexes are much slower, suggesting that the reactions with the aquo ions are inner sphere. This indicates that a third reactive species must be considered: the complex between the reduced metal ion and hydrogen peroxide, e.g.,

$$\mathrm{Fe(II)} + H_2O_2 \rightarrow FeO_2H^+ + H^+$$

This complex is then the precursor to the hydroxyl radical or the superoxidized metal ion.

Twenty years ago, with the accumulation of a sufficient database of hydroxyl radical kinetics, the similarity between the reactive species arising from Fenton chemistry and the hydroxyl radical appeared to settle the issue of the identity of this species in favor of $\cdot OH$ (Walling, 1975). Several recent studies also support the formation of free hydroxyl radicals with Fe^{2+} (Eberhardt and Colina, 1988), $\mathrm{Fe(EDTA)}^{2-}$ (Pogozelski *et al.*, 1995), Cu^+ (Gunther *et al.*, 1995), and both Cr(IV) and Cr(V) (Shi *et al.*, 1994a). A number of other studies, however, conclude that free $\cdot OH$ is not formed in the reactions of H_2O_2 with Fe^{2+} (Sawyer *et al.*, 1993; Wink *et al.*, 1994b), $\mathrm{Fe(EDTA)}^{2-}$ (Koppenol, 1985; Rush and Koppenol, 1986) and other polyaminocarboxylate complexes of iron (Rush and Koppenol, 1987), $Fe(DTPA)^-$ (Rahhal and Richter, 1988), or Cu^+ (Johnson *et al.*, 1988). Similarly, a reactive intermediate containing iron was observed in the reaction of a synthetic bleomycin with hydrogen peroxide (Guajardo and Mascharak, 1995).

More interesting are the observations that the oxidizing species may be somewhat variable. In one study, evidence was presented indicating that whereas $Cr^{2+} + H_2O_2$ always produced ·OH in the presence of > 0.1 M alcohol scavenger, $Cu^+ + H_2O_2$ formed a $Cu^{2+}{\cdot}HO_2^-$ complex that reacted with the alcohol (Masarwa *et al.*, 1988). In the absence of a scavenger, however, this complex also decomposed into Cu^{2+} + ·OH. Other work involving various iron complexes indicated that the extent of free ·OH formation depended on the nature of the chelating agent and on the H_2O_2 concentration (Yamazaki and Piette, 1991).

To further complicate this matter, the binding of ligands to metal ions in the body is probably highly variable. Indeed, recent work suggests that there are three distinct forms of iron associated with Fenton-induced DNA damage (Luo *et al.*, 1994). The distinctions were suggested to be related to the extent and nature of the binding to DNA.

An intriguing perturbation on Fenton chemistry has been suggested recently. Under appropriate conditions, Fe(II) complexes interact with H_2O_2 and HCl to bring about the chlorohydroxylation of alkenes (Sawyer *et al.*, 1995). The prevalence of Cl^- in the physiological milieu suggests that this chemistry should be investigated further.

3. THE HYDROXYL RADICAL

A large number of reactions of ·OH radicals in aqueous solutions have been studied. Indeed, rate constants have been reported for almost 1800 reactions (Buxton *et al.*, 1988; Ross *et al.*, 1997). The ·OH radical reacts by addition to unsaturated bonds, hydrogen abstraction, and electron transfer. The latter process may occur via addition, e.g., to metal ions. For the hydrogen-abstraction reactions, there is a linear correlation between the results obtained in the gas phase and those obtained in the aqueous phase (Wallington *et al.*, 1988). Polar compounds, such as ethers, alcohols, esters, acids, and nitriles, have similar reactivities in the two phases. Simple alkanes, however, react faster in the aqueous phase, by a factor of 20 for methane and less for the higher alkanes. On the basis of these findings, rate constants for H-abstraction by ·OH radicals measured in aqueous solutions are expected to be a good representation for rate constants in biological media. Addition reactions of ·OH radicals are very rapid, close to the diffusion-controlled limit for most double bonds and aromatic rings, and the rate constants in aqueous solutions should be similar to those in biological media.

The reduction potential for the couple $\cdot OH/OH^-$ is $E° = 1.90$ V (Stanbury, 1989). The derived potential for the couple $(H^+, \cdot OH)/H_2O$ is 2.72 V. Therefore, the hydroxyl radical can oxidize substrates with reduction potentials greater than 1.9 V by forming an addition complex that undergoes a subsequent reaction with a proton. This is of significant importance in understanding the role of ·OH in acidic media. This mechanism, however, is less important in biological systems. For example, ·OH can react with halide ions by electron transfer

$$\cdot OH + X^- \rightarrow OH^- + \cdot X$$

or by addition

$$\cdot OH + X^- \rightleftharpoons HOX^-$$

This latter reaction is particularly important for Cl^-, which has a reduction potential of 2.41 V (Stanbury, 1989). Subsequent reactions lead to the halogen atom and to the dihalide radical anion

$$HOX^- \rightleftharpoons OH^- + \cdot X$$

$$HOX^- + H^+ \rightarrow H_2O + \cdot X$$

$$HOX^- + X^- \rightarrow OH^- + \cdot X_2^-$$

$$\cdot X + X^- \rightleftharpoons \cdot X_2^-$$

Because the chlorine atom reacts rapidly with OH^- ($k = 1.8 \times 10^{10}$ liters mole^{-1} s^{-1}) and with H_2O ($k = 1.6 \times 10^5$ s^{-1}) (Klaning and Wolff, 1985), the rapid formation of the dichloride radical anion is restricted to acidic solutions and is unimportant in neutral media.

Hydroxyl radicals react very rapidly with most organic materials present in biological systems (Buxton *et al.*, 1988; Ross *et al.*, 1997). The rate constants for individual amino acids range from ~ 10^8 liters mole^{-1} s^{-1} for the aliphatic acids and approach ~ 10^{10} liters mole^{-1} s^{-1} for the aromatic and sulfur-containing acids. Rate constants for complete proteins were reported to be 10^{10} to 10^{11} liters mole^{-1} s^{-1} and are clearly diffusion controlled. All of the monomeric constituents of DNA react with OH radicals with rate constants between 10^9 and 10^{10} liters mole^{-1} s^{-1}, but when they are bound in the DNA, because of restrictions on diffusion, the rate constant per base unit is found to be between 4×10^8 and 8×10^8 liters mole^{-1} s^{-1}; the whole molecule reacting at the diffusion-controlled limit. The constituents of lipids also react very rapidly; a rate constant of 2×10^9 liters mole^{-1} s^{-1} has been measured for glycerol and values close to 10^{10} liters mole^{-1} s^{-1} were reported for unsaturated fatty acids. Glucose and other sugars react with rate constants on the order of 10^9 liters mole^{-1} s^{-1}. Antioxidants such as ascorbate and tocopherol react with rate constants close to 10^{10} liters mole^{-1} s^{-1}. Only urea, among bioorganic compounds, is known to react slowly. On the basis of the rate constants mentioned above, it has been argued that ·OH radicals react immediately at the site of their formation with little selectivity toward the various possible sites of attack (Samuni *et al.*, 1983).

4. HYDROPEROXYL AND SUPEROXIDE RADICALS

The hydroperoxyl radical, $HO_2\cdot$, is a weak acid, with a pK_a of 4.8 (Bielski *et al.*, 1985).

$$HO_2\cdot \rightleftharpoons H^+ + \cdot O_2^-$$

Rate constants for the reactions of both $HO_2\cdot$ and $\cdot O_2^-$ toward a large number of organic and inorganic reactants, and even some free radicals, have been reported (Bielski *et al.*, 1985; Ross *et al.*, 1997). Either radical can serve both as an oxidant

$$\cdot O_2^- + 2H^+ + e^- \rightarrow H_2O_2$$

$$\cdot HO_2 + H^+ + e^- \rightarrow H_2O_2$$

and as a reductant

$$O_2 + e^- \rightarrow \cdot O_2^-$$

$$O_2 + H^+ + e^- \rightarrow \cdot HO_2$$

Their reduction potentials, however, are quite different (Sawyer and Valentine, 1981). $\cdot O_2^-$ is a moderately strong reductant ($E°$ = –0.16 V at pH 7) but a weak oxidant ($E°$ = 0.89 V at pH 7), while $HO_2\cdot$ is a moderately strong oxidant ($E°$ = 1.44 V at pH 0) but a weak reductant ($E°$ = 0.12 V at pH 0), all potentials are given for 1 M O_2 standard state versus NHE [A recent publication reports a slightly higher $E°$ of –0.14 V for $\cdot O_2^-$ (Dohrmann and Bergmann, 1995).] The difference in reduction potentials between $HO_2\cdot$ and $\cdot O_2^-$ is manifested in the distinct chemical behavior of the radicals.

The reactions of $HO_2\cdot/\cdot O_2^-$ with bioorganic compounds are generally quite slow (Bielski *et al.*, 1985; Cabelli, 1997; Ross *et al.*, 1997). Indeed, the slowness of these reaction has led to the suggestion that the formation of superoxide acts as a sink for intracellularly generated radicals, with the superoxide ultimately removed by superoxide dismutase (Winterbourn, 1993). Only upper limits for the rate constants have been reported for the reactions of various amino acids with superoxide, although oxidation of cysteine at pH 7 had a measurable rate constant, on the order of 10^4 liters mole^{-1} s^{-1}. Unsaturated fatty acids react with $HO_2\cdot$ with rate constants around 10^3 liters mole^{-1} s^{-1}, but their reactions with $\cdot O_2^-$ are about five orders of magnitude slower. Only the antioxidants, ascorbate, tocopherol, and thiols react relatively rapidly. The rate constants are higher for $HO_2\cdot$ than for $\cdot O_2^-$ but the important values are those measured in neutral solutions for the mixture of radicals, 3×10^5 liters mole^{-1} s^{-1} for ascorbate (Cabelli and Bielski, 1983), 2×10^4 liters mole^{-1} s^{-1} for Trolox (Bielski, 1983), and $1.2 - 4.8 \times 10^5$ liters mole^{-1} s^{-1} for a series of thiols (Dikalov *et al.*, 1996).

Transition-metal ions may be reduced or oxidized by $HO_2\cdot/O_2^-$ (Bielski *et al.*, 1985). In some cases, transient complexes of the metal ions with $HO_2\cdot$ have been observed as intermediates in these reactions, with varying degrees of stability (Samuni and Czapski, 1970a,b; Samuni, 1972; Meisel *et al.*, 1974). Certain metal ions undergo both reduction and oxidation by $\cdot O_2^-$ of their oxidized and reduced states, respectively, resulting in superoxide dismutase-like activity. A number of metal complexes have been shown to exhibit such activity (Goldstein *et al.*, 1990; Bull *et al.*, 1983; Peretz *et al.*, 1982; Amar *et al.*, 1982; Butler and Halliwell, 1982; Solomon *et al.*, 1982; Ilan *et al.*, 1981; Pasternack and Halliwell, 1979; Pasternack and Skowronek, 1979; Brigelius *et al.*, 1974; Rabani *et al.*, 1973).

5. SINGLET OXYGEN

The first excited state of molecular oxygen, $O_2{}^1\Delta_g$, is located 94.3 kJ mole^{-1} above the ground state and has a radiative lifetime of 2.7×10^3 s; the next excited state, $O_2{}^1\Sigma_g^+$ has a lifetime of 7.1 s and is located at 156.9 kJ mole^{-1} (Kasha and Khan, 1970). This latter species quenches very rapidly and often to the $^1\Delta_g$ state (Ogryzlo, 1979), resulting in an estimated lifetime of 10^{-11} s in water, obviously too fast to allow this state to be important in physiology (Kearns, 1979). On the other hand, the lifetime of $O_2{}^1\Delta_g$ varies from 2 μs in water to 1,000 μs in $CFCl_3$ (Kearns, 1979), so that it has time to interact with other species in solution. A reduction potential for the couple $O_2{}^1\Delta_g/\cdot O_2^-$ has been calculated to be 0.83 V at 1 M O_2 standard state, assuming that the free energy of hydration is the same as for ground-state oxygen (Stanbury, 1989).

Singlet oxygen can be produced in a number of ways. A common method is by the photoexcitation of an organic molecule to its first excited singlet state, which can then undergo intersystem crossing to a lower triplet state. This triplet may then interact with ground state oxygen to produce singlet oxygen:

$$S + h\nu \rightarrow {}^1S$$

$$^1S \rightarrow {}^3S$$

$$^3S + O_2{}^3\Sigma_g^- \rightarrow S + O_2{}^1\Delta_g$$

This mechanism is likely to be important in photodynamic therapy and in photosensitivity.

There are several chemical sources of $O_2{}^1\Delta_g$ known, and probably many more that have not been established because of the weakness of the singlet oxygen emission. The earliest known and best established is the reaction of hydrogen peroxide with hypochlorite (Murray, 1979).

$$ClO^- + H_2O_2 \rightarrow Cl^- + H_2O + O_2{}^1\Delta_g$$

Labeling experiments have demonstrated that the oxygen is derived from the hydrogen peroxide, suggesting the intermediate formation of a chloroperoxide ion (Kasha and Khan, 1970). Similar to the hypochlorite/peroxide system, singlet oxygen formation has also been observed in the reaction of bromine with hydrogen peroxide (Howard and Ingold, 1968).

The disproportionation of $HO_2\cdot/\cdot O_2^-$ involves two doublet radicals, which, according to spin conservation rules, could result in the formation of singlet products (Allen, 1994),

$$HO_2\cdot + \cdot O_2^- + H^+ \rightarrow H_2O_2 + O_2{}^1\Delta_g$$

Several studies have sought evidence for such a pathway in this reaction (Barlow *et al.*, 1979; Foote *et al.*, 1980; Aubry *et al.*, 1981; Arudi *et al.*, 1984; Nagano and Fridovich, 1985; Kanofsky, 1986b). These studies have established clearly, however, that singlet oxygen production from the dismutation of superoxide is unimportant in aqueous solutions. The emission observed in halocarbon solutions containing superoxide probably results from the reductive dehalogenation of the halocarbon by $\cdot O_2^-$ and a subsequent self-reaction of the halocarbon peroxyl radical or reaction of superoxide with a halocarbon peroxyl radical (Kanofsky, 1986a).

Singlet oxygen production is also expected in the self-reactions of primary and secondary peroxyl radicals by the Russell mechanism

$$2 >CHOO\cdot \rightarrow >C{=}O + >CH(OH) + O_2{}^1\Delta_g$$

The production of $O_2{}^1\Delta_g$ was confirmed by chemical trapping (Howard and Ingold, 1968) and chemiluminescence (Kanofsky, 1986b) studies. The yields were low, which may be the result of the more facile formation of $O_2{}^1\Sigma_g^+$ in this reaction (Bogan *et al.*, 1984). Formation of $O_2{}^1\Delta_g$ was not observed from the self-reaction of *t*-butyl peroxyl radicals in the chemical trapping experiments, but a low yield was observed by chemiluminescence for this reaction and for the self-reaction of cumyl peroxyl radicals. This may arise from a reaction of the type

$$2(CH_3)_3COO \rightarrow (CH_3)_3COOC(CH_3)_3 + O_2{}^1\Delta_g$$

or, for cumyl peroxyl radicals, may involve fragmentation and subsequent reaction of methyl peroxyl radicals (Kanofsky, 1986b).

Several recent studies have proposed reactions producing singlet oxygen of possibly physiological importance. These reaction systems, however, may be considerably more complex than the postulated mechanisms. One such system is the Haber–Weiss reaction, which has been reported to produce singlet oxygen by spin-trap (Mao *et al.*, 1995) and chemiluminescence experiments (Khan and Kasha, 1994b). The initial step in the Haber-Weiss reaction

$$\cdot O_2^- + H_2O_2 \rightarrow O_2 + \cdot OH + OH^-$$

is known to be very slow, however, and the reaction is normally thought to proceed with the involvement of transition-metal ion catalysis. In a reaction similar to that of ClO^- with H_2O_2, $ONOO^-$ has been found to react with H_2O_2 to produce emission from $O_2{}^1\Delta_g$ (Di Mascio *et al.*, 1994). Another study indicates that the acidification of hypochlorite leads directly to singlet oxygen production (Khan and Kasha, 1994a). Chemiluminescence appears to commence at about the pK_a of HOCl and the yield was about 30% that of the ClO^-/H_2O_2 reaction. Acidification of peroxynitrite, ^-OONO, was also observed to generate $O_2{}^1\Delta_g$ chemiluminescence, but with 100% yield (Khan, 1995). Finally, the reaction of $\cdot NO$ with H_2O_2 has been proposed as a source of $O_2{}^1\Delta_g$ (Noronha-Dutra *et al.*, 1993). The reaction between these two species was investigated in the gas phase and an upper limit to the rate constant of 30 liters $mole^{-1}$ s^{-1} derived (Gray *et al.*, 1972). This suggests that the observed chemistry in the aqueous phase could involve other factors, such as transition-metal ion catalysis and, indeed, a second study did not observe emission from $O_2{}^1\Delta_g$ via this reaction (Di Mascio *et al.*, 1994).

The most common reactions of singlet oxygen are additions to alkenes and conjugated dienes (Schaap and Zaklika, 1979; Bartlett and Landis, 1979; Gollnick and Kuhn, 1979). With 1,3-dienes, there is a 1,4-cycloaddition reaction leading to peroxide formation, which can be viewed as a Diels–Alder reaction. 1,2-Cycloaddition across a double bond can result in dioxetane formation, which, in turn, may fragment to carbonyl products. Finally, addition to alkenes containing allylic hydrogen atoms can take place by an ene-type reaction, in which a hydrogen atom is transferred to the oxygen, leading to hydroperoxides.

Rate constants have been measured for the reactions of $O_2{}^1\Delta_g$ with alkenes and dienes both in the gas phase and in solution (Gollnick and Kuhn, 1979; Wilkinson and Brummer, 1981). In these studies, care had to be taken to separate physical quenching and chemical reaction. In methanolic solution, the rate constants increase from 5×10^3 liters mole^{-1} s^{-1} for cyclohexene, to 7×10^5 liters mole^{-1} s^{-1} for methylcyclohexene, to 7×10^6 liters mole^{-1} s^{-1} for 1,2-dimethylcyclohexene. The rate constants are about 20 times lower in the gas phase. Similar substituent effects are observed for linear alkenes. Among the fastest reactions of singlet oxygen are the cycloaddition reaction to 2,5-dimethylfuran, which has been reported to be $1–4 \times 10^8$ liters mole^{-1} s^{-1} in methanol and 1.5×10^7 liters mole^{-1} s^{-1} in the gas phase, the ene-type reaction with 2,3-dimethyl-2-butene, which is 4.4×10^7 liters mole^{-1} s^{-1} in methanol and 7.6×10^5 liters mole^{-1} s^{-1} in the gas phase, and the reaction with *N,N*-dimethylisobutenylamine, which is 4.9×10^8 liters mole^{-1} s^{-1} in solution (Schaap and Zaklika, 1979) and 1.5×10^7 liters mole^{-1} s^{-1} in the gas phase (Huie and Herron, 1973). This latter class of reactions with enamines has been proposed to occur by an electron transfer or charge transfer mechanism (Schaap and Zaklika, 1979).

Of considerable importance are the reactions of singlet oxygen with fatty acids. Rate constants for the reactive interaction of $O_2{}^1\Delta_g$ with methyl oleate, methyl linoleate, and ethyl linolenate of 2.4, 3.8, and 5.2×10^4 liters mole^{-1} s^{-1} were recently obtained in $CHCl_3$, calibrating the quantum yield of oxygen consumption with 2,3-dimethyl-2-butene (Tanielian and Mechin, 1994). The rate constants for physical quenching were all about 1×10^4 liters mole^{-1} s^{-1}.

Only recently have rate constants for reactions of $O_2{}^1\Delta_g$ in aqueous solutions begun to be measured. The 1,4-cycloaddition reaction with dimethylfuran has a rate constant of 8.2×10^8 liters mole^{-1} s^{-1} and the ene-type reaction with 2,4-dimethyl-2-butene is 1.1×10^8 liters mole^{-1} s^{-1}, both about three times greater than the results in methanol (Scully and Hoigné, 1987). Phenols show increasing reactivity with increasing pH, with the rate constant for the protonated phenol at least two orders of magnitude lower than the deprotonated form (Tratnyek and Hoigné, 1991). Good correlations were found between the rate constants and the half-wave potentials of the phenols, with the exception of those with substituents that are good through-resonance acceptors. Correlations were also tried with substituent constants and the best fits found with the use of σ^-.

The effect of solvent on the physical and chemical interaction of $O_2{}^1\Delta_g$ with 1,4-dimethylnaphthalene and derivatives was investigated recently (Aubury *et al.*, 1995). Physical quenching, k_q, and chemical reaction, k_r, rate constants were determined with 28 different solvents. The values increased by two orders of magnitude from cyclohexane to formamide. Physical quenching was always faster than chemical reaction and the two were found to be closely correlated, and the intercept of a plot of k_r against k_q led to a value of 0.29 for the ratio $k_r/(k_r + k_q)$. These results were interpreted to indicate that physical quenching and chemical reaction proceed through a common intermediate.

Singlet oxygen is quenched to the triplet state by various physiological species. The most efficient mechanism is probably an energy-transfer reaction with an organic singlet to produce an organic triplet

$$O_2{}^1\Delta_g + S \rightarrow O_2{}^3\Sigma_g^- + {}^3S$$

This, of course, is essentially the reverse of the key singlet oxygen formation reaction, but by a molecule with a lower triplet excitation energy. An important class of molecules that quench singlet oxygen very rapidly are the carotenoids, e.g., β-carotene and crocetin. The carotenoids quench singlet oxygen with rate constants on the order of 10^{10} liters mole^{-1} s^{-1}, near the diffusion limit. The rate constants for chemical reaction are generally less than 10^7 liters mole^{-1} s^{-1}, except in water where the rate constants are $>10^8$ liters mole^{-1} s^{-1} (Lissi *et al.*, 1993). This mechanism for quenching of singlet oxygen also takes place for sensitisers where the exothermicity of the formation reaction is not too large. An important example is bilirubin, which quenches with a rate constant of $1–2 \times 10^9$ liters mole^{-1} s^{-1}, depending on solvent, but also reacts with a rate constant of $1–4 \times 10^8$ liters mole^{-1} s^{-1} (Lissi *et al.*, 1993).

Another major class of quenching agents are the phenolic compounds, particularly α-tocopherol and its derivatives. The quenching rate constant for α-tocopherol increases with solvent polarity from about 3 to 70×10^7 liters mole^{-1} s^{-1} (Lissi *et al.*, 1993). The rate constant for chemical reaction increases also, from about 1 to 46×10^6 liters mole^{-1} s^{-1}. The quenching rate constants for these phenols and for tocopheramines have been found to correlate well with their reduction potentials (Scurlock *et al.*, 1989; Mukai *et al.*, 1991; Itoh *et al.*, 1994).

6. ORGANIC PEROXYL RADICALS

In addition to the hydroperoxyl radical discussed above, organic peroxyl radicals are of importance in biological systems, being produced either from natural components or from external contaminants. The mechanism of toxicity of CCl_4 and other polyhalogenated compounds is related to the high reactivity of the peroxyl radicals produced from these compounds (Brault *et al.*, 1985; Cheeseman *et al.*, 1985). Following attack by these peroxyl radicals or by other radicals on biological targets, additional radicals may be produced that may also form peroxyl radicals. The abstraction of a hydrogen atom from an organic compound or the addition of a radical to a double bond leads to the formation of a carbon-centered radical, which, in the presence of O_2, forms a peroxyl radical

$$\cdot R + O_2 \rightarrow RO_2\cdot$$

Rate constants for the reactions of carbon-centered radicals with oxygen are typically $>10^9$ liters mole^{-1} s^{-1} and irreversible (Neta *et al.*, 1990). In a few cases, however, the reaction is reversible at room temperature. An important example of this is the hydroxycyclohexadienyl peroxyl radical, produced in the reaction of ·OH with benzene in the presence of O_2 (Pan and von Sonntag, 1990). The forward rate constant for this reaction is relatively slow, 3.1×10^8 liters mole^{-1} s^{-1}, and the reverse reaction was found to have a rate constant of 1.2×10^4 s^{-1}.

Organic peroxyl radicals play an important role in physiological chemistry, particularly in the damage induced by ionizing radiation, possibly in aging, and in ischemia. As a result, there have been a number of studies of their reactions, with a particular emphasis on molecules of biological importance, and the kinetics (Howard and Scaiano, 1984; Neta *et al.*, 1990) and mechanisms (von Sonntag and Schuchmann, 1991) have been reviewed recently (Alfassi, 1997). Under physiological conditions, a number of different types of reactions of peroxyl radicals are known to be important, including electron transfer, addition to double bonds, and hydrogen abstraction.

Peroxyl radicals react with various compounds via different mechanisms. With saturated aliphatic compounds they may react by hydrogen abstraction,

$$RO_2\cdot + R'H \rightarrow RO_2H + \cdot R'$$

This reaction plays a major role when the compound contains weak C–H bonds or other weakly bonded hydrogens, such as those in thiols:

$$RO_2\cdot + RSH \rightarrow RO_2H + \cdot RS$$

With unsaturated compounds the reaction may be via addition to the unsaturated bond,

$$RO_2\cdot + {>}C{=}C{<} \rightarrow RO_2{-}\overset{|}{\underset{|}{C}}{-}C\cdot{<}$$

or by abstraction from allylic or bis-allylic C–H bonds. With compounds that are readily oxidized the reaction is generally via electron transfer, e.g., oxidation of phenolate ions to phenoxyl radicals.

$$RO_2^- + PhO\cdot \rightarrow RO_2\cdot + PhO^-$$

In the following sections we discuss the results for various mechanisms of peroxyl radical reactions.

6.1. Unimolecular Decomposition

An extensive review of the unimolecular reactions of peroxyl radicals is included in a recent article (von Sonntag and Schuchmann, 1991), and only a few of the findings will be summarized here. In the aqueous phase, a number of peroxyl radicals are able to undergo unimolecular decomposition to yield $\cdot O_2^-$. The rate constant for the reaction and the specific nature of the process is strongly dependent on the substituents. The concerted loss of a proton accelerates the reaction. The decomposition of α-amino- and α-hydroxyalkylperoxyl radicals and hydroxycyclohexadienylperoxyl radicals have been assumed to be internal abstraction reactions involving five-membered transition states. If weak C–H bonds are located β to the peroxyl carbon, internal abstraction reactions with six-membered transition states are possible. There is also evidence that the peroxyl group can add intramolecularly to double bonds.

6.2. Radical–Radical Reactions

The self-reactions of $RO_2\cdot$ also have been the subject of a number of investigations. There is a wide variation in the reactivity of primary, secondary, and tertiary peroxyl radicals toward self-reaction. Primary peroxyl radicals react very rapidly, near the diffusion limit; rate constants for these reactions have been measured primarily at room temperature and in aqueous solutions or the gas phase. Secondary peroxyl radicals react somewhat slower and tertiary peroxyl radicals react very slowly. Rate constants have been measured for a large number of secondary and tertiary radicals over a wide temperature range, particularly in non-aqueous solutions (Howard and Scaiano, 1984).

The temperature dependence observed in nonaqueous solvents for the second-order loss of secondary peroxyl radicals, when measured over a wide temperature range, often seems quite complex. This is because at low temperatures, these radicals exist in equilibrium with their dimer, the tetroxides. The temperature dependence of this equilibrium leads to the measured temperature dependence of the rate constant. As the temperature is raised, a disproportionation reaction becomes more important. For example, for the 2-propylperoxyl radical (Bennett, 1987; Bennett *et al.*, 1987a,b):

$$2(CH_3)_2CHOO\cdot \rightarrow (CH_3)_2CHOH + (CH_3)_2CO + O_2$$

This reaction has a relatively low temperature dependence. As the temperature is raised further, formation of alkoxyl radicals increases in importance

$$2(CH_3)_2CHOO\cdot \rightarrow 2(CH_3)_2CHO\cdot + O_2$$

This reaction has a higher temperature dependence than the disproportionation reaction. At intermediate temperatures, product analysis suggests that a third mechanism is also operative:

$$2(CH_3)_2CHOO\cdot \rightarrow 2(CH_3)_2CO + H_2O_2$$

The first two reaction pathways are similar for the reactions of primary and secondary peroxyl radicals both in the gas phase and in nonaqueous solutions; only the second pathway is possible for tertiary peroxyl radicals. This results in the much slower rate constants observed for the self-reactions of these radicals and their higher temperature dependence.

The reaction path leading to the formation of two carbonyl products and hydrogen peroxide probably occurs via a transition state containing a pair of five-membered rings. In water, a much less strained transition state with six-membered rings is possible. Indeed, this mechanism is very important for reactions in aqueous solutions; in the reaction of the hydroxymethylperoxyl radical, it is the major path, yielding hydrogen peroxide and formic acid (Bothe and Schulte-Frohlinde, 1978).

6.3. Abstraction of Hydrogen Atoms

The hydrogen-abstraction reactions of alkyl peroxyl radicals are slow and highly selective. For abstraction from saturated alkanes, the rate constants are typically less than 1 liter $mole^{-1}$ s^{-1} (Howard and Scaiano, 1984). The abstraction rates for allylic hydrogens are faster, with a rate constant of less than 10^4 liters $mole^{-1}$ s^{-1} for abstraction of the bis-allylic hydrogen in linoleic acid. Although these reactions are relatively slow, they constitute the chain-propagation steps in the autoxidation of unsaturated organic compounds (Porter, 1986; Porter *et al.*, 1994). The increased reactivity of bis-allylic C–H bonds reflects their much lower bond strengths (~372 kJ $mole^{-1}$) relative to primary, secondary, or tertiary C–H bonds (422, 413, and 402 kJ $mole^{-1}$, respectively) (Tsang, 1996). The bond strength for the O–H bond in *t*-butyl hydroperoxide was determined to be 359 kJ mole^{-1} (Tsang, 1996) and, from this value and a comparison of the activation energies for abstraction by secondary and tertiary peroxyl radicals, the bond strength in secondary hydroperoxides was estimated to be 365 kJ mole^{-1} (Denisov and Denisova, 1993). Halogenation at the α-carbon increases the reactivity of peroxyl radicals toward hydrogen abstraction enormously. By considering the increased reactivity of $CCl_3CCl_2COO\cdot$ and $CHCl_2CCl_2COO\cdot$ toward n-C_8H_{18}, the O–H bond dissociation energy of chlorinated hydroperoxides was estimated to be 407 kJ mole^{-1} (Denisov and Denisova, 1993). Apparent cellular lipid peroxidation increases exponentially with the number of bis-allylic C–H bonds (Wagner *et al.*, 1994), in contrast to the behavior in homogeneous solutions of polyunsaturated fatty acids, where a linear increase in autoxidation rate is observed (Cosgrove *et al.*, 1987). This difference may reflect the highly ordered arrangement of lipids within cell membranes.

Rate constants for hydrogen abstraction by $CCl_3O_2\cdot$ radicals were measured (Mosseri *et al.*, 1987) by competition kinetics. The rate constant for cyclohexane was found to be 1.0×10^3 liters mole^{-1} s^{-1}, whereas that for cyclohexene was found to be 1.0×10^5 liters mole^{-1} s^{-1}. Because the rate constant for the two CH_2 groups that are remote from the double bond is very low, the main contribution must be the abstraction of the activated allylic hydrogens of the other two CH_2 groups and/or the addition to the double bond. The same ambiguity exists in the reactions with fatty acids, where both addition to double bonds and abstraction of allylic and doubly allylic hydrogen atoms may contribute to the overall reactivity. Rate constants for reactions of $CCl_3O_2\cdot$ with oleic, linoleic, and linolenic acids were measured by two groups (Forni *et al.*, 1983; Brault *et al.*, 1985) under somewhat different conditions. The absolute rate constants reported vary by a factor of 5–10, possibly reflecting differences in solvent and pH. The order of reactivity, however, was oleic < linoleic < linolenic in both studies. The increasing reactivity may be ascribed to the increasing number of allylic and doubly allylic hydrogens and to the increasing number of double bonds; thus, the reaction may involve H-abstraction and/or addition.

Other hydrogens that may undergo abstraction readily are phenolic and thiolic. In these cases, however, the same products may be generated by a different mechanism, namely oxidation via electron transfer and subsequent protonation. It is not always clear which mechanism prevails. It has been suggested that reactions of peroxyl radicals that are weak oxidants, in low-polarity solvents, may occur by H-abstraction whereas strongly oxidizing radicals, particularly in more polar solvents, react by electron transfer (see below).

6.4. Addition to Double Bonds

The addition of alkylperoxyl radicals to carbon–carbon double bonds is also generally very slow and strongly dependent on the substituents about the double bond (Howard and Scaiano, 1984). Often, both addition and abstraction reactions can take place. For example, the rate constant at 393 K for the addition of $(CH_3)_3CO_2\cdot$ to 2,3-dimethyl-2-butene was reported to be 22 liters mole^{-1} s^{-1} in benzyl chloride and the abstraction rate constant was reported to be 28 liters mole^{-1} s^{-1} (Koelewijn, 1972). Rate constants for reactions of several peroxyl radicals with a series of unsaturated hydrocarbons and alcohols were determined by competition kinetics. For compounds of low reactivity, it was difficult to distinguish between H-abstraction and addition. Without product analysis, it is frequently difficult to ascertain if a reported rate constant refers to an addition or an abstraction reaction. For more complex compounds, however, as the reactivity increased, it became clear that this increase was ascribable to more reactive double bonds rather than to an increase in the number or the reactivity of abstractable hydrogen atoms.

Reactions of halogenated peroxyl radicals with alkenes are much faster than the corresponding reactions of alkylperoxyl radicals. The rate constants for a number of such reactions have been reported (Nahor and Neta, 1991; Alfassi *et al.*, 1993a; Shoute *et al.*, 1994). For each radical, the rate constants for the different alkenes varied by about two orders of magnitude. These variations do not follow the pattern expected for hydrogen abstraction, i.e., diallyl > allyl > alkyl and, within each type, tertiary > secondary > primary (March, 1985), and do not increase with the number of reactive hydrogens. Therefore, hydrogen abstraction is ruled out as a major route for these reactions. The rate constant for the addition reaction should depend on the substituents about the double bond. In fact, a good correlation was found between the measured rate constants and Taft's σ^* substituent constants (Taft, 1956; Weiberg, 1964), which reflect the inductive effect of the substituents. The negative slope found suggests that the mechanism involves an electrophilic addition to the double bond.

The reactivity of the alkylperoxyl radicals is also affected by the substituents attached to the alkyl group, mainly those at the α-carbon. The trihaloalkylperoxyl radicals exhibit increased reactivity with increasing electron affinity of the halogen, i.e., $CBr_3O_2\cdot$ < $CCl_3O_2\cdot$ << $CF_3O_2\cdot$.

6.5. Electron-Transfer Reactions

Over the past decade, there have been a large number of rate constants reported for the one-electron oxidation of substrates by peroxyl radicals. This has come about because of an increasing awareness of the role of peroxyl radicals in physiological processes, including aging, radiation-induced damage, and ischemia. Because of the involvement of $CCl_3O_2\cdot$ in the toxicity of CCl_4 (Brault, 1985; Cheeseman *et al.*, 1985), much of the work has focused on this radical, but there has also been a significant amount of interest in the behavior of other peroxyl radicals.

The rate constants for the electron-transfer reactions of peroxyl radicals depend strongly on the substituents on the radical. Rate constants increase greatly as the electron-withdrawing ability of the substituents increases. A study (Neta *et al.*, 1989b) of the reactions of various substituted methylperoxyl radicals with ascorbate ions found that log k correlated well with the polar substituent constant, σ^*. For each reductant, the points did not all fit on the same line but it appeared that the behavior of halogenated peroxyl radicals is somewhat different than for the nonhalogenated peroxyl radicals, with the former being somewhat more reactive but having a lower increase in reactivity with σ^* than the latter.

The electron-transfer reactions of halogenated peroxyl radicals are strongly solvent dependent (Alfassi *et al.*, 1987, 1993b; Neta *et al.*, 1989a). The behavior of these reactions in mixed water–alcohol solvents is somewhat complex, with maxima in the rate constants often observed at an alcohol mole fraction of 0.1–0.2. The temperature dependence of some reactions of $CCl_3O_2\cdot$ has been determined in some water–alcohol mixtures (Alfassi *et al.*, 1992). Both the Arrhenius preexponential factor and the activation energy increase as the proportion of water in the mixture increases, suggesting that the effect of solvent polarity on the rate constant is the result of two compensating effects. When the reactions of $CCl_3O_2\cdot$ with chlorpromazine and Trolox were studied in a range of different solvents (Alfassi *et al.*, 1993b), log k was found to correlate better with the Hildebrand solubility parameter, or cohesive energy density (Hildebrand *et al.*, 1970; Barton, 1983), than with any other single parameter. For Trolox, which is a phenolic compound, the removal of a proton from the compound may also be involved. In this case, the correlation was improved considerably by inclusion of a term to take into account the basicity of the solvent. The above results and other findings indicate that electron transfer to the peroxyl radical is concerted with transfer of a proton to the incipient hydroperoxide anion, and that the solvent assists in this process.

It has not yet been possible to make direct determinations of the reduction potentials of the peroxyl radicals. It is, of course, clear that electron-withdrawing substituents increase the reduction potential and that electron-donating substituents decrease it, so that there is some information about the relative reduction potentials. From the kinetic behavior of $CH_3O_2\cdot$ in aqueous solution, a potential of 0.6–0.7 V at pH 7 was estimated, as compared with an estimated potential for $CCl_3O_2\cdot$ of >1 V (Huie *et al.*, 1986). In more recent studies (Merenyi *et al.*,1994; Jonsson, 1996), reduction potentials have been estimated for a number of peroxyl radicals, varying from 0.4 to 1.3 V versus NHE.

Although most oxidation reactions by peroxyl radicals are via one-electron transfer mechanisms, several cases have been suggested to involve the transfer of more than one electron in a single reaction step. This was particularly observed for the oxidation of I^- and of organic sulfides. For example, although most of the reaction of dimethyl sulfide with $CCl_3O_2\cdot$ appeared to take place by electron transfer to form ultimately the dimer radical cation, about 25% of the reaction led to the formation of the sulfoxide (Schöneich *et al.*, 1991):

$$CCl_3O_2\cdot + Me_2S \rightarrow (CCl_3O{-}O{-}SMe_2)\cdot$$

$$(CCl_3O{-}O{-}SMe_2)\cdot + Me_2S \rightarrow CCl_3OO^- + (Me_2S\therefore SMe_2)^+$$

$$(CCl_3O{-}O{-}SMe_2)\cdot \rightarrow CCl_3O\cdot + OSMe_2$$

The relative contributions of the two paths depend on the electron affinity of the peroxyl radical.

The reaction of $CCl_3O_2\cdot$ with I^- also was shown to involve multielectron transfer (Schöneich *et al.*, 1991). Pulse conductometric and spectrophotometric measurements indicated that the reaction leads to simultaneous formation of I_3^- and I_2^-. This was interpreted in terms of a concerted two-electron transfer to form I_2 and an alkoxyl radical, the latter oxidizes another I^- to an I atom, and finally the I_2 and the I atom each bind another I^- ion to give the more stable complexes observed:

$$CCl_3O_2\cdot + I^- + H_2O \rightleftharpoons [CCl_3O{-}O(H){-}I]\cdot + OH^-$$

$$[CCl_3O{-}O(H){-}I]\cdot + I^- \rightarrow CCl_3O\cdot + OH^- + I_2$$

$$CCl_3O\cdot + I^- \rightarrow CCl_3O^- + I\cdot$$

$$\cdot I + I^- \rightleftharpoons \cdot I_2^-$$

$$I_2 + I^- \rightleftharpoons I_3^-$$

6.6. The Fate of Organic Peroxyl Radicals

From the above discussion it is clear that various organic peroxyl radicals exhibit moderate or low reactivities toward a number of biological components and may have the capacity to diffuse and attack certain targets selectively. In considering these attacks we can divide these radicals into two groups: the very reactive halogenated radicals formed from exogenous materials (e.g., anesthetics, pollutants) and the less reactive peroxyl radicals derived from biological targets (e.g., lipids). Radicals of the first group generally are formed following reductive dehalogenation of polyhalogenated compounds by cytochrome P450. They are strong oxidants capable of attacking the heme of cytochrome P450. Alternatively, they may diffuse and attack unsaturated fatty acids to initiate the lipid-peroxidation chain reaction. The latter reaction is not very rapid and may be readily prevented by antioxidants, such as ascorbate and tocopherol, which are oxidized by halogenated peroxyl radicals quite rapidly (to yield less reactive oxygen species). Peroxyl radicals of the second group, those not containing halogens or other strongly electron withdrawing substituents, are much less reactive and are capable of diffusing and attacking more selectively. Their reactions with antioxidants are relatively slow but their reactions with other biomolecules are even slower and thus they may decay by radical–radical reactions. For such radicals, however, the reactivity may not be the most important parameter that determines their fate; rather, their diffusion may be restricted by their large molecular weight or their position within a membrane structure so that they may react locally even if the rate of reaction is very low. This is the case in the propagation step of lipid peroxidation. After the peroxyl radicals react by any of the above mechanisms, the possibility still exists for producing the more reactive alkoxyl radicals, either by radical–radical decay or following reduction of the hydroperoxide formed.

7. ALKOXYL AND AROXYL RADICALS

Alkoxyl radicals are formed on reduction of dialkyl peroxides or alkyl hydroperoxides, the latter being the products of some peroxyl radical reactions. Fenton-type reactions with alkyl hydroperoxides form alkoxyl radicals (Kochi, 1962; Gilbert *et al.*, 1976). Alkoxyl radicals may react with organic molecules via H-abstraction, addition, or oxidation (Howard, 1984; Erben-Russ *et al.*, 1987). Most of these reactions are more rapid than those of alkylperoxyl radicals. In particular, H-abstraction from unsaturated fatty acids is about three orders of magnitude more rapid with alkoxyl than with alkylperoxyl; this is related to the fact that the RO–H bond is considerably stronger than the ROO–H bond (Tsang, 1996). However, alkoxyl radicals also undergo rapid monomolecular rearrangements or decompositions (Gilbert *et al.*, 1976, 1977, 1981; Paul *et al.*, 1978; Baignee *et al.*, 1983), which often compete with or practically prevent their reactions with other molecules. Primary and secondary alkoxyl radicals undergo 1,2- or 1,5-hydrogen shifts, converting the alkoxyl into 1-hydroxy or 4-hydroxyalkyl radicals. Tertiary alkoxyl radicals undergo β-scission to yield a ketone and an alkyl radical. These various alkyl radicals, in turn, will react with oxygen to yield peroxyl radicals.

Aroxyl radicals include those formed from various antioxidants and are generally less reactive than alkoxyl or peroxyl radicals; that is the essence of their antioxidant action. Lacking other targets, aroxyl radicals undergo radical–radical reactions to form dimeric products, generally biphenols (Ye and Schuler, 1989), which also act as antioxidants. Aroxyl radicals, however, are capable of oxidizing a "better antioxidant." For example, the aroxyl radical derived from tocopherol can oxidize ascorbate (Packer *et al.*, 1979), resulting in repair of tocopherol and in ascorbate serving as the ultimate antioxidant. A recent study (Böhm *et al.*, 1997) has shown that the tocopherol radical can also oxidize various carotenoids and that the carotenoid radical cations formed in such a process can, in turn, oxidize ascorbate. This may help explain the role of the carotenoids in the body, and why many studies appear to give contradictory results as to the efficacy of these species as antioxidants (Liebler *et al.*, 1997).

8. REACTIVE SPECIES INVOLVING NITRIC OXIDE

8.1. The Autoxidation of Nitric Oxide

Until recently, there had been only one study of the autoxidation of nitric oxide in water (Pogrebnaya *et al.*, 1975) and one in carbon tetrachloride (Nottingham and Sutter, 1986). The discovery that NO is an important mediator of a variety of physiological processes brought forth a burst of activity on the kinetics and mechanism of this reaction in water. These studies have shown clearly that the overall reaction, in the absence of any additives, is

$$4\cdot NO + O_2 + 2H_2O \rightarrow 4HNO_2$$

The kinetics of the reaction has been investigated by a number of methods, including flow studies with conductivity (Pogrebnaya *et al.*, 1975), stopped-flow spectrophotometry (Eguchi *et al.*, 1989; Wink *et al.*, 1993a; Awad and Stanbury, 1993; Pires *et al.*, 1994; Goldstein and Czapski, 1995b), stopped-flow spectrophotometry with a phenol red indicator to monitor H^+ formation (Kharitonov *et al.*, 1994), and a stirred reactor in which $\cdot NO$ and NO_2^- could be monitored simultaneously (Lewis and Deen, 1994).

The reaction kinetics have been found to be second order in $\cdot NO$ and first order in O_2, with the rate law

$$-\frac{d[\cdot NO]}{dt} = d\frac{[NO_2^-]}{dt} = 4k[\cdot NO]^2[O_2]$$

One study did report that the reaction is zero order in $\cdot NO$ and first order in O_2, but a nonstandard kinetic analysis of questionable validity was employed (Taha *et al.*, 1992).

All of the kinetic results are basically in agreement (Figure 1). The most thorough study of the kinetics was that of Awad and Stanbury (1993), and the solid line in Figure 1 is a fit to their data over the temperature range 10–40°C, and corresponds to $k = (2.2 \pm$

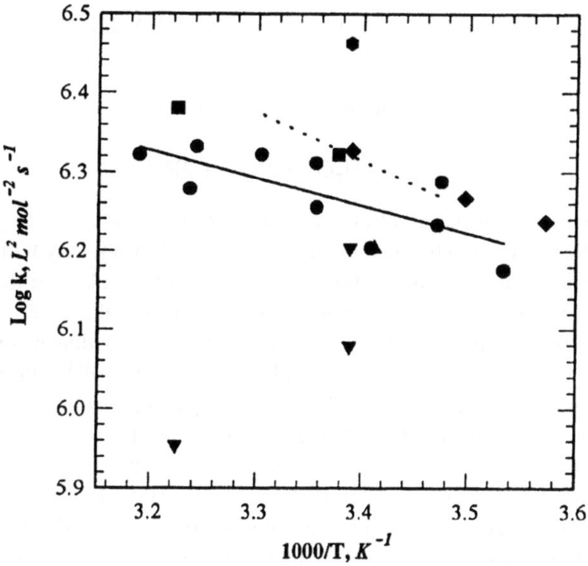

FIGURE 1. Arrhenius plot of the data on the reaction $4{\cdot}NO + O_2 = 4HNO_2$. ●, Awad and Stanbury (1993); ■, Lewis and Deen (1994); ▲, Kharitonov *et al.* (1994); ▼, Wink *et al.* (1993a); ♦ Pires *et al.* (1994); ⬢, Goldstein and Czapski (1995b); - - - - , Pogrebnaya *et al.* (1975); ——, fit to the data of Awad and Stanbury (1993).

1.8) $\times 10^7$ exp($-740 \pm 250/T$) liters^2 mole^{-2} s^{-1}. The results from most other studies essentially scatter around these data and probably are a realistic illustration of the uncertainty associated with the measurement of this rate constant.

There has been considerable interest in the autoxidation of ·NO in the gas phase, as it is probably the best known reaction that is nominally termolecular. Recent theoretical calculations have clarified the details of the reaction (McKee, 1995), supporting a mechanism in which a complex of ·NO and O_2 is first formed and then reacts with a second ·NO in the rate-determining step:

$$\cdot NO + O_2 \rightleftharpoons NO{\cdot}O_2$$

$$NO{\cdot}O_2 + \cdot NO \rightarrow ONOONO$$

$$ONOONO \rightarrow 2{\cdot}NO_2$$

Assuming a steady state in both $NO{\cdot}O_2$ and ONOONO, this mechanism results in a rate law proportional to the O_2 concentration and the square of the ·NO concentration, as observed in both the gas and liquid phases. In the aqueous-phase studies, N_2O_4 (O_2NNO_2), which is the dimer of $\cdot NO_2$, is not formed in the reaction either by rearrangement of ONOONO or by dimerization of $\cdot NO_2$. Otherwise, NO_3^- produced in the hydrolysis of N_2O_4, would be a major product, not NO_2^- as observed (Pogrebnaya *et al.*, 1975; Awad and Stanbury, 1993; Ignarro *et al.*, 1993; Lewis and Deen, 1994). At the high concentration of ·NO in a normal kinetic experiment, the formation of $\cdot NO_2$ is followed immediately by a second reaction with ·NO:

$$\cdot NO + \cdot NO_2 \rightleftharpoons N_2O_3$$

$$N_2O_3 + H_2O \rightarrow 2NO_2^- + 2H^+$$

In this mechanism, the formation of nitrite is irreversible, in agreement with experiment (Pires *et al.*, 1994). Although this mechanism is supported by analogy with the gas phase and by the kinetic results, a spin trapping experiment did not observe the formation of $\cdot NO_2$ during the autoxidation of ·NO in aqueous solution (Reszka *et al.*, 1994).

There have been a number of studies that have found evidence for an oxidizing intermediate in the autoxidation of NO (Wink *et al.*, 1993a, 1994a; Pires *et al.*, 1994; Lewis *et al.*, 1995; Goldstein and Czapski, 1995b; Kharitonov *et al.*, 1995; Hogg *et al.*, 1996). There are three potential oxidizing intermediates formed in the autoxidation scheme: ONOONO, $\cdot NO_2$, and N_2O_3. Generally, the results have appeared to support N_2O_3 as the active agent (Pires *et al.*, 1994; Lewis *et al.*, 1995; Kharitonov *et al.*, 1995). Pires *et al.* (1994) suggest that there may be two isomers of N_2O_3: $ONNO_2$ and ONONO, the latter of which nitrosates more rapidly than the former. This suggestion is based on the differences observed in nitrosation reactions by gaseous N_2O_3 dissolved into water compared with that formed in solution from HNO_2. Wink and co-workers (Wink *et al.*, 1993a, 1994a) have observed, however, that their results are not adequately explained by N_2O_3 or $\cdot NO_2$ alone and Goldstein and Czapski (1995b) have suggested that all of the intermediates, ONOONO, $\cdot NO_2$, and N_2O_3, may be important, depending on the oxidizable substrate and its concentration. Clearly, more information is needed on the behavior of this complicated system and more independent data are needed on the chemistry of the key intermediates, particularly N_2O_3.

8.2. The Reaction of Nitric Oxide with Superoxide

One of the most important reactions of nitric oxide is with superoxide, $\cdot O_2^-/HO_2\cdot$, which yields peroxynitrite, $OONO^-$ (Blough and Zafiriou, 1985):

$$\cdot NO + \cdot O_2^- \rightarrow OONO^-$$

An early study in which a mixture of ·NO, O_2, and formate was pulse irradiated found a rate constant that was clearly too low (Saran *et al.*, 1990). Three subsequent studies, all of which produced ·NO *in situ*, have shown that the reaction is quite fast. In two of these studies (Kobayashi *et al.*, 1995; Goldstein and Czapski, 1995c), employing pulse radiolysis, ·NO was produced by the reaction of the hydrated electron with NO_2^- :

$$e_{aq}^- + NO_2^- \rightarrow NO_2^{2-}$$

$$NO_2^{2-} + H_2O \rightarrow \cdot NO + 2OH^-$$

From the pulse, e_{aq}^- and ·OH are formed in essentially equal concentrations. In these experiments, the ·OH was allowed to react with formate, leading to $\cdot O_2^-$:

$$\cdot OH + HCO_2^- \rightarrow OH^- + \cdot CO_2^-$$

$$\cdot CO_2^- + O_2 \rightarrow CO_2 + \cdot O_2^-$$

and e_{aq}^- and any ·H produced can also react with O_2:

$$e_{aq}^- + O_2 \rightarrow \cdot O_2^-$$

The relative concentrations of ·NO and $\cdot O_2^-$ are then set by the relative concentrations of NO_2^- and O_2, and the absolute concentrations are set by the pulse intensity. These two studies were carried out under first-order conditions with $\cdot O_2^-$ in excess and resulted in rate constants of 3.8×10^9 liters $mole^{-1}$ s^{-1} (Kobayashi *et al.*, 1995) and 4.3×10^9 liters $mole^{-1}$ s^{-1} (Goldstein and Czapski, 1995c).

In the third study, laser-flash photolysis of NO_2^- resulted in the production of equal concentrations of ·OH and ·NO (Huie and Padmaja, 1993)

$$NO_2^- + h\nu \rightarrow \cdot NO + O^-$$

$$O^- + H^+ \rightarrow OH$$

Formate was used to convert the OH to $\cdot O_2^-$, as above. Because equal concentrations of ·NO and $\cdot O_2^-$ were produced, second-order kinetics was followed and the derived k/ε converted to an absolute rate constant by using the known absorption coefficients of $\cdot O_2^-$ and $OONO^-$. In this study, a rate constant of 6.7×10^9 liters $mole^{-1}$ s^{-1}, 65% higher than the average of the two pulse radiolysis studies, was obtained.

8.3. Reactions of Organic Peroxyl Radicals with Nitric Oxide

The fastest reaction of an organic peroxyl radical is likely to be its reaction with nitric oxide. The reactions of $RO_2\cdot$ with ·NO are fast, with rate constants of $1–3 \times 10^9$ liters $mole^{-1}$ s^{-1} for the peroxyl radicals derived from MeOH, 2-PrOH, and *t*-BuOH (Padmaja and Huie, 1993). These reactions lead to the formation of an intermediate species that absorbs light at around 300 nm, similar to the peroxynitrite species, ^-OONO, formed in the reaction of $\cdot O_2^-$ with ·NO:

$$RO_2\cdot + \cdot NO \rightarrow ROONO$$

This intermediate decays with a rate constant of 0.1–0.3 s^{-1} in water and about 200 times faster in alcoholic solutions. Although rate constants have not been reported for the reaction of any other peroxyl radicals with ·NO in solution, a large number of these reactions have been investigated in the gas phase (Wallington *et al.*, 1992). The reactions are all fast and, typically, involve oxygen-atom transfer, probably through a peroxynitrite intermediate:

$$RO_2\cdot + \cdot NO \rightarrow RO\cdot + \cdot NO_2$$

In the liquid phase, the organic peroxynitrite may also decompose to radical products, in which case ·NO will act as a potent prooxidant, or it may rearrange to an organic nitrate,

$$ROONO \rightarrow RONO_2$$

and ·NO would be an antioxidant. This behavior might be a function of solvent, as is the decomposition rate. In the chemical system in which the rates of formation and decay of the organic peroxynitrite were investigated (Padmaja and Huie, 1993), it was not possible to determine the mechanism of the decay of the organic peroxynitrite, nor have any other direct determinations in the liquid phase been reported. There have been, however, a few reports of the impact of ·NO on peroxidation reactions, which have shed some light on this reaction.

Once it was demonstrated that peroxynitrite is formed from ·NO, most of the emphasis has been on oxidation initiated by this species. Various studies, however, suggested that ·NO plays a protective role in cells. In an experiment in which a liposome suspension was subjected to a constant dose of $\cdot O_2^-$, low rates of infusion of ·NO enhanced lipid peroxidation, whereas excess ·NO resulted in inhibition (Rubbo *et al.*, 1994). Analysis of the products showed various NO_3-containing oxidation adducts. In other experiments, it was shown that infusion of ·NO into lung fibroblasts provided protection from the cytotoxicity of hypoxanthine/xanthine oxidase, hydrogen peroxide, or *t*-butyl hydroperoxide (Wink *et al.*, 1993b, 1995). In an extension of this work, employing several ·NO donor compounds and similar compounds that do not release ·NO, it was demonstrated that free ·NO was required as the protective agent in hydrogen peroxide-mediated cytotoxicity (Wink *et al.*, 1996). Nitric oxide was also found to inhibit the peroxynitrite-induced oxidation of phosphatidylcholine liposomes (Laskey and Mathews, 1996). Well-characterized oxidation markers were used to quantify the extent of peroxidation.

These various studies all point to a reaction of ·NO with the chain-propagating peroxyl radicals to terminate the autoxidation. This suggests that, in lipids at least, ROONO rearranges to $RONO_2$ and does not decompose to RO· and $\cdot NO_2$.

8.4. Peroxynitrite

The reaction between ·NO and $\cdot O_2^-$ results in the formation of the peroxynitrite anion, $OONO^-$ (Blough and Zafiriou, 1985). This species can be produced in several different ways and its properties have been investigated for many years (Edwards and Plumb, 1994). Because of the interest in its physiological behavior, there have been numerous studies recently on its chemical reactions. The peroxynitrite anion absorbs light in the near UV with a maximum absorption at 302 nm and an absorptivity of 1670 liters $mole^{-1}$ cm^{-1} (Hughes and Nicklin, 1968). The absorption spectrum of the acid, ONOOH, is blueshifted and has a maximum absorptivity of 770 liters $mole^{-1}$ cm^{-1} at 240 nm (Løgager and Sehested, 1993). There have been several determinations of the pK_a of ONOOH, both by absorption measurements and from the effect of pH on the rate of decomposition. Generally, we expect pK_a values based on kinetic measurements to be less certain because of possible effects of buffers or catalytic impurities, so we recommend the recent value of $pK_a = 6.5 \pm 0.1$ based on optical absorption following the pulse radiolysis of nitrite solutions (Løgager and Sehested, 1993).

Solutions of peroxynitrite decay with a rate constant that relates to the hydrogen ion concentration in a manner suggesting that only the acid form undergoes reaction

$$k = \frac{k_{HA}[H^+]}{[H^+] + K_a}$$

where k_{HA} is the first-order rate constant for the decay of the acid and K_a is the acid dissociation constant (Koppenol *et al.*, 1992; Løgager and Sehested, 1993). A rate constant of $k_{HA} = 1.2 \pm 0.2\ s^{-1}$ would include most of the values that have been reported at 298 K (Yang *et al.*, 1992; Koppenol *et al.*, 1992; Huie and Padmaja, 1993; Løgager and Sehested, 1993). The temperature dependence of the reaction indicates an activation energy of 77.4 kJ $mole^{-1}$ (Koppenol *et al.*, 1992). The apparent pK_a from the kinetic studies does not change significantly from 5 to 37°C.

The mechanism of the decay of peroxynitrous acid has been considerably more contentious. In acid solutions, there is quantitative production of nitrite (Plumb *et al.*, 1992). Several groups reported evidence that the mechanism involves homolytic decomposition (Mahoney, 1970; Beckman *et al.*, 1990; King *et al.*, 1992; Hogg *et al.*, 1992; Yang *et al.*, 1992).

$$ONOOH \rightarrow \cdot NO_2 + \cdot OH$$

This interpretation was challenged, however, and it was suggested that a reactive isomer is formed during the decomposition of ONOOH (Koppenol *et al.*, 1992). This conclusion has been supported by several recent publications (Crow *et al.*, 1994; Shi *et al.*, 1994b; Lemercier *et al.*, 1995; Goldstein and Czapski, 1995a; Padmaja *et al.*, 1996). The reactive intermediate apparently is derived from the *trans* conformer rather than the lower-energy *cis* conformer (McGrath and Rowland, 1994; Tsai *et al.*, 1994; Krauss, 1994).

Peroxynitrite can react with organic and inorganic compounds by a number of mechanisms, including one- and two-electron transfer oxidations, oxygen-atom transfer, and electrophilic nitrations. Iodide reacts with peroxynitrite with a rate constant proportional to $[H^+]$, indicating that ONOOH is the reactive species (Hughes *et al.*, 1971; Goldstein and Czapski, 1995a). The derived rate constant for the reaction of I^- with ONOOH is about 2.3×10^5 liters $mole^{-1}\ s^{-1}$ (Goldstein and Czapski, 1995a), which is about 100 times larger than the corresponding rate constants for HSO_5^- or for peroxycarboxylic acids (Edwards and Plumb, 1994). Cyanide and thiocyanate react with peroxynitrite in an alkaline solution with a rate constant that is independent of pH, indicating that the reaction is with $ONOO^-$ (Hughes *et al.*, 1971). $ONOO^-$ oxidizes cysteine to cystine with a rate constant of 5.9×10^3 liters $mole^{-1}\ s^{-1}$ at pH 7.4, which is 1000 times greater than the corresponding reactions of H_2O_2 (Radi *et al.*, 1991b). Ferrocyanide, $Fe(CN)_6^{4-}$, is oxidized by peroxynitrite with a rate independent of the concentration of $Fe(CN)_6^{4-}$ and equal to the rate constant for decomposition of $ONOO^-$ (Goldstein and Czapski, 1995a). Thiyl radicals have been observed in the reactions of cysteine and other thiols (glutathione and penicillamine) with peroxynitrite (Shi *et al.*, 1994b). Methionine is oxidized by peroxynitrite by both one- and two-electron reactions (Pryor *et al.*, 1994). The results were interpreted to suggest that the reaction took place through the steady-state formation of an activated peroxynitrous acid. The ascorbate anion reacts with ONOOH with a rate constant of 2.4×10^2 liters $mole^{-1}\ s^{-1}$ (Bartlett *et al.*, 1995; Squadrito *et al.*, 1995), producing ascorbyl radicals (Shi *et al.*, 1994b; Bartlett *et al.*, 1995). It was suggested that ascorbyl radical formation resulted from an electron-transfer reaction of an excited

ONOOH species, whereas ground-state ONOOH reacted by a two-electron, S_N2 displacement reaction (Squadrito *et al.*, 1995). Peroxynitrite reacts with α-tocopherol in methanol or acetonitrile, and with Trolox in water, by what appears to be two-electron oxidation without the formation of significant tocopherol radical (Hogg *et al.*, 1994). On the other hand, the nitration of tyrosine residues appears to take place with the intermediate formation of tyrosyl radicals (van der Vliet *et al.*, 1995). The kinetics of the reaction of peroxynitrite with the phenolic compounds phenol, tyrosine, and salicylate were first order in peroxynitrite but zero order in the phenolic compound, with the total rate constant approximately the same as the rate constant for the isomerization of peroxynitrite to nitrate (Ramezanian *et al.*, 1996). Both hydroxyphenolic and nitrophenolic products are formed, but the yields of these products are affected differently by pH and ·OH scavengers. L-Tryptophan reacts with $k = 1.3 \times 10^2$ liters mole^{-1} s^{-1} with a maximum rate constant at pH 5.1 (Padmaja *et al.*, 1996). Hydroxytryptophans were not observed, but at least two nitrotryptophans were formed.

The formation of peroxynitrite was also observed to promote superoxide-induced luminol chemiluminescence (Castro *et al.*, 1996). In this case, nitric oxide played a dual role: promoting the chemiluminescent reaction via formation of peroxynitrite and promoting a dark reaction by intercepting the luminol radical.

The decomposition of $ONOO^-$ in an alkaline solution is much faster in the presence of a transition-metal catalyst (Plumb *et al.*, 1992). In the presence of 2×10^{-4} mole liter^{-1} Cu(II), the rate of peroxynitrite decomposition over the pH range 11 to 13 was increased by a factor of 1000 over the rate predicted from the uncatalyzed rates at lower pH. Of particular importance is the observation that the product of the catalyzed reaction is NO_2^-, not NO_3^-. Peroxynitrite also reacts rapidly with the porphyrin (5,10,15,20)–tetrakis(N-methyl-4′-pyridyl)porphinato manganese(III) to produce an oxoMn(IV)TMPyP species (Groves and Marla, 1995). This species was found capable of oxidatively cleaving plasmid DNA. Fe(III)EDTA appears to efficiently catalyze the nitration of the phenols p-hydroxyphenylacetic acid, glycyl-tyrosine, phenol, and *p*-cresol, but superoxide dismutase was a relatively inefficient catalyst (Beckman *et al.*, 1992). The difference in behavior was ascribed to the ability of Fe(III)EDTA to react with the *cis*-peroxynitrite, whereas superoxide dismutase required the transformation of *cis*- to *trans*-peroxynitrite prior to reaction. Subsequent work showed that while the nitration of phenols was enhanced in the presence of Fe(III) complexes, the hydroxylation yield was decreased (Ramezanian *et al.*, 1996). Further, Cu(II) complexes were less efficient than the Fe(III) complexes. Contrary to these observations, iron did not appear to strongly influence peroxynitrite-mediated lipid peroxidation (Radi *et al.*, 1991a).

The study of the chemistry of peroxynitrite was jolted recently with evidence that its reactions may be mediated by CO_2. It had previously been shown that bicarbonate greatly enhanced the luminol chemiluminescence induced by reaction with peroxynitrite (Radi *et al.*, 1993). Then it was demonstrated that the decay of peroxynitrite was accelerated by carbonate in a manner that indicated either an interaction between CO_2 and $ONOO^-$ or between HCO_3^- and ONOOH. By taking advantage of the slow hydration–dehydration transformation of CO_2/H_2CO_3, it was possible to show that CO_2 was the important reactant (Lymar and Hurst, 1995). The rate constant for the reaction was determined to be 2.9×10^4 liters mole^{-1} s^{-1}, which is comparable to the faster reactions of peroxynitrite with organic reductants. The enthalpy of activation for the reaction was determined to be

$\Delta H^{\#} = 45$ kJ mole^{-1}, which is similar to the temperature dependences reported for several other reactions of peroxynitrite, suggesting a common mechanism (Denicola *et al.*, 1996).

In subsequent work, the reaction of this adduct with tyrosine was investigated (Lymar *et al.*, 1996). It was shown that the adduct was more reactive than $ONOO^-$, with $k \gg 2 \times 10^5$ liters mole^{-1} s^{-1}, and produced 3-nitrotyrosine and 3,3′-dinitrotyrosine as the major products. The results were consistent with the reaction mechanism

$$CO_2 + ONOO^- \rightarrow ONO_2CO_2^-$$

$$ONO_2CO_2^- + \text{tyrosine} \rightarrow \text{Tyr}\cdot + \cdot NO_2 + HCO_3^-$$

$$\text{Tyr}\cdot + \cdot NO_2 \rightarrow \text{nitrotyrosine}$$

$$\text{Tyr}\cdot + \text{Tyr}\cdot \rightarrow \text{dityrosine}$$

$$\cdot NO_2 + \text{tyrosine} \rightarrow NO_2^- + \text{Tyr}\cdot + H^+$$

$$\cdot NO_2 + \cdot NO_2 + H_2O \rightarrow NO_2^- + NO_3^- + 2H^+$$

Alternately, it was suggested that $ONO_2CO_2^-$ directly attacks the *ortho* position of tyrosine, forming nitrotyrosine (Gow *et al.*, 1996).

Other researchers explored the reaction of CO_2 with $ONOO^-$ with experiments designed to determine the products of the reaction, to establish the effect of hydration–dehydration of carbonate by use of carbonic anhydrase, and also to prevent catalysis by trace metals by the addition of DTPA (Uppu *et al.*, 1996). These results supported the earlier conclusion and established that the product of the decomposition was NO_3^-, not NO_2^- as might be expected by metal ion catalysis, nor were hydrogen peroxide or other hydroperoxidic products formed. It was shown that the adduct formed in the reaction of CO_2 and $ONOO^-$ was highly reactive toward phenol, 4-hydroxyphenylacetic acid, and ABTS. Based on these studies, Uppu *et al.* (1996), suggested that the initial $O{=}N{-}OOCO_2^-$ rearranges to $O_2N{-}OCO_2^-$, which is the oxidant in these systems. Subsequent theoretical calculations, however, indicated that this intermediate was unstable toward decomposition to $\cdot NO_2$ and $\cdot CO_3^-$ (Houk *et al.*, 1996).

These various studies suggest that many of the kinetic studies on peroxynitrite could have been influenced by the presence of trace amounts of CO_2, particularly because peroxynitrite is normally prepared in alkaline solutions that almost always are contaminated with carbonate. The decomposition rate constant for peroxynitrite recommended above would still be valid, however, because the peroxynitrite in that experiment was not from an alkaline solution.

Although the arguments that there is an important interaction between $ONOO^-$ and CO_2 are very convincing, a similar interaction between CO_2 and H_2O_2 was proposed some time ago (Navarro *et al.*, 1984). This was criticized in a subsequent publication (Csányi and Galbács, 1985) where the observed effect was explained to be the result of trace metal catalysis. The CO_2 brought about the disaggregation of the oxohydroxo metal complexes and the formation of carbonate complexes. Although the experiments on the effect of CO_2 on $ONOO^-$ chemistry were designed to minimize the influence of metal-ion catalysis, the fact that metal ions can have such a strong catalytic effect and that they can behave synergistically, suggests caution in interpreting the results on the interaction of CO_2 with $ONOO^-$. On the other hand, theoretical calculations indicate that the reaction of CO_2 with ^-OONO is 108 kJ mole^{-1} exothermic and that there is no barrier to the formation of $ONO_2CO_2^-$ (Houk *et al.*, 1996).

9. NITROGEN DIOXIDE

Nitrogen dioxide was thought of previously only as an exogenous free radical, but the discovery of the importance of nitric oxide as a key intermediate in many physiological processes has increased the possibility that $\cdot NO_2$ might be formed endogenously. Nitrogen dioxide is a free radical capable of reacting via H-atom abstraction, addition to unsaturated bonds, oxygen-atom transfer, addition to free radicals, and electron transfer. Nitrogen dioxide reacts rapidly with other free radicals, reasonably fast with one-electron reductants, but reacts much more slowly by addition or H-abstraction. Quantitative kinetic information on these latter reactions has proven difficult to obtain in water where the lifetime of $\cdot NO_2$ is short as a result of disproportionation reactions.

There are few rate data on $\cdot NO_2$ H-abstraction reactions in solution. Under normal conditions, *n*-alkanes fail to react, even after several months, in the presence of $\cdot NO_2/N_2O_4$; branched-chain alkanes, however, do react in a reasonable time (Titov, 1963). The reaction of $\cdot NO_2$ with toluene has an activation energy of 120–140 kJ mole^{-1} and the nitration of toluene by $\cdot NO_2$ goes to completion. The reaction of $\cdot NO_2$ with diphenylmethane has an activation energy of 71–92 kJ mole^{-1} and diphenylmethane is nitrated in 2 days at room temperature, whereas triphenylmethane has an activation energy of 21–42 kJ mole^{-1} and is nitrated quickly by $\cdot NO_2$ upon warming (Titov, 1963). Phenols also can be nitrated by a reaction that is initiated by the abstraction of the phenolic hydrogen in nonaqueous solvents.

$\cdot NO_2$ adds to the double bond of alkenes in the liquid phase (Titov, 1963). Rate constants of $\sim 2 \times 10^5$ liters mole^{-1} s^{-1} at pH 9.5 have been reported for the reaction of $\cdot NO_2$ with linoleate and $\sim 10^6$ liters mole^{-1} s^{-1} for arachidonate at pH 9.0 (Prütz *et al.*, 1985). In another study (Forni *et al.*, 1986), however, the rate constant for the reaction of $\cdot NO_2$ with linoleate at pH 6.5 was reported to be $< 5 \times 10^4$ liters mole^{-1} s^{-1}. It was also noted that linoleate did not appear to inhibit $\cdot NO_2$-induced tyrosine nitration, suggesting the rate constant is not as high as the pulse radiolysis measurements indicated (Prütz *et al.*, 1985).

There has been a thorough study of the mechanism of the reaction of $\cdot NO_2$ with alkenes and polyunsaturated fatty acids (Pryor *et al.*, 1982). The results were interpreted to suggest that two concurrent reaction mechanisms are operative, the first involving addition of $\cdot NO_2$ to the double bond and the second involving abstraction of the allylic hydrogen by $\cdot NO_2$. A third possible mechanism would have $\cdot NO_2$ first add to the double bond and then, via a five- or six-member ring, abstract the allylic hydrogen. In the presence of an adequate amount of $\cdot NO_2$ or O_2, the adduct radical can be scavenged to form a dinitro compound or a nitro peroxyl radical. Without sufficient scavenger, the nitro-alkyl radical dissociates to the initial reactants, or undergoes internal H-abstraction to form nitrous acid and an alkyl radical. The dissociation of the nitro-alkyl radical can lead to geometric isomerization of the alkene (Sprung *et al.*, 1974), which could have potential consequences for membrane structure.

Nitrogen dioxide is a moderately strong one-electron oxidant with a reduction potential for the $\cdot NO_2/NO_2^-$ couple of 1.04 V (Stanbury, 1989). There have been many direct determinations of the rate constants for the one-electron oxidation of organic, and a few inorganic, compounds by $\cdot NO_2$, primarily by pulse radiolysis (Neta *et al.*, 1988). $\cdot NO_2$ reacts with substituted phenolate ions with a rate constant that increases as the Hammett substituent constant, σ, becomes more negative (Alfassi *et al.*, 1986). The reaction of $\cdot NO_2$ with neutral phenols appears to be too slow to be measured by pulse radiolysis. The temperature dependence for the reactions of $\cdot NO_2$ with phenolate, *p*-methoxyphenolate, and ascorbate have also been reported (Alfassi *et al.*, 1990). Activation energies of 34, 26, and 36 kJ $mole^{-1}$, respectively, were determined for these reactions, values somewhat higher than generally derived for the other radicals.

Nitrogen dioxide reacts very rapidly with most other free radicals. A rate constant of 4×10^9 liters $mole^{-1}$ s^{-1} has been reported for the reaction of $HO_2\cdot$ with $\cdot NO_2$ (Sutton, 1975). This result was supported by subsequent pulse-radiolysis experiments in which a rate constant of 0.014 s^{-1} for the decomposition of the peroxynitrate product was derived (Sutton *et al.*, 1978). A different study, however, suggested a rate constant of ~10^7 liters $mole^{-1}$ s^{-1} (Warneck and Wurzinger, 1988). The rate constant for the electron-transfer reaction of $\cdot NO_2$ with $\cdot O_2^-$ is also reported in that work as ~10^8 liters $mole^{-1}$ s^{-1}. The higher rate constant is clearly more consistent with the gas-phase reaction, for which the recommended rate constant is 2.8×10^9 liters $mole^{-1}$ s^{-1} (Atkinson *et al.*, 1992).

$\cdot NO_2$ reacts rapidly ($k = 1 \times 10^9$ liters $mole^{-1}$ s^{-1}) with the carbon-centered radicals derived from methanol and 2-propanol (Elliot and Simsons, 1984), in which the radical center is located on the same carbon atom as the hydroxyl group, to yield nitrate and the corresponding aldehyde or ketone:

$$\cdot NO_2 + \cdot CH_2OH \rightarrow NO_2^- + CH_2O + H^+$$

The reaction of $\cdot NO_2$ with the radical derived from 2-methyl-2-propanol, in which the radical center is located away from the hydroxyl group, is not observed to produce nitrite ($k < 5 \times 10^4$ liters $mole^{-1}$ s^{-1}) but might take place by simple addition.

$$\cdot NO_2 + \cdot R \rightarrow RONO$$

10. HYPOCHLOROUS ACID

Hypochlorous acid is produced by the reaction of chloride ion with hydrogen peroxide, catalyzed by haloperoxidases such as myeloperoxidase or eosinophil peroxidase (Allen, 1994):

$$Cl^- + H_2O_2 \rightarrow HOCl + OH^-$$

Hypochlorous acid is a weak acid, with a $pK_a = 7.40$ at 25°C and 7.30 at 35°C (Adam *et al.*, 1992). Therefore, it will exist as both the anion and the acid at physiological pH. HOCl is in equilibrium with Cl_2O

$$2HOCl \rightleftharpoons Cl_2O{\cdot}H_2O$$

but the equilibrium constant has been estimated to be only 0.030 liters mole^{-1} at 50°C. It is also in equilibrium with Cl_2

$$HOCl + Cl^- + H^+ \rightleftharpoons Cl_2 + H_2O$$

with k_f = 1.6 × 10^5 liters^2 mole^{-2} s^{-1} and k_r = 12 s^{-1} at 50°C (Adam *et al.*, 1992). The forward reaction is too slow to be important under physiological conditions.

The two-electron reduction potential for ClO^-/Cl^-, $2OH^-$ is 0.890 V in basic solution (Kuhn and Rice, 1985); at pH 7, the potential (HClO, H^+/Cl^- is 1.07 V (Koppenol and Butler, 1985). For comparison, hydrogen peroxide has a two-electron reduction potential of 1.74 V at pH 0 and 1.32 V at pH 7; the potential for the couple HO_2^-/OH^- is 0.77 V at pH 14 (Koppenol and Butler, 1985). Therefore, hypochlorite is a relatively mild oxidant. Calculated one-electron reduction potentials for HOCl are very low, suggesting that this species is not likely to react by this mechanism (Koppenol and Butler, 1985).

Hypochlorite can react with hydrogen peroxide to produce singlet oxygen. The rate constant for the reaction reaches a maximum of 3 × 10^3 liters mole^{-1} s^{-1} at about pH 10, suggesting that either the specific reaction

$$HClO + HO_2^- \rightarrow Cl^- + H_2O + O_2\,{}^1\Delta_g$$

or

$$ClO^- + H_2O_2 \rightarrow Cl^- + H_2O + O_2\,{}^1\Delta_g$$

takes place (Held *et al.*, 1978). From the ionization constants, rate constants of 4.4 × 10^7 or 3.4 × 10^3 liters mole^{-1} s^{-1}, respectively, can be calculated for these two reactions. It has been argued that the most likely path is the first reaction, the nucleophilic attack of HO_2^- on the electrophilic chlorine of HOCl (Held *et al.*, 1978). Because the rate of production of HClO in the body depends on the H_2O_2 concentration, the reaction of HOCl with H_2O_2 is likely to limit the accumulation of hypochlorite.

Hypochlorite has been widely employed for the syntheses of halogenated organic compounds (Chakrabartty, 1978; Fontana *et al.*, 1989). Oxidation reactions can also occur, often concurrently. HOCl adds to double bonds to yield chlorohydrins and the formation of these species has been confirmed for unsaturated fatty acids (Winterbourn *et al.*, 1992). The formation of cholesterol chlorohydrins has been observed in membrane lipids, even from whole cells treated with HOCl (Carr *et al.*, 1996).

Chemiluminescence, with a maximum emission ~580 nm, was observed in the reaction of hypochlorous acid with tryptophan (Fornier de Violet *et al.*, 1984). The reaction is complex, with a square dependence of the chemiluminescence intensity on the square of the HOCl concentration and a clear threshold effect of the tryptophan concentration. It was suggested that this reaction is the cause of the weak luminescence observed during phagocytosis.

Relative rate constants have been measured for the reactions of hypochlorous acid with a series of biological compounds (Winterbourn, 1985). The most reactive compounds were cysteine, glutathione, and methionine followed by ascorbic acid, which was about 20 times less reactive (at pH 7.3). A further four to five times less reactive were taurine, NADH, histidine, and uric acid. Slower was serine and even slower was leucine. Dimethylsulfoxide was also investigated and found to be far less reactive than the other compounds. More recently, absolute rate constants have been measured by stopped-flow spectrophotometry for several HOCl reactions in two different laboratories (Folkes *et al.*, 1995; Prütz, 1996). The two most important water-soluble antioxidants, ascorbate and glutathione, react very rapidly with HOCl, with $k > 10^6$ liters mole^{-1} s^{-1}. Other thiols and amines also react rapidly. Several of the compounds exhibited slower secondary reactions when HOCl was in excess (Prütz, 1996). NADH also reacts very rapidly with HOCl, $k > 2 \times 10^5$ liters mole^{-1} s^{-1}, and there is a slower secondary reaction of the product with $k = 3.5 \times 10^3$ liters mole^{-1} s^{-1} (Prütz, 1996). This, apparently, is not a one-electron oxidation of NADH, but an addition to the nicotinamide group.

There was no evidence of the formation of free radicals in the rapid reactions of HOCl with amino or thiol groups (Folkes *et al.*, 1995). These reactions are probably electrophilic substitution reactions of the types

$$\text{R–NH}_2 + \text{HOCl} \rightarrow \text{R–NHCl} + \text{H}_2\text{O}$$

$$\text{GSH} + \text{HOCl} \rightarrow \text{GSCl} + \text{H}_2\text{O}$$

Some of the reactants, however, did form products capable of inducing the oxidation of $Fe(CN)_6^{4-}$, but the mechanism is unclear (Prütz, 1996).

The rate constants for the reactions of hypochlorous acid solutions with a number of amines increased strongly with pH, reached a maximum, and then decreased (Antelo *et al.*, 1995b). This behavior is consistent with either the reaction of the acid with the neutral amine as above, or the reaction of the anion with the protonated amine.

$$\text{R–NH}_3^+ + \text{ClO}^- \rightarrow \text{R–NHCl} + \text{H}_2\text{O}$$

These studies, and additional studies on several substituted butylamines (Antelo *et al.*, 1995a), indicated that the rate constant correlates with the Taft substituent constant, σ^*. This was interpreted to suggest a relationship between the rate constant and the pK_a of the amine.

Ferrous ion at pH 4 and a ferrous citrate complex at pH 5.0 reacted with similar rate constants of 1.7 and 1.3×10^4 liters mole^{-1} s^{-1} (Folkes *et al.*, 1995). In the reactions of HOCl with Fe^{2+}, a reactive species was formed that was capable of hydroxylating benzoate. The intermediate was not identified, but it was probably not ·OH or ·Cl.

The reaction of hypochlorous acid with the substitution-inert complex $Fe(CN)_6^{4-}$ was investigated with HOCl in excess. The decreasing rate constant with increasing pH from 5 to 9 was consistent with HOCl being the more reactive oxidant, with a rate constant of 2.2×10^2 liters mole^{-1} s^{-1} (Candeias *et al.*, 1994). An intermediate was formed in this reaction that hydroxylated benzoate, giving the same product distribution as ·OH, but with a yield of only 27%. In another study at pH 6.9, but with $Fe(CN)_6^{4-}$ in excess, the rate constant was 2.1 liters mole^{-1} s^{-1} (Prütz, 1996). The oxidation of two molecules of $Fe(CN)_6^{4-}$ per molecule of HOCl was also consistent with the production of a reactive species in the primary reaction. It was suggested that the explanation for the discrepancy was the catalytic effect of trace amounts of copper.

An additional rapid reaction of HOCl merits attention, and that is its reaction with superoxide

$$\cdot O_2^- + HOCl \rightarrow \cdot OH + Cl^- + O_2$$

which has a rate constant, determined by pulse radiolysis, of 7.5×10^6 liters $mole^{-1}$ s^{-1} (Long and Bielski, 1980). The production of ·OH in this reaction has been confirmed by the hydroxylation of benzoate (Candeias, 1993).

11. THE CARBONATE RADICAL

The increase in the radiolytic inactivation of bacteria when the medium contains carbonate has been ascribed to the formation of carbonate radicals, $\cdot CO_3^-$ (Wolcott *et al.*, 1994). The physiological importance of the carbonate radical, is uncertain because the bicarbonate anion is oxidized by ·OH radicals very slowly ($k = 8.5 \times 10^6$ liters $mole^{-1}$ s^{-1}) (Buxton *et al.*, 1988)

$$\cdot OH + HCO_3^- \rightarrow H_2O + \cdot CO_3^-$$

This may be offset, however, by the long lifetime of this radical, reflecting its slow self-reaction rate constant and its low reactivity toward most extracellular components. This will allow $\cdot CO_3^-$ to diffuse much farther than the much more reactive ·OH. It has also been pointed out that the HCO_3^- concentration at infection sites is probably enhanced by the production of CO_2 during the respiratory burst of phagocytic cells. Carbonate radicals might also be formed in the reaction of CO_2 with $ONOO^-$.

The pK_a of the carbonate radical is uncertain, with a value that lies between 7.0 and 8.2 (Eriksen *et al.*, 1985). This means that the exact mix of the radical between $\cdot HCO_3$ and $\cdot CO_3^-$ is also uncertain under physiological conditions. The reduction potential of the couple $\cdot CO_3^-/CO_3^{2-}$ has been determined to be 1.59 V at pH 12 (Huie *et al.*, 1991a), but will increase as the pH is decreased. There have been a large number of studies of the electron-transfer reactions of $\cdot CO_3^-$ (Neta *et al.*, 1988). The rate constants for the reactions with substituted phenols (Moore *et al.*, 1977) and anilines (Elango *et al.*, 1984) correlate well with the Hammet substituent constant, σ, with a slope of -1, suggesting that the radical is strongly electrophilic. Rate constants for a number of reactions of $\cdot CO_3^-$ with organic and inorganic reductants have been determined (Huie *et al.*, 1991b). The reactions of the carbonate radical with several alcohols and tetrahydrofuran were investigated as a function of temperature so as to establish the reactivity of this radical toward H-abstraction (Clifton and Huie, 1993). The rate constants increased as the C–H bond strengths decreased and as the number of reactive bonds increased.

Of particular importance is the reactivity exhibited by $\cdot CO_3^-$ toward biological molecules in neutral aqueous media. The reactivity is highest for indole heterocyclic compounds ($k > 10^8$ liters $mole^{-1}$ s^{-1}), moderate for sulfur-containing compounds ($k = 10^6$–10^7 liters $mole^{-1}$ s^{-1}), and low for aliphatic amino acids and peptides ($k < 10^5$ liters $mole^{-1}$ s^{-1}) (Chen and Hoffman, 1973). The radical is significantly more reactive with a deprotonated sulfhydryl group or imidazole group than the protonated forms (Chen and Hoffman, 1975).

In light of the general selectivity of $\cdot CO_3^-$, it is important to note that this radical reacts rapidly, $k = 5 \times 10^8$ liters mole^{-1} s^{-1} (Eriksen *et al.*, 1985), with $\cdot O_2^-$.

12. CONCLUSION

This chapter has outlined some of the main reactions of reactive oxygen species, with emphasis on reactions of biological relevance. Other chapters in this volume will discuss in detail specific biological ramifications of this chemistry. The chemistry of reactive oxygen species is a vast subject; only a brief introduction could be given here. Some of the specific data cited are from the biochemistry literature but a vast array of information is available in the physical, organic, and inorganic chemistry literature. This information includes rate constants and mechanisms for many reactions of reactive oxygen species with model compounds that parallel those of biological importance, both in the aqueous phase and in nonaqueous solvents. This information, along with the various structure–activity relationships that have been established, may be utilized to predict rate constants or mechanisms of reactions of biological importance that have not been studied directly. Solvent effects are also becoming better understood, so that cautious extrapolations may be possible from one medium to another.

13. REFERENCES

Adam, L. C., Fábián, I., Suzuki, K., and Gordon, G., 1992, Hypochlorous acid decomposition in the pH 5–8 region, *Inorg. Chem.* **31**:3534–3541.

Alfassi, Z. B. (ed), 1997, *Peroxyl Radicals*, Wiley, New York.

Alfassi, Z. B., Huie, R. E., and Neta, P., 1986, Substituent effects on rates of one-electron phenols by the radical ClO_2, NO_2, and SO_3^-, *J. Phys. Chem.* **90**:4156–4158.

Alfassi, Z. B., Mosseri, S., and Neta, P., 1987, Halogenated alkylperoxyl radicals as oxidants: Effects of solvents and substituents on rates of electron transfer, *J. Phys. Chem.* **91**:3383–3385.

Alfassi, Z. B., Huie, R. E., Neta, P., and Shoute, L. C. T., 1990, Temperature dependence of the rate constants for reactions of inorganic radicals with organic reductants, *J. Phys. Chem.* **94**:8800–8805.

Alfassi, Z. B., Huie, R. E., Kumar, M., and Neta, P., 1992, Temperature dependence on the rate constants for oxidation of organic compounds by peroxyl radicals in aqueous alcohol solutions, *J. Phys. Chem.* **96**:767–770.

Alfassi, Z. B., Huie, R. E., and Neta, P., 1993a, Rate constants for reactions of perhaloalkylperoxyl radicals with alkenes, *J. Phys. Chem.* **97**:6835–6838.

Alfassi, Z. B., Huie, R. E., and Neta, P., 1993b, Solvent effects on the rate constants for reaction of trichloromethylperoxyl radicals with organic reductants, *J. Phys. Chem.* **97**:7253–7257.

Alfassi, Z. B., Huie, R. E., and Neta, P., 1997, Kinetics studies of organic peroxyl radicals in aqueous solutions and mixed solvents, in *Peroxyl Radicals* (Z. B. Alfassi, ed.), pp. 235–281, Wiley, New York.

Allen, R. C., 1994, Role of oxygen in phagocyte microbicidal action, *Environ. Health Perspect.* **102**:201–208.

Amar, C., Vilkas, E., and Foos, J., 1982, Catalytic activity studies of some copper(II)-histidine-containing dipeptide complexes on aqueous superoxide ion dismutation, *J. Inorg. Biochem.* **17**:313–323.

Antelo, J. M., Arce, F., Castro, M. C., Crugeiras, J., Perez-Moure, J. C., and Rodriguez, P., 1995a, Kinetics of the formation, decomposition, and disproportionation reactions of N-chlorobutylamines, *Int. J. Chem. Kinet.* **27**:703–717.

Antelo, J. M., Arce, F., and Parajo, M., 1995b, Kinetic study of the formation of N-chloramines, *Int. J. Chem. Kinet.* **27**:637–647.

Arudi, R. L., Bielski, B. H., and Allen, A. O., 1984, Search for singlet oxygen luminescence in the disproportionation of HO_2/O_2^- *Photochem. Photobiol.* **39**:703–706.

Atkinson, R., Baulch, D. L., Cox, R. A., Hampson, R. F., Kerr, J. A., and Troe, J., 1992, Evaluated kinetic and photochemical data for atmospheric chemistry. Supplement IV, *J. Chem. Phys. Ref. Data* **21**:1125–1568.

Aubry, J.-M., Rigaudy, J., Ferrandini, C., and Pucheault, J., 1981, Search for singlet oxygen in the disproportionation of superoxide anion, *J. Am. Chem. Soc.* **103**:4965–4966.

Aubry, J.-M., Mandard-Cazin, B., Rougee, M., and Bensasson, R. V., 1995, Kinetic studies of singlet oxygen [4 + 2]-cycloadditions with cyclic 1,3-dienes in 28 solvents, *J. Am. Chem. Soc.* **117**:9159–9164.

Awad, H. H., and Stanbury, D. M., 1993, Autoxidation of NO in aqueous solution, *Int J. Chem. Kinet.* **25**:375–381.

Babior, B. M., 1994, Activation of the respiratory burst oxidase, *Environ. Health Perspect.* **102**:53–56.

Baignee, A., Howard, J. A., Scaino, J. C., and Stewart, L. C., 1983, Absolute rate constants for reactions of cumyloxy in solution, *J. Am. Chem. Soc.* **105**:6120–6123.

Bakac, A., and Espenson, J. H., 1983, Kinetics of the oxidation of chromium(II) by hydrogen peroxide: Flash-photolytic and stopped-flow studies based on radical-trapping reactions, *Inorg. Chem.* **22**:779–783.

Barlow, G. E., Bisby, R. H., and Cundall, R. B., 1979, Does disproportionation of superoxide produce singlet oxygen? *Radiat. Phys. Chem.* **13**:73–75.

Bartlett, D., Church, D. F., Bounds, P. L., and Koppenol, W. H., 1995, The kinetics of the oxidation of L-ascorbic acid by peroxynitrite, *Free Radical Biol. Med.* **18**:85–91.

Bartlett, P. D., and Landis, M. E., 1979, The 1,2-dioxetanes, in *Singlet Oxygen* (H. H. Wasserman and R. W. Murray, eds.), pp. 243–286, Academic Press, New York.

Barton, A. F. M., 1983, *CRC Handbook of Solubility Parameters and Other Cohesion Parameters,* CRC Press, Boca Raton, FL.

Beckman, J. S., Beckman, T. W., Chen, J., Marshall, P. A., and Freeman, B. A., 1990, Apparent hydroxyl radical production by peroxynitrite: Implication for endothelial injury from nitric oxide and superoxide, *Proc. Natl. Acad. Sci. USA* **87**:1620–1624.

Beckman, J. S., Ischiropoulos, H., Zhu, L., Woerd, M. v. d., Smith, C., Chen, J., Harrison, J., Martin, J. C., and Tsai, M., 1992, Kinetics of superoxide dismutase- and iron-catalyzed nitration of phenolics by peroxynitrite, *Arch. Biochem. Biophys.* **298**:438–445.

Bennett, J. E., 1987, A kinetic study of the self-reaction of prop-2-ylperoxyl radicals in solution using ultraviolet absorption spectroscopy, *J. Chem. Soc. Faraday Trans. 1* **83**:1805–1813.

Bennett, J. E., Brunton, G., Smith, J. R. L., Salmon, T. M. F., and Waddington, D. J., 1987a, Reactions of alkylperoxyl radicals in solution. A kinetic and product study of self-reactions of 2-propylperoxyl radicals between 253 and 323 K, *J. Chem. Soc. Faraday Trans. 1* **83**:2433–2447.

Bennett, J. E., Brunton, G., Smith, J. R. L., Salmon, T. M. F., and Waddington, D. J., 1987b, Reactions of alkylperoxyl radicals in solution. A kinetic study of self-reactions of 2-propylperoxyl radicals between 135 and 300 K, *J. Chem. Soc. Faraday Trans. 1* **83**:2421–2432.

Biaglow, J. E., Held, K. D., Manevich, Y., Tuttle, S., Kachur, A., and Uckun, F., 1996, Role of guanosine triphosphate in ferric ion-linked fenton chemistry, *Radiat. Res.* **145**:554–562.

Bielski, B. H. J., 1983, Evaluation of the reactivities of HO_2/O_2 with compounds of biological interest, in *Oxy Radicals and their Scavenger Systems* (G. Cohen and R. A. Greenwald, eds.), pp. 1–7, Elsevier, Amsterdam.

Bielski, B. H. J., Cabelli, D. E., Arudi, R. L., and Ross, A. B., 1985, Reactivity of HO_2/O_2^- radicals in aqueous solution, *J. Phys. Chem. Ref. Data* **14**:1041–1100.

Blough, N. V., and Zafiriou, O. C., 1985, Reaction of superoxide with nitric oxide to form peroxonitrite in alkaline aqueous solution, *Inorg. Chem.* **24**:3500–3504.

Bogan, D. J., Celii, F., Sheinson, R. S., and Coveleskie, R. A., 1984, Observation of $O_2(b^1\Sigma_g^+ \rightarrow X^3\Sigma_g^-)$ chemiluminescence from the self-reaction of isopropylperoxy radicals, *J. Photochem.* **25**:409–417.

Böhm, F., Edge, R., Land, E. J., McGarvey, D. J., and Truscott, T. G., 1997, Carotenoids enhance vitamin E antioxidant efficiency, *J. Am. Chem. Soc.* **119**:621–622.

Bothe, E., and Schulte-Frohlinde, D., 1978, The bimolecular decay of α-hydroxymethylkperoxyl radicals in aqueous solution, *Z. Naturforsch.* **33B**:786–788.

Brault, D., 1985, Model studies in cytochrome P-450-mediated toxicity of halogenated compounds: Radical processes involving iron porphyrins, *Environ. Health Perspect.* **64**:53–60.

Brault, D., Neta, P., and Patterson, L. K., 1985, The lipid peroxidation model for halogenated hydrocarbon toxicity. Kinetics of peroxyl radical processes involving fatty acids and Fe(III)-porphyrins, *Chem.-Biol. Interact.* **54**:289–297.

Brigelius, R., Spoettl, R., Bors, W., Lengfelder, E., Saran, M., and Weser, U., 1974, Superoxide dismutase activity of low molecular weight Cu^{2+}-chelates studied by pulse radiolysis, *FEBS. Lett.* **47**:72–75.

Buettner, G. R., and Jurkiewicz, B. A., 1996, Catalytic metals, ascorbate and free radicals: Combinations to avoid, *Radiat. Res.* **145**:532–541.

Bull, C., McClune, G. J., and Fee, J. A., 1983, The mechanisms of Fe-EDTA catalyzed superoxide dismutation, *J. Am. Chem. Soc.* **105**:5290–5300.

Butler, J., and Halliwell, B., 1982, Reaction of iron-EDTA chelates with the superoxide radical, *Arch. Biochem. Biophys.* **218**:174–178.

Buxton, G. V., Greenstock, C. L., Helman, W. P., and Ross, A. B., 1988, Critical review of rate constants for reactions of hydrated electrons, hydrogen atoms and hydroxyl radicals ($\cdot OH/\cdot O^-$) in aqueous solution, *J. Phys. Chem. Ref. Data* **17**:513–886.

Buxton, G. V., Mulazzani, Q. G., and Ross, A. B., 1995, Critical review of rate constants for reactions of transients from metal ions and metal complexes in aqueous solution, *J. Phys. Chem. Ref. Data* **24**(3):1055–1349.

Cabelli, D. E., 1997, The reactions of HO_2/O_2^- radicals in aqueous solution, in *Peroxyl Radicals* (Z. B. Alfassi, ed.), pp. 407–437, Wiley, New York.

Cabelli, D. E., and Bielski, B. H. J., 1983, Kinetics and mechanism for the oxidation of ascorbic acid by HO_2/O_2^- radicals. A pulsed radiolysis and stopped-flow photolysis study, *J. Phys. Chem.* **87**:1809–1812.

Candeias, L. P., 1993, Free hydroxyl radicals are formed on reaction between the neutrophil-derived species superoxide anion and hypochlorous acid, *FEBS Lett.* **333**:151–153.

Candeias, L. P., Stratford, M. R. L., and Wardman, P., 1994, Formation of hydroxyl radicals on reaction of hypochlorous acid with ferrocyanide, a model iron (II) complex, *Free Radical Res.* **20**:241–249.

Carr, A. C., van der Berg, J. J. M., and Winterbourn, C. C., 1996, Chlorination of cholesterol in cell membranes by hypochlorous acid, *Arch. Biochem. Biophys.* **332**:63–69.

Castro, L., Alvarez, M. N., and Radi, R., 1996, Modulatory role of nitric oxide on superoxide-dependent luminol chemiluminescence, *Arch. Biochem. Biophys.* **333**:179–188.

Chakrabartty, S. K., 1978, Alkaline hypohalite oxidations, in *Oxidation in Organic Chemistry* (W. S. Trahanovsky, ed.), pp. 343–370, Academic Press, New York.

Cheeseman, K. H., Albano, E. F., Tomasi, A., and Slater, T. F., 1985, Biochemical studies on the metabolic activation of halogenated alkanes, *Environ. Health Perspect.* **64**:85–101.

Chen, S. N., and Hoffman, M. Z., 1973, Rate constants for the reaction of the carbonate radical with compounds of biochemical interest in neutral aqueous solution, *Radiat. Res.* **56**:40–47.

Chen, S. N., and Hoffman, M. Z., 1975, Effect of pH on the reactivity of the carbonate radical in aqueous solution, *Radiat. Res.* **62**:18–27.

Clifton, C. L., and Huie, R. E., 1993, Rate constants for some hydrogen abstraction reactions of the carbonate radical, *Int. J. Chem. Kinet.* **25**:199–203.

Cosgrove, J. P., Church, D. F., and Pryor, W. A., 1987, The kinetics of the autoxidation of polyunsaturated fatty acids, *Lipids* **22**:299–304.

Crow, J. P., Spruell, C., Chen, J., Gunn, C., Ischiropoulos, H., Tsai, M., Smith, C. D., Radi, R., Koppenol, W. H., and Beckman, J. S., 1994, On the pH-dependent yield of hydroxyl radical products from peroxynitrite, *Free Radical Biol. Med.* **16**:331–338.

Csányi, L., and Galbács, Z. M., 1985, Carbon dioxide-mediated decomposition of hydrogen peroxide in alkaline solutions, *J. Chem. Soc. Faraday Trans. I* **81**:113-116.

Denicola, A., Freeman, B. A., Trujillo, M., and Radi, R., 1996, Peroxynitrite reaction with carbon dioxide/bicarbonate: Kinetics and influence on peroxynitrite-mediated oxidations, *Arch. Biochem. Biophys.* **333**:49–58.

Denisov, E. T., and Denisova, T. G., 1993, Kinetic parameters of the reactions $RO_2 + RH$ in the framework of the parabolic model of transition state, *Kinet. Catal.* **34**:173–179.

Dikalov, S., Khramtsov, V., and Zimmer, G., 1996, Determination of rate constants of the reactions of thiols with superoxide radical by electron paramagnetic resonance: Critical remarks on spectrophotometric approaches, *Arch. Biochem. Biophys.* **326**:207–218.

Di Mascio, P., Bechara, E. J. H., Medeiros, M. H. G., Briviba, K., and Sies, H., 1994, Singlet molecular oxygen production in the reaction of peroxynitrite with hydrogen peroxide, *FEBS Lett.* **355**:287–289.

Dohrmann, J., and Bergmann, B., 1995, Equilibria and rates of redox reactions involving the 2-*tert*-butyl-1,4-benzosemiquinone radical in aqueous solution: An investigation by potentiometry, ESR, and pulse radiolysis, *J. Phys. Chem.* **99:**1218–1227.

Eberhardt, M. K., and Colina, R., 1988, The Reaction of OH radicals with dimethyl sulfoxide. A comparative study of Fenton's reagent and the radiolysis of aqueous dimethyl sulfoxide solutions, *J. Org. Chem.* **53:**1071–1074.

Edwards, J. O., and Plumb, R. C., 1994, The chemistry of peroxonitrites, in *Progress in Inorganic Chemistry* (K. D. Karlin, ed.), pp. 599-635, Wiley, New York.

Eguchi, W., Tanigaki, M., Mutoh, K., and Tsuchiya, H., 1989, *Kagaku Kogaku Ronbunshu* **15:**1115.

Elango, T. P., Ramakrishnan, V., Vancheesan, S., and Kuriacose, J. C., 1984, Reaction of the carbonate radical with substituted anilines, *Proc. Indian Acad. Sci. (Chem. Sci.)* **93:**47–52.

Elliot, A. J., and Simsons, A. S., 1984, Reactions of NO_2 and nitrite ion with organic radicals, *Can. J. Chem.* **62:**1831–1834.

Erben-Russ, M., Michel, C., Bors, W., and Saran, M., 1987, Absolute rate constants of alkoxyl radical reactions in aqueous solution, *J. Phys. Chem.* **91:**2362–2365.

Eriksen, T. E., Lind, J., and Merényi, G., 1985, On the acid-base equilibrium of the carbonate radical, *Radiat. Phys. Chem.* **26:**197–199.

Folkes, L. K., Candeias, L. P., and Wardman, P., 1995, Kinetics and mechanisms of hypochlorous acid reactions, *Arch. Biochem. Biophys.* **323:**120–126.

Fontana, F., Minisci, F., Vismara, E., Faraci, G., and Platone, E., 1989, Chlorination by hypochlorous acid. Free-radical versus electrophilic reactions, in *Free Radicals in Synthesis and Biology* (F. Minisci, ed.), pp. 269–282, Kluwer, Dordrecht.

Foote, C. S., Shook, F. C., and Abakerli, R. A., 1980, Chemistry of superoxide ion. 4. Singlet oxygen is not a major product of dismutation, *J. Am. Chem. Soc.* **102:**2503–2504.

Forni, L. G., Packer, J. E., Slater, T. F., and Willson, R. L., 1983, Reaction of the trichloromethyl and halothane-derived peroxy radicals with unsaturated fatty acids: A pulse radiolysis study, *Chem.-Biol. Interact.* **45:**171–177.

Forni, L. G., Mora-Arellano, V. O., Packer, J. E., and Willson, R. L., 1986, Nitrogen dioxide and related free radicals: electron-transfer reactions with organic compounds in solutions containing nitrite or nitrate, *J. Chem. Soc. Perkin Trans. 2* **1986:**1–6.

Fornier de Violet, P., Veyret, B., Vincendeau, P., and Caristan, A., 1984, Chemiluminescence induced by oxidation of tryptophan by singlet oxygen and by hypochlorous acid. Implications in the luminescence emitted in phagocytosis, *Photochem. Photobiol.* **39:**707–712.

Gilbert, B. C., Holmes, R. G. G., Laue, H. A. H., and Norman, R. O. C., 1976, Electron spin resonance studies. Part L. Reactions of alkoxyl radicals generated from alkyl hydroperoxides and titanium (III) ion in aqueous solution, *J. Chem. Soc. Perkin Trans. 2* **1976:**1047–1052.

Gilbert, B. C., Marshall, P. D. R., Norman, R. O. C., Pineda, N., and Williams, P. S., 1977, Electron spin resonance studies. Part LII. Reactions of secondary alkoxyl radicals, *J. Chem. Res.* **1977:**101–113.

Gilbert, B. C., Marshall, P. D. R., Norman, R. O. C., Pineda, N., and Williams, P. S., 1981, Electron spin resonance studies. Part 61. The generation and reactions of the *t*-butoxyl radical in aqueous solution, *J. Chem. Soc. Perkin Trans. 2* **1981:**1392–1400.

Goldstein, S., and Czapski, G., 1995a, Direct and indirect oxidations by peroxynitrite, *Inorg. Chem.* **34:**4041–4048.

Goldstein, S., and Czapski, G., 1995b, Kinetics of nitric oxide autoxidation in aqueous solution in the absence and presence of various reductants. The nature of the oxidizing intermediates, *J. Am. Chem. Soc.* **117:**12078–12084.

Goldstein, S., and Czapski, G., 1995c, The reaction of NO with O_2^- and HO_2: A pulse radiolysis study, *Free Radical Biol. Med.* **19:**505–510.

Goldstein, S., Czapski, G., and Meyerstein, D., 1990, A mechanistic study of the copper(II)-peptide-catalyzed superoxide dismutation. A pulse radiolysis study, *J. Am. Chem. Soc.* **112:**6489–6492.

Goldstein, S., Meyerstein, D., and Czapski, G., 1993, The Fenton reagents, *Free Rad. Biol. Med.* **15:**435–445.

Gollnick, K., and Kuhn, H. J., 1979, Ene-reactions with singlet oxygen, in *Singlet Oxygen* (H. H. Wasserman, and R. W. Murray, eds.), pp. 287–427, Academic Press, New York.

Gow, A., Duran, D., Thom, S. R., and Ischiropoulos, H., 1996, Carbon dioxide enhancement of peroxynitrite-mediated protein tyrosine nitration, *Arch. Biochem. Biophys.* **333:**42–48.

Gray, D., Lissi, E., and Heicklen, J., 1972, The reaction of hydrogen peroxide with nitrogen dioxide and nitric oxide, *J. Phys. Chem.* **76:**1919.

Groves, J. T., and Marla, S. S., 1995, Peroxynitrite-induced DNA strand scission mediated by a manganese porphyrin, *J. Am. Chem. Soc.* **117:**9578–9579.

Guajardo, R. J., and Mascharak, P. K., 1995, Lipid peroxidation by synthetic analogues of iron bleomycin: Possible role of a low-spin {hydroperoxo}iron(III) intermediate in lipid peroxidation induced by bleomycin, *Inorg. Chem.* **34:**802–808.

Gunther, M. R., Hanna, P. M., Mason, R. P., and Cohen, M. S., 1995, Hydroxyl radical formation from cuprous ion and hydrogen peroxide: A spin-trapping study, *Arch. Biochem. Biophys.* **316:**515–522.

Halliwell, B., and Cross, C. E., 1994, Oxygen-derived species: Their relation to human disease and environmental stress, *Environ. Health Perspect.* **102:**5–12.

Halliwell, B., and Gutteridge, J. M. C., 1989, *Free Radicals in Biology and Medicine*, Oxford Univeristy Press (Clarendon), London.

Held, A. M., Halko, D. J., and Hurst, J. K., 1978, Mechanisms of chlorine oxidation of hydrogen peroxide, *J. Am. Chem. Soc.* **100:**5732–5740.

Held, K. D., Sylvester, F. C., Hopcia, K. L., and Biaglow, J. E., 1996, Role of Fenton chemistry in thiol-induced toxicity and apoptosis, *Radiat. Res.* **145:**542–553.

Hildebrand, J. H., Prausnitz, J. M., and Scott, R. L., 1970, *Regular and Related Solutions*, Van Nostrand–Reinhold, Princeton, NJ.

Hogg, N., Darley-Usmar, V. M., Eilson, M. T., and Moncada, S., 1992, Production of hydroxyl radicals from the simultaneous generation of superoxide and nitric oxide, *Biochem. J.* **281:**419–424.

Hogg, N., Joseph, J., and Kalyanaraman, B., 1994, The oxidation of α-tocopherol and Trolox by peroxynitrite, *Arch. Biochem. Biophys.* **314:**153–158.

Hogg, N., Singh, R. J., Goss, S. P. A., and Kalyanaraman, B., 1996, The reaction between nitric oxide and α-tocopherol: A reappraisal, *Biochem. Biophys. Res. Commun.* **224:**696–702.

Houk, K. N., Condroski, K. R., and Pryor, W. A., 1996, Radical and concerted mechanisms in oxidations of amines, sulfides, and alkenes by peroxynitrite, peroxynitrous acid, and the peroxynitrite–CO_2 adduct: Density functional theory transition state structures and energetics, *J. Am. Chem. Soc.* **118:**13002–13006.

Howard, J. A., 1984, The application of kinetic electron spin resonance spectroscopy to some reactions of t-butylperoxide in solution, *Rev. Chem. Intermed.* **5:**1–19.

Howard, J. A., and Ingold, K. U., 1968, The self-reaction of *sec*-butylperoxy radicals. Confirmation of the Russell mechanism, *J. Am. Chem. Soc.* **90:**1056–1058.

Howard, J. A., and Scaiano, J. C., 1984, Radical reaction rates in liquid. Oxyl, peroxyl, and related radicals, in *Landolt-Börnstein. Numerical Data on Functional Relationships in Science and Technology. New Series, Group II; Atomic and Molecular Physics* (K.-H. Hellwege and O. Madelung, eds.), Vol. 13, Part D, Springer-Verlag, Berlin (see also update in Vol. 18, Part D2, 1997).

Hughes, M. N., and Nicklin, H. G., 1968, The chemistry of pernitrites. Part 1. Kinetics of decomposition of pernitrous acid, *J. Chem. Soc. A* 450–452.

Hughes, M. N., Nicklin, H. G., and Sackrule, W. A. C., 1971, The chemistry of peroxonitrites. Part III. The reaction of peroxonitrite with nucleophiles in alkali, and other nitrite producing reactions, *J. Chem. Soc. A* **1971:**3722–3725.

Huie, R. E., and Herron, J. T., 1973, Kinetics of the reactions of singlet molecular oxygen ($O_2{}^1\Delta_g$) with organic compounds in the gas phase, *Int. J. Chem. Kinet.* **5:**197–211.

Huie, R. E., and Padmaja, S., 1993, The reaction of NO with superoxide, *Free Radical Res. Commun.* **18:**195–199.

Huie, R. E., Alfassi, Z. B., and Neta, P., 1986, Rate constants for one-electron oxidation by methylperoxyl radicals in aqueous solutions, *Int. J. Chem. Kinet.* **18:**1185–1191.

Huie, R. E., Clifton, C. L., and Neta, P., 1991a, Electron transfer reaction rates and equilibria of the carbonate and sulfate radical anions, *Radiat. Phys. Chem.* **38:**477–481.

Huie, R. E., Shoute, L. C. T., and Neta, P., 1991b, Temperature dependence of the rate constants for reactions of the carbonate radical with organic and inorganic reductants, *Int. J. Chem. Kinet.* **23:**541–552.

Ignarro, L. J., Fukuto, J. M., Griscavage, J. M., and Rogers, N. E., 1993, Oxidation of nitric oxide in aqueous solution to nitrite but not nitrate: Comparison with enzymatically formed nitric oxide from L-arginine, *Proc. Natl. Acad. Sci. USA* **90:**8103–8107.

Ilan, Y., Rabani, J., and Fridovich, I., 1981, Superoxide dismuting activity of an iron porphyrin, *Inorg. Nucl. Chem. Lett.* **17**:93–96.

Itoh, S., Nagaoka, S.-I., Mukai, K., Ikesu, S., and Kaneko, Y., 1994, Kinetic study of quenching reactions of singlet oxygen and scavenging reactions of free radicals by α-, β-, τ-, and δ- tocopheramines in ethanol solution and micellar dispersion, *Lipids* **29**:799–802.

Johnson, G. R. A., Nazhat, N. B., and Saadalla-Nazhat, R. A., 1988, Reaction of the aquacopper(I) ion with hydrogen peroxide, *J. Chem. Soc. Faraday Trans. 1* **84**:501–510.

Jonsson, M., 1996, Thermochemical properties of peroxides and peroxyl radicals, *J. Phys. Chem.* **100**:6814–6818.

Kanofsky, J. R., 1986a, Singlet oxygen production in superoxide ion–halocarbon systems, *J. Am. Chem. Soc.* **108**:2977–2979.

Kanofsky, J. R., 1986b, Singlet oxygen production from the reactions of alkylperoxy radicals. Evidence from 1268-nm chemiluminescence, *J. Org. Chem.* **51**:3386–3388.

Kasha, M., and Khan, A. U., 1970, The physics, chemistry, and biology of singlet molecular oxygen, in *International Conference on Singlet Molecular Oxygen and its Role in Environmental Sciences* (A. M. Trozzolo, ed.), pp. 5–23, Annals of the New York Academy of Sciences, New York.

Kawanishi, S., Inoue, S., and Yamamoto, K., 1994, Active oxygen species in DNA damage induced by carcinogenic metal compounds, *Environ. Health Perspect.* **102**:17–20.

Kearns, D. R., 1979, Solvent and solvent isotope effects on the lifetime of singlet oxygen, in *Singlet Oxygen* (H. H. Wasserman, and R. W. Murray, eds.), pp. 115-137, Academic Press, New York.

Khan, A. U., 1995, Quantitative generation of singlet ($^1\Delta_g$) oxygen from acidified aqueous peroxynitrite produced by the reaction of nitric oxide and superoxide anion, *J. Biolumin. Chemilumin.* **10**:329–333.

Khan, A. U., and Kasha, M., 1994a, Singlet molecular oxygen evolution upon simple acidification of aqueous hypochlorite: Application to studies on the deleterious health effects of chlorinated drinking water, *Proc. Natl. Acad. Sci. USA* **91**:12362–12364.

Khan, A. U., and Kasha, M., 1994b, Singlet molecular oxygen in the Haber–Weiss reaction, *Proc. Natl. Acad. Sci. USA* **91**:12365–12367.

Kharitonov, V. G., Sundquist, A. R., and Sharma, V. S., 1994, Kinetics of nitric oxide autoxidation in aqueous solution, *J. Biol. Chem.* **269**:5881–5883.

Kharitonov, V. G., Sundquist, A. R., and Sharma, V. S., 1995, Kinetics of nitrosation of thiols by nitric oxide in the presence of oxygen, *J. Biol. Chem.* **270**:28158–28164.

King, P. A., Anderson, V. E., Edwards, J. O., Gustafson, G., Plumb, R. C., and Suggs, J. W., 1992, A stable solid that generates hydroxyl radical upon dissolution in aqueous solutions: Reaction with proteins and nucleic acid, *J. Am. Chem. Soc.* **114**:5430–5432.

Klaning, U. K., and Wolff, T., 1985, Laser flash photolysis of HClO, ClO^-, HBrO, and BrO^- in aqueous solution. Reactions of Cl- and Br-atoms, *Ber. Bunsenges. Phys. Chem.* **89**:243–245.

Kobayashi, K., Miki, M., and Tagawa, S., 1995, Pulse-radiolysis study of the reaction of nitric oxide with superoxide, *J. Chem. Soc. Dalton Trans.* **1995**:2885–2889.

Kochi, J. K., 1962, Chemistry of alkoxy radicals: Cleavage reactions, *J. Am. Chem. Soc.* **84**:1193–1197.

Koelewijn, P., 1972, Epoxidation of olefins by alkylperoxyl radicals, *Recl. Trav. Chim. Pays-Bas* **91**:759–779.

Koppenol, W. H., 1985, The reaction of ferrous EDTA with hydrogen peroxide: Evidence against hydroxyl radical formation, *Free Radical Biol. Med.* **1**:281–285.

Koppenol, W. H., 1993, The centennial of the Fenton reaction, *Free Radical Biol. Med.* **15**:645–651.

Koppenol, W. H., and Butler, J., 1985, Energies of interconversion reactions of oxyradicals, *Adv. Free Radical Biol. Med.* **1**:91–131.

Koppenol, W. H., Moreno, J. J., Pryor, W. A., Ischiropoulos, H., and Beckman, J. S., 1992, Peroxynitrite, a cloaked oxidant formed by nitric oxide and superoxide, *Chem. Res. Toxicol.* **5**:834–842.

Krauss, M., 1994, Electronic structure and spectra of the peroxynitrite anion, *Chem. Phys. Lett.* **222**:513–516.

Kuhn, A. T., and Rice, C. L., 1985, The Halogens, in: *Standard Potentials in Aqueous Solution* (A. J. Bard, R. Parsons, and J. Jordan, eds.), pp. 67–92, Dekker, New York.

Laskey, R. E., and Mathews, W. R., 1996, Nitric oxide inhibits peroxynitrite-induced production of hydroxyeicosatetraenoic acids and F_2-isoprostanes in phosphatidylcholine liposomes, *Arch. Biochem. Biophys.* **330**:193–198.

Lemercier, J.-N., Squadrito, G. L., and Pryor, W. A., 1995, Spin trap studies on the decomposition of peroxynitrite, *Arch. Biochem. Biophys.* **321**:31–39.

Lewis, R. S., and Deen, W. M., 1994, Kinetics of the reaction of nitric oxide with oxygen in aqueous solutions, *Chem. Res. Toxicol.* **7**:568–574.

Lewis, R. S., Tannenbaum, S. R., and Deen, W. M., 1995, Kinetics of N-nitrosation in oxygenated nitric oxide solutions at physiological pH: Role of nitrous anhydride and effects of phosphate and chloride, *J. Am. Chem. Soc.* **117**:3933–3939.

Liebler, D. C., Stratton, S. P., and Kaysen, K. L., 1997, Antioxidant actions of β-carotene in liposomal and microsomal membranes: Role of carotenoid-membrane incorporation and α-tocopherol, *Arch. Biochem. Biophys.* **338**:244–250.

Lissi, E. A., Encinas, M. V., Lemp, E., and Rubio, M. A., 1993, Singlet oxygen $O_2(^1\Delta_g)$ bimolecular processes. Solvent and compartmentalization effects, *Chem. Rev.* **93**:699–723.

Løgager, T., and Sehested, K., 1993, Formation and decay of pernitrous acid: A pulse radiolysis study, *J. Phys. Chem.* **97**:6664–6669.

Long, C. A., and Bielski, B. H. J., 1980, Rate of reaction of superoxide radical with chlorine-containing species, *J. Phys. Chem.* **84**:555–557.

Luo, Y., Han, Z., Chin, S. M., and Linn, S., 1994, Three chemically distinct types of oxidants formed by iron-mediated Fenton reactions in the presence of DNA, *Proc. Natl. Acad. Sci. USA* **91**:12438–12442.

Lymar, S. V., and Hurst, J. K., 1995, Rapid reaction between peroxonitrite ion and carbon dioxide: Implications for biological activity, *J. Am. Chem. Soc.* **117**:8867–8868.

Lymar, S. V., Jiang, Q., and Hurst, J. K., 1996, Mechanism of carbon dioxide-catalyzed oxidation of tyrosine by peroxynitrite, *Biochemistry* **35**:7855–7861.

Mahoney, L. R., 1970, Evidence for the formation of hydroxyl radicals in the isomerization of pernitrous acid in aqueous solution, *J. Am. Chem. Soc.* **92**:5262–5263.

Mao, Y., Zang, L., and Shi, X., 1995, Singlet oxygen generation in the superoxide reaction, *Biochem. Mol. Biol. Interact.* **36**:227–232.

March, J., 1985, *Advanced Organic Chemistry*, 3rd ed., Wiley, New York.

Marcus, R. A., and Sutin, N., 1985, Electron transfer in chemistry and biology, *Biochim. Biophys. Acta* **811**:265–322.

Masarwa, M., Cohen, H., Meyerstein, D., Hickman, D. L., Bakac, A., and Espenson, J. H., 1988, Reactions of low-valent transition-metal complexes with hydrogen peroxide. Are they "Fenton-like" or not? 1. The case of Cu^+_{aq} and Cr^{2+}_{aq}, *J. Am. Chem. Soc.* **110**:4293–4297.

McGrath, M. P., and Rowland, F. S., 1994, Determination of the barriers of internal rotation in ONOOX (X=H, Cl) and characterization of the minimum energy conformers, *J. Phys. Chem.* **98**:1061–1067.

McKee, M. L., 1995, Ab initio study of the N_2O_4 potential energy surface. Computational evidence for a new N_2O_4 isomer, *J. Am. Chem. Soc.* **117**:1629–1637.

Meisel, D., Levanon, H., and Czapski, G., 1974, Hydroperoxyl radical reactions. II. Cupric ions in modulated photolysis. Electron paramagnetic resonance experiments, *J. Phys. Chem.* **78**:779–782.

Merenyi, G., Lind, J., and Engman, L., 1994. One- and two electron reduction potentials of peroxyl radicals and related species. *J. Chem. Soc. Perkin Trans. 2* **1994**:2551–2553.

Moffett, J. W., and Zika, R. G., 1987, Reaction kinetics of hydrogen peroxide with copper and iron in seawater, *Environ. Sci. Technol.* **21**:804–810.

Moore, J. S., Phillips, G. O., and Sosnowski, A., 1977, Reaction of the carbonate radical anion with substituted phenols, *Int. J. Radiat. Biol.* **31**:603–605.

Mosseri, S., Alfassi, Z. B., and Neta, P., 1987, Absolute rate constants for hydrogen abstraction from hydrocarbons by the trichloromethylperoxyl radical, *Int. J. Chem. Kinet.* **19**:309–317.

Mukai, K., Daifuku, K., Okabe, K., Tanigaki, T., and Inoue, K., 1991, Structure–activity relationship in the quenching reaction singlet oxygen by tocopherol (vitamin E) derivatives and related phenols. Finding of linear correlation between the rates of quenching of singlet oxygen and scavenging of peroxyl radicals in solution, *J. Org. Chem.* **56**:4188–4192.

Murray, R. W., 1979, Chemical sources of singlet oxygen, in *Singlet Oxygen* (H. H. Wasserman and R. W. Murray, eds.), pp. 59–114, Academic Press, New York.

Nagano, T., and Fridovich, I., 1985, Does the xanthine oxidase reaction generate singlet oxygen? *Photochem. Photobiol.* **41**:33–37.

Nahor, G. S., and Neta, P., 1991, Rate constants for reactions of perfluorobutylperoxyl radical with alkenes, *Int. J. Chem. Kinet.* **23**:941–946.

Navarro, J. A., Rosa, M. A. d. l., Roncel, M., and Rosa, F. F. d. l., 1984, Carbon dioxide-mediated decomposition of hydrogen peroxide in alkaline solutions, *J. Chem. Soc. Faraday Trans. 1* **80:**249–253.

Neta, P., Huie, R. E., and Ross, A. B., 1988, Rate constants for reactions of inorganic radicals in aqueous solution, *J. Phys. Chem. Ref. Data* **17:**1027–1284.

Neta, P., Huie, R. E., Maruthamuthu, P., and Steenken, S., 1989a, Solvent effects in the reactions of peroxyl radicals with organic reductants. Evidence for proton transfer mediated electron transfer, *J. Phys. Chem.* **93:**7654–7659.

Neta, P., Huie, R. E., Mosseri, S., Shastri, L. V., Mittal, J. P., Maruthamuthu, P., and Steenken, S., 1989b, Rate constants for reaction of substituted methylperoxyl radicals with ascorbate ions and TMPD, *J. Phys. Chem.* **93:**4099–4104.

Neta, P., Huie, R. E., and Ross, A. B., 1990, Rate constants for reactions of peroxyl radicals in fluid solutions, *J. Phys. Chem. Ref. Data* **19:**413–513.

Neta, P., Grodkowski, J., and Ross, A. B., 1996, Rate constants for reactions of aliphatic carbon-centered radicals in aqueous solution, *J. Phys. Chem. Ref. Data* **24:**709–1068.

Noronha-Dutra, A. A., Epperlein, M. M., and Woolf, N., 1993, Reaction of nitric oxide with hydrogen peroxide to produce potentially cytotoxic singlet oxygen as a model for nitric oxide-mediated killing, *FEBS Lett.* **321:**59–62.

Nottingham, W. C., and Sutter, J. R., 1986, Kinetics of the oxidation of nitric oxide by chlorine and oxygen in nonaqueous media, *Int. J. Chem. Kinet.* **18:**1289–1302.

Ogryzlo, E. A., 1979, Gaseous singlet oxygen, in *Singlet Oxygen* (H. H. Wasserman and R. W. Murray, eds.), pp. 35–58, Academic Press, New York.

Packer, J. E., Slater, T. F., and Willson, R. L., 1979, Direct observation of a free radical interaction between vitamin E and vitamin C, *Nature* **278:**737–738.

Padmaja, S., and Huie, R. E., 1993, The reaction of nitric oxide with organic peroxyl radicals, *Biochem. Biophys. Res. Commun.* **195:**539–544.

Padmaja, S., Ramazenuan, M. S., Bounds, P. L., and Koppenol, W. H., 1996, Reaction of peroxynitrite with L-tryptophan, *Redox Rep.* **2:**173–177.

Pan, X.-M., and von Sonntag, C., 1990, OH-radical-induced oxidation of benzene in the presence of oxygen. A pulse radiolysis study, *Z. Naturforsch.* **B45:**1337–1344.

Pasternack, R. F., and Halliwell, B., 1979, Superoxide dismutase activities of an iron porphyrin and other iron complexes, *J. Am. Chem. Soc.* **101:**1026–1031.

Pasternack, R. F., and Skowronek, W. R., 1979, Catalysis of the disproportionation of superoxide by metalloporphyrins, *J. Inorg. Biochem.* **11:**261–267.

Paul, H., Small, R. D., and Scaiano, J. C., 1978, Hydrogen abstraction by *tert*-butoxy radicals. A laser photolysis and electron spin resonance study, *J. Am. Chem. Soc.* **100:**4520–4527.

Peretz, P., Solomon, D., Weinraub, D., and Faraggi, M., 1982, Chemical properties of water-soluble porphyrins 3. The reaction of superoxide radicals with some metalloporphyrins, *Int. J. Radiat. Biol.* **42:**449–456.

Pires, M., Rossi, M. J., and Ross, D. S., 1994, Kinetic and mechanistic aspects of the NO oxidation by O_2 in aqueous phase, *Int. J. Chem. Kinet.* **26:**1207–1227.

Plumb, R. C., Edwards, J. O., and Herman, M. A., 1992, Problem of concurrent measurements of peroxonitrite and nitrite contents, *Analyst* **117:**1639–1641.

Pogozelski, W. K., McNeese, T. J., and Tullius, T. D., 1995, What species is responsible for strand scission in the reaction of $[Fe^{II}EDTA]^{2-}$ and H_2O_2 with DNA? *J. Am. Chem. Soc.* **117:**6428–6433.

Pogrebnaya, V. L., Usov, A. P., Baranov, A. V., Nesterenko, A. I., and Bez"yazychnyi, P. I., 1975, Oxidation of nitric oxide by oxygen in the liquid phase, *Zh. Prikl. Khim. (English Trans.)* **48:**1004–1007.

Porter, N. A., 1986, Mechanisms for the autoxidation of polyunsaturated lipids, *Acc. Chem. Res.* **19:**262–268.

Porter, N. A., Mills, K. A., and Carter, R. L., 1994, A mechanistic study of oleate autoxidation: Competing peroxyl H-atom abstraction and rearrangement, *J. Am. Chem. Soc.* **116:**6690–6696.

Prütz, W. A., 1996, Hypochlorous acid interactions with thiols, nucleotides, DNA and other biological substrates, *Arch. Biochem. Biophys.* **332:**110–120.

Prütz, W. A., Mönig, H., Butler, J., and Land, E. J., 1985, Reactions of nitrogen dioxide in aqueous model systems: Oxidation of tyrosine units in peptides and proteins, *Arch. Biochem. Biophys.* **243:**125–134.

Pryor, W. A., Lightsey, J. W., and Church, D. F., 1982, Reaction of nitrogen dioxide with alkenes and polyunsaturated fatty acids: Addition and hydrogen abstraction mechanisms, *J. Am. Chem. Soc.* **104:**6685–6692.

Pryor, W. A., Jin, X., and Squadrito, G. L., 1994, One- and two-electron oxidations of methionine by peroxynitrite, *Proc. Natl. Acad. Sci. USA* **91:**11173–11177.

Rabani, J., Klug-Roth, D., and Lilie, J., 1973, Pulse radiolytic investigations of the catalyzed diproportionation of peroxy radicals. Aqueous cupric ions, *J. Phys. Chem.* **77:**1169–1175.

Radi, R., Beckman, J. S., Bush, K. M., and Free, B. A., 1991a, Peroxynitrite-induced membrane lipid peroxidation: The cytotoxic potential of superoxide and nitric oxide, *Arch. Biochem. Biophys.* **288:**481–487.

Radi, R., Beckman, J. S., Bush, K. M., and Freeman, B. A., 1991b, Peroxynitrite oxidation of sulfhydryls, *J. Biol. Chem.* **266:**4244–4250.

Radi, R., Cosgrove, T. P., Beckman, J. S., and Freeman, B. A., 1993, Peroxynitrite-induced luminol chemiluminescence, *Biochem. J.* **290:**51–57.

Rahhal, S., and Richter, H. W., 1988, Reduction of hydrogen peroxide by the ferrous iron chelate of diethylenetriamine-*N, N, N′,N″, N‴*-pentaacetate, *J. Am. Chem. Soc.* **110:**3126–3133.

Ramezanian, M. S., Padmaja, S., and Koppenol, W. H., 1996, Nitration and hydroxylation of phenolic compounds by peroxynitrite, *Chem. Res. Toxicol.* **9:**232–240.

Reszka, K. J., Chignell, C. F., and Bilski, P., 1994, Spin trapping of nitric oxide (NO) by acinitromethane in aqueous solution, *J. Am. Chem Soc.* **116:**4119–4120.

Ross, A. B., Mallard, W. G., Hellman, W. P., Bielski, B. H. J., Buxton, G. V., Cabelli, D. E., Greenstock, C. L., Huie, R. E., and Neta, P., 1997, NDRL/NIST solution kinetics database. Ver. 3.0, *NIST Standard Reference Database* **40.**

Rubbo, H., Radi, R., Trujillo, M., Telleri, R., Kalyanaraman, B., Barnes, S., Kirk, M., and Freeman, B. A., 1994, Nitric oxide regulation of superoxide and peroxynitrite-dependent lipid peroxidation, *J. Biol. Chem.* **269:**26066–26075.

Rush, J. D., and Koppenol, W. H., 1986, Oxidizing intermediates in the reaction of ferrous EDTA with hydrogen peroxide. Reactions with organic molecules and ferrocytochrome, *J. Biol. Chem.* **261:**6730–6733.

Rush, J. D., and Koppenol, W. H., 1987, The reaction between ferrous polyaminocarboxylate complexes and hydrogen peroxide: An investigation of the reaction intermediates by stopped flow spectrophotometry, *J. Inorg. Biochem.* **29:**199–215.

Samuni, A., 1972, The precursors of the metal-complexed hydroperoxyl radical, *J. Phys. Chem.* **76:**2207–2213.

Samuni, A., and Czapski, G., 1970a, Complexes of peroxy radical with transition metal ions, *J. Phys. Chem.* **74:**4592–4594.

Samuni, A., and Czapski, G., 1970b, Oxidation of Ce^{3+} by HO_2 radical and Ce^{3+}–HO_2 complex formation, *Israel J. Chem.* **8:**551–561.

Samuni, A., Aronovitch, J., Chevion, M., and Czapski, G., 1983. Metal-mediated hydroxyl radical damage. A site-specific mechanism, in *Oxidative Damage and Related Enzymes*. Life Chemistry Reports, Suppl. 2, pp. 39–47.

Saran, M., Michel, C., and Bors, W., 1990, Reaction of NO with $O_2^{\cdot-}$. Implications for the action of endothelium-derived relaxing factor (EDRF), *Free Radical Res. Commun.* **10:**221–226.

Sawyer, D. T., and Valentine, J. S., 1981, How super is superoxide? *Acc. Chem. Res.* **14:**393–400.

Sawyer, D. T., Hage, J. P., and Sobkowiak, A., 1995, Iron(II)-induced activation of 1:1 HOOH/HCl for the chlorohydroxylation of olefins and the chlorination of hydrocarbons: Chlorinated Fenton chemistry, *J. Am. Chem. Soc.* **117:**106–109.

Sawyer, D. T., Knag, C., Llobet, A., and Redman, C., 1993, Fenton reagents (1:1 $Fe^{II}L_x$/HOOH) react via $[L_xFe^{II}OOH(BH^+)]$ (1) as hydroxylases (RH-ROH), not as generators of free hydroxyl radicals (HO·), *J. Am. Chem. Soc.* **115:**5817–5818.

Schaap, A. P., and Zaklika, K. A., 1979, 1,2-Cycloaddition reactions of singlet oxygen, in *Singlet Oxygen* (H. H. Wasserman and R. W. Murray, eds.), pp. 173–242, Academic Press, New York.

Schöneich, C., Aced, A., and Asmus, K.-D., 1991, Halogenated peroxyl radicals as two-electron-transfer agents. Oxidation of organic sulfides to sulfoxides, *J. Am. Chem. Soc.* **113:**376–377.

Scully, F. E., and Hoigné, J., 1987, Rate constants for reactions of singlet oxygen with phenols and other compounds in water, *Chemosphere* **16:**681–694.

Scurlock, R., Rougee, M., and Bensasson, R. V., 1989, Redox properties of phenols, their relationships to singlet oxygen quenching and to their inhibitory effects on benzo(a)pyrene-induced neoplasia, *Free Radical Res. Commun.* **8:**251–258.

Shi, X., Mao, Y., Knapton, A. D., Ding, M., Rojanasakul, Y., Gannett, P. M., Dalal, N., and Liu, K., 1994a, Reaction of Cr(VI) with ascorbate and hydrogen peroxide generates hydroxyl radicals and causes DNA damage: Role of a Cr(IV)-mediated Fenton-like reaction, *Carcinogenesis* **15:**2475–2478.

Shi, X., Rojanasakul, Y., Gannett, P., Liu, K., Mao, Y., Daniel, L. N., Ahmed, N., and Saffiotti, U., 1994b, Generation of thiyl and ascorbyl radicals in the reaction of peroxynitrite with thiols and ascorbate at physiological pH, *J. Inorg. Biochem.* **56:**77–86.

Shoute, L. C. T., Alfassi, Z. B., Neta, P., and Huie, R. E., 1994, Rate constants for reactions of (perhaloalkyl)peroxyl radicals with alkenes in methanol, *J. Phys. Chem.* **98:**5701–5704.

Solomon, D., Peretz, P., and Faraggi, M., 1982, Chemical properties of water-soluble prophyrins. 2. The reaction of iron(III) tetrakis(4-N-methylpyridyl) prophyrin with the superoxide radical dioxygen couple, *J. Phys. Chem.* **86:**1842–1849.

Spinks, J. W. T., and Woods, R. J., 1990, *Introduction to Radiation Chemistry*, 3rd ed., Wiley, New York.

Sprung, J. L., Akimota, H., and Pitts, J. N., 1974, Nitrogen dioxide catalyzed geometric isomerization of olefins. Isomerization kinetics of the 2-butenes and the 2-pentenes, *J. Am. Chem. Soc.* **96:**6549–6554.

Squadrito, G. L., Jin, X., and Pryor, W. A., 1995, Stopped-flow kinetic study of the reaction of ascorbic acid with peroxynitrite, *Arch. Biochem. Biophys.* **322:**53–59.

Stanbury, D. M., 1989, Reduction potentials involving inorganic free radicals in aqueous solution, *Adv. Inorg. Chem.* **33:**69–138.

Stohs, S. J., and Bagchi, D., 1995, Oxidative mechanisms in the toxicity of metal ions, *Free Radical Biol. Med.* **18:**321–336.

Sutton, H. C., 1975, Reactions of the hydroperoxyl radical (HO_2) with nitrogen dioxide and tetranitromethane in aqueous solution, *J. Chem. Soc. Faraday Trans. I* **71:**2142–2147.

Sutton, H. C., Seddon, W. A., and Sopchyshyn, F. C., 1978, Pulse radiolysis: Evidence for the reaction of HO_2 with NO_2 in aqueous solution, *Can. J. Chem.* **56:**1961–1964.

Taft, R. W., 1956, Separation of polar, steric, and resonance effects in reactivity, in *Steric Effects in Organic Chemistry* (M. S. Newman, ed.), p. 619, Wiley, New York.

Taha, Z., Kiechle, F., and Malinski, T., 1992, Oxidation of nitric oxide by oxygen in biological systems monitored by porphyrinic sensor, *Biochem. Biophys. Res. Commun.* **188:**734–739.

Tanielian, C., and Mechin, R., 1994, Reaction and quenching of singlet molecular oxygen with esters of polyunsaturated fatty acids, *Photochem. Photobiol.* **59:**263–268.

Titov, A. I., 1963, The free radical mechanism of nitration, *Tetrahedron* **19:**557–580.

Tratnyek, P. G., and Hoigné, J., 1991, Oxidation of substituted phenols in the environment: A QSAR analysis of rate constants for reaction with singlet oxygen, *Environ. Sci. Technol.* **25:**1596–1604.

Tsai, J.-H. M., Harrison, J. G., Martin, J. C., Hamilton, T. P., van der Woerd, M., Jablonsky, M. J., and Beckman, J. S., 1994, Role of conformation of peroxynitrate anion ($ONOO^-$) in its stability and toxicity, *J. Am. Chem. Soc.* **116:**4115–4116.

Tsang, W., 1996, Heats of formation of organic free radicals by kinetic methods, in *Energetics of Free Radicals* (A. Greenberg and J. Liebman, eds.), pp. 22–58, Chapman & Hall, London.

Uppu, R. M., Squadrito, G. L., and Pryor, W. A., 1996, Acceleration of peroxynitrite oxidants by carbon dioxide, *Arch. Biochem. Biophys.* **327:**335–343.

van der Vliet, A., Eiserich, J. P., O'Neill, C. A., Halliwell, B., and Cross, C. E., 1995, Tyrosine modification by reactive nitrogen species: A closer look, *Arch. Biochem. Biophys.* **319:**341–349.

von Sonntag, C., and Schuchmann, H.-P., 1991, The elucidation of peroxyl radical reactions in aqueous solution with the help of radiation-chemical methods, *Angew. Chem. Int. Ed. Engl.* **30:**1229–1253.

Wagner, B. A., Buettner, G. R., and Burns, C. P., 1994, Free radical-mediated lipid peroxidation in cells: Oxidizability is a function of cell lipid bis-allylic hydrogen content, *Biochemistry* **33:**4449–4453.

Walling, C., 1975, Fenton's reagent revisited, *Acc. Chem. Res.* **8:**125–131.

Wallington, T. J., Dagaut, P., and Kurylo, M. J., 1988, Correlation between gas-phase and solution-phase reactivities of hydroxyl radicals toward saturated organic compounds, *J. Phys. Chem.* **92:**5024–5028.

Wallington, T. J., Dagaut, P., and Kurylo, M. J., 1992, Ultraviolet absorption cross sections and reaction kinetics and mechanisms for peroxy radicals in the gas phase, *Chem. Rev.* **92:**667–710.

Wardman, P., and Candeias, L. P., 1996, Fenton chemistry: An introduction, *Radiat. Res.* **145:**523–531.

Warneck, P., and Wurzinger, C., 1988, Product quantum yields for the 305-nm photodecomposition of NO_3^- in aqueous solution, *J. Phys. Chem.* **92:**6278–6283.

Weiberg, K. B., 1964, *Physical Organic Chemistry*, Wiley, New York.

Wilkinson, F., and Brummer, J. G., 1981, Rate constants for the decay and reactions of the lowest electronically excited singlet state of molecular oxygen in solution, *J. Phys. Chem. Ref. Data* **10:**809–999.

Wink, D. A., Darbyshire, J. F., Nims, R. W., Saavedra, J. E., and Ford, P. C., 1993a, Reactions of the bioregulatory agent nitric oxide in oxygenated aqueous media: Determination of the kinetics for oxidation and nitrosation by intermediates generated in the NO/O_2 reaction, *Chem. Res. Toxicol.* **6:**23–27.

Wink, D. A., Hanbauer, I., Krishna, M. C., DeGraff, W., Gamson, J., and Mitchell, J. B., 1993b, Nitric oxide protects against cellular damage and cytotoxicity from reactive oxygen species, *Proc. Natl. Acad. Sci. USA* **90:**9813–9817.

Wink, D. A., Nims, R. W., Darbyshire, J. F., Christodoulou, D., Hanbauer, I., Cox, G. W., Laval, F., Laval, J., Cook, J. A., Krishna, M. C., DeGraff, W. G., and Mitchell, J. B., 1994a, Reaction kinetics for nitrosation of cysteine and glutathione in aerobic nitric oxide solutions at neutral pH. Insights into the fate and physiological effects of intermediates generated in the NO/O_2 reaction, *Chem. Res. Toxicol.* **7:**519–525.

Wink, D. A., Wink, C. B., Nims, R. W., and Ford, P. C., 1994b, Oxidizing intermediates generated in the fenton reagent: Kinetic arguments against the intermediacy of the hydroxyl radical, *Environ. Health Perspect.* **102:**11–15.

Wink, D. A., Cook, J. A., Krishna, M. C., Hanbauer, I., DeGraff, W., Gamson, J., and Mitchell, J. B., 1995, Nitric oxide protects against alkyl peroxide-mediated cytotoxicity: Further insights into the role nitric oxide plays in oxidative stress, *Arch. Biochem. Biophys.* **319:**402–407.

Wink, D. A., Cook, J. A., Pacelli, R., DeGraff, W., Gamson, J., Liebmann, J., Krishna, M. C., and Mitchell, J. B., 1996, The effect of various nitric oxide-donor agents on hydrogen peroxide-mediated toxicity: A direct correlation between nitric oxide formation and protection, *Arch. Biochem. Biophys.* **331:**241-248.

Winterbourn, C. C., 1985, Comparative reactivities of various biological compounds with myeloperoxidase–hydrogen peroxide–chloride, and similarity of the oxidant to hypochlorite, *Biochim. Biophys. Acta* **840:**204–210.

Winterbourn, C. C., 1993, Superoxide as an intercellular radical sink, *Free Radical Biol. Med.* **14:**85–90.

Winterbourn, C. C., Berg, J. J. M. v. d., Roitman, E., and Kuypers, F. A., 1992, Chlorohydrin formation from unsaturated fatty acids reacted with hypochlorous acid, *Arch. Biochem. Biophys.* **296:**547–555.

Wolcott, R. G., Franks, B. S., Hannum, D. M., and Hurst, J. K., 1994, Bactericidal potency of hydroxyl radical in physiological environments, *J. Biol. Chem.* **269:**9721–9728.

Yamazaki, I., and Piette, L. H., 1991, EPR Spin-trapping study on the oxidizing species formed in the reaction of the ferrous ion with hydrogen peroxide, *J. Am. Chem. Soc.* **113:**7588–7593.

Yang, G., Candy, T. E. G., Boaro, M., Wilkin, H. E., Jones, P., Nazhat, N. B., Saadalla-Nazhat, R. A., and Blake, D. R., 1992, Free radical yields from the homolysis of peroxynitrous acid, *Free Radical Biol. Med.* **12:**327–330.

Ye, M., and Schuler, R. H., 1989, Second-order combination reactions of phenoxyl radicals, *J. Phys. Chem.* **93:**1989–1902.

Part II

General Biochemistry and Molecular Biology

Chapter 3

The Steady-State Concentrations of Oxygen Radicals in Mitochondria

Cecilia Giulivi, Alberto Boveris, and Enrique Cadenas

Eres nube, eres mar, eres olvido.
Eres también aquello que se ha perdido.
Nubes, J. L. Borges

1. INTRODUCTION

The damaging effects of O_2 on living organisms have been known since the 1800s (Bert, 1878; Priestley, 1794; Lavoisier, 1783; Scheele, 1782), although, the molecular understanding of "O_2 toxicity" remained uncertain until 1954, when Gerschman, Gilbert, and collaborators proposed a common mechanism for X-ray and O_2 toxicities involving O_2 free radicals (Gerschman *et al.*, 1954). Later on, Chance *et al.* (1965) observed increased formation of hydrogen peroxide (H_2O_2) and inhibition of the mitochondrial reversed electron transfer *in vitro* and *in vivo* on hyperbaric oxygenation. These were the earliest contributions dealing with the molecular, subcellular, and cellular mechanism of O_2 toxicity. In the following years, the generation of hydrogen peroxide (H_2O_2) was observed in mitochondria isolated from different sources under normoxic (Boveris and Chance, 1973; Loschen *et al.*, 1973, 1971; Boveris *et al.*, 1972) and hyperoxic conditions (Boveris and Chance, 1973). Both NAD- and FAD-linked substrates are able to support H_2O_2 production (Boveris and Chance, 1973) modulated by different metabolic states (Boveris and Chance, 1973; Loschen *et al.*, 1971). This effect, along with the production of

Cecilia Giulivi and Enrique Cadenas Department of Molecular Pharmacology and Toxicology, School of Pharmacy, University of Southern California, Los Angeles, California 90033. *Present address for C.G.*: Department of Chemistry, University of Minnesota, Duluth, Minnesota 55812. **Alberto Boveris** Department of Physical Chemistry, School of Pharmacy and Biochemistry, University of Buenos Aires, Buenos Aires, Argentina.

Reactive Oxygen Species in Biological Systems, edited by Gilbert and Colton. Kluwer Academic / Plenum Publishers, New York, 1999.

superoxide anion ($\cdot O_2^-$) and H_2O_2 by submitochondrial particles (SMP), devoid of matrical components, suggested that a member of the respiratory chain or a H_2O_2 generator in redox equilibrium with a member of the respiratory chain was responsible for H_2O_2 generation. Studies performed with inhibitors of the mitochondrial respiratory chain pointed to an electron carrier on the substrate side of the antimycin-sensitive site (Loschen *et al.*, 1974a; Boveris and Chance, 1973) as the generator of H_2O_2. The midpotential ($E_0' = -50$ to $+50$ mV) pool acting between the rotenone- and the antimycin-sensitive sites, namely, (1) the flavoprotein succinate dehydrogenase, (2) ubiquinone, (3) iron–sulfur proteins, and (4) cytochromes *b*, were thermodynamically potential sources of $\cdot O_2^-$ and H_2O_2. The kinetic main role of ubiquinone in mitochondrial H_2O_2 generation was supported by the linear relationship between substrate-reducible ubiquinone and H_2O_2 production in mitochondria depleted and supplemented with variable amounts of ubiquinone (Boveris *et al.*, 1976; Boveris and Chance, 1973) and by the active production of H_2O_2 by the isolated complexes I (NADH-ubiquinone reductase) and III (ubiquinol-cytochrome *c* reductase) of beef heart mitochondria that share ubiquinone as a common component (Ragan *et al.*, 1977). Later on, a second site of $\cdot O_2^-$ and H_2O_2 production in mitochondrial membranes, although quantitatively less significant, was found at the NADH dehydrogenase segment (Turrens and Boveris, 1980). A production of $\cdot O_2^-$, amounting to about one-third of that observed from the succinate dehydrogenase-ubiquinone segment at physiological pH, was found in NADH-supplemented SMP in the presence of rotenone, antimycin, or KCN. Similar rates of $\cdot O_2^-$ production were observed during reversed electron transfer in the presence of succinate and ATP (Turrens and Boveris, 1980).

Several other subcellular fractions have been identified as intracellular sources of H_2O_2, namely, endoplasmic reticulum, peroxisomes, and soluble enzymes (Boveris *et al.*, 1972). The intracellular pO_2 (about 20 μM; Oshino *et al.*, 1975a) is certainly rate limiting for intracellular H_2O_2 production by mitochondria (rate at 20 μM equals 0.25 × rate at 240 μM O_2; Boveris and Cadenas, 1982), as well as by peroxisomes [K_m (O_2) = 100 μM; de Duve, 1965], endoplasmic reticulum [K_m (O_2) = 50 μM; Thurman *et al.*, 1972], and cytosolic enzymes [K_m (O_2) = 800 μM for xanthine oxidase; Fridovich and Handler, 1961]. Calculations estimating the importance of the intracellular sources to cytosolic H_2O_2 in rat liver (Table I) set the relative contributions of endoplasmic reticulum to about 57%, mitochondria and peroxisomes each to about 20%, and cytosolic enzymes to about 3%. When total cellular H_2O_2 is considered, the full peroxisomal oxygen uptake comes into consideration and peroxisomal H_2O_2 accounts for about 73%. The difference of the peroxisomal contribution to total and cellular H_2O_2 productions reflects the close association of peroxisomal oxidases and catalase in the organelle. If the total H_2O_2 production in rat liver is considered, mitochondria, endoplasmic reticulum, and cytosolic enzymes contribute about 7, 19, and 1%, respectively (Table I). In agreement with these values, most of the H_2O_2 production, obtained with perfused rat liver at the physiological concentrations of substrates, originates in the peroxisomes (72%) followed by the mitochondrial (27%) and the cytosolic (1%) fractions. Considering the cytosolic H_2O_2, because most of the peroxisomal H_2O_2 is consumed intraorganelle by catalase, mitochondria contribute about 78% (peroxisomes and cytosolic enzymes 19 and 3%, respectively; Table I). In contrast to the H_2O_2 contributions deduced from subcellular fractions, the microsomal production of H_2O_2 in perfused liver seems to be negligible. This is based on

Table I
Contribution of Intracellular Compartments to the Cellular Hydrogen Peroxide Production[a]

	$+d[H_2O_2]/dt$ (nmole/min/g liver)	Relative importance (%)	
		Cytosolic	Total
From subcellular fractions			
Mitochondria (succinate)	18.4	20	7
Microsomes (NADPH)	51.0	57	19
Peroxisomes (urate)	198.0		73
Intravesicular	180.0		
Diffusing[b]	18.0	20	
Soluble enzymes (xanthine)	2.6	3	1
From perfused liver			
Mitochondria (pyruvate-octanoate)	75	78	27
Microsomes (NADPH)	N.S.[c]	N.S.	N.S.
Peroxisomes (urate)	365		72
Intravesicular	185		
Diffusing[b]	18	19	
Soluble enzymes (hypoxanthine)	3	3	1

[a]The rates of H_2O_2 production at 240 μM, at 22 °C, and substrate-saturated (indicated in parentheses) were taken from Boveris *et al.* (1972) and corrected by the intracellular pO_2 estimated as 20 μM (Oshino *et al.*, 1975b) assuming Michaelis–Menten kinetics for cytochrome P450 reductase (rate at 20 μM = 0.34 V_{max}, and K_m for O_2 = 50 μM; Thurman *et al.*, 1972), urate oxidase (rate at 20 μM = 0.28 V_{max} and K_m for O_2 = 100 μM; de Duve, 1965), and xanthine oxidase (rate at 20 μM = 0.34 V_{max} and K_m for O_2 = 800 μM; Fridovich and Handler, 1961), and considering the pO_2 dependence of mitochondrial $\cdot O_2^-$ generation (rate at 20 μM O_2 = 0.26 × rate at 220 μM; Boveris and Cadenas, 1982). The rates at 37 °C were calculated from the rates at 22 °C using the Q_{10} rule. The rates of H_2O_2 production from perfused liver were taken from Oshino *et al.* (1975a) and corrected with the physiological substrate concentration for each fraction.

[b]Diffusing H_2O_2 was calculated as ~10% (Boveris *et al.*, 1972).

[c]N.S., not significant.

two experimental observations: (1) the lower rate of oxidized glutathione—as an index of H_2O_2 production—in perfused liver than that expected from the rates of microsomes *in vitro* and (2) the lack of the increase of H_2O_2 on pretreatment of animals with phenobarbital (Oshino *et al.*, 1975a; although microsomes isolated from these animals showed an increased rate of H_2O_2). Hence, H_2O_2 formation by microsomes *in vivo* does not occur as rapidly as would be expected from experiments with microsomes. Perhaps during the fragmentation and membrane vesicle formation that occurs on cell disruption to produce microsomes the arrangement of the components of the P450 system within the membrane is altered so that the electrons escape more easily to O_2.

The product of the bivalent reduction of O_2 accounts for about 13% of the physiological O_2 uptake in rat liver, a revisit to Wieland's concepts of about 70 years ago (Chance *et al.*, 1979). The mitochondrial productions of $\cdot O_2^-$ and H_2O_2 account for about 2 and 1% of the total O_2 uptake in rat liver, considering the O_2 molecule produced back in the dismutation reaction. In organs other than liver, in which peroxisomes are not so proliferated, such as heart, brain, and muscle, the mitochondrial production of $\cdot O_2^-$ and H_2O_2 is expected to be the most important physiological source of O_2 radicals.

On the basis that univalent and divalent reduction products of O_2 are produced in the oxidative metabolism under physiological conditions, it follows that either an increase in the generation of these species or a decrease in the antioxidant defenses will equally lead to increased intracellular steady-state concentrations of $\cdot O_2^-$, H_2O_2, and related reactive O_2 species and to the expression of O_2 poisoning. Considering the manifestations of O_2 toxicity, it is likely that the different effects observed could be understood in terms of (1) the primary reactions of $\cdot O_2^-$, H_2O_2, and related reactive O_2 species, considered as the primary responsible reactants, with essential biomolecules with loss of biological function, and (2) the exhaustion of the mechanisms that respond stoichiometrically (vitamin E, NADPH, reduced glutathione) to the reactive O_2 species.

The increased steady-state concentrations of reactive O_2 species occurring in acute-type situations or the slow cumulative effects of oxidative damage in cellular targets are both likely to occur under physiological conditions. Indeed, it has been frequently speculated that continuous oxidation at the normal O_2 tension may be significant in aging (Menzel, 1970; Gerschman, 1964).

Among the possible cellular targets, oxidative damage to mitochondrial DNA may be considered a relevant process (Giulivi and Cadenas, 1998; Giulivi *et al.*, 1995; Richter *et al.*, 1988) supported by the continuous exposure of this biomolecule to a steady-state concentration of O_2 radicals in the mitochondrial matrix, the presence of redox labile metals (Massa and Giulivi, 1993), and the redox cycling of these metal complexes by either $\cdot O_2^-$ or the mitochondrial respiratory chain (Massa and Giulivi, 1993). A correlation between the steady-state concentration of matrical H_2O_2 and the level of 8-hydroxy-desoxyguanosine, a product of DNA oxidative damage, has been reported (Figure 1; Giulivi and Cadenas, 1998; Giulivi *et al.*, 1995). The primary responsible reactant is likely to be the hydroxyl radical ($\cdot OH$)—as a product of the metal-catalyzed reaction of $\cdot O_2^-$ and H_2O_2—and the target, mitochondrial DNA. This biomolecule contains the information for several proteins of the electron-transfer chain and different types of RNA, not included in the nuclear genome (Wallace, 1992a; Tzagoloff and Myers, 1986). Accumulation of oxidative damage to mtDNA is a process favored by the lack of histones, high turnover rate, lack of introns, and lower repair capacity (Trounce *et al.*, 1989). This results in a higher mutation rate of mtDNA in comparison with nDNA (Wallace, 1992a). Because mitochondria, even with a mutated genome, have an intact capacity to replicate, a generation of dysfunctional mitochondria may proliferate with time leading to a lower energy availability in a tissue. Indeed, a decline in the activity of certain mitochondrial complexes has been observed with age (Trounce *et al.*, 1989) that could eventually establish the basis for the aging process (Wallace, 1992b).

Lately, a series of pathological processes have been linked to "oxidative stress" situations (Halliwell and Gutteridge, 1989). Although it is clear that the biological effects are produced by the increased concentrations of reactive O_2 species, no unambiguous correlations had been drawn because the quantification of the steady-state concentrations of these species is lacking. Therefore, it is becoming increasingly important to quantify adequately the steady-state levels of O_2 radicals to correlate them to oxidative stress situations and to understand the consequences of their ultimate actions.

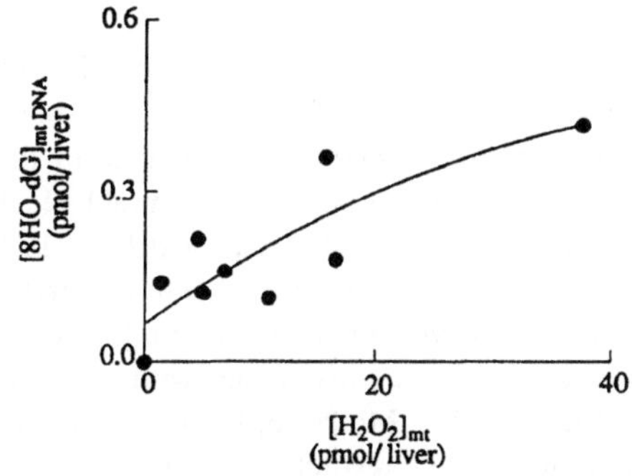

FIGURE 1. Dependence of the level of 8-hydroxy-desoxyguanosine in mitochondrial DNA on the steady-state concentration of hydrogen peroxide in rat liver mitochondria. Control or GSH-depleted isolated rat liver mitochondria were incubated with endogenous substrate, 5 mM succinate, or succinate plus antimycin. The steady-state concentrations of hydrogen peroxide were measured as described in Section 3.1.3. The mitochondrial DNA were extracted, purified, hydrolyzed, and the level of 8-hydroxy-desoxyguanosine measured by HPLC with electrochemical detection. The values were calculated from Giulivi *et al.* (1995) and Giulivi and Cadenas (1998).

The purpose of this chapter is to survey critically the methodology available for measuring the rates as well as the steady-state concentrations of H_2O_2 and $\cdot O_2^-$ in mitochondria, the main intracellular source of O_2 radicals. In addition, the modulation of the concentrations of O_2 radicals by metabolic states and by certain physiologic situations are discussed in terms of a continuous (not constant) production of O_2 radicals by mitochondria, thus emphasizing the dynamic character of this process.

2. ESTIMATION OF THE STEADY-STATE CONCENTRATION OF $\cdot O_2^-$ IN MITOCHONDRIA

The measurement of $\cdot O_2^-$ is a difficult task based on the low intracellular concentration (10^{-11}–10^{-10} M), the instability (it can be easily oxidized or reduced), and the poor reaction selectivity. Two methods for estimating the steady-state concentration of $\cdot O_2^-$ have been reported. The aconitase method is based on the dynamic state of inactivation–reactivation of the enzyme where aconitase activity serves as a measurement of the steady-state concentration of $\cdot O_2^-$. The second method is based on measurements of the rate of $\cdot O_2^-$ or H_2O_2 production from mitochondrial preparations from which the steady-state concentration of O_2^- is calculated.

2.1. Methods for Estimating the Steady-State Concentration of $\cdot O_2^-$ in the Mitochondrial Compartment

2.1.1. The Aconitase Method

Aconitase catalyzes the interconversion of tricarboxylic acids: citric acid, (+)-isocitric acid, and *cis*-aconitic acid; it functions both as an isomerase and as a hydratase. It is found in all respiring tissues of animals from protozoa to mammals. There are at least two types of aconitases: one in the cytoplasm and another in the mitochondria. The former catalyzes the formation of citrate needed for the fatty acid biosynthesis and feedback regulation of glycolysis (Glusker, 1971). The percentage of the cytoplasmic enzyme varies with the function of the tissues, being 70–80% in rat liver, 49% in pig liver (Eanes and Kun, 1971), and only 2% in pig heart (Eanes and Kun, 1971). The mitochondrial isoenzyme catalyzes mainly the conversion of citrate to isocitrate, which is degraded by the Krebs cycle. Both enzymes have a prosthetic group formed of the [4Fe-4S] cluster: Enzyme inactivation occurs on exposure to oxidants with release of Fe^{2+} (Kennedy *et al.*, 1983; Villafranca and Mildvan, 1971; Morrison, 1954), whereas reduction of the enzyme followed by the incorporation of Fe^{2+} restitutes the activity.

The method for estimating the steady-state concentration of $\cdot O_2^-$, initially developed by Gardner and Fridovich (1992) in *Escherichia coli* and later applied to rat 1a fibroblasts (Gardner and White, 1995), is based on the rapid inactivation of aconitase by $\cdot O_2^-$.

Aconitase is inactivated by $\cdot O_2^-$ (endogenously produced or facilitated by paraquat addition) and reactivated on reduction and addition of Fe^{2+}. This reversible inactivation is thought to be related to an oxidative attack of O_2^- on the $[4Fe\text{-}4S]^{2+}$ cluster based on the protective role of SOD, but not of catalase:

$$[4Fe\text{-}4S]^{2+} + 2H^+ + \cdot O_2^- \rightarrow [4Fe\text{-}4S]^{3+} + H_2O_2 \tag{1}$$

$$[4Fe\text{-}4S]^{3+} \rightarrow [3Fe\text{-}4S]^+ + Fe^{2+} \tag{2}$$

$$\underset{\text{active}}{[4Fe\text{-}4S]^{2+}} + 2H^+ + \cdot O_2^- \underset{b}{\overset{a}{\rightleftharpoons}} \underset{\text{inactive}}{[3Fe\text{-}4S]^+} + H_2O_2 + Fe^{2+} \tag{3}$$

The degree of aconitase inactivation was related to the level of $\cdot O_2^-$: thus, to calculate the steady-state concentration of $\cdot O_2^-$ in a given biological system, the following equations may be used:

$$-d[\text{aconitase}_{\text{act}}]/dt = k_{3a}[\cdot O_2^-]\,[\text{aconitase}_{\text{act}}] \tag{4}$$

$$+d[\text{aconitase}_{\text{act}}]/dt = k_{3b}\,[\text{aconitase}_{\text{inact}}] \tag{5}$$

Under steady-state conditions ($d[\text{aconitase}_{\text{act}}]/dt = 0$), therefore,

$$k_{3a}[\cdot O_2^-]\,[\text{aconitase}_{\text{act}}] = k_{3b}[\text{aconitase}_{\text{inact}}] \tag{6}$$

$$[\cdot O_2^-]_{ss} = k_{3b}[\text{aconitase}_{\text{inact}}]/k_{3a}[\text{aconitase}_{\text{act}}] \tag{7}$$

The pseudo-first-order rate constant (k_{3b}) can be calculated from the half time for the reactivation of aconitase in the biological system (15 and 3–5 min for *E. coli* and rat 1a fibroblasts, respectively), the percentage of active aconitase can be experimentally measured, and the rate constant for aconitase inactivation (k_{3a}) is $2 \times 10^7\ M^{-1}\ s^{-1}$ at 25 °C.

Using this approach, $\cdot O_2^-$ concentrations on the order of 20–40 and 300 pM were estimated in *E. coli* and SOD-deficient strains, respectively. In rat 1a fibroblasts, the $\cdot O_2^-$ concentration increased from 5.5 pM to 14.9 pM on addition of antimycin, an inhibitor of the mitochondrial chain (Gardner and White, 1995).

This technique seems to have the potential to be used in mitochondria as well as in other biological systems based on the ubiquitous distribution of aconitase, and the major advantage will be that no external manipulations are involved, allowing a direct measurement of matrical $\cdot O_2^-$. However, several points have to be considered:

1. The rate constant for aconitase inactivation by $\cdot O_2^-$ was taken from *in vitro* experiments performed with fluorocitrate-saturated aconitase. Although it is expected that aconitase *in vivo* will be fully saturated with citrate, the rate constant may actually differ from that operating in a biological setting (e.g., ionic strength, pH, buffer composition).
2. The sole participation of $\cdot O_2^-$ in the inactivation of aconitase is controversial based on the partial protection of aconitase inactivation by catalase, and the broad effect of oxidants that elicit the same effect (e.g., hydroxyl radical). The cluster conversion for the inactivation process (that entails the oxidation of the cluster) will depend on the relative compartmentalization and concentration of possible oxidants/reductants, besides $\cdot O_2^-$, determining the value of [aconitas-e_{act}]/[aconitase$_{inact}$] ratios.
3. The inactivation of aconitase might reflect the steady-state concentration of $\cdot O_2^-$ within the surroundings of the enzyme, based on the poor diffusibility of $\cdot O_2^-$ through plasma membranes and the lack of microhomogeneity of biological systems. In this regard, it is difficult to ascribe the inactivation of aconitase, present in mitochondria as well as in cytosol, as a sensor for specific changes in mitochondrial $[\cdot O_2^-]_{ss}$.

2.1.2. The Steady-State Approach Using the Rate of $\cdot O_2^-$ Formation

This approach allows the calculations of the steady-state concentrations of both $\cdot O_2^-$ and H_2O_2 in the mitochondrial compartment. It is important for these estimations to consider the extent of the univalent and bivalent reduction of oxygen (i.e., the primary production of $\cdot O_2^-$ or H_2O_2) and the existence of one or more mitochondrial $\cdot O_2^-$ and H_2O_2 generators. Concerning the first consideration, it is generally accepted that $\cdot O_2^-$ is the stoichiometric precursor of H_2O_2 (Boveris and Cadenas, 1982, 1975; Forman and Boveris, 1982; Chance *et al.*, 1979; Dionisi *et al.*, 1975). The experimentally determined $[\cdot O_2^-]/[H_2O_2]$ ratios in SMP are usually about 1.4 to 1.6 (Table II). The consistent underestimation (about 20 to 35%) in the rates of $\cdot O_2^-$ generation compared with the mitochondrial one (estimated from the rate of H_2O_2) can be explained in terms of the "wrong sideness" of the SMP preparations, in which usually 25 to 35% have the *c*-side facing the reaction medium, generating about 20–30% of the total $\cdot O_2^-$ production to the intravesicular space from which diffusible H_2O_2 reaches the reaction medium. Regarding the second consideration, it is accepted that $\cdot O_2^-$ is primarily produced in mitochondrial respiratory chain at two sites: (1) the autoxidation of the ubisemiquinone (Cadenas *et al.*,

Table II
Succinate-Dependent Production of $\cdot O_2^-$ and H_2O_2 in Submitochondrial Particles Supplemented with Antimycin (Rates at 22 °C and pO_2 = 220 μM)

Source	$+d[\cdot O_2^-]/dt$	$+d[H_2O_2]/dt$ (nmole/min/mg protein)	$[\cdot O_2^-]/[H_2O_2]$
Rat heart SMP[a]	5.3	3.2	
	4.9	3.2	
Mean value	5.1 ± 0.2	3.2 ± 0.3	1.6 ± 0.5
Beef heart mitochondria	4.4[a]	2.5[a]	
	4.0[b]	3.7[c]	
	2.6[c]	2.6[c]	
	3.8[c]	0.9[c]	
	2.8[d]		
	2.1[e]		
Mean value	3.3 ± 0.5	2.4 ± 0.5	1.4 ± 0.7

[a]Boveris and Cadenas (1975).
[b]Loschen *et al.* (1974b).
[c]Boveris *et al.* (1976).
[d]Cadenas and Boveris (1980).
[e]Turrens and Boveris (1980).

1977; Boveris *et al.*, 1976) and (2) the autoxidation of the flavin semiquinone of NADH dehydrogenase. Because both sources generate $\cdot O_2^-$, they are considered jointly for the steady-state approach with a common O_2 dependence (Boveris and Cadenas, 1982).

Under steady-state conditions, the rate of $\cdot O_2^-$ is equal to the rate of consumption ($d[\cdot O_2^-]/dt = 0$). Assuming that the main decay of $\cdot O_2^-$ in the mitochondrial matrix is through the dismutation catalyzed by MnSOD (based on the high enzyme concentration and the almost diffusional controlled dismutation rate) and that the rate of $\cdot O_2^-$ production measured from SMP truly reflects the production of this radical in intact mitochondria, the steady-state concentration of $\cdot O_2^-$ can be calculated from Eqs. (8) and (9) by measuring the rate of $\cdot O_2^-$ production, and using $k_{SOD} = 4.7 \times 10^9\ M^{-1}\ s^{-1}$ ($1.2 \times 10^9\ M^{-1}\ s^{-1}$ in molar subunits; Forman and Fridovich, 1973), and [MnSOD] = 11 μM (44 μM in Mn-containing subunits; Tyler, 1975):

$$+ d[\cdot O_2^-]/dt = - d[\cdot O_2^-]/dt = k_{SOD}\,[SOD]\,[\cdot O_2^-] \tag{8}$$

$$(+d[\cdot O_2^-]/dt)/k_{SOD}[SOD] = [\cdot O_2^-]_{ss} \tag{9}$$

The lack of diffusibility of $\cdot O_2^-$ through the mitochondrial membranes imposes the measurement of the rate of $\cdot O_2^-$ production in SMP using either the SOD-sensitive cytochrome *c* assay or the adrenochrome method. Alternatively, the rate of $\cdot O_2^-$ formation in mitochondria can be estimated from the rates of mitochondrial H_2O_2 production measured with any of the methods discussed in the next section.

Of note, the assumptions on which this method is based might not be valid under certain conditions; for example, other decay pathways for $\cdot O_2^-$ in mitochondria (e.g., oxidation of Fe/Cu pools; Massa and Giulivi, 1993) might become relevant under particular conditions, in which the steady-state equations need to be rewritten to illustrate the new situation.

The intramitochondrial steady-state concentrations of $\cdot O_2^-$ have been estimated from the rates of $\cdot O_2^-$ or H_2O_2 production in mitochondrial preparations from various tissues (Table III). The steady-state concentrations of $\cdot O_2^-$ estimated from the rates of $\cdot O_2^-$ from SMP are consistently lower (about 50%) than those calculated from the rates of H_2O_2 production from intact mitochondria. Of note, the rates of $\cdot O_2^-$ and H_2O_2 production by heart SMP are also lower (ca. 50%) than the values obtained with intact heart mitochondria, which can be attributed to partial denaturation during the sonication used to prepare SMP. Besides, this can be attributed to several factors that affect the measured rates in SMP, as follows.

1. Possible contamination of the SMP particles with endogenous SOD or with cytochrome *c* oxidase or reductase activities. Indeed, some SOD might still have been trapped in the intravesicular space or nonspecifically attached to the membranes based on the stimulatory effect of KCN (an inhibitor of the CuZn isoenzyme) on $\cdot O_2^-$ production

Table III
Steady-State Concentration of $\cdot O_2^-$ in Mitochondria Using the Steady-State Approach

	Source	Substrate/Inhibitor	$[\cdot O_2^-]_{ss}$ (pM)
Estimation from the rate of H_2O_2 production in mitochondria or SMP [a]	Pigeon heart mt	Succinate/ antimycin	18.0 [b]
	Rat heart mt	Succinate/ antimycin	10.8 [b]
	Beef heart SMP	Succinate/ antimycin	40.5 [c]
Estimation from the rate of $\cdot O_2^-$ production in SMP [d]	Beef heart	NADH/rotenone	1.1 [e]
	Beef heart	Succinate/ antimycin	2.1[f]
			3.1[c]
			4.6[c]
	Rat heart	Succinate/ antimycin	1.6[g]
	Rat brain	Dihydroorotate /antimycin	1.3[h]

[a]The rate of $\cdot O_2^-$ was considered double the rate of H_2O_2 production in agreement with the dismutation stoichiometry. The rates were corrected using 54 mg mitochondrial protein/g heart, 0.5 mg IM protein/mg mitochondrial protein (Costa *et al.*, 1988), and 0.7 ml/g wet organ (Tyler, 1975).

[b]Boveris *et al.* (1977).

[c]Boveris *et al.* (1976).

[d]The steady-state concentrations of $\cdot O_2^-$ were calculated from the rates of $\cdot O_2^-$ in SMP, using 25, 30 (Boveris *et al.*, 1972), and 54 mg mitochondrial protein (Costa *et al.*, 1988) per g brain, liver, and heart, respectively. To calculate the rate of $\cdot O_2^-$ in mitochondria from those of SMP, 21 (Schnaitman and Greenawalt, 1968) and 50% (Costa *et al.*, 1988) of the total mitochondrial protein was attributed to the inner membrane protein in liver and heart, respectively.

[e]Loschen *et al.* (1974b).

[f]Turrens and Boveris (1980).

[g]Dionisi *et al.* (1975).

[h]Forman and Kennedy (1976).

by the NADH-dehydrogenase segment (Turrens and Boveris, 1980). The underestimation of the rate of $\cdot O_2^-$ is the result of a competition between the detection system (cytochrome *c* or adrenaline) and endogenous contamination of SOD. For example, using cytochrome *c* as the detection system, it can be formulated that the rate of cytochrome *c* reduction is proportional to the rate of $\cdot O_2^-$ production (v_p):

$$+d[\text{cyt. } c^{2+}]/dt = f \times v_p \times (k_{\text{cyt. } c} / k_{\text{SOD}}) ([\text{cyt. } c]/[\text{SOD}]) \quad (10)$$

where f is the proportionality factor, $k_{\text{cyt. } c}$ is the rate constant for the reduction of cytochrome *c* by $\cdot O_2^-$ (2.6×10^5 M^{-1} s^{-1}), and k_{SOD} is the rate constant for the SOD-catalyzed $\cdot O_2^-$ dismutation.

Based on the definition of 1 unit of SOD activity and the 1.5-fold increase in $\cdot O_2^-$ production by SMP on addition of KCN, it can be estimated that this preparation of particles might contain about 5.2 nM SOD, which will result in a 50–75% underestimation using 10–20 μM cytochrome *c* [Eq. (10)]. Evidently, under these conditions, the SOD contamination can be overcome by raising the concentration of cytochrome *c* to 40 μM, or by using 1 mM KCN in the assay mixture. In the former case, i.e., when using higher concentrations of cytochrome *c*, it is essential to check that the cytochrome *c* is SOD-free (Fridovich, 1985). In the latter case, using KCN, not only CuZnSOD is inhibited but also cytochrome *c* oxidase; the latter by consuming ferrocytochrome *c* leads to an underestimation of the rate of $\cdot O_2^-$ production. The cytochrome *c* oxidase responsible for this interference and accessible to ferrocytochrome *c* is present on the "right-side-in" particles; normally, most of the particles (65 to 75%) have the "right-side-out" conformation—the matrical side facing the reaction medium—although, a fraction (25–35%) retains the "right-side-in" conformation depending on each batch. The interferences by cytochrome *c* oxidase and reductase activities can be overcome by using acetylated or succinylated cytochrome *c*. These modified cytochrome *c* still are reduced by $\cdot O_2^-$ albeit at a slower rate, while no longer are a substrate for these activities.

2. The most likely factors interfering with the assay for H_2O_2 [the cytochrome *c* peroxidase (CCP) assay] under these experimental conditions are reduced cytochrome *c* and the presence of catalase (trapped inside the particles or unspecifically attached to the membranes). A direct contact of cytochrome *c* (located in the "right-side-in" particles) with the CCP–H_2O_2 complex is expected to occur with the "right-side-in" particles that the preparation might contain. However, the underestimation in the rate of H_2O_2 by SMP seems to be larger than that attributable to a "cytochrome *c* reductase activity." Catalase seems to be the major interference in this assay by competing with CCP for H_2O_2. Assuming that the rate of antimycin-resistant respiration in SMP reflects the physiological rate of H_2O_2 production in intact mitochondria (10.8 nmole O_2/min per mg protein; Boveris *et al.*, 1976), it can be calculated (Appendix in Boveris *et al.*, 1972) that these preparations might contain a contamination of 2 μM heme catalase (0.5 μM catalase) that would result in the observed 35% of the rate of formation of cytochrome *c* peroxidase ES complex (3.78 nmole H_2O_2/ min per mg protein; Boveris *et al.*, 1976).

3. The H_2O_2 production by SMP (or by ubiquinone-reconstituted membranes) never exceeded 35% of the antimycin-insensitive O_2 uptake (Boveris *et al.*, 1976), a result contrasting with that observed in intact pigeon heart mitochondria, in which H_2O_2 production matches the antimycin-insensitive O_2 utilization (5.7 and 6 nmole H_2O_2/min

per mg protein, respectively). The antimycin-insensitive electron leak through cytochromes b and c_1 is consequently much more active in the sonicated (SMP) and acetone-treated (ubiquinone-reconstituted membranes) preparations than in the intact mitochondria (Boveris *et al.*, 1976).

4. Previously, we discussed the possible interference by catalase and/or SOD in the measurement of H_2O_2 and/or $\cdot O_2^-$ production. However, in mitochondrial preparations virtually free of contaminants, still the rate of H_2O_2 production is lower in SMP than in intact mitochondria (Table III; Boveris *et al.*, 1976). An interesting hypothesis that explains this apparent discrepancy is based on the interactions between components of the inner membrane and the matrix that take place in intact mitochondria and are disrupted in mitochondrial subfractions. In intact mitochondria, the high matrical concentration of SOD will favor the oxidation of the ubisemiquinone by O_2 and will prevent the participation of $\cdot O_2^-$ as a reductant of ubiquinol (Figure 2).

In SMP the role of $\cdot O_2^-$ as a reductant of ubiquinol seems unlikely because addition of 0.3 μM SOD increased (2-fold) the formation of H_2O_2 (Boveris *et al.*, 1976). This may indicate that matrical SOD, by consuming $\cdot O_2^-$, is kinetically favoring the oxidation of ubisemiquinone, and therefore, higher rates of production of H_2O_2 would be expected in intact mitochondria than in SMP.

5. Antimycin or other inhibitors of the bc_1 site promote $\cdot O_2^-$ production when electrons are fed into Complex III via the Q_O site [cytosolic side of the mitochondrial inner membrane (c-side)] when cytochrome oxidase is active (Ksenzenko *et al.,* 1983) and cytochrome c is in the oxidized state (Turrens and Boveris, 1980; Figure 3). There are two main possibilities regarding the sidedness of the ubisemiquinone oxidation, namely, that it occurs either at the c-side or at the m-side of the mitochondrial inner membrane. The work with inhibitors, following the studies performed by von Jagow and

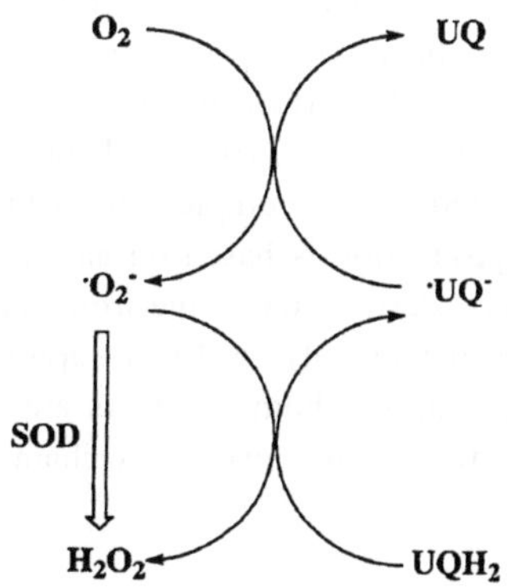

FIGURE 2. Redox transitions of ubiquinone and the effect of superoxide dismutase. The autoxidation of ubisemiquinone ($\cdot UQ^-$) produces superoxide anion ($\cdot O_2^-$), which in turn may oxidize ubiquinol (UQH_2) to the corresponding semiquinone. In this process, superoxide anion is a propagating species because it increases the steady-state concentration of the ubisemiquinone. Superoxide dismutase (SOD), exogenously added or present within the organelle, will prevent the latter reaction and will pull the oxidation of the ubisemiquinone by consuming superoxide anion.

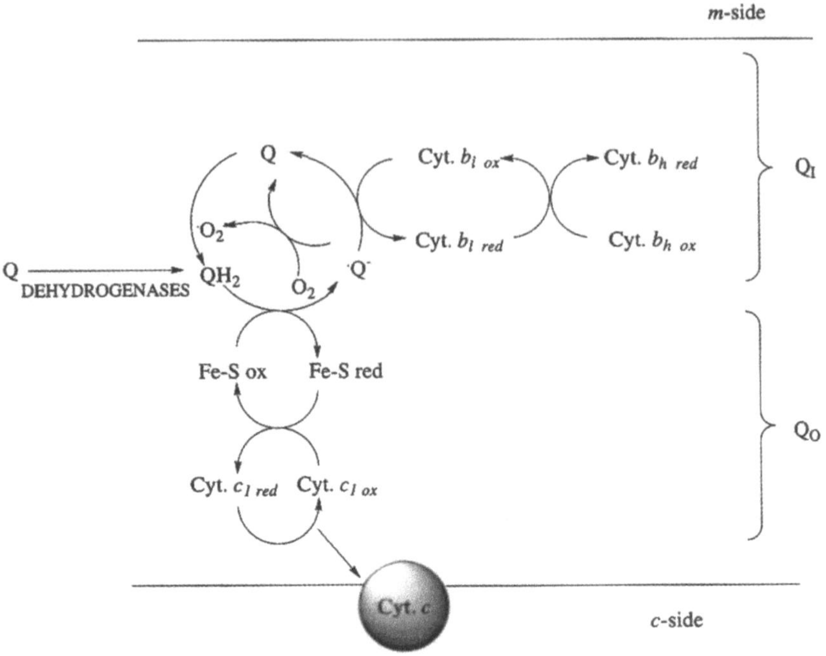

FIGURE 3. Scheme of the mitochondrial bc_1 complex.

Link (1986) and von Jagow *et al.* (1986), appears to indicate that Group III inhibitors (e.g., antimycin) compete with each other but bind independently of Group I and II inhibitors*; this is related to the fact that their point of action is at the Q_I site (*m*-side), which is most probably close to the middle of the membrane, compared with the Q_O site, which is located closer to the cytosolic surface of the mitochondrial inner membrane, and where the Group I and II inhibitors react. Group III inhibitors generally induce $\cdot O_2^-$ production when electrons are fed into the complex via the Q_O site and when cytochrome oxidase is active. The $\cdot O_2^-$ production is based on an autoxidation of the $\cdot UQ_O^-$. The presence of Group I inhibitors, destruction of the iron–sulfur cluster by British anti-lewisite (BAL or 2,3-dimercaptoethanol), or KCN suppress the $\cdot O_2^-$ generation. In contrast to inhibition by myxothiazol, both *b* centers are reduced in the presence of antimycin when electrons are fed into the respiratory chain of SMP on the route of the

*Inhibitors of the *bc* segment can be classified into four categories. Group I: bind at the Q_O site and block simultaneously two reactions, namely, electron transfer from ubiquinol to the iron–sulfur center and electron transfer onto the heme b_l center, e.g., various β-methoxyacrylates. Group II: bind at the Q_O site and block electron transfer from the iron–sulfur center to cytochrome c_1 and onto heme b_l center, e.g., hydroxyquinone analogues. Group III: bind at Q_I site and block electron transfer from the heme b_h center to ubiquinone, e.g., antimycin, funiculosin, and certain quinone analogues. Group IV: complete block of the Q_O center showing different binding properties than those of Group I and II inhibitors, e.g., stigmatellins.

various dehydrogenases, e.g., succinate- or NADH dehydrogenase, as well as when electrons are delivered to an isolated bc_1 complex by ubiquinol-9 or a homologue thereof (Figure 3). However, autoxidation of the ubisemiquinone at the *c*-side will release $\cdot O_2^-$ at the intermembrane space, leaving matrical MnSOD without physiological function. The rates of $\cdot O_2^-$ production by SMP (Table II) are not conclusive, as they account for about 30 to 75% of the H_2O_2 production by intact mitochondria depending on the preparations [3.90–4.20 nmole H_2O_2/min per mg protein for pigeon heart (Boveris and Chance, 1973) and 3.70 nmole H_2O_2/min per mg protein for rat heart (Cadenas and Boveris, 1980) in succinate-supplemented mitochondria in the presence of antimycin and uncoupler].

In general, the rates of H_2O_2 production in intact mitochondria and SMP and the antimycin-resistant respiration in intact mitochondria are higher (2- to 3-fold) than those based on the $\cdot O_2^-$ production. This can be explained in terms of a partial contamination by MnSOD, the intramatrical SOD isoenzyme, which is KCN-resistant. It can be estimated [Eq. (10)] that ca. 3.6 nM MnSOD (14.4 nM in subunits) could be present in these preparations and responsible for the $\cdot O_2^-$ underestimation.

2.2. Conclusions

By comparing the two methods available for estimating the steady-state concentration of $\cdot O_2^-$ in mitochondria, it seems that the aconitase method is technically easier than the steady-state approach. Contamination of SOD isoenzymes in SMP, and catalase in SMP and mitochondria may interfere with an accurate measurement of the rates of $\cdot O_2^-$ or H_2O_2. However, those interferences can be experimentally checked and quantitatively evaluated to correct the underestimation. Evidently, the preparation of mitochondria devoid of contaminating enzymes (usually from lysed erythrocytes) will be the best choice: For this purpose, perfusion of the organ (from which the mitochondrial preparation will be obtained) with saline solution to remove blood, and to avoid red blood cell lysis during the handling of the biological preparation are indicated.

For comparison purposes, the $[\cdot O_2^-]_{ss}$ obtained in rat 1a fibroblasts using the aconitase method, and those from rat liver mitochondria using the steady-state approach are listed in Table IV. Most of the values are within 10^{-11}–10^{-10} M, regardless of the method or approach used. The steady-state concentrations of $\cdot O_2^-$ obtained with the steady-state approach, clearly follow the changes in the rate of $\cdot O_2^-$ production, which, in turn, are modulated by the mitochondrial metabolic states: The $[\cdot O_2^-]_{ss}$ are higher in State 4 than in States 1, 3, or 3μ (see Table VI; Boveris and Chance, 1973; Loschen *et al.*, 1971). However, addition of an uncoupler to rat 1a fibroblasts did not decrease significantly the $[\cdot O_2^-]_{ss}$. This discrepancy may have different explanations: (1) The aconitase method may follow the activity changes modulated by the redox environment where the relatively major aconitase isoenzyme is located. This means that the redox environment will change according to the production of $\cdot O_2^-$ but not necessarily indicative of the sole participation of this species. (2) Aconitase reactivation in the presence of an uncoupler might be modulated by other processes not considered, e.g., limited availability of a reductant. (3) Because the aconitase method seems to have a lower limit of 5 pM (Table IV), changes in $[\cdot O_2^-]_{ss}$ below this threshold could not be detected.

Table IV
Steady-State Concentration of Superoxide Anion Estimated by the Aconitase Method and the Steady-State Approach

Method	Source	Substrate/inhibitor	$[\cdot O_2^-]_{ss}$ (pM)
Aconitase	Rat 1a fibroblasts[a]	F12K medium	5.0
		F12K medium/antimycin	13.5
		F12K/FCCP	5.5
		F12K/antimycin/FCCP	55.5
		F12K/KCN	9.0
Steady-state approach	Rat liver mitochondria	None	4.1[b]
		None/rotenone	2.6[b]
		Succinate	14.0[b]
		Succinate/antimycin	18.0[c]
		Succinate/ PCP	0.9[b]
		Succinate/antimycin/FCCP	95.0[d]
		Lactate/KCN	27.6[e]

[a]Gardner and White (1995).
[b]Boveris *et al.* (1972).
[c]Boveris *et al.* (1977).
[d]Cadenas and Boveris (1980).
[e]Oshino *et al.* (1973a).

3. ESTIMATION OF THE STEADY-STATE CONCENTRATION OF H_2O_2 IN MITOCHONDRIA

The catabolism of H_2O_2 is mainly performed by the enzymatic catalysis of catalase, glutathione peroxidase, and other peroxidases:

$$-d[H_2O_2]/dt = k_{cat}\,[\text{catalase}]\,[H_2O_2] + k_{GP}[\text{glutathione peroxidase}]\,[H_2O_2] \quad (11)$$

Under steady-state conditions, the rate of H_2O_2 production is equal to the rate of consumption ($d[H_2O_2]/dt = 0$); thus, to calculate the concentration of H_2O_2, measurement of (1) the rate of H_2O_2 production is a first requisite condition as well as (2) the consideration of the complex pathways for the H_2O_2 consumption. The cellular compartmentalization of catalase and glutathione peroxidase in different subcellular organelles emphasizes the uniqueness of each enzyme function, but allows them to cooperate effectively.

It was shown in perfused liver that the rate of GSSG release induced by H_2O_2 was about 40–50% of that induced by *t*-butyl-hydroperoxide at a similar infusion rate; furthermore, H_2O_2-induced release is saturated at ~ 50 nmole GSH/min per g liver. This result indicates that catalase decomposes about 50–60% of the infused H_2O_2 when the infusion rate is less than 2 μmole/min per g liver. The saturation of the H_2O_2-dependent GSSG release emphasizes the ability of catalase to decompose effectively unphysiological concentrations of H_2O_2, although this might lead to the incorrect conclusion that the catalatic activity of catalase is ineffective under physiological conditions. A number of factors, such as the compartmentalization of enzymes and the site of H_2O_2 production in the cell, must be taken into account when considering the physiological functions of catalase and glutathione peroxidase. It seems likely that the unique ability of glutathione peroxidase to decompose hydroperoxides may operate effectively only when the system is exposed to a minimal input of H_2O_2. Such conditions are afforded and maintained by the first-order nature of the catalase reaction with respect to the H_2O_2 concentration. Therefore, when calculating the steady-state concentration of H_2O_2 in a particular organ/tissue, it is critical to understand the synergistic catalysis of these two enzymes that deals with the natural production of hydroperoxides.

The catabolism of H_2O_2 in mitochondria is mainly achieved by the matrical glutathione peroxidase–reductase system because of the lack of mitochondrial catalase in most of the tissues. Thus, the first term of Eq. (11) becomes negligible, considering only the catabolism by glutathione peroxidase:

$$-d[H_2O_2]/dt = k_{GP}\,[\text{glutathione peroxidase}]\,[H_2O_2] \qquad (12)$$

The concentration of glutathione peroxidase can be calculated from the activity of the enzyme (calculated from its specific activity and molecular weight), and considering that 1 mg mitochondrial protein = 0.8 μl mitochondrial volume (Tyler, 1975). Using this approach, glutathione peroxidase concentrations of 1.0–1.5 μM are obtained, in good agreement with reported liver values of 2–4 μM. The rate constant of the catabolism of H_2O_2 by glutathione peroxidase is $2 \times 10^8\ M^{-1}\ s^{-1}$ (taking glutathione peroxidase molar and $5 \times 10^7\ M^{-1}\ s^{-1}$ in glutathione peroxidase subunit concentrations). Thus, the equation is reduced to the following expression, where $240\ s^{-1}$ is the product of $2 \times 10^8\ M^{-1}\ s^{-1}$ times 1.2×10^{-6} M, and the rate of H_2O_2 needs to be measured under specified conditions:

$$+d[H_2O_2]/dt = -d[H_2O_2]/dt = 240\ s^{-1}\ [H_2O_2]_{ss} \qquad (13)$$

Most of the assays used for the determination of H_2O_2-generation rates are based on enzyme–substrate compounds of catalase and peroxidases, which have a characteristic high affinity for H_2O_2. These assays are based on the free diffusibility of H_2O_2 across cellular membranes, the inaccessibility of catalase/peroxidase to the organelle, and the fact that H_2O_2 diffuses readily in favor of a gradient toward the extramitochondrial space, essentially free of peroxide. However, the assumption that *all* H_2O_2 produced diffuses to the extraorganelle space as well as the presence of other peroxidases (or peroxidaselike activity) may introduce an underestimation of the rate of H_2O_2 production. Another important consideration is implied in Eq. (13). Although mathematically it appears that the steady-state concentration of H_2O_2 follows a linear relationship with the rate of consumption or generation of H_2O_2, it is worth noting that Eq. (13) can only be valid under conditions in which a high steady-state concentration of NADPH, and GSH in mitochondria are maintained, i.e., the mitochondrial NADH-$NADP^+$ transhydrogenase activity is not the rate-limiting step when supplying reducing equivalents to the glutathione peroxidase/reductase system.

Several experimental approaches were used to calculate the rates of formation of H_2O_2, namely, the production of the catalase intermediate in subcellular suspensions (Chance and Oshino, 1971), the steady-state concentration of catalase intermediate in subcellular fractions (Chance and Oshino, 1971) or in perfused liver (Sies and Chance, 1970), peroxidase (cytochrome *c* or horseradish peroxidase)-coupled assays linked to absorption (Boveris *et al.*, 1972), or fluorescence changes (Loschen *et al.*, 1971). Each method has advantages and drawbacks that need to be evaluated in terms of the biological system chosen and/or the experimental design.

A brief description of the abovementioned methods is given below.

Methods for Estimating the Steady-State Concentration of H_2O_2 in the Mitochondrial Compartment

Monitoring Catalase Compound I. The direct optical measurement ($\varepsilon_{660-640} = 4$ mM^{-1} cm^{-1}) of the steady-state concentration of catalase Compound I is based on the dependence of the formation of Compound I on the rate of H_2O_2 and the hydrogen donor present (Oshino *et al.*, 1973b, 1975a,b; Sies *et al.*, 1973). *In vitro* experiments with purified rat liver catalase revealed a rather simple relationship between the steady state of catalase Compound I ([Compound I]$_{ss}$), the rate of H_2O_2 production ($+d[H_2O_2]/dt$), and the concentration of hydrogen donor, ethanol, ($a_{1/2}$) that produces a decrease in the [Compound I]$_{ss}$ value to half its saturating value; theoretical consideration of the kinetics of the catalase reactions provided the following equation for ethanol as hydrogen donor at room temperature:

$$a_{1/2} = 2.3 \times 10^{-3}\,[\text{catalase heme}]^{-1} \times (+d[H_2O_2]/dt) \qquad (14)$$

This equation indicates that the rate of H_2O_2 may be estimated when $a_{1/2}$ and [catalase heme] are known. To detect the mitochondrial production of H_2O_2 in perfused liver, Compound I was titrated with methanol in the absence and presence of octanoate. The obtained $a_{1/2}$ (0.12 mM in the control system and 0.40 mM in the octanoate-supplemented liver) are interpolated in plots of log (nmole H_2O_2/min per mole catalase) versus log $a_{1/2}$, obtained with perfused liver and glucose oxidase. These results indicate that the rate of H_2O_2 generation increases from 49 to 170 nmole/min per g wet weight liver (1.6 to 5.7 nmole/min per mg mitochondrial protein) on infusion with octanoate. The advantage of this technique is clearly that the rate of H_2O_2 is measured at quasi-physiological conditions (mitochondria within an organ), nondestructive, sensitive, and specific for H_2O_2, whereas the techniques performed with subcellular fractions lack the series of biological regulatory devices and are hampered by unphysiological metabolite accumulation.

Methods Based on Peroxidases: Horseradish and Cytochrome *c* Peroxidases. The measurement of the rate of H_2O_2 using horseradish peroxidase (HRP) or cytochrome *c* peroxidase (CCP) is based on the rate of formation of a relatively stable enzyme–substrate complex [H_2O_2–peroxidase; Eqs. (15)–(17)], characterized by a fairly large absorption change ($\varepsilon = 50$ mM^{-1} cm^{-1} at 419–407 nm for HRP and 417–402 nm for CCP). In addition, this assay offers a high affinity for H_2O_2 ($K \leq 10^{-8}$ M; Chance *et al.*, 1967), and in the case of CCP, its relative specificity for ferrocytochrome *c* as hydrogen donor (Yonetani and Ray, 1965). Reduced cytochrome *c* in intact mitochondria does not seem to interact with CCP, because of the impermeability of the mitochondrial outer membrane to the peroxidase. Also, the low protein concentration used in these assays decreases the optical interference from cytochromes that occurs immediately on addition of mitochondrial substrates/inhibitors.

$$\underset{(\lambda_{max} = 402\ nm)}{HRP} + H_2O_2 \rightarrow \underset{(\lambda_{max} = 410\ nm)}{HRP\text{–}H_2O_2\ (Compound\ I)} \quad (15)$$

$$\underset{(\lambda_{max} = 410\ nm)}{HRP\text{–}H_2O_2\ (Compound\ I)} + AH_2 \rightarrow {}^{\bullet}AH + \underset{(\lambda_{max} = 417\ nm)}{HRP\text{–}H_2O_2\ (Compound\ II)} \quad (16)$$

$$\underset{(\lambda_{max} = 417\ nm)}{HRP\text{–}H_2O_2\ (Compound\ II)} + {}^{\bullet}AH \rightarrow A + 2\ H_2O + HRP \quad (17)$$

The rate of H_2O_2 production can then be evaluated by following the rate of H_2O_2–peroxidase complex formation using a double-dual-wavelength spectrophotometer at the adequate λ_{max} pairs. The advantages of HRP over CCP is that the former is commercially available whereas the latter has to be purified from baker's yeast. However, interferences in the HRP method are to be expected because of the broad specificity of the enzyme–H_2O_2 complex toward endogenous and exogenous hydrogen donors. Consequently, controls are necessary such as the stability of the enzyme–H_2O_2 complex (i.e., spontaneous catalytic decomposition), the presence of interfering hydrogen donors that may halt the complex stability, the recovery of H_2O_2 (usually 95–103 and 60–80% for CCP and HRP, respectively), and the presence of catalase unspecifically attached to the mitochondrial membrane. To overcome some of these problems, the use of fluorescent exogenous hydrogen donors of the enzyme–H_2O_2 complex was introduced. The advantages are the easy availability of the HRP and the hydrogen donors, along with a simpler instrumentation than that of other methods employed. The technique is based on the peroxidase-catalyzed oxidation of a hydrogen donor to its corresponding derivative. Among the hydrogen donors so far used (e.g., guaiacol, benzidine, *o*-dianisidine, homovanillic acid, *p*-hydroxyphenylacetic acid, 7-hydroxy-6-methoxycoumarin, scopoletin), the most frequently employed is scopoletin. The HRP–scopoletin assay is carried out in a spectrofluorometer with excitation and emission wavelengths of 365 and 450 nm. The use of scopoletin may improve the sensitivity of the method considering the advantages of fluorometric versus absorption spectrophotometric techniques. However, the HRP–scopoletin assay gives lower values in different mitochondrial preparations (2–60%) as compared with the CCP method. The underestimation of the actual H_2O_2 production rate by the scopoletin method is likely related to the presence of hydrogen donors in mitochondrial preparations. It is advisable in such cases to estimate the concentration of endogenous hydrogen donors using the same HRP–scopoletin system supplemented with H_2O_2 (Rich *et al.*, 1976). Another possible interference is contamination with catalase, which can be avoided by extensive washing of the mitochondrial pellet or by quantifying the catalase/HRP heme ratio (Boveris *et al.*, 1977).

Measurement of the Steady-State Concentration of H_2O_2 in Intact Mitochondria. Based on the free diffusibility of H_2O_2, actual measurement of the steady-state concentration of H_2O_2 can be evaluated by allowing the sample to reach a diffusion equilibrium between the intra- and extraorganelle spaces. The extraorganelle concentration of H_2O_2 is then evaluated with the HRP or CCP methods or the HRP–scopoletin assay. Using this approach, the steady-state concentration of H_2O_2 in isolated mitochon-

Table V
Steady-State Concentrations of H_2O_2 in Rat Liver Hepatocytes and Isolated Mitochondria

Biological System	Method used for evaluation	$[H_2O_2]$ (M)
Hepatocytes	Rate of H_2O_2 production[a,b]	1.5×10^{-8}
	HRP–H_2O_2 complex[b]	1.5×10^{-7}
	Methanol titration of the catalase–H_2O_2 complex[c]	10^{-7}–10^{-9}
Rat liver mitochondria	HRP–scopoletin[d]	1.5×10^{-8}
		2.3×10^{-8} (GSH-depleted)

[a]This study.
[b]Giulivi (1989).
[c]Oshino *et al.* (1975a).
[d]Giulivi and Cadenas (1997).

dria (Giulivi and Cadenas, 1997), and in rat hepatocytes (Giulivi, 1989) were successfully determined (Table V).

It is important to note that the time when the diffusion equilibrium is reached will essentially depend on two factors: (1) the rate of H_2O_2 production by mitochondria, which is modulated by the metabolic state, and (2) the volume in which the mitochondria are incubated. For example, a mitochondrial suspension (1 mg protein/ml) incubated in 5 ml of adequate buffer will take 70 or 3 s to reach a diffusion equilibrium in State 1 or 4, respectively.

A similar approach consists of exogenous addition of H_2O_2 (5–10 μM) to the intact mitochondria, and measuring the remaining H_2O_2 at different time points. At relatively long times, the $[H_2O_2]$ does not change with time indicating that a new steady-state concentration has been reached. Using this technique, similar values of $[H_2O_2]$ have been obtained; however, this procedure may be complicated by the fact that H_2O_2 diffuses into mitochondria from outside, in contrast to the physiological situation, where most of the peroxide is generated inside the mitochondria.

4. HOW "STEADY" IS THE STEADY-STATE CONCENTRATION OF H_2O_2?

The production of H_2O_2 is mainly accomplished by the ubiquinol-cytochrome *c* segment (Complex III) and secondarily by the NADH-dehydrogenase flavoprotein (Complex I). The production of H_2O_2 by the former complex was envisaged as a nonenzymatic oxidation of a reduced form of the quinone by molecular O_2 [Eqs. (18) and (19); Boveris *et al.*, 1976] (similar equations may be written using the flavin moiety present in the NADH-dehydrogenase segment):

$$UQH_2 + O_2 \rightarrow \cdot UQ^- + \cdot O_2^- + 2\,H^+ \tag{18}$$

$$\cdot UQ^- + O_2 \rightarrow UQ + \cdot O_2^- \tag{19}$$

Then, the rate of $\cdot O_2^-$ production may be represented by

$$+ d[\cdot O_2^-]/dt = k_{18}[UQH_2]\,[O_2] + k_{19}[\cdot UQ^-]\,[O_2] \tag{20}$$

Based on the reduction potentials, Reaction (19) is thermodynamically favored ($\Delta G^{o\prime}$ = −7 kJ/mole) over Reaction (18) ($\Delta G^{o\prime}$ = 50 kJ/mole). Thus, assuming that Reaction (18) is negligible, which also agrees with the requirement of cytochrome c^{3+} for the univalent oxidation of ubiquinol (Ksenzenko *et al.*, 1983; Turrens and Boveris, 1980 [Eq. 21]),

$$UQH_2 + \text{cyt. } c^{3+} \rightarrow \cdot UQH + \text{cyt. } c^{2+} + H^+ \tag{21}$$

Equation (22) allows prediction of which factors will actually increase the rate of $\cdot O_2^-$ and H_2O_2, considering $\cdot O_2^-$ as the chemical precursor of H_2O_2:

$$+d[\cdot O_2^-]/dt = 2 \times (+d[H_2O_2]/dt) = k_{19}[\cdot UQ^-]\,[O_2] \tag{22}$$

Isolated mitochondria produce H_2O_2 at rates that depend primarily on their metabolic state. Based on the nonenzymatic nature of the production of H_2O_2 by mitochondria, it is likely that conditions that increase the reduction of components of the electron-transfer chain (e.g., the rate of electron flow through the chain is determined by the presence of uncouplers, substrate, mitochondrial inhibitors, O_2 concentration) will lead to higher rates of H_2O_2 production. In other words, considering the ubiquinol-cytochrome *b* segment as the most important source of O_2 radicals in mitochondria (60–80%), the level of O_2 reactive species is modulated by the steady-state concentration of the ubisemiquinone. However, a small pool of ubisemiquinone is expected to participate in the production of radicals, while most of it will decay by transferring electrons to the b-c_1 segment because the electron transfer rates between intramolecular redox groups (Ruzicka and Crane, 1970) exceed the autoxidation rate by a factor of 10^3–10^4 (Turrens and Boveris, 1980). To estimate the rate constant of the oxidation of the ubisemiquinone by O_2, the values for the rate of $\cdot O_2^-$ by mitochondria (2 × 4 nmole H_2O_2 /min per mg protein; Boveris and Chance, 1973), the O_2 concentration (220 μM), and the ubisemiquinone content (1% of the total ubiquinone; 0.1 nmole/ mg IM protein; Backstrom *et al.*, 1970) are used in Eq. (22), where k_{19} is found to be $1 \times 10^4\ M^{-1}\ s^{-1}$. Previous calculations gave a k_{19} value of $40\ M^{-1}\ s^{-1}$ (Boveris *et al.*, 1976) using the production of $\cdot O_2^-$ by ubiquinol-reconstituted membranes (0.24 nmole/min per mg IM protein) and an O_2 concentration of 1 mM. The discrepancy between the rate of O_2 radicals produced by SMP and mitochondria was thoroughly discussed before in terms of interferences and the intactness of organelles against subfractions. The high O_2 concentration used was based on its higher solubility in a lipid phase (5- to 10-fold) where it was assumed this reaction took place. With the value of k_{19}, it is possible to calculate the ubisemiquinone pool [Eq. (22)] that is actively involved in the production of O_2 radicals under different mitochondrial metabolic states (Table VI). Under State 1 or 3μ, negligible amounts of the ubisemiquinone are producing O_2 radicals (0–0.12%), whereas that level increases to 0.2–1.7% when substrate (State 4) or an inhibitor of the respiratory chain (State 3μ plus antimycin) are available (Table VI). It can be concluded from the data in Table VI that only a small percentage of the total ubiquinone content (0.01–1.7%) or of the reduced ubiquinone pool (0.1–1.5%) is actively involved in the production of O_2 radicals.

Finally, this physicochemical analysis may be extended to more concrete biological situations that illustrate the modulation of O_2 radicals by mitochondria. In the next section,

Table VI
Modulation of the Ubisemiquinone Levels by Metabolic States

Metabolic State	$+d[H_2O_2]/dt$[a] (nmole/min/mg protein)	$\cdot O_2^-$-producing ubisemiquinone pool[b] pmole/mg protein	%	Reduced UQ pool[c] (%)
State 1	0.04	0.6	0.012	18
State 4	0.7	11.0	0.22	38
State 3μ	0.0	0.0	0.0	18
State 3μ + antimycin	5.7	86.0	1.72	125

[a]Values taken from Chance *et al.* (1971).

[b]Values calculated using Eq. (22), $k_{19} = 1 \times 10^4\ M^{-1}\ s^{-1}$, and the total content of ubiquinone equal to 5 nmole/mg protein.

[c]Values interpolated from those of the steady-state reduction of different mitochondrial chain components from Chance and Williams (1956).

several conditions that affect the rate of O_2 species, some obvious and others implicit in Eq. (22), are presented and evaluated within a biological frame. Among the possible factors, we focus on the specific topics of organ specificity of mitochondria, changes in pO_2, substrate availability, and hormonal/metabolic regulation.

Factors that Modulate the Steady-State Concentration of Oxygen Radicals

Organ Specificity of Mitochondria. Mitochondria from different mammalian tissues have the ability to generate H_2O_2. For example, mitochondria isolated from pigeon heart (Chance *et al.*, 1971; Loschen *et al.*, 1971), rat liver (Boveris *et al.*, 1972), rat and ox heart, rat kidney, yeast, *Ascaris* muscle, and *Crithidia fasciculata* have all been shown to be active sources of H_2O_2. However, detailed studies on the effects of different metabolic conditions and of mitochondrial inhibitors have been carried out in pigeon heart and rat liver mitochondria. Although formation of H_2O_2 shows generally similar characteristics, such as a decrease of the rate for State 4–State 3 transition, different features are also observed: In pigeon and rat heart mitochondria, uncoupler and antimycin A are required to obtain maximal formation of H_2O_2, which is not the case for rat liver mitochondria. Addition of an uncoupler and antimycin A to mitochondria in State 4 increased 15- to 20-fold the formation of H_2O_2 in pigeon heart mitochondria but only 1.5-fold in rat liver mitochondria. Another difference is observed in the response to hyperbaric O_2: An increase in pO_2 up to 1.92 MPa increases the formation of H_2O_2 by a factor of 4 in pigeon heart mitochondria and by a factor of 15–20 in rat liver mitochondria (Boveris and Chance, 1973). These differences may originate in the metabolic adaptation of mitochondria to best suit the requirements of a particular tissue. For example, tightly coupled liver mitochondria exhibit a P/O ratio of 3 during NADH oxidation; however, brown fat mitochondria normally exhibit a P/O ratio below unity. It is suggested that the function of these mitochondria is to produce heat by uncoupled phosphorylation so as to maintain the temperature of newborn or hibernating animals. If the metabolic state of these mitochondria is comparable to State 3μ, then the rate of O_2 radicals under *normal* conditions is expected to be lower than in the *normal* liver mitochondria.

Changes in pO_2. The intracellular concentration of O_2 may be estimated to be 10–20 μM; thus, under normal physiological conditions, tissues are under hypoxic environments. A gradient of O_2 as high as 1000-fold (Sugano *et al.*, 1974) is expected from the capillaries toward mitochondria, and the steady-state concentration of O_2 in the latter compartment may be lower than that either in the peroxisomal and cytosolic spaces (Oshino *et al.*, 1975a) or in the milieu surrounding mitochondria. [The critical O_2 concentration for bioenergetic function of mitochondria corresponds approximately to 50% reduction of pyridine nucleotide being 60 and 80 nM in State 4 and 3, respectively (Sugano *et al.*, 1974).]

More severe hypoxic conditions, like those that trigger anaerobic glycolysis, occur in most vertebrates during short bursts of extreme muscular activity, e.g., in a 100-m sprint, during which O_2 cannot be carried to the muscles fast enough to oxidize pyruvate for generating ATP. Instead, the muscles use their stores of glycogen as fuel to generate ATP by anaerobic glycolysis. In this case, the steady-state level of reduction of each electron carrier in the respiratory chain is high, although the production of free radicals is expected to be low because O_2 is limiting. In the recovery period, when O_2 supply is restored, a burst of O_2 radicals is expected followed by a decrease because ATP synthesis is needed to regenerate the liver and muscle glycogen "borrowed" to carry out intense muscular activity in the sprint.

Hyperbaric O_2 exposure results in a marked increase in H_2O_2 production by isolated pigeon heart and rat liver mitochondria, in agreement with the almost linear dependence of mitochondrial $\cdot O_2^-$ production on O_2 tension. Hyperoxia and hyperbaric O_2 enhance H_2O_2 generation at the subcellular level and in isolated liver cells from 60 to 200%. However, the hyperbaric response appears to be greatly diminished at the organ level *in vivo*. The tissue O_2 level may be limited by the microvascular response and the endogenous rate of H_2O_2 generation may be substrate limited.

Substrate Availability. Rat liver mitochondria supplemented with FAD- or NAD-linked substrates produce H_2O_2 at 0.6–0.8 nmole/min per mg protein. However, it is important to remember that in State 3, i.e., when mitochondria are providing energy in the form of ATP to drive most of the biosynthetic reactions and energy-requiring reactions that constitute the living systems, the mitochondrial production of H_2O_2 drops to almost negligible values. Living organisms evolved complicated but efficient control processes to precisely regulate fuel oxidation and closely couple this to ATP synthesis, and also to ensure that these systems can respond rapidly to changing demands without compromising cellular energy homeostasis or misusing fuel reserves. Accordingly, the rate of O_2 radicals in mitochondria will be modulated by the dynamic changes characteristic of energy homeostasis. To evaluate the relative importance of the different mitochondrial states, it is thus necessary to know the particular conditions of the mitochondria within a tissue.

Another important aspect is to consider the substrate utilized by the mitochondria. The rate of H_2O_2 production will be greatly modulated by the type of oxidizable substrate available, and by the biochemical features of the mitochondria within a tissue. For example, during conditions of high lipid mobilization, such as starvation or diabetes, heart and kidney can obtain the bulk of their energy from the oxidation of ketone bodies, and even the brain, which normally uses only glucose as a fuel, will adapt to prolonged starvation by replacing up to 75% of its glucose requirements with ketone bodies. Although these mitochondria in State 4 (high substrate availability) show high rates of H_2O_2 production, these rates will vary with the availability of different substrates, such as octanoate, malate–glutamate, and succinate (Table VII).

Hormonal/ Metabolic Changes. Usually between 1 and 3% of the O_2 consumed "leaks" to the production of O_2 free radicals (Chance *et al.*, 1979). Then, it is expected that in conditions where the total tissue O_2 consumption is higher, such as hyperthyroidism (Pedersen *et al.*, 1963) or exercise (Davies *et al.*, 1982), the rates of H_2O_2 could be increased accordingly. Other metabolic conditions may include those with an increased level of ubiquinone, such as fasting, chronic treatment with dinitrophenol, cortisone treatment (Beyer *et al.*, 1962) or diabetes (Boveris *et al.*, 1969). These situations may lead to tissue damage linked to oxidative stress if the rates of production are not accompanied by an adequate increase in antioxidant defenses. This may account for the damaging effects (i.e., necrotic myopathy, focal necrosis) produced by acute heavy exercise as compared with the effects observed after daily exercise of low intensity (Salminen and Vihko, 1983).

5. CONCLUDING REMARKS

Research in free radicals in biology and biochemistry is becoming increasingly important because of its possible link to different pathophysiological states, with emphasis on mitochondrial diseases. This chapter has described several methods available to measure (or estimate) the steady-state concentration of $\cdot O_2^-$ and H_2O_2 in mitochondria which will be primarily responsible for increasing the physiological rate of free radical reactions. However, to interpret correctly the impact of these concentrations within a biological milieu, the limitations of these methods should be recognized: (1) The lack of homogeneity in microscopic systems may lead to under- or overestimation of metabolites

Table VII
Modulation of the Rate of Hydrogen Peroxide Production in Rat Liver Mitochondria by Different Substrates[a]

Substrate	$+d[H_2O_2]/dt$ (nmole/min/mg protein)
Endogenous	0.16
Malate–glutamate	
State 4	0.50
State 3	0.08
Succinate	
State 4	0.40
State 3	0.10
Octanoate	0.22
Palmitoylcarnitine	0.23

[a]Values taken from Boveris *et al.* (1972) and Boveris and Cadenas (1982).

or the occurrence of reactions. For example, the localized and transient increase of H_2O_2, which, in turn, may lead to site-specific damage, may go undetected by the available methodology. (2) The rate constants used in the calculations are extrapolated from chemical systems in dilute aqueous solutions; however, the intracellular milieu conditions may promote situations that do not occur in dilute solutions. For example, the mitochondrial matrix constitutes a unique environment constituted by a high protein concentration, on the order of 500 mg/ml (Hackenbrock, 1968), and a concomitant relative scarcity of water (Srere, 1980, 1982). The mitochondrial inner membrane surface is so extensive (the total mitochondrial inner membrane surface in one liver cell is 16,000 μm^2, or fivefold that of the cell membrane surface; Lehninger, 1964) that it has been calculated that a major proportion of matrix proteins must be adjacent to it (Srere, 1982). Thus, these conditions can favor protein–protein interactions involving both soluble and membrane proteins that might not take place in diluted solutions (McConkey, 1982). (3) The main consumption of H_2O_2 and $\cdot O_2^-$ is assumed to be of enzymatic nature, neglecting alternative pathways that may take place at the site where these metabolites are produced. (4) Although the estimation of the steady-state concentration of H_2O_2 using isolated organelles—which lack interorgan or interorganelle regulatory devices—or a monitoring system (i.e., the catalase intermediate) within the organ led to values (ca. 1×10^{-7} M, Table V; Chance *et al.,* 1979), the contribution of each source to the maintenance of this steady-state concentration varies according to the starting biological material used, i.e., subcellular fractions or perfused organs.

6. REFERENCES

Backstrom, D., Norling, B., Eherenberg, A., and Ernster, L., 1970, Electron spin resonance measurement on ubiquinone-depleted and ubiquinone-replenished submitochondrial particles, *Biochim. Biophys. Acta* **197:**108–111.

Bert, P., 1878, *Barometric Pressure Researches in Experimental Pathology* [translated from the 1878 French edition by M. A. Hitchcock and F. A. Hitchcock, Columbus, Ohio, College Book Co., 1943].

Beyer, R. E., Noble, W. M., and Hirschfeld, T. J., 1962, Alterations of rat-tissue coenzyme Q (ubiquinone) levels by various treatments, *Biochim. Biophys. Acta* **57:**376–379.

Boveris, A., and Cadenas, E., 1975, Mitochondrial production of superoxide anion and its relationship to the antimycin insensitive respiration, *FEBS Lett.* **54:**311–314.

Boveris, A., and Cadenas, E., 1982, Production of superoxide radicals and hydrogen peroxide in mitochondria, in *Superoxide Dismutase* (L. W. Oberley, ed.), pp. 15–30, CRC Press, Boca Raton.

Boveris, A., and Chance, B., 1973, The mitochondrial generation of hydrogen peroxide: general properties and effect of hyperbaric oxygen. *Biochem. J.* **134:**707–716.

Boveris, A., Peralta Ramos, M. C. de, Stoppani, A. O. M., and Foglia, V. G., 1969, Phosphorylation, oxidation, and ubiquinone content in diabetic mitochondria, *Proc. Soc. Exp. Biol. Med.* **132:**170–174.

Boveris, A., Oshino, N., and Chance, B., 1972, The cellular production of hydrogen peroxide, *Biochem. J.* **128:**617–630.

Boveris, A., Cadenas, E., and Stoppani, A. O. M., 1976, Role of ubiquinone in the mitochondrial generation of hydrogen peroxide, *Biochem. J.* **156:**435–444.

Boveris, A., Martino, E., and Stoppani, A. O. M., 1977, Evaluation of the horseradish peroxidase-scopoletin method for the measurements of hydrogen peroxide formation in biological systems. *Anal. Biochem.* **80:**145–158.

Cadenas, E., and Boveris, A., 1980, Enhancement of hydrogen peroxide formation by protophores and ionophores in antimycin-supplemented mitochondria, *Biochem. J.* **188:**31–37.

Cadenas, E., Boveris, A., Ragan, C. I., and Stoppani, A. O. M., 1977, Production of superoxide radicals and hydrogen peroxide by NADH-ubiquinone reductase and ubiquinol-cytochrome *c* reductase from beef heart mitochondria, *Arch. Biochem. Biophys.* **80:**248–257.

Chance, B., and Oshino, N., 1971, Kinetics and mechanisms of catalase in peroxisomes of the mitochondrial fraction, *Biochem. J.* **122:**225–233.

Chance, B., and Williams, G. R., 1956, The respiratory chain and oxidative phosphorylation, *Adv. Enzymol.* **17:**65–137.

Chance, B., Jamieson, D., and Coles, H., 1965, Energy-linked pyridine nucleotide reduction: Inhibitory effects of hyperbaric oxygen *in vitro* and *in vivo*, *Nature* **206:**257–263.

Chance, B., De Vault, D., Legallais, V., Mela, L., and Yonetani, T., 1967, Kinetics of electron transfer reactions in biological systems, in *Fast Reactions and Primary Processes in Chemical Kinetics* (S. Claesen, ed.), p. 437, Interscience, New York.

Chance, B., Boveris, A., Oshino, N., and Loschen, G., 1971, The nature of the catalase intermediate in its biological function, in *Oxidases and Related Redox Systems*, Vol. I (T. E. King, H. S. Mason, and M. Morrison, eds.), pp. 350–353, University Park Press, Baltimore.

Chance, B., Sies, H., and Boveris, A., 1979, Hydroperoxide metabolism in mammalian organs, *Physiol. Rev.* **59:**527–605.

Costa, L. E., Boveris, A., Koch, O. R., and Taquini, A. C., 1988, Liver and heart mitochondria in rats submitted to chronic hypobaric hypoxia, *Am. J. Physiol.* **255:**C123–C129.

Davies, K. J. A., Quintanilha, A. T., Brooks, G. A., and Packer, L., 1982, Free radicals and tissue damage produced by exercise, *Biochem. Biophys. Res. Commun.* **107:**1198–1205.

de Duve, C., 1965, The separation and characterization of subcellular particles, *Harvey Lect. Ser.* **59:**48–87.

Dionisi, O., Terranova, T., and Azzi, A., 1975, Superoxide radicals and hydrogen peroxide formation in mitochondria from normal and neoplastic tissues, *Biochim. Biophys. Acta* **403:**292–301.

Eanes, R. Z., and Kun, E., 1971, Separation and characterization of aconitase hydratase isoenzymes from pig tissues, *Biochim. Biophys. Acta* **227:**204–210.

Forman, H. J., and Boveris, A., 1982, Superoxide and hydrogen peroxide in mitochondria, in *Free Radicals in Biology*, Vol. 5 (W. B. Pryor, ed.), pp. 65–90, Academic Press, New York.

Forman, H., and Fridovich, I., 1973, Superoxide dismutase: A comparison of rate constants, *Arch. Biochem. Biophys.* **158:**396–400.

Forman, H. J., and Kennedy, J., 1976, Dihydroorotate-dependent superoxide production in rat brain and liver. A function of the primary dehydrogenase, *Arch. Biochem. Biophys.* **173:**219.

Fridovich, I., 1985, Cytochrome *c*, in *CRC Handbook of Methods for Oxygen Radicals Research* (R. A. Greenwald, ed.), pp.213–215, CRC Press, Boca Raton.

Fridovich, I., and Handler, P., 1961, Detection of free radicals generated during enzymic oxidation by the initiation of sulfite oxidation, *J. Biol. Chem.* **236:**1836–1840.

Gardner, P. R., and Fridovich, I., 1992, Inactivation–reactivation of aconitase in *Escherichia coli*, *J. Biol. Chem.* **267:**8757–8763.

Gardner, P., and White, C. W., 1995, Application of the aconitase method to the assay of superoxide in the mitochondrial matrices of cultured cells: Effects of oxygen, redox-cycling agents, TNF-α, IL-1, LPS, and inhibitors of respiration. in *The Oxygen Paradox* (K. J. A. Davies and F. Ursini, eds.), pp. 33–50, Cleup University Press, Padova, Italy.

Gerschman, R., 1964, Biological effects of oxygen. in *Oxygen in the Animal Organism* (F. Dickens, and E. Neil, eds.), pp. 475–494, Pergamon Press, Elmsford, NY.

Gerschman, R., Gilbert, D. L., Nye, S. W., Dwyer, P., and Fenn, W. O., 1954, Oxygen poisoning and X-irradiation: A mechanism in common, *Science* **119:**623–626.

Giulivi, C., 1989, Metabolism of hydroperoxides in eukaryotic cells, Ph.D. dissertation, University of Buenos Aires, Argentina.

Giulivi, C., and Cadenas, E., 1998, The role of mitochondrial glutathione in DNA base oxidation, *Biochem. Biophys. Acta*, in press.

Giulivi, C., Boveris, A., and Cadenas, E., 1995, Hydroxyl radical generation during mitochondrial electron transfer and the formation of 8-hydroxydesoxyguanosine in mitochondrial DNA, *Arch. Biochem. Biophys.* **316:**909–916.

Glusker, J. P., 1971, Aconitase, in *The Enzymes*, Vol. V (P. D. Boyer, ed.), pp. 413–439, Academic Press, New York.

Hackenbrock, C. R., 1968, Chemical and physical fixation of isolated mitochondria in low-energy and high-energy states, *Proc. Natl. Acad. Sci. USA* **61:**598–605.

Halliwell, B., and Gutteridge, J. M. C., 1989, in *Free Radicals in Biology and Medicine*, Oxford University Press (Clarendon), London.

Kennedy, M. C., Emptage, M. H., Dreyer, J.-L., and Beinert, H., 1983, The role of iron in the activation-inactivation of aconitase. *J. Biol. Chem.* **258:**11098–11105.

Ksenzenko, M., Konstantinov, A., Khomutov, G. B., Tikhonov, A. N., and Ruuge, E., 1983, Effect of electron transfer inhibitors on superoxide generation in the cytochrome bc_1 site of the mitochondrial respiratory chain, *FEBS Lett.* **155:**19–24.

Lavoisier, A. L., 1783, *Memoires de Médecine et de Physique Médicale*, Vol. 5, p. 569, Societe Royal de Medicine, Paris.

Lehninger, A. L., 1964, Mitochondria in the intact cell, in *The Mitochondrion: Molecular Basis of Structure and Function*, pp. 17–40, Benjamin, New York.

Loschen, G., Flohé, L., and Chance, B., 1971, Respiratory chain linked H_2O_2 production in pigeon heart mitochondria, *FEBS Lett.* **18:**261–264.

Loschen, G., Azzi, A., and Flohé, L., 1973, Mitochondrial H_2O_2 formation: Relationship with energy conservation. *FEBS Lett.* **33:**84–88.

Loschen, G., Azzi, A., Richter, C., and Flohé, L., 1974a, Superoxide radicals as precursors of mitochondrial hydrogen peroxide, *FEBS Lett.* **42:**68–72.

Loschen, G., Azzi, A., and Flohé, L., 1974b, Mitochondrial hydrogen peroxide formation, in *Alcohol and Aldehyde Metabolizing Systems* (R. G. Thurman, T. Yonetani, J. R. Williamson, and B. Chance, eds.), pp. 215–229, Academic Press, New York.

Massa, E.M., and Giulivi, C., 1993, Alkoxyl and methyl radical formation during cleavage of *tert*-butylhydroperoxide by a mitochondrial membrane-bound, redox active pool of copper: An EPR study, *Free Radical Biol. Med.* **14:**559–565.

McConkey, E. H., 1982, Molecular evolution, intracellular organization, and the quinary structure of proteins, *Proc. Natl. Acad. Sci. USA* **79:**3236–3240.

Menzel, D. B., 1970, Toxicity of ozone, oxygen, and radiation, *Annu. Rev. Pharmacol.* **10:**379–394.

Morrison, J. F., 1954, The purification of aconitase, *Biochem. J.* **56:**99–105.

Oshino, N., Chance, B., Sies, H., and Bücher, T., 1973a, The role of H_2O_2 generation in perfused rat liver and the reaction of catalase compound I and hydrogen donors, *Arch. Biochem. Biophys.* **154:**117–131.

Oshino, N., Oshino, R., and Chance, B., 1973b, The characteristics of the peroxidatic reaction of catalase in ethanol oxidation, *Biochem. J.* **131:**555–563.

Oshino, N., Jamieson, D., Sugano, T., and Chance, B., 1975a, Optical measurement of the catalase–hydrogen peroxide intermediate (compound I) in the liver of anaesthesized rats and its implication to hydrogen peroxide production *in situ*, *Biochem. J.* **146:**67–77.

Oshino, N., Jamieson, D., and Chance, B., 1975b, The properties of hydrogen peroxide production under hyperoxic and hypoxic conditions of perfused rat liver, *Biochem. J.* **146:**53–65.

Pedersen, S., Tata, J. R., and Ernster, E., 1963, Ubiquinone (coenzyme Q) and the regulation of basal metabolic rate by thyroid hormones, *Biochim. Biophys. Acta* **69:**407–409.

Priestley, J., 1794, *The Discovery of Oxygen, Part 1*, Alembic Club Reprints, No. 7, Simpkin, Marshall, Hamilton and Kent, London.

Ragan, C. I., Cadenas, E., Boveris, A., and Stoppani, A. O. M., 1977, Production of superoxide radicals and hydrogen peroxide by NADH-ubiquinone reductase and ubiquinol-cytochrome *c* reductase from beef heart mitochondria, *Arch. Biochem. Biophys.* **180:**248–257.

Rich, P. A., Boveris, A., Bonner, W. D., Jr., and Moore, A. L., 1976, Hydrogen peroxide generation by the alternate oxidase of higher plants, *Biochem. Biophys. Res. Commun.* **71:**695–703.

Richter, C., Park, J.-W., and Ames, B., 1988, Normal oxidative damage to mitochondrial and nuclear DNA is extensive, *Proc. Natl. Acad. Sci. USA* **85:**6465–6467.

Ruzicka, F. J., and Crane, F. L., 1970, Quinone interaction with the respiratory chain-linked NADH dehydrogenase of beef heart mitochondria, *Biochim. Biophys. Acta* **223:**71–85.

Salminen, A., and Vihko, V., 1983, Lipid peroxidation in exercise myopathy, *Exp. Mol. Pathol.* **38:**380–388.

Scheele, C. W., 1782, *Chemische Abhandlungen von Luft und Feuer*, 2nd ed., p. 132.

Schnaitman, C., and Greenawalt, J. W., 1968, Enzymic properties of the inner and outer membranes of rat liver mitochondria, *J. Cell Biol.* **38:**158–168.

Sies, H., and Chance, B., 1970, The steady-state level of catalase compound I in isolated hemoglobin-free perfused rat liver, *FEBS Lett.* **11**:172–176.
Sies, H., Bücher, T., Oshino, N, and Chance, B., 1973, Heme occupancy of catalase in hemoglobin-free perfused rat liver and isolated rat liver catalase, *Arch. Biochem. Biophys.* **154**:106–116.
Srere, P. A., 1980, The infrastructure of the mitochondrial matrix, *Trends Biochem. Sci.* **5**:120–121.
Srere, P. A., 1982, The structure of the mitochondrial inner membrane-matrix compartment, *Trends Biochem. Sci.* **7**:375–378.
Sugano, T., Oshino, N., and Chance, B., 1974, Mitochondrial functions under hypoxic conditions. The steady-states of cytochrome *c* reduction and of energy metabolism, *Biochim. Biophys. Acta* **347**:340–358.
Thurman, R. G., Ley, H. G., and Scholz, R., 1972, Hepatic microsomal ethanol oxidation, hydrogen peroxide formation, and the role of catalase, *Eur. J. Biochem.* **25**:420.
Trounce, I., Byrne, E., and Marzuki, S., 1989, Decline in skeletal muscle mitochondrial respiratory chain function: possible factor in ageing, *Lancet* **1**:637–639.
Turrens, J. F., and Boveris, A., 1980, Generation of superoxide anion by the NADH-dehydrogenase of bovine heart mitochondria, *Biochem. J.* **191**:421–427.
Tyler, D. D., 1975, Polarographic assay and intracellular distribution of superoxide dismutase in rat liver, *Biochem. J.* **147**:493–504.
Tzagoloff, A., and Myers, A.M., 1986, Genetics of mitochondrial biogenesis, *Annu. Rev. Biochem.* **55**:249–285.
Villafranca, J. J., and Mildvan, A. S., 1971, The mechanism of aconitase action, *J. Biol. Chem.* **246**:772–779.
von Jagow, G., and Link, T. A., 1986, Use of specific inhibitors on the mitochondrial bc_1 complex, *Methods Enzymol.* **126**:253–271.
von Jagow, G., Link, T. A., and Ohnishi, T., 1986, Organization and function of cytochrome *b* and ubiquinone in the cristae membrane of beef heart mitochondria, *J. Bioenerg. Biomembr.* **18**:157–179.
Wallace, D.C., 1992a, Diseases of the mitochondrial DNA, *Annu. Rev. Biochem.* **61**:1175–1212.
Wallace, D.C., 1992b, Mitochondrial genetics: A paradigm for aging and degenerative diseases? *Science* **256**:628–632.
Yonetani, T., and Ray, G. S., 1965, Studies on cytochrome *c* peroxidase. I. Purification and some properties. *J. Biol. Chem.* **240**:4503–4508.

Chapter 4

The Role of Transition Metal Ions in Free Radical-Mediated Damage

Mordechai Chevion, Eduard Berenshtein, and Ben-Zhan Zhu

1. INTRODUCTION

Molecular oxygen (O_2) probably appeared on the Earth's surface about 2×10^9 years ago as a result of photosynthetic microorganisms acquiring the ability to split water. Oxygen is now the most abundant element in the biosphere. Its concentration in dry air has risen to 21%. Iron is the second most abundant metal in the Earth's crust whereas copper is more scarce.

A free radical is an atom or a molecule with one or more unpaired electrons. This definition includes the oxygen molecule (a biradical), a hydrogen atom, and most of the transition metal ions. To avoid a spin, and possibly, orbital restriction, oxygen accepts electrons one at a time. This considerably slows down the reaction of oxygen with the majority of covalent molecules, which are nonradicals, and can be considered an advantage for life in oxygen. A major disadvantage is, however, that electrons when added singly to oxygen lead to the formation of reactive intermediates, two of which are free radicals. The four-electron reduction of oxygen to water gives rise sequentially to the superoxide anion radical ($\cdot O_2^-$), hydrogen peroxide (H_2O_2), hydroxyl radical ($\cdot OH$), and water (Halliwell and Gutteridge, 1989).

It has been recognized since the late nineteenth century that oxygen, which is essential for most living systems, is also inherently toxic (Frank and Massaro, 1980; Cadenas,

---

Mordechai Chevion, Eduard Berenshtein, and Ben-Zhan Zhu Department of Cellular Biochemistry, Hebrew University–Hadassah Schools of Medicine and Dental Medicine, Jerusalem 91120, Israel.

Reactive Oxygen Species in Biological Systems, edited by Gilbert and Colton. Kluwer Academic / Plenum Publishers, New York, 1999.

1989). In fact, long before this time, one of the two discoverers of oxygen, Joseph Priestley, suggested that "as a candle burns much faster in dephlogisticated than in common air, so we might, as may be said, live out too fast, and the animal powers be too soon exhausted in this pure kind of air"(Priestley, 1906).

It was not until 1954 (Gerschman *et al.*, 1954; Gerschman, 1981) that it was proposed that most of the damaging effects of elevated oxygen concentrations in living organisms could be attributed to the formation of free radicals. However, this idea did not capture the interest of many biologists and clinicians until the discovery in 1969 of an enzyme specific for the catalytic removal of $\cdot O_2^-$, the superoxide dismutase (SOD) (McCord and Fridovich, 1969). This was a major turning point in the research and understanding of free radicals in biology and medicine. Soon after the discovery of SOD, it was found that $\cdot OH$ appeared to be produced in the superoxide-generating system (Beauchamp and Fridovich, 1970). The mechanism of $\cdot OH$ production was proposed through the following Haber–Weiss reaction:

$$\cdot O_2^- + H_2O_2 \rightarrow \cdot OH + OH^- + O_2$$

Studies of the kinetics of this reaction found it to be very slow at physiological conditions (Ferradini *et al.*, 1978). So, the question of how $\cdot OH$ is formed in $\cdot O_2^-$ -generating systems remained to be answered.

The catalytic role of transition metals, particularly iron and copper, was not appreciated until the Pinawa meeting in 1977, when it was presented that adventitious levels of iron in buffer solutions were on the order of 1 μM, and that this level of iron would change the results observed in $\cdot O_2^-$ -generating systems (Czapski and Ilan, 1978; Koppenol *et al.*, 1978). Just as important, it was shown that chelating agents will alter the reactivity of iron in $\cdot O_2^-$ -generating systems. It was demonstrated that EDTA enhances the reactivity of iron toward $\cdot O_2^-$ whereas diethylenetriaminepentaacetic acid (DTPA) and desferrioxamine (DFO) dramatically slow it down (Buettner *et al.*, 1978; Gutteridge *et al.*, 1979). Based on the above considerations, it was suggested that traces of soluble iron or copper can catalyze the transformation of $\cdot O_2^-$ to the highly reactive $\cdot OH$, via the metal-catalyzed Haber–Weiss reaction (or $\cdot O_2^-$ -driven Fenton reaction) (Haber and Weiss, 1934; Anbar and Levitzki, 1966; Borg *et al.*, 1978; Halliwell, 1978; Koppenol *et al.*, 1978; McCord and Day, 1978; Borg and Schaich, 1984; Czapski, 1984; Aust *et al.*, 1985; Aronovitch *et al.*, 1986; Goldstein and Czapski, 1986; Chevion, 1988; Halliwell and Gutteridge, 1989; Stadtman, 1990):

$$\cdot O_2^- + \mathrm{Fe(III)/Cu(II)} \rightarrow O_2 + \mathrm{Fe(II)/Cu(I)} \quad (1)$$

$$\mathrm{Fe(II)/Cu(I)} + H_2O_2 \rightarrow \mathrm{Fe(III)/Cu(II)} + OH^- + \cdot OH \quad (2)$$

These metal ions are rather insoluble under physiological conditions, and remain in solution only by becoming complexed to low- or high-molecular weight cellular components. Consequently, they serve as catalytic centers for free radical production. For example, copper forms stable complexes with many proteins (Shinar *et al.*, 1983), whereas iron makes stable complexes with nucleotide diphosphates and triphosphates. For iron, partial displacement of organic ligand by hydroxide anions renders $Fe(OH)^+$

suitable for the Fenton reaction. This is on both a kinetic account, where its rate with H_2O_2 is 2×10^6 M^{-1} s^{-1}, and on a thermodynamic basis where $E°$ for $Fe(OH)^{2+}/Fe(OH)^+$ = 0.304 V, rather than ~ 0.77 V, which is usually cited for Fe^{3+}/Fe^{2+}. Once these transition metal ions are completely displaced from such complexes and become aquo-complexes in the bulk solution, they polymerize and precipitate out as unreactive polynuclear structures (Chevion, 1988).

Using the Marcus theory and the experimental data in the literature, it has been known that in most cases the reaction of these metal complexes with H_2O_2 is unlikely to occur via an outer-sphere electron-transfer mechanism. It is suggested that the first step in this process is the formation of a transient complex, which may decompose to a ·OH radical or a higher oxidation state of the metal. Alternatively, it may yield an organic free radical in the presence of organic substrates. Thus, the question of whether ·OH radicals are being formed or not, via the Fenton reaction, depends on the relative rates of the decomposition reactions of the metal–peroxide complex, and that of its reaction with organic substrates (Goldstein *et al.*, 1993).

2. THE SITE-SPECIFIC MECHANISM OF METAL-MEDIATED PRODUCTION OF FREE RADICALS

2.1. General

The site-specific mechanism of free radical formation and the subsequently induced damage stem from the assumption that copper or iron complexes are the essential mediators of the damage (Samuni *et al.*, 1980, 1981, 1983, 1984; Gutteridge and Wilkins, 1983; Kohen and Chevion, 1986; Aronovitch *et al.*, 1986; Yamamoto and Kawanishi, 1989). These biologically bound metal ions can undergo redox cycling. Reducing agents such as $\cdot O_2^-$, ascorbate, the favism-inducing agents, thiols, phenolic compounds, hydrazines, as well as other molecules could reduce the metal ion within its complex yielding the cupro or ferro states, and concurrently form $\cdot O_2^-$ and/or H_2O_2. Subsequently, these reduced metal ions can react with hydrogen peroxide (the Fenton reaction) yielding the ·OH [Reaction (2)]. The site-specific metal-mediated mechanism is characterized as follows.

1. It explains the funneling of free radical damage to the specific metal-binding sites. Relatively low-reactive reducing agents such as $\cdot O_2^-$ or ascorbate, whose life spans are comparatively long, can migrate a relatively long way until they encounter a redox-active metal ion and react with it.
2. It explains the transformation of rather low-reactive species, such as $\cdot O_2^-$ or ascorbate, to the highly reactive species such as ·OH, which is known to cause a variety of molecular disruptions (Walling, 1982). Generally, the ·OH is characterized by very high kinetic rate constants for its reaction with a variety of biological molecules or residues. This radical can act by causing breaks in the polymeric backbone of a macromolecule, by abstracting a hydrogen, or by addition to a double bond.
3. It explains the possible "multihit" effects that are often experimentally observed. If we assume that the off-rate of copper or iron is slow enough that its spatial

locus is fixed, then the metal can undergo more than a single redox cycle. Alternatively, repeated redox cycles can take place at the same locus by rapidly exchanging transition metal. In each such cycle, one ·OH is released and the repeated redox reactions account for the production of multiple ·OH. Because of its high reactivity, it is likely that the ·OH will inflict its damage within a few encounters, i.e., near the site of its formation—the metal-binding site. Thus, this mechanism demonstrates how multihits by ·OH will take place near the metal-binding site.

These features indicate that the site-specific mechanism could explain either an amplification or a dampening of the damage caused by a given number of ·OH radicals. If the metal-binding site is important for function or activity, or for spatial orientation or conformation, then the funneling effect will cause marked amplification of the biological damage, because all of the deleterious radicals will be produced and will react at this pertinent site. In contrast, if the metal-binding site is relatively unimportant, the funneling effect will direct the damage to the redundant site, and the recorded biological damage will be noticeably reduced.

The free-radical-induced DNA break is an illustrative example of the amplification for such a mechanism (Chevion, 1988). DNA damage can take place in a variety of modes, including breaks, depurination–depyrimidation, and chemical modification of the bases or the sugar. When a ·OH radical hits the DNA polymer, it could result in a single-strand break (SSB). The mechanism of such a SSB involves the addition of ·OH to C=C or an abstraction of H atoms from the sugar moiety. The rate determining step of this process is the rate of heterolytic cleavage of the phosphate bond. SSBs can be repaired, in most cells, by a variety of repair mechanisms. Thus, a low incidence of SSBs on the DNA is usually efficiently tolerated. In contrast, when the two strands are broken at spatially nearby sites, a double-strand break (DSB) is formed and cannot be properly corrected. This will lead to a loss of genetic information and eventually to death of the cell. In the presence of copper and ascorbate, the site-specific mechanism of DNA breaks will lead to multihit effects at the copper-binding sites. By using an *in vitro* system of purified DNA, repair processes have been eliminated, allowing precise quantitation of the total number of strand breaks produced during exposure to an ascorbate/copper mixture. In this mechanism, ascorbate could play a dual role: to produce the necessary hydrogen peroxide in a copper-catalyzed reaction, and to reduce DNA-bound copper which would then serve as a center for repeated and site-specific production of ·OH. This has resulted in a high incidence of DSBs, compared with DSBs that were formed by ionizing radiation. It has been shown that more than 100 SSBs are necessary to produce one DSB by γ–irradiation (Kohen *et al.*, 1986). In contrast, when DNA was exposed to ascorbate and copper and underwent cleavage by the site-specific mechanism, one DSB for every 2.3 SSBs was observed. This high yield of DSBs, representing an increase by a factor of more than 40, can be directly attributed to the multihits of ·OH. This increase in probability of DSBs may provide a predictor for the potential hazards to DNA from various xenobiotics.

It should be noted that in some cases, hydroxyl radicals can be produced by organic Fenton reactions, which are metal-independent (Koppenol and Butler, 1985; Nohl and Jordan, 1987).

2.2. The Role of Transition Metal Ions in Converting Low Reactive Molecules to Highly Reactive Species

A number of enzymatic systems and chemical oxidants have been implicated in the activation of carcinogens. Studies have also provided evidence suggesting that transition metals, particularly copper, but also iron and manganese, are capable of directly mediating the metabolic activation of xenobiotics (Table I). This can be explained by virtue of the site-specific formation of more reactive species such as ·OH, organic free radicals, and other electrophiles (Emerit and Cerutti, 1981; Birnboim, 1982, 1992; Rao, 1991; Rumyantseva *et al.*, 1991; Swauger *et al.*, 1991; Yourtee *et al.*, 1992; Li and Trush, 1993a,b, 1994).

Copper is a redox-active essential element, playing important roles in redox-active centers in a variety of metalloenzymes, such as in cytochrome *c* oxidase, which is critical for oxidative phosphorylation. Other copper-containing enzymes include lysyl oxidase, which is involved in the cross-linking of elastin and collagen; tyrosinase, which is needed for melanin pigment formation; dopamine β-hydroxylase, necessary for catecholamine production and therefore nerve and metabolic function; SOD, specializing in the disposal of potentially damaging $\cdot O_2^-$ produced in normal metabolic reactions; ceruloplasmin, a potential extracellular free radical scavenger as well as a ferroxidase in blood plasma and other extracellular fluids; and tryptophan oxygenase, ascorbate oxidase, polyphenol oxidase, and lactase. Copper-containing proteins without known enzymatic function have also been identified, and the association of copper with amino acids, small peptides, and perhaps other metabolites has also been reported (Howell and Gawthorne, 1987; Beinert, 1991; Linder, 1991). The physiological roles of copper in mammals and/or humans are associated with erythropoiesis, the development of connective tissues, the development of bones, the development of central nervous system, immune competence, and pigmentation. Studies on the interaction of copper with biomolecules have also demonstrated that copper exists in chromosomes (Wacker and Vallee, 1959; Bryan *et al.*, 1981) and is closely associated with DNA bases, particularly G–C sites (Pezzano and Podo, 1980;

Table I
Transition-Metal-Mediated Conversion of Low-Reactive Molecules into Highly Reactive Species

Low-reactive molecules	Metals involved	Highly reactive species
Superoxide	Cu/Fe	·OH (possibly hypervalent metal ions)
Paraquat	Cu/Fe	·OH, $\cdot O_2^-$
Phorbol esters	Cu/Fe	·OH, $\cdot O_2^-$
Ascorbate	Cu/Fe	·OH, $\cdot O_2^-$, and ascorbyl radical
Flavins	Cu/Fe	·OH, $\cdot O_2^-$, carbon-centered or other radicals
Pyrimidines	Cu/Fe	·OH, $\cdot O_2^-$, carbon-centered or other radicals
Lipid peroxides	Cu/Fe	Peroxy and alkoxy radicals, cytotoxic aldehydes
Phenolic compounds	Cu	·OH, $\cdot O_2^-$, semiquinone, and quinone
Hydrazines	Cu/Fe/Mn	·OH, $\cdot O_2^-$, nitrogen-centered and carbon-centered radicals
Thiols	Cu	·OH, $\cdot O_2^-$, and sulfur-centered radical
Bleomycin	Fe	·OH, cytotoxic propenals

Agarwal *et al.*, 1989; Geierstanger *et al.*, 1991). DNA-associated copper may be involved in maintaining normal chromosome structure and in gene regulatory processes (Prutz *et al.*, 1990; O'Halloran, 1993). Besides the physiological roles of copper, it is also involved in pathological processes. The involvement of copper in xenobiotic activation and damages has been the subject of recent toxicological studies.

2.3. Natural and Xenobiotic Molecules Participating in Site-Specific Damage

2.3.1. Superoxide Radical

Important biological roles of $\cdot O_2^-$ are associated with its reaction with nitric oxide to produce peroxynitrite (for details, see Part III of this volume). There is no compelling evidence that this reaction is mediated by transition metals.

Purified solutions of the enzyme penicillinase were γ-irradiated in the presence of formate, a highly effective scavenger of ·OH and solvated electrons. In the presence of formate, most of the primary radicals produced by radiation are transformed to the formate radical $\cdot CO_2^-$ that, in the presence of oxygen, is efficiently converted to $\cdot O_2^-$. Copper ions dramatically enhanced the $\cdot O_2^-$-induced inactivation of the enzyme, whereas EDTA completely eliminated this effect. It has also been shown that in the absence of metal ions, $\cdot O_2^-$ is harmless toward phages and bacteria (Samuni *et al.*, 1980, 1981, 1983, 1984).

In these studies, the $\cdot O_2^-$ could play two roles: reducing the copper in its complex with a biological molecule [Reaction (1)], and providing H_2O_2 (by spontaneous or SOD-driven dismutation) that is required for the production of the ultimate damaging species, ·OH [Reaction (2)]. It should be mentioned, however, that there are cases in which $\cdot O_2^-$ per se has been shown to exert its toxicity (without the requirement of adventitious metals). In addition, $\cdot O_2^-$ has been shown to react with variable rapidity with epinephrine, 6-hydroxydopamine, catechols, vitamin E, and many other biologically relevant compounds (Chevion, 1988).

2.3.2. Ascorbate

Ascorbate plays a key role in protecting cells against oxidative damage (Cameron *et al.*, 1979; Frei *et al.*, 1989). Paradoxically, in the presence of traces of ferric or cupric ions, ascorbate can promote the generation of the same reactive species it is known to destroy. This prooxidant activity derives from the ability of ascorbate to reduce Fe^{3+} or Cu^{2+} to Fe^{2+} or Cu^+, respectively, and to reduce O_2 to $\cdot O_2^-$ and H_2O_2.

The prooxidant activity of ascorbate has been extensively used to cause biological damage, in particular in the presence of adventitious copper, to enzymes, proteins, and nucleic acids (reviewed by Chevion, 1988). An example of enzyme inactivation that is in accord with the site-specific mechanism is the mechanism of inhibition of catalase by ascorbate (Davison *et al.*, 1986). Ascorbate reversibly inhibits catalase, and this inhibition is enhanced and rendered irreversible by the prior addition of Cu(II)-bis-histidine. The failure of conventional scavengers of active forms of oxygen to protect the activity of catalase has led the investigators to conclude that the mechanism might involve the site-specific generation of damaging intermediates, probably ·OH. An additional example is that when albumin is exposed to copper and ascorbate (Marx and Chevion, 1986), it is site-specifically cleaved, yielding peptide fragments of specific lengths.

Ascorbate has also been extensively used in combination with copper or iron complexes to generate SSBs and DSBs in DNA (Stoewe and Prutz, 1987). It is assumed that the metal or the chelate binds to the DNA either at phosphate residue or sugar on the backbone, or at the purine/pyrimidine bases. There the metal or the chelate serves as a center for repetitive formation of oxygen active species. The addition of copper ions and ascorbate to DNA causes a dramatic increase in DSBs (efficiency of about 85%), compared with random hits. A high incidence of DSBs has also been identified when the Cu^{2+}–1,10-phenanthroline complex was used together with ascorbate to induce DNA scission (Sigman, 1986; Thederahn *et al.*, 1989; Dizdaroglu *et al.*, 1990). The Cu^{+}–1,10-phenanthroline complex binds to the minor groove of DNA, and the activated species, whether ·OH or a metal-oxo species or another metal-bound species, abstracts a hydrogen atom specifically from C-1′ of the DNA sugar. This site-specific DNA cleavage is also in accord with the site-specific mechanism. Also, by using copper-2,2′-bipyridyl, where the majority of ·OH are formed in the bulk solution, and not on the DNA, mainly SSBs have been generated (Chevion, 1988).

2.3.3. Pyrimidines

Isouramil and divicine are two naturally occurring pyrimidines found in broad beans (*Vicia faba*). They are incriminated as the agents responsible for the acute hemolytic crisis (favism) following the ingestion of broad beans in glucose-6-phosphate dehydrogenase-deficient human subjects (Chevion *et al.*, 1983; Navok and Chevion, 1984).

The chemical structure of these two pyrimidines is very similar to that of dialuric and ascorbic acids. All of these compounds are chemical reductones and contain an "ene-diol" or "enolamine" group which is activated by a carbonyl group at the α-position. They can undergo a two-electron redox cycle, yielding their oxidized form, $\cdot O_2^-$ and H_2O_2 (Chevion, 1988).

Isouramil could cause enzyme inactivation in a mechanism that is identical to that suggested for ascorbate. As the pathogenesis of favism is not yet completely understood, the level of trace elements, and in particular labile and redox-active iron and copper, may have a significant bearing on the incidence and severity of the onset of the favic crisis.

2.3.4. Paraquat

Paraquat (PQ^{2+}) (1,1′-dimethyl-4,4′-bipyridylium dichloride) is a widely used herbicide that was first synthesized at the end of the nineteenth century. It was primarily used as a redox indicator (also known as methyl viologen). Since 1966, with its increasing agricultural use throughout the world, there have been a large number of cases of systemic intoxication resulting from ingestion of paraquat. Today, paraquat is known to be highly lethal to people and animals, and its use has become more limited.

Studies of the chemistry of the paraquat radical showed that it reacts quickly with oxygen, yielding $\cdot O_2^-$. Thus, paraquat toxicity can be regarded as a bona fide model for

$\cdot O_2^-$ toxicity. The mechanism of paraquat toxicity has been extensively studied in bacterial model and mammalian systems (reviewed by Chevion, 1988).

The essential mediatory role of copper and iron in paraquat toxicity could be clearly demonstrated. Several chelators of iron and copper minimized the effect of paraquat toxicity. There is only a small amount of endogenously available metal ions in the cells, and paraquat can utilize them to catalyze its toxic effect. Comparison between the corresponding killing rate constant shows that on a molar basis, copper is a 24-fold more efficient mediator of paraquat toxicity than iron.

The different efficiencies of catalysis of the paraquat-induced deleterious effects by either iron or copper may reflect artifactual as well as mechanistic reasons. Because of low solubility, it is not easy to introduce Fe^{3+} into cells. Thus, the level of available iron in the system might be only a small faction of the total added ions. Copper, on the other hand, is much more soluble so that the concentration of redox-active copper could equal to the total added amount. Mechanistically, the differences between the effects of iron and copper may be related to different rates of reduction of the metal complex by either $\cdot O_2^-$ or paraquat radical, or to different rate constants for the reduced metal complex in the Fenton reaction (Koppenol and Butler, 1985; Stern, 1985) yielding the highly reactive $\cdot OH$.

It has been shown that DFO protects against paraquat-induced toxicity both in mice (Kohen and Chevion, 1985) and in plants (Peleg *et al.*, 1992), DFO was also tried out successfully on severely paraquat-intoxicated human subjects in Holland, and moderately intoxicated subjects in Israel.

2.3.5. Peroxides

A major portion of the toxicity of H_2O_2 in *E. coli* is attributed to DNA damage mediated by the Fenton reaction, which generates $\cdot OH$ from H_2O_2, DNA-bound iron, and a constant source of reducing equivalents. Kinetic peculiarities of DNA damage produced by H_2O_2 *in vivo* can be reproduced by DNA in an *in vitro* Fenton reaction system in which iron catalyzes the univalent reduction of H_2O_2 by NADH. Iron chelators such as DFO and 1,10-phenanthroline were found to protect both bacterial and mammalian cells against H_2O_2-induced toxicity (Mello Filho and Meneghini, 1985; Imlay *et al.*, 1988). Besides H_2O_2, transition metals have also been implicated in the activation of several organic peroxides, such as benzoyl peroxide and *t*-butyl hydroperoxide (Swauger *et al.*, 1991; Massa and Giulivi, 1993). Benzoyl peroxide is both a tumor promoter and a progressor in mouse skin (O'Connell *et al.*, 1986). Studies have demonstrated that benzoyl peroxide can be activated to DNA-damaging intermediates via copper-catalyzed cleavage of the peroxide bond, formation of the benzoyloxyl radical, which may then produce sensitive sites in DNA through hydrogen-abstraction reactions. Further investigations on the effects of Cu^+ on the mutagenesis of benzoyl peroxide in the *supF* gene of a reporter plasmid have shown that the mutation frequency caused by benzoyl peroxide alone is enhanced by the presence of Cu^+. Furthermore, DNA-sequencing studies showed that 22 of 25 point mutations examined occurred at G-C to A-T transversions. These results demonstrate that the interaction between benzoyl peroxide and copper produces site-specific promutagenic DNA damage. A more recent study has demonstrated that reaction of benzoyl peroxide and Cu^+ generates $\cdot OH$ or a similar reactive intermediate that causes oxidative DNA base modification, including the formation of 8-hydroxyguanosine. This oxidative DNA base modification may be responsible for benzoyl peroxide/Cu^+-mediated mutagenesis (Akman *et al.*, 1992, 1993).

2.3.6. Phenolic Compounds

Phenolic compounds are widely distributed in nature. They are also prevalent as environmental pollutants and drugs. Some phenolic compounds are mutagenic and/or carcinogenic (Irons, 1985; Nohl *et al.*, 1986). Two enzymatic systems, the cytochrome P450 family and the peroxidases, as well as transition metals have been implicated in the activation of various phenolic compounds. The activation of phenolic compounds via a Cu^{2+}/Cu^+ redox cycle has been demonstrated (Kawanishi *et al.*, 1989; Rahman *et al.*, 1989; Inoue *et al.*, 1990; Ahmad *et al.*, 1992; Calderaro *et al.*, 1993; Li and Trush, 1993a,b, 1994; Naito *et al.*, 1994). These compounds include carcinogenic benzene-derived metabolites such as 1,4-hydroquinone (HQ), catechol, and 1,2,4-benzenetriol (Kari *et al.*, 1992; Snyder *et al.*, 1993), polychlorophenol-derived metabolites such as tetrachloro-1,4-hydroquinone (TCHQ), and flavonoids such as quercetin, and *O*-phenylphenol-derived metabolites.

Recent studies have demonstrated that Cu^{2+} strongly catalyzes the oxidation of HQ to 1,4-benzoquinone and H_2O_2, through a semiquinone anion radical intermediate, and a Cu^{2+}/Cu^+ redox-cycle mechanism is critical for such an activation process. The Cu^{2+}-mediated oxidation of HQ is biologically relevant. For example, the presence of copper in primary bone marrow stromal cell cultures significantly enhances HQ-induced cytotoxicity (Li and Trush, 1993a). Furthermore, oxidation of HQ by Cu^{2+} has been demonstrated to result in the formation of DNA strand breaks (Li and Trush, 1993b). The DNA cleaving activity of the HQ/Cu^{2+} system is dependent on a Cu^{2+}/Cu^+ redox cycle, the presence of oxygen, and the generation of H_2O_2. Experiments using scavengers demonstrated that the site-specific generation of reactive oxygen intermediates could account for the HQ/Cu^{2+}-induced DNA damage.

Dopa, dopamine, adrenaline, and noradrenaline are derivatives of the aromatic amino acid phenylalanine, and they are often referred to collectively as *catecholamines*. Catecholamines are important physiological regulators of cardiac contractility and metabolism, and their oxidation products could contribute to myocardial damage, as the heart is sensitive to oxidant stress. The oxidation of catecholamines can produce $\cdot O_2^-$ and H_2O_2 in a complicated series of reactions. The rate of oxidation is greatly accelerated by the presence of transition metal ions, and it produces not only oxygen-derived species, but also quinones and semiquinones (Halliwell and Gutteridge, 1989; Rao, 1991). These can combine with various cellular constituents to form covalent bonds, usually with thiol groups. Hence, they can deplete cellular GSH concentrations. GSH plays a key role in drug metabolism and antioxidant defense systems, so its depletion can render the cell more sensitive to oxidant stress.

Most quinone antibiotics undergo redox cycling to produce $\cdot O_2^-$ and H_2O_2 *in vivo* (Powis, 1989). Rifamycin SV, an antibiotic often used in the treatment of tuberculosis, can oxidize in the presence of transition metal ions to give a quinone (called rifamycin S), with intermediate formation of a semiquinone, and $\cdot O_2^-$ (Halliwell and Gutteridge,

1989). It has been suggested that the bactericidal action of rifamycin SV may involve increased oxidant generation within the bacteria, as well as prevention of bacterial RNA synthesis. Rifamycin has many biological and chemical properties similar to those of the transition metal chelator 1,10-phenanthroline. Both have been shown to mediate ·OH dependent damage to DNA, lipids, or carbohydrates in the presence of transition metal ions, especially copper ion.

Other phenolic compounds, such as 1,2,4-benzenetriol, catechol, caffeic acid, and 2-hydroxyestradiol, can also undergo copper-dependent activation, generating reactive oxygen species, which participate in DNA damage (Li and Trush, 1994). Structure–activity analysis of various phenolic compounds has shown that in the presence of copper, the DNA cleaving activity for phenolic compounds with a 1,4-hydroquinone structure, such as 1,2,4-benzenetriol and TCHQ, is greater than those with a catechol group including caffeic acid and 2-hydroxyestradiol. Compounds having one phenol group such as eugenol, 2-acetaminophenol, and acetaminophen are the least active.

It is reasonable to propose that the DNA-associated copper in cells may have the potential to activate phenolic compounds via a copper-redox reaction, producing reactive oxygen species and electrophilic phenolic intermediates on or close to the DNA. The interaction of these reactive intermediates with DNA may result in a spectrum of lesions, including oxidative DNA base modifications, DNA adducts of phenolic intermediates, and DNA strand breakage. These lesions might contribute to site-specific mutations and to the initiation of carcinogenesis.

2.3.7. Hydrazines

Most hydrazines tested are carcinogens and mutagens. Large amounts of carcinogenic hydrazines are present in edible mushrooms. The widely eaten false morel (*Gyromitra esculenta*) contains 11 hydrazines, 3 of which are known carcinogens. One of these, *N*-methyl-*N*-formylhydrazine, is present at a concentration of 50 mg/100 g and causes lung tumors in mice at an extremely low dietary level of 20 μg per mouse per day. The most common commercial mushroom, *Agaricus bisporus*, contains about 300 mg of agaritine, the δ-glutamyl derivative of the mutagen 4-hydroxy-methylphenylhydrazine, per 100 g of mushrooms, as well as smaller amounts of the closely related carcinogen *N*-acetyl-4-hydroxy-methylphenylhydrazine (Ames, 1983).

A number of hydrazines are able to penetrate to the O_2-binding crevice of the hemoglobin molecule. Two of the most studied hydrazines are phenylhydrazine and its derivative acetylphenylhydrazine. Phenylhydrazine and its derivatives slowly oxidize in aqueous solution to form $\cdot O_2^-$ and H_2O_2, a reaction catalyzed by trace transition metal ions (Rumyantseva *et al.*, 1991; Yamamoto and Kawanishi, 1991a,b).

It was discovered that methemoglobin, acting as a peroxidase, oxidizes phenylhydrazine in the presence of H_2O_2 (Halliwell and Gutteridge, 1989). Oxyhemoglobin also oxidizes phenylhydrazine, but H_2O_2 is not required. These oxidase and peroxidase reactions of hemoglobin form a phenylhydrazine product that can react with O_2 to give $\cdot O_2^-$, and can also lead to formation of the phenyldiazine radical and phenyl radical. Of these various species, the most damaging seems to be the phenyldiazine radical, which can denature the hemoglobin molecule, and stimulate peroxidation of membrane lipids, causing eventual hemolysis.

2.3.8. Thiols

Thiol-containing compounds are generally good antioxidants. Under certain conditions, various thiols, including many radioprotective agents, can be toxic, causing death of cultured mammalian cells, in a biphasic fashion. These thiols are toxic at intermediate (0.2–1.0 mM) drug levels, but not at high or low concentrations. The role of Fenton chemistry in thiol-induced toxicity and apoptosis was clearly demonstrated using dithiothreitol (DTT) as a model thiol (Van Steveninck *et al.*, 1985; Held and Biaglow, 1993a,b). The toxicity of DTT in V79 cells has several characteristics: It is very dependent on the culture medium; the toxicity is decreased or prevented by addition of exogenous catalase, but SOD has no effect; the toxicity is increased by addition of copper (either free or derived from ceruloplasmin); and the toxicity can be modified intracellularly by altering glucose availability or pentose cycle activity. These findings are consistent with a mechanism whereby DTT oxidation produces H_2O_2 in a reaction catalyzed by metals, predominantly copper, followed by the reaction of H_2O_2 in Fenton reaction to produce the ultimate toxic species, ·OH.

Studies comparing 12 thiols have shown that the magnitude of cell killing and pattern of dependence on thiol concentration vary among different agents, with the toxicity depending on the interplay between the rates of two reactions: thiol oxidation and the reaction between the thiol and H_2O_2 produced during the thiol oxidation. The addition of other metals, such as Zn^{2+}, and metal chelators, such as EDTA, can also alter DTT toxicity by altering the rates of either thiol oxidation or the Fenton reaction.

2.4. Involvement of Iron and Copper in Tissue Injury Associated with Ischemia and Reperfusion

Reactive oxygen-derived species (ROS), both nonradical and free radical, have been implicated in tissue injury following ischemia and reperfusion of the heart (McCord, 1985; Garlick *et al.*, 1987; Arroyo *et al.*, 1987; Zweier *et al.*, 1987; Opie, 1989; Vandeplassche *et al.*, 1989; Shlafer *et al.*, 1990), lung (Ayene *et al.*, 1992; Basoglou *et al.*, 1992; Kinnula *et al.*, 1995; Marx and Van Asbeck, 1996), brain (Cao *et al.*, 1988), and kidney (Gamelin and Zager, 1988; Weight *et al.*, 1996), as well as in a variety of other pathologies (Halliwell and Gutteridge, 1984, 1985, 1989, 1990; Aruoma *et al.*, 1991), including apoptotic processes (Oberhammer *et al.*, 1992; Gottlieb *et al.*, 1994). The outcome of the stress is associated with exacerbation of cellular injury, and with immediate and/or programmed cell death. Reperfusion following ischemia occasionally enhances release of intracellular enzymes, excessive influx of Ca^{2+}, breakdown of sarcolemmal phospholipids, and disruption of cell membranes, resulting ultimately in cell death (Das, 1993). These events are known as *reperfusion injury.* The *free radical hypothesis,* the *calcium overloading hypothesis*, and the *loss of sarcolemmal phospholipid hypothesis* have been proposed, among others, to explain this phenomenon. Many lines of evidence have been published in recent years, supporting the free radical hypothesis.

The free radical hypothesis is based on the original work of Granger *et al.* (1981), McCord (1985), and others (Borg *et al.*, 1978; Manson *et al.*, 1983; Halliwell and Gutterridge, 1984). During ischemia, there is a buildup of adenosine resulting from the catabolism of intracellular ATP. It is converted to hypoxanthine and at the same time, during ischemia, a Ca^{2+}-activated protease is thought to catalyze the conversion of xanthine dehydrogenase to xanthine oxidase. This conversion can also occur under the influence of tumor necrosis factor, interleukins 1 and 3, neutrophil elastase, complement system (C5), and chemotactic peptide *N*-formyl-Met-Leu-Phe (Friedl *et al.*, 1989; Balkley, 1993). The ischemic duration required for the conversion of the dehydrogenase to the oxidase varies from organ to organ. It takes about 30 min for kidney (McCord, 1985) and dog heart (Hearse *et al.*, 1986). According to the original proposal, reperfusion results in a burst of superoxide anion radicals generated from the oxidation of hypoxanthine to xanthine and uric acid (McCord, 1985). There are several other pathways that could lead to the production of superoxide radical following ischemia and during the early phase of reperfusion. These include oxidation products of catecholamines, through semiquinones and quinone structures, cytochromes, cytokines, and oxidases (see Figure 1).

Considering that superoxide radical is a relatively low-reactive species, it is conceivable that the $\cdot O_2^-$ produced during the early phase of reperfusion is converted to ·OH through the catalysis of traces of redox-active labile metal pools.

Circumstantial evidence supporting the causative role of newly mobilized redox-active iron in tissue injury has been reported (Nayini *et al.*, 1985; Holt *et al.*, 1986; Gower *et al.*, 1989). Iron chelation was shown to provide protection against tissue injury following ischemia (Aust *et al.*, 1985; Mayers *et al.*, Appelbaum *et al.*, 1985; Ferreira *et al.*, 1990; DeBoer and Clark, 1992; Morita *et al.*, 1995), whereas the addition of iron or copper to the perfusate facilitated injury to reperfused hearts (Bernier *et al.*, 1986; Powell *et al.*, 1991; Karwatowska-Prokopczuk *et al.*, 1992). Using the isolated rat heart model the involvement of endogenous iron and copper in reperfusion injury was shown (Chevion *et al.*, 1993). Elevated levels of low-molecular-weight iron (LMWI) have also been reported (Voogd *et al.*, 1992, 1994).

Labile iron pool (LIP) of cells (Breuer *et al.*, 1995, 1996; Cabantchik *et al.*, 1996) constitutes the primary source of metabolic and catalytically reactive iron in the cytosol. LIP is homeostatically regulated in cells, and can be markedly altered following massive mobilization of iron. LIP has been shown to sharply rise following brief exposure to Fe(II) or to oxidative stress. The mobilized iron following ischemia is comprised of two fractions: The major one contains iron in macromolecular structures such as iron containing proteins, and the minor fraction is LIP. LIP can be measured directly by the extent of quenching of calcein fluorescence (see below). The LIP method, as well as direct measurement of the labile copper pool, is awaiting further development. It is important to note that the intracellular levels of *redox-active* iron and copper, rather than the *total* level of these transition metals, are important indicators of the susceptibility of the cells and the tissue to the oxidative stress.

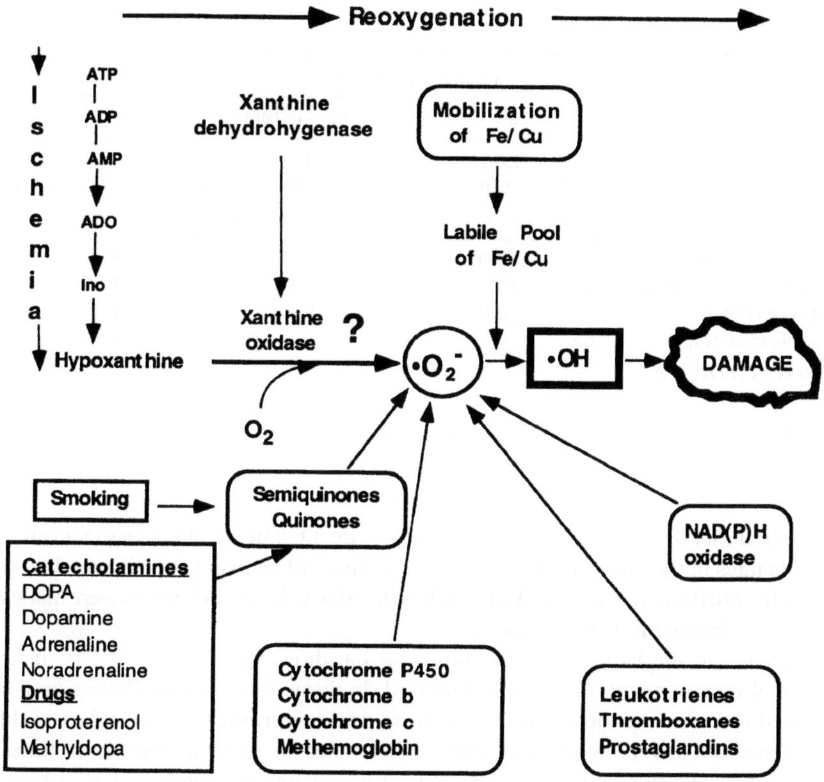

FIGURE 1. Possible sources of superoxide radical during ischemia and reperfusion, and the mobilization of iron and copper during ischemia which serve as catalyst in the production of hydroxyl radicals.

2.4.1. Heart

Recent efforts have been made to identify and quantitate labile pools of iron in tissues subjected to ischemia and reperfusion. Voogd *et al.* (1992, 1994) have used the DFO-detectable iron pool. These investigators disrupted the tissue under study in the cold and in the presence of DFO, and subsequently quantitated the Fe/DFO by HPLC. They found that the DFO-detectable labile iron pool is 2.1 μM in the nonischemic heart and increases to 55 μM following 45 min of global cardiac ischemia (see Table II). Both values seem rather high for physiological levels and even for injured tissues. It is plausible that during the sample processing, which includes tissue homogenization and cell rupture, even on ice, newly released proteases (and possibly nucleases) digest iron-containing structures and iron is released as a low-molecular-weight component.

In view of these obvious difficulties, trials to quantitate iron in coronary flow were undertaken (Nohl *et al.*, 1991; Chevion *et al.*, 1993). These studies were based on the

Table II
Iron Levels in Isolated Rat Heart (H) and Its Coronary Flow (CF) and in K562 Cells in Tissue Culture (K)

	Time of ischemia						
	0 min	15–18 min	30–35 min	45 min	60 min	System	Other treatment
	Reported LMWI (μM)						
Chevion *et al.* (1993)	0.36	0.48	0.90	0.86	0.89	CF	
Voogd *et al.* (1992, 1994)	2.1	13.0	35.0	55.1		H	
Nohl *et al.* (1991)		0.16			0.34	CF	
Ambrosio *et al.* (1987)	6.8		7.2			H	
Breuer *et al.* (1995, 1996)	0.35					K	Iron loading 0.77

assumption that the levels in the coronary flow reflect the intracellular tissue levels, and large changes in the intracellular distribution of iron will be manifested in metal leakage from cells into the coronary flow. Both studies found that the total Fe levels were markedly increased following ischemia (Table II).

Because of the hardships associated with the direct quantitation of LIP, indirect functional measurements have been employed. In these, the coronary flow is used as a source of redox-active metals in the ascorbate-driven conversion of salicylate to dihydroxybenzoate derivatives (DHBA). These indirect measurements also show that the early fraction of reperfusion is rich in redox-active iron and copper (Chevion *et al.*, 1993).

The causative roles of iron and copper in reperfusion injury could also be demonstrated via myocardial preconditioning. Preconditioning is a protective procedure against extended ischemia, which is attained by a series of short episodes of ischemia/reperfusion prior to the extended ischemia.

In a series of unpublished studies we found a significant reduction in the mobilization of both iron and copper following extended ischemia in preconditioned hearts, as compared with non-preconditioned groups (Figure 2).

Hemodynamic recovery was markedly higher in the preconditioned hearts compared with controls. The recovery of the "working index" of the preconditioned hearts was markedly higher than for the non-preconditioned hearts, substantiating the correlation between the degree of metal mobilization and the extent of cardiac damage. The data further support the notion that the mobilized metals play a causative role in myocardial reperfusion injury.

What are the cellular sources of iron and copper found in the coronary flow? While these metal ions in the free state are found only in minute concentrations within the cell, their total levels are in the micromolar range. In addition, we as well as others have shown that the newly mobilized metals are mostly attached to low-molecular-weight structures. Thus, their source is evidently degraded iron and/or copper-containing proteins. Of special interest was the SOD, which is an enzyme associated with protection against free

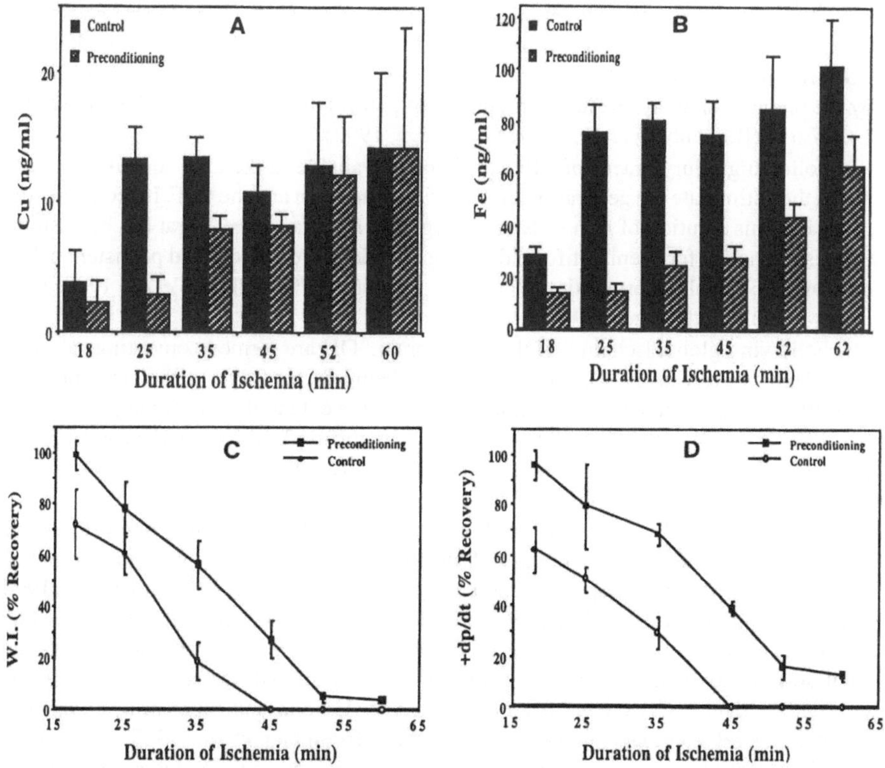

FIGURE 2. Hemodynamic parameters of hearts and levels of iron and copper in the coronary flow as a function of ischemic duration. Hearts were subjected to ischemia with and without prior preconditioning. (A) Copper levels in the first coronary flow fraction of reperfusion. (B) Iron Levels in the first coronary flow fraction of reperfusion. (C) Working Index (W.I.) of the hearts subjected to ischemia and reperfusion. (D) +*dp*/*dt* of hearts subjected to ischemia and reperfusion.

radicals. We have tried to identify any SOD activity in the coronary flow, but were unsuccessful.

2.4.2. Brain

Ischemic or traumatic injuries to the brain or spinal cord often result in more extensive tissue damage than equivalent insults to other tissues. The brain and central nervous system (CNS) may be especially prone to radical damage for a number of reasons, including the fact that several areas of the brain such as the globus pallidus, the substantia nigra, and the red nucleus are rich in iron and copper (Hallgren and Sowander, 1958; Harrison *et al.*, 1968; Hill and Switzer, 1984; Youdim, 1988a,b). Brain tissue contains nonheme iron at a level of 0.074 μg/mg protein (Youdim and Green, 1978).

The iron content of cerebrospinal fluid (CSF) has been determined to range between 0.2 and 1.1 μM. Because normal transferrin levels are around 0.24 μM, it is likely that CSF transferrin is completely saturated with iron. When the level of iron was examined by the bleomycin assay, the level of redox-active (free) iron was found to be around 0.55 ± 0.27 μM (Bleijenberg *et al.*, 1971; Gutteridge, 1992).

Following injury, traces of LIP can become available in the CNS and ascorbic acid might then stimulate the generation of ·OH within the brain and the CSF. Indeed, injection of an aqueous solution of iron salts or hemoglobin into the cortex of rat has been shown to cause transient focal epileptiform discharges, lipid peroxidation, and persistent behavioral and electrical abnormalities (Rosen and Frumin, 1979; Willmore *et al.*, 1980, 1986; Halliwell and Gutteridge, 1990).

Following global ischemia of the gerbil brain, ·OH are formed converting salicylate to its DHBA derivatives. The burst of ·OH is being formed during the early phase of reperfusion (less than 5 min). Correlation between the extent of brain damage, as evident from the changes in the locomotor activity of the reperfused animals, and the level of ·OH burst was clearly shown. This is in accordance with the hypothesized role of redox-active metals in brain reperfusion injury (Cao *et al.*, 1988).

Damage to the CNS is a major clinical problem. There is a deep interest in novel treatment modalities, such as the use of antioxidants, chelating agents, and free radical scavengers in arresting the spread of tissue injury from its original site (Chow *et al.*, 1994; Clemens and Panetta, 1994; Bar *et al.*, 1995; Tynecka, 1995; Chan *et al.*, 1996). α-Lipoic acid (Packer *et al.*, 1997), α-tocopherol, Trolox (a vitamin E analogue), SOD, DFO, and catalase were studied. In addition, a series of 21 amino steroids (referred to as *lazaroids*) with antioxidant activity have been developed by Upjohn and other drug companies. Indeed, several members of the lazaroid drugs (Braughler *et al.*, 1987) were found effective in minimizing neurological damage and protecting CNS against reperfusion injury and trauma.

2.4.3. Eye

Free radicals have been considered as possible causative agents in a number of ocular disorders. Some of these disorders are associated with light exposure (such as photic damage to the retina and perhaps age-related macular degenerations). Others are associated with ischemic injury, the classic example being retinopathy of prematurity (ROP) and the extreme conditions of retinal artery or vein occlusions. Free radicals may also play a role in conditions where ischemia and reperfusion are more subtle, but none the less critical, such as diabetic retinopathy, sickle-cell disease, and perhaps glaucoma. In these conditions, repeated events of transient ischemia may well occur. The retina, with its high content of long-chain polyunsaturated fatty acids and high rate of oxygen consumption, is especially vulnerable to free-radical-induced damage.

It is therefore not surprising that in both clinical and experimental systems, attempts to curb free-radical-induced damage have been made. At the experimental level, a number of studies showed that SOD reduced retinal injury following ischemia and reperfusion in rats and rabbits (Szabo *et al.*, 1991a,b; Nayak *et al.*, 1993). DFO and allopurinol reduced retinal damage in the isolated cat eye model subjected to ischemia/reperfusion, as measured by electroretinographic (ERG) recordings (Gehbach and Purple, 1994). At the clinical level, vitamin E was given to babies with ROP. Different studies yielded controversial results and the use of vitamin E is not advocated. A large, randomized clinical trial is currently being conducted by the National Eye Institute to evaluate the role of various antioxidants and mineral supplements in the development of macular degeneration. At present, the use of antioxidants or free radical scavengers in clinical settings of ischemia/reperfusion is not a standard procedure, neither in ocular disorders nor in other organs (such as the heart) where such treatment may be even more relevant.

Using the conversion of salicylate to DHBA, it was possible to show that ·OH are being formed in the early phase of the reperfused retina (Ophir *et al.*, 1993). Trials to protect the ischemic retina using different antioxidants showed varying degrees of success. Treatment by combinations of the chelator DFO and zinc or DFO and gallium, a non-redox active metal, which often neutralize the prooxidant effects of iron and copper, has proven highly effective in minimizing the production of ·OH and injury to the retina. This was measured by ERG and histological evaluation, and maintenance of the endogenous levels of antioxidants and energy metabolites (Ophir *et al.*, 1994; Berenshtein *et al.*, 1996).

3. INTERVENTION AND PREVENTION

Intervention and prevention of biological damage caused by the site-specific mechanism can be accomplished by various approaches.

One classical mode of protection is use of specific scavengers for ·OH. Initial predictions for efficient scavenging of homogeneously produced and distributed ·OH could not be met experimentally in biological systems. These findings are in accord with the fact that often the mechanism of damage is the site-specific mechanism. The lack of efficient scavenging could mean that scavengers usually do not enjoy free access to the metal-binding site, because of spatial restrictions and coordination requirements. Thus, the effective concentration of the scavenger at the ·OH formation site is low, leading to reduced efficiency of scavenging. It should be noted that several ·OH scavengers such as mannitol, salicylate, and β-mercaptopropionylglycine have variable degree of transition metal binding ability as well, which is more likely to account for their protective effects than the direct scavenging of ·OH (Chevion, 1988; Halliwell and Gutteridge, 1989).

Another classical mode of protection is the use of specific enzymes to remove reactive oxygen-derived species, H_2O_2 and $\cdot O_2^-$. H_2O_2 is an essential component in the site-specific mechanism, and is often produced from $\cdot O_2^-$ by spontaneous or enzymatic dismutation. Enzymes such as catalase and glutathione peroxidase can remove H_2O_2 at different concentrations and with different mechanisms, inhibiting the formation of ·OH, and minimizing molecular and cellular damage (Halliwell, 1990; Halliwell *et al.*, 1992; Rice-Evans and Diplock, 1993). $\cdot O_2^-$ can act as a reducing agent for the transition metals. In this regard, $\cdot O_2^-$ could be an important reductant in cells even though there are other reducing species at higher concentrations. Because of its size, this small diatomic molecule can penetrate into condensed structures. SOD would generally offer protection in such a system by removing superoxide that serves as a reducing agent for transition metals. $\cdot O_2^-$ could also be a precursor for H_2O_2. Therefore, in some cases where the rate-limiting step for the observed biological damage is the Fenton reaction and H_2O_2 is the limiting substrate, SOD may enhance the biological damage rather than protect against it.

The interaction between $\cdot O_2^-$ and nitric oxide could also affect the end results of these free radicals. SOD could influence this reaction, thereby altering the end results.

3.1. "Pull" Mechanism of Protection

Addition of a specific chelator for iron and/or copper, such as DFO or DTPA, can remove or "pull" out labile metal ions from biological binding sites, thus protecting these sites (Graf *et al.*, 1984; Menasche *et al.*, 1988). The metal may become inaccessible to reducing agents or H_2O_2. DFO was used extensively to treat iron-overload patients (Halliwell, 1989; Gabutti and Piga, 1996; Marx and Van Asbeck, 1996) and found to protect cells against the poisonous effects of paraquat (Kohen and Chevion, 1985, 1986). DFO was also found to protect against anthracycline cardiotoxicity by iron chelation (Hershko *et al.*, 1996). Likewise, 1,10-phenanthroline protected bacterial cells and mammals against hydrogen peroxide-induced toxicity (Mello Filho and Meneghini, 1985; Imlay *et al.*, 1988). Using the retrogradely perfused isolated rat heart, the copper-specific chelator neocuproine provided protection against hydrogen peroxide-induced cardiac damage and against ischemia/reperfusion-induced arrhythmias (Appelbaum *et al.*, 1990). Also, TPEN [*N,N,N′,N′*-tetrakis(2-pyridylmethyl)-ethylenediamine], a heavy metal chelator, provided nearly complete (>90%) protection against ischemia/reperfusion-induced arrhythmias (Appelbaum *et al.*, 1988).

It is known that circulating free iron can be lethal (Eaton, 1996). Humans developed two iron-binding proteins to soak up free iron so as to prevent the generation of toxic free radicals. One of these proteins is transferrin, a high-affinity, low-capacity circulating protein (2 atoms of ferric iron per molecule of transferrin). Transferrin receptors on the surface of iron-requiring cells recognize iron-loaded transferrin. The other is ferritin, a lower-affinity, high-capacity iron-binding protein (maximum of 4500 atoms of iron per molecule of ferritin), for which there are receptors only on the surface of iron-storage cells, such as reticuloendothethelial cells. Iron is trapped inside the ferritin protein shell as harmless Fe^{3+} (Herbert *et al.*, 1996). Hence, the evolution of iron storage and transport proteins not only provides a convenient way of moving iron around the body, but may also be regarded as an antioxidant defense. For example, in normal individuals, per milliliter of serum, there are approximately 300,000 molecules of transferrin per molecule of ferritin, so the concentration of non-transferrin-bound iron ions is negligible (Halliwell and Gutteridge, 1989).

Most plasma copper is attached to the protein ceruloplasmin, which has antioxidant properties. These are partly related to the ability of ceruloplasmin to oxidize Fe^{2+} to Fe^{3+}, which decreases Fe^{2+}-dependent production of $\cdot OH$ from H_2O_2, and Fe^{2+}-dependent lipid peroxidation. The small amount of plasma copper not attached to ceruloplasmin is apparently attached to either histidine, small peptides, or albumin. None of these forms of copper can apparently generate reactive oxidants in solution, and $\cdot OH$ or Cu(III) that are formed by reaction of bound copper ions with H_2O_2, appear to attack the copper-binding site. Indeed, albumin tightly binds copper ions, preventing their binding to more important sites. Thus, albumin is acting as an antioxidant in this context (Marx and Chevion, 1986; Halliwell and Gutteridge, 1989).

3.2. "Push" Mechanism of Protection

There is a variable degree of similarity between the coordination chemistry of copper and iron, on the one hand, and zinc, on the other. This particularly relates to the nature of their ligands (Bray and Bettger, 1990). Thus, zinc, a redox-inactive metal, could compete and displace ("push" out) Cu^{2+} and Fe^{2+} from various biological sites (Har-El and Chevion, 1989). Once zinc replaces copper, it will divert the site of formation and reaction of free radicals, and would provide an effective protection of this formerly copper-specific site. We have applied this strategy to paraquat-induced cell killing. Using a tenfold excess of Zn over adventitious copper, an effective mediator for paraquat toxicity, the synergistic killing was prevented (Korbashi *et al.*, 1989). Similarly, zinc complexes proved protective against ischemia-induced arrhythmia, using rat heart in the Langendorf configuration (Powell *et al.*, 1990). It appears that in cell damage caused by oxidative stress, endogenous or exogenous copper, and to a lesser degree iron, in the presence of a reducing environment, induces a cyclic Fenton-like reaction which is most deleterious to the cell. The most susceptible sites are the metal-binding sites.

3.3. "Pull–Push" Mechanism of Protection

It is proposed that the complex zinc–desferrioxamine (Zn/DFO) ($\log \beta = 14$) would be the ultimate protector against free-radical-induced transition-metal-mediated injury. Zn/DFO could act by both "pull" and "push" mechanisms. Metal-free DFO is a randomly oriented linear molecule, consisting of several subunits, which does not penetrate easily into cells. On metal binding the DFO within its complex with zinc assumes a well-defined and organized globular structure (Chevion, 1988, 1991). This structural alteration is expected to render the Zn/DFO complex more permeable into cells than the metal-free species. Once within the cell, Zn/DFO would readily exchange with available iron or copper ("pull" mechanism) to yield ferrioxamine or copper–DFO complex, respectively. In this way, controlled levels of zinc would be liberated within the cell and could act to further protect by "pushing" out additional redox-active metal ions from their binding sites, and act as a secondary antioxidant. Thus, this combination of "pull" and "push" mechanisms could efficiently prevent the site-specific metal-mediated oxidative damage.

We have investigated the protective effect of another complex that acts via the "push–pull" mechanism. Gallium–DFO complex (Ga/DFO) at low concentrations caused a nearly complete protection against reperfusion injury following global cardiac ischemia of up to 25 min. A nearly complete inhibition of mobilization of copper and iron into the coronaries of the perfused heart was also observed (Figure 3).

The ability of Ga/DFO (unpublished data) and Zn/DFO (Ophir *et al.*, 1994) to curb free radical formation in the ischemic/reperfused retina was also evaluated. This was accomplished by monitoring the biochemical profile of retinal tissue subjected to ischemia/reperfusion, with or without the drugs, by ERG recordings of retinal function, and by measuring the levels of conversion of salicylate to its hydroxylated products—2,5-

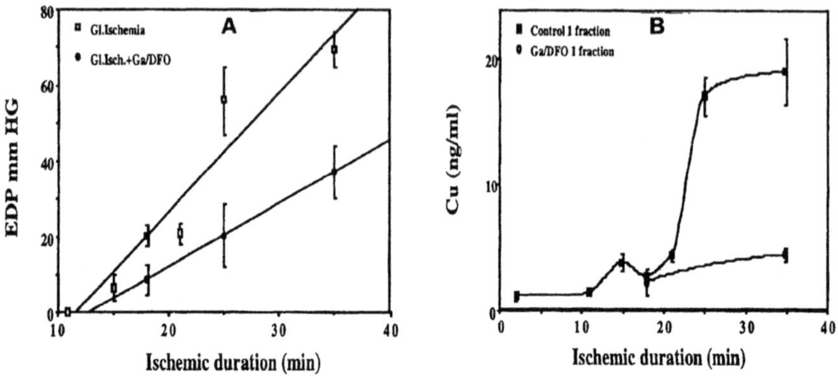

FIGURE 3. (A) The end diastolic pressure (EDP) of hearts subjected to ischemia and reperfusion ± Ga/DFO. (B) The level of Cu in the first coronary flow fraction of hearts subjected to ischemia and reperfusion ± Ga/DFO.

and 2,3-DHBA. The results obtained were in accord with the proposed high efficiency of the "push–pull" mechanism.

It should be noted that protecting agents can act in more than one way. For example, low concentrations of histidine produce a marked decrease in paraquat-induced cellular killing. This protective effect may be related to a combination of the following modes: histidine is (1) an effective ·OH scavenger, (2) an efficient chelator for copper and iron, (3) an efficient donor of hydrogen atoms, and (4) histidine forms a tight complex with H_2O_2 that might be less active in the Fenton reaction. Similarly, while DFO exerts its protection mainly through its chelation of iron, it also acts through scavenging free radicals such as ·OH, $\cdot O_2^-$, peroxyl radicals, and ferryl radicals (Halliwell, 1989; Darley-Usmar *et al.*, 1989; Green *et al.*, 1993; Denicola *et al.*, 1995; Van Reyk and Dean, 1996). Two new modes of action of DFO have recently been demonstrated: scavenging of semiquinone radical and stimulation of hydrolysis of chlorinated phenols to the corresponding hydroxy derivatives (Zhu *et al.*, 1998).

4. METHODS FOR THE DETECTION OF REDOX-ACTIVE LABILE POOLS OF TRANSITION METALS

It should be reiterated that only a very small fraction of the transition metals found in the biological systems is labile and redox-active. Thus, the quantitation of this fraction is important. The methods employed to detect and measure these fractions of iron and copper are summarized below.

4.1. The Bleomycin Assay for Iron

This assay is based on the fact that degradation of DNA by the antibiotic bleomycin requires the presence of ferrous iron. If other reagents are present in excess, the extent of the DNA degradation is proportional to the amount of iron in the sample that is available to bleomycin (Halliwell and Gutteridge, 1989; Evans and Halliwell, 1994). Bleomycin has a fairly low affinity for iron, and it cannot remove iron ions at pH 7.4 from pure iron proteins. Thus, the extent of bleomycin-mediated DNA degradation indicates that bleomycin-detectable iron presumably represents labile iron ions.

4.2. The Phenanthroline Assay for Copper

This assay is based on the fact that 1,10-phenanthroline, in the presence of copper ions, oxygen, and a suitable reducing agent, can degrade DNA. DNA degradation results in the release of a product from DNA that reacts, on heating with thiobarbituric acid at acidic pH, to form a pink chromogen (Halliwell and Gutteridge, 1989; Evans and Halliwell, 1994). Phenanthroline-detectable copper has been found in some sweat samples and in CSF. The phenanthroline assay detects copper bound to histidine, but not to ceruloplasmin. It has been observed that CSF from patients with Parkinson's disease has a high level of phenanthroline-detectable copper relative to controls, although there is no rise in the total Fe or Mn content of the fluid.

4.3. ESR/Ascorbate Assay

This assay is based on the fact that Fe-EDTA is an excellent catalyst of ascorbate oxidation, whereas Cu-EDTA is a very poor catalyst. The steady-state concentration of ascorbyl free radical can be determined by ESR, at a limit as low as ~ 5 nM. An analogous spectrophotometric method has been in use, following the rate of dissipation of ascorbate at 265 nm (Buettner, 1988, 1990).

4.4. ESR/DFO–Nitric Oxide Assay

This assay is based on the fact that DFO and nitric oxide form characteristic paramagnetic complexes with intracellular free iron, which can be measured by ESR (Kozlov *et al.*, 1992).

4.5. LIP (Labile Iron Pool) Assay

This assay is based on the rapid and stoichiometric quenching of the fluorescence of the probe calcein by divalent, "chelatable" metal ions such as Fe(II) and Cu(II) (Cabantchik *et al.*, 1996). This is the only available method for direct quantitation of low concentrations of the cellular pool of labile iron.

4.6. Ascorbate-Driven DNA Breakage and Ascorbate-Driven Conversion of Salicylate to Its Hydroxylated Metabolites

This method is based on the fact that in the presence of redox-active forms of copper and iron, ascorbate acts as a potent prooxidant promoting degradation of DNA and the extent of degradation is analyzed by agarose gel electrophoresis. The conversion of salicylate to its DHBA and catechol derivatives, which are analyzed by HPLC coupled with electrochemical detection (HPLC-ECD), is a complementary method (Chevion *et al.*, 1993).

4.7. DFO-Available LMWI

This method is based on the fact that DFO-available iron in biological tissues yielding Fe/DFO can be measured by HPLC (Gower *et al.*, 1989). Tissue fractions are disrupted in the cold in the presence of DFO, the parent compound, and ferrioxamine; they are then extracted using solid-phase cartridges and quantitated by reversed-phase HPLC using UV detection. Calculation of the ferrioxamine:DFO ratio and comparison with a standard curve using a series of known iron concentrations allows determination of the micromolar amounts of DFO-available iron in biological samples.

5. REFERENCES

Agarwal, K., Sharma, A., and Talukder, G., 1989, Effects of copper on mammalian cell components, *Chem. Biol. Interact.* **69:**1–16.

Ahmad, M. S., Fazal, F., Rahman, A., Hadi, S. M., and Parish, J. H., 1992, Activities of flavonoids for the cleavage of DNA in the presence of Cu(II): Correlation with generation of reactive oxygen species, *Carcinogenesis* **13:**605–608.

Akman, S. A., Doroshow, J. H., and Kensler, T. W., 1992, Copper-dependent site-specific mutagenesis by benzoyl peroxide in the sup F gene to the mutation reporter plasmid pS189, *Carcinogenesis* **13:**1783–1787.

Akman, S. A., Kensler, T. W., Doroshow, J. H., and Dizdaroglu, M., 1993, Copper ion-mediated modification of bases in DNA in vitro by benzoyl peroxide, *Carcinogenesis* **14:**1971–1974.

Ambrosio, G., Zweier, J. L., Jacobus, W. E., Weisfeldt, M. L., and Flaherty, J. T., 1987, Improvement of postischemic myocardial function and metabolism induced by administration of deferoxamine at the time of reflow: The role of iron in the pathogenesis of reperfusion injury, *Lab. Invest.* **76:**906–915.

Ames, B. N., 1983, Dietary carcinogens and anticarcinogens, *Science* **221:**1256–1264.

Anbar, M., and Levitzki, B., 1966, Copper-induced radiolytic deactivation of α-amylase and catalase, *Radiat. Res.* **27:**32–34.

Appelbaum, Y. J., Kuvin, J., Chevion, M., and Uretzky, G., 1988, TPEN, a heavy metal chelator, protected the isolated perfused rat heart from reperfusion induced arrhythmias, *J. Mol. Cell. Cardiol.* **20**(Suppl. V):Abstract #32.

Appelbaum, Y. J., Kuvin, J., Borman, J. B., Uretzky, G., and Chevion, M., 1990, The protective role of neocuproine against cardiac damage in isolated perfused rat heart, *Free Radical Biol. Med.* **8:**133–143.

Aronovitch, J., Samuni, A., Godinger, D., and Czapski, G., 1986, In vivo degradation of bacterial DNA by H_2O_2 and *O*-phenanthroline, in *Superoxide and Superoxide Dismutase in Chemistry, Biology and Medicine* (G. Rotilio, ed.), pp. 364–348, Elsevier, Amsterdam.

Arroyo, C., Kramer, J., Dickens, B., and Weglicki, W., 1987, Identification of free radicals in myocardial ischemia/reperfusion by spin trapping with nitrone DMPO, *FEBS Lett.* **221:**101–104.

Aruoma, O., Kaur, H., and Halliwell, B., 1991, Oxygen free radicals in human diseases, *J. R. Soc. Health.* **111:**172–177.

Aust, S. D., Morehouse, L. A., and Thomas, C. E., 1985, Role of metals in oxygen radical reaction, *Free Radical Biol. Med.* **1:**3–25.

Ayene, I. S., Dodia, C., and Fisher, A. B., 1992, Role of oxygen in oxidation of lipid and protein during ischemia/reperfusion in isolated perfused rat lung, *Arch. Biochem. Biophys.* **296:**183–189.

Balkley, J. B., 1993, Endothelial xanthine oxidase: A radical transducer of inflammatory signals for reticuloendothelial activation, *Br. J. Surg.* **80:**684–686.

Bar, P. R., 1995, Neuroprotective effects of radical scavengers in an intact dorsal root ganglion hypoxia model, *Ann. N. Y. Acad. Sci.* **765:**111–118.

Basoglu, A., Kocak, H., Pac, M., Cerrahogly, M., Bakan, N., Yekeler, I., Yuksek, M. S., and Goksu, S., 1992, Oxygen free radical scavengers and reperfusion injury in dog lung preserved in cold ischemia, *Thorac. Cardiovasc. Surg.* **40:**144–147.

Beauchamp, C., and Fridovich, I., 1970, A mechanism of the production of ethylene from ethanol, *J. Biol. Chem.* **245:**4641–4646.

Beinert, H., 1991, Copper in Biological Systems: A Report from "The 6th Manziana Conference," September 23–27, *J. Inorg. Biochem.* **44:**173–218.

Berenshtein, E., Banin, E., Pe'er, J., Kitrossky, N., and Chevion, M., 1996, Ga/DFO protect retina against reperfusion injury, VIII Biennial Meeting International Society for Free Radical Research, Barcelona, p. 157.

Bernier, M., Hearse, D. J., and Manning, A. S., 1986, Reperfusion-induced arrhythmias and oxygen-derived free radicals. Studies with "anti-free radical" interventions and a free radical-generating system in the isolated perfused rat heart, *Circ. Res.* **58:**331–340.

Birnboim, H. C., 1982, DNA strand breakage in human leukocytes exposed to a tumor promoter phorbol-myristate acetate, *Science* **215:**1247–1249.

Birnboim, H. C., 1992, Effect of lipophilic chelators on oxyradical-induced DNA strand breaks in human granulocytes: Paradoxical effect of 1,10-phenanthroline, *Arch. Biochem. Biophys.* **294:**17–21.

Bleijenberg, B. G., Eljk, H. G. V., and Eijnse, B., 1971, The determination of non-heme iron and transferrin in cerebrospinal fluid, *Clin. Chim. Acta* **31:**277–281.

Borg, D. C., Schaich, K. M., Elmore, J. J., and Bell, J. A., 1978, Cytotoxic reaction of free radical species of oxygen, *Photochem . Photobiol.* **28:**887–907.

Borg, D. C., and Schaich, K. M., 1984, Cytotoxicity from coupled redox cycling of autoxidizing xenobiotics and metals, *Israel J. Chem.* **24:**38–53.

Braughler, J. M., Pregenzer, J. F., Chase, R. L., Duncan, L. A., Jacobsen, E. J., and McCall, J. M., 1987, Novel 21-amino steroids as potent inhibitors of iron-dependent lipid peroxidation, *J. Biol. Chem.* **262:**10438–10440.

Bray, T. M., and Bettger, W. J., 1990, The physiological role of zinc as an antioxidant, *Free Radical Biol. Med.* **8:**281–291.

Breuer, W., Epsztein, S., Milligram, P., and Cabantchik, Z. I., 1995, Transport of iron and other transition metals into cells revealed by a fluorescent probe, *Am. J. Physiol.* **268:**C1354–C1361.

Breuer, W., Epsztein, S., and Cabantchik, Z. I., 1996, Dynamics of the cytosolic chelatable iron pool of K562 cells, *FEBS Lett.* **382:**304–308.

Bryan, S. E., Vizard, D. L., Beary, D. A., La Biche, R. A., and Hardy, K. J., 1981, Partitioning of zinc and copper within subnuclear nucleoprotein particles, *Nucleic Acids Res.* **9:**5811–5823.

Buettner, G. R., 1988, In the absence of catalytic metals ascorbate does not autoxidize at pH 7: Ascorbate as a test for catalytic metals, *J. Biochem. Biophys. Methods* **16:**27–40.

Buettner, G. R., 1990, Ascorbate oxidation: UV absorbance of ascorbate and ESR spectroscopy of the ascorbyl radical as assays for iron, *Free Radical Res. Commun.* **10:**5–9.

Buettner, G. R., Oberley, L. W., and Leuthauser, S. W. H. C., 1978, The effect of iron on the distribution of superoxide and hydroxyl radicals as seen by spin trapping and on the superoxide dismutase assay, *Photochem. Photobiol.* **28:**693–695.

Cabantchik, Z. I., Glickstein, H., Milgram, P., and Breuer, W., 1996, A fluorescence assay for assessing chelation of intracellular iron in a membrane model system and in mammalian cells, *Anal. Biochem.* **233:**221–227.

Cadenas, E., 1989, Biochemistry of oxygen toxicity, *Annu. Rev. Biochem.* **58:**79–110.

Calderaro, M., Martins, E. A. L., and Meneghini, R., 1993, Oxidative stress by menadione affects cellular copper and iron homeostasis, *Mol. Cell. Biochem.* **126:**17–23.

Cameron, E., Pauling, L., and Leibovitz, B., 1979, Ascorbic acid and cancer, *Cancer Res.* **39:**663–681.

Cao, W., Carney, J. M., Duchon, A., Floyd, R. A., and Chevion, M., 1988, Oxygen free radical involvement in ischemia and reperfusion injury to brain, *Neurosci. Lett.* **88:**233–238.

Chan, P. H., Charles, J., Epstein, H., Kinouchi, Kamii, H., Chen, S. F., Carlson, E., Gafni, J., Yang, G., and Reola, L., 1996, Neuroprotective role of Cu,Zn-superoxide dismutase in ischemic brain damage, *Ad. Neurol.* **71:**271–280.

Chevion, M., 1988, A site specific mechanism for free radical induced biological damage: The essential role of redox-active transition metals, *Free Radical Biol. Med.* **5:** 27–37.

Chevion, M., 1991, Protection against free radical-induced and transition metal-mediated damage: The use of "pull" and "push" mechanisms, *Free Radical Res. Commun.* **12–13:**691–696.

Chevion, M., Mager, J., and Glaser, G., 1983, Favism-inducing agents, in *Handbook of Naturally Occurring Food Toxicants* (M. Rechigl, Jr., ed.), pp. 63–79, CRC Press, Boca Raton.

Chevion, M., Jiang, Y., Har-El, R., Berenshtein, E., Uretzky, G., and Kitrossky, N., 1993, Copper and iron are mobilized following myocardial ischemia: Possible criteria for tissue injury, *Proc. Natl. Acad. Sci. USA* **90:**1102–1106.

Chow, H. S., Lynch-3rd, J. J., Rose, K., and Choi, D. W., 1994, Trolox attenuates cortical neuronal injury induced by iron, ultraviolet light, glucose deprivation, or AMPA, *Brain Res.* **639**(1)**:**102–108.

Clemens, J. A., and Panetta, J. A., 1994, Neuroprotection by antioxidants in model of global and focal ischemia, *Ann. N. Y. Acad. Sci.* **738:**250–256.

Czapski, G., 1984, On the use of ·OH scavengers in biological systems, *Israel J. Chem.* **24:**29–32.

Czapski, G., and Ilan, Y. A., 1978, On the generation of the hydroxyl agent from the superoxide radical: Can the Haber–Weiss reaction be the source of ·OH radicals? *Photochem. Photobiol.* **28:**651–654.

Darley-Usmar, V. M., Hersey, A., and Garland, L. G., 1989, A method for comparative assessment of antioxidant as peroxyl radical scavengers, *Biochem. Pharmacol.* **38:**1645–1649.

Das, D. K., 1993, *Pathophysiology of Reperfusion Injury*, CRC Press, Boca Raton.

Davison, A. J., Kettle, A. J., and Fatur, D. J., 1986, Mechanism of the inhibition of catalase by ascorbate, *J. Biol. Chem.* **261:**1193–1200.

DeBoer, D. A., and Clark, R. E., 1992, Iron chelation in myocardial preservation after ischemia–reperfusion injury: The importance of pretreatment and toxicity, *Ann. Thorac. Surg.* **53:**412–418.

Denicola, A., Souza, J. M., Gatti, R. M., Augusto, O., and Radi, R., 1995, Desferrioxamine inhibition of the hydroxyl radical-like reactivity of peroxynitrite: Role of hydroxamic groups, *Free Radical Biol. Med.* **19:**11–19.

Dizdaroglu, M., Aruoma, O. I., and Halliwell, B., 1990, Modification of bases in DNA by copper ion-1,10-phenanthroline complexes, *Biochemistry* **29:**8447–8451.

Eaton, J. W., 1996, Iron: The essential poison, *Redox Rep.* **2:**215.

Emerit, I., and Cerutti, P., 1981, Tumor promoter phorbol-12-myristate-13-acetate induces chromosomal damage via indirect action, *Nature* **293:**144–146.

Evans, P. J., and Halliwell, B., 1994, Measurement of iron and copper in biological systems: Bleomycin and copper-phenanthroline assays, *Methods Enzymol.* **233:**83–92.

Ferradini, C., Foos, J., and Houee, C., 1978, The reaction between superoxide anion and hydrogen peroxide, *Photochem. Photobiol.* **28:**697–700.

Ferreira, R., Burgos, M., Milei, J., Llesuy, S., Molteni, L., Hourquebie, H., and Boveris, A., 1990, Effect of supplementing cardioplegic solution with deferoxamine on reperfused human myocardium, *J. Thorac. Cardiovasc. Surg.* **100:**708–714.

Frank, L., and Massaro, D., 1980, Oxygen toxicity, *Am. J. Med.* **69:**117–126.

Frei, B., England, L., and Ames, B. N., 1989, Ascorbate is an outstanding antioxidant in human blood plasma, *Proc. Natl. Acad. Sci. USA* **86:**6377–6381.

Friedl, H. P., Till, G. O., Ryan, U. S., and Ward, P. A., 1989, Mediator-induced activation of xanthine oxidase in endothelial cells, *FASEB J.* **3:**2512–2518.

Gabutti, V., and Piga, A., 1996, Results of long-term iron-chelating therapy, *Acta Haematol.* **95:**26–36.

Gamelin, L. M., and Zager, R. A., 1988, Evidence against oxidant injury as a critical mediator of postischemic acute renal failure, *Am. J. Physiol.* **255:**F450–F460.

Garlick, P. B., Davies, M. J., Hearse, M. J., and Slater, T. F., 1987, Direct detection of free radicals in the reperfused rat heart using electron spin resonance spectroscopy, *Circ. Res.* **61:**757–760.

Gehbach, P. L., and Purple, R. L., 1994, Enhancement of retinal recovery by conjugated deferoxamine after ischemia–reperfusion, *Invest. Ophthalm. Vis. Sci.* **35**(2):669–676.

Geierstanger, B. H., Kagawa, T. F., Chen, S. L., Quigley, G. T., and Ho, P. S., 1991, Base-specific binding of copper(II) to Z-DNA: The 1.3 angstrom single crystal structure of d(m5CGUAm5CG) in the presence of $CuCl_2$, *J. Biol. Chem.* **266:**20185–20191.

Gerschman, R., 1981, Historical introduction to the "free radical theory" of oxygen toxicity, in *Oxygen and Living Processes, an Interdisciplinary Approach* (D. Gilbert, ed.), pp. 44–46, Springer-Verlag, Berlin.

Gerschman, R., Gilbert, D., Nye, S. W., Dwyer, P., and Fenn, W. O., 1954, Oxygen poisoning and X-irradiation: A mechanism in common, *Science* **119:**623–626.

Goldstein, S., and Czapski, G., 1986, The role and mechanism of metal ions and their complexes in enhancing damage in biological systems or in protecting these systems from the toxicity of O_2^- , *Free Radical Biol. Med.* **2:**3–11.

Goldstein, S., Meyerstein, D., and Czapski, G., 1993, The Fenton reagents, *Free Radical Biol. Med.* **15:**435–445.

Gottlieb, R. A., Burleson, K. O., and Cloner, R. A., 1994, Reperfusion injury induced apoptosis in rabbit cardiomyocytes, *J. Clin. Invest.* **94:**1621–1628.

Gower, J., Healing, G., and Green, C., 1989, Measurement by HPLC of desferrioxamine-available iron in rabbit kidneys to assess the effect of ischemia on the distribution of iron within the total pool, *Free Radical Res. Commun.* **5:**291–299.

Graf, E., Mahoney, J. R., Bryant, R. G., and Eaton, J. W., 1984, Iron-catalyzed hydroxyl radical formation. Stringent requirement for free iron coordination site, *J. Biol. Chem.* **259:**3620–3624.

Granger, D. N., Rutili, G., and McCord, J. M., 1981, Superoxide radical in feline intestinal ischemia, *Gastroenterology* **81:**22–29.

Green, E. S. R., Rice-Evans, H., Rice-Evans, P., Davies, M. J., Salah, N., and Rice-Evans, C. A., 1993, The efficacy of mono-hydroxamates as free radical scavenging agents compared with di- and tri-hydroxamates, *Biochem. Pharmacol.* **45:**357–366.

Gutteridge, J. M. C., 1992, Ferrous ions detected in cerebrospinal fluid by using bleomycin and DNA damage, *Clin. Sci.* **82:**315–320.

Gutteridge, J. M. C., and Wilkins, S., 1983, Copper salt-dependent hydroxyl radical formation damage to proteins acting as antioxidants, *Biochim. Biophys. Acta* **759:**38–41.

Gutteridge, J. M. C., Richmond, R., and Halliwell, B., 1979, Inhibition of the iron-catalyzed formation of hydroxyl radicals from superoxide and of lipid peroxidation by desferrioxamine, *Biochem. J.* **184:**469–472.

Haber, F., and Weiss, J., 1934, The catalytic decomposition of hydrogen peroxide by iron salts, *Proc. R. Soc. London Ser. A* **147:**332–351.

Hallgren, B., and Sowander, P., 1958, The effect of age on the non-haemin iron in the human brain, *J. Neurochem.* **3:**41–51.

Halliwell, B., 1978, Superoxide-dependent formation of hydroxyl radicals in the presence of iron chelates, *FEBS Lett.* **92:**321–326.

Halliwell, B., 1989, Protection against tissue damage in vivo by desferrioxamine: What is its mechanism of action? *Free Radical Biol. Med.* **7:**645–651.

Halliwell, B., 1990, How to characterize a biological antioxidant, *Free Radical Res. Commun.* **9:**1–32.

Halliwell, B., and Gutteridge, J. M. C., 1984, Oxygen toxicity, oxygen radicals, transition metals and disease, *Biochem. J.* **219:**1–14.

Halliwell, B., and Gutteridge, J. M. C., 1985, The importance of free radicals and catalytic metal ions in human disease, *Mol. Aspects Med.* **8:**89–193.

Halliwell, B., and Gutteridge, J. M. C., 1989, *Free Radicals in Biology and Medicine*, Oxford University Press (Clarendon), London.

Halliwell, B., and Gutteridge, J. M. C., 1990, Role of free radicals and catalytic metal ions in human diseases: An overview, *Methods Enzymol.* **186:**1–85.

Halliwell, B., Gutteridge, J. M., and Cross, C. E., 1992, Free radicals, antioxidants, and human disease: Where are we now? *J. Lab. Clin. Med.* **119:**598–620.

Har-El, R., and Chevion, M., 1989, Zn(II) protect against free radical-induced damage: Studies on single and double-strand DNA breakage, *Free Radical Res. Commun.* **12–13:**509–515.

Harrison, W. W., Netsky, M. G., and Brown, M. D., 1968, Trace elements in human brain: Copper, zinc, iron and magnesium, *Clin. Chim. Acta* **21:**55–60.

Hearse, D. J., Manning, A. S., Downey, J. M., and Yellon, D. M., 1986, Xanthine oxidase: A critical mediator of myocardial injury during ischemia and reperfusion? *Acta Physiol. Scand. Suppl.* **548:**65–78.

Held, K. D., and Biaglow, J. E., 1993a, Role of copper in the oxygen radical-mediated toxicity of the thiol-containing radioprotector dithiothreitol in mammalian cells, *Radiat. Res.* **134:**375–382.

Held, K. D., and Biaglow, J. E., 1993b, Mechanisms of the oxygen radical-mediated toxicity of various thiol-containing compounds in cultured mammalian cells, *Radiat. Res.* **139:**152–123.

Herbert, V., Shaw, S., and Jayatilleke, E., 1996, Vitamin C-driven free radical generation from iron, *J. Nutr.* **126:**1213S–1220S.

Hershko, C., Pinson, A., and Link, G., 1996, Prevention of anthracycline cardiotoxicity by iron chelation, *Acta Haematol.* **95:**87–92.

Hill, J. M., and Switzer, R. C., 1984, The regional distribution and cellular localization of iron in the rat brain, *Neuroscience* **11:**595–603.

Holt, S., Gunderson, M., Joyce, K., Nayini, N. R., Eyster, G. F., Garitano, A. M., Zonia, C., Krause, G. S., Aust, S. D., and White, B. C., 1986, Myocardial tissue iron delocalization and evidence for lipid peroxidation after two hours of ischemia, *Ann. Emerg. Med.* **15:**1155–1159.
Howell, J. M., and Gawthorne, J. M., 1987, *Copper on Animal and Man*, CRC Press, Boca Raton.
Imlay, J. A., Chin, S. M., and Linn, S., 1988, Toxic DNA damage by hydrogen peroxide through the Fenton reaction in vivo and in vitro, *Science* **240:**640–642.
Inoue, S., Yamamoto, K., and Kawanishi, S., 1990, DNA damage induced by metabolites of *O*-phenylphenol in the presence of copper(II) ion, *Chem. Res. Toxicol.* **3:**144–149.
Irons, R. D., 1985, Quinones as toxic metabolites of benzene, *J. Toxicol. Environ. Health* **16:**673–678.
Kari, F. W., Bucher, J., Eustis, S. L., Haseman, J. K., and Huff, J. E., 1992, Toxicity and carcinogenicity of hydroquinone in F344/N rats and B6csF1 mice, *Food Chem. Toxicol.* **30:**737–747.
Karwatowska-Prokopczuk, E., Czarnowska, E. and Beresewicz, A., 1992, Iron availability and free radical induced injury in the isolated ischemic/reperfused rat heart, *Cardiovasc. Res.* **26:**58–66.
Kawanishi, S., Inoue, S., and Kawanishi., M., 1989, Human DNA damage induced by 1,2,4-benzenetriol, a benzene metabolite, *Cancer Res.* **49:**164–168.
Kinnula, V. I., Crapo, J. D., and Raivio, K. O., 1995, Biology of disease. Generation and disposal of reactive oxygen metabolites in the lung, *Lab Invest.* **73**(1)**:**3–19.
Kohen, R., and Chevion, M., 1985, Paraquat toxicity is enhanced by iron and inhibited by desferrioxamine in laboratory mice, *Biochem. Pharmacol.* **34:**1841–1843.
Kohen, R., and Chevion, M., 1986, Transition metals potentiate paraquat toxicity, *Free Radical Res. Commun.* **1:**79–88.
Kohen, R., Szyf, M., and Chevion, M., 1986, Quantitation of single and double strand DNA breaks in vitro and in vivo, *Anal. Biochem.* **154:**485–491.
Koppenol, W. H., and Butler, J., 1985, Energetics in interconversion reactions of oxyradicals, *Adv. Free Radical Biol.* **1:**91–131.
Koppenol, W. H., Butler, J., and Leenwen, J. W V., 1978, The Haber–Weiss cycle, *Photochem. Photobiol.* **28:**655–660.
Korbashi, P., Katzhendler, J., Saltman, P., and Chevion, M., 1989, Zinc protects *Escherichia coli* against copper-mediated paraquat-induced damage, *J. Biol. Chem.* **264:**8479–8482.
Kozlov, A. V., Yegorov, D. Y., Vladimirov, Y. A., and Azizova, O. A., 1992, Intracellular free iron in liver tissue and liver homogenate: Studies with electron paramagnetic resonance on the formation of paramagnetic complexes with Desferal and nitric oxide, *Free Radical Biol. Med.* **13:**9–16.
Li, Y., and Trush, M. A., 1993a, Oxidation of hydroquinone by copper: Chemical mechanism and biological effects, *Arch. Biochem. Biophys.* **300:**346–355.
Li, Y., and Trush, M. A., 1993b, DNA damage resulting from the oxidation of hydroquinone by copper: Role for a Cu(II)/Cu(I) redox cycle and reactive oxygen generation, *Carcinogenesis* **14:**1303–1311.
Li, Y., and Trush, M. A., 1994, Reactive oxygen-dependent DNA damage resulting from the oxidation of phenolic compounds by a copper-redox cycle mechanism, *Cancer Res.* **54:**1895–1898.
Linder, M. C., 1991, *Biochemistry of Copper*, Plenum Press, New York.
Manson, P. N., Antheneli, R. N., Jim, M., Bulkley, G. B., and Hoopes, J. E., 1983, The role of oxygen-free radicals in ischemic tissue injury in island skin flaps, *Ann. Surg.* **198:**87–90.
Marx, G., and Chevion, M., 1986, Site-specific modification of albumin by free radicals, *Biochem. J.* **236:**397–400.
Marx, J. J. M., and Van Asbeck, B. S., 1996, Use of iron chelators in preventing hydroxyl radical damage: Adult respiratory distress syndrome as an experimental model for the pathophysiology and treatment of oxygen-radical-mediated tissue damage, *Acta Haematol.* **95:**49–62.
Massa, E. M., and Giulivi, C., 1993, Alkoxyl and methyl radical formation during cleavage of *tert*-butyl hydroperoxide by a mitochondrial membrane-bound, redox active copper pool: An ESR study, *Free Radical Biol. Med.* **14:**559–565.
Mayers, C. L., Weiss, S. J., Krish, M. M., and Shlafer, M., 1985, Involvement of hydrogen peroxide and hydroxyl radical in the oxygen paradox: Reduction of creatine kinase release by catalase, allopurinol or deferoxamine, but not by superoxide dismutase, *J. Mol. Cell. Cardiol.* **17:**675–684.
McCord, J. M., 1985, Oxygen-derived free radicals in postischemic tissue injury, *N. Engl. J. Med.* **312:**159–163.
McCord, J. M., and Day, E. D., 1978, Superoxide-dependent production of hydroxyl radical catalyzed by iron–EDTA complex, *FEBS Lett.* **86:**139–143.

McCord, J. M., and Fridovich, I., 1969, Superoxide dismutase: An enzyme function for erythrocytes, *J. Biol. Chem.* **244**:6049–6055.

Mello Filho, A. C. D., and Meneghini, R., 1985, Protection of mammalian cells by *O*-phenanthroline from lethal and DNA-damaging effects produced by active oxygen species, *Biochim. Biophys. Acta* **847**:82–89.

Menasche, P., Pasquier, C., Bellucci, S., Lorente, P., Jaillon, P., and Piwnica, A., 1988, Deferoxamine reduced neutrophil mediated free radical production during cardiopulmonary bypass in man, *J. Thorac. Cardiovasc. Surg.* **96**:582–589.

Naito, S., Ono, Y., Somiya, I., Inoue, S., Ito, K., Yamamoto, K., and Kawanishi, S., 1994, Role of active oxygen species in DNA damage by pentachlorophenol metabolites, *Mutat. Res.* **310**:79–88.

Navok, T., and Chevion, M., 1984, Transition metals mediate enzymatic inactivation caused by favism-inducing agents, *Biochem. Biophys. Res. Commun.* **122**:297–303.

Nayak, M. S., Kita, M., and Marmor, M. F., 1993, Protection of the rabbit retina from ischemic injury by superoxide dismutase and catalase, *Invest. Ophthalmol. Vis. Sci.* **34**:2018–2022.

Nayini, N. R., White, B. C., Aust, S. D., Huang, R. R., Indrieri, R. J., Evans, A. T., Bialek, H., Jacobs, W. A., and Komara, J., 1985, Post resuscitation iron delocalization and malondialdehyde production in the brain following prolonged cardiac arrest, *Free Radical Biol. Med.* **1**:111–116.

Nohl, H., and Jordan, W., 1987, The involvement of biological quinones in the formation of hydroxyl radicals via the Haber–Weiss reaction, *Bioorg. Chem.* **15**:374–382.

Nohl, H., Jordan, W., and Youngman, R. J., 1986, Quinones in biology: Function in electron transfer and oxygen activation, *Adv. Free Radical Biol. Med.* **2**:211–279.

Nohl, H., Stolze, K., Napetschnig, S., and Ishikawa, T., 1991, Is oxidative stress primarily involved in reperfusion injury of the ischemic heart? *Free Radical Biol. Med.* **11**:581–588.

Oberhammer, F. A., Pavalka, M., and Sharma, S., 1992, Induction of apoptosis in cultured hepatocytes and in regressing liver by transforming growth factor b1, *Proc. Natl. Acad. Sci. USA* **89**:5408–5412.

O'Connell, J. F., Klein-Szanto, A. P., DiGiovanni, D. M., Fries, J. W., and Slaga, T. J., 1986, Enhanced malignant progression of mouse skin tumors by the free-radical generator benzoyl peroxide, *Cancer Res.* **46**:2863–2865.

O'Halloran, T. V., 1993, Transition metals in control of gene expression, *Science* **261**:715–725.

Ophir, A., Berenshtein, E., Kitrossky, N., Berman, E. R., Photiou, S., Rothman, Z., and Chevion, M., 1993, Hydroxyl radical generation in the cat retina during reperfusion following ischemia, *Exp. Eye Res.* **57**:351–357.

Ophir, A., Berenshtein, E., Kitrossky, N., and Averbukh, E., 1994, Protection of the transiently ischemic cat retina by zinc-desferrioxamine, *Invest. Ophthalmol. Vis. Sci.* **35**:1212–1222.

Opie, L. H., 1989, Reperfusion injury and its pharmacological modification, *Circulation* **80**:1049–1062.

Packer, L., Tritschler, H. J., and Wesel, K., 1997, Neuroprotection by the metabolic antioxidant α-lipoic acid, *Free Radical Biol. Med.* **22**(1/2):359–378.

Peleg, I., Zer, H., and Chevion, M., 1992, Paraquat toxicity in *Pisum sativum*: Effects on soluble and membrane-bound proteins, *Physiol. Plant.* **86**:131–135.

Pezzano, H., and Podo, F., 1980, Structure of binary complexes of mono- and polynucleotides with metal ions of the first transition group, *Chem. Rev.* **80**:365–399.

Powell, S., Saltmann, P., Uretzky, G., and Chevion, M., 1990, The effect of zinc on reperfusion arrhythmias in the isolated perfused rat heart, *Free Radical Biol. Med.* **8**:33–46.

Powell, S. R., Hall, D., and Shih, A., 1991, Copper loading of hearts increases postischemic reperfusion injury, *Circ. Res.* **69**:881–885.

Powis, G., 1989, Free radical formation by antitumor quinones, *Free Radical Biol. Med.* **6**:63–101.

Priestley, J., 1906, *The Discovery of Oxygen*, University of Chicago Press, Chicago.

Prutz, W. A., Butler, J., and Land, E. J., 1990, Interaction of Cu(I) with nucleic acids, *Int. J. Radiat. Biol.* **58**:215–234.

Rahman, A., Shahabuddin, Hadi, S. M., Parish, J. H., and Ainley, K., 1989, Strand scission in DNA induced by quercetin and Cu(II): Role of Cu(I) and oxygen free radicals, *Carcinogenesis* **10**:1833–1839.

Rao, G. S., 1991, Release of 2-thiobarbituric acid reactive products from glutamate, deoxypuridine or DNA during autoxidation of dopamine in the presence of copper ions, *Pharmacol. Toxicol.* **69**:164–166.

Rice-Evans, C. A., and Diplock A. T., 1993, Current status of antioxidant therapy, *Free Radical Biol. Med.* **15**:77–96.

Rosen, A. D., and Frumin, N. V., 1979, Focal epileptogenesis following intracortical haemoglobin injection, *Exp. Neurol.* **66**:277–284.

Rumyantseva, G. V., Kennedy, C. H., and Mason, R. P., 1991, Trace transition metal-catalyzed reactions in the microsomal metabolism of alkyl hydrazines to carbon-centered free radicals, *J. Biol. Chem.* **266**:21422–21427.

Samuni, A., Kalkstein, A., and Czapski, G., 1980, Does oxygen enhance the radiation-induced inactivation of penicillinase? *Radiat. Res.* **82**:65–70.

Samuni, A., Chevion, M., and Czapski, G., 1981, Unusual copper-induced sensitization of the biological damages due to superoxide radicals, *J. Biol. Chem.* **256**:12632–12635.

Samuni, A., Aronovitch, J., Godinger, D., Chevion, M., and Czapski, G., 1983, On the cytotoxicity of vitamin C and metal ions: A site-specific Fenton mechanism, *Eur. J. Biochem.* **137**:119–124.

Samuni, A., Chevion, M., and Czapski, G., 1984, Roles of copper and O_2^- in the radiation induced inactivation of T7 bacteriophage, *Radiat. Res.* **99**:562–572.

Shinar, E., Navok, T., and Chevion, M., 1983, The analogous mechanisms of enzymatic inactivation induced by ascorbate and superoxide in the presence of copper, *J. Biol. Chem.* **258**:14778–14783.

Shlafer, M., Brosamer, K., Forder, J. R., Simon, R. H., Ward, P. A., and Grum, C. M., 1990, Cerium chloride as a histochemical marker of hydrogen peroxide in reperfused ischemic hearts, *J. Mol. Cell. Cardiol.* **22**:83–97.

Sigman, D., 1986, Nuclease activity of 1,10-phenanthroline-copper-ion, *Acc. Chem. Res.* **19**:180–186.

Snyder, R., Witz, G., and Goldstein, D., 1993, The toxicology of benzene, *Environ. Health Perspect.* **100**:293–306.

Stadtman, E. R., 1990, Metal ion-catalyzed oxidation of proteins: Biochemical mechanism and biological consequences, *Free Radical Biol. Med.* **9**:315–325.

Stern, A., 1985, *Oxidative Stress*, pp. 331–349, Academic Press, New York.

Stoewe, R., and Prutz, W. A., 1987, Copper-catalyzed DNA damage by ascorbate and hydrogen peroxide, *Free Radical Biol. Med.* **3**:97–105.

Swauger, J. E., Dolan, P. M., Zweier, J. L., Kuppusamy, P., and Kensler, T. W., 1991, Role of benzoyoxyl radical in DNA damage mediated by benzoyl peroxide, *Chem. Res. Toxicol.* **4**:233–228.

Szabo, M. E., Dray-Lefaix, M. T., Doly, M., and Braquest, P., 1991a, Free radical-mediated effects in reperfusion injury: A histologic study with SOD and EGB761 in rat retina, *Ophthalmic Res.* **23**:225–234.

Szabo, M. E., Dray-Lefaix, M. T., Doly, M., Carre, C., and Braquest, P., 1991b, Ischemia and reperfusion-induced histologic changes in the rat retina, *Invest. Ophthalmol. Vis. Sci.* **32**:1471–1478.

Thederahn, T. B., Kuwabara, M. C., Larsen, T. A., and Sigman, D. S., 1989, Nuclease activity of 1,10-phenanthroline-copper: Kinetic mechanism, *J. Am. Chem. Soc.* **111**:4941–4946.

Tynecka, M., 1995, [The role of iron in the pathology of nervous system], *Neurol. Neurochir. Pol.* **29**(3):409–417.

Vandeplassche, G., Hermans, C., Thone, F., and Borgers, M., 1989, Mitochondrial hydrogen peroxide generation by NADH-oxidase activity following regional myocardial ischemia in the dog, *J. Mol. Cell. Cardiol.* **21**:383–392.

Van Reyk, D. M., and Dean, R. T., 1996, The iron-selective chelator desferal can reduce chelated copper, *Free Radical Res.* **24**:55–60.

Van Steveninck, J., Van der Zee, J. and Dubbelman, T. M. A. R., 1985, Site-specific and bulk-phase generation of hydroxyl radicals in the presence of cupric irons and thiol compounds, *Biochem. J.* **232**:309–311.

Voogd, A., Sluiter, W., Eijk, H. G. v., and Koster, J. F., 1992, Low molecular weight iron and the oxygen paradox in isolated rat hearts, *J. Clin. Invest.* **90**:2050–2055.

Voogd, A., Sluiter, W., and Koster, J. F., 1994, The increased susceptibility to hydrogen peroxide of the (post-)ischemic rat heart is associated with the magnitude of the low molecular weight iron pool, *Free Radical Biol. Med.* **16**:453–458.

Wacker, W. E. C., and Vallee, B. L., 1959, Nucleic acids and metals, *J. Biol. Chem.* **234**:3257–3262.

Walling, C., 1982, The nature of the primary oxidants in oxidation mediated by metal ions, in *Oxidase and Related Redox Systems.* (T. E. King, H. S. Mason, and M. Morrison, eds.)pp. 85–97, Pergamon Press, Elmsford, NY.

Weight, S. C., Bell, P. R. F., and Nicholson, M. L., 1996, Renal ischemia–reperfusion injury, *B. J. Surg.* **83**:162–170.

Willmore, L. J., Ballinger, W. E., Boggs, W., and Rubin, G. W. S. J. J., 1980, Dendritic alteration in rat isocortex within an iron-induced chronic epileptic focus, *Neurosurgery* **7**:142–146.

Willmore, L. J., Triggs, W. J., and Grey, J. D., 1986, The role of iron-induced hippocampal peroxidation in acute epileptogenesis, *Brain Res.* **382**:422–426.

Yamamoto, K., and Kawanishi, S., 1989, Hydroxyl free radical is not the main active species in site-specific DNA damage induced by copper(II) ion and hydrogen peroxide, *J. Biol. Chem.* **264**:15435–15440.

Yamamoto, K., and Kawanishi, S., 1991a, Site-specific DNA damage induced by hydrazine in the presence of manganese and copper ion, *J. Biol. Chem.* **266**:1509–1515.

Yamamoto, K., and Kawanishi, S., 1991b, Site-specific DNA damage induced by phenylhydrazine and phenelzine in the presence of Cu(II) and Fe(III) complexes: Roles of active oxygen species and carbon radicals, *Chem. Res. Toxicol.* **5**:440–446.

Youdim, M. B. H., 1988a, *Brain Iron. Neurochemical and Behavioral Aspects*, Taylor & Francis, London.

Youdim, M. B. H., 1988b, Iron in the brain: implication for Parkinson's and Alzheimer's diseases, *Mt. Sinai J. Med.* **55**:97–101.

Youdim, M. B. H., and Green, A. R., 1978, Iron deficiency and neurotransmitter synthesis and function, *Proc. Nutr. Soc.* **37**:173–179.

Yourtee, D. M., Elkins, L. L., Nalvarte, E. L., and Smith, R. E., 1992, Amplification of doxorubicin mutagenicity by cupric ion, *Toxicol. Appl. Pharmacol.* **116**:57–65.

Zhu, B. Z., Har-El, R., Kitrossky, N., and Chevion, M., 1998, New modes of action of desferrioxamine: Scavenging of semiquinone radical and stimulation of hydrolysis of tetrachlorohydroquinone, *Free Radical Biol. Med.* **24**:360–369.

Zweier, J. L., Flaherty, J. T., and Weisfeldt, M. L., 1987, Direct measurement of free radicals generation following reperfusion of ischemic myocardium, *Proc. Natl. Acad. Sci. USA* **84**:1404–1407.

Chapter 5

Biochemistry of Redox Signaling in the Activation of Oxidative Stress Genes

Beatriz González-Flecha and Bruce Demple

1. INTRODUCTION

The term *oxidative stress* has become widely used (and abused) in biological science, thanks to the apparent involvement of reactive oxygen species in a wide variety of biological processes. Many examples are presented below and in the other chapters of this volume. A versatile definition of the term is that oxidative stress is an imbalance between the production and disposal of reactive oxygen (Sies, 1991), which we will adopt for this chapter [see Figure 1 for a general summary of H_2O_2 and $\cdot O_2^-$ (superoxide) metabolism]. Thus, excessive generation of superoxide, as in cells exposed to redox-cycling agents (Kappus and Sies, 1981) such as paraquat (PQ), and genetic deficiency in superoxide dismutase (SOD) (Imlay and Fridovich, 1991), may both be considered as forms of oxidative stress related to superoxide. Elevated levels of H_2O_2 or nitric oxide can constitute different forms of oxidative stress. Thus, different biochemical forms of oxidative stress may occur depending on the nature of the agent(s) involved and the kinetics with which the stress develops (abrupt or acute versus gradual).

Adopting the above definition of oxidative stress underscores the importance of obtaining a quantitative understanding of the rates of reactive oxygen production and elimination under different physiological conditions. This necessity is beginning to be

Beatriz González-Flecha Physiology Program, Department of Environmental Health, Harvard School of Public Health, Boston, Massachusetts 02115. **Bruce Demple** Division of Toxicology, Department of Cancer Cell Biology, Harvard School of Public Health, Boston, Massachusetts 02115.

Reactive Oxygen Species in Biological Systems, edited by Gilbert and Colton. Kluwer Academic / Plenum Publishers, New York, 1999.

filled by studies in the past few years. A relatively complete picture of reactive oxygen metabolism in *Escherichia coli* can now be envisioned (Figure 1).

Diverse species seem to have evolved genetic mechanisms to cope with oxidative stress. Enteric bacteria (*E. coli* and *Salmonella typhimurium*) exhibit differential responses to superoxide-generating agents or H_2O_2, as discussed below. Separate responses to H_2O_2 and superoxide have also been described in the yeast *Saccharomyces cerevisiae* (Moradas-Ferreira *et al.*, 1996). This ability of cells to distinguish different types of oxidative stress implies the existence of sensing molecules that respond differentially. In *E. coli*, where molecular analysis has proceeded furthest, different sensing molecules have indeed been identified (Hidalgo and Demple, 1996b). OxyR protein is triggered in cells exposed to H_2O_2 or nitrosothiols; SoxR protein is activated on superoxide or nitric oxide

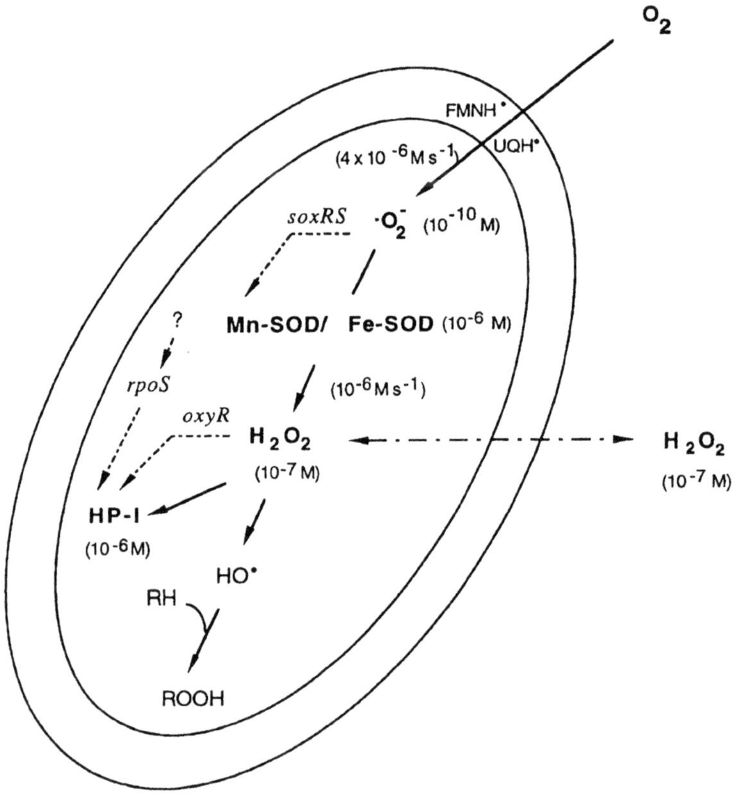

FIGURE 1. Metabolism of superoxide anion and hydrogen peroxide in *E. coli*. FMNH˙, NADH dehydrogenase flavin semiquinone; UQH˙, ubisemiquinone; Mn-SOD and Fe-SOD, manganese- and iron-containing superoxide dismutases; HP-I; catalase hydroperoxidase-I; *rpoS*, σ^{38} subunit of RNA polymerase (stationary phase). The numbers correspond to experimental values for steady-state concentrations or production rates (Imlay and Fridovich, 1991; González-Flecha and Demple, 1995, 1997). O_2 and probably H_2O_2 pass freely through the membrane. Genetic systems in bacteria that control oxidative metabolism enzymes are italicized.

stress; Fnr protein is active only during anaerobic growth. The mechanisms by which such differential sensing is achieved are discussed in separate sections below. Additionally, eukaryotic examples of systems for response to oxidative stress are presented.

2. HYDROGEN PEROXIDE: THE CELLULAR SIGNAL FOR OxyR

2.1. Genetic and Physiological Evidence

The OxyR protein (34 kDa) is a redox sensor and transcription regulator (Storz and Altuvia, 1994). The *oxyR* locus was identified in *E. coli* and *S. typhimurium* (Christman *et al.*, 1985) as the site of mutations that confer increased resistance to H_2O_2, cumene hydroperoxide, and *t*-butyl hydroperoxide, but not to $\cdot O_2^-$ -generating systems such as menadione (MD) or PQ. Two-dimensional electrophoresis of cell extracts from the H_2O_2-resistant mutants showed overexpression of eight or nine proteins that are H_2O_2-inducible in wild-type cells (Christman *et al.*, 1985; Greenberg and Demple, 1989). Deletion of the *oxyR* gene prevents the induction of these proteins and confers increased sensitivity to H_2O_2 (Christman *et al.*, 1985). The *oxyR*-regulated proteins include catalase-hydroperoxidase I (HP-I), the *katG* gene product, an alkylhydroperoxide reductase, glutathione reductase, and the Dps DNA binding protein (Christman *et al.*, 1985; Altuvia *et al.*, 1994).

The ability of OxyR to sense changes in H_2O_2 levels *in vivo* has been shown in studies that follow the expression of the *oxyR*-regulated *katG* gene (encoding catalase-hydroperoxidase I) after H_2O_2 treatment. Gene expression was monitored by measuring either mRNA levels or the β-galactosidase activity from a *katG'::lacZ* operon fusion. A wide variety of stimuli were assayed ranging from the standard single-dose administration of 5–100 μM H_2O_2 (Demple and Halbrook, 1983; Christman *et al.*, 1985; Morgan *et al.*, 1986; Tao *et al.*, 1989; González-Flecha and Demple, 1995) to the use of H_2O_2-generating systems in the extracellular compartment (González-Flecha and Demple, 1995). These experiments have shown that the magnitude of the response depends on the type of stimulus as much as it does on its severity. Pulse-type stimuli give a maximal response at a H_2O_2/catalase ratio of ~15,000. The response is a transient increase in the *oxyR*-dependent activities which peaks at ~10 min after treatment (González-Flecha and Demple, 1995). Ramp-type (gradually increasing) stimuli produce much higher maximal responses for the same H_2O_2/catalase ratio (González-Flecha and Demple, 1995), indicating a dynamic interplay between the rates of production and elimination of H_2O_2, on the one side, and the rates of activation and inactivation of OxyR, on the other. For each particular treatment, a different steady state of activated OxyR would be achieved, which will in turn produce a different increase in the expression of the downstream genes. Note that, to date (1997), no specific mechanism of OxyR inactivation has been identified; conceivably the activity is shut off through reduction or proteolysis.

Interestingly, even metabolic changes in the rate of H_2O_2 generation that occur when cells progress from lag phase to exponential growth in rich medium, are sensed by OxyR (González-Flecha and Demple, 1997). This physiological activation of OxyR is responsible for the fine tuning of the catalase and other activities that control the cellular H_2O_2 concentration homeostatically (González-Flecha and Demple, 1997).

The relevance of OxyR in regulating antioxidant activities under physiological conditions is also supported by the fact that *oxyR*-deficient *S. typhimurium* and *E. coli* strains are hypersensitive to H_2O_2 and have a high spontaneous mutation rate during aerobic growth (Storz *et al.*, 1987; Greenberg and Demple, 1988; González-Flecha and Demple, 1997).

2.2. *In Vitro* Experiments

The role of OxyR as the sensor in the response to H_2O_2 was indicated by the finding that the steady-state level of *oxyR* mRNA (Christman *et al.*, 1989), and amount of OxyR protein (Storz *et al.*, 1990) are not increased after H_2O_2 treatment. This indicates that posttranslational modification of the OxyR protein is the first event in the cascade leading to induction of *oxyR*-dependent activities.

The mechanism by which OxyR senses H_2O_2 was first studied by evaluating the ability of purified OxyR to activate the transcription of the *oxyR*-regulated *ahpFC* and *katG* genes *in vitro* (Storz *et al.*, 1990). Unexpectedly, the OxyR protein *in vitro* activated transcription of the *katG* gene, even though the protein was isolated from untreated cells (Storz *et al.*, 1990). This activity could be eliminated by treating OxyR with high levels of the reducing agent dithiothreitol aerobically, or lower levels anaerobically. This reductive inactivation was reversed on removal of the reducing agent and exposure to air. Thus, OxyR is apparently activated on isolation from the cell into O_2-equilibrated buffers.

The behavior of OxyR suggested that its transcriptional activity might be controlled by a redox-active amino acid residue such as cysteine. The OxyR residue involved in sensing H_2O_2 was studied by site-directed or random mutagenesis, followed by screening for strains with the *oxyR*-regulated proteins either noninducible by H_2O_2 or overproduced. This analysis revealed that cysteine-199 (of six cysteines in total) was critical for activity (Kullik *et al.*, 1995). Random mutagenesis of the entire *oxyR* gene and screening for overexpression of *oxyS* (a strong *oxyR*-dependent gene) showed eight different mutations causing single amino acid exchanges. Four of these mutations (histidine-198 to tyrosine, histidine-198 to arginine, arginine-201 to cysteine, and cysteine-208 to tyrosine) mapped close to the proposed redox-active cysteine-199 residue (Kullik *et al.*, 1995). Two other mutations affected histidine residues (histidine-114 and histidine-198) that are speculated to form hydrogen bonds with cysteine-199 (Kullik *et al.*, 1995). Although these experiments point to cysteine-199 as the redox-active center, it was still unclear what chemical modification this residue undergoes on reaction with H_2O_2. The existence of intermolecular disulfide bridges is disfavored (Kullik *et al.*, 1995), and one model (Storz and Altuvia, 1994) proposes the reversible oxidation of cysteine-199 to a sulfenic acid ($-SO^-$), as reported for streptococcal NADH peroxidase (Poole and Claiborne, 1989; Claiborne *et al.*, 1993). However, recent data indicate that this model should be revised. Specifically, cysteine-208 seems to be essential for activation of OxyR, and a peptide with a disulfide linking residues 199 and 208 has been isolated from the active protein (M. Zheng and G. Storz, personal communication).

In addition to H_2O_2, OxyR has recently been reported to be activated by *S*-nitrosothiols (Hausladen *et al.*, 1996). Because *S*-nitrosylation of OxyR induces *katG* transcription *in vitro*, and denitrosylation inactivates the protein, the mechanism of activation by *S*-nitrosothiols seems to be mediated by nitrosation rather than oxidation (Hausladen *et al.*, 1996).

Regardless of the mechanism of activation, the transduction of the oxidative stress signal to RNA polymerase seems to be related to a conformational change in OxyR after oxidation (Storz *et al.*, 1990). Activated (oxidized) OxyR is a tetramer that binds to specific promoter regions increasing transcription of *oxyR*-dependent genes (Toledano *et al.*, 1994) (see Figure 2).

3. THE CELLULAR SIGNAL FOR SoxR

3.1. Genetic and Physiological Evidence

In *E. coli*, Mn-containing SOD (MnSOD) and the DNA repair enzyme endonuclease IV are induced in cells exposed to PQ during aerobic growth, but not on H_2O_2 exposure (Hassan and Fridovich, 1977; Chan and Weiss, 1987). Large-scale analyses by two-dimensional gel electrophoresis (Morgan *et al.*, 1986; Greenberg and Demple, 1989) revealed that PQ or MD induce the synthesis of ~40 proteins in addition to the ~40 proteins induced by H_2O_2 directly—a total of nearly 100 proteins responding to oxidative stress! This level of complexity dictated genetic analysis as the best initial approach.

Two independent observations were exploited in different laboratories to identify a regulatory locus governing a key group of the superoxide-responsive proteins. B. Weiss and colleagues had demonstrated the induction of endonuclease IV by PQ (Chan and

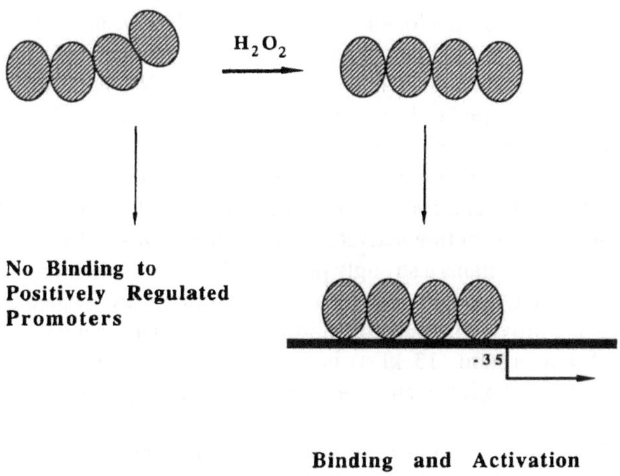

FIGURE 2. A model for OxyR sensing and regulation. OxyR (hatched ovals) is a tetramer in both the reduced and oxidized forms; nonactivated (reduced) OxyR binds only the *oxyR* promoter, which it represses (Storz and Altuvia, 1994; Toledano *et al.*, 1994). Reduced OxyR does not bind to the positively regulated promoters. Oxidation by hydrogen peroxide (proposed to target cysteine-199; Kullik *et al.*, 1995) changes the protein conformation and allows binding and activation of the OxyR-dependent genes (*ahpC*, *katG*, *oxyS*, and *dpS*) (Toledano *et al.*, 1994).

Weiss, 1987), and had cloned the gene (*nfo*) encoding this enzyme (Cunningham *et al.*, 1986). They generated a reporter fusion (*nfo::lac*) and used this to screen for bacterial strains with increased expression of the reporter resulting from mutations in other genes (Tsaneva and Weiss, 1990). Our work took advantage of an adaptive response to MD (Greenberg and Demple, 1989), in which cells exposed to low levels of this compound developed resistance to otherwise toxic MD concentrations. We sought mutant strains in which such resistance was expressed without prior exposure, i.e., constitutive mutants (Greenberg *et al.*, 1990). Both searches identified the same locus at 92 min on the *E. coli* genomic map, which we named *soxR* (for *s*uper*ox*ide *r*esponse or *r*esistance). Cloning *soxR* and the generation of null (deletion) mutations was facilitated by the prior cloning of genes in the same region (Greenberg *et al.*, 1990; Tsaneva and Weiss, 1990). These studies showed that *soxR* acts in an overall positive fashion, because the locus is required for activation of the system.

Two-dimensional gel analysis showed *soxR* to control at least eight of the inducible proteins (Greenberg *et al.*, 1990), including MnSOD (encoded by *sodA*), endonuclease IV, and glucose-6-phosphate dehydrogenase (Greenberg *et al.*, 1990; Tsaneva and Weiss, 1990). Additional regulated gene products have since been identified: fumarase C protein, aconitase, and ferredoxin:NADPH oxidoreductase (Hidalgo and Demple, 1996b). All of the above activities can be considered to have some relation to coping with oxidative stress. However, activation of *soxR* also mediates elevated resistance to multiple antibiotics, and this is mediated through induction of an antisense RNA encoded by *micF*, which interferes with synthesis of the OmpF porin (Chou *et al.*, 1993), and the AcrAB efflux pumps that help eliminate some antibiotics (Ma *et al.*, 1996). Still other *soxR*-regulated genes have been identified but their functions are unknown (Hidalgo and Demple, 1996b). Together, the group of genes coregulated under *soxR* control constitutes a *regulon*.

Molecular analysis of the cloned *soxR* locus quickly revealed a new level of complexity: The locus actually contained two genes required to functionally complement deletion mutations (Amábile-Cuevas and Demple, 1991; Wu and Weiss, 1991). These genes, *soxR* and *soxS* (named to follow *soxR*), are arranged head to head in the chromosome, with transcription of *soxR* initiated within the *soxS* structural genes, whereas *soxS* transcription has a more conventionally located start site in the intergenic region (Wu and Weiss, 1991). The *soxR* gene encodes a 17-kDa protein related to the MerR family of mercury-responsive transcription activators (Amábile-Cuevas and Demple, 1991). Near the N-terminus, SoxR contains a strongly predicted helix–turn–helix motif that probably mediates DNA binding; the four cysteine residues of SoxR are located in a cluster near the C-terminus, and provide ligands for an iron–sulfur center essential for SoxR activity. The predicted SoxS protein (13 kDa) is homologous to the C-terminal one-third of AraC/XylS-type transcription factors, and also contains a predicted helix–turn–helix motif.

Transcriptional analysis showed that the *soxS* transcript is dramatically induced in PQ-treated cells, while the level of *soxR* RNA was unchanged (Wu and Weiss, 1991). Complementary experiments (Amábile-Cuevas and Demple, 1991) showed that high-level expression of SoxS protein in the absence of SoxR was sufficient to activate expression of regulon genes (such as *nfo* or *sodA*) and to confer resistance phenotypes against PQ or antibiotics; treatment with PQ had no further effect on gene expression of phenotypes (Amábile-Cuevas and Demple, 1991). On the other hand, expression of SoxR in the absence of SoxS did not activate any of the regulon genes, even in cells exposed to PQ. Collectively, these observation suggested a novel type of two-stage regulation in which SoxR regulates *soxS* expression in response to a redox signal, and SoxS is the direct activator of the regulon genes (Demple and Amábile-Cuevas, 1991).

Experimental evidence for this two-stage model was provided through the use of reporter fusions containing the soxS promoter (*soxS'::lacZ*). Expression of such reporters was found to be strongly inducible by PQ in a *soxR*-dependent fashion (Nunoshiba *et al.*, 1992; Wu and Weiss, 1992). Systematic analysis showed that the induction response of *soxS'::lacZ* is restricted to superoxide-generating compounds, whereas H_2O_2, heat shock, UV light, and various metals did not induce transcription (Nunoshiba *et al.*, 1992). Most notably, genetic deficiency in SOD activated the stress response during normal growth (Nunoshiba *et al.*, 1992), consistent with the burden of excess $\cdot O_2^-$ expected in such circumstances (Imlay and Fridovich, 1991) and suggesting a specific signaling role for superoxide.

Biochemical experiments have substantiated the regulatory role of SoxS protein (Li and Demple, 1994). Purified SoxS binds specifically to the promoter regions of *soxRS* regulon genes such as *sodA* and *nfo*. This binding serves to enhance the subsequent binding of RNAP and stimulates *in vitro* transcription. *In vitro* transcription has also been reconstituted with a MalE–SoxS fusion protein (Jair *et al.*, 1996). At least one close relative of SoxS exists in *E. coli*, the 14-kDa MarA protein (Cohen *et al.*, 1993a), which mediates the induction of many of the same genes (Greenberg *et al.*, 1991; Chou *et al.*, 1993; Ariza *et al.*, 1994). MarA synthesis is regulated in response to salicylate and some antibiotics (Cohen *et al.*, 1993b). Thus, both SoxS and MarA are regulatory proteins that themselves are unresponsive to biological signals.

3.2. *In Vitro* Analysis of SoxR Transcriptional Activity

The genetic and physiological analysis showed SoxR to be the redox-sensing regulator of the *soxRS* regulon (Nunoshiba *et al.*, 1992; Wu and Weiss, 1992). These observations directed attention on the properties of this protein, which was predicted to be a transcription activator. The first direct evidence for this point was provided by DNA binding experiments in which cell extracts overexpressing SoxR were shown to have specific, high-affinity binding activity for the *soxS* promoter (Nunoshiba *et al.*, 1992). Subsequent purification of SoxR confirmed this binding specificity and affinity (apparent $K_d \sim 10^{-10}$ M), and demonstrated that SoxR binds a region of perfect dyad symmetry centered between the −10 and −35 promoter elements (Hidalgo and Demple, 1994). Such positioning for binding is more typical of repressors than activators, although the homologous MerR (mercury response) protein interacts in a similar manner with its target for transcription activation (Summers, 1992). For both MerR and SoxR, the target promoters have −10 and −35 elements separated by 19-bp spacers (compared with the optimal 17 bp), and deleting 1 or 2 bp from the spacer region of *soxS* dramatically increases the constitutive activity of these promoters (Hidalgo and Demple, 1997). Thus, activated SoxR (like activated MerR) seems to compensate for the overwinding between −10 and −35 by remodeling the promoter structure (Hidalgo and Demple, 1997) (Figure 3).

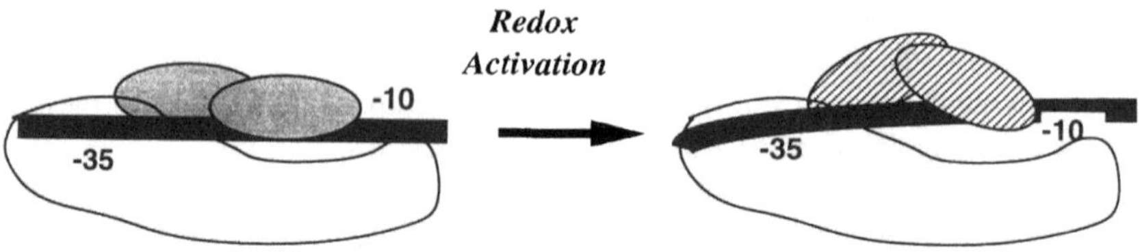

FIGURE 3. A model for SoxR sensing and regulation. Both resting SoxR [left, protein with reduced [2Fe-2S] centers (Ding *et al.*, 1996; Gaudu and Weiss, 1996)] and activated SoxR (right, protein with oxidized [2Fe-2S] centers) bind in the center of the RNA polymerase recognition elements (–35 and –10) of the *soxS* promoter (Hidalgo and Demple, 1994). A conformational change in SoxR is proposed to occur on oxidation (Hidalgo *et al.*, 1995; Hidalgo and Demple, 1997). Nonactivated SoxR has no detectable effect on transcription. Activated SoxR compensates for the suboptimal spacing between the promoter elements, remodeling the structure for more rapid initiation by the polymerase.

Purification of SoxR immediately revealed one of the protein's most obviously striking properties: a reddish-brown color (Hidalgo and Demple, 1994). Such color implied a prosthetic group bound to the protein, and the details of the absorbance spectrum suggested that this might be an iron–sulfur center. Potential for iron binding was also indicated by the cysteine cluster near the SoxR C-terminus (Amábile-Cuevas and Demple, 1991; Wu and Weiss, 1991). Analysis of the purified protein for 20 different metals showed significant amounts only of Fe, at a stoichiometry approaching 2 per SoxR monomer (Hidalgo and Demple, 1994). In attempts to prevent oxidation of the protein during purification, SoxR was purified in buffers containing 1 mM 2-mercaptoethanol, which yielded a colorless form of the protein lacking detectable Fe (Hidalgo and Demple, 1994).

Both the Fe-containing (Fe-SoxR) and the iron-free (apo-SoxR) forms of the protein bound the *soxS* promoter, in the same position and with the same affinity. However, only Fe-SoxR was able to stimulate *in vitro* transcription of *soxS* at least 100-fold (Hidalgo and Demple, 1994). This activation is evidently not related to recruitment of RNAP, because Fe-SoxR enhanced subsequent RNAP binding only ~2-fold; instead, the activation results from a step subsequent to DNA binding, such as an allosteric change in the protein–DNA complex (Figure 3), dependent on the Fe in the protein.

Detailed analysis of the properties of Fe-SoxR showed that the protein is a homodimer containing a pair of [2Fe-2S] centers (Hidalgo *et al.*, 1995; Wu *et al.*, 1995). In the protein isolated under aerobic conditions, these centers are in the oxidized form (formally, Fe^{3+}-Fe^{3+}), and this is the form of SoxR first found to be transcriptionally active (Hidalgo and Demple, 1994). Chemical reduction of these centers with dithionite (to yield the Fe^{2+}-Fe^{3+} form) generated an electron paramagnetic resonance spectrum typical of [2Fe-2S] proteins (Hidalgo *et al.*, 1995; Wu *et al.*, 1995), consistent with the four cysteines of SoxR acting as the ligands for the metal center. More recent work confirms this by showing that substitution of any of the SoxR cysteines with alanine eliminates detectable Fe from the protein (Bradley *et al.*, 1997). Such cysteine-to-alanine derivatives also lack detectable activation *in vivo* in response to PQ, even though these proteins are stable and bind the *soxS* promoter with apparently normal affinity. This analysis has provided an *in vivo* demonstration of the critical regulatory role played by the SoxR [2Fe-2S] centers.

The key question, of course, was the mechanism by which SoxR activity is regulated. Two possibilities have been considered: oxidation–reduction or assembly–disassembly of the iron–sulfur centers. Consistent with the former hypothesis was the isolation of SoxR with intact [2Fe-2S] centers even from cells not treated with activating agents such as PQ (Hidalgo and Demple, 1994). The latter hypothesis was supported by the ease with which apo-SoxR could be generated and the stable folding and normal DNA binding by this protein. Indeed, biological activities that mediate disassembly and assembly of the SoxR [2Fe-2S] clusters have been identified. Glutathione destroys these [2Fe-2S] clusters in an oxygen-dependent reaction that probably involves sulfur-based radicals, and glutathione-deficient bacteria activate SoxR more readily and to a greater extent than wild-type cells (Ding and Demple, 1996). This biological thiol could therefore operate *in vivo* to limit SoxR activity. Rapid assembly of [2Fe-2S] centers to generate fully active SoxR is mediated by NifS-type proteins (Hidalgo and Demple, 1996a), which may be involved in the assembly of many types of iron–sulfur centers (Flint, 1996).

Although assembly–disassembly mechanisms could play a role in controlling SoxR function, as they do in Fnr and the mammalian iron-response proteins (Klausner *et al.*, 1993; Khoroshilova *et al.*, 1995; Hentze and Kuhn, 1996), the best current evidence is that the critical reactions are oxidation and reduction. Controlling the oxidation state of SoxR in *in vitro* transcription reactions proved difficult, owing to the ready oxidation of SoxR by oxygen (Hidalgo and Demple, 1994). This sensitivity arises from a redox potential of about –285 mV for the SoxR [2Fe-2S] centers *in vitro* (Ding *et al.*, 1996; Gaudu and Weiss, 1996). The development of redox equilibrium conditions for *in vitro* transcription reactions allowed assessment of SoxR activity as a function of oxidation state. These studies showed that fully reduced SoxR loses most of its *soxS*-specific transcriptional activity (Ding *et al.*, 1996; Gaudu and Weiss, 1996) but is still able to bind the *soxS* promoter (Gaudu and Weiss, 1996). Moreover, prior binding of SoxR to *soxS* DNA did not prevent its inactivation by chemical reduction (Gaudu and Weiss, 1996). Reduced SoxR could be readily reactivated by exposure to air or the chemical oxidant potassium ferricyanide (Ding *et al.*, 1996). This reversibility could confer sensitivity in redox sensing by SoxR.

Key support for the *in vitro* model of redox-regulated SoxR activity has been provided by analysis of mutant forms of the protein. Constitutively activated forms of SoxR were generated by the genetic studies that identified this system (Greenberg *et al.*, 1990). Alterations affecting three different regions of the protein (Nunoshiba and Demple, 1994) were examined: an arginine-to-cysteine substitution in the N-terminal helix–turn–helix motif; a serine-to-leucine change in the center of the polypeptide; and a glycine-to-aspartate change near the C-terminus, in a short region affected by many other constitutive mutations (Nunoshiba and Demple, 1994). Curiously, all three mutant proteins displayed Fe content, *soxS* DNA binding, and transcriptional activation similar to wild-type SoxR, and all three could be inactivated *in vitro* by reduction or removal of the [2Fe-2S] clusters (Hidalgo *et al.*, 1997). Thus, the three mutations had not generated proteins that were simply locked in an active conformation; these proteins are still redox-regulated. However, *in vivo* analysis of the status of the [2Fe-2S] centers showed that they were present in only 2–4% as the reduced form even in cells not treated with activating agents, in contrast to wild-type SoxR ranging from 42 to 95% reduced (Hidalgo *et al.*, 1997). This result suggested that the mutant proteins might be hypersensitive to oxidation. Indeed, one protein had an *in vitro* redox potential shifted by –65 mV (Hidalgo *et al.*, 1997), consistent with an increased rate of oxidation during normal aerobic growth.

The [2Fe-2S] centers of SoxR are probably maintained in the reduced state enzymatically, although the responsible reductase(s) have yet to be identified (Nunoshiba and Demple, 1994; Gaudu and Weiss, 1996; Hidalgo *et al.*, 1997). Thus, mutations in other regions of the protein could affect the rate of this inactivating reaction. The cluster of constitutive mutations near the C-terminus may identify a region for interaction with a reductase (Nunoshiba and Demple, 1994; Gaudu and Weiss, 1996). In this model, SoxR is positioned in a dynamic equilibrium of reduction, maintaining the resting protein, and oxidation, generating the active form that triggers the defense regulon (Hidalgo *et al.*, 1997). The biochemistry of iron–sulfur reduction is well established for many other proteins, but the nature of the reducing enzyme(s) will be of great interest.

The oxidants that activate SoxR remain to be established. Oxygen itself may suffice (Hidalgo and Demple, 1994; Wu *et al.*, 1995), but superoxide may also be present in air-exposed buffers. Oxidation of the SoxR [2Fe-2S] centers by superoxide is expected to yield H_2O_2, but this is the most abundant product of $\cdot O_2^-$ in any case and so would be hard to identify as a specific product of SoxR [2Fe-2S] oxidation. As noted above, other oxidants, such as H_2O_2, are ineffective in activating SoxR (Nunoshiba *et al.*, 1992).

Redox-cycling agents that generate superoxide also deplete NAD(P)H pools (Kappus and Sies, 1981). This effect could limit reductase activity and allow oxidized SoxR to accumulate, and such a mechanism of SoxR activation has been proposed (Liochev and Fridovich, 1992). However, SoxR activation occurs rapidly (Nunoshiba *et al.*, 1992; Wu and Weiss, 1992; Nunoshiba and Demple, 1993; H. Ding and B. Demple, unpublished data), so that proposed reductase effects and SoxR oxidation would both have to occur in a short time. Direct action of physiological signals by oxidation of the [2Fe-2S] centers of SoxR therefore seems likely. Such a mechanism is consistent with the activating effect of SOD deficiency (Nunoshiba *et al.*, 1992). The chemistry by which nitric oxide activates SoxR must also be explained (Nunoshiba *et al.*, 1993). Activation by ·NO occurs most efficiently under anaerobic conditions (Nunoshiba *et al.*, 1993, 1995), thus excluding oxidants derived from oxygen. Therefore, the possibility that ·NO itself or oxygen-independent derivatives of ·NO oxidize the SoxR [2Fe-2S] clusters needs to be investigated.

4. OXYGEN: AN INACTIVATING SIGNAL FOR Fnr

4.1. Genetic and Physiological Evidence

The Fnr protein (30 kDa) is a global transcriptional regulator that, when oxygen is limiting, facilitates the switch in *E. coli* from aerobic respiration to fermentation (Guest *et al.*, 1996; Lynch and Lin, 1996). The *fnr* gene mediates the upregulation of over 70 genes involved in cellular adaptation to anoxic growth (Guest *et al.*, 1996; Lynch and Lin, 1996).

The ability of Fnr to sense anaerobiosis *in vivo* has been shown in studies of the expression of operon or protein fusions of different respiratory genes to the *lacZ* gene in wild-type and *fnr*-deficient strains grown in aero- or anaerobiosis. Fnr has been shown to mediate both transcriptional activation of genes encoding aerobic respiratory activities and repression of the genes of the aerobic pathway. The *narGHJI*, *dmsABC*, and *frdABCD* genes that encode nitrate reductase, trimethylamine *N*-oxide:dimethyl sulfoxide reductase, and fumarate reductase, respectively, have been shown to be under positive control of Fnr in response to anaerobiosis (Stewart, 1982; Jones and Gunsalus, 1987; Cotter and Gunsalus, 1989). The *cyoABCDE* and *cydAB*, and *ndh* genes, encoding the cytochrome *o* and *d* oxidase complexes, and the NADH dehydrogenase II, respectively, are negatively regulated by Fnr under anaerobic conditions (Spiro *et al.*, 1989; Cotter *et al.*, 1990). Thus, Fnr protein appears to be active in DNA binding anaerobically but not aerobically.

4.2. *In Vitro* Experiments

As in the case of OxyR, the role of Fnr as the sensor of the modulon was indicated by the finding that the level of Fnr protein does not differ between aerobically and anaerobically grown *E. coli* (Trageser *et al.*, 1990).

A proposed mechanism of oxygen sensing by Fnr was developed through sequence analysis, which revealed that Fnr protein is homologous to the catabolite regulatory protein (Crp) (Shaw *et al.*, 1983). Because Crp is not involved in sensing O_2, it was speculated that the specificity of Fnr could be attributed to the additional 26 amino acids at the N-terminus of Fnr that are not present in Crp.

Site-directed mutagenesis of Fnr has shown that three of the four cysteine residues in the N-terminal cluster (cysteine-20, -23, and -29, but not cysteine-16) and the only other cysteine residue of this protein (cysteine-122), are essential for the normal triggering of Fnr activity *in vivo* (Melville and Gunsalus, 1990; Sharrocks *et al.*, 1990; Kiley and Reznikoff, 1991). Depletion of Fe in the growth medium was also reported to inhibit Fnr activity *in vivo* (Spiro *et al.*, 1989; Green and Guest, 1993), which indicated an important role for metal binding. A key hypothesis was that the metal ion coordination state would be O_2 sensitive (Spiro *et al.*, 1989). On this basis, Unden and Guest (1985) suggested the presence of an O_2-sensing iron–sulfur cluster anchored to the N-terminal region of Fnr.

The mechanism by which Fnr senses anaerobiosis was also approached by analyzing purified mutant proteins active under aerobic conditions: the Fnr152 protein (two amino acid substitutions, aspartic acid-22 to serine and glutamic acid-27 to arginine, plus the insertion of a serine after position 17; Melville and Gunsalus, 1996), the DA154 protein (one substitution, aspartic acid-154 to alanine; Lazazzera *et al.*, 1993), and the LH28-DA154 protein (two substitutions, leucine-28 to histidine and aspartic acid-154 to alanine; Lazazzera *et al.*, 1996). The analysis of the LH28-DA154 protein identified an Fe-S cluster associated with the aerobically purified mutant protein resulting in an increased DNA binding over the aerobically purified wild-type protein (Khoroshilova *et al.*, 1995). Further analysis indicated that the type of cluster is $[4Fe\text{-}4S]^{2+}$ in the LH28-DA154 protein, and that the loss of the Fe-S cluster by exposure to oxygen leads to conversion to the monomeric form and decreased DNA binding (Lazazzera *et al.*, 1996).

Independent evidence for the role of the Fe-S cluster in sensing oxygen was reported by Green *et al.* (1996), who were able to reconstitute the active form of wild-type Fnr by treatment of the purified apoprotein with ferrous ions, cysteine, and the NifS protein of *Azotobacter vinelandii* in anaerobiosis. The incorporation of two $[4Fe\text{-}4S]^{2+}$ clusters per Fnr dimer increased DNA binding about sevenfold compared with apo-Fnr (Green *et al.*, 1996).

The current mechanistic model (Figure 4), based on the above evidence, suggests that Fnr protein is present during aerobic growth as a monomeric apoprotein with weak DNA binding activity. In anaerobiosis, formation of the [4Fe-4S] clusters is proposed to facilitate dimerization and consequent binding to DNA at the target regulatory sites.

The transduction of the signal sensed by Fnr to DNA seems to be related to changes in the oligomeric state of the Fnr protein which increase DNA binding at Fnr-dependent promoters. Two alternative mechanisms have been suggested to date. From studies with purified Fnr152 protein, sequential binding to Fnr recognition sites of two Fnr monomers and subsequent assembly of a stable dimer has been suggested (Melville and Gunsalus, 1996). In contrast, studies on DA154 and LH28-DA154 proteins purified in the presence of oxygen suggest that anaerobiosis would lead to dimerization of free Fnr in solution, followed by binding of the dimers at the Fnr-dependent promoters (Lazazzera *et al.*, 1993).

5. EUKARYOTIC TRANSCRIPTION FACTORS IN REDOX SIGNALING

5.1. Signaling Cascades for NF-κB and AP-1

The genetic responses to oxidative stress in eukaryotic cells are less well understood than the responses in bacteria. Transcription factors that are exclusively activated by reactive oxygen species (ROS) or that selectively control the expression of antioxidant and repair enzymes have not been identified. On the contrary, the response to increased

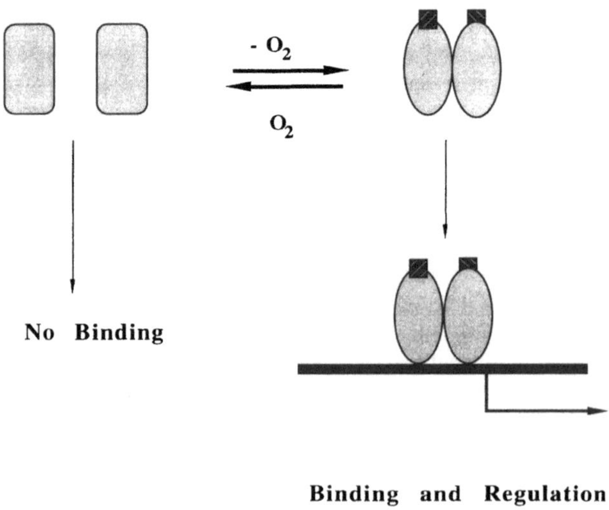

FIGURE 4. A model for Fnr sensing and regulation. In the presence of oxygen, Fnr is predominantly present as free monomers and does not bind DNA (left). At low oxygen concentrations (right), a [4Fe-4S] center (dark squares) is assembled in each monomer to stabilize the dimeric form of the protein. Fnr dimers can bind DNA at specific sites and regulate (increase or diminish) the transcription of specific sets of genes (Beinert and Kiley, 1996; Lazazzera *et al.*, 1996).

concentrations of ROS involves the activation of many functionally unrelated genes associated with signal transduction, proliferation, and immunological defense reactions (reviewed in Schulze-Osthoff *et al.*, 1995, and Chapter 6). Most of the genes induced by oxidative stress can be equally activated by other physiological signals, such as growth factors or cytokines, suggesting that ROS may act as second messengers that integrate a diverse set of gene-inducing signals into a common genetic response. Several conditions inducing oxidative stress result in activation of NF-κB and AP-1, two widely used transcription factors in eukaryotic cells.

5.1.1. Nuclear Factor-κB (NF-κB)

5.1.1a. Cellular Signal. NF-κB is one of the eukaryotic transcription factors shown to respond to oxidative stress (Schreck *et al.*, 1991; Toledano and Leonard, 1991). A variety of extracellular stimuli, such as bacterial lipopolysaccharide, viral infection, UV or γ-irradiation, the cytokines IL-1 and TNF, and virtually all inducers of T-cell activation, trigger NF-κB within minutes (Schulze-Osthoff *et al.*, 1995). Measurement of different indicators of oxidative stress has shown that most inducers of NF-κB also induce a transient increase in ROS concentrations (Schulze-Osthoff *et al.*, 1995). Conversely, activation of NF-κB by most inducers is blocked by antioxidants such as glutathione precursors (*N*-acetylcysteine), metal chelators, thiols, dithiocarbamates, and vitamin E (Meyer *et al.*, 1992; Schreck *et al.*, 1992a,b). Direct treatment of human T-cell (Jurkat) and cervical carcinoma (HeLa) lines with 30 to 250 μM H_2O_2 rapidly induced NF-κB DNA binding and transcriptional activation (Schreck *et al.*, 1991; Meyer *et al.*, 1993) whereas superoxide-generating compounds (PQ, MD, and doxorubicin) and singlet oxygen-generating agents [3,3′-(1,4-naphthylidene)dipropionate] failed to increase NF-κB DNA binding *in vitro* (Ignarro, 1990; Schreck *et al.*, 1992a), although the possibility remains that these compounds are not being effectively internalized or metabolized within the cell. Cell lines stably overexpressing catalase alone, but not SOD alone, had diminished NF-κB activation in response to TNF or okadaic acid. Inhibition by aminotriazole of catalase in the overexpressing cells restored the NF-κB inducibility (Schmidt *et al.*, 1995). These data point to H_2O_2 as a critical intermediate in the activation of NF-κB by diverse agents.

5.1.1b. Transcriptional Activation. In all of these cases, NF-κB activation does not require new protein synthesis (Baeuerle and Baltimore, 1988). Activated NF-κB is a heterodimer of a 50-kDa DNA-binding protein (p50) and a 65-kDa DNA-binding and transcription-activating subunit (p65 or RelA). In most cell types, inactive NF-κB is present in the cytoplasm bound to a third protein, the inhibitory subunit I-κB. Activation of NF-κB is associated with release and proteolytic degradation of I-κB (Henkel *et al.*, 1993; Palombella *et al.*, 1994; Traenckner *et al.*, 1994). The degradation of I-κB is preceded by phosphorylation of serines 32 and 36, which increases its turnover (Traenckner *et al.*, 1995). The phosphorylation of I-κB is prevented by dithiocarbamate (Traenckner *et al.*, 1995), suggesting that active oxygen species would be involved in the regulation of the steady-state level of I-κB, either by activation of an I-κB kinase or by inactivation of an I-κB phosphatase (Schulze-Osthoff *et al.*, 1995). Finally, activated NF-κB is translocated to the nucleus and binds specific recognition sites in NF-κB-regulated promoters initiating transcription of the target genes. NF-κB does not seem to sense an oxidative signal directly, because H_2O_2 did not activate the NF-κB–I-κB complex for DNA binding *in vitro*. Indeed, oxidation of a conserved cysteine in the RelA protein inactivates DNA binding by NF-κB.

5.1.2. Activator Protein-1 (AP-1)

Cellular Signal. AP-1 is another important redox-responsive transcription factor. In contrast to NF-κB, AP-1 is always present in the nucleus (Karin, 1995). AP-1 is rapidly and transiently induced by a variety of extracellular agents such as mitogenic signals, phorbol esters, differentiation-inducing stimuli, UV irradiation, and heavy metals, which may have oxidative stress in common (Schulze-Osthoff *et al.*, 1995). On the other hand, AP-1 binding activity is also stimulated by treatment of cells with various antioxidants, including thiocarbamates, *N*-acetyl-L-cysteine, butylated hydroxyanisole, 2-mercaptoethanol, and dithiothreitol (Meyer *et al.*, 1993; Schenk *et al.*, 1994), and also after transient overexpression of human thioredoxin (Amstad *et al.*, 1991). As a result, activation of AP-1 was proposed to reflect an *antioxidant* response. However, it was reported recently that semiquinone radicals and H_2O_2 are intermediates in the metabolism of both pyrrolidine dithiocarbamate and butylated hydroxyanisole in hepatoma cells (Pinkus *et al.*, 1996), indicating that these agents might in fact activate AP-1 via an oxidative stress pathway.

The activity called AP-1 consists of heterodimers of any of several Jun proteins (c-Jun, JunB, and JunD) and any of several Fos proteins (c-Fos, FosB, Fra1, and Fra2). Although the c-Fos·c-Jun is the more stable heterodimer, c-Jun homodimers can also bind AP-1 sites (Angel and Karin, 1991). Several pathways and mechanisms, including *c-fos* and *c-jun* induction as well as the posttranslational phosphorylation of c-Fos or c-Jun, contribute to AP-1 activity (Karin, 1995). In this connection, DNA binding activity can be restored to air-oxidized c-Jun or c-Fos *in vitro* through reduction of key cysteine residues (Abate *et al.*, 1990). This reductive reactivation might be linked to the antioxidant response pathway cited above, and can be mediated by thioredoxin and thioredoxin reductase (Abate *et al.*, 1990), or by the "Ref-1" activity of a DNA repair enzyme (Xanthoudakis *et al.*, 1992).

Three distinct mitogen-activated kinases (MAPKs) contribute to induction of AP-1 activity (Karin and Hunter, 1995). Some MAPKs are involved in increasing the amount of AP-1 complexes, whereas other MAPKs increase the activity of existing proteins. Extracellular-regulated kinases (ERKs) stimulate AP-1 activity through induction of c-Fos synthesis, which leads to formation of stable Fos–Jun heterodimers. The Jun N-terminal kinases (JNKs), also called *stress-activated kinases* (SAPKs), stimulate c-Jun and ATF2 transcriptional activities and can thereby also enhance *c-jun* transcription. Finally, activation of the novel Ras-activated protein kinase FRK (phosphoregulating kinase) results in a further increase in AP-1 activity by enhancing the transcription-activating function of c-Fos. The differential activation of these protein kinases and their distinct effects on *c-fos* and *c-jun* transcription and on AP-1 composition may affect the selection for AP-1 target genes. As for NF-κB, the redox-sensitive component(s) of the AP-1 pathway remains to be identified.

5.2. Regulation via Redox-Sensitive Thiols

NF-κB and AP-1 are not the only transcription factors sensitive to changes in the intracellular levels of ROS. Several transcription factors contain cysteine residues whose oxidation decreases DNA binding (Sun and Oberley, 1996). Redox-sensitive thiols have been found in zinc finger (Egr-1), helix–loop–helix (USF), and leucine zipper proteins (c-Fos and c-Jun) (Bandyopadhyay and Gronostajski, 1994; Wu *et al.*, 1996). *In vitro* experiments have shown that oxidation prevents DNA binding of these proteins. The nuclear protein Ref-1, which reduces thiol groups and can use thioredoxin as a cofactor, restores DNA-binding activity of oxidized Fos and Jun proteins (Xanthoudakis *et al.*, 1992). Because the reduced state of the cysteine residues is essential for DNA binding, and oxidation and reduction processes have been described, it is suggested that the redox modification of the DNA-binding domain could be a mechanism to regulate transcriptional activity (Table I).

6. PERSPECTIVES

The redox-signaling systems described here employ sophisticated molecular devices to discriminate among different growth conditions and types of oxidative stress. We have focused on systems of redox-regulated transcriptional control, especially in prokaryotes, but related biochemical control mechanisms could operate to modulate other activities. For example, the mammalian iron response protein can be activated by various types of oxidative stress through the disassembly of its [4Fe-4S] center (Klausner *et al.*, 1993; Hentze and Kuhn, 1996). This is formally similar to the inactivation mechanism of Fnr (Khoroshilova *et al.*, 1995), but the activated (apo) form of the iron response protein acts in posttranscriptional control (Klausner *et al.*, 1993; Hentze and Kuhn, 1996). In another example, the oxidation state of the *A. vinelandii* nitrogenase protein controls its ATPase activity allosterically (Georgiadis *et al.*, 1992). This may be formally related to the mechanism operating in SoxR, but in nitrogenase an enzyme activity is affected. The redox status of key cysteine residues in mammalian *N*-methyl-D-aspartate receptors has been proposed to regulate their signal transduction properties in response to nitric oxide

Table I
Summary of Redox Signals and Regulators[a]

Proposed redox-signaling mechanisms in transcription factors				
Protein	Proposed signal	Proposed sensor in protein	Proposed active form(s)	Resting form
OxyR	H_2O_2, RS-NO	Cys-SH	Cys-S-S-Cys, Cys-S-NO	Cys-SH
SoxR	$\cdot O_2^-$, ·NO	[2Fe-2S]	Oxidized	Reduced
Fnr	Oxygen	[4Fe-4S]	Fe-protein	Apoprotein
Fos, Jun type	Oxidation	Cysteine-SH	Cys-S-S-Cys, Cys-SOH, or $Cys-SO_2H$	Cys-SH

[a]See text for critical discussion of individual systems.

(Lipton *et al.*, 1993). This mechanism could be related to that proposed for OxyR (Storz *et al.*, 1990; Hausladen *et al.*, 1996), although the detailed chemistries remain to be established.

It should not be surprising if the redox signaling mechanisms described here are employed widely in biology. The ability to discriminate aerobic from anaerobic conditions, or H_2O_2 stress from $\cdot O_2^-$ stress, could be useful in many contexts. The chemistry of these sensing systems is well adapted in each case, and may represent the repeated turning of deleterious reactions to the advantage of cells during evolution. Although the oxygen- or oxidant-sensitivity of [4Fe-4S] clusters can lead to problematic inactivation of some hydratase enzymes (Gardner and Fridovich, 1992; Liochev and Fridovich, 1993, 1994; Keyer and Imlay, 1996), this same property has been adapted to advantage in Fnr. In contrast, the stability of [2Fe-2S] clusters in both the reduced and the oxidized state makes them well suited to their role in sensing oxidative stress in the SoxR protein. Structural effects of the oxidation state of a [2Fe-2S] center are beginning to be elucidated for putidaredoxin, whose association with a cytochrome P450 is modulated by oxidation–reduction (Lyons *et al.*, 1996). The reactivity of cysteines toward oxidants has been made useful rather than damaging in the case of OxyR.

Caution should be exercised in interpreting the behavior of proteins *in vitro*, in the absence of genetic or other *in vivo* evidence. Numerous transcription factors have now been described that lose activity on exposure to air; often these proteins have conserved cysteine residues in a DNA-binding motif (Sun and Oberley, 1996) (see Table I for comparison with other regulators). Reducing agents *in vitro* often restore binding activity in these cases, but such effects do not prove actual redox regulation. For some proteins, such behavior may instead be artifacts of their removal from the reducing environment of the cell. In this context it is worth noting that protein cysteines *in vivo* can be surprisingly difficult to oxidize to cystine—this seems to require enzymatic intervention, even in the relatively unprotected periplasmic space of gram-negative bacteria (Grauschopf *et al.*, 1995). The other side of this issue is that cellular activities exist that may help maintain proteins in the reduced, active forms. These include glutathione (or derivatives of it) and thioredoxin in virtually all cell types, while mammalian cells harbor a protein that may have dual roles in DNA repair and redox regulation (Xanthoudakis *et al.*, 1992).

ACKNOWLEDGMENTS. We are grateful to all of our colleagues for many insightful discussions. Work in the authors' laboratory was supported by grants from the U.S. National Institutes of Health (grant CA37831) and the Amyotrophic Lateral Sclerosis Association. B.G.-F. acknowledges support from the Pew Foundation.

7. REFERENCES

Abate, C., Patel, L., Rauscher, F. J., III, and Curran, T., 1990, Redox regulation of Fos and Jun DNA-binding activity in vitro, *Science* **249:**1157–1161.

Altuvia, S., Almiron, M., Huisman, G., Kolter, R., and Storz, G., 1994, The *dps* promoter is activated by OxyR during growth and by IHF and sigma S in stationary phase, *Mol. Microbiol.* **13:**265–272.

Amábile-Cuevas, C. F., and Demple, B., 1991, Molecular characterization of the *soxRS* genes of *Escherichia coli*: Two genes control a superoxide stress regulon, *Nucleic Acids Res.* **19:**4479–4484.

Amstad, P., Peskin, A., Shah, G., Mirault, M. E., Moret, R., Zbinden, I., and Cerutti, P., 1991, The balance between Cu,Zn-superoxide dismutase and catalase affects the sensitivity of mouse epidermal cells to oxidative stress, *Biochemistry* **30:**9305–9313.

Angel, P., and Karin, M., 1991, The role of Jun, Fos and the AP-1 complex in cell-proliferation and transformation, *Biochim. Biophys. Acta* **1072:**129–157.

Ariza, R. R., Cohen, S. P., Bachhawat, N., Levy, S. B., and Demple, B., 1994, Repressor mutations in the *marRAB* operon that activate oxidative stress genes and multiple antibiotic resistance in *Escherichia coli*, *J. Bacteriol.* **176:**143–148.

Baeuerle, P. A., and Baltimore, D., 1988, I kappa B: A specific inhibitor of the NF-kappa B transcription factor, *Science* **242:**540–546.

Bandyopadhyay, S., and Gronostajski, R. M., 1994, Identification of a conserved oxidation-sensitive cysteine residue in the NFI family of DNA-binding proteins, *J. Biol. Chem.* **269:**29949–29955.

Beinert, H., and Kiley, P., 1996, Redox control of gene expression involving iron-sulfur proteins. Change of oxidation-state or assembly/disassembly of Fe-S clusters? [comment], *FEBS Lett.* **382:**218–219; discussion 220–221.

Bradley, T. M., Hidalgo, E., Leautaud, V., Ding, H., and Demple, B., 1997, Cysteine-to-alanine replacements in the *E. coli* SoxR protein and the role of the [2Fe-2S] centers in transcriptional activation, *Nucleic Acids Res.* **25:**1469–1475.

Chan, E., and Weiss, B., 1987, Endonuclease IV of *Escherichia coli* is induced by paraquat, *Proc. Natl. Acad. Sci. USA* **84:**3189–3193.

Chou, J. H., Greenberg, J. T., and Demple, B., 1993, Posttranscriptional repression of *Escherichia coli* OmpF protein in response to redox stress: Positive control of the *micF* antisense RNA by the *soxRS* locus, *J. Bacteriol.* **175:**1026–1031.

Christman, M. F., Morgan, R. W., Jacobson, F. S., and Ames, B. N., 1985, Positive control of a regulon for defenses against oxidative stress and some heat-shock proteins in *Salmonella typhimurium*, *Cell* **41:**753–762.

Christman, M. F., Storz, G., and Ames, B. N., 1989, OxyR, a positive regulator of hydrogen peroxide-inducible genes in *E. coli* and *S. typhimurium*, is homologous to a family of bacterial regulatory proteins, *Proc. Natl. Acad. Sci. USA* **86:**3484–3488.

Claiborne, A., Miller, H., Parsonage, D., and Ross, R. P., 1993, Protein-sulfenic acid stabilization and function in enzyme catalysis and gene regulation, *FASEB J.* **7:**1483–1490.

Cohen, S. P., Hächler, H., and Levy, S. B., 1993a, Genetic and functional analysis of the multiple antibiotic resistance (*mar*) locus in *Escherichia coli*, *J. Bacteriol.* **175:**1484–1492.

Cohen, S. P., Levy, S. B., Foulds, J., and Rosner, J. L., 1993b, Salicylate induction of antibiotic resistance in *Escherichia coli*: Activation of the *mar* operon and a *mar*-independent pathway, *J. Bacteriol.* **175:**7856–7862.

Cotter, P. A., and Gunsalus, R. P., 1989, Oxygen, nitrate, and molybdenum regulation of *dmsABC* gene expression in *Escherichia coli*, *J. Bacteriol.* **171:**3817–3823.

Cotter, P. A., Chepuri, V., Gennis, R. B., and Gunsalus, R. P., 1990, Cytochrome o (*cyoABCDE*) and d (*cydAB*) oxidase gene expression in *Escherichia coli* is regulated by oxygen, pH, and the *fnr* gene product, *J. Bacteriol.* **172:**6333–6338.

Cunningham, R. P., Saporito, S. M., Spitzer, S. G., and Weiss, B., 1986, Endonuclease IV (*nfo*) mutant of *Escherichia coli*, *J. Bacteriol.* **168:**1120–1127.

Demple, B., and Amábile-Cuevas, C. F., 1991, Redox redux: The control of oxidative stress responses, *Cell* **67:**837–839.

Demple, B., and Halbrook, J., 1983, Inducible repair of oxidative DNA damage in *Escherichia coli*, *Nature* **304:**466–468.

Ding, H., and Demple, B., 1996, Glutathione-mediated destabilization in vitro of [2Fe-2S] centers in the SoxR regulatory protein, *Proc. Natl. Acad. Sci. USA* **93:**9449–9453.

Ding, H., Hidalgo, E., and Demple, B., 1996, The redox state of the [2Fe-2S] clusters in SoxR protein regulates its activity as a transcription factor, *J. Biol. Chem.* **271:**33173–33175.

Flint, D. H., 1996, *Escherichia coli* contains a protein that is homologous in function and N-terminal sequence to the protein encoded by the *nifS* gene of *Azotobacter vinelandii* and that can participate in the synthesis of the Fe-S cluster of dihydroxy-acid dehydratase, *J. Biol. Chem.* **271:**16068–16074.

Gardner, P. R., and Fridovich, I., 1992, Inactivation–reactivation of aconitase in *Escherichia coli*. A sensitive measure of superoxide radical, *J. Biol. Chem.* **267**:8757–8763.

Gaudu, P., and Weiss, B., 1996, SoxR, a [2Fe-2S] transcription factor, is active only in its oxidized form, *Proc. Natl. Acad. Sci. USA* **93**:10094–10098.

Georgiadis, M. M., Komiya, H., Chakrabarti, P., Woo, D., Kornuc, J. J., and Rees, D. C., 1992, Crystallographic structure of the nitrogenase iron protein from *Azotobacter vinelandii*, *Science* **257**:1653–1659.

González-Flecha, B., and Demple, B., 1995, Metabolic sources of hydrogen peroxide in aerobically growing *Escherichia coli*, *J. Biol. Chem.* **270**:13681–13687.

González-Flecha, B., and Demple, B., 1997, Homeostatic regulation of the intracellular concentration of hydrogen peroxide in aerobically growing *Escherichia coli*, *J. Bacteriol.* **179**:382–388.

Grauschopf, U., Winthe, R. J. R., Korber, P., Zander, T., Dallinger, P., and Bardwell, J. C., 1995, Why is DsbA such an oxidizing disulfide catalyst? *Cell* **83**:947–955.

Green, J., and Guest, J. R., 1993, A role for iron in transcriptional activation by FNR, *FEBS Lett.* **329**:55–58.

Green, J., Bennett, B., Jordan, P., Ralph, E. T., Thomson, A. J., and Guest, J. R., 1996, Reconstitution of the [4Fe-4S] cluster in FNR and demonstration of the aerobic–anaerobic transcription switch *in vitro*, *Biochem. J.* **316**:887–892.

Greenberg, J. T., and Demple, B., 1988, Overproduction of peroxide-scavenging enzymes in *Escherichia coli* suppresses spontaneous mutagenesis and sensitivity to redox-cycling agents in *oxyR*-mutants, *EMBO J.* **7**:2611–2617.

Greenberg, J. T., and Demple, B., 1989, A global response induced in *Escherichia coli* by redox-cycling agents overlaps with that induced by peroxide stress, *J. Bacteriol.* **171**:3933–3939.

Greenberg, J. T., Monach, P., Chou, J. H., Josephy, P. D., and Demple, B., 1990, Positive control of a global antioxidant defense regulon activated by superoxide-generating agents in *Escherichia coli*, *Proc. Natl. Acad. Sci. USA* **87**:6181–6185.

Greenberg, J. T., Chou, J. H., Monach, P. A., and Demple, B., 1991, Activation of oxidative stress genes by mutations at the *soxQ/cfxB/marA* locus of *Escherichia coli*, *J. Bacteriol.* **173**:4433–4439.

Guest, J. R., Green, J., Irvine, A. S., and Spiro, S., 1996, The FNR modulon and Fnr-regulated gene expression, In *Regulation of Gene Expression in Escherichia coli* (E. C. C. Lin and A. S. Lynch, eds.), pp. 317–342, R. G. Landes, Austin, TX.

Hassan, H. M., and Fridovich, I., 1977, Regulation of the synthesis of superoxide dismutase in *Escherichia coli*. Induction by methyl viologen, *J. Biol. Chem.* **252**:7667–7772.

Hausladen, A., Privalle, C. T., Keng, T., Deangelo, J., and Stamler, J. S., 1996, Nitrosative stress—Activation of the transcription factor OxyR, *Cell* **86**:719–728.

Henkel, T., Machleidt, T., Alkalay, I., Kronke, M., Ben-Neriah, Y., and Baeuerle, P. A., 1993, Rapid proteolysis of I kappa B-alpha is necessary for activation of transcription factor NF-kappa B, *Nature* **365**:182–185.

Hentze, M. W., and Kuhn, L. C., 1996, Molecular control of vertebrate iron metabolism: mRNA-based regulatory circuits operated by iron, nitric oxide, and oxidative stress, *Proc. Natl. Acad. Sci. USA* **93**:8175–8182.

Hidalgo, E., and Demple, B., 1994, An iron-sulfur center essential for transcriptional activation by the redox-sensing SoxR protein, *EMBO J.* **13**:138–146.

Hidalgo, E., and Demple, B., 1996a, Activation of SoxR-dependent transcription in vitro by noncatalytic or NifS-mediated assembly of [2Fe-2S] clusters into apo-SoxR, *J. Biol. Chem.* **271**:7269–7272.

Hidalgo, E., and Demple, B., 1996b, Adaptive responses to oxidative stress: The *soxRS* and *oxyR* regulons, In *Regulation of Gene Expression in Escherichia coli* (E. C. C. Lin and A. S. Lynch, eds.), pp. 435–452, R. G. Landes, Austin, TX.

Hidalgo, E., and Demple, B., 1997, Spacing of promoter elements regulates the basal expression of the *soxS* gene and converts SoxR from a transcriptional activator to a repressor, *EMBO J.* **16**:1056–1067.

Hidalgo, E., Bollinger, J. M., Jr., Bradley, T. M., Walsh, C. T., and Demple, B., 1995, Binuclear [2Fe-2S] clusters in the *Escherichia coli* SoxR protein and role of the metal centers in transcription, *J. Biol. Chem.* **270**:20908–20914.

Hidalgo, E., Ding, H., and Demple, B., 1997, Redox signal transduction: Mutations shifting [2Fe-2S] centers of the SoxR sensor-regulator to the oxidized form, *Cell* **88**:121–129.

Ignarro, L. I., 1990, Biosynthesis and metabolism of endothelial-derived nitric oxide, *Annu. Rev. Pharmacol. Toxicol.* **30**:535–560.

Imlay, J. A., and Fridovich, I., 1991, Assay of metabolic superoxide production in *Escherichia coli*, *J. Biol. Chem.* **266**:6957–6965.

Jair, K. W., Fawcett, W. P., Fujita, N., Ishihama, A., and Wolf, R. E. J., 1996, Ambidextrous transcriptional activation by SoxS: Requirement for the C-terminal domain of the RNA polymerase alpha subunit in a subset of the *Escherichia coli* superoxide-inducible genes, *Mol. Microbiol.* **19**:306–317.

Jones, H. M., and Gunsalus, R. P., 1987, Regulation of *Escherichia coli* fumarate reductase (*frdABCD*) operon expression by respiratory electron acceptors and the *fnr* gene product, *J. Bacteriol.* **169**:3340–3349.

Kappus, H., and Sies, H., 1981, Toxic drug effects associated with oxygen metabolism: Redox cycling and lipid peroxidation, *Experientia* **37**:1233–1241.

Karin, M., 1995, The regulation of AP-1 activity by mitogen-activated protein kinases, *J. Biol. Chem.* **270**:16483–16486.

Karin, M., and Hunter, T., 1995, Transcriptional control by protein phosphorylation: Signal transmission from the cell surface to the nucleus, *Curr. Biol.* **5**:747–757.

Keyer, K., and Imlay, J. A., 1996, Superoxide accelerates DNA damage by elevating free-iron levels, *Proc. Natl. Acad. Sci. USA* **93**:13635–13640.

Khoroshilova, N., Beinert, H., and Kiley, P. J., 1995, Association of a polynuclear iron-sulfur center with a mutant FNR protein enhances DNA binding, *Proc. Natl. Acad. Sci. USA* **92**:2499–2503.

Kiley, P. J., and Reznikoff, W. S., 1991, Fnr mutants that activate gene expression in the presence of oxygen, *J. Bacteriol.* **173**:16–22.

Klausner, R. D., Rouault, T. A., and Harford, J. B., 1993, Regulating the fate of mRNA: The control of cellular iron metabolism, *Cell* **72**:19–28.

Kullik, I., Toledano, M. B., Tartaglia, L. A., and Storz, G., 1995, Mutational analysis of the redox-sensitive transcriptional regulator OxyR: Regions important for oxidation and transcriptional activation, *J. Bacteriol.* **177**:1275–1284.

Lazazzera, B. A., Bates, D. M., and Kiley, P. J., 1993, The activity of the *Escherichia coli* transcription factor FNR is regulated by a change in oligomeric state, *Genes Dev.* **7**:1993–2005.

Lazazzera, B. A., Beinert, H., Khoroshilova, N., Kennedy, M. C., and Kiley, P. J., 1996, DNA binding and dimerization of the Fe-S-containing FNR protein from *Escherichia coli* are regulated by oxygen, *J. Biol. Chem.* **271**:2762–2768.

Li, Z., and Demple, B., 1994, SoxS, an activator of superoxide stress genes in *Escherichia coli*. Purification and interaction with DNA, *J. Biol. Chem.* **269**:18371–18377.

Liochev, S. I., and Fridovich, I., 1992, Fumarase C, the stable fumarase of *Escherichia coli*, is controlled by the *soxRS* regulon, *Proc. Natl. Acad. Sci. USA* **89**:5892–5896.

Liochev, S. I., and Fridovich, I., 1993, Modulation of the fumarases of *Escherichia coli* in response to oxidative stress, *Arch. Biochem. Biophys.* **301**:379–384.

Liochev, S. I., and Fridovich, I., 1994, The role of $O_2^-\cdot$ in the production of $HO^\cdot$: *In vitro* and *in vivo*, *Free Radical Biol. Med.* **16**:29–33.

Lipton, S. A., Choi, Y. B., Pan, Z. H., Lei, S. Z., Chen, H. S., Sucher, N. J., Loscalzo, J., Singel, D. J., and Stamler, J. S., 1993, A redox-based mechanism for the neuroprotective and neurodestructive effects of nitric oxide and related nitroso-compounds, *Nature* **364**:626–632.

Lynch, A. S., and Lin, E. C. C., 1996, Regulation of aerobic and anaerobic metabolism by the Arc system, in *Regulation of Gene Expression in Escherichia coli* (E. C. C. Lin and A. S. Lynch, eds.), pp. 361–381, R. G. Landes, Austin, TX.

Lyons, T. A., Ratnaswamy, G., and Pochapsky, T. C., 1996, Redox-dependent dynamics of putidaredoxin characterized by amide proton exchange, *Protein Sci.* **5**:627–639.

Ma, D., Alberti, M., Lynch, C., Nikaido, H., and Hearst, J. E., 1996, The local repressor AcrR plays a modulating role in the regulation of *acrAB* genes of *Escherichia coli* by global stress signals, *Mol. Microbiol.* **19**:101–112.

Melville, S. B., and Gunsalus, R. P., 1990, Mutations in *fnr* that alter anaerobic regulation of electron transport-associated genes in *Escherichia coli*, *J. Biol. Chem.* **265**:18733–18736.

Melville, S. B., and Gunsalus, R. P., 1996, Isolation of an oxygen-sensitive FNR protein of *Escherichia coli*: Interaction at activator and repressor sites of FNR-controlled genes, *Proc. Natl. Acad. Sci. USA* **93**:1226–1231.

Meyer, M., Caselmann, W. H., Schluter, V., Schreck, R., Hofschneider, P. H., and Baeuerle, P. A., 1992, Hepatitis B virus transactivator MHBst: Activation of NF-kappa B, selective inhibition by antioxidants and integral membrane localization, *EMBO J.* **11**:2991–3001.

Meyer, M., Schreck, R., and Baeuerle, P. A., 1993, H_2O_2 and antioxidants have opposite effects on activation of NF-kappa B and AP-1 in intact cells: AP-1 as secondary antioxidant-responsive factor, *EMBO J.* **12**:2005–2015.

Moradas-Ferreira, P., Costa, V., Piper, P., and Mager, W., 1996, The molecular defences against reactive oxygen species in yeast, *Mol. Microbiol.* **19**:651–658.

Morgan, R. W., Christman, M. F., Jacobson, F. S., Storz, G., and Ames, B. N., 1986, Hydrogen peroxide-inducible proteins in *Salmonella typhimurium* overlap with heat shock and other stress proteins, *Proc. Natl. Acad. Sci. USA* **83**:8059–8063.

Nunoshiba, T., and Demple, B., 1993, Potent intracellular oxidative stress exerted by the carcinogen 4-nitroquinoline-N-oxide, *Cancer Res.* **53**:3250–3252.

Nunoshiba, T., and Demple, B., 1994, A cluster of constitutive mutations affecting the C-terminus of the redox-sensitive SoxR transcriptional activator, *Nucleic Acids Res.* **22**:2958–2962.

Nunoshiba, T., Hidalgo, E., Amábile Cuevas, C. F., and Demple, B., 1992, Two-stage control of an oxidative stress regulon: The *Escherichia coli* SoxR protein triggers redox-inducible expression of the *soxS* regulatory gene, *J. Bacteriol.* **174**:6054–6060.

Nunoshiba, T., deRojas-Walker, T., Wishnok, J. S., Tannenbaum, S. R., and Demple, B., 1993, Activation by nitric oxide of an oxidative-stress response that defends *Escherichia coli* against activated macrophages, *Proc. Natl. Acad. Sci. USA* **90**:9993–9997.

Nunoshiba, T., deRojas-Walker, T., Tannenbaum, S. R., and Demple, B., 1995, Roles of nitric oxide in inducible resistance of *Escherichia coli* to activated murine macrophages, *Infect. Immun.* **63**:794–798.

Palombella, V. J., Rando, O. J., Goldberg, A. L., and Maniatis, T., 1994, The ubiquitin–proteasome pathway is required for processing the NF-κB1 precursor protein and the activation of NF-κB, *Cell* **78**:773–785.

Pinkus, R., Weiner, L. M., and Daniel, V., 1996, Role of oxidants and antioxidants in the induction of AP-1, NF-κB and glutathione S-transferase gene expression, *J. Biol. Chem.* **271**:13422–13429.

Poole, L. B., and Claiborne, A., 1989, The non-flavin redox center of the streptococcal NADH peroxidase. I. Thiol reactivity and redox behavior in the presence of urea, *J. Biol. Chem.* **264**:12322–12329.

Schenk, H., Klein, M., Erdbrugger, W., Droge, W., and Schulze-Osthoff, K., 1994, Distinct effects of thioredoxin and antioxidants on the activation of transcription factors NF-kappa B and AP-1, *Proc. Natl. Acad. Sci. USA* **91**:1672–1676.

Schmidt, K. N., Amstad, P., Cerutti, P., and Baeuerle, P. A., 1995, The roles of hydrogen peroxide and superoxide as messengers in the activation of transcription factor NF-κB, *Chem. Biol.* **2**:13–22.

Schreck, R., Rieber, P., and Baeuerle, P. A., 1991, Reactive oxygen intermediates as apparently widely used messengers in the activation of the NF-kappa B transcription factor and HIV-1, *EMBO J.* **10**:2247–2258.

Schreck, R., Grassmann, R., Fleckenstein, B., and Baeuerle, P. A., 1992a, Antioxidants selectively suppress activation of NF-kappa B by human T-cell leukemia virus type I Tax protein, *J. Virol.* **66**:6288–6293.

Schreck, R., Meier, B., Mannel, D. N., Droge, W., and Baeuerle, P. A., 1992b, Dithiocarbamates as potent inhibitors of nuclear factor kappa B activation in intact cells, *J. Exp. Med.* **175**:1181–1194.

Schulze-Osthoff, K., Los, M., and Baeuerle, P. A., 1995, Redox signalling by transcription factors NF-kappa B and AP-1 in lymphocytes, *Biochem. Pharmacol.* **50**:735–741.

Sharrocks, A. D., Green, J., and Guest, J. R., 1990, In vivo and in vitro mutants of FNR, the anaerobic transcriptional regulator of *E. coli*, *FEBS Lett.* **270**:119–122.

Shaw, D., Rice, D. W., and Guest, J. R., 1983, Homology between CAP and Fnr, a regulator of anaerobic respiration in *Escherichia coli*, *J. Mol. Biol.* **166**:241–247.

Sies, H., 1991, Oxidative stress: Introduction, In *Oxidative Stress: Oxidants and Antioxidants* (H. Sies, ed.), pp. xv–xxii, Academic Press, London.

Spiro, S., Roberts, R. E., and Guest, J. R., 1989, FNR-dependent repression of the *ndh* gene of *Escherichia coli* and metal ion requirement for FNR-regulated gene expression, *Mol. Microbiol.* **3**:601–608.

Stewart, V., 1982, Requirement of Fnr and NarL functions for nitrate reductase expression in *Escherichia coli* K-12, *J. Bacteriol.* **151**:1320–1325.

Storz, G., and Altuvia, S., 1994, OxyR regulon, *Methods Enzymol.* **234**:217–223.

Storz, G., Christman, M. F., Sies, H., and Ames, B. N., 1987, Spontaneous mutagenesis and oxidative damage to DNA in *Salmonella typhimurium*, *Proc. Natl. Acad. Sci. USA* **84**:8917–8921.

Storz, G., Tartaglia, L. A., and Ames, B. N., 1990, Transcriptional regulator of oxidative stress-inducible genes: Direct activation by oxidation, *Science* **248:**189–194.

Summers, A. O., 1992, Untwist and shout: A heavy metal-responsive transcriptional regulator, *J. Bacteriol.* **174:**3097–3101.

Sun, Y., and Oberley, L. W., 1996, Redox regulation of transcriptional activators, *Free Radical Biol. Med.* **21:**335–348.

Tao, K., Makino, K., Yonei, S., Nakata, A., and Shinagawa, H., 1989, Molecular cloning and nucleotide sequencing of *oxyR*, the positive regulatory gene of a regulon for an adaptive response to oxidative stress in *Escherichia coli*: Homologies between OxyR protein and a family of bacterial activator proteins, *Mol. Gen. Genet.* **218:**371–376.

Toledano, M. B., and Leonard, W. J., 1991, Modulation of transcription factor NFκB binding activity by oxidation-reduction *in vitro*, *Proc. Natl. Acad. Sci. USA* **88:**4328–4332.

Toledano, M. B., Kullik, I., Trinh, F., Baird, P. T., Schneider, T. D., and Storz, G., 1994, Redox-dependent shift of OxyR–DNA contacts along an extended DNA-binding site: A mechanism for differential promoter selection, *Cell* **78:**897–909.

Traenckner, E. B., Wilk, S., and Baeuerle, P. A., 1994, A proteasome inhibitor prevents activation of NF-kappa B and stabilizes a newly phosphorylated form of I kappa B-alpha that is still bound to NF-kappa B, *EMBO J.* **13:**5433–5441.

Traenckner, E. B., Pahl, H. L., Henkel, T., Schmidt, K. N., Wilk, S., and Baeuerle, P. A., 1995, Phosphorylation of human I kappa B-alpha on serines 32 and 36 controls I kappa B-alpha proteolysis and NF-kappa B activation in response to diverse stimuli, *EMBO J.* **14:**2876–2883.

Trageser, M., Spiro, S., Duchene, A., Kojro, E., Fahrenholz, F., Guest, J. R., and Unden, G., 1990, Isolation of intact FNR protein (Mr 30,000) of *Escherichia coli*, *Mol. Microbiol.* **4:**21–27.

Tsaneva, I. R., and Weiss, B., 1990, *soxR*, a locus governing a superoxide response regulon in *Escherichia coli* K-12, *J. Bacteriol.* **172:**4197–4205.

Unden, G., and Guest, J. R., 1985, Isolation and characterization of the Fnr protein, the transcriptional regulator of anaerobic electron transport in *Escherichia coli*, *Eur. J. Biochem.* **146:**193–199.

Wu, J., and Weiss, B., 1991, Two divergently transcribed genes, *soxR* and *soxS*, control a superoxide response regulon of *Escherichia coli*, *J. Bacteriol.* **173:**2864–2871.

Wu, J., and Weiss, B., 1992, Two-stage induction of the *soxRS* (superoxide response) regulon of *Escherichia coli*, *J. Bacteriol.* **174:**3915–3920.

Wu, J., Dunham, W. R., and Weiss, B., 1995, Overproduction and physical characterization of SoxR, a [2Fe-2S] protein that governs an oxidative response regulon in *Escherichia coli*, *J. Biol. Chem.* **270:**10323–10327.

Wu, X., Bishopric, N. H., Discher, D. J., Murphy, B. J., and Webster, K. A., 1996, Physical and functional sensitivity of zinc finger transcription factors to redox change, *Mol. Cell. Biol.* **16:**1035–1046.

Xanthoudakis, S., Miao, G., Wang, F., Pan, Y. C., and Curran, T., 1992, Redox activation of Fos-Jun DNA binding activity is mediated by a DNA repair enzyme, *EMBO J.* **11:**3323–3335.

Chapter 6

Regulation of Mammalian Gene Expression by Reactive Oxygen Species

Dana R. Crawford

1. INTRODUCTION

It is now known that the modulation of gene expression by reactive oxygen species is a widespread occurrence. This modulation has been found for genes from organisms ranging from bacteria and yeast to human, and in response to a wide range of oxidant stress agents, such as superoxide, hydrogen peroxide, nitric oxide, redox-active quinones, glutathione depletion, and others (Crawford *et al.*, 1995). Most of these genes have been identified in bacteria, most notably those under the control of OxyR and SoxRS (Storz and Tartaglia, 1992; Greenberg *et al.*, 1990). As expected, many of the modulated genes in bacteria code for antioxidant enzymes. In fact, the first studies in this field were undertaken at the protein level, where it was reported that oxidative stress induces the expression of the antioxidant enzymes superoxide dismutase (Hassan and Fridovich, 1977) and catalase (Yashpe-Purer *et al.*, 1977). Subsequent studies demonstrated that dozens of proteins are induced by hydrogen peroxide and superoxide in bacteria, including the antioxidant enzyme genes for catalase, alkyl hydroperoxide reductase, glutathione reductase, and manganese superoxide dismutase.

Reports on the modulation of gene expression by oxidative stress in mammalian cells have lagged behind those in bacteria, perhaps reflecting the surprising lack of a strong modulation of mammalian antioxidant gene expression by reactive oxygen species.

Dana R. Crawford Department of Biochemistry and Molecular Biology, The Albany Medical College, Albany, New York 12208.

Reactive Oxygen Species in Biological Systems, edited by Gilbert and Colton. Kluwer Academic / Plenum Publishers, New York, 1999.

Initially, this was a surprising result because there was already a precedent for the modulation of expression of these genes by reactive oxygen species in bacteria. Furthermore, it simply made intuitive sense that cells would increase their expression of superoxide dismutase when exposed to superoxide, or that catalase and glutathione peroxidase would increase in response to hydrogen peroxide. This lack of response suggested that reactive oxygen species might not be important in regulating gene expression in mammals. Subsequent studies, however, have revealed that oxidative stress modulates the expression of a number of mammalian genes, and the identification and regulation of these genes is the focus of this chapter.

2. EXPERIMENTAL APPROACHES

There are two obvious considerations for studying the modulation of gene expression by reactive oxygen species: the model system to be studied, and the technique used to identify modulated genes. To date, *in vitro* cell culture has been the most popular system in the study of oxidant-modulated genes. This approach has the major advantages of ease of use and homogeneity. Homogeneity is especially important when using sensitive and time-consuming methods of analysis, such as subtractive hybridization and differential display. Differences between treated and control preparations that are not specific to the oxidant stress will lead to a great many false positives. For these types of studies, the biological cells, tissue, or organism of choice is exposed to a wide choice of reactive oxygen species, including reagent hydrogen peroxide and potassium superoxide; oxidant-generating chemical and enzymatic systems; ionizing and ultraviolet radiation; hyperbaric oxygen; glutathione depletion, and many more.

Preferably, the system is also one in which a cellular response to oxidative stress has been demonstrated. For this reason, adaptive response model systems have often been used. Adaptive response refers to the ability of cells or organisms to better resist the damaging effects of a toxic agent when first preexposed to a lower dose. It is a universal phenomenon, having been observed in prokaryotes, yeast, mammals, and plants to a number of different types of damaging agents, including alkylating agents, heat stress, oxidant stress, radiation, and heavy metals (Crawford and Davies, 1994). In general, adaptation appears to involve the modulation of expression of many genes. A lower, initial dose of oxidant elicits a protective response in the cells. This approach therefore represents a method by which protective genes can be identified. It has been used successfully to identify bacterial responses, including modulated gene expression, to hydrogen peroxide and superoxide anion (Kullik and Storz, 1994). The same approach is now being used in mammalian cells, as discussed below.

There are a number of techniques that can and have been used to identify oxidant-modulated genes. For our discussion, we are assuming that the intent is to identify as yet unknown sequences. Earlier studies used polyacrylamide gel electrophoresis to assess modulated proteins. These analyses have led to two important discoveries in the field: the identification of 30–40 induced bacterial proteins by hydrogen peroxide as assessed by 2D gel electrophoresis, and the identification of induced heme oxygenase protein in cultured mammalian cells as assessed by 1D gel electrophoresis (Keyse and Tyrrell, 1989; Christman *et al.*, 1985). Although these were landmark studies in the field, neither protein-based approach is very sensitive, especially when applied to mammalian studies. Molecular biology techniques have therefore, for the most part, supplanted protein gel analyses as the technique of choice in mammalian cells.

There are three primary molecular biology techniques for analyzing the modulation of gene expression, all at the level of mRNA: differential hybridization, subtractive hybridization, and differential display. Differential hybridization is the most commonly used of these techniques. However, although it is significantly easier to use than the other two techniques, it is also much less sensitive. This means that it is only able to detect higher abundant mRNAs. It has been used successfully to identify CL100, a hydrogen peroxide-inducible mRNA that was originally identified in human fibroblasts (Keyse and Emslie, 1992). Subtractive hybridization is a much more sensitive technique, and is able to detect modulated sequences up to 50 times less abundant than necessary for differential hybridization. This technique was successfully used to detect the growth arrest and DNA damage (*gadd*) genes that are induced by a number of types of stress, including oxidative (Fargnoli *et al.*, 1990). Differential display is a relatively new technique that has rapidly gained popularity in the field, and has now been used to identify many oxidant-modulated genes (Liang and Pardee, 1992; Crawford *et al.*, 1994). Its sensitivity approaches that of subtractive hybridization. Unlike the other two techniques, differential display has the benefit of being able to detect reduced as well as induced sequences.

3. MODULATION OF NUCLEAR GENE EXPRESSION BY OXIDANT STRESS

The expression of an increasing number of mammalian genes is now known to be modulated by oxidative stress. Most of these modulations involve inductions, and are summarized in Table I. Surprisingly, few of these involve the "classical" antioxidant enzymes superoxide dismutase, catalase, or glutathione peroxidase. This is in marked contrast to the observed response in bacteria, where a number of antioxidant enzymes are induced by hydrogen peroxide and superoxide (Kullik and Storz, 1994; Storz and Tartaglia, 1992; Greenberg *et al.*, 1990; Chapter 5). In mammalian cells, modest inductions at best have been observed for antioxidant enzyme mRNAs, and in most cases, no modulation at all. In fact, this lack of response probably set the field back a number of years, as it appeared that oxidative stress was unimportant in the modulation of mammalian gene expression. However, several important studies changed this. The first had to do with cellular protooncogene expression. Crawford and Cerutti (1987) found that mRNA to the protooncogenes c-*fos* and c-*myc* but not h-*ras* was induced in mouse epidermal JB6 cells by superoxide and hydrogen peroxide generated by xanthine/xanthine oxidase. Parallel analysis of antioxidant gene expression did not detect any modulation in antioxidant gene expression. Additional studies not only confirmed these inductions, but also demonstrated the induction of other protooncogene and immediate early gene mRNAs including c-*jun*, *egr*, and JE (Shibanuma *et al.*, 1988; Muhlematter *et al.*, 1989; Nose *et al.*, 1991). In addition, other sources of oxidants induced expression, such as stock hydrogen peroxide and *t*-butylhydroperoxide (Nose *et al.*, 1991; Muhlematter *et al.*, 1989). Thus, not only were reactive oxygen species able to modulate mammalian gene expression, but they did so for a group of genes that are considered to be among the most

Table I
mRNAs Whose Steady-State Levels Are Induced by Oxidative Stress in Mammalian Cells

Genes	Modulating agent	References
c-*fos*, c-*myc*	Xanthine/xanthine oxidase	Crawford and Cerutti (1987)
c-*fos*, c-*jun*, *egr*-1, JE	Hydrogen peroxide	Nose *et al.* (1991)
c-*fos*	*t*-Butylhydroperoxide	Muhlematter *et al.* (1989)
Heme oxygenase	Multiple oxidants	Keyse and Tyrrell (1989)
gadd45 and *gadd153*	Hydrogen peroxide	Fornace *et al.* (1989)
CL100	Hydrogen peroxide	Keyse and Emslie (1992)
Interleukin 8	Hydrogen peroxide	DeForge *et al.* (1993)
γ-Glutamyl transpeptidase	Menadione	Kugelman *et al.* (1994)
Vimentin, cytochrome IV, RP-L4	Diethylmaleate	Ammendola *et al.* (1994)
c-*fos*, c-*jun*, c-*myc*, HSP	Xanthine/xanthine oxidase	Yamamoto *et al.* (1993)
Glucose-regulated proteins	Singlet oxygen, other oxidants	Gomer *et al.* (1991)
c-*fos* and zif/268	Nitric oxide	Morris (1995)
adapts	Hydrogen peroxide	Crawford *et al.* (1996a,b), Wang *et al.* (1996)
Catalase, MnSOD, GPx	Hydrogen peroxide	Shull *et al.* (1991)
MnSOD	Xanthine/xanthine oxidase	Shull *et al.* (1991)
	Hyperbaric oxygen	Stevens and Autor (1977)
	Multiple oxidants	Stralin and Marklund (1994)

important in the mammalian genome. Cellular protooncogenes are critical to cellular growth and differentiation, as evidenced by the observation that their aberrant expression leads to cancer (Anderson *et al.*, 1992). Therefore, reactive oxygen species must be important mediators of mammalian gene expression.

A second important study in the field was made by Keyse and Tyrrell (1989). Again using a mammalian cell culture model, these investigators discovered that the protein and mRNA to heme oxygenase are strongly induced by oxidative stress. Subsequent studies have demonstrated induction of heme oxygenase protein as well by reactive oxygen species. In fact, heme oxygenase is now the prototype mRNA for assessing cellular oxidative stress at the mRNA level, and it has been employed in a number of clinically related studies as a marker of oxidative stress-related disease. There are a number of reasons for its popularity as a marker, including its ready detection, its specificity of induction by oxidative stress, its sustained induction (as compared with immediate early gene mRNAs such as c-*fos*), and its inducibility in many tissues and mammalian species.

A third study important to the modulation of gene expression by oxidative stress field was made by Fornace *et al.* (1989). They discovered a number of genes that were induced by growth arrest and DNA damage (*gadd*). These genes were also found to be induced by oxidative stress. The two most studied of the *gadd* genes are *gadd45* and *gadd153*, and they have been implicated in growth arrest, cellular protection, apoptosis, and DNA repair (Zhan *et al.*, 1994; Smith *et al.*, 1994; Fornace *et al.*, 1992).

In addition to the above, a growing number of mRNAs have now been reported to be modulated by reactive oxygen species. These include CL100, interleukin 8, γ-glutamyl transpeptidase, vimentin, cytochrome IV, ribosomal protein L4, heat shock protein, and glucose-regulated proteins (Kugelman *et al.,* 1994; Gomer *et al.*, 1991; Yamamoto *et al.*, 1993; DeForge *et al.*, 1993; Keyse and Emslie, 1992). Nitric oxide is one of these reactive oxygen species, and it modulates the mRNA levels of the immediate early genes c-*fos* and zif/268 (Morris, 1995). An interesting caveat has been found for heat shock. In general, the effect of oxidative stress has been modest, if any, on the induction of heat shock sequences (Tacchini *et al.*, 1995; Bruce *et al.*, 1993). However, Bruce *et al.* (1993) have found that under conditions where no increases in the steady-state levels of heat shock mRNA are observed, oxidative stress activates heat shock transcription factor. It is also important to note that the modulation of expression of a number of genes appears to be mediated by reactive oxygen species, even though the inducing agent is not itself a classic oxidant. Examples include the induction of vascular cell adhesion molecule-1 by interleukin 1 β (IL-1β) (Marui *et al.*, 1993), and the induction of cyclooxygenase-2 by lipopolysaccharide, IL-1, and tumor necrosis factor (TNF) (Feng *et al.*, 1995).

We have used an adaptive response system to identify genes that may be protective against oxidative stress. In our system, HA-1 hamster fibroblasts exposed to a minimally toxic level of hydrogen peroxide ("pretreatment") undergo a protective response that allows these cells to better withstand the damaging effects of a subsequent higher dose of peroxide (Crawford *et al.*, 1996a; Wiese *et al.*, 1995; Spitz *et al.*, 1987). This adaptive response is dependent on the *de novo* synthesis of protein and mRNA (Wiese *et al.*, 1995). Thus, the exposure of HA-1 fibroblasts to a pretreatment concentration of hydrogen peroxide allows us to identify genes whose modulation may be important in the protection of HA-1 cells against the damaging effects of oxidative stress. Using the technique of differential display, we have identified 13 RNAs that are modulated following exposure to a pretreatment concentration of peroxide (Crawford *et al.*, 1996a–c; Wang *et al.*, 1996). Seven of these are induced, and six reduced. The induced mRNAs have been designated *adapts*. Interestingly, one of these *adapts*, *adapt15*, is expressed coordinately with *gadd45* and *gadd153*, and appears to be a member of the same *gadd* family. Other *adapts* that we have studied include *adapt33*, a novel and apparently untranslated RNA; *adapt66*, a homologue of the transcriptional regulator *mafG*; *adapt73*, a cardiogenic-shock inducible homologue (Crawford and Davies, 1997); and *adapt78*, whose gene maps to a region of human chromosome 21 thought to be important in dementia (Crawford *et al.*, 1997). It is interesting to note that we also see the induction of *gadd153*, *gadd45,* and heme oxygenase mRNAs during adaptive response.

As already indicated, most investigators have observed minimal modulation of classical antioxidant gene expression in response to oxidative stress. However, there are enough reports of modulation of these genes to warrant continued interest and study in this area. In tracheobronchial epithelia, catalase and, to a lesser extent, MnSOD and glutathione peroxidase, but not copper/zinc-superoxide dismutase (Cu/Zn-SOD) mRNA levels are elevated in response to hydrogen peroxide. However, only MnSOD mRNA levels are elevated in these same cells in response to xanthine/xanthine oxidase (Shull *et al.*, 1991). MnSOD mRNA is also elevated in neonatal rat lung following hyperbaric lung treatment (Stevens and Autor, 1977). No modulation of Cu/Zn-SOD mRNA was observed in this study. In fact, Cu/Zn-SOD is so rarely modulated by oxidative stress that it is often used as an unmodulated internal marker to normalize the levels of other genes that are modulated by oxidative stress. It appears that MnSOD is the most frequently modulated antioxidant gene mRNA. However, even in this case, the modulation of MnSOD by reactive oxygen species is significantly less than that reported for other agents not considered to be classic oxidants, most notably IL-1, endotoxin, and TNF. Typical of, and consistent with, these observations are the results of recent studies by Stralin and Marklund (1994). These investigators examined the induction of both MnSOD and Cu/Zn-SOD in human fibroblasts using a wide variety of oxidative agents. They observed no induction of Cu/Zn-SOD and modest (2- to 3-fold) induction of MnSOD. This MnSOD induction is well below the sometimes 20- to 100-fold induction reported for MnSOD mRNA induction by IL-1, endotoxin, and TNF (Gwinner *et al.*, 1995; Suzuki *et al.*, 1993). It should be pointed out that most analyses in this area of gene expression have centered on superoxide dismutase, glutathione peroxidase, and catalase. More detailed analyses of other scavengers and related enzymes such as glucose-6-phosphate dehydrogenase and glutathione-S-transferase may reveal modulation by oxidative stress under the right conditions and cell type.

4. MODULATION OF MITOCHONDRIAL GENE EXPRESSION BY OXIDANT STRESS

Mitochondria are known to be especially sensitive to oxidative damage. The types of mitochondrial damage sustained in response to oxidative stress include the release of calcium, loss of membrane potential, depletion of ATP, lipid peroxidation, protein oxidation, DNA damage, and loss of electron transport capacity, among others (Zhang *et al.*, 1990; Farber *et al.*, 1990; Bindoli, 1988; Shay and Werbin, 1987). It is now known that, like nuclear genes, the expression of mitochondrial genes is modulated by reactive oxygen species. By and large, these modulations usually involve the downregulation of mitochondrial RNA levels. Decreases in overall mitochondrial transcription have been reported in various experimental systems, in response to xanthine/xanthine oxidase (Vincent *et al.*, 1994) and peroxyl radicals (Kristal *et al.*, 1994a,b). One of these reports identified two specific mitochondrial genes that were involved in this downregulation: 12 S ribosomal RNA and NADH dehydrogenase subunit 4 mRNA (Kristal *et al.*, 1994a). We have observed a dramatic downregulation in the steady-state levels of at least six specific mitochondrial RNAs in HA-1 hamster cells exposed to hydrogen peroxide (Crawford *et al.*, 1996c). These RNAs include 16 S rRNA, 12 S rRNA, and mRNAs to NADH dehydrogenase subunit 6, ATPase subunit 6, and cytochrome oxidase subunits I and III. Furthermore, these downregulations were preferential for mitochondrial RNA as compared with cytoplasmic RNAs. This latter observation further underscores the hypersensitivity of mitochondria to oxidative stress. There are also examples of mitochondrial RNA inductions, in response to transformation (Coral *et al.*, 1989) and hypoxemia (Coral-Debrinski *et al.*, 1991).

5. MODES OF REGULATION

There are three major mechanisms by which cells can modulate their levels of mRNA: transcription, mRNA stability, and precursor processing. Because the identification of oxidant-modulated genes is a relatively recent observation, even less information is known about their mechanism of regulation. However, enough analyses have been done in this area to indicate that all of the major mechanisms of mRNA level regulation also apply to oxidant-modulated mRNAs.

To date, most analyses in this area have assessed the contribution of transcription. These studies have revealed that the induction of both c-*fos* and c-*jun* mRNA levels is mediated, at least in part, by increased gene transcription (Devary *et al.*, 1991; Crawford *et al.*, 1988). The induction of heme oxygenase by direct oxidant exposure or by depletion of intracellular glutathione in human fibroblasts, and both glutathione S-transferase Ya subunit and NAD(P)H:quinone oxidoreductase by hydrogen peroxide, also involve increased transcription (Tyrrell *et al.*, 1993; Rushmore *et al.*, 1991; Rushmore and Pickett, 1990). These results indicate the presence of oxidant-responsive elements somewhere in these genes. In fact, these genes have served as valuable models for this important area of oxidant research. Oxidant-responsive elements have now been identified for several genes whose inductions depend, at least in part, on transcription. These include c-*fos*, c-*jun*, heme oxygenase, glutathione S-transferase Ya subunit, and NAD(P)H:quinone. Probably the most work in this area has been done on the glutathione S-transferase Ya subunit gene by Pickett and co-workers (Rushmore *et al.*, 1991; Rushmore and Pickett, 1990). Oddly enough, the identified element in this gene has been designated the antioxidant-responsive element (ARE), as it is bound and activated by a series of compounds classically considered to be antioxidants. However, it has been determined that their ability to stimulate transcription is based on their ability to produce a certain amount of reactive oxygen species. The ARE has been found to be present in the promoter regions of both glutathione S-transferase Ya subunit and NAD(P)H:quinone reductase genes (Pinkus *et al.*, 1995; Bergelson *et al.*, 1994; Rushmore and Pickett, 1990). Importantly, both genes are transcriptionally stimulated at this site by hydrogen peroxide (Rushmore and Pickett, 1990). Induction of transcription of the c-*fos* gene by hydrogen peroxide depends on the serum response element (Amstad *et al.*, 1992). It appears that an AP-1 response element is important in the stimulation of c-*jun* transcription (Devary *et al.*, 1991). Several oxidant-responsive elements reside in the heme oxygenase promoter close to the TATA box. However, a critical enhancer region with an AP-1-like sequence is also present in this gene more than 4 kb upstream of the TATA box (Alam and Den, 1992). This would suggest that the induction of c-*fos* and c-*jun* mRNA, as mentioned above, and their subsequent protein products, may also be involved in the transcriptional induction of heme oxygenase during oxidative stress.

In addition to the identification of oxidant-responsive elements, several important transcription factors have been identified that are mediators of oxidative stress. These factors are induced or activated by reactive oxygen species, then bind and activate nuclear genes involved in overall cellular response to oxidative stress. Most notably, they include AP-1 and NF-κB. Oxidative stress induces the mRNA levels of both c-*jun* and c-*fos*, and activates NF-κB (Crawford and Cerutti, 1987; Nose *et al.*, 1991; Schreck *et al.*, 1992). In addition, there is yet another level of redox regulation of these transcription factors: DNA binding. The binding and subsequent activity of a number of transcription factors including c-Fos, c-Jun, SP1, v-ETS, v-Rel, v-Myb, and p53 is affected by their redox state (Burdon, 1995; Ammendola *et al.*, 1994). In almost all cases, reduction favors binding, with NF-κB being the exception. Cysteine residues are especially important in the reduction and subsequent activation of most of these transcription factors. This introduces yet another level of regulation, that of the reductase molecule. This has been extensively studied for AP-1. It has been determined that a DNA repair enzyme designated Ref-1 stimulates the DNA binding activity of Fos–Jun dimers by reducing a critical cysteine residue (Xanthoudakis *et al.,* 1994).

We have found that *mafG* homologue, designated *adapt66* in our studies, is induced by hydrogen peroxide (Crawford *et al.*, 1996b). *mafG* is a member of the *maf* family of genes encoding nuclear proteins that recognize AP-1-like response elements. MafB, MafK, MafF, and MafG are all able to heterodimerize with each other, c-Fos, and erythroid cell-specific transcription factor NF-E2, to affect the transcription of target genes. We speculate that the induction of *mafG* homologue/*adapt66* by hydrogen peroxide, and its known interaction with oxidant-inducible c-*fos*, may represent important mechanisms by which oxidative stress can modulate gene expression.

The effect of oxidative stress on mRNA stabilization has not been reported. We have recently assessed the contribution of mRNA stability to the induction of *adapt15*, *adapt33,* c-*fos*, c-*jun*, and *gadd45* mRNA following exposure to hydrogen peroxide. Peroxide induced a significant increase in the stability of *adapt33* and *gadd45* but not *adapt15*, c-*fos*, or c-*jun* mRNAs. Calcium is involved in this stabilization effect, as intracellular chelation of calcium dramatically inhibits the induction of *adapt33* mRNA by hydrogen peroxide. It has previously been shown that increased cytosolic calcium stabilizes certain cytokines in lymphoid cells. We also observe a significant acceleration of the processing of *adapt15* precursor RNA following the exposure of HA-1 cells to hydrogen peroxide. This apparently accounts for some of the observed increase in the steady-state levels of mature *adapt15* mRNA that we see in these cells following oxidative stress. We also observe an apparent degradation of mitochondrial precursor RNAs under the same conditions which may also involve precursor processing (Crawford *et al.*, 1996c).

6. SIGNAL TRANSDUCTION

The identification of genes that are transcriptionally induced by reactive oxygen species has led, logically, to studies aimed at understanding the signal tranduction pathway leading up to their induction. Detailed reports are presented elsewhere in this volume. Many of the signal transduction studies in the field to date have analyzed the pathways leading up to the activation of AP-1 and Nf-κB, because, thus far, these are the major transcription factors known to be activated by oxidative stress. Both AP-1, which is comprised of c-Fos and c-Jun proteins, and NF-κB are activated by phosphorylation, and therefore indicate that a signal transduction pathway involving phosphorylation is important in their activations. Although most studies to date have focused on oxidative stress, it is now clear that low concentrations of reactive oxygen species also elicit important cellular responses, especially proliferation (Burdon, 1995). The following studies demonstrate the importance of oxidation–reduction in signal transduction, and presumably reflect cellular proliferative, stress, and overlapping responses to reactive oxygen species.

Larsson and Cerutti (1989) demonstrated the translocation and enhancement of the phosphotransferase activity of protein kinase C (PKC) in mouse JB6 epidermal cells treated with hydrogen peroxide. In addition, they found that ribosomal protein S6, a mediator of cellular mitogenic response, was activated by phosphorylation following several types of oxidative stress (Larsson and Cerutti, 1988). Kuo *et al.* (1995) found that both superoxide and hydrogen peroxide generated by paraquat activate PKC. Whisler *et al.* (1994) demonstrated activation of PKC by hydrogen peroxide in human Jurkat T cells.

Other mediators of signal transduction also appear to be involved in oxidant response. Guyton *et al.* (1996) identified several cellular mitogenic kinases that are activated by hydrogen peroxide, including ERK2, JNK, and p38. CHOP/Gadd153 protein, whose mRNA is strongly induced by reactive oxygen species, is activated by phosphorylation of p38 MAP kinase following stress (Wang and Ron, 1996). Its expression is p38 independent. Stevenson *et al.* (1994) used MAP kinase and its subsequent activation of a c-Fos activating complex of transcription factors to assess the role of MAP kinase in c-*fos* induction by hydrogen peroxide, X-irradiation, and phorbol ester. They found that all three agents activated MAP kinase, and importantly, that these activations were mediated by reactive oxygen species. Hydrogen peroxide has been shown to increase the levels of ceramide, which in turn activates c-*jun* through SAPK/JNK in U937 cells (Verheij *et al.*, 1996). Nitric oxide is also involved in signal transduction, as *S*-nitrosylation of $p21^{ras}$ triggers downstream signaling (Lander *et al.*, 1995). Similar to the modulation of nuclear gene expression by oxidative stress, agents that are not considered to be classic oxidants can also activate signal transduction in a redox-dependent manner. An example of this is platelet-derived growth factor (PDGF)-induced tyrosine phosphorylation and MAP kinase stimulation, which has been found to be dependent on PDGF-generated hydrogen peroxide in rat vascular smooth muscle cells (Sundaresan *et al.*, 1995).

Calcium appears to be an important cellular mediator of gene expression following oxidative stress. The induction of c-*fos*, *gadd153*, *adapt15*, *adapt33*, and *adapt66* mRNAs by oxidative stress is strongly dependent on calcium, as chelation of calcium prior to exposure of the cells to reactive oxygen species dramatically inhibits their induction. Furthermore, calcium ionophores by themselves induce the mRNA levels of these genes (Crawford *et al.*, 1996a,c; Trump and Berezesky, 1995; Bartlett *et al.*, 1992; Wang *et al.*, 1996). It has already been demonstrated that the induction of some of these genes involves PKC, which is activated by calcium. Calmodulin kinase is another important mediator of signal transduction that is activated by calcium.

7. GENE EXPRESSION AND OXIDANT STRESS-RELATED DISEASE

The modulation of gene expression by oxidative stress is not limited to cell culture or experimental animal-related research studies. A number of these genes have now been associated with oxidative stress-related diseases and disorders, and therefore have potential usefulness as clinical therapeutic targets. This is not surprising, as a number of chronic mammalian disorders have been associated with oxidative stress, and long-term pathological effects often correlate with stable alterations in gene expression (Anderson *et al.*, 1992). Examples of pathologies for which the aberrant expression of oxidant-modulated genes has been implicated are shown in Table II.

Cancer was one of the earliest studied mammalian diseases in the oxidative stress field. The studies of Slaga *et al.* (1981) and, separately, Weitzman *et al.* (1985) revealed

Table II
Examples of Pathologies for which the Aberrant Expression of Oxidant-Modulated Genes Has Been Implicated

Pathology	Gene	Expression	References
Cancer	MnSOD	Decreased	St. Clair and Oberley (1991), Li *et al.* (1995), Yan *et al.* (1996)
Down syndrome	Cu/Zn-SOD	Increased	Schickler *et al.* (1989)
Neurodegeneration	Heme oxygenase-1	Increased	Castellani *et al.* (1995)
Brain trauma	c-*fos*	Increased	Sharp and Sagar (1994)
Inflammation	Cyclooxygenase-2, IL-8	Increased	DeForge *et al.* (1993), Feng *et al.* (1995)
Cataractogenesis	c-*fos*	Increased	Spector (1995)
HIV infection	MnSOD	Decreased	Flores *et al.* (1993)
Aging protection	Cu/Zn-SOD, catalase	Increased	Orr and Sohal (1994)

that chronic oxidant treatment generated tumors in rodents, and that reactive oxygen species from activated neutrophils were able to transform target fibroblasts in cell culture. Subsequent studies have attempted to identify important genes that mediate these oxidant-related transformations, most likely protooncogenes and/or tumor suppressor genes. The observation that cellular protooncogene mRNA levels are induced by oxidative stress was an important contribution to these transformation studies. It has been shown that oxidative stress can act as a tumor promoter (Cerutti, 1985). Classic tumor promotion involves the modulation of gene expression (Cerutti, 1985). Therefore, the observation that oxidant tumor promoters such as hydrogen peroxide and *t*-butyl hydroperoxide modulate cellular protooncogenes, which are mediators of tumor promotion, was consistent with a role for oxidative stress in tumor development (Sun *et al.*, 1993). It has also been demonstrated that a decreased level of MnSOD is associated with a number of tumor cell lines (St. Clair and Oberley, 1991). This and recent MnSOD overexpression studies suggest that MnSOD acts as a cellular tumor suppressor gene (Li *et al.*, 1995; Yan *et al.*, 1996).

Neurodegenerative diseases and disorders also have a significant association with oxidative stress. As with cancer, the aberrant expression of oxidant-related genes is also associated with neurodegeneration. The overexpression of Cu/Zn-SOD is associated with Down syndrome, and mutations in the gene for this enzyme are associated with familial amyotrophic lateral sclerosis. Overexpression of Cu/Zn-SOD produces Down-like characteristics in transgenic mice, and is associated with increased lipid peroxidation (Schickler *et al.*, 1989; Avraham *et al.*, 1988). Paradoxically, brain nigral cells, the target of neuron degeneration in patients with Parkinson's disease, survive better when Cu/Zn-SOD is overexpressed in a transgenic mouse model (Nakao *et al.*, 1995). In addition, increased levels of heme oxygenase-1 protein, an oxidant-inducible sequence, have been associated with the neurofibrillary pathology of Alzheimer's disease as well as progressive supranuclear palsy, subacute sclerosing panencephalitis, Pick's disease, and corticobasal degeneration (Castellani *et al.*, 1995). c-*fos* induction is associated with a number of brain pathological stimuli including ischemia, hypoxia, seizures, and cortical injury (Sharp and

Sagar, 1994). The oxidant-mediated modulation of gene expression has also been associated with excitotoxity, which is thought to play an important role in neuronal degeneration, and is associated with the generation of reactive oxygen species (Bondy and LeBel, 1993).

In addition to the above, a number of other genes have been identified whose expressions are modulated by reactive oxygen species in association with a disease state. For example, reactive oxygen species are strongly involved in the induction of vascular cell adhesion molecule-1 (VCAM-1) by IL-1β (Marui *et al.*, 1993). Both oxidative stress and VCAM-1 expression are early features in the pathogenesis of atherosclerosis and other inflammatory diseases. Reactive oxygen species also induce another mediator of inflammatory response, cyclooxygenase-2, and it is speculated that this induction contributes to the deleterious amplification of prostanoids during inflammation (Feng *et al.*, 1995). IL-8 is also induced by reactive oxygen species, perhaps acting as an early signal to recruitment of neutrophils during inflammation (DeForge *et al.*, 1993). c-*fos* mRNA levels are induced under conditions of cataractogenesis (Spector, 1995). Infection of cells with the human immunodeficiency virus results in suppressed MnSOD expression (Flores *et al.*, 1993), possibly leading to increased oxidative stress, which has been suggested to promote viral replication. Finally, the levels of antioxidant gene mRNAs appear to be important in aging (Orr and Sohal, 1994). There are many more diseases and disorders associated with oxidative stress, and undoubtedly, oxidant-modulated genes will continually be found to play critical roles.

Finally, the observation that mitochondrial RNA levels are modulated by reactive oxygen species may prove to have important clinical significance. The mitochondrion is the most important organelle for intracellular production of oxidants in resting-state cells. A number of human diseases have now been linked to aberrations in the mitochondrial genome (Wallace, 1992; Taylor *et al.*, 1994; Bandy and Davison, 1990). Reactive oxygen species have been implicated in these alterations. This is not surprising because reactive oxygen species are continuously generated in the mitochondrion proximal to its genome, and the mitochondrial DNA is not protected by histones. Some of these alterations may occur in genes whose rRNA and/or mRNA products are modulated by oxidative stress, and may lead to an improper modulation during oxidative and perhaps other stress. We speculate that in HA-1 fibroblasts, the degradation of mitochondrial RNAs represents an early stage "shut down" of mitochondria, which may be important in combating the damaging effects of oxidative stress.

8. CONCLUDING REMARKS

Our understanding of the modulation of gene expression by reactive oxygen species has expanded dramatically over the last decade. Sensitive techniques able to detect low-abundant mRNA species; the elucidation of signal transduction pathways; and more sophisticated transcriptional analyses have combined to produce a great amount of insightful, exciting, and sometimes puzzling information. The observed redox regulation in itself is far more complicated than originally imagined. This is a remarkable progression, given that not so long ago it was uncertain as to whether reactive oxygen species were even important in the modulation of gene expression in mammals.

What is the function of mammalian oxidant-inducible genes? To date, there is no consensus answer to this question. However, it is often assumed that the induced genes serve a protective function, especially because a number of them have been identified as part of a protective "adaptive response." This would suggest that at least some of the modulated species are involved in either directly preventing oxidative damage to the cell, such as the antioxidant enzymes, or in removing and repairing oxidatively damaged macromolecules. In fact, antioxidant and repair genes are already known to be modulated in bacteria in the adaptive response studies described in the Introduction. There are also adaptive response model systems for mammalian cells whereby a pretreatment concentration of reactive oxygen also elicits a cellular protective response. In bovine vascular endothelial cells, activities of total SOD, catalase, and glutathione peroxidase were all reported to increase following exposure to adaptive response pretreatment levels of hydrogen peroxide (Lu *et al.*, 1993). The *adapt* series of oxidant-induced mRNAs were also identified under conditions where a protective cellular response occurred, in HA-1 hamster cells (Crawford *et al.*, 1996a; Wiese *et al.*, 1995). In addition, overexpression of heme oxygenase has been implicated in a protective adaptive response to UVA irradiation that protects cells against exposure to oxidative stress (Vile *et al.*, 1994). Ferritin appears to play an important role in this protective effect. Furthermore, heme oxygenase converts heme to biliverdin, a scavenger of reactive oxygen species. Myocardial cells express apparent protective genes in response to several oxidant-based stresses (Das *et al.*, 1995). Heat shock protein and mRNA are known to to be protective to mammalian cells, and their modest inductions by reactive oxygen species have been reported (Tacchini *et al.*, 1995; Bruce *et al.*, 1993). Adaptation can also involve constitutive overexpression of protective genes, and it has been found that Chinese hamster cells adapt to chronic exposure to high levels of hydrogen peroxide and atmospheric oxygen by dramatically overexpressing catalase as well as several antioxidants (Spitz *et al.*, 1992; Sullivan *et al.*, 1992).

Clearly, much more study will be needed to elucidate the exact details of oxidant-modulated gene expression and the genes involved. Important areas of future research include the identification and characterization of the binding proteins to mRNAs modulated at the mRNA stability level; the continued identification and characterization of redox-sensitive transcription factors; the sequence identification of oxidant-response elements; the further elucidation of signal transduction pathways that are triggered by reactive oxygen species; identification of the cellular reductases and oxidases that act on gene regulatory factors; and the grouping of mediators into those involved in protective stress response, proliferation, differentiation, and apoptosis. The continued identification of oxidant-modulated genes is also important.

The final elucidation of cellular response to reactive oxygen species, to a large extent, may ultimately depend on how intracellular phosphorylation/dephosphorylation events translate into intracellular redox mechanisms governing expression of genes associated with reestablishing the intracellular balance between prooxidant production and antioxidant capacity. These different systems are interrelated and dependent on the actions of each other to express an integrated function. Already, significant inroads have been made in this area. Overall, these studies will not only provide new insight into our understanding of cellular responses to reactive oxygen species, but should also lead to new directions in the areas of detection and therapy of oxidative stress-related diseases and disorders.

9. REFERENCES

Alam, J., and Den, Z., 1992, Distal AP-1 binding sites mediate basal level enhancement and TPA induction of the mouse heme oxygenase-1 gene, *J. Biol. Chem.* **267**:21894–21900.

Ammendola, R., Mesuraca, M., Russo, T., and Cimino, F., 1994, The DNA-binding efficiency of Sp1 is affected by redox changes, *Eur. J. Biochem.* **225**:483–489.

Amstad, P. A., Krupitza, G., and Cerutti, P. A., 1992, Mechanism of c-fos induction by active oxygen, *Cancer Res.* **52**:3952–3960.

Anderson, M. W., Reynolds, S. H., You, M., and Maronpot, R. M., 1992, Role of proto-oncogene activation in carcinogenesis, [Review], *Environ. Health Perspect.* **98**:13–24.

Avraham, K. B., Schickler, M., Sapoznikov, D., Yarom, R., and Groner, Y., 1988, Down's syndrome: Abnormal neuromuscular junction in tongue of transgenic mice with elevated levels of human Cu/Zn-superoxide dismutase, *Cell* **54**:823–829.

Bandy, B., and Davison, A. J., 1990, Mitochondrial mutations may increase oxidative stress: Implications for carcinogenesis and aging? *Free Radical Biol. Med.* **8**:523–539.

Bartlett, J. D., Luethy, J. D., Carlson, S. G., Sollott, S. J., and Holbrook, N.J., 1992, Calcium ionophore A23187 induces expression of the growth arrest and DNA damage inducible CCAAT/enhancer-binding protein (C/EBP)-related gene, gadd153. Ca^{2+} increases transcriptional activity and mRNA stability, *J. Biol. Chem.* **267**:20465–20470.

Bergelson, S., Pinkus, R., and Daniel, V., 1994, Intracellular glutathione levels regulate Fos/Jun induction and activation of glutathione S-transferase gene expression, *Cancer Res.* **54**:36–40.

Bindoli, A., 1988, Lipid peroxidation in mitochondria, *Free Radical Biol. Med.* **5**:247–261.

Bondy, S. C., and LeBel, C. P., 1993, The relationship between excitotoxicity and oxidative stress in the central nervous system, *Free Radical Biol. Med.* **14**:633–642.

Bruce, J. L., Price, B. D., Coleman, C. N., and Calderwood, S. K., 1993, Oxidative injury rapidly activates the heat shock transcription factor but fails to increase levels of heat shock proteins, *Cancer Res.* **53**:12–15.

Burdon, R. H., 1995, Superoxide and hydrogen peroxide in relation to mammalian cell proliferation, *Free Radical Biol. Med.* **18**:775–794.

Castellani, R., Smith, M. A., Richey, P. L., Kalaria, R., Gambetti, P., and Perry, G., 1995, Evidence for oxidative stress in Pick disease and corticobasal degeneration, *Brain Res.* **696**:268–271.

Cerutti, P. A., 1985, Prooxidant states and tumor promotion, *Science* **227**:375–381.

Christman, M. F., Morgan, R. W., Jacobson, F. S., and Ames, B. N., 1985, Positive control of a regulon for defenses against oxidative stress and some heat-shock proteins in Salmonella typhimurium, *Cell* **41**:753–762.

Coral, M., Paris, B., Baffet, G., Tichonicky, L., Guguen-Guillouzo, C., Kruh, J., and Defer, N., 1989, Increased levels of the mitochondrial ND5 transcript in chemically induced rat hepatomas, *Exp. Cell Res.* **184**:158–166.

Coral-Debrinski, M., Stepien, G., Shoffner, J. M., Lott, M. T., Kanter, K., and Wallace, D. C., 1991, Hypoxemia is associated with mitochondrial DNA damage and gene induction. Implications for cardiac disease, *J. Am. Med. Assoc.* **266**:1812–1816.

Crawford, D. R., and Cerutti, P. A., 1987, Expression of oxidant stress-related genes in tumor promotion of mouse epidermal cells JB6, in *Proceedings of the 2nd Conference on Anticarcinogens and Radioprotectors* (O. F. Nygaard, P. Cerutti, O. Nygaard, and M Simic, eds.), pp. 183–190, Plenum Press, New York.

Crawford, D. R., and Davies, K. J. A., 1994, Adaptive response and oxidative stress. *Environ. Health Perspect.* **102**:25–28.

Crawford, D. R., and Davies, K. J. A., 1997, *adapt73* is a chemical stress and cardiogenic shock response-associated RNA, *Surgery* **121**:581–587.

Crawford, D., Zbinden, I., Amstad, P., and Cerutti, P., 1988, Oxidant stress induces the proto-oncogenes c-fos and c-myc in mouse epidermal cells, *Oncogene* **3**:27–32.

Crawford, D. R., Edbauer-Nechamen, C. A., Lowry, C. V. , Salmon, S.L., Kim, Y. K., Davies, J. M.S., and Davies, K. J. A., 1994, Assessing gene-expression during oxidative stress, *Methods Enzymol.* **234**:175–217.

Crawford, D. R., Edbauer-Nechamen, C. A., Schools, G.P., Salmon, S. L., Davies, J. M. S., and Davies, K. J. A., 1995, Oxidant-modulated gene expression, in *The Oxygen Paradox* (K. J. A. Davies, ed.), pp. 327–336, C. L.E. U. P. Press, Padova, Italy.

Crawford, D. R., Schools, G. P., Salmon, S. L., and Davies, K. J. A., 1996a, Hydrogen peroxide induces the expression of *adapt15*, a novel RNA associated with polysomes in hamster HA-1 cells, *Arch. Biochem. Biophys.* **325**:256–264.

Crawford, D.R., Leahy, K. P., Wang, Y., Schools, G. P., Kochheiser, J. C., and Davies, K. J. A., 1996b, Oxidative stress induces the levels of a mafG homolog in hamster HA-1 cells, *Free Radical Biol. Med.* **21**:521–525.

Crawford, D. R., Wang, Y., Schools, G. P., Kochheiser, J., and Davies, K. J. A., 1996c, Down-regulation of mammalian mitochondrial RNAs during oxidative stress, *Free Radical Biol. Med.* **22**:551–559.

Crawford, D. R., Leahy, K. P., Abramova, N., Lan Lan, Wang, Y., and Davies, K. J.A., 1997, Hamster *adapt78* mRNA is a Down syndrome critical region-homologue that is inducible by oxidative stress, *Arch. Biochem. Biophys.* **342**:6–12.

Das, D. K., Maulik, N., and Moraru, I. I., 1995, Gene expression in acute myocardial stress. Induction by hypoxia, ischemia, reperfusion, hyperthermia and oxidative stress [Review], *J. Mol. Cell. Cardiol.* **27**:181–193.

DeForge, L. E., Preston, A. M., Takeuchi, E., Kenney, J., Boxer, L. A., and Remick, D. G., 1993, Regulation of interleukin 8 gene expression by oxidant stress, *J. Biol. Chem.* **268**:25568–25576.

Devary, Y., Gottlieb, R.A., Lau, L. F., and Karin, M., 1991, Rapid and preferential activation of the c-jun gene during the mammalian UV response, *Mol. Cell. Biol.* **11**:2804–2811.

Farber, J. L., Kyle, M. E., and Coleman, J. B., 1990, Mechanisms of cell injury by activated oxygen species, *Lab. Invest.* **62**:670–679.

Fargnoli, J., Holbrook, N. J., and Fornace, A. J., Jr., 1990, Low-ratio hybridization subtraction, *Anal. Biochem.* **187**:364–373.

Feng, L., Xia, Y., Garcia, G.E., Hwang, D., and Wilson, C. B., 1995, Involvement of reactive oxygen intermediates in cyclooxygenase-2 expression induced by interleukin-1, tumor necrosis factor-alpha, and lipopolysaccharide, *J. Clin. Invest.* **95**:1669–1675.

Flores, S. C., Marecki, J. C., Harper, K. P., Bose, S. K., Nelson, S. K., and McCord, J. M., 1993, Tat protein of human immunodeficiency virus type 1 represses expression of manganese superoxide dismutase in HeLa cells, *Proc. Natl. Acad. Sci. USA* **90**:7632–7636.

Fornace, A. J., Jr., Nebert, D. W., Hollander, M. C., Luethy, J. D., Papathanasiou, M., Fargnoli, J., and Holbrook, N. J., 1989, Mammalian genes coordinately regulated by growth arrest signals and DNA-damaging agents, *Mol. Cell. Biol.* **9**:4196–4203.

Fornace, A. J., Jr., Jackman, J., Hollander, M. C., Hoffman-Liebermann, B., and Liebermann, D. A., 1992, Genotoxic-stress-response genes and growth-arrest genes. gadd, MyD, and other genes induced by treatments eliciting growth arrest, *Ann. N. Y. Acad. Sci.* **663**:139–153.

Gomer, C. J., Ferrario, A., Rucker, N., Wong, S., and Lee, A. S., 1991, Glucose regulated protein induction and cellular resistance to oxidative stress mediated by porphyrin photosensitization, *Cancer Res.* **51**:6574–6579.

Greenberg, J. T., Monach, P., Chou, J.H., Josephy, P. D., and Demple, B., 1990, Positive control of a global antioxidant defense regulon activated by superoxide-generating agents in Escherichia coli, *Proc. Natl. Acad. Sci. USA* **87**:6181–6185.

Guyton, K. Z., Liu, Y., Gorospe, M., Xu, Q., and Holbrook, N. J., 1996, Activation of mitogen-activated protein kinase by H_2O_2. Role in cell survival following oxidant injury, *J. Biol. Chem.* **271**:4138–4142.

Gwinner, W., Tisher, C. C., and Nick, H. S., 1995, Regulation of manganese superoxide dismutase in glomerular epithelial cells: Mechanisms for interleukin 1 induction, *Kidney Int.* **48**:354–362.

Hassan, H. M., and Fridovich, I., 1977, Regulation of the synthesis of superoxide dismutase in Escherichia coli, *J. Biol. Chem.* **252**:7667–7672.

Keyse, S. M., and Emslie, E. A., 1992, Oxidative stress and heat shock induce a human gene encoding a protein-tyrosine phosphatase, *Nature* **359**:644–647.

Keyse, S. M., and Tyrrell, R. M., 1989, Heme oxygenase is the major 32-kDa stress protein induced in human skin fibroblasts by UVA radiation, hydrogen peroxide, and sodium arsenite, *Proc. Natl. Acad. Sci. USA* **86**:99–103.

Kristal, B. S., Chen, J., and Yu, B. P., 1994a, Sensitivity of mitochondrial transcription to different free radical species, *Free Radical Biol. Med.* **16**:323 329.

Kristal, B. S., Kim, J. D., and Yu, B. P., 1994, Tissue-specific susceptibility to peroxyl radical-mediated inhibition of mitochondrial transcription, *Redox Rep.* **1**:51–55.

Kugelman, A., Choy, H. A., Liu, R., Shi, M.M., Gozal, E., and Forman, H. J., 1994, gamma-glutamyl transpeptidase is increased by oxidative stress in rat alveolar L2 epithelial cells, *Am. J. Respir. Cell Mol. Biol.* **11**:586–592.

Kullik, I., and Storz, G., 1994, Transcriptional regulators of the oxidative stress response in prokaryotes and eukaryotes, *Redox Rep.* **1**:23–29.

Kuo, M. L., Lee, K. C., Lin, J. K., and Huang, T. S., 1995, Pronounced activation of protein kinase C, ornithine decarboxylase and c-jun proto-oncogene by paraquat-generated active oxygen species in WI-38 human lung cells, *Biochim. Biophys..Acta* **1268**:229–236.

Lander, H. M., Ogiste, J. S., Pearce, S. F., Levi, R., and Novogrodsky, A., 1995, Nitric oxide-stimulated guanine nucleotide exchange on p21ras, *J. Biol. Chem.* **270**:7017–7020.

Larsson, R., and Cerutti, P., 1988, Oxidants induce phosphorylation of ribosomal protein S6, *J. Biol. Chem.* **263**:17452–17458.

Larsson, R., and Cerutti, P., 1989, Translocation and enhancement of phosphotransferase activity of protein kinase C following exposure in mouse epidermal cells to oxidants [published erratum appears in *Cancer Res.* 1990 Jan. 1; **50**(1):212], *Cancer Res.* **49**:5627–5632.

Li, J. J., Oberley, L. W., St. Clair, D. K., Ridnour, L. A., and Oberley, T. D., 1995, Phenotypic changes induced in human breast cancer cells by overexpression of manganese-containing superoxide dismutase, *Oncogene* **10**:1989–2000.

Liang, P., and Pardee, A. B., 1992, Differential display of eukaryotic messenger RNA by means of the polymerase chain reaction, *Science* **257**:967–971.

Lu, D., Maulik, N., Moraru, I. I., Kreutzer, D. L., and Das, D. K., 1993, Molecular adaptation of vascular endothelial cells to oxidative stress, *Am. J. Physiol.* **264**:C715–C722.

Marui, N., Offermann, M. K., Swerlick, R., Kunsch, C., Rosen, C. A., Ahmad, M., Alexander, R. W., and Medford, R. M., 1993, Vascular cell adhesion molecule-1 (VCAM-1) gene transcription and expression are regulated through an antioxidant-sensitive mechanism in human vascular endothelial cells, *J. Clin. Invest.* **92**:1866–1874.

Morris, B. J., 1995, Stimulation of immediate early gene expression in striatal neurons by nitric oxide, *J. Biol. Chem.* **270**:24740–24744.

Muhlematter, D., Ochi, T., and Cerutti, P., 1989, Effects of tert-butyl hydroperoxide on promotable and non-promotable JB6 mouse epidermal cells, *Chem. Biol. Interact.* **71**:339–352.

Nakao, N., Frodl, E. M., Widner, H., Carlson, E., Eggerding, F. A., Epstein, C.J., and Brundin, P., 1995, Overexpressing Cu/Zn superoxide dismutase enhances survival of transplanted neurons in a rat model of Parkinson's disease, *Nature Med.* **1**:226–231.

Nose, K., Shibanuma, M., Kikuchi, K., Kageyama, H., Sakiyama, S., and Kuroki, T., 1991, Transcriptional activation of early-response genes by hydrogen peroxide in a mouse osteoblastic cell line, *Eur. J. Biochem.* **201**:99–106.

Orr, W. C., and Sohal, R. S., 1994, Extension of life-span by overexpression of superoxide dismutase and catalase in Drosophila melanogaster, *Science* **263**:1128–1130.

Pinkus, R., Weiner, L. M., and Daniel, V., 1995, Role of quinone-mediated generation of hydroxyl radicals in the induction of glutathione S-transferase gene expression, *Biochemistry* **34**:81–88.

Rushmore, T. H., and Pickett, C. B., 1990, Transcriptional regulation of the rat glutathione S-transferase Ya subunit gene. Characterization of a xenobiotic-responsive element controlling inducible expression by phenolic antioxidants, *J. Biol. Chem.* **265**:14648–14653.

Rushmore, T. H., Morton, M. R., and Pickett, C. B., 1991, The antioxidant responsive element. Activation by oxidative stress and identification of the DNA consensus sequence required for functional activity, *J. Biol. Chem.* **266**:11632–11639.

Schickler, M., Knobler, H., Avraham, K. B., Elroy-Stein, O., and Groner, Y., 1989, Diminished serotonin uptake in platelets of transgenic mice with increased Cu/Zn-superoxide dismutase activity, *EMBO J.* **8**:1385–1392.

Schreck, R., Albermann, K., and Baeuerle, P. A., 1992, Nuclear factor kappa B: An oxidative stress-responsive transcription factor of eukaryotic cells, *Free Radical Res. Commun.* **17**:221–237.

Sharp, F. R., and Sagar, S. M., 1994, Alterations in gene expression as an index of neuronal injury: Heat shock and the immediate early gene response, *Neurotoxicology* **15**:51–59.

Shay, J. W., and Werbin, H., 1987, Are mitochondrial DNA mutations involved in the carcinogenic process? *Mutat. Res.* **186**:149–160.

Shibanuma, M., Kuroki, T., and Nose, K., 1988, Induction of DNA replication and expression of proto-oncogene c-myc and c-fos in quiescent Balb/3T3 cells by xanthine/xanthine oxidase, *Oncogene* **3:**17–21.

Shull, S., Heintz, N. H., Periasamy, M., Manohar, M., Janssen, Y. M., Marsh, J. P., and Mossman, B. T., 1991, Differential regulation of antioxidant enzymes in response to oxidants, *J. Biol. Chem.* **266:**24398–24403.

Slaga, T. J., Klein-Szanto, A. J., Triplett, L. L., Yotti, L. P., and Trosko, K. E., 1981, Skin tumor-promoting activity of benzoyl peroxide, a widely used free radical-generating compound, *Science* **213:**1023–1025.

Smith, M. L., Chen, I. T., Zhan, Q. , Bae, I., Chen, C. Y., Gilmer, T. M., Kastan, M. B., O'Connor, P. M., and Fornace, A. J., Jr., 1994, Interaction of the p53-regulated protein Gadd45 with proliferating cell nuclear antigen, *Science* **266:**1376–1380.

Spector, A., 1995, Oxidative stress-induced cataract: Mechanism of action, *FASEB J.* **9:**1173–1182.

Spitz, D. R., Dewey, W. C., and Li, G. C., 1987, Hydrogen peroxide or heat shock induces resistance to hydrogen peroxide in Chinese hamster fibroblasts, *J. Cell. Physiol.* **131:**364–373.

Spitz, D. R., Adams, D. T., Sherman, C. M., and Roberts, R. J., 1992, Mechanisms of cellular resistance to hydrogen peroxide, hyperoxia, and 4-hydroxy-2-nonenal toxicity: The significance of increased catalase activity in H_2O_2-resistant fibroblasts, *Arch. Biochem. Biophys.* **292:**221–227.

St. Clair, D. K., and Oberley, L. W., 1991, Manganese superoxide dismutase expression in human cancer cells: A possible role of mRNA processing, *Free Radical Res. Commun.* **12–13**(Pt.2)**:**771–778.

Stevens, J. B., and Autor, A. P., 1977, Induction of superoxide dismutase by oxygen in neonatal rat lung, *J. Biol. Chem.* **252:**3509–3514.

Stevenson, M. A., Pollock, S. S., Coleman, C. N., and Calderwood, S. K., 1994, X-irradiation, phorbol esters, and H_2O_2 stimulate mitogen-activated protein kinase activity in NIH-3T3 cells through the formation of reactive oxygen intermediates, *Cancer Res.* **54:**12–15.

Storz, G., and Tartaglia, L. A., 1992, OxyR: A regulator of antioxidant genes, *J.Nutr.* **122:**627–630.

Stralin, P., and Marklund, S. L., 1994, Effects of oxidative stress on expression of extracellular superoxide dismutase, CuZn-superoxide dismutase and Mn-superoxide dismutase in human dermal fibroblasts, *Biochem. J.* **298:**347–352.

Sullivan, S. J., Oberley, T. D., Roberts, R. J., and Spitz, D. R., 1992, A stable $O_2^{\cdot-}$ resistant cell line: Role of lipid peroxidation byproducts in $O_2^{\cdot-}$ mediated injury, *Am. J. Physiol.* **262:**L748–L756.

Sun, Y., Colburn, N. H., and Oberley, L. W., 1993, Decreased expression of manganese superoxide dismutase mRNA and protein after immortalization and transformation of mouse liver cells, *Oncol. Res.* **5:**127–132.

Sundaresan, M., Yu, Z. X., Ferrans, V. J., Irani, K., and Finkel, T., 1995, Requirement for generation of H_2O_2 for platelet-derived growth factor signal transduction, *Science* **270:**296–299.

Suzuki, K., Tatsumi, H., Satoh, S., Senda, T., Nakata, T., Fujii, J., and Taniguchi, N., 1993, Manganese-superoxide dismutase in endothelial cells: Localization and mechanism of induction, *Am. J. Physiol.* **265:**H1173–H1178.

Tacchini, L., Pogliaghi, G., Radice, L., Anzon, E., and Bernelli-Zazzera, A., 1995, Differential activation of heat-shock and oxidation-specific stress genes in chemically induced oxidative stress, *Biochem. J.* **309:**453–459.

Taylor, R. W., Birch-Machin, M. A., Bartlett, K., Lowerson, S.A., and Turnbull, D. M., 1994, The control of mitochondrial oxidations by complex III in rat muscle and liver mitochondria. Implications for our understanding of mitochondrial cytopathies in man, *J. Biol. Chem.* **269:**3523–3528.

Trump, B. F., and Berezesky, I. K., 1995, Calcium-mediated cell injury and cell death, *FASEB J.* **9:**219–228.

Tyrrell, R. M., Applegate, L. A., and Tromvoukis, Y., 1993, The proximal promoter region of the human heme oxygenase gene contains elements involved in stimulation of transcriptional activity by a variety of agents including oxidants, *Carcinogenesis* **14:**761–765.

Verheij, M., Bose, R., Lin, X. H., Yao, B., Jarvis, W. D., Grant, S., Birrer, M.J., Szabo, E., Zon, L. I., Kyriakis, J. M., Haimovitz-Friedman, A., Fuks, Z., and Kolesnick, R. N.,1996, Requirement for ceramide-initiated SAPK/JNK signalling in stress-induced apoptosis, *Nature* **380:**75–79.

Vile, G. F., Basu-Modak, S., Waltner, C., and Tyrrell, R. M., 1994, Heme oxygenase 1 mediates an adaptive response to oxidative stress in human skin fibroblasts, *Proc. Natl. Acad. Sci. USA* **91:**2607–2610.

Vincent, F., Corral-Debrinski, M., and Adolphe, M., 1994, Transient mitochondrial transcript level decay in oxidative stressed chondrocytes, *J. Cell. Physiol.* **158:**128–132.

Wallace, D. C., 1992, Mitochondrial genetics: A paradigm for aging and degenerative diseases? *Science* **256:**628–632.

Wang, X. Z., and Ron, D., 1996, Stress-induced phosphorylation and activation of the transcription factor CHOP (GADD153) by p38 MAP kinase, *Science* **272:**1347–1349.

Wang, Y., Crawford, D. R., and Davies, K. J. A., 1996, adapt33, a novel oxidant-inducible RNA from hamster HA-1 cells, *Arch. Biochem. Biophys.* **332:**225–260.

Weitzman, S. A., Weitberg, A. B., Clark, E. P., and Stossel, T. P., 1985, Phagocytes as carcinogens: Malignant transformation produced by human neutrophils, *Science* **227:**1231–1233.

Whisler, R. L., Newhouse, Y. G., Beiqing, L., Karanfilov, B. K., Goyette, M. A., and Hackshaw, K. V., 1994, Regulation of protein kinase enzymatic activity in Jurkat T cells during oxidative stress uncoupled from protein tyrosine kinases: Role of oxidative changes in protein kinase activation requirements and generation of second messengers, *Lymph. Cyt. Res.* **13:**399–410.

Wiese, A. G., Pacifici, R. E., and Davies, K. J. A., 1995, Transient adaptation to oxidative stress in mammalian cells, *Arch. Biochem. Biophys.* **318:**231–240.

Xanthoudakis, S., Miao, G. G., and Curran, T., 1994, The redox and DNA-repair activities of Ref-1 are encoded by nonoverlapping domains, *Proc. Natl. Acad. Sci. USA* **91:**23–27.

Yamamoto, N., Maki, A., Swann, J. D., Berezesky, I. K., and Trump, B. F., 1993, Induction of immediate early and stress genes in rat proximal tubule epithelium following injury: The significance of cytosolic ionized calcium, *Renal Failure* **15:**163–171.

Yan, T., Oberley, L. W., Zhong, W., and St. Clair, D. K., 1996, Manganese-containing superoxide dismutase overexpression causes phenotypic reversion in SV40-transformed human lung fibroblasts, *Cancer Res.* **56:**2864–2871.

Yashpe-Purer, Y., Henis, Y., and Yashpe, J., 1977, Regulation of catalase level in Escherichia coli K12, *Can. J. Microbiol.* **23:**84–91.

Zhan, Q., Lord, K. A., Alamo, I., Jr., Hollander, M. C., Carrier, F., Ron, D., Kohn, K. W., Hoffman, B., Liebermann, D. A., and Fornace, A. J., Jr., 1994, The gadd and MyD genes define a novel set of mammalian genes encoding acidic proteins that synergistically suppress cell growth, *Mol. Cell Biol.* **14:**2361–2371.

Zhang, Y., Marcillat, O., Giulivi, C., Ernster, L., and Davies, K. J., 1990, The oxidative inactivation of mitochondrial electron transport chain components and ATPase, *J. Biol. Chem.* **265:**16330–16336.

Chapter 7

Inflammatory Regulation of Manganese Superoxide Dismutase

John F. Valentine and Harry S. Nick

1. REACTIVE OXYGEN SPECIES

Reactive oxygen species (ROS) including the superoxide anion ($\cdot O_2^-$), hydroxyl radical ($\cdot OH$) or their metabolites (H_2O_2, HOCl) have been implicated in over 100 human clinical conditions (Cross *et al.*, 1987). Exposure to hyperoxia (Freeman and Crapo, 1981), xenobiotics (Burger *et al.*, 1980; Lieber, 1988), cigarette smoke (Cross *et al.*, 1987), or mineral dust (Janssen *et al.*, 1994) has been shown to result in oxygen radical production and tissue injury followed by both an acute and possibly a chronic inflammatory response. Any physiologic event that results in an inflammatory response and thus causes tissue damage can be associated with ROS. For example, the early tissue damage in inflammatory diseases such as lupus, glomerulonephritis, asthma, rheumatoid arthritis, ulcerative colitis, and Crohn's disease is most likely a consequence of ROS release.

ROS are normally and continuously produced during aerobic respiration by the mitochondrial electron transport chain. As a result of normal respiration, estimates of reactive oxygen radical leak from the mitochondrial electron transport chain range from 1 to 5% of the oxygen utilized during oxidative phosphorylation (Harris *et al.*, 1992). These highly reactive compounds must be metabolically detoxified to prevent cellular injury. Other enzymatic processes such as those carried out by xanthine oxidase, cyclooxygenase, and lipoxygenase (Cross *et al.*, 1987) result in the intracellular production

---

John F. Valentine Department of Medicine, University of Florida, and Gainesville VA Medical Center, Gainesville, Florida, 32610. **Harry S. Nick** Department of Biochemistry and Molecular Biology, University of Florida, Gainesville, Florida 32610.

Reactive Oxygen Species in Biological Systems, edited by Gilbert and Colton. Kluwer Academic / Plenum Publishers, New York, 1999.

of oxygen free radicals. ROS are also generated as a consequence of exposure to ionizing radiation and ischemia reperfusion injury (Cross *et al.*, 1987).

ROS damage cells by oxidation and inactivation of enzymes (Vlessis *et al.*, 1991), DNA base hydroxylation, nicking, and cross-linking (Imlay and Linn, 1988; Burger *et al.*, 1980) and by peroxidation of polyunsaturated lipids in the plasma membrane (Grisham and Granger, 1988) which may then act as chemoattractants and perpetuate the inflammatory response.

2. ROS AND THE INFLAMMATORY RESPONSE

ROS are also a major component of the immune response (Weiss, 1989). The neutrophilic acute component results in an increased production of ROS through the action of the neutrophil's membrane-bound NADPH oxidase. This extracellular production of oxygen free radicals serves as the center of the neutrophil's tissue destructive capacity (Grisham and Granger, 1988). $\cdot O_2^-$ secretion accounts for greater than 90% of the oxygen consumed by activated neutrophils, and if there were no dismutation, the concentration of superoxides secreted into the activated neutrophil-substrate cleft would reach 100–1000 mM/liter (Weiss, 1989). The production of intracellular oxygen radicals is also a consequence of proinflammatory cytokine production by both tissue resident cells and infiltrating immune cells. The cytotoxic mechanisms of tumor necrosis factor α (TNFα) and interleukin 1 (IL-1) have been directly linked to intracellular oxygen radical production (Wong *et al.*, 1989; Schulze-Osthoff *et al.*, 1992; Hirose *et al.*, 1993). In specific cellular systems, intracellular ROS production contributes directly to the mechanism of apoptosis (Wong and Goeddel, 1994a). Schulze-Osthoff *et al.* (1992) have demonstrated that exposure of cells to TNFα results in increased production of ROS at the ubisemiquinone step (complex III) of the mitochondrial electron transport chain and that the toxic effect of TNFα can be overcome by the addition of antioxidants. In addition, the generation of mitochondrion-derived oxygen free radicals induced by IL-1 or TNFα may be a mechanism through which these agents activate transcription factor NF-κB (Schreck and Baeuerle, 1994) as well as other redox-sensitive transcription factors, thus further propagating the inflammatory response.

3. ANTIOXIDANT DEFENSE MECHANISMS

Even though excessive oxygen radical production during an inflammatory response may result in extensive tissue damage, oxygen radical production is a normal consequence of aerobic life. As such, all aerobic organisms from bacteria to humans have developed mechanisms to detoxify ROS. Many of these mechanisms have remained highly conserved owing to their vital importance to the survival of the organism. These biochemical defenses include chemical antioxidants such as β-carotene, vitamin E, and vitamin C (Harris *et al.*, 1992) as well as the antioxidant enzymes superoxide dismutase (SOD), catalase, and glutathione peroxidase (Harris *et al.*, 1992). The first line of cellular defense are the superoxide dismutases. These enzymes catalyze the reaction $2 \cdot O_2^- + 2H^+ \rightarrow H_2O_2 + O_2$ (Bannister *et al.*, 1987). Hydrogen peroxide is then inactivated by catalase and glutathione peroxidase to form water. In mammals, three SODs have been identified.

Manganese superoxide dismutase (MnSOD) is a nucleus-encoded mitochondrial matrix protein (Bannister *et al.*, 1987) located on human chromosome 6. This enzyme is a homotetramer with individual subunits of 23 kDa. Unrelated to the MnSOD is copper/zinc superoxide dismutase (Cu/ZnSOD), a cytoplasmic protein (Bannister *et al.*, 1987) located on human chromosome 21. Cu/ZnSOD is a homodimer with a subunit size of 32 kDa. Extracellular SOD (EC-SOD) possesses 22.6% amino acid sequence homology with Cu/ZnSOD (Hjalmarsson *et al.*, 1987). EC-SOD is tetrameric, glycosylated, and bound to heparin moieties on the surface of cells resulting in low serum levels, which can be increased by injection of heparin (Karlsson and Marklund, 1987).

4. STIMULUS-DEPENDENT REGULATION OF THE SODs

Despite the potential cytoprotective importance of the SODs, little is documented about the molecular mechanisms that regulate basal tissue levels of antioxidant enzymes as well as levels during the inflammatory response. However, most recently, abundant evidence has accumulated demonstrating that MnSOD and EC-SOD are regulated by agents that are tightly linked to the inflammatory process.

Of the SODs, stimulus-dependent regulation of gene expression has been most extensively studied for the nucleus-encoded, mitochondrially localized antioxidant enzyme, MnSOD. Table I summarizes the cytokine responsiveness of MnSOD in a variety of cell types. We and others have found a 5- to 25-fold induction of MnSOD mRNA levels following cellular exposure to IL-1, TNFα or bacterial endotoxin (LPS) in pulmonary epithelial (Visner *et al.*, 1990) and endothelial cell lines (Visner *et al.*, 1991), small intestinal epithelial cell lines (Valentine and Nick, 1992), primary cultures of myenteric neurons (Valentine *et al.*, 1996), primary cultures of colonic smooth muscle cells (Tannahill *et al.*, 1997), primary cultures of mesangial (Stephanz *et al.*, 1996) and glomerular epithelial cells (Gwinner *et al.*, 1995) as well as primary cultures of neuronal and glial cells (Kifle *et al.*, 1996; Mokuno *et al.*, 1994). In rat intestinal and pulmonary epithelial cells, the induction of MnSOD mRNA is rapid and can be detected as early as 1 hr after treatment with LPS, TNFα, or IL-1. The induction peaks at approximately 8 hr with MnSOD mRNA levels decreasing to basal levels within 36 to 48 hr. IL-6 did not induce MnSOD in any of the intestinal, lung, or endothelial cells, although it was found to be a potent inducer of MnSOD levels in primary cultures of rat hepatocytes, resulting in a 15-fold induction of MnSOD mRNA (Dougall and Nick, 1991). The cell-specific responsiveness of MnSOD to IL-6 in hepatocytes has not been investigated but may result from any number of possibilities including the absence of IL-6 receptor expression, specific signal transduction pathways, or tissue-specific transcription factors.

Similar inductions of MnSOD by interferon-γ were observed in several cell types (Harris *et al.*, 1991), but interferon-γ did not induce MnSOD in A549 cells, a human lung cancer cell line (Wong and Goeddel, 1988), or in T84 cells, a human colon cancer cell line (Valentine, unpublished data).

Surprisingly, glucocorticoids repress both basal and LPS- or cytokine-stimulated levels of MnSOD mRNA and protein in the adult rat intestinal epithelial cell line IEC-6 but not in the fetal rat intestinal epithelial cell line IRD-98 (Valentine and Nick, 1994), or in adult rat lung epithelial L2 cells (Nick, unpublished data). It is not known if the

Table I
Cytokine Responsiveness of MnSOD in a Variety of Cell Types

Cell type	Stimulus	Reference
Murine astrocytes	LPS, TNF, and IL-1	Mokuno *et al.* (1994)
Primary cultures of rat neurons and astrocytes	LPS, TNF, and IL-1	Kifle *et al.* (1996)
Human fetal brain	TNF or LT	Wong (1995)
Rat golmerular epithelial	IL-1α & β, LPS	Gwinner *et al.* (1995)
Rat kidney mesangial	IL-1β	Stephanz *et al.* (1996)
Human embryonic kidney 293	TNF	Wong *et al.* (1989)
Hepatocytes	IL-6	Dougall and Nick (1991)
Rat pulmonary artery and microvascular endothelial	LPS, TNF, and IL-1	Visner *et al.* (1992)
Porcine pulmonary artery endothelial	LPS	Visner *et al.* (1991)
Human airway epithelial	IFNγ	Jacoby and Choi (1994)
L2 (rat pulmonary epithelial)	LPS, TNF, and IL-1	Visner *et al.* (1990)
IEC-6 (adult rat small intestine)	LPS, TNF, IL-1	Valentine and Nick (1992)
IRD-98 (fetal rat small intestine)		
Primary cultures of colonic smooth muscle	LPS, TNF, and Il-1	Tannahill *et al.* (1997)
Myenteric plexus neurons	TNF and IL-β	Valentine *et al.* (1996)
A549 (human lung carcinoma)	TNF and IFNγ	Harris *et al.* (1991)
HS153 (human fibroblasts)		
ACHN (renal carcinoma)		
A375 (human melanoma)		
A549 (human lung carcinoma)	TNF, IL-1α & β	Wong and Goeddel (1988)
HeLa (human cervical)		
HT-29 (human colon cancer)		
KB (oral epidermoid carcinoma)		
RD (rhabdosarcoma)		
U373 & U87MG (glioblastoma)		
WI-38 (normal fetul lung fibroblasts)		
MCF-7 (breast carcinoma)		
U-937 (human monoblastic leukemic)	TNF + vitamin D_3	Iwamoto *et al.* (1990)
Insulin-producing RIN	IL-1	Bedoya *et al.* (1995)
C57BL/6 MBE (microvascular endothelial)	TNF	Shaffer *et al.* (1990)
Human keratinocytes	IL-1 and TNF	Lontz *et al.* (1995)
A375 (human melanoma)	IL-1	Masuda *et al.* (1988)
A375 (human melanoma)	IL-1	Eastgate *et al.* (1993)
TF-1 (factor-dependent leukemic)		
A549 (human lung carcinoma)		
MCF-7 (human breast cancer)	TNF	Asoh *et al.* (1989)
KG-1 (myeloblasts)	TNF	Kizaki *et al.* (1993)
HL-60 (promyelocytes)		
ML-3 (early myelomonoblasts)		
THP-1 & U-937 (human monoblastic leukemic)		
K562 & HEL (early myeloid/erythroid)		
WI-38 (normal fetal lung fibroblasts)		
Lu-CSF-1 (human adenocarcinoma of lung)		
A172 (human glioblastoma)	TNF	Melendez and Baglioni (1993)
HT-1080 (fibrosarcoma)		
HOS (osteosarcoma)		
SK-MEL-109 (melanoma)		

difference in response to glucocorticoids in the two rat intestinal epithelial cell lines is related to the fetal versus adult origin of the cells or differences in differentiation between the two cell lines. In primary cultures of rat hepatocytes cultured longer than 16 hr, MnSOD mRNA levels are unchanged when treated with glucocorticoids alone. However, cotreatment with glucocorticoids and IL-6 results in a greater induction of MnSOD mRNA levels than treatment with IL-6 or dexamethasone alone (Dougall and Nick, 1991). The induction of IL-6 receptor expression in hepatocytes by glucocorticoids (Snyers *et al.*, 1990) may explain this effect. There have been no *in vivo* studies on the effects of glucocorticoid treatment on MnSOD levels in animals or humans.

Most studies have found that Cu/ZnSOD is constitutively expressed with no direct response to free radical levels, acute-phase cytokines, or LPS. A twofold induction of Cu/ZnSOD mRNA levels is observed in primary cultures of rat colonic smooth muscle cells in response to IL-1 treatment for 8 hr (Tannahill *et al.*, 1997). However, levels of both Cu/ZnSOD and MnSOD are dramatically induced during the process of differentiation of 3T3 fibroblasts into adipocytes (Nick, unpublished data). One hypothesis to explain this result is that the abundant fat deposition in the cytoplasm of adipocytes is highly susceptible to oxygen radical-mediated oxidation and that enhanced cytoplasmic antioxidants are required for cell survival. It is not known if the induction of Cu/ZnSOD is the result of metabolic factors, adipocyte-specific response elements, or some other mechanism.

5. MnSOD: A POTENT CYTOPROTECTIVE ANTIOXIDANT ENZYME

As indicated, there are abundant data that MnSOD levels can be induced by proinflammatory cytokines; however, of equal importance is the finding that elevated levels of this enzyme are protective against cytokine-induced cytotoxicity. The induction of MnSOD by TNFα (Wong and Goeddel, 1988; Visner *et al.*, 1990; Valentine and Nick, 1992) results in protection of the cell from TNFα-mediated cytotoxicity. Elegant experiments by Wong *et al.* (1989) using sense and antisense human MnSOD expression vectors have clearly demonstrated that overexpression of MnSOD markedly reduced the cytotoxic effects of TNFα exposure; conversely, inhibition of MnSOD expression with an antisense vector rendered the cells more susceptible to TNFα cytotoxicity. This effect is likely the result of mitochondrial protection from TNFα-dependent oxygen radical production. MnSOD is also induced by and protects against the damaging effects of IL-1 in multiple cell lines (Hirose *et al.*, 1993).

The cytoprotective effects of MnSOD have also been demonstrated in models of radiation damage. The induction of MnSOD in bone marrow stem cells results in reduced radiation toxicity (Hirose *et al.*, 1993; Eastgate *et al.*, 1993), a free-radical-mediated event. Similarly, IL-1 pretreatment of mice results in radioprotection of small intestinal crypt cells (Wu and Miyamoto, 1990). This effect may be the result of increased MnSOD levels as IL-1 induces MnSOD in the rat small intestinal crypt cell line IEC-6 (Valentine and Nick, 1994). In an animal model of radiation injury and fibrosis, Lefaix *et al.* (1996) demonstrated a 70% reduction in the area and volume of fibrosis in the skin of γ-irradiated pigs treated with either liposomal bovine Cu/ZnSOD or free human MnSOD. The comparable results for Cu/ZnSOD and MnSOD in this study are quite interesting and raise questions on the intracellular and extracellular localization of the administered Cu/ZnSOD and MnSOD that were not answered in these experiments. Experiments by Wong (1995) revealed that transfection of cells with a human MnSOD expression vector augmented cellular resistance to radiation and that overexpression of antisense MnSOD RNA, reduce resistance. Furthermore, these studies demonstrated that overexpression of Cu/ZnSOD or EC-SOD did not result in radiation protection. However, transfection of cells with either Cu/ZnSOD or EC-SOD containing a mitochondrial leader sequence did enhance cellular radiation resistance, thus clearly demonstrating the requirement for intramitochondrial SOD expression for radiation resistance in cell culture.

Protective effects for administered MnSOD but not Cu/ZnSOD have been observed in a model of adjuvant arthritis and bleomycin-induced lung fibrosis (Parizada *et al.*, 1991). However, the localization of administered SODs must be investigated for a complete understanding of the protective effects. Others have shown that cytokine (Tsan *et al.*, 1991) or 85% oxygen (Crapo and Tierney, 1974) pretreatment results in MnSOD induction, which is coupled to the subsequent prolonged survival in normally lethal 100% oxygen environments. To further explore the cytoprotective effects of enhanced SOD levels, investigators have developed transgenic mice overexpressing Cu/ZnSOD or MnSOD. Transgenic mice overexpressing Cu/ZnSOD in the lung experienced less toxicity and survived significantly longer when exposed to 99% oxygen (White *et al.*, 1991). Other investigators developed MnSOD transgenic mice in which the surfactant promoter was used to drive MnSOD overexpression in the lung (Wispe *et al.*, 1992). These animals also demonstrated resistance and improved survival to the toxic effects of 95% oxygen. Yen *et al.* (1996) used the β-actin promoter to drive overexpression of human MnSOD in the heart of transgenic mice. These mice have documented mitochondrial expression of the transgene that results in protection from adriamycin-induced cardiac toxicity.

6. MnSOD AND ONCOGENESIS

In addition to its cytoprotective and oxygen scavenging role, MnSOD has been strongly associated with the suppression of oncogenesis (Church *et al.*, 1993) and promotion of cellular differentiation in some models (St. Clair *et al.*, 1994). MnSOD expression is frequently reduced and often resistant to regulation in many tumor cell lines (Borrello *et al.*, 1993). Striking experiments by Church *et al.* (1993) demonstrated that overexpression of MnSOD in the human melanoma cell line UACC-903 results in loss of the malignant phenotype characterized by the ability to grow in soft agar or formation of tumors in nude mice. Most interestingly, this cell line has a deletion of chromosome 6 often characteristic of human melanoma cells in a region harboring the MnSOD locus. It is hypothesized that increased levels of MnSOD alter the redox state of the cells and may change the activity of redox-sensitive transcription factors. Similar findings were observed in the human breast cancer cell line MCF-7 (J. J. Li *et al.*, 1995). In these cells, overexpression of MnSOD reduced plating efficiency, reduced colony formation in soft agar, and reduced tumor development in nude mice. As a result, the authors have implicated MnSOD as a potential tumor suppressor protein. The cause of altered MnSOD regulation and expression in some tumor cells has not been identified. Along the same lines, we have observed that MnSOD is not induced by LPS, TNFα, IL-1, or interferon-γ in the human colon cancer cell line T84, even though receptor expression and function has been established for interferon-γ as treatment with this cytokine results in increased expression of FAS mRNA levels (Valentine, unpublished data). These data therefore implicate aberrant gene regulation as an additional explanation for reduced levels of MnSOD in cells derived from various malignancies.

7. MnSOD GENE ABLATION

In inflammatory states, the production of oxygen free radicals by cytokines and activated neutrophils may inundate a tissue's defense mechanisms. Tissues such as the intestine may be particularly susceptible to radical-mediated damage because the colonic mucosa, submucosa, and muscularis mucosa have been found to contain low levels of SOD, catalase, and glutathione peroxidase activity (Grisham *et al.*, 1990). SOD-deficient mutants have been described in *E. coli* (Farr *et al.*, 1986), yeast (Van Loon *et al.*, 1986), and *Drosophila* (Phillips *et al.*, 1989) that express increased sensitivity to oxidant stress manifested by increased mutation rates, oxygen and paraquat hypersensitivity, and reduced life span. Most strikingly, Y. Li *et al.* (1995) have reported that mice harboring a homozygous ablation of MnSOD do not survive longer than 10 days after birth, with death a consequence of severe dilated cardiomyopathy. In this model, animals heterozygotic for the MnSOD mutated gene did not display an abnormal phenotype even though MnSOD activity was reduced 49–55%. Homozygotic (–/–) animals were normal size at birth, but showed severe growth retardation over the next several days and by day 10 nearly all had died. Autopsy revealed a dilated cardiomyopathy, marked steatosis of the liver, and submucosal calcifications in the small intestine. Elevation of lipid peroxides was not observed, although succinate dehydrogenase and aconitase were severely reduced in the heart and other tissues, a likely consequence of oxygen radical-mediated damage. The absence of MnSOD in this model demonstrates the necessity of free radical scavenging mechanisms for normal aerobic respiration. The susceptibility of heterozygotic mutants to inflammatory challenge is currently being investigated.

8. SIGNAL TRANSDUCTION

TNFα induces MnSOD through the binding of the P55 receptor (Wong, 1995). Furthermore, antibodies that activate the p55 TNFα receptor induce MnSOD expression whereas those that activate the p75 TNFα receptor do not (Tartaglia *et al.*, 1991). The signal transduction pathway through which activation of the TNFα receptor results in the induction of MnSOD is not clear. It is unlikely that oxygen radicals serve as a second messenger because oxygen radical generators such as hyperoxia, ionizing radiation, paraquat, adriamycin, and H_2O_2 do not lead to MnSOD induction (Wong *et al.*, 1992). Furthermore, protein kinase C and NF-κB do not appear to be involved in the induction of MnSOD by TNFα or IL-1 because inhibitors of protein kinase C (Wong, 1995; Gwinner *et al.*, 1995) and NF-κB (Wong, 1995; Bedoya *et al.*, 1995) do not block the induction of MnSOD. Furthermore, we have convincing evidence from *in vivo* footprinting experiments that clearly demonstrate that a potential NF-κB DNA-binding site in the MnSOD promoter is not occupied (Kuo and Nick, unpublished data).

Conflicting data exist on the role of the lipoxygenase pathway in the induction of MnSOD gene expression by proinflammatory mediators. NDGA, an inhibitor of the lipoxygnease pathway, inhibits the induction of MnSOD by TNFα or an agonist antibody to the TNFα p55 receptor (Wong and Goeddel, 1994b). However, experiments by Gwinner *et al.* (1995) reveal that NDGA does not inhibit the induction of MnSOD by IL-1. In pulmonary epithelial and endothelial cells, antimycin A, an inhibitor of complex III in the mitochondrial electron transport chain, blocks the induction of MnSOD by TNFα, but has no effect on the induction of MnSOD by LPS or IL-1 (Rogers and Nick, unpublished data). These experiments indicate that TNFα and IL-1 utilize different pathways to induce MnSOD and that at present no consistent signal transduction mechanism can be associated with MnSOD gene activation by proinflammatory mediators.

9. MOLECULAR MECHANISMS CONTROLLING MnSOD GENE EXPRESSION

The molecular mechanism of induction of SOD mRNA levels by LPS, TNFα, or IL-1 has been the most thoroughly investigated for rat MnSOD. The induction of MnSOD by these agents is dependent at least in part on *de novo* transcription. Nuclear run-off assays demonstrate an increase in newly transcribed MnSOD message following treatment with IL-1, TNFα, or LPS in both lung L2 cells (Nick, unpublished data) and intestinal IEC-6 cells (Valentine, unpublished data). However, the inductions seen with nuclear run-off assays are not as great as those observed by Northern analysis. Experiments evaluating the effect of MnSOD mRNA sequences on the half-life of the rabbit globin gene driven by a c-fos promoter have identified a coding region determinant of instability that reduces the globin mRNA half-life from 12–15 hr to 1–2 hr (Davis and Nick, unpublished data). The destabilizing effect is inhibited by actinomycin D; thus, the protein required for function of the MnSOD coding region determinant of instability may also be under the regulation of stimuli of MnSOD induction and contribute to enhanced stimulus-dependent MnSOD mRNA stabilization.

Stimulus-dependent changes in rat MnSOD chromatin structure have been evaluated by DNase I hypersensitive studies (Kuo and Nick, unpublished data). DNase I cuts DNA at a higher frequency in the areas of open chromatin structure which have been shown to house binding sites for *trans*-activating factors. Seven hypersensitive sites have been identified in the promoter region and within the MnSOD gene. High-resolution analysis of site I in the MnSOD promoter revealed that it is made up of five subsites, and a unique sixth site is observed only in LPS- and TNFα-treated cells.

To determine the functional characteristics of the hypersensitive sites, promoter deletion analysis has been performed using a human growth hormone (hGH) reporter gene in rat lung pulmonary epithelial cells (Rogers and Nick, unpublished data). LPS and TNFα treatment results in increased expression of hGH mRNA and secreted protein. Deletion of the promoter fragment to 77 bp or less eliminates both basal and stimulated expression. The inductions observed with the promoter growth hormone constructs are much less than for MnSOD mRNA observed by Northern analysis. Evaluation of the hypersensitive sites within the MnSOD gene in conjunction with the promoter fragments has identified an enhancer element in exon II that results in dramatic amplification of the induction of growth hormone by LPS, TNFα, and IL-1 to levels comparable to those observed for MnSOD by Northern analysis.

In vivo footprinting has been used to determine MnSOD DNA–protein contacts of the basal and stimulus-dependent *trans*-activating factors at single nucleotide resolution within the MnSOD promoter (Kuo and Nick, unpublished data). Analysis of approximately 700 bp of the MnSOD promoter has identified the binding sites of 10 *trans*-activating factors which are potentially involved in basal gene expression. As predicted by the DNase I hypersensitivity analysis, a factor has been identified that is only present after LPS, TNFα, or IL-1 treatment. The sequence of the putative binding site and the specific contacts implicate nuclear factor IL6 (NF-IL6) as the *trans*-activating factor. Further analysis of the DNase I hypersensitive sites outside the promoter region is under way.

10. MnSOD LEVELS IN INFLAMMATORY MODELS

Data from cell culture have demonstrated that cytokines and inflammatory mediators can induce MnSOD mRNA and protein levels in an assortment of cell types. The induction is at least partially transcriptional. An enhancer element and a stimulus-dependent *trans*-activating factor binding site have been identified. Both cell culture and animal studies have documented a cytoprotective role for MnSOD in response to oxidant stress from a range of mechanisms. To determine whether the cytokine induction of MnSOD observed in cell culture also occurs in tissue inflammation, animal models of inflammatory responses have been evaluated.

Most animal inflammatory models have examined inhalation of particulate irritants such as chrysotile asbestos (Quinlan *et al.*, 1995), silica (Janssen *et al.*, 1992), cristobalite, and titanium dioxide (Janssen *et al.*, 1994). Inhalation of silica, asbestos, cristobalite, and fibrogenic titanium dioxide all resulted in increased MnSOD mRNA levels. The study by Janssen *et al.* (1994) also examined the inflammatory response by bronchoalveolar lavage and found a correlation between the levels of MnSOD induction and the inflammatory response. Conversely, inhalation of nonfibrogenic titanium dioxide-F did not increase MnSOD levels or the number of inflammatory cells in the bronchoalveolar lavage. The induction of MnSOD by inhalation of mineral dust is likely the result of cytokine release by exposure of alveolar macrophages to mineral dust (Driscoll and Maurer, 1991).

From a historical standpoint, an early study by Frank *et al.* (1978) demonstrated that endotoxin treatment of adult rats dramatically improved survival following 72 hr of hyperoxia. Survival increased from 33% in control animals to 97% in the endotoxin-treated animals. Furthermore, the endotoxin-treated rats had highly significant reductions in pulmonary edema and hemorrhage compared with controls. The endotoxin-treated rats demonstrated significant increases in lung catalase, glutathione peroxidase, and SOD activity. Crapo and Tierney (1974) determined that exposure of rats to 85% oxygen for 7 days resulted in a 50% increase in SOD activity which correlated with survival in normally lethal 100% oxygen. Similarly, Forman and Fisher (1981) determined that exposure of rats to 80% O_2 for 7 days resulted in an increase in MnSOD activity in whole lung of 154% of controls. No changes were found for Cu/ZnSOD or glutathione peroxidase activity. Another study examined the induction of MnSOD by hyperoxia in the rat lung (Chang *et al.*, 1995). Rats were exposed to 85% oxygen for 7 and 14 days. Intracellular concentration of Cu/ZnSOD did not increase, but total lung Cu/ZnSOD did increase as a result of cellular hypertrophy. MnSOD levels increased 1.6- and 3-fold by immunogold labeling at the 7-day time point in type II cells and in the interstitial fibroblast, respectively. Western analysis of rat lung homogenates revealed a 3.7-fold increase at 7 days and 5.22-fold increase in MnSOD protein levels at 14 days. Lesser but significant increases were also found for MnSOD activity. The long duration of hyperoxia required for increases in MnSOD indicates that hyperoxia likely causes tissue damage that results in cytokine release and induction of MnSOD. Similarly, chronic alcohol administration (50% of calories as alcohol) to monkeys results in mild steatosis and enlarged mitochondria with distorted cristae. Associated with these changes are a nonsignificant reduction in Cu/ZnSOD activity and a significant 115% increase in MnSOD activity (Keen *et al.*, 1985). These differences may have been the result of changes in the tissue levels of zinc and manganese, or the result of alcohol-induced hepatic inflammation. Hepatic ethanol metabolism results in the generation of reactive oxygen species in humans (Lieber, 1988) and baboons (Shaw *et al.*, 1981), whereas chronic alcohol intake results in steatosis and alcoholic hepatitis in humans.

Inflammatory regulation of MnSOD in the rat colon has been studied by Tannahill *et al.* (1995) who demonstrated the most rapid and pronounced inflammatory induction of MnSOD yet observed. Adult rats were given 5% acetic acid enemas to produce a mild to moderate colitis and the levels of MnSOD mRNA and protein were evaluated at various time points within the first 24 hr following administration of the enema. Increases in MnSOD mRNA levels were observed as early as 4 hr after initiation of colitis. The induction of MnSOD mRNA was preceded by an induction of IL-1β mRNA observed at 15 min. By 12 hr, MnSOD mRNA levels increased 14-fold in the epithelial layer and 96-fold in the muscle layer of the colon. Western analysis demonstrated a 40-fold increase in the epithelial layer MnSOD protein content by 24 hr and a 22-fold increase in MnSOD protein content of the muscle layer of the colon. Immunocytochemistry demonstrated that the increased MnSOD protein levels were confined to the epithelial cells at the base of the glands, diffusely in the smooth muscle cells with a dramatic induction in the neurons of the myenteric plexus. Furthermore, MnSOD was shown to be inducible by TNFα, IL-1, and LPS in primary cultures of colonic smooth muscle cells (Tannahill *et al.*, 1997) and in primary cultures of myenteric neurons (Valentine *et al.*, 1996).

The induction observed following acetic acid-induced colitis is not simply a response to injury. Superficial injury such as observed by acetic acid-induced colitis would not explain the induction of MnSOD in the myenteric plexus neurons. Additionally, in intestinal epithelial cell culture, injury produced by *Clostridium difficile* toxin, dilute acetic acid, hyperoxia, or mechanical disruption does not induce MnSOD (Valentine, unpublished data).

A model has been proposed to consolidate the cell culture and animal data showing that cytokines are cytotoxic but also induce MnSOD protective against cytokine toxicity (Nick and Valentine, 1994; Figure 1). The initial tissue injury results in cytokine release from the injured cells as well as from the resident inflammatory cells, yielding various cellular responses. The release of cytokines may serve at least three functions: (1) They act though autocrine and paracrine mechanisms to increase levels of MnSOD in surround-

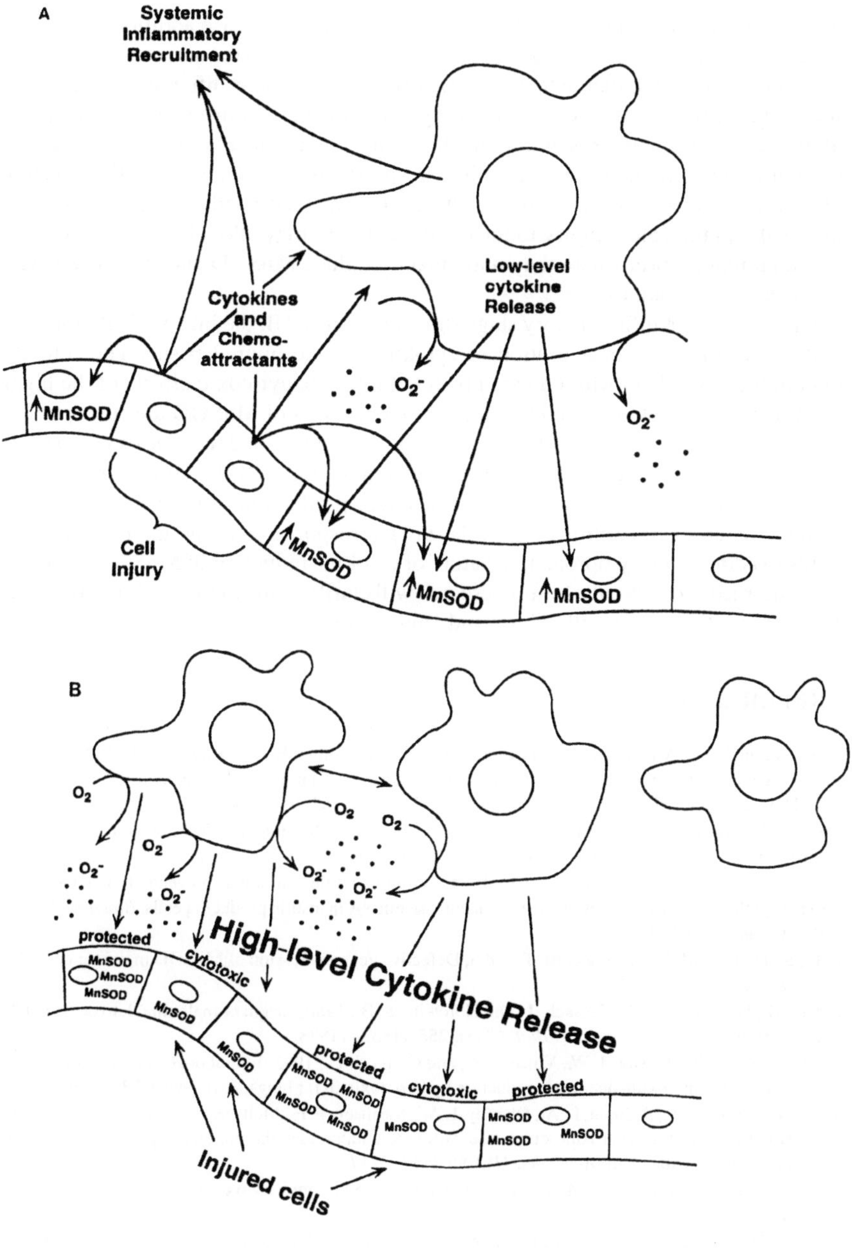

FIGURE 1. A proposed model of the role of MnSOD in an acute inflammatory response. (A) The initial tissue injury results in local cytokine release and induction of MnSOD. (B) The acute inflammatory phase produces an increase of cytotoxic cytokines resulting in damage to those cells that have low levels of MnSOD. See text for discussion. (Used with permission from the Crohn's and Colitis Foundation of America, Inc., *Progress in Inflammatory Bowel Disease* 1994;**15:**8–11.)

ing cells. (2) They stimulate resident inflammatory cells to release additional cytokines and chemoattractants leading to further increases in MnSOD in the surrounding cells. (3) The cytokines and chemoattractants released from the resident inflammatory cells and injured cells initiate systemic inflammatory recruitment. If during this initial phase, cytokines exist at a low concentration, they may not produce significant toxicity in surrounding cells. If, however, the initial response is intense, high initial levels of cytokines may overwhelm the antioxidant mechanisms and result in significant injury. Additionally, in the colon, injury to the epithelial lining may allow bacterial products to enter the lamina propria where they may directly induce MnSOD and stimulate macrophage release of cytokines.

During the acute inflammatory stage shown in Figure 1B, the influx of inflammatory cells leads to a higher concentration of cytokines in the injured tissue. The cells that contain induced levels of MnSOD are protected from the cytotoxic effects of the higher concentrations of cytokines. Cells with insufficient levels of MnSOD are susceptible to cytokine-mediated injury. Extremely high levels of cytokines and reactive oxygen species may still overcome the cell defenses. The induction of EC-SOD by similar mechanisms could provide cellular protection from extracelluar production of oxygen free radicals. This model is an oversimplification of the *in vivo* response to tissue injury. The model attempts to address the cytoprotective nature of cytokine induction of MnSOD in the face of cytokine-mediated toxicity. Future research will clarify many of the confounding issues in the interpretation of the cell culture and animal model data.

11. REFERENCES

Asoh, K., Watanabe, Y., Mizoguchi, H., Mawatari, M., Ono, M., Kohno, K., and Kuwano, M., 1989, Induction of manganese superoxide dismutase by tumor necrosis factor in human breast cancer MCF-7 cell line and its TNF-resistant variant, *Biochem. Biophys. Res. Commun.* **162**:794–801.

Bannister, J. V., Bannister, W. H., and Rotilio, G., 1987, Aspects of the structure, function and applications of superoxide dismutase, *CRC Crit. Rev. Biochem.* **2**:111–180.

Bedoya, F. J., Flodstrom, M., and Eizirik, D. L., 1995, Pyrrolidine dithiocarbamate prevents IL-1-induced nitric oxide synthase mRNA, but not superoxide dismutase mRNA in insulin producing cells, *Biochem. Biophys. Res. Commun.* **210**:816–822.

Borrello, S., De Leo, M. E., and Galeotti, T., 1993, Defective gene expression of MnSOD in cancer cells, *Mol. Aspects Med.* **14**:253–258.

Burger, R. M., Berkowitz, A. R., Peisach, J., and Horowitz, S. B., 1980, Origin of malondialdehyde from DNA degraded by Fe (II)-bleomycin, *J. Biol. Chem.* **255**:11832–11838.

Chang, L. Y., Kang, B. H., Slot, J. W., Vincent, R., and Crapo, J. D., 1995, Immunocytochemical localization of the sites of superoxide dismutase induction by hyperoxia in rat lungs, *Lab. Invest.* **73**:29–39.

Church, S. L., Grant, J. W., Ridnour, L. A., Oberley, L. W., Swanson, P. E., Meltzer, P. S., and Trent, J. M., 1993, Increased manganese superoxide dismutase expression suppresses the malignant phenotype of human melanoma cells, *Proc. Natl. Acad. Sci. USA* **90**:3113–3117.

Crapo, J. D., and Tierney, D. F., 1974, Superoxide dismutase and pulmonary oxygen toxicity, *Am. J. Physiol.* **226**:1402–1407.

Cross, C. E., Halliwell, B., Borish, E. T., Pryor, W. A., Ames, B. N., Saul, R. L., McCord, J. M., and Harman, D., 1987, Oxygen radical and human disease, *Ann. Intern. Med.* **107**:526–545.

Dougall, W. C., and Nick, H. S., 1991, Manganese superoxide dismutase: A hepatic acute phase protein regulated by interleukin 6 and glucocorticoids, *Endocrinology* **129**:2376–2384.

Driscoll, K. E., and Maurer, J. K., 1991, Cytokine and growth factor release by alveolar macrophages: Potential biomarkers of pulmonary toxicity, *Toxicol. Pathol.* **19**:389–405.

Eastgate, J., Moreb, J., Nick, H. S., Keiichiro, S., Taniguichi, N., and Zucali, J. R., 1993, A role for manganese superoxide dismutase in radioprotection of hematopoietic stem cells by interleukin-1, *Blood* **81**:639–646.

Farr, S. B., D'Ari, R., and Touati, D., 1986, Oxygen-dependent mutagenesis in Escherichia coli lacking superoxide dismutase, *Proc. Natl. Acad. Sci. USA* **83**:8268–8272.

Forman, H. J., and Fisher, A. B., 1981, Antioxidant enzymes of rat granular pneumocytes; constitutive levels and effect of hyperoxia, *Lab Invest.* **45**:1–6.

Frank, L., Yam, J., and Roberts, R. J., 1978, The role of endotoxin in protection of adult rats from oxygen-induced lung toxicity, *J. Clin. Invest.* **61**:269–275.

Freeman, B. A. and Crapo, J. D., 1981, Hyperoxia increases oxygen radical production in rat lungs and lung mitochondria, *J. Biol. Chem.* **256**:10986–10992.

Grisham, M. B., and Granger, D. N., 1988, Neutrophil-mediated mucosal injury, *Dig. Dis. Sci.* **53**:6S–15S.

Grisham, M. B., MacDermott, R., and Deitch, E., 1990, Oxidant defense mechanisms in the human colon, *Inflammation* **13**:669–680.

Gwinner, W., Tisher, C. C., and Nick, H. S., 1995, Regulation of manganese superoxide dismutase in glomerular epithelial cells: Mechanisms for interleukin-1 induction, *Kidney Int.* **48**:354–362.

Harris, C. A., Derbin, K. S., Hunte-McDonough, B., Krauss, M. R., Chen, K. T., Smith, D. M., and Epstein, L. B., 1991, Manganese superoxide dismutase is induced by IFN-γ in multiple cell types, *J. Immunol.* **147**:149–154.

Harris, M. L., Schiller, H. J., Reilly, P. M., Donowitz, M., Grisham, M. B., and Bulkley, G. B., 1992, Free radicals and other reactive oxygen metabolites in inflammatory bowel disease: Cause, consequence or epiphenomenon? *Pharmacol. Ther.* **53**:375–408.

Hirose, K., Longo, D. L., Oppenheim, J. J., and Matsushima, K., 1993, Overexpression of mitochondrial manganese superoxide dismutase promotes the survival of tumor cells exposed to interleukin-1, tumor necrosis factor, selected anticancer drugs, and ionizing radiation, *FASEB J.* **7**:361–368.

Hjalmarsson, K., Marklund, S. L., Engstrom, A., and Edlund, T., 1987, Isolation and sequence of complementary DNA encoding human extracellular superoxide dismutase, *Proc. Natl. Acad. Sci. USA* **84**:6340–6344.

Imlay, J. A., and Linn, S., 1988, DNA damage and oxygen radical toxicity, *Science* **240**:1302–1309.

Iwamoto, S., Takeda, K., Kamijo, R., and Konno, K., 1990, Induction of resistance to TNF cytotoxicity and mitochondrial superoxide dismutase on U-937 cells by 1,25-dihydroxyvitamin D3, *Biochem. Biophys. Res. Commun.* **170**:73–79.

Jacoby, D. B., and Choi, A. M., 1994, Influenza virus induces expression of antioxidant genes in human epithelial cells, *Free Radical Biol. Med.* **16**:821–824.

Janssen, Y. M. W., Marsh, J. P., Absher, M. P., Memenway, D., Vacek, P. M., Leslie, K. O., Borm, P. J. A., and Mossman, B. T., 1992, Expression of antioxidant enzymes in rat lungs after inhalation of asbestos or silica, *J. Biol. Chem.* **15**:10625–10630.

Janssen, Y. M. W., Marsh, J. P., Driscoll, K. E., Borm, P. J. A., Obersorster, G., and Mossman, B. T., 1994, Increased expression of manganese-containing superoxide dismutase in rat lungs after inhalation of inflammatory and fibrogenic minerals, *Free Radical Biol. Med.* **16**:315–322.

Karlsson, K., and Marklund, S. L., 1987, Heparin-induced release of extracellular superoxide dismutase to human blood plasma, *Biochem. J.* **242**:55–59.

Keen, C. L., Tamura, T., Lonnerdal, B., Hurley, L. S., and Halsted, C. H., 1985, Changes in hepatic superoxide dismutase activity in alcoholic monkeys, *J. Clin. Nutr.* **41**:929–932.

Kifle, Y., Monnier, J., Chesrown, S. E., Raizada, M. K., and Nick, H. S., 1996, Regulation of the manganese superoxide dismutase and inducible nitric oxide synthase gene in rat neuronal and glial cells, *J. Neurochem.* **66**:2128–2135.

Kizaki, M., Sakashita, A., Karmakar, A., Lin, C. W., and Koeffler, H. P., 1993, Regulation of manganese superoxide dismutase and other antioxidant genes in normal and leukemic hematopoeitic cells and their relationship to cytotoxicity by tumor necrosis factor, *Blood* **82**:1142–1150.

Lefaix, J. L., Delanian, S., Leplat, J. J., Tricaud, Y., Martin, M., Nimrod, A., Baillet, F., and Daburon, F., 1996, Successful treatment of radiation-induced fibrosis using Cu/Zn-SOD and Mn-SOD: An experimental study, *Int. J. Radiat. Oncol. Biol. Phys.* **35**:305–312.

Li, J. J., Oberley, L. W., St. Clair, D. K., Ridnour, L. A., and Oberley, T. D., 1995, Phenotypic changes induced in human breast cancer cells by overexpression of manganese-containing superoxide dismutase, *Oncogene* **10**:1989–2000.

Li, Y., Huang, T. T., Carlson, E. J., Melov, S., Ursell, P. C., Olson, J. L., Noble, L. J., Yoshimura, M. P., Berger, C., and Chan, P. H., 1995, Dilated cardiomyopathy and neonatal lethality in mutant mice lacking manganese superoxide dismutase, *Nature Genet.* **11:**376–381.

Lieber, C. S., 1988, Biochemical and molecular basis of alcohol-induced injury to liver and other tissues, *N. Engl. J. Med.* **319:**1639–1650.

Lontz, W., Sirsjo, A., Liu, W., Lindberg, M., Rollman, O., and Torma, H., 1995, Increased mRNA expression of manganese superoxide dismutase in psoriasis skin lesions and in cultured human keratinocytes exposed to IL-1 and TNF-alpha, *Free Radical Biol. Med.* **18:**349–355.

Masuda, A., Longo, D. L., Kobayashi, Y., Appella, E., Oppenheim, J. J., and Matsushima, K., 1988, Induction of mitochondrial manganese superoxide dismutase by interleukin 1, *FASEB J.* **2:**3087–3091.

Melendez, J. A., and Baglioni, C., 1993, Differential induction and decay of manganese superoxide dismutase mRNAs, *Free Radical Biol. Med.* **14:**601–608.

Mokuno, K., Ohtani, K., Suzumura, A., Kiyowasa, K., Hirose, Y., Kasai, K., and Kato, K., 1994, Induction of manganese superoxide dismutase by cytokines and lipopolysaccharide in cultured mouse astrocytes, *J. Neurochem.* **63:**612–616.

Nick, H. S., and Valentine, J. F., 1994, A potential cytoprotective mechanism in the colon: Regulation of manganese superoxide dismutase, *Prog. Inflam. Bowel Dis.* **15:**8–11.

Parizada, B., Werber, M. M., and Nimrod, A., 1991, Protective effects of human recombinant MnSOD in adjuvant arthritis and bleomycin-induced lung fibrosis, *Free Radical Res. Commun.* **15:**297–301.

Phillips, J. P., Campbell, S. D., Michaud, D., Charbonneau, M., and Hilliker, A. J., 1989, Null mutation of copper/zinc superoxide dismutase in Drosophila confers hypersensitivity to paraquat and reduced longevity, *Proc. Natl. Acad. Sci. USA* **86:**:2761–2765.

Quinlan, T. R., BeruBe, K. A., Marsh, J. P., Janssen, Y. M. W., Taishi, P., Leslie, K. O., Hemenway, D., O'Shaughnessy, P. T., Vacek, P., and Mossman, B. T., 1995, Patterns of inflammation, cell proliferation, and related gene expression in lung after inhalation of chrysotile asbestos, *Am. J. Pathol.* **147:**728–739.

Schreck, R., and Baeuerle, P. A., 1994, Assessing oxygen radicals as mediators in activation of inducible eukaryotic transcription factor NF-κB, *Methods Enzymol.* **234:**151–163.

Schulze-Osthoff, K., Bakker, A.C., Vanhaesebroeck, B., Beyaert, R., Jacob, W.A., Fiers, W., 1992, Cytotoxic activity of tumor necrosis factor is mediated by early damage of mitochondrial functions, *J. Biol. Chem.* **267:**5317–5323.

Shaffer, J. B., Treanor, C. P., and Del Vecchio, P. J., 1990, Expression of bovine and mouse endothelial cell antioxidant enzymes following TNF-alpha exposure, *Free Radical Biol. Med.* **8:**487–502.

Shaw, S., Jayatilleke, E., Ross, W. A., and Gordon, E. R., 1981, Ethanol-induced lipid peroxidation: Potentiation by long-term alcohol feeding and attenuation by methionine, *J. Lab. Clin. Med.* **98:**417–424.

Snyers, L., DeWit, L., and Content, J., 1990, Glucocorticoid up-regulation of high-affinity interleukin-6 receptors on human epithelial cells, *Proc. Natl. Acad. Sci. USA* **87:**2364–2368.

St. Clair, D. K., Oberley, T. D., Muse, K. E., and St. Clair, W. H., 1994, Expression of manganese superoxide dismutase promotes cellular differentiation, *Free Radical Biol. Med.* **16:**275–282.

Stephanz, G. B., Gwinner, W., Cannon, J. K., Tisher, C. C., and Nick, H. S., 1996, Heat-aggregated IgG and interleukin-1-beta stimulate manganese superoxide dismutase in mesangial cells, *Exp. Nephrol.* **4:**151–158.

Tannahill, C. L., Stevenot, S. A., Eaker, E. Y. , Sallustio, J. E., Nick, H. S., and Valentine, J. F., 1997, Regulation of superoxide dismutase in primary cultures of rat colonic smooth muscle cells, *Am. J. Physiol.* **272:**G1230–G1235.

Tannahill, C. L., Stevenot, S. A., Campbell-Thompson, M., Nick, H. S., and Valentine, J. F., 1995, Induction and immunolocalization of manganese superoxide dismutase in acute acetic acid colitis, *Gastroenterology* **109:**800–811.

Tartaglia, L. A., Weber, R. F., Figari, I. S., Reynolds, C., Palladino, M. A., Jr., and Goeddel, D. V., 1991, The two different receptors for tumor necrosis factor mediate distinct cellular responses, *Proc. Natl. Acad. Sci. USA* **88:**9292–9296.

Tsan, M., Lee, C. Y., and Terada, L. S., 1991, Interleukin 1 protects rats against oxygen toxicity, *J. Appl. Physiol.* **71:**688–697.

Valentine, J. F., and Nick, H. S., 1992, Acute-phase induction of manganese superoxide dismutase in intestinal epithelial cell lines, *Gastroenterology* **103:**905–912.

Valentine, J. F., and Nick, H. S., 1994, Glucocorticoid repression of basal and stimulated manganese superoxide dismutase expression in rat intestinal epithelial cells, *Gastroenterology* **107:**1662–1670.

Valentine, J. F., Tannahill, C. L., Stevenot, S. A., Nick, H. S., Sallustio, J. E., and Eaker, E. Y., 1996, Colitis and interleukin-1β up-regulate inducible nitric oxide synthase and superoxide dismutase in rat myenteric neurons, *Gastroenterology* **111:**56–64.

Van Loon, A. P., Pesold-Hurt, B., and Schatz, G., 1986, A yeast mutant lacking mitochondrial manganese-superoxide dismutase is hypersensitive to oxygen, *Proc. Natl. Acad. Sci. USA* **83:**3820–3824.

Visner, G. A., Dougall, W. C., Wilson, J. M., Burr, I. M., and Nick, H. S., 1990, Regulation of manganese superoxide dismutase by lipopolysaccharide, interleukin-1 and tumor necrosis factor, *J. Biol. Chem.* **265:**2856–2864.

Visner, G. A., Block, E. R., Burr, I. M., and Nick, H. S., 1991, Regulation of manganese superoxide dismutase in porcine pulmonary artery endothelial cells, *Am. J. Physiol.* **260:**L444–L449.

Visner, G. A., Chesrown, S. E., Monnier, J., Ryan, U. S., and Nick, H. S., 1992, Regulation of manganese superoxide dismutase: IL-1 and TNF induction in pulmonary artery and microvascular endothelial cells, *Biochem. Biophys. Res. Commun.* **188:**453–462.

Vlessis, A. A., Muller, P., Bartos, D., and Trunkey, D., 1991, Mechanism of peroxide-induced cellular injury in cultured adult cardiac myocytes, *FASEB J.* **5:**2600–2605.

Weiss, S. J., 1989, Tissue destruction by neutrophils, *N. Engl. J. Med.* **320:**365–376.

White, C. W., Avraham, K. B., Shanley, P. F., and Groner, Y., 1991, Transgenic mice with expression of elevated levels of copper-zinc superoxide dismutase in the lungs are resistant to pulmonary oxygen toxicity, *J. Clin. Invest.* **87:**2162–2168.

Wispe, J. R., Warner, B. B., Clark, J. C., Dey, C. R., Neuman, J., Classer, S. W., Crapo, J. D., Chang, L. Y., and Whisett, J. A., 1992, Human Mn-superoxide dismutase in pulmonary epithelial cells of transgenic mice confers protection from oxygen injury, *J. Biol. Chem.* **267:**23937–23941.

Wong, G. H. W., 1995, Protective roles of cytokines against radiation: Induction of mitochondrial MnSOD, *Biochim. Biophys. Acta* **1271:**205–209.

Wong, G. H. W., and Goeddel, D. V., 1988, Induction of manganous superoxide dismutase by tumor necrosis factor: Possible protective mechanism, *Science* **242:**941–944.

Wong, G. H. W., and Goeddel, D. V., 1994a, Fas antigen and p55 TNF receptor signal apoptosis through distinct pathways, *J. Immunol.* **152:**1751–1755.

Wong, G. H. W., and Goeddel, D. V., 1994b, One-day Northern blotting for detection of mRNA: NDGA inhibits the induction of MnSOD mRNA by agonists of type 1 TNF receptor, *Methods Enzymol.* **234:**244–252.

Wong, G. H. W., Elwell, J. H., Oberley, L. W., and Goeddel, D. V., 1989 Manganous superoxide dismutase is essential for cellular resistance to cytotoxicity of tumor necrosis factor, *Cell* **58:**923–931.

Wong, G. H. W., Kamb, A., and Goeddel, D. V., 1992, Cell and Molecular Biology (J. Scandalios, ed.), pp. 69–96, Cold Spring Harbor Laboratory Press, Cold Spring Harbor, NY.

Wu, S. G., and Miyamoto, T., 1990, Radioprotection of the intestinal crypts of mice by recombinant human interleukin-1 alpha, *Radiat. Res.* **123:**112–115.

Yen, H. C., Oberley, T. D., Vichitbandha, S., Ho, Y. S., and St. Clair, D. K., 1996, The protective role of manganese superoxide dismutase against adriamycin-induced acute cardiac toxicity in transgenic mice, *J. Clin. Invest.* **98:**1253–1260.

Chapter 8

Antioxidant Protection and Oxygen Radical Signaling

John M. C. Gutteridge and Barry Halliwell

1. REACTIVE OXYGEN, NITROGEN, IRON, AND COPPER SPECIES

1.1. Oxygen and Reactive Oxygen Species (ROS)

The element oxygen (O) exists in air as a diatomic molecule (O_2), and was first isolated and characterized between 1772 and 1779 by Priestley, Lavoisier, and Scheele (see Chapter 1). Oxygen appeared in significant amounts on the surface of the Earth more than 2.5 billion years ago (Gilbert, 1996), and geological evidence supports a biological origin from the photosynthetic activity of the blue-green bacteria (cyanobacteria). The percentage of oxygen in dry air is now around 21%, the major component of air being nitrogen (78%). Oxygen in the air is a negligible amount of the total present on the Earth, most of which is in water molecules and mineral reservoirs of the Earth's crust, where it is by far the most abundant element. The stable nature of oxygen with its characteristically poor reactivity is a result of its unusual chemistry. O_2 contains two unpaired electrons each in their own orbital but with the same spin quantum number. This makes it difficult for O_2 to accept two electrons from another molecule, e.g., in a covalent bond, because they will have opposite spins. Only when this spin restriction is overcome can the potential oxidizing capacity of oxygen be expressed. To bypass this spin restriction, oxygen prefers to accept electrons one at a time to complete its four-electron reduction to water:

John M. C. Gutteridge Oxygen Chemistry Laboratory, Critical Care Unit, Royal Brompton Hospital, London SW3 6NP, England. **Barry Halliwell** Biochemistry Department, National University of Singapore, Singapore 119260.

Reactive Oxygen Species in Biological Systems, edited by Gilbert and Colton. Kluwer Academic / Plenum Publishers, New York, 1999.

$$O_2 + e^- + H^+ \rightarrow HO_2\cdot \tag{1}$$

$$HO_2\cdot \rightarrow H^+ + \cdot O_2^- \tag{2}$$

(pH 7.4)

$$HO_2\cdot + e^- + H^+ \rightarrow H_2O_2 \tag{3}$$

$$H_2O_2 + e^- \rightarrow OH^- + \cdot OH \tag{4}$$

$$\cdot OH + e^- + H^+ \rightarrow H_2O \tag{5}$$

The problem with this strategy is that the addition of single electrons produces free radicals such as superoxide ($\cdot O_2^-$) and the hydroxyl radical ($\cdot OH$) whose reactivity is considerably greater than O_2. To achieve greater reactivity, oxygen uses iron ions and to a lesser extent copper ions, as catalysts for oxidase, oxygenase, and antioxidant reactions, and for oxygen and electron transport by proteins. The simple reason is that metals undergo facile one-electron transfer, i.e.,

$$Fe^{2+} \leftrightharpoons Fe(III) + e^- \tag{6}$$

$$Cu^+ \leftrightharpoons Cu^{2+} + e^- \tag{7}$$

It is well established that a range of molecules loosely classified as *reactive oxygen species* (ROS) and *reactive nitrogen species* (RNS) (Table I) is generated in the human body (Babior, 1978; Fridovich, 1989; Moncada and Higgs, 1991; Sies, 1991; Frei, 1994; Chapter 10). Some ROS/RNS arise by "accidents of chemistry." For example, $\cdot O_2^-$ and H_2O_2 are produced by the chemical reaction of O_2 with autoxidizable biomolecules such as adrenaline, dopamine, tetrahydrofolates, and some components of mitochondrial and P450-associated electron transport chains. In fact, many "autoxidation" reactions are very slow because of the spin restriction and they are usually catalyzed by traces of transition metal ions in the reaction system. Such $\cdot O_2^-$ generation is usually regarded as the unavoidable consequence of having these "autoxidizable" molecules in a body that needs O_2 and transition metal ions (Fridovich, 1989; Moncada and Higgs, 1991; Babior, 1991; Sies, 1991; Frei, 1994; Halliwell, 1996b). Low-wavelength electromagnetic radiation (e.g., γ rays) splits water to generate the viciously reactive $\cdot OH$ whereas UV light can cleave the O–O bond in H_2O_2 to give $2\cdot OH$.

1.2. Nitrogen and Reactive Nitrogen Species (RNS)

Nitrogen was first recognized by Scheele and Lavoisier, first prepared by Rutherford, and later named by Chaptal in 1823. The free gas accounts for 78% by volume of the Earth's atmosphere, and its inert properties constitute a major global antioxidant for organic matter, deterring combustion and other oxidation reactions. For nitrogen to participate in biochemical processes, it must first be "fixed" into nitrogen-containing compounds. This is achieved by nitrogen moving through a worldwide cycle, the *nitrogen cycle*, that involves biochemical, geological, and atmospheric processes. Three simple oxides of nitrogen are of current biomedical interest: (1) nitrous oxide (N_2O), a colorless gas with a sweet taste, used as an anesthetic (laughing gas); (2) nitrogen dioxide ($\cdot NO_2$), a toxic brown paramagnetic gas (free radical) that exists in equilibrium with its dimer dinitrogen tetroxide (peroxide) N_2O_4; and (3) nitric oxide ($\cdot NO$). Nitrogen dioxide is an environmental pollutant, and may be produced *in vivo* from reactions of $\cdot NO$. It is a powerful initiator of lipid peroxidation, and can damage proteins and DNA. When human blood is exposed to $\cdot NO_2$, there is a rapid depletion of both ascorbate and urate and the onset of lipid peroxidation (Halliwell *et al.*, 1992b). Nitric oxide is a colorless gas and a weak reducing agent, which was first recognized as a distinct gas by Joseph Priestley in 1772.

Biological interest in nitric oxide and other RNS (Table I) has exploded since the recent observation that the vascular endothelium and other cells in the body produce small amounts of it from the amino acid L-arginine (reviewed in Moncada and Higgs, 1991). Nitric oxide is poorly reactive with most molecules in the human body (nonradicals), but as a free radical it can react extremely rapidly with other free radicals such as superoxide,

Table I
Reactive Species[a,b]

Radicals	Nonradicals
Reactive Oxygen Species (ROS)	
Superoxide, $\cdot O_2^-$	Hydrogen peroxide, H_2O_2
Hydroxyl, $\cdot OH$	Hypochlorous acid, HOCl
Peroxyl, $RO_2\cdot$	Hypobromous acid, HOBr
Alkoxyl, $RO\cdot$	Ozone, O_3
Hydroperoxyl, $HO_2\cdot$	Singlet oxygen $^1\Delta_g$
Reactive Nitrogen Species (RNS)	
Nitric oxide (nitrogen monoxide), $\cdot NO$	Nitrous acid, HNO_2
Nitrogen dioxide, $\cdot NO_2$	Nitrosyl cation, NO^+
	Nitroxyl anion, NO^-
	Dinitrogen tetroxide, N_2O_4
	Dinitrogen trioxide, N_2O_3
	Peroxynitrite, $ONOO^-$
	Peroxynitrous acid, ONOOH
	Nitryl (nitronium), NO_2^+
	(e.g., as nitryl chloride, NO_2Cl)
	Alkyl peroxynitrites, ROONO

[a]Adapted from Halliwell (1996b).

[b]ROS is a collective term that includes both oxygen radicals and certain nonradicals that are oxidizing agents and/or are easily converted into radicals (HOCl, O_3, $ONOO^-$, $^1O_2$, H_2O_2). RNS is also a collective term including nitric oxide and nitrogen dioxide radicals, as well as such nonradicals as HNO_2 and N_2O_4. $ONOO^-$ is often included in both categories. *Reactive* is not always an appropriate term: H_2O_2, $\cdot NO$, and $\cdot O_2^-$ react quickly with few molecules whereas $\cdot OH$ reacts quickly with almost all molecules. $RO_2\cdot$, $RO\cdot$, HOCl, $\cdot NO_2$, $ONOO^-$, and O_3 have intermediate reactivities. HOCl and NO_2Cl could also be regarded as "reactive chlorinating species," HOBr as a "reactive brominating species"

amino acid radicals, and certain transition metal ions. The reaction between nitric oxide and superoxide (k_2 6.7 × 10^9 M^{-1} s^{-1}) (Huie and Padmaja, 1993) produces peroxynitrite ($ONOO^-$). Peroxynitrous acid, formed by its protonation, is a powerful oxidant itself, but can also decompose to yield further oxidants with the chemical reactivities of $\cdot NO_2$, $\cdot OH$, and NO_2^+. The exact chemistry of damage by $ONOO^-$ is a matter of considerable current debate (van der Vliet *et al.,* 1994; Pryor and Squadrito, 1995; Kaur *et al.*, 1997).

In addition to putative "accidental" generation of ROS/RNS *in vivo*, some are made deliberately. For example, $\cdot NO$ has multiple physiological roles (Moncada and Higgs, 1991). Activated phagocytes generate $\cdot O_2^-$, H_2O_2, and (in the case of neutrophils) HOCl as one of their many mechanisms for killing foreign organisms (Babior, 1978). They can also generate RNS, for a similar reason. This essential defense mechanism can go wrong: Several diseases (such as rheumatoid arthritis and inflammatory bowel diseases) are accompanied by inappropriate phagocyte activation and resulting tissue damage, to which ROS/RNS contribute (Grisham, 1994; Halliwell, 1996b; Gutteridge and Quinlan, 1996). Other useful roles of ROS are known; e.g., H_2O_2 is generated in the thyroid gland to allow functioning of a peroxidase enzyme in thyroid hormone biosynthesis (Deme *et al.*, 1994).

It seems likely that if ROS fulfill important additional biological functions, it will be the selectively reactive ones such as $\cdot O_2^-$ and H_2O_2 that are used, rather than the indiscriminately reactive $\cdot OH$ (although H_2O_2 might exert effects by causing $\cdot OH$ formation at a specific site, such as H_2O_2-induced cardiovascular reflexes (Stahl *et al.*, 1992). Similarly, $\cdot NO$ and possibly its derivatives (e.g., NO^+) are widely used physiologically, whereas reactive RNS such as $\cdot NO_2$ and $ONOO^-$ may be too indiscriminately damaging. The interactions of RNS and ROS are also important. H_2O_2 activates NF-κB (Schreck *et al.*, 1992) but $\cdot NO$ inhibits activation (Peng *et al.*, 1995). In vascular endothelium, $\cdot O_2^-$ antagonizes the action of $\cdot NO$ and causes vasoconstriction (Laurindo *et al.*, 1991) and it has been suggested that this is a physiological mechanism for regulating vascular tone (Halliwell, 1989). Unfortunately, the rapid reaction of $\cdot NO$ with $\cdot O_2^-$ generates $ONOO^-$ (Huie and Padmaja, 1993), a species possibly responsible for several of the cytotoxic effects attributed to excess $\cdot NO$, such as destruction of Fe-S clusters in certain enzymes (Castro *et al.*, 1994). To add to the complexity, $\cdot NO$ modulates damage by $ONOO^-$. Peroxynitrite aggravates lipid peroxidation, but $\cdot NO$ reacts rapidly with the peroxyl radicals that propagate this process (Rubbo *et al.*, 1996). If the resulting ROONO species can be metabolized without the release of free radicals, then $\cdot NO$ effectively inhibits lipid peroxidation. Clearly, this $\cdot NO$/ROS ratio is all important. Diminished availability of $\cdot NO$ and increased ROS formation may be key events in the pathology of atherosclerosis (Beckman *et al.*, 1994; Rubbo *et al.,* 1996; Darley-Usmar and Halliwell, 1996).

1.3. Iron and Reactive Iron Species (RIS)

Iron is the fourth most abundant element in the Earth's crust and the second most abundant metal (after aluminum). It is the metallic iron at the Earth's center that accounts for its magnetic field as well as for its overall mass density. This metallic iron was exploited many centuries ago for navigation purposes with the pioneering development of the magnetic compass.

Aerobic life forms had to use iron, but it was only available to them as insoluble ferric complexes because O_2 oxidized Fe^{2+} to Fe(III) oxides. Siderophores were evolved by microorganisms, which permitted them to survive in the environment where there is poor bioavailability of iron; the siderophores allowed the capture and assimilation of iron, which could then be used as a catalyst for oxygen utilization. Aerobes eventually came to dominate the planet, although it appears that earlier primitive anaerobic life forms may also have used iron–sulfur redox chemistry for energy capture because soluble Fe^{2+} was easily available to them in the anaerobic world to donate electrons. The move of life forms to land from the seas and tidal pools occurred when a protective screen of oxygen and ozone was able to filter out most of the damaging solar UVC radiation (wavelengths between 100 and 280 nm) (Chapter 12). Part of the marine environment is still reflected in blood and body fluids with Na^+, K^+, Ca^{2+}, Mg^{2+}, and Cl^- ions predominating. These ions are present in sea water at concentrations some 10^5–10^6 times greater than those of trace metals such as Fe, Cu, Zn, Mn, Mo, Sn, and V. Despite aluminum and iron being the most abundant metals in the Earth's crust, they do not appear as major ions in surface waters, reflecting their poor solubility at neutral pH values. Iron is present in sea water mainly as colloidal particles of hydrated ferric oxide (finely dispersed rust) representing a true solution concentration of iron ions of around 10^{-9} M. Ever increasing industrialization is leading to acidification of surface waters, with a consequential rise in the solubility of both iron and aluminum (Martin, 1994), allowing both to enter biological ecochains with unknown long-term consequences.

1.3.1. Reactive Iron Species in Biological Systems

Within cells there normally exists a pool of low-molecular-mass redox active iron which is essential for the synthesis of iron-requiring enzymes and proteins, including enzymes essential to the replication of DNA (Breuer *et al.*, 1996). This pool of iron is the target of iron chelators and is also a form of iron sensed by iron regulatory proteins. The amount and nature of the ligands attached to this iron remain unknown. However, a recently introduced fluorescence assay based on calcein may enhance our knowledge of intracellular iron pools (Cabantchik *et al.*, 1996). In contrast to the intracellular environment, extracellular compartments do not require, or normally contain, a low-molecular-mass iron pool. Iron-binding proteins such as transferrin and lactoferrin do not even remotely approach iron saturation in healthy subjects; indeed, they retain a considerable iron-binding capacity, and are able to remove mononuclear forms of iron that enter cellular fluids. The differences between intracellular and extracellular compartments, and their requirements for low-molecular-mass iron deserve special comment, as it is iron in this form that is the most likely catalyst of biological free radical reactions (Halliwell and Gutteridge, 1990).

Inside the cell, low-molecular-mass iron need not pose a serious threat as a free radical catalyst, provided that the cell has specific defenses to safely and speedily remove all of the $\cdot O_2^-$, H_2O_2, and organic peroxides (such as lipid peroxides) that could react with such iron. This is achieved by intracellular enzymes such as the superoxide dismutases, catalase, and glutathione peroxidase and possibly also by thioredoxin-dependent H_2O_2 removal systems (Netto *et al.*, 1996). In the extracellular space, however, we see a different pattern of protection against free radical chemistry. Here, proteins bind, conserve, transport, and recycle iron, and while doing so keep it in non- or poorly reactive forms that do not react with H_2O_2 or organic peroxides (for a review of Fenton chemistry see Chapter 2). Proteins such as transferrin and lactoferrin bind mononuclear iron, whereas haptoglobins bind hemoglobin, and hemopexin binds heme. In addition, plasma contains a ferrous ion oxidizing protein (ferroxidase), ceruloplasmin (see Section 3.3.1c). By keeping iron in a poorly reactive state, molecules such as $\cdot O_2^-$, H_2O_2, ·NO, and lipid peroxides can survive long enough to perform important and useful functions as signal, trigger, and intercellular messenger molecules, as we shall elaborate in this chapter.

During situations of iron overload, plasma transferrin can become fully loaded with iron (100% iron saturation) and allow low-molecular-mass iron to accumulate in the plasma. Such iron, when present in micromolar concentrations, can bind to various added chelating agents (such as EDTA, desferrioxamine, and bleomycin) that cannot abstract iron from transferrin. This nontransferrin bound iron can be associated with several ligands including citrate (Grootveld *et al.*, 1989) and other organic acids and possibly albumin. Low-molecular-mass ligands for iron inside the cell are also a subject of considerable debate. ATP (Gurgueira and Meneghini, 1996), ADP, GTP, pyrophosphates, inositol phosphates, amino acids, and polypeptides have all been proposed.

1.3.2. Iron Chelates as Catalysts and Stimulators of Free Radical Reactions

In 1981 the authors introduced the "bleomycin assay" as a first attempt to detect and measure chelatable redox active iron that could participate in free radical reactions (Gutteridge *et al.*, 1981b). The assay procedure is based on the ability of the metal-ion binding glycopeptide antitumor antibiotic bleomycin to degrade DNA in the presence of an iron salt, oxygen, and a suitable iron reducing agent. The ternary bleomycin–iron–oxygen complex binds tightly to DNA, and during the redox cycling of iron, releases base propenals from the DNA molecule. These are unstable and rapidly degrade to release malondialdehyde (MDA), which is derived from the deoxyribose sugar. MDA can be accurately measured by reacting it with thiobarbituric acid. Binding of the ternary complex to DNA makes the reaction site-directed and prevents most biological antioxidants from interfering. Ceruloplasmin is an exception because it appears to be able to catalyze the oxidation of certain ferrous complexes as well as ferrous ions (see Section 3.3.1c). To prevent interference from ceruloplasmin, the concentration of ascorbate added to the reaction is sufficient to inhibit its ferroxidase activity (Gutteridge, 1991a). The bleomycin assay can be directly applied to most biological fluids (see Table II), and if it is positive, it is reasonable to assume that bleomycin has been able to chelate a low-molecular-mass form of iron from the sample and that such iron can be redox cycled. Meticulous removal of adventitious iron (known to be present in all laboratory glassware, reagents, and chemicals) is essential to achieve a sensitivity of detection in the low micromolar range. Blood serum, or plasma, from normal healthy individuals does not contain detectable levels of iron in the bleomycin assay. Indeed, such samples usually give assay values less than the reagent controls, because their iron-binding capacity allows them to remove traces of iron contaminating the reagents used. The latter iron is still present despite cleaning all reagents by treatments with Chelex resin or by dialysis against the iron-binding protein ovotransferrin (Gutteridge, 1987a). The most likely explanation as to why plasma transferrin can remove iron from the reagents but ovotransferrin dialysis cannot, is that iron is present in the reagents in a polynuclear form and not readily chelatable until ascorbate is added to the reaction. Ascorbate releases mononuclear iron, which can then bind to ovotransferrin. Bleomycin, with a binding constant of 10^{15} for ferric ions, is not a strong enough chelator to remove iron correctly loaded onto transferrin, or lactoferrin, or into ferritin or heme proteins. Using the bleomycin assay, low-molecular-mass iron has been detected in a variety of biological fluids as well as in tissue homogenates (see Table II). By not including ascorbate in the reagents, ferrous salts plus the effect of any endogenous reducing agents can be speciated using the bleomycin assay (Gutteridge, 1991b). Other approaches to the measurement of low-molecular-mass iron in biological fluids have been based on alternative ways of measuring chelatable iron. A useful example is the binding of iron by desferrioxamine followed by separation of desferrioxamine and ferrioxamine (Green *et al.*, 1989).

Table II
Bleomycin-Detectable Iron in Biological Samples

	Bleomycin-chelatable iron, μmole/liter (unless otherwise noted)	Comments	Refs.
Blood serum/plasma			
Normal healthy controls (25 in ref. cited, 10^3 in studies combined)	0	Never present. Transferrin has a large iron-binding capacity	*Biochim. Biophys Acta* **869**:119, 1986
Arthritis (osteo-, rheumatoid) (31)	0	Never present unless arthropathy related to iron-overload disease	*Biochim. Biophys. Acta* **869**:119, 1986
Hemochromatosis (7)	4.3 ± 6.7	Much of the iron bound to citrate	*Clin. Sci.* **68**:463, 1985; *J. Biol. Chem.* **264**:4417, 1989
Fulminant hepatic failure (20)	2.2 ± 2.2	2/20 patients did not show BLM-iron	*Free Radical Res.* **20**:139, 1994
Myeloid leukemia			
Before chemotherapy (6)	1.0 ± 1.8	2/6 patients showed BLM-iron	*FEBS Lett.* **241**:202, 1988
After chemotherapy (4)	9.5 ± 8.5	All showed BLM-iron	
Nonlymphocytic leukemia			
Before chemotherapy (9)	0.4 ± 1.0	2/9 patients showed BLM-iron	*FEBS Lett.* **308**:4, 1992
1–6 days after chemotherapy (9)	2.6 ± 2.6	6/9 patients showed BLM-iron	
Adult respiratory distress syndrome, patients with multiorgan failure (5)	0.46 ± 0.22	Iron present in 33% of ARDS patients managed during the period of study	*Thorax* **49**:707, 1994
Meningism related due to treatment with iron-dextran (1)	36	Sample showed an iron-binding capacity	*Lancet* **2**:435, 1983; *Q. J. Med.* **219**:345, 1985
Cerebral and ocular toxicity related to treatment with desferrioxamine and prochlorperazine (1)	0		

(continued)

Table II (*Continued*)

	Bleomycin-chelatable iron, μmole/liter (unless otherwise noted)	Comments	Refs.
Cardiopulmonary bypass extra-corporeal blood circulation (4)	0.27 ± 0.9	Mock bypass. BLM-iron present in 2 samples, BLM-iron present in 27% of patients receiving warm blood cardioplegia	*FEBS Lett.* **328:**103, 1993; *Free Radical Res.* **21:**53, 1994
Hypotransferrinemic mice (8)	16.0 ± 5.0	Both the control and hpx/hpx mice had NTA-detectable iron (1.6 and 1.3 μM) present, suggesting problems with this assay when compared with the bleomycin assay	*Biochim. Biophys. Acta* **1156:**19, 1992
Control mice (5)	0		
Newborn infants (cord) (25)	0.27 ± 0.56	6/25 samples showed BLM-iron	*FEBS Lett.* **303:**201, 1992
Newborn infants 1–5 days (13)	0.33 ± 0.72	3/13 samples showed BLM-iron	*Free Radical Res. Commun.* **16:**285, 1992
Kwashiokor patients (50)	1.0–19.5 (range)	BLM-iron present in 58% of the Kwashiokor patients	*Eur. J. Clin. Nutr.* **49:** 208, 1995
Marasmus patients (6)	0		
Sweat fluids			
Arm fluid (12)	0.72 ± 2.5	BLM-iron present in only one sample	*Clin. Chim. Acta.* **145:**267, 1985
Trunk fluid (10)	4.62 ± 2.9	BLM-iron present in all samples	
Synovial fluids			
Rheumatoid arthritis (21)	3.61 ± 1.9	Approximately 40% of synovial fluids from patients show BLM-iron when incubated at pH 5.3	*Biochem. J.* **245:**415, 1987; *Clin. Sci.* **66:**691–695, 1985
Pleural exudate			
Rat (inflammatory) (2)	0.85[a]	Collection conditions unknown	Unpublished
Colostrum			
Bovine (4)	4.1 ± 8.2[a]	BLM-iron present in only one sample. Collection conditions unknown	Unpublished
Eye vitreous humor			
Rabbit (3 pooled samples)	4.8[a]	Collection conditions unknown	Unpublished

Table II (*Continued*)

	Bleomycin-chelatable iron, μmole/liter (unless otherwise noted)	Comments	Refs.
Cerebrospinal fluids			
Normal human (12)	1.8 ± 1.3	Unbuffered assay reagents	*Lancet* **2**:459, 1982
Normal human (15)	0.55 ± 0.27	pH 7.0	*Clin. Sci.* **82**:315, 1992
Neuronal ceroid lipofuscinoses (15)	5.37 ± 2.0	Unbuffered assay reagents	*Lancet* **2**:459, 1982
Neuronal ceroid lipofuscinoses (7)	1.6 ± 0.8	pH 5.3	*Am. J. Med. Genet.* **5**:193, 1988
Encephalopathy (9)	1.0 ± 0.4	pH 5.3	*Am. J. Med. Genet.* **5**:193, 1988
Multiple sclerosis (6)	1.75 ± 0.8	Unbuffered assay reagents	*Lancet* **2**:459, 1982
Epilepsy (5)	2.48 ± 0.8	Unbuffered assay reagents	*Lancet* **2**:459, 1982
Lung lining fluids (recovered as bronchoalveolar lavage)			
Normal human controls (6)	0.132 ± 0.02	BLM-iron present in all samples	*Biochem. Biophys. Res. Commun.* **220**:1024, 1996
Adult respiratory distress syndrome (survivors) (13)	0.30 ± 0.09	BLM-iron present in 12 of 13 samples	As above
Adult respiratory distress syndrome (nonsurvivors) (11)	0	No BLM-iron detected but high level of transferrin present	As above
Urine			
Renal reperfusion injury			
Control (8)	138 ± 54 pmole/30 min		*Kidney Int.* **34**:474, 1988
15–45 min reperfusion (8)	1590 ± 776 pmole/30 min		
45–75 min reperfusion (6)	1249 ± 506 pmole/30 min		
Cellular Ultrafiltrates			
10,000-Da exclusion			
Brain tissue, gerbil (8)	20.5 ± 3.5 nmole/g wet wt		*Free Radical Res. Commun.* **11**:317, 1991
Red blood cells, human (1)	0.65 nmole/liter		Unpublished
Red blood cells, human + parasite material (*P. falciparum*) (1)	1.29 nmole/liter		Unpublished
Red blood cells, rat (10)	0		Unpublished
Red blood cells, rat + *P. bergheii* (8)	0.91 ± 0.39 nmole/liter		

(*continued*)

Table II (*Continued*)

	Bleomycin-chelatable iron, μmole/liter (unless otherwise noted)	Comments	Refs.
30,000-Da exclusion			
Kidney tissue control (15)	18 nmole/g wet wt		*Biochem. J.* **291:**901, 1993
Kidney tissue after ischemia–reperfusion (15)	57 nmole/g wet wt		
Cellular homogenates			
Bacteria (5 different strains)	0.003–1.046 μmole/mg protein		*Biochem. Int.* **8:**89, 1984
Homogenized tissue, human			
Cerebellum	21 ± 10 nmole/mg tissue	Levels of Cu released on tissue homogenization broadly comparable to those of BDI (*FEBS Lett.* **353:**246, 1994; *Free Radical Res.* **23:**465, 1995)	
Substantia nigra	53 ± 18 nmole/mg tissue		
Arterial wall	4.9 ± 2.8		
Atherosclerotic material		Copper ions are powerful catalysts of low-density lipoprotein (LDL) oxidation, which may contribute to atherosclerosis (*Biochem. J.* **286:**901, 1992; *FEBS Lett.* **368:**513, 1995). Lesion extracts will stimulate *in vitro* LDL oxidation	
"Gruel from advanced lesions"	~ 4.0		
Endarterectomy samples (13)	0.7 ± 0.3 nmole/mg protein		
Aneurysms (10)	0.3 ± 0.2 nmole/mg protein		
Hepatic cells (female mice)			
C3H/HeJ (fatty streak-resistant)			
Control	120 ± 2 nmole/mg	In response to an atherogenic diet, fatty streak-resistant mice exhibited higher levels of liver apoferritin and low levels of BLM-iron than did fatty streak-susceptible mice, implicating IRE-BP signaling through iron (*J. Clin. Invest.* **95:**2104, 1995)	
Atherogenic diet	78 ± 9 nmole/mg		
C57BL/6J (fatty streak-susceptible)			
Control	93 ± 9 nmole/mg		
Atherogenic diet	86 ± 9 nmole/mg		

[a]Unpublished data samples provided by Drs. Armstrong, Griffiths, and Peto. Mean values are shown (± SEM or SD as derived in appropriate references).

Bleomycin-Detectable Ferrous Ions in Biological Samples

	Bleomycin iron, μmole/liter	Comments	Ref.
Blood plasma			
Preterm and term cord plasma (13)	3.6 ± 0.8	No ascorbate added. BLM-iron-positive samples selected	*Free Radical Res.* **22:**555, 1995
Cerebrospinal fluids			
Normal fluids (16)	0.55 ± 0.28	No ascorbate added	*Clin. Sci.* **82:**315, 1992

1.4. Copper and Reactive Copper Species (RCS)

Copper is an essential trace element and distinct metabolic changes occur when dietary intake or intestinal absorption of copper is inadequate. The total copper content of an average human is about 80 mg. Absorption of copper ions takes place in the stomach and small intestine, and subsequent transport to the liver is thought to be facilitated by complexing with albumin, which in humans has one high-affinity copper-binding site. In the liver, copper is incorporated into apoceruloplasmin to form the active protein ceruloplasmin, which is released into the plasma.

Copper ions readily attach to amino groups of proteins and to amino acids where they may still react with H_2O_2 or organic peroxides to form ·OH, RO·, and RO_2· radicals. The reactive oxidant formed appears to attack the ligand to which the copper is bound and is not released into free solution. Thus, the binding of copper ions to albumin may be a biologically significant protective mechanism (Gutteridge and Wilkins, 1983; Gutteridge, 1986b; Halliwell, 1988). An assay that measures chelatable copper in biological fluids has been developed by one of the authors (Gutteridge, 1984). In this assay the chelating agent 1,10-phenanthroline degrades DNA in the presence of copper ions, oxygen, and a suitable reducing agent. Degradation of DNA results in the release of products that on heating with thiobarbituric acid at acidic pH form a pink chromogen. The assay detects copper bound to the high-affinity binding site of albumin and to histidine but not to ceruloplasmin. Because most of the plasma copper is associated with ceruloplasmin, release of copper ions from this protein during periods of oxidative stress is most likely to account for the presence of chelatable copper in plasma. Peroxynitrite appears to be particularly effective at releasing copper from ceruloplasmin (Swain *et al.*, 1994) as does superoxide generated in a cell line producing ceruloplasmin (Mukhopadhyay *et al.*, 1996). Some of the biological fluids, and conditions under which chelatable copper has been detected are summarized in Table III.

Table III
Phenanthroline-Detectable Copper in Some Biological Materials

Material	Level (μM)	Comments
Plasma/serum, healthy humans	0	Never detected in fresh samples. Amount of nonceruloplasmin copper must be <0.1 μM. Detected in stored samples, probably related to breakdown of ceruloplasmin on storage.
Plasma, untreated Wilson's disease	2.7	A "copper-overload" disease (*Pediatr. Res.* **37**:219, 1995). Variably present in patients treated with copper chelating agents.
Plasma, patients with liver failure	1–2	May be related to metal release from liver and/or failure of liver to synthesize ceruloplasmin (*Free Radical Res.* **20**:139, 1994).
Sweat, human		
Arm	1.8 ± 3.9	Sweat also contains iron (Table II). It may be an excretory route for both metals. Presence of these metals facilitates bacterial growth (*Clin. Chim. Acta* **145**:287, 1985).
Trunk	27 ± 3	

2. ANTIOXIDANT DEFENSES: ESSENTIAL BUT INCOMPLETE

2.1. Biological Antioxidants

The term *antioxidant* is frequently used in the biomedical literature, but rarely defined, often implying that it refers to chemicals with chain-breaking properties such as vitamin E (α-tocopherol) and vitamin C (ascorbic acid). The authors (Halliwell and Gutteridge, 1998) take a much wider view and define an antioxidant as "any substance that when present at low concentrations, compared to those of an oxidizable substrate, significantly delays, or inhibits, oxidation of that substrate."

Antioxidants can act at many different stages in an oxidative sequence such as:

- Removing oxygen or decreasing local O_2 concentrations
- Removing catalytic metal ions
- Removing key ROS/RNS such as $\cdot O_2^-$, H_2O_2, HOCl, singlet O_2, or $ONOO^-$
- Scavenging initiating radicals such as ·OH, RO·, $RO_2\cdot$
- Breaking the chain of an initiated sequence

Many antioxidants have more than one mechanism of action. For example, propyl gallate, a partially water-soluble phenolic antioxidant used by the food industry, is a chain-breaking antioxidant, a powerful scavenger of ·OH radicals, and an iron-binding agent. Cells have formidable defenses against oxidative damage, many of which may at first sight not seem to be antioxidants. Antioxidant protection can operate at different levels within cells, such as

- Preventing radical formation
- Intercepting formed radicals
- Repairing oxidative damage
- Increasing elimination of damaged molecules
- Promoting the death of cells with excessively damaged DNA so as to prevent transformed cells from arising

The different antioxidant strategies used within cells, inside membranes, and in extracellular fluids prompted us to propose in 1986 that such marked differences were important for humoral signaling (Halliwell and Gutteridge, 1986).

2.2. Antioxidant Defenses

The deleterious effects of ROS, RNS, RIS, and RCS are controlled by antioxidant defenses. Some antioxidants are synthesized in the human body, and include enzymes, certain other proteins, and low-molecular-mass scavengers. Examples are superoxide dismutases, catalases, peroxidases, thiol-specific antioxidants, metallothioneins, other metal ion-binding and storage proteins, urate, GSH, and ubiquinol (Fridovich, 1989; Sies, 1991; Chubatsu and Meneghini, 1993; Frei, 1994; Gutteridge and Halliwell, 1994; Ernster and Dallner, 1995; Netto *et al.*, 1996; Chapter 17). These defenses protect aerobes against most of the toxic effects of ambient O_2 (21%), but all aerobes (including humans) are injured if exposed to excess O_2 (Balentine, 1982).

It follows that, in general, aerobes have not evolved an excess of antioxidant defenses, although often defenses are inducible by elevated O_2 if sufficient time for adaptation is allowed (Balentine, 1982). Indeed, antioxidants do not even prevent damage by RNS/ROS at ambient O_2 (Table IV). Animals thus rely on a second line of defense in the form of repair systems, of which the most important may be those that remove mutagenic lesions in DNA induced by ROS/RNS (Demple and Harrison, 1994). Superimposed on such defenses are inducible proteins such as heme oxygenase-1. Heme oxygenases remove the prooxidant heme and in the process produce the antioxidant bilirubin (Stocker, 1990). However, it should be noted that another prooxidant, RIS, is formed which is potentially more redox-active than the heme iron that is removed. The constant assault by ROS/RNS on DNA throughout the long human life span may contribute to the age-related development of cancer (Demple and Harrison, 1994). Indeed, aging itself may involve the cumulative effects of oxidative damage over a life span (Ames *et al.*, 1993; Orr and Sohal, 1994), through the constant triggering of oxidative reactions that evolved to ensure survival in the short term (e.g., by killing foreign invaders).

Additional protection is provided by dietary antioxidants. The physiological role of some of these is well established (e.g., vitamin E, ascorbate) whereas the role of others (e.g., flavonoids, carotenoids) is currently uncertain. Dietary antioxidants appear to be important in delaying/preventing certain human diseases, especially cardiovascular dis-

Table IV
Damage by RNS and ROS Occurs in the Human Body

Target of damage	Evidence
DNA	Low levels of oxidative base damage products are present in DNA isolated from human cells; levels often increase in subjects with chronic inflammatory diseases or subjected to oxidative stress, e.g., smoking. Some base damage products are excreted in urine, presumably resulting from DNA repair processes. Smokers and rheumatoid arthritis patients excrete more 8-hydroxydeoxyguanosine.
Protein	Attack of ROS on proteins produces carbonyls and other amino acid modifications (e.g., methionine sulfoxide, 2-oxo-histidine, protein peroxides, hydroxylation of tyrosine to dopa, formylkynurenine). Low levels of carbonyls and certain other products (e.g., *ortho*-tyrosine) have been detected in healthy human tissues and body fluids. Nitrotyrosines, products of tyrosine attack by RNS, have been detected in healthy human body fluids (nanomolar levels) and at greater levels in atherosclerotic lesions, human urine, and body fluids/tissues from patients with chronic inflammatory diseases. Bityrosine and chlorotyrosine have also sometimes been detected.
Lipid	Accumulation of "age pigments" in tissues. Lipid peroxidation in atherosclerotic lesions. Presence of specific end products of peroxidation (e.g., isoprostanes) in body fluids from healthy humans; levels increase during oxidative stress, e.g., smoking.
Uric acid	Attacked by ROS/RNS to generate allantoin, cyanuric acid, parabanic acid, oxonic acid, and other products, which are present in human body fluids. Levels increase in chronic inflammatory/metal overload diseases.

[a]Adapted from Halliwell (1996b).

ease and some types of cancer (for reviews see Block *et al.*, 1992; Gutteridge and Halliwell, 1994; Gey, 1995).

2.3. Oxidative Stress

Some oxidative damage occurs constantly in the human body (Table IV). If damage increases because of antioxidant depletion and/or production of more ROS, RNS, RIS, and RCS, oxidative stress is said to occur (Sies, 1991). Mild oxidative stress often results in upregulation of the synthesis of antioxidant defenses to restore the balance, e.g., if rats are gradually acclimatized to elevated O_2, they can tolerate pure O_2 for much longer, apparently because of increased synthesis of antioxidant defense enzymes and of GSH in the lung (Iqbal *et al.*, 1989). Exposure of *E. coli* to excess $\cdot O_2^-$ or H_2O_2 activates a complex adaptive response system (Storz and Toledano, 1994; Chapter 5). H_2O_2 (Schreck *et al.*, 1992) and other peroxides (e.g., those in atherosclerotic lesions; Lusis and Navab, 1993) activate NF-κB by causing displacement of an inhibitory subunit.

However, severe oxidative stress can produce major interdependent derangements of cell metabolism, including DNA strand breakage (often an early event), increases in intracellular "free" Ca^{2+}, decompartmentalization of "catalytic" iron and copper ions (i.e., increases in RIS/RCS), damage to membrane ion transporters and/or other specific proteins, and lipid peroxidation. Cell injury and death may result, the latter by apoptosis or necrosis (Halliwell, 1987; Orrenius *et al.*, 1989; Cochrane, 1993; Sarafian and Bredesen, 1994). Oxidative stress can lead to damage to all types of biomolecule. By reversibly oxidizing specific "detector" proteins (e.g., oxyR) in *E. coli*, ROS, RIS, and RCS (and possibly RNS) can alter gene expression. In excess, ROS, RIS, RNS, and RCS can damage DNA and proteins. Damage to proteins affects the function of receptors, enzymes, and transport proteins, e.g., the inability to phosphorylate tyrosine residues in proteins after nitration by RNS (Kong *et al.*, 1996). Repair enzymes may also be damaged, and polymerases show decreased fidelity of nucleic acid replication (reviewed by Wiseman and Halliwell, 1996). Lipid peroxidation end products (e.g., such cytotoxic aldehydes as MDA and 4-hydroxynonenal) cause damage to proteins and to DNA, e.g., by forming DNA base adducts (El Ghissassi *et al.*, 1995). Hypochlorous acid in the presence of Fe^{2+} ions can generate ·OH radicals (Candeias *et al.*, 1994; Folkes *et al.*, 1995), and as a powerful oxidizing agent can also directly damage lipids, proteins, and DNA, leaving a chlorinated product on proteins as a marker of its reactivity (Kettle, 1996; Eiserich *et al.*, 1996). Hydroxyl radical attacks all four DNA bases (reviewed by Wiseman and Halliwell, 1996); products derived from ·NO (such as HNO_2 and $ONOO^-$) but not ·NO itself, cause nitrosation and deamination, e.g., guanine to xanthine and 8-nitroguanine, adenine to hypoxanthine (de Rojas-Walker *et al.*, 1995; Spencer *et al.*, 1996). The chemical pattern of DNA damage can be used to help identify the damaging species, e.g., DNA fragmentation in cell cultures by H_2O_2 probably involves ·OH formation in the nucleus, whereas damage by cigarette smoke and $ONOO^-$ seems to be mainly related to RNS (Spencer *et al.*, 1995).

3. WHY HAVE WE NOT EVOLVED BETTER ANTIOXIDANT DEFENSES? IS THERE A NORMAL PHYSIOLOGICAL ROLE FOR OXIDATIVE STRESS?

If ROS, RIS, RCS, and RNS were fearsomely cytotoxic species, one would expect evolutionary pressure to evolve an excess of antioxidant defenses. Instead, our antioxidants seem to control levels of ROS, RIS, RCS, and RNS rather than eliminate them completely (Table IV). Of course, maintaining excess antioxidant defenses would have an energy cost—it could be "cheaper" to repair (DNA) or destroy (proteins, lipids) damaged biomolecules (Halliwell, 1996b). The ·OH that arises from exposure to "background" radiation will probably inevitably cause damage. The diseases that appear to result from continuing biological insult by residual ROS/RNS occur after the reproductive period, and thus might not be selected against. The importance of diet-derived antioxidants in maintaining health is increasingly appreciated (Block *et al.*, 1992; Gutteridge and Halliwell, 1994; Gey, 1995).

Another possibility is that ROS and RNS are physiologically useful in controlled amounts (just as RIS and RCS are known to be). Phagocyte killing is a clear example (Babior, 1978; Chapter 19), but it may be one of many (Halliwell, 1996b). Lymphocytes can release $\cdot O_2^-$, and this might influence their function (Maly, 1990; Morikawa and Morikawa, 1996). Fibroblasts and vascular endothelial cells release $\cdot O_2^-$ *in vitro*, but whether they do so *in vivo* is as yet uncertain. Cells in culture respond to ROS (Burdon, 1994); low levels of ROS have been repeatedly shown to stimulate cell proliferation, whereas higher levels exert (cytotoxic) inhibitory actions, i.e., the "redox balance" seems to be critical in affecting cell behavior through signaling (Figure 1). ROS affect cell–cell communication, an action perhaps important in carcinogenesis (Ruch and Klaunig, 1988; Hu *et al.*, 1995).

Hence, it has repeatedly been suggested that certain ROS (and perhaps by extension RCS, RIS, and RNS) might be used as "signal, messenger, and trigger molecules" (Halliwell and Gutteridge, 1986; Saran and Bors, 1989; Schreck *et al.*, 1992; Barja, 1993; Sarafian and Bredesen, 1994; Burdon, 1994; Krieger-Brauer and Kather, 1995), e.g., mediating the effects of PDGF on smooth muscle cells (Sundaresan *et al.*, 1995), activating adenylate cyclase (Tan *et al.*, 1995), and iron regulatory proteins (Rouault and Klausner, 1996). We are seeing increasing examples of redox regulation of gene expression; not only oxyR and NF-κB but also the role of thioredoxin and of AP-1, which is

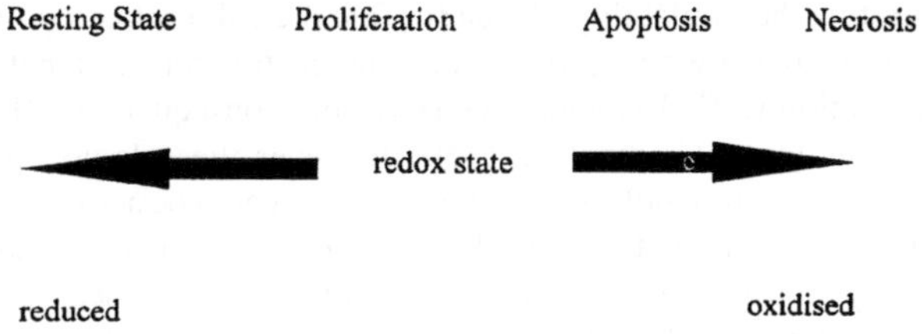

FIGURE 1. Cellular activity.

composed of the *jun* and *fos* gene products, whose transcription is enhanced by ROS (Schreck *et al.*, 1992; Staal *et al.*, 1994; Schulze-Osthoff *et al.*, 1995).

3.1. Antioxidants and Intracellular Signaling

Oxygen metabolism occurs within cells, and it is here that we expect to find antioxidants evolved to deal speedily and specifically with ROS/RNS. Enzymes such as the superoxide dismutases rapidly promote the dismutation of superoxide into hydrogen peroxide and oxygen at a rate considerably faster than it occurs uncatalyzed (Fridovich, 1989). Hydrogen peroxide, a product of the dismutation reaction, can be destroyed by at least two enzyme systems, namely, catalase and the glutathione peroxidases. During normal aerobic metabolism, these enzymes function in concert to eliminate toxic reduction intermediates of oxygen inside the cell, thereby allowing a small pool of RIS (the low-molecular-mass intracellular iron pool) to safely exist to provide iron for the manufacture of iron-containing proteins.

Intracellular Proteins Controlling RIS

Ferritin and Hemosiderin. Iron is stored in cells within two major proteins: ferritin and hemosiderin. Ferritin is a soluble protein located in the cytosol, whereas hemosiderin is insoluble and found mainly in lysosomes. There is probably some ferritin in every human cell but the largest amounts are present in the liver parenchyma. Ferritin consists of a hollow protein shell, 2.5 nm thick and enclosing a cavity of 8 nm wide. The shell is composed of 24 subunits each of which is about 20 kDa giving a protein of ~ 480 kDa, which can contain up to 4500 ions of iron per molecule of protein. Iron enters the protein as Fe^{2+}, but is stored as Fe(III) in the central core in structures similar to the mineral ferrihydrite. For iron release to occur, a reductive conversion back to Fe^{2+} is required. Ferritin is a relatively safe storage form of iron in the body, as little of its iron can be mobilized by $\cdot O_2^-$, H_2O_2, and lipid hydroperoxides (LOOH) to participate in radical chemistry (Bolann and Ulvik, 1987). The small amount of iron released under stress by ROS does not appear to be that loaded into the central core (Bolann and Ulvik, 1990). Findings that show linear relationships between the iron loading of ferritin and its ability to drive ·OH formation and lipid peroxidation may result from the pool of iron at the surface of the core, or the use of degraded commercial preparations.

Hemosiderin is considerably less inclined than ferritin to release iron (Gutteridge and Hou, 1986) that can stimulate free radical reactions essentially because it is insoluble. Thus, conversion of ferritin to hemosiderin during conditions of iron overload may be protective, by limiting the availability of iron for free radical reactions (O'Connell *et al.*, 1986). When ferrous ions are being loaded into the central core of ferritin, it has been proposed that oxidation to Fe(III) generates H_2O_2 and consequently ·OH radicals as a by-product of core building (Grady *et al.*, 1989); however, these do not appear to escape from the interior of the ferritin molecule. Hence, ferritin would be acting as a preventative antioxidant during iron loading of its central core, by not allowing ·OH radicals to escape into solution, and by limiting the availability of iron that can be released by $\cdot O_2^-$-generating systems. Recently, copper ions have been found to be associated with ferritin *in vitro,* and these are mainly responsible for the weak ability of ferritin to cause the catalytic decay of $\cdot O_2^-$ in solution (Bolann and Ulvik, 1993). If copper is found to be associated with native ferritin *in vivo*, a novel antioxidant function may be suggested. Further support for an antioxidant role for ferritin comes from studies in which oxidative damage to the endothelium, mediated by heme, caused induction of mRNAs for both heme oxygenase and ferritin (Balla *et al.*, 1992), of which the latter appeared to be the major cytoprotectant.

Intracellular Iron Signaling. Cells normally accumulate iron via the binding of transferrin to high-affinity surface receptors (TfR) followed by endocytosis. There is also a transferrin-independent pathway of cellular iron uptake that is said to involve a ferrireductase and an Fe^{2+} transmembrane transport system (reviewed in DeSilva *et al.*, 1996). The reductase is proposed to provide iron in a soluble form to the membrane transporter. When nontransferrin bound iron appears in plasma, related to iron overload or lack of transferrin (apotransferrinemia), it is rapidly cleared by the membrane-bound transport system constitutively present on parenchymal cells of organs, particularly those of liver, heart, pancreas, and the adrenals. This latter system does not require endocytosis of a protein for iron delivery. It may function to absorb RIS generated by damage to other tissues. For example, the low-molecular-mass iron and copper seen in plasma of fulminant hepatic failure patients was quickly lost when they were given a new liver (Evans *et al.*, 1994). As already mentioned, the intracellular environment can cope better with low-molecular-mass iron than can the extracellular compartments.

The rate of synthesis of TfR and ferritin is regulated at the posttranscriptional level by cellular iron, and coordinated by the iron-dependent binding of cytosolic proteins called the *iron responsive element binding proteins* (IRE-BP) [now known as *iron regulatory proteins* (IRP)] which bind to specific sequences on their mRNAs (Klausner *et al.*, 1993). It appears that low-molecular-mass iron is capable of acting as a signal to regulate ferritin and TfR synthesis in this way. In the absence of iron, IRE-BP binds to the iron-responsive element (IRE) (the 3′-untranslated region of the TfR message contains a set of stem-loop structures termed IRE) stabilizing the transcript. When iron is present, the protein dissociates from the IRE and degradation of the mRNA occurs. Recent work has shown that one of the IRE-BPs is identical to the cytosolic enzyme aconitase (reviewed in Kennedy *et al.*, 1992; Haile *et al.*, 1992a,b). The protein functions as an active aconitase when it has an Fe-S cluster present or as an RNA-binding protein when iron is absent. Switching between these two forms depends on cellular iron status such that when iron is replete it is an active aconitase, whereas when deprived of iron it has only RNA-binding activity. The ability of low-molecular-mass iron to activate iron-requiring acontitase has recently been used to detect and measure RIS in plasma (Gutteridge *et al.*, 1996). It has been suggested that superoxide, nitric oxide, and ascorbate can inactivate aconitase switching it to an RNA-binding protein with upregulation of TfR, although challenges to this interpretation have been made (Drapier *et al.*, 1994; Hentze and Kuhn, 1996). Intracellular low-molecular-mass iron may also be regulated by the oxidative stress protein heme oxygenase, which can lead to increases in intracellular levels of ferritin (Vile and Tyrrell, 1993).

3.2. Antioxidants and Membrane Signaling

Within the hydrophobic interior of membranes, lipophilic radicals are formed that are usually different from those seen in the intracellular aqueous space. Lipophilic radicals require hydrophobic antioxidants for their removal. α-Tocopherol, a fat-soluble vitamin, is a poor antioxidant outside a membrane but is extremely effective when incorporated into the membrane bilayer (Gutteridge, 1977). Membrane stability and protection against oxidative insult depend very much on the way in which the membrane is assembled from its lipid components. Structural organization requires the "correct" ratios of phospholipids and that their fatty acids be esterified (reviewed in Gutteridge and Halliwell, 1988).

When a cell is damaged, or dies, it is highly likely that its lipids will undergo peroxidation (Halliwell and Gutteridge, 1984). Tissue damage releases RIS/RCS and activates enzymes that catalyze peroxidation of polyunsaturated fatty acids, leading to a buildup of lipid peroxides (Herold and Spiteller, 1996). Peroxidation of membrane polyunsaturated fatty acids produces a plethora of reactive primary peroxides and secondary carbonyls and it was suggested many years ago by one of the authors that lipid oxidation products such as these, resulting from cell death, could act as triggers for new cell growth (Gutteridge and Stocks, 1976). Through the detailed work of Esterbauer *et al.*, (1988), we now have clearer insights into the biological reactivity of lipid oxidation products. 4-Hydroxy-2-nonenal (HNE), a peroxidation product of (*n*-6) fatty acids (when RIS or RCS are present), is a potent trigger for chemotaxis, can inactivate thiol-containing molecules, and activate certain enzymes (reviewed in Esterbauer *et al.*, 1991). As a general rule, low levels of ROS, and possibly reactive carbonyls, activate cellular processes, whereas higher levels turn them off. The resting cell is normally in a reduced state and is progressively activated as oxidation increases up to a maximum. Too much oxidation depresses cell function (Burdon, 1994; McConkey *et al.*, 1996) until eventually apoptosis or necrosis is triggered (see Figure 1).

3.3. Antioxidants and Extracellular Signaling: Basic Principles

Human extracellular fluids contain little or no catalase activity, and extremely low levels of superoxide dismutase. Glutathione peroxidases, in both selenium-containing and non-selenium-containing forms, are present in plasma but there is little glutathione in plasma (1–2 μM) to satisfy an enzyme with a K_m for GSH in the millimolar range. "Extracellular" superoxide dismutases (EC-SOD) have been identified (Marklund *et al.*, 1982) and shown to contain copper, zinc, and attached carbohydrate groups. By allowing the limited survival of $\cdot O_2^-$, H_2O_2, lipid peroxides (LOOH), and possibly other ROS/RNS in extracellular fluids, the body can utilize these molecules, and others such as nitric oxide (·NO), as useful messenger, signal, or trigger molecules (Halliwell and Gutteridge, 1986). A key feature of such a proposal is that $\cdot O_2^-$, H_2O_2, LOOH, LOOH, and HOCl do not encounter reactive iron or copper, and that extracellular antioxidant protection has evolved to keep iron and copper in poorly or nonreactive forms (Halliwell and Gutteridge 1986; Gutteridge, 1995).

The major copper-containing protein of human plasma is ceruloplasmin, unique for its intense blue coloration. This protein's ferroxidase activity makes a major contribution to extracellular antioxidant protection by decreasing ferrous ion-driven lipid peroxidation and Fenton chemistry (Gutteridge *et al.*, 1980; Gutteridge and Stocks, 1981; see Section 3.3.1c).

3.3.1. Extracellular Proteins Controlling RIS

3.3.1a. Transferrin and Lactoferrin. The transferrin proteins, which include human plasma transferrin (M_r 79,000), egg white ovotransferrin, sometimes called *conalbumin* (M_r 77,700), lactoferrin from body secretions and milk (M_r 82,400), melanotransferrin (M_r 97,000) from melanoma cells, and uteroferrin (M_r 35,000), are unique mononuclear ferric ion-binding proteins. Human apotransferrin (Tf) has two metal-binding lobes, the C- and N-termini, which when fully loaded with iron give a ratio of 2 moles of iron per mole of protein (Fe_2 Tf). Two monoferric forms of transferrin can, therefore, exist with one ferric ion bound to the N-terminus (Fe_N Tf) or to the C-terminus (Tf Fe_C). For iron to bind to transferrin, an anion, usually bicarbonate (HCO_3^-), is required (Kojiman and Bates, 1981). The bicarbonate coordinates with the iron forming a bridge between the metal and a cationic group on the protein. Transferrin, ovotransferrin, and lactoferrin have been ascribed antimicrobial properties through their ability to deprive bacteria of iron essential for growth (reviewed in Ward *et al.*, 1996). Melanotransferrin, on the other hand, may function *in vivo* to actively acquire iron for rapid cell (tumor) growth (Rose *et al.*, 1986).

Under normal physiological conditions, plasma transferrin has a binding constant for ferric ions of around 10^{22}, with the C-lobe having a higher affinity for iron than the N-lobe. In normal healthy individuals, transferrin is only up to one-third loaded with iron and retains a considerable iron-binding capacity. This can be calculated in plasma by measuring the total nonheme iron content both before and after saturating the protein with an iron salt. At the lower iron saturation (i.e., around 25%), it is the weaker more acid-labile "N" site of transferrin that is predominantly occupied by iron. The reduction potential of transferrin-bound iron has been reported to be around –500 mV.

Lactoferrin is considerably more stable under acid conditions than is transferrin, and is thought to hold onto its iron down to pH values as low as 4.0. Iron can be released from transferrin by lowering the pH value; and such iron release is greatly amplified if an iron-reducing molecule and an iron chelator are also present. Transferrin is mainly synthesized in the liver and secreted into the circulation as the apoprotein, although other tissues can also synthesize transferrin at much lower levels. Transferrin binds ferric ions in the circulation, keeping levels of mononuclear iron in the plasma at effectively zero. Iron-loaded transferrin enters cells by binding to high-affinity transferrin receptors (TfR) on the surface of cells. The expression of TfRs is determined not only by cellular iron requirements but also by cell growth and differentiation. Changes in TfR gene expression are posttranscriptionally regulated, in response to variations in intracellular iron levels, through the IRP (Section 3.1.1b).

Inside the cell, iron is released from transferrin by acidification, through a proton-pumping ATPase, within endosomes (reviewed in Crichton and Charloteaux-Wauters, 1987). The released iron is used for storage, and for the synthesis of DNA and of iron-containing proteins. The remaining apotransferrin is returned to the extracellular environment to continue the cycle of iron binding and cell delivery.

3.3.1b. Transferrin as a Free Radical Catalyst. Claims that lactoferrin and transferrin were efficient Fenton catalysts producing ·OH radicals from H_2O_2 have not been substantiated. Careful studies by the authors and others showed that these claims were unlikely for both lactoferrin and transferrin (Gutteridge *et al.*, 1981a; Aruoma and

Halliwell, 1987) at neutral pH values. Addition of a ferrous salt to apotransferrin or apolactoferrin at an acid pH value (4.5–5.0), however, does appear to generate ·OH radicals (Klebanoff *et al.*, 1989). But, as proposed for hemoglobin (Gutteridge, 1986a), site-specifically generated ·OH radicals are not likely to escape into free solution from the iron–transferrin complex. In many of the early studies (showing generation of ·OH from H_2O_2 by iron-loaded transferrin and iron-loaded lactoferrin), it seems highly likely that ferric ions were not correctly loaded onto the high-affinity-binding sites and became detached during the assay procedures. Transferrin reacts poorly with nonchelated ferric ions at neutral pH values, leading to ineffective iron loading of the protein (Aisen *et al.*, 1978). Another problem noted in such experiments was the presence of contaminating iron chelators used commercially to prepare the apoprotein, e.g., EDTA. Iron–EDTA chelates are highly redox-active.

Proteolytic degradation of transferrin leads to fragments retaining an iron-binding capacity (Williams, 1974). Proteolysis is more rapid with aged or low iron-loaded transferrin, whereas iron saturation confers resistance (Williams, 1974). Elastase from the bacterium *Pseudomonas aeruginosa*, an infecting organism of human tissue, however, appears to be able to degrade diferric-transferrin and diferric-lactoferrin to produce iron-containing fragments that stimulate ·OH formation from H_2O_2, perhaps because they release iron more easily (Britigan and Edeker, 1991). Neutrophil myeloperoxidase, in the presence of H_2O_2 and halide ions, decreases the iron-binding capacity of transferrin and lactoferrin, although the latter was more resistant (Winterbourn and Molloy, 1988). Loss of iron-binding capacity at sites of inflammation would be a serious loss of antioxidant protection against iron-driven free radical reactions. However, the secretion of lactoferrin by activated phagocytes may be an attempt to maintain this ability (Gutteridge *et al.*, 1981a).

3.3.1c. Haptoglobins and Hemopexin. Haptoglobins are glycoproteins, M_r about 86,000, found in the α_1-globulin fraction of human serum. They form stable complexes with hemoglobin, but not with heme, both *in vivo* and *in vitro*. The resulting bond is one of the strongest noncovalent protein interactions known, with a stoichiometry of 1 mole of haptoglobin to 1 mole of hemoglobin. The amount of haptoglobin present in the total plasma volume has been calculated to be sufficient to bind and conserve 3 g of hemoglobin, ensuring that no prooxidant hemoglobin is normally present in the plasma. Hemoglobin can stimulate the process of lipid peroxidation by at least two mechanisms: (1) the heme ring reacts with peroxides to form active oxo-iron species and amino acid radicals (McArthur and Davis, 1993) and (2) when a large excess of peroxide is present it causes fragmentation of the pyrrole rings with the release of RIS (Gutteridge, 1986a). Binding of hemoglobin to haptoglobin greatly decreases its prooxidant properties (Gutteridge, 1987b). The concentration of haptoglobins in normal plasma ranges from 0.5 to 2.0 g/liter, and one study has suggested that hypohaptoglobinemia may be associated with familial epilepsy (Panter *et al.*, 1985).

Hemopexin is a plasma β-glycoprotein with a M_r of around 60,000, which tightly binds heme but not hemoglobin. Heme is transported to liver parenchymal cells by a receptor-mediated process involving endocytosis of hemopexin (Smith and Morgan, 1979). Like the iron transport protein transferrin, hemopexin is not degraded when delivering heme to cells, returning to the circulation as an intact protein (Smith and Morgan, 1979). The normal plasma concentration of hemopexin ranges from 0.15 to 1.3

g/liter. Albumin can also tightly bind heme to form a complex known as *methemalbumin*. The prooxidant properties of heme are greatly diminished when it is bound to hemopexin, thereby conserving iron as well as decreasing its reactivity (Gutteridge and Smith, 1988).

3.3.1d. Ferrous Oxidases. Enzymes that catalyze the oxidation of ferrous ions, or ferrous complexes, to their ferric state are called *ferrous oxidases* or *ferroxidases*. Iron chelators can have ferroxidaselike activities when they displace the equilibrium between ferrous and ferric ions in solution. By binding ferric ions they pull the reaction strongly to the right, i.e.,

$$Fe^{2+} \rightarrow Fe(III) \rightarrow \text{Ferric chelate}$$

Many iron chelators show a ferroxidaselike activity, examples being transferrin, bleomycin, EDTA, and desferrioxamine (Goodwin and Whitten, 1965; Harris and Aisen, 1973; Caspary *et al.*, 1979).

Ceruloplasmin, the Major Biological Ferroxidase. Ceruloplasmin is the major copper-containing protein of extracellular fluids. It has a relative molecular mass of approximately 132 kDa with six or seven copper ions per molecule. Six of these coppers are tightly bound to ceruloplasmin and can only be released at low pH in the presence of a reducing agent, while the seventh copper is labile and chelatable (reviewed in Gutteridge and Stocks, 1981). It has been suggested that native ceruloplasmin, carrying this seventh copper ion, can under abnormal circumstances interact with cells to become prooxidant and contribute to the development of atherosclerosis by causing oxidation of low-density lipoprotein (LDL) (Ehrenwald *et al.*, 1994; Ehrenwald and Fox, 1996).

However, there is no evidence that the protein itself is prooxidant to LDL. Ceruloplasmin may be able to supply copper to cells via receptor-mediated delivery for incorporation into other copper-containing proteins such as Cu/ZnSOD and cytochrome oxidase. This copper-donor role is sometimes referred to as a copper "transport" function. However, ceruloplasmin does not specifically bind and transport copper in the way that transferrin binds and transports mononuclear iron. Ceruloplasmin catalyzes the oxidation of a wide variety of polyamine and polyphenol substrates *in vitro*. However, with the possible exception of certain bioamines, these oxidations appear to have no biological significance *in vivo*. A role for ceruloplasmin as a ferroxidase enzyme *in vivo* was first proposed by Frieden and his colleagues (Osaki *et al.*, 1966). The protein can catalyze oxidation of ferrous ions to the ferric state, the electrons being passed onto oxygen to form water. It has been proposed that this ferroxidase activity is essential *in vivo* for normal iron metabolism, and for the incorporation of ferric ions into transferrin and into ferritin although the latter has recently been suggested as unlikely (Treffry *et al.*, 1995). Several scientists have pointed out that ferrous ions readily autoxidize in neutral aerobic solutions without the requirement for an enzyme:

$$4Fe^{2+} + O_2 \rightleftharpoons 4Fe(III) + 4{\cdot}O_2^- \tag{8}$$

$$4{\cdot}O_2^- + 4H^+ \rightarrow 2H_2O_2 + 2O_2 \tag{9}$$

$$2H_2O_2 + 2Fe^{2+} \rightarrow 2{\cdot}OH + 2OH^- + 2Fe(III) \tag{10}$$

Ceruloplasmin, however, rapidly removes ferrous ions from solutions at neutral pH values, with the complete reduction of oxygen to water:

$$4Fe^{2+} + O_2 + 4H^+ \rightarrow 2H_2O + 4Fe(III) \quad (11)$$

By transferring four electrons at the enzyme's active (ferroxidase) center, no reactive forms of oxygen are released into solution (Gutteridge and Stocks, 1981). Compare Eqs. (8) and (9) with (11).

Further, in Wilson's disease, where serum ceruloplasmin levels are often extremely low, there is no gross abnormality in iron metabolism. To explain the near-normal iron metabolism seen in patients with Wilson's disease, it has been suggested that a second ferroxidase enzyme exists. This enzyme, ferroxidase II, is a cupro-lipoprotein complex (Lykins *et al.*, 1977) that is almost completely absent from freshly taken normal human plasma (Gutteridge *et al.*, 1985). Whenever plasma is stored or mishandled, however, a metalloproteinase closely associated with ceruloplasmin rapidly degrades the native protein (132 kDa) into fragments of 116, 50, and 19 kDa. Ceruloplasmin has structural similarities and sequence homologies with factors V and VIII of the coagulation cascade, and with the iron-binding protein lactoferrin. The blood clotting factors V and VIII are normally activated by serine protease cleavage, and the similarity of ceruloplasmin to these clotting factors may in part explain why it is so sensitive to proteolysis. In separated plasma, some of these fragments release copper that is chelatable to 1,10-phenanthroline, and that can cause the oxidation of plasma lipoproteins (Gutteridge *et al.*, 1985). The fact that fresh plasma does not catalyze LDL oxidation argues against the alleged prooxidant effect of the intact ceruloplasmin molecule. As fragmentation occurs, ceruloplasmin ferroxidase I activity is lost, but a new ferroxidase activity (ferroxidase II) appears which cannot be inhibited by adding azide. Ferroxidase II in human plasma, therefore, appears to be an artifact (Gutteridge *et al.*, 1985).

Ceruloplasmin accounts for most (96% or more) of the total plasma copper, with a normal adult concentration of 0.35 g/liter. Ceruloplasmin is a minor acute-phase protein induced in response to tissue damage. Levels will, therefore, change, to some extent, in a wide variety of diseases. The ferroxidase activity of ceruloplasmin was reported by the authors to be of importance as an antioxidant *in vivo* (Gutteridge, 1978; Halliwell and Gutteridge, 1990), by rapidly removing Fe^{2+} capable of reacting with H_2O_2 and organic peroxides, to give $\cdot OH$ and alkoxyl radicals ($RO\cdot$), respectively. Ceruloplasmin can also react stoichiometrically with $\cdot O_2^-$, H_2O_2, and $\cdot OH$ and nonspecifically bind ferric and cupric ions (reviewed in Gutteridge and Quinlan, 1996).

The ferroxidase activity of ceruloplasmin competes effectively for ferrous ions in the Fenton reaction. The second-order rate constant for the reaction of ceruloplasmin with ferrous ions has been reported as $27 \times 10^3\ M^{-1}s^{-1}$, which is considerably faster than Fe^{2+} reacts with H_2O_2 ($76\ M^{-1}s^{-1}$). It is likely that ceruloplasmin would function *in vivo* in a similar way to prevent hydroxyl radical formation.

Xanthine oxidase. Xanthine oxidoreductase is a cytosolic enzyme containing molybdenum and iron at its active center, with a M_r of around 275,000. Most of the enzyme present *in vivo* is the NAD^+-dependent "D" (dehydrogenase) form. However, conversion to the "O" (oxidase) form can occur during oxidative stress, whereby the enzyme becomes NAD^+- independent and transfers electrons directly onto dioxygen to form $\cdot O_2^-$ and H_2O_2.

"D"-to-"O" conversions can occur during periods of ischemia and contribute to the oxygen toxicity of reoxygenation injury (reviewed in Bulkley, 1994). Its substrate specificity is low because apart from hypoxanthine and xanthine it can catalyze the oxidation of many different products including aldehydes. Indeed, aldehydes produced during the peroxidation of lipids (Bounds and Winston, 1991) and DNA (Gutteridge *et al.*, 1990) can act as substrates for xanthine oxidase, with products of the reaction being superoxide and hydrogen peroxide.

The intestinal mucosa is reported to contain a "ferroxidase activity" that promotes the incorporation of iron into transferrin. The enzyme responsible for this activity was described as xanthine oxidase acting on xanthine as a substrate (Topham *et al.*, 1986). The "O" form appears to be responsible for the observed ferroxidase activity, the "D" form having no such activity. It was reported that xanthine oxidase has 1000 times the ferroxidase activity of ceruloplasmin, and can operate at very low concentrations of ferrous ions and dioxygen.

3.3.2. Extracellular Proteins Controlling RCS

As already mentioned, there are no *specific* proteins in human plasma for the binding of copper ions in the way that apotransferrin binds iron ions, although it could be argued that albumin fulfills this role. Copper ions bind nonspecifically to all plasma proteins and amino acids, although human albumin has one high-affinity copper-binding site. Copper ions are more reactive with H_2O_2 than iron ions, and by our limited ability to detect them in biological fluids during oxidative stress (Table III) we conclude that they are very carefully controlled. Metallothioneins are low-molecular-mass proteins (around 6500) found in the cytosol of eukaryotic cells, as well as in plasma in an extracellular form. They are rich in sulfur and possess the ability to bind a variety of metal ions (reviewed in Cousins, 1985). Copper binds avidly to metallothionein in multiple stoichiometries (Chen *et al.*, 1996). Indeed, metallothionein has been proposed to act as an antioxidant (Thornalley and Vasak, 1985).

4. HOW IMPORTANT IS REDOX CONTROL OF CELL SIGNALING?

How important is cell signalling by ROS, RIS, RCS, and RNS (other than the obvious case of ·NO)? It is, as yet, hard to say (Halliwell, 1996b). Many transport proteins/receptors/signal transduction systems contain essential –SH groups. Attack on these by ROS, RIS, RCS, and RNS may oxidize them, often reversibly, and "send a message." This does not, of course, mean that this is the normal mechanism by which the message is sent. Thus, some cells do not activate NF-κB in response to ROS (Brennan and O'Neill, 1995). Human adipocytes contain a membrane-bound H_2O_2-generating system responsive to insulin (Krieger-Brauer and Kather, 1996), but there is no evidence as yet that the effects of insulin on fat metabolism are mediated via H_2O_2. We urgently need to move away from studies of isolated cell systems (often affected by exposure to unphysiological O_2 levels, and complex pro-/antioxidant reactions involving such constituents of growth media as iron ions, copper ions, fetal sera, histidine, and thiols) to the *in vivo* situation (Halliwell, 1996b).

ACKNOWLEDGMENTS. J.M.C.G. thanks the British Lung Foundation, the British Oxygen Group, the British Heart Foundation, and the Dunhill Medical Trust for their generous support. B.H. thanks the British Heart Foundation, World Cancer Research Fund, Medical Research Council, Wellcome Trust, Parkinson's Disease Research Foundation, and the Ministry of Agriculture, Fisheries and Food for their generous support.

5. REFERENCES

Aisen, P., Leibman, A., and Zwier, J., 1978, Stoichiometric and site characteristics of the binding of iron to human transferrin, *J. Biol. Chem.* **253**:1930–1937.

Ames, B. N., Shigenaga, M. K., and Hagen, T. M., 1993, Oxidants, antioxidants, and the degenerative disease of aging, *Proc. Natl. Acad. Sci. USA* **90**:7915–7922.

Aruoma, O., and Halliwell, B., 1987, Superoxide-dependent and ascorbate-dependent formation of hydroxyl radicals from hydrogen peroxide in the presence of iron. Are lactoferrin and transferrin promoters of hydroxyl-radical generation? *Biochem. J.* **241**:273–278.

Babior, B. M., 1978, Oxygen-dependent microbial killing by phagocytes, *N. Engl. J. Med.* **298**:659–668.

Balentine, J., 1982, *Pathology of Oxygen Toxicity*, Academic Press, New York.

Balla, G., Jacob, H. S., Balla, J., Rosenberg, M., Nath, K., Apple, F., Eaton, J. W., and Vercellotti, G. M., 1992, Ferritin: A cytoprotective antioxidant strategem of endothelium, *J. Biol. Chem.* **267**:18148–18153.

Barja, G., 1993, Oxygen radicals, a failure or a success of evolution? *Free Radical Res. Commun.* **18**:63–70.

Beckman, J. S., Ye, Y. Z., Anderson, P., Chen, J., Accavitti, M., Tarpey, M. M., and White, C. R., 1994, Extensive nitration of protein tyrosines in human atherosclerosis detected by immunohistochemistry, *Biol. Chem. Hoppe-Seyler* **375**:81–88.

Block, G., Patterson, B., and Subar, A., 1992, Fruit, vegetables and cancer prevention—A review of the epidemiological evidence, *Nutr. Cancer* **18**:1–29.

Bolann, B. J., and Ulvik, R. J., 1987, Release of iron from ferritin by xanthine oxidase. Role of the superoxide radical, *Biochem. J.* **243**:55–59.

Bolann, B. J., and Ulvik, R. J., 1990, On the limited ability of superoxide to release iron from ferritin, *Eur. J. Biochem.* **193**:899–904.

Bolann, B. J., and Ulvik, R. J., 1993, Stimulated decay of superoxide caused by ferritin-bound copper, *FEBS* **328**:263–267.

Bounds, P. L., and Winston, G. W., 1991, The reaction of xanthine oxidase with aldehydic products of lipid peroxidation, *Free Radical Biol. Med.* **11**:447–454.

Brennan, P., and O'Neill, L. A. J., 1995, Effects of oxidants and antioxidants on nuclear factor kβ activation in three different cell lines: Evidence against a universal hypothesis involving O_2 radicals, *Biochim. Biophys. Acta.* **1260**:167–175.

Breuer, W., Epsztein, S., and Cabantchik, Z. I., 1996, Dynamics of the cytosolic chelatable iron pool of K562 cells, *FEBS Lett.* **384**:304–308.

Britigan, B. E., and Edeker, B. K., 1991, Pseudomonas and neutrophil products modify transferrin and lactoferrin to create conditions that favour hydroxyl radical formation, *J. Clin. Invest.* **88**:1092–1102.

Bulkley, G. B., 1994, Reactive oxygen metabolites and reperfusion injury: Aberrant triggering of reticuloendothelial function, *Lancet* **344**:934–936.

Burdon, R. H., 1994, Superoxide and H_2O_2 in relation to mammalian cell proliferation, *Free Radical Biol. Med.* **18**:775–794.

Cabantchik, Z. I., Glickstein, H., Milgram, P., and Breuer, W., 1996, A fluorescence assay for assessing chelation of intracellular iron in a membrane model system and mammalian cells, *Anal. Biochem.* **233**:221–227.

Candeias, L. P., Stratford, M. R., and Wardman, P., 1994, Formation of hydroxyl radicals on reaction of HOCl with ferrocyanide, a model iron (II) complex, *Free Radical Res.* **20**:241–249.

Caspary, W. J., Lanzo, D. A., Niziak, C., Friedman, R., and Bachur, N. R., 1979, Bleomycin A_2: A ferrous oxidase, *Mol. Pharmacol.* **16**:256–290.

Castro, L., Rodriguez, I. N., and Radi, R., 1994, Aconitase is readily inactivated by peroxynitrite, but not by its precursor, nitric oxide, *J. Biol. Chem.* **269**:29409–29415.

Chen, P., Munoz, A., Nettesheim, D., Shaw, C., and Petering, D., 1996, Stoichiometry and cluster specificity of copper binding to metallothionein: Homogeneous metal clusters, *Biochem. J.* **317**:395–402.

Chubatsu, L. S., and Meneghini, R., 1993, Metallothionein protects DNA from oxidative damage, *Biochem. J.* **291**:193–198.

Cochrane, C. G., 1993, Cellular injury by oxidants, *Am. J. Med.* **91**:235–305.

Cousins, R. J., 1985, Absorption, transport and hepatic metabolism of copper and zinc: Special reference to metallothionein and ceruloplasmin, *Physiol. Rev.* **65**:238–309.

Crichton, R. R., and Charloteaux-Wauters, M., 1987, Iron transport and storage, *Eur. J. Biochem.* **164**:485–506.

Darley-Usmar, U., and Halliwell, B., 1996, Blood radicals—reactive nitrogen species, reactive oxygen species, transition metal ions and the vascular system, *Pharm. Res.* **5**:649–662.

Deme, D., Daissiere, J., De Sandro, V., and Dupuy, C. J., 1994, The Ca^{2+}-NADPH-dependent H_2O_2 generator in thyroid plasma membranes—inhibition by diphenyleneindonium, *Biochem. J.* **301**:75–81.

Demple, B , and Harrison, L., 1994, Repairs of oxidative damage to DNA. Enzymology and biology, *Annu. Rev. Biochem.* **63**:915–948.

de Rojas-Walker, T., Tamir, S., Ji, H., Wishnok, J. S., and Tannenbaum, S. R., 1995, NO induces oxidative damage in addition to deamination in macrophage DNA, *Chem. Res. Toxicol.* **8**:473–477.

DeSilva, D. M., Askwith, C. C., and Kaplan, J., 1996, Molecular mechanisms of iron uptake in eukaryotes, *Physiol. Rev.* **76**:31–47.

Drapier, J. C., Hirling, H., Wietzerbin, J., Kaldy, P., and Kuhn, L. C., 1994, Reciprocal modulation of aconitase activity and RNA-binding activity of iron regulatory factor by NO, *Adv. Exp. Med. Biol.* **356**:141–148.

Ehrenwald, E., and Fox, P. L., 1996, Role of endogenous ceruloplasmin in low density lipoprotein oxidation by human U937 monocytic cells, *J. Clin. Invest.* **97**:884–890.

Ehrenwald, E., Chisolm, G. M., and Fox, P. L., 1994, Intact human ceruloplasmin oxidatively modifies low density lipoprotein, *J. Clin. Invest.* **93**:1493–1501.

Eiserich, J. P., Cross, C. E., Jones, A. D., Halliwell, B., and Van der Vliet, A., 1996, Formation of nitrating and chlorinating species by reaction of nitrite with HOCl. A novel mechanism for NO-mediated protein modification, *J. Biol. Chem.* **211**:19199–19208.

El Ghissassi, F., Barbin, A., Nair, J., and Bartsch, H., 1995, Formation of 1,N6-ethenoadenine and 3,N4-ethenocytosine by lipid peroxidation products and nucleic acid bases, *Chem. Res. Toxicol.* **8**:278–283.

Ernster, E., and Dallner, G., 1995, Biochemical, physiological and medical aspects of ubiquinone function, *Biochim. Biophys. Acta.* **1271**:195–204.

Esterbauer, H., Zollner, J., and Schaur, R. J., 1988, Hydroxyl alkenals: Cytotoxic products of lipid peroxidation, *ISI Atlas Sci.* **1**:311–317.

Esterbauer, H., Schaur, R. J., and Zollner, H., 1991, Chemistry and biochemistry of 4-hydroxynonenal, malonaldehyde and related aldehydes, *Free Radical Biol. Med.* **11**:81–128.

Evans, P. J., Evans, R. W., Bomford, A., Williams, R., and Halliwell, B., 1994, Metal ions catalytic for free radical reactions in the plasma of patients with fulminant hepatic failure, *Free Radical Res.* **20**:139–144.

Folkes, L. K., Candeias, L. P., and Wardman, P., 1995, Kinetics and mechanisms of HOCl reactions, *Arch. Biochem. Biophys.* **323**:120–126.

Frei, B., (ed.), 1994, *Natural Antioxidants in Human Health and Disease*, Academic Press, San Diego.

Fridovich, I., 1989, Superoxide dismutase. An adaption to a paramagnetic gas, *J. Biol. Chem.* **264**:7761–7764.

Gey, F., 1995, Ten-year retrospective on the antioxidant hypothesis of arteriosclerosis—Threshold plasma levels of antioxidant micronutrients related to minimum cardiovascular risk, *J. Nutr. Biochem.* **6**:206–236.

Gilbert, D. L., 1996, Evolutionary aspects of atmospheric oxygen and organisms, in *Handbook of Physiology: Section 4. Environmental Physiology: Vol. 2* (M. J. Fregly and C. M. Bratteis, eds.), pp. 1059–1094, Oxford University Press, London.

Goodwin, J. F., and Whitten, C. F., 1965, Chelation of ferrous sulphate solutions by desferrioxamine B, *Nature* **205**:281–283.

Grady, J. K., Chen, Y., Chasteen, D., and Harris, D. C., 1989, Hydroxyl radical production during oxidation deposition of iron in ferritin, *J. Biol. Chem.* **264**:20224–20229.

Green, C. J., Gaver, J. D., Healing, G., Cotterill, L., Fuller, B. J., Simpkin, S., 1989, The importance of iron, calcium and free radicals in reperfusion injury: An overview of studies in ischaemic rabbit kidney, *Free Radical Res. Commun.* **7**:255–264.

Grisham, M. B., 1994, Oxidants and free radicals in inflammatory bowel disease, *Lancet* **344**:859–861.

Grootveld, M., Bell, J. D., Halliwell, B., Aruoma, O. I., Bomford, A., and Sadler, P. J., 1989, Non-transferrin-bound iron in plasma or serum from patients with idiopathic hemochromatosis, *J. Biol. Chem.* **264**:4417–4422.

Gunther, M. R., Kelman, D. J., Corbett, J. T., and Mason, R. P., 1995, Self-peroxidation of metmyoglobin results in formation of an oxygen-reactive tryptophan radical, *J. Biol. Chem.* **270**:16075–16081.

Gurgueira, S. A., and Meneghini, R., 1996, An ATP-dependent iron transport system in isolated rat liver nuclei, *J. Biol. Chem.* **271**:13616–13620.

Gutteridge, J. M. C., 1977, The membrane effects of vitamin E, cholesterol and their acetates on peroxidative susceptibility, *Res. Commun. Chem. Pathol. Pharmacol.* **77**:379–386.

Gutteridge, J. M. C., 1978, Caeruloplasmin: A plasma protein, enzyme and antioxidant, *Ann. Clin. Biochem.* **15**:293–296.

Gutteridge, J. M. C., 1984, Copper-phenanthroline induced site specific oxygen radical damage to DNA. Detection of loosely bound trace copper in biological fluids, *Biochem. J.* **218**:983–985.

Gutteridge, J. M. C., 1986a, Iron promoters of the Fenton reaction and lipid peroxidation can be released from haemoglobin by peroxides, *FEBS Lett.* **201**:291–925.

Gutteridge, J. M. C., 1986b, Antioxidant properties of the proteins caeruloplasmin, albumin and transferrin. A study of their activity in serum and synovial fluid from patients with rheumatoid arthritis, *Biochim. Biophys. Acta* **917**:119–127.

Gutteridge, J. M. C., 1987a, A method for removal of trace iron contamination from biological buffers, *FEBS Lett.* **214**:362–364.

Gutteridge, J. M. C., 1987b, The antioxidant activity of haptoglobin towards haemoglobin stimulated lipid peroxidation, *Biochim. Biophys. Acta* **917**:219–223.

Gutteridge, J. M. C., 1987c, Bleomycin-detectable iron in knee-joint synovial fluid from arthritic patients and its relationship to the extracellular antioxidant activities of caeruoplasmin, transferrin, and lactoferrin, *Biochem. J.* **245**:415–421.

Gutteridge, J. M. C., 1991a, Plasma ascorbate levels and inhibition of the antioxidant activity of caeruloplasmin, *Clin. Sci.* **81**:313–317.

Gutteridge, J. M. C., 1991b, Ferrous ions detected in cerebrospinal fluid by using bleomycin and DNA damage, *Clin. Sci.* **82**:315–320.

Gutteridge, J. M. C., 1995, Signal messenger and trigger molecules from free radical reactions, and their control by antioxidants, in *Signalling Mechanisms—from Transcription Factors to Oxidative Stress* (L. Packer and K. Wirtz, eds.), pp. 157–164, Springer-Verlag, Berlin.

Gutteridge, J. M. C., and Halliwell B., 1988, The antioxidant proteins of extracellular fluids, in *Cellular Antioxidant Defense Mechanisms* (C. K. Chow, ed.), pp. 1–23, CRC Press, Boca Raton.

Gutteridge, J. M. C., and Halliwell, B., 1994, *Antioxidants in Nutrition, Health and Disease.* pp. 1–143, Oxford University Press, London.

Gutteridge, J. M. C., and Hou, Y. Y., 1986, Iron complexes and their reactivity in the bleomycin assay for radical promoting loosely bound iron, *Free Radical Res.Commun.* **2**:143–151.

Gutteridge, J. M. C., and Quinlan, G. J., 1996, Reactive oxygen species, antioxidant protection and lung injury, in *Acute Respiratory Distress Syndrome in Adults* (T. W. Evans and C. Haslett, eds.), pp. 167–196, Chapman & Hall, London.

Gutteridge, J. M. C., and Smith, A., 1988, Antioxidant protection by hemopexin of haem stimulated lipid peroxidation, *Biochem. J.* **256**:861–865.

Gutteridge, J. M. C., and Stocks, J., 1976, Peroxidation of cell lipids, *J. Med. Lab. Sci.* **53**:281–285.

Gutteridge, J. M. C., and Stocks, J., 1981, Caeruloplasmin: Physiological and pathological perspectives, *CRC Crit. Rev. Clin. Lab. Sci.* **14**:257–329.

Gutteridge, J. M. C., and Wilkins, S., 1983, Copper dependent hydroxyl radical formation. Damage to proteins acting as antioxidants, *Biochim. Biophys. Acta* **759**:38–41.

Gutteridge, J. M. C., Richmond, R., and Halliwell, B., 1980, Oxygen free radicals and lipid peroxidation: Inhibition by the protein caeruloplasmin, *FEBS Lett*, **112**:269–272.

Gutteridge, J. M. C., Paterson, S. K., Segal, A. W., and Halliwell, B., 1981a, Inhibition of lipid peroxidation by the iron-binding protein lactoferrin, *Biochem. J.* **199**:259–261.

Gutteridge, J. M. C., Rowley, D. A., and Halliwell, B., 1981b, Superoxide-dependent formation of hydroxyl radicals in the presence of iron salts. Detection of "free" iron in biological systems by using bleomycin-dependent degradation of DNA, *Biochem. J.* **199**:263–265.

Gutteridge, J. M. C., Winyard, P., Blake, D. R., Lunec, J., Brailsford, S., and Halliwell, B., 1985, The behaviour of caeruloplasmin in stored human extracellular fluids in relation to ferroxidase II activity, lipid peroxidation and phenanthroline-detectable copper, *Biochem. J.* **230:**517–523.

Gutteridge, J. M. C., West, M., Eneff, K., and Floyd, R. A., 1990, Bleomycin-iron damage to DNA with formation of 8-hydroxyl-deoxyguanosine and base propenals. Indications that xanthine oxidase generates superoxide from DNA degradation products, *Free Radical Res. Commun.* **10:**159–165.

Gutteridge, J. M. C., Mumby, S., Koizumi, M., and Taniguchi, N., 1996, "Free" iron in neonatal plasma activates aconitase: Evidence for biologically reactive iron, *Biochem. Biophys. Res. Commun.* **229:**806–809.

Haile, D. J., Rouault, T. A., Harford, J. B., Kennedy, M. C., Blondin, G., Beinert, H., and Klausner, R. D., 1992a, Cellular regulation of the iron-responsive element binding protein: Disassembly of the cubane iron-sulfur cluster results in high affinity RNA binding, *Proc. Natl. Acad. Sci. USA* **89:**11735–11739.

Haile, D. J., Rouault, T. A., Tang, C. K., Chin, J., Hartford, J. B., and Klausner, R. D., 1992b, Reciprocal control of RNA-binding and aconitase activity in the regulation of the iron-responsive element binding protein: Role of the iron-sulfur cluster, *Proc. Natl. Acad. Sci. USA* **89:**7536–7540.

Halliwell, B., 1987, Oxidants and human disease: Some new concepts, *FASEB J.* **1:**358–364.

Halliwell, B., 1988, Albumin—An important extracellular antioxidant? *Biochem. Pharmacol.* **37:**569–571.

Halliwell, B., 1989, Superoxide, iron, vascular endothelium and reperfusion injury, *Free Radical Res. Commun.* **5:**315–318.

Halliwell, B., 1996a, Oxidative stress, nutrition and health. Experimental strategies for optimization of nutritional antioxidant intake in humans, *Free Radical Res.* **25:**57–74.

Halliwell, B., 1996b, Free radicals, proteins and DNA—oxidative damage versus redox regulation, *Biochem. Soc. Trans.* **24:**1023–1027.

Halliwell, B., and Gutteridge, J. M. C., 1984, Lipid peroxidation, oxygen radicals, cell damage and antioxidant therapy, *Lancet* **1:**1396–1397.

Halliwell, B., and Gutteridge, J. M. C., 1986, Oxygen free radicals and iron in relation to biology and medicine: Some problems and concepts, *Arch. Biochem. Biophys.* **246:**501–514.

Halliwell, B., and Gutteridge, J. M. C., 1990, The antioxidants of human extracellular fluids, *Arch. Biochem. Biophys.* **280:**1–8.

Halliwell, B., Gutteridge, J. M. C., and Cross, C. E., 1992a, Free radicals, antioxidants, and human disease: Where are we now? *J. Lab. Clin. Med.* **119:**598–620.

Halliwell, B., Hu, M. -L., Louise, S., Duval, T. R., Tarkington, B. K., Motchnik, P., and Cross, C. E., 1992b, Interaction of nitrogen dioxide with human plasma. Antioxidant depletion and oxidative damage, *FEBS Lett.* **313:**62–66.

Halliwell, B., and Gutteridge, J. M. C., 1998, *Free Radicals in Biology and Medicine*, third edition, Oxford University Press, London.

Harris, D. C., and Aisen, P., 1973, Facilitation of Fe(II) autoxidation by Fe(III) complexing agents, *Biochim. Biophys. Acta* **329:**156–158.

Hentze, N. W., and Kuhn, L. C., 1996, Molecular control of vertebrate iron metabolism: mRNA-based regulatory circuits operated by iron, NO, and oxidative stress, *Proc. Natl. Acad. Sci. USA* **93:**8175–8182.

Herold, M., and Spiteller, G., 1996, Enzymic production of hydroperoxides of unsaturated fatty acids by injury of mammalian cells, *Chem. Phys. Lipids* **79:**113–121.

Hu, J., Speisty, H., and Cotgreave, I. A., 1995, The inhibitory effects of boldine, glaucine and probucol on TPA-induced down regulation of gap junction function. Relationships to intracellular peroxides, protein kinase C translocation and connexin 43 phosphorylation, *Biochem. Pharmacol.* **50:**1635–1643.

Huie, R. E., and Padmaja, S., 1993, The reaction of NO with superoxide, *Free Radical Res. Commun.* **18:**195–199.

Iqbal, J., Clerch, L. B., Hass, M. A., Frank, L., and Massaro, D., 1989, Endotoxin increases lung Cu, Zn superoxide dismutase mRNA, O_2 raises enzyme synthesis, *Am. J. Physiol.* **257:**L61–L64.

Kaur, H., Whiteman, M., and Halliwell, B., 1997, Peroxynitrite-dependent aromatic hydroxylation and nitration of salicylate and phenylalanine. Is hydroxyl radical involved? *Free Radical Res.* **26:**71–82.

Kennedy, M. N., Mende-Mueller, L., Blondin, G. A., and Beinert, H., 1992, Purification and characterisation of cytosolic aconitase from beef liver and its relationship to the iron-responsive element binding protein (IRE-BP), *Proc. Natl. Acad. Sci. USA* **89:**11730–11734.

Kettle, A. J., 1996, Neutrophils convert tyrosyl residues in albumin to chlorotyrosine, *FEBS Lett.* **379:**103–106.

Klausner, R. D., Rouault, T. A., and Harford, J. B., 1993, Regulating the fate of mRNA: The control of cellular iron metabolism, *Cell* **72:**19–28.

Klebanoff, S. J., Waltersdorph, A. M., Michel, B. R., and Rosen, H., 1989, Oxygen-based free radical generation by ferrous ions and deferoxamine, *J. Biol. Chem.* **264:**19765–19771.

Kojiman, N., and Bates, G. W., 1981, The formation of Fe^{3+}–transferrin–CO_3^{2-} via the binding and oxidation of Fe^{2+}, *J. Biol. Chem.* **256:**12034–12039.

Kong, S. K., Yim, M. B., Stadtman, E. R., and Chock, P. B., 1996, Peroxynitrite disables the tyrosine phosphorylation regulatory system: Lymphocyte-specific tyrosine kinase fails to phosphorylate nitrated cdc2 (60-20) NH_2 peptide, *Proc. Natl. Acad. Sci. USA* **93:**3377–3382.

Krieger-Brauer, H. I., and Kather, H., 1995, The stimulus-sensitive H_2O_2-generating system present in human fat-cell plasma membranes is multireceptor-linked and under antagonistic control by hormones and cytokines, *Biochem. J.* **307:**543–548.

Laurindo, F. R., da Luz, P. L., Vint, L., Rocha, T. F., Jaeger, R. G., and Lopes, E. A., 1991, Evidence for superoxide radical-dependent coronary vasospasm after angioplasty in intact dogs, *Circulation* **83:**1705–1715.

Lusis, A. J., and Navab, M., 1993, Lipoprotein oxidation and gene expression in the artery wall. New opportunities for pharmacologic intervention in atherosclerosis, *Bio. Pharm.* **46:**2119–2126.

Lykins, L. F., Akey, C. W., Christian, E. G., Duval, G. W., and Topham, R. W., 1977, Dissociation and reconstitution of human ferroxidase II, *Biochemistry* **16:**693–698.

Maly, F. E., 1990, The B lymphocyte: A newly recognized source of reactive oxygen species with immunoregulatory potential, *Free Radical Res. Commun.* **8:**143–148.

Marklund, S. L., Holme, E., and Hellner, L., 1982, Superoxide dismutase in extracellular fluids, *Clin. Chim. Acta* **126:**41–51.

Martin, R. B., 1994, Aluminium: A neurotoxic product of acid rain, *Acc. Chem. Res.* **27:**204–210.

McArthur, K. M., and Davies, M. J., 1993, Detection and reactions of the globin radical in haemoglobin, *Biochim. Biophys. Acta* **1202:**172–181.

McConkey, D. J., Orrenius, S., and Jondal, M., 1996, The regulation of apoptosis in thymocytes, *Biochem. Soc. Trans.* **22:**606–610.

Moncada, S., and Higgs, E. A., 1991, Endogenous nitric oxide: Physiology, pathology and clinical relevance, *Eur. J. Clin. Invest.* **21:**361–374.

Morikawa, K., and Morikawa, S., 1996, Immunomodulatory effect of recombinant human SOD on human B lymphocyte function in vitro, *Cell. Immunol.* **172:**70–76.

Mukhopadhyay, C. K., Ehrenwald, R., and Fox, P. L., 1996, Ceruloplasmin enhances smooth muscle cell and endothelial cell-mediated low density lipoprotein oxidation by a superoxide-dependent mechanism, *J. Biol. Chem.* **371:**14773–14778.

Netto, L. E. S., Chae, H. Z., Kang, S. W., Rhee, S. G., and Stadtman, E. R., 1996, Removal of H_2O_2 by thiol-specific antioxidant enzyme (TSA) is involved with its antioxidant properties. TSA possesses thiol peroxidase activity, *J. Biol. Chem.* **271:**727–731.

O'Connell, M., Halliwell, B., Moorhouse, C. P., Aruoma, O. I., Baum, H., and Peters, T. J., 1986, Formation of hydroxyl radicals in the presence of ferritin and haemosiderin, *Biochem. J.* **234:**727–731.

Orr, W. C., and Sohal, R. S., 1994, Extension of life span by over expression of superoxide dismutase and catalase in Drosophila melanogaster, *Science* **263:**1128–1130.

Orrenius, S., McConkey, D. J., Bellomo, G., and Nicotera, P., 1989, Role of Ca^{2+} in toxic cell killing, *Trends Pharm. Sci.* **10:**281–285.

Osaki, S., Johnson, D. A., and Frieden, E., 1966, The possible significance of the ferrous oxidase activity of ceruloplasmin in normal human serum, *J. Biol. Chem.* **241:**2746–2751.

Panter, S. S., Sadrzadeh, S. M. H., Hallaway, P. E., Haines, J. L., Anderson, V. W., and Eaton, J. W., 1985, Hypohaptoglobinemia associated with familial epilepsy, *J. Exp. Med.* **161:**748–784.

Peng, H. B., Libby, P., and Liao, J. K., 1995, Induction and stabilization of I kappa B alpha by NO mediates inhibition of NF-kappa B, *J. Biol. Chem.* **270:**14214–14219.

Pryor, W. A., and Squadrito, G. L., 1995, The chemistry of peroxynitrite: A product from the reaction of nitric oxide with superoxide, *Am. J. Physiol.* **268:**L699–L722.

Rose, T. M., Plowman, G. D., Teplow, D. B., Dreyer, W. J., Hellstrom, K. E., and Brown, J. P., 1986, Primary structure of the human melanoma-associated antigen p97 (melanotransferrin) deduced from the mRNA sequence, *Proc. Natl. Acad. Sci. USA* **83:**1261–1265.

Rouault, R. A., and Klausner, R. O., 1996, Iron-sulfur clusters as biosensors of oxidants and iron, *Trends Biochem. Sci.* **21**:174–177.

Rubbo, H., Darley-Usmar, V., and Freeman, B., 1996, Nitric oxide regulation of tissue free radical activity, *Chem. Res. Toxicol.* **9**:809–820.

Ruch, R. J., and Klaunig, J. E., 1988, Inhibition of mouse hepatocyte intercellular communication by paraquat-generated oxygen free radicals, *Toxicol. Appl. Pharmacol.* **94**:427–436.

Sarafian, T. A., and Bredesen, D. E., 1994, Is apoptosis mediated by reactive oxygen species? *Free Radical Res.* **21**:1–8.

Saran, M., and Bors, W., 1989, Oxygen radicals acting as chemical messengers: A hypothesis, *Free Radical Res. Commun.* **7**:213–220.

Schreck, R. A., Albermann, K., and Baeuerle, P. A., 1992, Nuclear factor kappa B: An oxidative stress-responsible transcription factor of eukaryotic cells (a review), *Free Radical Res. Commun.* **17**:221–237.

Schulze-Osthoff, K., Schenk, H., and Droge, W., 1995, Effects of thioredoxin on activation of transcription factor NFκB, *Methods Enzymol.* **252**:253–264.

Sies, H. (ed.), 1991, *Oxidative Stress, Oxidants and Antioxidants*, Academic Press, San Diego.

Smith, A., and Morgan, W. T., 1979, Haem transport to the liver by haemopexin: Receptor-mediated uptake with recycling of the protein, *Biochem. J.* **182**:47–54.

Spencer, J. P. E., Jenner, A., Chimel, K., Aruoma, O. I., Cross, C. E., Wu, R., and Halliwell, B., 1995, DNA damage in human respiratory tract epithelial cells: Damage by gas phase cigarette smoke apparently involves attack by reactive nitrogen species in addition to oxygen radicals, *FEBS Lett.* **375**:179–182.

Spencer, J. P. E., Wong, J., Jenner, A., Aruoma, O. I., Cross, C. E., and Halliwell, B., 1996, Base modification and strand breakage in isolated cell thymus DNA and in DNA from human skin epidermal keratinocytes exposed to peroxynitrite or 3-morpholinosydnonimine, *Chem. Res. Toxicol.* **9**:1152–1158.

Staal, F. J., Anderson, M. T., Staal, G, E., Herzenberg, L. A., Gitler, C., and Herzenberg, L. A., 1994, Redox regulation of signal transduction: Tyrosine phosphorylation and calcium influx, *Proc. Natl. Acad. Sci. USA* **91**:3619–3622.

Stahl, G. L., Halliwell, B., and Longhurst, J. C., 1992, Hydrogen peroxide-induced cardiovascular reflexes. Role of hydroxyl radicals, *Circ. Res.* **71**:295–302.

Stocker, R., 1990, Induction of haem oxygenase as a defense against oxidative stress, *Free Radical Res. Commun.* **9**:101–112.

Storz, G., and Toledano, M. B., 1994, Regulation of bacterial gene expression in response to oxidative stress, *Methods Enzymol.* **236**:196–207.

Sundaresan, M., Yu, Z. X., Ferrans, V. J., Irtani, K., and Fintrel, T., 1995, Requirement for generation of H_2O_2 for PDGF signal transduction, *Science* **270**:296–299.

Swain, J., Darley-Usmar, V., and Gutteridge, J. M. C., 1994, Peroxynitrite can release copper from caeruloplasmin: Implications for atherosclerosis, *FEBS Lett.* **342**:49–52.

Tan, C. M., Xenoyannis, Y., and Feldman, R. D., 1995, Oxidant stress enhances adenylyl cyclase activation, *Circ. Res.* **77**:710–717.

Thornalley, P. J., and Vasak, M., 1985, Possible role for metallothionein in protection against radiation-induced oxidative stress. Kinetics and mechanisms of its reaction with superoxide and hydroxyl radicals, *Biochim. Biophys. Acta.* **827**:36–44.

Topham, R. W., Jackson, M. R., Joslin, S. A., and Walker, M. C., 1986, Studies of the ferroxidase activity of native and chemically modified xanthine oxidase, *Biochem. J.* **235**:39–44.

Treffry, A., Gelvan, D., Konijn, A. M., and Harrison, P. M., 1995, Ferritin does not accumulate iron oxidized by caeruloplasmin, *Biochem. J.* **305**:21–23.

van der Vliet, A., O'Neill, C. A., Halliwell, B., Cross, C. E., and Kaur, H., 1994, Aromatic hydroxylation and nitration of phenylalanine and tyrosine by peroxynitrite. Evidence for hydroxyl radical production from peroxynitrite, *FEBS. Lett.* **339**:89–92.

van der Vliet, A., Eiserich, J. P., O'Neill, C. A., Halliwell, B., and Cross, C. E., 1995, Tyrosine modification by reactive nitrogen species: A closer look, *Arch. Biochem. Biophys.* **319**:341–349.

Vile, G. F., and Tyrrell, R. M., 1993, Oxidative stress resulting from ultraviolet A irradiation of human skin fibroblasts leads to a heme oxygenase-dependent increase in ferritin, *J. Biol. Chem.* **268**:14678–14681.

Ward, C. G., Bullen, J. J., and Rogers, H. J., 1996, Iron and infection: New developments and their implications, *J. Trauma, Injury Infect. Crit. Care* **41**:356–364.

Williams, J., 1974, The formation of iron-binding fragments of ovotransferrin by limited proteolysis, *Biochem. J.* **141**:745–752.

Winterbourn, C. C., and Molloy, A. L., 1988, Susceptibilities of lactoferrin and transferrin to myeloperoxidase-dependent loss of iron-binding capacity, *Biochem. J.* **250**:613–616.

Wiseman, H., and Halliwell, B., 1996, Damage to DNA by reactive oxygen and nitrogen species: Role in inflammatory disease and progression to cancer, *Biochem. J.* **313**:17–29.

Part III

Nitrogen Reactive Species

Chapter 9

Nitric Oxide Synthase

Nicolas J. Guzman and Bismark Amoah-Apraku

1. INTRODUCTION

Few biological observations have generated so much interest in modern times as the discovery of the mammalian L-arginine:nitric oxide (NO) pathway. This interest stems from the realization that NO, the end product of the activation of this pathway, is a major regulator of important physiologic functions as diverse as the maintenance of vascular tone and neurotransmission, as well as a mediator of pathophysiological disorders such as septic shock and allograft rejection (Moncada *et al.*, 1991).

The generation of oxides of nitrogen by mammalian cells was first described early in this century by Mitchell and collaborators (Mitchell *et al.*, 1916). However, there was very little interest in the potential role of these compounds in human biology during the six decades following this early discovery. Most of the studies performed during this period focused on the role of nitrogen oxides in the exacerbation of pulmonary diseases as a result of exposure to environmental "smog," and on the ability of these compounds to cure meat. Interestingly, pharmacological compounds that release NO had been in clinical use with great success for over 150 years even though their mechanism of action was until recently completely obscure.

The independent discoveries in the late 1970s and 1980s of the biological formation of NO in macrophages, endothelium, and neurons, and the corresponding physiological roles of this gas as modulator of immune function, vascular tone, and neurotransmission (Forstermann *et al.*, 1995a; Moncada *et al.*, 1991; Nathan and Xie, 1994a; Stuehr and

Nicolas J. Guzman and Bismark Amoah-Apraku Division of Nephrology and Hypertension, Georgetown University Medical Center, Washington, D.C. 20007.

Reactive Oxygen Species in Biological Systems, edited by Gilbert and Colton. Kluwer Academic / Plenum Publishers, New York, 1999.

Griffith, 1992), respectively, opened one of the most exciting and productive eras of scientific research in mammalian biology.

2. BIOCHEMISTRY OF NO FORMATION

Nitric oxide synthase (NOS; EC 1.14.13.39) is a hemeprotein that catalyzes the calmodulin-dependent oxidation of one of the terminal guanidino groups of L-arginine resulting in the stoichiometric formation of L-citrulline and NO (Figure 1) (Stuehr and Griffith, 1992). This reaction involves the transfer of five electrons from molecular oxygen (O_2) onto the substrate guanidino nitrogen and requires NADPH as cosubstrate, and tetrahydrobiopterin (BH_4), FAD, and FMN as cofactors (Stuehr and Griffith, 1992). The oxidation of L-arginine occurs in two phases (Figure 1). Phase I involves the hydroxylation of L-arginine to the intermediate product N^{ω}-L-hydroxyarginine (OHArg) (Marletta, 1993, 1994b). This reaction resembles a classical cytochrome P450 hydroxylation consuming 1 mole of NADPH and possibly involving the heme iron of NOS. In phase II, OHArg is oxidized to L-citrulline in a BH_4-dependent reaction that consumes another mole of NADPH as reducing equivalent and releases NO (Marletta, 1993, 1994b). In the absence of L-arginine and BH_4, the activation of molecular O_2 by NOS results in a divalent reduction of O_2 to yield superoxide anions ($\cdot O_2^-$) and hydrogen peroxide (H_2O_2) (Zweier *et al.*, 1988).

3. ISOFORMS OF NOS

The NOS enzyme family is comprised of three distinct isoforms each originating from a different gene (Table I) (Forstermann *et al.*, 1995a; Nathan and Xie, 1994b). According to their mode of activation, these can be grouped into the constitutively expressed, Ca^{2+}-calmodulin dependent NOS isoforms (cNOS) originally found in neurons (nNOS, bNOS, or NOS1) and endothelial cells (eNOS or NOS3), and the immunologically inducible, Ca^{2+}-independent NOS originally discovered in macrophages and whose expression is regulated by DNA transcription (iNOS or NOS2) (Forstermann *et al.*, 1995a; Stuehr and Griffith, 1992).

All NOS isoforms have two functionally distinct domains: The oxygenase domain resides in the N-terminus and contains heme, and the L-arginine and BH_4 binding sites;

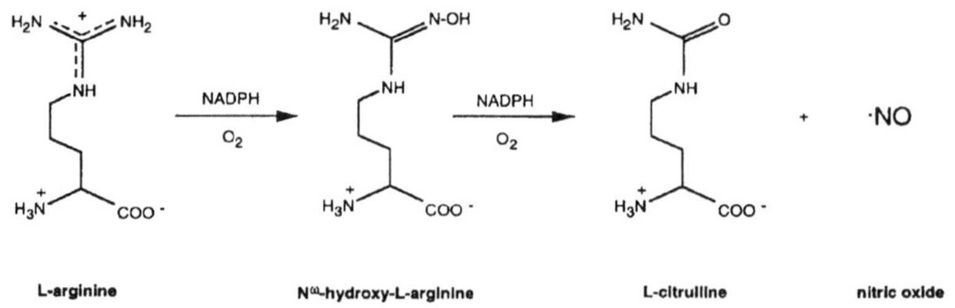

FIGURE 1. Two-phase reaction catalyzed by NOS for the conversion of L-arginine to L-citrulline.

the reductase domain located in the C-terminus is highly homologous to P450 reductase and contains FAD, FMN, and the NADPH binding site (Chen *et al.*, 1996; Ghosh *et al.*, 1996; Lowe *et al.*, 1996). The heme group has been identified as a cytochrome P450-type iron-protoporphyrin IX (Fe-PPIX) prosthetic group that functions in the turnover of L-arginine. This characteristic places NOSs in a unique group by being the first examples of mammalian P450s that have both the reductase and heme domains as part of the same molecule (Marletta, 1993, 1994b).

In their active state, all three NOS isoforms are homodimers of subunits ranging from 130 to 160 kDa (Bredt *et al.*, 1991; Schmidt *et al.*, 1991; Xie *et al.*, 1992). Dimerization activates NO synthesis by enabling electrons to transfer between the reductase and oxygenase domains (Siddhanta *et al.*, 1996). The heme group and BH_4 contained within the oxygenase domain appear to be essential for enzyme dimerization to occur (Cho *et al.*, 1995; Klatt *et al.*, 1996). Interestingly, L-arginine alters the enzyme's heme iron spin equilibrium, increases its NADPH oxidation, and promotes assembly of active dimeric NOS from inactive monomers, all of these actions that would provide a positive feedback for NO synthesis (Sennequier and Stuehr, 1996). Conversely, NO appears to interfere with intracellular assembly of dimeric NOS by preventing heme insertion and decreasing heme availability, an effect that could represent an inhibitory feedback mechanism to limit NOS activity (Albakri and Stuehr, 1996).

A calmodulin (CaM) binding region is present in all NOS isoforms between the oxygenase and reductase domains (Anagli *et al.*, 1995; Bredt *et al.*, 1992; Venema *et al.*, 1996). Binding of CaM to NOS appears to bring about a conformational change that allows the flow of electrons from NADPH to FAD, FMN, and heme (Marletta, 1994b; Su *et al.*, 1995). Whereas CaM binding to cNOS is Ca^{2+}-dependent and regulates enzyme activity, iNOS forms a very tight complex with CaM at low levels of Ca^{2+}, which makes its activity largely Ca^{2+}-independent (Marletta, 1994b; Xie and Nathan, 1994).

3.1. Neuronal NOS (nNOS or NOS1)

3.1.1. Distribution and Function

Neuronal NOS is a 150- to 160-kDa protein originally isolated from cerebellum (Table I) (Bredt and Snyder, 1989, 1990). nNOS is also expressed in synaptic terminals of nonadrenergic noncholinergic (NANC) nerves throughout the body including the gastrointestinal, urogenital and respiratory tracts, in certain areas of the spinal cord, in sympathetic ganglia and adrenal glands, in a variety of epithelial cells, in kidney macula densa cells, in pancreatic islet cells, and in human and rat skeletal muscle (T. M. Dawson and Dawson, 1996; Forstermann *et al.*, 1995a; Nathan and Xie, 1994b; Vanderwinden *et al.*, 1993; Yamamoto *et al.*, 1993; Yoshida *et al.*, 1993; X. Zhang *et al.*, 1993). In humans, nNOS is highly abundant in skeletal muscle, its levels exceeding those in brain (Bredt, 1996; Reid, 1996). Its activity in that organ correlates closely with fast twitch (type II) muscle fibers. nNOS also colocalizes with a number of neurotransmitters including somatostatin in the hippocampus, and choline acetyltransferase and dopamine β-hydroxylase in pre- and postganglionic sympathetic neurons, respectively (Forstermann *et al.*, 1995a; Hirsch *et al.*, 1993).

Table I
NOS Isozymes

Isozyme	Other names[a]	Enzyme characteristics
Neuronal NOS	nNOS	150–160 kDa
	bNOS	Cytosolic and membrane-bound
	cNOS	Constitutive expression
	ncNOS	Ca^{2+}/calmodulin-dependent
	bcNOS	Low-output NO production
	NOS1	N-terminus PDZ domain
	NOS I	Cell prototype: neurons
Inducible NOS	iNOS	125–135 kDa
	macNOS	Cytosolic (predominantly)
	hepNOS	Inducible expression
	NOS2	Ca^{2+}/calmodulin-independent
	NOS II	High-output NO production
		Cell prototype: macrophage
Endothelial NOS	eNOS	135 kDa
	cNOS	Membrane-bound (caveolae)
	ecNOS	Constitutive expression
	NOS3	Ca^{2+}/calmodulin-dependent
	NOS III	Low-output NO production
		Cotranslational myristylation
		Posttranslational palmitoylation
		Cell prototype: endothelium

[a]n, neuronal; b, brain; e, endothelial; c, constitutive; i, inducible; mac, macrophage; hep, hepatocyte.

nNOS is constitutively expressed and its activity is regulated by Ca^{2+} via CaM activation (Stuehr and Griffith, 1992). The primary stimulus for NO synthesis in neurons is activation of NMDA-type glutamate receptors (Bredt and Snyder, 1989). Overactivation of these receptors leads to excessive production of NO, which mediates excitotoxicity and neuronal death in conditions such as cerebral ischemia, and in degenerative disorders such as Alzheimer's disease, Huntington's chorea, and amyotrophic lateral sclerosis (Brenman *et al.*, 1996; V. L. Dawson and Dawson, 1996a; Hunot *et al.*, 1996; Knowles and Moncada, 1994; Lowenstein *et al.*, 1994; Zhang *et al.*, 1994). In NANC nerve terminals, NO acts as a neurotransmitter mediating smooth muscle relaxation (Jaffrey and Snyder, 1995; Lowenstein *et al.*, 1994). NO also functions as a retrograde messenger in the central nervous system possibly mediating certain aspects of hippocampal long-term potentiation (important in memory and learning) and cerebellar long-term depression (important in motor learning) (Doyle *et al.*, 1996).

3.1.2. Regulation of Expression

The human nNOS gene resides at chromosome 12q24.2 and comprises 28 exons that span over 150 kb (Deng *et al.*, 1995). Eight unique exons 1 resulting in eight different mRNAs have been isolated (Deng *et al.*, 1995; Forstermann *et al.*, 1995a). These variants are localized to different cell populations suggesting that cell-type-specific transcription/splicing factors may control nNOS expression. Mice with a disrupted nNOS gene ($nNOS^{-/-}$ mutant) suffered from hypertrophic pyloric stenosis (Huang *et al.*, 1993), a condition associated in humans with gastric sphincter NANC nerve dysfunction caused by nNOS deficiency. Indeed, recent studies suggest that nNOS is a susceptibility locus for infantile pyloric stenosis in humans (Chung *et al.*, 1996). $nNOS^{-/-}$ mutant mice did not initially exhibit any other obvious neurological abnormalities (Huang *et al.*, 1993). This was later attributed to the fact that the mutation resulted in residual nNOS catalytic activity in the brain at levels up to 8% of those found in wild-type mice (Bredt, 1996; Brenman *et al.*, 1996). $nNOS^{-/-}$ mutant mice are resistant to brain injury following cerebral ischemia, which confirms the role of NO in this form of excitotoxicity (Huang *et al.*, 1994). Male $nNOS^{-/-}$ mutants also exhibit an increase in aggressive behavior and excess and inappropriate sexual behavior (Nelson *et al.*, 1995).

Transcription of the human nNOS gene results in a 10-kb mRNA encoding for a protein comprised of 1433 amino acids (Figure 2). Over 90% of NOS activity in skeletal muscle is membrane associated (Bredt, 1996). nNOS immunoreactivity in neurons is largely associated with rough endoplasmic reticulum and synaptic membrane structures. Thus, over 60% of NOS activity in cortical extracts is in the particulate fraction, and over 85% of nNOS immunoreactivity in monkey visual cortex is associated with axonal or dendritic structures (Bredt, 1996; Brenman *et al.*, 1996).

In contrast to eNOS, nNOS lacks the myristoylation site necessary for membrane association (see below) (Bredt *et al.*, 1991). However, the extended N-terminus domain of nNOS contains 230 amino acids that are not present in eNOS and are not required for catalytic activity. Within this unique N-terminal domain, nNOS contains a PDZ/GLGF motif of about 100 amino acids that is shared by a diverse group of cytoskeletal proteins and enzymes commonly found concentrated at specialized cell–cell junctions, such as neuronal synapses, epithelial zona occludens, and septate junctions (Bredt, 1996). Recent studies indicate that nNOS binds to α1-syntrophin in the dystrophin-associated glycoprotein complex (DAGC) of the sarcolemma of skeletal muscle fast fibers (Bredt, 1996;

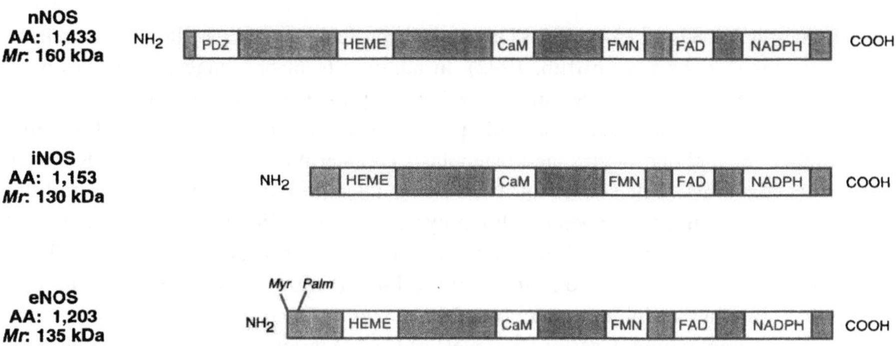

FIGURE 2. Schematic representation of NOS isozymes showing relationships of consensus binding sequences for various cofactors and other proteins. nNOS, neuronal NOS; iNOS, inducible NOS; eNOS, endothelial NOS; AA, amino acids; *Mr*, molecular mass; PDZ, region containing "GLGF" anchoring motif; CaM, calmodulin binding sequence; *Myr*, myristylation site; *Palm*, palmitoylation site.

Brenman *et al.*, 1996). The functional significance of this association is not entirely clear at this time but binding of nNOS to the DAGC beneath skeletal muscle sarcolemma may provide a mechanism for signal transduction mediated by the DAGC. NO causes cGMP-mediated relaxation of skeletal muscle (Bredt, 1996; Reid, 1996). Interestingly, patients with Duchenne muscular dystrophy exhibit a disruption of the DAGC and have markedly reduced levels of nNOS in dystrophic muscle (Grozdanovic *et al.*, 1996). In brain, nNOS reportedly binds to a PDZ/GLGF motif of postsynaptic density-95 protein (PSD-95) and a related protein PSD-93. nNOS and PSD-95 are coexpressed in numerous populations of developing and mature neurons (Brenman *et al.*, 1996). It appears that PDZ domain interactions are important in organizing proteins at synaptic membranes. Thus, the unique N-terminal domain of nNOS containing the PDZ/GLGF motif appears to be important in determining the subcellular distribution of the enzyme and may provide a mechanism for cell-type-specific nNOS functions based on PDZ domain interactions.

Upregulation of nNOS expression has been reported during fetal development in rat lung (North *et al.*, 1994). Hyperosmotic stimulation upregulates the expression of nNOS mRNA in the supraoptic and paraventricular nuclei of rat hypothalamus (Kadowaki *et al.*, 1994). Rapid upregulation of nNOS activity and mRNA has been observed in ischemic lesions during focal cerebral ischemic injury in the rat (Zhang *et al.*, 1994). Interestingly, the expression of nNOS by neurons that retained their morphological structure in the area of the infarct suggests that nNOS containing neurons are more resistant to the ischemic insult (Zhang *et al.*, 1994). It is not clear whether the mechanisms responsible for the upregulation of nNOS are transcriptional or posttranscriptional in nature.

Phosphorylation of nNOS by Ca^{2+}/calmodulin kinase II and protein kinases A and G have been reported to reduce enzyme activity *in vitro* (Forstermann *et al.*, 1995a).

3.2. Inducible NOS (iNOS or NOS2)

3.2.1. Distribution and Function

Inducible NOS is a 125- to 135-kDa cytosolic protein originally isolated from murine macrophages (Stuehr and Griffith, 1992). In addition to macrophages, iNOS has been found in many other cell types including hepatocytes, vascular smooth muscle cells, cardiac myocytes, renal mesangial and epithelial cells, brain microglia, Kupffer cells, endothelial cells, chondrocytes, and osteoblasts (Forstermann *et al.*, 1995a,b; Knowles and Moncada, 1994). The activity of iNOS is regulated primarily by DNA transcription, which can be induced by bacterial lipopolysaccharide (LPS), various cytokines and peptides, hypoxia/ischemia, and other heterogeneous inducers (Forstermann *et al.*, 1995a; Nathan and Xie, 1994b; Xie and Nathan, 1994). Table II shows a partial list of agents that can either stimulate or inhibit induction of this enzyme. Once synthesized, iNOS forms a very tight complex with CaM at low levels of Ca^{2+} (Marletta, 1994b; Xie and Nathan, 1994). As a result of this interaction, its activity is Ca^{2+}-independent and leads to sustained high-output NO production (Table I). Cytokines such as IL-1β and tumor necrosis factor-α can also stimulate the synthesis of the Na^+-independent System y^+ cationic amino

Table II
Agents that Stimulate or Inhibit iNOS Expression

Stimulatory	Inhibitory
• Bacterial lipopolysaccharide • Cytokines: IL-1β, IL-4, γ-interferon, tumor necrosis factor-α • Angiotensin II • Cyclic AMP-elevating agents • Plasmin • Hypoxia/ischemia • Growth factors: basic and acidic fibroblast growth factor, platelet-derived growth factor, epidermal growth factor • Lipid A and lipid A analogues • Glia-derived neurotrophic protein S100-β • Oxidative stress • Silica • Ultraviolet light	• Cytokines: IL-4, IL-6, IL-10 • Transforming growth factor-β • NF-κB inhibitors: NO, glucocorticoids, aspirin, proteasome inhibitors, pyrrolidine dithiocarbamate, antioxidants • Steroids: aldosterone, progesterone, glucocorticoids (posttranscriptional) • Nonsteroidal anti-inflammatory drugs • Peptides: angiotensin II, thrombin, osteopontin • Growth factors: basic and acidic fibroblast growth factor, platelet-derived growth factor, epidermal growth factor • Tumor suppressor gene product p53 • P_{2Y} purinoreceptor agonists • Protein kinase C activators • Tyrosine kinase inhibitors • Protein phosphatase 1 and 2A inhibitors • Antioxidant: rotenone (posttranscriptional) • L-Canavanine

acid transporter *cat-2* (Gill *et al.*, 1996). This enhances the uptake of L-arginine necessary to sustain NO production by iNOS (Durante *et al.*, 1996; Gill *et al.*, 1996).

The large amounts of NO synthesized as a result of iNOS activation can have both cytoprotective and cytotoxic effects. NO synthesized by iNOS can be protective against microbial infections such as malaria, viral encephalitis, and leishmaniasis (Cattell and Jansen, 1995; Nussler and Billiar, 1993). Similarly, it can prevent glomerular thrombosis during endotoxemia (Shultz and Raij, 1992) and pregnancy (Raij, 1994). On the other hand, NO lowers blood pressure and causes cardiovascular collapse when produced in large amounts, such as during bacterial endotoxemia (Moncada *et al.*, 1991). Moreover, NO can be cytotoxic by reacting with oxygen radicals and forming peroxynitrite and other reactive species (V. L. Dawson and Dawson, 1996b), or by inactivating enzymes that contain critical iron-sulfur centers such as ribonucleotide reductase, mitochondrial aconitase, and succinate:ubiquinone reductase (Knowles and Moncada, 1994). NO has been implicated as a mediator of tissue injury in several inflammatory disorders including ulcerative colitis, arthritis, allograft rejection, and glomerulonephritis (see references in Cattell and Jansen, 1995). Transgenic mice lacking the iNOS gene (iNOS$^{-/-}$ "knockout" mutant) are more susceptible to infection by the protozoan parasite *Leishmania major* than wild-type mice, which are highly resistant to this parasite (Wei *et al.*, 1995), and to *Listeria monocytogenes* infection (Macmicking *et al.*, 1995). iNOS$^{-/-}$ mutant mice also show reduced nonspecific inflammatory responses to carrageenan (Wei *et al.*, 1995). Interestingly, although iNOS$^{-/-}$ mutant mice appear to be protected from the cardiovascular collapse and early death induced by bacterial endotoxin, their overall survival following LPS administration does not appear to be significantly different than that of wild-type mice (Laubach *et al.*, 1995; Macmicking *et al.*, 1995). Extensive LPS-induced liver and lung damage has been found in both iNOS$^{-/-}$mutant and wild-type mice and probably accounts for the lack of differences in mortality (Laubach *et al.*, 1995; Macmicking *et al.*, 1995).

3.2.2. Regulation of Expression

The expression of iNOS appears to be subject to strict and highly tissue-specific transcriptional regulatory control. The complexity of the mechanisms governing its induction can be appreciated by contrasting the effects of certain agents in different cell types (Table II). For example, angiotensin II has been reported to inhibit IL-1β-mediated NOS II induction in rat vascular smooth muscle (Nakayama *et al.*, 1994). Conversely, it augments induction of iNOS by the same cytokine in rat cardiac myocytes (Ikeda *et al.*, 1995). Moreover, in rat astroglia angiotensin II inhibits iNOS induction by LPS but fails to inhibit its induction by IL-1β, γ-interferon (IFNγ), and tumor necrosis factor-α (Chandler *et al.*, 1995). Growth factors such as basic fibroblast growth factor and platelet-derived growth factor also appear to have both stimulatory and inhibitory effects on iNOS induction in different cell types (Forstermann *et al.*, 1995b). Thus, expression of iNOS appears to occur as a result of extensive cell-type-specific cross talk between complex networks of intracellular signaling mediators.

The human iNOS gene has been localized to chromosome 17cen-q11.2 where it comprises 26 exons that span 37 kb (Chartrain *et al.*, 1994; Marsden *et al.*, 1994). Interestingly, the chromosomal position of human iNOS is syntenic to a region of rat chromosome 10 implicated in the development of spontaneous hypertension (Marsden *et al.*, 1994). Moreover, alleles of the iNOS locus cosegregate highly significantly with blood pressure in the Dahl salt-sensitive rat (Deng and Rapp, 1995), which is a model of hypertension responsive to L-arginine, the substrate for NOS, and to dexamethasone, an inhibitor of iNOS (Chen and Sanders, 1993). Transcription of the human iNOS gene results in a 4.5-kb mRNA that encodes for a protein comprised of 1153 amino acids (Forstermann *et al.*, 1995a).

Cloning of the human, rat, and murine iNOS genes demonstrated the presence of promoter/enhancer regions in their 5′-flanking segments (Adachi *et al.*, 1993; Chartrain *et al.*, 1994; Lowenstein *et al.*, 1993; Marsden *et al.*, 1994; Xie *et al.*, 1993). The mouse iNOS promoter region studied consists of a 1749-bp fragment containing a TATA box 30 bp upstream of the mRNA transcription initiation site, along with at least 24 oligonucleotide elements homologous to consensus sequences for the binding of several trans-activating proteins (Lowenstein *et al.*, 1993; Xie *et al.*, 1993). These include ten copies of γ-interferon response element; three copies of γ-activated site; two copies each of nuclear factor-κB (NF-κB), interferon-α-stimulated response element, activator protein 1 (AP-1), and tumor necrosis factor-α response element sites; one nuclear factor interleukin 6 (NF-IL6) site, one Oct site, and one AP-1 site (Lowenstein *et al.*, 1993; Xie *et al.*, 1993).

The proximal NF-κB sequence comprised by nucleotides –85 to –76 of the murine iNOS promoter has been found to be essential for LPS inducibility (Xie *et al.*, 1994). NF-κB proteins are a ubiquitous group of transcription factors that exist mainly as inactive dimers in the cytoplasm of cells bound to inhibitory proteins known as inhibitors of kappa-B (I-κB) (Muller *et al.*, 1993). On cell activation, IκB dissociates from the NF-κB dimer allowing this to migrate to the nucleus where it binds to the iNOS promoter and initiates transcription (Muller *et al.*, 1993). Activation of NF-κB appears to mediate iNOS induction by many of the known inducers (Table II) (Adcock *et al.*, 1994; Amoah-Apraku *et al.*, 1995a; Flodstrom *et al.*, 1996; Kleinert *et al.*, 1996b; Spink *et al.*, 1995; Xie *et al.*, 1994). It also appears that the ability of cells to form active NF-κB dimers with transactivating potential may depend on the presence or activation of specific NF-κB proteins. Therefore, the composition of NF-κB dimers may contribute to impart tissue specificity to iNOS responses (Amoah-Apraku *et al.*, 1995a; Xie *et al.*, 1994).

A variety of NF-κB inhibitors have been reported to prevent iNOS induction. Glucocorticoids inhibit cytokine-stimulated iNOS induction by preventing NF-κB activation probably through the induction of I-κB synthesis (Auphan *et al.*, 1995; Kleinert *et al.*, 1996a; Scheinman *et al.*, 1995). Aspirin also inhibits iNOS induction by an as yet unknown mechanism (Farivar *et al.*, 1996); however, it is likely that inhibition of NF-κB is also, at least in part, responsible for aspirin's inhibitory effect (Kopp and Ghosh, 1994). Antioxidants inhibit iNOS expression probably by preventing NF-κB activation although posttranscriptional inhibitory effects have also been postulated (Hecker *et al.*, 1996; Tetsuka *et al.*, 1996). Proteasome inhibitors prevent ubiquitin-dependent I-κB degradation, which blocks NF-κB activation (Palombella *et al.*, 1994) and macrophage iNOS induction (Griscavage *et al.*, 1996). Lastly, NO itself appears to inhibit iNOS expression by inducing and stabilizing I-κB thereby preventing NF-κB activation (Colasanti *et al.*, 1995; Peng *et al.*, 1995).

Although initial cloning of the proximal 400 bp of the 5′-flanking region of the human iNOS gene showed 66% homology with its murine counterpart (Chartrain *et al.*, 1994), functional evidence of enhancer activity was not obtained until recently. Analysis of the first 3.8 kb upstream of the human iNOS gene demonstrated basal promoter activity but failed to show cytokine inducibility (Devera *et al.*, 1996). This is in contrast with the murine promoter where cytokine-responsive elements are located within the proximal 1.0-kb region (Xie *et al.*, 1993). In the human promoter, cytokine-responsive elements were only identified in three distinct regions located substantially farther upstream: –3.8 to –5.8, –5.8 to –7.0, and –7.0 to –16 kb (Devera *et al.*, 1996). These observations emphasize the enormous complexity that appears to be involved in the regulation of iNOS gene expression, and the major differences that exist between species.

Activation of tyrosine kinases appears to be necessary for iNOS expression in various cell types including human pancreatic islet cells (Corbett *et al.*, 1996), human chondrocytes (Geng *et al.*, 1995), rat brain glial cells (Feinstein *et al.*, 1994), and mouse macrophages (Dong *et al.*, 1993; Eason and Martin, 1995). The mechanisms of the interaction between tyrosine kinases and iNOS have not been elucidated. A recent report indicates that ERK1/ERK2 activation appears to be necessary for the induction of iNOS by IL-1β and IFNγ in rat ventricular myocytes and cardiac microvascular endothelial cells (Singh *et al.*, 1996). Activation of specific cellular kinases in response to cytokines may serve to target iNOS gene expression to specific cells or regions within an organ.

In addition to transcriptional control, iNOS can also be regulated at the posttranscriptional and posttranslational levels. LPS appears to prolong iNOS mRNA half-life in addition to increasing iNOS mRNA transcription in mouse macrophages (Weisz *et al.*,

1994). Similarly, IFNγ stabilizes iNOS mRNA in these cells (Vodovotz *et al.*, 1993). TGF-β enhances the degradation of iNOS mRNA and iNOS protein, and reduces iNOS mRNA translation (Vodovotz *et al.*, 1993). IL-4 has been reported to decrease iNOS mRNA several hours after the latter has been induced in mouse peritoneal macrophages (Bogdan *et al.*, 1994).

Although serine phosphorylation of iNOS can be demonstrated *in vitro*, there is currently no evidence to suggest a direct effect of phosphorylation on enzymatic function (Nathan and Xie, 1994b).

3.3. Endothelial NOS (eNOS or NOS3)

3.3.1. Distribution and Function

Endothelial NOS is a 135-kDa membrane-bound protein originally isolated from endothelial cells (Table I) (Forstermann *et al.*, 1995a). It is found in various types of arterial and venous endothelial cells in different tissues, including human tissues (Pollock *et al.*, 1993). In addition to endothelial cells, eNOS is also found in rat skeletal muscle in association with mitochondria (Bates *et al.*, 1996; Kobzik *et al.*, 1995), in rat hippocampal pyramidal cells and other regions of the brain (Dinerman *et al.*, 1994; Tomimoto *et al.*, 1994), in syncytiotrophoblasts of human placenta (Myatt *et al.*, 1993a, 1993b), and in canine kidney epithelium (Tracey *et al.*, 1994) and colonic interstitial cells (Xue *et al.*, 1994).

eNOS is constitutively expressed in cells and its activity is regulated by Ca^{2+}, which causes calmodulin to bind and activate the enzyme (Stuehr and Griffith, 1992). Activation of eNOS results in low-output NO production (Moncada *et al.*, 1991). eNOS can be activated by hemodynamic forces such as shear stress or via receptor activation by Ca^{2+}-mobilizing agents such as bradykinin, acetylcholine, and purinergic agonists (Moncada *et al.*, 1991). NO produced by eNOS plays an important role in the control of vascular tone and blood pressure, and in the modulation of platelet function (Moncada *et al.*, 1991).

3.3.2. Regulation of Expression

The human eNOS gene has been localized to chromosome 7q35-7q36 and comprises 26 exons that span 21 kb (Nathan and Xie, 1994b; Robinson *et al.*, 1994). Transcription of this gene results in a 4.3-kb mRNA coding for a protein comprised of 1203 amino acids (Forstermann *et al.*, 1995a). During its synthesis, eNOS undergoes *N*-myristoylation of Gly-2, a necessary step for membrane association (Sessa *et al.*, 1993). Subsequently, eNOS undergoes posttranslational palmitoylation of cysteines 15 and/or 26, which targets the enzyme to caveolae (Garciacardena *et al.*, 1996; Robinson and Michel, 1995). This processing is likely to provide endothelial cells with an efficient mechanism to locally produce NO in response to hemodynamic forces and receptor activation. Interestingly, bradykinin treatment promotes eNOS depalmitoylation resulting in cytosolic translocation (Robinson *et al.*, 1995). This would remove the enzyme from its site of action in the membrane and expose it to phosphorylation by cytosolic protein kinases (Robinson *et al.*, 1995). *In vitro*, phosphorylation of eNOS by protein kinase C appears to reduce its activity

(Davda *et al.*, 1994; Michel *et al.*, 1993). However, the physiological significance of eNOS phosphorylation remains to be determined. Like the other isoforms, eNOS activity is also downregulated by NO possibly through an interaction with the prosthetic heme group (Buga *et al.*, 1993).

Analysis of the proximal ~1.6 kb of the 5′-flanking region of the human eNOS gene demonstrated a "TATA-less" region with several oligonucleotide consensus sequences for the binding of transcription factors. These include AP-1, AP-2, sterol regulatory element (SRE), retinoblastoma control element (RCE), shear stress response element (SSRE), nuclear factor 1 (NF-1), cAMP response element (CRE), and Sp1 binding sites (Zhang *et al.*, 1995). It was observed that the Sp1 binding site at −103 is absolutely required for basal expression of eNOS, whereas a GATA element present at −230 exerts a modulatory influence on the level of expression (Zhang *et al.*, 1995). Very little else is known about possible transcriptional regulation of the human eNOS gene. Flow-induced shear stress has been shown to increase eNOS mRNA expression and eNOS activity in human endothelial cells (Noris *et al.*, 1995). However, it is not known whether this effect is transcriptionally mediated. Tumor necrosis factor-α decreases the expression of eNOS in human endothelial cells by destabilizing eNOS and shortening its half-life (Yoshizumi *et al.*, 1993).

Several factors regulate eNOS expression in experimental animal models and cultured cells. Hypoxia increases eNOS gene expression in bovine endothelial cells by augmenting its transcription (Arnet *et al.*, 1996). The transcription factors that may be involved in this effect are still unknown. Chronic exercise increases endothelial cell eNOS expression in dogs (Sessa *et al.*, 1994). Chronic ethanol exposure augments eNOS activity in bovine pulmonary endothelial cells (Davda *et al.*, 1993). Focal cerebral ischemia increases cerebral eNOS expression in rats (Z. G. Zhang *et al.*, 1993). eNOS expression is developmentally regulated in rat lung and increases maximally near term in pregnant animals (North *et al.*, 1994). Pregnancy and estrogens increase eNOS expression in several tissues of guinea pigs (Weiner *et al.*, 1994). Agents that increase intracellular cAMP reportedly decrease cardiac myocyte eNOS expression in rats (Belhassen *et al.*, 1996). With the exception of hypoxia, it is not clear whether the mechanisms responsible for the abovementioned effects are transcriptional or posttranscriptional in nature.

4. INHIBITORS OF NOS ACTIVITY

The complexity of the biochemical reaction catalyzed by NOS and its requirements for various cofactors and prosthetic groups offer a multitude of potential targets for selective pharmacological inhibition of the different NOSs. The availability of isoform-selective NOS inhibitors would be a major breakthrough because of their potential impact in the therapy of various inflammatory and neurodegenerative disorders. Several classes of NOS inhibitors have been developed that have different degrees of selectivity for the three isoforms.

4.1. L-Arginine Analogues and Other Amino Acid-Based Inhibitors of NOS

This is the largest and probably most useful group of NOS inhibitors. These compounds are N^{ω}-substituted analogues of L-arginine and include N^{ω}-nitro-L-arginine

(L-NNA), N^{ω}-nitro-L-arginine methyl ester (L-NAME), N^{ω}-methyl-L-arginine (L-NMA), and N^{ω}-amino-L-arginine (L-NAA), among others (Marletta, 1994a; Southan and Szabo, 1996; Stuehr and Griffith, 1992). L-NAME is an order of magnitude less potent than L-NNA and requires the presence of esterase activity to fully exhibit its inhibitory action. These compounds have limited selectivity for individual NOS isoforms and relatively low inhibitory potency. However, some L-arginine analogues exhibit a certain degree of selectivity mostly toward the constitutive isoforms. For example, L-NNA and L-NAME show selectivity toward eNOS. Conversely, L-NMA and L-NAA have no significant preference for either isoform.

The mechanisms of inhibition of NOS activity by these compounds vary but all involve occupation of the L-arginine binding site, which prevents the metabolism of this amino acid to NO (Marletta, 1994a; Southan and Szabo, 1996; Stuehr and Griffith, 1992). Thus, they are competitive inhibitors and their effects can be largely reversed by L-arginine. However, L-NMA also causes slow irreversible inhibition of iNOS via a mechanism-based process in which the inhibitor is enzymatically converted to *N*-hydroxy-*N*-methyl-L-arginine by NADPH-dependent hydroxylation. Similarly, several of these analogues have been reported to inhibit NOS activity by mechanisms that are independent of their ability to compete with L-arginine for binding. For example, L-NMA causes loss of heme from the enzyme and inhibits L-arginine uptake by cells (a rate-limiting step in NO production by iNOS). In addition, L-NMA can be metabolized to NO by NOS. L-NAME and other esters of L-arginine have been reported to act as muscarinic receptor antagonists. More general nonspecific effects of L-arginine analogues include inhibition of activity of iron-containing enzymes such as catalase and interference with various iron-containing systems (Marletta, 1994a; Southan and Szabo, 1996).

L-NMA and dimethylarginines such as symmetric and asymmetric dimethylarginine are endogenously produced inhibitors of NOS that can accumulate to significant levels in the plasma of patients with chronic renal insufficiency and hypercholesterolemia (Aneman *et al.*, 1994; Marletta, 1994a).

In addition to L-arginine analogues, derivatives of L-citrulline and L-lysine are also being studied as potential inhibitors of NOS. L-Thiocitrulline is a very potent nonselective inhibitoı of NOS that acts by binding to heme and reducing its redox capacity (Frey *et al.*, 1994; Joly *et al.*, 1995; Narayanan and Griffith, 1994). On the other hand, *S*-ethyl- and *S*-methyl-thiocitrulline appear to be substantially more potent inhibitors of bNOS than of iNOS and eNOS (Furfine *et al.*, 1994). However, their clinical usefulness may be limited because of their poor uptake into cells. The inhibitory effects of all thiocitrulline derivatives can be reversed by L-arginine (Narayanan *et al.*, 1995). Derivatives of L-lysine are being studied as potentially selective inhibitors of iNOS (Moore *et al.*, 1994; Southan and Szabo, 1996).

4.2. Non-Amino-Acid-Based Nitrogen-Containing Inhibitors of NOS

A large number of nitrogen-containing compounds are being studied as inhibitors of NOS. Of these, only aminoguanidines, isothioureas, and some imidazoles and indazoles appear to have isoform-selective inhibitory effects (Fukuto and Chaudhuri, 1995; Griffiths *et al.*, 1995; Moore *et al.*, 1993; Nakane *et al.*, 1995; Southan *et al.*, 1995a,b).

Aminoguanidine is a selective inhibitor of iNOS that has been used successfully in several models of inflammation and endotoxic shock (Cross *et al.*, 1994; Griffiths *et al.*, 1995; Weigert *et al.*, 1995). This compound has low toxicity and significant potential for clinical use. Although the selectivity of aminoguanidine for iNOS may vary in different tissues, it is as potent as L-NMA in its ability to inhibit iNOS in macrophages and significantly less potent than this arginine analogue in inhibiting cNOS (Southan and Szabo, 1996). Aminoguanidine appears to inhibit NOS by binding as ligand to the heme iron at the catalytic site (Southan and Szabo, 1996). L-Arginine attenuates this inhibition and therefore competitive binding with this amino acid appears to be involved in the mechanism of NOS inhibition. Recent reports suggest that aminoguanidine is a mechanism-based inhibitor, its inhibitory effect increasing with time of exposure to NOS (Sennequier and Stuehr, 1996; Southan and Szabo, 1996).

S-substituted isothioureas constitute a new group of NOS inhibitors that are reported to be 10- to 30-fold more potent than L-NMA (Southan *et al.*, 1995b). The selectivity of these agents for NOS isoforms is variable. *S*-methylisothiourea and *S*-(2-aminoethyl)isothiourea appear to be relatively selective toward iNOS whereas *S*-ethylisothiourea and *S*-isopropylisothiourea are potent inhibitors of eNOS and iNOS with little selectivity toward either isoform (Aranow *et al.*, 1996; Garvey *et al.*, 1994; Nakane *et al.*, 1995; Seo *et al.*, 1996; Szabo *et al.*, 1994). Their mechanism of action is unclear but the NOS inhibition is dose-dependently prevented by excess of L-arginine, suggesting that isothioureas are competitive inhibitors of NOS at the L-arginine binding site (Nakane *et al.*, 1995; Southan *et al.*, 1995b). These agents have also been used successfully in the treatment of experimental endotoxic shock (Szabo *et al.*, 1994).

Of the imidazoles and indazoles, 7-nitroindazole is the most important NOS inhibitor developed to date. This compound is preferentially taken up by neurons, making it highly selective for bNOS over eNOS (Southan and Szabo, 1996). In cell homogenates, 7-nitroindazole is also a potent inhibitor of eNOS (Southan and Szabo, 1996). In experimental animals, 7-nitroindazole has potent antinociceptive effects, inhibits hippocampal long-term potentiation, causes distinct changes in regional cerebral blood flow, and protects against neurotoxicity from middle cerebral artery occlusion, all of these effects related to selective bNOS inhibition (Doyle *et al.*, 1996; Moore *et al.*, 1993; Southan and Szabo, 1996). The mechanism of NOS inhibition by 7-nitroindazole appears to be related to binding of this compound to the heme group of the enzyme affecting both the L-arginine and BH_4 binding sites (Mayer *et al.*, 1994).

4.3. Compounds that Interfere with Cofactor Availability

As reviewed above, NOS requires CaM and BH_4 for its activity. Agents that inhibit CaM binding or synthesis of BH_4 are therefore important candidates for use as NOS inhibitors.

Compounds that bind to CaM thereby impeding its binding to NOS will inhibit enzyme activity. These inhibitors are therefore specific for cNOS, which is dependent on Ca^{2+}-calmodulin for activity (Bredt and Snyder, 1990). CaM inhibitors such as calcineurin, trifluoperazine, chlorpromazine, calmidazolium, W-7, and fendiline have been used extensively for *in vitro* studies of NOS (Marletta, 1994a; Stuehr and Griffith, 1992). However, the usefulness of these agents for *in vivo* studies is limited by their lack of selectivity among the various CaM-dependent systems. Recent studies suggest that the interaction of CaM with NOS is unique and, therefore, the design of agents that will selectively interfere with this process may eventually be possible (Marletta, 1994a; Zoche *et al.*, 1996).

Tetrahydrobiopterin synthesis can occur via two distinct pathways (Masada *et al.*, 1990): One is a *de novo* pathway that utilizes guanosine triphosphate (GTP) as a substrate and in which GTP cyclohydrolase I (GTPCH) is the rate-limiting enzyme. This pathway appears to be the main source of BH_4 during induced NO synthesis and can be blocked by the selective GTPCH inhibitor 2,4-diamino-6-hydroxy pyrimidine. The second is a salvage pathway that utilizes dihydropteridines such as dihydrobiopterin and sepiapterin, which are subsequently converted to BH_4 by dihydrofolate reductase. This pathway can be blocked by the sepiapterin reductase inhibitors *N*-acetyl-serotonin, phenprocoumon, and dicumarol. NO production can be reduced by blockade of either pathway (Amoah-Apraku *et al.*, 1995b; Gross and Levi, 1992). However, the use of inhibitors of BH_4 synthesis to manipulate NOS activity is limited by the fact that BH_4 is also a cofactor for important aromatic amino acid hydroxylases that participate in the biosynthesis of norepinephrine, epinephrine, dopamine, and serotonin.

4.4. Agents that Inhibit NOS Expression

These inhibitors are obviously directed to interfere with the expression of iNOS. As mentioned above, iNOS requires activation of the transcription factor NF-κB for its expression. NF-κB inhibitors such as glucocorticoids, aspirin, pyrrolidine dithiocarbamate, and antioxidants including vitamin E have all been shown to inhibit iNOS expression. This inhibition may explain some of the beneficial effects of glucocorticoids and aspirin in various inflammatory disorders in which NO is implicated as an injurious factor.

5. ASSAYS FOR MEASURING NOS ACTIVITY

NOS activity can be determined by measuring the products of its reaction (L-citrulline and NO) or the effects of NO on target systems (guanylate cyclase activity, bioassays) (Hevel and Marletta, 1994; Knowles and Moncada, 1994).

5.1. L-Citrulline Conversion Assay

Measurement of the generation of ^{14}C- or ^{3}H-labeled L-citrulline from labeled L-arginine is a simple and widely used method for the determination of NOS activity. This assay takes advantage of the polar nature of L-arginine, which can be easily separated from nonpolar L-citrulline by ion-exchange chromatography (Bredt and Snyder, 1989; Davda *et al.*, 1994). It is commonly used to measure NOS activity in crude cell or tissue extracts (Bredt and Snyder, 1989; Davda *et al.*, 1994) and in cultured cells *in situ* (Davda *et al.*, 1993, 1994). The L-citrulline conversion assay is very sensitive detecting less than 100 nM of product and is unaffected by the presence of extraneous factors that affect spectroscopic NO measurements (e.g., hemoglobin, opalescence in solutions). For maximal accuracy in obtaining quantitative results it is important that all L-arginine present in the sample be removed or accounted for (e.g., 100 μM in tissues; 500–1000 μM in cell culture medium), and that unlabeled L-citrulline (1 mM) be added to prevent recycling of the labeled product.

5.2. Assays for the Measurement of NO in Biological Fluids

The half-life of NO in biological fluids is approximately 6 s (Moncada *et al.*, 1991), making direct measurement of this compound very difficult. The generation of NO can be measured semiquantitatively using spectroscopic and electrochemical methods. Spectroscopic methods include chemiluminescence, ultraviolet (UV)–visible spectroscopy, and electron spin resonance (paramagnetic resonance). Electrochemical methods make use of modified glass microelectrodes or specially designed porphyrinic microsensors that permit direct *in situ* measurement of NO in biological samples.

The chemiluminescence method measures the intensity of the fluorescent radiation emitted by nitrogen dioxide ($\cdot NO_2$) formed after chemical oxidation of NO by ozone (Hevel and Marletta, 1994; Kiechle and Malinski, 1993; Knowles and Moncada, 1994). This method requires initial reduction of the oxygenation products of NO, NO_2^- and NO_3^-, back to NO, and passage of NO from the liquid to the gas phase using an inert gas as carrier. This method has a detection limit of approximately 20 nM and is probably the best one for determination of the total amount of NO released by a system (Kiechle and Malinski, 1993).

Two different UV–visible spectroscopic methods are widely used for the determination of NO. The oxygenation products of NO, NO_2^- and NO_3^-, can be measured by a colorimetric assay using the Griess reagent, which is a mixture of sulfanilic acid and *N*-(1-naphthyl)ethylenediamine (Ding *et al.*, 1988; Hevel and Marletta, 1994; Kiechle and Malinski, 1993). First, NO_3^- needs to be converted to NO_2^- by nitrate reductase or metallic catalysts. The spectrum of the product of the reaction of NO with *N*-(1-naphthyl) ethylenediamine shows a band at 548 nm. The absorbance of this peak is proportional to the concentration of NO. The limit of detection of this method is approximately 50 nM.

NO oxidizes oxyhemoglobin to methemoglobin, which yields a high absorbance band at 406 nm that is used as the analytical signal for spectroscopy (Ding *et al.*, 1988; Hevel and Marletta, 1994; Kiechle and Malinski, 1993; Knowles and Moncada, 1994). This method is very sensitive with a detection limit of 2 nM. However, a number of factors can interfere with this assay including color or opalescence in samples and the presence of blood resulting in hemoglobin concentrations above 5 μM.

With a detection limit of approximately 4 μM, the electron spin resonance method is the least sensitive of all. This method is useful primarily for the study of interactions between NO and other molecules or components of biological systems (Kiechle and Malinski, 1993).

Electrochemical detection methods using NO-sensitive microsensors are very sensitive and can be useful for monitoring *in situ* NO levels on the surface of cell membranes and measuring the kinetics of NO release (Kiechle and Malinski, 1993). Their limit of detection ranges from 500 nM for the modified glass (Clark) electrode to 10 nM for the porphyrinic microsensor. One potential problem with these methods is the difficulty in obtaining adequate calibration consistently especially in studies measuring NO production under flow conditions.

5.3. Biological Assays of NO Production

The most commonly used bioassay for NO production is the measurement of guanylate cyclase activity both in soluble form and within cells. This assay is very sensitive and has been extremely useful in investigations of the L-arginine:NO pathway in a variety of tissues (Bredt and Snyder, 1989; Hevel and Marletta, 1994; Knowles and Moncada, 1994). However, it does not permit accurate quantitation of NOS activity or NO production and is limited by potential interference from exogenous factors such as oxyhemoglobin, superoxide anion, and other compounds that react with NO and prevent it from reacting with guanylate cyclase.

6. CONCLUSIONS

The discovery of the L-arginine:NO pathway has proven to be one of the most important and exciting scientific developments of the last decade. This pathway is involved in a myriad of physiologic roles that range from the control of vascular tone and blood pressure to the modulation of skeletal muscle function. An increasing number of pathological conditions are now being recognized to be the result of either excessive or insufficient NO production. Intensive efforts are being dedicated to the development of selective NOS inhibitors with the hope of targeting specific tissue sources of excessive NO production. Similarly, genetic approaches to treat conditions resulting from reduced NOS expression are likely to be explored in the near future. New developments in the field of NO research are likely to continue at the current rapid pace. This should bring important additional benefits to most areas of clinical medicine.

7. REFERENCES

Adachi, H., Iida, S., Oguchi, S., Ohshima, H., Suzuki, H., Nagasaki, K., Kawasaki, H., Sugimura, T., and Esumi, H., 1993, Molecular cloning of a cDNA encoding an inducible calmodulin-dependent nitric-oxide synthase from rat liver and its expression in COS-1 cells, *Eur. J. Biochem.* **217**:37–43.

Adcock, I. M., Brown, C. R., Kwon, O., and Barnes, P. J., 1994, Oxidative stress induces NF kappa b DNA binding and inducible NOS messenger RNA in human epithelial cells, *Biochem. Biophys. Res. Commun.* **199**:1518–1524.

Albakri, Q. A., and Stuehr, D. J., 1996, Intracellular assembly of inducible NO synthase is limited by nitric oxide-mediated changes in heme insertion and availability, *J. Biol. Chem.* **271**:5414–5421.

Amoah-Apraku, B., Chandler, L. J., Harrison, J., Tang, S. S., Ingelfinger, J. R., and Guzman, N. J., 1995a, NF-κB and transcriptional control of renal epithelial inducible nitric oxide synthase, *Kidney Int.* **48**:674–682.

Amoah-Apraku, B., Tang, S. S., Ingelfinger, J. R., and Guzman, N. J., 1995b, Guanosine triphosphate cyclohydrolase I regulates nitric oxide synthesis in renal proximal tubules, *J. Am. Soc. Nephrol.* **5**:1630–1633.

Anagli, J., Hofmann, F., Quadroni, M., Vorherr, T., and Carafoli, E., 1995, The calmodulin-binding domain of the inducible (macrophage) nitric oxide synthase, *Eur. J. Biochem.* **233**:701–708.

Aneman, A., Backman, V., Snygg, J., Vonbothmer, C., Fandriks, L., and Pettersson, A., 1994, Accumulation of an endogenous inhibitor of nitric oxide synthase during graded hemorrhagic shock, *Circ. Shock* **44**:111–114.

Aranow, J. S., Zhuang, J., Wang, H. L., Larkin, V., Smith, M., and Fink, M. P., 1996, A selective inhibitor of inducible nitric oxide synthase prolongs survival in a rat model of bacterial peritonitis: Comparison with two nonselective strategies, *Shock* **5**:116–121.

Arnet, U. A., McMillan, A., Dinerman, J. L., Ballermann, B., and Lowenstein, C. J., 1996, Regulation of endothelial nitric-oxide synthase during hypoxia, *J. Biol. Chem.* **271:**15069–15073.

Auphan, N., DiDonato, J. A., Rosette, C., Helmberg, A., and Karin, M., 1995, Immunosuppression by glucocorticoids: Inhibition of NF-κB through induction of I-κB synthesis, *Science* **270:**286–290.

Bates, T. E., Loesch, A., Burnstock, G., and Clark, J. B., 1996, Mitochondrial nitric oxide synthase: A ubiquitous regulator of oxidative phosphorylation? *Biochem. Biophys. Res. Commun.* **218:**40–44.

Belhassen, L., Kelly, R. A., Smith, T. W., and Balligand, J. L., 1996, Nitric oxide synthase (NOS3) and contractile responsiveness to adrenergic and cholinergic agonists in the heart—Regulation of NOS3 transcription in vitro and in vivo by cyclic adenosine monophosphate in rat cardiac myocytes, *J. Clin. Invest.* **97:**1908–1915.

Bogdan, C., Vodovotz, Y., Paik, J., Xie, Q., and Nathan, C., 1994, Mechanism of suppression of nitric oxide synthase expression by interleukin-4 in primary mouse macrophages, *J. Leukocyte Biol.* **55:**227–233.

Bredt, D. S., 1996, Targeting nitric oxide to its targets, *Proc. Soc. Exp. Biol. Med.* **211:**41–48.

Bredt, D. S., and Snyder, S. H., 1989, Nitric oxide mediates glutamate-linked enhancement of cGMP levels in the cerebellum, *Proc. Natl. Acad. Sci. USA* **86:**9030–9033.

Bredt, D. S., and Snyder, S. H., 1990, Isolation of nitric oxide synthase, a calmodulin-requiring enzyme, *Proc. Natl. Acad. Sci. USA* **87:**682–685.

Bredt, D. S., Ferris, C. D., and Snyder, S. H., 1992, Nitric oxide synthase regulatory sites. Phosphorylation by cyclic AMP-dependent protein kinase, protein kinase C, and calcium/calmodulin protein kinase; identification of flavin and calmodulin binding sites, *J. Biol. Chem.* **267:**10976–10981.

Bredt, D. S., Hwang, P. M., Glatt, C. E., Lowenstein, C., Reed, R. R., and Snyder, S. H., 1991, Cloned and expressed nitric oxide synthase structurally resembles cytochrome P-450 reductase, *Nature* **351:**714–718.

Brenman, J. E., Chao, D. S., Gee, S. H., McGee, A. W., Craven, S. E., Santillano, D. R., Wu, Z. Q., Huang, F., Xia, H. H., Peters, M. F., Froehner, S. C., and Bredt, D. S., 1996, Interaction of nitric oxide synthase with the postsynaptic density protein PSD-95 and alpha 1-syntrophin mediated by PDZ domains, *Cell* **84:**757–767.

Buga, G. M., Griscavage, J. M., Rogers, N. E., and Ignarro, L. J., 1993, Negative feedback regulation of endothelial cell function by nitric oxide, *Circ. Res.* **73:**808–812.

Cattell, V., and Jansen, A., 1995, Inducible nitric oxide synthase in inflammation, *Histochem. J.* **27:**777–784.

Chandler, L. J., Kopnisky, K., Richards, E., Crews, F. T., and Sumners, C., 1995, Angiotensin II decreases inducible nitric oxide synthase expression in rat astroglial cultures, *Am. J. Physiol. Cell. Physiol.* **37:**C700–C707.

Chartrain, N. A., Geller, D. A., Koty, P. P., Sitrin, N. F., Nussler, A. K., Hoffman, E. P., Billiar, T. R., Hutchinson, N. I., and Mudgett, J. S., 1994, Molecular cloning, structure, and chromosomal localization of the human inducible nitric oxide synthase gene, *J. Biol. Chem.* **269:**6765–6772.

Chen, P. F., Tsai, A. L., Berka, V., and Wu, K. K., 1996, Endothelial nitric-oxide synthase—Evidence for bidomain structure and successful reconstitution of catalytic activity from two separate domains generated by a baculovirus expression system, *J. Biol. Chem.* **271:**14631–14635.

Chen, P. Y., and Sanders, P. W., 1993, Role of nitric oxide synthesis in salt-sensitive hypertension in Dahl/Rapp rats, *Hypertension* **22:**812–818.

Cho, H. J., Martin, E., Xie, Q. W., Sassa, S., and Nathan, C., 1995, Inducible nitric oxide synthase: Identification of amino acid residues essential for dimerization and binding of tetrahydrobiopterin, *Proc. Natl. Acad. Sci. USA* **92:**11514–11518.

Chung, E., Curtis, D., Chen, G., Marsden, P. A., Twells, R., Xu, W. M., and Gardiner, M., 1996, Genetic evidence for the neuronal nitric oxide synthase gene (NOS1) as a susceptibility locus for infantile pyloric stenosis, *Am. J. Hum. Genet.* **58:**363–370.

Colasanti, M., Persichini, T., Menegazzi, M., Mariotto, S., Giordano, E., Caldarera, C. M., Sogos, V., Lauro, G. M., and Suzuki, H., 1995, Induction of nitric oxide synthase mRNA expression—Suppression by exogenous nitric oxide, *J. Biol. Chem.* **270:**26731–26733.

Corbett, J. A., Kwon, G., Marino, M. H., Rodi, C. P., Sullivan, P. M., Turk, J., and McDaniel, M. L., 1996, Tyrosine kinase inhibitors prevent cytokine-induced expression of iNOS and COX-2 by human islets, *Am. J. Physiol. Cell. Physiol.* **39:**C1581–C1587.

Cross, A. H., Misko, T. P., Lin, R. F., Hickey, W. F., Trotter, J. L., and Tilton, R. G., 1994, Aminoguanidine, an inhibitor of inducible nitric oxide synthase, ameliorates experimental autoimmune encephalomyelitis in SJL mice, *J. Clin. Invest.* **93:**2684–2690.

Davda, R. K., Chandler, L. J., Crews, F. T., and Guzman, N. J., 1993, Ethanol enhances the endothelial nitric oxide synthase response to agonists, *Hypertension* **21:**939–943.

Davda, R. K., Chandler, L. J., and Guzman, N. J., 1994, Protein kinase C modulates receptor-independent activation of endothelial nitric oxide synthase, *Eur. J. Pharmacol. Mol. Pharmacol.* **266:**237–244.

Dawson, T. M., and Dawson, V. L., 1996, Nitric oxide synthase: Role as a transmitter/mediator in the brain and endocrine system, *Annu. Rev. Med.* **47:**219–227.

Dawson, V. L., and Dawson, T. M., 1996a, Nitric oxide actions in neurochemistry, *Neurochem. Int.* **29:**97–110.

Dawson, V. L., and Dawson, T. M., 1996b, Nitric oxide neurotoxicity, *J. Chem. Neuroanat.* **10:**179–190.

Deng, A. Y., and Rapp, J. P., 1995, Locus for the inducible, but not a constitutive, nitric oxide synthase cosegregates with blood pressure in the Dahl salt-sensitive rat, *J. Clin. Invest.* **95:**2170–2177.

Deng, A. Y., Rapp, J. P., Kato, H., and Bihoreau, M. T., 1995, Linkage mapping of the neuronal nitric oxide synthase gene (Nos1) to rat chromosome 12, *Mamm. Genome* **6:**824.

Devera, M. E., Shapiro, R. A., Nussler, A. K., Mudgett, J. S., Simmons, R. L., Morris, S. M., Billiar, T. R., and Geller, D. A., 1996, Transcriptional regulation of human inducible nitric oxide synthase (NOS2) gene by cytokines: Initial analysis of the human NOS2 promoter, *Proc. Natl. Acad. Sci. USA* **93:**1054–1059.

Dinerman, J. L., Dawson, T. M., Schell, M. J., Snowman, A., and Snyder, S. H., 1994, Endothelial nitric oxide synthase localized to hippocampal pyramidal cells: Implications for synaptic plasticity, *Proc. Natl. Acad. Sci. USA* **91:**4214–4218.

Ding, A. H., Nathan, C. F., and Stuehr, D. J., 1988, Release of reactive nitrogen intermediates and reactive oxygen intermediates from mouse peritoneal macrophages, *J. Immunol.* **141:**2407–2412.

Dong, Z., Qi, X., Xie, K., and Fidler, I. J., 1993, Protein tyrosine kinase inhibitors decrease induction of nitric oxide synthase activity in lipopolysaccharide-responsive and lipopolysaccharide-nonresponsive murine macrophages, *J. Immunol.* **151:**2717–2724.

Doyle, C., Holscher, C., Rowan, M. J., and Anwyl, R., 1996, The selective neuronal NO synthase inhibitor 7-nitro-indazole blocks both long-term potentiation and depotentiation of field EPSPs in rat hippocampal CA1 in vivo, *J. Neurosci.* **16:**418–424.

Durante, W., Liao, L., Iftikhar, I., O'Brien, W. E., and Schafer, A. I., 1996, Differential regulation of L-arginine transport and nitric oxide production by vascular smooth muscle and endothelium, *Circ. Res.* **78:**1075–1082.

Eason, S., and Martin, W., 1995, Involvement of tyrosine kinase and protein kinase C in the induction of nitric oxide synthase by lipopolysaccharide and interferon-gamma in J774 macrophages, *Arch. Int. Pharmacodyn. Ther.* **330:**225–240.

Farivar, R. S., Chobanian, A. V., and Brecher, P., 1996, Salicylate or aspirin inhibits the induction of the inducible nitric oxide synthase in rat cardiac fibroblasts, *Circ. Res.* **78:**759–768.

Feinstein, D. L., Galea, E., Cermak, J., Chugh, P., Lyandvert, L., and Reis, D. J., 1994, Nitric oxide synthase expression in glial cells: Suppression by tyrosine kinase inhibitors, *J. Neurochem.* **62:**811–814.

Flodstrom, M., Welsh, N., and Eizirik, D. L., 1996, Cytokines activate the nuclear factor kappa B (NF-kappa B) and induce nitric oxide production in human pancreatic islets, *FEBS Lett.* **385:**4–6.

Forstermann, U., Gath, I., Schwarz, P., Closs, E. I., and Kleinert, H., 1995a, Isoforms of nitric oxide synthase—Properties, cellular distribution and expressional control, *Biochem. Pharmacol.* **50:**1321–1332.

Forstermann, U., Kleinert, H., Gath, I., Schwarz, P., Closs, E. I., and Dun, N. J., 1995b, Expression and expressional control of nitric oxide synthases in various cell types, in *Nitric Oxide* (L. Ignarro and F. Murad, eds.) pp. 171–186, San Diego, Academic Press.

Frey, C., Narayanan, K., McMillan, K., Spack, L., Gross, S. S., Masters, B. S., and Griffith, O. W., 1994, L-Thiocitrulline—a stereospecific, heme-binding inhibitor of nitric-oxide synthases, *J. Biol. Chem.* **269:**26083–26091.

Fukuto, J. M., and Chaudhuri, G., 1995, Inhibition of constitutive and inducible nitric oxide synthase: Potential selective inhibition, *Annu. Rev. Pharmacol. Toxicol.* **35:**165–194.

Furfine, E. S., Harmon, M. F., Paith, J. E., Knowles, R. G., Salter, M., Kiff, R. J., Duffy, C., Hazelwood, R., Oplinger, J. A., and Garvey, E. P., 1994, Potent and selective inhibition of human nitric oxide synthases—Selective inhibition of neuronal nitric oxide synthase by S-methyl-L-thiocitrulline and S-ethyl-L-thiocitrulline, *J. Biol. Chem.* **269:**26677–26683.

Garciacardena, G., Oh, P., Liu, J. W., Schnitzer, J. E., and Sessa, W. C., 1996, Targeting of nitric oxide synthase to endothelial cell caveolae via palmitoylation: Implications for nitric oxide signaling, *Proc. Natl. Acad. Sci. USA* **93**:6448–6453.

Garvey, E. P., Oplinger, J. A., Tanoury, G. J., Sherman, P. A., Fowler, M., Marshall, S., Harmon, M. F., Paith, J. E., and Furfine, E. S., 1994, Potent and selective inhibition of human nitric oxide synthases—inhibition by non-amino acid isothioureas, *J. Biol. Chem.* **269**:26669–26676.

Geng, Y., Maier, R., and Lotz, M., 1995, Tyrosine kinases are involved with the expression of inducible nitric oxide synthase in human articular chondrocytes, *J. Cell. Physiol.* **163**:545–554.

Ghosh, D. K., Abusoud, H. M., and Stuehr, D. J., 1996, Domains of macrophage NO synthase have divergent roles in forming and stabilizing the active dimeric enzyme, *Biochemistry* **35**:1444–1449.

Gill, D. J., Low, B. C., and Grigor, M. R., 1996, Interleukin-1 beta and tumor necrosis factor-alpha stimulate the cat-2 gene of the L-arginine transporter in cultured vascular smooth muscle cells, *J. Biol. Chem.* **271**:11280–11283.

Griffiths, M. J. D., Messent, M., Curzen, N. P., and Evans, T. W., 1995, Aminoguanidine selectively decreases cyclic CMP levels produced by inducible nitric oxide synthase, *Am. J. Respir. Crit. Care Med.* **152**:1599–1604.

Griscavage, J. M., Wilk, S., and Ignarro, L. J., 1996, Inhibitors of the proteasome pathway interfere with induction of nitric oxide synthase in macrophages by blocking activation of transcription factor NF-kappa B, *Proc. Natl. Acad. Sci. USA* **93**:3308–3312.

Gross, S. S., and Levi, R., 1992, Tetrahydrobiopterin synthesis: An absolute requirement for cytokine-induced nitric oxide generation by vascular smooth muscle, *J. Biol. Chem.* **267**:25722–25729.

Grozdanovic, Z., Gosztonyi, G., and Gossrau, R., 1996, Nitric oxide synthase I (NOS-I) is deficient in the sarcolemma of striated muscle fibers in patients with Duchenne muscular dystrophy, suggesting an association with dystrophin, *Acta Histochem.* **98**:61–69.

Hecker, M., Preiss, C., Schinikerth, V. B., and Busse, R., 1996, Antioxidants differentially affect nuclear factor kappa B-mediated nitric oxide synthase expression in vascular smooth muscle cells, *FEBS Lett.* **380**:224–228.

Hevel, J. M., and Marletta, M. A., 1994, Nitric-oxide synthase assays, in *Methods Enzymol.* **233**:250–258.

Hirsch, D. B., Steiner, J. P., Dawson, T. M., Mammen, A., Hayek, E., and Snyder, S. H., 1993, Neurotransmitter release regulated by nitric oxide in PC-12 cells and brain synaptosomes, *Curr. Biol.* **3**:749–754.

Huang, P. L., Dawson, T. M., Bredt, D. S., Snyder, S. H., and Fishman, M. C., 1993, Targeted disruption of the neuronal nitric oxide synthase gene, *Cell* **75**:1273–1286.

Huang, Z. H., Huang, P. L., Panahian, N., Dalkara, T., Fishman, M. C., and Moskowitz, M. A., 1994, Effects of cerebral ischemia in mice deficient in neuronal nitric oxide synthase, *Science* **265**:1883–1885.

Hunot, S., Boissiere, F., Faucheux, B., Brugg, B., Mouattprigent, A., Agid, Y., and Hirsch, E. C., 1996, Nitric oxide synthase and neuronal vulnerability in Parkinson's disease, *Neuroscience* **72**:355–363.

Ikeda, U., Maeda, Y., Kawahara, Y., Yokoyama, M., and Shimada, K., 1995, Angiotensin II augments cytokine-stimulated nitric oxide synthesis in rat cardiac myocytes, *Circulation* **92**:2683–2689.

Jaffrey, S. R., and Snyder, S. H., 1995, Nitric oxide: A neural messenger, *Annu. Rev. Cell. Dev. Biol.* **11**:417–440.

Joly, G. A., Narayanan, K., Griffith, O. W., and Kilbourn, R. G., 1995, Characterization of the effects of two new arginine/citrulline analogues on constitutive and inducible nitric oxide synthases in rat aorta, *Br. J. Pharmacol.* **115**:491–497.

Kadowaki, K., Kishimoto, J., Leng, G., and Emson, P. C., 1994, Up-regulation of nitric oxide synthase (Nos) gene expression together with Nos activity in the rat hypothalamo-hypophysial system after chronic salt loading—evidence of a neuromodulatory role of nitric oxide in arginine vasopressin and oxytocin secretion, *Endocrinology* **134**:1011–1017.

Kiechle, F. L., and Malinski, T., 1993, Nitric oxide: Biochemistry, pathophysiology, and detection, *Am. J. Clin. Pathol.* **100**:567–575.

Klatt, P., Pfeiffer, S., List, B. M., Lehner, D., Glatter, O., Bachinger, H. P., Werner, E. R., Schmidt, K., and Mayer, B., 1996, Characterization of heme-deficient neuronal nitric-oxide synthase reveals a role for heme in subunit dimerization and binding of the amino acid substrate and tetrahydrobiopterin, *J. Biol. Chem.* **271**:7336–7342.

Kleinert, H., Euchenhofer, C., Ihrigbiedert, I., and Forstermann, U., 1996a, Glucocorticoids inhibit the induction of nitric oxide synthase II by down-regulating cytokine-induced activity of transcription factor nuclear factor-kappa B, *Mol. Pharmacol.* **49**:15–21.

Kleinert, H., Euchenhofer, C., Ihrigbiedert, I., and Forstermann, U., 1996b, In murine 3T3 fibroblasts, different second messenger pathways resulting in the induction of NO synthase II (iNOS) converge in the activation of transcription factor NF-kappa B, *J. Biol. Chem.* **271:**6039–6044.

Knowles, R. G., and Moncada, S., 1994, Nitric oxide synthases in mammals, *Biochem. J.* **298:**249–258.

Kobzik, L., Stringer, B., Balligand, J. L., Reid, M. B., and Stamler, J. S., 1995, Endothelial-type nitric oxide synthase in skeletal muscle fibers: Mitochondrial relationships, *Biochem. Biophys. Res. Commun.* **211:**375–381.

Kopp, E., and Ghosh, S., 1994, Inhibition of NF-kappa B by sodium salicylate and aspirin, *Science* **265:**956–959.

Laubach, V. E., Shesely, E. G., Smithies, O., and Sherman, P. A., 1995, Mice lacking inducible nitric oxide synthase are not resistant to lipopolysaccharide-induced death, *Proc. Natl. Acad. Sci. USA* **92:**10688–10692.

Lowe, P. N., Smith, D., Stammers, D. K., Riverosmoreno, V., Moncada, S., Charles, I., and Boyhan, A., 1996, Identification of the domains of neuronal nitric oxide synthase by limited proteolysis, *Biochem. J.* **314:**55–62.

Lowenstein, C. J., Alley, E. W., Raval, P., Snowman, A. M., Snyder, S. H., Russell, S. W., and Murphy, W. J., 1993, Macrophage nitric oxide synthase gene—Two upstream regions mediate induction by interferon-gamma and lipopolysaccharide, *Proc. Natl. Acad. Sci. USA* **90:**9730–9734.

Lowenstein, C. J., Dinerman, J. L., and Snyder, S. H., 1994, Nitric oxide—A physiologic messenger, *Ann. Intern. Med.* **120:**227–237.

Macmicking, J. D., Nathan, C., Hom, G., Chartrain, N., Fletcher, D. S., Trumbauer, M., Stevens, K., Xie, Q. W., Sokol, K., Hutchinson, N., Chen, H., and Mudgett, J. S., 1995, Altered responses to bacterial infection and endotoxic shock in mice lacking inducible nitric oxide synthase, *Cell* **81:**641–650.

Marletta, M. A., 1993, Nitric oxide synthase: Function and mechanism, *Adv. Exp. Med. Biol.* **338:**281–284.

Marletta, M. A., 1994a, Approaches toward selective inhibition of nitric oxide synthase, *J. Med. Chem.* **37:**1899–1907.

Marletta, M. A., 1994b, Nitric oxide synthase: Aspects concerning structure and catalysis, *Cell* **78:927–930.**

Marsden, P. A., Heng, H., Duff, C. L., Shi, X. M., Tsui, L. C., and Hall, A. V., 1994, Localization of the human gene for inducible nitric oxide synthase (NOS2) to chromosome 17q11.2-q12, *Genomics* **19:**183–185.

Masada, M., Matsumoto, J., and Akino, M., 1990, Biosynthetic pathways of pteridines and their association with phenotypic expression in vitro in normal and neoplastic pigment cells from goldfish, *Pigment Cell Res.* **3:**61–70.

Mayer, B., Klatt, P., Werner, E. R., and Schmidt, K., 1994, Molecular mechanisms of inhibition of porcine brain nitric oxide synthase by the antinociceptive drug 7-nitro-indazole, *Neuropharmacology* **33:**1253–1259.

Michel, T., Li, G. K., and Busconi, L., 1993, Phosphorylation and subcellular translocation of endothelial nitric oxide synthase, *Proc. Natl. Acad. Sci. USA* **90:**6252–6256.

Mitchell, H. H., Shonle, H. A., and Grindley, H. S., 1916, The origin of nitrate in the urine, *J. Biol. Chem.* **24:**461.

Moncada, S., Palmer, R. M. J., and Higgs, E. A., 1991, Nitric oxide: Physiology, pathophysiology and pharmacology, *Pharmacol. Rev.* **43:**109–142.

Moore, P. K., Wallace, P., Gaffen, Z., Hart, S. L., and Babbedge, R. C., 1993, Characterization of the novel nitric oxide synthase inhibitor 7-nitro indazole and related indazoles: Antinociceptive and cardiovascular effects, *Br. J. Pharmacol.* **110:**219–224.

Moore, W. M., Webber, R. K., Jerome, G. M., Tjoeng, F. S., Misko, T. P., and Currie, M. G., 1994, L-N-6-(1-iminoethyl)lysine: A selective inhibitor of inducible nitric oxide synthase, *J. Med. Chem.* **37:**3886–3888.

Muller, J. M., Ziegler-Heitbrock, H. W., and Baeuerle, P. A., 1993, Nuclear factor kappa B, a mediator of lipopolysaccharide effects, *Immunobiology* **187:**233–256.

Myatt, L., Brockman, D. E., Eis, A., and Pollock, J. S., 1993a, Immunohistochemical localization of nitric oxide synthase in the human placenta, *Placenta* **14:**487–495.

Myatt, L., Brockman, D. E., Langdon, G., and Pollock, J. S., 1993b, Constitutive calcium-dependent isoform of nitric oxide synthase in the human placental villous vascular tree, *Placenta* **14:**373–383.

Nakane, M., Klinghofer, V., Kuk, J. E., Donnelly, J. L., Budzik, G. P., Pollock, J. S., Basha, F., and Carter, G. W., 1995, Novel potent and selective inhibitors of inducible nitric oxide synthase, *Mol. Pharmacol.* **47:**831–834.

Nakayama, I., Kawahara, Y., Tsuda, T., Okuda, M., and Yokoyama, M., 1994, Angiotensin-II inhibits cytokine-stimulated inducible nitric oxide synthase expression in vascular smooth muscle cells, *J. Biol. Chem.* **269**:11628–11633.

Narayanan, K., and Griffith, O. W., 1994, Synthesis of L-thiocitrulline, L-homothiocitrulline, and S-methyl-L-thiocitrulline—A new class of potent nitric oxide synthase inhibitors, *J. Med. Chem.* **37**:885–887.

Narayanan, K., Spack, L., McMillan, K., Kilbourn, R. G., Hayward, M. A., Masters, B. S. S., and Griffith, O. W., 1995, S-alkyl-L-thiocitrullines—Potent stereoselective inhibitors of nitric oxide synthase with strong pressor activity in vivo, *J. Biol. Chem.* **270**:11103–11110.

Nathan, C., and Xie, Q. W., 1994a, Nitric oxide synthases: Roles, tolls, and controls, *Cell* **78**:915–918.

Nathan, C., and Xie, Q. W., 1994b, Regulation of biosynthesis of nitric oxide, *J. Biol. Chem.* **269**:13725–13728.

Nelson, R. J., Demas, G. E., Huang, P. L., Fishman, M. C., Dawson, V. L., Dawson, T. M., and Snyder, S. H., 1995, Behavioural abnormalities in male mice lacking neuronal nitric oxide synthase, *Nature* **378**:383–386.

Noris, M., Morigi, M., Donadelli, R., Aiello, S., Foppolo, M., Todeschini, M., Orisio, S., Remuzzi, G., and Remuzzi, A., 1995, Nitric oxide synthesis by cultured endothelial cells is modulated by flow conditions, *Circ. Res.* **76**:536–543.

North, A. J., Star, R. A., Brannon, T. S., Ujiie, K., Wells, L. B., Lowenstein, C. J., Snyder, S. H., and Shaul, P. W., 1994, Nitric oxide synthase type I and type III gene expression are developmentally regulated in rat lung, *Am. J. Physiol.* **266**:L635–L641.

Nussler, A. K., and Billiar, T. R., 1993, Inflammation, immunoregulation and inducible nitric oxide synthase, *J. Leukocyte Biol.* **54**:171–178.

Palombella, V. J., Rando, O. J., Goldberg, A. L., and Maniatis, T., 1994, The ubiquitin–proteasome pathway is required for processing the NF-kappa B1 precursor protein and the activation of NF-kappa B, *Cell* **78**:773–785.

Peng, H. B., Libby, P., and Liao, J. K., 1995, Induction and stabilization of I kappa B alpha by nitric oxide mediates inhibition of NF-kappa b, *J. Biol. Chem.* **270**:14214–14219.

Pollock, J. S., Nakane, M., Buttery, L., Martinez, A., Springall, D., Polak, J. M., Förstermann, U., and Murad, F., 1993, Characterization and localization of endothelial nitric oxide synthase using specific monoclonal antibodies, *Am. J. Physiol. Cell. Physiol.* **265**:C1379–C1387.

Raij, L., 1994, Glomerular thrombosis in pregnancy: Role of the L-arginine–nitric oxide pathway, *Kidney Int.* **45**:775–781.

Reid, M. B., 1996, Reactive oxygen and nitric oxide in skeletal muscle, *News Physiol. Sci.* **11**:114–119.

Robinson, L. J., and Michel, T., 1995, Mutagenesis of palmitoylation sites in endothelial nitric oxide synthase identifies a novel motif for dual acylation and subcellular targeting, *Proc. Natl. Acad. Sci. USA* **92**:11776–11780.

Robinson, L. J., Weremowicz, S., Morton, C. C., and Michel, T., 1994, Isolation and chromosomal localization of the human endothelial nitric oxide synthase (NOS3) gene, *Genomics* **19**:350–357.

Robinson, L. J., Busconi, L., and Michel, T., 1995, Agonist-modulated palmitoylation of endothelial nitric oxide synthase, *J. Biol. Chem.* **270**:995–998.

Scheinman, R. I., Cogswell, P. C., Lofquist, A. K., and Baldwin, A. S., Jr., 1995, Role of transcriptional activation of IκBα in mediation of immunosuppression by glucocorticoids, *Science* **270**:283–286.

Schmidt, H. H. H. W., Pollock, J. S., Nakane, M., Gorsky, L. D., Förstermann, U., and Murad, F., 1991, Purification of a soluble isoform of guanylyl cyclase-activating factor synthase, *Proc. Natl. Acad. Sci. USA* **88**:365–369.

Sennequier, N., and Stuehr, D. J., 1996, Analysis of substrate-induced electronic, catalytic, and structural changes in inducible NO synthase, *Biochemistry* **35**:5883–5892.

Seo, H. G., Fujiwara, N., Kaneto, H., Asahi, M., Fujii, J., and Taniguchi, N., 1996, Effect of a nitric oxide synthase inhibitor, S-ethylisothiourea, on cultured cells and cardiovascular functions of normal and lipopolysaccharide-treated rabbits, *J. Biochem.* **119**:553–558.

Sessa, W. C., Barber, C. M., and Lynch, K. R., 1993, Mutation of N-myristoylation site converts endothelial cell nitric oxide synthase from a membrane to a cytosolic protein, *Circ. Res.* **72**:921–924.

Sessa, W. C., Pritchard, K., Seyedi, N., Wang, J., and Hintze, T. H., 1994, Chronic exercise in dogs increases coronary vascular nitric oxide production and endothelial cell nitric oxide synthase gene expression, *Circ. Res.* **74**:349–353.

Shultz, P. J., and Raij, L., 1992, Endogenously synthesized nitric oxide prevents endotoxin-induced glomerular thrombosis, *J. Clin. Invest.* **90:**1718–1725.

Siddhanta, U., Wu, C. Q., Abusoud, H. M., Zhang, J. L., Ghosh, D. K., and Stuehr, D. J., 1996, Heme iron reduction and catalysis by a nitric oxide synthase heterodimer containing one reductase and two oxygenase domains, *J. Biol. Chem.* **271:**7309–7312.

Singh, K., Balligand, J. L., Fischer, T. A., Smith, T. W., and Kelly, R. A., 1996, Regulation of cytokine-inducible nitric oxide synthase in cardiac myocytes and microvascular endothelial cells—Role of extracellular signal-regulated kinases 1 and 2 (ERK1/ERK2) and STAT1 alpha, *J. Biol. Chem.* **271:**1111–1117.

Southan, G. J., and Szabo, C., 1996, Selective pharmacological inhibition of distinct nitric oxide synthase isoforms, *Biochem. Pharmacol.* **51:**383–394.

Southan, G. J., Szabo, C., O'Connor, M. P., Salzman, A. L., and Thiemermann, C., 1995a, Amidines are potent inhibitors of nitric oxide synthases: Preferential inhibition of the inducible isoform, *Eur. J. Pharmacol. Mol. Pharm.* **291:**311–318.

Southan, G. J., Szabo, C., and Thiemermann, C., 1995b, Isothioureas: Potent inhibitors of nitric oxide synthases with variable isoform selectivity, *Br. J. Pharmacol.* **114:**510–516.

Spink, J., Cohen, J., and Evans, T. J., 1995, The cytokine responsive vascular smooth muscle cell enhancer of inducible nitric oxide synthase—Activation by nuclear factor-kappa B, *J. Biol. Chem.* **270:**29541–29547.

Stuehr, D. J., and Griffith, O. W., 1992, Mammalian nitric oxide synthases, *Adv. Enzymol.* **65:**287–346.

Su, Z. Z., Blazing, M. A., Fan, D., and George, S. E., 1995, The calmodulin–nitric oxide synthase interaction—Critical role of the calmodulin latch domain in enzyme activation, *J. Biol. Chem.* **270:**29117–29122.

Szabo, C., Southan, G. J., and Thiemermann, C., 1994, Beneficial effects and improved survival in rodent models of septic shock with S-methylisothiourea sulfate, a potent and selective inhibitor of inducible nitric oxide synthase, *Proc. Natl. Acad. Sci. USA* **91:**12472–12476.

Tetsuka, T., Baier, L. D., and Morrison, A. R., 1996, Antioxidants inhibit interleukin-1-induced cyclooxygenase and nitric-oxide synthase expression in rat mesangial cells—Evidence for post-transcriptional regulation, *J. Biol. Chem.* **271:**11689–11693.

Tomimoto, H., Akiguchi, I., Wakita, H., Nakamura, S., and Kimura, J., 1994, Histochemical demonstration of membranous localization of endothelial nitric oxide synthase in endothelial cells of the rat brain, *Brain Res.* **667:**107–110.

Tracey, W. R., Pollock, J. S., Murad, F., Nakane, M., and Forstermann, U., 1994, Identification of an endothelial-like type III NO synthase in LLC-PK1 kidney epithelial cells, *Am. J. Physiol.* **266:**C22–C28.

Vanderwinden, J. M., De, L. M., Schiffmann, S. N., Mailleux, P., Lowenstein, C. J., Snyder, S. H., and Vanderhaeghen, J. J., 1993, Nitric oxide synthase distribution in the enteric nervous system of Hirschsprung's disease, *Gastroenterology* **105:**969–973.

Venema, R. C., Sayegh, H. S., Kent, J. D., and Harrison, D. G., 1996, Identification, characterization, and comparison of the calmodulin-binding domains of the endothelial and inducible nitric oxide synthases, *J. Biol. Chem.* **271:**6435–6440.

Vodovotz, Y., Bogdan, C., Paik, J., Xie, Q.-W., and Nathan, C., 1993, Mechanisms of suppression of macrophage nitric oxide release by transforming growth factor-β, *J. Exp. Med.* **178:**605–613.

Wei, X. Q., Charles, I. G., Smith, A., Ure, J., Feng, C. J., Huang, F. P., Xu, D. M., Muller, W., Moncada, S., and Liew, F. Y., 1995, Altered immune responses in mice lacking inducible nitric oxide synthase, *Nature* **375:**408–411.

Weigert, A. L., Higa, E. M. S., Niederberger, M., McMurtry, I. F., Raynolds, M., and Schrier, R. W., 1995, Expression and preferential inhibition of inducible nitric oxide synthase in aortas of endotoxemic rats, *J. Am. Soc. Nephrol.* **5:**2067–2072.

Weiner, C. P., Lizasoain, I., Baylis, S. A., Knowles, R. G., Charles, I. G., and Moncada, S., 1994, Induction of calcium-dependent nitric oxide synthases by sex hormones, *Proc. Natl. Acad. Sci. USA* **91:**5212–5216.

Weisz, A., Oguchi, S., Cicatiello, L., and Esumi, H., 1994, Dual mechanism for the control of inducible-type NO synthase gene expression in macrophages during activation by interferon-gamma and bacterial lipopolysaccharide—Transcriptional and post-transcriptional regulation, *J. Biol. Chem.* **269:**8324–8333.

Xie, Q. W., and Nathan, C., 1994, The high-output nitric oxide pathway: role and regulation, *J. Leukocyte. Biol.* **56:**576–582.

Xie, Q., Cho, H. J., Calaycay, J., Mumford, R. A., Swiderek, K. M., Lee, T. D., Ding, A., Troso, T., and Nathan, C., 1992, Cloning and characterization of inducible nitric oxide synthase from mouse macrophages, *Science* **256:**225–228.

Xie, Q., Whisnant, R., and Nathan, C., 1993, Promoter of the mouse gene encoding calcium-independent nitric oxide synthase confers inducibility by interferon-γ and bacterial lipopolysaccharide, *J. Exp. Med.* **177**:1779–1784.

Xie, Q., Kashiwabara, Y., and Nathan, C., 1994, Role of transcription factor NF-kB/Rel in the induction of nitric oxide synthase, *J. Biol. Chem.* **269**:4705–4708.

Xue, C., Pollock, J., Schmidt, H. H. H. W., Ward, S. M., and Sanders, K. M., 1994, Expression of nitric oxide synthase immunoreactivity by interstitial cells of the canine proximal colon, *J. Auton. Nerv. Syst.* **49**:1–14.

Yamamoto, R., Bredt, D. S., Snyder, S. H., and Stone, R. A., 1993, The localization of nitric oxide synthase in the rat eye and related cranial ganglia, *Neuroscience* **54**:189–200.

Yoshida, K., Okamura, T., Kimura, H., Bredt, D. S., Snyder, S. H., and Toda, N., 1993, Nitric oxide synthase-immunoreactive nerve fibers in dog cerebral and peripheral arteries, *Brain Res.* **629**:67–72.

Yoshizumi, M., Perrella, M. A., Burnett, J. C., and Lee, M., 1993, Tumor necrosis factor downregulates an endothelial nitric oxide synthase mRNA by shortening its half-life, *Circ. Res.* **73**:205–209.

Zhang, R., Min, W., and Sessa, W. C., 1995, Functional analysis of the human endothelial nitric oxide synthase promoter—SP1 and GATA factors are necessary for basal transcription in endothelial cells, *J. Biol. Chem.* **270**:15320–15326.

Zhang, X., Verge, V., Wiesenfeld, H. Z., Ju, G., Bredt, D., Snyder, S. H., and Hökfelt, T., 1993, Nitric oxide synthase-like immunoreactivity in lumbar dorsal root ganglia and spinal cord of rat and monkey and effect of peripheral axotomy, *J. Comp. Neurol.* **335**:563–575.

Zhang, Z. G., Chopp, M., Zaloga, C., Pollock, J. S., and Forstermann, U., 1993, Cerebral endothelial nitric oxide synthase expression after focal cerebral ischemia in rats, *Stroke* **24**:2016–2021.

Zhang, Z. G., Chopp, M., Gautam, S., Zaloga, C., Zhang, R. L., Schmidt, H. H. H. W., Pollock, J. S., and Forstermann, U., 1994, Upregulation of neuronal nitric oxide synthase and mRNA, and selective sparing of nitric oxide synthase-containing neurons after focal cerebral ischemia in rat, *Brain Res.* **654**:85–95.

Zoche, M., Bienert, M., Beyermann, M., and Koch, K. W., 1996, Distinct molecular recognition of calmodulin-binding sites in the neuronal and macrophage nitric oxide synthases: A surface plasmon resonance study, *Biochemistry* **35**:8742–8747.

Zweier, J. L., Kuppasamy, P., and Lutty, G. A., 1988, Measurement of endothelial cell free radical generation: Evidence for a central mechanism of free radical injury in post ischemic tissues, *Proc. Natl. Acad. Sci. USA* **85**:4046–4056.

Chapter 10

The Chemical Biology of Nitric Oxide

David A. Wink, Martin Feelisch, Yoram Vodovotz, Jon Fukuto, and Matthew B. Grisham

1. INTRODUCTION

It has become apparent in the last 20 years that reactive chemical species play important roles in the etiology of many pathophysiological conditions. Reactive oxygen species (ROS), including oxygen-derived free radicals, have been invoked as causative agents in numerous diseases (Halliwell and Gutteridge, 1989a). In the late 1980s, it was discovered that another radical, nitric oxide (·NO), is formed *in vivo* and plays crucial roles in the regulation of numerous physiological processes ranging from blood pressure control to neurotransmission, as well as being a key component of the immune system (for reviews see Moncada *et al.*, 1991; Ignarro, 1990; Furchgott and Vanhoutte, 1989; Feldman *et al.*, 1993). Research has shown that this diatomic radical is produced in nearly every tissue *in vivo*, prompting the journal *Science* to choose ·NO as "molecule of the year" in 1992 (Culotta and Koshland, 1992).

The biological role of ·NO was discovered via two different lines of research. In the late 1970s, it was shown that the innermost cellular layer of a blood vessel, the endothelium, produces a labile substance that can migrate to the underlying smooth muscle, to activate soluble guanylate cyclase and lead to the relaxation of vascular smooth muscle (Furchgott and Zawadzki, 1980). As the chemical nature of this messenger molecule remained unknown, it was termed descriptively *endothelium-derived relaxation factor*

David A. Wink and Yoram Vodovotz Tumor Biology Section, Radiation Biology Branch, National Cancer Institute, Bethesda, Maryland 20892. **Martin Feelisch** Wolfson Institute for Biomedical Research, London W1P 9LN, England. **Jon Fukuto** Department of Molecular Pharmacology, University of California, Los Angeles, California 90269. **Matthew B. Grisham** Department of Physiology, Louisiana State University Medical Center, Shreveport, Louisiana 71130.

Reactive Oxygen Species in Biological Systems, edited by Gilbert and Colton. Kluwer Academic / Plenum Publishers, New York, 1999.

(EDRF). Through the 1980s, substantial work carried out by several groups demonstrated that the action of nitrovasodilators was mediated by ·NO (Feelisch and Noack, 1987; Feelisch *et al.*, 1988; Ignarro, 1989). In 1987, Ignarro and co-workers as well as Moncada and co-workers independently showed that ·NO was produced by the endothelium and, in fact, was EDRF (Ignarro *et al.*, 1987; Palmer *et al.*, 1987). Since these initial findings, substances other than ·NO, such as *S*-nitrosothiols, have been proposed to account for the biological activity of EDRF (Myers *et al.*, 1990; Stamler *et al.*, 1992). A recent study comparing different proposed chemical candidates for EDRF demonstrated that ·NO, and not another derivative or nitrogenous product, is the active principle of EDRF (Feelisch *et al.*, 1994). Concurrently with these activities in cardiovascular research, Hibbs and co-workers had discovered that cytotoxic activity of activated macrophages largely depends on supply with the amino acid, L-arginine, which has since been demonstrated to be the precursor for the biosynthesis of ·NO (Hibbs *et al.*, 1987). Further pioneering work showed that iron nitrosyl complexes were formed in activated macrophages (Lancaster and Hibbs, 1990; Bastian *et al.*, 1994; Drapier *et al.*, 1991; Vanin *et al.*, 1992; Pellat *et al.*, 1990). Tannenbaum and co-workers showed that infection resulted in a tenfold elevation of plasma nitrate and nitrite (Green *et al.*, 1981), and Stuehr and Marletta demonstrated that macrophages generate nitrate and nitrite via an enzyme now known as *nitric oxide synthase* (NOS) (Stuehr and Marletta, 1985; Stuehr and Nathan, 1989). Taken together, these observations showed that ·NO is a critical part of the immune response.

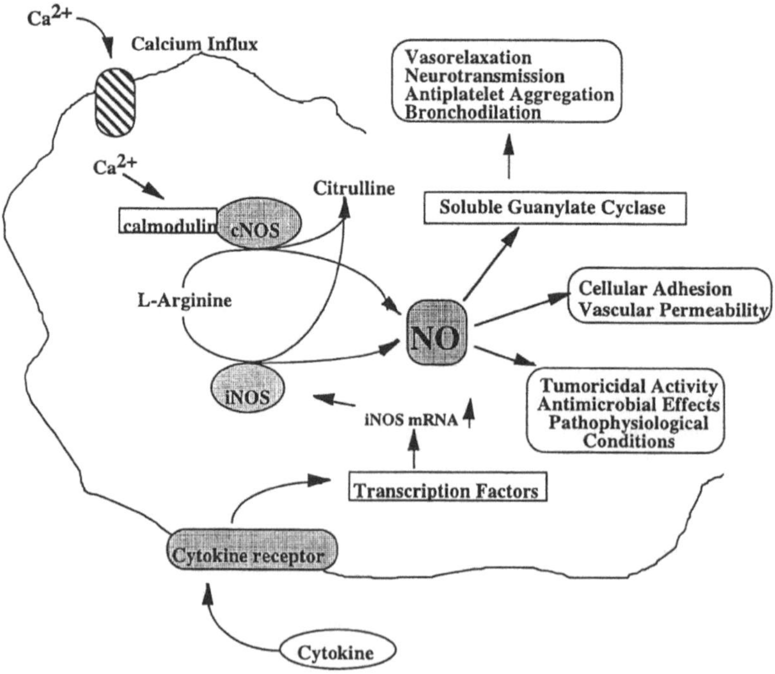

FIGURE 1. Regulation of cellular production of NO.

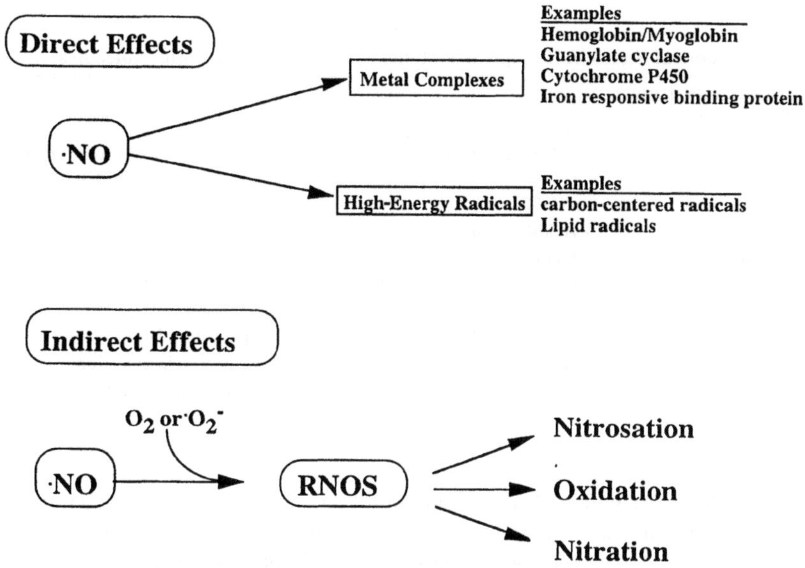

FIGURE 2. Direct and indirect effects of NO.

These initial observations in the immune and cardiovascular systems were extended by the work of Garthwaite and co-workers who discovered that ·NO is produced by neurons stimulated with glutamate (Garthwaite *et al.*, 1988). Development of an antibody to NOS has allowed a mapping of the distribution of this enzyme in the brain (Forstermann *et al.*, 1990; Bredt *et al.*, 1990, 1991). Since then, NOS has been found to be expressed throughout the body.

Nitric oxide is derived *in vivo* from the enzyme NOS (Chapter 9; Griffith and Stuehr, 1995; Nathan and Xie, 1994; Marletta, 1993, 1994; Stuehr *et al.*, 1995) (Figure 1). There are three isoforms of this enzyme: type I (neuronal nitric oxide synthase, nNOS; NOS1), type II (inducible nitric oxide synthase, iNOS; NOS2), and type III (endothelial nitric oxide synthase, eNOS; NOS3). Types I and III are usually constitutively present in the cell (cNOS); however, under certain conditions, their expression can also be induced. These isoenzymes are activated by an increase in intracellular calcium, which facilitates the binding of calmodulin to NOS, thus activating the enzyme (Stuehr *et al.*, 1995). The second class of NOS, the so-called inducible form, is expressed in cells after exposure to certain cytokines; however, some tissues express this isoform of NOS even under basal conditions. The major difference between the "constitutive" and "inducible" isoforms is the amount and duration of ·NO produced by either NOS. All three isoforms have similar specific activities when purified to homogeneity; however, the total amount of ·NO generated per cell by cNOS is low relative to that generated by iNOS. The flux of ·NO generated by cNOS is of short duration, whereas iNOS generates considerably higher concentrations of ·NO for periods of hours to days. Therefore, physiological versus potentially toxic actions of ·NO might be dictated by the presence and activity of specific

Table I
Chemical Stress in Biological Systems Mediated by Diatomic Molecules

Type of stress mediated by the indirect effects	Source	Chemistry
Nitrosative Stress	$\cdot NO/O_2$ reaction	Amine and thiol nitrosation to form nitrosamines and *S*-nitrosothiols
	$\cdot NO/\cdot O_2^-$ reaction	
	Acidic nitrite	
Oxidative Stress	Metal/peroxide ($E > 1.2$ V)	Hydroxylation
	$\cdot NO/\cdot O_2^-$ reaction ($E < 1.2$ V)	DNA base oxidation and strand breakage
	$\cdot NO/O_2$ reaction ($E < 0.7$ V)	Lipid peroxidation
	Acidic nitrite ($E < 0.7$ V)	Heme protein oxidation
		Catecholamine oxidation

isoforms of NOS. As will be described in the following text, this difference in fluxes of ·NO is critical to the understanding of the chemical reactions involving ·NO *in vivo*.

The concept of the "chemical biology of ·NO" takes into account the reactions involving this radical that can occur *in vivo*, and discusses them in the context of relevant biological effects (Wink *et al.*, 1996c). Conceptually, this concept shall provide a guide to the location and timing of the chemical reactions that ·NO can undergo *in vivo*. This chemical biology of ·NO encompasses two distinct categories, consisting of direct and indirect effects (Figure 2). The former are defined as those chemical reactions in which ·NO will interact directly with a biological target. The latter are defined as the chemistry mediated by reactive nitrogen oxide species (RNOS) derived from the interaction of ·NO with superoxide ($\cdot O_2^-$) or oxygen (O_2) (Table I). A variety of RNOS can be formed by these interactions, leading to specific types of chemistry such as nitrosation, oxidation, nitration, and hydroxylation. Because this chemistry can be rather complex in biological systems, it is useful to unmask the predominating chemistry by the array of metabolites that are actually detectable *in vivo*. For instance, nitrosation chemistry leads to formation of *N*-nitrosamines and *S*-nitrosothiols. These nitrosated adducts have a defined spectrum of biological effects in their own rights. However, a variety of reactions can lead to formation of the same products. By understanding the sources of these reactions, it will be easier to evaluate the relevance of a particular reaction for the biological effect of ·NO in the target tissue.

2. DIRECT EFFECTS

In the absence of oxygen, ·NO in biological systems is a relatively unreactive chemical species. With the exception of oxygen and superoxide, which are discussed in the following sections, ·NO reacts primarily with metals and highly carbon, nitrogen, and oxygen reactive radicals generated under extreme conditions in biological environments. The most important interaction of ·NO with respect to biology is with metal complexes. Like many other reactions, the nature and extent of the interaction between ·NO and different metal complexes varies, and not all metal complexes react with ·NO. There are two important aspects to consider in this context: the rate of the reaction and the stability of the product. One of the most important reactions in biological systems is the interaction of ·NO with iron complexes.

To better understand the significance of iron for its interactions with ·NO, the kinetics and types of reactions need to be discussed. ·NO can react with some transition metals to form a metal nitrosyl; however, not every metal will react with ·NO, nor will they form a stable nitrosyl adduct (Cotton and Wilkinson, 1988). The reaction rate and stability of these products depend on the oxidation state, as well as on the other ligands coordinated to the metal. For example, ferric ion (Fe^{3+}) in aqueous solution does not form a stable nitrosyl product, yet reduction to the ferrous state yields an iron nitrosyl complex (Epstien *et al.*, 1980):

$$(H_2O)_6Fe(II)^{2+} + \cdot NO \rightarrow (H_2O)_5Fe(II)NO^{2+}$$

$$k_1 = 6.2 \times 10^5\ M^{-1}\ s^{-1} \tag{1}$$

If the aqueous coordination sphere is replaced by the metal chelator EDTA, the rate of reaction with ·NO increases to > $10^7\ M^{-1}\ s^{-1}$ (Zang *et al.*, 1988):

$$EDTA(Fe(II) + \cdot NO \rightarrow EDTA(Fe(II)\text{–}NO$$

$$k_2 = 1 \times 10^7\ M^{-1}\ s^{-1} \tag{2}$$

In contrast, the ferric EDTA complex does not react rapidly with ·NO to form a nitrosyl adduct. Hence, the oxidation state and the ligands coordinated to the metal determine the rate of formation as well as the stability of iron nitrosyls.

Similar trends are observed when examining the reactivity of ·NO with biologically relevant heme complexes. Nitrosyl metal complexes are readily formed from both ferrous and ferric heme complexes. As discussed above, ferric ion complexes with ligands such as water or EDTA do not readily bind ·NO, yet the ferric heme Fe(III)TPPS [TPPS = *meso*-tetrakis(4-sulfonatophenyl)porphyrin complex reacts with ·NO to form a detectable nitrosyl complex (Hoshino *et al.*, 1993):

$$Fe(III)(TPPS) + \cdot NO \rightarrow Fe(III)(TPPS)NO$$

$$k_3 = 7.2 \times 10^6\ M^{-1}\ s^{-1} \tag{3}$$

Compared with the ferric complex, the ferrous complex has a rate that is more than two orders of magnitude higher:

$$Fe(II)(TPPS) + \cdot NO \rightarrow Fe(II)(TPPS)NO$$

$$k_4 = 1.8 \times 10^9\ M^{-1}\ s^{-1} \tag{4}$$

Heme complexes have the highest association rate constant for ·NO in forming a metal nitrosyl complex. As before, the ferrous state has a higher on-rate for ·NO than does the ferric state.

Another important reaction of ·NO in biological systems is its influence on Fenton-type reactions (Huie and Neta, 1998). This type of reaction occurs between ferrous iron and hydrogen (or alkyl) peroxide, producing powerful oxidants such as metallo-oxo or hydroxyl radicals (Halliwell and Gutteridge, 1989a; Wink *et al.*, 1994c). This reaction has been proposed to increase damage to biomolecules, ultimately leading to oxidative stress in cells. Kanner and co-workers showed that Fenton-type oxidation was abated by ·NO (Kanner *et al.*, 1991). Furthermore, the oxidation of dihydrorhodamine by Fe-EDTA and hydrogen peroxide (H_2O_2) is also prevented by ·NO (Miles *et al.*, 1996). Dopamine is oxidized by Fenton-type reactions, and the resultant oxidized metabolites, which may play a role in some neurodegenerative diseases, are reduced in the presence of ·NO (Cook *et al.*, 1996). Furthermore, DNA strand breaks, as well as 8-oxo-deoxyguanine mutations result from Fenton chemistry and are abated by ·NO (Pacelli *et al.*, 1994, Oshima, personal communication). These findings suggest that ·NO is good antioxidant for Fenton-type oxidation reactions.

One explanation for these results is that ·NO can scavenge powerful oxidants, such as hydroxyl radical (·OH) or the metallo-oxo species generated by this reaction. Though at first this seems an attractive explanation, the competing reactions of other biological molecules for oxidants such as ·OH suggest a limited importance for ·NO *in vivo*. However, in the presence of H_2O_2, the ferrous nitrosyl complex reacts to form ferric iron and nitrite (NO_2^-) without the generation of the powerful oxidants. The most likely mechanism for the protective effect of ·NO is the reaction of ·NO with the ferrous iron, which decreases the production of powerful oxidants from H_2O_2.

Another possible influence ·NO may have on the Fenton reaction is to reduce the ferric iron to ferrous iron (Farias-Eisner *et al.*, 1996). The ferrous state is the valence state that facilitates the formation of powerful oxidants when reacted with peroxides. Under these conditions, ·NO would provide electrons to drive the Haber–Weiss reaction (Huie and Neta, 1998). (The biological significance of this process, however, remains to be demonstrated.) ·NO can also scavenge $\cdot O_2^-$ to form peroxynitrite ($ONOO^-$), thereby inhibiting the Haber–Weiss reaction and limiting the Fenton reaction. Therefore, the valence state of the binding transition metal must be taken into consideration when evaluating the role of ·NO in biological processes.

The two basic types of iron complexes that ·NO interacts with are heme and nonheme complexes. As discussed above, heme-containing complexes, especially in the ferrous state, have a strong affinity for ·NO. Below we discuss the various reactions of different heme and nonheme complexes and their importance in biology.

2.1. Heme Complexes

2.1.1. Hemoglobin and Myoglobin

Hemoglobin and myoglobin were among the first biological complexes shown to form Fe–NO adducts (Antonini and Brunori, 1971). In fact, HbNO has been used in some cases as a "dosimeter" for ·NO (for review see Singel and Lancaster, 1996). Oxyhemoglobin was shown to react quantitatively with ·NO to form methemoglobin (Doyle and Hoekstra, 1981):

$$Hb(Fe(II)O_2) + \cdot NO \rightarrow Hb(Fe(III)) + NO_3^- \tag{5}$$

This has been developed into a method for detecting ·NO (Feelisch, 1991). This reaction, which occurs at 3×10^7 M^{-1} s^{-1} (Doyle and Hoekstra, 1981), represents the primary mechanism by which ·NO is detoxified *in vivo* (Lancaster, 1994). Diffusion of ·NO to the blood results in the destruction of ·NO, thereby controlling this molecule's spreading *in vivo*. Also important is the reaction of ·NO with the hypervalent state of hemoglobin or myoglobin. The reaction of methemoglobin or myoglobin with H_2O_2 results in a ferryl pi cation radical species (Grisham and Everse, 1991):

$$\text{Fe(III) or Fe(II)} + H_2O_2 \rightarrow \text{high-valent metal-oxo complexes} \quad (6)$$

Exposure of high-valent metal-oxo complexes to ·NO results in a return to their original oxidation state (Kanner *et al.*, 1991; Wink *et al.*, 1994c; Gorbunov *et al.*, 1995):

$$\text{High-valent metal-oxo complexes} + \cdot NO \rightarrow \text{Fe(II)/Fe(III)} + NO_2^-/NO_3^- \quad (7)$$

The rereduction of these hypervalent heme complexes could prevent the breakdown of these reactive complexes, which otherwise could lead to potentially deleterious effects mediated by ROS.

2.1.2. Soluble Guanylate Cyclase

The best example of a direct effect is the activation by ·NO of soluble guanylate cyclase (sGC), leading to its translocation to the plasma membrane (Murad, 1994). This critical regulatory enzyme participates in the regulation of vascular tone, neurotransmission, platelet aggregation, and numerous other processes throughout the body. This enzyme is a heterodimer (Murad, 1994) that contains a protoporphyrin IX cofactor (Ignarro *et al.*, 1984) and that catalyzes the conversion of GTP into cGMP. Both ·NO and CO bind to the heme moiety of guanylate cyclase (DeRubertis *et al.*, 1978; Gerzer *et al.*, 1981a) (Figure 3). Binding of ·NO to this prosthetic group was required for the activation of guanylate cyclase by ·NO and nitrovasodilators (Edwards *et al.*, 1981; Gerzer *et al.*, 1981b; Ignarro *et al.*, 1982 a,b, 1984; Ohlstein *et al.*, 1982).

Recent studies have examined the mechanism by which ·NO activates guanylate cyclase. Several studies have shown that contained within the heme pocket are two histidines available for coordination, one at the distal and one at the proximal position (Yu *et al.*, 1994). Burstyn and colleagues suggest that the ferrous heme protein contains either one or both histidines bound to the ferrous metal (Burstyn *et al.*, 1995). These histidine groups are displaced when ·NO binds to the heme site (Yu *et al.*, 1994; Stone and Marletta, 1994; Deinum *et al.*, 1996). The mechanism by which ·NO activates sGC is presumably through binding to the heme and displacing the proximal histidine ligand to form a pentacoordinate complex. From the literature, the rate constant for the reaction of ·NO with sGC is estimated to be > 10^5 M^{-1} s^{-1} (Stone and Marletta, 1995).

The ferrous heme complexes discussed above have a high affinity for ·NO. It appears that the formation of a pentacoordinate ferrous nitrosyl is crucial for activation of this enzyme (Yu *et al.*, 1994; Stone and Marletta, 1994). One unique feature of the ferrous state of guanylate cylcase is its low affinity for O_2 relative to ·NO. Generally, heme proteins in the ferrous state will bind ·NO, CO, and also O_2. However, it appears that the

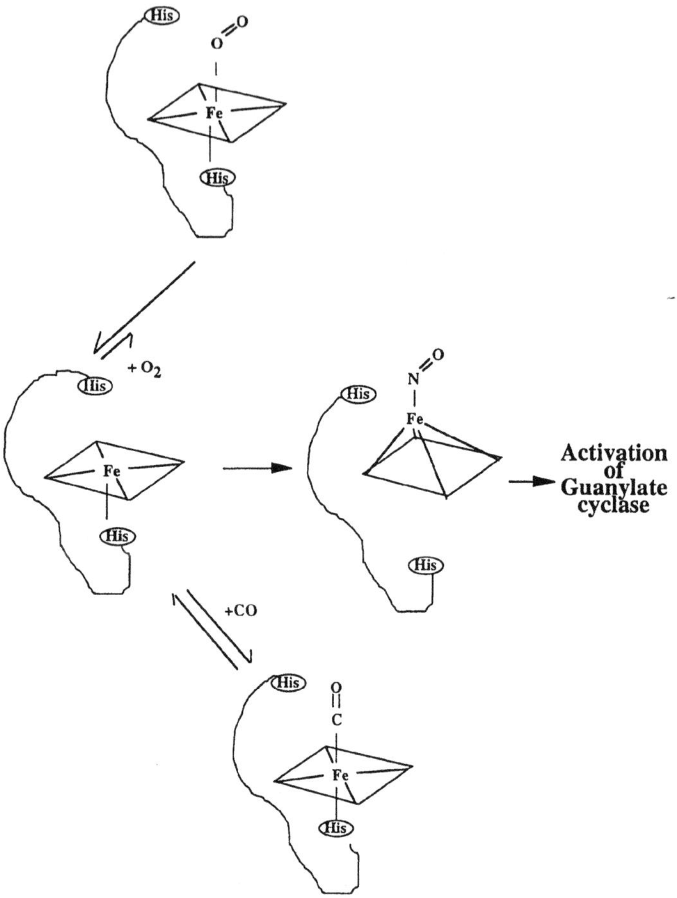

FIGURE 3. Mechanisms of acitivation of soluble guanylate cyclase of NO and CO.

heme pocket in sGC is unique in minimizing the binding of oxygen, either because of the polarity of the heme pocket or because of the inability of the proximal histidine ligand to adequately bind and stabilize the hexacoordinate ferrous oxy complex (Deinum *et al.*, 1996). It is this structural perturbation that activates sGC to convert GTP to cGMP. Additionally, the presence of the porphyrin is required for the activation of sGC by CO. This is thought to be related to the formation by CO of a hexacoordinate complex, as opposed to the pentacoordinate product formed by ·NO. Comparison of the activation by ·NO and CO suggests that ·NO forms a complex 200 times more stable than that formed by CO (Stone and Marletta, 1994; Burstyn *et al.*, 1995). It was proposed that the lability of the CO complex was responsible for the poor activity of CO (Stone and Marletta, 1995).

2.1.3. Cytochrome P450 and Other Monooxygenases

The cytochrome P450s are a family of enzymes that are involved in the synthesis and catabolism of such bioproducts as fatty acids, steroids, prostaglandins, and leukotrienes. There are numerous isoforms of cytochrome P450 that catalyze the synthesis of hormones, metabolism of drugs, and mediate the production of certain carcinogens. A report in 1992 postulated that endogenously generated ·NO could play a role in regulating the synthesis of testosterone, suggesting a role of ·NO derived from NOS in the regulation of hormone metabolism (Adams *et al.*, 1992). Cytochrome P450 can be inhibited by exogenous and endogenous sources of ·NO (Wink *et al.*, 1993c; Khatsenko *et al.*, 1993; Stadler *et al.*, 1994), both reversibly and irreversibly (Wink *et al.*, 1993c) (Figure 4). Reversible inhibition is a direct effect that occurs when ·NO binds to the heme, preventing the binding of oxygen and thus inhibiting catalysis (Wink *et al.*, 1993c). Irreversible inhibition is an indirect effect mediated by RNOS formed by the autoxidation of ·NO, and the inhibition is abated in the presence of bovine serum albumin and glutathione (Wink *et al.*, 1993c). Kim and colleagues proposed that the irreversible inhibition was related to the removal of heme from the protein (Kim *et al.*, 1995a). A plausible mechanism for this process is that ·NO binds to the heme to form a pentacoordinate adduct

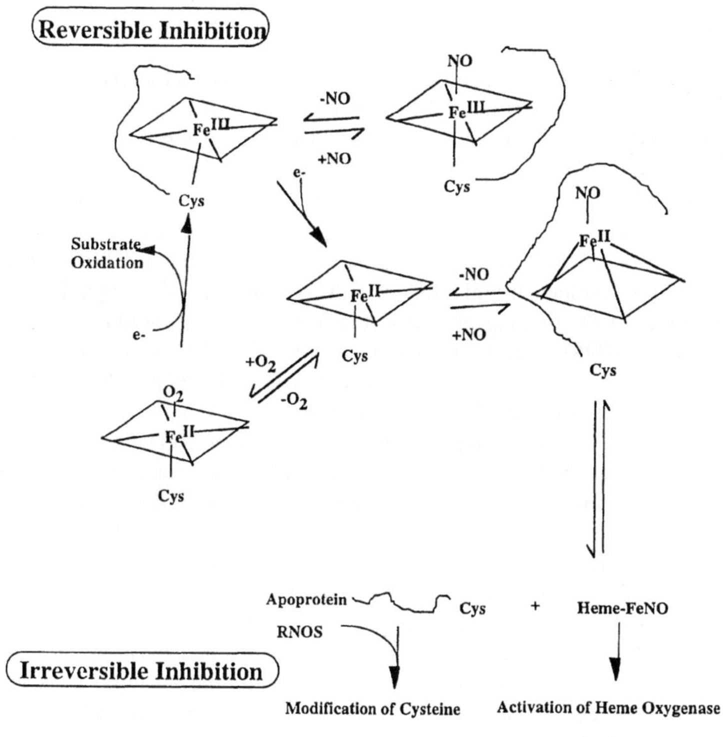

FIGURE 4. Mechanism of inhibition of cytochrome P450 by NO.

with the axial cysteine ligand removed, analogous to the binding of ·NO to sGC. If the pocket is sufficiently opened, an RNOS can nitrosate or oxidize this cysteine ligand, preventing reattachment to the heme and resulting in the irreversible inhibition of activity. It should be noted that different isoforms have different thresholds of inhibition. For instance, the 2B1 isoform is more susceptible than the 1A1 isoform with respect to both reversible and irreversible inhibition (Wink *et al.*, 1993c). Therefore, different fluxes of ·NO, which depend on distance from the source, as well as scavengers, may dictate the profile of inhibition of monooxygenases in tissues.

The inhibition of P450 has been thought to regulate hormone metabolism (Adams *et al.*, 1992) as well as to explain the inhibition of drug metabolism during chronic infection (Khatsenko *et al.*, 1993). Stadler and co-workers showed that hepatocytes, in which iNOS was activated, exhibited a dramatic decrease not only in the activity of P450 but also in the expression of P450 mRNA (Stadler *et al.*, 1994). Another important aspect of the binding of ·NO to P450 followed by the removal of heme is the activation of heme oxygenase in hepatocytes (Kim *et al.*, 1995a,b). These reactions with P450 may be important in pathophysiological mechanisms, as cytokine-induced formation of ·NO inhibited the activity of P450 associated with arachidonic acid, thereby increasing renal vasodilation by this pathway (Oyekan, 1995). The interaction of ·NO with P450 can therefore play both a regulatory and a pathophysiological role.

2.1.4. Cyclooxygenase

Cyclooxygenase (COX) is a key enzyme involved in the metabolism of arachidonic acid (AA) to prostaglandins (Figure 5). Kanner and co-workers first showed that COX can be inhibited by ·NO (Kanner *et al.*, 1992). Comparing the hemoglobin-mediated oxidation of linoleic acid, they concluded that the peroxidation of linoleate was inhibited by the binding of ·NO to the heme moiety within COX (Kanner *et al.*, 1992). Another report examined the binding of ·NO with prostaglandin H synthase, which is analogous to COX. These workers found that the ferric state does not efficiently bind ·NO, yet the ferrous state binds ·NO and strongly displaces the proximal histidine ligand, in a manner analogous to that of the binding of ·NO to the ferrous state of P450 or sGC described above (Tsai *et al.*, 1994). Because microsomes exposed to ·NO donors and ·NO gas up to 1 mM did not inactivate COX activity, these authors concluded that ·NO did not inhibit COX by direct binding to heme (Tsai *et al.*, 1994).

Examination of *in vivo* and cellular systems showed that ·NO can either attenuate or augment COX activity. ·NO formed in Kupffer cells inhibits the formation of the products of COX, suggesting that in this case ·NO might directly inhibit this enzyme (Stadler *et al.*, 1993). However, several other studies have shown that ·NO derived from NOS markedly enhances the activity of prostaglandin synthase (Salvemini *et al.*, 1993, 1994; McDaniel *et al.*, 1996; Corbett *et al.*, 1993; Sautebin and Di Rosa, 1994; Laszlo *et al.*, 1994). It was proposed that ·NO activates COX by a pathway independent of cGMP (Davidge *et al.*, 1995). The assay employed in this study involved the monitoring of PGE_2, which is converted from PGH_2 via isomerase activity or thermal decomposition. However, PGH_2, in addition to being converted to PGE_2, can also be converted to PGI_2 (prostacyclin) by prostaglandin H synthase (PHS). This assay, therefore, measures the cumulative

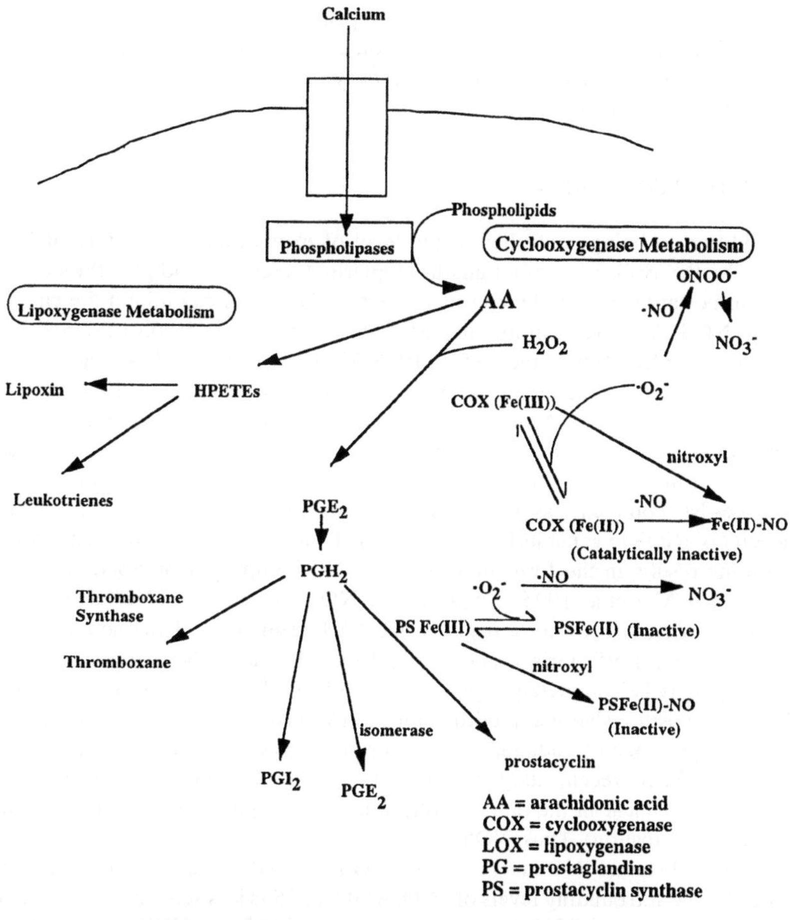

FIGURE 5. Influence of NO on arachidonic acid metabolism.

influence of ·NO on a number of enzymes involved in this pathway. It is not clear if the enhancement by ·NO of the formation of end products is related to a direct effect on the activity of COX, or to the reaction of PGH_2 with isomerase or PHS. Several reports have suggested that superoxide inhibits the synthesis of prostacyclin in endothelial cells (Davidge *et al.*, 1995) as well as that of PGE_2 in vascular smooth muscle (Inoue *et al.*, 1993; Kelner and Uglik, 1994). Furthermore, $·O_2^-$ has been shown to be involved in the inhibition of PGH_2/PGE_2 isomerase (Kelner and Uglik, 1994) and PHS (Davidge *et al.*, 1995). A study reported that 5,5-dimethyl-1-pyrroline-*N*-oxide (DMPO), a free radical spin trap, could inhibit the synthesis of PHS-2 stimulated by LPS, suggesting that either $·O_2^-$ or oxidants derived from Fenton-type chemistry may be involved (Hempel *et al.*, 1994). These results imply that the enhancement of the synthesis of PGE_2 by ·NO may be related to the scavenging of superoxide via the $·NO/·O_2^-$ reaction. In neutrophils and endothelial cells, the source of superoxide appears to be COX (Chen *et al.*, 1994; Cosentino *et al.*, 1994). These reports may indicate that both the interaction of $\cdot O_2^-$ and ·NO and the modulation by superoxide of enzymatic activity may be important in prostaglandin metabolism.

2.1.5. Nitric Oxide Synthase

In addition to being controlled at the level of expression, the activity of NOS is controlled by the availability of tetrahydrobiopterin, L-arginine, and glutathione. One of the most important factors to be considered is that ·NO itself can inhibit the enzymatic activity of NOS. As NOS is similar in part of its structure to P450 (Griffith and Stuehr, 1995; Nathan and Xie, 1994; Stuehr *et al.*, 1995; Marletta, 1993, 1994), it is not surprising that ·NO can actually attenuate the activity of NOS (Griscavage *et al.*, 1994; Abu-Soud *et al.*, 1995; Hurshman and Marletta, 1995; Griscavage *et al.*, 1995). A comparison of the different isoforms of NOS shows that eNOS and nNOS are more susceptible than iNOS (Griscavage *et al.*, 1995). This is apparently related to the different stabilities of the resulting Fe–NO complexes as well as electron fluxes. These findings are consistent with the capacity of iNOS to generate higher amounts of ·NO in localized areas. The oxidation of L-arginine results in the formation of an Fe–NO complex (Abu-Soud *et al.*, 1995; Hurshman and Marletta, 1995). Exposure of NOS to authentic ·NO also decreases enzymatic activity, suggesting a negative feedback regulation of the activity of NOS (Abu-Soud *et al.*, 1995; Griscavage *et al.*, 1995). Notably, the activity of iNOS in macrophages may be completely suppressed in cells producing high levels of ·NO by as yet undefined posttranslational modifications(s) (Vodovotz *et al.*, 1994), in some cases requiring the presence of endogenous transforming growth factor-β1 (Vodovotz *et al.*, 1996). Results from a recent study indicate that the formation of Fe–NO by nNOS makes the enzyme's K_m linear in the range of physiological oxygen concentration (D. Stuehr, personal communication). Thus, NOS may serve as a sensor for oxygen and possibly also for H_2O_2. The fluxes of ·NO derived from nNOS and eNOS are likely to be feedback-controlled by the surrounding levels of ·NO, while iNOS is less sensitive to these effects. Significant formation of RNOS may not occur via nNOS or eNOS, even under high intracellular calcium concentrations, and suggests that the source of RNOS and their attendant indirect effects *in vivo* must be largely derived from iNOS.

2.1.6. Catalase

This is a heme-containing enzyme that is critical in protecting cells against damage mediated by H_2O_2. Hoshino and colleagues showed that ·NO could bind to the heme moiety, forming a ferric nitrosyl with an on-rate of $3 \times 10^7\ M^{-1}\ s^{-1}$ and K_d of $1 \times 10^5\ M^{-1}$ (Hoshino *et al.*, 1993). Kim and co-workers demonstrated that hepatocytes stimulated with cytokines had reduced catalase activity that correlated with the levels of nitrite produced (Kim *et al.*, 1995a,b). Farias-Eisner and co-workers examined the toxicity of ·NO and H_2O_2 and concluded that the inhibition by ·NO of catalase could play a role in the tumoricidal activity of macrophages. The authors of this report calculated that 15–20 μM ·NO would be capable of inhibiting the activity of catalase by as much as 80% when the Fe–NO inhibitory mechanism applies (Farias-Eisner *et al.*, 1996). An examination of the effect of ·NO donors on cytotoxicity mediated by H_2O_2 demonstrated that ·NO inhibited the consumption of H_2O_2 by cells in culture (Wink *et al.*, 1996a). Observations by Li and colleagues suggest that catalase and peroxide can attenuate the concentrations and fluxes of ·NO, thereby reducing its bioavailability (Li *et al.*, 1992). It was shown that ·NO can partially inhibit the consumption of H_2O_2 with a K_i of 0.18 μM (Brown, 1995b). Importantly, the combination of catalase and hydrogen peroxide can consume ·NO (Brown, 1995b; Brunelli *et al.*, 1994).

There appears to be two possible direct mechanisms for the inhibition by ·NO of catalase (Figure 6). The first reaction would involve H_2O_2 reacting with catalase to form complex I. In the absence of ·NO, complex I can react with H_2O_2 to form O_2. However, ·NO can also react with complex I, forming complex II, which may further react with another molecule of ·NO and thereby convert 2 moles of ·NO and 1 mole of hydrogen peroxide to 2 moles of NO_2^-. An examination of the reaction of ·NO with complex I formed from H_2O_2 (Kanner *et al.*, 1991; Wink *et al.*, 1994c; Gorbunov *et al.*, 1995) suggests that the rate of this reaction is > 10^7 M^{-1} s^{-1} (D. Wink, unpublished observation). This reaction could well be a source of nitrite *in vivo*. Though this mechanism may slow down the consumption of hydrogen peroxide, it does provide a mechanism by which ·NO is consumed and by which its bioavailability is regulated. Increases in glutathione peroxidase, which *per se* does not react with ·NO, increase the bioavailability of ·NO (from eNOS) (Loscalzo *et al.*, 1996). This suggests that H_2O_2 may play a crucial role in

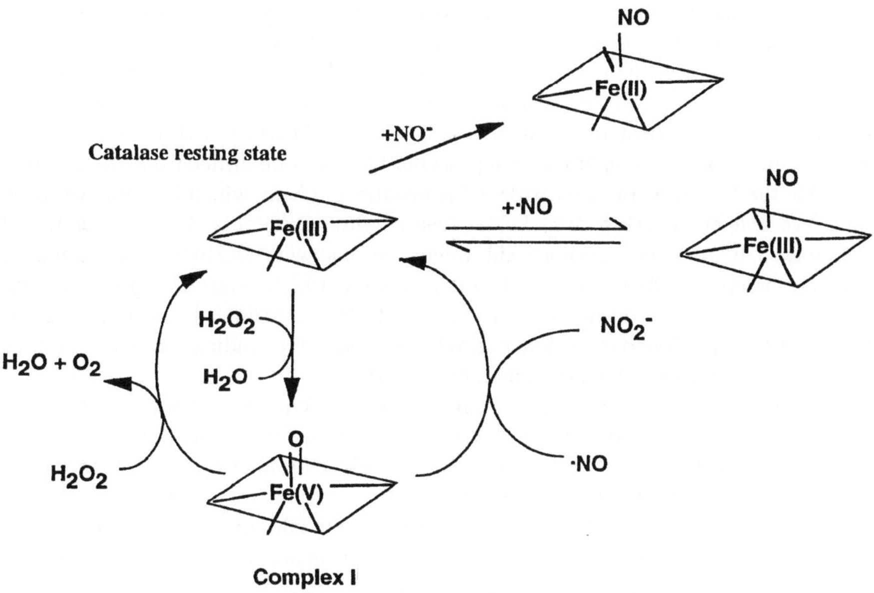

FIGURE 6. Nitric oxide and catalase activity.

regulation of the levels of ·NO *in vivo* via chemistry analogous to that described for catalase/peroxide/·NO.

The second mechanism would require that ·NO bind to catalase. Under these conditions, ·NO would be in excess and may compete directly for the catalase heme center. This may be important in certain pathophysiological conditions in which ·NO and H_2O_2 are formed sequentially.

2.1.7. Cytochrome aa_3 (Mitochondria)

One of the primary cellular targets for the cytotoxic action of ·NO has been proposed to be the mitochondrion (Hibbs *et al.*, 1987; Lancaster and Hibbs, 1990; Moncada *et al.*, 1991). Dinitrosyl adducts of aconitase are formed in cells after exposure to ·NO and may be a key element in the inhibition of mitochondrial activity via disruption of the citric acid cycle (Lancaster and Hibbs, 1990). At low doses, ·NO deenergizes mitochondria (Kurose *et al.*, 1993), and one report suggests that the influence of ·NO on the mitochondria may play a role in the relaxation of smooth muscle cells both physiologically and pathophysiologically (Geng *et al.*, 1992; Roberts *et al.*, 1992; Szabo *et al.*, 1996).

As discussed below, there are several components of the mitochondria that ·NO and RNOS might influence. Under inflammatory conditions, complex I (NADH:ubiquinone oxidoreductase) and complex II (succinate:ubiquinone oxidoreductase) are both inhibited by ·NO, although it has been suggested that cytochrome *c* oxidase is the primary target at high concentrations of ·NO (Cleeter *et al.*, 1994; Brown *et al.*, 1995; Brown, 1995a; Cassina and Radi, 1996; Moro *et al.*, 1996; Lisdero *et al.*, 1996) (Figure 7).

Cytochrome *c* oxidase can form a nitrosyl adduct (Rousseau *et al.*, 1988). It appears that the binding of ·NO to cytochrome oxidase may influence the activity of other mitochondrial enzymes. Interestingly, ·NO under anaerobic conditions is consumed at a rate four times slower than oxygen. Under these conditions, it was proposed that ·NO directly binds to the cytochrome oxidase site, where it is converted to reduced nitrogenous products by a mechanism analogous to that of nitrite reductase (Clarkson *et al.*, 1995). However, under aerobic conditions, it appears that electrons are shifted from the reduction of ·NO to the formation of superoxide in the respiratory chain, which is an energetically more favorable process (Figure 7). Under these conditions, the flux of superoxide would be increased. It is thought that superoxide might then react with ·NO to form peroxynitrite. However, MnSOD, which exists in the mitochondria at high concentrations, may limit the formation of peroxynitrite by converting $\cdot O_2^-$ to H_2O_2. Another interesting aspect is that eNOS has been found in mitochondria (Bates *et al.*, 1996), indicating that this source of ·NO may modulate oxygen tension intracellularly.

A recent report has suggested that mitochondrial function can be inhibited by a mechanism that does not involve the formation of peroxynitrite (Knowles *et al.*, 1996). The reversible inhibition described in this report involved the binding of ·NO to the heme, whereas irreversible inhibition involved the modification of components of this organelle by RNOS. This is considered a direct effect, whereas the inhibition by peroxynitrite and other RNOS would be considered an indirect effect. There are clear differences, however, regarding the inhibition of mitochondria following the induction of iNOS *in vitro* and *in vivo*. Whereas the stimulation with interferon-γ and LPS of hepatocytes in culture inhibits

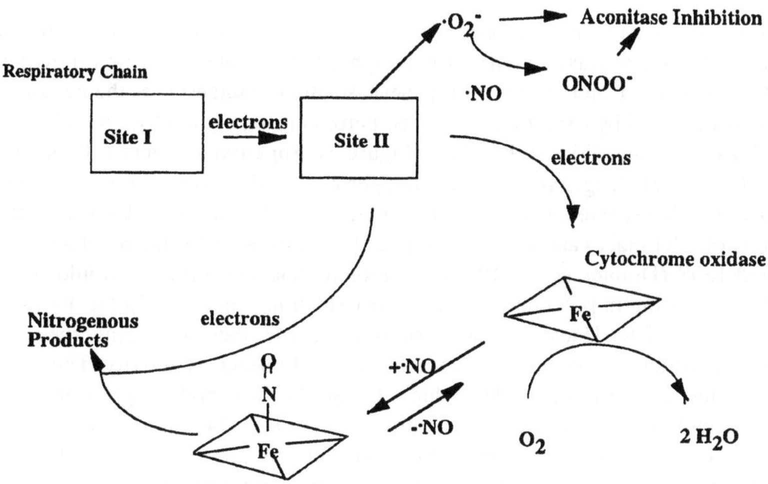

FIGURE 7. Nitric oxide and mitochondrial respiration.

respiration, the same stimuli had no such effect *in vivo* (Stadler *et al.*, 1991). This discrepancy may suggest that oxyhemoglobin and diffusion may play an important role in mitochondrial inhibition, limiting the formation of RNOS and resulting in reversible inhibition.

2.2. Nonheme Iron Proteins

Nonheme proteins include the iron–sulfur (FeS) proteins as well as iron complexes bound to amino acids such as histidine. ·NO may activate, inhibit, or not affect nonheme proteins. For example, proteins that bind DNA respond differently to oxidative stress. SoxR (Chapter 5), which has two Fe_2S_2 clusters, responds to both O_2^- and ·NO leading to the induction of SoxS, which in turn activates a defense response to oxidative stress (Nunoshiba *et al.*, 1993). However, another enzyme that contains an Fe_2S_2 moiety, ferrochelatase, is inhibited by ·NO (Kim *et al.*, 1995a,b). Yet other enzymes that have FeS centers, such as endonuclease III which contains an Fe_3S_4 cluster, are unaffected by as much as 10 mM ·NO (Wink and Laval, 1994). Here, we will discuss the direct effects of ·NO on various nonheme iron proteins.

2.2.1. Aconitase, Reactive Chemical Species, and Mitochondrial Function

This enzyme catalyzes the conversion of citrate to isocitrate, which can facilitate the formation of NADH by isocitrate dehydrogenase and form α-ketoglutarate in the Krebs cycle. Aconitase is a protein that contains an Fe_4S_4 center in which three iron molecules are bound via cysteine to the protein, while apical iron binds the substrate. The three iron molecules bound to the protein are in a tetrahedral environment, while the apical iron is hexacoordinate. The conversion of isocitrate to α-ketoglutarate is facilitated by the binding of citrate to the ferrous form of the apical iron in aconitase. One of the first targets of ·NO in the tumoricidal activity of macrophages to be described was mitochondrial aconitase (Lancaster and Hibbs, 1990). Several studies suggest that the activity of the enzyme is modified by oxidation of $\cdot O_2^-$ and peroxynitrite, but not by ·NO (Castro *et al.*, 1994; Hausladen and Fridovich, 1994) (Figure 7). Superoxide reacts with aconitase at $3 \times 10^7\ M^{-1}\ s^{-1}$. Hydrogen peroxide and oxygen can also inactivate this enzyme, but with considerably slower rate constants on the order of $2 \times 10^2\ M^{-1}\ s^{-1}$. However, anaerobic solutions of ·NO inactivate aconitase, suggesting a still undetermined balance between ·NO and ROS (Drapier *et al.*, 1993). Superoxide and peroxynitrite would react with aconitase, resulting in potentially irreversible oxidation (Figure 7). In the above studies (Castro *et al.*, 1994; Hausladen and Fridovich, 1994), these oxidizing species could oxidize the cluster by one-electron transfer, hence limiting its activity. These reactive intermediates may form disulfides, which disrupt the structural integrity of the cluster and irreversibly inhibit the enzyme (Castro *et al.*, 1994). The binding of ·NO to apical clusters may be important and reversible, and probably plays an important role in the regulation of iron levels. Also, a balance may exist between the binding of citrate and ·NO to the apical iron, suggesting that aconitase may be a switch for citrate metabolism controlled reversibly by ·NO.

2.2.2. Aconitase, Iron Response Binding Protein, and Iron Homeostasis

Another important aconitaselike protein is the iron responsive binding protein (IRB), which contains an aconitase active site and is found in the cytosol. This protein is critical in regulating the transcription of iron responsive elements (IRE; RNA structures), which regulate the transferrin receptor (TfR) or ferritin posttranscriptionally (Klausner *et al.*, 1993) (Figure 8). The IRB exist in two forms: the holoprotein, which contains Fe_4S_4 and possesses aconitase activity but cannot bind to the IRE; and Fe_3S_4 (apoprotein), which has no aconitase activity but can bind to the IRE. Stimulation of iNOS activity increases iron uptake (Weiss *et al.*, 1993; Drapier *et al.*, 1993). Furthermore, treatment of cortical neurons with ·NO derived from either 100 μM *S*-nitroso-*N*-acetylpenicillamine (SNAP) or from stimulation of NMDA receptors resulted in the binding of IRB to IRE (Jaffrey *et al.*, 1994). Like the mitochondrial aconitase, $\cdot O_2^-$ and $ONOO^-$ inhibit the aconitase activity of IRB, whereas ·NO does not (Bouton *et al.*, 1996). However, ·NO does stimulate the binding of IRB to the IRE, unlike $\cdot O_2^-$ or $ONOO^-$ (Bouton *et al.*, 1996). Superoxide and peroxynitrite may modify the IRB such that it cannot effectively bind to the IRE, probably by oxidation of a critical thiol group (reviewed in Drapier and Bouton, 1996; Hentze and Kuhn, 1996). Therefore, these reactive species may simply inhibit the protein irreversibly, so that neither aconitase activity nor IRE binding can take place. However, ·NO can bind to the apical iron in the ferrous state, forming a Fe–NO which may labilize the apical iron, thus inducing binding to the IRE. Because hexacoordinate ferrous complexes bind ·NO reversibly, this would provide a switch for the IRE. An important element in the role of ·NO in iron homeostasis may be the reaction of ·NO with ferritin to form Fe–NO complexes. As discussed above, cellular exposure to ·NO results in the formation of a Fe–NO complex (Lee *et al.*, 1994). An iron nitrosyl can be formed from ferritin, resulting in labilization of iron from the iron stores (Lee *et al.*, 1994; Bastian *et al.*, 1994).

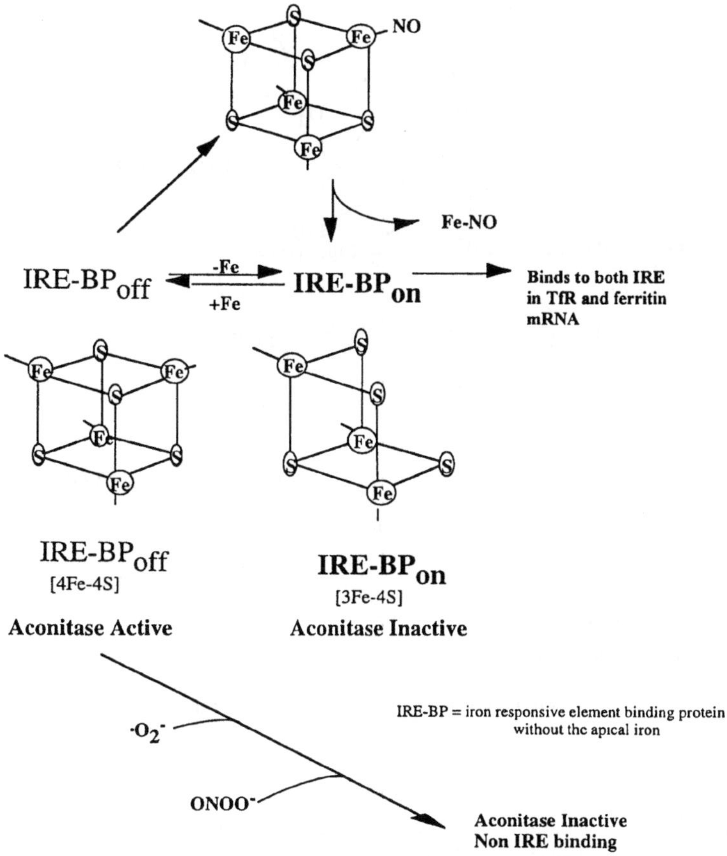

FIGURE 8. Nitric oxide and iron response elements.

2.2.3. Lipoxygenase

Lipoxygenase is a nonheme iron enzyme that catalyzes the peroxidation of polyunsaturated acids such as AA to hydroperoxyeicosa-6,8,11,14-tetraenoic acid (5-HPETE). This is the precursor for a number of leukotrienes, most notably LTB_4, which is generated by platelets, is a strong chemoattractant, and can increase vascular permeability making it a strong systemic vasodilator. Products of lipoxygenase are thought to be important in the induction of iNOS in macrophages (Imai *et al.*, 1993; Ryoyama *et al.*, 1993). Another report has correlated the tumoricidal activity of a macrophage with the inhibition of lipoxygenase products, as well as glutathione *S*-transferase (GST) (Hubbard and Erickson, 1995). This enzyme has also been shown to play an important role in atherosclerosis and other inflammatory diseases. In endotoxemia, it was suggested that ·NO derived from

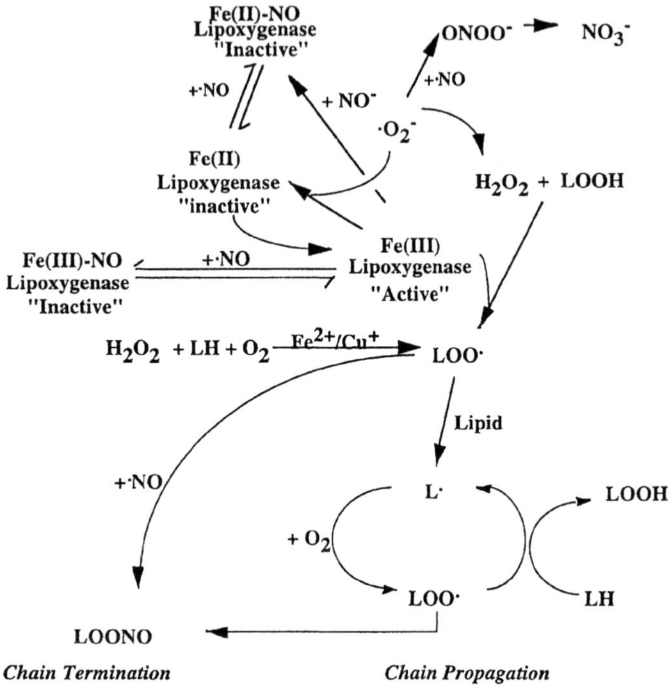

FIGURE 9. Nitric oxide and lipid metabolism.

cNOS inhibits the formation of lipoxygenase products and leads to loss of vascular integrity in the colon (Laszlo and Whittle, 1995).

Kanner and colleagues demonstrated that ·NO inhibited formation of lipoxygenase products (Kanner *et al.*, 1992). In platelets and neuroblastoma cells, ·NO derived from sodium nitroprusside (SNP) and SNAP inhibits the formation of lipoxygenase products (Nakatsuka and Osawa, 1994; Maccarrone *et al.*, 1996). Iron in the active site of lipoxygenase is in the ferric state where it can bind H_2O_2, resulting in lipid oxidation and subsequent lipid peroxidation (Figure 9). There are two proposed mechanisms of inhibition which will be discussed below, one involving the radical coupling of ·NO to prevent the formation of products and the second involving the modification by ·NO of the iron center. Maccarrone and colleagues suggested that ·NO mediates the reduction of ferric to the inactive ferrous form of iron, thereby inhibiting the enzyme (Maccarrone *et al.*, 1996). Another report suggests that the ferrous form can be oxidized by ·NO to activate the enzyme. Rubbo and colleagues examined the lipoxygenase-induced lipid oxidation of linoleic acid. These workers demonstrated that the inhibition by ·NO of product formation occurs via radical scavenging and limiting lipid peroxidation, and not through the inactivation of the Fe center (discussed further below) (Rubbo *et al.*, 1995). These reports suggest that different modes of inhibition can occur at different fluxes of ·NO .

2.3. Radical Reactions

2.3.1. Reaction with Free Radicals

Because ·NO has an unpaired electron, rapid reactions with other radical species are generally the rule. For instance, ·NO reacts with hydroxyl radical at a rate very close to diffusion control in aqueous solution (Buxton *et al.*, 1988):

$$\cdot NO + OH \rightarrow HNO_2$$

$$k_8 = 1 \times 10^{10}\ M^{-1}\ s^{-1} \quad (8)$$

·NO also reacts with $\cdot NO_2$ to form the RNOS, N_2O_3, with a rate constant of $2 \times 10^9\ M^{-1}\ s^{-1}$ (Schwartz and White, 1983):

$$\cdot NO + \cdot NO_2 \rightarrow N_2O_3$$

$$k_9 = 2 \times 10^9\ M^{-1}\ s^{-1} \quad (9)$$

$$N_2O_3 + H_2O \rightarrow 2\ NO_2^- \quad (10)$$

$$k_{10} = 1000\ s^{-1}$$

It has been further shown that alkyl, alkoxy, and alkylperoxide radicals rapidly react with ·NO (Padmaja and Huie, 1993):

$$\cdot ROO + \cdot NO \rightarrow ROONO$$

$$k_{11} = 2 \times 10^9\ M^{-1}\ s^{-1} \quad (11)$$

Reactions like those in Eq. (11) may help in understanding the effect of ·NO on lipid peroxidation (Rubbo *et al.*, 1994) and the ability of ·NO to sensitize hypoxic mammalian cells to radiation (Mitchell *et al.*, 1993).

2.3.2. Lipid Peroxidation

The Fenton-catalyzed peroxidation of low-density lipoproteins and other lipids critical in diseases such as atherosclerosis can be abated by ·NO (Struck *et al.*, 1995; Hogg *et al.*, 1995; Cayatte *et al.*, 1994; Seccia *et al.*, 1996; Hayashi *et al.*, 1995; Laskey and Mathews, 1996; Rubbo *et al.*, 1995). In addition, lipid peroxidation in membranes is thought to play a role in the cytotoxic mechanism of hydrogen peroxides (Halliwell and Gutteridge, 1986). It was shown that ·NO also abates the toxicity mediated by alkylperoxide (Wink *et al.*, 1995b). This report suggested that chain termination of lipid peroxidation may be playing a protective role.

Lipid peroxidation is initiated by the formation of a Fenton-derived oxidant that results in the production of a lipid radical. In the presence of oxygen, this radical is converted to a hydroperoxide radical, which then abstracts a hydrogen from other lipids.

These steps result in chain propagation and, consequently, enhanced lipid peroxidation. Nitric oxide rapidly reacts (2×10^9 M^{-1} s^{-1}) with these lipid peroxyl radicals, resulting in chain termination by the formation of LOONO products (Padmaja and Huie, 1993). This chain termination thus limits the extent of lipid peroxidation and appears to be protective. However, it is not clear what effect LOONO and other lipid derivatives of ·NO may have under physiological conditions (Figure 9).

2.3.3. Radicals Induced by Ionizing Radiation

Another process where ·NO can react with high-energy radicals is the radiosensitization of hypoxic cells. It has been long known that hypoxia induces radioresistance. In 1957, Howard-Flanders demonstrated that anaerobic bacteria were radiosensitized by ·NO (Howard-Flanders, 1957). A similar phenomenon occurs in mammalian cells, with ·NO derived from a variety of agents (Mitchell *et al.*, 1993). ·NO may sensitize hypoxic cells to radiation because it can react with radicals formed on the DNA to yield adducts that cannot be repaired by the cells' DNA repair machinery (Figure 10).

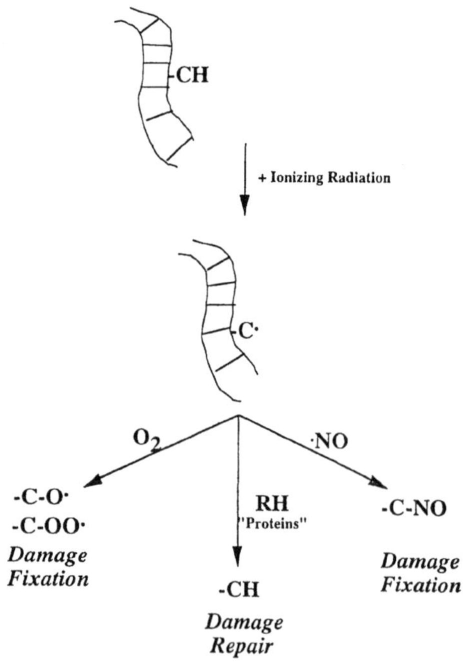

FIGURE 10. Radiosensitization by NO.

3. INDIRECT EFFECTS

The indirect effects of ·NO involve the formation of RNOS, resulting in derivatives that are either oxidized or reduced redox states of ·NO. The activation of ·NO to these reactive intermediates generally requires the presence of either oxygen or superoxide, leading to the formation of RNOS such as N_2O_3, $\cdot NO_2$, or $ONOO^-$. In addition to these RNOS, nitroxyl (NO^-) can also have some biological effects and might play a significant role in the metabolism of ·NO. To understand the biological implications of these complexes, which are often intermixed chemically, we have to further catalogue the indirect effects of ·NO into nitrosation and oxidation chemistry. Oxidation chemistry includes the removal of one or two electrons from a given substrate, as well as hydroxylation reactions. Nitrosation reactions refer to the addition of an equivalent of NO^+ added to amine, thiol, or hydroxy groups. For instance, intermediates in the $\cdot NO/O_2$ reaction convert thiol peptides to *S*-nitrosothiol peptides. Another type of chemistry that is often discussed in the literature is nitration, which is the donation of an equivalent of an NO_2^+ group to a nucleophile such as a hydroxy group or an activated aromatic ring. The formation of nitrotyrosine from different RNOS is a good example of this type of chemistry (Ischiropoulos *et al.*, 1992, 1996; Beckman *et al.*, 1992). An antibody developed to nitrated tyrosines in various proteins has served as an indicator of the presence of RNOS *in vivo*, providing a method to determine whether or not these RNOS are involved in a given biological system (Beckman *et al.*, 1994b). Nitrated products have been proposed to originate not only from peroxynitrite but also from other RNOS (van der Vliet *et al.*, 1994, 1995). A recent report has examined different mechanisms by which nitrotyrosine might be formed and concluded that $ONOO^-$ in the presence of carbon dioxide should be the primary source (Gow *et al.*, 1996). It appears that the presence of nitrated proteins can serve as a good "dosimeter" of the presence of ·NO and RNOS *in vivo*. As will be shown below, nitrosation and oxidation reactions are mutually independent and their chemistry is very distinct. Thus, using dosimeters of nitrosation and oxidation, the chemistry responsible for some of the biological effects of ·NO can be determined.

3.1. $\cdot NO/O_2$ Chemistry

In biological systems, nitrosation results in the formation of *N*-nitrosamines and *S*-nitrosothiols. These adducts have been detected in biological systems and in some cases have been directly correlated with the activity of NOS (Gaston *et al.*, 1993). The formation of *N*-nitrosamines *in vivo* has been linked to the activity of stimulated macrophages (Marletta, 1988; Lewis *et al.*, 1995), hepatocytes (Liu *et al.*, 1991, 1992), and neutrophils (Miles *et al.*, 1995) which express iNOS (Figure 11). A dramatic increase in the formation of *S*-nitrosothiols is also observed under stimulated conditions (Gaston *et al.*, 1993). There are three possible chemical mechanisms by which nitrosation chemistry can result from RNOS: autoxidation of ·NO, acidification of nitrite, or the reaction of peroxynitrite with ·NO. In all of these reactions, the major chemical species responsible for nitrosation are isomers of N_2O_3.

The most familiar reaction of ·NO is the autoxidation reaction, which is partly responsible for the instability of ·NO in aerobic aqueous solutions and is the source of

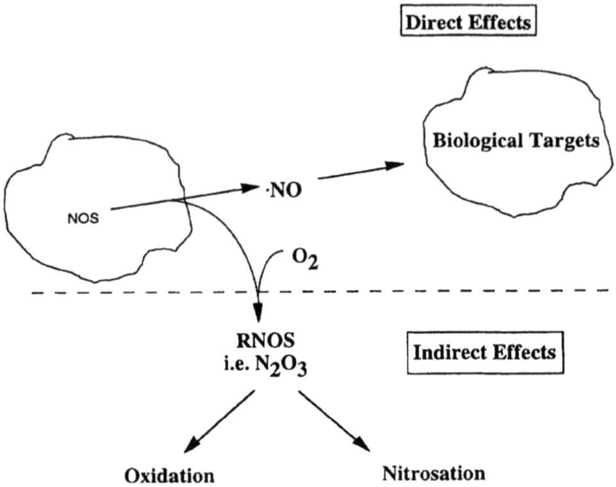

FIGURE 11. The $\cdot NO/O_2$ reaction in biological systems.

the RNOS formed in air pollution and cigarette smoke. This reaction has been under intensive investigation over the past few years. The reaction in aqueous solution is

$$4\cdot NO + O_2 + 2\,H_2O \rightarrow 4\,HNO_2 \tag{12}$$

In a variety of condensed phase conditions, acid, base, neutral, and hydrophobic organic solvents, the rate equation is third order:

$$-d[O_2]/dt = 1/4 * -d[\cdot NO]/dt = k_1[\cdot NO]^2[O_2]$$

being second order with respect to ·NO (see summary in Wink *et al.*, 1996b,d). The k_1 has been reported to be 1.6–2.0 × 10^6 M^{-2} s^{-1} (though one report suggests that it is closer to 2.9 × 10^6 M^{-2} s^{-1}, probably because of the lack of direct measurement of ·NO in solution; Goldstein and Czapski, 1995). Under biological conditions, the most important rate constant is one corresponding to the situation in which ·NO is limiting and oxygen is in excess. Because ·NO produced *in vivo* is formed from oxygen, ·NO is most likely to be the limiting reagent; therefore, k_{obs} for biologically relevant conditions is 6–8 × 10^6 M^{-2} s^{-1}(Ford *et al.*, 1993):

$$-d[\cdot NO]/dt = k_{obs}[\cdot NO]^2[O_2]$$

The kinetics of this reaction gives some insight into how ·NO can produce toxic RNOS yet regulate numerous physiological functions (Wink *et al.*, 1993a). The second-order dependency on ·NO means that the lifetime of ·NO, even in a large excess of oxygen, is determined by its concentration. For instance, at 1 μM · NO, the half-life of ·NO is about 13 min whereas at 100 μM its half-life is 8 s. *In vivo*, cells generate micromolar amounts of ·NO. The ·NO will diffuse from the source, thereby becoming diluted and increasing its half-life. This allows ·NO to directly react with biological targets such as the heme moiety in sGC without interference from the NO/O_2 reaction. However, as the concentration of ·NO increases, autoxidation increases dramatically, and exponentially increases the formation of RNOS.

It has been shown that in the gas phase and in hydrophobic media, the ·NO/O_2 reaction initially results in ·NO_2 after the rate-limiting step:

$$2\cdot NO + O_2 \rightarrow 2\cdot NO_2 \tag{13}$$

Nitrogen dioxide can further react to give N_2O_4 and N_2O_3:

$$2\cdot NO_2 \Leftrightarrow N_2O_4 \tag{14}$$

$$\cdot NO_2 + \cdot NO \Leftrightarrow N_2O_3 \tag{15}$$

·NO_2 is thought to be responsible for the toxicity attributed to air pollution and cigarette smoke. ·NO_2 is a highly reactive radical that can initiate lipid peroxidation, oxidize thiols, and nitrate aromatic groups (Williams, 1988). In addition, dinitrogen trioxide and dinitrogen tetroxide can nitrosate and nitrate substrates (Williams, 1988).

The RNOS formed from the ·NO/O_2 reaction in aqueous solution have been suggested to differ from that in the gas phase (Wink *et al.*, 1993a). A series of experiments examining the oxidation of ferrocyanide by RNOS formed in aqueous media suggested that ·NO_2 is not formed in aqueous solution. Further experiments involving the quenching of this oxidation reaction by azide (N_3) yielded further evidence for such a process. Furthermore, it was concluded from these studies that the chemical intermediate responsible for the observed chemistry was not N_2O_3 formed from acidified nitrite (Wink *et al.*, 1993a). Pires and co-workers examined the nitrosation of phenols and came to a similar conclusion (Pires *et al.*, 1994). These results were reexamined and suggested that Eqs. (13) and (14) did apply to the aqueous solution and that there had been a simple misinterpretation (Goldstein and Czapski, 1995). However, these latter studies provided limited data using a concentration range of 1.4 and 1.7 mM ·NO, and did not address the results obtained in the presence of azide, thereby invalidating the authors' discussion. An extension of the original studies (Wink *et al.*, 1993a) has shown that there were no selectivity differences between nitrosation and oxidation over a 100-fold range of concentration of ·NO, and thus the chemistry of the intermediates is independent of ·NO (Wink *et al.*, unpublished observations). Because the involvement of ·NO_2 depends on ·NO, the intermediates associated with gas-phase autoxidation are unlikely to be involved in the aqueous solution. An examination of the stoichiometry shows that the conclusion of Pires and colleagues proposing an isomer of N_2O_3 as the primary reactant in aqueous solution is correct.

The chemistry of N_2O_3 and its potential isomers has been examined over a number of years. This intermediate can nitrosate as well as oxidize substrates (Wink *et al.*, 1993a). The oxidizing potential of this species is 0.7 V, considerably lower than that for the activated isomer of peroxynitrous acid and hydroxyl radical (Grisham and Miles, 1994). The primary chemistry N_2O_3 carries out in biological systems is the nitrosation of thiols and amines. In the case of thiols, N_2O_3 forms *S*-nitrosothiols whereas with amines it forms *N*-nitrosamines:

$$RSH + N_2O_3 \rightarrow RSNO + HNO_2 \tag{16}$$

$$R'R''NH + N_2O_3 \rightarrow R'R''NNO + HNO_2 \tag{17}$$

Both *N*-nitrosamines and *S*-nitrosothiols can be generated in the presence of iNOS (Marletta, 1988; Lewis *et al.*, 1995; Liu *et al.*, 1991, 1992; Miles *et al.*, 1995).

One of the important aspects of these reactions is that bioorganic functional groups have the highest affinity for N_2O_3. A series of selectivity studies was carried out using a variety of nucleic and amino acids, and it was found that at neutral pH thiol-containing peptides had a higher affinity for N_2O_3 than for any other bioorganic molecule (Wink *et al.*, 1994a). Tyrosine had an affinity 20 times lower than GSH, whereas the rest of the amino acids and nucleotides had > 1000-fold lower affinities (Wink *et al.*, 1994a). GSH has a high affinity for N_2O_3 and plays a critical role in abating the toxicity of these RNOS. Indeed, depletion of GSH by L-buthionine sulfoximine (BSO) renders cells considerably more susceptible to cytotoxicity caused by ·NO. Subsequent studies have also probed the effect of GSH on cell survival following treatment with ·NO gas and ·NO donors, arriving at similar conclusions (Walker *et al.*, 1995). Increasing metallothionein concentration can also protect cells from toxicity induced by ·NO (Schwarz *et al.*, 1995).

Is the autoxidation of ·NO relevant *in vivo,* and where and when is it most likely to occur? As discussed above, the autoxidation reaction is slow in an aqueous solution at 1 μM ·NO. However, in hydrophobic layers in which the rate constant is similar, the concentrations of both ·NO and O_2 may be some 30-fold higher. Therefore, the rate of the ·NO/O_2 reaction increases it by more than a factor of 1000. This means that for 1 μM levels of ·NO detected in aqueous solution, areas such as membranes may have 10–30 μM ·NO with 200–500 μM oxygen, reducing the half-life of ·NO from 13 min to under 10 s. This suggests that the first part of the ·NO/O_2 reaction would take place in the hydrophobic layers, and the intermediates associated with the gas-phase reaction, i.e., nitrogen dioxide, would be important.

3.2. ·NO/O_2^- Chemistry

The reaction between superoxide and ·NO is very important in the fundamental behavior of ·NO in biological systems (Beckman *et al.*, 1990; Pryor and Squadrito, 1996). Huie and Padmaja showed that ·NO and ·O_2^- reacted at near diffusion control to form peroxynitrite (Huie and Padmaja, 1993):

$$\cdot NO + \cdot O_2^- \rightarrow ONOO^- \tag{18}$$

One of the first observations suggesting that this reaction could be important in biological systems was the enhancement by SOD of the effect of EDRF (Furchgott and Zawadzki, 1980). It was thought that concentrations of ·NO might in part be controlled by superoxide. Because of the high rate constant for the formation of the powerful oxidant, $ONOO^-$, this reaction was hypothesized to play an important role in the contribution of ·NO to various pathophysiological conditions (Beckman *et al.*, 1990). As will be discussed below, there is much more to this reaction than simple formation of $ONOO^-$.

The reaction rate constant suggests that $ONOO^-$ could be formed *in vivo*. The question is, when and where? One of the most important considerations for whether a chemical reaction occurs *in vivo* is the relative pseudo-first-order rate constants and not just the rate constant itself. The amount of participation of a given reaction is determined by the concentrations of the reactants as much as by the rate constant (for summary of evaluating RNOS see Wink *et al.*, 1996c). The cellular concentrations of $\cdot O_2^-$ and ·NO under normal conditions are thought to be in the range of 5 pM (Chapter 3) and 0.1–1 μM, respectively. This suggests that the production of $\cdot O_2^-$ determines the reaction between these radicals as well as the amount of $ONOO^-$ formed. The concentration of ·NO determines whether the reaction occurs at any specific location. The occurrence of this reaction depends on the competing reaction of $\cdot O_2^-$ with SOD. For example, SOD reacts with superoxide at a rate constant similar to that of the reaction of $\cdot O_2^-$ with ·NO, and is thus the most important competing reaction to consider. For 10% of the $\cdot O_2^-$ formed to be converted to $ONOO^-$, the concentration of ·NO would be have to be 0.4 to 5 μM. In a cell, the cytosolic concentration of SOD is thought to be between 4 and 10 μM, while the mitochondrial SOD is probably as high as 20 μM in the organelle in which most of the superoxide is produced (estimated from Nikano *et al.*, 1990). In addition, other enzymes that react with $\cdot O_2^-$, such as aconitase ($3 \times 10^7\ M^{-1}\ s^{-1}$) and ferricytochrome *c* ($5 \times 10^6\ M^{-1}\ s^{-1}$), could also participate in abating the reaction of ·NO with $\cdot O_2^-$. Therefore, the reaction of $\cdot O_2^-$ and NO·, despite the high rate constant, is probably confined to specific sites.

Peroxynitrite in neutral solution is a powerful oxidant. It can oxidize thiols such as GSH and thioethers such as methionine, nitrate tyrosine residues, nitrate and oxidize guanosine, initiate lipid peroxidation, and cleave DNA. The oxidant responsible for this is an excited state of peroxynitrous acid (HOONO; Koppenol *et al.*, 1992). Peroxynitrite is in equilibrium with peroxynitrous acid ($pK_a = 6.6$; Pryor and Squadrito, 1996). In the absence of adequate substrate, the protonated form simply rearranges to nitrate. This can be thought of as a detoxification pathway. However, at high enough concentrations, substrates such as tyrosine (1 mM tyrosine to yield 50%) can react to give nitrotyrosine. It is thought that most of the nitration and oxidative chemistry proceeds through the $HOONO^*$ species (Koppenol *et al.*, 1992). There are some species that can directly react with anions rather than requiring protonation. With respect to metals, $ONOO^-$ can react at $6 \times 10^3\ M^{-1}\ s^{-1}$ with FeEDTA (Beckman *et al.*, 1992). In the case of Cu/ZnSOD and FeEDTA, the metal species enhance nitration reactions. Also, heme-containing enzymes such as myeloperoxidase ($6 \times 10^6\ M^{-1}\ s^{-1}$), lactoperoxidase ($3.3 \times 10^5\ M^{-1}\ s^{-1}$), and horseradish peroxidase ($3.2 \times 10^6\ M^{-1}\ s^{-1}$) react directly with $ONOO^-$ (Floris *et al.*, 1993). Peroxynitrite can react with ferrihorseradish peroxidase to form a potent nitrating agent (Ischiropoulos *et al.*, 1996). Only compound II was formed in the case of myeloperoxidase and lactoperoxidase, but compound I was formed in the case of horseradish peroxidase. In fact, this report proposes an interesting mechanism by which compound I is formed initially and then rapidly oxidizes nitrite to nitrogen dioxide. We postulate that $\cdot NO_2$ formation in the presence of ·NO leads to N_2O_3, which carries out a number of nitrosation reactions. Another important reaction is that with CO_2. Peroxynitrite has been shown to react with CO_2 to form $ONOO(CO_2)^-$ (Lymar and Hurst, 1995), which has been shown to dramatically enhance nitration and oxidation reactions (Gow *et al.*, 1996; Lymar *et al.*, 1996; Uppu *et al.*, 1996).

Another important consideration of the NO/$\cdot O_2^-$ reaction is that $\cdot$NO and $\cdot O_2^-$ can react with peroxynitrite to give nitrogen dioxide (Miles *et al.*, 1996; Koppenol *et al.*, 1992; Beckman *et al.*, 1994a):

$$ONOO^- + \cdot NO \xrightarrow{H^+} \cdot NO_2 + NO_2^- \tag{19}$$

$$ONOO^- + \cdot O_2^- \xrightarrow{H^+} \cdot NO_2 + OH^- + O_2 \tag{20}$$

This places further restrictions on the chemistry mediated directly by $ONOO^-$. To examine the contribution of $\cdot$NO and $\cdot O_2^-$ on nitrosative and oxidative chemistry, several studies have examined the reactions of $\cdot$NO donors with xanthine oxidase (Miles *et al.*, 1996). For instance, $\cdot$NO does not alter the oxidation of xanthine by xanthine oxidase (XO) but does affect the production of superoxide (Wink *et al.*, 1993b; Rubbo *et al.*, 1994; Clancy *et al.*, 1992; Miles *et al.*, 1995). XO generates both $\cdot O_2^-$ and H_2O_2 and is thus a convenient way of simulating oxidative stress. It has been proposed that $\cdot$NO reacts with the superoxide formed from XO to yield peroxynitrite, which then isomerizes to nitrate (Wink *et al.*, 1993b; Rubbo *et al.*, 1994; Clancy *et al.*, 1992; Miles *et al.*, 1995):

$$ONOO^- \xrightarrow{H^+} NO_3^- \tag{21}$$

If the same conditions were used in the presence of dihydrorhodamine (DHR; Miles *et al.*, 1996), an increase in oxidation is observed in the presence of $\cdot$NO/XO:

$$ONOO^- + DHR \rightarrow \text{oxidation} \tag{22}$$

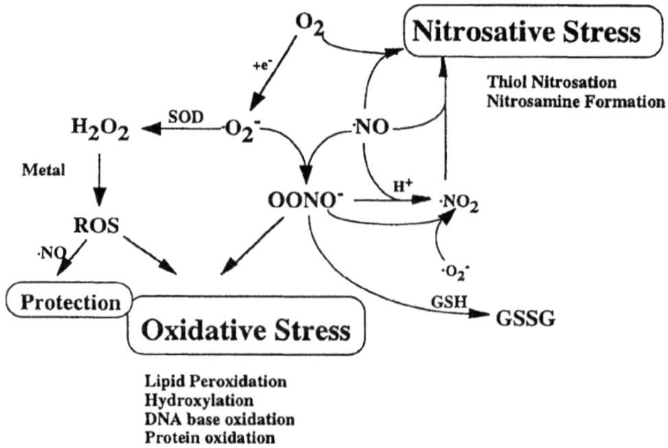

FIGURE 12. Chemistry of continuous and simultaneous fluxes of $\cdot$NO and $\cdot O_2^-$.

Peroxynitrite is known to oxidize DHR [Eq. (22)] (Kooy *et al.*, 1994), further supporting the hypothesis that peroxynitrite is an RNOS generated from ·NO/XO.

What is intriguing about this study is that the fluxes of ·NO and $\cdot O_2^-$ were varied relative to each other, and that maximal oxidation through peroxynitrite was only achieved at a molar ratio of 1 to 1. An excess of either radical quenched the oxidation chemistry of $ONOO^-$. In the presence of excess ·NO or $\cdot O_2^-$, peroxynitrite is converted to $\cdot NO_2$ (Beckman *et al.*, 1994a; Koppenol *et al.*, 1992; Miles *et al.*, 1996), which in turn can rapidly react with ·NO to form the nitrosating species, N_2O_3:

$$\cdot NO_2 + \cdot NO \rightarrow N_2O_3 \tag{23}$$

Thus, although nitrosation does *not* occur directly through peroxynitrite or peroxynitrous acid, there are several mechanisms by which it can occur through the $\cdot NO/O_2^-$ reaction, which may be important in the biology of ·NO (Figure 12).

4. THE BIOCHEMICAL TARGETS OF RNOS

As discussed above, there are various mechanisms and reactions by which ·NO can lead to oxidizing and nitrosating species. Furthermore, nitrosative and oxidative stresses involve distinct types of chemistry and leave different signatures on pathophysiological processes. Here, we will discuss the biological targets of both oxidizing and nitrosating species.

4.1. Nitrosative Stress

Nitrosative stress refers to processes in which the fluxes of ·NO become high enough to result in nitrosation of amines and thiols (Figure 13). This can lead to the formation of

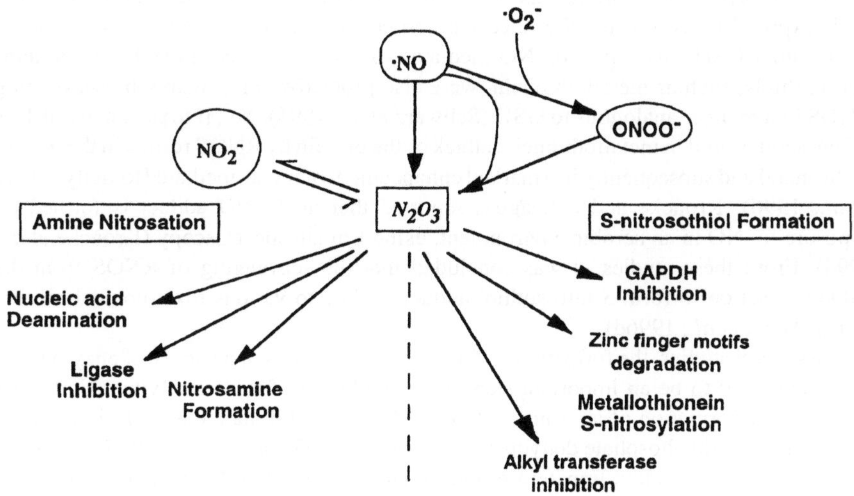

FIGURE 13. The chemistry of nitrosative stress.

protein thionitrosyls, often modulating the functions of these proteins. Alternatively, the nitrosation of amines can potentially lead to the formation of carcinogenic nitrosamines or result in deamination. Here, we will discuss the targets of thiol and amine nitrosation.

4.1.1. Thiol Nitrosation

S-nitrosothiol adducts (RSNO) can be readily formed under nitrosative stress conditions. These complexes have been suggested to play important roles in signal transduction (Stamler, 1994). Ignarro and colleagues showed that these adducts could stimulate the conversion of GTP to cGMP by sGC (Ignarro *et al.*, 1980b), and suggested they were key intermediates in the action of various nitrovasodilator compounds such as sodium nitroprusside and nitroglycerin (Ignarro *et al.*, 1980a; Ignarro, 1989). Although EDRF was reported to be ·NO (Ignarro *et al.*, 1987; Palmer *et al.*, 1987), a later report suggested that RSNO adducts of various peptides, in particular *S*-nitrosocysteine, could be involved in the mechanisms modulating vascular tone (Myers *et al.*, 1990). Stamler and co-workers have suggested that *S*-nitrosothiol protein adducts might serve as a circulating source of ·NO (Stamler *et al.*, 1992). Further studies have revealed that *S*-nitrosothiol protein adducts are formed as a consequence of various immune responses (Gaston *et al.*, 1993).

S-nitrosothiols form from nitrosating agents derived from acidified nitrite (Williams, 1988). Pryor and co-workers showed that *S*-nitrosothiols cannot be formed from direct interaction with ·NO, but rather through the formation of RNOS (Pryor *et al.*, 1982). Formation of *S*-nitrosothiol peptides can occur by reaction with $\cdot NO_x$ intermediates of the $\cdot NO/O_2$ reaction (Wink *et al.*, 1994a; Walker *et al.*, 1995; DeMaster *et al.*, 1995). These reactive RNOS have a high affinity for thiol-containing peptides (Wink *et al.*, 1994a).

The formation of these adducts could play two roles in cell toxicology. First, GSH is critical in cellular protection against the toxicity of RNOS (Wink *et al.*, 1994a; Walker *et al.*, 1995), presumably through the initial formation of *S*-nitrosoglutathione (GSNO). Cells exposed to ·NO or to ·NO-releasing compounds showed little toxicity, yet when intracellular GSH was depleted, ·NO-mediated toxicity increased. In addition, proteins rich in thiols, such as metallothionein, were also protective, presumably by scavenging RNOS in a manner analogous to GSH (Schwarz *et al.*, 1995). Yet, if toxic metals such as Cd are sequestered in metallothionein, attack of the protein by RNOS results in the release of the metal and subsequently in a marked enhancement of metal-mediated toxicity (Misra *et al.*, 1996). Kroncke and colleagues reported that an RSNO adduct is formed on exposure to ·NO in an aerobic environment, using Raman spectroscopy (Kroncke *et al.*, 1994). From these studies, it was concluded that the scavenging of RNOS from the $\cdot NO/O_2$ reaction to form *S*-nitrosothiol adducts is key to various toxicological mechanisms (Wink *et al.*, 1996d).

In several studies, the formation of *S*-nitrosothiol adducts of a variety of enzymes has been suggested to be an important step in the inhibition of their catalytic activity. For example, the formation of an *S*-nitrosothiol adduct may stimulate the ADP ribosylation of glyceraldehyde phosphate dehydrogenase (Molina y Vedia *et al.*, 1992). ·NO inhibits the activity of DNA alkyl transferase both *in vitro* and *in vivo* (Laval and Wink, 1994) (Figure 14). This protein has critical –SH residues that on exposure to aerobic solutions of ·NO form a S–NO adduct. This adduct inhibits the transfer of an alkyl group from the

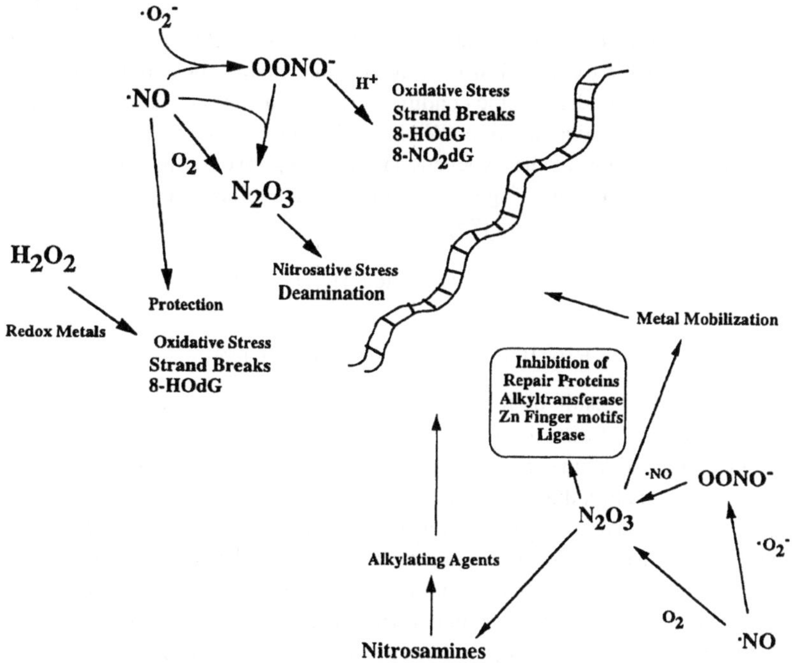

FIGURE 14. Potential mechanisms for ·NO-mediated DNA damage.

O^6 position of guanine to the thiol residue within the protein, resulting in potentiation of the toxicity of alkylating agents (Laval and Wink, 1994) (Figure 14). Proteins that contain zinc finger motifs (which involve Zn sulfur ligation) lose their structural integrity on exposure to ·NO, resulting in the inhibition of enzymatic activity (Wink and Laval, 1994; Kroncke *et al.*, 1994).

4.1.2. Amine Nitrosation

In the presence of secondary amines, macrophages expressing iNOS can form nitrosamines (Marletta, 1988). It was proposed that nitrosamines formed by an activated immune system might play a role in the carcinogenesis associated with inflammation. Lui and colleagues demonstrated that nitrosamines could be generated from woodchuck liver chronically infected with hepatitis, suggesting that nitrosative stress sufficient to form carcinogenic nitrosamines can occur *in vivo* (Liu *et al.*, 1991; 1992).

Though ·NO *per se* does not interact with bioorganic molecules such as DNA or proteins, RNOS such as N_2O_3 and $ONOO^-$ can alter DNA, resulting in a variety of lesions (Wink *et al.*, 1996b,d) (Figure 14). Bacterial and mammalian cells exposed to ·NO exhibit single-strand breaks in their DNA (Arroyo *et al.*, 1992; Nguyen *et al.*, 1992). Further studies showed that deamination also occurs under these conditions (Wink *et al.*, 1991; Nguyen *et al.*, 1992). It was proposed that the formation of RNOS via the autoxidation of ·NO was responsible for these lesions. Several other studies showed in addition that in DNA exposed to ·NO under aerobic conditions, nucleic acids such as cytosine, adenine, and guanine underwent deamination (Wink *et al.*, 1991; Nguyen *et al.*, 1992). It was proposed that nitrosation of the exocyclic amine group resulted in the formation of a primary nitrosamine [Eq. (24)] (Figure 14). This was then followed by rapid deamination, resulting in a hydroxyl group [Eq. (25)].

$$NH_2\text{-}R + N_2O_3 \rightarrow R\text{-}NHNO + NO_2^- \quad (24)$$

$$R\text{-}NHNO \rightarrow R\text{-}NNOH \rightarrow R\text{-}OH + N_2 \quad (25)$$

This chemistry would result, by formal exchange of an NH_2 group against an OH group, in the conversion of cytosine to uracil, guanine to xanthine, methylcytosine to thymine, and adenine to hypoxanthine. This mechanism involving ·NO may thus contribute to the spontaneous deamination that occurs *in vivo*.

Further studies have examined the resultant mutations of a shuttle vector transfected and retrieved from bacterial and mammalian cells treated with nitrosative agents to determine the types of mutations present after repair. Plasmids treated with nitrosating agents derived from either ·NO gas under aerobic conditions (Routledge *et al.*, 1993), ·NO donor compounds (Routledge *et al.*, 1994b), or acidified nitrite (Routledge *et al.*, 1994a) were transfected into different cells. The plasmids were recovered and the mutations determined. The results were primarily GC-to-AT transitions consistent with the proposed deamination mechanisms.

4.2. Oxidative Stress

The primary oxidizing species derived from ·NO is $ONOO^-$, whereas nitrosation is mediated by isomers of N_2O_3. The $\cdot NO/O_2$ reaction to form N_2O_3 can also lead to oxidation of substrates, but only with substances with lower oxidation potential such as metals (Wink *et al.*, 1996d) and catecholamines (Cook *et al.*, 1996). Species such as N_2O_3 may help keep cells in an oxidized state. Nitrogen dioxide may also play a role, but its oxidative chemistry is likely to be limited *in vivo* by virtue of its rapid reaction with ·NO. In a recent report comparing agents that can undergo either nitrosation or one-electron oxidation, it was found that substances whose potential was > 0.7 V underwent nitrosation of the exocyclic amine group, whereas substances below this potential underwent one-electron oxidation when exposed to ·NO under aerobic conditions (Grisham and Miles, 1994).

As discussed above, the oxidation of substrates with $ONOO^-$ is dependent on the flux of ·NO and the concentration of the substrate. If the fluxes of ·NO are considerably higher than those of O_2^-, then $ONOO^-$ is rapidly converted to a nitrosating species. However, if the fluxes of ·NO and $\cdot O_2^-$ are equivalent, $ONOO^-$ is formed and can react with different substrates. Pryor and co-workers showed that $ONOO^-$ can oxidize methionine by one- and two-electron mechanisms (Pryor *et al.*, 1994), whereas intermediates from the ·NO/O_2 reaction do not (Wink *et al.*, 1994a). Also, proteins such as α-1-proteinase are oxidized by peroxynitrite (Moreneo and Pryor, 1992).

Radi and co-workers reported that $ONOO^-$ can intitate lipid peroxidation in membranes (Radi *et al.*, 1991). Along the same lines, Hogg and colleagues suggested that $ONOO^-$ may play a role in initiating the peroxidation of low-density lipoproteins, thus affecting atherosclerosis (Hogg *et al.*, 1993). However, both groups have gone on to demonstrate that ·NO can also abate these processes, reinforcing the importance of the interaction among the various chemical species (Radi *et al.*, 1991; Hogg *et al.*, 1993).

Several studies have shown that synthetically generated $ONOO^-$ can cause various lesions to DNA . Concentrations of $ONOO^-$ ranging from 0.05 to 8 mM were used to induce DNA strand breaks *in vitro* (King *et al.*, 1992; Salgo *et al.*, 1995). Oxidation of guanine to form HOdG was observed in the presence of 3-morpholinosydnonimine (SIN-1), an ·NO donor thought to generate both ·NO and superoxide (Inoue and Kawanishi, 1995). However, another study suggested that $ONOO^-$ does not increase the levels of HOdG in DNA (Yermilov *et al.*, 1995a). In addition to oxidation products, 8-nitroguanine has also been detected as a product of the reaction of $ONOO^-$ with guanine, suggesting that oxidation of nucleotides can also be associated with nitration (Yermilov *et al.*, 1995a,b). deRojas-Walker and colleagues have suggested that oxidative damage to DNA in activated macrophages is related to the formation of $ONOO^-$ (deRojas-Walker *et al.*, 1995). Analogously to the treatment of the plasmid pSP189 with ·NO described in their Table 1, Juedes and Wogan treated plasmids with $ONOO^-$ and transfected them into *E. coli* and AD293 cells (Juedes and Wogan, 1996). Treatment with 2.5 mM $ONOO^-$ resulted in GC → TA (65%) transversions, GC → CG (28%) transversions, and GC--AT transitions (11%), suggesting a spectrum of mutations different from that produced by agents that give rise to nitrosative stress.

Most of the studies involving $ONOO^-$ were conducted in the presence of large concentrations of the synthetically generated compound. These preparations are usually contaminated with excessive NO_2^- and H_2O_2, possibly contributing to the observed result.

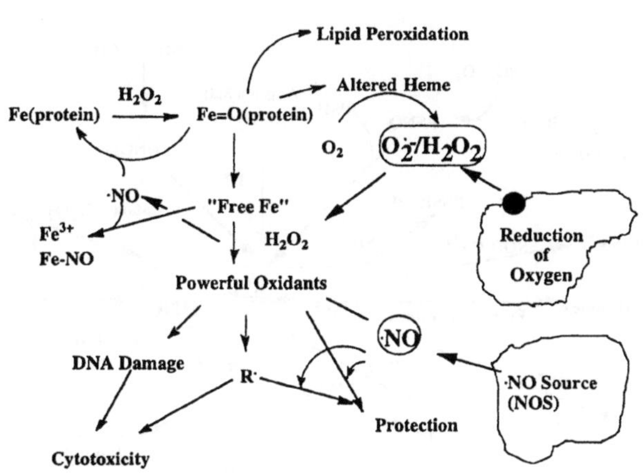

FIGURE 15. The chemistry of nitric oxide with reactive oxygen species.

Contrary to its delivery in boluses, the chemistry of $ONOO^-$ formed *in vivo* depends on the relative fluxes of ·NO and $\cdot O_2^-$ in any given microenvironment (Miles *et al.*, 1996). The amount of $ONOO^-$ that can react directly with the biological target is minimal if the flux of ·NO is in excess. In fact, damage to DNA mediated by XO and H_2O_2 is abolished by the presence of ·NO (Pacelli *et al.*, 1994). It is assumed that Fenton chemistry mediates the DNA strand breaks that are abated by ·NO (Pacelli *et al.*, 1994). Furthermore, hydroxylation reactions are similarly quenched by the presence of ·NO (Miles *et al.*, 1996). These results suggest that the presence of ·NO could abate the oxidation chemistry mediated by oxidants generated in the classical Haber–Weiss chemistry (Figure 15). Taken together, these protective effects indicate that the involvement of RNOS in modifying DNA directly might be limited *in vivo,* while ·NO can protect against oxidative stress.

5. NITROXYL CHEMISTRY

Nitroxyl (HNO/NO^-) is a RNOS that may also contribute to the biology of ·NO. Nitroxyl may be a product formed through the activity of NOS (Hobbs *et al.*, 1994; Schmidt *et al.*, 1996), or through the subsequent oxidation of HOArg (Pufahl *et al.*, 1995), a product of NOS (Figure 16). Nitroxyl is the one-electron reduction product of ·NO. This chemical species is isoelectronic to oxygen, occurring in both singlet and triplet states. Stanbury has calculated the reduction potential to form either the triplet or singlet state as +0.34 and −0.33 V, respectively, using energies of formation values (Stanbury, 1989). Based on thermodynamics, it might be expected that the triplet state would dominate and thereby ·NO would be reduced faster than oxygen. However, as discussed below, this is not the case.

Gratzel and co-workers first reported the direct observation of this RNOS using pulse radiolysis techniques, and they determined an absorption maximum at 260 nm with a molar extinction coefficient of 1200 M^{-1} s^{-1}(Gratzel *et al.*, 1970). The rate constant for

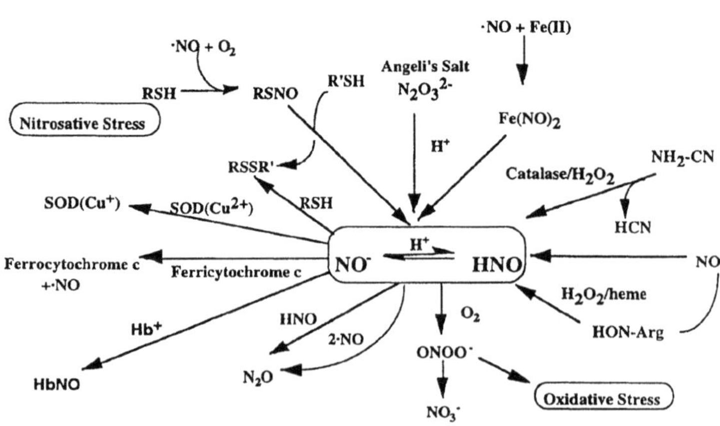

FIGURE 16. Sources and reactivity of nitroxyl (NO^-) *in vivo.*

reduction of ·NO by an aquated electron (e^-_{aq}) derived from ionizing radiation is nearly diffusion controlled (10^{10} M^{-1} s^{-1}):

$$\cdot NO + e^-_{aq} \rightarrow NO^- \quad (26)$$

Nitroxyl can be formed from ·NO by weaker reductants. But to understand the reduction of ·NO, a brief description of its electron configuration is necessary. The higher energy of the singlet state is expected and is analogous to that of oxygen. One might predict that the triplet state would dominate its chemistry; however, the extent of reduction of ·NO at an electrode surface is higher than that of oxygen (Wink *et al.*, 1995a). Furthermore, preliminary results show that reduced methyl viologen (MV^+) is oxidized by ·NO at a rate of 10^4 M^{-1} s^{-1}. If we compare the rate constant for oxidation of MV^+ by oxygen to form superoxide ($k = 10^6$ M^{-1} s^{-1}), the formation of superoxide is 100 times greater than that of nitroxyl (unpublished results). The more facile oxidation of MV^+ by oxygen supports the notion that the redox potential of $O_2/\cdot O_2^-$ is lower than that of $\cdot NO/NO^-$ at neutral pH. Because in the course of its biological formation by NOS ·NO is derived, in part, from oxygen, and because the latter is always in large excess over ·NO, superoxide would be > 10^4 times more likely to be formed than nitroxyl via an outer-sphere electron transfer. Thus, it appears unlikely that ·NO will ever have the chance to competitively inhibit the formation of $\cdot O_2^-$. To further support the notion that NO^- is unlikely to be derived from the same sources as $\cdot O_2^-$, the interactions of ·NO with XO were examined. The biological reduction of O_2 via flavin cofactors results in the formation of both $\cdot O_2^-$ and H_2O_2 (Halliwell and Gutteridge, 1989a). Yet, under an atmosphere of ·NO, no formation of NADH or ureate was observed, implying that ·NO does not accept electrons from this enzyme (Wink *et al.*, 1993b, 1994c). Therefore, the formation of NO^- from direct one-electron reduction of ·NO would not be expected, under physiological conditions.

It is conceivable that NO^- may be formed *in vivo* by the interaction of adjacent thiol groups with *S*-nitrosothiols (Stamler, 1994), by metal-catalyzed reduction of ·NO (Bonner and Stedman, 1996), or by the metabolism of exogenously introduced or endogenously generated compounds. Because ·NO does *not* directly react with thiols in biological systems, RNOS must account for the formation of *S*-nitrosothiols (Wink *et al.*, 1994a; Pryor *et al.*, 1982) prior to release of NO^-. For instance, the intermediates formed in the $\cdot NO/O_2$ reaction preferentially react with thiols to form *S*-nitrosothiols:

$$4\cdot NO + O_2 \rightarrow 2\,N_2O_3 \quad (27)$$

$$N_2O_3 + RSH \rightarrow RSNO + HNO_2 \quad (28)$$

$$RSNO \rightarrow \cdot RS + \cdot NO \quad (29)$$

$$RSNO + R'S^- \rightarrow RS\text{-}SR' + NO^- \quad (30)$$

$$2\,NO^- + 2H^+ \rightarrow N_2O + H_2O \quad (31)$$

$$NO^- + 2{\cdot}NO \rightarrow N_2O + NO_2^- \tag{32}$$

$$^3NO^- + O_2 \rightarrow ONOO^- \rightarrow NO_3^- \tag{33}$$

These *S*-nitrosothiols can then decompose either homolytically to yield thiyl radical (·RS) and ·NO, or heterolytically to yield N_2O, nitrite, and nitrate. Therefore, under conditions of high local concentrations of ·NO in an aerobic environment where nitrosation occurs, reactions such as those described in Eqs. (29)–(33) may take place.

Another potential source of NO^- is the metal-catalyzed reduction of ·NO. Bonner and co-workers investigated the reduction by Fe(II) (Bonner and Pearsall, 1982). They postulated that ferrous ion (Fe(II)) reduced ·NO to $\cdot NO^-$ via the intermediacy of $Fe(NO)_2$. Interestingly, these species have been detected in cells and tissues exposed to activated immune cells (Lancaster and Hibbs, 1990). Because such dinitrosyl species are formed *in vivo*, this chemistry may play a role in the biology of ·NO (Bonner and Stedman, 1996).

There are several ways to generate NO^- from exogenous donors. The most notable is sodium trioxodinitrate ($N_2O_3^{2-}$; Angeli's salt). Though this salt is stable in alkaline solution ($pH > 11$), at neutral pH it decomposes to form NO_2^- and N_2O (Bonner and Ravid, 1975). Bonner and co-workers suggested that the monoanion of trioxodinitrate, $HN_2O_3^-$, decomposes by a simple heterolytic cleavage of the N–N bond to yield nitroxyl and nitrite (Bonner and Akhtar, 1981; Bonner and Ravid, 1975):

$$HN_2O_3^- \rightarrow NO^- + HNO_2 \tag{34}$$

However, Doyle and Mahapatro suggested a more complex mechanism, according to which the N–N bond is cleaved homolytically, and which involved the intermediacy of NO_2^{2-} (Doyle and Mahapatro, 1984):

$$HN_2O_3^- \rightarrow NO_2^{2-} + {\cdot}NO + H^+ \tag{35}$$

$$2H^+ + NO_2^{2-} \rightarrow {\cdot}NO + H_2O \tag{36}$$

$$NO_2^{2-} + N_2O_3^{2-} \rightarrow N_2O_3^{3-} + NO_2^- \tag{37}$$

$$N_2O_3^{3-} + {\cdot}NO + 2\,H^+ \rightarrow N_2O + NO_2^- + H_2O \tag{38}$$

In a later paper, it was proposed that NO^- from Angeli's salt could be trapped by metmyoglobin (Mb^+) (Bazylinkski and Hollocher, 1985). However, Doyle and colleagues later discussed that NO^- may be formed in the decomposition mechanism of trioxodinitrate(II), but probably not via Eqs. (35)–(38) (Doyle *et al.*, 1988). Though the exact mechanism is still to be determined, the formation of N_2O is likely to originate from NO^-. The spin state of NO^- released from trioxodinitrate(II) is thought to be singlet (Bonner and Stedman, 1996). Other chemical sources of NO^- are derived from the decomposition of benzenesulfohydroxylamic acid ($PhSO_2NHOH$; Piloty's acid) or its derivatives (Feelisch and Stamler, 1996). Though NO^- is not likely to result from simple one-electron reduction of NO *in vivo*, decomposition of various chemical complexes could result in the formation of $\cdot NO^-$. As this would generally be in the singlet state, it should *not* react with O_2 to give peroxynitrite.

The pK_a of HNO is 4.7, meaning that at physiological pH it exists in the unprotonated form:

$$NO^- + H^+ \Leftrightarrow HNO \tag{39}$$

It has been reported that the dimerization of HNO to form N_2O has a rate constant of 10^9 $M^{-1}\ s^{-1}$ (Doyle *et al.*, 1988):

$$2\ HNO \rightarrow N_2O + H_2O \tag{40}$$

Nitroxyl is thought to react with oxygen [see Eq. (33)]. Huie and Padmaja reported that the rate constant for a related reaction, i.e., that between $\cdot O_2^-$ and $\cdot NO$, was at near diffusion control (Huie and Padmaja, 1993). This rate constant [Eq. (18)] was 100 times greater than previously reported (Saran and Bors, 1994), and it is thought that the previous value might actually represent the reaction rate between nitroxyl and oxygen ($5 \times 10^7\ M^{-1}\ s^{-1}$) to form peroxynitrite [Eq. (33)]. However, Fukuto and colleagues had suggested that $\cdot NO^-$ derived from hydroxybenzenesulfonamide and trioxodinitrate(II) is oxidized by molecular oxygen to form $\cdot NO$ and $\cdot O_2^-$ (Fukuto *et al.*, 1993). They reported that under anaerobic conditions a small amount of $\cdot NO$ is produced, as was previously reported (Maragos *et al.*, 1991). Yet, on addition of oxygen, a dramatic increase in the production of $\cdot NO$ was observed, which was accompanied by the consumption of O_2 and the formation of H_2O_2. Thus, it was proposed that NO^- reduced oxygen to form superoxide. Superoxide would then dismutate to give hydrogen peroxide [Eq. (43)]:

$$N_2O_3^{2-} \rightarrow NO^- + NO_2^- \tag{41}$$

$$^1NO^- + O_2 \rightarrow \cdot NO + O_2^- \qquad \text{(for singlet)} \tag{42}$$

$$2\ O_2^- + 2H^+ \rightarrow H_2O_2 \tag{43}$$

$$^1NO^- + O_2^- + 2\ H^+ \rightarrow \cdot NO + H_2O_2 \tag{44}$$

There appears to be a balance between the formation of superoxide and that of $\cdot NO$, through complex mechanisms. Yet, when done under different conditions, i.e., the presence of oxygen, no increased $\cdot NO$ production was observed (Zamora *et al.*, 1995). One possible explanation is that under conditions where $\cdot NO$ was detected, the solution was rigorously purged, which may have resulted in some $\cdot NO$ escaping the complex radical chemistry.

Nitroxyl can reduce metal ions via inner- and outer-sphere electron-transfer mechanisms. For instance, ferricytochrome *c* is reduced by trioxodinitrate(II), presumably through the intermediacy of NO^- (Doyle *et al.*, 1988). Yet, when methemoglobin (Hb^+) and metmyoglobin (Mb^+) are exposed to trioxodinitrate(II), formation of the corresponding nitrosyl adducts is observed. These reactions are postulated to occur through an inner-sphere electron-transfer mechanism and are thought to proceed at rates greater than

$10^4 M^{-1} s^{-1}$. Nitroxyl anion can also oxidize thiol-containing compounds, resulting in the formation of disulfide and hydroxylamine (Doyle *et al.*, 1988). These rapid interactions of nitroxyl with oxygen, metals, and thiols may limit the lifetime of NO^- in biological systems. Its half-life *in vivo* will thus largely depend on the relative concentration of the reactants near the site of its formation.

Preliminary studies have shown that Angeli's salt can enhance the toxicity of H_2O_2 (Wink *et al.*, 1996a). Nitroxyl oxidizes thiols, which might have deleterious effects on cellular processes. Nitroxyl may also influence the function of heme-containing enzymes. As discussed above, the ferric state of lipoxygenase and COX (which is the active form) has relatively little affinity for ·NO. Compounds that release $\cdot NO^-$ readily form Fe(II)–NO complexes, which could inactivate lipoxygenase and/or COX and possibly influence a number of other signal transduction pathways as well. One of the most important roles of $\cdot NO^-$ may involve its ability to scavenge ·NO (e.g., see Gratzel *et al.*, 1970). The formation of nitroxyl *in vivo* may act analogously to $\cdot O_2^-$ in attenuating ·NO, as we have observed that nitroxyl donors can reduce basal production of ·NO (unpublished observations). This may have ramifications for cellular adhesion processes which are controlled by the endothelium-derived release of ·NO.

6. PERSPECTIVE

As discussed above, ·NO can influence a variety of cellular events. Because of its broad implications for physiology, there have been intense discussions as to the mechanisms by which ·NO and its chemistry may influence pathophysiological conditions. As with all of the above chemical reactions, the most important factors are timing, concentration of the reactive species generated, rate constants of the reaction in question, and location of the reactions of ·NO and RNOS in biological systems. Most if not all of the direct effects of ·NO will occur in the concentrations range from 0.001 to 1 μM, where the primary interactions will be with metals and radical species. These low concentrations allow ·NO to interact with different biological complexes and to migrate away from the cellular source without formation of RNOS. Therefore, physiologically important roles such as the activation of sGC, reversible inhibition of mitochondrial respiration (resulting in increased local pO_2 levels), attenuation of COX/lipoxygenase products, and other processes probably occur at rather modest fluxes of ·NO. Under pathophysiological conditions, these same fluxes of ·NO may be "out of balance," leading to abnormal blood flow and unwanted cellular signaling. Yet these same direct effects can protect cells from oxidative stress induced by peroxides, and prevent lipid peroxidation, Fenton-type oxidation of DNA and protein, as well as leukocyte adhesion. However, if locally ·NO levels rise to micromolar concentrations, RNOS formation starts to become important. These indirect reactions can lead to chemical modification and alteration of the tertiary structure of macromolecules, either by oxidation or by nitrosation, thus affecting cellular responses. Although such reactions may be important in fighting various pathogens, they may well also contribute to pathophysiological processes such as cancer or neurological disorders if left uncontrolled.

In many cases, the most important factors are the fluxes of ·NO and the duration of ·NO production. The term *flux* is defined as the concentrations of ·NO to which a target is exposed per unit of time. For example, 100 μM ·NO delivered by an ·NO donor with a half-life of 2 min might result in a concentration of ·NO of >5 μM for less than a minute. Yet, the same 100 μM ·NO delivered by an ·NO donor with a half-life of 1 hr may only reach a flux of 0.5 μM but last for hours. Therefore, *in vivo*, the compound with the longer half-life may not generate sufficiently high conentration of ·NO to result in formation of RNOS, whereas the compound with the shorter half-life will. Therefore, temporal considerations are as important as stoichiometric ones when evaluating whether or not RNOS can be formed and whether the effects of ·NO will be most likely direct or indirect.

Another important factor to consider *in vivo* is the spatial distribution of ·NO. As it diffuses away from the source, ·NO dilutes proportionally to the distance it travels. Therefore, RNOS will most likely be present near the site of formation while direct effects of ·NO probably occur much farther away. As discussed above, ·NO can regulate its own production from the constitutive isoforms of NOS, yet iNOS is not readily inhibited directly by ·NO. Therefore, it is unlikely that RNOS, with perhaps the possible exception of $\cdot NO^-$, would be formed from nNOS or eNOS. This suggests that the major source of RNOS is iNOS. Therefore, analysis of the expression of iNOS may allow an evaluation of the location of the indirect effects, while studying the distribution of nNOS and eNOS may allow the localization of nearly exclusively direct effects. With these concepts in mind, one can evaluate where and when direct actions of ·NO might apply and when indirect ones are mediated by RNOS. Another important factor may be the location of point sources of $\cdot O_2^-$, leading to highly localized areas of RNOS formation. Yet another consideration for an indirect effect is the relationship of a specific protein or other biomolecule to the proximity of lipid or other hydrophobic layers. The $\cdot NO/O_2$ reaction and the production of RNOS associated with nitrosative stress will most likely occur in these regions. Hence, proteins that are membrane-bound are the ones most likely to be nitrosated.

7. REFERENCES

Abu-Soud, H. M., Wang, J., Rousseau, D. L., Fukuto, J. M., Ignarro, L. J., and Stuehr, D. J. 1995, Neuronal nitric oxide synthase self-inactivates by forming a ferrous–nitrosyl complex during aerobic catalysis, *J. Biol. Chem.* **270:**22997–23006.

Adams, M. L., Nock, B., Truong, R., and Cicero, T. J., 1992, Nitric oxide control of steroidogenesis: Endocrine effects of NG-nitro-L-arginine and comparisons to alcohol, *Life Sci.* **50:**PL35–40.

Antonini, E., and Brunori, M., 1971, The derivatives of ferrous hemoglobin and myoglobin, *Frontiers Biol.* **21:**31–33.

Arroyo, P. L., Hatch-Pigott, V., Mower, H. F., and Cooney, R. V., 1992, Mutagenicity of nitric oxide and its inhibition by antioxidants, *Mutat. Res.* **281:**193–202.

Bastian, N., Yim, C. Y., Hibbs, J. B., and Samlowski, W. E., 1994, Induction of iron-derived ESR signals in murine cancers by nitric oxide: Evidence for multiple intracellular targets, *J. Biol. Chem.* **269:**5127–5131.

Bates, T. E., Loesch, A., Burnstock, G., and Clark, J. B., 1996, Mitochondrial nitric oxide synthase: A ubiquitous regulator of oxidative phosphorylation? *Biochem. Biophys. Res. Commun.* **218:**40–44.

Bazylinkski, D. A., and Hollocher, T. C., 1985, Metmyoglobin and methemoglobin as efficient traps for nitrosyl hydride (nitroxyl) in neutral aqueous solution, *J. Am. Chem. Soc.* **107:**7982–7986.

Beckman, J. S., Beckman, T. W., Chen, J., Marshall, P. H., and Freeman, B. A., 1990, Apparent hydroxyl radical production by peroxylnitrites: Implications for endothelial injury from nitric oxide and superoxide. *Proc. Natl. Acad. Sci. USA* **87:**1620–1624.

Beckman, J. S., Ischiropoulos, H., Zhu, L., van der Woerd, M., Smith, C., Chen, J., Harrison, J., Martin, J. C., and Tsai, M., 1992, Kinetics of superoxide dismutase- and iron-catalyzed nitration of phenolics by peroxynitrite, *Arch. Biochem. Biophys.* **298:**438–445.

Beckman, J. S., Chen, J., Ischiropoulos, H., and Crow, J. P., 1994a, Oxidative chemistry of peroxynitrite, *Methods Enzymol.* **233:**229–240.

Beckman, J. S., Ye, Y. Z., Anderson, J., Chen, M. A., Accavitti, M. A., Tarpey, M. M., and White, C. R., 1994b, Extensive nitration of protein tyrosines in human atherosclerosis detected by immunohistochemistry, *Biol. Chem. Hoppe-Seyler* **375:**81–88.

Bonner, F. T., and Akhtar, M. J., 1981, Formation of nitrosyltricyanonickelate in a direct NO-displacement reaction, *Inorg. Chem.* **20:**3155–3160.

Bonner, F. T., and Pearsall, K. A., 1982, Aqueous nitrosyliron(II) chemistry. 1. Reduction of nitrite and nitric oxide by iron(II) and (trioxodinitrato)iron(II) in acetate buffer. Intermediacy of nitrosyl hydride, *Inorg. Chem.* **21:**1973–1978.

Bonner, F. T., and Ravid, B., 1975, Thermal decomposition of oxyhyponitrite (sodium trioxodinitrate(II)) in aqueous solution, *Inorg. Chem.* **14:**558–563.

Bonner, F. T., and G. Stedman, 1996, The chemistry of nitric oxide and redox-related species, in *Methods in Nitric Oxide Research* (M. Feelisch and J. Stamler, eds.), pp. 13–18, Wiley, New York.

Bouton, C., Raveau, M., and Drapier, J. C., 1996, Modulation of iron regulatory protein function. Further roles of nitrogen and oxygen-derived reactive species, *J. Biol. Chem.* **271:**2300–2306.

Bredt, D. S., Hwang, P. M., and Snyder, S. H., 1990, Localization of nitric oxide synthase indicating a neural role for nitric oxide. *Nature* **347:**768–770.

Bredt, D. S., Glatt, C. E., Hwang, P. M., Fotuhi, M., Dawson, T. M., and Snyder, S. H., 1991, Nitric oxide synthase protein and mRNA are discretely localized in neuronal populations of the mammalian CNS together with NADPH diaphorase, *Neuron* **7:**615–624.

Brown, G. C., 1995a, Nitric oxide regulates mitochondrial respiration and cell functions by inhibiting cytochrome oxidase, *FEBS Lett.* **369:**136–139.

Brown, G. C., 1995b, Reversible binding and inhibition of catalase by nitric oxide, *Eur. J. Biochem.* **232:**188–191.

Brown, G. C., Bolanos, J. P., Heale, S. J., and Clark, J. B., 1995, Nitric oxide produced by activated astrocytes rapidly and reversibly inhibits cellular respiration, *Neurosci. Lett.* **193:**201–204.

Brunelli, L., Koppenol, W. H., Bertolini, A., and Beckman, J. S., 1994, Catalase can scavenge nitric oxide, *Circulation* **90:**A2461.

Burstyn, J. N., Yu, A. E., Dierks, E. A., Hawkins, B. K., and Dawson, J. H. 1995, Studies of the heme coordination and ligand binding properties of soluble guanylyl cyclase (sGC): Characterization of Fe(II)sGC and Fe(II)sGC(CO) by electronic absorption and magnetic circular dichroism spectroscopies and failure of CO to activate the enzyme, *Biochemistry* **34:**5896–5903.

Buxton, G. V., Greenstock, C. L., Helman, W. P., and Ross, A. B., 1988, Critical review of rate constants for reactions of hydrated electrons, hydrogen atoms and hydroxyl radicals in aqueous solution, *J. Phys. Chem. Ref. Data* **17:**513–886.

Cassina, A., and Radi, R., 1996, Differential inhibitory action of nitric oxide and peroxynitrite on mitochondrial electron transport, *Arch. Biochem. Biophys.* **328:**309–316.

Castro, L., Rodrigue, M., and Radi, R., 1994, Aconitase is readily inactivated by peroxynitrite, but not by its precursor, nitric oxide, *J. Biol. Chem.* **269:**29409–29415.

Cayatte, A. J., Palacino, J. J., Horten, K., and Cohen, R. A., 1994, Chronic inhibition of nitric oxide production accelerates neointima formation and impairs endothelial function in hypercholesterolemic rabbits, *Arterioscler. Thromb.* **14:**753–759.

Chen, L. Y., Lawson, D. L., and Mehta, J. L., 1994, Reduction in human neutrophil superoxide anion generation by n-3 polyunsaturated fatty acids: Role of cyclooxygenase products and endothelium-derived relaxing factor, *Thromb. Res.* **76:**317–322.

Clancy, R. M., Leszczynska-Piziak, J., and Abramson, S. B., 1992, Nitric oxide, an endothelial cell relaxation factor, inhibits neutrophil superoxide anion production via a direct action on the NADPH oxidase, *J. Clin. Invest.* **90:**1116–1121.

Clarkson, R. B., Norby, S. W., Boyer, S., Vahdi, N., Smirnov, A., Nims, R. W., and Wink, D. A., 1995, Direct observation of the kinetics of accumulation and disappearance of nitric oxide within the Chinese hamster ovary cells using a novel intracellular electron paramagnetic resonance technique, *Biochim. Biophys. Acta* **1243**:496–502.

Cleeter, M. W., Cooper, J. M., Darley-Usmar, V. M., Moncada, S., and Schapira, A. H., 1994, Reversible inhibition of cytochrome c oxidase, the terminal enzyme of the mitochondrial respiratory chain, by nitric oxide. Implications for neurodegenerative diseases, *FEBS Lett.* **345**:50–54.

Cook, J. A., Wink, D. A., Blount, V., Krishna, M. C., and Hanbauer, I., 1996, Effects of nitric oxide on the uptake mechanisms of dopamine. Insights into potential mechanisms of NO-mediated neurotoxicity, *Neurochem. Int.* **516**:609–617.

Corbett, J. A., Kwon, G., Turk, J., and McDaniel, M. L., 1993, IL-1 beta induces the coexpression of both nitric oxide synthase and cyclooxygenase by islets of Langerhans: Activation of cyclooxygenase by nitric oxide, *Biochemistry* **32**:13767–13770.

Cosentino, F., Sill, J. C., and Katusic, Z. S., 1994, Role of superoxide anions in the mediation of endothelium-dependent contractions, *Hypertension* **23**:229–235.

Cotton, F. A., and Wilkinson, G., eds., 1988, *Advanced Inorganic Chemistry*, Wiley, New York.

Culotta, E., and Koshland, D. E., 1992, NO, news is good news, *Science* **258**:1862–1865.

Davidge, S. T., Baker, P. N., Laughlin, M. K., and Roberts, J. M., 1995, Nitric oxide produced by endothelial cells increases production of eicosanoids through activation of prostaglandin H synthase, *Circ. Res.* **77**:274–283.

Deinum, G., Stone, J. R., Babcock, G. T., and Marletta, M. A., 1996, Binding of nitric oxide and carbon monoxide to soluble guanylate cyclase as observed with resonance Raman spectroscopy, *Biochemistry* **35**:1540–1547.

DeMaster, E. G., Quast, B. J., Redfern, B., and Nagasawa, H. T., 1995, Reaction of nitric oxide with the free sulfhydryl group of human serum albumin yields a sulfenic acid and nitrous oxide, *Biochemistry* **34**:11494–11499.

deRojas-Walker, T., Tamir, S., Ji, H., Wishnok, J. S., and Tannenbaum, S. R., 1995, Nitric oxide induces oxidative damage in addition to deamination in macrophage DNA, *Chem. Res. Toxicol.* **8**:473–477.

DeRubertis, F. R., Craven, P. A., and Pratt, D. W., 1978, Electron spin resonance study of the role of nitrosyl-heme in the activation of guanylate cyclase by nitrosoguanidine and related agonists, *Biochem. Biophys. Res. Commun.* **83**:158–167.

Doyle, M. P., and Hoekstra, J. W., 1981, Oxidation of nitrogen oxide by bound dioxygen in hemoproteins, *J. Inorg. Biochem.* **14**:351–356.

Doyle, M. P., and Mahapatro, S. N., 1984, Nitric oxide dissociation from trioxodinitrate(II) in aqueous solution, *J. Am. Chem. Soc.* **106**:3678–3679.

Doyle, M. P., Mahapatro, S. N., Broene, R. D., and Guy, J. K., 1988, Oxidation and reduction of hemoproteins by trioxodinitrate(II). The role of nitrosyl hydride and nitrite, *J. Am. Chem. Soc.* **110**:593–599.

Drapier, J. C., and Bouton, C., 1996, Modulation by nitric oxide of metalloprotein regulatory activities, *BioEssays* **18**:1–8.

Drapier, J. C., Pellat, C., and Henry, Y., 1991, Generation of EPR detectable nitrosyl-iron complexes in tumor target cells cocultured with activated macrophage, *J. Biol. Chem.* **269**:5127–5131.

Drapier, J. C., Hirling, H., Wietzerbin, J., Kaildy, P., and Kuhn, L. C., 1993, Biosynthesis of nitric oxide activates iron regulatory factor in macrophages, *EMBO J.* **12**:3643–3649.

Edwards, J. C., Barry, B. K., Gruetter, D. Y., Ohlstein, E. H., Baricos, W. H., and Ignarro, L. J., 1981, Activation of hepatic guanylate cyclase by nitrosyl-heme complexes. Comparison of unpurified and partially purified enzyme, *Biochem. Pharmacol.* **30**:2531–2538.

Epstien, I. R., Kustin, K., and Warshaw, L. J., 1980, A kinetic study of the oxidation of iron(II) by nitric acid, *J. Am. Chem. Soc.* **102**:3751–3757.

Farias-Eisner, R., Chaudhuri, G., Aeberhard, E., and Fukuto, J. M., 1996, The chemistry and tumoricidal activity of nitric-oxide hydrogen-peroxide and the implications to cell resistance susceptibility, *J. Biol. Chem.* **271**:6144–6151.

Feelisch, M., 1991, The biochemical pathways of nitric oxide formation from nitrovasodilators: Appropriate choice of exogenous NO donors and aspects of preparation and handling of aqueous NO solutions, *J. Cardiovasc. Pharmacol.* **17**:S25–S33.

Feelisch, M., and Noack, H., 1987, Correlation between nitric oxide formation during degradation of organic nitrates and activation of guanylate cyclase, *Eur. J. Pharmacol.* **139**:19–30.

Feelisch, M., and Stamler, J. S., 1996, Donors of nitrogen oxides, in *Methods in Nitric Oxide Research* (M. Feelisch and J. Stamler, eds.), pp. 71–118, Wiley, New York.

Feelisch, M., Noack, E., and Schroder, H., 1988, Explanation of the discrepancy between the degree of organic nitrate decomposition, nitrite formation and guanylate cyclase stimulation, *Eur. Heart J.* **Suppl A:**57–62.

Feelisch, M., te Pool, M., Zamora, R., Deussen, A., and Moncada, S., 1994, Understanding the controversy over the identity of EDRF, *Nature* **368:**62–64.

Feldman, P. L., Griffith, O. W., and Stuehr, D. J., 1993, The surprising life of nitric oxide, *Chem. Eng. News* **71:**26–38.

Floris, R., Piersma, S. R., Yang, G., Jones, P., and Wever, R., 1993, Interaction of myeloperoxidase with peroxynitrite. A comparison with lactoperoxidase, horse radish peroxidase and catalase, *Eur. J. Biochem.* **215:**767–775.

Ford, P. C., Wink, D. A., and Stanbury, D. M., 1993, Autoxidation kinetics of aqueous nitric oxide, *FEBS Lett.* **326:**1–3.

Forstermann, U., Gorsky, L. D., Pollock, J. S., Schmidt, H. H. H. W., Heller, M., and Murad, F., 1990, Regional distribution of EDRF/NO-synthesizing enzyme(s) in rat brain, *Biochem. Biophys. Res. Commun.* **168:**727–732.

Fukuto, J. M., Hobbs, A. J., and Ignarro, L. S., 1993, Coversion of nitroxyl (HNO) to nitric oxide (NO) in biological systems: The role of physiological oxidants and relevance to the biological activity of HNO, *Biochem. Biophys. Res. Commun.* **196:**707–713.

Furchgott, R. F., and Vanhoutte, P. M., 1989, Endothelium-derived relaxing and contracting factors, *FASEB J.* **3:**2007–2018.

Furchgott, R. F., and Zawadzki, J. V., 1980, The obligatory role of endothelial cells in the relaxation of arterial smooth muscle by acetylcholine, *Nature* **288:**373–376.

Garthwaite, J., Charles, S. L., and Chess-Williams, R., 1988, Endothelium-derived relaxing factor release on activation of NMDA receptors suggests a role as intercellular messenger in the brain, *Nature* **336:**385–388.

Gaston, B., Reilly, J., Drazen, J. M., Fackler, J., Ramdev, P., Arnelle, D., Mullins, M. E., Sugarbaker, D. J., Chee, C., Singel, D. J., Loscalzo, J., and Stamler, J. S., 1993, Endogenous nitrogen oxides and bronchodilator S-nitrosothiols in human airways, *Proc. Natl. Acad. Sci. USA* **90:**10957–10961.

Geng, Y., Hansson, G. K., and Holme, E., 1992, Interferon-gamma and tumor necrosis factor synergize to induce nitric oxide production and inhibit mitochondrial respiration in vascular smooth muscle cells, *Circ. Res.* **71:**1268–1276.

Gerzer, R., Bohme, E., Hofmann, F., and Schultz, G., 1981a, Soluble guanylate cyclase purified from bovine lung contains heme and copper, *FEBS Lett.* **132:**71–74.

Gerzer, R., Hofmann, F., and Schultz, G., 1981b, Purification of a soluble, sodium-nitroprusside-stimulated guanylate cyclase from bovine lung, *Eur. J. Biochem.* **116:**479–486.

Goldstein, S., and Czapski, G., 1995, Kinetics of nitric oxide autoxidation in aqueous solution in absence and presence of various reductants. The nature of the oxidizing intermediates, *J. Am. Chem. Soc.* **117:**12078–12088.

Gorbunov, N. V., Osipov, A. N., Day, B. W., Zayas-Rivera, B., Kagan, V. E., and Elsayed, N. M., 1995, Reduction of ferrylmyoglobin and ferrylhemoglobin by nitric oxide: A protective mechanism against ferryl hemoprotein-induced oxidations, *Biochemistry* **34:**6689–6699.

Gow, A., Duran, D., Thom, S. R., and Ischiropoulos, H., 1996, Carbon dioxide enhancement of peroxynitrite-mediated protein tyrosine nitration, *Arch. Biochem. Biophys.* **333:**42–48.

Gratzel, M., Taniguchi, S., and Henglein, A., 1970, Pulsradiolytische Untersuchung kurzlebiger zwischenprodukte der NO-reduktion in wassriger losung, *Ber. Bunsenges. Phys. Chem.* **74:**292–297.

Green, L. C., Ruiz de Luzuriaga, K., Wagner, D. A., Rand, W., Istfan, N., Young, V. R., and Tannenbaum, S. R., 1981, Nitrate biosynthesis in man, *Proc. Natl. Acad. Sci. USA* **78:**7764–7768.

Griffith, O. W., and Stuehr, D. J., 1995, Nitric oxide synthases: Properties and catalytic mechanism, *Annu. Rev. Physiol.* **57:**707–736.

Griscavage, J. M., Fukuto, J. M., Komori, Y., and Ignarro, L. J., 1994, Nitric oxide inhibits neuronal nitric oxide synthase by interacting with the heme prosthetic group. Role of tetrahydrobiopterin in modulating the inhibitory action of nitric oxide, *J. Biol. Chem.* **269:**21644–21649.

Griscavage, J. M., Hobbs, A. J., and Ignarro, L. J., 1995, Negative modulation of nitric oxide synthase by nitric oxide and nitroso compounds, *Adv. Pharmacol.* **34:**215–234.

Grisham, M. B., and Everse, J., 1991, Prooxidant activity of hemoglobin and myoglobin, in *Peroxidases in Chemistry and Biology,* Vol. I (J. Evers, K. E. Evers, and M. B. Grisham, eds.), pp. 335–344, Wiley, New York.

Grisham, M. B., and Miles, A. M., 1994, Effects of aminosalicylates and immunosuppressive agents on nitric oxide dependent N-nitrosation reactions, *Biochem. Pharmacol.* **47:**1897–1902.

Halliwell, B., and Gutteridge, J. M. C., 1986, Oxygen free radicals and iron in relation to biology and medicine: Some problems and concepts, *Arch. Biochem. Biophys.* **246:**501–514.

Halliwell, B., and Gutteridge, J. M. C., 1989a, in *Free Radicals in Biology and Medicine,* pp. 210–211, Oxford University Press (Clarendon), London.

Halliwell, B., and Gutteridge, J. M. C., 1989b, in *Free Radicals in Biology and Medicine,* pp. 416–509, Oxford University Press (Clarendon), London.

Hausladen, A., and Fridovich, I., 1994, Superoxide and peroxynitrite inactivate aconitases, but nitric oxide does not, *J. Biol. Chem.* **269:**29405–29408.

Hayashi, K., Noguchi, N., and Niki, E., 1995, Action of nitric oxide as an antioxidant against oxidation of soybean phosphatidylcholine liposomal membranes, *FEBS Lett.* **370:**37–40.

Hempel, S. L., Monick, M. M., He, B., Yano, T., and Hunninghake, G. W., 1994, Synthesis of prostaglandin H synthase-2 by human alveolar macrophages in response to lipopolysaccharide is inhibited by decreased cell oxidant tone, *J. Biol. Chem.* **269:**32979–32984.

Hentze, M. W., and Kuhn, L. C., 1996, Molecular control of vertebrate iron metabolism:mRNA-based regulatory circuits operated by iron, nitric oxide, and oxidative stress, *Proc. Natl. Acad. Sci. USA* **93:**8175–8182.

Hibbs, J. B., Vavrin, Z., and Taintor, R. R., 1987, L-Arginine is required for the expression of the activated macrophage effector mechanism causing selective metabolic inhibition in target cells, *J. Immunol.* **138:**550–565.

Hobbs, A. J., Fukuto, J. M., and Ignarro, L. J., 1994, Formation of free nitric oxide from L-arginine by nitric source, *Proc. Natl. Acad. Sci. USA* **91:**10992–10999.

Hogg, N., Darley-Usmar, V. M., Wilson, M. T., and Moncada, S., 1993, The oxidation of a human low-density lipoprotein by the simultaneous generation of superoxide and nitric oxide, *FEBS Lett.* **326:**199–203.

Hogg, N., Struck, A., Goss, S. P., Santanam, N., Joseph, J., Parthasarathy, S., and Kalyanaraman, B., 1995, Inhibition of macrophage-dependent low density lipoprotein oxidation by nitric-oxide donors, *J. Lipid Res.* **36:**1756–1762.

Hoshino, M., Ozawa, K., Seki, H., and Ford, P. C., 1993, Photochemistry of nitric oxide adducts of water-soluble iron(III) porphyrin and ferrihemoproteins studied by nanosecond laser photolysis, *J. Am. Chem. Soc.* **115:**9568–9575.

Howard-Flanders, P., 1957, Effect of nitric oxide on the radiosensitivity of bacteria, *Nature* **180:**1191–1192.

Hubbard, N. E., and Erickson, K. L., 1995, Role of 5′-lipoxygenase metabolites in the activation of peritoneal macrophages for tumoricidal function, *Cell. Immunol.* **160:**115–122.

Huie, R. E., and Padmaja, S., 1993, The reaction of NO with superoxide, *Free Radical Res. Commun.* **18:**195–199.

Hurshman, A. R., and Marletta, M. A., 1995, Nitric oxide complexes of inducible nitric oxide synthase: Spectral characterization and effect on catalytic activity, *Biochemistry* **34:**5627–5634.

Ignarro, L. J., 1989, Endothelium-derived nitric oxide:Pharmacology and relationship to the actions of organic esters, *Pharm. Res.* **6:**651–659.

Ignarro, L. J., 1990, Biosynthesis and metabolism of endothelium-derived nitric oxide, *Annu. Rev. Pharmacol. Toxicol.* **30:**535–560.

Ignarro, L. J., Barry, B. K., Gruetter, D. Y., Edwards, J. C., Ohlstein, E. H., Gruetter, C. A., and Baricos, W. H., 1980a, Guanylate cyclase activation by nitroprusside and nitroguanidine is related to formation of S-nitrosothiol intermediates, *Biochem. Biophys. Res. Commun.* **94:**93–100.

Ignarro, L. J., Edwards, J. C., Gruetter, D. Y., Barry, B. K., and Gruetter, C. A., 1980b, Possible involvement of S-nitrosothiols activation of guanylate cyclase nitroso compounds, *FEBS Lett.* **110:**275–278.

Ignarro, L. J., Degnan, J. N., Baricos, W. H., Kadowitz, P. J., and Wolin, M. S., 1982a, Activation of purified guanylate cyclase by nitric oxide requires heme: Comparison of heme-deficient, heme-reconstituted, and heme-containing forms of soluble enzyme from bovine lung, *Biochim. Biophys. Acta* **718:**49–59.

Ignarro, L. J., Wood, K. S., and Wolin, M. S., 1982b, Activation of purified soluble guanylate cyclase by protoporphyrin IX, *Proc. Natl. Acad. Sci. USA* **79:**2870–2873.

Ignarro, L. J., Ballot B., and Wood, K. S., 1984, Regulation of soluble guanylate cyclase activity by porphyrins and metalloporphyrins, *J. Biol. Chem.* **259**:6201–6207.

Ignarro, L. J., Buga, G. M., Wood, K. S., Byrns, R. E., and Chaudhuri, G., 1987, Endothelium-derived relaxing factor produced and released from artery and vein is nitric oxide, *Proc. Natl. Acad. Sci. USA* **84**:9265–9269.

Imai, Y., Kolb, H., and Burkart, V., 1993, Nitric oxide production from macrophages is regulated by arachidonic acid metabolites, *Biochem. Biophys. Res. Commun.* **197**:105–109.

Inoue S., and Kawanishi, S., 1995, Oxidative DNA damage induced by simultaneous generation of nitric oxide and superoxide, *FEBS Lett.* **371**:86–88.

Inoue, T., Fukuo, K., Morimoto, S., Koh, E., and Ogihara, T., 1993, Nitric oxide mediates interleukin-1-induced prostaglandin E2 production by vascular smooth muscle cells, *Biochem. Biophys. Res. Commun.* **194**:420–424.

Ischiropoulos, H., Zhu, L., Chen, J., Tsai, M., Martin, J. C., Smith, C. D., and Beckman, J. S., 1992, Peroxynitrite-mediated tyrosine nitration catalyzed by superoxide dismutase, *Arch. Biochem. Biophys.* **298**:431–437.

Ischiropoulos, H., Nelson, J., Duran, D., and Al-Mehdi, A., 1996, Reactions of nitric oxide and peroxynitrite with organic molecules and ferrihorseradish peroxidases: Interference with the determination of hydrogen peroxide, *Free Radical Biol. Med.* **20**:373–381.

Jaffrey, S. R., Cohen, N. A., Rouault, T. A., Klausner, R. D., and Synder, S. H., 1994, The iron-responsive elements binding protein: A target for synaptic actions of nitric oxide, *Proc. Natl. Acad. Sci. USA* **91**:12994–12998.

Juedes, M. J., and Wogan, G. N., 1996, Peroxynitrite-induced mutation spectra of pSP189 following replication in bacteria and in human cells, *Mutat. Res.* **349**:51–61.

Kanner, J., Harel, S., and Granit, R., 1991, Nitric oxide as an antioxidant, *Arch. Biochem. Biophys.* **289**:130–136.

Kanner, J., Harel, S., and Granit, R., 1992, Nitric oxide, an inhibitor of lipid oxidation by lipoxygenase, cyclooxygenase and hemoglobin, *Lipids* **27**:46–49.

Kelner, M. J., and Uglik, S. F., 1994, Mechanism of prostaglandin E2 release and increase in PGH2/PGE2 isomerase activity by PDGF: Involvement of nitric oxide, *Arch. Biochem. Biophys.* **312**:240–243.

Khatsenko, O. G., Gross, S. S., Rifkind, A. B., and Vane, J. R., 1993, Nitric oxide is a mediator of the decrease in cytochrome P450-dependent metabolism caused by immunostimulants, *Proc. Natl. Acad. Sci. USA* **90**:11147–11151.

Kim, Y. -M., Bergonia, H. A., Muller, C., Pitt, B. R., Watkins, W. D., and Lancaster, J. R., 1995a, Loss and degradation of enzyme-bound heme induced by cellular nitroxide synthesis, *J. Biol. Chem.* **270**:5710–5713.

Kim, Y. -M., Bergonia, H. A., Muller, C., Pitt, B. R., Watkins, W. D., and Lancaster, J. R., 1995b, Nitric oxide and intracellular heme, *Adv. Pharmacol.* **34**:277–291.

King, P. A., Anderson, V. E., Edwards, J. O., Gustafson, G., Plumb, R. C., and Suggs, J. W., 1992, A stable solid that generates hydroxyl radical upon dissolution in aqueous solution: Reactions with proteins and nucleic acids, *J. Am. Chem. Soc.* **114**:5430.

Klausner, R. D., Rouault, T. A., and Harford, J. B., 1993, Regulating the fate of mRNA: The control of cellular iron metabolism, *Cell* **72**:19–28.

Knowles, R. G., Darley-Usmar, V., and Moncada, S., 1996, Nitric oxide and peroxynitrite exert distinct effects on mitochondrial respiration which are differentially blocked by glutathione or glucose, *Biochem. J.* **314**:877–880.

Kooy, N. W., Royall, J. A., Ischiropoulos, H., and Beckman, J. S., 1994, Peroxynitrite-mediated oxidation of dihydrorhodamine 123, *Free Radical Biol. Med.* **16**:149–156.

Koppenol, W. H., Moreno, J. J., Pryor, W. A., Ischiropoulos, H., and Beckman, J. S., 1992, Peroxynitrite, a cloaked oxidant formed by nitric oxide and superoxide, *Chem. Res. Toxicol.* **5**:834–842.

Kroncke, K.-D., Fechsel, K., Schmidt, T., Zenke, F. T., Dasting, I., Wesener, J. R., Bettermann, H., Breunig, K. D., and Kolb-Bachofen, V., 1994, Nitric oxide destoys zinc-finger clusters inducing zinc release from metallothionein and inhibition of the zinc finger-type yeast transcription activator LAC9, *Biochem. Biophys. Res. Commun.* **200**:1105–1110.

Kurose, I., Miura, S., Fukumura, D., Yonei, Y., Saito, H., Tada, S., Suematsu, M., and Tsuchiya, M., 1993, Nitric oxide mediates Kupffer cell-induced reduction of mitochondrial energization in hepatoma cells: A comparison with oxidative burst, *Cancer Res.* **53**:2676–2682.

Lancaster, J., 1994, Simulation of the diffusion and reaction of endogenously produced nitric oxide, *Proc. Natl. Acad. Sci. USA* **91:**8137–8141.

Lancaster, J. R., and Hibbs, J. B., 1990, EPR demonstration of iron-nitrosyl complex formation by cytotoxic activated macrophages, *Proc. Natl. Acad. Sci. USA* **87:**1223–1227.

Laskey, R. E., and Mathews, W. R., 1996, Nitric oxide inhibits peroxynitrite-induced production of hydroxyeicosatetraenoic acids and F2-isoprostanes in phosphatidylcholine liposomes, *Arch. Biochem. Biophys.* **330:**193–198.

Laszlo, F., and Whittle, B. J., 1995, Colonic microvascular integrity in acute endotoxaemia: Interactions between constitutive nitric oxide and 5-lipoxygenase products, *Eur. J. Pharmacol.* **277:**R1–3.

Laszlo, F., Whittle, B. J., and Moncada, S., 1994, Interactions of constitutive nitric oxide with PAF and thromboxane on rat intestinal vascular integrity in acute endotoxaemia, *Br. J. Pharmacol.* **113:**1131–1136.

Laval, F., and Wink, D. A., 1994, Inhibition by nitric oxide of the repair protein O6-methylguanine-DNA-methyltransferase, *Carcinogenesis* **15:**443–447.

Lee, M., Arosio, P., Cozzi, A., and Chasteen, N. D., 1994, Identification of the EPR-active iron-nitrosyl complexes in mammalian ferritins, *Biochemistry* **33:**3679–3687.

Lewis, R. S., Tamir, S., Tannenbaum, S. R., and Deen, W. H., 1995, Kinetic analysis of the fate of nitric oxide synthesized by macrophages in vitro, *J. Biol. Chem.* **270:**29350–29355.

Li, Y., Severn, A., Rogers, M. V., Palmer, R. M., Moncada, S., and Liew, E. Y., 1992, Catalase inhibits nitric oxide synthesis and the killing of intracellular Leishmania major in murine macrophages, *Eur. J. Immunol.* **22:**441–446.

Lisdero, C., Riobo, N., Schopfer, F., and Boveris, A., 1996, Nitric oxide inhibits electron transfer and increases superoxide radical production in rat heart mitochondria and submitochondrial particles, *Arch. Biochem. Biophys.* **328:**85–92.

Liu, R. H., Baldwin, B., Tennant, B. C., and Hotchkiss, J. H., 1991, Elevated formation of nitrate and N-nitrosodimethylamine in woodchucks (Marmota monax) associated with chronic woodchuck hepatitis virus infection, *Cancer Res.* **51:**3925–3929.

Liu, R. H., Jacob, J. R., Tennant, B. C., and Hotchkiss, J. H., 1992, Nitrite and nitrosamine synthesis by hepatocytes isolated from normal woodchucks (Marmota monax) and woodchucks chronically infected with woochuck hepatitis virus, *Cancer Res.* **52:**4139–4143.

Loscalzo, J., Benoit, S. E., Valeri, C. R., Barnard, M. R., and Michelson, A. D., 1996, Decreased platelet inhibition by nitric oxide in two brothers with a history of arterial thrombosis, *J. Clin. Invest.* **97:**979–987.

Lymar, S. V., and Hurst, J. K., 1995, Rapid reaction between peroxynitrite ion and carbon dioxide: Implications for biological activity, *J. Am. Chem. Soc.* **290:**52–57.

Lymar, S. V., Jiang, Q., and Hurst, J. K., 1996, Mechanism of carbon dioxide-catalyzed oxidation of tyrosine by peroxynitrite, *Biochemistry* **35:**7855–7861.

Maccarrone, M., Corasaniti, M. T., Guerrieri, P., Nistico, G., and Finazzi Agro, A., 1996, Nitric oxide donor compounds inhibit lipoxygenase activity. *Biochem. Biophys. Res. Commun.* **219:**128–133.

Maragos, C. M., Morley, D., Wink, D. A., Dunams, T. M., Saavedra, J. E., Hoffman, A., Bove, A. A., Isaac, L., Hrabie, J. A., and Keefer, L. K., 1991, Complexes of ·NO with nucleophiles as agents for the controlled biological release of nitric oxide. Vasorelaxant effects, *J. Med. Chem.* **34:**3242–3247.

Marletta, M. A., 1988, Mammalian synthesis of nitrite, nitrate, nitric oxide and N-nitrosating agents, *Chem. Res. Toxicol.* **1:**249–257.

Marletta, M. A., 1993, Nitric oxide synthase structure and mechanism, *J. Biol. Chem.* **268:**12231–12234.

Marletta, M. A., 1994, Nitric oxide synthase: Aspects concerning structure and catalysis, *Cell* **78:**927–930.

McDaniel, M. L., Kwon, G., Hill, J. R., Marshall, C. A., and Corbett, J. A., 1996, Cytokines and nitric oxide in islet inflammation and diabetes, *Proc. Soc. Exp. Biol. Med.* **211:**24–32.

Miles, A. M., Gibson, M., Krishna, M., Cook, J. C., Pacelli, R., Wink, D. A., and Grisham, M. B., 1995, Effects of superoxide on nitric oxide-dependent N-nitrosation reactions, *Free Radical Res.* **233:**379–390.

Miles, A. M., Bohle, D. S., Glassbrenner, P. A., Hansert, B., Wink, D. A., and Grisham, M. B., 1996, Modulation of superoxide-dependent oxidation and hydroxylation reactions by nitric oxide, *J. Biol. Chem.* **271:**40–47.

Misra, R. R., Hochadel, J. F., Smith, G. T., Waalkes, M. P., and Wink, D. A., 1996, Evidence that nitric oxide enhances cadmium toxicity by displacing the metals from metallothionein, *Chem. Res. Toxicol.* **10:**326–332.

Mitchell, J. B., Wink, D. A., DeGraff, W., Gamson, J., Keefer, L. K., and Krishna, M. C., 1993, Hypoxic mammalian cell radiosensitization by nitric oxide, *Cancer Res.* **53:**5845–5848.

Molina y Vedia, L., McDonald, B., Reep, B., Brune, B., DiSilvio, M., Billiar, T. R., and Lapetina, E. G., 1992, Nitric oxide-induced S-nitrosylation of glyceraldehyde-3-phosphate dehydrogenase inhibits enzymatic activity and increases endogenous ADP-ribosylation, *J. Biol. Chem.* **267**:24929–24932.

Moncada, S., Palmer, R. M. J., and Higgs, E. A., 1991, Nitric oxide: Physiology, pathophysiology, and pharmacology, *Pharmacol. Rev.* **43**:109–142.

Moreneo, J. J., and Pryor, W. A., 1992, Inactivation of α-1-proteinase inhibitor by peroxynitrite, *Chem. Res. Toxicol.* **5**:425–431.

Moro, M. A., Knowles, R. G., Darley-Usmar, V., and Moncada, S., 1996, Nitric oxide and peroxynitrite exert distinct effects on mitochondrial respiration which are differentially blocked by glutathione or glucose, *Biochem. J.* **314**:877–880.

Murad, F., 1994, The nitric oxide–cyclic GMP signal transduction system for intracellular and intercellular communication, *Recent Prog. Horm. Res.* **49**:239–248.

Myers, P. R., Minor, R. L., Guerra, R., Bates, J. N., and Harrison, D. G., 1990, Vasorelaxant properties of the endothelium derived relaxing factor more closely resemble S-nitrosocysteine than nitric oxide, *Nature* **345**:161–163.

Nakatsuka, M., and Osawa, Y., 1994, Selective inhibition of the 12-lipoxygenase pathway of arachidonic acid metabolism by L-arginine or sodium nitroprusside in intact human platelets, *Biochem. Biophys. Res. Commun.* **200**:1630–1634.

Nathan, C., and Xie, Q., 1994, Regulation of biosynthesis of nitric oxide, *J. Biol. Chem.* **269**:13725–13728.

Nguyen, T., Brunson, D., Crespi, C. L., Penman, B. W., Wishnok, J. S., and Tannenbaum, S. R., 1992, DNA damage and mutation in human cells exposed to nitric oxide, *Proc. Natl. Acad. Sci. USA* **89**:3030–3034.

Nikano, M., Kimura, H., Hara, M., Kuroiwa, M., Kato, M., Totsune, K., and Yoshikawa, T., 1990, A highly sensitive method for determining both Mn- and Cu-Zn superoxide dismutase activities in tissue and blood cells, *Anal. Biochem.* **187**:277–280.

Nunoshiba, T., DeRojas-Walker, T., Wishnok, J. S., Tannenbaum, S. R., and Demple, B., 1993, Activation by nitric oxide of an oxidative-stress response that defends Escherichia coli against macrophages. *Proc. Natl. Acad. Sci. USA* **90**:9993–9997.

Ohlstein, E. H., Wood, K. S., and Ignarro, L. J., 1982, Purification and properties of heme-deficient hepatic soluble guanylate cyclase: Effects of heme and other factors on enzyme activation by NO, NO-heme, and protoporphyrin IX, *Arch. Biochem. Biophys.* **218**:187–198.

Oyekan, A. O., 1995, The suppression by lipopolysaccharide of cytochrome P450-dependent renal vasodilation in the rat is mediated by nitric oxide, *Eur. J. Pharmacol.* **277**:123–132.

Pacelli, R., Krishna, M. C., Wink, D. A., and Mitchell, J. B., 1994, Nitric oxide protects DNA from hydrogen peroxide-induced double strand cleavage, *Proc. Am. Assoc. Cancer Res.* **35**:540.

Padmaja, S., and Huie, R. E., 1993, The reaction of nitric oxide with organic peroxyl radicals, *Biochem. Biophys. Res. Commun.* **195**:539–544.

Palmer, R. M. J., Ferrige, A. G., and Moncada, S., 1987, Nitric oxide release accounts for the biological activity of endothelium-derived relaxing factor, *Nature* **327**:524–526.

Pellat, C., Henry, Y., and Drapiers, J. C., 1990, IFN-α activated macrophages: Detection by electron paramagnetic resonance of complexes between L-arginine-derived nitric oxide and non-heme iron, *Biochem. Biophys. Res. Commun.* **166**:119–125.

Pires, M., Ross, D. S., and Rossi, M. J., 1994, Kinetic and mechanistic aspects of the NO oxidation by O_2 in aqueous phase, *Int. J. Chem. Kinet.* **26**:1207–1227.

Pryor, W. A., and Squadrito, G. L., 1996, The chemistry of peroxynitrite and peroxynitrous acid: Products from the reaction of nitric oxide with superoxide, *Am. J. Physiol.* **268**:L699–721.

Pryor, W. A., Church, D. F., Govindan, C K., and Crank, G , 1982, Oxidation of thiols by nitric oxide and nitrogen dioxide: Synthetic utility and toxicological implications, *J. Org. Chem.* **47**:156–159.

Pryor, W. A., Jin, X., and Squadrito, G., 1994, One and two electron oxidations of methionine by peroxynitrite, *Proc. Natl. Acad. Sci. USA* **92**:11173–11177.

Pufahl, R. A., Wishnok, J. S., Marletta, M. A., 1995, Hydrogen peroxide-supported oxidation of NG-hydroxy-L-arginine by nitric oxide synthase, *Biochemistry* **34**:1930–1941.

Radi, R., Beckman, J. S., Bush, K. M., and Freeman, B. A., 1991, Peroxynitrite-induced membrane lipid peroxidation: Cytotoxic potential of superoxide and nitric oxide, *Arch. Biochem. Biophys.* **288**:481–487.

Roberts A. B., Vodovotz, Y., Roche, N. S., Sporn, M. B., and Nathan, C. F., 1992, Role of nitric oxide in antagonistic effects of transforming growth factor-beta and interleukin-1 beta on the beating rate of cultured cardiac myocytes, *Mol. Endocrinol.* **6:**1921–1930.

Rousseau, D. L., Sing, S., Ching, Y. C., and Sassoroli, M., 1988, Nitrosyl cytochrome c oxidase. Formation and properties of mixed valcnce enzyme, *J. Biol. Chem.* **263:**5681–5685.

Routledge, M. N., Wink, D. A., Keefer, L. K., and Dipple, A., 1993, Mutations induced by saturated aqueous nitric oxide in the pSP189 supF gene in human Ad293 and E. coli MBM7070 cells, *Carcinogenesis* **14:**1251–1254.

Routledge, M. N., Mirsky, F. J., Wink, D. A., Keefer, L. K., and Dipple, A., 1994a, Nitrite-induced mutations in a forward mutation assay: Influence of nitrite concentration and pH, *Mutat. Res.* **322:**341–346.

Routledge, M. N., Wink, D. A., Keefer, L. K., and Dipple, A., 1994b, DNA sequence changes induced by two nitric oxide donor drugs in the supF assay, *Chem. Res. Toxicol.* **7:**628–632.

Rubbo, H., Radi, R., Trujillo, M., Telleri, R., Kalyanaraman, B., Barnes, S., Kirk, M., and Freeman, B. A., 1994, Nitric oxide regulation of superoxide and peroxynitrite dependent lipid peroxidation: Formation of novel nitrogen containing oxidized lipid derivatives, *J. Biol. Chem.* **269:**26066–26075.

Rubbo, H., Parthasarathy, S., Barnes, S., Kirk, M., Kalyanaraman, B., and Freeman, B. A., 1995, Nitric oxide inhibition of lipoxygenase-dependent liposome and low-density lipoprotein oxidation: Termination of radical chain propagation reactions and formation of nitrogen-containing oxidizcd lipid derivatives, *Arch. Biochem. Biophys.* **324:**15–25.

Ryoyama, K., Nomura, T., and Nakamura, S., 1993, Inhibition of macrophage nitric oxide production by arachidonate-cascade inhibitors, *Cancer Immunol. Immunother.* **37:**385–389.

Salgo, M. G., Stone, K., Squadrito, G. L., Battista, J. R., and Pryor, W. A., 1995, Peroxynitrite causes DNA nicks in plasmid pBR322, *Biochem. Biophys. Res. Commun.* **210:**1025–1030.

Salvemini, D., Misko, T. P., Masferrer, J. L., Seibert, K., Currie, M. G., and Needleman, P., 1993, Nitric oxide activates cyclooxygenase enzymes, *Proc. Natl. Acad. Sci. USA* **90:**7240–7244.

Salvemini, D., Seibert, K., Masferrer, J. L., Misko, T. P., Currie, M. G., and Needleman, P., 1994, Endogenous nitric oxide enhances prostaglandin production in a model of renal inflammation, *J. Clin. Invest.* **93:**1940–1947.

Saran, M., and Bors, W., 1994, Pulse radiolysis for investigation of nitric oxide-related reactions, *Methods Enzymol.* **233:**20–34.

Sautebin, L., and Di Rosa, M., 1994, Nitric oxide modulates prostacyclin biosynthesis in the lung of endotoxin-treated rats, *Eur. J. Pharmacol.* **262:**193–196.

Schmidt, H. H. M., Hofmann, H., Schindler, U., Shutenko, Z. S., Cunningham, D. D., and Feelisch, M., 1996, No · NO from NO synthase, *Proc. Natl. Acad. Sci. USA* **93:**14492–14497.

Schwartz, S. E., and White, W. H., 1983, *Trace Atmospheric Constituents. Properties, Transformation and Fates*, pp. 1–117, Wiley, New York.

Schwarz, M. A., Lazo, J. S., Yalowich, J. C., Allen, W. P., Whitmore, M., Bergonia, H. A., Tzeng, E., Billiar, T. R., Robbins, P. D., Lancaster, J. R., and Pitt, B. R., 1995, Metallothionein protects against the cytotoxic and DNA damaging effects of nitric oxide, *Proc. Natl. Acad. Sci. USA,* **92:**4452–4456.

Seccia, M., Perugini, C., Albano, E., and Bellomo, G., 1996, Inhibition of Cu^{2+}-induced LDL oxidation by nitric oxide: A study using NO donors with different half-time of NO release, *Biochem. Biophys. Res. Commun.* **220:**306–309.

Singel, D. J., and Lancaster, J. R., 1996, Electron paramagnetic resonance spectroscopy and nitric oxide biology, in *Methods in Nitric Oxde Research* (M. Feelisch and J. Stamler, eds.), pp. 341–346, Wiley, New York.

Stadler, J., Billiar, T. R., Curran, R. D., Stuehr, D. J., Ochoa, J. B., and Simmons, R. L., 1991, Effect of exogenous and endogenous nitric oxide on mitochondrial respiration of rat hepatocytes, *Am. J. Physiol.* **260:**C910–C916.

Stadler, J., Harbrecht, B. G., Di Silvio, M., Curran, R. D., Jordan, M. L., Simmons, R. L., and Billiar, T. R., 1993, Endogenous nitric oxide inhibits the synthesis of cyclooxygenase products and interleukin-6 by rat Kupffer cells, *J. Leukocyte Biol.* **53:**165–172.

Stadler, J., Trockfeld, J., Shmalix, W. A., Brill, T., Siewert, J. R., Greim, H., and Doehmer, J., 1994, Inhibition of cytochromes P450 1A by nitric oxide, *Proc. Natl. Acad. Sci. USA* **91:**3559–3563.

Stamler, J. S., 1994, Redox signaling: Nitrosylation and related target interactions of nitric oxide, *Cell* **78:**931–936.

Stamler, J. S., Jaraki, O., Osbourne, J., Simon, D. I., Keaney, J., Vita, J., Singel, D., Valeri, R., and Loscalzo, J., 1992, Nitric oxide circulates in mammalian plasma primarily as an S-nitroso adduct of serum albumin, *Proc. Natl. Acad. Sci. USA* **89:**7674–7677.

Stanbury, D. M., 1989, Reduction potentials involving inorganic free radicals in aqueous solution, *Adv. Inorg. Chem.* **33:**69–136.

Stone, J. R., and Marletta, M. A., 1994, Soluble guanylate cyclase from bovine lung: Activation with nitric oxide and carbon monoxide and spectral characterization of the ferrous and ferric state, *Biochemistry* **33:**5636–5640.

Stone, J. R., and Marletta, M. A., 1995, The ferrous heme of guanylate cyclase: Formation of hexacoordinate complexes with carbon monoxide and nitrosomethane, *Biochemistry* **34:**16397–16403.

Struck, A. T., Hogg, N., Thomas, J. P., and Kalyanaraman, B., 1995, Nitric oxide donor compounds inhibit the toxicity of oxidized low-density lipoprotein to endothelial cells, *FEBS Lett.* **361:**291–294.

Stuehr, D. J., and Marletta, M. A., 1985, Mammalian nitrate biosynthesis: Mouse macrophages produce nitrite and nitrate in response to *Escherichia coli* lipopolysaccharide, *Proc. Natl. Acad. Sci. USA* **82:**7738–7742.

Stuehr, D. J., and Nathan, C. F., 1989, A macrophage product responsible for cytostasis and respiratory inhibition in tumor target cells, *J. Exp. Med.* **169:**1543–1555.

Stuehr, D. J., Abu-Soud, H. M., Rousseau, D. L., Feldman, P. L., and Wang, J., 1995, Control of electron transfer in neuronal nitric oxide synthase by calmodulin, substrate, substrate analogs, and nitric oxide, *Adv. Pharmacol.* **34:**207–213.

Szabo, C., Zingarelli, B., and Salzman, A. L., 1996, Role of poly-ADP ribosyltransferase activation in the vascular contractile and energetic failure elicited by exogenous and endogenous nitric oxide and peroxynitrite, *Circ. Res.* **78:**1051–1063.

Tsai, A. L., Wei, C., and Kulmacz, R. J., 1994, Interaction between nitric oxide and prostaglandin H synthase, *Arch. Biochem. Biophys.* **31:**367–372.

Uppu, R. M., Squadrito, G. L., and Pryor, W. A., 1996, Acceleration of peroxynitrite oxidants by carbon dioxide, *Arch. Biochem. Biophys.* **327:**335–343.

van der Vliet, A., O'Neill, C. A., Halliwell, B., Cross, C. E., and Kaur, H., 1994, Aromatic hydroxylation and nitration of phenylalanine and tyrosine by peroxynitrite. Evidence for hydroxyl radical production from peroxynitrite, *FEBS Lett.* **339:** 89–92.

van der Vliet, A., Eiserich, J. P., O'Neill, C. A., Halliwell, B., and Cross, C. E., 1995, Tyrosine modification by reactive nitrogen species: A closer look, *Arch. Biochem. Biophys.* **319:**341–349.

Vanin, A. F., Mordvintcev, P. I., Hausschildt, S., and Mulsch, A., 1992, The relationship between L-arginine nitric oxide synthases, nitrite release, and dinitrosyl-iron complex formation by activated macrophage, *Biochim. Biophys. Acta* **1177:**37–42.

Vodovotz, Y., Kwon, N., Popischil, M., Manning, J., Paik, J., and Nathan, C., 1994, Inactivation of nitric oxide synthase following prolonged incubation of mouse macrophages with interferon-gamma and bacterial lipopolysaccharide, *J. Immunol* **152:**4110–4118.

Vodovotz, Y., Letterio, J. J., Geiser, A. G., Chesler, L., Roberts, A. B., and Sparrow, J., 1996, Control of nitric oxide production by endogenous transforming growth factor-b1 and systemic nitric oxide in retinal pigment epithelial cells and peritoneal macrophages, *J. Leukocyte. Biol.* **60:**261–270.

Walker, M. W., Kinter, M. T., Roberts, R. J., and Spitz, D. R., 1995, Nitric oxide-induced cytotoxicity: Involvement of cellular resistance to oxidative stress and the role of glutathione in protection, *Pediatr. Res.* **37:**41–47.

Weiss, G., Goossen, B., Doppler, W., Pantopoulos, K., Werner-Felmayer, G., and Hentze, M. W., 1993, Translational regulation via iron-responsive elements by the nitric oxide/NO synthase pathway, *EMBO J.* **12:**3651–3657.

Williams, D. L. H., 1988, *Nitrosation*, Cambridge University Press, London.

Wink, D. A., and Laval, J., 1994, The Fpg protein, a DNA repair enzyme, is inhibited by the biomediator nitric oxide in vitro and in vivo, *Carcinogenesis* **15:**2125–2129.

Wink, D. A., Kasprzak, K. S., Maragos, C. M., Elespuru, R. K., Misra, M., Dunams, T. M., Cebula, T. A., Koch, W. H., Andrews, A. W., Allen, J. S., and Keefer, L. K., 1991, DNA deaminating ability and genotoxicity of nitric oxide and its progenitors, *Science* **254:**1001–1003.

Wink, D. A., Darbyshire, J. F., Nims, R. W., Saveedra, J. E., and Ford, P. C., 1993a, Reactions of the bioregulatory agent nitric oxide in oxygenated aqueous media: Determination of the kinetics for oxidation and nitrosation by intermediates generated in the NO/O_2 reaction, *Chem. Res. Toxicol.* **6:**23–27.

Wink, D. A., Hanbauer, I., Krishna, M. C., DeGraff, W., Gamson, J., and Mitchell, J. B., 1993b, Nitric oxide protects against cellular damage and cytotoxicity from reactive oxygen species, *Proc. Natl. Acad. Sci. USA* **90:**9813–9817.

Wink, D. A., Osawa, Y., Darbyshire, J. F., Jones, C. R., Eshenaur, S. C., and Nims, R. W., 1993c, Inhibition of cytochromes P450 by nitric oxide and a nitric oxide-releasing agent, *Arch. Biochem. Biophys.* **300:**115–123.

Wink, D. A., Nims, R. W., Darbyshire, J. F., Christodoulou, D., Hanbauer, I., Cox, G. W., Laval, F., Laval, J., Cook, J. A., Krishna, M. C., DeGraff, W., and Mitchell, J. B., 1994a, Reaction kinetics for nitrosation of cysteine and glutathione in aerobic nitric oxide solutions at neutral pH. Insights into the fate and physiological effects of intermediates generated in the NO/O_2 reaction, *Chem. Res. Toxicol.* **7:**519–525.

Wink, D. A., Wink, C. B., Nims, R. W., and Ford, P. C., 1994b, Oxidizing intermediates generated in the Fenton reagent: Kinetic arguments against the intermediacy of the hydroxyl radical, *Environ. Health Perspect.* **102**(Suppl. 3):11–15.

Wink, D. A., Hanbauer, I., Laval, F., Cook, J. A., Krishna, M. C., and Mitchell, J. B., 1994c, Nitric oxide protects against the cytotoxic effects of reactive oxygen species, *Ann. N. Y. Acad. Sci.* **738:**265–278.

Wink, D. A., Christodoulou, D., Ho, M., Krishna, M. C., Cook, J. A., Haut, H., Randolph, J. K., Sullivan, M., Coia, G., Murray, R., and Meyer, T., 1995a, A discussion of electrochemical techniques for the detection of nitric oxide, *Methods: A Companion to Methods in Enzymology* **7:**71–77.

Wink, D. A., Cook, J. A., Krishna, M. C., Hanbauer, I., DeGraff, W., Gamson, J., and Mitchell, J. B., 1995b, Nitric oxide protects against alkyl peroxide-mediated cytotoxicty: Further insights into the role nitric oxide plays in oxidative stress, *Arch. Biochem. Biophys.* **319:**402–407.

Wink, D. A., Cook, J., Pacelli, R., DeGraff, W., Gamson, J., Liebmann, J., Krishna, M., and Mitchell, J. B., 1996a, Effect of various nitric oxide-donor agents on peroxide mediated toxicity. A direct correlation between nitric oxide formation and protection, *Arch. Biochem. Biophys.* **331:**241–248.

Wink, D. A., Grisham, M., Mitchell, J. B., and Ford, P. C., 1996b, Direct and indirect effects of nitric oxide. Biologically relevant chemical reactions in biology of NO, *Methods Enzymol.* **268:**12–31.

Wink, D. A., Grisham, M. B., Miles, A. M., Nims, R. W., Krishna, M. C., Pacelli, R., Teague, D., Poore, C. M. B., and Cook, J. C., 1996c, Methods for the determination of selectivity of the reactive nitrogen oxide species for various substrates, *Methods Enzymol.* **268:**120–130.

Wink, D. A., Hanbauer, I., Grisham, M. B., Laval, F., Nims, R. W., Laval, J., Cook, J. C., Pacelli, R., Liebmann, J., Krishna, M. C., Ford, M. C., and Mitchell, J. B., 1996d, The chemical biology of NO. Insights into regulation, protective and toxic mechanisms of nitric oxide, *Curr. Top. Cell. Regul.* **34:**159–187.

Yermilov, V., Rubio, J., Becchi, M., Friesen, M. D., Pignatelli, B., and Ohshima, H., 1995a, Formation of 8-nitroguanine by the reaction of guanine with peroxynitrite in vitro, *Carcinogenesis* **16:**2045–2050.

Yermilov, V., Rubio, J., and Ohshima, H., 1995b, Formation of 8-nitroguanine in DNA treated with peroxynitrite in vitro and its rapid removal from DNA by depurination, *FEBS Lett.* **376:**207–210.

Yu, A. E., Hu, S., Spiro, T. G., and Burstyn, J. N., 1994, Resonance Raman spectroscopy of soluble guanylyl cyclase reveals displacement of distal and proximal heme ligand by NO, *J. Am. Chem. Soc.* **116:**4117–4118.

Zamora R., Grzesiok, A., Weber, H., and Feelisch, M., 1995, Oxidative release of nitric oxide accounts for guanylyl cyclase stimulating, vasodilator and anti-platelet activity of Piloty's acid: A comparison with Angeli's salt, *Biochem. J.* **312** (Pt. 2):333–339.

Zang, V., Kotowski, M., and van Eldik, R., 1988, Kinetics mechanism of the formation of Fe(II)(edta)NO in the system Fe(II)edta/NO/HONO/NO_2^- in aqueous solution, *Inorg. Chem.* **27**:3279–3283.

Chapter 11

Nitroxides as Protectors against Oxidative Stress

James B. Mitchell, Murali C. Krishna, Amram Samuni, Angelo Russo, and Stephen M. Hahn

1. INTRODUCTION

The term *oxidative stress* has emerged to encompass a broad variety of biological stresses, some of which have obvious implications for health care. Many modalities used in cancer treatment including X rays, and some chemotherapy drugs, exert their cytotoxicity by producing oxygen-related free radicals, thereby imposing added burden to normal cellular detoxification systems. A variety of toxic oxygen-related species including superoxide ($\cdot O_2^-$), hydrogen peroxide (H_2O_2), and hydroxyl radical ($\cdot OH$) can be produced by a diverse group of initiating agents and diseases. When left unchecked, these redox-active species can undoubtedly damage cells and tissues. There is obvious interest in discovering and implementing additional approaches, apart from inherent intracellular detoxification systems, to protect cells, tissues, animals, and humans against oxidative stress.

Stable nitroxide free radicals are employed to probe various biophysical and biochemical processes including paramagnetic contrast agents in MRI (Bennett *et al.*, 1987a,b), probes of membrane structure (Berliner, 1979), and as sensors of oxygen in biological systems (Strzalka *et al.*, 1990). Nitroxides are known to react with several biologically relevant free radicals (Mehlhorn and Packer, 1984; Chateauneuf *et al.*, 1988;

James B. Mitchell, Murali C. Krishna, and Angelo Russo Radiation Biology Branch, National Cancer Institute, National Institutes of Health, Bethesda, Maryland 20892. **Amram Samuni** Molecular Biology, School of Medicine, Hebrew University, Jerusalem 91010, Israel. **Stephen M. Hahn** Department of Radiation Oncology, University of Pennsylvania, Philadelphia, Pennsylvania 19104.

Reactive Oxygen Species in Biological Systems, edited by Gilbert and Colton. Kluwer Academic / Plenum Publishers, New York, 1999.

Nilsson *et al.*, 1989; Belkin *et al.*, 1987). The metabolism of stable nitroxides and the corresponding hydroxylamines has been studied in various cellular and subcellular fractions (Sentjurc *et al.*, 1989; Iannone *et al.*, 1989, 1990a,b) and was found to involve one-electron redox processes. Our initial observation was that a persistent nitroxide spin adduct produced in the reaction between ·OH and the commonly used spin-trapping agent DMPO, reacted with superoxide. Our conclusion was that monitoring the concentration of the DMPO–OH spin adduct does not reflect the hydroxyl radical concentrations in the presence of a high flux of superoxide radicals (Samuni *et al.*, 1989). We extended this initial observation to show that a related stable free radical, 2-ethyl-2,5,5-trimethyl-3-oxazolidine-1-oxyl (OXANO), and its corresponding hydroxylamine (OXANOH) react with $\cdot O_2^-$ in a metal-independent manner. By both reducing and oxidizing $\cdot O_2^-$, the OXANO/OXANOH couple exhibited a superoxide dismutase (SOD)-like activity (Samuni *et al.*, 1988). To determine if other nitroxides had similar SOD-like properties, we used electron paramagnetic resonance (EPR) spectrometry to investigate a series of nitroxides for their reactions with $\cdot O_2^-$ (Samuni *et al.*, 1988, 1989). The five-membered ring nitroxides were reduced by $\cdot O_2^-$ to the corresponding hydroxylamines, which in turn were reoxidized to the parent nitroxides by $\cdot O_2^-$. Furthermore, we found that alicyclic nitroxides may function as SOD mimics by being first oxidized by superoxide to the oxoammonium intermediate and subsequently reduced to the nitroxide by another superoxide. In other words, our results suggested that SOD-like activity is not limited to OXANO but is shared by stable nitroxide spin labels in general (Samuni *et al.*, 1990).

In this chapter, we will review studies evaluating the protective effects of stable nitroxides in mammalian cells, isolated organs, and whole animals, subjected to various types of oxidative damage. Nitroxides have been shown to protect biological systems both *in vitro* and *in vivo* by several modes of actions and the chemical mechanism(s) underlying these observations will be discussed.

2. CHEMISTRY OF NITROXIDES

Nitroxides described in this chapter belong to several ring types such as (A) the five-membered oxazolidine ring, (B) the five-membered saturated pyrrolidine ring, (C) the five-membered unsaturated pyrroline ring, and (D) the six-membered piperidine ring. Four oxidation states exist for a given ring structure, namely, amine, hydroxylamine, nitroxide, and oxoammonium, as shown below:

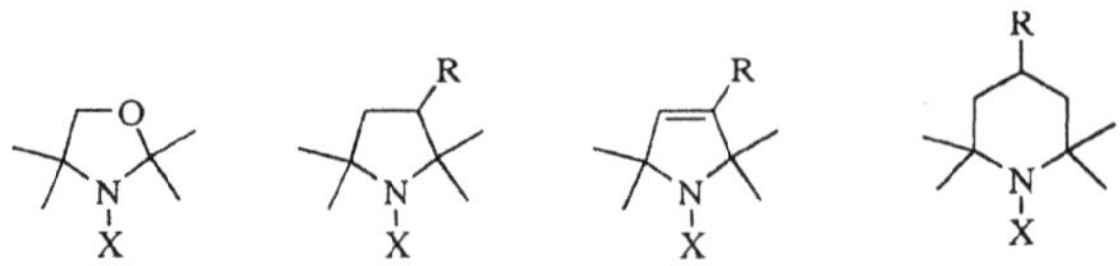

A B C D

X = –H, amine
X = –OH, hydroxylamine
X = –O·, nitroxide radical
X = =O$^+$, oxoammonium cation

While, in general, the amines and the hydroxylamines are chemically stable, stability to the nitroxide radicals is conferred by the bulky substituents at the α-position of the ring structure. The observed antioxidant effects of the nitroxides can be explained in terms of the shuttling between three oxidation states as given below, where the nitroxide (RR′NO·) acts as a reducing agent to provide detoxification by being oxidized by one electron to the corresponding oxoammonium cation ($RR'NO^+$):

$$RR'NO \text{ (oxidation)} \rightarrow RR'NO^+ \tag{1}$$

The oxoammonium cation can undergo reduction either to the nitroxide by a one-electron reducing agent,

$$RR'NO^+ \text{ (one-electron reduction)} \rightarrow RR'NO\cdot \tag{2}$$

or to the hydroxylamine by two-electron reducing agents,

$$RR'NO^+ \text{ (two-electron reduction)} \rightarrow RR'NOH \tag{3}$$

or undergo comproportionation to give the nitroxides,

$$RR'NOH + RR'NO^+ \text{ (comproportionation)} \rightarrow 2\ RR'NO\cdot \tag{4}$$

These redox reactions facilely provide reducing equivalents for detoxification, and more importantly, allow a dynamic equilibrium to be established for the corresponding nitroxide. The hydroxylamine also is a weak reducing agent and can donate an H-atom like a classic antioxidant and in the process is converted to the nitroxide,

$$RR'NOH \text{ (one-electron oxidation)} \rightarrow RR'NO\cdot \tag{5}$$

Nitroxides and hydroxylamines have been shown to inhibit lipid peroxidation:

$$X\cdot + RR'NO\cdot \rightarrow RR'NOX \tag{6}$$

$$X\cdot + RR'NOH \rightarrow RR'NO\cdot + XH \tag{7}$$

$$L\cdot + RR'NO\cdot \rightarrow RR'NOL \tag{8}$$

$$L\cdot + RR'NOH \rightarrow RR'NO\cdot + LH \tag{9}$$

$$LO\cdot + RR'NOH \rightarrow RR'NO\cdot + LOH \tag{10}$$

Because nitroxides and hydroxylamines are interconvertible [e.g., Reactions (7), (9), and (10)], effective protection against lipid peroxidative damage is observed even at low concentrations of either the hydroxylamine or the nitroxide. Miura *et al.* (1993) suggest that hydrophilic nitroxides and hydroxylamines act as preventive antioxidants by inhibiting initiation reactions in lipid peroxidation such as Reactions (6) and (7), where X· is the initiating species. The lipophilic nitroxides and hydroxylamines interrupt chain-propagating lipid peroxidation reactions as shown in Reactions (8)–(10).

Cyclic voltammetric techniques have shown that unlike the nitroxide/hydroxylamine couple, the nitroxide/oxoammonium redox couple acts as a reversible redox couple to mediate the catalytic dismutation of superoxide in a pH-dependent manner (Krishna *et al.*, 1992, 1996a):

$$RR'NO\cdot + HO_2\cdot + H^+ \rightarrow RR'NO^+ + H_2O_2 \tag{11}$$

$$RR'NO^+ + \cdot O_2^- \rightarrow RR'NO\cdot + O_2 \tag{12}$$

Summing Eqs. (11) and (12) gives

$$2\,\cdot O_2^- + 2\,H^+ \rightarrow H_2O_2 + O_2 \tag{13}$$

which is the $\cdot O_2^-$ dismutation reaction. Because nitroxides facilitate this reaction, they represent the first class of metal-independent SOD mimics. In the SOD mimetic reactions, the univalent oxidation of the nitroxide, primarily by the protonated form of superoxide [Eq. (11)], is a pH-dependent reaction. The reduction of the oxoammonium cation by superoxide is diffusion-controlled and the reaction rate constant has been estimated to be $1.2 \times 10^{10}\ M^{-1}\ s^{-1}$. For several piperidine nitroxides examined at pH values in the range of 5.5–8, the catalytic rate constants were found to be between 10^5 and $1.2 \times 10^8\ M^{-1}\ s^{-1}$; consistent with Eqs. (11) and (12), the reaction is efficient at lower pH (Krishna *et al.*, 1996a).

In addition to their direct catalytic role as SOD mimics, nitroxides can act as pro-catalysts by stimulating catalaselike activities in heme proteins and provide antioxidant defense by detoxifying oxidants such as hydrogen peroxide and organic peroxides. Organic peroxides and hydrogen peroxide can react with heme proteins such as hemoglobin, myoglobin, and cytochrome *c* and convert the heme iron into the ferryl state where the valence state of the heme iron is +4. In this valence state, the altered heme protein possesses oxidizing capabilities similar to peroxidase enzymes such as horseradish peroxidase. The higher valence oxidation states of heme iron can inflict significant biological damage by oxidizing critical targets, particularly at the membrane. Nitroxides can facilitate the antioxidant catalytic properties in the heme iron by reducing the ferryl heme and directly detoxify the hypervalent heme iron (Krishna *et al.*, 1996b).

Oxidation of Transition Metal Ions

Low-molecular-weight complexes of transition metals such as copper and iron are thought to mediate peroxide-associated free radical generation that subsequently results in biological damage via Fenton reactions [Eq. (15)] or metal-catalyzed Haber–Weiss reactions [Eqs. (14) and (15)].

$$M^n + \cdot O_2^- \rightarrow M^{n-1} + O_2 \tag{14}$$

$$M^{n-1} + H_2O_2 \rightarrow OH^- + \cdot OH + M^n \tag{15}$$

When these transition metal ions are associated with critical targets such as DNA, site-specific free radical generation can cause significant toxicity. Metal chelators such as desferrioxamine are used to inhibit Reaction (14) and thereby limit the generation of hydroxyl radical or similar oxidants. Metal-catalyzed free radical generation can also be prevented by oxidizing the reduced metals to preempt Reaction (15), i.e.,

$$M^{n-1} + RRNO\cdot \rightarrow M^{n} + RRNOH \tag{16}$$

Nitroxides have been effective in facilitating Reaction (16) (Mitchell *et al.*, 1990; Samuni *et al.*, 1991a), when the metal is associated with either DNA or protein, thereby preventing the Fenton reaction (15) and its consequences.

3. NITROXIDE-MEDIATED PROTECTION AGAINST SUPEROXIDE-, HYDROGEN PEROXIDE-, AND ORGANIC HYDROPEROXIDE-INDUCED CYTOTOXICITY

The initial observation that stable nitroxides function chemically as SOD mimics (Samuni *et al.*, 1988, 1990) led to their evaluation *in vitro* to determine the protective effects against damage induced by several oxidative processes. First, it was found that a variety of nitroxides were not toxic to mammalian cells over the concentration range from 0.1 to 10.0 mM for exposure times of 1–2 hr at 37°C (Mitchell *et al.*, 1990; Hahn *et al.*, 1992b). Further, it was shown that treatment of mammalian cells with nitroxide concentrations of 25–100 mM for 30 min was not cytotoxic (Mitchell *et al.*, 1991). Second, using EPR spectrometry, it was found that nitroxides partition into cells instantaneously (Mitchell *et al.*, 1990; Samuni *et al.*, 1991a). Because low-molecular weight nitroxides can readily enter cells, they would be expected to better protect intracellular oxidative damage than exogenously added SOD, which may be restricted from intracellular entry because of its size.

3.1. Exposure to Superoxide-Generating System

When V79 cells were exposed to $\cdot O_2^-$ generated by the hypoxanthine/xanthine oxidase (HX/XO) reaction, it was found that many stable nitroxides, including the six-membered piperidine derivatives, conferred cytoprotection when applied immediately before the induction of oxidative stress as shown in Figure 1 (Mitchell *et al.*, 1990). Both the oxidized and reduced form of Tempol (4-hydroxy-2,2,6,6-tetramethylpiperidine-1-oxyl) provided substantial protection against HX/XO-mediated cytotoxicity. Catalase, which detoxifies H_2O_2, and Desferal (DF), which is a ferric ion chelator, also provided protection, whereas SOD had no effect. Pretreating these particular cells with SOD for 6 hr did not result in cytoprotection. Based on these experiments, the following conclusions were drawn. First, extracellular $\cdot O_2^-$ generated by the HX/XO reaction is in and of itself not toxic to cells, as SOD, which in V79 cells can only localize in the extracellular space, had no protective effects. Second, even though $\cdot O_2^-$ is generated by the HX/XO reaction, H_2O_2 mediates the damage because $\cdot O_2^-$ dismutates to produce H_2O_2, which freely diffuses through the cell membrane to localize at critical sites. Catalase completely prevented the damage induced by the HX/XO reaction by intercepting H_2O_2. Third, the damage mediated by H_2O_2 is transition metal dependent, as DF completely prevented the cytotoxicity induced by the HX/XO reaction. There was no change in the metabolism of H_2O_2

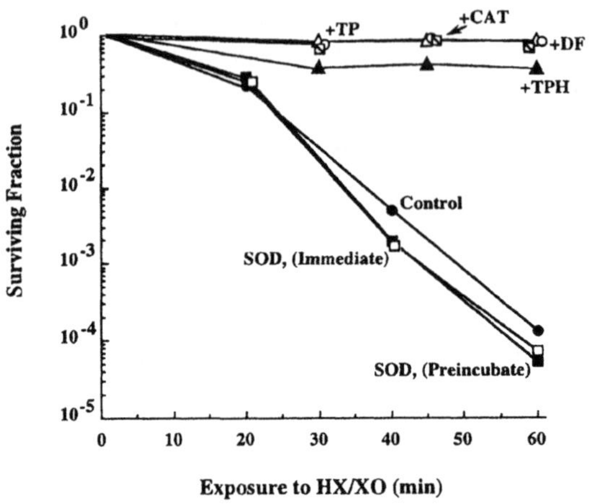

FIGURE 1. Protection from hypoxanthine/xanthine oxidase (HX/XO)-induced cell killing. Chinese hamster V79 cells in full medium at 37°C were exposed to 0.05 unit/ml XO + 0.5 mM HX for various time periods in the presence of various additives: ●, no additives; □, 100 units/ml SOD added simultaneously with XO; ■, 100 units/ml SOD, preincubated for 6 hr with cells prior to addition of XO; △, 5 mM Tempol (TP); ▲, 5 mM Tempol-H (TPH); ○, 500 μM Desferal, preincubated for 2 hr with the cells prior to the addition of XO; ⧅, 100 units/ml catalase added simultaneously with XO. Data adapted from Mitchell *et al.* (1990), with permission.

by V79 cells in the presence and absence of nitroxides (Mitchell *et al.*, 1990). Lastly, the mode of nitroxide protection against the cytotoxicity induced by HX/XO is by oxidizing reduced transition metals or terminating intracellular free radical chain reactions (Mitchell *et al.*, 1990). The dismutation of $\cdot O_2^-$ by nitroxides, though a potential mode of action, is not be the major detoxification pathway in cells exposed to extracellularly generated $\cdot O_2^-$ radicals.

3.2. Hydrogen Peroxide

In studies designed to evaluate the protective effects of stable nitroxides in V79 cells exposed to nonenzymatically generated H_2O_2, Tempol was found to inhibit completely the damage as shown in Figure 2A. In this study a range of Tempol concentrations provided protection against H_2O_2-mediated cytotoxicity, with concentrations of 1–5 mM providing near complete protection. Tempol-H and DF were also protective. Therefore, stable nitroxides can interrupt transition-metal-mediated H_2O_2-induced damage, presumably by inhibiting the generation of site-specific hydroxyl radicals (Mitchell *et al.*, 1990). Similar protective effects against H_2O_2-induced damage have been observed in beating cardiomyocytes as shown in Figure 2B (Samuni *et al.*, 1991c) and mutant bacterial cells that are repair deficient and also hypersensitive to H_2O_2 (Samuni *et al.*, 1991a). Reddan and colleagues evaluated the ability of Tempol to block hydrogen peroxide toxicity of cultured rabbit lens epithelial cells and found that Tempol blocked or decreased hydrogen peroxide-induced inhibition of cell growth and decreased induction of single-strand DNA breaks (Reddan *et al.*, 1992, 1993).

3.3. Organic Hydroperoxides

Nitroxides have also been shown to protect monolayered cells exposed to organic hydroperoxides such as *t*-butyl hydroperoxide (TBH) (Samuni *et al.*, 1991b). Damage to

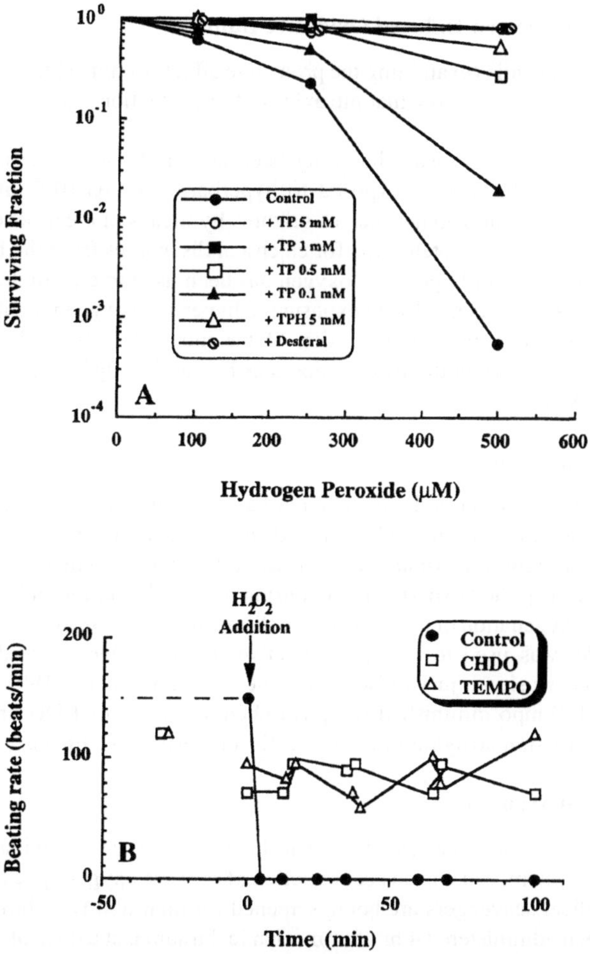

FIGURE 2. (A) Survival of Chinese hamster V79 cells following exposure to varying concentrations of H_2O_2 in the absence or presence of Tempol, Tempol-H, or Desferal. Data adapted from Mitchell *et al.* (1990) with permission. (B) Effect of nitroxides on the H_2O_2-induced loss of cell contractility. The beating rate of isolated rat heart cells after incubation with 250 μM H_2O_2 in the absence (●) and the presence of 0.5 mM cyclohexane doxyl (□) or 0.5 mM Tempo (△). The plated cells were kept at 37°C in Ham's F-10 culture medium under an atmosphere of 5% CO_2 and 95% air. Data from Samuni *et al.* (1991c), with permission.

cells by TBH is independent of both $\cdot O_2^-$ and H_2O_2. This protection has been proposed to result from the reaction of nitroxides with the alkyl, alkoxyl, and alkyl peroxyl radicals formed by the decomposition of TBH (Samuni *et al.*, 1991b). The reaction of these radicals with nitroxides has been documented by others (Belkin *et al.*, 1987; Chateauneuf *et al.*, 1988; Mehlhorn and Packer, 1984; Nilsson *et al.*, 1989).

3.4. Hyperbaric Oxygen-Induced Oxidative Damage

Several model studies examining the protective effects of nitroxides at isolated organ and whole body level also show that nitroxides offer protection against diverse types of oxidative damage.

Hyperbaric oxygen exposure has long been associated with free radical toxicity (Gerschman *et al.*, 1954). Rats exposed to hyperbaric oxygen (0.5 mPa O_2) can be assessed for neurological damage such as seizures by means of electroencephalography (EEG). The duration of the latent time for electrical discharges from the brain to appear after the administration of hyperbaric oxygen has been used to quantitate neurotoxicity. Neurological toxicity associated with exposure to hyperbaric oxygen in rats was reversed by the intraperitoneal (i.p.) administration of the nitroxides, Tempol and Tempo, as measured by the changes in the EEG at doses as low as 2.5 mg/kg body wt (Bitterman and Samuni, 1995).

3.5. Reperfusion Injury

Reoxygenation after periods of ischemia has been shown to generate free radicals such as superoxide, which ostensibly can mediate cardiac tissue injury (McCord, 1987). Low-molecular-weight antioxidants as well as antioxidant enzymes such as SOD and catalase have been applied experimentally during reperfusion to minimize injury associated with the reoxygenation of ischemic tissue (Simpson *et al.*, 1987). Tempo, a lipophilic stable nitroxide, was used to test the protective role of stable nitroxides in regional postischemic reperfusion injury in isolated rat hearts. Gelvan *et al.* (1991) demonstrated that 0.4–1 mM Tempo diminished the postischemic release of LDH and reduced the duration of reperfusion arrhythmias, which reflect reperfusion damage.

3.6. Mechanical Trauma

Mechanical trauma to the brain has been shown to alter the antioxidant status of the central nervous system, causing an accumulation of reactive oxygen species (Siesjo *et al.*, 1989). Free radical scavengers are being screened for their ability to limit this damage. Nitroxides, when administered 4 hr after mechanical trauma, at a dose of 50 mg/kg body wt, were shown to provide protection in rat brains, as measured by neurological severity scores as well as the integrity of the blood–brain barrier. These results suggest that the nitroxides may be useful clinically for head trauma (Yannai *et al.*, 1996).

3.7. Experimental Colitis and Gastric Mucosal Injury

Evidence supporting the association between free radicals and the pathogenesis of colitis and gastric mucosal injury is growing. The effect of intravenous (i.v.) or i.p. administration of Tempol on the gastric mucosal damage induced by ethanol, indomethacin, and aspirin was studied by Rachmilewitz *et al.* (1994), who measured the size of mucosal lesions as well as the levels of mediators of inflammatory response such as myeloperoxidase and leukotrienes. Tempol, given at a dose of 100 mg/kg body wt in rats 5 min prior to the administration of ethanol or indomethacin, completely prevented the mucosal lesions and decreased the levels of myeloperoxidase and leukotrienes (Rachmilewitz *et al.*, 1994).

In another study, Karmeli *et al.* (1995) examined the effect of antioxidants in a rat model of experimental colitis mediated by a free radical mechanism with acetic acid or trinitrobenzene sulfonic acid. Nitroxides, administered intragastrically at a dose of 0.5 g/kg body weight, were studied in this model. Nitroxides led to a decrease in the size of the ulcerative lesion and a decrease in the levels of leukotrienes (Karmeli *et al.*, 1995).

4. NITROXIDE-MEDIATED PROTECTION AGAINST IONIZING RADIATION

4.1. *In Vitro* Radioprotection by Nitroxides

Radiation-induced cellular damage and cytotoxicity are critically influenced by oxygen free radicals. Hence, cells exposed to ionizing radiation under aerobic conditions are more sensitive to radiation-induced cytotoxicity when compared with cells exposed to radiation under hypoxic conditions (Hall, 1994). Efforts over the years have been directed toward modulating the effects of radiation by enhancing or suppressing cytotoxicity mediated by free radicals (Alexander and Charlesby, 1954; Bacq, 1965). Sulfhydryl compounds are excellent biological scavengers of free radicals (Bacq *et al.*, 1953) and, by extension, the classic group of radiation protectors are the aminothiol compounds including cysteine, cysteamine, and WR-2721 (Amifostine) (Patt *et al.*, 1949; Yuhas and Storer, 1969).

Chinese hamster V79 cells in the presence of the nitroxide Tempol were protected against the aerobic lethal damage induced by X rays in a dose-dependent manner as shown in Figure 3 (Mitchell *et al.*, 1991). From the data presented in Figure 3 and subsequent studies, several points regarding the effects of Tempol on the radiation response can be made (Mitchell *et al.*, 1991). First, Tempol-mediated protection of V79 cells was concentration dependent requiring millimolar concentrations to exert significant radioprotection. It should be noted that the requirement for millimolar concentrations of Tempol to provide radioprotection is similar to that for other radioprotectors such as aminothiols; however, aminothiols in general are more efficient radioprotectors on a concentration basis compared with Tempol. Second, the hydroxylamine form of Tempol did not provide radioprotection. Third, Tempol modestly radiosensitized cells irradiated under hypoxic conditions, in agreement with previous studies (Millar *et al.*, 1977, 1978, 1983). Fourth, Tempol treatment did not increase intracellular glutathione concentration, or induce intracellular SOD mRNA. Lastly, neither exogenous SOD nor DF altered the radiation response. This last point suggests that it is unlikely (although it cannot be ruled out) that the SOD mimetic action of nitroxides is responsible for aerobic radioprotection. Previous reports suggested that SOD protected bone marrow progenitor cells against radiation cytotoxicity *in vitro* and *in vivo* (Petkau *et al.*, 1975; Petkau and Chelack, 1984; Petkau, 1987); however, not all workers have been able to reproduce these findings (Abe *et al.*, 1981). We have repeatedly demonstrated *in vitro* that exogenously applied SOD does not afford aerobic radioprotection. Likewise, it is unlikely that Tempol radioprotection results from its effect on metal reduction because the metal chelator DF does not provide *in vitro* radioprotection.

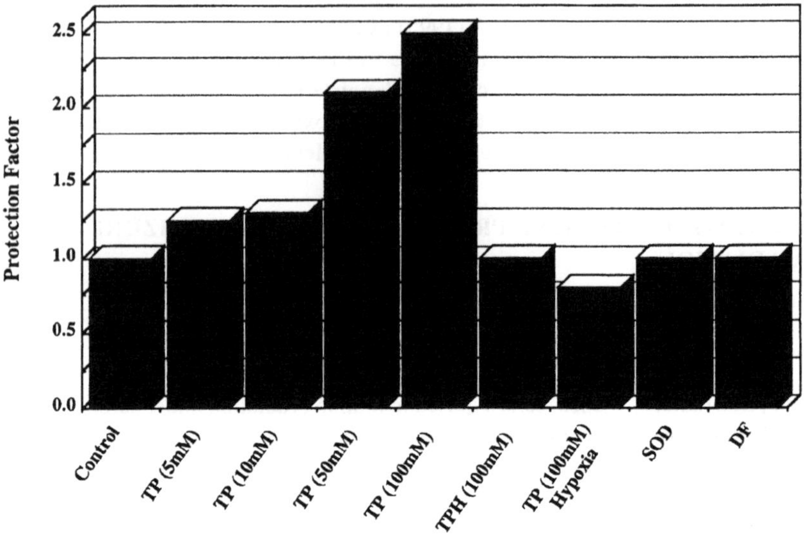

FIGURE 3. Protection of Chinese hamster V79 cells exposed to X rays treated with Tempol (TP) at concentrations of 5, 10, 50, and 100 mM (aerobic and hypoxic); Tempol-H (TPH) at 100 mM; SOD (100 μg/ml); or Desferal (500 μM). Protection factors were calculated at 10% survival by dividing the radiation dose of control cells by the radiation dose of treated cells. Data adapted from Mitchell *et al.* (1991) with permission).

Nitroxides could provide biological radioprotection by inhibiting the damage mediated by radiation-induced reactive species (X·) to biologically important molecules (BIM) such as DNA by at least two modes.

Damage

$$X\cdot + \text{BIM-H} \rightarrow XH + \text{BIM}\cdot \tag{17}$$

Protection by radical scavenging

$$H^+ + X\cdot + RR'NO\cdot \rightarrow HX + RR'NO^+ \tag{18}$$

$$X\cdot + RR'NO\cdot \rightarrow RR'NO\text{-}X \tag{19}$$

Protection by repair

$$X\cdot + \text{BIM-H} \rightarrow XH + \text{BIM}\cdot \tag{20}$$

$$H^+ + \text{BIM}\cdot + RR'NO\cdot \rightarrow \text{BIM-H} + RR'NO^+ \tag{21}$$

Radiation exposure can result in the formation of carbon-centered radicals on BIM (in the case of radiation, the most likely intracellular target is the DNA) as shown in Eq. (17). As proposed in Eqs. (18) and (19), nitroxides could scavenge radiation-induced reactive species. Likewise, Eqs. (20) and (21) suggest that nitroxides have the capability to reduce a carbon-centered radical on BIM to normal by donating an electron (BIM-H). A combination of Eqs. (18)–(19) and (20)–(21) is also feasible.

Radiation-induced unrepaired DNA double-strand breaks and subsequent production of chromosome aberrations have been closely linked with radiation-induced cell killing (Puck, 1958; Carrano, 1973; Bedford *et al.*, 1978). DeGraff *et al.* (1992b) showed a direct correlation between Tempol-mediated *in vitro* radioprotection (cell survival) and reduced DNA double-strand breaks. Johnstone *et al.* (1995) showed that Tempol significantly reduced the frequency of radiation-induced chromosome aberrations in human peripheral blood lymphocytes. Both studies strongly suggest that Tempol-mediated radioprotection is accompanied by a reduction in DNA damage.

4.2. *In Vivo* Radioprotection by Nitroxides

The protective effects of nitroxides in *in vitro* experiments, prompted the study of nitroxides using *in vivo* models. So as to screen stable nitroxides as *in vivo* radioprotectors, the toxicity, pharmacology, and *in vivo* radioprotective effects of Tempol were studied in C3H mice (Hahn *et al.*, 1992a). Tempol was injected into mice at the maximally tolerated dose (275 mg/kg i.p.), and whole blood levels of nitroxide were measured by EPR spectroscopy as shown in Figure 4A. The peak total Tempol concentration was observed 10 min after the i.p. injection. No detectable nitroxide was found in whole blood 24 hr after injection. Because the peak Tempol concentration occurred 10 min after injection, mice were injected with Tempol (275 mg/kg i.p.) 10 min prior to whole-body radiation (7.0–13.0 Gy). Tempol provided a dose modification factor of 1.3 at the $LD_{50/30}$ level as shown in Figure 4B. *In vivo*, Tempol is rapidly reduced to the hydroxylamine, a form shown not to be radioprotective *in vitro* (see above). Despite the rapid reduction of Tempol *in vivo*, significant radioprotection was observed. It is expected that *in vivo* radioprotection would be increased if higher concentrations of the oxidized form of the nitroxide were available at the time of irradiation. Currently our laboratory has initiated a search for other nitroxides that are bioreduced at a slower rate and hence would yield a higher concentration of the oxidized form at the time of irradiation.

The ability to selectively protect normal tissues in cancer patients receiving radiation treatment would be most advantageous. If selective protection of normal tissues were possible, higher radiation doses could be delivered to the tumor accompanied with higher local control rates. The key, however, is *selective* normal tissue protection because if a systemic radioprotector also protects the tumor, no advantage would be realized. To test whether systemic administration of Tempol would protect local irradiation delivered to a tumor, Hahn *et al.* (1997b) used a RIF-1 transplantable rodent tumor to evaluate the effect of Tempol on local tumor control by radiation. As shown in Figure 5A, the administration of Tempol to tumor bearing mice at the same concentration and timing as shown in Figure 4 resulted in no protection of tumor. To identify a mechanism for the apparent differential protection of the hematopoietic system (see above) and tumor tissue, pharmacological studies revealed that the percentage of oxidized compound was approximately two-fold greater in the bone marrow compartment compared with RIF-1 tumor (see Figure 5B) at the time of irradiation. Greater bioreduction of Tempol occurs in RIF-1 tumor and may provide at least a partial explanation for the absence of tumor radioprotection, as it

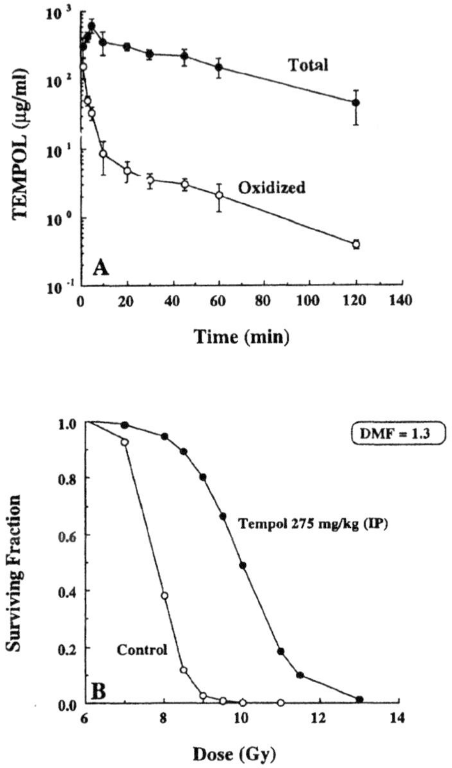

FIGURE 4. *In vivo* studies of Tempol in C3H mice. (A) The time dependence of Tempol in the whole blood of mice. Mice were injected with Tempol (275 mg/kg, i.p.) and at various times after injection, whole blood levels were measured by EPR spectroscopy. The closed symbols represent the total Tempol concentration (oxidized plus reduced). The open symbols represent only the oxidized form. (B) Survival of mice 30 days after whole-body radiation. Control mice received saline i.p. and the treated mice received Tempol (275 mg/kg, i.p.) 10 min prior to radiation. The dose modification factor (DMF) is 1.3. Data from Hahn *et al.* (1992a) with permission.

previously had been demonstrated that the oxidized form of Tempol is the active radioprotector (see above). These preliminary data imply that a potential difference exists between normal tissues and tumor with respect to their ability for bioreduction. Future studies will focus on other normal tissues and different tumor systems to determine the generality of the findings.

Another application of nitroxides as radioprotectors that may have benefit in a clinical setting would be to protect against radiation-induced alopecia, a common radiotherapeutic problem. Hair loss as a result of irradiation especially from whole brain irradiation often leads to cosmetic, social, and psychological problems for the radiotherapy patient. Clinically, no successful interventions are available. Topical application of Tempol was evaluated for possible protective effects against radiation-induced alopecia using guinea pig skin as a model. For single acute doses up to 30 Gy, Tempol, when topically applied 15 min prior to irradiation, provided a marked increase in the rate and extent of new hair recovery when compared with untreated skin (Goffman *et al.*, 1992). By EPR spectroscopy Tempol was detected in treated skin specimens. EPR measurements of blood samples failed to show any systemic nitroxide signal resulting from topical application, nor could Tempol be detected in brain tissue after application on the scalp. Recently these studies have been extended to evaluate fractionated radiation delivery with multiple

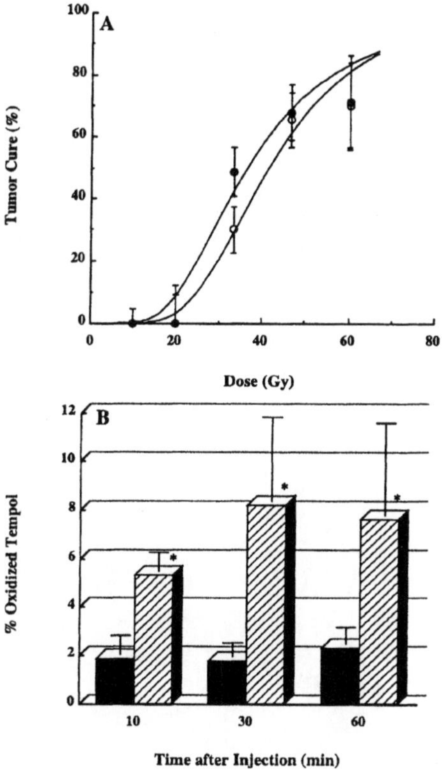

FIGURE 5. (A) Radiation tumor control curves for animals treated in the absence or presence of Tempol. (●) Tempol-treated mice; (○): PBS-treated mice. There was no statistical difference between the two curves ($p = 0.54$). The calculated $TCD_{50/30}$ values for Tempol-treated and PBS-treated mice were 36.7 and 41.8 Gy, respectively. There was no statistical difference between these values ($p = 0.32$). The error bars represent one standard error above and below the value shown. (B) The percent oxidized Tempol with time in RIF-1 tumor versus bone marrow cells harvested simultaneously from C3H mice. Hatched bars: bone marrow; solid bars: RIF-1 tumor. Error bars represent the standard error of the mean. A greater percentage of oxidized Tempol was present in bone marrow compared with RIF-1 tumor cells (*$p \leq 0.004$, as analyzed by Student's *t* test). Data adapted from Hahn *et al.* (1997b) with permission.

applications of Tempol or Tempo (Cuscela *et al.*, 1996). Topical administration of Tempol or Tempo 15 min before each radiation treatment (daily fractions of 7 Gy, for a total of eight treatments over 10 days) resulted in statistically significant radioprotection with respect to hair loss and regrowth of hair in the treatment field (Cuscela *et al.*, 1996). A representative example of Tempo-mediated radioprotection 9 weeks after treatment is shown in Figure 6. Histological evaluation showed that radiation treatment alone resulted in a marked decrease in the number of hair follicles and poor development of remaining follicles; however, nitroxide pretreatment resulted in no appreciable decrease in hair follicles and hair follicles appeared mature, similar to unirradiated controls (Cuscela *et al.*, 1996). These studies suggest that the topical application of nitroxides may be useful clinically to reduce the undesirable toxicity of radiation-induced alopecia. Likewise, topical application of nitroxides may have utility in other sites such as rectum and bladder, two normal tissues at risk in radiation treatment of prostate and/or cervix cancer. The advantage of topical application is that high concentrations of the nitroxide could be used directly on the tissue at risk. Systemic levels of the drug would hopefully be inadequate to protect the tumor or cause systemic toxicity.

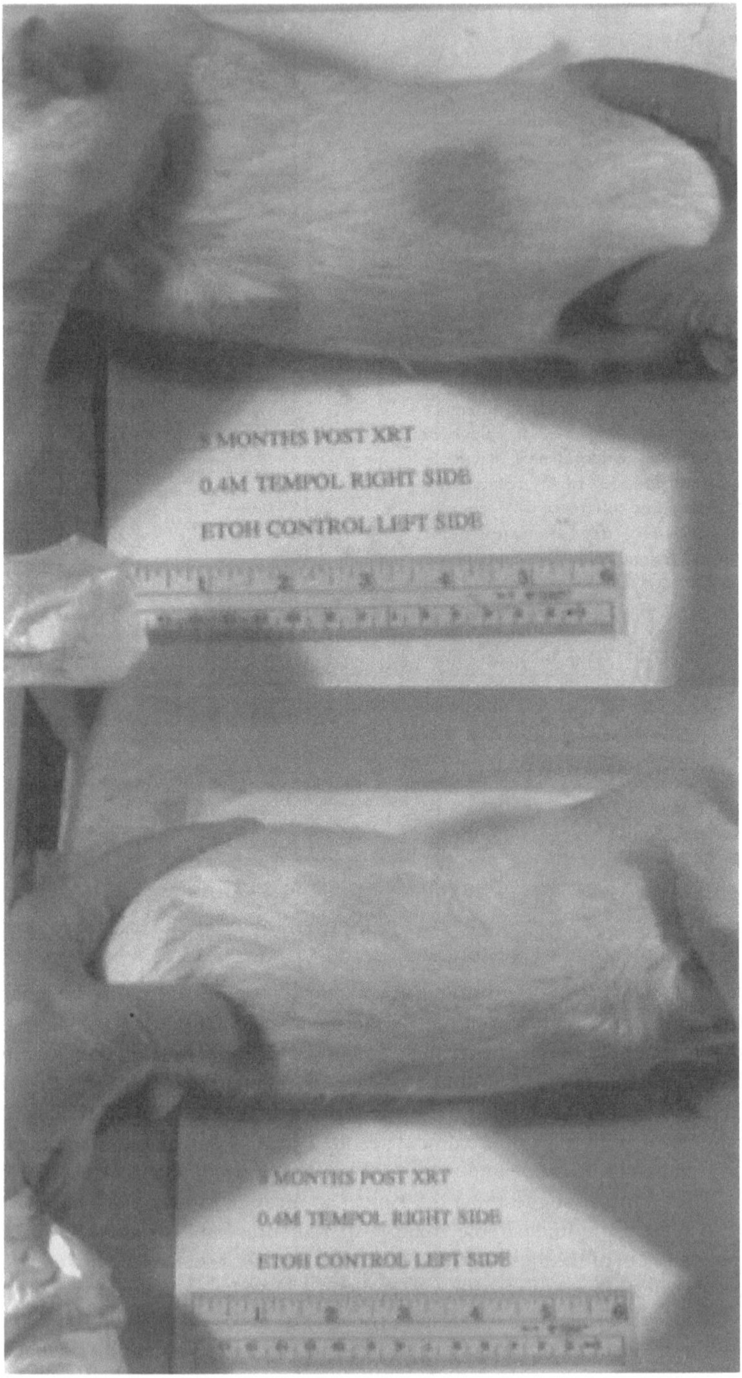
MONTHS POST XRT
0.4M TEMPOL RIGHT SIDE
ETOH CONTROL LEFT SIDE
MONTHS POST XRT
0.4M TEMPOL RIGHT SIDE
ETOH CONTROL LEFT SIDE

5. NITROXIDE-MEDIATED PROTECTION AGAINST REDOX-CYCLING CHEMOTHERAPY DRUGS

5.1. Oxidation of Semiquinones

Some chemotherapeutic agents undergo bioreductive activation by enzymatic processes to produce intermediates that can damage DNA, thereby exerting tumoricidal effects. Quinone-based drugs (Q) such as adriamycin, mitomycin C, and the benzotriazine-based drug SR 4233 are used clinically and, not surprisingly, the efficacy of the drugs is limited by normal tissue toxicity. In addition to these agents, another quinone-based antibiotic, streptonigrin, is known to exert cytotoxic effects by free radical generation via bioreductive activation to the semiquinone ($SQ^{-}\cdot$):

$$Q \text{ (bioreductive activation)} \rightarrow SQ^{-}\cdot \quad (22)$$

$$SQ^{-}\cdot + O_2 \rightarrow Q + \cdot O_2^{-}$$

$$2\,\cdot O_2^{-} + 2\,H^{+} \rightarrow H_2O_2 + O_2 \quad (23)$$

$$H_2O_2 + \text{transition metals} \rightarrow \text{Oxidants} \quad (24)$$

For many of these drugs, a free radical mechanism for the antiproliferative effects as well as the side effects is gaining acceptance. In the case of adriamycin, however, the mechanism of tumoricidal action is less clear. While inhibition of topoisomerase II has been suggested as its mechanism of tumor cell cytotoxicity, oxyradicals generated by the redox cycling of the semiquinone may be responsible for the cardiac toxicity of adriamycin. Because nitroxides are cell permeable, can participate in radical–radical recombination reactions, and have been shown to react with semiquinone radicals (DeGraff *et al.*, 1994) as well as autoxidation products such as superoxide, these compounds are useful research tools for evaluating free radical modes of chemotherapy drug cytotoxicity. The effect of Tempol on the cytotoxicity of adriamycin and streptonigrin in mammalian cells was studied in an effort to determine the importance of free radical mechanisms on the cytotoxicity of these drugs (DeGraff *et al.*, 1994). Whereas Tempol was effective in completely inhibiting the cytotoxicity and the DNA double-strand breaks induced by streptonigrin, it had no effect on either the cytotoxicity or DNA double-strand breaks induced by adriamycin. Although Tempol is cell permeable and reacts with chemically generated semiquinone radicals of streptonigrin and adriamycin, the lack of protection against adriamycin-induced cytotoxicity and DNA double-strand breaks suggests that free radical pathways do not play a major role in the antiproliferative effects of adriamycin.

FIGURE 6. Illustration of radioprotection by Tempol of skin 8 months postirradiation. The animal's left side (top panel) shows hair loss as a result of fractionated irradiation treatment alone (7 Gy fractions over 10 days, eight total treatments). The animal's right side (bottom panel) shows that topical application of Tempol 15 min before each radiation fraction afforded significant protection. See Cuscela *et al.* (1996).

5.2. Adriamycin-Induced Cardiotoxicity

The cumulative dose-limiting effect of adriamycin, clinically, is related to irreversible damage to the myocardium. Oxygen radicals generated by the redox cycling of the drug have been proposed as the cause of cardiotoxicity and agents to inhibit redox cycling are being investigated (Monti *et al.*, 1991). In a recent study, Tempol was evaluated for protective effects against acute cardiotoxicity induced by adriamycin in isolated rat hearts (Monti *et al.*, 1996). When rat hearts were perfused with 100 μg/ml of adriamycin for 60 min, significant impairment of the contractile function as well as increased lipid peroxidation were observed. Coperfusion of adriamycin with Tempol (2.5 mM) significantly inhibited the contractile impairment of the cardiac tissue and caused a decrease in lipid peroxidation indices. The chemical basis for the observed protective effects was proposed to involve free radical scavenging by nitroxides.

5.3. Mitomycin C

Mitomycin C (MMC) is a quinone-containing antineoplastic agent whose semiquinone radical intermediate has been implicated in DNA interstrand cross-links leading to cytotoxicity. Nitroxides have been shown to be effective in reversing the hypoxic cytotoxicity of MMC *in vitro* (Krishna *et al.*, 1991). The cytoprotective effects provided by the nitroxides presumably result from reoxidizing the semiquinone of MMC back to the quinone.

$$SQ^{-}\cdot + RR'NO \rightarrow Q + RR'NOH \tag{25}$$

The semiquinone has been implicated to cause DNA interstrand cross-links.

An example of the use of nitroxides to prevent damage mediated by MMC was recently evaluated in a swine model of chemotherapy extravasation (Hahn *et al.*, 1997a). Extravasation tissue injury from chemotherapeutic drugs such as MMC is a serious clinical problem with few antidotes available. Several nitroxides were screened as protectors of MMC-induced skin necrosis. 3-Carbamoyl-proxyl (3-CP) was the lone nitroxide that protected when given 5 min after MMC extravasation. Histological sections of the 3-CP- and MMC-treated pig skin showed a marked reduction in the degree of acute inflammation and the absence of deep dermal scarring when compared with MMC alone.

5.4. 6-Hydroxydopamine

An analogue of the neurotransmitter dopamine, 6-hydroxydopamine, is used clinically in the treatment of neuroblastoma. The neurotransmitter analogues take advantage of localizing in tumor cells by the dopamine uptake system on the cell membrane.

FIGURE 7. (A) Mutagenicity as assessed by the XPRT forward mutation assay in AS52 cells exposed to ionizing radiation in the absence or presence of Tempol. Tempol treatment alone did not increase the mutation frequency above control. Tempol provided near-complete protection from the cytotoxic effects (top) and mutagenic effects (bottom) of radiation treatment. (B) Analysis of a field inversion gel electrophoresis (FIGE) assay of DNA extracted from cells treated either with radiation or radiation plus Tempol. Tempol protected against radiation-induced DNA damage (the FIGE assay provides a measure of DNA double-strand break damage). Data from DeGraff *et al.* (1992b) with permission.

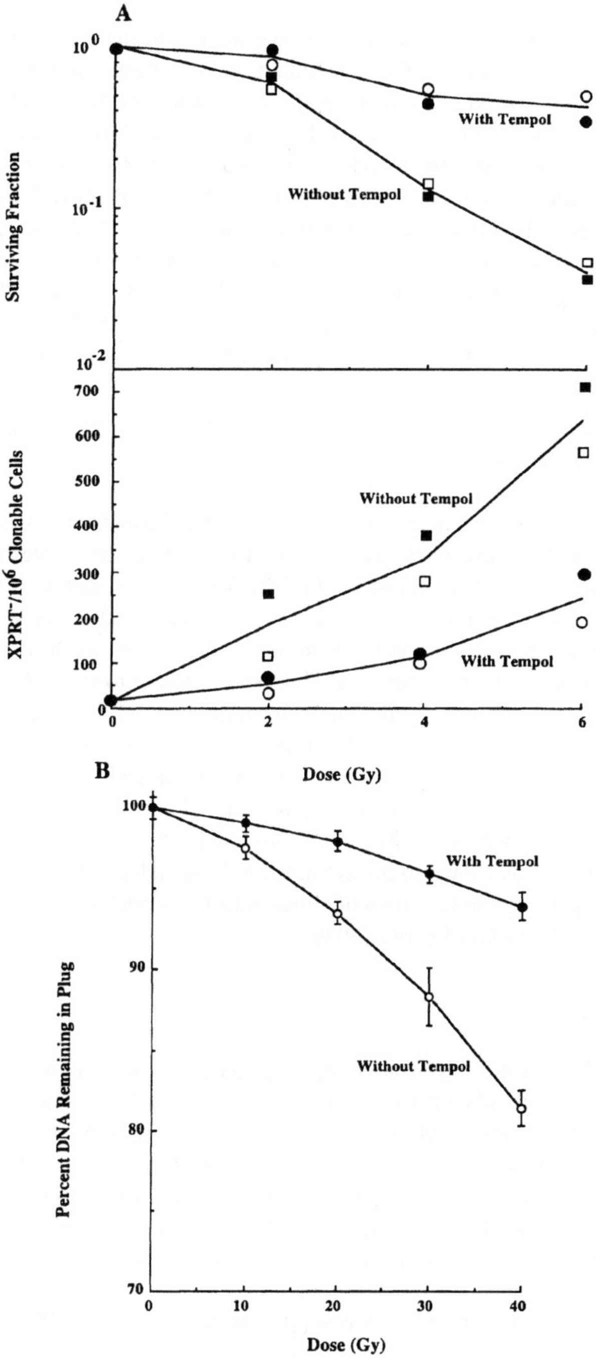
A
Surviving Fraction
10^0
10^{-1}
10^{-2}
With Tempol
Without Tempol
XPRT⁻/10⁶ Clonable Cells
700
600
500
400
300
200
100
0
Without Tempol
With Tempol
0
2
4
6
Dose (Gy)
B
Percent DNA Remaining in Plug
100
90
80
70
With Tempol
Without Tempol
0
10
20
30
40
Dose (Gy)

However, systemic toxicity, particularly of the sympathetic nervous system, limits its therapeutic usefulness. Reactive oxygen species generated by the autoxidation of the drug have been shown to mediate the systemic toxicity (Sachs and Johnson, 1975). Adjunctive use of antioxidants such as the radioprotector WR-2721, to limit the systemic toxicity, have not been successful. A recent study (Purpura *et al.*, 1996) using a murine model showed that the mortality observed on administration of toxic doses of 6-hydroxydopamine is decreased by prior i.p. administration of Tempol. In mice bearing experimental neuroblastoma tumors, Tempol alone had no effect on tumor growth; however, treatment of these animals with a combination of 6-hydroxydopamine and Tempol resulted in an enhanced tumor response relative to treatment with 6-hydroxydopamine alone. The protective effects of the nitroxide were attributed to its ability to scavenge radicals generated by the autoxidation of 6-hydroxydopamine.

6. NITROXIDE PROTECTION AGAINST MUTAGENIC REACTIVE OXYGEN SPECIES

Reactive oxygen species generated by both chemical agents and X rays have been implicated in mutation induction (Hsie *et al.*, 1986). With the impressive degree of nitroxide-mediated protection against cytotoxicity from superoxide (HX/XO) and hydrogen peroxide as shown in Figures 1 and 2, we questioned whether Tempol would also afford protection against mutation induction to cells exposed to these oxidants. Two important observations were made. First, Tempol treatment to cells alone was neither mutagenic nor toxic at 5 or 10 mM, the concentration used in the cytoprotection of mammalian cells (DeGraff *et al.*, 1992a). Second, Tempol was found to act as an antimutagenic agent in cells exposed to HX/XO and hydrogen peroxide. Tempol treatment provided near-complete protection against mutations at the *XPRT* locus in AS52 cells induced by either HX/XO or hydrogen peroxide treatment. In addition, as shown in Figure 7A, Tempol was shown to protect against X ray-induced mutations in the same cell system, suggesting that the protection results from Tempol protection against DNA damage (Figure 7B) (DeGraff *et al.*, 1992b).

7. SUMMARY

The use of nitroxides has been greatly expanded over the past few years with the discovery of their antioxidant activity. The ability of nitroxides to function as mimics of antioxidant enzymes such as SOD and catalase and their ability to act as efficient radical scavengers are unique protective capabilities. The observed protective effects of nitroxides at the cellular and animal level against diverse oxidative insults have made them potentially useful as agents to assess free radical modes of cytotoxicity and as a new class of antioxidants that may have utility in clinical biomedical research.

ACKNOWLEDGMENT. This research was supported in part by grant 95-00287 from the USA-Israel Binational Science Foundation (BSF).

8. REFERENCES

Abe, M., Nishidai, T., Yukawa, Y., Takahashi, M., Ono, K., Hiraoka, M., and Ri, N., 1981, Studies on the radioprotective effects of superoxide dismutase in mice, *Int. J. Radiat. Oncol. Biol. Phys.* **7**:205–209.

Alexander, P., and Charlesby, A., 1954, Physico-chemical methods of protection against ionizing radiations, in *Radiobiology Symposium* (Z. M. Bacq and P. Alexander, eds.), pp. 49–56, Butterworths, London.

Bacq, Z. M., 1965, *Chemical Protection Against Ionizing Radiation*, Thomas, Springfield, IL.

Bacq, Z. M., Dechamps, G., Fischer, P., Herve, A., LeBihan, H., Lecomte, J., Pirotte, M., and Rayet, P., 1953, Protection against X-rays and therapy of radiation sickness with β-mercaptoethylamine, *Science* **117**:633–636.

Bedford, J. S., Mitchell, J. B., Griggs, H. G., and Bender, M. A., 1978, Radiation-induced cellular reproductive death and chromosome aberrations, *Radiat. Res.* **76**:573–586.

Belkin, S., Mehlhorn, R. J., Hideg, K., Hankovsky, O., and Packer, L., 1987, Reduction and destruction of nitroxide spin probes, *Arch. Biochem. Biophys.* **256**:232–243.

Bennett, H. F., Brown R. D., III, Koenig, S. H., and Swartz, H. M., 1987a, Effects of nitroxides on the magnetic field and temperature dependence of 1/T1 of solvent water protons, *Magn. Reson. Med.* **4**:93–111.

Bennett, H. F., Swartz, H.M ., Brown R. D., III, and Koenig, S. H., 1987b, Modification of relaxation of lipid protons by molecular oxygen and nitroxides, *Invest. Radiol.* **22**:502–507.

Berliner, L. J ., 1979, *Spin Labelling II: Theory and Applications*, Academic Press, New York.

Bitterman, N., and Samuni, A., 1995, Nitroxide stable radicals protect against hyperoxic-induced seizures in rats, *Undersea Hyperb. Med.* **22**:47–48.

Carrano, A. V., 1973, Chromosome aberrations and radiation-induced cell death: II. Predicted and observed cell survival, *Mutat. Res.* **17**:355–366.

Chateauneuf, J., Lusztyk, J., and Ingold, K. U., 1988, Absolute rate constants for the reactions of some carbon-centered radicals with 2,2,6,6-tetramethylpiperidine-N-oxyl, *J. Org. Chem.* **53**:1629–1632.

Cuscela, D., Coffin, D., Lupton, G., Cook, J. A., Glass, J., Krishna, M. C., Muldoon, R., Bonner, R. F., and Mitchell, J.B., 1996, Protection from radiation-induced alopecia with topical application of nitroxides: Fractionated studies, *Cancer J. Sci. Am.* **2**:273–278.

DeGraff, W.G., Krishna, M.C., Russo, A., and Mitchell, J.B., 1992a, Antimutagenicity of a low molecular weight superoxide dismutase mimic against oxidative mutagens, *Environ. Mol. Mutagen.* **19**:21–26.

DeGraff, W. G., Krishna, M. C., Kaufman, D., and Mitchell, J. B., 1992b, Nitroxide-mediated protection against x-ray- and neocarzinostatin-induced DNA damage, *Free Radical Biol. Med.* **13**:479–487.

DeGraff, W., Hahn, S. M., Mitchell, J. B., and Krishna, M. C., 1994, Free radical modes of cytotoxicity of adriamycin and streptonigrin, *Biochem. Pharmacol.* **48**:1427–1435.

Gelvan, D., Saultman, P., and Powell, S., 1991, Cardiac reperfusion damage prevented by a stable nitroxide free radical, *Proc. Natl. Acad. Sci. USA* **88**:4680–4684.

Gerschman, R., Gilbert, D. L., Nye, S.W., Dwyer, P., and Fenn, W. O., 1954, Oxygen poisoning and X-irradiation: A mechanism in common, *Science* **119**:623–626.

Goffman, T., Cuscela, D., Glass, J., Hahn, S., Krishna, M. C., Lupton, G., and Mitchell, J. B., 1992, Topical application of nitroxide protects radiation induced alopecia in guinea pigs, *Int. J. Radiat. Oncol. Biol. Phys.* **22**:803–806.

Hahn, S. M., Tochner, Z., Krishna, M. C., Glass, J., Wilson, L., Samuni, A., Sprague, M., Venzon, D., Glatstein, E., Mitchell, J. B., and Russo, A., 1992a, Tempol, a stable free radical, is a novel murine radiation protector, *Cancer Res.* **52**:1750–1753.

Hahn, S. M., Wilson, L., Krishna, M. C., Liebmann, J., DeGraff, W., Gamson, J., Samuni, A., Venzon, D., and Mitchell, J. B., 1992b, Identification of nitroxide radioprotectors, *Radiat. Res.* **132**:87–93.

Hahn, S. M., Sullivan, F. J., DeLuca, A. M., Sprague, M., Hampshire, V. A., Krishna, M. C., Russo, A., and Mitchell, J. B., 1997a, Protection of mitomycin C induced skin extravasation with the nitroxide, 3-carbamoyl-proxyl (3-CP), *Int. J. Oncol.* **10**:119–123.

Hahn, S. M., Sullivan, F. J., DeLuca, A. M., Krishna, M. C., Wersto, N., Venzon, D., Russo, A., and Mitchell, J. B., 1997b, Evaluation of tempol radioprotection in a murine tumor model, *Free Radical Biol. Med.* **22**:1211–1216.

Hall, E. J., 1994, The oxygen effect and reoxygenation, in *Radiobiology for the Radiologist*, pp. 133–153, Lippincott, Philadelphia.

Hsie, A. W., Recio, L., Katz, D .S., Lee, C. Q., Wagner, M., and Schenley, R. L., 1986, Evidence for reactive oxygen species inducing mutations in mammalian cells, *Proc. Natl. Acad. Sci. USA* **83:**9616–9620.

Iannone, A., Bini, A., Swartz, H. M., Tomasi, A., and Vannini, V., 1989, Metabolism in rat liver microsomes of the nitroxide spin probe Tempol, *Biochem. Pharmacol.* **38:**2581–2586.

Iannone, A., Tomasi, A., Vannini, V., and Swartz, H. M., 1990a, Metabolism of nitroxide spin labels in subcellular fractions of rat liver. I. Reduction in the cytosol, *Biochim. Biophys. Acta* **1034:**290–293.

Iannone, A., Tomasi, A., Vannini, V., and Swartz, H. M., 1990b, Metabolism of nitroxide spin labels in subcellular fractions of rat liver. II. Reduction by microsomes, *Biochim. Biophys. Acta* **1034:**285–289.

Johnstone, P. A. S., DeGraff, W. G., and Mitchell, J. B., 1995, Protection of radiation-induced chromosomal aberrations by the nitroxide Tempol, *Cancer* **75:**2323–2327.

Karmeli, F., Eliakim, R., Okon, E., Samuni, A., and Rachmilewitz, D., 1995, A stable nitroxide radical effectively decreases mucosal damage in experimental colitis, *Gut* **37:**386–393.

Krishna, M.C., DeGraff, W., Tamura, S., Gonzalez, F., Samuni, A., Russo, A., and Mitchell, J.B., 1991, Mechanisms of hypoxic and aerobic cytotoxicity of mitomycin C in Chinese hamster V79 cells, *Cancer Res.* **51:**6622–6628.

Krishna, M.C., Grahame, D.A., Samuni, A., Mitchell, J.B., and Russo, A., 1992, Oxoammonium cation intermediate in the nitroxide-catalyzed dismutation of superoxide, *Proc. Natl. Acad. Sci. USA* **89:**5537–5541.

Krishna, M. C., Russo, A., Mitchell, J. B., Goldstein, S., Dafni, H., and Samuni, A., 1996a, Do nitroxide antioxidants act as scavengers of superoxide or as SOD mimics? *J. Biol. Chem.* **271:**26026–26031.

Krishna, M. C., Samuni, A., Taira, J., Goldstein, S., Mitchell, J. B., and Russo, A., 1996b, Stimulation by nitroxides of catalase-like activity of hemeproteins, *J. Biol. Chem.* **271:**26018–26025.

McCord, J. M., 1987, Oxygen-derived radicals: A link between reperfusion injury and inflammation, *FED Proc.* **46:**2402–2406.

Mehlhorn, R. J., and Packer, L., 1984, Electron paramagnetic resonance spin destruction methods for radical detection, *Methods Enzymol.* **105:**215–220.

Millar, B. C., Fielden, E. M., and Smithen, C. E., 1977, Polyfunctional radiosensitizers III. Effect of the biradical Ro-03-6061) in combination with other radiosensitizers on the survival of hypoxic V-79 cells, *Radiat. Res.* **69:**489–499.

Millar, B. C., Fielden, E. M., and Smithen, C. E., 1978, Polyfunctional radiosensitizers IV. The effect of contact time and temperature on sensitization of hypoxic Chinese hamster cells in vitro by bifunctional nitroxyl compounds, *Br. J. Cancer* **37:**73–79.

Millar, B. C., Jenkins, T. C., Fielden, E. M., and Jinks, S., 1983, Polyfunctional radiosensitizers. VI. Dexamethasone inhibits shoulder modification by uncharged nitroxyl biradicals in mammalian cells irradiated in vitro, *Radiat. Res.* **96:**160–172.

Mitchell, J. B., Samuni, A., Krishna, M. C., DeGraff, W. G., Ahn, M. S., Samuni, A., and Russo, A , 1990, Biologically active metal-independent superoxide dismutase mimics, *Biochemistry* **29:**2802–2807.

Mitchell, J. B., DeGraff, W., Kaufman, D., Krishna, M. C., Samuni, A., Finkelstein, E., Ahn, M. S., Hahn, S. M., Gamson, J., and Russo, A., 1991, Inhibition of oxygen-dependent radiation-induced damage by the nitroxide superoxide dismutase mimic, Tempol, *Arch. Biochem. Biophys.* **289:**62–70.

Miura, Y., Utsumi, H., and Hamada, A., 1993, Antioxidant activity of nitroxide radicals in lipid peroxidation of rat liver microsomes, *Arch. Biochem. Biophys.* **300:**148–156.

Monti, E., Paracchini, L., Perletti, G., and Piccinni, F., 1991, Protective effects of spin-trapping agents on adriamycin-induced cardiotoxicity in isolated rat atria, *Free Radical Res. Commun.* **14:**41–45.

Monti, E., Cova, D., Guido, E., Morelli, R., and Oliva, C., 1996, Protective effects of the nitroxide Tempol against the cardiotoxicity of adriamycin, *Free Radic. Biol. Med.* **21:**463–470.

Nilsson, U. A., Olsson, L. I., Carlin, G., and Bylund-Fellenius, A. C., 1989, Inhibition of lipid peroxidation by spin labels. Relationships between structure and function, *J. Biol. Chem.* **264:**11131–11135.

Patt, H. M., Tyree, E. B., Staube, R. L., and Smith, D. E., 1949, Cysteine protection against X-irradiation, *Science* **110:**213–214.

Petkau, A., 1987, Role of superoxide dismutase in modification of radiation injury, *Br. J. Cancer Suppl.* **8:**87–95.

Petkau, A., and Chelack, W. S., 1984, Radioprotection by superoxide dismutase of macrophage progenitor cells from mouse bone marrow, *Biochem. Biophys. Res. Commun.* **119:**1089–1095.

Petkau, A., Chelack, W. S., Pleskach, S. D., Meeker, B. E., and Brady, C. M., 1975, Radioprotection of mice by superoxide dismutase, *Biochem. Biophys. Res. Commun.* **65:**886–893.

Puck, T. T., 1958, Action of radiation on mammalian cells: III. Relationships between reproductive death and induction of chromosome anomalies by X-irradiation of euploid human cells in vitro, *Proc. Natl. Acad. Sci. USA* **44**:772–780.

Purpura, P., Westman, L., Will, P., Eidelman, A., Kagan, V. E., Osipov, A. N., and Schor, N. F., 1996, Adjunctive treatment of murine neuroblastoma with 6-hydroxydopamine and Tempol, *Cancer Res.* **56**:2336–2342.

Rachmilewitz, D., Karmeli, F., Okon, E., and Samuni, A., 1994, A novel antiulcerogenic stable radical prevents gastric mucosal lesions in rats, *Gut* **35**:1181–1188.

Reddan, J., Sevilla, M., Giblin, F., Padgaonkar, V., Dziedzic, D., and Leverenz, V., 1992, Tempol and deferoxamine protect cultured rabbit lens epithelial cells from H_2O_2 insult: Insight into the mechanism of H_2O_2-induced injury, *Lens Eye Toxic. Res.* **9**:385.

Reddan, J. R., Sevilla, M. D., Giblin, F. J., Padgaonkar, V., Dziedzic, D. C., Leverenz, V., Misra, I. C., and Peters, J. L., 1993, The superoxide dismutase mimic Tempol protects cultured rabbit lens epithelial cells from hydrogen peroxide insult, *Exp. Eye Res.* **56**:543.

Sachs, A., and Johnson, G., 1975, Mechanisms of action of 6-hydroxydopamine, *Biochem. Pharmacol.* **24**:1–8.

Samuni, A., Krishna, M. C., Riesz, P., Finkelstein, E., and Russo, A., 1988, A novel metal-free low molecular weight superoxide dismutase mimic, *J. Biol. Chem.* **263**:17921–17924.

Samuni, A., Krishna, C. M., Riesz, P., Finkelstein, E., and Russo, A., 1989, Superoxide reaction with nitroxide spin-adducts, *Free Radical Biol. Med.* **6**:141–148.

Samuni, A., Krishna, M. C., Mitchell, J. B., Collins, C. R., and Russo, A., 1990, Superoxide reaction with nitroxides, *Free Radical Res. Comms.* **9**:241–249.

Samuni, A., Godinger, D., Aronovitch, J., Russo, A., and Mitchell, J. B., 1991a, Nitroxides block DNA scission and protect cells from oxidative damage, *Biochemistry* **30**:555–561.

Samuni, A., Mitchell, J. B., DeGraff, W., Krishna, M. C., Samuni, A., and Russo, A., 1991b, Nitroxide SOD-mimics: Modes of action, *Free Radical Res. Comms.* **12–13**:187–194.

Samuni, A., Winkelsberg, D., Pinson, A., Hahn, S. M., Mitchell, J. B., and Russo, A., 1991c, Nitroxide stable radicals protect beating cardiomyocytes against oxidative damage, *J. Clin. Invest.* **87**:1526–1530.

Sentjurc, M., Pecar, S., Chen, K., Wu, M., and Swartz, H. M., 1989, Cellular metabolism of proxyl nitroxides and hydroxylamines, *Biochim. Biophys. Acta* **1073**:329–335.

Siesjo, B. K., Agardh, C. D., and Bengtsson, F., 1989, Free radicals and brain damage, *Cereb. Brain Metab. Rev.* **1**:165–211.

Simpson, P. J., Mickelson, J. K., and Luchesi, B. R., 1987, Free radical scavengers in myocardial ischemia, *FED. Proc.* **46**:2413–2421.

Strzalka, K., Walczak, T., Sarna, T., and Swartz, H. M., 1990, Measurement of time-resolved oxygen concentration changes in photosynthetic systems by nitroxide-based EPR oximetry, *Arch. Biochem. Biophys.* **281**:312–318.

Yannai, E. B., Zhang, R., Trembovler, V., Samuni, A., and Shohami, E., 1996, Cerebroprotective effect of stable nitroxide radicals in closed head injury in the rat, *Brain Res.* **717**:22–28.

Yuhas, J. M., and Storer, V. B., 1969, Differential chemoprotection of normal and malignant tissues, *J. Natl. Cancer Inst.* **42**:331–335.

Part IV

Environmental Pro- and Antioxidants

Chapter 12

Stratospheric Ozone and Its Effects on the Biosphere

Sasha Madronich

1. INTRODUCTION

Ultraviolet (UV) photons have enough energy to break or modify bonds of some molecules, often creating photofragments that are highly reactive. Solar UV radiation thus drives the photochemical oxidation of natural and pollutant compounds in the atmosphere. These atmospheric gases and particles in turn influence the flow of UV radiation to the Earth's biosphere, where numerous detrimental effects are possible (UNEP, 1994).

The most important UV-absorbing atmospheric compound, ozone (O_3), is subject to reductions that raise concern for potential consequences of increased UV irradiation of animals, plants, outdoor materials, soils, waters, and air. Ozone controls the availability of photons in the UV-B (280–320 nm) wavelength range, and therefore of the shortest wavelengths to which the environment is exposed. Many target molecules (e.g., DNA) have pronounced absorptions in this wavelength region. UV stress is already recognized in some communities and ecosystems, and further increases are generally deemed undesirable.

This chapter provides a brief review of some basic relationships between biologically effective UV radiation and the atmosphere. A quantitative emphasis is put on the role of stratospheric ozone, as here much is known with good certainty, at least compared with effects such as from clouds or pollutants. Estimation of long-term UV trends is still problematic and current estimates given here will likely require updating or revision.

Sasha Madronich Atmospheric Chemistry Division, National Center for Atmospheric Research, Boulder, Colorado 80307.

Reactive Oxygen Species in Biological Systems, edited by Gilbert and Colton. Kluwer Academic / Plenum Publishers, New York, 1999.

2. BIOLOGICALLY EFFECTIVE RADIATION

UV radiation is implicated in numerous photobiological processes, such as the induction of skin cancer and cataracts among humans. These biological responses, however defined, are generally wavelength dependent, and are often most sensitive at the shortest UV-B wavelengths. Figure 1 illustrates some known biological spectral sensitivity functions (action spectra), for which the sensitivities increase by several orders of magnitude from longer UV-A (320–400 nm) to shorter UV-B wavelengths. Over this same wavelength range, the environmental UV radiation is usually increasing in the opposite direction (toward longer wavelengths), so that the largest contributions are from intermediate wavelengths having both sufficient radiation levels and notable biological sensitivity.

A useful measure of biologically effective radiation is the weighted irradiance (or dose rate or exposure),

$$W \equiv \int F(\lambda) B(\lambda) d\lambda \tag{1}$$

Here λ is wavelength (nm), $F(\lambda)$ is the spectral UV irradiance (W m^{-2} nm^{-1}), and $B(\lambda)$ is a function describing the wavelength dependence of the biological sensitivity (the action spectrum).

$F(\lambda)$ may be either calculated or measured directly (see Section 3) so that for any given action spectrum $B(\lambda)$ (e.g., erythema induction), it is possible to estimate the weighted irradiance W. Usually, only the spectral shape of $B(\lambda)$ is known and the action spectrum is arbitrarily normalized to unity at a specific wavelength (e.g., 300 nm); in this case, W has units of radiant energy incident on a unit area per unit time (e.g., W m^{-2}), but with the understanding that its absolute value is as arbitrary as the normalization of the action spectrum. Thus, values of W are most useful for computing relative (percent) changes in biologically effective exposure, for specified changes in environmental UV

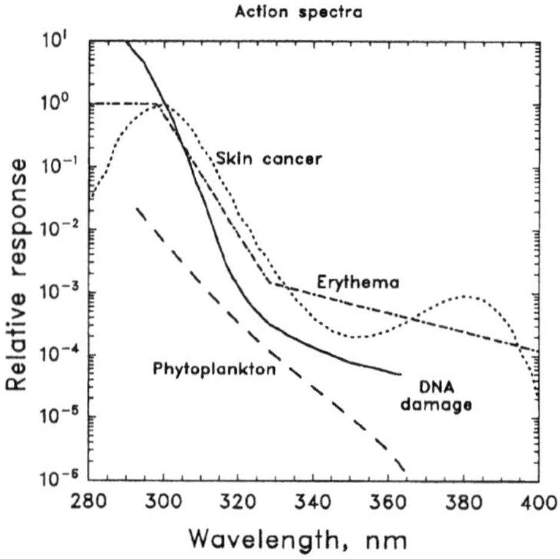

FIGURE 1. Biological spectral sensitivity functions (action spectra) for damage to *in vitro* DNA (Setlow, 1974), erythema (skin reddening) induction in humans (McKinlay and Diffey, 1987), squamous cell carcinoma induction in mice adjusted for human skin transmission (de Gruijl and van der Leun, 1994), and *in situ* inhibition of photosynthesis in Antarctic phytoplankton (Boucher *et al.*, 1994).

levels (e.g., seasonal and geographic variations, increases related to stratospheric ozone depletion changes). The instantaneous weighted irradiance, W, may also be integrated over various time intervals (e.g., hourly, daily, yearly) to yield the corresponding cumulative UV exposure doses (e.g., J m^{-2}).

The response of the weighted irradiance W to atmospheric ozone changes is of special interest. Radiation amplification factors (RAFs) are a convenient measure of UV/ozone sensitivity,

$$RAF \cong \delta W/(-\delta O_3) \tag{2}$$

where δ implies relative (or percent) change. Values are given in Table I for a selection of different processes. Action spectra that decay most steeply with increasing wavelength have the largest RAFs. If significant response exists at UV-A wavelengths, even when falling below the detection threshold in some studies (see note *b* in Table I), the sensitivity

Table I
Radiation Amplification Factors (RAFs) at 30°N[a]

Effect	RAF January (290 DU)	RAF July (305 DU)	Reference
Animals			
Erythema reference	1.1	1.2	McKinlay and Diffey (1987)
SKH-1 corrected for human skin transmission	1.2	1.1	de Gruijl and van der Leun (1994)
Melanoma induction in fish	0.1	0.1	Setlow *et al.* (1993)
Generalized DNA damage	2.0	1.9	Setlow (1974)
Damage to cornea	1.2	1.1	Pitts *et al.* (1977)
Damage to lens (cataract)	0.8	0.7	Pitts *et al.* (1977)
Occupational exposure limit	1.4	1.5	ACGIH (1992)
Immune suppression	[0.4][b] 1.0	[0.4] 0.8	De Fabo and Noonan (1983)
Plants			
Generalized plant spectrum	2.0	1.6	Caldwell *et al.* (1986)
Inhibition of growth of cress seedlings	[3.6] 3.8	3.0	Steinmetz and Wellmann (1986)
DNA damage in alfalfa	0.5	0.6	Quaite *et al.* (1992)
Inhibition of phytoplankton photosynthesis, in Antarctic community	0.8	0.8	Boucher *et al.* (1994)
Tropospheric photolysis			
$O_3 + h\nu \rightarrow O(^1D) + O_2$	2.1	1.8	Madronich and Granier (1994)
Materials damage			
Yellowness induction in polyvinylchloride	0.2	0.2	Andrady *et al.* (1989)
Yellowness induction in polycarbonate	0.4	0.4	Andrady *et al.* (1989)
Other weighting functions			
Solar Light Robertson–Berger meter (Model 501)	1.2	1.1	M. Morys (personal communication, 1994)
UV-B (280–320 nm)	0.87	0.71	

[a]RAFs computed on basis of daily integral. Adapted from Madronich *et al.* (1995).

[b]Values in brackets show effect of extrapolating original data to 400 nm with an exponential tail, for cases where the effect is larger than 0.2 RAF unit.

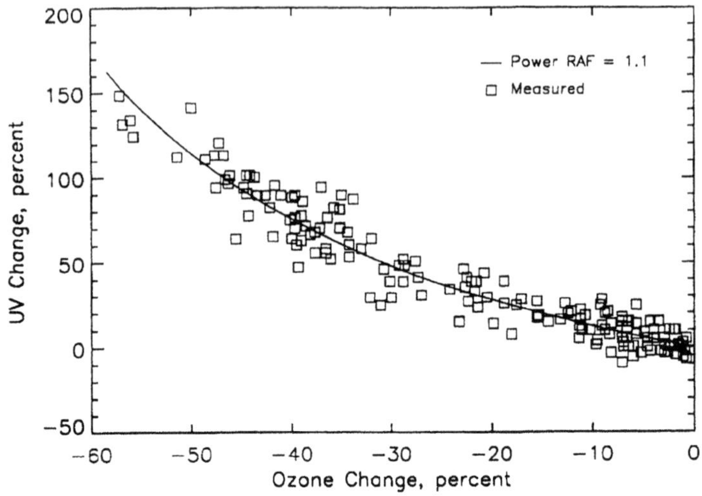

FIGURE 2. Dependence of erythemally weighted UV radiation on ozone column changes, measured at the South Pole between 1 February 1991 and 12 December 1992. Adapted from Booth and Madronich (1994).

to ozone is smaller. Uncertainties in action spectra are usually large, and standard reference spectra (e.g., the erythemal induction sensitivity shown in Figure 1) are often used for some comparison purposes.

For a given action spectrum, direct measurements generally confirm the predicted sensitivities to ozone changes. Figure 2 shows the dependence of erythemal UV radiation on ozone at the South Pole, with the solid line given by Eq. (2) (after integration over the large ozone change). The knowledge of such dependences is key to estimating the UV radiation at different locations and times, or under scenarios of future ozone depletion.

3. UV RADIATION AND THE ATMOSPHERE

3.1. Atmospheric Ozone

Ozone is a naturally occurring atmospheric gas and an effective filter of solar UV-B radiation. Although its known history spans over 150 years, our quantitative understanding of the processes controlling its atmospheric amounts is still evolving. First found in the 1840s by German chemist Schönbein and identified chemically a few decades later, ozone was by the 1880s detected routinely in the natural atmosphere. By the 1920s, accurate determinations of the total amount of overhead atmospheric ozone were made by British physicist G. Dobson, via spectroscopic measurement of solar radiation at several wavelengths differentially absorbed by ozone. Since then, numerous additional techniques have been used to characterize the climatology of ozone, including vertical profiles, seasonal variations, geographical distributions, and long-term trends. A typical mid-latitude profile is shown in Figure 3. The large amount of ozone in the stratosphere absorbs sufficient UV (and visible) radiation to heat the locally thin air, resulting in the

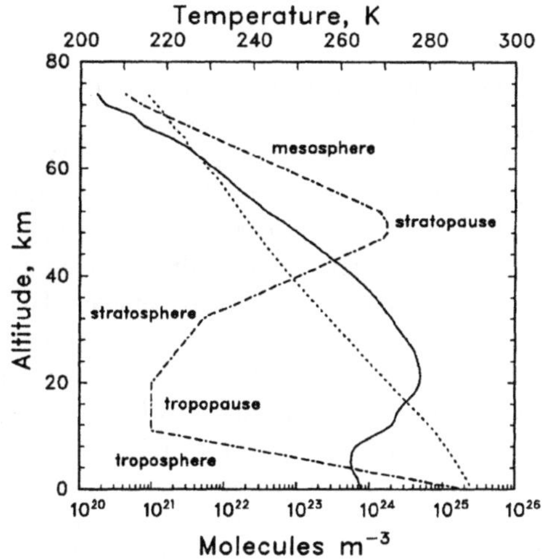

FIGURE 3. Typical atmospheric vertical profiles of air (dotted curve), ozone (solid, multiplied by 10^6), and temperature (dot-dash, top scale). Values from the US Standard Atmosphere (1976).

temperature inversions (tropopause and stratopause) that delimit the stratosphere from ca. 10 to 50 km in Figure 3.

The amount of atmospheric ozone is determined by a dynamical balance between production and destruction. Chapman (1930) first proposed that the main source is the photolysis of oxygen molecules (O_2) at UV-C (100–280 nm) wavelengths,

$$O_2 + h\nu\ (\lambda < 240\ \text{nm}) \rightarrow O + O$$

followed by addition of oxygen atoms (O) to O_2,

$$O + O_2 \rightarrow O_3$$

The maximum ozone production occurs in the stratosphere: At higher altitudes (mesosphere and above) the O_2 concentrations are small, while at lower altitudes (troposphere) the necessary UV-C photons are no longer available because of absorption by the large overhead amounts of O_2 and O_3. The destruction of ozone occurs via direct photolysis at visible and UV wavelengths,

$$O_3 + h\nu \rightarrow O_2 + O$$

and by reaction with O,

$$O_3 + O \rightarrow 2\ O_2$$

The Chapman scheme overpredicts measured ozone, and a number of other destruction pathways have been identified. Because these rely on reactions with compounds present in much smaller amount than the ozone, they are effective only when operating in catalytic cycles, generally of the type

$$O_3 + X \rightarrow XO + O_2$$

$$O + XO \rightarrow X + O_2 \quad \text{or} \quad O_3 + XO \rightarrow X + 2\,O_2$$

Natural cycles were identified by Bates and Nicolet (1950) for hydrogen compounds (X = OH, generated photochemically from stratospheric water vapor and methane), and by Crutzen (1970) for nitrogen species (X = NO, generated from stratospheric N_2O). A similar chlorine cycle (X = Cl) was identified by Stolarski and Cicerone (1974), and Rowland and Molina (1975) proposed that stratospheric Cl concentrations could be increasing as a result of chlorofluorocarbons (CFCs) released by human activities. The resulting coupled chemistry is complex, but it should be noted that normally only a fraction of these minor species is present in the ozone destroying forms (X, XO), with the balance existing as relatively unreactive "reservoir" species (e.g., HNO_3, $ClONO_2$, HCl). Crutzen, Molina, and Rowland were recognized for their seminal contributions with the 1995 Nobel prize for chemistry.

A major revision of understanding occurred with the observation, beginning in the early 1980s and continuing to date, of extremely low ozone over Antarctica during springtime (Farman *et al.*, 1985). This drastic seasonal ozone destruction is believed to result from reactions taking place on the surface of cloud particles formed in the extremely cold polar winter stratosphere. Chlorine reservoir species (HCl, $ClONO_2$) are transformed by gas–particle reactions to Cl_2 and HOCl, which are then easily photolyzed, on polar sunrise, to ozone-destroying Cl and ClO (Solomon, 1990). Similar reactivation of chlorine is thought to occur on the surface of volcanic aerosol particles (Hoffman and Solomon, 1989; Brasseur *et al.*, 1990) and may explain the particularly low ozone levels observed for several years after the eruption in Mt. Pinatubo in June 1991 (WMO, 1994a).

CFCs are rich sources of stratospheric chlorine, and have shown disturbing increases in the past few decades (WMO, 1994a). An important aspect of the ozone–CFC problem is the time delay between industrial CFC production and ultimate removal from the atmosphere: Several decades are required for their release to the troposphere, transport

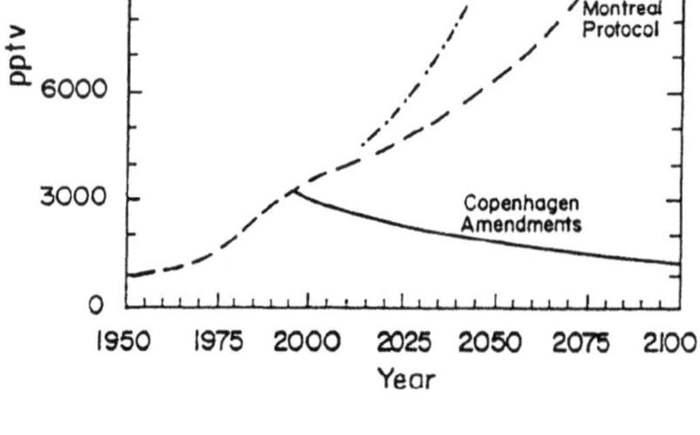

FIGURE 4. Projected levels of stratospheric chlorine under different emission scenarios resulting from international agreements. From WMO (1994a).

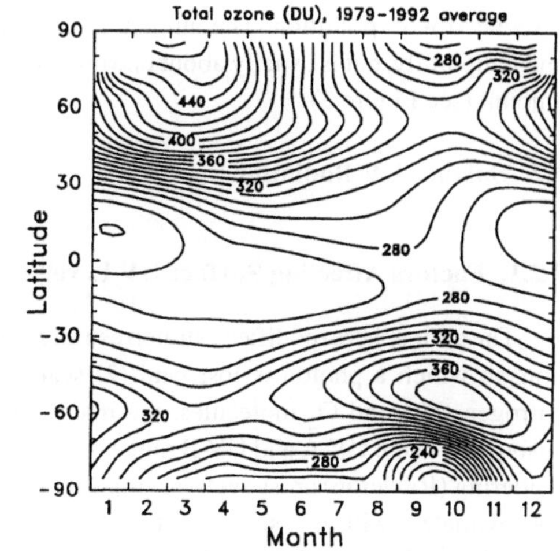

FIGURE 5. Seasonal and latitudinal distribution of the vertical ozone column amount, averaged over 1979–1992, measured by the Total Ozone Mapping Spectrometer (TOMS, version 6) aboard the Nimbus 7 satellite (data courtesy of R. Stolarski, NASA).

from the lower troposphere to the stratosphere, photochemical conversions within the stratosphere, slow removal from the stratosphere by downward transport, and eventual rainout of soluble species. The production of CFCs is now restricted by international agreements (the 1987 Montreal Protocol with its periodic reassessments). Figure 4 shows projected levels of stratospheric chlorine reaching a maximum near the year 2000, decreasing to 1980 levels by about 2040. Slowing or reversal of the growth of atmospheric concentrations of some CFCs has now been detected (WMO, 1994a). Studies combining atmospheric change scenarios with epidemiological data suggest that without such international agreements, that is, with continuing ozone depletion and concomitant UV radiation increases, skin cancer incidence in the United States would increase about fourfold by the year 2100, to about 1.5 million new cases per year (Slaper *et al.*, 1996).

The geographical and seasonal distribution of ozone, measured from a satellite instrument, is shown in Figure 5. The amount of ozone in a vertical column (from surface to space) is often expressed as the vertical thickness it would occupy if collected as a pure gas at standard temperature and pressure (STP: 1 atm, 273.15 K), in units of millicentimeters or Dobson units (DU). According to the ideal gas law, the number density of molecules at STP is ca. 2.69×10^{19} molecules cm^{-3}, so that the Dobson unit is alternatively expressed as 1 DU = 2.69×10^{16} molecules cm^{-2}. The geographical distribution seen in Figure 5 reflects the fact that ozone is made primarily in the upper levels of the tropical stratosphere, then transported poleward in both hemispheres. The largest values therefore occur at high latitudes when descending air motion brings ozone-rich air to the lower stratosphere and the troposphere.

Ozone also occurs in the troposphere, related in part to downward transport from the stratosphere, and in part to local photochemical production from

$$NO_2 + h\nu\ (\lambda < 420\ \text{nm}) \rightarrow NO + O$$

followed as before by the oxygen atom recombination reaction, $O + O_2 \rightarrow O_3$. The production depends on the presence of nitrogen oxides and hydrocarbons, and is often associated with pollution. The amount of tropospheric ozone is quite variable spatially and temporally, but is usually about one order of magnitude smaller than the stratospheric amount (see Figure 3).

3.2. Surface UV Radiation

3.2.1. Factors Affecting Surface UV Levels

The atmosphere modifies extraterrestrial solar radiation by two processes: (1) absorption, in which a photon is lost, and (2) scattering, in which the photon is redirected. Nitrogen (N_2) and O_2 molecules, the main constituents of air (see Figure 3), have no absorption in the UV-A and UV-B ranges but their large abundances make even their small scattering (Rayleigh) cross sections important. The molecular scattering efficiency scales approximately with λ^{-4}, so that diffuse (scattered) sky radiation is much more important at the shorter UV wavelengths than, e.g., in the visible range (and, within the visible range, near the blue end of the spectrum). The main gaseous absorber of atmospheric UV-B radiation is ozone. Cloud and aerosol particles are efficient scatterers, and may be absorbing as well, depending on composition.

The transmission of the atmosphere is illustrated in Figure 6. The solid curve (computed for one set of reference conditions, see figure caption) shows the importance of ozone absorption in the UV-B range, as well as the weaker wavelength dependence from scattering by air molecules. The other curves show the impacts of changing (one at the time) different environmental factors, as discussed in more detail below.

3.2.1a. Solar Zenith Angle. The angle between the sun and the zenith (overhead) direction is the factor that determines the effective atmospheric optical path traversed by solar radiation. A solar zenith angle change from 0 to 60° is approximately equivalent to

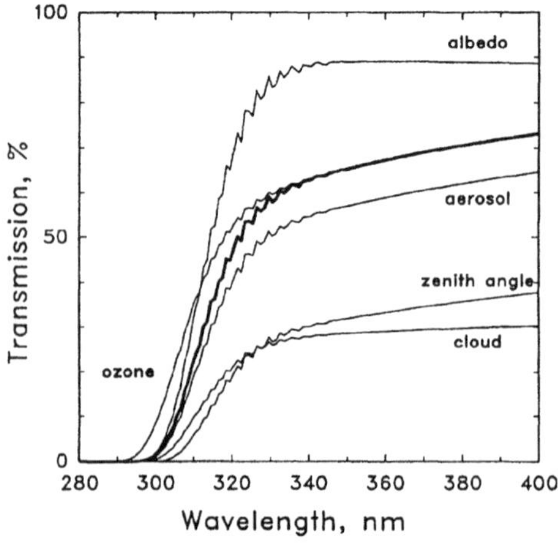

FIGURE 6. Atmospheric spectral transmission. The thick solid curve is for one set of reference conditions: no clouds, ozone column = 300 DU, no aerosols (unlimited horizontal visibility), solar zenith angle = 30°, ground albedo = 0.1. The thin curves show the effect of individually changing conditions: cloud optical depth = 20, ozone column = 150 DU, aerosol visible range = 10 km, solar zenith angle = 60°, albedo = 0.9.

doubling the amount of ozone traversed by a solar ray. Scattering by air molecules, haze, or clouds is also more important for larger solar zenith angles (low sun).

3.2.1b. Ozone. Absorption by ozone only affects radiation at wavelengths shorter than about 330 nm, as can be seen from Figure 6. At these wavelengths, even small changes in atmospheric ozone can be important to the radiation reaching the surface, as shown in more detail in Figure 7. Relative (percent) changes are largest at the shortest UV-B wavelengths, while absolute energy increments are greatest in the range from 300 to 320 nm, with diminishing contributions from shorter and longer wavelengths. Stratospheric and tropospheric ozone are both efficient absorbers, but have slightly different interactions with the direct solar beam, which dominates in the stratosphere, and with the diffuse radiation, which is more important in the troposphere (Brühl and Crutzen, 1989).

3.2.1c. Clouds. These usually reduce surface UV levels, although temporary enhancements are possible under broken cloud conditions. Realistic clouds have complex microphysical and morphological properties, which usually are not well known for any specific situation. In one highly simplified view, the cloud properties are parameterized by three quantities: the optical depth (τ), which depends on the total amount of liquid water as well as on cloud particle sizes; the single scattering albedo (ω_0), describing the extent of absorption (e.g., by droplet impurities); and the asymmetry factor (g), which accounts for the preferential forward scattering by the larger particles. Also useful is knowledge of frequency of cloud occurrence, of percent horizontal coverage, of vertical stratification, and so forth. The effects of clouds at visible and UV wavelengths are somewhat different. Figure 8 shows the wavelength-dependent transmission through idealized clouds of various optical depths, relative to cloud-free skies. Cloud scattering can enhance photon paths through tropospheric ozone, resulting in more absorption at the shortest UV-B wavelengths. However, the interaction between cloud and Rayleigh scattering by air is also stronger at shorter wavelengths, so that effective cloud UV transmissions can be somewhat larger than at visible wavelengths. Many measurements of UV reductions by clouds are available, though usually without much simultaneous information about cloud properties. The effects can vary from near-total darkness to some

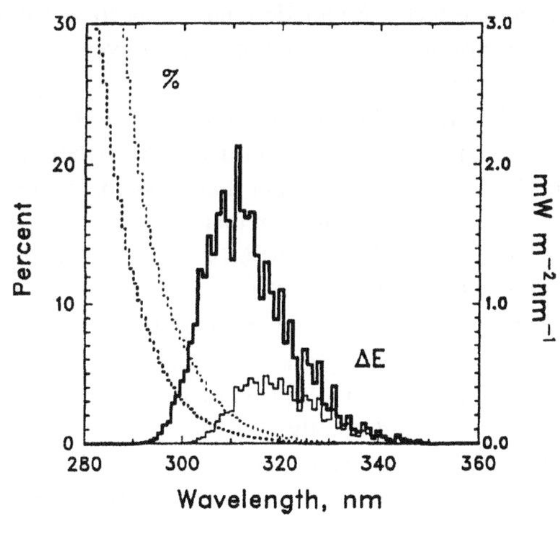

FIGURE 7. Increases of UV radiation at the surface for a 1% decrease in the total ozone column (near 300 DU). Solid curves and right scale are for energy increments, dotted curves and left scale are for percent increases. Thick and thin curves are for overhead sun and 60° solar zenith angle, respectively. From Madronich *et al.* (1995).

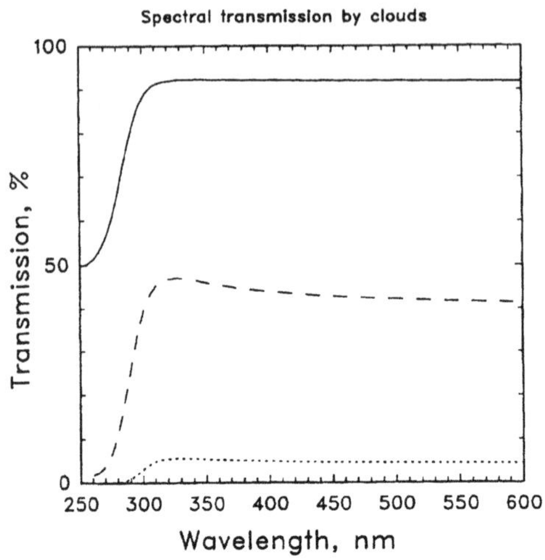

FIGURE 8. Spectral transmission relative to cloud-free conditions, for thin cloud (solid curve, optical depth = 2), moderate cloud (dashed curve, optical depth = 20), and very thick cloud (dotted curve, optical depth = 200). For overhead sun, no aerosol, ozone column = 300 DU, surface albedo = 0.1.

enhancement, but long-term average reductions by 10–50% are common (e.g., Frederick and Snell, 1990; Schafer *et al.*, 1996).

3.2.1d. Aerosols. The frequent haziness of the lower atmosphere is caused by airborne particles of various natural and pollution-related origins, e.g., mineral dust, sea salt particles, soot, ammonium sulfate, and sulfuric acid. Reductions in surface UV radiation can be significant, especially if the particles are good light absorbers (e.g., the carbonaceous particles derived from combustion). Sulfate particles are widespread in industrial regions, where by some estimates they have reduced UV levels by 5–20% since preindustrial days (Liu *et al.*, 1991), similar to the reductions seen in Figure 6.

3.2.1e. Surface Elevation. Terrains above sea level have less atmosphere overhead, and thus may be expected to allow higher transmission at all wavelengths. For a cloud-free, pollution-free atmosphere, this increase is about 5–6% km^{-1}. In practice, the altitude effect can include other, more local variables, e.g., higher elevations are often above pollution layers and sometimes even above low clouds, resulting in larger vertical UV variations.

3.2.1f. Surface Reflections. Scattering from surfaces provides another potential source of UV exposure. Surface reflectivities are generally wavelength dependent, and in the UV are often smaller than at visible wavelengths. For vegetation, the UV albedo (or fraction of incident light that is reflected) is typically 1–5%, while soils and sands fall in the range from 5 to 30% (see Madronich, 1993a,b, for a compilation of measured values). For clean snow and ice, values can approach 100%, though more commonly fall in 50–80% range, and lower values are possible (e.g., old snow). High surface reflectivity over an extended region can also enhance the radiation downwelling from the sky (see Figure 6), through multiple scattering/reflection events. This so-called "photon trapping" can be observed readily at visible wavelengths, e.g., between snow and cloud, but for clear skies it is even more pronounced at UV wavelengths because of wavelength dependence of atmospheric molecular (Rayleigh) scattering.

3.2.1g. Target Geometry. The angle of incidence between the light and a target surface is clearly of importance. For example, reflections from snow, or radiation from horizon directions can both contribute or even dominate, depending on the angle of the exposed surface. It should be noted that most standard UV measurements report the radiation incident on an upward-facing horizontal surface.

3.2.2. Clear Sky Erythemal Doses and the UV Index

A typical diurnal profile of erythemal radiation under cloud-free skies is illustrated in Figure 9. The radiation peaks sharply at solar noon, and about three-fourths of the daily total dose is incurred during these central 5–6 hr. Seasonal differences are related to the angle of the sun, the annual natural cycle of ozone, and to a lesser extent the cycle in the sun–Earth distance. Integrated daily doses, presented in Figure 10, also show a geographical dependence, with subsolar tropical maxima and summer/winter cycles at middle and polar latitudes.

The standard UV Index disseminated to the public at some locations is defined as the noontime erythemal irradiance divided by 0.025 W m^{-2} (WMO, 1994b; Long *et al.*, 1996). Values of the index range over 0–2 (minimal exposure risk), 3–4 (low), 5–6 (moderate), 7–9 (high), and 10 or greater (very high). For the case given in Figure 10, the index values are 1.8 for 21 December, 5.5 for 21 March, and 9.9 for 21 June.

3.2.3. Techniques for Measurement of Environmental UV

Direct measurement of UV radiation levels is possible using a variety of instruments. Spectroradiometers provide the most complete information on the wavelength dependence, but often at the cost of complexity. Simpler instruments feature single or multiple detector/filter combinations (broad or narrow band), but the results can be more difficult

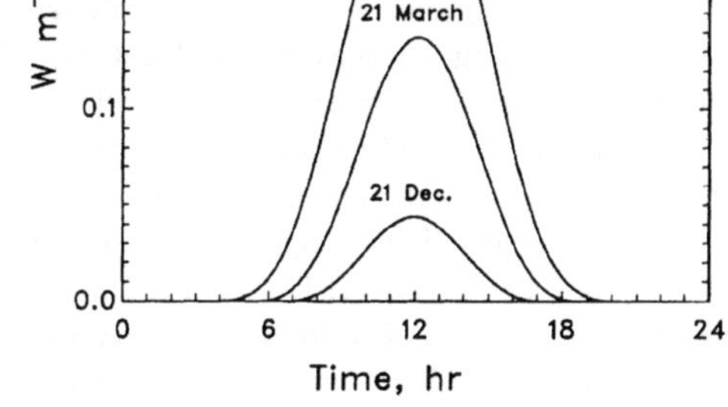

FIGURE 9. Erythemal dose rates at 40° N, calculated for cloud-free skies, sea level, ground albedo = 0.1. Climatological values of the total ozone column are 360 DU for 21 March, 340 DU for 21 June, and 320 DU for 21 December.

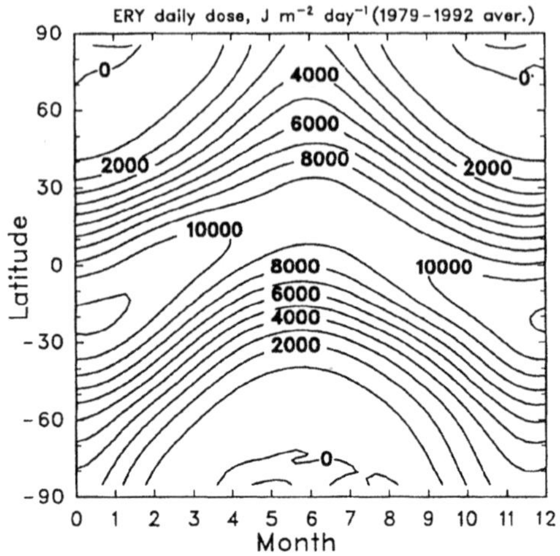

FIGURE 10. Erythemal daily doses, computed for cloud-free skies, based on ozone measurements from the TOMS satellite instrument.

to interpret (e.g., because of the strong wavelength-dependence of UV-B radiation). Chemical and biological dosimeters have also been developed, and may be useful for their small size and ease of handling.

All measurements are subject to uncertainties, and the choice of an instrument depends on the intended use. Many instruments are capable of detecting large UV differences (e.g., broad geographical distributions, cloud effects), and detailed spectral information may be particularly useful to unravel the controlling factors. However, the monitoring of long-term trends is much more difficult, with the inevitable instrumental drift requiring frequent absolute calibrations.

4. TRENDS IN ENVIRONMENTAL LEVELS OF UV RADIATION

Atmospheric O_3 amounts have declined significantly in the past two decades, according to observations from satellite-based instruments and ground monitoring stations (WMO, 1994a). The largest decreases (the so-called "ozone hole") are observed over Antarctica each spring (September–December), with O_3 column values dipping to near 100 DU in recent years, compared with 300 DU during the 1950s and 1960s.

Smaller but statistically significant ozone depletion trends are also found in the Northern polar regions and at middle latitudes of both hemispheres, with reductions of 5–20% depending on season and location. The midlatitude ozone depletions were particularly noticeable for several years following the 1991 eruption of Mt. Pinatubo, with volcanic aerosol providing the surfaces needed to reactivate chlorine reservoir species. Tropical regions, on the other hand, have not shown any significant ozone decreases.

There are no reliable long-term measurements of UV changes corresponding to the above ozone trends. Direct UV measurements spanning the 1970s and 1980s exist for only a few locations. The high variability of UV radiation at the surface makes trend estimation very difficult, and the UV monitoring instruments must be kept stable over

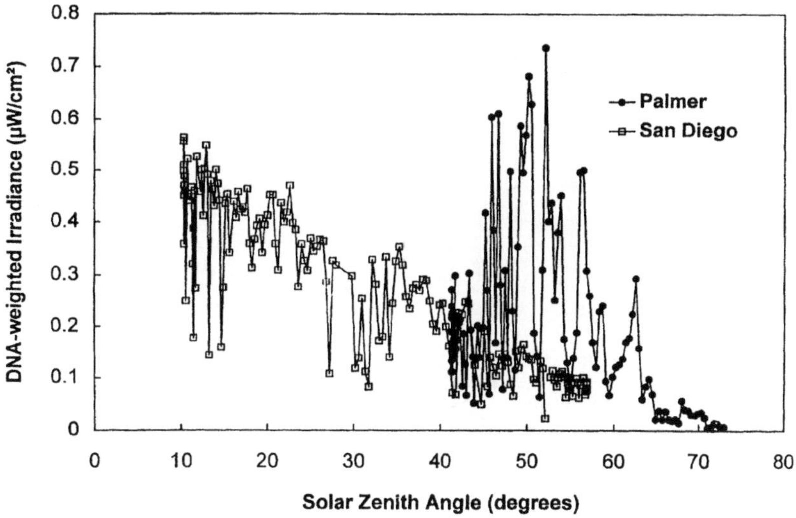

FIGURE 11. DNA-damage weighted irradiances, measured in 1993 at solar noon in San Diego (California) and Palmer Station (Antarctica). From Madronich *et al.* (1995), data courtesy of R. C. Booth.

periods of decades. For example, some reported trends (Scotto *et al.*, 1988) have been shown to be unreliable because of instrument drift and calibration shifts (DeLuisi, 1993; Weatherhead *et al.*, 1997). Thus, direct knowledge of the natural (pre-ozone depletion) UV levels has been lost. Many more UV monitoring stations have been deployed in recent years, using improved instrumentation and calibration procedures, but the data record is still far too short for reliable trend estimation. Direct UV measurements do confirm high UV levels under drastic ozone reductions. For example, the DNA-damaging irradiance at Palmer Station (Antarctica) can exceed summertime maximum values observed in Southern California, as shown in Figure 11; the high values are associated with the ozone hole. Measurements at the South Pole (e.g., see Figure 2) and other locations show good agreement with theory, over a wide range of ozone changes.

The experimental validation of the ozone–UV relationship provides the basis for inferring global distributions of surface UV radiation from satellite ozone data and atmospheric radiative transfer modeling. The derived clear sky geographical and seasonal distribution of erythemal radiation was shown in Figure 10. UV trends may also be estimated from the ozone trends, as shown in Figure 12. UV increases associated with the ozone hole are clearly evident (springtime austral latitudes), with smaller but still significant increases at all other locations except the tropics.

The implications of enhanced UV levels can be estimated, at least roughly, for some biological impacts. Table II shows projected increases in the incidence of two types of skin cancer related to the 1979–1992 increases in UV radiation. These are steady-state estimates because they do not consider any induction time for skin cancer, nor any additional changes in UV beyond the 1992 levels, but of course do assume unchanging human exposure patterns and behavior. Although percent increases are largest in the polar

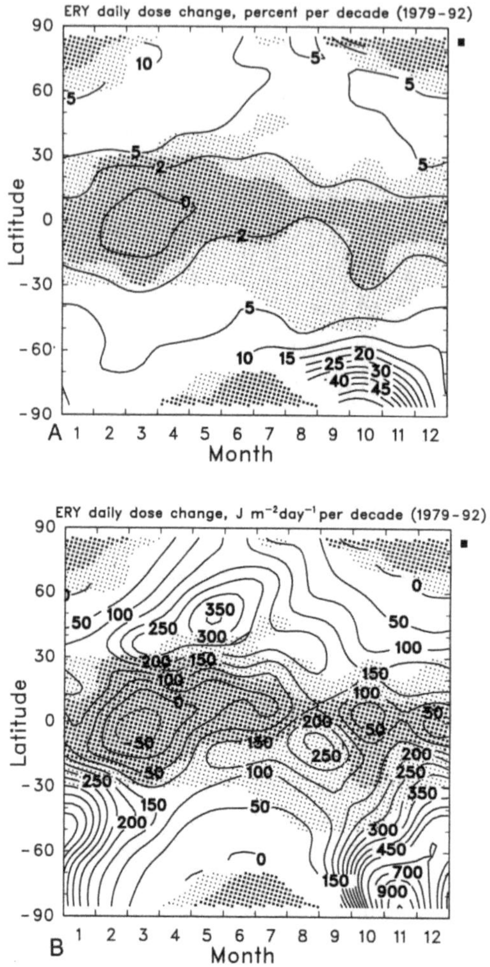

FIGURE 12. Estimated trends in erythemal daily doses, based on TOMS (version 6) ozone column measurements over 1979–1992. Panel (a) shows percent changes, while panel (b) gives energy increments. Shading marks regions where trends differ from zero by less than one (heavy) or two (light) standard deviations.

regions, the enhancements are most important at middle latitudes where the incidence of skin cancer is already high.

Another biological impact is evident under the large UV enhancements associated with the Antarctic ozone hole. The edge of the hole fluctuates, much like weather patterns, and it is possible to experience alternating high and low UV-B exposures over the span of several days. Studies have focused on phytoplankton growth in the Southern Ocean, where *in situ* photosynthetic carbon fixation was found to be reduced by 6–12% on days with enhanced UV-B radiation (Smith *et al.*, 1992).

UV radiation trends can also be caused by factors other than stratospheric ozone depletion. In industrialized regions, high amounts of gaseous and particulate pollutants may attenuate UV radiation substantially, thus offsetting partly the UV increases related

Table II
Expected Increases (Percent)[a] in Annual UV Doses and Skin Cancer Incidence Related to Stratospheric Ozone Depletion from 1979 through 1992

Latitude	Total ozone	Erythema induction dose	DNA damage dose	Skin cancer dose	Basal cell carcinoma incidence[b]	Squamous cell carcinoma incidence[b]
85 N	−8.8±3.2	7.1±1.7	14.8±3.6	10.6±2.5	15.1±5.6	28.5±11.2
75 N	−9.0±2.9	7.6±1.7	14.9±3.3	10.8±2.4	15.4±5.8	29.1±11.5
65 N	−7.4±1.7	7.6±1.7	14.1±3.2	10.3±2.3	14.7±5.6	27.7±11.0
55 N	−7.4±1.3	7.2±1.7	12.9±3.2	9.5±2.3	13.5±5.3	25.4±10.3
45 N	−6.6±1.2	6.5±1.7	10.9±3.0	8.1±2.2	11.6±4.7	21.6±9.0
35 N	−4.8±1.4	5.0±1.8	8.2±3.0	6.1±2.2	8.6±4.0	16.0±7.6
25 N	−2.7±1.5	3.0±1.9	4.8±3.1	3.5±2.2	5.0±3.5	9.0±6.4
15 N	−1.5±1.1	1.7±1.4	2.6±2.3	1.9±1.6	2.7±2.4	4.8±4.4
5 N	−0.6±1.6	0.7±1.9	1.2±3.0	0.8±2.2	1.2±3.1	2.1±5.5
5 S	−1.1±1.4	1.2±1.7	2.0±3.0	1.4±2.0	2.0±2.8	3.6±5.2
15 S	−1.9±1.3	2.2±1.5	3.5±2.4	2.5±1.7	3.6±2.6	6.5±4.8
25 S	−2.6±1.6	3.1±1.7	5.0±2.4	3.6±1.9	5.1±3.1	9.2±5.8
35 S	−4.0±1.6	4.8±1.6	7.9±2.6	5.7±1.9	8.1±3.6	14.9±6.8
45 S	−5.6±1.4	7.2±1.6	12.5±2.7	8.9±1.9	12.7±4.8	23.9±9.2
55 S	−9.0±1.5	10.9±2.0	19.7±3.6	14.2±2.6	20.4±7.4	39.3±15.1
65 S	−15.0±2.0	16.3±3.0	30.5±5.8	21.9±4.1	31.9±12.2	64.0±26.6
75 S	−19.5±2.6	24.1±5.4	49.8±12.0	34.0±7.7	50.6±21.4	107.7±52.0
85 S	−21.1±3.0	31.0±6.8	72.0±17.6	46.5±10.5	70.6±31.2	159.6±83.6

[a]Evaluated over the 14-year data record, expressed as percent relative to the 1979 intercept. Uncertainties are one standard deviation. From Madronich and de Gruijl (1993).

[b]Based on skin cancer dose increase and relevant biological amplification factors.

to stratospheric ozone depletion (Brühl and Crutzen, 1989; Liu *et al.*, 1991). Long-term changes in cloud cover can also induce corresponding UV radiation trends, although satellite cloud images show no evidence for such systematic cloudiness variations over 1979–1992 (Herman *et al.*, 1996).

The trends given here are likely to undergo future updates and revisions. For example, a recent reanalysis of satellite ozone measurements (Herman *et al.*, 1996) gives ozone reductions somewhat smaller than reported earlier (e.g., Stolarski *et al.*, 1991; WMO, 1994a), with corresponding UV radiation increases also smaller than earlier estimates (e.g., Madronich and de Gruijl, 1993), by ca. 30% on average, with the exact values depending on latitude and season. Furthermore, future changes in stratospheric ozone are unlikely to be simple monotonic extrapolations of past and current trends, but will instead depend on both natural fluctuations and the extent of continuing human impacts on the atmosphere.

5. CONCLUSIONS

Exposure of biota to UV radiation is now recognized as largely detrimental even at levels typical of the natural atmosphere. The main factors controlling atmospheric UV radiation (solar zenith angle, ozone column amount, clouds and aerosols, surface reflections) are fairly well understood in principle, but usually at least some of these are not known for an arbitrary location and time of interest. Reductions in stratospheric ozone have so far caused increases in biologically effective UV radiation of some 5–15% at middle latitudes, with larger percent increases in polar regions, but no significant increases in the tropics. Such increases, while difficult to detect directly, are estimated to have substantial public health impacts, e.g., on skin cancer incidence.

International agreements to phase out the production and release of ozone-destroying chemicals are intended to prevent further depletion of stratospheric ozone, and lead to full recovery by the middle of the next century. The success of such agreements is contingent on strict global compliance, and assumes that our current understanding of ozone chemistry, which in many details is still evolving, is at least sufficient today to allow prediction of future atmospheric responses to anthropogenic emissions.

6. REFERENCES

ACGIH, 1992, *1991–1992 Threshold Limit Values*, American Conference of Governmental and Industrial Hygienists.

Andrady, A. L., Torikai, A., and Fueki, K., 1989, Photodegradation of rigid PVC formulations III. Wavelength sensitivity to light-induced yellowing by monochromatic light, *J. Appl. Polym. Sci.* **37**:935–946.

Bates, D. R., and Nicolet, M., 1950, Atmospheric hydrogen, *Publ. Astron. Soc. Pac.* **62**:106.

Booth, R. C., and Madronich, S., 1994, Radiation amplification factors—improved formulation accounts for large increases in ultraviolet radiation associated with Antarctic ozone depletion, in *Ultraviolet Radiation and Biological Research in Antarctica* (C. S. Weiler and P. A. Penhale, eds.), pp. 39–42, American Geophysical Union Antarctic Research Series, Washington, DC.

Boucher, N., Prezelin, B. B., Evens, T., Jovine, R., Kroon, B., Moline, M. A., and Schofield, O., 1994, Icecolors '93: Biological weighting function for the ultraviolet inhibition of carbon fixation in a natural antarctic phytoplankton community, *Antarct. J.-Review 1994*, pp. 272–275.

Brasseur, G. P., Granier, C., and Walters, S., 1990, Future changes in stratospheric ozone and the role of heterogeneous chemistry, *Nature* **348**:626–628.

Brühl, C., and Crutzen, P. J., 1989, On the disproportionate role of tropospheric ozone as a filter against solar UV-B radiation, *Geophys. Res. Lett.* **16**:703–706.

Caldwell, M. M., Camp, L. B., Warner, C. W., and Flint, S. D., 1986, Action spectra and their key role in assessing biological consequences of solar UV-B radiation change, in *Stratospheric Ozone Reduction, Solar Ultraviolet Radiation and Plant Life* (R. C. Worrest and M. M. Caldwell, eds.), pp. 87–111, Springer-Verlag, Berlin.

Chapman, S., 1930, On ozone and atomic oxygen in the upper atmosphere, *Philos. Mag.* **10**:369.

Crutzen, P. J., 1970, The influence of nitrogen oxide concentrations on the atmospheric ozone content, *Q. J. R. Meteorol. Soc.* **96**:320.

De Fabo, E. C., and Noonan, F. P., 1983, Mechanism of immune suppression by ultraviolet radiation *in vivo*. I. Evidence for the existence of a unique photoreceptor in skin and its role in photoimmunology, *J. Exp. Med.* **158**:84–98.

de Gruijl, F. R., and van der Leun, J. C., 1994, Estimate of the wavelength dependency of ultraviolet carcinogenesis and its relevance to the risk assessment of a stratospheric ozone depletion, *Health Phys.* **4**:317–323.

DeLuisi, J. J., 1993, Possible calibration shift in the U.S. surface UV network instrumentation 1979 to 1985, Paper presented at the U.S. Department of Energy UV-B Critical Issues Workshop, Cocoa Beach, FL, 24–26 February.

Farman, J. C., Gardiner, B. G., and Shanklin, J. D., 1985, Large losses of total ozone in Antarctica reveal seasonal ClO_x/NO_x interaction, *Nature* **315**:207–210.

Frederick, J. E. and Snell, H. E., 1990, Tropospheric influence on solar ultraviolet radiation: The role of clouds, *J. Climate* **3**:373–381.

Herman, J. R., Bhartia, P. K., Ziemke, J., Ahmad, Z., and Larko, D., 1996, UV-B increases (1979–1992) from decreases in total ozone, *Geophys. Res. Lett.* **23**:2117–2120.

Hoffman, D. J., and Solomon, S., 1989, Ozone destruction through heterogeneous chemistry following the eruption of El Chichon, *J. Geophys. Res.* **94**:5029–5041.

Liu, S. C., McKeen, S. A., and Madronich, S., 1991, Effects of anthropogenic aerosols on biologically active ultraviolet radiation, *Geophys. Res. Lett.* **18**:2265–2268.

Long, C. S., Miller, A. J., Lee, H.-T., Wild, J. D., Przywarty, R. C., and Hufford, D., 1996, Ultraviolet index forecasts issued by the National Weather Service, *Bull. Am. Meteorol. Soc.* **77**:729–748.

Madronich, S., 1993a, The atmosphere and UV-B radiation at ground level, in *Environmental UV Photobiology* (L. O. Björn and A. R. Young, eds.), pp. 1–39, Plenum Press, New York.

Madronich, S., 1993b, UV radiation in the natural and perturbed atmosphere, in *Environmental Effects of UV (Ultraviolet) Radiation* (M. Tevini, ed.), pp. 17–69, Lewis Publisher, Boca Raton, FL.

Madronich, S., and de Gruijl, F. R., 1993, Skin cancer and UV radiation, *Nature* **366**:23.

Madronich, S. and Granier, C., 1994, Tropospheric chemistry changes due to increases in UV-B radiation, in *Stratospheric Ozone Depletion/UV-B Radiation in the Biosphere* (H. R. Biggs and M. E. B. Joyner, eds.), pp. 3–10, NATO Advanced Research Workshop, Springer-Verlag.

Madronich, S., McKenzie, R. L., Björn, L., and Caldwell, M. M., 1995, Changes in ultraviolet radiation reaching the Earth's surface, *Ambio* **24**:143–152.

McKinlay, A. F., and Diffey, B. L., 1987, A reference action spectrum for ultraviolet induced erythema in human skin, in *Human Exposure to Ultraviolet Radiation: Risks and Regulations* (W. R. Passchler and B. F. M. Bosnajokovic, eds.), pp. 83–87, Elsevier, Amsterdam.

Pitts, D. G., Cullen, A. P., and Hacker, P. D., 1977, Ocular effects of ultraviolet radiation from 295 to 365 nm, *Invest. Ophthalmol. Visual Sci.* **16**:932–939.

Quaite, F. E., Sutherland, B. M., and Sutherland, J. C., 1992, Action spectrum for DNA damage in alfalfa lowers predicted impact of ozone depletion, *Nature* **358**:576–578.

Rowland, F. S., and Molina, M. J., 1975, Chlorofluoromethanes in the environment, *Rev. Geophys. Space Phys.* **13**:1–35.

Schafer, J. S., Saxena, V. K., Wenny, B. N., Barnard, W., and DeLuisi, J. J., 1996, Observed influence of clouds on ultraviolet-B radiation, *Geophys. Res. Lett.* **23**:2625–2628.

Scotto, J., Cotton, G., Urbach, F., Berger, D., and Fears, T., 1988, Biologically effective ultraviolet radiation: Surface measurements in the United States, 1974 to 1985, *Science* **239**:762–764.

Setlow, R. B., 1974, The wavelengths in sunlight effective in producing skin cancer: A theoretical analysis, *Proc. Natl. Acad. Sci. USA,* **71**:3363–3366.

Setlow, R. B., Grist, E., Thompson, K., and Woodhead, A. D., 1993, Wavelengths effective in induction of malignant melanoma, *Proc. Natl. Acad. Sci. USA* **90**:6666–6670.

Slaper, H., Velders, G. J. M., Daniel, J. S., de Gruijl, F. R., and van der Leun, J. C., 1996, Scenario study on ozone depletion and skin cancer incidence illustrating the Vienna Convention achievements, *Nature* **384**:256–258.

Smith, R. C., Prezelin, B. B., Baker, K. S., Bidigare, R. R., Boucher, N. P., Coley, T., Karentz, D., MacIntyre, S., Matlick, H. A., Menzies, D., Ondrusek, M., Wan, Z., and Waters, K. J., 1992, Ozone depletion: Ultraviolet radiation and phytoplankton biology in Antarctic waters, *Science* **255**:952–959.

Solomon, S., 1990, Progress towards a quantitative understanding of Antarctic ozone depletion, *Nature* **347**:347–354.

Steinmetz, V., and Wellmann, E., 1986, The role of solar UV-B in growth regulation of cress (Lepidium sativum L.) seedlings, *Photochem. Photobiol.* **43**:189–193.

Stolarski, R. S., and Cicerone, R. J., 1974, Stratospheric chlorine: A possible sink for ozone, *Can. J. Chem.* **52**:1610.

Stolarski, R. S., Bloomfield, P., McPeters, R. D., and Herman, J. R., 1991, Total ozone trends deduced from Nimbus 7 TOMS data, *Geophys. Res. Lett.* **18**:1015–1018.

UNEP, 1994, *Environmental Effects of Stratospheric Ozone Depletion—1994 Update* (J. van der Leun, M. Tevini, and X. Tang, eds.), United Nations Environmental Programme, Nairobi, Kenya.

US Standard Atmosphere, 1976, National Oceanic and Atmospheric Administration, National Aeronautics and Space Administration, United States Air Force, Washington, October.

Weatherhead, E. C., Tiao, G. C., Reinsel, G. C., Frederick, J. E., DeLuisi, J. J., Choi, D., and Tam, W., 1997, Analysis of long-term behavior of ultraviolet radiation measured by Robertson-Berger meters at 14 sites in the United States, *J. Geophys. Res.* **102**:8737–8754.

WMO, 1994a, *Scientific Assessment of Stratospheric Ozone: 1994* (D. Albritton and R. Watson, eds.), World Meteorological Organization, Global Ozone Research and Monitoring Project, Report No. 37.

WMO, 1994b, *Report of the WMO Meeting of Experts on UV-B Measurements, Data Quality and Standardization of UV Indices*, World Meteorological Organization Global Atmosphere Watch, Report No. 95.

Chapter 13

Ozone and Nitrogen Dioxide

Daniel B. Menzel and Dianne M. Meacher

1. INTRODUCTION

Ozone (O_3) and nitrogen dioxide ($\cdot NO_2$) are the most plentiful photochemical oxidants in ambient air. These gases also are the most potent toxicants to the respiratory tract. O_3 and $\cdot NO_2$ cause epithelial cytotoxicity and increased cell permeability, produce intermittent and persistent inflammation, degrade the primary structure of the lung, and exacerbate chronic lung diseases such as asthma and bronchitis. The thesis we explore is that both O_3 and $\cdot NO_2$ produce these effects through the production of free radicals, which leads to peroxidation of plasma membranes of cells lining the respiratory tract.

The serious nature of the public health threat from O_3, in particular, has been recognized by the U.S. Environmental Protection Agency (EPA). The World Health Organization and the European Community also have declared that O_3 and $\cdot NO_2$ are pulmonary toxicants of major importance. A wealth of data exists on the adverse health effects of these two pollutants. This review cannot be as exhaustive as the U.S. EPA reference works on O_3 and $\cdot NO_2$ (Air Quality Criteria for Ozone and Related Photochemical Oxidants, 1996; Air Quality Criteria for Oxides of Nitrogen, 1993). Readers interested in more detail should refer to these multivolume works, which have been prepared by manifold experts and reviewed extensively.

Much of the data included in the EPA Criteria documents, however, demonstrate the toxicity of $\cdot NO_2$ and O_3 in humans and multiple experimental animal species at concentrations two to ten times those in polluted urban air. Questions have been raised about the relevance to human disease of experiments using high concentrations of the oxidants.

Daniel B. Menzel and Dianne M. Meacher Department of Community and Environmental Medicine, University of California, Irvine, Irvine, California 92697-1825.

Reactive Oxygen Species in Biological Systems, edited by Gilbert and Colton. Kluwer Academic / Plenum Publishers, New York, 1999.

The partial combustion of fossil fuels by mobile vehicles and electric power plants is responsible for the current levels of O_3 and $\cdot NO_2$ in urban areas around the world. As discussed in Section 2, O_3 is formed in the troposphere from nitrogen oxides. $\cdot NO_2$ is one of the products of the conversion process. The gas-phase reactions are dependent on light. The O_3 concentrations in urban air in the United States have been diminishing mainly as a result of vigorous efforts to reduce the emissions of nitrogen oxides and hydrocarbons (Figure 1), yet further reductions in the emissions from autos of nitrogen oxides are offset by the increase in the number of autos being driven. Even the most modern internal combustion engine emits substantial nitrogen oxides. Reduction in O_3 and $\cdot NO_2$ levels below current levels will require stringent enforcement of measures designed to reduce overall usage of fossil fuels. Developing countries demand the same use of fossil fuels as is now occurring in the developed countries. Consequently, the toxicity of $\cdot NO_2$ and O_3 is becoming more widespread demanding world cooperation for a solution. In a nutshell, $\cdot NO_2$ and O_3 toxicity is likely to be the most common environmentally related cause of disease aside from tobacco smoke in the next 50 years.

Before stringent efforts to curb precursor emissions can be taken, the relationship between the exposure to $\cdot NO_2$ and O_3 and human health effects must be demonstrated in a quantitative manner. Because the research agenda has been driven by essentially descriptive toxicology, a clear understanding of the mechanisms of action of O_3 and $\cdot NO_2$ still lags behind a need to regulate. Sound regulations require a defensible and biologically relevant dose–response relationship. To have a solid basis for regulations, government leaders have come to understand that mechanisms of actions are of vital import.

In this review, we focus on lipid peroxidation as the main mechanism of toxicity of both $\cdot NO_2$ and O_3. At first glance, the problem seems to be simple but several factors

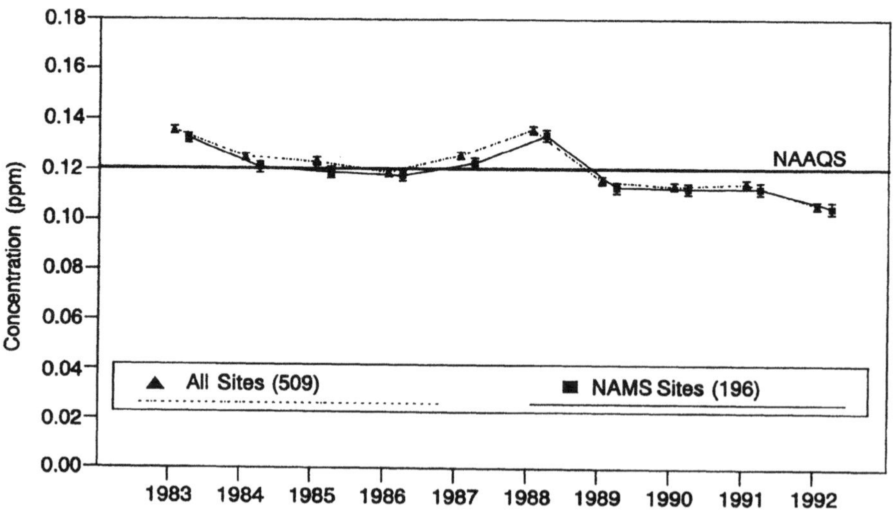

FIGURE 1. The composite averages of the second highest maximum 1-hr ozone concentration with 95% confidence intervals at National Air Monitoring Stations (NAMS) and at all U.S. sites, 1983 to 1992, in relation to the U.S. National Ambient Air Quality Standard (NAAQS). Taken from U.S. EPA (1996a).

make the analysis complex. One complexity is that humans are exposed in free-living populations to episodes of oxidant air pollution that occur with greater frequency in areas with sunny weather, and the episodes occur more so in spring and summer than in winter. Experimental animal studies, on the other hand, often are carried out at constant levels of the toxicants during both the active and sleep cycles of the animals. Such protocols are easy to carry out but have little in common with the lifelong exposures of humans living in polluted air. Human epidemiological data on the effects of oxidant air pollutants reflect the sum of exposures to both O_3 and $\cdot NO_2$ at varying concentrations. No one breathing polluted outdoor air is exposed only to $\cdot NO_2$ or to O_3. Most epidemiological studies are hampered by the covariance of the two gases in the atmosphere related to the chemistry of the reactions in the gas phase (see Section 2). Epidemiological studies associate certain health effects with O_3 or $\cdot NO_2$ by various statistical maneuvers, but no hard data exist to demonstrate causation of lung disease in humans by these toxicants alone.

The anatomical effects of O_3 and $\cdot NO_2$ also are complicated by the chemical reactivities of the two pollutants and their transfer from the gas phase to the liquid phase lining the respiratory tract or to the underlying cells. Thus, the dose to cells changes among different regions of the lung.

Most investigators would agree that the most important aspect of studies of $\cdot NO_2$ and O_3 is the remodeling of the lung during lifetime exposures. Assuming that the mechanisms of action of both $\cdot NO_2$ and O_3 are the same in experimental animals and humans, J. A. Graham, F. J. Miller, and D. B. Menzel proposed the paradigm shown in Figure 2 as the basis for an experimental approach investigating mechanisms of toxic action (unpublished data).

To resolve the sociopolitical issues regarding the adverse health effects of $\cdot NO_2$ and O_3 exposure, quantitative relationships between the exposure dose and the molecular dose delivered to the lung are needed. Experimental animals are important to this type of research because they can be confined, have short lifetimes, and can be examined by highly invasive procedures.

Differences exist in biochemistry between humans and experimental animals. Although the primary composition of lung cell membranes is similar in both humans and experimental animals, the expression of genes involved in defense or in regulation of the cell cycle may be different. The extrapolation parallelogram attempts to model species differences in biochemistry by studying effects on isolated lung cells *in vitro* versus whole

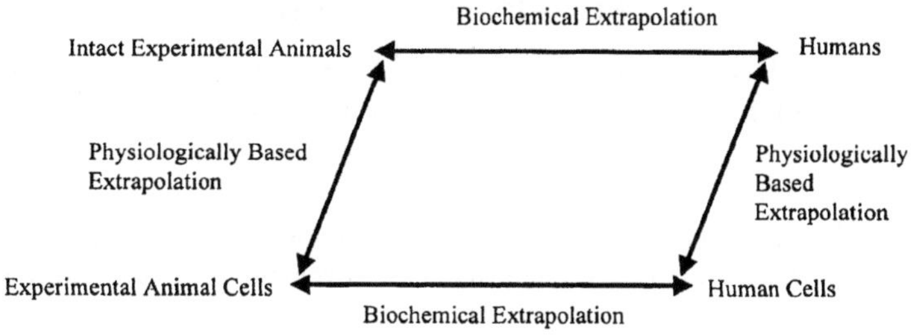

FIGURE 2. The extrapolation modeling paradigm. The parallelogram illustrates how relationships between *in vitro* and *in vivo* effects and between experimental animal studies and human studies are interrelated.

animal exposures. Complex mathematical models have been developed to relate the exposure dose to the molecular dose at the airway. $\cdot NO_2$ and O_3 research has attempted to quantify the responses in humans and experimental animals for many endpoints such as production of inflammatory hormones, induction of GSH-dependent protective pathways, and depletion of lung antioxidant reserves. One can appreciate the problem of trying to do these experiments quantitatively. So far this aspect has not progressed as well as has the mathematical modeling of the deposition of O_3 and $\cdot NO_2$ in the airways.

The extrapolation parallelogram paradigm will collapse if the mechanisms of pulmonary toxicity of O_3 and $\cdot NO_2$ are different between experimental animals and humans. Thus, mechanisms of action are the foundation for the entire experimental air pollution health effects enterprise!

2. BACKGROUND

2.1. Gas-Phase Chemistry of O_3 and $\cdot NO_2$

O_3 and $\cdot NO_2$ are the primary constituents of photochemical smog. Nitric oxide ($\cdot NO$) is formed from nitrogen in air during combustion processes (Boström, 1993):

$$N_2 + O \rightarrow \cdot NO + N \tag{1}$$

and

$$N + O_2 \rightarrow \cdot NO + O \tag{2}$$

$\cdot NO$ is oxidized to $\cdot NO_2$ in the atmosphere. The burning of fossil fuels by motor vehicles and stationary sources for heating and electric power generation is chiefly responsible for the $\cdot NO_2$ in the atmosphere (Mohsenin, 1994). In cities, motor vehicles are the main source of nitrogen oxides in outdoor air (Berglund, 1993).

In the clean troposphere, $\cdot NO_2$ is formed primarily from the reaction of $\cdot NO$ with O_3 and secondarily from the reaction of $\cdot NO$ with the hydroperoxyl radical ($\cdot HO_2$) (Boström, 1993; Huie, 1994):

$$\cdot NO + O_3 \rightarrow \cdot NO_2 + O_2 \tag{3}$$

and

$$\cdot NO + \cdot HO_2 \rightarrow \cdot NO_2 + \cdot OH \tag{4}$$

$\cdot NO_2$ also is a product of the reaction of $\cdot NO$ and peroxy radicals ($\cdot RO_2$) formed from volatile organic compounds in the atmosphere (Boström, 1993):

$$\cdot RO_2 + \cdot NO \rightarrow \cdot RO + \cdot NO_2 \tag{5}$$

Volatile organic compounds are emitted from industries and transportation vehicles. The compounds also are released during house painting, vehicle refueling, and road paving (Breslin, 1995).

$\cdot NO_2$ contributes to the formation of photochemical smog when it is photolyzed to $\cdot NO$ and atomic oxygen:

$$\cdot NO_2 + h\nu \rightarrow O + \cdot NO \quad (6)$$

Atomic oxygen reacts with O_2 to form O_3 (Boström, 1993):

$$O + O_2 \rightarrow O_3 \quad (7)$$

Almost all of the O_3 in ambient air is formed secondarily from photochemical reactions that depend on volatile organic compounds and $\cdot NO_2$ (Lippmann, 1992) [Reactions (5)–(7)]. In addition to precursors, weather conditions including hot temperatures and inversion layers have a major impact on O_3 formation.

O_3 is nearly insoluble in water (Sandström, 1995). It decomposes in aqueous solutions to produce hydrogen peroxide, superoxide, and hydroxyl radicals (Victorin, 1992). Though not itself a free radical, O_3 causes radical formation in biological systems (Pryor, 1994).

$\cdot NO_2$ is a nitrogen-centered free radical and is partly soluble in water (Mohsenin, 1994). $\cdot NO_2$ can initiate free-radical chain reactions, yet its stability allows it to reach high levels in ambient air. $\cdot NO_2$ is solubilized to nitrous and nitric acids as well as nitric oxide (Elsayed, 1994). Roughly equal amounts of nitric and nitrous acid are formed because only the $\cdot NO_2$ dimer is soluble in water at the concentrations of $\cdot NO_2$ reached in the lung. Some reactions involving nitrite have been reported, but nitrite does not appear to be a major biological product on $\cdot NO_2$ exposure. $\cdot NO_2$ gas reactions dominate the biological effects.

2.2. Exposure

The public health importance of $\cdot NO_2$ and O_3 can be appreciated from the number of people exposed in the United States. The U.S. National Ambient Air Quality Standard promulgated in 1979 for O_3 is 0.12 ppm (averaged over 1 hr) which can be exceeded once per year. Nonetheless, greater than half of U.S. residents are estimated to live in areas where O_3 levels are near or above the standard (Kleeberger, 1995). O_3 concentrations as high as 0.3 to 0.4 ppm may occur in urban areas such as the Los Angeles basin that have heavy mobile vehicle traffic and sunny weather (Bromberg and Koren, 1995). Comparable levels occur in Japan, Europe, and South America.

O_3 produced in metropolitan areas can be transported downwind long distances. Diurnal patterns of O_3 levels (Figure 3) show a rise to an early afternoon peak, and the highest O_3 levels occur in late spring or in summer in urban areas (U.S. EPA, 1996a). Outdoor levels of O_3 usually are higher than indoor levels so that infiltration of outdoor air is the sole source of indoor O_3 (Zhang and Lioy, 1994). The time and site of exercising outdoors are the main determinants of exposure of an individual to a short peak of O_3 (Vostal, 1994).

$\cdot NO_2$ is the major driving force in the formation of O_3. Because $\cdot NO_2$ arises mostly from autos, geographic and diurnal patterns occur (Figure 4). The range of annual average outdoor levels of $\cdot NO_2$ in many parts of the United States is from 0.015 to 0.035 ppm, concentrations below the U.S. National Ambient Air Quality Standard of 0.053 ppm (annual arithmetic mean) (Schlesinger, 1992). Nevertheless, polluted urban air can contain levels from 0.05 to 0.2 ppm $\cdot NO_2$ with peaks of 0.5 ppm (Mayorga, 1994). The highest levels of ambient $\cdot NO_2$ occur, on average, in the late afternoon and evening hours

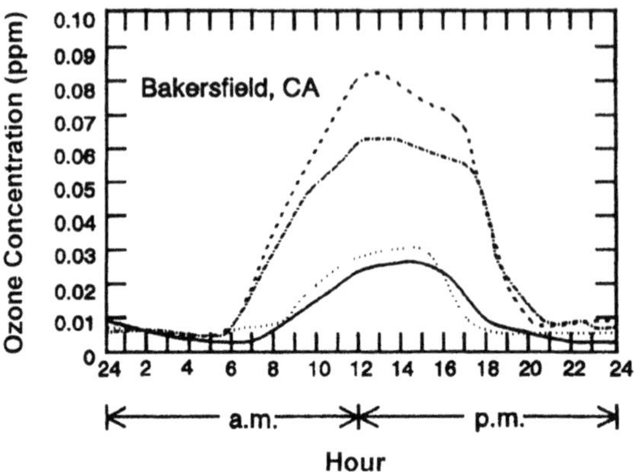

FIGURE 3. Typical diurnal cycle of ozone in polluted air. Solid line represents first quarter; alternating dashed and dotted line represents second quarter; dashed line represents third quarter; dotted line represents fourth quarter. Taken from U.S. EPA (1996a).

(U.S. EPA, 1993a). In urban areas with hourly outdoor $\cdot NO_2$ levels greater than 0.2 ppm, an additional diurnal episode can occur midmorning (U.S. EPA, 1993a). The highest monthly levels of ambient $\cdot NO_2$ occur in the fall and winter months (U.S. EPA, 1993a).

$\cdot NO_2$ is generated indoors as well as outdoors whereas O_3 indoors occurs almost exclusively through infiltration of outdoor air. Microenvironments that have high levels of $\cdot NO_2$ are homes with gas cooking stoves and gas-fueled space heaters and water heaters. Tobacco-smoke-filled rooms and inside autos and buildings in heavy traffic areas are also common exposure locales (Berglund, 1993). Several studies indicate that $\cdot NO_2$ levels inside motor vehicles may be greater than those measured at closeby outdoor locations (American Thoracic Society, 1996). Furthermore, indoor air may contain higher levels of $\cdot NO_2$ than outdoor air when gas appliances are used (Berglund, 1993). Concentrations of $\cdot NO_2$ up to 0.6 ppm for 45 min have been measured during cooking with a gas stove (Goldstein *et al.*, 1988). In general, humans are seldom exposed to outdoor levels greater than 0.5 ppm $\cdot NO_2$ or indoor levels greater than 1 ppm (Schlesinger, 1992). Thus, it is questionable whether experimental animal studies above 2 ppm $\cdot NO_2$ are of use in understanding the biological mechanism of action, and most are omitted from this review. Mechanisms found at such high levels are suspect and probably of little biological utility.

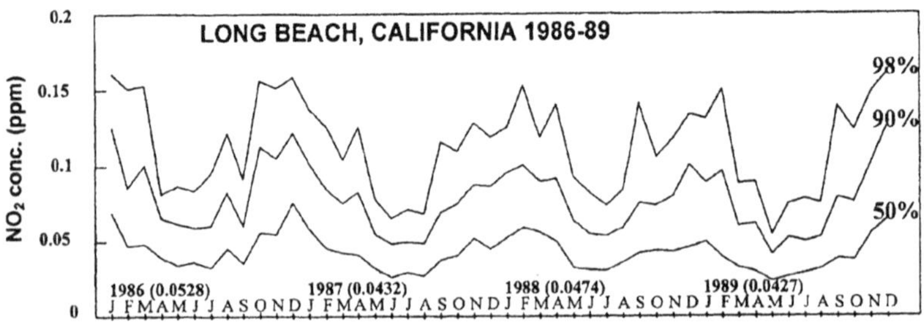

FIGURE 4. Monthly 50th, 90th, and 98th percentiles of 1-hr nitrogen dioxide concentrations at Long Beach, CA, 1986 to 1989. The episodic and cyclic nature of $\cdot NO_2$ concentrations occurs in many other locations as well. Taken from U.S. EPA (1993a).

2.3. Dosimetry Modeling to Estimate the Regional Deposition of O_3 and $\cdot NO_2$ in the Lungs of Mammals

2.3.1. Regional O_3 Uptake in the Lung

Quantification of O_3 and $\cdot NO_2$ effects has been a major objective of oxidant air pollutant research. The quantification aspect is driven by the need to justify public health measures aimed at reducing O_3 generation. From a theoretical view, the search for quantitative relationships between exposure concentrations and biological effects has generated much data useful in examining hypotheses on modes of action. The first monograph that attempted to relate biological effect to exposure concentration by mathematical modeling of the biochemical reactions of $\cdot NO_2$ and O_3 and the regional uptake of these gases in the lung was by Miller and Menzel (1984). Mathematical modeling predicts that the maximum dose of O_3 to lung tissue is to the centriacinar region, the junction between the conducting airways and the pulmonary region (Miller *et al.*, 1978). Subsequent morphological measurements support this prediction.

One apparent anomaly in O_3 toxicity is the susceptibility of the cells lining the junction between the respiratory bronchiole and the alveolus. Intuitively, one would suppose that the regions of the respiratory tract first exposed to O_3 during breathing would receive the greatest O_3 dose. Despite its high reactivity, only about 50% of inhaled O_3 is removed from the mucus-lined nasal and pharyngeal regions of the lung (Gerrity *et al.*, 1988). O_3 absorption occurs in each of the three main divisions (extrathoracic, tracheobronchial, and pulmonary) of the respiratory tract (Miller, 1995). O_3 reaches the pulmonary region of the lung because of its poor aqueous solubility.

For the same exposure, adult humans receive approximately twice the dose of O_3 as do adult rats and children receive slightly greater doses than do adults (Lippmann, 1992). Children under the age of 6 y receive greater O_3 doses than do older children and adults (Kleinman, 1991). (For added details see U.S. EPA, 1996b.)

In controlled human exposures, the total respiratory uptake of O_3 in humans is 78 to 97% compared with an average of 47% in rats and guinea pigs. The upper respiratory uptake efficiency of O_3 is approximately 40% during either steady unidirectional or cyclical flow whereas lower respiratory uptake efficiency is about 90% for steady unidirectional flow and 65% for cyclical flow. The O_3 uptake in the pulmonary region is increased with exercise, at least in part, because of the augmentation in gas transport caused by convection in that region. The increased O_3 uptake may occur even though minute ventilation in exercising and resting humans exposed to O_3 may be fairly constant (see Miller, 1995, for additional details).

The mathematical dosimetry models of O_3 deposition are based on the concept of transfer of material across a physical boundary by chemical reaction (Miller and Menzel, 1984; Overton, 1984). O_3 removal from the air space to the cells lining the respiratory tract depends on the chemical reactivity of O_3 and the amount of biological compounds present to react with the O_3 (Van der Vliet *et al.*, 1995). The respiratory tract is lined with a fluid layer throughout its length. In those regions where the fluid lining layer is aqueous rather than lipophilic, little O_3 would be transferred because of the low aqueous solubility of O_3 were it not for the presence of easily oxidized compounds such as vitamin C

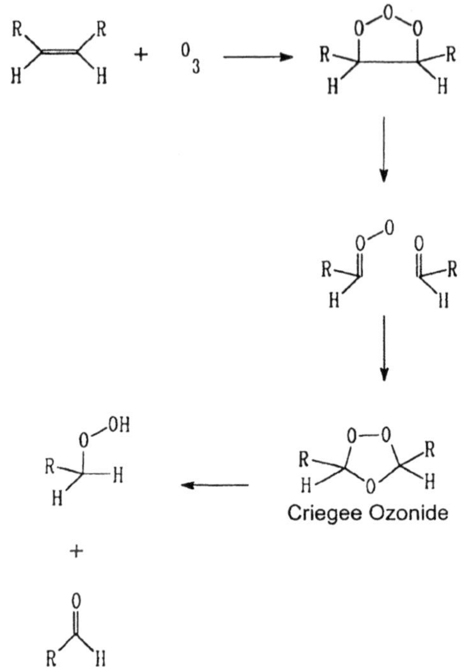

FIGURE 5. Classic ozone addition to the unsaturations in olefins to form the Criegee ozonide.

(ascorbic acid), unsaturated fatty acids, and thiols including GSH. The tissues underlying the lung lining fluid are protected from O_3 by the easily oxidized components.

The mathematical models assume that most of the chemical reaction of O_3 in the lung is through the Criegee ozonide formation mechanism (Figure 5). Note that the formation of the Criegee ozonide is *not* a free-radical-mediated reaction (Pryor, 1994). Nevertheless, Criegee oxonides can lead to the formation of free radicals. A hydroperoxide produced from the Criegee ozonide can decompose thermally to peroxyl and hydroxyl free radicals. Peroxyl radicals can initiate lipid peroxidation. Another pathway of radical formation from the reaction of O_3 with olefins is the decomposition of the initial ozonide via β-scission to a peroxyl radical and an aldehyde (Figure 6). Other pathways by which O_3 generates free radicals could exist as suggested by Pryor (1994). An additional important pathway of O_3 toxicity could be by direct oxidation of protein intercalated in the plasma membrane (Banerjee and Mudd, 1992).

2.3.2. Regional $\cdot NO_2$ Uptake in the Lung

$\cdot NO_2$ deposition in the lung is in some ways simpler than that of O_3. The total respiratory uptake of $\cdot NO_2$ is 90% during maximum ventilation by healthy subjects and asthmatic adults (Ewetz, 1993). In humans at rest, about 50% is absorbed in the nasal cavity; during exercise, however, less than 20% of $\cdot NO_2$ is absorbed by nasal surfaces (Mohsenin, 1994). Uptake of $\cdot NO_2$ in the upper respiratory tracts of dogs, rabbits, and

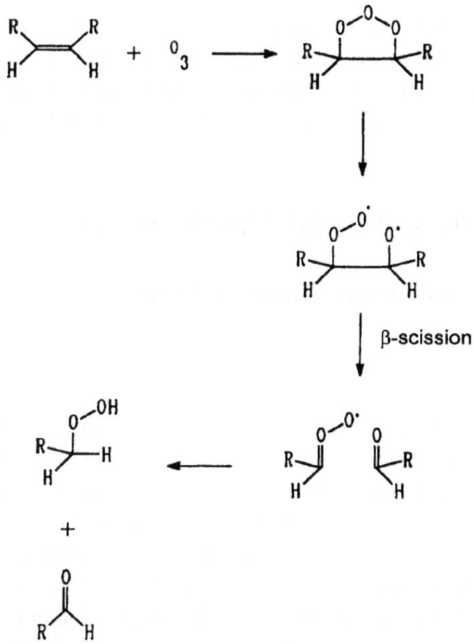

FIGURE 6. β-Scission of the Criegee ozonide to form the peroxyl radical.

rats is 25 to 85% (Ewetz, 1993). Absorption of inhaled $\cdot NO_2$ in the lungs of rhesus monkeys is 31 to 50% (Ewetz, 1993). The nasopharyngeal region is an important barrier to $\cdot NO_2$.

Slightly more $\cdot NO_2$ than O_3 is deposited in the airways because of the greater aqueous solubility of $\cdot NO_2$ (American Thoracic Society, 1996). Dosimetry models predict that $\cdot NO_2$ is deposited evenly in the airways with increased amounts deposited in the terminal bronchioli of human and animal lungs (Sandström, 1995). Little $\cdot NO_2$ is predicted to be deposited in the alveoli during resting breathing unless very high levels are inhaled. Both the percentage of $\cdot NO_2$ absorbed by the lower respiratory tract and the total uptake of $\cdot NO_2$ by the respiratory tract are estimated to increase in exercising humans (Overton, 1984).

The absorption of $\cdot NO_2$, like that of O_3, occurs by chemical reaction in the lung lining fluid (Postlethwait *et al.*, 1995) even though $\cdot NO_2$ is less reactive than O_3. The antioxidants GSH and ascorbic acid may be the compounds that drive $\cdot NO_2$ absorption whereas unsaturated fatty acids may be responsible for no greater than 20% of the absorption (Postlethwait *et al.*, 1995). As noted above, $\cdot NO_2$ is hydrated as the dimer producing equal amounts of nitrite and nitrate. It is not surprising that much more nitrate than nitrite has been found in urine (Ewetz, 1993).

In summary, the dose (defined as the moles of $\cdot NO_2$ or O_3 reaching the lung cells) varies throughout the respiratory tract even though both gases are highly reactive. Much of the $\cdot NO_2$ and O_3 appear to be destroyed by the protective fluid layer lining the lung, in which ascorbic acid and GSH normally are found. The mucus forms a viscous layer that is continually renewed and excreted after it is moved to the mouth and swallowed. The amounts of $\cdot NO_2$ and O_3 required for biological effects are very small compared with the exposure concentrations. These conclusions reflect the need to conduct experiments at low relevant levels of both gases.

3. MOLECULAR MECHANISMS OF TOXIC ACTION

3.1. Initiation of Peroxidation of Membrane Lipids

3.1.1. Ozone

Initiation of peroxidation of membrane lipids by O_3 and $\cdot NO_2$ is an attractive mechanism of toxic action (Pryor, 1994; Menzel, 1976, 1970; Roehm *et al.*, 1971b). O_3 generates free radicals both *in vitro* and *in vivo*. The observation that vitamin E (α-tocopherol) prevents O_3-initiated lipid peroxidation is evidence that O_3 exposure leads to the formation of free radicals. Vitamin E is the single most potent protective agent against $\cdot NO_2$ and O_3 toxicity and is more effective with $\cdot NO_2$ than with O_3 (Menzel *et al.*, 1972; Roehm *et al.*, 1971a). Probably this reflects the decomposition of ozonation products into cytotoxic compounds that are not produced by $\cdot NO_2$. O_3 is inherently more toxic than $\cdot NO_2$.

Protection by vitamin E against O_3 and $\cdot NO_2$ damage has not been demonstrated in humans as yet because no long-term studies have been carried out. Experimental animal studies show that vitamin E supplementation reduces the lifetime mortality in mice exposed to O_3 (Donovan *et al.*, 1977) and rats exposed to O_3 and $\cdot NO_2$ (Roehm *et al.*, 1971a). The reduction in mortality of the supplemented animals is observed in the last third of the life spans of the animals. Few noninvasive measures of the effects of O_3 or $\cdot NO_2$ exposure have been developed. Most such studies in humans have relied on pulmonary function tests, which are crude, imprecise, and difficult to replicate even in the same subject. Thus, a lifetime clinical trial will be needed to demonstrate the efficacy of vitamin E supplementation for prevention of injury caused by O_3 and $\cdot NO_2$.

Vitamin E protects by scavenging peroxyl radicals produced from O_3-induced peroxidation rather than by reacting directly with O_3 (Pryor, 1994). Vitamin E and olefins have similar rates of reactivity with O_3. Nevertheless, in lung lining fluid and cellular membranes, the greater amount of unsaturated fatty acids compared with that of vitamin E (less than 1–2 mole% as α-tocopherol in synthetic bilayers; see Shoaf *et al.*, 1989b) implies that much more O_3 reacts with olefins than with vitamin E. Bilayers enriched with vitamin E at greater than 2 mole% are physically unstable and thus the amount of vitamin E intercalated in the membrane is limited (Shoaf *et al.*, 1989b). In contrast to vitamin E, vitamin C and thiols such as GSH react sufficiently fast with O_3 to prevent the oxidant from reacting with critical cellular target molecules.

The fate of O_3 and it biological effects can be explained via the Criegee mechanism for the addition of O_3 to unsaturated fatty acids (Roehm *et al.*, 1971b). In the hypothesis proposed by Menzel (1970), the addition of O_3 results in a 1,2,3-trioxolane (Figure 6), which is the same as Criegee's original hypothesis. Pryor (1994) also begins with this initial ozonide but proposes two mechanisms that are involved in the production of radicals by O_3. In the first, a carbon-centered radical is separated from an olefin when O_3 reacts with the olefin to form a 1,2,3-trioxolane, which generates a diradical by O–O bond homolysis. The diradical then breaks down into alkyl and peroxyl radicals following β-scission. This conclusion was based on the findings that the yield of radicals is uninfluenced by water, iron, or hydrogen peroxide. In the second mechanism of radical production by O_3, the oxidant gas reacts with electron donors such as GSH or its anion to produce the thiyl radical and the O_3 radical anion. The O_3 radical anion then reacts with a proton to form the hydroxyl ion and molecular oxygen. The lung lining fluid or cellular membranes contain olefinic compounds such as unsaturated fatty acids, cholesterol, and tryptophan, as well as GSH, phenolic compounds, and vitamin C.

Radicals are produced in roughly 10% of reactions of O_3 with compounds in the lung (Pryor, 1994). More commonly, a carbonyl oxide and a carbonyl compound are formed either directly from the 1,2,3-trioxolane or following the generation of the diradical. In a lipophilic environment such as a cell membrane, the carbonyl oxide and the carbonyl compound reunite to form a Criegee ozonide. This pathway is the same as the original Criegee mechanism and is that proposed by Menzel. In other words, the intermediates from the 1,2,3-trioxolane are kept together by the solvent effects of the highly ordered lipid bilayer.

In an aqueous medium, water adds to the carbonyl compound to form a hydroxyhydroperoxide. Hydroxyhydroperoxides can produce hydrogen peroxide and aldehydes. Fatty acid ozonides also can decompose to aldehydes. Almost all aldehydes generated from monounsaturated fatty acids by O_3 must come from ozonation because monounsaturated fatty acids do not autoxidize readily. Aldehydes from the ozonation of olefinic fatty acids have been detected in both rats and humans (Pryor, 1994). Nevertheless, Pryor (1994) proposes that radical- and non-radical-mediated pathways may cause approximately equivalent amounts of lung damage by O_3 because one radical, via the autocatalytic nature of the peroxidative chain reaction mechanism, can cause lipid peroxidation of numerous unsaturated fatty acids. This conclusion is, however, in opposition to the known biological result that large doses of vitamin E prevent both $\cdot NO_2$ and O_3 toxicity (Menzel *et al.*, 1972). The production of peroxyl radicals is important to the biological outcome because vitamin E reacts preferentially with peroxyl radicals.

Ozonides are not benign compounds and initiate peroxidation of polyunsaturated fatty acids and isolated cell membranes (Menzel, unreported results). Vitamin E prevents the ozonide-initiated peroxidation (Menzel, unreported experiments). Because vitamin E as α-tocopherol reacts mostly with peroxyl radicals and poorly with hydrogen peroxide, it seems likely that the typical inhibited autoxidation reaction kinetics are related to peroxyl radicals.

Pryor (1992) has suggested that the highly reactive O_3 molecule is consumed in the lung lining fluid or, where the fluid is missing or scant, in the membranes of the lung epithelial cells. Therefore, cellular changes resulting from O_3 exposures could depend on by-product molecules. Pryor (1992) has predicted that O_3 itself is unable to permeate through fluid thicker than about 0.1 μm. As reported by Miller (1995), Robert Mercer has estimated that only about 2% of the pulmonary region has a surfactant layer that is between approximately 0.1 and 0.2 μm thick whereas about 98% of the region has a surfactant layer on average approximately 0.02 μm thick. Miller (1995) emphasizes that the lung lining fluid varies considerably in thickness throughout the respiratory tract and may be patchy in distal conducting airways. In mathematical models of O_3 deposition, simulation of patchy or no lung lining fluid enhances O_3 reaction with lung cell lipids.

Peroxidation initiated by O_3 is a complex event (see Menzel, 1970). Endoperoxides as well as hydroperoxides are formed during the O_3-initiated peroxidation of thin layers of polyunsaturated fatty acids (Roehm *et al.*, 1971b). Endoperoxides formed during ozonation of arachidonic acid resemble prostaglandin H_2 (Roycroft *et al.*, 1977). Probably endoperoxide formation is favored by the nature of the oxygen addition (Menzel *et al.*, 1976). The question is then still open whether O_3 itself or some ozonation product is responsible for the observed toxicity.

3.1.2. Nitrogen Dioxide

Like the molecular mechanisms of toxic action of O_3, the mechanisms of toxicity of $\cdot NO_2$ are incompletely understood. The results of a number of investigations suggest that pulmonary damage caused by $\cdot NO_2$ inhalation results from lipid peroxidation of cellular membranes (Elsayed, 1994; Moldéus, 1993). For example, in rats exposed to 0.04 to 4 ppm $\cdot NO_2$ for 9 to 27 months, lipid peroxidation as determined by ethane exhalation increased in a dose-dependent manner (Sagai and Ichinose, 1987).

Hydrogen abstraction has been proposed as the mechanism of reaction of low concentrations of $\cdot NO_2$ with fatty acids that have doubly allylic hydrogen atoms (Gallon and Pryor, 1994). Postlethwait *et al.* (1995) suggested that inhaled $\cdot NO_2$ is likely to react with components of the lung lining fluid before it reaches the lung surface. These investigators propose that free radicals produced from the antioxidants GSH and vitamin C may initiate $\cdot NO_2$ cytotoxicity through lipid peroxidation. Lipid peroxidation clearly occurs when lipid bilayers are exposed to $\cdot NO_2$ (Shoaf *et al.*, 1989a). The peroxidation is inhibited by vitamin E in the bilayer phase. Vitamin C sealed in the liquid phase within the lipid bilayers is readily oxidized, which suggests that $\cdot NO_2$ easily penetrates the bilayer and, therefore, probably the cell membrane.

The mechanism of $\cdot NO_2$ free-radical-mediated toxicity appears simpler than that of O_3. $\cdot NO_2$ is a classical initiator generating carbon-centered free radicals, which start the chain reaction within lipids. Vitamin C occurs in the lung lining fluids of humans and experimental animals. There is controversy over the amount of vitamin C oxidized by atmospheres containing nitrogen oxides including $\cdot NO_2$. Cigarette smoke nitric oxide (the concentration of which can be as high as 100 ppm) is oxidized to $\cdot NO_2$ with a half time of 8 to 60 min (Norman and Keith, 1965). Some studies report a decline in lung lavage fluid content of ascorbate in smokers, whereas others do not. It is not at all clear that prophylactic treatment with either vitamin E or C protects smokers against noncancerous toxicity (exacerbation of asthma and bronchitis and initiation of emphysema) nor from cancer produced by substances contained in cigarette smoke.

3.2. Oxidation of Membrane Proteins

An alternate hypothesis to the lipid peroxidation hypothesis of the biochemical basis of O_3 and $\cdot NO_2$ toxicity is that the gases oxidize proteins. Freeman and Mudd (1981) proposed that the primary macromolecules oxidized within cell membranes by O_3 are proteins because no oxidation of membrane lipids was detected. More recent studies clearly show that O_3 and $\cdot NO_2$ initiate oxidation of membrane fatty acid as discussed previously.

Oxidation of intracellular GSH to GSSG in human red blood cells (RBCs) treated with O_3 was reported as evidence that O_3 is able to cross cell membranes. Mixed disulfides with both cytoplasmic and membrane proteins, as well as GSSG, were thought to be formed following O_3 treatment although no effects on hemoglobin were observed. This is in contrast to the strong reaction of synthetic ozonides to cause Heinz bodies in human RBCs (Menzel *et al.*, 1975a). Heinz bodies are mixed disulfides between hemoglobin and membrane-linked proteins. Recent methods have now been developed to analyze for mixed disulfides in oxidatively stressed cells; readers should consult the review by Reed (1990) for details on the formation of mixed disulfides. There is some question about the results reported by Freeman and Mudd (1981) in light of current methods. In more recent studies, Banerjee and Mudd (1992) found that only amino acid residues external to the membrane are oxidized by O_3 in RBC ghosts.

In protein-containing reverse micelles, O_3 caused oxidation of tryptophan residues possibly forming a Criegee ozonide (Pryor and Uppu, 1993). Only about 0.07 mole of hydrogen peroxide per mole of tryptophan was formed when lysozyme was treated with O_3. Hydrogen peroxide is not destroyed during ozonation, nor is it consumed by protein, and protein structure has no effect on the reaction. Results of experiments of Uppu *et al.* (1995) with human blood cell membranes indicated that proteins and unsaturated lipids react simultaneously and competitively when limiting amounts of O_3 are present.

In sum, oxidation, either directly by O_3 and $\cdot NO_2$ or indirectly by peroxides or hydrogen peroxide generated particularly by O_3, remains a viable and stimulating hypothesis by Mudd. More research is needed into this hypothesis using modern biochemical methods.

3.3. Mixtures of O_3 and $\cdot NO_2$

$\cdot NO_2$ and O_3 can act synergistically to cause pulmonary injury when animals are exposed to mixtures of the two gases (Elsayed, 1994; Last *et al.*, 1994b). Chemical species of greater toxicity than $\cdot NO_2$ and O_3 may be formed (Elsayed, 1994). Despite the fact that O_3 and $\cdot NO_2$ have been found to have synergistic effects on morbidity and mortality in rodents, the oxidants acted antagonistically on components in human plasma (O'Neill *et al.*, 1995). Previously, the individual gases had been shown to consume antioxidants and alter proteins and lipids in plasma (O'Neill *et al.*, 1995).

4. POSSIBLE MEDIATORS OF O_3 AND $\cdot NO_2$ TOXICITY

4.1. Ozonides

The toxic effects of O_3 may be mediated by fatty acid ozonides generated when O_3 reacts with polyunsaturated fatty acid moieties of membrane phospholipids. The *in vitro* cytotoxicity of a model ozonide, methyl linoleate-9,10-ozonide, is three times greater than that of a model peroxidative agent, cumene hydroperoxide, when phagocytizing ability of rat alveolar macrophages is used as an endpoint (de Vries *et al.*, 1994). Vitamins

E and C were protective against the effect of the ozonide whereas GSH depletion increased the effect of the ozonide, results similar to those seen with O_3 treatment.

The detoxification of polyunsaturated fatty acid ozonides with GSH can be catalyzed by glutathione *S*-transferases (Hempenius *et al.*, 1992; Vos *et al.*, 1989; Rietjens *et al.*, 1987). Based on semiempirical molecular orbital calculations, Hempenius *et al.* (1992) proposed that GSH interacts with one of the carbon atoms of the ozonide ring. This suggested mechanism differs from that of the glutathione *S*-transferase-mediated reaction of GSH with one of the oxygen atoms of a hydroperoxide (Hempenius *et al.*, 1992). The role of glutathione *S*-transferases in O_3 toxicity remains to be investigated. Humans express a number of forms of glutathione *S*-transferases and gene deletion occurs in the Mu class glutathione *S*-transferases. Deletion of the glutathione *S*-transferase Mu gene has been associated with increased risk of lung cancer in cigarette smokers (Nazar-Stewart *et al.*, 1993).

4.2. 4-Hydroxynonenal

Other possible mediators of the toxic effects associated with exposure to O_3 and $\cdot NO_2$ are 4-hydroxyalkenals. These reactive aldehydes are produced during lipid peroxidation. Because 4-hydroxyalkenals are amphipathic, the chemicals can be present both in the cytosol and in membranes (Danielson *et al.*, 1987). Aldehydes have longer half-lives than do free radicals and could therefore diffuse to reach distant targets (Esterbauer *et al.*, 1991).

The most cytotoxic aldehyde produced from peroxidation of rat liver microsomal lipid is 4-hydroxynonenal, which is generated in a relatively large amount (Esterbauer *et al.*, 1991). 4-Hydroxynonenal is produced from the ω-6-polyunsaturated fatty acids linoleic acid and arachidonic acid (Esterbauer *et al.*, 1991). Strohmaier *et al.* (1995) suggest that 4-hydroxynonenal is a specific biomarker of lipid peroxidation.

A physiological steady-state level of 4-hydroxynonenal ranging from 0.1 to 1.4 μM occurs in venous blood plasma from human subjects (Strohmaier *et al.*, 1995). The concentrations are of the same range as those that block proliferation of endothelial cells and fibroblasts *in vitro*. These findings suggest that lipid peroxidation and specifically 4-hydroxynonenal are involved in the regulation of cell division *in vivo*.

Inhibition of cell proliferation by concentrations of 4-hydroxyalkenals that had no effect on cell viability occurs with a number of cell types *in vitro* (Esterbauer *et al.*, 1991). Treatment of rat neutrophils with 4-hydroxynonenal concentrations less than 0.1 μM (levels in the physiological range) stimulated chemotaxis, modulated adenylate cyclase activity, and increased the activity of phospholipase C, an enzyme important in cell membrane signal transduction by G-proteins (Esterbauer *et al.*, 1991). In isolated rat hepatocytes, micromolar concentrations of 4-hydroxynonenal increase the intracellular calcium ion level probably through phospholipase C (Carini *et al.*, 1996).

4-Hydroxynonenal may regulate phospholipase A_2 activity to increase the release of arachidonic acid from platelets (Selley *et al.*, 1988). Concentrations of 4-hydroxynonenal less than 100 μM potentiate aggregation and stimulate thromboxane A_2 production in platelets (Selley *et al.*, 1988). In cultured bovine aortic endothelial cells treated with 10 μM 4-hydroxynonenal, prostacyclin synthesis was markedly inhibited while thromboxane A_2 synthesis was unchanged (Moldovan *et al.*, 1994). Therefore, the balance between thromboxane A_2 and prostacyclin production in aortic endothelial cells is altered by 4-hydroxynonenal. Similar effects might occur in the lung on O_3 or $\cdot NO_2$ exposure because of the generation of 4-hydroxynonenal. Platelet aggregation in the capillary and changes in vessel caliber could result in the edema and fluid accumulation in the lung observed with O_3 primarily.

Thiols including GSH conjugate readily with 4-hydroxyalkenals at physiological pH to form very stable products (Esterbauer *et al.*, 1975). In addition to cysteine, the 4-hydroxyalkenals react with other amino acid residues of proteins (Esterbauer *et al.*, 1991). Glutathione *S*-transferases increase the rates of GSH conjugation with 4-hydroxyalkenals about 300- to 600-fold (Alin *et al.*, 1985). The relative activity of glutathione *S*-transferases with 4-hydroxyalkenals is M1–1 > A1–1 > P1–1 (Berhane *et al.*, 1994).

4-Hydroxynonenal protein adducts are increased in lavage fluid proteins and alveolar macrophages from subjects exposed to 0.25 ppm O_3 (Hamilton *et al.*, 1996), which suggests that O_3 exposure does produce biologically active alkenals. Inflammation and cytotoxicity following O_3 exposure could result from the release of aldehydes, including 4-hydroxynonenal, during O_3-initiated lipid peroxidation.

Aldehydes including 4-hydroxynonenal are released from rat alveolar macrophages treated with low $\cdot NO_2$ concentrations (Robison *et al.*, 1995). Alveolar macrophage phagocytic and bactericidal activity are decreased by O_3 exposure. Aldehydes may be responsible for this inhibition. Apoptosis is induced by 4-hydroxynonenal in murine alveolar macrophages (Li *et al.*, 1996). Low concentrations (10 to 25 μM) of 4-hydroxynonenal also induce the stress protein heme oxygenase 1 in the alveolar macrophages. 4-Hydroxynonenal could mediate apoptosis that results from oxidative stress.

In summary, aldehydes, particularly 4-hydroxynonenal, are likely to play an important role in the cytotoxicity, increased permeability of the lung, and recruitment of inflammatory cells following O_3 exposure. Additional research is needed to sort out the relationship between the effects noted in isolated systems and those in the human lung under chronic exposure to O_3.

5. PROTEIN INDUCTION

5.1. Ozone

Two-dimensional gel electrophoresis of proteins from alveolar macrophages obtained from humans exposed to 0.4 ppm O_3 for 2 hr shows that the rate of synthesis had increased for 45 proteins but decreased for 78 proteins (Devlin and Koren, 1990). Levels of mRNA for clotting factors (tissue factor and factor VII) increased in alveolar macrophages of humans exposed to O_3 under these same conditions (McGee *et al.*, 1990). A 45-kDa protein of unknown function is induced by O_3 in the BEAS-2B human bronchial epithelial cell line in a dose-related manner (Sun *et al.*, 1994). Extracellular GSH inhibited the induction.

In rat lung, O_3 induces the expression of macrophage inflammatory protein-2 (Haddad *et al.*, 1995). Heat shock protein 70 increased 277% in rat lung following chronic inhalation of 0.15 ppm O_3 (Wong *et al.*, 1996).

O_3 exposure (0.1 ppm) increases the expression of interleukin (IL) 8 in human A549 alveolar type 2-like cells (Jaspers *et al.*, 1997). Treatment with hydrogen peroxide, to simulate general oxidative stress, did not increase IL-8 mRNA levels. The DNA-binding activity of the transcription factors nuclear factor (NF)-κB, NF-IL-6, and activator protein-1 (AP-1) increased with the O_3 exposure. NF-κB, NF-IL-6, and perhaps AP-1 regulate the expression of IL-8. Together with the observations on protein levels, these data suggest that O_3 alters gene expression, perhaps via transcription factors.

The antioxidant responsive element in the regulatory region of rat glutathione *S*-transferase class Alpha Ya subunit gene is responsive to hydrogen peroxide (Rushmore *et al.*, 1991). Although an antioxidant responsive element has not been located in the regulatory regions of human glutathione *S*-transferase class Alpha subunits, perhaps human class Alpha glutathione *S*-transferases are induced by hydrogen peroxide produced during lipid peroxidation following O_3 exposure. Glutathione *S*-transferase A1-1, which is active with 4-hydroxyalkenals (Berhane *et al.*, 1994), could play a role in cellular detoxification of lipid peroxidation products.

5.2. Nitrogen Dioxide

Most studies of the effects of $\cdot NO_2$ on proteins have analyzed changes in enzyme activities (see Section 6.8). Only one recent report of proteins induced by $\cdot NO_2$ could be found in the literature. Increased expression of the adhesion molecule Mac-1 occurs in granulocytes following exposure of subjects with mild asthma to 0.26 ppm $\cdot NO_2$ for 30 min with intermittent exercise (Strand *et al.*, 1996). This study may indicate a link between air pollution and the near-epidemic increases in asthma. Much needs to be done to make a closely connected case between $\cdot NO_2$ action and effect on inflammation.

6. RELATING MECHANISMS TO TOXIC EFFECTS

6.1. Lung Inflammation

Mechanisms involved in the development of lung inflammation following O_3 exposure are only beginning to be understood. Platelet-activating factor, as well as other mediators from epithelial cells, may contribute to O_3 toxicity (Wright *et al.*, 1994). Treatment of guinea pig tracheal epithelial cells in primary culture with 0.05 to 1.0 ppm O_3 for 1 hr increased the release of platelet activating factor, perhaps through a mechanism that involves the activation of phospholipases A_2, C, and D.

Hazbun *et al.* (1993) proposed that the mechanism of action of O_3 for lung inflammation and damage after short-term exposure may be via inactivation by oxidation of neutral endopeptidase in the airways, which results in increased levels and activity of substance P. In support of their hypothesis, acute exposure of exercising human subjects to 0.25 ppm O_3 increases levels in airway lavage of substance P and of 8-epi-prostaglandin $F_{2\alpha}$.

As discussed previously, O_3 induces mRNA expression of macrophage inflammatory protein-2 in rat lung, which may recruit neutrophils to the lung (Haddad *et al.*, 1995). Dexamethasone prevents the induction as well as the influx of neutrophils into the lung. Another neutrophil chemotactic cytokine that was previously mentioned to be induced by O_3 is IL-8 (Jaspers *et al.*, 1997). IL-8 also may mediate lung inflammation. The roles of cytokines in O_3 toxicity beyond acute inflammation are unclear but interesting.

6.2. Lung Function

Tsukagoshi *et al.* (1995) suggested that superoxide anions liberated from airway inflammatory cells may contribute to airway hyperresponsiveness. O_3-induced airway responsiveness to bradykinin but not acetylcholine was blocked by apocynin administered prior to exposure of Brown Norway rats to 3 ppm O_3 for 3 or 6 hr. Apocynin is an inhibitor of superoxide anion-generating NADPH oxidase. The antioxidants allopurinol and deferoxamine blocked O_3-induced airway responsiveness to both bradykinin and acetylcholine. Neither neutral endopeptidase activity nor the number of bradykinin receptors in the lung were changed significantly by O_3 exposure. Unfortunately, the exposure concentrations were unrealistic and difficult to relate to humans.

6.3. Lung Structure

Lung protein metabolism has been examined to investigate how structural changes in the lung are caused by O_3 and $\cdot NO_2$. Winters *et al.* (1994) suggested that O_3 may cause lung pathology by directly altering elastin and by enhancing its degradation. O_3, unlike $\cdot NO_2$, caused fragmentation of elastin *in vitro* and increased its susceptibility to proteolysis. Furthermore, vitamin C, EDTA, and uric acid protected O_3-treated elastin from proteolysis.

Rats chronically exposed to O_3 at 0.5 or 1.0 ppm developed mild to moderate lung fibrosis as determined histologically (Last *et al.*, 1994a). In female rats exposed to these levels, the increased fibrotic collagen was measurable biochemically. No effect on collagen deposition was observed in the small group of animals exposed to 0.12 ppm O_3.

Exposure of human subjects to 3 to 4 ppm $\cdot NO_2$ for 3 hr decreased activity in bronchoalveolar lavage fluid of α-1-protease inhibitor, important in blocking leukocyte elastase activity (Mohsenin and Gee, 1987). Elastase can cause proteolytic damage in the lung. Nevertheless, $\cdot NO_2$ at a tenfold higher concentration failed to affect α-1-protease inhibitor activity *in vitro* (Mohsenin and Gee, 1989). The investigators examined whether $\cdot NO_2$ might cause its effects via lipid peroxidation. The *in vitro* activity of the α-1-protease inhibitor was reduced by $\cdot NO_2$ treatment in the presence of linoleic acid but not stearic acid. In addition, GSH and glutathione peroxidase were effective in preventing the inhibition of α-1-protease inhibitor activity by $\cdot NO_2$ and linoleic acid treatment whereas α-tocopherol was partially effective. Because dithiothreitol and methionine sulfoxide peptide reductase reactivated α-1-protease inhibitor that was oxidized in the presence of $\cdot NO_2$ and linoleic acid, the investigators suggested that a critical methionine residue may have been oxidized. Nonetheless, Johnson *et al.* (1990) failed to find an effect of $\cdot NO_2$ on α-1-protease inhibitor activity on the lungs of normal human subjects exposed short-term to low levels of $\cdot NO_2$.

6.4. Host Defenses

The result of O_3 and $\cdot NO_2$ toxicity using the infectivity method, one of the earliest types of mechanistic studies of O_3 and $\cdot NO_2$, was the increased susceptibility of experimental animals to pathogens delivered as aerosols (Coffin and Gardner, 1972). Some O_3 and $\cdot NO_2$ effects have been traced to inhibition of alveolar macrophage defense functions (see U.S. EPA, 1996b; U.S. EPA, 1993b, for details). Most rodents are maintained as specific respiratory-pathogen-free stocks whereas humans frequently have episodes of infections. The inflammatory response to O_3 and $\cdot NO_2$ could be related to the generation of cytokines secondary to the primary oxidative damage. The effects of O_3 on alveolar macrophages, however, appear specific. Alveolar macrophages isolated from rats exposed to 0.8 ppm showed increased chemotaxis and greater adhesion to unexposed lung type II cells (Bhalla, 1996). Antibodies to the leukocyte adhesion protein CD11b or the epithelial adhesion molecule ICAM-1 reversed the adhesion of O_3-treated cells. O_3 effects may be specific for different cell-to-cell interactions that, in turn, could modulate the infiltration of proteins into the airways.

The effects of both $\cdot NO_2$ and O_3 on host defenses could be initiated through free-radical mechanisms involving the cell membrane or gene expression. The time between exposure and appearance of edema is about 18–20 hr. It does not seem likely that free radicals could survive that long. More stable intermediates including ozonides could be involved. The observations on macrophages and the function of the mucociliary escalator illustrate the huge time difference between the initial reaction of unsaturated fatty acids in the cell membrane with O_3 and $\cdot NO_2$ and the biological response.

Chronic exposure of rats to 0.15 ppm O_3 caused a significant delay in particle clearance in the upper respiratory tract (Mannix *et al.*, 1996). Additionally, a trend toward accelerated clearance in the deep lung of the O_3-exposed rats was observed. The results of the investigators match those observed following acute exposures to higher levels of O_3 (about 0.8 ppm). Alveolar macrophages are sensitive to $\cdot NO_2$ (Moldéus, 1993). In experimental animals, the alveolar macrophage functions of phagocytic capacity and mobility are diminished at $\cdot NO_2$ levels as low as 0.3 ppm. Both O_3 and $\cdot NO_2$ caused immature macrophages to accumulate in the airways of rats exposed for 11 weeks to 0.2 ppm O_3 or 4.0 ppm $\cdot NO_2$ (Mochitate *et al.*, 1992).

6.5. Lung Permeability

Airway permeability is affected by exposure to less than 1.0 ppm O_3. Bhalla *et al.* (1992) suggest that inflammatory cells are involved in causing increased airway permeability following exposure of rats to O_3. These investigators demonstrated that three agents that affect leukocytes or their products—cyclophosphamide, the leukotriene D4 antagonist FPL 55712, and indomethacin—lessened airway permeability induced by exposure of rats to 0.8 ppm O_3 for 2 hr. Furthermore, IL-1 and tumor necrosis factor-α, cytokines involved in producing changes in epithelial permeability, appear to mediate macrophage activation and adherence to epithelial cells following an acute exposure of rats to O_3 (Pearson and Bhalla, 1997; Bhalla *et al.*, 1996).

O_3 may enhance airway permeability by impairing cytoskeletal function in epithelial cells (Yu *et al.*, 1994). Increases in paracellular permeability of canine bronchial epithelial cells following *in vitro* exposure to 0.2, 0.5, or 0.8 ppm O_3 for 3 hr was inhibited by vitamin A, vitamin E, or the actin polymerizing agent phalloidin. At 18 hr after treatment with O_3, the permeability diminished in cells treated with the lower concentrations of O_3 but increased in cells treated with the highest concentration.

Rasmussen *et al.* (1992) recommend monitoring subjects exposed to $\cdot NO_2$ for prolonged durations in order to observe delayed responses. Normal human subjects at rest exposed to 2.3 ppm $\cdot NO_2$ for 5 hr showed a reduction in alveolar permeability 11 hr after the beginning of the exposure.

The changes in airway permeability are important to the clinical outcome of exposure to O_3 and $\cdot NO_2$. The changes in airway permeability occur 18–20 hr after a single acute exposure. Because the permeability increases in a dose-response-related manner, it has been tempting to some to assume that these effects are the underlying mechanism of action. On the other hand, the inflammatory response and alteration in airway permeability may be far down the chain of events. Important though these effects may be, it is likely that the initial event is peroxidation of the cell membrane. Unfortunately, no group has made a systematic time study of the wide variety of effects of $\cdot NO_2$ or O_3 to determine the order of events.

6.6. Membrane Fluidity

Rietjens *et al.* (1987) concluded that increased membrane fluidity in rat alveolar macrophages enriched with polyunsaturated fatty acids did not cause the enhanced *in vitro* sensitivity of the cells to O_3 and $\cdot NO_2$ that was observed. Cells had been preincubated with arachidonic acid complexed to bovine serum albumin. The increased sensitivity of the macrophages occurred at a concentration of the arachidonic acid higher than the concentration that had caused a maximum increase in membrane fluidity. Nevertheless, the investigators stated that their results support a role for lipid oxidation in the toxicity of O_3 and $\cdot NO_2$.

One characteristic of free-radical-mediated reactions is polymerization of the monomers to high polymers. Annihilation of two radicals is kinetically favored and results in dimers, trimers, and so forth. If lipid peroxidation involving peroxyl radicals occurs shortly after exposure to O_3 and $\cdot NO_2$, then the fluidity of the plasma membrane ought to be decreased. Transmembrane proteins should not be able to diffuse as readily as in cells not exposed to O_3. Nonetheless, changes in membrane fluidity should be reinvestigated using more sensitive and quantitative methods. A number of fluorescent reporter groups are available for use.

6.7. Arachidonic Acid Metabolism

Augmentation of metabolism of eicosanoids (arachidonic acid metabolites) has been observed following exposure of lung cells to O_3 or $\cdot NO_2$. Eicosanoids can cause changes in airway inflammation, permeability, and smooth-muscle reactivity. Acute exposure to O_3 or $\cdot NO_2$ modified pulmonary arachidonic acid metabolism in rabbits (Schlesinger *et al.*, 1990). Radioimmunoassay of bronchopulmonary lavage following exposure of rabbits to 1 ppm O_3 for 2 hr revealed increased amounts of prostaglandins E_2 and $F_{2\alpha}$. Whereas thromboxane B_2 levels were elevated after exposure to 1 ppm $\cdot NO_2$, the levels were decreased after 3 and 10 ppm $\cdot NO_2$. Levels of 6-keto-prostaglandin $F_{1\alpha}$ were diminished following exposure to 10 ppm $\cdot NO_2$. Exposure to a mixture of 0.3 ppm O_3 and 3 ppm $\cdot NO_2$ led to synergistic increases in prostaglandins E_2 and $F_{2\alpha}$.

Structure–activity relationships of O_3-fatty acid degradation products on eicosanoid metabolism in cultured human airway epithelial cells were examined by Leikauf *et al.*

(1993). The stimulatory effect of the aldehydes and hydroxyhydroperoxides (three-, six-, and nine-carbon molecules in each series) increased with greater chain length. The hydroxyhydroperoxides were more potent than the corresponding aldehydes.

6.8. Enzyme Activities

A unique profile of induced antioxidant enzymes in the lung seems to be obtained following exposure to a particular oxidant (Quinlan *et al.*, 1994) (see Part II of this volume). Nevertheless, manganese superoxide dismutase is induced in pulmonary cells by a number of oxidants that elicit inflammation or pulmonary fibrosis (Quinlan *et al.*, 1994).

Antioxidant enzyme activities increase following short-term exposure of laboratory animal to O_3. In rats exposed to 0.7 ppm O_3 for 5 days, the antioxidant enzymes copper-zinc superoxide dismutase, manganese superoxide dismutase, catalase, and glutathione peroxidase are induced in the lung (Rahman *et al.*, 1991). The GSH antioxidant system of bronchoalveolar lavage fluid cells is stimulated by exposure of female rats to 0.8 ppm O_3 for 6 hr daily for 3 or 7 days (Boehme *et al.*, 1992).

Plopper *et al.* (1994) determined that long-term exposure of rats to O_3 causes changes in antioxidant enzyme activities in the lung that are site specific. The activities of glutathione S-transferase, glutathione peroxidase, and superoxide dismutase, normalized for DNA content, were increased significantly in the distal bronchioles. Changes occurred in the activities of some of these enzymes in distal trachea and major and minor daughter bronchi, but no changes were found in lobar bronchi and parenchyma.

Sagai and Ichinose (1987) studied the relationship between lipid peroxidation and the activities of antioxidative enzymes after acute and chronic exposure of rats to $\cdot NO_2$. They found that the activities of antioxidative enzymes increased and then slowly decreased. Lipid peroxides also increased, but after returning to the control level, the peroxides increased again slowly. These results may be related to inhibition of antioxidative enzymes with chronic exposure to $\cdot NO_2$. Lipid peroxidation is likely even as these enzymes are synthesized.

Glycolytic enzyme activities in the lung also are increased by oxidant gas inhalation. Following an 11-week exposure of rats to 0.2 ppm O_3, the specific activities of glucose-6-phosphate dehydrogenase, glutathione peroxidase, pyruvate kinase, and hexokinase were increased 40 to 70% in alveolar macrophages (Mochitate *et al.*, 1992). Environmentally relevant levels of O_3 increase xenobiotic metabolism in the lungs of rats (Sindhu *et al.*, 1996). Lung microsomes from rats chronically exposed to 0.15 ppm O_3 showed a 72% increase in benzphetamine *N*-demethylase activity and a threefold increase in [^{3}H]benzo[*a*]pyrene metabolism compared with controls.

Again, no studies have been done on a molecular level to determine the time of induction and the relation of one enzyme to another in the process. No studies of mRNA levels for the antioxidative enzymes have appeared, for example. Controversy surrounds the role of free radicals as mediators for the GSH-dependent antioxidative enzymes. It is not clear if a free-radical responsive element exists in humans for regulation of glutathione *S*-transferases. In light of the destruction of ozonides by glutathione *S*-transferases, this area of research would be a fruitful one. At this stage, there is not enough information to know if there are any interrelations between the enzymes or if a common O_3 or $\cdot NO_2$ intermediate is responsible for activation of all of the enzymes.

7. ROLES OF ANTIOXIDANTS

Reactive oxygen species can interact with and alter the functions of polyunsaturated cell membrane lipids, DNA, and proteins with thiol groups. Nevertheless, free radicals also play a role in normal physiological functions. Antioxidants are discussed here in terms of their functions with $\cdot NO_2$ and O_3 exposure. (See Chapter 14 for a comprehensive discussion of antioxidants.)

7.1. Ozone

GSH, present in the lung lining fluid of humans at approximately 0.4 mM (Cantin *et al.*, 1987), may be a direct O_3 scavenger. Thiols including GSH react with O_3 at greater rates than do olefins (Pryor *et al.*, 1984). Thiol groups are oxidized by O_3 to the sulfone and sulfoxide (Menzel, 1971). Only GSSG can be reduced by glutathione reductase. Depletion of intracellular GSH could be related to the inability of glutathione reductase to reduce these higher oxidation products back to GSH.

Using human plasma as a substitute for lung lining fluid, O_3 was found to react primarily with vitamin C and urate (Van der Vliet *et al.*, 1995). The reactions were more efficient at low (2 ppm) concentrations than at high (16 ppm) concentrations. Oxidative injury to plasma proteins and lipids, which was observed only after extended O_3 treatment, was not inhibited by GSH or dihydrolipoic acid. Neither did addition of these thiols (at ≤ 1 mM) diminish the depletion rates of vitamin C or urate. The investigators suggested that thiol supplementation may protect the lower respiratory tract from O_3 damage by enhancing reactive absorption of O_3 in the upper respiratory tract. However, the concentration of O_3 used far exceeds those likely to occur in humans breathing polluted air. The efficiency of scavenging O_3 may be quite different at low O_3 concentrations.

In rat alveolar macrophages *in vitro,* α-tocopherol was equally protective against O_3 and $\cdot NO_2$ (Rietjens *et al.*, 1986). Earlier studies with rats and mice found that the toxicity of $\cdot NO_2$ was more greatly reduced by vitamin E than was that of O_3 (see Section 3.1.1).

Pretreatment of rats with the antioxidant dimethylthiourea prevented increased airway permeability caused by exposure to 0.6 to 0.8 ppm O_3 for 2 hr (Bhalla, 1994). Vitamin C deficiency did not influence significantly lung permeability, lung function, or lung pathology following a 1-week recovery after a continuous 1-week exposure of guinea pigs to 0.2, 0.4, or 0.8 ppm O_3 (Kodavanti *et al.*, 1995a,b). In contrast to rats and mice, guinea pigs do not synthesize vitamin C and can be made deficient with diets free of vitamin C.

Taurine, which is present in leukocytes at high intracellular concentrations, has been proposed as a protective agent against oxidative damage from O_3 (Schuller-Levis *et al.*, 1994). Taurine traps hypochlorous acid and may block formation of nitrite and tumor necrosis factor-α, which have been shown to cause tissue damage.

7.2. Nitrogen Dioxide

Supplementation of human subjects with vitamins C and E for 1 month prior to acute exposure of up to 4 ppm $\cdot NO_2$ reduced the levels of conjugated dienes and prevented a reduction in elastase inhibitory capacity in alveolar lining fluid (Mohsenin, 1991). Conjugated dienes were the main lipid peroxidation products generated in the subjects when the $\cdot NO_2$ exposure was conducted without vitamin supplementation.

Tu *et al.* (1995) emphasize the cooperative roles of GSH and vitamin C in the protection of cells against the oxidant effects of $\cdot NO_2$. These investigators found that GSH depletion increased cytotoxicity caused by $\cdot NO_2$ in human umbilical vein endothelial cells and increased the rate of vitamin C consumption. Moreover, vitamin C completely protected even GSH-depleted cells from $\cdot NO_2$-dependent toxicity and significantly decreased GSH depletion by $\cdot NO_2$ in control cells (cells not depleted of GSH).

As stated earlier, $\cdot NO_2$ is a free radical. β-Carotene rapidly scavenges $\cdot NO_2$ as well as thiyl and sulfonyl radicals (Everett *et al.*, 1996). Different mechanisms are involved, however, for the various reactions. For the $\cdot NO_2$ radical, the mechanism involves electron abstraction to produce a radical cation. For the GSH thiyl radical, the mechanism is radical addition to produce an adduct radical. Sulfonyl radicals react with $\cdot NO_2$ by both mechanisms. The β-carotene radical cations and adduct radicals are stable and decay to nonradical products slowly.

8. SYSTEMIC EFFECTS OF O_3

8.1. Hematological Effects

A reduction in the RBC deformability was observed following exposure to 1.0 ppm for 4 hr (Morgan *et al.*, 1985). RBCs must maintain a concave disk conformation to traverse the capillaries and supply oxygen to the distal organs. The spleen carries out a careful surveillance process removing any defects in RBCs that might cause them to lodge in capillaries and not carry out their vital oxygenation of the tissues. Several measures of deformability have been developed, but the simplest is filtration of RBCs through membranes with a pore size just smaller than the diameter of the RBC. Exposure of human and mouse RBCs to O_3 causes the RBCs to become rigid and thus less able to deform under physiological pressures and pass through membranes having the same pore size as the fenestra of the spleen. Although Morgan *et al.* (1985) clearly demonstrated the occurrence of this rigidity, the mechanism is unknown. Speculatively, the O_3-induced rigidity could be related to cross-linking reactions in the plasma membrane or to disruption of the cytoskeleton of the RBCs. One should recall that cross-linking reactions are common between fatty acid alkyl free radicals. Commercially, vegetable oils unsuited for human consumption are converted to factus, a peroxidative polymer used in the dry-cleaning industry to absorb solvents following cleaning. A common product containing factus is the old-fashioned "art gum" eraser. The cross-linking reaction is only about 2–5 mole%, yet liquid vegetable oil is converted to a solid by this reaction.

Heinz body formation in mouse and human RBCs (Menzel *et al.*, 1975a) from O_3 and fatty acid ozonides is another example of the cross-linking initiated by O_3. Heinz bodies were first noted following the exposure of workers to aniline dye precursors. The

Heinz body is an insoluble aggregation of hemoglobin bound to the inner membrane of the RBC. The type of Heinz body can be revealed by redox dyes to determine the oxidation state of the heme. The important point is that ingestion of α-tocopheryl acetate (vitamin E) by the volunteers in the study prevented Heinz body formation in a dose–response manner (Menzel *et al.*, 1975b). Thus, vitamin E is clearly protective against major oxidative effects of O_3 by-products such as ozonides.

Margination has been noted in numerous studies of experimental animals exposed to O_3 and $\cdot NO_2$. Adhesion of cells through cell-to-cell interactions is mediated by membrane-bound adhesion molecules. The mechanism of activation of these molecules by O_3 is not clear. Platelets undergo aggregations via the prostaglandin cascade (Roycroft *et al.*, 1977). O_3 peroxidation of arachidonic acid leads to a cycloperoxide similar to PGH_2 that initiates platelet aggregation. The precise structure of the cycloperoxide is not known except that it is not a substrate for the human platelet enzymes and leads to other prostanoids. These early results do, however, support the concept that O_3 can react with membrane lipids to produce hydroperoxides with biological activity. In this sense, O_3 does lead to a "cascade" of biologically active compounds.

8.2. Effects on Hepatic Drug Metabolism

Pentobarbital is metabolized in the liver by cytochrome P450. The duration of anesthesia produced by pentobarbital in several animal species is proportional to the metabolism of pentobarbital in the liver (Graham *et al.*, 1981). Increases in the pentobarbital sleeping time are proportional to the inhibition of the cytochrome P450 activity of the liver (Graham *et al.*, 1982a,b). Cytochrome P450 is differentially expressed according to sex, being greater in males than in females. Pentobarbital-induced sleeping time in female mice was increased by ambient concentrations of either O_3 (Graham *et al.*, 1981) or $\cdot NO_2$ (Miller *et al.*, 1980). The concentration of pentobarbital in the brain on awakening is generally not affected by O_3 exposure if no direct central nervous system effects have occurred (Graham *et al.*, 1985). Thus, both O_3 and $\cdot NO_2$ apparently cause changes in metabolism in organs other than the lung and may affect the permeability of the blood–brain barrier. The mechanism of action is unknown.

9. SUMMARY

9.1. The Physical Perspective

The health effects of $\cdot NO_2$ and O_3 are interconnected because $\cdot NO_2$ and O_3 are produced in the troposphere from the same precursors and occur together. The precursors are mostly man-made and come from the incomplete combustion of fossil fuels. The production of O_3 from $\cdot NO_2$ and hydrocarbons is a light-dependent reaction. The photochemical nature of O_3 production results in increased levels of O_3 as sunlight increases. At night, little O_3 or $\cdot NO_2$ is formed. The diurnal cycle of human activity, particularly the heavy usage of mobile vehicles in the morning hours, contributes to the cyclic pattern of exposure to a mixture of $\cdot NO_2$ and O_3. Awareness of the cyclic, as well as the episodic, nature of exposures to O_3 and $\cdot NO_2$ is important for investigators attempting to elucidate the health effects of these toxicants. Experimental animals are useful in the study of O_3 and $\cdot NO_2$ because the cyclic exposures that occur with free-living humans can be changed to constant levels of exposure, which eliminate the complex kinetics of exposure and effect that occur in humans.

Another aspect that makes O_3 and $\cdot NO_2$ toxicity difficult to understand is the structure of the lung. The human lung is an equally bifurcating structure. O_3 can reach the deepest parts of the lung because, even though O_3 is highly reactive, it is poorly soluble in water and therefore does not react sufficiently with the aqueous lung lining fluid to be completely eliminated in the upper respiratory tract or the airways. $\cdot NO_2$, on the other hand, is fairly soluble in water and the deposition of $\cdot NO_2$ is relatively uniform in the airways. $\cdot NO_2$ does not exhibit the regional toxicity in the lung to the extent of that found with O_3.

$\cdot NO_2$ and O_3 covary in the atmosphere. Humans living in urban areas are exposed to both $\cdot NO_2$ and O_3 in ambient air. Further, $\cdot NO_2$ exposure may occur indoors for short periods whereas O_3 exposure almost always occurs outdoors. $\cdot NO_2$ or O_3 effects are quantified by statistical models of nonspecific measures such as lung function. Controlled human exposures to $\cdot NO_2$ or O_3 alone are reliable indicators of the toxicity of a single gas but the effects observed in such studies have not been compared quantitatively with the effects observed in epidemiological studies.

9.2. Ozone

At first glance, the increased susceptibility to O_3 of lung cells at the junction between the alveoli and the conducting airways would seem to suggest a special sensitivity of these cells to O_3. The Miller–Overton mathematical model of O_3 deposition in the lung predicts that this region, the centriacinar region, receives a greater O_3 dose than do the surrounding regions. The fluid lining the airways and the lipid surfactant layer lining the alveolus are not continuous. In fact, the surfactant layer is very patchy. When cells lining the lumen of the lung have no overlying lung-lining layer, O_3 is deposited in the cell membrane directly from the gas phase. The very rapid reaction of O_3 with biomolecules may limit O_3 from diffusing from the lung cells into the bloodstream even when there is no overlying fluid layer. However, the effects of O_3 exposure on RBCs suggest that if O_3 has not diffused into the blood, then some reactive product has. Similarly, effects on the central nervous system, the blood–brain barrier, and the hepatic metabolism of foreign compounds suggest a systemic effect of O_3 related to a diffusible by-product.

Three chemical classes seem most likely to produce the effects observed in the blood and organs other than the lung. Aldehydes produced by the decomposition of the 1,2,3-trioxolane are potential agents that could diffuse into the blood stream. However, only α, β-unsaturated aldehydes are likely to have sufficient reactivity to cause cytotoxicity. 4-Hydroxynonenal, an α, β-unsaturated aldehyde, is produced from lipid peroxidation rather than directly from ozonides.

Other products formed mainly from the peroxidation of polyunsaturated fatty acids in lung membranes are endoperoxides, which normally are generated by cyclization of fatty acid hydroperoxide radicals (Menzel, 1980; Pryor *et al.*, 1976). So far, no studies have reported endoperoxides circulating in blood. No comparisons between the biological effects of O_3 and other peroxidation initiators have been reported. This class of membrane peroxidation by-products needs further study.

A likely candidate for effects of O_3 on organs other than the lung is the ozonide. Polyunsaturated fatty acid ozonides are relatively stable, can be isolated in good yield from polyunsaturated fatty acids, and have inflammatory effects. When ozonides are injected subcutaneously, a classic immediate wheal response occurs. Ozonides are converted by glutathione *S*-transferases to unknown products. Probably, products from the reduction of the ozonide are formed, but addition of GSH to the ozonide is possible. Ozonides and endoperoxides resemble the prostanoids and may be responsible for the initial inflammatory response of the lung on O_3 exposure.

Biologically, vitamin E is the most potent protector of experimental animals against O_3 toxic effects. The decrease in O_3-caused mortality in mice and rats by vitamin E implies that peroxidation is biologically more important to the pulmonary toxicity of O_3 than are other metabolic pathways. Studies of protection by vitamin E from O_3 toxicity in humans have produced mixed results. Few experiments in humans have used the high doses of vitamin E found to be effective with *in vitro* studies of cells treated with O_3 or ozonides. Roughly 400 μg of synthetic vitamin E preparations would have to be consumed twice a day by humans to achieve a blood level similar to that found protective in experimental animals. The kinetics and the limits of loading of the human lung with vitamin E are not known, however.

9.3. Nitrogen Dioxide

$\cdot NO_2$ deposition in the airways is relatively uniform according to mathematical models such as those of Overton and Miller. Anatomic lesions occur throughout the respiratory tract. The relatively uniform response to $\cdot NO_2$ suggests that regional deposition is less important in the toxicity of $\cdot NO_2$ than O_3, where regional differences have been predicted and found.

Starting in the lumen of the lung, $\cdot NO_2$ diffuses to the lung lining fluid. The lung lining fluid contains many biochemicals likely to reduce $\cdot NO_2$, most notably vitamin C, GSH, and protein thiols. The reaction of $\cdot NO_2$ with vitamin C is much faster than is the initiation of lipid peroxidation in artificial lipid bilayers. Probably, the secretion of vitamin C into the lung lining fluid is a major protective factor against $\cdot NO_2$ pulmonary toxicity. In humans, studies of the depletion of the lung lining fluid content of ascorbic acid have produced mixed results. Some of the uncertainty may be related to different protocols and measurements done by various investigators. A number of experiments have been done with cigarette smoke rather than $\cdot NO_2$. From the standpoint of public health, clarifying whether vitamin C has a protective role in reducing the health effects of passive cigarette smoking is important. Nevertheless, between the patches of lung lining fluid, cells are exposed directly to $\cdot NO_2$. Therefore, regardless of the $\cdot NO_2$ dose or vitamin C dose, some cells will be affected. Protection will never be 100%.

Once $\cdot NO_2$ reaches the cell membrane, because it is water-soluble, it is hydrated to form one molecule of nitrite and one molecule of nitrate. Both nitrite and nitrate are found in the blood and urine of humans or animals exposed to $\cdot NO_2$. Nitrosation reactions with exogenous amines, given to trap nitrite formed in the lung, do not indicate that nitrosation as a consequence of $\cdot NO_2$ inhalation is a major pathway. On the other hand, $\cdot NO_2$ gas solvated in water does react with polyunsaturated fatty acids to abstract a hydrogen from the polyunsaturated fatty acid to produce nitrous acid and a carbon-centered free radical.

Lipid peroxidation of membrane polyunsaturated fatty acids clearly occurs in both humans and experimental animals on inhalation of $\cdot NO_2$ at concentrations found in polluted air. Like its preventive effects on O_3 toxicity, vitamin E is the most protective agent so far found against mortality from chronic $\cdot NO_2$ exposure of experimental animals. Biologically, peroxidation seems to be the key to understanding $\cdot NO_2$ toxicity as well as that of O_3.

9.4. Oxidation of Lung Proteins and Amino Acids

Mudd's lab has demonstrated that protein oxidation occurs while lipid peroxidation goes on. It is not clear if supplemental vitamin E or C prevents the oxidation of lung proteins or amino acids. So, it is not possible to contrast this insightful hypothesis of protein oxidation with that of lipid peroxidation. Ozonation of tryptophan produces a potent inducer of cytochrome P450. More work is needed here. No doubt additional reactions that compete with lipid peroxidation occur especially when humans inhale the complex mixture of O_3, $\cdot NO_2$, and other chemicals present in polluted air.

9.5. Inflammation and $\cdot NO_2$ and O_3 Toxicity

Inflammation occurs with both O_3 and $\cdot NO_2$ exposures of experimental animals and humans. Changes in levels of inflammatory mediators are detected in the lung lavage fluid of humans and experimental animals exposed to either $\cdot NO_2$ or O_3. The recruitment of inflammatory cells to the lung seems to result from the release of some of these cytokines. Bhalla's group continues to probe this aspect of toxicity and may be able to shed light on why some asthmatics may be more sensitive to O_3 and $\cdot NO_2$ than are asymptomatic subjects.

The consequences of recurring bouts of inflammation in the lung are not known, but surely are not beneficial. Recurring infections in children alter lung development and disrupt the oxygenation of blood with all of the consequences. The cytotoxic effects of $\cdot NO_2$ and O_3 on the alveolar macrophage reduce the ability of the host to defend itself against adventitious respiratory infections. Because repeated inflammation may be particularly important in children, Menzel (1992, 1993) suggested administration of supplemental vitamin E.

9.6. Future Research

The most pressing gap in our knowledge of the health effects of $\cdot NO_2$ and O_3 is the quantitative relation between the dose to lung cells versus the toxic action on these lung cells. It seems that the patchy nature of the lung lining fluid hinders direct experiments to measure the influence of the lung lining fluid. Experiments *in vivo* are unlikely to give a clear-cut answer to the question of whether the health effects result from the direct reaction of either $\cdot NO_2$ or O_3 on lung cells or from indirect effects from products formed in the lung lining fluid. An innovative paradigm is needed here.

Better mapping of the time sequence of events within lung cells during and after exposure to both $\cdot NO_2$ and O_3 is needed. Bhalla's group has done a time sequence of O_3 effects on lung permeability and cell adhesion molecules. Together with a time sequence of cytokine elaboration, these studies point to inflammation as a late event compared with the initial reaction of O_3 with cell membranes. More studies of this type, especially in human volunteers with controlled clinical exposure conditions, are needed.

The complex mixture of products formed by O_3 and $\cdot NO_2$ initiation of lipid peroxidation in the lung provides a rich source for the generation of new ideas of mechanisms of action of these pollutants, as well as of other chemicals that may produce free-radical-mediated toxicity. Few experiments have exploited the concept of endoperoxide formation during lipid peroxidation.

Studies of $\cdot NO_2$ and O_3 have demonstrated that mathematical models of biological effects of chemicals are powerful tools. Mathematical modeling of active oxygen reactions is expected to clarify the many questions of the relative importance of competing chemical reactions. There is a desperate need to place *in vitro* models into an *in vivo* context.

Lastly, profound morphological changes occur in the lungs of experimental animals exposed continuously to $\cdot NO_2$ and O_3. Morphological changes have been found at the National Ambient Air Quality Standard of 0.12 ppm, which had been thought to provide a margin of safety. Whereas the lipid peroxidation theory satisfies many of the questions about the actions of $\cdot NO_2$ and O_3, there are large areas concerning the effects of $\cdot NO_2$ and O_3 on cell-to-cell interactions where mechanisms and effects are only poorly understood. A fertile area of research should be on the activation of nuclear transcription factors such as NF-κB and AP-1. $\cdot NO_2$ or O_3 also may activate other transcription factors. Do these pathways of activation depend on free radicals? Do peroxidation by-products activate nuclear transcription factors? These are questions that deserve intense study and that will be applicable to other oxidants in addition to O_3 and $\cdot NO_2$.

10. REFERENCES

Alin, P., Danielson, U. H., and Mannervik, B., 1985, 4-Hydroxyalk-2-enals are substrates for glutathione transferase, *FEBS Lett.* **179**:267–270.

American Thoracic Society, 1996, Committee of the Environmental and Occupational Health Assembly of the American Thoracic Society. Health effects of outdoor air pollution. Part 2, *Am. J. Respir. Crit. Care Med.* **153**:477–498.

Banerjee, S. K., and Mudd, J. B., 1992, Reaction of ozone with glycophorin in solution and in lipid vesicles, *Arch. Biochem. Biophys.* **295**:84–89.

Berglund, M., 1993, Exposure, *Scand. J. Work Environ. Health* **19**:14–20.

Berhane, K., Widersten, M., Engström, Å., Kozarich, J. W., and Mannervik, B., 1994, Detoxication of base propenals and other α,β-unsaturated aldehyde products of radical reactions and lipid peroxidation by human glutathione transferases, *Proc. Natl. Acad. Sci. USA* **91**:1480–1484.

Bhalla, D. K., 1994, Alteration of ozone-induced airway permeability by oxygen metabolites and antioxidants, *Toxicol. Lett.* **73**:91–101.

Bhalla, D. K., 1996, Alteration of alveolar macrophage chemotaxis, cell adhesion, and cell adhesion molecules following ozone exposure of rats, *J. Cell. Physiol.* **169**:429–438.

Bhalla, D. K., Daniels, D. S., and Luu, N. T., 1992, Attenuation of ozone-induced airway permeability in rats by pretreatment with cyclophosphamide, FPL 55712, and indomethacin, *Am. J. Respir. Cell Mol. Biol.* **7**:73–80.

Bhalla, D. K., Hoffman, L. A., and Pearson, A. C., 1996, Modification of macrophage adhesion by ozone: Role of cytokines and cell adhesion molecules, *Ann. N. Y. Acad. Sci.* **796**:38–46.

Boehme, D. S., Hotchkiss, J. A., and Henderson, R. F., 1992, Glutathione and GSH-dependent enzymes in bronchoalveolar lavage fluid cells in response to ozone, *Exp. Mol. Pathol.* **56**:37–48.

Boström, C. E., 1993, Nitrogen oxides in ambient air—Properties, sources and concentrations, *Scand. J. Work Environ. Health* **19** (Suppl. 2):9–13.

Breslin, K., 1995, The impact of ozone [news], *Environ. Health Perspect.* **103:**660–664.

Bromberg, P. A., and Koren, H. S., 1995, Ozone-induced human respiratory dysfunction and disease, *Toxicol. Lett.* **82–83:**307–316.

Cantin, A. M., North, S. L., Hubbard, R. C., and Crystal, R. G., 1987, Normal alveolar epithelial lining fluid contains high levels of glutathione, *J. Appl. Physiol.* **63:**152–157.

Carini, R., Bellomo, G., Paradisi, L., Dianzani, M. U., and Albano, E., 1996, 4-Hydroxynonenal triggers Ca^{2+} influx in isolated rat hepatocytes, *Biochem. Biophys. Res. Commun.* **218:**772–776.

Coffin, D. L., and Gardner, D. E., 1972, Interaction of biological agents and chemical air pollutants, *Ann. Occup. Hyg.* **15:**219–235.

Danielson, U. H., Esterbauer, H., and Mannervik, B., 1987, Structure–activity relationships of 4-hydroxyalkenals in the conjugation catalysed by mammalian glutathione transferases, *Biochem. J.* **247:**707–713.

Devlin, R. B., and Koren, H. S., 1990, The use of quantitative two-dimensional gel electrophoresis to analyze changes in alveolar macrophage proteins in humans exposed to ozone, *Am. J. Respir. Cell Mol. Biol.* **2:**281–288.

de Vries, J., Hempenius, R. A., and Rietjens, I. M., 1994, Ozone-induced lung toxicity: Mediated by ozonides? *Toxicol. Lett.* **72:**243–246.

Donovan, D. H., Williams, S. J., Charles, J. M., and Menzel, D. B., 1977, Ozone toxicity: Effect of dietary vitamin E and polyunsaturated fatty acids, *Toxicol. Lett.* **1:**135–139.

Elsayed, N. M., 1994, Toxicity of nitrogen dioxide: An introduction, *Toxicology* **89:**161–174.

Esterbauer, H., Zollner, H., and Scholz, N., 1975, Reaction of glutathione with conjugated carbonyls, *Z. Naturforsch. C* **30:**466–473.

Esterbauer, H., Schaur, R. J., and Zollner, H., 1991, Chemistry and biochemistry of 4-hydroxynonenal, malonaldehyde and related aldehydes, *Free Radical Biol. Med.* **11:**81–128.

Everett, S. A., Dennis, M. F., Patel, K. B., Maddix, S., Kundu, S. C., and Willson, R. L., 1996, Scavenging of nitrogen dioxide, thiyl, and sulfonyl free radicals by the nutritional antioxidant beta-carotene, *J. Biol. Chem.* **271:**3988–3994.

Ewetz, L., 1993, Absorption and metabolic fate of nitrogen oxides, *Scand. J. Work Environ. Health* **19**(Suppl. 2):21–27.

Freeman, B. A., and Mudd, J. B., 1981, Reaction of ozone with sulfhydryls of human erythrocytes, *Arch. Biochem. Biophys.* **208:**212–220.

Gallon, A. A., and Pryor, W. A., 1994, The reaction of low levels of nitrogen dioxide with methyl linoleate in the presence and absence of oxygen, *Lipids* **29:**171–176.

Gerrity, T. R., Weaver, R. A., Berntsen, J., House, D. E., and O'Neil, J. J., 1988, Extrathoracic and intrathoracic removal of O_3 in tidal-breathing humans, *J. Appl. Physiol.* **65:**393–400.

Goldstein, I. F., Lieber, K., Andrews, L. R., Kazembe, F., Foutrakis, G., Huang, P., and Hayes, C., 1988, Acute respiratory effects of short-term exposures to nitrogen dioxide, *Arch. Environ. Health* **43:**138–142.

Graham, J. A., Menzel, D. B., Miller, F. J., Illing, J. W., and Gardner, D. E., 1981, Influence of ozone on pentobarbital-induced sleeping time in mice, rats, and hamsters, *Toxicol. Appl. Pharmacol.* **61:**64–73.

Graham, J. A., Menzel, D. B., Miller, F. J., Illing, J. W., and Gardner, D. E., 1982a, Effect of ozone on drug-induced sleeping time in mice pretreated with mixed-function oxidase inducers and inhibitors, *Toxicol. Appl. Pharmacol.* **62:**489–497.

Graham, J. A., Miller, F. J., Gardner, D. E., Ward, R., and Menzel, D. B., 1982b, Influence of ozone and nitrogen dioxide on hepatic microsomal enzymes in mice, *J. Toxicol. Environ. Health* **9:**849–856.

Graham, J. A., Menzel, D. B., Mole, M. L., Miller, F. J., and Gardner, D. E., 1985, Influence of ozone on pentobarbital pharmacokinetics in mice, *Toxicol. Lett.* **24:**163–170.

Haddad, E. B., Salmon, M., Sun, J., Liu, S., Das, A., Adcock, I., Barnes, P. J., and Chung, K. F., 1995, Dexamethasone inhibits ozone-induced gene expression of macrophage inflammatory protein-2 in rat lung, *FEBS Lett.* **363:**285–288.

Hamilton, R. F., Hazbun, M. E., Jumper, C. A., Eschenbacher, W. L., and Holian, A., 1996, 4-Hydroxynonenal mimics ozone-induced modulation of macrophage function ex vivo, *Am. J. Respir. Cell Mol. Biol.* **15:**275–282.

Hazbun, M. E., Hamilton, R., Holian, A., and Eschenbacher, W. L., 1993, Ozone-induced increases in substance P and 8-epi-prostaglandin F2 alpha in the airways of human subjects, *Am. J. Respir. Cell Mol. Biol.* **9**:568–572.

Hempenius, R. A., de Vries, J., and Rietjens, I. M., 1992, Molecular orbital study on the glutathione-dependent detoxication of ozonides, *Eur. J. Pharmacol.* **228**:207–211.

Huie, R. E., 1994, The reaction kinetics of NO_2, *Toxicology* **89**:193–216.

Jaspers, I., Flescher, E., and Chen, L. C., 1997, Ozone-induced IL-8 expression and transcription factor binding in respiratory epithelial cells, *Am. J. Physiol.* **272**:L504–511.

Johnson, D. A., Frampton, M. W., Winters, R. S., Morrow, P. E., and Utell, M. J., 1990, Inhalation of nitrogen dioxide fails to reduce the activity of human lung alpha-1-proteinase inhibitor, *Am. Rev. Respir. Dis.* **142**:758–762.

Kleeberger, S. R., 1995, Genetic susceptibility to ozone exposure, *Toxicol. Lett.* **82–83**:295–300.

Kleinman, M. T., 1991, Effects of ozone on pulmonary function: The relationship of response to dose, *J. Exp. Anal. Environ. Epidemiol.* **1**:309–325.

Kodavanti, U. P., Costa, D. L., Dreher, K. L., Crissman, K., and Hatch, G. E., 1995a, Ozone-induced tissue injury and changes in antioxidant homeostasis in normal and ascorbate-deficient guinea pigs, *Biochem. Pharmacol.* **50**:243–251.

Kodavanti, U. P., Hatch, G. E., Starcher, B., Giri, S. N., Winsett, D., and Costa, D. L., 1995b, Ozone-induced pulmonary functional, pathological, and biochemical changes in normal and vitamin C-deficient guinea pigs, *Fundam. Appl. Toxicol.* **24**:154–164.

Last, J. A., Gelzleichter, T. R., Harkema, J., and Hawk, S., 1994a, Consequences of prolonged inhalation of ozone on Fischer-344/N rats: Collaborative studies. Part I: Content and cross-linking of lung collagen, *Res. Rep. Health Eff. Inst.* 1–29; discussion 31–40.

Last, J. A., Sun, W. M., and Witschi, H., 1994b, Ozone, NO, and NO_2: Oxidant air pollutants and more, *Environ. Health Perspect.* **102**(Suppl. 10):179–184.

Leikauf, G. D., Zhao, Q., Zhou, S., and Santrock, J., 1993, Ozonolysis products of membrane fatty acids activate eicosanoid metabolism in human airway epithelial cells, *Am. J. Respir. Cell Mol. Biol.* **9**:594–602.

Li, L., Hamilton, R. F., Kirichenko, A., and Holian, A., 1996, 4-Hydroxynonenal-induced cell death in murine alveolar macrophages, *Toxicol. Appl. Pharmacol.* **139**:135–143.

Lippmann, M., 1992, Ozone, in *Environmental Toxicants: Human Exposures and Their Health Effects* (M. Lippmann, ed.), pp. 465–519, Van Nostrand–Reinhold, Princeton, NJ.

Mannix, R. C., Phalen, R. F., Oldham, M. J., Mautz, W. J., and Kleinman, M. T., 1996, Effects of repeated exposure to nitric acid vapor and ozone on respiratory tract clearance in the rat, *Inhal. Toxicol.* **8**:595–605.

Mayorga, M. A., 1994, Overview of nitrogen dioxide effects on the lung with emphasis on military relevance, *Toxicology* **89**:175–192.

McGee, M. P., Devlin, R., Saluta, G., and Koren, H., 1990, Tissue factor and factor VII messenger RNAs in human alveolar macrophages: Effects of breathing ozone, *Blood* **75**:122–127.

Menzel, D. B., 1970, Toxicity of ozone, oxygen, and radiation, *Annu. Rev. Pharmacol.* **10**:379–394.

Menzel, D. B., 1971, Oxidation of biologically active reducing substances by ozone, *Arch. Environ. Health* **23**:149–153.

Menzel, D. B., 1976, Proceedings: Oxidants and human health, *J. Occup. Med.* **18**:342–345.

Menzel, D. B., 1980, Protection against environmental toxicants, in *Vitamin E: A Comprehensive Treatise* (L. F. Machlin, ed.), pp. 474–494, Dekker, New York.

Menzel, D. B., 1992, Antioxidant vitamins and prevention of lung disease, *Ann. N. Y. Acad. Sci.* **669**:141–155.

Menzel, D. B., 1993, Antioxidants in lung disease, *Toxicol. Ind. Health* **9**:323–336.

Menzel, D. B., Roehm, J. N., and Lee, S. D., 1972, Vitamin E: The biological and environmental antioxidant, *J. Agric. Food Chem.* **20**:481–486.

Menzel, D. B., Slaughter, R. J., Bryant, A. M., and Jauregui, H. O., 1975a, Heinz bodies formed in erythrocytes by fatty acid ozonides and ozone, *Arch. Environ. Health* **30**:296–301.

Menzel, D. B., Slaughter, R. J., Bryant, A. M., and Jauregui, H. O., 1975b, Prevention of ozonide-induced Heinz bodies in human erythrocytes by vitamin E, *Arch. Environ. Health* **30**:234–236.

Menzel, D. B., Roycroft, J. H., Nixon, J. R., Isaac, S. R., and Porter, N. A., 1976, Monocyclic peroxides as inhibitors of arachidonic acid and prostaglandin endoperoxide analog initiated aggregation of human platelets, *Res. Commun. Chem. Pathol. Pharmacol.* **15**:767–785.

Miller, F. J., 1995, Uptake and fate of ozone in the respiratory tract, *Toxicol. Lett.* **82–83**:277–285.

Miller, F. J., and Menzel, D. B. (eds.), 1984, *Fundamentals of Extrapolation Modeling of Inhaled Toxicants: Ozone and Nitrogen Dioxide*, Hemisphere Publishing, Washington, DC.

Miller, F. J., Menzel, D. B., and Coffin, D. L., 1978, Similarity between man and laboratory animals in regional pulmonary deposition of ozone, *Environ. Res.* **17:**84–101.

Miller, F. J., Graham, J. A., Illing, J. W., and Gardner, D. E., 1980, Extrapulmonary effects of NO_2 as reflected by pentobarbital-induced sleeping time in mice, *Toxicol. Lett.* **6:**267–274.

Mochitate, K., Ishida, K., Ohsumi, T., and Miura, T., 1992, Long-term effects of ozone and nitrogen dioxide on the metabolism and population of alveolar macrophages, *J. Toxicol. Environ. Health* **35:**247–260.

Mohsenin, V., 1991, Lipid peroxidation and antielastase activity in the lung under oxidant stress: Role of antioxidant defenses, *J. Appl. Physiol.* **70:**1456–1462.

Mohsenin, V., 1994, Human exposure to oxides of nitrogen at ambient and supra-ambient concentrations, *Toxicology* **89:**301–312.

Mohsenin, V., and Gee, J. B., 1987, Acute effect of nitrogen dioxide exposure on the functional activity of alpha-1-protease inhibitor in bronchoalveolar lavage fluid of normal subjects, *Am. Rev. Respir. Dis.* **136:**646–650.

Mohsenin, V., and Gee, J. L., 1989, Oxidation of alpha 1-protease inhibitor: Role of lipid peroxidation products, *J. Appl. Physiol.* **66:**2211–2215.

Moldéus, P., 1993, Toxicity induced by nitrogen dioxide in experimental animals and isolated cell systems, *Scand. J. Work Environ. Health* **19**(Suppl. 2):28–36.

Moldovan, N. I., Lupu, F., Moldovan, L., and Simionescu, N., 1994, 4-Hydroxynonenal induces membrane perturbations and inhibition of basal prostacyclin production in endothelial cells, and migration of monocytes, *Cell Biol. Int.* **18:**985–992.

Morgan, D. L., Dorsey, A. F., and Menzel, D. B., 1985, Erythrocytes from ozone-exposed mice exhibit decreased deformability, *Fundam. Appl. Toxicol.* **5:**137–143.

Nazar-Stewart, V., Motulsky, A. G., Eaton, D. L., White, E., Hornung, S. K., Leng, Z. T., Stapleton, P., and Weiss, N. S., 1993, The glutathione S-transferase mu polymorphism as a marker for susceptibility to lung carcinoma, *Cancer Res.* **53:**2313–2318.

Norman, V., and Keith, C. H., 1965, Nitrogen oxides in tobacco smoke, *Nature* **205:**915–916.

O'Neill, C. A., Van der Vliet, A., Eiserich, J. P., Last, J. A., Halliwell, B., and Cross, C. E., 1995, Oxidative damage by ozone and nitrogen dioxide: Synergistic toxicity in vivo but no evidence of synergistic oxidative damage in an extracellular fluid, *Biochem. Soc. Symp.* **61:**139–152.

Overton, J. H., 1984, Physicochemical processes and the formulation of dosimetry models, *J. Toxicol. Environ. Health* **13:**273–294.

Pearson, A. C., and Bhalla, D. K., 1997, Effects of ozone on macrophage adhesion in vitro and epithelial and inflammatory responses in vivo: The role of cytokines, *J. Toxicol. Environ. Health* **50:**143–157.

Plopper, C. G., Duan, X., Buckpitt, A. R., and Pinkerton, K. E., 1994, Dose-dependent tolerance to ozone. IV. Site-specific elevation in antioxidant enzymes in the lungs of rats exposed for 90 days or 20 months, *Toxicol. Appl. Pharmacol.* **127:**124–131.

Postlethwait, E. M., Langford, S. D., Jacobson, L. M., and Bidani, A., 1995, NO_2 reactive absorption substrates in rat pulmonary surface lining fluids, *Free Radical Biol. Med.* **19:**553–563.

Pryor, W. A., 1992, How far does ozone penetrate into the pulmonary air/tissue boundary before it reacts?, *Free Radical Biol. Med.* **12:**83–88.

Pryor, W. A., 1994, Mechanisms of radical formation from reactions of ozone with target molecules in the lung, *Free Radical Biol. Med.* **17:**451–465.

Pryor, W. A., and Uppu, R. M., 1993, A kinetic model for the competitive reactions of ozone with amino acid residues in proteins in reverse micelles, *J. Biol. Chem.* **268:**3120–3126.

Pryor, W. A., Stanley, J. P., and Blair, E., 1976, Autoxidation of polyunsaturated fatty acids: II. A suggested mechanism for the formation of TBA-reactive materials from prostaglandin-like endoperoxides, *Lipids* **11:**370–379.

Pryor, W. A., Giamalva, D. H., and Church, D. F., 1984, Kinetics of ozonation. 2. Amino acids and model compounds in water and comparison to rates in nonpolar solvents, *J. Am. Chem. Soc.* **106:**7094–7100.

Quinlan, T., Spivack, S., and Mossman, B. T., 1994, Regulation of antioxidant enzymes in lung after oxidant injury, *Environ. Health Perspect.* **102**(Suppl. 2):79–87.

Rahman, I., Clerch, L. B., and Massaro, D., 1991, Rat lung antioxidant enzyme induction by ozone, *Am. J. Physiol.* **260:**L412–418.

Rasmussen, T. R., Kjaergaard, S. K., Tarp, U., and Pedersen, O. F., 1992, Delayed effects of NO_2 exposure on alveolar permeability and glutathione peroxidase in healthy humans, *Am. Rev. Respir. Dis.* **146:**654–659.

Reed, D. J., 1990, Glutathione: Toxicological implications, *Annu. Rev. Pharmacol. Toxicol.* **30:**603–631.

Rietjens, I. M., Poelen, M. C., Hempenius, R. A., Gijbels, M. J., and Alink, G. M., 1986, Toxicity of ozone and nitrogen dioxide to alveolar macrophages: Comparative study revealing differences in their mechanism of toxic action, *J. Toxicol. Environ. Health* **19:**555–568.

Rietjens, I. M., van Tilburg, C. A., Coenen, T. M., Alink, G. M., and Konings, A. W., 1987, Influence of polyunsaturated fatty acid supplementation and membrane fluidity on ozone and nitrogen dioxide sensitivity of rat alveolar macrophages, *J. Toxicol. Environ. Health* **21:**45–56.

Robison, T. W., Forman, H. J., and Thomas, M. J., 1995, Release of aldehydes from rat alveolar macrophages exposed in vitro to low concentrations of nitrogen dioxide, *Biochim. Biophys. Acta* **1256:**334–340.

Roehm, J. N., Hadley, J. G., and Menzel, D. B., 1971a, Antioxidants vs lung disease, *Arch. Intern. Med.* **128:**88–93.

Roehm, J. N., Hadley, J. G., and Menzel, D. B., 1971b, Oxidation of unsaturated fatty acids by ozone and nitrogen dioxide. A common mechanism of action, *Arch. Environ. Health* **23:**142–148.

Roycroft, J. H., Gunter, W. B., and Menzel, D. B., 1977, Ozone toxicity: Hormone-like oxidation products from arachidonic acid by ozone-catalyzed autoxidation, *Toxicol. Lett.* **1:**75–82.

Rushmore, T. H., Morton, M. R., and Pickett, C. B., 1991, The antioxidant responsive element. Activation by oxidative stress and identification of the DNA consensus sequence required for functional activity, *J. Biol. Chem.* **266:**11632–11639.

Sagai, M., and Ichinose, T., 1987, Lipid peroxidation and antioxidative protection mechanism in rat lungs upon acute and chronic exposure to nitrogen dioxide, *Environ. Health Perspect.* **73:**179–189.

Sandström, T., 1995, Respiratory effects of air pollutants: Experimental studies in humans [see comments], *Eur. Respir. J.* **8:**976–995.

Schlesinger, R. B., 1992, Nitrogen oxides, in *Environmental Toxicants: Human Exposures and Their Health Effects* (M. Lippmann, ed.), pp. 412–453, Van Nostrand–Reinhold, Princeton, NJ.

Schlesinger, R. B., Driscoll, K. E., Gunnison, A. F., and Zelikoff, J. T., 1990, Pulmonary arachidonic acid metabolism following acute exposures to ozone and nitrogen dioxide, *J. Toxicol. Environ. Health* **31:**275–290.

Schuller-Levis, G., Quinn, M. R., Wright, C., and Park, E., 1994, Taurine protects against oxidant-induced lung injury: Possible mechanism(s) of action, *Adv. Exp. Med. Biol.* **359:**31–39.

Selley, M. L., McGuiness, J. A., Jenkin, L. A., Bartlett, M. R., and Ardlie, N. G., 1988, Effect of 4-hydroxy-2,3-trans-nonenal on platelet function, *Thromb. Haemost.* **59:**143–146.

Shoaf, C. R., Wolpert, R. L., and Menzel, D. B., 1989a, Antioxidant effects of α-tocopherol and ascorbate in liposomes exposed to nitrogen dioxide, *Inhal. Toxicol.* **1:**315–329.

Shoaf, C. R., Wolpert, R. L., and Menzel, D. B., 1989b, Nitrogen dioxide-initiated peroxidation of liposomal membrane systems, *Inhal. Toxicol.* **1:**301–314.

Sindhu, R. K., Mautz, W. J., Fujita, I., Wang, N., and Kikkawa, Y., 1996, Effect of chronic exposure to ozone and nitric acid on cytochrome P-450 monooxygenase system of rat lung and liver, *Inhal. Toxicol.* **8:**695–708.

Strand, V., Salomonsson, P., Lundahl, J., and Bylin, G., 1996, Immediate and delayed effects of nitrogen dioxide exposure at an ambient level on bronchial responsiveness to histamine in subjects with asthma, *Eur. Respir. J.* **9:**733–740.

Strohmaier, H., Hinghofer-Szalkay, H., and Schaur, R. J., 1995, Detection of 4-hydroxynonenal (HNE) as a physiological component in human plasma, *J. Lipid Mediat. Cell Signal.* **11:**51–61.

Sun, W., Wu, R., and Last, J. A., 1994, Coordinated expression of a 45 kD protein and ozone toxicity in a human bronchial epithelial cell line, *Am. J. Respir. Cell Mol. Biol.* **10:**673–682.

Tsukagoshi, H., Haddad, E. B., Sun, J., Barnes, P. J., and Chung, K. F., 1995, Ozone-induced airway hyperresponsiveness: Role of superoxide anions, NEP, and BK receptors, *J. Appl. Physiol.* **78:**1015–1022.

Tu, B., Wallin, A., Moldeus, P., and Cotgreave, I., 1995, The cytoprotective roles of ascorbate and glutathione against nitrogen dioxide toxicity in human endothelial cells, *Toxicology* **98:**125–136.

Uppu, R. M., Cueto, R., Squadrito, G. L., and Pryor, W. A., 1995, What does ozone react with at the air/lung interface? Model studies using human red blood cell membranes, *Arch. Biochem. Biophys.* **319:**257–266.

Urban, T., Hurbain, I., Urban, M., Clement, A., and Housset, B., 1995, [Oxidants and antioxidants. Biological effects and therapeutic perspectives] Oxydants et antioxydants. Effets biologiques et perspectives therapeutiques, *Ann. Chir.* **49**:427–434.

U.S. Environmental Protection Agency, 1993a, Air Quality Criteria for Oxides of Nitrogen, Vol. I, Report no. EPA/600/8-91/049aF, U.S. EPA, Research Triangle Park, NC.

U.S. Environmental Protection Agency, 1993b, Air Quality Criteria for Oxides of Nitrogen, Vol. III, Report no. EPA/600/8-91/049cF, U.S. EPA, Research Triangle Park, NC.

U.S. Environmental Protection Agency, 1996a, Air Quality Criteria for Ozone and Related Photochemical Oxidants, Vol. I, Report no. EPA/600/P-93/004aF, U.S. EPA, Research Triangle Park, NC.

U.S. Environmental Protection Agency, 1996b, Air Quality Criteria for Ozone and Related Photochemical Oxidants, Vol. III, Report no. EPA/600/P-93/004cF, U.S. EPA, Research Triangle Park, NC.

Van der Vliet, A., O'Neil, C. A., Eiserich, J. P., and Cross, C. E., 1995, Oxidative damage to extracellular fluids by ozone and possible protective effects of thiols, *Arch. Biochem. Biophys.* **321**:43–50.

Victorin, K., 1992, Review of the genotoxicity of ozone, *Mutat. Res.* **277**:221–238.

Vos, R. M., Rietjens, I. M., Stevens, L. H., and van Bladeren, P. J., 1989, Methyl linoleate ozonide as a substrate for rat glutathione S-transferases: Reaction pathway and isoenzyme selectivity, *Chem. Biol. Interact.* **69**:269–278.

Vostal, J. J., 1994, Physiologically based assessment of human exposure to urban air pollutants and its significance for public health risk evaluation, *Environ. Health Perspect.* **102**(Suppl. 4):101–106.

Winters, R. S., Burnette-Vick, B. A., and Johnson, D. A., 1994, Ozone, but not nitrogen dioxide, fragments elastin and increases its susceptibility to proteolysis, *Am. J. Respir. Crit. Care Med.* **150**:1026–1031.

Wong, C. G., Bonakdar, M., Mautz, W. J., and Kleinman, M. T., 1996, Chronic inhalation exposure to ozone and nitric acid elevates stress-inducible heat shock protein 70 in the rat lung, *Toxicology* **107**:111–119.

Wright, D. T., Adler, K. B., Akley, N. J., Dailey, L. A., and Friedman, M., 1994, Ozone stimulates release of platelet activating factor and activates phospholipases in guinea pig tracheal epithelial cells in primary culture, *Toxicol. Appl. Pharmacol.* **127**:27–36.

Yu, X. Y., Takahashi, N., Croxton, T. L., and Spannhake, E. W., 1994, Modulation of bronchial epithelial cell barrier function by in vitro ozone exposure, *Environ. Health Perspect.* **102**:1068–1072.

Zhang, J., and Lioy, P. J., 1994, Ozone in residential air: Concentrations, I/O ratios, indoor chemistry, and exposures, *Indoor Air* **4**:95–105.

Chapter 14

Dietary Antioxidants and Nutrition

Catherine Rice-Evans and Saimar Arif

Compelling chemical, biochemical, clinical, and epidemiological evidence supports the contention that the antioxidant nutrients exert vital contributions toward the prevention or delayed onset of cancer and cardiovascular disease. Antioxidants are considered to exert their effects by attenuating oxidative events that contribute to the pathophysiology of these diseases. A summary of some of the evidence is presented in this review.

1. INTRODUCTION

Many epidemiological studies have led to the observation that diets rich in vegetables and fruit, and the antioxidants they contain, are protective against a wide range of chronic diseases and degenerative conditions associated with aging, including coronary heart disease, certain cancers, and cataract (reviewed in Diplock, 1994; Steinmetz and Potter, 1993a,b). For example, insufficient fruit and vegetable consumption increases the rate of most types of cancer about twofold as shown by about 200 epidemiological studies that are remarkably consistent (Ames *et al.*, 1993; Block *et al.*, 1992; Henderson *et al.*, 1991).

WHO has advised the intake of a lower limit of 400 g/day of mixed fresh fruits and vegetables (excluding tubers) and including at least 30 g of pulses, nuts, and seeds—a recommendation that has been endorsed by government agencies. This recommendation was based primarily on the apparently healthy fruit and vegetable intakes of populations living in Mediterranean countries with low rates of coronary heart disease.

There is considerable evidence for a role for the dietary antioxidant nutrients, vitamins E, C, and β-carotene, in the maintenance of health, in contributing to the

---

Catherine Rice-Evans and Saimar Arif International Antioxidant Research Centre, UMDS–Guy's Hospital, London SE1 9RT, England.

Reactive Oxygen Species in Biological Systems, edited by Gilbert and Colton. Kluwer Academic / Plenum Publishers, New York, 1999.

decreased incidence of disease (cancer, cataract, cardiovascular disease), and in protecting against the recurrence of pathological events. Extended international dietary surveys reveal that the calculated intake of essential antioxidants such as vitamins C, E, and β-carotene is inversely related to the risk of ischemic heart disease, as it is for certain forms of cancer.

Table I
Hierarchy of Levels of Antioxidant Nutrients (Based on Amounts/100 g Raw) Present in Selected Fruits and Vegetables[a]

		Cartenoids (μg)		Vitamin C (mg)		Vitamin E (mg)
Fruits	Mango	1800[b]	Black currant	200[c]	Avocado	3.2
	Cantaloupe melon	1000	Guava	230[d]	Blackberry	2.37
	Paw paw	810	Strawberry	77	Olive (green)	1.99
	Passion fruit	750	Lemon	58	Mango	1.05
	Apricot	405[e]	Orange	54[f]	Black currant	1
	Plum	295	Lynchee	45		
	Watermelon	230	Mango	37		
	Pink grapefruit	280	Nectarine	37		
	Black currant	100	Peach	31		
	Peach	58	Apple	3–20		
	Nectarine	58				
	Apple	18				
Vegetables	Carrots	8115[g]	Spring greens	180	Sweet potato	4.56
	Sweet potato	3930	Red pepper	140	Spinach	1.71
	Red pepper	3840	Green pepper	120	(Curly kale)	[1.7]
	Spinach	3535	Curly kale	110	Watercress	1.46
	Curly kale	3145	Brussels sprout	115	(Broccoli)	[1.30]
	Spring greens	2630	Broccoli	87	Tomato	1.22
	Watercress	2520	Watercress	62	Asparagus	1.16
	Leek	735	Mange-tout	54	Parsnip	1
	Mange-tout	695	Cabbage	49	Leek	0.92
	Tomato[h]	640	Cauliflower	43		
	Courgette	610	Tomato	17		
	Broccoli	575				

[a]Data from McCance and Widdowson (1992).
[b]Levels range from 300 to 3000 μg.
[c]Levels range from 150 to 230 mg.
[d]Levels range from 9 to 410 mg.
[e]Levels range from 200 to 3370 μg.
[f]Levels range from 44 to 79 mg.
[g]Levels range from 4300 to 11,000 μg.
[h]Tomato is an important source of the carotenoid lycopene.
**ca. 50% vitamin C lost on cooking

Table II
Cartenoid Composition of Fruits and Vegetables

Cartenoid	Source
Canthaxanthin	Mushroom
Capsanthin	Red and yellow peppers, oranges
α-Carotene, β-carotene	Carrots, red palm oil, sweet potato, oranges, berries, grapes, black currants
α-Carotenc	Oranges, banana
β-Carotene	Banana, mango, papaya, watermelon, artemia, star fruit, paprika, cassava, potato, taro, red pepper
α-Crptoxanthin	Papaya, watermelon
β-Crptoxanthin	Mango, papaya, peaches, oranges, mandarin, star fruit, cassava, taro
Lutein	Banana, oranges, mandarin, star fruit, cassava, yellow peppers, potato, berries, grapes, black currants, peas, sprouts, greens, broccoli
Lycopene	Tomato, papaya, watermelon, cassava, taro
Phytoene	Pumpkin, mango, papaya
Phytofluene	Apricots, peaches
Zeaxanthin	Paprika, red and yellow peppers

Fruits and vegetables are the principal sources of two of the major dietary antioxidants, namely, vitamin C and the carotenoids (Table I). Vitamin C occurs in citrus fruits, green peppers, cabbage, strawberries, and green leafy vegetables; β-carotene is found in carrots, dark leafy greens, such as spinach, greens, and broccoli, orange and yellow vegetables, and orange fruits. Fruits and vegetables contain not only vitamins C, E, and β-carotene but also other antioxidants, as well as other constituents that may act as agents of other anticarcinogenic and cardioprotective mechanisms independent of their antioxidant properties. These other antioxidants include other carotenoids (Table II) such as lycopene, the major red constituent of tomatoes, and capsanthin, the red constituent of peppers (Khackik *et al.*, 1992; Camara and Moneger, 1978).

One of the most consistent epidemiological findings has been the association between higher intakes of vitamin C-rich foods with high blood plasma vitamin C levels and a reduction in risk for stomach cancer. It is of interest that ascorbic acid is secreted into gastric juice in concentrations that often exceed those in plasma, and a major other effect that could also be the basis of protection is its ability to inhibit nitrosamine formation, procarcinogens that occur in food and can be formed *in vivo* by reaction of nitrite with other dietary or endogenous amines or amides. Smoking has a dramatic influence on vitamin C levels. Cigarette smokers have lower blood vitamin C levels than nonsmokers (Pelletier, 1975). Cigarette smoke contains an abundance of oxidants and prooxidants that stress the antioxidant defense systems of the body (Pryor, 1986). Smoking increases the turnover of vitamin C as shown by a study describing that, in comparison with nonsmokers who consume on average 60 mg vitamin C/day (U.S. recommended dietary allowance), smokers would need 150 mg/day to achieve the same blood levels (Schectman *et al.*, 1991). Following adjustment for the potentially confounding variables, blood serum vitamin C levels remain 21% lower in smokers compared with nonsmokers. Other data show that the association between smoking and blood levels of vitamin C is independent of effects of poor dietary intake, suggesting that cigarette smoking decreases plasma vitamin C levels.

Although carotenoids are synthesized only by higher plants (as well as algae and some bacteria), they are widely distributed in animals, which acquire them via their diet. Thus, many crustaceans and finned fish contain carotenoids, which color the animals pink. Carotenoid function is usually linked to their role in pro-vitamin A activity. In addition, they have a second protective function in plants through scavenging of free radicals that would otherwise lead to damage of sensitive cellular constituents such as DNA. Their presence in leafy vegetables is masked by the green chlorophylls but in other foodstuffs the carotenoids are more visually evident, contributing to the red, yellow, and orange colors of vegetables and fruits. Levels of a range of individual carotenoids in commonly consumed fruits and vegetables in the United Kingdom from a more recent detailed study are listed in Table III (Scott and Hart, 1995). There is a wide variation in carotenoid content in individual fruits and vegetables as well as a seasonal variation. The predominant dietary source of lycopene is tomatoes. Good dietary sources of lutein are peas, sprouts, greens, broccoli, spinach, and peppers. Mangoes and apricots are major sources of β-carotene and oranges of cryptoxanthin in fruits.

Table III
The Cartenoid Content of Vegetables μg/100 g "Wet Weight" as Eaten[a,b]

		β-Carotene	Lutein	Zeaxanthin	β-Cryptoxanthin	Lycopene	α-Carotene
Bean green†	Raw	376	494				70
	Cooked	373	548				26
Beans baked	Canned	30	25			1659	
Beans broad	Raw	261	506				
	Cooked	406	620	34			
Beans french	Raw	383	479				
	Cooked	443	577				
Beans runner	Raw	343	555				33
	Cooked	538	632				tr
Broccoli fresh	Raw	919	1614				
	Cooked	1381	1949				
Brussels sprout†	Raw	553	610				
	Cooked	555	621				
Cabbage green	Raw	59	80				
	Cooked	71	111				
Cabbage Savoy	Raw	50	103				
	Cooked	273	341				
	Outside leaves	11849	14457				
Carrots†	Raw	8538	268				3268
	Cooked	9907	300				3851
Carrots (May)	Raw	8597	170				2660
	Cooked	9059	149				2838
Carrots (Sept.)	Raw	10800	283				3610
	Cooked	12302	313				3767
Cauliflower	Raw	nd	tr				
	Cooked	nd	tr				

Table III (*Continued*)

		β-Carotene	Lutein	Zeaxanthin	β-Cryptoxanthin	Lycopene	α-Carotene
Cucumber	Raw	295	670				
Greens	Raw	1949	3046				
	Cooked	3105	4437				
Leeks	Raw	87	161				
	Cooked	84	164				
Lettuce broadhead	Raw	1988	1611				
Lettuce iceberg	Raw	91	110				
Marrow	Raw	48	128				
	Cooked	53	135				
Mixed vegetables†	Raw	4195	882	84			1045
	Cooked	4505	976	59			1045
Parsley	Raw	4523	5812				
Peas†	Raw	438	1633				
	Cooked	660	1991				
Pepper green	Raw	298	660				
	Fried	543	1023				
Pepper green*	Raw	714	1116				
Pepper orange	Raw	416	503	1608	90		167
Pepper orange*	Raw	1128	2492	8480	779		636
Spinach	Raw	4021	5869				
	Cooked	5501	7410				
Spring onions	Raw	142	255				
Sweet corn†	Raw	59	522	437			60
	Cooked	24	444	375			33
Tomato fresh	Raw	439	78			2937	
	Cooked	648	120			3703	
Tomatoes tin	In sauce	423	105			6205	
	Reheated	258	116				
Watercress	Raw	5919	10713				
			Fruits				
Apple peel*	1.5 mm	229	434	16	20		nd
Apples*	Eating	35	84	13	11		nd
Apricots*	Dried	1014	44	23	75		nd
	Fresh	1766	101	31	231		37
	Tinned	882	59	16	32		nd
Bananas*		39	33	4	nd		50
Mandarins*		285	50	142	1774		12
Oranges*	Valencia	14	64	50	83		nd
Peach*	Fresh	103	78	42	86		tr
Satsumas*		23	44	41	1180		nd
Strawberry*	Fresh	11	6	nd	nd		nd

[a]Modified from Scott and Hart (1995).

[b]nd, not detectible; tr, trace; *, saponified; †, frozen.

There is also a strong and consistent association between a high consumption of carotene-containing vegetables and fruits and a concomitant high β-carotene status, with a lower risk of cancer. The relationship is most evident for lung cancer. The U.K. Government's Dietary and Nutritional Survey of British Adults shows men to have an average daily intake of 2.4 mg of carotene compared with 2.1 mg for women (Gregory *et al.*, 1990). The higher intake among males can be attributed to their higher intake of calories generally. In the United Kingdom there is currently no official recommendation on the optimum daily intake of β-carotene or indeed antioxidants generally. However, various expert groups including the U.S. National Cancer Institute are recommending a diet that would provide at least 6 mg of β-carotene per day. This is far higher than the current British intake. The U.K. diet therefore appears to fall short of the level of these vitamins that the body may need to protect itself from chronic disease. According to the Dietary and Nutritional Survey of British Adults, the level of carotene intake is lower among the 16–24 age group, falling to mean intakes of 1.6 mg in women and 1.9 mg in men, compared with 2.4 and 2.8 mg, respectively in the 50–64 age group. The low intakes of carotene among the younger age groups reflect a lower intake of fruits and vegetables. It is of some concern that the "age differential" in β-carotene intake reflects different eating habits between generations; subjects currently aged 16–24 may not change their diets as they get older. If the correlation between low β-carotene and a high incidence of certain cancers and heart disease proves correct, then this suggests an increase in chronic disease rates when today's 16- to 24-year-olds reach 50 and 60 years of age.

Only certain fruits and vegetables contain significant amounts of vitamin E, major sources being cereal oils, olive oil, and nuts. In the context of fruits and vegetables, vegetarians have a lower risk of atherosclerosis, in which the vitamin E/cholesterol ratio is enhanced. In addition, like the carotenoids and vitamin C, vitamin E has been reported to stimulate the immune system and protect against the development of cancer by enhancing immune surveillance (Dorgan and Schatzkin, 1991; Mergens and Bhagavan, 1990; Mirvish, 1986). Vitamins E and C reduce nitrite, inhibiting the production of nitrosamines and nitrosamides (Mirvish, 1986), compounds that might induce tumors in humans (Bostick *et al.*, 1993). Selenium, an essential constituent of the enzyme glutathione peroxidase (which reduces peroxides and thus prevents damage to intracellular membranes), may also stimulate the immune system and inhibit DNA synthesis and cell proliferation (Medina, 1986).

A relatively low intake of antioxidant nutrients, and especially in smokers, probably contributes to the higher incidence of coronary heart disease in northern Europe, especially Scotland. Cigarette smokers eat fewer cereals and cereal products (the main dietary source of vitamin E) and fruits and vegetables (the sources of vitamin C and the carotenoids). Furthermore, in countries where the incidence of coronary heart disease is very high, such as Scotland, Northern Ireland, and Finland, the measured blood plasma levels of vitamin E are significantly lower than in Mediterranean countries where the incidence is low and consumption of vitamin E, mostly from olive oil and cereal oil products, and vitamin C, from fresh citrus fruits and specific vegetables, is high. In addition to trapping free radicals, these antioxidants may also have other anticarcinogenic and cardioprotective effects.

Recent work is also highlighting the additional antioxidant role of the "flavonoid" constituents of fruits and vegetables; this role may contribute to the inverse correlation

Table IV
Some Dietary Sources of Flavonoids

Flavan-3-ol	
Epicatechin	
Catechin	
Epigallocatechin	Green and black teas
Epicatechin gallate	Red wine
Epigallocatechin gallate	
Flavanon	
Naringin	Peel of citrus fruits
Taxifolin	Citrus fruits
Flavon-3-ol	
Kaempferol	Endive, leek, broccoli, radish, grapefruit, black tea
Quercetin	Onion, lettuce, broccoli, cranberry, apple skin, berries, olive, tea, red wine
Myricetin	Cranberry, grapes, red wine
Flavone	
Chrysin	Fruit skin
Apigenin	Celery, parsley
Anthocyanins	
Malvidin	Red grapes, red wine
Cyanidin	Cherry, raspberry, strawberry, grapes
Apigenidin	Colored fruit and peels
Phenyl propanoids	
Ferulic acid	Wheat, corn, rice, tomatoes, spinach, cabbage, asparagus
Caffeic acid	White grapes, white wine, olives, olive oil, spinach, cabbage, asparagus, coffee
p-Coumaric acid	White grapes, white wine, tomatoes, spinach, cabbage, asparagus
Chlorogenic acid	Apples, pears, cherries, plums, peaches, apricots, blueberries, tomatoes, anis

noted between indices of antioxidants found in fresh fruits and vegetables and decreased risk of coronary heart disease and certain cancers. Flavonoids are also effective antioxidants (reviewed in Rice-Evans and Packer, 1997) but do not occur at such high concentrations as vitamin C and carotenoids in certain foods. Table IV shows the flavonoid constituents of some beverages, fruits, and vegetables and the antioxidant activities and specific polyphenolic constituents relative to vitamins C and E. In support of this are the findings from a Dutch epidemiological study of elderly males showing that those with high dietary intake of tea, onions, and apples particularly (and red wine) (i.e., higher intake of flavonoids) have a lower mortality from coronary heart disease and decreased incidence of heart attack (Hertog *et al.*, 1993). This is further emphasized by the "French paradox" (Renaud and de Lorgeril, 1992)—the fact that the Southern French have a low incidence of coronary heart disease despite a high-fat diet and smoking tendencies. Their high intake of fresh fruits and vegetables, oil from olives, and high red wine consumption may be important determinants. The Mediterranean diet is rich in antioxidants and flavonoids from intake of fresh fruits and vegetables, olive oil, tomatoes, garlic, and red wine.

2. CANCER

It has been proposed that cancer-inducing damage to DNA might be prevented by vitamins E, C, and β-carotene through the scavenging of excess free radicals that are by-products of many normal metabolic functions or the effects of carcinogens or radiation (Cerutti, 1994). Inverse correlations have been demonstrated between cancer risk and indices of certain micronutrients found in fresh fruits and vegetables, especially vitamin C and β-carotene. There is now good evidence linking high intake of carotenoid-containing fresh fruits and vegetables with low incidence of certain cancers (Block, 1992) and analysis of data for specific antioxidant nutrients indicates a strong and consistent association between poor plasma β-carotene levels and increased risk for developing lung cancer and between poor plasma ascorbate and gastrointestinal cancers. Studies have also indicated that poor β-carotene status may be associated with the risk for developing cancer at other sites. However, it is not clear whether an antioxidant response is involved. Less is known for vitamin E and cancer. Out of 200 studies the relationship between cancer risk and fruit and vegetable intake is exceptionally strong and consistent. The evidence is particularly strong for cancer sites in respiratory and upper digestive tracts and substantial for other sites. The effect is not only statistically significant, it also is clinically significant. In general, people with low dietary intakes of fruits and vegetable have double the risk for cancer seen in those with high intakes of these foods. Few other risk factors except cigarette smoking (and alcohol intake for certain cancers) confer risks of this magnitude (Block *et al.*, 1992). However, the evidence is weak for hormonal cancers, breast, prostate, cervix, and other cancers. When the epidemiological data are analyzed in terms of individual antioxidant nutrients, the evidence, compared with low dietary intake of fruits and vegetables doubling the risk of cancer, is not quite as definite but it is still strong both for vitamin C (Block, 1992) and for carotene (Van Poppel, 1993). Less is known about vitamin E. Studies that relate blood vitamin E levels to cancer risk have yielded inconsistent results (Knekt, 1992). However, two recent epidemiological studies, one of lung cancer (Mayne *et al.*, 1994) and one of oral cancer (Gridley *et al.*, 1992), found reduced cancer risks in individuals who took relatively high-dose single-entity vitamin E supplements.

2.1. Dietary/Habitual Intakes

A wide variety of international studies have investigated the relationship between intake of fruits and vegetables and risk of or mortality from various cancers, or between levels of antioxidant nutrients in the blood, reflecting dietary intake, and disease risk in healthy populations as well as in patients. Looking in detail at some of the more recent studies in healthy and disease populations (Torun *et al.*, 1995; Yuan *et al.*, 1995; Bunin *et al.*, 1994; Ferraroni *et al.*, 1994; Lavecchia *et al.*, 1994; Mayne *et al.*, 1994; Hunter *et al.*, 1993; Liu *et al.*, 1993; Malvy *et al.*, 1993; Tavani *et al.*, 1993; Buckley *et al.*, 1992; Comstock *et al.*, 1992, 1991; Eicholzer *et al.*, 1992; Freudenheim *et al.*, 1992; Longnecker *et al.*, 1992; Shibata *et al.*, 1992; Tominaga *et al.*, 1992; Vena *et al.*, 1992; Boeing *et al.*, 1991; Boeing and Frentzelbeyme, 1991; Bravo *et al.*, 1991; Demesquita *et al.*, 1991; Devet *et al.*, 1991; Gerber *et al.*, 1991; Herrero *et al.*, 1991; Knekt *et al.*, 1991, 1990, 1988; Olsen *et al.*, 1991; Palan *et al.*, 1991; Potischman *et al.*, 1991; Riboli *et al.*, 1991;

Richardson *et al.*, 1991; Smith and Waller, 1991; Stahelin *et al.*, 1991; Vaneenwyk *et al.*, 1991; West *et al.*, 1991; Zatonski *et al.*, 1991; Zaridze and Filipchenko, 1991; Ziegler *et al.*, 1991; Zaridze *et al.*, 1991; de Vries and Snow, 1990; Kalandidi *et al.*, 1990; LeGardeur *et al.*, 1990; Basu *et al.*, 1989; Connett *et al.*, 1989; Le Marchand *et al.*, 1989; Mills *et al.*, 1989; Verreault *et al.*, 1989; Brock *et al.*, 1988; Fontham *et al.*, 1988; Knekt, 1988; Koo, 1988; Rohan *et al.*, 1988; Russell *et al.*, 1988; Wald *et al.*, 1988, 1987, 1984; Burr *et al.*, 1987; Heinonen *et al.*, 1987, 1985; Miyamoto *et al.*, 1987; Paganini-Hill *et al.*, 1987; Menkes *et al.*, 1986; Hirayama, 1985; Colditz *et al.*, 1985; Ghosh and Das, 1985; Kolonel *et al.*, 1985; Nomura *et al.*, 1985; Salonen *et al.*, 1985; Hinds *et al.*, 1984; Willett *et al.*, 1984), it is clear that increased intake of dark green/yellow vegetables, carotenoid-containing vegetables, cruciferi, as well as tomatoes and fruit decreases risk of and mortality from lung, stomach and gastrointestinal, colorectal, oral, and head and neck cancers. This is apparently related to the carotenoid and vitamin C content of the fruits and vegetables.

In a study of 11,580 elderly subjects initially free of cancer (Shibata *et al.*, 1992) dietary vitamin C and fruit and vegetable intake was associated with decreased cancer risk at all sites in females—median tertiles for dietary vitamin C being 105, 176, and 256 mg/day and median tertiles for fruits and vegetables being 4.4, 6.64, and 9.66 servings /day. In a Dutch study of 164 patients with pancreatic cancer versus 480 controls, age range 35–79 years, vitamin C intake was significantly protective in women only and β-carotene intake especially derived from legumes and tomatoes (Demesquita *et al.*, 1991) was inversely associated with risk for cancer.

A higher dietary consumption of leafy greens, carrots, and fresh fruits (and fresh fish) was shown to be highly effective in protection against lung cancer in a Hong Kong population of Chinese female nonsmokers, 88 lung cancer patients, and 137 controls (Koo, 1988). A higher consumption of fruits being inversely related to lung cancer was also the outcome of the Athens study (Kalandidi *et al.*, 1990): 154 females were admitted to hospital with lung cancer versus 145 controls (admitted for orthopedic noncarcinogenic reasons). The effect is only partly explained in terms of vitamin C. A Hawaiian study of dietary fruit and vegetable intake and decreased risk for lung cancer revealed an especially strong association with β-carotene derived from dark green vegetables, cruciferous vegetables, and tomatoes (Le Marchand *et al.*, 1989).

Considering the β-carotene levels in the blood as a marker of dietary intake, the BUPA study of 22,000 males attending a London screening clinic showed that those developing cancer had a considerably decreased plasma β-carotene level, the strongest association being for lung cancer (Wald *et al.*, 1988). A New Zealand study measured blood serum levels of β-carotene in 2073 participants consisting of 389 cancer cases, 391 controls (hospital patients without cancer), 618 family members of cancer patients, and 675 controls (Smith and Waller, 1991). The study found that low levels of β-carotene were associated particularly with cancers of sites for which smoking is a strong risk factor, i.e., lung, gastrointestinal tract, cervix, and uterus. Furthermore, family members of lung cancer patients had lower blood levels of β-carotene than family members of controls. No differences were apparent in levels in patients with cancer of the breast, colon, prostate, or skin and their family members.

The Milan women's study conducted in 1984–1991 reveals an inverse relationship between β-carotene intake and cancer, highlighting the protective effects against esophageal cancer of high intakes of fresh fruits (Tavani *et al.*, 1993). In the Maryland study (Comstock *et al.*, 1991), the 436 subsequent cancers were studied in comparison with 765 controls. The prediagnostic β-carotene levels in the blood serum of the volunteers indicate strong protective associations with lung cancer and protective associations with melanoma and bladder cancer. A Finnish study of 117 cases of lung cancer in 4538 males initially cancer-free (Knekt *et al.*, 1991b) found a correlation between dietary intake of vitamin C and carotenoids and lung cancer occurrence. A similar follow-up study on β-carotene levels of 284 lung cancer cases diagnosed in 6800 males from Hawaii of Japanese ancestry was even more significant than the aforementioned.

A large range of studies on habitual intake of fresh fruits and vegetables have shown protective association of vitamins C and β-carotene for colorectal cancer (Ferraroni *et al.*, 1994; Zaridze and Filipchenko, 1991) pancreatic cancer (Olsen *et al.*, 1991) especially in relation to legumes and tomatoes (Demesquita *et al.*, 1991), gastric cancer (Lavecchia *et al.*, 1994), lung cancer (Morris *et al.*, 1994; LeGardeur *et al.*, 1990; Fontham *et al.*, 1988; Hinds *et al.*, 1984), cancer of the urinary tract for carotenoid intake only (Vena *et al.*, 1992), and cervical cancer (Bravo *et al.*, 1991).

Overall, it should be noted that the evidence in relation to cancer protection is less compelling in general for vitamin E intake, compared with vitamin C and β-carotene, although some strong data emerged from the Iowa women's study (Bostick *et al.*, 1993) (see Section 2.2) and the Basel study (Eicholzer *et al.*, 1992) showing increased risk for subsequent cancer at low vitamin E levels.

2.2. Supplementation

Antioxidants in fruits and vegetables may account for part of their beneficial effect. However, the effects of dietary intakes of the antioxidants vitamins C, E, and β-carotene are difficult to delineate from the other antioxidant constituents of fruits and vegetables such as flavonoids as well as other important anticarcinogenic components that are not antioxidants. These might include indoles that induce detoxifying enzymes, phytoestrogens that act as antiestrogens (and that may also have an antioxidant role as some of these compounds are isoflavonoids), and low-intake folic acid that has been associated with several cancers (Glynn and Albanes, 1994; Giovannucci *et al.*, 1993; Freudenheim *et al.*, 1991). Supplementation studies have helped to provide evidence pointing to the nature of the specific antioxidant nutrients in fruits and vegetables contributing toward the protective effects (Omenn *et al.*, 1996; Greenberg *et al.*, 1994; Heinonen *et al.*, 1994; Kirkpatrick *et al.*, 1994; Lamm *et al.*, 1994; Lockwood *et al.*, 1994; Mayne *et al.*, 1994; Wang *et al.*, 1994; Wei *et al.*, 1994; Benner *et al.*, 1993; Blot *et al.*, 1993; Bostick *et al.*, 1993; Hunter *et al.*, 1993; Rohan *et al.*, 1993; Barone *et al.*, 1992; Enstrom *et al.*, 1992; Gridley *et al.*, 1992; London *et al.*, 1992; Paganelli *et al.*, 1992; Shibata *et al.*, 1992; Cameron and Campbell, 1991; Nomura *et al.*, 1991; Ziegler *et al.*, 1991; Sundstrom *et al.*, 1989; Verreault *et al.*, 1989; Lu *et al.*, 1986; Moertel *et al.*, 1985).

A clinical trial of antioxidant supplements has demonstrated clinical improvement in the treatment of oral leukoplakia (Kaugars *et al.*, 1994). Although use of tobacco is often associated with the precancerous oral leukoplakia, there appears to an increased risk for malignant transformation in patients with oral leukoplakia who do not use tobacco. In previous studies, 15–71% of human oral leukoplakias were partially or completely resolved with oral β-carotene administration. This present study investigated the effectiveness of an antioxidant supplement (30 mg β-carotene, 800 IU vitamin E, 1000 mg vitamin C per day for 9 months) in 79 patients with oral leukoplakia.

Clinical improvement of oral lesions was demonstrated in 55% of patients after antioxidant supplementation. Improvement was more likely to occur in patients who had decreased their use of tobacco or alcohol (90% showed clinical improvement). However, approximately one-half of the patients who did not change their habits of alcohol and tobacco use also showed clinical improvement of oral lesions. On the basis of the data from the current study of 79 patients, it appears that a reduction in risk-factor exposures (which were mainly tobacco products) and the use of combined antioxidant supplements (vitamins E, C, and β-carotene) was successful in achieving clinical improvement of oral leukoplakia in a majority of patients.

The Linxian intervention study (Wang *et al.*, 1994; Blot *et al.*, 1993) also investigated combined supplements. This prospective study for subsequent cancer mortality demonstrates the positive health benefits of the antioxidant nutrients. Linxian, a rural country in north central China, has one of the highest rates of cancer of the esophagus in the world. The Linxian population has a low intake of dietary vitamins/minerals and higher cancer rates particularly of the stomach and esophagus, the death rate being >130/100,000 versus 13/100,000 in the United States. From four Linxian communes, 29,584 individuals aged 40–69 were recruited in 1985. The results are striking for gastric cancer, and especially a combined daily dose of vitamin E (30 mg, twice the U.S. recommended dietary allowance), β-carotene (15 mg), and selenium (50 μg), which for 5 1/4 years significantly reduced rates of cancer death (13% reduction in mortality) and overall mortality (9% reduction) in a population of healthy adults, especially gastric cancer (21% reduction in mortality). No improvement was observed in those receiving the combination containing 120 mg/day vitamin C.

Other supplementation studies with vitamins C and E have shown protective effects against oral and esophageal cancers. In one study, high-dosage supplementation for 18 months in breast cancer patients showed partial remission (Lockwood *et al.*, 1994). Megadose supplementation (2 g vitamin C, 400 IU vitamin E, plus other components) in patients with bladder cancer showed a significant decrease in recurrence versus those on the recommended dietary allowance (60 mg vitamin C, 30 IU vitamin E, plus other components) (Lamm *et al.*, 1994) .

The Iowa women's study consisted of 35,215 participants aged 55–69, free of cancer at the outset of the study. Of the 212 subsequent cases of colon cancer, vitamin E intake was inversely associated with risk especially in women age younger than 65 (Bostick *et al.*, 1993). A U.S. study of 1103 patients with oral and pharyngeal cancer and 1262 controls over four areas of the United States shows a reduced risk of oral cancer with supplementary vitamin E use, estimated at 100 IU/day (Gridley *et al.*, 1992).

A study of vitamin C supplementation (Enstrom *et al.*, 1992) has shown an inverse correlation between death caused by cancer and total vitamin C intake (for men but not for women) based on 11,348 subjects aged 25–74 individually examined during 1971 to 1974 and followed up in 1984. The mean intake of the whole study population was 83 mg dietary vitamin C per day (16% of the population being regular supplement users). Supplemented incurable cancer patients (294 versus 1532 not supplemented) had a median overall survival time double that of the controls (Cameron and Campbell, 1991).

Many studies have reported beneficial effects of β-carotene supplementation and cancer prevention, but the necessity for reappraisal was the outcome of the Finnish ATBC trial (Heinonen *et al.*, 1994). This was a randomized double-blind primary prevention trial involving approximately 29,000 male smokers, aged 50–69, supplementing with 20 mg β-carotene and/or 50 mg vitamin E daily for an average of 5 1/4 years. The participants had smoked a mean of 20 cigarettes daily for a mean of >30 years. Lung cancer was significantly increased in those supplementing with β-carotene. The adverse effects of high β-carotene supplementation in such high-risk populations are also substantiated by the CARET study (Omenn *et al.*, 1996). This multicenter, double-blind, randomized primary prevention trial involved 18,300 smokers, former smokers, and asbestos workers supplementing with 30 mg β-carotene and 25,000 IU retinol. After 4 years of supplementation, an increase in lung cancer risk was observed in the supplemented group. The trial was stopped prematurely.

3. CARDIOVASCULAR DISEASE

3.1. Dietary Intake

Evidence suggests that the presence of the antioxidant vitamins C, E, and β- carotene in the blood may have a protective role against cardiovascular disease and many recent studies are showing an inverse correlation between vitamin E levels in the blood and risk of or mortality from coronary heart disease, and in several instances for β-carotene, as well as for vitamin C and stroke (Kee *et al.*, 1995; Luoma *et al.*, 1995; Duthie *et al.*, 1994; Elinder *et al.*, 1994; Hensrud *et al.*, 1994; Knekt *et al.*, 1994; Leng *et al.*, 1994; Levy *et al.*, 1994; Morris *et al.*, 1994; Singh *et al.*, 1994; Street *et al.*, 1994; Tse *et al.*, 1994; Torun *et al.*, 1994; Van Lente *et al.*, 1994; Donnan *et al.*, 1993; Hense *et al.*, 1993; Kardinaal *et al.*, 1993; Adamscampbell *et al.*, 1992; Bolton-Smith *et al.*, 1992, 1991; Okamoto, 1992; Pronczuk *et al.*, 1992; Trout, 1992 (review); Gey *et al.*, 1991; Riemersma *et al.*, 1991, 1990; Gey and Puska, 1989; Salonen *et al.*, 1988, 1987; Jacques *et al.*, 1987; Kok *et al.*, 1987). Selected studies are highlighted below. The WHO European cross-cultural epidemiological study (Gey, 1994; Gey *et al.*, 1991) showed an inverse correlation between diet-derived plasma vitamin E levels (lipid-standardized) and mortality from ischemic heart disease in 100 apparently healthy males aged 40–49 from each of 16 different population groups. Twelve of the groups had a common plasma cholesterol level (5.7–6.2 mM) and the same blood pressure. The results also reveal the "ischemic heart disease" slope across Europe, those in southern Europe having higher blood vitamin E levels and lower mortality from ischemic heart disease compared with northern Europeans, with lower vitamin E levels in the blood and higher mortalities. Median levels of vitamin E were 28 μM in Spain (highest) and 19.5 μM in Edinburgh (lowest).

From the point of view of northern Europe, the coronary heart disease capital of the world, a low intake of antioxidant nutrients, especially in smokers, is likely to be a contributory factor to the higher incidence of coronary heart disease in Scotland. Cigarette smokers eat less fruits and vegetables and few cereals (Fulton *et al.*, 1988). A Scottish case–control study (Riemersma *et al.*, 1991) of a heavy-smoking population (with a high incidence of coronary heart disease) investigated relationships between plasma concentrations of vitamins C, E, and β-carotene and risk for angina pectoris. This case–control study of 6000 males aged 35–54 found that plasma concentrations of vitamin E (α-tocopherol)/cholesterol in 110 angina patients and 394 controls were inversely related to angina risk, after adjustment for age, smoking, blood pressure, lipids, and relative weight. The relationship with plasma β-carotene declined and that with vitamin C disappeared after adjustment for smoking. No correlation was found for vitamin A.

The Scottish Heart Health study (Bolton-Smith *et al.*, 1991) revealed that 17% of males with the highest mortality from coronary heart disease (age group 40–59) ate no green vegetables and 30% no fruit. The results suggested that high-risk populations might benefit from eating diets rich in antioxidants. Furthermore, the study (Bolton-Smith *et al.*, 1992) of 5123 males and 5236 females aged 40–59 assessing dietary intake of vitamins C, E, and β-carotene from fruits, vegetables, and so forth by food frequency questionnaires revealed a significantly higher risk of undiagnosed coronary heart disease in men lower in the higher quintiles; the opposite trend was found for diagnosed coronary heart disease (possibly because of changes in diet on diagnosis).

A more recent study has undertaken comparison of antioxidant status and free-radical-induced peroxidation of plasma lipoproteins (see later) in healthy young persons from Naples and Bristol (Parfitt *et al.*, 1994). The research found that ischemic heart disease mortality is much lower in southern Italy than in the United Kingdom. This is not completely explained by differences in classical risk factors. Because the "Mediterranean diet" includes large amounts of fresh fruits and vegetables, intake of antioxidant vitamins is likely to be high. There is increasing evidence that increased oxidation of plasma low-density lipoprotein (LDL) may be an important mechanism in the development of heart disease. The present study compared dietary intake, antioxidant status, and plasma lipid peroxidation in subjects on typical regional diets from southern Italy and the United Kingdom. The subjects from southern Italy consumed more fresh tomatoes and more monounsaturated fat (from olive oil) and had higher plasma levels of vitamin E and β-carotene than subjects from the United Kingdom. Intakes of vitamin C and total uncooked fruits and vegetables were similar in both groups, as were plasma levels of vitamin A, selenium, and copper. All indices of plasma lipid peroxidation were significantly lower in the subjects from southern Italy. The most striking finding was the much lower levels of plasma lipid peroxidation in the Naples group. The higher levels of the lipid antioxidants vitamin E and β-carotene in their lipoproteins are most likely to have accounted for this by conferring greater protection against free radical damage. The results suggest that high plasma concentrations of lipid antioxidants, particularly vitamin E and β-carotene, in southern Italians, as a consequence of characteristic features of their diet, such as high consumption of olive oil and fresh tomatoes, may contribute to the relatively low mortality from ischemic heart disease in this region by reducing free radical damage to plasma lipoproteins.

Knekt *et al.* (1994) undertook an observational study of the relationship between dietary intake of antioxidant vitamins and subsequent coronary heart disease in 5133 Finnish men and women aged 30–69, initially free from disease. During the 14-year follow-up there were 186 (males) and 58 (females) new fatal coronary disease cases in which the inverse relationships between vitamin E intake and relative risk were most significant for both men and women. Interestingly, there was a significant difference for men versus controls regarding fruit and vegetable intake, but the difference was not significant for women.

Investigating the question as to whether low levels of antioxidants are risk factors for myocardial infarction per se (Street *et al.*, 1994), 25,802 subjects aged 23–58 (who had donated blood samples in 1974) were studied during 1981 to 1988: 123 were diagnosed with myocardial infarction and, in comparison with the 246 selected controls, the findings indicated clearly that low blood levels of carotenoids especially β-carotene are associated with increased risk for subsequent myocardial infarction among smokers and a suggestive trend for decreasing lutein levels. The protective effects of higher vitamin E levels were suggested only in persons with high cholesterol levels.

The Euramic study (Kardinaal *et al.*, 1993), an observational multicenter European effort, involved analysis and comparison of data from 683 male patients, with a first myocardial infarction, and 727 controls. Lower β-carotene levels (in adipose tissue) were found in the myocardial infarction cases compared with controls. In addition, a significant inverse relationship between adipose tissue β-carotene levels and the development of the first myocardial infarction was found in nearly 1500 previously healthy males. There was also a graded association between levels of β-carotene and myocardial infarction risk according to smoking status. Vitamin E levels were not associated with myocardial infarction risk.

Clinical investigations of 29 patients with cardiovascular disease and its extent assessed in terms of at least 70% obstruction of the coronary artery versus 73 controls showed lower levels of vitamin E and β-carotene in patients with coronary heart disease (Levy *et al.*, 1994). In other surgical studies (Mezzetti *et al.*, 1995) on 48 patients undergoing coronary bypass surgery, 24 nonsmokers and 24 smokers, vitamin E levels in blood serum and arterial tissue were measured. Tissue vitamin E levels were almost 50% lower in the smokers, who also had increased oxidative damage of the tissue, the level being inversely related to tissue vitamin E. The severity of the coronary atherosclerotic lesions was inversely related to arterial vitamin E for both nonsmokers and smokers but much more significant for the latter. Vitamin C levels were also lower in tissue (by 50%) and in plasma (by 40%) in smokers. All data are highly significant (Mezzetti *et al.*, 1995).

3.2. Supplementation

More recent studies reveal that it is now becoming possible to suggest the intake of antioxidant nutrients is associated with the low subsequent incidence of disease (Stephens *et al.*, 1996; Hodis *et al.*, 1995; Paolisso *et al.*, 1995; Bellizzi *et al.*, 1994; Bostom *et al.*, 1994; Demaat *et al.*, 1994; Manson *et al.*, 1993; Rimm *et al.*, 1993; Stampfer *et al.*, 1993; Cerna *et al.*, 1992; Luostarinen *et al.*, 1992; Trout, 1992; Ringer *et al.*, 1991; Chernomorets *et al.*, 1990). A good example is the Health Professionals Study, which included self-supplementing groups (Rimm *et al.*, 1993; Stampfer *et al.*, 1993). The results show that supplementation with ≥100 IU of vitamin E daily for at least 2 years gives an approximately 37% reduction in coronary heart disease risk compared with those who did not supplement. The data do not prove a causal relationship but provide evidence of an association between a higher intake of vitamin E and a lower risk of coronary heart disease. This is an important study because the levels associated with these intakes would not be achieved from diet alone. β-Carotene was not associated with a lower risk among those who never smoked but was inversely associated with risk among current and former smokers. The study (Rimm *et al.*, 1993) involved 39,910 U.S. male health professionals aged 40–75, free of diagnosed coronary heart disease, diabetes, and hypercholesterolemia. During the 4-year follow-up, there were 667 cases of coronary heart disease with maximal reduction in men supplementing with 100–249 IU of vitamin E for at least 2 years, with no further decrease in risk with higher doses. This may relate to the fact that blood levels plateau and probably also the tissue levels. In addition, 87,245 female nurses aged 34–59 were involved in the study. During the 8-year follow-up, there were 522 cases of major coronary disease with 40% lower risk among regular users of vitamin E supplements for at least 2 years; <100 IU had little or no apparent benefit (Stampfer *et al.*, 1993).

The Food and Agriculture Organisation recalculation study for 19 western European and 5 non-European countries assessed premature mortality from coronary heart disease in males under the age of 65. Premature mortality was strongly related to vitamin E intake in all countries (Bellizzi *et al.*, 1994). Vitamin C and β-carotene moderately correlated with coronary heart disease.

The levels of vitamin E required for protection in the health professionals studies are consistent with the results of a clinical study of 156 males aged 40–59 with previous coronary artery bypass graft surgery (Hodis *et al.*, 1995). By supplementing four groups of patients with >100 IU vitamin E per day, <100 IU vitamin E per day, >250 mg vitamin C per day, or <250 mg vitamin C per day, subjects taking >100 IU vitamin E per day were demonstrated to have less coronary artery lesion progression than subjects with intake <100 IU/day for all lesions. Vitamin C intake had no effect on lesion progression (Hodis *et al.*, 1995).

The Iowa women's study (Kushi *et al.*, 1996) investigated the effects of vitamin C, E, and A intake both from food sources and from supplements on coronary heart disease deaths in a population of 34,000 postmenopausal women aged 55–69. The results showed an association between a higher coronary heart disease risk and a lower vitamin E intake from food but not from supplements. A recent secondary prevention trial (Stephens *et al.*, 1996) examined the consequences of vitamin E supplementation for myocardial infarction risk in 2002 patients with angiographically confirmed atherosclerosis. The results showed significantly fewer fatal myocardial infarctions in the supplemented group but there was an adverse trend toward cardiovascular deaths.

The Physicians Health Study (Hennekens *et al.*, 1996), one of the largest and longest ongoing intervention trials, investigated the effects of β-carotene supplementation on coronary dysfunction. The study commenced in 1982 with 22,000 male physicians aged 40–80; a preliminary analysis was done 6 years later of 333 subjects, all of whom had a previous coronary event, angina pectoris, and/or coronary revascularization. At this interim stage it was reported that the administration of 50 mg β-carotene on alternate days provided 44% reduction in all major coronary events defined as myocardial infarction, revascularization, or death, as reported in Gaziano *et al.*, (1990). After the 12-year follow-up, there were no differences in the incidence of cardiovascular disease or cancer, or death from all causes (Hennekens *et al.*, 1996).

3.3. Effects of Antioxidant Nutrients in Protecting LDL against Oxidation

Considerable evidence supports a role for vitamin E in protecting LDL (the cholesterol-carrying molecule of the body) from oxidation (oxidized LDL is implicated in the pathogenesis of coronary heart disease and is involved in the formation of the cholesterol-laden foam cells and the fatty streak *in vivo*). Several human studies report beneficial effects (Gaziano *et al.*, 1995; Jialal *et al.*, 1995; Mezzetti *et al.*, 1995; Princen *et al.*, 1995; Suzukawa *et al.*, 1995; Allard *et al.*, 1994; Brown *et al.*, 1994; Gilligan *et al.*, 1994; McDowell *et al.*, 1994; Nyyssonen *et al.*, 1994; Belcher *et al.*, 1993; Frei and Gaziano, 1993; Jialal and Grundy, 1993; Reaven *et al.*, 1993; Reaven and Witztum, 1993; Princen *et al.*, 1992; Dieber-Rotheneder *et al.*, 1991).

Some individuals have LDL with increased susceptibility (or lowered resistance) to oxidation and the strongest predictor appears to be a decreased vitamin E/cholesterol ratio (Frei and Gaziano, 1993). A wide range of human supplementation studies in normal populations have been undertaken to investigate the effects of vitamin E on the resistance of LDL to oxidation.

All of the studies show that vitamin E protects LDL against oxidative modification in supplemental levels ranging from 60 IU (40 mg)/day to 1600 mg/day (Suzukawa *et al.*, 1995; Brown *et al.*, 1994; Reaven and Witztum, 1993; Belcher *et al.*, 1993; Princen *et al.*, 1992; Dieber-Rotheneder *et al.*, 1991). Others have shown similar results in patients with hyperlipidemia. Recent results have proposed that supplementation of at least 400 IU is required to decrease the susceptibility of LDL to oxidation. However, β-carotene does not apparently enhance the oxidation resistance of LDL on supplementation (Gaziano *et al.*, 1995; Witztum, 1994), although others have shown effectiveness in smokers (Allard *et al.*, 1994). Combination supplementation apparently supports the contention that only vitamin E contributes to enhancing the resistance of LDL to oxidation (Gilligan *et al.*, 1994; Nyyssonen *et al.*, 1994; Jialal and Grundy, 1993; Reaven *et al.*, 1993).

3.4. Vitamin C and Stroke

Studies on hypertensive patients showed that those with uncontrolled and controlled hypertension have lower blood vitamin C levels than control normotensives (Tse *et al.*, 1994). Hensrud *et al.* (1994) studying 65,000 subjects from 65 counties in the People's Republic of China have shown that blood levels of β-carotene and vitamin C are inversely related to mortality rate for coronary heart disease with the relationship of low vitamin C levels and stroke being particularly significant. In the United Kingdom, Gale *et al.*, (1995) reported that vitamin C status in the elderly is a strong predictor of death from stroke. The 20-year survival of 730 men and women who had taken part in the DHSS's nutritional survey in 1973 was followed. The subjects had originally no history of stroke, cerebral arteriosclerosis, or coronary heart disease. During follow-up there were 634 deaths of which 124 were caused by stroke. Both plasma vitamin C levels and dietary vitamin C intake were inversely related to death from stroke. The association remained after adjustment for age, gender, other dietary components, and cardiovascular risk factors. Those subjects in the highest third of vitamin C intake (≥45 mg vitamin C—blood plasma level 28 μM) had a 60% reduced rate of stroke and a 20% reduced risk of mortality from coronary heart diseases compared with those in the lowest third (<28 mg vitamin C intake—blood plasma vitamin C level 12 μM). The findings of the relationships between low vitamin C levels and mortality from stroke were barely modified after exclusion of either all subjects who died during the first 2 years follow-up or during the first 10 years, an exercise carried out to ascertain the influence on the outcome of those who may have had undetectable symptoms at the start of the study.

Another study supporting the benefits of vitamin C and reduced risk of cerebrovascular disease for the elderly examined the association between a low vitamin C concentration in the elderly and a high fibrinogen concentration (Khaw and Woodhouse, 1995). The researchers hypothesized that vitamin C may protect against cardiovascular disease through an effect on blood clotting and response to infection. Maintaining adequate vitamin C, for example by eating an extra orange daily (around 60 mg), may reduce some of the seasonal variation in mortality. During the winter, dietary vitamin C intake is reduced, leading to lower serum ascorbate concentrations. This may be related to raised blood clotting factors (plasma fibrinogen), increased susceptibility to infection, and ultimately to increased levels of mortality from CVD during the winter months. In the United Kingdom, CVD mortality is 20–30% higher in the winter than in the summer. In the study, 96 men and women aged 65–74 were followed every 2 months for a 1-year period. Their dietary intakes, vitamin C concentration, clotting factors, acute-phase proteins (related to infection), and respiratory function were measured. Blood vitamin C concentration was inversely related to the clotting factors fibrinogen and factor VIIC, as well as to acute-phase proteins. The strong inverse relationship remained after adjustment for age, sex, and smoking status. There was significant seasonal variation in vitamin C intake from a mean intake of 65 mg/24 hr in the winter up to 90 mg/24 hr in the summer.

Subjects who reported taking vitamin C supplements had significantly higher serum ascorbate concentrations than subjects who did not take vitamin C (70.6 versus 49.1 μmole/liter) and lower plasma fibrinogen (2.68 versus 2.87 g/liter). The relationship between plasma ascorbate and clotting factors remained after supplement takers were excluded from the analysis. This indicates that vitamin C intakes within the normal dietary range have beneficial effects on clotting factors. Estimates from the study indicate that increasing vitamin C intake could have a measurable impact on risk factors. An extra 60 mg vitamin C per day (i.e., the amount contained in one orange) is associated with an increase in serum ascorbate concentration from 50 μmole/liter to 60 μmole/liter and a decline in fibrinogen concentration of 0.15 g/liter, equivalent to a decline of about 10% in ischemic heart disease risk. The study also found that smoking was associated with lower concentrations of plasma ascorbate and higher concentrations of blood clotting factors, despite smokers having similar vitamin C intakes to nonsmokers.

Although dietary intakes of vitamin E and carotene were not related to blood clotting factors, dietary vitamin E was low at 2.6 mg/24 hr. Other research indicates that the benefits of vitamin E on cardiovascular health may only be apparent at levels in excess of this, which are usually only achieved by supplements. In summary, the researchers emphasize the importance of maintaining vitamin C intakes, especially during the winter months when fruit and vegetable consumption is often reduced. They speculate that vitamin C may reduce the incidence of infections and thus lower plasma fibrinogen concentrations and there is experimental evidence that large doses of vitamin C increase fibrinolytic activity. Raised fibrinogen and factor VIIC are well-recognized risk factors for myocardial infarction and stroke. Evidence also exists that chronic and acute infection, and raised white cell count are risk factors for CVD as infection may contribute to the inflammatory processes occurring in atherosclerosis. Increased concentration of fibrinogen, which occurs during inflammation, is thought to be the mechanism by which acute and chronic illness could increase CVD.

Several other investigations have revealed that vitamin C levels are inversely associated with blood pressure in healthy persons (Trout, 1992; Salonen *et al.*, 1987) and that vitamin C intake is lower in individuals who subsequently suffer from peripheral vascular disease (Leng *et al.*, 1994). Furthermore, supplementation with vitamin C was shown to lower blood pressure (Trout, 1992) and in other studies to decrease blood cholesterol levels.

4. CONCLUSIONS

In recent years the potential for the antioxidants vitamins C, E, and β-carotene in disease prevention has been highlighted. The degenerative diseases of aging such as cancer, cardiovascular disease, cataracts, and brain dysfunction are increasingly found to have, in good part, an oxidative origin (Ames *et al.*, 1993). Much evidence is available to support the notion that dietary antioxidants such as vitamins C, E, and the carotenoids play a major role in minimizing this damage. Insufficient fruit and vegetable consumption increases the rate of most types of cancer about twofold. As a result, fruits and vegetables that contain these antioxidants are taking a prominent place in contemporary recommendations to promote good health and prevent disease. Indeed, evidence for increasing the recommended dietary allowance for vitamin C from 60 to 200 mg has been provided by the studies of Levine *et al.*, (1996), and it is possible to attain these levels from fruit and vegetable consumption, given the correct daily diet.

Epidemiological evidence indicates that avoidance of smoking and increased consumption of fruits and vegetables will have major effects in reducing rates of coronary heart disease and many cancers; in addition, diets rich in vitamins E and C have been shown to be particularly important in protection against heart disease, while the additional role of fruits and vegetables (particularly those high in β-carotene and vitamin C), as well as the control of infection, are significant in reducing cancer rates. Other factors include avoidance of intense sun exposure, increased participation in physical exercise, and reduction in alcohol consumption (Ames *et al.*, 1995).

5. REFERENCES

Adamscambell, L. L., Nwankwo, M. U., Ukoli, F. A., Omene, J. A., and Kuller, L.H., 1992, Serum retinol, carotenoids, vitamin E, and cholesterol in Nigerian women, *J. Nutr.* **3**:58–61.

Allard, J. P., Royall, D, Kurian, R., Muggli, R., and Jeefeebhoy, K. N., 1994, Effects of beta-carotene supplementation on lipid peroxidation in humans, *Am. J. Clin. Nutr.* **59**:884–890.

Ames, B. N., Shigenega, M. K., and Hagen, T. M., 1993, Oxidants, antioxidants and the degenerative diseases of ageing, *Proc. Nat. Acad. Sci. USA* **90**:7915–7922.

Ames, B. N., Golf, L. S., and Willett, W. C., 1995, The causes and prevention of cancer, *Proc. Nat. Acad. Sci. USA* **92**:5258–5265.

Barone, J., Taioli, E., Hebert, J. R., and Wynder, E. L., 1992, Vitamin supplement use and risk for oral and esophageal cancer, *Nutr. Cancer* **18**:31–41.

Basu, T. K., Hill, G. B., Ng, D., Abdi, E., and Temple, N., 1989, Serum vitamins A and E, beta-carotene, and selenium in patients with breast cancer, *J. Amer. Coll. Nutr.* **8**:524–529.

Belcher, J. D., Balla, J., Balla, G., Jacobs, D. R., Gross, M., Jacob, H. S., and Vercellotti, G. M., 1993, Vitamin E, LDL, and endothelium—Brief oral vitamin supplementation prevents oxidised LDL-mediated vascular injury in vitro, *Arterioscler. Thromb.* **13**:1779–1789.

Bellizzi, M. C., Franklin, M. F., Duthie, G. G., and James, W. P., 1994, Vitamin E and coronary heart disease: The European paradox, *Eur. J. Clin. Nutr.* **48**:822–831.

Benner, S. E., Winn, R. J., Lippman, S. M., Poland, J., Hansen, K. S., Luna, M. A., and Hong, W. K., 1993, Regression of oral leukoplakia with alpha-tocopherol—A community clinical oncology program chemoprevention study, *J. Natl. Cancer Inst.* **85**:44–47.

Block, G., 1992, A role for antioxidants in reducing cancer risk, *Nutr. Rev.* **50**:207–213.

Block, G., Patterson, B., and Subar, A., 1992, Fruit, vegetables and cancer prevention. A review of the epidemiological evidence, *Nutr. Cancer* **18**:1–29.

Blot, W. J., Li, J. Y., Taylor, P. R., Guo, W., Dawsey, S., Wang, G. Q., Yang, C. S., Zheng, S. F., Gail, M., Li, G. Y., Yu, Y., Liu, B. Q., Tangrea, J., Sun, Y. H., Liu, F. S., Fraumeni, J. F., Zhang, Y. H., and Li, B., 1993, Nutrition intervention trials in Linxian, China: Supplementation with specific vitamin/mineral combinations. Cancer incidence and disease-specific mortality in the general population, *J. Nat. Cancer Inst.* **85**:1483–1492

Boeing, H., and Frentzelbeyme, R., 1991, Regional risk factors for stomach cancer in the FRG, *Environ. Health Perspect.* **94**:83–89.

Boeing, H., Frentzelbeyme, R., Berger, M., Berndt, V., Gores, W., Kornder, M., Lohmeier, R., Menarcher, A., Mannl, H. F. K., Meinhardt, M., Muller, R., Ostermeier, H., Paul, F., Schwemmle, K., Wagner, K. H., and Wahrendorf, J., 1991, Case–control study on stomach cancer in Germany, *Int. J. Cancer* **47**:858–864.

Bolton-Smith, C., Woodward, M., Smith, W. C., and Tunstall-Pedoe, H., 1991a, Dietary and non-dietary predictors of serum total and HDL-cholesterol in men and women: Results from the Scottish Heart Health Study, *Int. J. Epidemiol.* **20**:95–104.

Bolton-Smith, C., Smith, W. C., Woodward, M., and Tunstall-Pedoe, H., 1991b, Nutrient intakes of different social-class groups: Results from Scottish Heart Health Study, *Br. J. Nutr.* **65**:321–335.

Bolton-Smith, C., Woodward, M., and Tunstall-Pedoe, H., 1992, The Scottish Heart Health Study, Dietary intake by food frequency questionnaire and odds ratios for coronary heart disease risk. II. The antioxidant vitamins and fibre, *Eur. J. Nutr.* **46**:85–93.

Bostick, R. M., Potter, J. D., McKenzie, D. R., Sellers, T. A., Kushi, L. H., Steinmetz, K. A., and Folsom, A. R., 1993, Reduced risk of colon cancer with high intake of vitamin E—The Iowa Womens Health Study, *Cancer Res.* **53**:4230–4237.

Bostom, A. G., Yanek, L., Hume, A. L., Eaton, C. B., McQuade, W., Nadeau, M., Perrone, G., Jacques, P. F., and Selhug, J., 1994, High dose ascorbate supplementation fails to affect plasma homocyst(e)ine levels in patients with coronary heart disease, *Atherosclerosis* **111**:267–270.

Bravo, M. P., Castellanos, E., and Calero, J. D., 1991, Dietary factors and prostatic cancer, *Urol. Int.* **46**:163–166.

Brock, K. E., Berry, G., Mock, P. A., MacLennan, R., Truswell, A. S., and Brinton, L. A., 1988, Nutrients in diet and plasma and risk of in situ cervical cancer, *J. Natl. Cancer Inst.* **80**:580–585.

Brown, K. M., Morrice, P. C., and Duthie, G. G., 1994, Vitamin E supplementation suppresses indexes of lipid peroxidation and platelet counts in blood of smokers and non-smokers but plasma lipoprotein concentrations remain unchanged, *Am. J. Clin. Nutr.* **60**:383–387.

Buckley, D. I., McPherson, R. S., North, C. Q., and Becker, T. M., 1992, Dietary micronutrients and cervical dysplasia in south-western American-Indian women, *Nutr. Cancer* **17**:179–185.

Bunin, G. R., Kuijten, R. R., Boesel, C. P., Buckley, J. D., and Meadows, A. T., 1994, Maternal diet and risk of astrocytic glioma in children—A report from the Childrens Cancer Group (United States and Canada), *Cancer Causes Control* **5**:177–187.

Burr, M. L., Samloff, I. M., Bates, C. J., and Holliday, R. M., 1987, Evaluation of vitamin A and C status in normal and malignant conditions and their possible role in cancer prevention, *Jpn. J. Cancer Res.* **76**:1174–1178.

Camara, B., and Moneger, R., 1978, Free and esterified carotenoids in green and red fruits and Capsicum annuum, *Phytochemistry* **17**:91–93.

Cameron, E., and Campbell, A., 1991, Innovation vs. quality control: An unpublishable' clinical trial of supplemental ascorbate in incurable cancer, *Med. Hypoth.* **36**:185–189.

Cerna, O., Ramacsay, L., and Ginter, E., 1992, Plasma lipids, lipoproteins and atherogenic index in men and women administered vitamin C, *Cor Vasa* **34**:246–254.

Cerutti, P. A., 1994, Oxyradicals and cancer, *Lancet* **344**:862–863.

Chernomorets, N. N., Kotlubei, G. V., Vatutin, N. T., Zhibotovskaia, I. A., Gnilitskaia, V. B., Alifanova, R. E., Loback, E. I., Mal'tseva, N. V., and Mitrofanov, A. N., 1990, The effect of tocopherol and nicotinic acid on the microcirculation and blood coagulability in patients with ischemic heart disease, [in Russian], *Vrach. Delo* **12**:6–8.

Colditz, G. A., Branch, L. G., Lipnick, R. J., Willett, W. C., Rosner, B., Posner, B. M., and Hennekens, C. H., 1985, Increased green and yellow vegetable intake and lowered cancer deaths in an elderly population, *Am. J. Clin. Nutr.* **41**:32–36.

Comstock, G. W., Helzlsouer, K. J., and Bush, T. L., 1991, Prediagnostic serum levels of carotenoids and vitamin E as related to subsequent cancer in Washington County, Maryland, *Am. J. Clin. Nutr.* **53** (1 Suppl.):260S–264S.

Comstock, G. W., Bush, T. L., and Helzlsouer, K., 1992, Serum retinol, beta-carotene, vitamin E, and selenium as related to subsequent cancer of specific sites, *Am. J. Epidemiol.* **135**:115–121.

Connett, J. E., Kuller, L. H., Kjelsberg, M. O., Polk, F. B., Collins, G., Rider, A., and Hulley, S. B., 1989, Relationship between carotenoids and cancer. The Multiple Risk Factor Intervention Trial (MRFIT) Study, *Cancer* **64**:126–134.

DeMaat, M. P. M., Princen, H. M. G., Dagneli, P. C., Kamerling, S. W. A., and Kluft, C., 1994, Effect of fish oil and vitamin E on the cardiovascular risk indicators fibrinogen, C-reactive protein and PAI activity in healthy young volunteers, *Fibrinolysis* **8**:50–52.

DeMesquita, H. B. B., Maisonneuve, P., Runia, S., and Moerman, C. J., 1991, Intake of foods and nutrients and cancer of the exocrine pancreas—A population-based case–control study in the Netherlands, *Int. J. Cancer* **48**:540–549.

DeVet, H. C. W., Knipschild, P. G., Grol, M. E. C., Schouten, H. J. A., and Sturmans, F., 1991, The role of beta-carotene and other dietary factors in the etiology of cervical dysplasia—Results of a case–control study, *Int. J. Epidemiol.* **20**:603–610.

De Vries, N., and Snow, G. B., 1990, Relationships of vitamins A and E and beta-carotene serum levels to head and neck cancer patients with or without second primary tumours, *Eur. Arch. Oto-Rhino-Laryngol.* **247**:368–370.

Dieber-Rotheneder, M., Puhl, H., Waeg, G., Striegl, G., and Esterbauer, H., 1991, Effect of oral supplementation with D-α-tocopherol on the vitamin E content of human low density lipoproteins and resistance to oxidation, *J. Lipid Res.* **32**:1325–1332.

Diplick, A., 1994, *Antioxidants and Disease Prevention*, Molecular Aspects of Medicine, **15**(4), Pergamon Press, New York.

Donnan, P. T., Thomson, M., Fowkes, F. G. R., Prescott, R. J., and Housely, E., 1993, Diet as a risk factor for peripheral arterial disease in the general population—The Edinburgh Artery Study, *Am. J. Clin. Nutr.* **57**:917–921.

Dorgan, A. F., and Schatzkin, A., 1991, Antioxidant micronutrients in cancer prevention, *Hematol. Oncol. Clin. North Am.* **5**:43–68.

Duthie, G. G., Beattie, J. A. G., Arthur, J. R., Franklin, M., Morrice, P. C., and James, W. P. T., 1994, Blood antioxidants and indexes of lipid peroxidation in subjects with angina pectoris, *Nutrition* **10**:313–316.

Eicholzer, M., Stahelin, H. B., and Gey, K. F., 1992, Inverse correlation between essential antioxidants in plasma and subsequent risk to develop cancer, ischemic heart disease and stroke respectively: 12-year follow-up of the Prospective Basel Study, *EXS* **62**:398–410.

Elinder, L. S., Hadell, K., Johansson, J., Molgaard, J., Olsson, A. G., and Walldius, G., 1994, Serum lipid soluble antioxidants in patients with hyperlipoproteinemia, *Nutr. Metab. Cardiovasc. Dis.* **4**:197–203.

Enstrom, J. E., Kanim, L. E., and Klein, M. A., 1992, Vitamin C intake and mortality among a sample of the United States population, *Epidemiology* **3**:194–202.

Ferraroni, M., LaVecchia, C., Davanzo, B., Negri, E., Franceschi, S., and DeCarli, A., 1994, Selected micronutrient intake and the risk of colorectal cancer, *Br. J. Cancer* **70**:1150–1155.

Fontham, E. T., Pickle, L. W., Haenszel, W., Correa, P., Lin, Y. P., and Falk, R. T., 1988, Dietary vitamins A and C and lung cancer risk in Louisiana, *Cancer* **62**:2267–2273.

Frei, B., and Gaziano, J. M., 1993, Content of antioxidants, preformed lipid hydroperoxides, and cholesterol as predictors of the susceptibility of human LDL to metal ion-dependent and ion-independent oxidation, *J. Lipid Res.* **34**:2135–2145.

Freudenheim, J. L., Graham, S., Marshall, J. R., Haughey, B. P., Cholewinski, S., and Wilkinson, G., 1991, Folate intake and carcinogenesis of the colon and rectum, *Int. J. Epidemiol.* **20**:368–374.

Freudenheim, J. L., Graham, S., Byers, T. E., Marshall, J. R., Haughey, B. P., Swanson, M. K., and Wilkinson, G., 1992, Diet, smoking, and alcohol in cancer of the larynx—A case–control study, *Nutr. Cancer* **17**:33–45.

Fulton, M., Thomson, M., Elton, R. A., Brown, S., Wood, D. A., and Oliver, M. F., 1988, Cigarette smoking, social class and nutrient intake: Relevance to coronary heart disease, *Eur. J. Clin. Nutr.* **42**:797–803.

Gale, C. R., Martyn, C. N., Winter, P. D., and Cooper, C., 1995, Vitamin C and risk of death from stroke and coronary heart disease in cohort of elderly people, *Br. Med. J.* **310**:1563–1566.

Gaziano, J. M., Manson, J. E., Ridker, P. M., Burng, J. E., and Hennekens, C. H., 1990, β-Carotene therapy for chronic stable angina, *Circulation* **82**(Suppl. III):201.

Gaziano, J. M., Hatta, A., Flynn, M., Johnson, E. J., Krinsky, N. I., Ridker, P. M., Hennekens, C. H., and Frei, B., 1995, Supplementation with beta-carotene in vivo and in vitro does not inhibit low density lipoprotein oxidation, *Atherosclerosis* **112**:187–195.

Gerber, M., Richardson, S., Salkeld, R., and Chappuis, P., 1991, Antioxidants in female breast cancer patients, *Cancer Invest.* **9**:421–428.

Gey, K. F., 1994, The relationship of antioxidant status and the risk of cancer and cardiovascular disease: A critical evaluation of observational data, in *Free Radicals in the Environment, Medicine and Toxicology* (H. Nohl, H. Esterbauer, and C. Rice-Evans, eds.), pp. 181–219, Richelieu Press, London.

Gey, K. F., Puska, P., 1989, Plasma vitamins E and A inversely correlated to mortality from ischemic heart disease in cross-cultural epidemiology, *Ann. N.Y. Acad. Sci.* **570**:268–282.

Gey, K. F., Puska, P., Jordan, P., and Moser, U. K., 1991, Inverse correlation between plasma vitamin E and mortality from ischemic heart disease in cross-cultural epidemiology, *Am. J. Clin. Nutr.* **53**(1 Suppl.):326S–334S.

Ghosh, J., and Das, S., 1985, Evaluation of vitamin A and C status in normal and malignant conditions and their possible role in cancer prevention, *Jpn. J. Cancer Res.* **76**:1174–1178.

Gilligan, D. M., Sack, M. N., Guetta, V., Casino, P. R., Quyyumi, A. A., Rader, D. J., Panza, J. A., and Cannon, R. O., 1994, Effect of antioxidant vitamins on low density lipoprotein oxidation and impaired endothelium-dependent vasodilation in patients with hypercholesterolemia, *J. Am. Coll. Cardiol.* **24**:1611–1617.

Giovanucci, E., Stampfer, M. J., Colditz, G. A., Rimm, E. B., Trichopoulos, D., Rosner, B. A., Speizer, F. E., and Willett, W. C., 1993, Folate, methionine, and alcohol intake and risk of colorectal cancer, *J. Nat. Cancer Inst.* **85**:875–884.

Glynn, S. A., and Albanes, D., 1994, Folate and cancer—A review of the literature, *Nutr. Cancer* **22**:101–119.

Greenberg, E. R., Baron, J. A., Tosteson, T. D., Freeman, D. H., Beck, G. J., Bond, J. H., Colachhio, T. A., Coller, J. A., Frankl, H. D., Haile, R. W., Mandel, J. S., Nierenberg, D. W., Rothstein, R., Snover, D. C., Stevens, M. M., Summers, R. W., and Vanstolk, R. U., 1994, Clinical trial of antioxidant vitamins to prevent colorectal adenoma, *N. Engl. J. Med.* **331**:141–147.

Gregory, J., 1990, *The Dietary and Nutritional Survey of British Adults*, HMSO, London.

Gridley, G., McLaughlin, J. K., Block, G., Blot, W. J., Gluch, M., and Fraumeni, J. F., 1992, Vitamin supplement use and reduced risk of oral and pharyngeal cancer, *Am. J. Epidemiol.* **135**:1083–1092.

Heinonen, P. K., Koskinen, T., and Tuimala, R., 1985, Serum levels of vitamins A and E in women with ovarian cancer, *Arch. Gynecol.* **237**:37–40.

Heinonen, P. K., Kuoppala, T., Koskinen, T., and Punnonen, R., 1987, Serum vitamins A and E and carotene in patients with gynecologic cancer, *Arch. Gynecol. Obstet.* **241**:151–156.

Heinonen, O. P., Huttunen, J. K., Albanes, D., *et al.*, 1994, Effect of vitamin E and beta-carotene on the incidence of lung cancer and other cancers in male smokers, *N. Eng. J. Med.* **330**:1029–1035.

Henderson, B. E., Ross, R. K., and Pike, M. C., 1991, Toward the primary prevention of cancer, *Science* **254**:1131–1138.

Hennekens, C. H., Buring, J. E., Manson, J. E., Stampfer, M., Rosner, B., Cook, N. R., Belanger, C., Lamotten, F., Gaziano, J. M., Ridker, P. M., Willett, W., and Peto, R., 1996, Lack of effect of long-term supplementation with β-carotene on the incidence of malignant neoplasms and cardiovascular disease, *N. Engl. J. Med.* **334**:1145–1149.

Hense, H. W., Stender, M., Bors, W., and Keil, U., 1993, Lack of an association between serum vitamin E and myocardial infarction in a population with high vitamin E levels, *Atherosclerosis* **103**:21–28.

Hensrud, D. D., Heimburger, D. C., Chen, J., and Parpia, B., 1994, Antioxidant status, erythrocyte fatty acids, and mortality from cardiovascular disease and Keshan disease in China, *Eur. J. Clin. Nutr.* **48:**455–64.

Herrero, R., Potischman, N., Brinton, L. A., Reeves, W. C., Brenes, M. M., Tenorio, F., DeBritton, R. C., and Gaitan, E., 1991, A case–control study of nutrient status and invasive cervical cancer. 1. Dietary indicators, *Am. J. Epidemiol.* **134:**1335–1346.

Hertog, M. G. L., Feskens, E. J. M., Hollman, P. C. H., Katan, M. B., and Kromhout, D., 1993, Dietary antioxidant flavonoids and risk of coronary heart disease: The Zutphen Elderly Study, *Lancet* **342:**1007–1011.

Hinds, M. W., Kolonel, L. N., Hankin, J. H., and Lee, J., 1984, Dietary vitamin A, carotene, vitamin C and risk of lung cancer in Hawaii, *Am. J. Epidemiol.* **119:**227–237.

Hirayama, T., 1985, A large scale cohort study on cancer risks by diet—With special reference to the risk reducing effects of green-yellow vegetable consumption, [review] *Princess Takamatsu Symp.* **16:**41–53.

Hodis, H. N., Mack, W. J., LaBree, L., Cashin-Hemphill, L., Sevanian, A., Johnson, R., and Azen, S. P., 1995, Serial coronary angiographic evidence that antioxidant vitamin intake reduces progression of coronary artery atherosclerosis, *J. Am. Med. Assoc.* **273:**1849–1854.

Hunter, D. J., Manson, J. E., Colditz, G. A., Stampfer, M. J., Rosner, B., Hennekens, C. H., Speizer, F. E., and Willett, W. C., 1993, A prospective study of the intake of vitamin C, vitamin E, and vitamin A and the risk of breast cancer, *N. Engl. J. Med.* **329:**234–240.

Jacques, P. F., Hartz, S. C., McGandy, R. B., Jacob, R. A., and Russell, R. M., 1987, Ascorbic acid, HDL, and total plasma cholesterol in the elderly, *J. Am. Coll. Nutr.* **6:**169–174.

Jialal, I., and Grundy, S. M., 1993, Effect of combined supplementation with alpha-tocopherol, ascorbate and beta-carotene on low density lipoprotein oxidation, *Circulation* **88:**2780–2786.

Jialal, I., Fuller, C. J., and Huet, B. A., 1995, The effect of alpha-tocopherol supplementation on LDL oxidation, *Arterioscler. Thromb. Vasc. Biol.* **15:**190–198.

Kalandidi, A., Katsouyanni, K., Voropoulou, N., Bastas, G., Saracci, R., and Trichopoulos, D., 1990, Passive smoking and diet in the etiology of lung cancer among non-smokers, *Cancer Causes Control* **1:**15–21.

Kardinaal, A. F. M, Kok, F. J., Ringstad, J., Gomezarecena, J., Mazaev, V. P., Kohlmeier, L., Martin, B. C., Aro, A., Kark, J. D., Delgadorodriguez, M., Riemersma, R. A., Vantveer, P., Huttunen, J. K., and Martinmoreno, J. M., 1993, Antioxidants in adipose tissue and risk of myocardial infarction—The Euramic Study, *Lancet* **342:**1379–1384.

Kaugars, G. E., Silverman, S., Lovas, J. G. L., Brandt, R. B., Riley, W. T., Dao, Q., Singh, V. N., and Gallo, J., 1994, A clinical trial of antioxidant supplements in the treatment of oral leukoplakia, *Oral Surg. Oral Med. Oral Pathol.* **78:**462–468.

Kee, F., Weston, J., McCrum, E. E., McMaster, D., Yarnell, J., Fruchart, J. C., Bard, J. M., Cambien, F., and Evans, A. E., 1995, The effect of diet on lipid apoprotein and lipoparticle variation in the ECTIM study in Belfast, *Rev. Epidemiol. Sante Publique* **43:**18–25.

Khackik, F., Beecher, G. R., Goli, M. B., and Lusby, W. R., 1992, Separation and quantitation of carotenoids in food, *Methods Enzymol.* **213:**336–346.

Khaw, K., and Woodhouse, P., 1995, Interrelation of vitamin C, infection, haemostatic factors and cardiovascular disease, *B. Med. J.* **310:**1559–1563.

Kirkpatrick, C. S., White, E., and Lee, J. A. H., 1994, Case–control study of malignant melanoma in Washington state. 2. Diet, alcohol, and obesity, *Am. J. Epidemiol.* **139:**869–880.

Knekt, P., 1988, Serum vitamin E level and risk of female cancers, *Int. J. Epidemiol.* **17:**281–286.

Knekt, P., Aromaa, A., Maatela, J., Aaran, R. K., Nikkari, T., Hakama, M., Hakulinen, T., Peto, R., Saxen, E., and Teppo, L., 1988a, Serum vitamin E and risk of cancer among Finnish men during a 10-year follow-up, [review] *Am. J. Epidemiol.* **127:**28–41.

Knekt, P., Aromaa, A., Maatela, J., Alfthan, G., Aaran, R. K., Teppo, L., and Hakama, M., 1988b, Serum vitamin E, serum selenium and the risk of gastrointestinal cancer, *Int. J. Cancer* **42:**846–850.

Knekt, P., Aromaa, A., Maatela, J., Aaran, R. K., Nikkari, T., Hakama, M., Hakulinen, T., Peto, R., and Teppo, L., 1990, Serum vitamin A and subsequent risk of cancer: Cancer incidence follow-up of the Finnish Mobile Clinic Health Examination Survey, *Am. J. Epidemiol.* **132:**857–858.

Knekt, P., Aromaa, A., Maatela, J., Alfthan, G., Aaran, R. K., Nikkari, T., Hakama, M., Hakulinen, T., and Teppo, L., 1991a, Serum micronutrients and risk of cancers of low incidence in Finland, *Am. J. Epidemiol.* **134:**356–361.

Knekt, P., Jarvinen, R., Seppanen, R., Rissanen, A., Aromaa, A., Heinonen, O. P., Albanes, D., Heinonen, M., Pukkala, E., and Teppo, L., 1991b, Dietary antioxidants and the risk of lung cancer, *Am. J. Epidemiol.* **134:**471–479.

Knekt, P., Reunanen, A., Jarvinen, R., Seppanen, R., Heliovaara, M., and Aromaa, A., 1994, Antioxidant vitamin intake and coronary mortality in a longitudinal population study, *Am. J. Epidemiol.* **139:**1180–1189.

Kok, F. J., De Bruijn, A. M., Vermeeren, R., Hofman, A., Van Laar, A., De Bruin, M., Hermus, R. J., and Valkenburg, H. A., 1987, Serum selenium, vitamin antioxidants, and cardiovascular mortality: A 9-year follow-up study in the Netherlands, *Am. J. Nutr.* **45:**462–468.

Kolonel, L. N., Hinds, M. W., Nomura, A. M., Hankin, J. H., and Lee, J., 1985, Relationship of dietary vitamin A and ascorbic acid intake to the risk for cancers of the lung, bladder and prostate in Hawaii, *Natl. Cancer Inst. Monogr.* **69:**137–142.

Koo, L. C., 1988, Dietary habits and lung cancer risk among Chinese females in Hong Kong who never smoked, *Nutr. Cancer* **11:**155–172.

Kushi, L. H., Folsom, A. R., Prineas, R. J., *et al.*, 1996, Dietary antioxidants and death from coronary heart disease in post-menopausal women, *N. Engl. J. Med.* **334:**1156–1152.

Lamm, D. L., Riggs, D. R., Shriver, J. S., Vangilder, P. F., Rach, J. F., and Dehaven, J. I., 1994, Megadose vitamins in bladder cancer—A double-blind clinical trial, *J. Urol.* **151:**21–26.

LaVecchia, C., Ferraroni, M., Davanzo, B., DeCarli, A., and Franceschi, S., 1994, Selected micronutrient intake and the risk of gastric cancer, *Cancer Epidemiol. Biomarkers Prev.* **3:**393–398.

LeGardeur, B. Y., Lopez, A., and Johnson, W. D., 1990, A case–control study of serum vitamins A, E, and C in lung cancer patients, *Nutr. Cancer* **14:**133–140.

Le Marchand, L., Yoshizawa, C. N., Kolonel, L. N., Hankin, J. H., and Gookman, M. T., 1989, Vegetable consumption and lung cancer risk: A population-based case–control study in Hawaii, *J. Natl. Cancer Inst.* **81:**1158–1164.

Leng, G. C., Horrobin, D. F., Fowkes, F. G. R., Smith, F. B., Lowe, G. D. O., Donnan, P. T., and Ells, K., 1994, Plasma essential fatty acids, cigarette smoking, and dietary antioxidants in peripheral arterial disease—A population-based case–control study, *Arterioscler. Thromb.* **14:**471–478.

Levine, M., Conry-Cantilena, C., Wang, Y., Welch, R. W., Washko, P. W., Dhariwal, K. R., Park, J. B., Lazarev, A., Graumlich, J. F., King, J., and Cantilena, L. R., 1996, Vitamin C pharmacokinetics in healthy volunteers: Evidence for a recommended dietary allowance, *Proc. Natl. Acad. Sci. USA* **93:**3704–3709.

Levy, Y., Nseir, W., Boulos, M., Omansky, T., Ben-Amotz, A., and Cohen, A., 1994, Relationship between plasma antioxidants and coronary artery disease [in hebrew], *Harefuah* **127:**154–157.

Liu, T. P., Soong, S. J., Wilson, N. P., Craig, C. B., Cole, P., Macaluso, M., and Butterworth, C. E., Jr., 1993, A case–control study of nutritional factors and cervical dysplasia, *Cancer Epidemiol. Biomarkers Prev.* **2:**525–530.

Lockwood, K., Moesgaard, S., Hanioka, T., and Folkers, K., 1994, Apparent partial remission of breast cancer in high risk patients supplemented with nutritional antioxidants, essential fatty acids and coenzyme Q, *Mol. Aspects Med.* **15:**231–240.

London, S. J., Stein, E. A., Henderson, I. C., Stampfer, M. J., Wood, W. C., Remine, S., Dmochowski, J. R., Robert, N. J., and Willett, W. C., 1992, Carotenoids, retinol, and vitamin E and risk of proliferative benign breast disease and breast cancer, *Cancer Causes Control* **3:**503–512.

Longnecker, M. P., Martinmoreno, J. M., Knekt, P., Nomura, A. M. Y., Schober, S. E., Stahelin, H. B., Wald, N. J., Gey, K. F., and Willett, W. C., 1992, Serum alpha-tocopherol concentration in relation to subsequent colorectal cancer—Pooled data from 5 cohorts, *J. Natl. Cancer Inst.* **84:**430–435.

Lu, S. H., Ohshima, H., Fu, H. M., Tian, Y., Li, F. M., Blettner, M., Wahrendorf, J., and Bartsch, H., 1986, Urinary excretion of N-nitrosoamino acids and nitrate by inhabitants of high- and low-risk areas for esophageal cancer in northern China: Endogenous formation of nitrosoproline and its inhibition by vitamin C, *Cancer Res.* **46:**1485–1491.

Luoma, P. V., Nayha, S., Sikkila, K., and Hassi, J., 1995, High serum alpha-tocopherol, albumin, selenium and cholesterol, and low mortality from coronary heart disease in northern Finland, *J. Intern. Med.* **237:**49–54.

Luostarinen, R., Siegbahn, A., and Saldeen, T., 1992, Effect of dietary fish oil supplemented with different doses of vitamin E on neutrophil chemotaxis in healthy volunteers, *Nutr. Res.* **12:**1419–1430.

Malvy, D. J., Burtschy, B., Arnaud, J., Sommelet, D., Leverger, G., Dostalova, L., Drucker, J., and Amedee-Manesme, O., 1993, Serum beta-carotene and antioxidant micronutrients in children with cancer. The 'Cancer in Children and Antioxidant Micronutrients' French Study Group, *Int. J. Epidemiol.* **22:**761–771.

Manson, J. E., Gaziano, J. M., Jonas, M. A., Hennekens, C. H., 1993, Antioxidants and cardiovascular disease—A review, *J. Am. Coll. Nutr.* **12:**426–432.

Mayne, S. T., Janerich, D. T., Greenwald, P., Chorost, S., Tucci, C., Zaman, M. B., Melamed, M. R., Kiely, M., and McKneally, M. F., 1994, Dietary beta carotene and lung cancer risk in US non-smokers, *J. Natl. Cancer Inst.* **86:**33–38.

McCance, R. A., and Widdowson, E. M., 1992, *The Composition of Foods*, Royal Society of Chemistry–MAFF, London.

McDowell, I. F. W., Brennan, G. M., McEneny, J., Young, I. S., Nicholls, D. P., McVeigh, G. E., Bruce, I., Trimble, E. R., and Johnston, G. D., 1994, The effect of probucol and vitamin E treatment on the oxidation of low density lipoprotein and forearm vascular responses in humans, *Eur. J. Clin. Invest.* **24:**759–765.

Medina, D., 1986, Mechanisms of selenium inhibition of tumorigenesis, *J. Am. Coll. Toxicol.* **5:**21–27.

Menkes, M. S., Comstock, G. W., Vuilleumier, J. P., Helsing, K. J., Rider, A. A., and Brookmeyer, R., 1986, Serum beta-carotene, vitamins A and E, selenium, and the risk of lung cancer, *N. Engl. J. Med.* **315:**1250–1254.

Mergens, W. J., and Bhagavan, H. N., 1990, α-Tocopherols (vitamin E). in *Nutrition and Cancer Prevention, Investigating the Role of Micronutrients* (T. E. Moon and M. S. Micozzis, eds.), pp. 305–340, Dekker, New York.

Mezzetti, A., Lapenna, D., Pierdomenico, S. D., Calafiore, A. M., Costantini, F., Riariosforza, F., Imbastaro, T., Neri, M., and Cuccurullo, F., 1995, Vitamin E, vitamin C and lipid peroxidation in plasma and arterial tissue of smokers and non-smokers, *Atherosclerosis* **112:**91–99.

Miller, N., Diplock, A. T., and Rice-Evans, C., 1995, Evaluation of the total antioxidant activity as a marker of the deterioration of apple juice on storage, *J. Agric. Food Chem.* **43:**1794–1801.

Mills, P. K., Beeson, W. L., Phillips, R. L., and Fraser, G. E., 1989, Cohort study of diet, lifestyle, and prostate cancer in Adventist men, *Cancer* **64:**598–604.

Mirvish, S. S., 1986, Effects of vitamins C and E on N-nitroso compound formation, carcinogenesis, and cancer, *Cancer (Philadelphia)* **58:**1842–1850.

Miyamoto, H., Araya, Y., Ito, M., Isobe, H., Dosaka, H., Shimizu, T., Kishi, F., Yamamoto, I., Honma, H., and Kawakami, Y., 1987, Serum selenium and vitamin E concentrations in families of lung cancer patients, *Cancer* **60:**1159–1162.

Moertel, C. G., Fleming, T. R., Creagan, E. T., Rubin, J., O'Connell, M. J., and Ames, M. M., 1985, High-dose vitamin C versus placebo in the treatment of patients with advanced cancer who have had no prior chemotherapy. A randomised double-blind comparison, *N. Engl. J. Med.* **312:**137–141.

Morris, D. L., Kritchevsky, S. B., and Davis, C. E., 1994, Serum carotenoids and coronary heart disease. The Lipid Research Clinics Coronary Primary Prevention Trial and Follow-up Study, *J. Am. Med. Assoc.* **272:**1439–1441.

Nomura, A. M., Stemmerman, G. N., Heilbrun, L. K., Salked, R. M., and Vuilleumier, J. P., 1985, Serum vitamin levels and the risk of cancer of specific sites in men of Japanese ancestry in Hawaii, *Cancer Res.* **45:**2369–2372.

Nomura, A. M. Y., Kolonel, L. N., Hankin, J. H., and Yoshizawa, C. N., 1991, Dietary factors in cancer of the lower urinary tract, *Int. J. Cancer* **48:**199–205.

Nyyssonen, K., Porkkala, E., Salonen, R., Korpela, H., and Salonen, J. T., 1994, Increase in oxidation resistance of atherogenic serum lipoproteins following antioxidant supplementation—A randomized double-blind placebo-controlled clinical trial, *Eur. J. Clin. Nutr.* **48:**633–642.

Okamoto, K., 1992, The relationship between dietary ascorbic acid intake and serum lipid concentration in the aged [in Japanese], *Nippon Ronen Igakkai Zasshi Jpn. J. Geriatr.* **29:**908–911.

Olsen, G. W., Mandel, J. S., Gibson, R. W., Wattenberg, L. W., and Schuman, L. M., 1991, Nutrients and pancreatic cancer—A population-based case–control study, *Cancer Causes Control* **2:**291–297.

Omenn, G. S., Goodman, G. E., Thornquist, M. D., Balmes, J., Cullen, M. R., Glass, A., Keough, J. P., Meyskens, F. L., Jr., Valanis, B., Williams, J. H., Jr., Barnhart, S., Chemiack, M. G., Brodkin, C. A., and Hammar, S., 1996, Effect of a combination of betacarotene and vitamin A on lung cancer and cardiovascular disease, *N. Engl. J. Med.* **334:**1150–1155.

Paganelli, G. M., Biasco, G., Brandi, G., Santucci, R., Gizzi, G., Villani, V., Cianci, M., Miglioli, M., and Barbara, L., 1992, Effect of vitamin A, vitamin C, and vitamin E supplementation on rectal cell proliferation in patients with colorectal adenomas, *J. Natl. Cancer Inst.* **84:**47–51.

Paganini-Hill, A., Chao, A., Ross, R. K., and Henderson, B. E., 1987, Vitamin A, beta-carotene, and the risk of cancer: A prospective study, *J. Natl. Cancer Inst.* **79**:443–448.

Palan, P. R., Mikhail, M. S., Basu, J., and Romney, S. L., 1991, Plasma levels of antioxidant beta-carotene and alpha-tocopherol in uterine cervix dysplasias and cancer, *Nutr. Cancer* **15**:13–20.

Paolisso, G., Gambardella, A., Giugliano, D., Galzerano, D., Amato, L., Volpe, C., Balbi, V., Varrocchio, M., and D'Onofrio, F., 1995, Chronic intake of pharmacological doses of vitamin E might be useful in the therapy of elderly patients with coronary heart disease, *Am. J. Clin. Nutr.* **61**:848–852.

Parfitt, V. J., Rubba, P., Bolton, C., Marotta, G., Hartog, M., and Mancini, M., 1994, Comparison of antioxidant status and free radical peroxidation of plasma lipoproteins in healthy young persons from Naples and Bristol, *Eur. Heart J.* **15**:871–876.

Pelletier, O., 1975, Vitamin C and cigarette smokers, *Ann. N.Y. Acad. Sci.* **258**:156–168.

Potischman, N., Herrero, R., Brinton, L. A., and Reeves, W. C., 1991, A case–control study of nutrient status and invasive cervical cancer. 2. Serologic indicators, *Am. J. Epidemiol.* **134**:1347–1355.

Princen, H. M. G., Vanpoppel, G., Vogelezang, C., Buytenhek, R., and Kok, F. J., 1992, Supplementation with vitamin E but not beta-carotene in vivo protects low density lipoprotein from lipid peroxidation in vitro—Effect of cigarette smoking, *Arterioscler. Thromb.* **12**:554–562.

Princen, H. M. G., Vanduyvenvoorde, W., Buytenhek, R., Vanderlaarse, A., Vanpoppel, G., Leuven, J. A. G., and Vanhinsbergh, V. W. M., 1995, Supplementation with low doses of vitamin E protects LDL from lipid peroxidation in men and women, *Arterioscler. Thromb. Vasc. Biol.* **15**:325–333.

Pronczuk, A., Kipervarg, Y., Hayes, K. C., 1992, Vegetarians have higher plasma alpha-tocopherol relative to cholesterol than do nonvegetarians, *J. Am. Coll. Nutr.* **11**:50–55.

Reaven, P. D., and Witzum, J. L., 1993, Comparison of supplementation of RRR-alpha-tocopherol and racemic alpha-tocopherol in humans—Effects on lipid levels and lipoprotein susceptibility to oxidation, *Arterioscler. Thromb.* **13**:601–608.

Reaven, P. D., Khouw, A., Beltz, W. F., Parthasarathy, S., and Witztum, J. L., 1993, Effect of dietary antioxidant combinations in humans—Protection of LDL by vitamin E but not by beta-carotene, *Arterioscler. Thromb.* **13**:590–600.

Renaud, S., and De Lorgeril, M., 1992, Wine, alcohol, platelets and the French paradox for coronary heart disease, *Lancet* **339**:1523–1526.

Riboli, E., Gonzalez, C. A., Lopezabente, G., Errezola, M., and Izarzugaza, I., 1991, Diet and bladder cancer in Spain — a multicenter case–control study, *Int. J. Cancer* **49**:214–219.

Rice-Evans, C., and Packer, L., eds., 1997, *Flavonoids in Health and Disease*, Dekker, New York.

Rice-Evans, C. A., Miller, N., Bolwell, G. P., Bramley, P. M., and Pridham, J. B., 1995, The relative antioxidant activities of plant-derived polyphenolic flavonoids, *Free Radical Res.* **22**:375–383.

Richardson, S., Ferber, M., and Cenee, S., 1991, The role of fat, animal protein and some vitamin consumption in breast cancer—A case-control study in southern France, *Int. J. Cancer* **48**:1–9.

Riemersma, R. A., Oliver, M., Elton, R. A., Alfthan, G., Vartiainen, E., Salo, M., Rubba, P., Mancini, M., Georgi, H., Vuilleumier, J. P., and Gey, K. F., 1990, Plasma antioxidants and coronary heart disease: Vitamins C and E, and selenium, *Eur. J. Clin. Nutr.* **44**:143–150.

Riemersma, R. A., Wood, D. A., Macintyre, C. C. A., Elton, R. A., Gey, K. F., and Oliver, M. F., 1991, Risk of angina pectoris and plasma concentrations of vitamin A, vitamin C, and vitamin E and carotene, *Lancet* **337**:1–5.

Rimm, E. B., Stampfer, M. J., Ascherio, A., Giovannucci, E., Colditz, G. A., and Willett, W. C., 1993, Vitamin E consumption and the risk of coronary heart disease in men [comment], *N. Engl. J. Med.* **328**:1450–1456.

Ringer, T. V., DeLoof, M. J., Winterrowd, G. E., Francom, S. F., Gaylor, S. K., Ryan, J. A., Sanders, M. E., and Hughes, G. S., 1991, Beta-carotene's effects on serum lipoproteins and immunologic indices in humans, *Am. J. Clin. Nutr.* **53**:688–694.

Rohan, T. E., McMichael, A. J., and Baghurst, P. A., 1988, A population-based case–control study of diet and breast cancer in Australia, *Am. J. Epidemiol.* **128**:478–489.

Rohan,T. E., Howe, G. R., Friedenreich, C. M., Jain, M., and Miller, A. B., 1993, Dietary fibre, vitamin A, vitamin C, and vitamin E, and risk of breast cancer—A cohort study, *Cancer Causes Control* **4**:29–37.

Russell, M. J., Thomas, B. S., and Bulbrook, R. D., 1988, A prospective study of the relationship between serum vitamins A and E and risk of breast cancer, *B. J. Cancer* **57**:213–215.

Salah, N., Miller, N. J., Paganga, G., Tijburg, L., Bolwell, G. P., and Rice-Evans, C., 1995, Polyphenolic flavanols as scavengers of aqueous phase radicals and as chain-breaking antioxidants, *Arch. Biochem. Biophys.* **322:**339–346.

Salonen, J. T., Salonen, R., Lappetelainen, R., Maenpaa, P. H., Alfthan, G., and Puska, P., 1985, Risk of cancer in relation to serum concentrations of selenium and vitamins A and E: Matched case–control analysis of prospective data, *B. Med. J. Clin. Res. Ed.* **290:**417–420.

Salonen, J. T., Salonen, R., Ihanainen, M., Parviainen, M., Seppanin, R., Seppanen, K., and Rauramaa, R., 1987, Vitamin C deficiency and low linolenate intake associated with elevated blood pressure: The Kuopio Ischaemic Heart Disease Risk Factor Study, *J. Hypertens. Suppl.* **5:**S521–S524.

Salonen, J. T., Salonen, R., Seppanen, K., Kantola, M., Parviainen, M., Alfthan, G., Maenpaa, P. H., Taskinen, E., and Rauramaa, R., 1988, Relationship of serum selenium and antioxidants to plasma lipoproteins, platelet aggregability and prevalent ischaemic heart disease in eastern Finnish men, *Atherosclerosis* **70:**155–160.

Schectman, G., Byrd, J. C., and Hoffmann, R., 1991, Ascorbic acid requirements for smokers: Analysis of a population survey, *Am. J. Clin. Nutr.* **53:**1466–1470.

Scott, K. J., and Hart, D. J., 1995, Development and evaluation of an HPLC method for the analysis of carotenoids in foods, and the measurement of the carotenoid content of vegetables and fruits commonly consumed in the UK, *Food Chem.* **54:**101–110.

Seddon, J. M., Ajani, U. A., Sperduto, R. D., Hiller, R., Blair, N., Burdon, T. C., Farker, M. D., Gragoudas, E. S., Haller, J , Miller, D. T., Yannuzzi, L. A., and Willett, W., 1994, Dietary carotenoids, vitamins A, C and E, and advanced age-related macular degeneration, *J. Am. Med. Assoc.* **272:**1413–1420.

Shibata, A., Paganini-Hill, A., Ros, R. K., and Henderson, B. E., 1992, Intake of vegetables, fruits, beta-carotene, vitamin C and vitamin supplements and cancer incidence among the elderly: A prospective study, *B. J. Cancer* **66:**673–679.

Singh, R. B., Niaz, M. A., Bishnoi, I., Sharma, J. P., Gupta, S., Rastogi, S. S., Singh, R., Begum, R., Chibo, H., and Shoumin, Z., 1994, Diet, antioxidant vitamins, oxidative stress and risk of coronary artery disease: The Peerzada Prospective Study, *Acta Cardiol.* **49:**453–467.

Smith, A. H., and Waller, K. D., 1991, Serum beta-carotene in persons with cancer and their intermediate families, *Am. J. Epidemiol.* **133:**661–671.

Stahelin, H. B., Gey, K. F., Eicholzer, M., Ludin, E., Bernasconi, F., Thurneysen, J., and Brubacher, G., 1991, Plasma antioxidant vitamins and subsequent cancer mortality in the 12-year follow-up of the prospective Basel Study, *Am. J. Epidemiol.* **133:**766–775.

Stampfer, M. J., Hennekens, C. H., Manson, J. E., Colditz, G. A., Rosner, B., and Willett, W. C., 1993, Vitamin E consumption and the risk of coronary disease in women, *N. Engl. J. Med.* **328:**1444–1449.

Steinmetz, K. A., and Potter, J. D., 1991, Vegetables, fruit, and cancer I. Epidemiology, *Cancer Causes Control* **2:**325–327.

Steinmetz, K. A., and Potter, J. D., 1993, Vegetables and fruit and cancer. I. Epidemiology, *Cancer Causes Control* **2:**325–357.

Steinmetz, K. A., and Potter, J. D., 1993b, Vegetables and fruit and cancer. II. Mechanisms, *Cancer Causes Control* **2:**427–442.

Stephens, N. G., Parsons, A., Schofield, P. M., Kelly, F., Cheesman, K., and Mitchinson, M. J., 1996, Randomised controlled trial of vitamin E in patients with coronary disease, Cambridge Heart Antioxidant Study (CHAOS), *Lancet* **347:**781–786.

Street, D. A., Comstock, F. W., Salkeld, R. M., Schuep, W., and Klag, M. J., 1994, Serum antioxidants and myocardial infarction—Are low levels of carotenoids and alpha-tocopherol risk factors for myocardial infarction? *Circulation* **90:**1154–1161.

Sundstrom, H., Korpela, H., Sajanti, E., and Kauppila, A., 1989, Supplementation with selenium, vitamin E and their combination in gynaecological cancer during cytotoxic chemotherapy, *Carcinogenesis* **10:**273–278.

Suzukawa, M., Ishikawa, T., Yoshida, H., and Nakamura, H., 1995, Effect of in vivo supplementation with low dose vitamin E on susceptibility of low density lipoprotein and high density lipoprotein to oxidative modification, *J. Am. Coll. Nutr.* **14:**46–52.

Tavani, A., Negri, E., Franceschi, S., and Lavacchia, C., 1993, Risk factors for esophageal cancer in women in northern Italy, *Cancer* **72:**2531–2536.

Tominaga, K., Saito, Y., Mori, K., Miyazawa, N., Yokoi, K., Koyama, Y., Shimamura, K., Imura, J., and Nagai, M., 1992, An evaluation of serum microelement concentrations in lung cancer and matched noncancer patients to determine the risk of developing lung cancer—A preliminary study, *Jpn. J. Clin. Oncol.* **22:**96–101.

Torun, M., Yardim, S., Sargin, H., and Simsek, B., 1994, Evaluation of serum beta-carotene levels in patients with cardiovascular diseases, *J. Clin. Pharm. Ther.* **19:**61–63.

Torun, M., Akgul, S., and Sargin, H., 1995, Serum vitamin E level in patients with breast cancer, *J. Clin. Pharm. Ther.* **20:**173–178.

Trout, D. L., 1992, Vitamin C and cardiovascular risk factors, [review], *Am. J. Clin. Nutr.* **53**(1 Suppl.):322S–325S.

Tse, W. Y., Maxwell, S. R. J., Thomason, H., Blann, A., Thorpe, G. H. G., Waite, M., and Holder, R., 1994, Antioxidant status in controlled and uncontrolled hypertension and its relationship to endothelial damage, *J. Hum. Hyperten.* **8:**843–849.

Vaneenwyk, J., Davis, F. G., and Bowen, P. E., 1991, Dietary and serum carotenoids and cervical intraepithelial neoplasia, *Int. J. Cancer* **48:**34–38.

Van Lente, F., Daher, R., and Waletsky, J. A., 1994, Vitamin E compared with other potential risk factor concentrations in patients with and without coronary artery disease: A case-matched study, *Eur. J. Clin. Chem. Clin. Biochem.* **32:**583–587.

Van Poppel, G., 1993, Carotenoids and cancer, *Eur. J. Cancer* **29A/9:**1335–1344.

Vena, J. E., Graham, S., Freudenheim, J., Marshall, J., Zielezny, M., Swanson, M., and Sufrin, G., 1992, Diet in the epidemiology of bladder cancer in western New York, *Nutr. Cancer* **18:**255–264.

Verreault, R., Chu, G., Mandelson, M. J., and Shy, K., 1989, A case–control study of diet and invasive cervical cancer, *Int. J. Cancer* **43:**1050–1054.

Wald, N. J., Boreham, J., Hayward, J. L., and Bulbrook, R. D., 1984, Plasma retinol, beta-carotene and vitamin E levels in relation to the future risk of breast cancer, *Br. J. Cancer* **49:**321–324.

Wald, N. J., Thompson, S. G., Densem, J. W., Boreham, J., and Bailey, A., 1987, Serum vitamin E and subsequent risk of cancer, *Br. J. Cancer* **56:**69–72.

Wald, N. J., Thompson, S. G., Densem, J. W., Boreham, J., and Bailey, A., 1988, Serum beta-carotene and subsequent risk of cancer: Results from the BUPA Study, *Br. J. Cancer* **57:**428–433.

Wang, G. Q., Dawsey, S. M., Li, J. Y., Taylor, P. R., Li, B., Blot, W. J., Weinstein, W. M., Liu, F. S., Lewin, K. J,, Wang, J., Wiggett, S., Gail, M. J., and Yang, C. S., 1994, Effects of vitamin mineral supplementation on the prevalence of histological dysplasia and early cancer of the esophagus and stomach—Results from the general population trial in Linxian, China, *Cancer Epidemiol. Biomarkers Prev.* **3:**161–166.

Wei, Q., Matanoski, G. M., Farmer, E. R., Strickland, P., and Grossman, L., 1994, Vitamin supplementation and reduced risk of basal cell carcinoma, *J. Clin. Epidemiol.* **47:**829–837.

West, D. W., Slattery, M. L., Robinson, L. M., French, T. K., and Mahoney, A. W., 1991, Adult dietary intake and prostate cancer risk in Utah—A case–control study with special emphasis on aggressive tumors, *Cancer Causes Control* **2:**85–94.

Willett, W. C., Polk, B. F., Underwood, B. A., Stampfer, M. J., Pressel, S., Rosner, B., Taylor, J. O., and Schneider, K., 1984, Relation of serum vitamins A and E and carotenoids to the risk of cancer, *N. Engl. J. Med.* **310:**430–434.

Witzum, J. L., 1994, The oxidation hypothesis of atherosclerosis, *Lancet* **344:**793–795.

Yuan, J. M., Wang, Q. S., Ross, R. K., Henderson, B. E., and Yu, M. C., 1995, Diet and breast cancer in Shanghai and Tianjin, China, *Br. J. Cancer* **71:**1353–1358.

Zaridze, D. G., and Filipchenko, V. V., 1991, The role of diet in etiology of colorectal cancer, *Vopr. Onkol.* **37:**152–158.

Zaridze, D., Lifanova, Y., Maximovitch, D., Day, N. E., and Duffy, S. W., 1991, Diet, alcohol consumption and reproductive factors in a case–control study of breast cancer in Moscow, *Int. J. Cancer* **48:**493–501.

Zatonski, W., Przewonziak, K., Howe, G. R., Maisonneuve, P., Walker, A. M., and Boyle P, 1991, Nutritional factors and pancreatic cancer, *Int. J. Cancer* **48:**390–394.

Ziegler, R. G., Jones, C. J., Brinton, L. A., Norman, S. A., Mallin, K., Levine, R. S., Lehman, H. F., Hamman, R. F., Trumble, A. C., Rosenthal, J. F., and Hoover, R. N., 1991, Diet and the risk of in situ cervical cancer among white women in the United States, *Cancer Causes Control* **2:**17–29.

Part V

Internal Pro- and Antioxidants

Chapter 15

Xanthine Oxidase in Biology and Medicine

Dale A. Parks, Kelly A. Skinner, Sidhartha Tan, and Henry B. Skinner

1. INTRODUCTION

Xanthine oxidoreductase (xanthine:NAD^+ oxidoreductase, EC 1.1.1.204) is an enzyme first discovered in milk and extensively characterized over a period of several decades (see Kooij, 1994, for historical overview). Because of the availability and relative stability of the enzyme, it was one of the first flavoproteins to be purified (Ball, 1939) and crystallized (Avis *et al.*, 1955). It was also the first mammalian enzyme found to contain molybdenum (De Renzo *et al.*, 1953) and iron (Fridovich and Handler, 1958). Xanthine oxidoreductase had been relegated to the role of a model protein for the study of oxidation–reduction reactions until relatively recently when Granger *et al.* (1981) proposed that xanthine oxidase (XO) was a significant biological source of reactive oxygen species and could play an integral role in the tissue injury associated with ischemia–reperfusion. Over the next decade, there was a literal explosion in the literature, with over 3000 publications characterizing the production of reactive oxygen species by XO (Kooij, 1994; Weinbroum *et al.*, 1995; Nielsen *et al.*, 1995; White *et al.*, 1996) and the roles of these XO-derived oxidants (Parks and Granger, 1986b; Yokoyama *et al.*, 1990; Kurose and Granger, 1994), which include an integral function in chemotaxis, regulation of nitric oxide (Baggiolini *et al.*, 1993; Miyamoto *et al.*, 1996), and in the etiology of a variety of pathological processes (Parks and Granger, 1986b). This review will summarize publications that have led to controversies as well as some of the seminal observations that have prompted interest in xanthine oxidase and related enzymes.

Dale A. Parks, Kelly A. Skinner, Sidhartha Tan, and Henry B. Skinner Department of Anesthesiology, University of Alabama at Birmingham, Birmingham, Alabama 35233.

Reactive Oxygen Species in Biological Systems, edited by Gilbert and Colton. Kluwer Academic / Plenum Publishers, New York, 1999.

2. STRUCTURAL INFORMATION

The molybdenum-containing hydroxylases (xanthine oxidoreductase, aldehyde oxidase, and sulfite oxidase) comprise a group of closely related metalloflavoproteins. This review will focus on XO, the most well characterized of the molybdenum (Mo) enzymes, but will refer to aldehyde oxidase (AO) and sulfite oxidase because of the similarity of the enzymes. All members of the family are proteins of M_r of ~300,000 and are comprised of two identical and catalytically independent subunits. Each subunit contains one molybdenum, as the pterin molybdenum cofactor, a flavin adenine nucleotide (FAD), and two iron–sulfur centers, FeSI and FeSII (Bray *et al.*, 1996; Hille and Nishino, 1995). The oxidative hydroxylation of reducing substrates (e.g., purines and aldehydes) occurs at the molybdenum center, reducing the molybdenum from Mo(VI) to Mo(IV). The reducing equivalents are transferred intramolecularly via the iron–sulfur centers to FAD, where reaction with physiological oxidizing substrates occurs. Molybdenum hydroxylases share a broad specificity for reducing substrates and the capability of producing reactive oxygen species, such as superoxide ($\cdot O_2^-$) and hydrogen peroxide (H_2O_2), from oxygen (Barber *et al.*, 1982; Hall and Krenitsky, 1986; Krenitsky *et al.*, 1974). XO produces $\cdot O_2^-$ predominantly from metabolism of purine substrates although various aldehyde substrates are also suitable (Hall and Krenitsky, 1986; Krenitsky *et al.*, 1974). AO shares extensive sequence similarity (84.3%) and cofactor requirements with XO (Ichida *et al.*, 1993; Wright *et al.*, 1995). Like XO, AO may utilize a wide variety of substrates but has a higher affinity for aldehydes than purines. Both AO and XO produce $\cdot O_2^-$ in a very similar manner, as both oxidases possess nearly identical internal electron transport systems (Barber *et al.*, 1982).

3. XDH-TO-XO CONVERSION

Xanthine oxidoreductase is the enzyme responsible for converting hypoxanthine to xanthine and xanthine to uric acid, an essential plasma antioxidant (Parks and Granger, 1986b; White *et al.*, 1996; Freeman and Crapo, 1982). Xanthine oxidoreductase exists in normal tissues predominantly as xanthine dehydrogenase (XDH) and is converted to the XO form during ischemia or hypoxia (Yokoyama *et al.*, 1990). Conversion of XDH to XO can occur reversibly by oxidation of essential sulfhydryl groups or through limited proteolytic cleavage of the amino terminus (Amaya *et al.*, 1990). The XDH form utilizes NAD^+ as an electron acceptor, whereas the XO form utilizes molecular oxygen as the electron acceptor, thereby producing $\cdot O_2^-$.

It had been suggested that XO was not responsible for ischemic liver injury because the duration of a typical ischemic insult is 2 hr whereas the rate of XDH-to-XO conversion in *ex vivo* perfused liver and hepatocytes is relatively slow, requiring 4–6 hr of ischemia for significant proteolytic conversion (Engerson *et al.*, 1987; Yokoyama *et al.*, 1990). However, XDH + XO can be released from ischemic liver or hypoxic hepatocytes and appear in the perfusate or media prior to detectable conversion (S. Tan *et al.*, 1993a, 1998; Kooij *et al.*, 1994). XDH + XO is also released into the systemic circulation following systemic shock or surgical intervention *in vivo* (S. Tan *et al.*, 1993a, 1995), where XDH-to-XO conversion, via sulfhydryl oxidation, may be extremely fast, requiring <1 min (Yokoyama *et al.*, 1990). Kupffer cells contain considerable XDH + XO activity and the rate of conversion of XDH to XO in these cells is much more rapid than in other cell types (Wiezorek *et al.*, 1994). Furthermore, ~15–20% of the XDH + XO exists in the XO form under normal conditions allowing significant $\cdot O_2^-$ production without the necessity of additional conversion (Parks and Granger, 1986b; Engerson *et al.*, 1987). Therefore, recent observations have reemphasized the importance of XO as a key source of oxidants following hepatic ischemia–reperfusion. It is possible that AO may exhibit similar toxicity comparable to XO because of similarities in the internal electron transport system (Barber *et al.*, 1982).

4. REGULATION AND GENE EXPRESSION

Although the molybdenum-containing hydroxylases have so far proven refractory to attempts at X-ray crystallography, insights have been gained through the cloning of cDNAs and genes for XDH and AO. The complete predicted amino acid sequence information is currently available for the human (Wright *et al.*, 1993; Rump *et al.*, 1993), rat (Amaya *et al.*, 1990), mouse (Terao *et al.*, 1992), *Drosophila* (Keith *et al.*, 1987), *Calliphora* (Houde *et al.*, 1989), and chicken (Sato *et al.*, 1995) xanthine oxidoreductase enzymes. The predicted amino acid sequence of AO is available for human (Wright *et al.*, 1993), bovine (Calzi *et al.*, 1995), and rabbit (Turner *et al.*, 1995). The mouse XDH gene has been cloned and analyzed (Cazzaniga *et al.*, 1994). The gene is composed of 36 exons and extends for ~70 kb.

Consensus binding sites for the transcriptional factors LBP-1, IBP-1, and a protein recognizing the TGGCA motif were revealed ~1 kb sequence upstream of exon 1. Type I and type II interferon (IFN) responsive elements were not identified. Several indirect and inverted repeats are evident although the functional significance of these sequences remains unclear.

Xanthine oxidoreductase is regulated through mechanisms acting at the level both of the gene (Falciani *et al.*, 1992; Dupont *et al.*, 1992) and of the protein (Phan *et al.*, 1989). Under normal conditions, xanthine oxidoreductase is constitutively expressed at high levels in the liver and intestine and at considerably lower levels in mammary gland epithelium and endothelium (Jarasch *et al.*, 1981, 1986; Ghezzi *et al.*, 1984; Parks and Granger, 1986b). Expression of xanthine oxidoreductase has been demonstrated to be dramatically upregulated by IFNα, a variety of viruses, bacterial lipopolysaccharide, and the interferon inducer, tilorone (Deloria *et al.*, 1985; Moochhala and Renton, 1991; Ghezzi *et al.*, 1984, 1985; Terao *et al.*, 1992). Endothelial cell XO activity may also be regulated at the transcriptional level by oxygen tension, with hypoxia causing an increase in XO (Hassoun *et al.*, 1995), and by IFNγ, which induces XO activity and mRNA expression in rat lung endothelial cells (Dupont *et al.*, 1992).

XO-derived oxidants can also affect transcription of proinflammatory cytokines, increasing mRNA levels for interleukin-1β (IL-1β), tumor necrosis factor-α (TNFα), and transforming growth factor (TGF-β1) by a mechanism that may include activation of cyclic AMP response element binding protein (CREB), a nuclear transcription regulatory factor (Shenkar and Abraham, 1996; Schwartz *et al.*, 1995). XO-derived oxidants also stimulate the induction of nitric oxide synthase (NOS) activity and inducible NOS mRNA in isolated rat hepatocytes. This process appears to be regulated by the cellular level of reduced glutathione (Duval *et al.*, 1995).

5. INTERACTION WITH NITRIC OXIDE

There are at least three ways that nitric oxide (·NO) has been postulated to interact with XO or XO-derived oxidants (Figures 1 and 2). First, ·NO has been proposed to posttranscriptionally regulate $\cdot O_2^-$ production by interacting with the XO or AO iron–sulfur moiety, sulfhydryl groups, or by reversible alteration of the flavin prosthetic site, thereby inhibiting activity (Fukahori *et al.*, 1994; Hassoun *et al.*, 1995). Second, as mentioned above, XO- and AO-derived oxidants have been demonstrated to stimulate the induction of NOS activity and mRNA, which could increase ·NO production. Finally, XO has been proposed as a regulator of ·NO based on the observation that XO inhibitors decrease $\cdot O_2^-$ generation and potentiate ·NO-mediated vasorelaxation (Miyamoto *et al.*, 1996). This regulation has been attributed to the almost diffusion-limited (6.7×10^9 M^{-1} s^{-1}) (Huie and Padmaja, 1993) interaction of ·NO with $\cdot O_2^-$ to form peroxynitrite ($ONOO^-$).

XO may modulate the pathological effects of peroxynitrite, which is a potent oxidizing agent that damages lipids, proteins, carbohydrates, deoxyribonucleic acid, and sulfhydryl groups (Beckman *et al.*, 1994; Rubbo *et al.*, 1994; Pryor and Squadrito, 1995; Freeman *et al.*, 1995). XO may also modulate the $ONOO^-$-induced inactivation of mitochondrial Mn superoxide dismutase (Ischiropoulos *et al.*, 1992), alterations in lipid aggregatory properties of surfactant protein A (Haddad *et al.*, 1994), inactivation of sodium transport (Hu *et al.*, 1994), inactivation of α1-antiproteinase (the major inhibitor of serine proteases in human body fluids; Moreno and Pryor, 1992), nitration of tyrosine (Beckman, 1996), and nitration of actin which may subsequently alter myocardial function (Beckman *et al.*, 1996).

6. TISSUE DISTRIBUTION AND CELLULAR LOCALIZATION

A definitive characterization of the localization of XO and AO in tissues and cells can enhance our understanding of the physiological and pathological role of these cytosolic oxidases. Tissue distribution and localization of XO has been accomplished

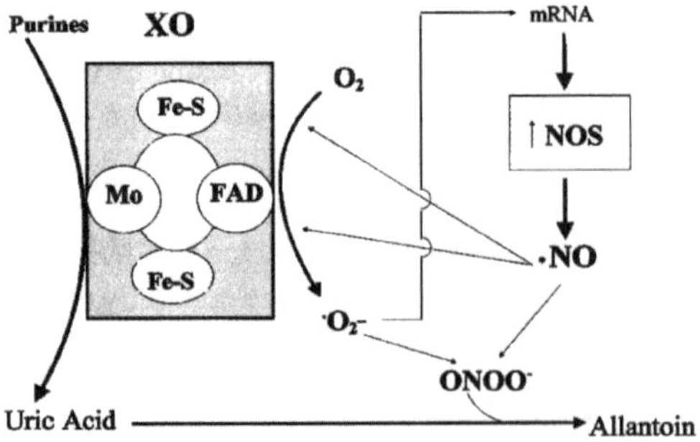

FIGURE 1. The potential role of XO in generation of oxidants, formation of antioxidants, and regulation of nitric oxide.

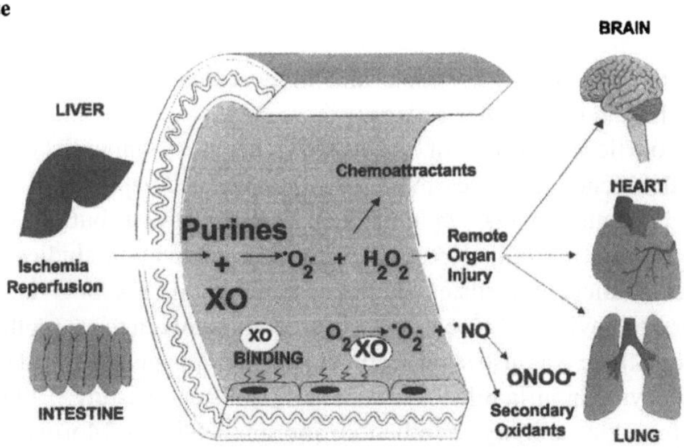

FIGURE 2. Proposed scheme for role of XO as a circulating mediator that binds to vascular cell surface of endothelium, produces reactive oxygen species, and directly or indirectly results in multisystem organ dysfunction.

utilizing both histochemical techniques and, more recently, immunolocalization techniques with XO antibodies.

6.1. Histochemical Localization

Many of the earlier attempts to localize XO histochemically were hampered by a lack of understanding of many of the properties of the enzyme, and have been extensively reviewed (Kooij, 1994). The failure to recognize that xanthine oxidoreductase exists predominately as XDH and not as XO is a major reason for the conflicting histochemical localization of XO and estimates of enzyme activity. A second reason for conflicting histochemical data was that tissues were often fixed, which inactivates enzyme activity (Sackler, 1966). A third reason for suboptimal localization is that soluble enzymes like XDH + XO "leak" from tissues (Sackler, 1966). Finally, nonspecific precipitation of the formazan reaction product often precluded a precise localization of XDH + XO (Kooij, 1994). Accounting for these potential complications, Kooij observed that XDH + XO was present in sinusoidal endothelial cells as well as hepatocytes, and that activity was higher in pericentral than periportal hepatocytes (Kooij, 1994).

6.2. Immunolocalization

We recently generated antibodies specific for cytosolic molybdenum-containing hydroxylases (Skinner *et al.*, unpublished observations) through expression of the carboxy-terminal 358 amino acids of human XO, subsequently identified as AO (Turner *et al.*, 1995; Wright *et al.*, 1993). Human XO and AO share 55% identity and 85% similarity over the selected 358 carboxy-terminal amino acids. The high degree of amino acid conservation resulted in generation of antibodies that recognized both XO and AO. The use of a recombinant antigen alleviated the previously observed cross-reactivity problems. By means of these monoclonal and polyclonal antibodies, XO/AO staining in normal tissues was found in (1) epithelial cells of liver, lung, intestine, and kidney, (2) smooth muscle cells of large vessels and myocardium, and, to a lesser extent, (3) endothelial cells and (4) Kupffer cells.

The hepatocytic localization of XO/AO (Skinner *et al.*, unpublished observations) is consistent with biochemical assessments of XO in isolated hepatocytes (Tan *et al.*, 1998; DeGroote and Littauer, 1988), while the zonal pattern of distribution is consistent with observations of Kooij (Kooij, 1994). Our immunolocalization studies were consistent with previous studies of Hellsten-Westing in which monoclonal antibodies against purified human or bovine milk localized XO predominately in the smooth muscle of large vessels although there was also significant XO in macrophages and mast cells, suggesting a role of the enzyme in inflammatory processes (Hellsten-Westing, 1993). Considerably less XO was found in capillary endothelium and small venules. Moriwaki and colleagues using polyclonal antibody against XO purified from human liver demonstrated the presence of the enzyme only in hepatic and intestinal epithelial cells and endothelial lining cells of liver, intestine, heart, kidney, aorta, lung, and brain (Moriwaki *et al.*, 1993). These observations are in marked contrast to the immunolocalization studies of XDH + XO by Jarasch *et al.* (1986) in which XO was localized exclusively in the capillary endothelial cells and epithelial cells of the mammary gland. Interestingly, XO was not detected in liver, intestine, kidney, or lung epithelial cells, Kupffer cells, or muscle cells, which is in marked contrast to histochemical localization studies, enzymatic activity assays, and other immunolocalization studies (Tan *et al.*, 1998; DeGroote and Littauer, 1988; Kooij, 1994; Moriwaki *et al.*, 1993; Hellsten-Westing, 1993). This apparent discrepancy may be attributed to the observation that IgG and lactoferrin may copurify with XDH + XO and that antibodies against purified XO may also detect IgG and lactoferrin (Sarnesto *et al.*, 1996; Clare and Lecce, 1991).

7. CIRCULATING XO

Hepatocellular damage has been demonstrated to increase XDH + XO in the circulation of humans (Giler *et al.*, 1975; Ramboer *et al.*, 1972). Circulating XDH + XO is also increased following thoracic aorta cross-clamping (S. Tan *et al.*, 1995), in adult respiratory disease (Grum *et al.*, 1991), and kidney disease (Giler *et al.*, 1975) with concomitant liver disease. Endothelial cells may also be a source of circulating XDH + XO, as interruption of the blood supply to an upper limb of human patients undergoing orthopedic procedure increased plasma levels of XDH + XO (Friedl *et al.*, 1990). Results in animal models confirm and extend these clinical observations. In animal models, there is evidence that a variety of conditions, including ischemia–reperfusion, sepsis, burns, acute viral infection, and hemorrhagic shock, release XO from liver and gut into the plasma of rat (Terada *et al.*, 1992; Oda *et al.*, 1989; Grum *et al.*, 1991; Akaike *et al.*, 1990; Anderson *et al.*, 1991; Repine *et al.*, 1987; Tan *et al.*, 1993; Weinbroum *et al.*, 1995; Yokoyama *et al.*, 1990) and rabbit (Nielsen *et al.*, 1994, 1995). *In vitro* studies indicate that it is likely that both hepatocytes and vascular endothelium are significant sources of circulating XDH + XO activity, as exposure of freshly isolated hepatocytes (Tan *et al.*, 1998; DeGroote and Littauer, 1988) or cultured endothelial cells to hypoxia results in release of XDH + XO into the media (Partridge *et al.*, 1992).

XO has also been detected in the plasma of patients undergoing surgical interventions that compromise hepatic and/or intestinal blood flow (Friedl *et al.*, 1990; S. Tan *et al.*, 1995). We have recently observed that XO activity was significantly increased in the plasma of patients undergoing liver transplantation and correlated well with aspartate transaminase (AST) levels, a clinically relevant enzymatic marker of tissue injury (unpublished observations). Plasma XO activity was also correlated with the uric acid oxidation product allantoin, an indicator of oxidant stress. These data indicate that preserved human liver releases XO into systemic circulation on reperfusion and that XO is associated with increased tissue damage. This finding also suggested that XO plasma half-life may be hours, not minutes as previously reported (Fridovich, 1986). We have also observed a twofold increase in circulating XO in patients undergoing thoracic aorta aneurysm repair, a procedure that renders liver, intestine, and all distal tissues ischemic (S. Tan *et al.*, 1995). Subsequent reperfusion of tissues below the right renal artery did not increase circulating XO, suggesting that the liver and gut were significant sources of the XO. This protracted elevation in plasma XO is also consistent with the enzyme's plasma half-life of hours. The appearance of an oxidant-generating enzyme in the circulation may have far-reaching clinical implications in damage to tissue remote from the ischemic organ (see Section 10.4 and Figure 2).

8. GLYCOSAMINOGLYCAN BINDING AND POTENTIAL RELOCALIZATION OF XO

Numerous proteins are synthesized and secreted by liver parenchymal cells, traverse the interstitial space, enter the intravascular compartment, and finally bind to glycosaminoglycans (GAGs) on endothelial cell surfaces (Hook *et al.*, 1984; Adachi and Marklund, 1989; Saxena *et al.*, 1990). It is postulated (Figure 2) that XO behaves in a manner similar to these GAG-binding proteins (Adachi *et al.*, 1993; Fukushima *et al.*, 1995). The binding of XO to GAGs is saturable, high affinity, and reversible by heparin (Radi *et al.*, 1997; Adachi *et al.*, 1993; Tan *et al.*, 1993; White *et al.*, 1996). The dissociation constant (K_d) is similar to the heparin reversible-binding of other GAG-binding proteins to vascular endothelium and concordant with noncovalent electrostatic interaction (Adachi *et al.*, 1993; Radi *et al.*, 1997). Interestingly, binding of XO to cell surfaces strongly influences the catalytic properties (increased K_m and K_i), oxidant-producing capacities (enhanced univalent reduction of oxygen to $\cdot O_2^-$), and increases the stability of XO (Radi *et al.*, 1997). Concentration of XO at the vascular cell surface may have significant pathological implications, as $\cdot O_2^-$ could then be produced at a site demonstrated to be inaccessible to antioxidants such as SOD (Radi *et al.*, 1997). Therefore, circulating XO and its subsequent binding to GAGs may cause oxidant injury to tissues with low endogenous XO activity, such as myocardium and lung. Consistent with this contention is the marked increase in XO specific activity of the lung (Weinbroum *et al.*, 1995) and myocardium (Nielsen *et al.*, 1997) following hepatoenteric ischemia–reperfusion and relocalization of XO following liver transplantation (see Section 10). These observations have significant implications for XO–GAG interactions, suggesting that XO can be (1) released from metabolically stressed cells, (2) bound to surface GAGs to become an external source of oxidant production, and (3) internalized by endocytosis to become an intracellular locus of oxidant production

9. PHYSIOLOGIC FUNCTIONS

9.1. Normal Physiologic Functions of XO and AO

Despite years of investigation and thousands of publications, the physiologic functions of XO and AO have remained poorly characterized. The primary functions of XO and AO can be categorized as (1) purine metabolism, (2) oxidant and, paradoxically, antioxidant production, (3) signal transduction, and (4) drug metabolism.

9.1.1. Purine Metabolism

The primary physiologic function of xanthine oxidoreductase is as the rate-limiting enzyme that catalyzes the terminal two steps in the degradation of purines, formation of xanthine from hypoxanthine and uric acid from xanthine. The final step in purine metabolism, formation of uric acid, determines whether purine nucleotides can be synthesized from the purine base by a purine salvage pathway or whether the purine base is irreversibly lost. In humans, uric acid is of special relevance because it is the terminal product of purine metabolism, reflecting the evolutionary loss of urate oxidase, which catabolizes uric acid to allantoin (Wu *et al.*, 1989). Consequently, human plasma contains up to 500 μM uric acid (Ames *et al.*, 1981), whereas most mammals excrete allantoin and urea as the major nitrogen-containing purine product (Mishra and Delivoria-Papadopoulos, 1988; Parks and Granger, 1986b). In addition to purines, XO is capable of oxidizing a wide variety of substrates, including pteridines, other heterocyclic compounds, and aldehydes. AO shares an equally broad substrate specificity and plays a role in the metabolism of purines, pyrimidines, and a variety of heterocyclic compounds (Hall and Krenitsky, 1986). Acetaldehyde can serve as a substrate for both AO and XO and thereby results in formation of $\cdot O_2^-$ (Shaw and Jayatilleke, 1990; Krenitsky *et al.*, 1986; Hall and Krenitsky, 1986).

9.1.2. Antioxidant Production by XO

Uric acid is a physiologically important plasma antioxidant (Jenkinson *et al.*, 1991) that is oxidized by various reactive species to relatively stable oxidation products (Mendez *et al.*, 1996). The levels of uric acid in plasma are considerably higher than other plasma antioxidants (Mize *et al.*, 1995). Uric acid is a critical water-soluble antioxidant (Ames *et al.*, 1981) that effectively protects biological target molecules against oxidation by hydroxyl radicals ($\cdot$OH), hypochlorous acid (HOCl), and peroxynitrite (van der Vliet *et al.*, 1994; Becker *et al.*, 1991; Hicks *et al.*, 1993; Kaur and Halliwell, 1990). Concurrent with the oxidation and depletion of uric acid in the plasma was the appearance of uric acid oxidation products, allantoin and parabanic acid, which could serve as indicators of the reaction of reactive species with endogenous uric acid in biological systems (van der Vliet *et al.*, 1994; Kaur and Halliwell, 1990). Uric acid and ascorbic acid are even more critical in early gestation, for the water-soluble antioxidants constitute the principal antioxidant defenses. In addition to functioning as an antioxidant, physiologic concentrations of uric acid significantly inhibit XO activity and the formation of $\cdot O_2^-$ in human blood (Tan *et al.*, 1993a). In newborn plasma, a similar pattern and extent of XO inhibition by uric acid was observed, except that $\cdot O_2^-$ formation was inhibited to an even greater extent than in adult plasma (Tan *et al.*, 1993a).

9.1.3. XO in Signal Transduction

Studies utilizing the generation of oxidants by XO plus substrate have demonstrated a role for XO in stimulation of macrophages (Mendez *et al.*, 1996), enhancement of adenylyl cyclase activation (probably via tyrosine kinase-mediated effects on the catalytic subunit of adenylyl cyclase; C. M. Tan *et al.*, 1995), induction of c-*myc* and c-*fos* expression in JB6 cells (Singh and Aggarwal, 1995), and attenuation of internal stores of calcium release and bradykinin-stimulated calcium influx in a time-dependent manner (Wesson and Elliott, 1994). In addition, XO in the presence of lumazine results in the closing of chloride channels in parietal cells similar to that observed with guanosine 5′-*O*-(3-thiotriphosphate) (Sakai and Takeguchi, 1994).

9.1.4. Drug Metabolism

The cytosolic enzymes, XO and AO, complement the microsomal cytochrome P450 system in the mixed-function oxidase metabolism of drugs (Beedham, 1985; Beedham *et al.*, 1995). AO plays a role in the metabolism of a variety of *N*-substituted heterocyclic xenobiotic compounds (Hall and Krenitsky, 1986) and ethanol (Shaw and Jayatilleke, 1990).

9.2. Deficiencies of XO: Xanthinuria and Molybdenum Deficiency

Hereditary xanthinuria is classified into three categories: classical xanthinuria types I and II, and molybdenum cofactor deficiency. In classical type I xanthinuria, only XDH activity is lacking, while in type II AO is also lacking (Simmonds *et al.*, 1995). In molybdenum cofactor deficiency, activity of all three molybdenum-containing enzymes (sulfite oxidase, XO, and AO) is lacking (Ichida *et al.*, 1997).

9.2.1. Classical Type I and II Xanthinuria

The classical type I and II xanthinurias are rare autosomal recessive disorders with a combined incidence of only 1:69,000 (Harkness *et al.*, 1986). Individuals suffering from the xanthinuria may develop urinary tract calculi, acute renal failure, or myositis related to the tissue depositions of xanthine although some subjects with homozygous mutations are asymptomatic (Simmonds *et al.*, 1995). It has recently been demonstrated (Ichida *et al.*, 1997) that point mutations in the XDH gene were the primary genetic defect responsible for classical type I xanthinuria. These deletion and nonsense mutations occurred in the flavin binding and molybdenum cofactor regions of XO and would result in synthesis of truncated peptides or peptides lacking flavin and/or molybdenum cofactor binding sites.

9.2.2. Molybdenum Cofactor Deficiency

Although type I and II deficiencies are rather benign, molybdenum cofactor deficiency is usually associated with severe neurological disorders, mental retardation, and ocular lens dislocation (Ichida *et al.*, 1997; Sardesai, 1993), and have been attributed to toxicity of sulfite and/or inadequate amounts of inorganic sulfate needed for formation of sulfated compounds in the brain. Molybdenum cofactor deficiency in the neonatal period has been reported to result in intractable seizures, profound failure to thrive, spastic quadriplegia, feeding difficulties, profound psychomotor retardation, and early death (Roesel *et al.*, 1986; Duran *et al.*, 1978; Slot *et al.*, 1993). In a single patient, a variant of molybdenum cofactor deficiency was clinically expressed as Marfan-like habitus with dislocated lenses, vertebral abnormality, learning disability, moderate hemiplegia, increased medial lentiform MRI signal, and intermittent microscopic hematuria. Despite biochemical parameters indicative of a severe deficiency state, the patient has survived into the third decade (Mize *et al.*, 1995). It is highly unlikely that molybdenum deficiency will occur in normal patients, as only trace amounts of molybdenum are required in the diet. However, molybdenum deficiency has been reported in a patient receiving protracted total parenteral nutrition with clinical signs characterized by tachycardia, headache, mental disturbances, coma, and very low levels of uric acid (low XO and AO activity) and inorganic sulfate (low sulfite oxidase activity).

9.3. Changes in XO during Fetal Development

XO activity in human fetal tissues has been reported by us (Winkler *et al.*, 1990) and others (Vettenranta and Raivio, 1990) to achieve adult levels as early as 20 weeks of gestation (i.e., midpoint of pregnancy). Enzymatic antioxidant systems are not fully developed until shortly before birth, making the water-soluble antioxidants, such as uric acid and ascorbic acid, even more critical in early gestation as they constitute the principal antioxidant defenses. In fetal rat and rabbit lung, the development of antioxidant enzymes, such as SOD, catalase, and glutathione peroxidase, parallels the development of surfactant. The developmental changes were characterized by a 150–200% elevation in fetal lung antioxidant enzyme levels during the final 10 to 15% of gestation (Frank and Sosenko, 1987). In guinea pig brain, most of the antioxidant enzyme systems increase and attain adult levels in the last 30% of the gestation period (Mishra and Delivoria-Papadopoulos, 1988). The temporal difference in development of oxidant-producing systems (i.e., XO) and antioxidant defense systems renders the premature infant extremely susceptible to oxidant-induced tissue injury. This is consistent with studies demonstrating that administration of antioxidants to the mother during this vulnerable period of development attenuates the myocardial damage associated with fetal hypoxia (Tan *et al.*, 1996).

10. PATHOLOGY

XO has been implicated to play a significant role in numerous pathological states, including ischemia–reperfusion, hypoxia–reoxygenation, multiple organ dysfunction, preservation–transplantation, vascular disease, respiratory distress syndrome, and alcohol metabolism (Table I).

10.1. Ischemia–Reperfusion

XO-derived oxidants were first implicated as mediators of ischemia-induced injury to the small intestine (Granger *et al.*, 1981; Parks *et al.*, 1982). It was demonstrated that ischemia–reperfusion increased intestinal vascular and mucosal permeability (Grogaard *et al.*, 1982; Parks and Granger, 1983), and produced morphological alterations that were attenuated by administration of the XO inhibitor allopurinol or the antioxidants SOD or catalase (Parks and Granger, 1983, 1986a). Subsequent studies revealed that ischemia–reperfusion induced the accumulation of neutrophils within the mucosa, as measured by myeloperoxidase (MPO) activity (Kurose and Granger, 1994). The increase in MPO activity was significantly attenuated by the administration of allopurinol or SOD, suggesting that XO-derived oxidants were either directly or indirectly responsible for the recruiting of neutrophils into the extravascular space. It has been subsequently demonstrated that XO plays an integral role in the etiology of ischemia–reperfusion injury to various tissues, including the liver, myocardium, lung, skeletal muscle, pancreas, stomach, kidney, and brain, and has been extensively reviewed (Granger and Korthuis, 1995; Marubayashi *et al.*, 1994; Abello *et al.*, 1994; Frederiks *et al.*, 1995; Bienvenu and Granger, 1993; Parks and Granger, 1986b).

Table I
Pathological States in which Xanthine Oxidase Involvement Has Been Implicated

Ischemia–reperfusion
Inflammation
Systemic shock
Organ preservation/transplantation
Multiple organ dysfunction
Atherosclerosis
Hypertension
Alcoholic liver disease
Surgical interventions
Fetal hypoxia and distress
Respiratory distress syndrome
Acute viral infections
Claudication and ischemia of lower extremities
Duchenne muscular dystrophy

10.2. Neonatal Cerebral Hypoxia

Allopurinol administration in neonatal rats ameliorated the increase in cerebral water content and neuropathological damage following brain hypoxia–ischemia (Palmer *et al.*, 1990). More interestingly, allopurinol administered after the cerebral hypoxic–ischemic insult also reduced brain damage in immature rats (Palmer *et al.*, 1993). However, allopurinol given to preterm newborn humans did not alter the incidence of periventricular leukomalacia (PVL). Four hundred infants of 24–32 weeks gestation were randomly enterally administered allopurinol or an equivalent dose of placebo. At admission, plasma hypoxanthine concentrations were significantly higher in infants who subsequently developed PVL, bronchopulmonary dysplasia, or retinopathy of prematurity, but there was no difference in the primary endpoint (PVL) between the treated and control groups (Russell and Cooke, 1995). In this study, it is unclear whether the timing or the dose was optimal for prevention of an endpoint, like PVL, that involves complex processes.

10.3. Premature Infants

In cord blood, an increase in plasma hypoxanthine levels has been observed in asphyxiated babies that correlated with the amount of acidosis (Pietz *et al.*, 1988). Plasma hypoxanthine levels at birth were also found to be significantly higher in premature infants with cavitating PVL than in unaffected infants (Russell *et al.*, 1992). At birth, significant XO activity has been detected in the plasma of cord blood of healthy, term infants by us (Tan *et al.*, 1993a) and others (Supnet *et al.*, 1994). Under normal conditions, XO is substrate limited (Parks and Granger, 1986b) but could react with the elevated hypoxanthine and xanthine to generate oxidants and produce cytotoxicity.

10.4. Multisystem Organ Dysfunction

XO has been demonstrated to be released from damaged liver and intestine (Terada *et al.*, 1992; Nielsen *et al.*, 1994; Weinbroum *et al.*, 1995; Nielsen *et al.*, 1997) and bind to GAGs on vascular cell surfaces (Radi *et al.*, 1997; Adachi *et al.*, 1993; Yokoyama *et al.*, 1990; Tan *et al.*, 1993b). XO is released from ischemic liver (Nielsen *et al.*, 1994; Weinbroum *et al.*, 1995) and subsequently increases pulmonary XO specific activity >80-fold and pulmonary permeability (Weinbroum *et al.*, 1995). Although the precise mechanisms of ischemia–reperfusion injury are complex, circulating XO plays a significant role, as inhibition of XO attenuates both the pulmonary and myocardial injury associated with liver and/or intestinal ischemia–reperfusion (Weinbroum *et al.*, 1995; Nielsen *et al.*, 1996, 1997). These findings indicate that XO released during reperfusion of ischemic liver binds to GAGs of pulmonary vascular endothelium or epithelium and the subsequent generation of vascular cell-derived $\cdot O_2^-$ may result in extrahepatic tissue damage, perhaps by reaction with $\cdot$NO to produce $ONOO^-$. Recent data indicate that circulating XO may actually exist as a glycosaminoglycan–XO complex, as treatment with endoglycosidases enhances target cell binding of XO >10-fold (unpublished observations). Consistent with this contention is the observation that heparin, a GAG that interferes with protein–GAG binding and has no direct oxidant scavenging properties, significantly decreases endothelial cell injury caused by XO (Hiebert and Liu, 1991).

10.5. Preservation–Transplantation

Preservation–transplantation is a special case of *ex vivo* and subsequent *in vivo* ischemia. During cold preservation, hepatocytes swell and develop surface protrusions, called *blebs*, which result in the release of hepatocellular enzymes including XO (Rosser and Gores, 1995; Lemasters *et al.*, 1995). The circulating XO could be responsible for pulmonary and vascular complications commonly observed following reperfusion of the preserved graft (Parks and Freeman, 1987; Carithers *et al.*, 1988; Wood *et al.*, 1985; Knights *et al.*, 1987). The most common preservation solution, University of Wisconsin, has been supplemented with allopurinol, a XO inhibitor, to minimize tissue injury (Sumimoto *et al.*, 1996; Toledo-Pereyra, 1991). Immunolocalization studies were performed using our polyclonal anti-XO/AO antibody, in biopsies of liver and lung collected 24 and 48 hr posttransplant of 6-hr preserved livers. XO/AO staining was predominately seen in hepatocytes, endothelial cells, and Kupffer cells of control tissues. Posttransplant (24 hr), immunoreactive XO/AO staining was intense in pericentral hepatocytes, intermediate in midzonal regions, and lowest adjacent to portal vein and to a lesser extent in Kupffer cells. XO/AO staining was most intense in degenerating hepatocytes 48 hr posttransplant. The regions of highest XO/AO staining correspond to regions where ischemia and hypoxia would be most severe, because oxygen is extracted as blood flows from the portal triads (Zone 1) to central vein branches (Zone 3). XO/AO and nitrotyrosine staining, consistent with the involvement of $ONOO^-$, in endothelial cells of thoracic aorta were intense and coincident 24 hr posttransplant, consistent with XO/AO binding to GAGs on vascular cell surfaces. To a lesser extent, XO/AO and nitrotyrosine staining were evident in endothelial cells of small vessels of the heart and alveolar epithelial cells. Following *ex vivo* hepatic ischemia, there was a dramatic increase of intracellular XO/AO in regions immediately adjacent to central vein branches (Zone 3), especially in hepatocytes and to a lesser extent Kupffer cells. In contrast to the XO/AO staining, there was little or no nitrotyrosine staining in the control livers. Following ischemia–reperfusion, the pattern of nitrotyrosine staining was coincident with XO/AO staining. These studies also indicate that XO/AO is released from hepatocytes and may bind to GAGs on the vascular cell surface. Coincident localization of XO/AO and nitrotyrosine coupled with our observation that iNOS was increased following transplantation supports the contention that $\cdot O_2^-$ produced by XO/AO could react with ·NO produced by NOS and nitrate tyrosine.

10.6. Vascular Disease

There are also data demonstrating that XO plays a role in systemic vascular diseases such as atherosclerosis and hypertension.

10.6.1. Atherosclerosis

XO has been found to be higher in the plasma of patients with atherosclerotic disease than in normal subjects (Mohacsi *et al.*, 1996). In addition, XO and uric acid have been identified in human atherosclerotic material (aneurysms and endarterectomies), suggesting that XO may play a role in the development of atherosclerotic lesions (Patetsios *et al.*, 1996; Swain and Gutteridge, 1995). This is consistent with immunohistochemical localization of XO/AO in human atherosclerotic lesions and the apparent association of XO/AO with the vascular surface (unpublished observations). This is also concordant with the observation that the 2.5-fold increase in plasma XO associated with hypercholesterolemia feeding in rabbits severely impaired acetylcholine (ACh)-mediated relaxation of the thoracic aorta (White *et al.*, 1996). Pretreatment of aortic vessels with either heparin, which competes with XO for binding to sulfated GAGs or allopurinol resulted in a partial restoration (36–40%) of ACh-dependent relaxation. Furthermore, $\cdot O_2^-$-dependent lucigenin chemiluminescence measured in intact ring segments from hypercholesterolemic rabbits was decreased by addition of heparin (White *et al.*, 1996), allopurinol (White *et al.*, 1996), or oxypurinol (Ohara *et al.*, 1993; Mugge *et al.*, 1994). Incubation of vascular rings from rabbits on a normal diet with purified XO (10 mU/ml) also impaired ·NO-dependent relaxation but only in the presence of purine substrate (Swain and Gutteridge, 1995). These data indicate that XO may play a significant role in the pathogenesis of atherosclerosis.

10.6.2. Hypertension

It has been demonstrated that XO and uric acid levels are increased in hypertensive patients and have a significant association with mean arterial blood pressure (Newaz *et al.*, 1996). Furthermore, XO activity is greater in mesenteric microvessels of spontaneously hypertensive rats (SHR) than normal rats (unpublished observations) and chronic inhibition of XO activity normalizes the increased arteriolar tone and mean arterial blood pressure. Myocardial XO activity was demonstrated to increase over 200% from 2 months to 18 months in SHR, but was unaltered in normotensive animals (Janssen *et al.*, 1993). These observations support the contention that XO may play a significant role in the pathogenesis of altered vascular function that results in hypertension.

10.7. Respiratory Distress Syndrome

The presence of multisystem organ dysfunction in many adult patients with respiratory distress syndrome (ARDS) implicates a role for a circulating cytotoxic mediator. High levels of circulating XO have been reported in patients with ARDS with concurrent hepatic dysfunction (Grum *et al.*, 1991). No elevation in circulating XO was observed in critically ill patients without ARDS or hepatic dysfunction. This suggests that XO may play a role in the development of ARDS following hepatic injury. In a randomized prospective study, premature infants in danger of hyperoxia were administered allopurinol, which resulted in a decreased mortality and improvement in renal function (Boda *et al.*, 1984). Treatment of hyperoxia-exposed premature baboons with allopurinol for the first 6 days of life resulted in amelioration of hyperoxia-associated pathological changes in the lung, improved ventilation, and induction of antioxidant defenses compared with vehicle-treated baboons exposed to 100% oxygen for the same time period (Frank and Sosenko, 1987). These findings suggest that XO may have a significant role in respiratory distress syndrome.

10.8. Alcohol Metabolism

Hepatic dysfunction is the primary manifestation of long-term chronic ethanol consumption although there are also strong links to cardiovascular effects such as cardiomyopathies, arrhythmias, coronary heart disease, stroke, hypertension, and respiratory distress syndrome (Moss *et al.*, 1996; Klatsky, 1995). There are three major pathways for ethanol metabolism all of which result in formation of acetaldehyde (Weiner *et al.*, 1994; Lieber, 1994). Acetaldehyde can serve as a substrate for both AO and XO and thereby result in the formation of $\cdot O_2^-$ (Shaw and Jayatilleke, 1990; Krenitsky *et al.*, 1986; Hall and Krenitsky, 1986). The K_m of XO for acetaldehyde (~30 mM) and the availability of preferred substrates, such as xanthine and hypoxanthine, may limit the *in vivo* significance (Fridovich, 1966). However, AO has a much lower K_m (~1 mM) for acetaldehyde and other aldehydes (Rajagopalan *et al.*, 1968; Barber *et al.*, 1982) and is a likely source of oxidants during ethanol metabolism (Mira *et al.*, 1995; Shaw and Jayatilleke, 1990). The metabolism of ethanol also results in formation of NADH, which may in turn serve as a substrate for AO, which is a significant source of oxidant formation during ethanol metabolism (Mira *et al.*, 1995).

We immunolocalized XO/AO in livers from patients diagnosed with fatty liver or significant alcoholic liver disease (unpublished observations). In both fatty and alcoholic liver, XO/AO was localized around the branches of the central vein, the region most susceptible to ethanol-induced tissue injury, while connective tissue in the advanced alcoholic liver did not stain positively. In advanced alcoholic liver disease, there was remarkable bile duct proliferation and intense XO/AO staining within the epithelium. Chronic alcohol consumption has a profound hypermetabolic effect on the liver that may result in hypoxia, especially in the centrivenous region where XO/AO activity is concentrated, and could ultimately contribute to liver damage and explain many of the similarities between ethanol-induced liver injury and ischemia-induced liver injury (Lieber, 1994; Nordmann, 1994). Damage to liver and other tissues associated with chronic ethanol consumption has been attributed to (1) altered metabolism and relative hypoxia at a microvascular level, (2) formation of reactive oxygen and nitrogen species, (3) formation of protein adducts and consequent alterations in function/structure, and (4) depletion in antioxidant defenses.

Chronic ethanol consumption, like ischemia, results in increased oxidant production as demonstrated indirectly by increased lipid peroxidation (Guerri *et al.*, 1994; Chen *et al.*, 1996; Lieber, 1994; Nordmann, 1994) and directly by EPR (Reinke *et al.*, 1987, 1991; McCay *et al.*, 1992). Chronic exposure to ethanol administration also depletes essential tissue and plasma antioxidants, overwhelms antioxidant defenses, and impairs critical biochemical processes rendering tissues more susceptible to oxidant injury (Nordmann, 1994; Grootveld and Halliwell, 1987). Chronic ethanol administration also increases $\cdot O_2^-$ production (Mira *et al.*, 1995; Shaw and Jayatilleke, 1990), production of ·NO (Wang *et al.*, 1995), and activates inflammatory cells, such as Kupffer cells (Rosser and Gores, 1995). The gene expression of oxidant-generating and antioxidant-producing enzymes is modulated by chronic ethanol administration (Nanji *et al.*, 1995) and may ultimately determine whether tissue damage is manifested.

10.9. XO Activity as a Predictor of Outcome

In global hypoxia, liver and intestinal injury releases XO into the circulation, causing multisystem oxidant-mediated damage. Because fetal hypoxia is usually global in origin, fetal brain damage following hypoxia–ischemia may result from circulating XO, or from XO-mediated heart failure. Serial measurements in premature newborn infants revealed that plasma XO was significantly increased in babies who had a poor outcome (defined by catastrophic clinical events), compared with babies with good outcomes, who actually experienced a drop in XO levels (Supnet *et al.*, 1994). Levels of lipid peroxidation in plasma in these infants showed an identical association. Bronchopulmonary dysplasia is a late condition affecting infants with respiratory distress syndrome or infants subjected to barotrauma. Infants with simple respiratory distress syndrome could be segregated from those infants who developed bronchopulmonary dysplasia by the magnitude of the epithelial lining fluid markers, including XO, leukocytes, elastase, and antioxidants. While infants developing bronchopulmonary dysplasia typically exhibited increased concentrations of these markers during the first week of life, those infants with simple respiratory distress syndrome displayed low, uniform, or decreasing values of these markers over this interval (Contreras *et al.*, 1996).

11. SUMMARY

Despite extensive study into the role of XO in biology (Yokoyama *et al.*, 1990; Parks and Granger, 1986b; Kurose and Granger, 1994), there remains considerable controversy as to the importance of this enzyme in physiology and pathology (Figure 3). It has long been recognized that XO can produce oxidants but the significance of XO-generated oxidants appeared limited because of the time required for conversion from the normally occurring dehydrogenase form to the oxidant-producing oxidase form (Engerson *et al.*, 1987; Yokoyama *et al.*, 1990). It has been subsequently demonstrated that XDH could be released from cells and tissues with high XO specific activity (Parks and Granger, 1986b) into the plasma following ischemia–reperfusion, sepsis, burns, acute viral infection, and hemorrhagic shock (Oda *et al.*, 1989; Grum *et al.*, 1991; Akaike *et al.*, 1990; Tan *et al.*, 1993a). Once in the plasma the conversion to XO was rapid (<1 min) and complete (Yokoyama *et al.*, 1990; Kooij *et al.*, 1994). This circulating XO could bind and

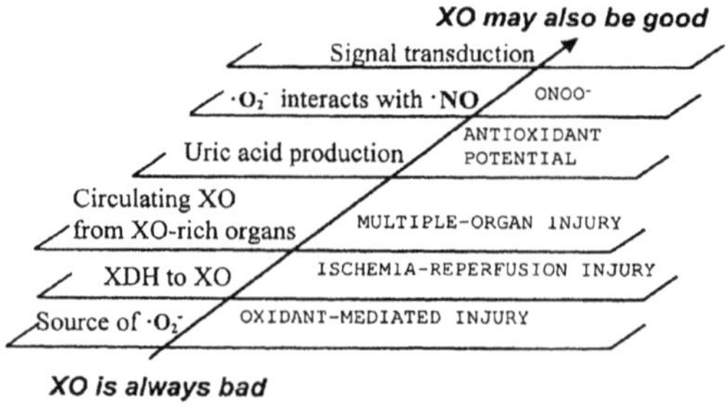

FIGURE 3. Evolution in the concepts for the role of xanthine oxidase in biology.

concentrate on vascular endothelial cell surfaces (White *et al.*, 1996; Tan *et al.*, 1993a; Adachi *et al.*, 1993), and involve tissues with low endogenous XO (e.g., myocardium, lung, and brain). It has also been demonstrated that the role of XO extends beyond the original concept of oxidant-induced tissue injury, with XO-derived oxidants stimulating transcription of proinflammatory cytokines (IL-1β, TNFα, and TGF-β1), in part, through activation of nuclear transcription regulatory factors (Shenkar and Abraham, 1996; Schwartz *et al.*, 1995). There also appears to be regulation of the vasoactive actions of ·NO by XO-generated oxidants to decrease rates of $\cdot O_2^-$ generation, thus potentiating (or preserving) ·NO-mediated vasorelaxation (Miyamoto *et al.*, 1996). This regulation may be related to the reaction of ·NO with $\cdot O_2^-$ to form $ONOO^-$. Nitric oxide may also regulate $\cdot O_2^-$ production posttranscriptionally by binding to the XDH + XO iron–sulfur moiety, sulfhydryl groups, or by reversible alteration of the flavin prosthetic site, thereby inhibiting activity (Fukahori *et al.*, 1994; Hassoun *et al.*, 1995). In summary, the simple concepts proposed over a decade ago for the role of XO in pathology are no longer sufficient and will require a reevaluation as to the importance of these molybdenum hydroxylases in biology.

ACKNOWLEDGMENTS. This work was supported in part by National Institutes of Health grant HL 48676.

12. REFERENCES

Abello, P. A., Buchman, T. G., and Bulkley, G. B., 1994, Shock and multiple organ failure, *Adv. Exp. Med. Biol.* **366:**253–268.

Adachi, T., and Marklund, S. L., 1989, Interactions between human extracellular superoxide dismutase C and sulfated polysaccharides, *J. Biol. Chem.* **264:**8537–8541.

Adachi, T., Fukushima, T., Usami, Y., and Hirano, K., 1993, Binding of human xanthine oxidase to sulphated glycosaminoglycans on the endothelial-cell surface, *Biochem. J.* **289**(Pt. 2):523–527.

Akaike, T., Ando, M., Oda, T., Ijiri, S., and Maeda, H., 1990, Dependence on $O_2^-\cdot$ generation by xanthine oxidase of pathogenesis of influenza virus infection in mice, *J. Clin. Invest.* **85:**739–745.

Amaya, Y., Yamazaki, K., Sato, M., Noda, K., and Nishino, T., 1990, Proteolytic conversion of xanthine dehydrogenase from the NAD-dependent type to the $O_2^-\cdot$ dependent type. Amino acid sequence of rat liver xanthine dehydrogenase and identification of the cleavage sites of the enzyme protein during irreversible conversion by trypsin, *J. Biol. Chem.* **265:**14170–14175.

Ames, B. N., Cathcart, R., Schwiers, E., and Hochstein, P., 1981, Uric acid provides an antioxidant defense in humans against oxidant and radical-caused aging and cancer: A hypothesis, *Proc. Natl. Acad. Sci. USA* **78:**6858–6862.

Anderson, B. O., Moore, E. E., Moore, F. A., Leff, J. A., Terada, L. S., Harken, A. H., and Repine, J. E., 1991, Hypovolemic shock promotes neutrophil sequestration in lungs by a xanthine oxidase-related mechanism, *J. Appl. Physiol.* **71:**1862–1865.

Avis, P. G., Bergel, F., and Bray, R. C., 1955, Cellular constituents. The chemistry of xanthine oxidase. Part I. The preparation of a crystalline xanthine oxidase from cow's milk, *Genetics* **129**(4):1099–1109.

Baggiolini, M., Boulay, F., Badwey, J. A., and Curnutte, J. T., 1993, Activation of neutrophil leukocytes: chemoattractant receptors and respiratory burst [review], *FASEB J.* **7**(11):1004–1010.

Ball, E. G., 1939, Xanthine oxidase: Purification and properties, *J. Biol. Chem.* **128:**51–67.

Barber, M. J., Coughlan, M. P., Rajagopalan, K. V., and Siegel, L. M., 1982, Properties of the prosthetic groups of rabbit liver aldehyde oxidase: A comparison of molybdenum hydroxylase enzymes, *Biochemistry* **21:**3561–3568.

Becker, B. F., Reinholz, D. N., Leipert, B., Raschke, P., Permanetter, B., and Gerlach, E., 1991, Role of uric acid as an endogenous radical scavenger and antioxidant, *Chest* **100**(Suppl. 3):176S–181S.

Beckman, J. S., 1996, Oxidative damage and tyrosine nitration from peroxynitrite, *Chem. Res. Toxicol.* **9**(5):836–844.

Beckman, J. S., Ye, Y. Z., Anderson, P. G., Chen, J., Accavitti, M. A., Tarpey, M. M., and White, C. R., 1994, Extensive nitration of protein tyrosines in human atherosclerosis detected by immunohistochemistry, *Biol. Chem. Hoppe Seyler* **375:**81–88.

Beckman, J. S., and Koppenol, W. H., 1996, Superoxide dismutase, toxicity, neuronal networks. Nitric oxide, superoxide, and peroxynitrite—the good, the bad, and the ugly, *Am. J. Physiol. Cell Physiol.* **40**(5):C1424–C1437.

Beedham, C., 1985, Molybdenum hydroxylases as drug-metabolizing enzymes [review], *Drug Metab. Rev.* **16**(1–2):119–156.

Beedham, C., Critchley, D. J., and Rance, D. J., 1995, Substrate specificity of human liver aldehyde oxidase toward substituted quinazolines and phthalazines: A comparison with hepatic enzyme from guinea pig, rabbit, and baboon, *Arch. Biochem. Biophys.* **319:**481–490.

Bienvenu, K., and Granger, D. N., 1993, Leukocyte adhesion in ischemia/reperfusion [review] [23 refs.], *Blood Cells* **19**(2):279–288; discussion 288–289.

Boda, D., Nemeth, I., Hencz, P., and Denes, K., 1984, Effect of allopurinol treatment in premature infants with idiopathic respiratory distress syndrome, *Dev. Pharmacol. Ther.* **7**(6):357–367.

Bray, R. C., Bennett, B., Burke, J. F., Chovnick, A., Doyle, W. A., Howes, B. D., Lowe, D. J., Richards, R. L., Turner, N. A., Ventom, A., and Whittle, J. R., 1996, Recent studies on xanthine oxidase and related enzymes [review], *Biochem. Soc. Trans.* **24**(1):99–105.

Calzi, M. L., Raviolo, C., Ghibaudi, E., de Gioia, L., Salmona, M., Cazzaniga, G., Kurosaki, M., Terao, M., and Garattini, E., 1995, Purification, cDNA cloning, and tissue distribution of bovine liver aldehyde oxidase, *J. Biol. Chem.* **270:**31037–31045.

Carithers, R. L., Fairman, R. P, Mendez-Picon, G., Posner, M. P., and Mills, A. S., 1988, Postoperative care, in *Transplantation of the Liver* (W. C. Maddrey, ed.), pp. 111–142, Elsevier, Amsterdam.

Cazzaniga, G., Terao, M., Lo Schiavo, P., Galbiati, F., Segalla, F., Seldin, M. F., and Garattini, E., 1994, Chromosomal mapping, isolation, and characterization of the mouse xanthine dehydrogenase gene, *Genomics* **23**(2):390–402.

Chen, L. H., Huang, C. Y., and Osio, Y., 1996, Large doses of ethanol and activities of liver enzymes which detoxify reactive oxygen species: Effects of antioxidants, *Biochem. Arch.* **12**(2):95–104.

Clare, D. A., and Lecce, J. G., 1991, Copurification of bovine milk xanthine oxidase and immunoglobulin, *Arch. Biochem. Biophys.* **286:**233–237.

Contreras, M., Hariharan, N., Lewandoski, J. R., Ciesielski, W., Koscik, R., and Zimmerman, J. J., 1996, Bronchoalveolar oxyradical inflammatory elements herald bronchopulmonary dysplasia, *Crit. Care Med.* **24**(1):29–37.

DeGroote, H., and Littauer, A., 1988, Reoxygenation injury in isolated hepatocytes: Cell death precedes conversion of xanthine dehydrogenase to xanthine oxidase, *Biochem. Biophys. Res. Commun.* **155:**278–282.

Deloria, L., Abbott, V., Gooderham, N., and Mannering, G. J., 1985, Induction of xanthine oxidase and degression of cytochrome P-450 by interferon inducers: Genetic difference in the responses of mice, *Biochem. Biophys. Res. Commun.* **131:**109–114.

De Renzo, E. C., Kaleita, E., Heyler, P., Oleson, J. J., Hutchings, B. L., and Williams, J. H., 1953, The nature of the xanthine oxidase factor, *J. Am. Chem. Soc.* **75:**753–758.

Dupont, G. P., Huecksteadt, T. P., Marshall, B. C., Ryan, U. S., Michael, J. R., and Hoidal, J. R., 1992, Regulation of xanthine dehydrogenase and xanthine oxidase activity and gene expression in cultured rat pulmonary endothelial cells, *J. Clin. Invest.* **89:**197–202.

Duran, M., Beemer, F. A., van de Heiden, C., Korteland, J., de Bree, P. K., Brink, M., Wadman, S. K., and Lombeck, I., 1978, Combined deficiency of xanthine oxidase and sulphite oxidase: A defect of molybdenum metabolism or transport? *J. Inherit. Metab. Dis.* **1**(4):175–178.

Duval, D. L., Sieg, D. J., and Billings, R. E., 1995, Regulation of hepatic nitric oxide synthase by reactive oxygen intermediates and glutathione, *Arch. Biochem. Biophys.* **316:**699–706.

Engerson, T. D., McKelvey, T. G., Rhyne, D. B., Boggio, E. B., Snyder, S. J., and Jones, H. P., 1987, Conversion of xanthine dehydrogenase to oxidase in ischemic rat tissues, *J. Clin. Invest.* **79:**1564–1570.

Falciani, F., Ghezzi, P., Terao, M., Cazzaniga, G., and Garattini, E., 1992, Interferons induce xanthine dehydrogenase gene expression in L929 cells, *Biochem. J.* **285**(Pt. 3):1001–1008.

Frank, L., and Sosenko, I. R., 1987, Prenatal development of lung antioxidant enzymes in four species, *J. Pediatr.* **110**(1):106–110.

Frederiks, W. M., Kooij, A., and Bosch, K. S., 1995, Role of xanthine oxidase activity in tissue damage of rat liver after ischemia, *Transplant. Proc.* **27**(5):2855–2856.

Freeman, B. A., and Crapo, J. D., 1982, Biology of disease: Free radicals and tissue injury, *Lab. Invest.* **47**:412–426.

Freeman, B. A., White, C. R., Gutierrez, H., Paler-Martinez, A., Tarpey, M. M., and Rubbo, H., 1995, Oxygen radical–nitric oxide reactions in vascular diseases [review], *Adv. Pharmacol.* **34**:45–69.

Fridovich, I., 1966, The mechanism of the enzymatic oxidation of aldehydes, *J. Biol. Chem.* **241**:3126–3128.

Fridovich, I., 1986, Biological effects of the superoxide radical [review], *Arch. Biochem. Biophys.* **247**:1–11.

Fridovich, I., and Handler, P., 1958, Xanthine oxidase. III. Sulfite oxidation as an ultrasensitive assay, *J. Biol. Chem.* **233**:1578–1580.

Friedl, H. P., Smith, D. J., Till, G. O., Thomson, P. D., Louis, D. S., and Ward, P. A., 1990, Ischemia–reperfusion in humans. Appearance of xanthine oxidase activity, *Am. J. Pathol.* **136**:491–495.

Fukahori, M., Ichimori, K., Ishida, H., Nakagawa, H., and Okino, H., 1994, Nitric oxide reversibly suppresses xanthine oxidase activity, *Free Radical Res.* **21**(4):203–212.

Fukushima, T., Adachi, T., and Hirano, K., 1995, The heparin-binding site of human xanthine oxidase, *Biol. Pharm. Bull.* **18**(1):156–158.

Ghezzi, P., Bianchi, M., Mantovani, A., Spreafico, F., and Salmona, M., 1984, Enhanced xanthine oxidase activity in mice treated with interferon and interferon inducers, *Biochem. Biophys. Res. Commun.* **119**(1):144–149.

Ghezzi, P., Bianchi, M., Gianera, L., Landolfo, S., and Salmona, M., 1985, Role of reactive oxygen intermediates in the interferon-mediated depression of hepatic drug metabolism and protective effect of N-acetylcysteine in mice, *Cancer Res.* **45**(8):3444–3447.

Giler, S., Sperling, O., Brosh, S., Urca, I., and DeVries, A., 1975, Serum xanthine oxidase in jaundice, *Clin. Chim. Acta* **63**:37–40.

Granger, D. N., and Korthuis, R. J., 1995, Physiologic mechanisms of postischemic tissue injury [review], *Annu. Rev. Physiol.* **57**:311–332.

Granger, D. N., Rutili, G., and McCord, J. M., 1981, Role of superoxide radicals in feline intestinal ischemia, *Gastroenterology* **81**:22–29.

Grogaard, B., Parks, D. A., Granger, D. N., McCord, J. M., and Forsberg, J. O., 1982, Effects of ischemia and oxygen radicals on mucosal albumin clearance in intestine, *Am. J. Physiol.* **242**(5):G448–454.

Grootveld, M., and Halliwell, B., 1987, Measurement of allantoin and uric acid in human body fluids. A potential index of free-radical reactions in vivo, *Biochem. J.* **243**:803–808.

Grum, C. M., Ragsdale, R. A., Ketai, L. H., and Simon, R. H., 1991, Plasma xanthine oxidase activity in patients with adult respiratory distress syndrome, *J. Crit. Care* **2**:22–26.

Guerri, C., Montoliu, C., and Renau-Piqueras, J., 1994, Involvement of free radical mechanism in the toxic effects of alcohol: Implications for fetal alcohol syndrome [review], *Adv. Exp. Med. & Biol.* **366**:291–305.

Haddad, I. Y., Beckman, J. S., Garver, R. I., and Matalon, S., 1994, Structural and functional alterations of surfactant protein A by peroxynitrite, *Chest* **105**(Suppl. 3):84S

Hall, W. W., and Krenitsky, T. A., 1986, Aldehyde oxidase from rabbit liver: Specificity toward purines and their analogs, *Arch. Biochem. Biophys.* **251**:36–46.

Harkness, R. A., McCreanor, G. M., Simpson, D., and MacFadyen, I. R., 1986, Pregnancy in and incidence of xanthine oxidase deficiency, *J. Inherit. Metab. Dis.* **9**(4):407–408.

Hassoun, P. M., Yu, F. S., Zulueta, J. J., White, A. C., and Lanzillo, J. J., 1995, Effect of nitric oxide and cell redox status on the regulation of endothelial cell xanthine dehydrogenase, *Am. J. Physiol.* **268**(5 Pt. 1):L809–817.

Hellsten-Westing, Y., 1993, Immunohistochemical localization of xanthine oxidase in human cardiac and skeletal muscle, *Histochemistry* **100**(3):215–222.

Hicks, M., Wong, L. S., and Day, R. O., 1993, Identification of products from oxidation of uric acid induced by hydroxyl radicals, *Free Radical Res. Commun.* **18**(6):337–351.

Hiebert, L. M., and Liu, J., 1991, Heparin protects cultured arterial endothelial cells from damage by toxic oxygen metabolites, *Atherosclerosis* **83**:47–51.

Hille, R., and Nishino, T., 1995, Flavoprotein structure and mechanism. 4. Xanthine oxidase and xanthine dehydrogenase [review], *FASEB J.* **9**(11):995–1003.

Hook, M., Kjellen, L., and Johansson, S., 1984, Cell-surface glycosaminoglycans, *Annu. Rev. Biochem.* **53:**847–869.

Houde, M., Tiveron, M. C., and Bregegere, F., 1989, Divergence of the nucleotide sequences encoding xanthine dehydrogenase in Calliphora vicina and Drosophila melanogaster, *Gene* **85**(2):391–402.

Hu, P., Ischiropoulos, H., Beckman, J. S., and Matalon, S., 1994, Peroxynitrite inhibition of oxygen consumption and sodium transport in alveolar type II cells, *Am. J. Physiol.* **266**(6 Pt. 1):L628–634.

Huie, R. E., and Padmaja, S., 1993, The reaction of NO with superoxide, *Free Radical Res. Commun.* **18**(4):195–199.

Ichida, K., Amaya, Y., Noda, K., Minoshima, S., Hosoya, T., Sakai, O., Shimizu, N., and Nishino, T., 1993, Cloning of the cDNA encoding human xanthine dehydrogenase (oxidase): Structural analysis of the protein and chromosomal location of the gene, *Gene* **133:**279–284.

Ichida, K., Amaya, Y., Kamatani, N., Nishino, T., Hosoya, T., and Sakai, O., 1997, Identification of two mutations in human xanthine dehydrogenase gene responsible for classical type I xanthinuria, *J. Clin. Invest.* **99**(10):2391–2397.

Ischiropoulos, H., Zhu, L., Chen, J., Tsai, M., Martin, J. C., Smith, C. D., and Beckman, J. S., 1992, Peroxynitrite-mediated tyrosine nitration catalyzed by superoxide dismutase, *Arch. Biochem. Biophys.* **298:**431–437.

Janssen, M., De Jong, J. W., Pasini, E., and Ferrari, R., 1993, Myocardial xanthine oxidoreductase activity in hypertensive and hypercholesterolemic rats, *Cardioscience* **4**(1):25–29.

Jarasch, E. D., Grund, C., Bruder, G., Heid, H. W., Keenan, T. W., and Franke, W. W., 1981, Localization of xanthine oxidase in mammary-gland epithelium and capillary endothelium, *Cell* **25**(1):67–82.

Jarasch, E. D., Bruder, G., and Heid, H. W., 1986, Significance of xanthine oxidase in capillary endothelial cells, *Acta Physiol. Scand.* **548:**39–46.

Jenkinson, S. G., Roberts, R. J., DeLemos, R. A., Lawrence, R. A., Coalson, J. J., King, R. J., Null, D. M., Jr., and Gerstmann, D. R., 1991, Allopurinol-induced effects in premature baboons with respiratory distress syndrome, *J. Appl. Physiol.* **70**(3):1160–1167.

Kaur, H., and Halliwell, B., 1990, Action of biologically-relevant oxidizing species upon uric acid. Identification of uric acid oxidation products, *Chem. Biol. Interact.* **73:**235–247.

Keith, T. P., Riley, M. A., Kreitman, M., Lewontin, R. C., Curtis, D., and Chambers, G., 1987, Sequence of the structural gene for xanthine dehydrogenase (rosy locus) in Drosophila melanogaster, *Genetics* **116**(1):67–73.

Klatsky, A. L., 1995, Cardiovascular effects of alcohol, *Sci. Am. Sci. Med.* **2**(2):28–37.

Knights, K. M., Gourlay, G. K., and Cousins, M. J., 1987, Changes in rat hepatic microsomal mixed function oxidase activity following exposure to halothane under various oxygen concentrations, *Biochem. Pharmacol.* **36:**897–906.

Kooij, A. 1994, A re-evaluation of the tissue distribution and physiology of xanthine oxidoreductase [review], *Histochem. J.* **26**(12):889–915.

Kooij, A., Schiller, H. J., Schijns, M., Van Noorden, C. F. J., and Frederiks, W. M., 1994, Conversion of xanthine dehydrogenase into xanthine oxidase in rat liver and plasma at the onset of reperfusion after ischemia, *Hepatology* **19:**1488–1495.

Krenitsky, T. A., Tuttle, J. V., Cattau, E. L., and Wang, P. A., 1974, A comparison of the distribution and electron acceptor specificities of xanthine oxidase and aldehyde oxidase, *Comp. Biochem. Physiol.* **49:**687–703.

Krenitsky, T. A., Spector, T., and Hall, W. W., 1986, Xanthine oxidase from human liver: Purification and characterization, *Arch. Biochem. Biophys.* **247:**108–119.

Kurose, I., and Granger, D. N., 1994, Evidence implicating xanthine oxidase and neutrophils in reperfusion-induced microvascular dysfunction [review], *Ann. N.Y. Acad. Sci.* **723:**158–179.

Lemasters, J. J., Bunzendahl, H., and Thurman, R. G., 1995, Reperfusion injury to donor livers stored for transplantation, *Liver Transplant. Surg.* **1**(2):124–138.

Lieber, C. S., 1994, Alcohol and the liver: 1994 update, *Gastroenterology* **106:**1085–1105.

Marubayashi, S., Dohi, K., and Kawasaki, T., 1994, Role of free radicals in hepatic reperfusion injury, *Ann. N.Y. Acad. Sci.* **723:**368–370.

McCay, P. B., Reinke, L. A., and Rau, J. M., 1992, Hydroxyl radicals are generated by hepatic microsomes during NADPH oxidation: Relationship to ethanol metabolism, *Free Radical Res. Commun.* **15**(6):335–346.

Mendez, C., Garcia, I., and Maier, R. V., 1996, Oxidants augment endotoxin-induced activation of alveolar macrophages, *Shock* **6**(3):157–163.

Mira, L., Maia, L., Barreira, L., and Manso, C. F., 1995, Evidence for free radical generation due to NADH oxidation by aldehyde oxidase during ethanol metabolism, *Arch. Biochem. Biophys.* **318:**53–58.

Mishra, O. P., and Delivoria-Papadopoulos, M., 1988, Anti-oxidant enzymes in fetal guinea pig brain during development and the effect of maternal hypoxia, *Brain Res.* **470**(2):173–179.

Miyamoto, Y., Akaike, T., Yoshida, M., Goto, S., Horie, H., and Maeda, H., 1996, Potentiation of nitric oxide-mediated vasorelaxation by xanthine oxidase inhibitors, *Proc. Soc. Exp. Biol. Med.* **211**(4):366–373.

Mize, C., Johnson, J. L., and Rajagopalan, K. V., 1995, Defective molybdopterin biosynthesis: Clinical heterogeneity associated with molybdenum cofactor deficiency, *J. Inherit. Metab. Dis.* **18**(3):283–290.

Mohacsi, A., Kozlovszky, B., Kiss, I., Seres, I., and Fulop, T., Jr., 1996, Neutrophils obtained from obliterative atherosclerotic patients exhibit enhanced resting respiratory burst and increased degranulation in response to various stimuli, *Biochim. Biophys. Acta* **1316**(3):210–216.

Moochhala, S. M., and Renton, K. W., 1991, A role for xanthine oxidase in the loss of cytochrome P-450 evoked by interferon, *Can. J. Physiol. Pharmacol.* **69**(7):944–950.

Moreno, J. J., and Pryor, W. A., 1992, Inactivation of alpha 1-proteinase inhibitor by peroxynitrite, *Chem. Res. Toxicol.* **5**(3):425–431.

Moriwaki, Y., Yamamoto, T., Suda, M., Nasako, Y., Takahashi, S., Agbedana, O. E., Hada, T., and Higashino, K., 1993, Purification and immunohistochemical tissue localization of human xanthine oxidase, *Biochim. Biophys. Acta* **1164**(3):327–330.

Moss, M., Bucher, B., Moore, F. A., Moore, E. E., and Parsons, P. E., 1996, The role of chronic alcohol abuse in the development of acute respiratory distress syndrome in adults, *J. Am. Med. Assoc.* **275**(1):50–54.

Mugge, A., Brandes, R. P., Boger, R. H., Dwenger, A., Bode-Boger, S., Kienke, S., Frolich, J. C., and Lichtlen, P. R., 1994, Vascular release of superoxide radicals is enhanced in hypercholesterolemic rabbits, *J. Cardiovasc. Pharmacol.* **24**(6):994–998.

Nanji, A. A., Griniuviene, B., Sadrzadeh, S. M., Levitsky, S., and McCully, J. D., 1995, Effect of type of dietary fat and ethanol on antioxidant enzyme mRNA induction in rat liver, *J. Lipid Res.* **36**(4):736–744.

Newaz, M. A., Adeeb, N. N., Muslim, N., Razak, T. A., and Htut, N. N., 1996, Uric acid, xanthine oxidase and other risk factors of hypertension in normotensive subjects, *Clin. Exp. Hypertens.* **18**(8):1035–1050.

Nielsen, V. G., Weinbroum, A., Tan, S., Samuelson, P. N., Gelman, S., and Parks, D. A., 1994, Xanthine oxidoreductase release after descending thoracic aorta occlusion and reperfusion in rabbits, *J. Thorac. Cardiovasc. Surg.* **107**(5):1222–1227.

Nielsen, V. G., McCammon, A. T., Tan, S., Kirk, K. A., Samuelson, P. N., and Parks, D. A., 1995, Xanthine oxidase inactivation attenuates postocclusion shock after descending thoracic aorta occlusion and reperfusion in rabbits, *J. Thorac. Cardiovasc. Surg* **110**(3):715–722.

Nielsen, V. G., Tan, S., Weinbroum, A., McCammon, A. T., Samuelson, P. N., Gelman, S., and Parks, D. A., 1996, Lung injury after hepatoenteric ischemia–reperfusion: Role of xanthine oxidase, *Am. J. Respir. Crit. Care Med.* **154**(5):1364–1369.

Nielsen, V. G., Tan, S., Baird, M. S., Samuelson, P. N., and Parks, D. A., 1997, Xanthine oxidase mediates myocardial injury after hepatoenteric ischemia–reperfusion, *Crit. Care Med.* **25**(6):1044–1050.

Nordmann, R., 1994, Alcohol and antioxidant systems [review], *Alcohol* **29**(5):513–522.

Oda, T., Akaike, T., Hamamoto, T., Suzuki, F., Hirano, T., and Maeda, H., 1989, Oxygen radicals in influenza-induced pathogenesis and treatment with pyran polymer-conjugated SOD, *Science* **244:**974–976.

Ohara, Y., Peterson, T. E., and Harrison, D. G., 1993, Hypercholesterolemia increases endothelial superoxide anion production, *J. Clin. Invest.* **91**(6):2546–2551.

Palmer, C., Vannucci, R. C., and Towfighi, J., 1990, Reduction of perinatal hypoxic–ischemic brain damage with allopurinol, *Pediatr. Res.* **27:**332–336.

Palmer, C., Towfighi, J., Roberts, R. L., and Heitjan, D. F., 1993, Allopurinol administered after inducing hypoxia–ischemia reduces brain injury in 7-day-old rats, *Pediatr. Res.* **33**(4 Pt. 1):405–411.

Parks, D. A., and Freeman, B. A., 1987, *Anesthesia and Organ Transplantation*, Saunders, Philadelphia.

Parks, D. A., and Granger, D. N., 1983, Ischemia-induced vascular changes: Role of xanthine oxidase and hydroxyl radicals, *Am. J. Physiol.* **245:**G285–G289.

Parks, D. A., and Granger, D. N., 1986a, Contributions of ischemia and reperfusion to mucosal lesion formation, *Am. J. Physiol.* **250:**G749–G753.

Parks, D. A., and Granger, D. N., 1986b, Xanthine oxidase: Biochemistry, distribution and physiology, *Acta Physiol. Scand.* **548:**87–99.

Parks, D. A., Bulkley, G. B., Granger, D. N., Hamilton, S. R., and McCord, J. M., 1982, Ischemic injury in the cat small intestine: Role of superoxide radicals, *Gastroenterology* **82:**9–15.

Partridge, C. A., Blumenstock, F. A., and Malik, A. B., 1992, Pulmonary microvascular endothelial cells constitutively release xanthine oxidase, *Arch. Biochem. Biophys.* **294:**184–187.

Patetsios, P., Rodino, W., Wisselink, W., Bryan, D., Kirwin, J. D., and Panetta, T. F., 1996, Identification of uric acid in aortic aneurysms and atherosclerotic artery, *Ann. N.Y. Acad. Sci.* **800:**243–245.

Phan, S. H., Gannon, D. E., Varini, J., Ryan, U. S., and Ward, P. A., 1989, Xanthine oxidase activity in rat pulmonary artery endothelial cells and its alteration by activated neutrophils, *Am. J. Pathol.* **134:**1201–1211.

Pietz, J., Guttenberg, N., and Gluck, L., 1988, Hypoxanthine: A marker for asphyxia, *Obstet. Gynecol.* **72**(5):762–766.

Pryor, W. A., and Squadrito, G. L., 1995, The chemistry of peroxynitrite: A product from the reaction of nitric oxide with superoxide, *Am. J. Physiol.* **268**(5 Pt. 1):L699–L722.

Radi, R., Rubbo, H., Bush, K., and Freeman, B. A., 1997, Xanthine oxidase binding to glycosaminoglycans—Kinetics and superoxide dismutase interactions of immobilized xanthine oxidase–heparin complexes, *Arch. Biochem. Biophys.* **339:**125–135.

Rajagopalan, K. V., Handler, P., Palmer, G., and Beinert, H., 1968, Studies of aldehyde oxidase by electron paramagnetic resonance spectroscopy. I. Spectra at equilibrium states, *J. Biol. Chem.* **243:**3784–3796.

Ramboer, C., Piessens, F., and DeGroote, J., 1972, Serum xanthine oxidase and liver disease, *Digestion* **7:**183–195.

Reinke, L. A., Lai, E. K., DuBose, C. M., and McCay, P. B., 1987, Reactive free radical generation in vivo in heart and liver of ethanol-fed rats: Correlation with radical formation in vitro, *Proc. Natl. Acad. Sci. USA* **84:**9223–9227.

Reinke, L. A., Kotake, Y., McCay, P. B., and Janzen, E. G., 1991, Spin-trapping studies of hepatic free radicals formed following the acute administration of ethanol to rats: In vivo detection of 1-hydroxyethyl radicals with PBN, *Free Radical Biol. Med.* **11**(1):31–39.

Repine, J. E., Cheronis, J. C., Rodell, T. C., Linas, S. L., and Patt, A., 1987, Pulmonary oxygen toxicity and ischemia–reperfusion injury: A mechanism in common involving xanthine oxidase and neutrophils, *Am. Rev. Respir. Dis.* **136:**483–485.

Roesel, R. A., Bowyer, F., Blankenship, P. R., and Hommes, F. A., 1986, Combined xanthine and sulphite oxidase defect due to a deficiency of molybdenum cofactor, *J. Inherit. Metab. Dis.* **9**(4):343–347.

Rosser, B. G., and Gores, G. J., 1995, Liver cell necrosis: Cellular mechanisms and clinical implications [review], *Gastroenterology* **108**(1):252–275.

Rubbo, H., Radi, R., Trujillo, M., Telleri, R., Kalyanaraman, B., Barnes, S., Kirk, M., and Freeman, B. A., 1994, Nitric oxide regulation of superoxide and peroxynitrite-dependent lipid peroxidation. Formation of novel nitrogen-containing oxidized lipid derivatives, *J. Biol. Chem.* **269:**26066–26075.

Rump, A. F. E., Rösen, R., and Klaus, W., 1993, Free radical scavenging properties of β-adrenoceptor blockers are not relevant for cardioprotection in isolated rabbit hearts, *Naunyn Schmiedebergs Arch. Pharmakol.* **348:**431–434.

Russell, G. A., and Cooke, R. W., 1995, Randomised controlled trial of allopurinol prophylaxis in very preterm infants, *Arch. Dis. Child Fetal Neonatal Ed.* **73**(1):F27–F31.

Russell, G. A., Jeffers, G., and Cooke, R. W., 1992, Plasma hypoxanthine: A marker for hypoxic–ischaemic induced periventricular leucomalacia? *Arch. Dis. Child* **67**(4 Spec. No.):388–392.

Sackler, M. L., 1966, Xanthine oxidase from liver and duodenum of the rat: Histochemical localization and electrophoretic heterogeneity, *J. Histochem. Cytochem.* **14**(4):326–333.

Sakai, H., and Takeguchi, N., 1994, A GTP-binding protein inhibits a gastric housekeeping chloride channel via intracellular production of superoxide, *J. Biol. Chem.* **269:**23426–23430.

Sardesai, V. M., 1993, Molybdenum: An essential trace element, *Nutr. Clin. Pract.* **8**(6):277–281.

Sarnesto, A., Linder, N., and Raivio, K. O., 1996, Organ distribution and molecular forms of human xanthine dehydrogenase/xanthine oxidase protein, *Lab. Invest.* **74**(1):48–56.

Sato, A., Nishino, T., Noda, K., and Amaya, Y., 1995, The structure of chicken liver xanthine dehydrogenase. cDNA cloning and the domain structure, *J. Biol. Chem.* **270:**2818–2826.

Saxena, U., Klein, M. G., and Goldberg, I. J., 1990, Metabolism of endothelial cell-bound lipoprotein lipase. Evidence for heparan sulfate proteoglycan-mediated internalization and recycling, *J. Biol. Chem.* **265:**12880–12886.

Schwartz, M. D., Repine, J. E., and Abraham, E., 1995, Xanthine oxidase-derived oxygen radicals increase lung cytokine expression in mice subjected to hemorrhagic shock, *Am. J. Respir. Cell Mol. Biol.* **12**(4):434–440.

Shaw, S., and Jayatilleke, E., 1990, The role of aldehyde oxidase in ethanol-induced hepatic lipid peroxidation in the rat, *Biochem. J.* **268**(3):579–583.

Shenkar, R., and Abraham, E., 1996, Plasma from hemorrhaged mice activates CREB and increases cytokine expression in lung mononuclear cells through a xanthine oxidase-dependent mechanism, *Am. J. Respir. Cell Mol. Biol.* **14**(2):198–206.

Simmonds, H. A., Reiter, S., and Nishino, T., 1995, *The Metabolic Basis of Inherited Disease*, 7th ed., McGraw–Hill, New York.

Singh, N., and Aggarwal, S., 1995, The effect of active oxygen generated by xanthine/xanthine oxidase on genes and signal transduction in mouse epidermal JB6 cells, *Int. J. Cancer* **62**(1):107–114.

Skinner, K. A., Crow, J. P., Skinner, H. B., Thompson, J. A., Chandler, R. T., and Parks, D. A., 1997, Free and protein associated nitrotyrosine formation following rat liver transplantation, *Arch. Biochem. Biophys.* **342**(2):282–288.

Slot, H. M., Overweg-Plandsoen, W. C., Bakker, H. D., Abeling, N. G., Tamminga, P., Barth, P. G., and van Gennip, A. H., 1993, Molybdenum-cofactor deficiency: An easily missed cause of neonatal convulsions, *Neuropediatrics* **24:**139–142.

Sumimoto, K., Matsura, T., Oku, J. I., Fukuda, Y., Yamada, K., and Dohi, K., 1996, Protective effect of UW solution on postischemic injury in rat liver: Suppression of reduction in hepatic antioxidants during reperfusion, *Transplantation* **62**(10):1391–1398.

Supnet, M. C., David-Cu, R., and Walther, F. J., 1994, Plasma xanthine oxidase activity and lipid hydroperoxide levels in preterm infants, *Pediatr. Res.* **36**(3):283–287.

Swain, J., and Gutteridge, J. M., 1995, Prooxidant iron and copper, with ferroxidase and xanthine oxidase activities in human atherosclerotic material, *FEBS Lett.* **368**(3):513–515.

Tan, C. M., Xenoyannis, S., and Feldman, R. D., 1995, Oxidant stress enhances adenylyl cyclase activation, *Circ. Res.* **77**(4):710–717.

Tan, S., Radi, R., Gaudier, F., Evans, R. A., Rivera, A., Kirk, K. A., and Parks, D. A., 1993a, Physiologic levels of uric acid inhibit xanthine oxidase in human plasma, *Pediatr. Res.* **34**(3):303–307.

Tan, S., Yokoyama, Y., Dickens, E., Cash, T. G., Freeman, B. A., and Parks, D. A., 1993b, Xanthine oxidase activity in the circulation of rats following hemorrhagic shock, *Free Radical Biol. Med.* **15**(4):407–414.

Tan, S., Gelman, S., Wheat, J. K., and Parks, D. A., 1995, Circulating xanthine oxidase in human ischemia reperfusion, *South. Med. J.* **88**(4):479–482.

Tan, S., Liu, Y. Y., Nielsen, V. G., Skinner, K., Kirk, K. A., Baldwin, S. T., and Parks, D. A., 1996, Maternal infusion of antioxidants (Trolox and ascorbic acid) protects the fetal heart in rabbit fetal hypoxia, *Pediatr. Res.* **39**(3):499–503.

Tan, S., Yokoyama, Y., Nielsen, V. G., Murdock, A. D., Adams, C., and Parks, D. A., 1998, Hypoxia–reoxygenation is as damaging as ischemia–reperfusion in the rat liver, *Crit. Care Med.* **26:**1089–1085.

Terada, L. S., Dormish, J. J., Shanley, P. F., Leff, J. A., Anderson, B. O., and Repine, J. E., 1992, Circulating xanthine oxidase mediates lung neutrophil sequestration after intestinal ischemia–reperfusion, *Am. J. Physiol.* **263:**L394–L401.

Terao, M., Cazzaniga, G., Ghezzi, P., Bianchi, M., Falciani, F., Perani, P., and Garattini, E., 1992, Molecular cloning of a cDNA coding for mouse liver xanthine dehydrogenase. Regulation of its transcript by interferons in vivo, *Biochem. J.* **283:**863–870.

Toledo-Pereyra, L. H., 1991, Liver transplantation reperfusion injury. Factors in its development and avenues for treatment, *Klin. Wochenschr.* **69:**1099–1104.

Turner, N. A., Doyle, W. A., Ventom, A. M., and Bray, R. C., 1995, Properties of rabbit liver aldehyde oxidase and the relationship of the enzyme to xanthine oxidase and dehydrogenase, *Eur. J. Biochem.* **232**(2):646–657.

van der Vliet, A., Smith, D., O'Neill, C. A., Kaur, H., Darley-Usmar, V., Cross, C. E., and Halliwell, B., 1994, Interactions of peroxynitrite with human plasma and its constituents: Oxidative damage and antioxidant depletion, *Biochem. J.* **303**(Pt. 1):295–301.

Vettenranta, K., and Raivio, K. O., 1990, Xanthine oxidase during human fetal development, *Pediatr. Res.* **27**(3):286–288.

Wang, J. F., Greenberg, S. S., and Spitzer, J. J., 1995, Chronic alcohol administration stimulates nitric oxide formation in the rat liver with or without pretreatment by lipopolysaccharide, *Alcoholism Clin. Exp. Res.* **19**(2):387–393.

Weinbroum, A., Nielsen, V. G., Tan, S., Skinner, K. A., Gelman, S., Matalon, S., Bradley, E., and Parks, D. A., 1995, Liver ischemia–reperfusion increases pulmonary permeability in the rat: Role of circulating xanthine oxidase, *Am. J. Physiol.* **268**(6):G988–G996.

Weiner, F. R., Esposti, S. D., and Zern, M. A., 1994, *The Liver: Biology and Pathobiology*, 3rd ed., Raven Press, New York.

Wesson, D. E., and Elliott, S. J., 1994, Xanthine oxidase inhibits transmembrane signal transduction in vascular endothelial cells, *J. Pharmacol. Exp. Ther.* **270**(3):1197–1207.

White, C. R., Darley-Usmar, V., Berrington, W. R., McAdams, M., Thompson, J. A., Parks, D. A., Tarpey, M. M., and Freeman, B. A., 1996, Circulating plasma xanthine oxidase contributes to vascular dysfunction in hypercholesterolemic rabbits, *Proc. Natl. Acad. Sci. USA* **93**:8745–8749.

Wiezorek, J. S., Brown, D. H., Kupperman, D. E., and Brass, C. A., 1994, Rapid conversion to high xanthine oxidase activity in viable Kupffer cells during hypoxia, *J. Clin. Invest.* **94**(6):2224–2230.

Winkler, C., Tan, S., Wheat, J. K., and Parks, D. A., 1990, Xanthine oxidase activity in human fetal tissues, *Pediatr. Res.* **27**(4):231A.

Wood, R. P., Shaw, B. W., Jr., and Starzl, T. E., 1985, Extrahepatic complications of liver transplantation, *Semin. Liver Dis.* **5**:377–384.

Wright, R. M., Vaitaitis, G. M., Wilson, C. M., Repine, T. B., Terada, L. S., and Repine, J. E., 1993, cDNA cloning, characterization, and tissue-specific expression of human xanthine dehydrogenase/xanthine oxidase, *Proc. Natl. Acad. Sci. USA* **90**:10690–10694.

Wright, R. M., Vaitaitis, G. M., Weigel, L. K., Repine, T. B., McManaman, J. L., and Repine, J. E. 1995, Identification of the candidate ALS2 gene at chromosome 2q33 as a human aldehyde oxidase gene, *Redox Rep.* **1**:313–321.

Wu, X., Lee, C. C., Muzny, D. M., and Caskey, C. T., 1989, Urate oxidase: Primary structure and evolutionary implications, *Proc. Natl. Acad. Sci. USA* **86**:9412–9416.

Yokoyama, Y., Beckman, J. S., Beckman, T. K., Wheat, J. K., Cash, T. G., Freeman, B. A., and Parks, D. A., 1990, Circulating xanthine oxidase: Potential mediator of ischemic injury, *Am. J. Physiol.* **258**:G564–G570.

Chapter 16

Melatonin

Antioxidative Protection by Electron Donation

Burkhard Poeggeler

The resistance to a new idea increases by the square of its importance.
Bertrand Russell

1. THE PRIMARY FUNCTIONS OF MELATONIN: ELECTRON DONATION, RADICAL SCAVENGING, AND ANTIOXIDATIVE PROTECTION

The neurohormone melatonin has demonstrated a remarkably broad spectrum of actions (Hardeland *et al.*, 1993). Until recently, however, melatonin was investigated mainly for its specific neurohormonal effects. The established role of this pineal indoleamine as a mediator of the signal darkness is mediated exclusively through specific membrane receptors localized within the brain. The participation of melatonin in the entrainment of circadian rhythms and the involvement of the hormone in regulating mammalian photoperiodism have been demonstrated conclusively. Certain actions of the neurohormone such as those related to the regulation and timing of reproduction and those associated with the entrainment of endogenous biological rhythms are very likely induced and mediated by the activation of specific membrane receptors (Dubocovich, 1995; Reiter, 1980).

Many actions of melatonin such as its potent chemopreventive and neuroprotective effects cannot be explained by the activation of specific membrane receptors (Hardeland *et al.*, 1993; Poeggeler *et al.*, 1993). The life-prolonging, antiaging effects of melatonin cannot be antagonized by melatonin receptor antagonists and are mimicked by compounds with the same electropotential and radical-scavenging ability as melatonin

Burkhard Poeggeler Department of Pathology, University of South Alabama, Mobile, Alabama 36617.

Reactive Oxygen Species in Biological Systems, edited by Gilbert and Colton. Kluwer Academic / Plenum Publishers, New York, 1999.

(Oakin-Bendahan *et al.*, 1995; Poeggeler *et al.*, 1993). Melatonin can cross the cell membrane with ease and can induce many intracellular effects that are clearly not mediated by membrane receptors localized on the surface of the cell (Reiter *et al.*, 1995). Melatonin elicits specific physiological responses, such as encystment in unicellular organisms, that do not require the presence of membrane receptors (Balzer and Hardeland, 1991, 1996). The indoleamine provides antioxidative protection in organs and organisms that do not possess specific membrane receptors for this endogenous indoleamine (Reiter *et al.*, 1995).

The presence of melatonin in unicellular organisms, plants, and invertebrates and the pleiotropic spectrum of physiological effects elicted by the indoleamine in different organs and organisms have been considered inconsistent with the classic neurohormonal concept of melatonin receptor activation and specific cellular response (Hardeland *et al.*, 1993). The rather limited distribution of specific melatonin receptors in certain brain areas cannot explain the broad spectrum of actions elicted by physiological levels of the endogenous indoleamine in tissues such as the liver, lung, and heart (Reiter *et al.*, 1995). Furthermore, the high solubility of the indole and its extremely high diffusibility infer that melatonin's actions are not limited to the activation of certain receptors on the surface of cells (Menendez-Peleaz *et al.*, 1993). Finally, the recent discovery that melatonin accumulates in high concentrations in the nucleus of cells intracellularly suggests that melatonin can elicit additional physiological responses that are not mediated by membrane receptors (Menendez-Peleaz *et al.*, 1993; Reiter *et al.*, 1995).

Melatonin and chemically related endogenous compounds with high resonance stability and electroreactivity are a class of chemoprotectants that act by several mechanisms, including radical scavenging, protection against thiol depletion and calcium overload as well as specific enzyme induction and inhibition (Reiter *et al.*, 1995). Melatonin can elicit highly specific genomic and neuronal responses, by activation of both intracellular receptors and membrane receptors localized on the surface of cells, but the indoleamine can also induce a broad range of effects directly without the mediation of certain specific receptors simply by the presence of the molecule and its ability to scavenge reactive oxygen species wherever and whenever they are generated (Dubocovich, 1995; Reiter *et al.*, 1995). Gene expression as well as enzyme induction, protection, and inhibition can be regulated by both mechanisms and it seems that melatonin acts in concert on both receptors and radicals to achieve antioxidative protection (Antolin *et al.*, 1996; Balzer and Hardeland, 1996; Barlow-Walden *et al.*, 1995; Reiter *et al.*, 1995).

Melatonin is now known to act as an endogenous electron donor and radical scavenger that provides antioxidative protection ubiquitously (Poeggeler, 1993; Poeggeler *et al.*, 1993, 1994, 1995, 1996). It is very fascinating that the structure of melatonin differs markedly from that of any other known antioxidant (Poeggeler *et al.*, 1993, 1994, 1996). Melatonin is a very stable, non-redox-active molecule (Poeggeler *et al.*, 1993). Neither free hydroxyl nor free sulfuryl groups are found (Hardeland *et al.*, 1993; Poeggeler *et al.*, 1993). The electron-rich, mesomery-stabilized indole ring system is shielded by an electron-donating, electron-inducing methoxy substituent in 5′ position and a resonance-stabilizing ethyl-*N*-acetyl side chain in 3′ position. The methyl group shields the melatonin molecule from reactions such as autoxidation, redox recycling, and dimerization or hydroxylation, presumably by preventing the formation of highly reactive phenoxy radicals (Poeggeler *et al.*, 1993). Radical scavenging by melatonin is always accompanied by a loss of indole fluorescence and the opening of the indole ring system. It has been demonstrated conclusively that the radical-mediated, nonenzymatic, oxidative cleavage of the pyrrole ring of the heterocyclic aromatic ring system is the mechanism by which melatonin is metabolized on exposure to highly reactive oxygen intermediates (Figure 1). The cleavage of the pyrrole ring is generally possible with all indole molecules and is found to occur to a significant extent with many indole compounds on exposure to reactive oxygen intermediates. However, melatonin exhibits an exceptionally high electroreactivity related to the resonance stability of its electron-rich, heterocyclic aromatic ring system and the presence of inductive, mesomery-stabilizing substituents (Hardeland *et al.*, 1993; Poeggeler *et al.*, 1993).

The ability of melatonin to cross all physiological barriers, to enter all cell compartments with ease, and to be metabolized, enzymatically or nonenzymatically in all of them, results in the broad spectrum of effects seen after the administration of the indole compound in many different organs and organisms (Hardeland *et al.*, 1993). It has been demonstrated conclusively that melatonin and structurally related indoleamines are substrates for enzymatic and nonenzymatic reactions involving the transfer of electrons that utilize reactive oxygen intermediates. The detection of melatonin and other bioactive indoleamines such as 5-methoxytryptamine in extrapineal tissues was only possible after

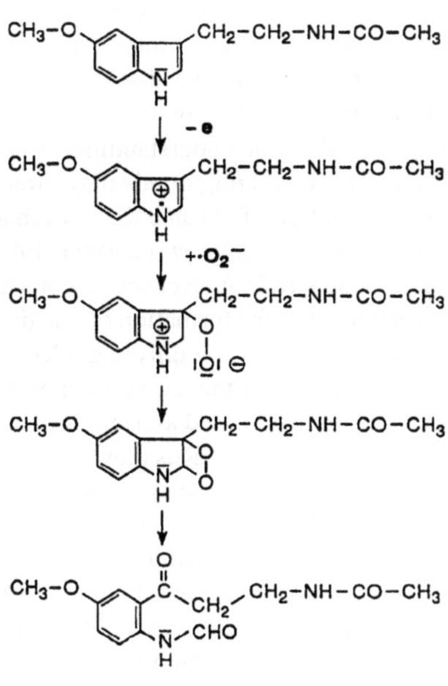

FIGURE 1. Electron donation and superoxide anion radical scavenging by melatonin. Mechanisms of electron donation and radical-mediated indole oxidation.

the development of preservative extraction procedures that avoided the nonenzymatic destruction of these compounds (Poeggeler and Hardeland, 1994).

The fact that melatonin and structurally related indoleamines were rapidly oxidized in extrapineal tissues during homogenization and extraction whenever conventional methods were applied, initiated experiments in which the radical scavenging activities of melatonin and 5-methoxytryptamine were demonstrated (Poeggeler and Hardeland, 1994). Thus, difficulty in recovering the indoleamines from tissues that are subject to oxidative stress mediated by free radicals during the extraction process has stimulated research on the electrochemical properties of melatonin (Poeggeler and Hardeland, 1994). The discovery of the potent electron-donating and radical-scavenging effects exerted by melatonin and 5-methoxytryptamine was accidental, as the oxidative chemistry of melatonin and other structurally related indoleamines was studied only because those compounds were lost on homogenization and extraction in tissues exposed to free radicals and oxidative stress (Hardeland *et al.*, 1993; Poeggeler and Hardeland, 1994). As is often the case, methodological difficulties in measuring endogenous compounds initiated research activities into the oxidative chemistry of these compounds and provided us with new insights into the physiological functions of these agents (Hardeland *et al.*, 1993; Poeggeler *et al.*, 1991). Comparative research on the occurrence of indoleamines in phylogenetically distinct organisms such as unicellular algae, plants, and invertebrates revealed the broad distribution of melatonin and structurally related tryptophan metabolites and many new physiological effects of these highly electroreactive indoleamines were detected and described (Balzer and Hardeland, 1996; Hardeland *et al.*, 1993; Poeggeler, 1993; Poeggeler *et al.*, 1991).

The specific vulnerability of certain cells and tissues to radical-mediated oxidative stress and damage has been at the center of interest in free radical research (Yu, 1994). Recently, many laboratories have begun to search for endogenous antioxidants that provide protection against oxidotoxicity and these compounds are now the target of intensive research efforts. The discovery of endogenous electron donors, which, like melatonin, are able to protect against radical-mediated damage and radical-based disease, has greatly extended our understanding of physiological antioxidative defense mechanisms. The use of these compounds may offer a pharmacological strategy for protection against and prevention of chronic diseases that have as their basis free-radical-mediated oxidative damage (Poeggeler *et al.*, 1993, 1994, 1995, 1996).

Oxygen is the terminal electron acceptor in all aerobic organisms and partial reduction of oxygen leads to the formation of reactive oxygen intermediates. Sulfur, in contrast, is used by all organisms as a universal electron donor and is present in many versatile nucleophilic antioxidants that serve as endogenous electron donors and reductants (Yu, 1994). L-Cystathionine is used as a storage and transport form of bioactive organic sulfur (Bender, 1975; Reed, 1995; Rusakow *et al.*, 1993). The sulfur amino acids L-cysteine, glutathione, and taurine can be synthesized from L-cystathionine (Bender, 1975; Ohlenschläger, 1991). Sulfuryl donors such as L-cysteine and glutathione are potent endogenous antioxidants and reductants (Meister, 1992; Ohlenschläger, 1991). They are, however, very reactive and often not stable (Meister, 1992; Ohlenschläger, 1991; Wada *et al.*, 1995, 1996). Autoxidation and redox recycling of reactive thiol compounds like L-cysteine and glutathione lead to the formation of reactive oxygen intermediates and the depletion of endogenous reductants such as ascorbic acid (Meister, 1992; Ohlenschläger, 1991).

Cystathionine does not contain a redox-reactive sulfuryl group. The sulfur atom of this thioether is shielded by large non-redox-reactive alkyl groups. Therefore, L-cystathionine is much more stable than L-cysteine and it has been suggested that this endogenous thioether be used as a nontoxic source of thiols (Itoh *et al.*, 1992; Kitamura *et al.*, 1989; Wada *et al.*, 1995, 1996). L-Cystathionine is converted to L-cysteine and homoserine under physiological conditions by the enzyme γ-cystathionase (Itoh *et al.*, 1992; Kitamura *et al.*, 1989; Wada *et al.*, 1995, 1996).

Thus, L-cystathionine has been used as a physiological prodrug, which slowly releases L-cysteine intracellularly, enhancing the endogenous thiol pool safely and without the prooxidant side effects seen after direct supplementation of L-cysteine (Itoh *et al.*, 1992). During the investigations of the putative antioxidative activity of L-cystathionine, researchers were surprised to find that L-cystathionine acts directly as a potent scavenger of free radicals *in vitro* and *in vivo* (Wada *et al.*, 1995, 1996). Even without conversion to L-cysteine and other endogenous thiols, potent antioxidative effects of L-cystathionine were demonstrated, which were a consequence of the direct radical-scavenging activity of the thioether itself (Wada *et al.*, 1995, 1996). Whereas L-cystathionine does not have a redox-reactive sulfuryl group, it retains a central sulfur atom, which is able to donate electrons and reduce free radicals (Wada *et al.*, 1995, 1996).

Melatonin contains an electron-rich aromatic indole ring system and can therefore act as an endogenous electron donor that reduces and repairs electrophilic radicals (Poeggeler *et al.*, 1996). The formation of a nitrogen-centered indolyl radical can be demonstrated when melatonin is oxidized by free radicals (Poeggeler *et al.*, 1993, 1994, 1996). The formation of a sulfur-centered radical can be observed when L-cystathionine is exposed to free radicals. The partially oxidized sulfur atom retains an unpaired electron and therefore reacts immediately with superoxide anion radicals to form a stable sulfoxide molecule with paired electrons in covalent bonds (Poeggeler *et al.*, unpublished findings; Wada *et al.*, 1996). Like melatonin, L-cystathionine is oxidized in a two-step process involving electron donation and radical scavenging (Poeggeler *et al.*, 1994; Wada *et al.*, 1996).

L-Cystathionine has been detected in all organisms screened for its presence or absence (Bender, 1975; Reed, 1995). The thioether is present in plants, fungi, and animals. It can be found in all microorganisms, including prokaryotes and archaebacteria (Bender, 1975; Reed, 1995). The ubiquitous presence of the compound suggests that it emerged early on, even before oxygen was used as the universal electron acceptor and oxidant (Bender, 1975; Reed, 1995). Thus, the electron-donating activity of the compound might be its most important physiological function. Radical-scavenging activity and antioxidative protection are important for the survival of all organisms and therefore electron donors appeared early in evolution (Poeggeler *et al.*, 1993).

Many indole compounds structurally similar to melatonin and sulfur compounds structurally related to L-cystathionine have been detected in prokaryotes, protozoans, plants, fungi, and animals, and all act as potent electron donors, radical scavengers, and antioxidants (Bender, 1975; Hamada and Nagase, 1996; Nakano *et al.*, 1991; Ohlenschläger, 1991; Poeggeler *et al.*, 1993; Reed, 1995). The primary function of these pigmentlike, electron-rich compounds is antioxidative protection (Poeggeler *et al.*, 1993). Recently, two new 5-methoxylated indole compounds, OPC 1560 and OPC 15161, have been isolated from the fungus *Thielavia minor* during a screening program that searched for radical scavengers (Nakano *et al.*, 1991). These indoles have a chemical structure very similar to melatonin and also act as very potent electron donors and radical scavengers providing antioxidative protection *in vitro* and *in vivo* (Hamada and Nagase, 1996; Nakano *et al.*, 1991).

2. THE EVOLUTION OF ENDOGENOUS ELECTRON DONORS: EVIDENCE FOR THE OXYGEN CONNECTION

The early occurrence of melatonin during evolution also argues against the notion that melatonin is simply another hormone acting through the specific activation of receptors located either outside or inside the cell (Poeggeler, 1993; Poeggeler *et al.*, 1993). Melatonin could rather be viewed as an ancient molecule, that originated as a radical scavenger providing antioxidative protection and later evolved into a cytoprotective mediator acting through specific binding sites (Poeggeler *et al.*, 1993). Because nonenzymatic photooxidation of the molecule generated a passive light–dark rhythm in primitive organisms early in evolution, the melatonin molecule might have been selected as a natural signal for the duration of darkness later on (Hardeland *et al.*, 1993; Poeggeler *et al.*, 1993).

Melatonin and other highly electroreactive indole compounds synthesized from the aromatic amino acid tryptophan appeared early in evolution when the first photosynthetic organisms produced large amounts of oxygen and aerobic cells began to rely on this highly reactive natural oxidant for respiration and metabolism (Poeggeler, 1993). Melatonin is an evolutionarily highly conserved molecule that is found in organisms as different as algae and humans and structurally related tryptophan metabolites may be present in all organisms (Hardeland *et al.*, 1993; Poeggeler, 1993; Poeggeler *et al.*, 1993). Melatonin immunoreactivity can be detected in the most primitive photosynthetic bacteria (Manchester *et al.*, 1995). The indoleamine has recently also been found in other prokaryotes (Tilden *et al.*, 1997). Thus, the evolution of the hormone can be traced back at least 3 billion years (Hardeland *et al.*, 1993; Manchester *et al.*, 1995; Poeggeler, 1993; Poeggeler *et al.*, 1991, 1993; Tilden *et al.*, 1997). Because the structure of the molecule has not changed, melatonin is one of the most ancient compounds shared by almost all living creatures (Poeggeler *et al.*, 1993).

In all organisms melatonin has the same chemical structure (Poeggeler, 1993). However, it does not act by a single mechanism and comparative studies of different organisms and organs show that the indoleamine is a versatile molecule that serves various functions in signaling and protection (Balzer and Hardeland, 1996; Hardeland *et al.*, 1993). The detection of a nonfluctuating intracellular pool of endogenous melatonin that approaches and sometimes exceeds micromolar concentrations in many different tissues argues in favor of the importance of intracellular actions of the indoleamine (Menendez-Peleaz *et al.*, 1993).

The tissue levels of the indoleamine are at least 1000 times higher than the circulating levels (Barlow-Walden *et al.*, 1995; Menendez-Peleaz *et al.*, 1993). The turnover of tissue melatonin is very high, and in contrast to circulating melatonin, which is primarily oxidized to 6-hydroxymelatonin in the liver, tissue melatonin is oxidized almost entirely to kynurenamines (Hardeland *et al.*, 1993; Hirata *et al.*, 1974; Kopin *et al.*,

1961; Poeggeler *et al.*, unpublished findings). 6-Halomelatonins are also almost entirely oxidized to 6-halokynurenamines (Flaugh *et al.*, 1979), being "suicide substrates" for the liver mixed-function oxidases that catalyze the enzymatic conversion of melatonin into 6-hydroxymelatonin (Kopin *et al.*, 1961). Using these compounds, the measurement of the radical-mediated turnover of melatonin unrelated to 6-hydroxylation by mixed-function oxidase activity might be possible. However, indoleamine-2,3-dioxygenase-catalyzed, superoxide anion radical-dependent oxidation of melatonin is a competitive, enzyme-based source of kynurenamines in the brain and pineal gland (Hirata *et al.*, 1974). The fact that antioxidants and radical scavengers protect against the nonenzymatic oxidation of melatonin in many different tissues argues in favor of the importance of radical-mediated melatonin destruction (Poeggeler and Hardeland, 1994). Melatonin indole fluorescence and immunoreactivity are lost whenever free radicals are generated and oxidative stress is induced (Hardeland *et al.*, 1993; Poeggeler *et al.*, 1994). Obviously, antioxidative protection is a primary action of this endogenous indoleamine (Poeggeler *et al.*, 1993).

Melatonin detoxifies only highly reactive free radicals, whereas the indolyl cation radical formed on the one-electron oxidation of the indoleamine selectively reacts with the omnipresent, weakly oxidizing superoxide anion radical to form a nontoxic, water-soluble kynuramine, 5-methoxy-*N*-acetyl-*N*-formyl-kynurenamine (Figure 2). Melatonin consumes one molecule of oxygen by scavenging the two most important oxygen-based

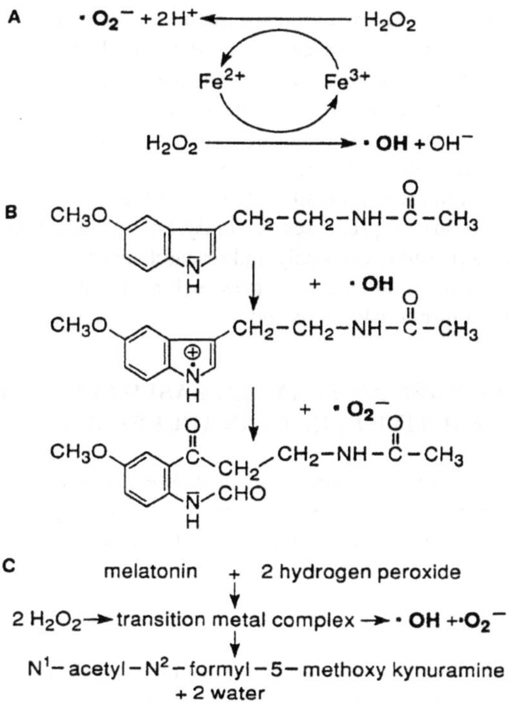

FIGURE 2. Oxidation of melatonin by electrophile intermediates. (A) Generation of electrophile intermediates with formation of superoxide anion radicals and hydroxyl radicals from hydrogen peroxide. (B) Oxidation of melatonin by hydroxyl radicals and superoxide anion radicals formed from hydrogen peroxide breakdown. (C) Reduction of hydrogen peroxide to water and oxidation of melatonin to the corresponding kynuramine by electron donation and radical scavenging.

free radicals and is irreversibly oxidized by the nonenzymatic cleavage of the pyrrole ring (Poeggeler *et al.*, 1994). The parent molecule reduces hydroxyl radicals to hydroxyl anions by electron donation, forming a cation radical that is oxidized on the reaction with another radical, the superoxide anion radical, generating the stable kynuramine.

The reaction of one radical, the indolyl cation radical, with another radical, the superoxide anion radical, terminates the chain of free radical reactions catalyzed by the breakdown of hydrogen peroxide into hydroxyl radicals and superoxide anion radicals. Whereas melatonin is irreversibly oxidized to the corresponding kynuramine, hydrogen peroxide, that is the precursor for both hydroxyl radicals and superoxide anion radicals, is reduced to water. In contrast to other antioxidants, melatonin is not a reducing agent that can be recycled. Therefore, melatonin does not act like a redox-active compound such as glutathione, nicotinamide, ascorbate, urate, and tocopherol and is devoid of any prooxidant effects (Chan and Tang, 1996; Poeggeler *et al.*, 1993).

Melatonin provides on-site protection against the oxidative damage mediated by highly reactive oxygen radicals. Melatonin is the most potent endogenous scavenger of hydroxyl radicals found to date (Poeggeler *et al.*, 1993; Tan *et al.*, 1993a). It is by far more effective than any other natural or synthetic antioxidant, radical scavenger, or spin trap tested so far (Poeggeler *et al.*, 1993, 1996; Tan *et al.*, 1993a). The indoleamine reacts with this highly reactive free radical at a diffusion-controlled rate whenever and wherever the hydroxyl radical is generated (Poeggeler *et al.*, 1996; Tan *et al.*, 1993a). Recent findings suggest an essential role for melatonin in chemoprevention and neuroprotection (Hardeland *et al.*, 1993; Poeggeler *et al.*, 1993; Reiter *et al.*, 1995).

Highly reactive oxygen-based free radicals such as hydroxyl radicals, alkoxy radicals, and peroxyl radicals cannot be detoxified enzymatically in contrast to other reactive oxygen intermediates such as superoxide anion radicals and hydrogen peroxide. Low-molecular-weight antioxidants such as melatonin are the only endogenous protective defense against oxidative damage induced by the former electrophilic intermediates (Poeggeler *et al.*, 1993; Yu, 1994). It has been proposed that indolic compounds such as melatonin are principally involved in the regulation of electron transfer and the control of peroxidative processes. Recently, knowledge about oxygen-based free radical reactions has increased enormously and research activities focusing on the antioxidative effects of endogenous radical scavengers such as melatonin and structurally related indoleamines have continuously increased.

3. OXYGEN AND OXYGEN-BASED FREE RADICALS: HIGHLY REACTIVE ELECTRON ACCEPTORS

Oxygen is required for many life-sustaining metabolic reactions. Oxygen and its activated intermediates, the reactive oxygen species (ROS), however, may react with cellular compounds with resultant degradation or inactivation of essential biomolecules:

$$B_{red} \xrightarrow{\cdot R} B_{ox}$$

Eukaryotic cells have to constantly cope with highly reactive, oxygen-derived free radicals. Their defense against free radicals is achieved by natural antioxidant molecules as well as endogenous antioxidant enzymes and repair systems. All organisms increase their physiological efficiency during development until they reach sexual maturity or divide. After they reproduce, physical performance gradually declines until the parent organism dies (Sohal and Weinbruch, 1996). The belief that a slow-acting endogenous toxin, oxygen, is to blame for the age-related deterioration of living systems with an aerobic metabolism is a simple and attractive concept (Shigenaga *et al.*, 1994). Natural and synthetic radical-scavenging compounds increase the life span of laboratory animals and antioxidants such as melatonin can prevent the detrimental biochemical, physiological, and behavioral changes that occur in senescent animals and humans (Poeggeler *et al.*, 1993).

Aged animals and humans are melatonin deficient and more sensitive to radical-mediated oxidative stress (Poeggeler *et al.*, 1993). Hydroxyl radical-mediated damage increases exponentially with age in animals and humans and can kill irreplaceable brain cells that regulate many important body functions (Agarwal and Sohal, 1996; Shigenaga *et al.*, 1994; Zhang *et al.*, 1993). The loss of these specific brain cells might be a pacemaker of the aging process (Poeggeler *et al.*, 1993). A melatonin replacement therapy that replenishes the melatonin levels to those seen in young, healthy humans could enhance life quality of elderly people and prevent or at least delay the onset of age-related chronic degenerative diseases, because the indoleamine exhibits remarkable life-prolonging effects in rodents (Oakin-Bendahan *et al.*, 1995; Pierpaoli and Regelson, 1994; Poeggeler *et al.*, 1993).

Phytochemicals with low toxicity but high electroreactivity and antioxidative potential are under active development as pharmaceutical tools to treat these conditions and at least some natural hydroxyl radical scavengers have been introduced successfully to cure those chronic diseases that have as their basis free radical damage to cells and organs (Poeggeler *et al.*, 1993; Shigenaga *et al.*, 1994). Recently, many electron-rich radical scavengers with potent antioxidative activity have been discovered that antagonize oxygen radical-mediated toxicity (Nakano *et al.*, 1991; Poeggeler *et al.*, 1993; Shigenaga *et al.*, 1994). All such compounds are very electroreactive and contain atoms that can easily donate electrons, e.g., oxygen, sulfur, and nitrogen (Poeggeler *et al.*, 1993; Shigenaga *et al.*, 1994).

In aerobic cells, oxygen is the terminal acceptor for electrons generated from reducing equivalents (Floyd, 1993). Melatonin safeguards electron transfer in the mitochondrial respiratory chain and attenuates mitochondrial dysfunction induced by excitotoxic agents such as kainate (Giusti *et al.*, 1995, 1996; Poeggeler, 1993). Phenolic antioxidants often act as mitochondrial toxins (Harman, 1983). Melatonin, however, improves efficiency of energy metabolism and decreases formation of superoxide anion radicals during mitochondrial respiration (Poeggeler, 1993, and unpublished observations). Oxygen serves as an electron acceptor in all aerobic organisms because it allows a high yield in energy production through respiration given its high electrochemical potential (Yu, 1994). However, because of its electronic structure, the reduction of oxygen is achieved through single electron transport steps that lead to the production of oxygen-based free radicals (Yu, 1994).

One-electron transfer reactions are difficult to control. Because oxygen serves as the universal electron acceptor in aerobic organisms, the generation of reactive oxygen intermediates is a consequence of respiration and energy metabolism and the accidental formation of highly reactive and damaging free oxygen radicals cannot be prevented entirely. Endogenous redox-active compounds can either amplify or attenuate radical generation during biochemical reactions that require electron transfer (Yu, 1994). Partially reduced reactive oxygen intermediates are very electrophilic and catalyze the oxidation of a broad range of substrates. In a free form, they can react spontaneously with a broad range of biomolecules (Yu, 1994).

Partially reduced oxygen species formed by one-electron reduction reactions are the source of a continuous flux of reactive oxygen intermediates (Yu, 1994). Paradoxically, it is the one-electron reduction of oxygen and its reactive intermediates that creates the most dangerous oxygen-derived radicals (Yu, 1994). Superoxide anion radicals, hydrogen peroxide, and the highly reactive and toxic hydroxyl radicals can be generated in the presence of redox-active transition metals and oxygen (Yu, 1994). They are the primary endogenous reactive oxygen species:

$$O_2 \rightarrow e^- \rightarrow \cdot O_2^- \rightarrow e^- \rightarrow H_2O_2 \rightarrow + e^- \rightarrow \cdot OH$$

One-electron reactions with biomolecules lead to secondary reactive oxygen species such as peroxyl radicals, alkoxy radicals, and other organic radicals:

$$R + \cdot OH \rightarrow \cdot R + OH^-$$

$$\cdot R = \cdot OOR,\ \cdot OR,\ \cdot R$$

These secondary reactive oxygen radicals are not as toxic and reactive as the highly electrophilic hydroxyl radical, but they can induce and promote radical chain reactions that lead to the cooxidation and peroxidation of biomolecules. Peroxidation as well as autoxidation of biomolecules occurs via free radical mechanisms that are catalyzed by transition metals and can be separated into three discrete reaction steps: initiation, propagation, and termination. The whole process consumes oxygen and produces a complex variety of peroxidized and thereby irreversibly damaged and rearranged biomolecules (Yu, 1994).

During oxidative stress, damage to biomolecules can exceed the scavenging and repair potential of the endogenous antioxidative control mechanisms of an organism. Stress and aging are characterized by a prolonged, dangerous, and life-threatening prooxidative state. If the damage to essential biomolecules and certain cell compartments reaches a critical level, cell dysdifferentiation, degeneration, and finally cell death are the consequence of accumulative oxygen toxicity. Organisms that rely on oxygen to maintain their organization and viability, fight incessantly, using their endogenous antioxidative scavenger systems against oxidative stress and damage induced by oxygen-based free radicals.

All multicellular life forms eventually die, because chronic oxygen toxicity mediated mainly by highly reactive and toxic oxygen radicals is accumulative and results in an exponential and irreversible damage to biomolecules and cells. The progressive, age-related decay of organ functions demonstrates that the endogenous antioxidative defense mechanisms cannot always cope successfully with the toxic by-products of oxygen metabolism and their devastating effects on living organisms (Shigenaga *et al.*, 1994). The free radical theory of aging suggests that the aging process and age-associated chronic degenerative conditions are the accumulative result of endogenous one-electron transfer reactions that involve the generation of highly reactive free radicals (Harman, 1956, 1983).

4. ONE-ELECTRON TRANSFER REACTIONS: THE MECHANISMS OF RADICAL FORMATION AND REDUCTION

Organisms have evolved successful strategies to simultaneously harvest and transfer energy in one-electron transfer reactions and build and protect their complex, delicate biostructures by two-electron transfer reactions. In general, a molecule that loses electrons is oxidized and a molecule that gains electrons is reduced. All radical (one-electron) and redox [*n*-electron(s)] reactions proceed because there is a difference in energy of transferable electrons between two molecules:

$$A_{red} - e^- \rightarrow A_{ox} + e^-$$

Nevertheless, loss of an electron is always accompanied by radical generation. Addition of an electron can reduce these electrophilic species and regenerate the partially oxidized molecule generated on one-electron transfer reactions. Free radicals are chemically very reactive, related mainly to the fact that they are not in a stable spin state (Floyd, 1993). Therefore, free radicals readily give up or accept an electron to stabilize their unpaired electron.

In general, all free radicals are electrophiles that accept free electrons, causing the production of another free radical. Although the newly produced radical is less reactive than the parent compound, in most cases it is still unstable and thus can also react with another molecule to produce yet another free radical. In this way, a chain of radical reactions can occur, leading to massive damage of biomolecules. Radical scavengers can react with free radicals, forming stable radicals that terminate this chain reaction.

Antioxidants can be defined as substances that significantly delay or prevent oxidation of other molecules (Halliwell, 1995). Radical scavengers can be defined as those compounds that react with free radicals carrying a single electron thereby reducing the radical and its reactivity toward other molecules. Electron donors are compounds that donate electrons to other substances thereby reducing them. Radical–radical reactions terminate one-electron transfer reactions by forming covalent bonds between the two electrophilic reactants:

$$R \rightarrow - e^- \rightarrow \cdot R$$

$$\cdot R \rightarrow + e^- \rightarrow R$$

$$\cdot R + \cdot R \rightarrow R\text{–}R$$

Early during evolution, aerobic cells developed mechanisms that avoid or achieve strict control over high-energy species, such as the highly reactive and toxic electron-deficient hydroxyl radicals. Agents with high redox activity such as flavins, porphyrins, and many xenobiotic agents, multivalent transition metals such as iron and copper, excited states of biomolecules such as singlet oxygen, and pigment molecules in the highly reactive singlet state would create havoc with their fragile biological organization if they could not be scavenged and detoxified by low-molecular-weight antioxidants.

Aerobic cells are equipped with highly efficient mechanisms to cope with oxidative stress and are loaded with endogenous compounds that protect them against oxidative damage. The susceptibility of any cell, organ, or organism to oxidative stress is a function of the overall balance between the factors that favor oxidative stress and those that provide antioxidative protection. Oxidative damage is the result of an insufficient antioxidative potential or an excessive oxidative stress.

Oxidative metabolism is always accompanied by one-electron transfer reactions that involve the generation of highly reactive oxygen-based free radicals. All one-electron transfer reactions create free radicals, highly reactive molecules with an unpaired electron, which generally induce and promote the peroxidation of biomolecules. Thus, one-electron reactions are necessary to reduce oxygen and allow for oxidative metabolism, but are inherently dangerous, because free radicals and other electrophilic molecules are generated as by-products.

In principle, one-electron transfer reactions are described as

$$A + D \rightarrow \cdot A^{-} + \cdot D^{+} \tag{1A}$$

$$R \rightarrow \cdot R^{+} + e^{-} \tag{1B}$$

Two- or multielectron transfer reactions are described as

$$A + D \rightarrow A\text{–}D \tag{2A}$$

$$R \rightarrow R^{2+} + 2e^{-} \tag{2B}$$

It is immediately evident that one-electron transfer reactions result in partially reduced and oxidized intermediates that are generally more reactive than their parent compounds and thus can undergo secondary reactions with other possible participants in an indiscriminate fashion. Electron-deficient radicals [as shown in Eq. (1B)] are highly reactive and unstable. They participate nonselectively in secondary one-electron reactions with other electron donors yielding more reactive intermediates with an unpaired electron.

On the contrary, the formation of a stable two-electron bond as symbolized in Eq. (2A) is a structure-promoting process, inherently much more safe and selective than the process described by Eq. (1A). The electron-deficient cation shown in Eq. (2B) is nonreactive and recombines selectively with anions carrying a negative charge of the same magnitude.

The antioxidative activity of melatonin is dependent on its ability to serve as a modulator and mediator of one-electron transfer processes and radical reactions. Because of its basic physiochemical properties—the highly electroreactive and resonance-stabilized, heteroaromatic indole ring system with side chains providing inducing and mesomerically stabilizing substituents with electron-donating properties—the melatonin molecule can detoxify reactive oxygen species, repair damaged biomolecules, and safeguard the electron flow in one-electron transfer reactions occurring during energy metabolism and oxygen utilization (Poeggeler *et al.*, 1993, 1996).

The important and unique position of this endogenous indoleamine in the overall antioxidative defense is related to the fact that melatonin is an extraordinarily potent electron donor that can detoxify electron-deficient, electrophilic intermediates and repair partially oxidized organic radicals and biomolecules. Melatonin plays a key role in scavenging highly reactive and toxic initiating radicals such as hydroxyl radicals, but detoxifies promoting radicals such as alkoxy, the peroxyl and organic radicals as well thereby breaking the chain of an initiated peroxidation sequence (Reiter *et al.*, 1995).

In contrast to nearly every other antioxidant, the melatonin molecule itself is irreversibly oxidized during the process. Melatonin does not transfer reducing equivalents to oxygen or other oxidants such as hydrogen peroxide and therefore does not generate partially reduced reactive oxygen species during radical-scavenging, one-electron transfer reactions. It is also completely devoid of autoxidative, redox-recycling activity associated with the highly redox-active phenolic and thiolic antioxidants. In contrast to these compounds, on electron transfer to an electrophilic radical, melatonin is partially oxidized to a highly stable and inert nitrogen-centered indolyl cation radical with an unpaired, delocalized electron that itself reacts selectively and irreversibly with the omnipresent superoxide anion radical to form a water-soluble and nontoxic metabolite, the kynuramine 5-methoxy-*N*-acetyl-*N*-formyl kynurenamine (Figure 3).

Melatonin detoxifies an electrophilic and a nucleophilic oxygen radical in a two-step reaction thereby removing two reactive oxygen intermediates simultaneously. However, the molecule reacts specifically only with highly reactive electrophilic intermediates and cannot be oxidized itself by compounds with a half wave oxidation potential lower than that of the methoxylated indoleamine at 780 mV (Poeggeler and Hardeland, 1994).

Therefore, melatonin is highly stable and nonreactive (Poeggeler *et al.*, 1993). It detoxifies selectively only highly reactive radicals and does not act on electrophiles with a low oxidation potential (Chan and Tang, 1996; Longoni *et al.*, 1995; Marshall *et al.*, 1996; Poeggeler *et al.*, 1993). In contrast to hydroxylated indoleamines and highly reactive and unstable phenolic antioxidants, melatonin and other 5-methoxylated indo-

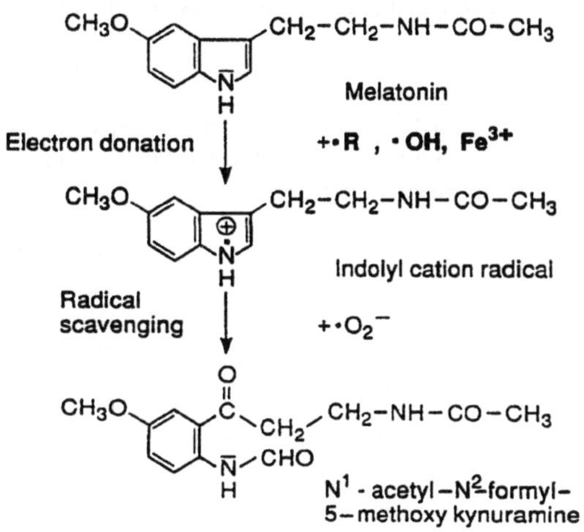

FIGURE 3. Irreversible oxidation of melatonin and formation of a stable kynuramine: a two-step process involving electron donation and radical scavenging.

leamines such as 5-methoxytryptamine do not autoxidize or reduce transition metals such as copper and iron (Chan and Tang, 1996; Marshall *et al.*, 1996; Poeggeler *et al.*, 1993, 1994).

In contrast, the melatonin molecule cannot be regenerated once it has reacted with those oxygen radicals, and thus cannot initiate redox-recycling radical reactions, which rely on electron donors that are reducible (Chan and Tang, 1996; Hardeland *et al.*, 1993; Marshall *et al.*, 1996; Poeggeler *et al.*, 1993, 1994). The long-lived indolyl cation radical is very stable and much less reactive than other organic radicals (Poeggeler and Tan, unpublished observations). Therefore, it cannot be reduced by other physiological antioxidants, even if they are extremely redox active (Poeggeler, unpublished observations). Melatonin acts as a highly selective and electroreactive endogenous electron donor that is sacrificed and irreversibly oxidized on reaction with reactive oxygen intermediates, thereby sparing and repairing other biomolecules (Poeggeler *et al.*, 1993, 1994, 1995, 1996).

5. ELECTRON DONATION: THE MOST POTENT AND VERSATILE ANTIOXIDATIVE PROTECTION AGAINST FREE RADICALS

Although antioxidants can act at several steps in the oxidative sequence and may provide antioxidative protection by many different mechanisms, principally only three different direct chemical reactions are available to control one-electron transfer processes that involve free radical reactions and to terminate the peroxidation of biomolecules, which is initiated and promoted by oxygen-based free radicals:

1. Atom transfer from a substrate to the reactive radical. These reactions are usually characterized by hydrogen atom donation:

$$SH + \cdot R \rightarrow RH + \cdot S$$

2. Charge transfer from a substrate to the reactive radical. These reactions are always characterized exclusively by electron donation:

$$D + \cdot R \rightarrow [\cdot D]^+ + R^-$$

3. Adduct formation between the substrate and the reactive radical. These reactions are characterized by the formation of a covalent bond between the spin trap and the reactive radical:

$$A + \cdot R \rightarrow \cdot AR$$

Adduct formation is a radical detoxification mechanism that has only been realized recently with the synthesis of highly specific spin-trap reagents (Floyd, 1993). Endogenously, certain thiol compounds form bioactive ·NO-radical adducts and act as physiological carriers and donors for this important endogenous radical species (Gilbert, 1994). The biological antioxidant defense network consists nearly exclusively of atom-transferring compounds that donate hydrogen, such as ascorbate, urate, nicotinamide, tocopherols, carotenoids, and glutathione (Yu, 1994). These compounds can also act as electron donors (Poeggeler *et al.*, 1995, 1996). However, in general, all such agents catalyze reductions of oxidants and radicals by transferring reducing equivalents in the form of hydrogen atoms (Yu, 1994).

Melatonin is the only known endogenous antioxidant that acts exclusively by donating electrons to highly electrophilic intermediates and partially reduced, electron-deficient biomolecules (Poeggeler *et al.*, 1996). Thus, melatonin catalyzes charge transfer reactions and reduces radicals to their corresponding anions. Electron donation is a very efficient way to terminate radical reactions, because it can both detoxify highly reactive initiating radicals and repair partially oxidized biomolecules. Melatonin itself is partially oxidized to the corresponding indolyl cation radical during the electron transfer reaction. The indolyl cation radical is irreversibly oxidized on reaction with the omnipresent superoxide anion radical to a stable kynurenamine, the 5-methoxy-*N*-acetyl-*N*-formyl kynurenamine. Therefore, melatonin acts as a true antioxidant by removing oxygen-based free radicals and repairing their oxidation products. It provides effective antioxidant protection to biomolecules by donating an electron to electron-deficient compounds and by removing electron-rich superoxide anion radicals, which would otherwise irreversibly oxidize those organic radicals according to the following equations:

$$M + \cdot R \rightarrow [\cdot M]^{+} + R^{-}$$

$$[\cdot M]^{+} + \cdot O_2^{-} \rightarrow M_{ox}$$

Only the second reaction, which is a reaction of one free radical, the highly resonance-stabilized indolyl cation radical, with another free radical, the weakly oxidizing superoxide anion radical, irreversibly terminates the chain of free radical reactions (Hardeland *et al.*, 1993; Poeggeler *et al.*, 1994). However, the indolyl cation radical is extremely stable and unreactive. This partially oxidized molecule is not reducible by natural occurring electron donors and reacts specifically and exclusively only with superoxide anion radicals. The oxidation of the melatonin molecule is an irreversible process, which can, however, be blocked by superoxide anion radical scavengers. Even under these conditions, which are characterized by the accumulation of the partially oxidized melatonin intermediate, the nitrogen-centered indolyl cation radical, further radical reactions can be completely inhibited and no adduct formation with other nucleophilic biomolecules such as ascorbate, glutathione, and cysteine occurs. Because the kynurenamine is the only degradation product formed on melatonin oxidation, no unspecific adduct formation is measurable. Thus, the indolyl radical cation is inert and not toxic and does not react with other biomolecules such as highly reactive thiol and phenoxy radicals formed on the radical-mediated, partial oxidation of other antioxidants with free hydroxyl or thiol groups such as ascorbate, urate, tocopherol, cysteine, *N*-acetylcysteine, and glutathione (Poeggeler *et al.*, 1993).

Melatonin acts differently than all common chain-breaking antioxidants. In contrast to hydroxylated indoleamines such as serotonin and *N*-acetylserotonin, melatonin cannot easily be oxidized and it does not autoxidize in the presence of oxygen or the ubiquitously present superoxide anion radical. Hydroxylated indoleamines can act as autoxidizable electron donors that transfer single electrons to oxygen-generating superoxide anion radicals and other partially reduced oxygen metabolites. These reactive oxygen species can promote redox recycling in the presence of the highly unstable indoleamines serotonin and *N*-acetylserotonin (Poeggeler *et al.*, 1993).

Hydroxylated indoleamines can be substrates of electron transfer reactions that participate in redox-recycling, radical-generating reactions. Serotonin and *N*-acetylserotonin are potent inducers of redox recycling in the presence of transition metals such as iron or copper. They catalyze the formation of radicals in the presence of redox-active compounds such as phenols and thiols. Hydroxylated indoleamines can do both: scavenge and generate radicals. Unlike melatonin, they act nonselectively and can therefore be either pro- or antioxidant. Serotonin and *N*-acetylserotonin are good substrates for flavin- and porphyrin-containing enzymes and produce reactive oxygen intermediates on oxidation by these peroxidases or mono- and dioxygenases. The enzymatic oxidation products of serotonin and *N*-acetylserotonin can be prooxidant and cytotoxic (Poeggeler *et al.*, 1993).

Methoxylated indoleamines such as melatonin and 5-methoxytryptamine are also a good substrate for specific mono- or dioxygenases and biologically active transition metal complexes such as heme proteins. In contrast to the unstable hydroxylated indoleamines that promote oxygen consumption and radical generation on incubation with redox-active enzyme systems, no increase in radical generation and oxygen consumption can be observed on incubation of methoxylated indoleamines in the presence of these enzymatic systems (Poeggeler *et al.*, 1993).

The protective methyl group in 5′ position completely prevents quinone–imine formation and radical-promoting activity associated with the hydroquinone–semiquinone–quinonimine redox cycle characteristic for hydroxylated indoleamines lacking this group. The formation of an excited molecule or a resonance-stabilized indolyl radical is the reason for the extremely efficient and powerful antioxidant activity of melatonin and the unique biophysical and biochemical properties of molecules that carry a methoxylated indole ring system. Melatonin and 5-methoxytryptamine act as highly specific and selective electron donors. Melatonin has been called a "miracle molecule" because of its extremely powerful antioxidative activity (Poeggeler *et al.*, 1993).

It is the unique chemical structure of the indole compound that makes the melatonin molecule capable of detoxifying the most damaging and dangerous electrophilic intermediates (Poeggeler *et al.*, 1994). Methoxylated tryptophan metabolites are the only other indole compounds with which melatonin shares its extremely efficient radical-scavenging ability (Poeggeler *et al.*, 1996). The kynurenamines generated on the enzymatic or nonenzymatic oxidation of melatonin and 5-methoxytryptamine, cannot be reduced and are therefore not redox active, unlike the quinonimines formed on the oxidation of hydroxylated indole compounds (Hardeland *et al.*, 1993).

Compounds that contain redox-active hydroxyl and sulfuryl groups are nonselective agents that can transfer reducing equivalents to other redox-active compounds in an indiscriminate fashion (Poeggeler *et al.*, 1993). These compounds transfer reducing equivalents to other electroreactive molecules thereby reducing them (Poeggeler *et al.*, 1993). In contrast to selective electron donors such as melatonin and 5-methoxytryptamine, which react exclusively with highly reactive radicals, these nonspecific agents can be either pro- or antioxidant (Yu, 1994). Very little research has been done on the potent prooxidant effects of redox-active compounds such as phenols, thiols, and indoles (Poeggeler *et al.*, 1993). It is, however, well documented that these compounds, which are generally regarded as protective antioxidants, can in fact increase oxidative stress and damage (Chan and Tang, 1996; Marshall *et al.*, 1996; Poeggeler *et al.*, 1993).

6. ENDOGENOUS ELECTRON DONORS: EXTREMELY POTENT HYDROXYL AND PEROXYL RADICAL SCAVENGERS

To demonstrate that melatonin scavenges hydroxyl radicals, the extremely reactive and short-lived oxygen radicals were generated *in vitro* by hydrogen peroxide exposed to UV light. This highly specific hydroxyl radical-generating system allowed the quantification of the hydroxyl radical-scavenging abilities of indoleamines such as melatonin in comparison with well-known scavengers. Using the photolysis of hydrogen peroxide, we reduced the probability of side reactions and were able to ensure that radical scavenging and not other reactions were responsible for the antioxidative actions exerted by the compounds tested (Tan *et al.*, 1993a).

The detection of hydroxyl radicals per se is virtually impossible because of the short half life and the extreme high reactivity of this oxygen radical. Therefore, the spin-trapping reagent 5,5-dimethyl-pyrroline-*N*-oxide (DMPO) was used to measure DMPO–OH adduct formation by high-performance liquid chromatography coupled to an electrochemical detector (HPLC-ECD). The results were validated by electron spin resonance (ESR). The fractions containing the DMPO–OH adduct collected by the HPLC system indicated the specific 1:2:2:1 ESR spectrum characteristic for hydroxyl radical adducts (Tan *et al.*, 1993a).

The hydroxyl radical-quenching effects of melatonin and structurally related indoles were quantified by measuring the amounts of specific DMPO–OH adducts generated on the photolysis of hydrogen peroxide and compared with those of other highly active radical scavengers (Tan *et al.*, 1993). The hydroxyl radical-scavenging activity of melatonin (IC_{50} = 21 μM) was found to be much greater than that of the most important endogenous radical scavenger, glutathione (IC_{50} = 123 μM), and the classical hydroxyl radical scavenger mannitol (IC_{50} = 286 μM) (Tan *et al.*, 1993a).

5-Methoxytryptamine proved to be also a fairly efficient hydroxyl radical scavenger in this system. However, its maximal inhibitory activity was found to be less than 40% that of melatonin and it is obvious that more hydroxyl radicals escape detoxification by the indoleamine than are trapped. *N*-Acetylserotonin was not active as a scavenger in this highly specific hydroxyl radical-generating system. Interestingly, serotonin even increased hydroxyl radical formation. The hydroxylated indoleamine showed a strong prooxidant effect in this specific hydroxyl radical-generating system. DMPO–OH adduct formation was increased twofold on incubation with the indoleamine. The prooxidant effect of serotonin might be related to the highly reactive unshielded hydroxyl group in 5 position, which can catalyze the photolysis of hydrogen peroxide initiated by the irradiation with UV light (Tan *et al.*, 1993a).

Of the naturally occurring indoleamines, melatonin was by far the most efficient hydroxyl radical scavenger tested. 5-Methoxytryptamine showed some minor free radical-quenching activity, but *N*-acetylserotonin was totally inactive. Surprisingly, serotonin acted as a potent prooxidant by increasing hydroxyl radical formation even in a system that did not include any transition metals such as iron. Thus, very minor changes in the chemical structure of the indole compounds have strong effects on the ability of indoleamines to react with free radicals. The data demonstrate that the hydroxyl radical-scavenging ability of melatonin is related to its specific molecular structure. The methylation of the hydroxyl group at C-5 of the aromatic indole ring system proved to be essential for hydroxyl radical-scavenging activity, whereas the *N*-acetyl-ethyl group in the side chain of the indoleamine provided the synergistic action required to detoxify the highly reactive radical efficiently probably by enhancing the resonance stability of the melatonin molecule. The data presented by Tan *et al.* (1993a) demonstrated clearly that melatonin is by far the most powerful and effective hydroxyl radical scavenger detected to date.

Melatonin does not react specifically only with hydroxyl radicals, but also detoxifies other highly reactive electrophilic species by electron donation (Chan and Tang, 1996; Miller *et al.*, 1996; Pierrefieche *et al.*, 1993; Poeggeler *et al.*, 1994; Sewerynek *et al.*, 1995a,c). The indoleamine quenches excited biomolecules and singlet oxygen as well (Cagnoli *et al.*, 1996; Poeggeler *et al.*, 1994). Peroxyl radicals are highly reactive and toxic secondary radicals. Shortly after the demonstration that melatonin is a highly effective hydroxyl radical scavenger, Pieri *et al.* (1994, 1995) reported that the indole exhibits similar scavenging action with regard to the peroxyl radical. Using a specific peroxyl radical generator and β-phycoerythrin as a fluorescent indicator protein, Pieri *et al.* (1994) demonstrated that melatonin scavenged the peroxyl radical twice as effectively as Trolox, a water-soluble vitamin E analogue. Melatonin was far superior to vitamin C and glutathione. Later the same research group confirmed and extended their findings by demonstrating that melatonin was more than twice as potent as vitamin E in neutralizing peroxyl radicals (Pieri *et al.*, 1995). Vitamin E is a well-known and very important endogenous chain-breaking antioxidant. Thus, melatonin can reduce the initiation of radical reactions by the highly reactive hydroxyl radical and terminate radical chain reactions by eliminating the peroxyl radical.

Recently, it has been confirmed that melatonin detoxifies selectively highly reactive radicals (Longoni *et al.*, 1995; Marshall *et al.*, 1996). Melatonin does not act like common chain-breaking antioxidants (Marshall *et al.*, 1996). Whereas melatonin is extremely potent in blocking hydroxyl radical-initiated lipid peroxidation, it does not block peroxidation induced by radicals with low reactivity (Longini *et al.*, 1995). Thus, even nanomolar concentrations of melatonin protect against hydroxyl radical-mediated lipid peroxidation, but micromolar concentrations are needed to protect against endogenous lipid peroxidation and even millimolar concentrations are required to block ascorbate-iron-induced lipid peroxidation (Longini *et al.*, 1995; Marshall *et al.*, 1996; Reiter *et al.*, 1995).

In contrast to the indolyl cation radical, melatonin does not react with superoxide anion radicals directly (Marshall *et al.*, 1996). The indoleamine does not chelate iron and does not exert any anti- or prooxidant effects by binding to this or to other transition metals (Chan and Tang, 1996; Marshall *et al.*, 1996; Poeggeler *et al.*, 1993). In contrast to highly reactive phenolic antioxidants, melatonin cannot reduce endogenous lipid peroxides (Poeggeler *et al.*, unpublished findings). Because of its high selectivity and specificity toward reactive radicals with high electrophilicity, melatonin is not as potent as chain-breaking phenolic antioxidants in inhibiting lipid peroxidation and its propagation

(Marshall *et al.*, 1996). It does, however, scavenge a broader range of endogenous reactive radicals than previously believed (Gilad *et al.*, 1997; Marshall *et al.*, 1996).

Very recently, it has been demonstrated that melatonin is an excellent scavenger of peroxynitrite (Gilad *et al.*, 1997). Melatonin caused a dose-dependent inhibition of the oxidation of dihydrorhodamine 123 by peroxynitrite and protected against peroxynitrite toxicity with an EC_{50} value of less than 10 μM (Gilad *et al.*, 1997). Thus, melatonin may protect against a broad range of highly toxic and reactive electrophilic intermediates. The molecule can detoxify not only free radicals, but also highly reactive electrophilic molecules such as peroxynitrite (Gilad *et al.*, 1997). To date, it is not known if melatonin reacts with the ·NO radical, although it is unlikely that melatonin does react because this radical species is not very reactive. However, melatonin has to be tested for its reactivity toward the ·NO radical, before any firm conclusions can be drawn. Melatonin does not react with radicals or electrophilic molecules of low reactivity (Marshall *et al.*, 1996). This specificity ensures that the indole is not wasted (Poeggeler *et al.*, 1993).

The selectivity of melatonin limits its direct antioxidative potential as measured *in vitro* (Longini *et al.*, 1995; Marshall *et al.*, 1996). Melatonin is 1000 times more potent than other radical scavengers in detoxifying hydroxyl and peroxyl radicals, but it also is 1000 times less active in inhibiting lipid peroxidation than chain-breaking antioxidants (Marshall *et al.*, 1996; Poeggeler *et al.*, 1996). These findings suggest that melatonin acts in a way unlike most other antioxidants. The synergistic effects of chain-breaking antioxidants and melatonin can be explained by the fact that the action of these compounds is different and complementary (Poeggeler *et al.*, 1995).

The potent hydroxyl and peroxyl radical-scavenging activities of melatonin encouraged further research on radical scavenging and repair mediated by the indole and structurally related compounds. The radical-scavenging properties of melatonin and structurally related, endogenous indoles were evaluated recently in kinetic competition studies using the specific radical trapping reagent 2,2′-azino-bis-(3-ethylbenz-thiazoline-6-sulfonic acid) (ABTS) and compared with those of other antioxidants. In the presence of highly reactive radicals, ABTS is oxidized to the stable thiazoline cation radical, $ABTS^{*+}$, which, given its intense green color, can be measured photometrically at an absorbance of 420 nm (Poeggeler *et al.*, 1996).

Melatonin and the structurally related indole compounds 5-methoxytryptophol, 5-methoxyindole acetic acid, and 5-methoxytryptamine, as well as the phenolic and thiolic antioxidants ascorbic acid, Trolox, and glutathione inhibited ABTS cation radical formation and catalyzed ABTS cation radical reduction. Melatonin was the most potent radical scavenger and electron donor when compared with the methoxylated indole analogues and the other antioxidants tested. Melatonin, the methoxylated indole analogues, and all other antioxidants tested acted as potent electron donors that scavenged initiating and propagating radicals and repaired partially oxidized organic radicals (Poeggeler *et al.*, 1995, 1996).

Reduction of ABTS cation radicals and the inhibition of their formation were assayed by coincubating with electron donors and antioxidants or vehicle with ABTS in the presence or absence of rat brain homogenate and the Fenton reagents hydrogen peroxide and iron sulfate. Melatonin, the structurally related indole compounds, and the other antioxidants reacted rapidly with hydroxyl, peroxyl, and ABTS cation radicals at a diffusion-controlled rate as shown by competition studies with the specific radical trapping agent ABTS. The second-order rate constants as determined by kinetic competition studies with ABTS were all extremely high, but the indole compounds were at least 1000 times more effective than the other radical scavengers (Poeggeler *et al.*, 1996).

Melatonin was by far the most potent electron donor and radical scavenger (Poeggeler *et al.*, 1996). Melatonin metabolites and other tryptophan metabolites showed very little radical-scavenging activity as measured by ABTS cation radical reduction. ABTS is a highly effective tool for the quantification of one-electron transfer reactions, the dye being sufficiently sensitive to allow the measurement of radical formation and reduction directly *in situ*. The second-order rate constant for the reaction of melatonin with the hydroxyl radical was 0.6×10^{11} M^{-1} s^{-1} and the second-order rate constant for the reaction of melatonin with peroxyl radicals was 0.75×10^{11} M^{-1} s^{-1} as determined by competition kinetics with ABTS in the absence or presence of brain homogenate (Poeggeler *et al.*, 1996).

7. PROTECTION AGAINST OXIDATIVE STRESS AND DAMAGE: THE IMPORTANT ROLE OF RADICAL REDUCTION AND REPAIR

Recent findings demonstrate that melatonin and other endogenous antioxidative agents not only detoxify highly reactive oxygen radicals, but also reduce and repair long-lived organic radicals such as the thiazoline cation radicals generated on the partial oxidation of ABTS:

$$\cdot ABTS^{+} + M \rightarrow ABTS + \cdot M^{+}$$

Chain-breaking antioxidants such as ascorbic acid, Trolox, and glutathione act synergistically with mediators of one-electron transfer reactions such as melatonin (Poeggeler *et al.*, 1995). Thus, antioxidative vitamins, endogenous thiolic radical scavengers, and indolic electron donors might act in concert to protect against oxidative damage induced by electrophilic intermediates (Poeggeler *et al.*, 1995). Radical scavenging and repair are complex processes even when measured *in vitro* and antioxidative compounds might exert multiple mechanisms catalyzing radical reduction and repair even in fairly simple chemical test tube systems:

$$\cdot ABTS^{+} + R \rightarrow ABTS + \cdot R^{+}$$

Very recently, we have applied the colorimetric assay of ABTS cation radical reduction to monitor the antioxidative status of biological fluids and tissues *ex vivo*:

$$[\cdot ABTS^{+}]/[R] = [\cdot R^{+}]/[ABTS]$$

This method is highly sensitive in detecting the total antioxidative protection available by endogenous radical scavengers present in tissues and body fluids and can be used to quantify radical-reducing activity of biological samples *ex vivo* (Rice-Evans and Miller, 1994). ABTS cation radicals are not substrates for P450 reductase and ABTS reduction catalyzed enzymatically by microsomes or mitochondria is insignificant (Poeggeler *et al.*, unpublished findings). Therefore, this assay can be used as proposed by Rice-Evans and Miller (1994) to measure total antioxidative protection mediated nonenzymatically by radical-scavenging activity of antioxidants in biological tissues and fluids.

We adopted this method developed by Rice-Evans and Miller (1994) to assay for "total antioxidative status" in plasma and in brain, lung, and liver tissue of rats. The highly sensitive method allows quantification of the protection that is endogenously available against free radical damage in any body tissue or fluid *in vivo* as soon as the sample is harvested. This is the first time a method allows for the measurement of total radical-scavenging capacity of body tissues and fluids *in vivo*, although the radical-reducing capacity is actually measured *ex vivo* (Poeggeler *et al.*, unpublished findings).

ABTS cation radicals can be generated by flash photolysis in high yield and purity. They must, however, be stabilized with ethanol or other organic modifiers. We and other laboratories are currently investigating the possibility of using microdialysis coupled to HPLC with on-line spectrophotometry to measure radical-scavenging activity *in vivo* as opposed to *ex vivo*. However, it is very difficult to implement this method, because all organic modifiers are toxic when infused at the concentrations required (3%) to stabilize ABTS cation radicals *in vivo* (Poeggeler, unpublished data).

Pretreatment of animals with pharmacological doses of melatonin, 0.5, 5.0, and 50.0 mg/kg i.p., enhanced the antioxidative potential of tissues and body fluids dose dependently as measured *ex vivo* by ABTS cation radical reduction. ABTS cation radical reduction as measured *ex vivo* was increased by 24, 37, and 78%, respectively, after administration of melatonin (Poeggeler, unpublished data).

Neither ascorbate, Trolox, nor glutathione, which, like melatonin, were all very active *in vitro*, had any significant effect on ABTS radical reduction as measured by the assay system developed by Rice-Evans and Miller (1994) *ex vivo*. Interestingly, other antioxidants, such as L-cysteine and *N*-acetyl-L-cysteine, that were effective as antioxidants *in vitro* proved to be strongly prooxidant *ex vivo* (Nath and Salahudeen, 1993; Poeggeler, unpublished findings; Puka-Sundvall *et al.*, 1995). We tested a broad range of natural and synthetic radical scavengers and antioxidants (Poeggeler *et al.*, unpublished findings). Unfortunately, all of them proved to be inactive or even acted as a prooxidant *ex vivo* (Poeggeler *et al.*, unpublished data). Interestingly, with the exception of melatonin, all of the 5-methoxylated indoleamines tested, which acted as highly potent electron donors and radical scavengers *in vitro*, did not show any significant antioxidative effects *in vivo* (Poeggeler, unpublished data).

Depletion of intracellular glutathione by fasting the animals proved to be the most efficient way to reduce antioxidative protection by body tissue samples. Depletion of plasma ascorbic acid by incubation of the plasma samples with ascorbic acid oxidase abolished antioxidative protection by rat plasma. Thus, whereas exogenous glutathione and ascorbate were inactive in enhancing radical-scavenging ability of body tissues and fluids, endogenous levels of the antioxidants were very important for antioxidative protection in both compartments. At 50 mg/kg, melatonin completely protected against the prooxidant effects of glutathione and ascorbate depletion (Poeggeler *et al.*, unpublished findings).

L-Cystathionine was the only other compound tested that enhanced the endogenous antioxidative potential *ex vivo*. The endogenous sulfur-containing compound proved to be effective at 5 and 50 mg/kg, increasing ABTS cation radical reduction by 23 and 54%, respectively. The lower dose of 0.5 mg/kg was ineffective in enhancing ABTS cation radical reduction and endogenous antioxidative potential (Poeggeler, unpublished data).

L-Cystathionine, an endogenous agent, is a sulfur ether that can act as a potent radical scavenger and electron donor *in vitro* (Wada *et al.*, 1995, 1996). It scavenged the hydroxyl radical with a second-order rate constant of 0.9×10^9 M^{-1} s^{-1} and the peroxyl radical with a second-order reaction constant of 7.5×10^9 M^{-1} s^{-1} (Poeggeler, unpublished data). As with melatonin, we could not find prooxidant effects of L-cystathionine *in vitro* (Poeggeler, unpublished data). However, all other phenolic and thiolic antioxidants, including L-cysteine and glutathione, proved to be highly prooxidant when incubated in the presence of transition metals such as iron and reducing agents such as ascorbate (Poeggeler, unpublished data). Both melatonin and L-cystathionine blocked the prooxidant effects of L-cysteine and *N*-acetyl-L-cysteine completely when given *ex vivo* (Poeggeler *et al.*, unpublished data).

Like melatonin, L-cystathionine reduced ABTS cation radicals *in vitro*. However, compared with melatonin, L-cystathionine exhibited a much lower ABTS cation radical-reducing activity *in vitro* (Poeggeler, unpublished observations). It has been suggested that L-cystathionine can scavenge both superoxide anion radicals and hydroxyl radicals (Wada *et al.*, 1996). The thioether reduced significantly superoxide anion radical-dependent chemiluminescence and superoxide anion radical-mediated nitroblue tetrazolium reduction (Wada *et al.*, 1996). L-Cystathionine was active in activated human leukocytes and in the xanthine–xanthine oxidase system (Wada *et al.*, 1996). However, the scavenging effect of L-cystathionine on hydroxyl radical-induced lipid peroxidation of erythrocyte membrane ghosts was not very strong (Wada *et al.*, 1996). Thus, the authors concluded that L-cystathionine preferentially scavenges superoxide anion radicals (Wada *et al.*, 1996).

When exposed to the hydroxyl radical-generating hydrogen peroxide/UV-light system, L-cystathionine was highly effective in reducing DMPO–OH adduct formation and ABTS cation radical generation when added at micromolar concentrations (Poeggeler *et al.*, unpublished findings). The inhibition of DMPO–OH adduct formation and ABTS cation radical generation was also observed when L-cystathionine was coincubated in the presence of the Fenton reagents hydrogen peroxide and iron sulfate (Poeggeler *et al.*, unpublished findings). In these assay systems, L-cystathionine proved to be a very potent hydroxyl radical scavenger. Using the specific peroxyl radical generator 2,2′-azo-bis(2-amidinopropane)dihydrochloride and β-phycoerythrin as a fluorescent indicator protein, we were able to demonstrate that L-cystathionine scavenges peroxyl radicals at least as effectively as Trolox, a water-soluble vitamin E analogue (Poeggeler *et al.*, unpublished findings).

L-Cystathionine was also extremely efficient in reducing hydroxyl and peroxyl radical-induced ABTS cation radical formation (measured as described in Poeggeler *et al.*, 1996) and lipid peroxidation (measured by the method of Sewerynek *et al.*, 1995a) in brain and liver homogenates (Poeggeler *et al.*, unpublished data). These data are in agreement with earlier findings of Wada *et al.* (1995) demonstrating that L-cystathionine can reduce endogenous lipid peroxidation and tissue damage induced by ischemia and reperfusion *in vivo*.

Using pulse radiolysis we have been able to demonstrate that L-cystathionine, L-cysteine, and *N*-acetyl-L-cysteine scavenge hydroxyl radicals, but not superoxide anion radicals (Poeggeler *et al.*, unpublished observations). Using the same sensitive and specific method of pulse radiolysis, Felix *et al.* (1996) demonstrated that *N*-acetyl-L-cysteine reacts selectively with hydroxyl radicals, but not with superoxide anion radicals.

The findings of Wada *et al.* (1996) can be explained by the fact that the sulfur-centered radical formed on the oxidation of the parent compound L-cystathionine by highly reactive hydroxyl and peroxyl radicals reacts rapidly with superoxide anion radicals, thereby removing these oxygen intermediates from the reaction medium (Poeggeler *et al.*, unpublished findings; Wada *et al.*, unpublished findings). It is well known that sulfur atoms with unpaired electrons react avidly and rapidly with superoxide anion radicals forming stable sulfoxides with two electron pairs shared by the sulfur and oxygen atoms (Wada *et al.*, 1996).

Activated leukocytes and the xanthine–xanthine oxidase system generate enough hydroxyl radicals to initiate the oxidation of the thioether forming the sulfur-centered L-cystathionine radical capable of removing the superoxide anion radicals (Wada *et al.*, 1996). Thus, L-cystathionine reacts very much like melatonin, detoxifying two radicals on its oxidation in a two-step process of electron donation and radical–radical reaction. Whereas the thioether itself seems to be inert against the oxidation by superoxide anion radicals, the immediate oxidation product, the sulfur-centered L-cystathionine radical, is a good substrate for the reaction with the omnipresent superoxide anion radical as proposed by Wada *et al.* (1996). In contrast to melatonin, two molecules of L-cystathionine are sacrificed in the process generating L-cystathioninesulfoxide, an endogenous metabolite of L-cystathionine (Machida *et al.*, 1996; Poeggeler *et al.*, 1994; Wada *et al.*, 1995, 1996).

8. ELECTRON DONATION: POTENT ANTIOXIDATIVE PROTECTION WITHOUT PROOXIDANT SIDE EFFECTS

Reactive radicals must be reduced and repaired by electron donation so as to avoid the initiation of radical chain reactions and the accumulation of oxidative damage. Endogenous radical scavengers provide antioxidative protection directly by reducing these electrophilic oxidants. However, they also reduce oxidative stress indirectly by reducing the accumulation of superoxide anion radicals. Both melatonin and L-cystathionine cannot react directly with superoxide anion radicals, but their metabolites do. Radical–radical reactions are crucial in terminating radical chain reactions, because only these reactions irreversibly remove electrophilic molecules with unpaired electrons and thereby block their ability to induce oxidative damage to other biomolecules.

Recently, we compared the effects of melatonin and L-cystathionine administration on the endogenous levels of lipid peroxides in brain tissue of rats as measured by a newly developed, highly specific and sensitive assay described by Sewerynek *et al.* (1995a). Whereas acute administration of either compound at a dose of 5 or 30 mg/kg i.p. had no effects on the endogenous levels of lipid peroxides in brain tissue, semichronic and chronic administration of either melatonin or L-cystathionine for 3 or 30 days orally in the drinking water at a dose of 30 mg/kg reduced endogenous lipid peroxidation significantly.

To our great surprise, L-cystathionine was by far the more potent compound, reducing endogenous lipid peroxides by 60% to 40% of control levels after administration for 30 days, compared with a reduction of only 30% to 70% of control levels after 30 days of administration of melatonin. Both compounds reduced endogenous lipid peroxidation significantly by 15% after 3 days of administration. It is difficult to explain why melatonin as the more potent radical scavenger was inferior in reducing endogenous peroxidation compared with L-cystathionine. However, it has been demonstrated that L-cystathionine can react with a broader spectrum of endogenous radicals involved in the propagation of lipid peroxidation and obviously reduces radicals with low reactivity more efficiently than melatonin (Poeggeler *et al.*, unpublished findings; Wada *et al.*, 1995, 1996).

It is important to note that L-cysteine and *N*-acetyl-L-cysteine, administered chronically for 30 days in the drinking water at a high dose of 300 mg/kg (because of low blood–brain barrier penetration), increased endogenous peroxidation significantly in brain tissue as measured by the method of Sewerynek *et al.* (1995a). Similar prooxidant effects were noted after the administration of 5-hydroxy-L-tryptophan and *N*-acetyl-5-hydroxy-L-tryptophan, the direct precursors of serotonin and *N*-acetylserotonin, at a dose of 300 mg/kg given with the drinking water (Poeggeler, unpublished findings). Thus, prooxidant effects of putative antioxidants may contribute to the toxicity seen after administration of these compounds to animals (Nath and Salahudeen, 1993; Puka-Sundvall *et al.*, 1995).

These unexpected findings do not argue in favor of the hypothesis that L-cystathionine provides antioxidative protection as a prodrug for L-cysteine or a precursor for glutathione as suggested recently (Itoh *et al.*, 1992; Kitamura *et al.*, 1989; Wada *et al.*, 1995, 1996). Rather the genuine radical-scavenging effects of the sulfur ether itself, which have been observed by others *in vitro* and *ex vivo* as well (Wada *et al.*, 1995, 1996), may be responsible for providing antioxidative protection *ex vivo*:

$$R_2S + \cdot R \rightarrow \cdot R_2S + R$$

$$\cdot R_2S + \cdot O_2^- \rightarrow R_2SO_2^-$$

$$R_2SO_2^- + R_2S \rightarrow 2\ R_2SO$$

Two molecules of the thioether L-cystathionine are oxidized to the corresponding sulfoxide, L-cystathionesulfoxide. The latter is an endogenous metabolite of L-cystathionine (Machida *et al.*, 1995). It can be generated nonenzymatically by radical-mediated oxidation of the parent compound or enzymatically by mixed-function oxidase and other heme proteins. L-Cystathioninesulfoxide can be further oxidized to L-cystathioninesulfone or be reduced to the parent compound L-cystathionine. L-Cystathionine is present in millimolar concentrations in the brain (Kodama *et al.*, 1985, 1988; Reed, 1995; Tallan *et al.*, 1957).

The pineal gland contains the highest concentrations of L-cystathionine found in the brain (Kodama *et al.*, 1988). The compound is present in millimolar concentrations in the mammalian pineal gland (Kodama *et al.*, 1988). The human pineal contains even higher concentrations of L-cystathionine (Kodama *et al.*, 1988; Reed, 1995; Tallan *et al.*, 1957). The extremely high endogenous concentrations of the thioether might compensate for the inferior radical-scavenging ability of L-cystathionine compared with melatonin and other endogenous indolic electron donors. Thus, the compound may act as an endogenous antioxidant much like melatonin and may do so by the same mechanism: electron donation and radical scavenging.

We have investigated the putative neuroprotective effects of L-cystathionine *in vivo*. The thioether protects against free-radical-mediated excitotoxicity in the central nervous system when given at a dose of 60 mg/kg i.p., that is comparable to the highest dose effective in enhancing endogenous antioxidative potential *in vivo* (Poeggeler *et al.*, unpublished data). So far, L-cystathionine is the only endogenous compound that matches the electron-donating activity of melatonin (Poeggeler *et al.*, unpublished findings). The thioether provides antioxidative protection without exerting any prooxidant effects and, much like melatonin, is metabolized to a nontoxic metabolite.

We also measured the effects of melatonin and L-cystathionine on hydroxyl radical-induced lipid peroxidation. We measured the extent of lipid peroxidation induced after exposure of brain homogenates to a hydroxyl radical-generating system as described in Poeggeler *et al.* (1996). Hydroxyl radical-induced lipid peroxidation was strongly decreased when either melatonin or L-cystathionine was added to the homogenate (Poeggeler *et al.*, unpublished findings; Reiter *et al.*, 1995; Sewerynek *et al.*, 1995a). These findings confirm observations made by Longini *et al.* (1995) showing that melatonin strongly reduces hydroxyl radical-induced lipid peroxidation.

Melatonin and L-cystathionine also reduce lipid peroxidation initiated by hydroxyl radicals when injected systemically at a dose of 30 mg/kg i.p. Lipid peroxidation was measured 30 min after the administration of these electron donors *ex vivo* in the brain homogenate by the method of Sewerynek *et al.* (1995a). Melatonin reduced hydroxyl radical-induced lipid peroxidation by 75%, whereas L-cystathionine reduced the levels of endogenous lipid peroxides by 60% (Poeggeler, unpublished data). These data demonstrate conclusively that electron donors and radical scavengers such as melatonin protect against hydroxyl radical-mediated peroxidation *in vitro* as well as *in vivo* (Longini *et al.*, 1995; Reiter *et al.*, 1995; Sewerynek *et al.*, 1995a,c).

The search for endogenous compounds that can provide antioxidative protection and the investigation of mechanisms that allow radical scavenging and quenching of reactive oxygen intermediates is a fascinating and fast progressing field of modern biomedicine. The screening for gerontoprotective agents yielded results suggesting that melatonin and other endogenous electron donors can exert potent life-prolonging effects in vertebrates, invertebrates, and even protozoans (Miquel and Economos, 1979; Oeiru and Vochitu, 1965; Pierpaoli and Regelson, 1994; Thomas and Smith-Sonneborn, 1996). Preliminary data from our laboratory have shown that both melatonin and L-cystathionine greatly increase life expectation of rotifers, genus *Philodina*, and of insects, *Drosophila melanogaster* (fruit flies), when fed at concentrations of 0.5 and 1.0 mM, respectively (Poeggeler, unpublished findings). Thus, melatonin acts as a potent gerontoprotective agent in rats, mice, fruit flies, rotifers, and ciliates and thereby prolongs the life span of vertebrates, invertebrates, and even unicellular organisms (Oakin-Bendahan *et al.*, 1995; Pierpaoli and Regelson, 1994; Poeggeler and Hardeland, in preparation; Thomas and Smith-Sonneborn, 1996). This review on melatonin and other structurally or functionally related electron donors may encourage further research on this important endogenous indoleamine and structurally or functionally related compounds.

More research on radical scavenging and repair is essential to gain a better understanding of mechanisms that are involved in antioxidative protection by endogenous compounds. The measurement of radical reactions has been made possible by the use of new and advanced methods. We are nevertheless only beginning to understand the complexity of radical reactions and their importance for biological processes. Research has focused too narrowly on highly reactive oxygen radicals that can be relatively easily detected indirectly by measuring adduct formation or oxidative damage. The process of "slow oxidotoxicity" that might be related to the generation of long-lived organic radicals has received less attention. Specific radical trapping agents such as ABTS may change that. I am convinced that the use of these specific radical trapping compounds will provide new and important insights into the endogenous mechanisms that enhance or reduce oxidative stress and damage.

9. MELATONIN: A POTENT ENDOGENOUS ANTIOXIDANT

Recently, *in vivo* studies have demonstrated that melatonin is an extremely potent antioxidant that protects against many chemical and physical stressors (Reiter *et al.*, 1995). Damage to DNA resulting from the exposure of rodents to radiation or carcinogens such as safrole is markedly reduced when melatonin is coadministered (Blinkenstaff *et al.*, 1994; Tan *et al.*, 1993b, 1994). The induction of cataracts in neonatal rat pups injected with a glutathione-depleting drug can be prevented by pretreating the animals with melatonin (Abe *et al.*, 1994). Paraquat-induced peroxidation in the lungs of adult rats is completely inhibited by melatonin administration and the mortality after administration of this highly toxic, radical-generating herbicide is strongly reduced in rodents after coinjection of melatonin (Melchiorri *et al.*, 1994; Reiter *et al.*, 1995). Likewise, bacterial endotoxin-induced free radical damage is significantly inhibited by melatonin pretreatment (Sewerynek *et al.*, 1995b). Finally, oxidative damage and neurodegeneration induced by the convulsant kainate can be inhibited very efficiently by melatonin coadministration (Guisti *et al.*, 1996). All of these potent antioxidative actions of melatonin have been reviewed and discussed in great detail (Reiter *et al.*, 1995).

Physiological concentrations of melatonin have been shown to inhibit the nitric oxide radical-generating enzyme, nitric oxide synthetase (Pozo *et al.*, 1994). The reduction of ·NO generation could contribute to melatonin's neuroprotective actions, as the radical can react with superoxide anion radicals to produce peroxynitrite, which in turn can degrade to form hydroxyl radicals (Reiter *et al.*, 1995). Melatonin also inhibits the radical-generating enzyme mixed-function oxidase in the liver; this effect might contribute to the hepatoprotective effects of the indoleamine (Kothari and Subramanian, 1992; Reiter *et al.*, 1995). The inhibition of porphyrin synthesis elicited by melatonin administration could mediate the protective effects of the endogenous indoleamine against phototoxicity (Antolin *et al.*, 1996).

Recently, it has been demonstrated that antioxidants can modulate the protein levels of Bcl-2 and Bax, two gene products involved in the regulation of cell death (Haendeler *et al.*, 1996). Bcl-2 inhibits cell death caused by necrosis and apoptosis, whereas Bax increases cell death (Golstein, 1997; Kane *et al.*, 1993). Bcl-2 strongly decreases the generation of reactive oxygen species and inhibits the release of cytochrome *c* from mitochondria (Golstein, 1997; Kane *et al.*, 1993). Mitochondrial superoxide anion radical generation is reduced and electron leakage is diminished by Bcl-2 (Golstein, 1997; Kane *et al.*, 1993). Bcl-2 protects cells against oxidative stress and damage and ensures survival even in a prooxidative environment. Preliminary data from our laboratory indicate that melatonin increases mitochondrial protein levels of Bcl-2 and reduces intracellular protein levels of Bax (Poeggeler *et al.*, unpublished findings). In contrast to vitamins C and E, melatonin is able to modulate protein levels of Bcl-2 and Bax not only *in vitro* but also *in vivo* (Haendeler *et al.*, 1996; Poeggeler *et al.*, unpublished findings). We are now investigating the effects of melatonin and pro- and antioxidative compounds on gene expression of Bcl-2 and Bax *in vitro* and *in vivo*. Our findings might explain why melatonin prevents apoptosis *in vitro* and *in vivo* (Sainz *et al.*, 1995). The pineal hormone may act as an antinecrotic and antiapoptotic agent by radical scavenging or through activation of specific binding sites (Sainz *et al.*, 1995).

Melatonin stimulates brain glutathione peroxidase activity (Barlow-Walden *et al.*, 1995). Glutathione peroxidase metabolizes reduced glutathione to its oxidized form and by doing so converts hydrogen peroxide to water, thereby reducing the generation of hydroxyl radicals by eliminating their precursor (Barlow-Walden *et al.*, 1995). Recently, research activity has focused on the effects of melatonin on glutathione peroxidase activity and gene expression.

Melatonin increases mRNA levels for manganese superoxide dismutase and mRNA levels for copper-zinc superoxide dismutase (Antolin *et al.*, 1996). The cytoprotective effects of melatonin might at least in part be mediated by the increase in gene expression for antioxidant enzymes (Antolin *et al.*, 1996). Recently, nuclear receptors for melatonin have been identified in the rat brain (Becker-Andre *et al.*, 1994). It is possible that activation of these binding sites mediates the genomic effects of melatonin and the powerful antiapoptotic and antinecrotic activity of this neurohormone (Sainz *et al.*, 1995).

For at least a decade it was assumed that the most important functions of melatonin are related to its actions on the circadian and neuroendocrine systems (Reiter, 1980). Furthermore, it has been demonstrated recently that these actions of melatonin are almost exclusively mediated by the activation of specific membrane receptors localized outside the cells (Dubocovich, 1995). In view of the data presented in this review, it is very likely that many effects of melatonin are related to the intracellular actions of this endogenous indoleamine. These intracellular functions may be receptor dependent, as demonstrated for the genomic effects of melatonin, or receptor independent, as shown for the radical-scavenging effects of the indoleamine. The multiple functions of melatonin at the cellular level are consistent with the widespread distribution of the indoleamine from unicellular prokaryotes to humans (Hardeland *et al.*, 1993; Poeggeler, 1993). Only future research will determine whether we can harvest the antioxidative potential of this extremely potent endogenous electron donor and use melatonin to treat and prevent chronic diseases that have as their basis free-radical-mediated oxidative stress and damage.

10. REFERENCES

Abe, M., Reiter, R. J., Orrhii, P. H., Hara, M., Poeggeler, B., and Barlow-Walden, L. R., 1994, Inhibitory effect of melatonin on cataract formation in newborn rats: Evidence of an antioxidative role for melatonin, *J. Pineal Res.* **17**:94–100.

Agarwal, S., and Sohal, R. S., 1994, DNA oxidative damage and life expectancy in houseflies, *Proc. Natl. Acad. Sci. USA* **91**:12332–12335.

Antolin, I., Rodriguez, C., Sainz, R. M., Mayo, J. C., Uria, H., Kotler, M. I., RodriguezColunga, M. J., Tolivia, D., and Menendez-Pelaez, A., 1996, Melatonin prevents cell damage: Effect on gene expression for antioxidant enzymes, *Fed. Am. Soc. Exp. Biol.* **10:**882–890.

Balzer, I., and Hardeland, R., 1991, Photoperiodism and effects of indoleamines in a unicellular alga, *Gonyaulax polyedra, Science* **253:**795–797.

Balzer, I., and Hardeland, R., 1996, Melatonin in algae and higher plants—Possible role as phytohormone and antioxidant, *Bot. Acta* **109:**180–183.

Barlow-Walden, L. R., Reiter, R. J., Abe, M., Pablos, M. I., Menendez-Pelaez, A., Chen, L.-D., and Poeggeler, B., 1995, Melatonin stimulates brain glutathione peroxidase activity, *Neurochem. Int.* **26:**497–502.

Becker-Andre, M., Wiesenthal, I., Schaeren-Wiemers, N., Andre, E., Missbach, M., Saurat, J. H., and Carlberg, C., 1994, Pineal hormone melatonin binds and activates an orphan of the nuclear receptor superfamily, *J. Biol. Chem.* **269:**28531–28534.

Bender, D. A., 1975, Methionine, in *Amino Acid Metabolism* (D. A. Bender, ed.), pp.126–132, Wiley, New York.

Blinkenstaff, R. T., Brandstadter, S. M., Reddy, S., and Witt, R., 1994, Potential radioprotective agents: 1. Homologs of melatonin, *J. Pharm. Sci.* **83:**216–218.

Cagnoli, C. M., Atabay, C., Kharlamov, E., and Manev, H., 1996, Melatonin protects neurons from singlet-oxygen-induced apoptosis, *J. Pineal Res.* **18:**222–226.

Chan, T. Y., and Tang, P. L., 1996, Characterization of the antioxidant effects of melatonin and related indoleamines in vitro, *J. Pineal Res.* **20:**187–191.

Dubocovich, M. L., 1995, Melatonin receptors: Are there multiple subtypes? *Trends Pharmacol.* **16:**50–56.

Felix, K., Pairet, M., and Zimmermann, R., 1996, The antioxidative activity of the mucoregulatory agents: Ambroxol, bromhexine and N-acetyl-L-cysteine. A pulse radiolysis study, *Life Sci.* **59:**1141–1147.

Flaugh, M. E., Crowell, T. A., Clemens, J. A., and Sawyer, B. D., 1979, Synthesis and evaluation of antiovulatory activity of a variety of melatonin analogues, *J. Med. Chem.* **22:**63–68.

Floyd, R. A., 1993, Basic free radical chemistry, in *Free Radicals in Aging* (B. P. Yu, ed.), pp. 39–55, CRC Press, Boca Raton.

Gilad, E., Cuzzocrea, S., Zingarelli, B., Salzman, A. L., and Szabo, D., 1997, Melatonin is a scavenger of peroxynitrite, *Life Sci.* **60:**PL 169–PL 174.

Gilbert, D., 1994, Keeping reactive oxygen species (ROS) in their proper place, *Ann. N.Y. Acad. Sci.* **738:**1–7.

Giusti, P., Guisella, M., Lipartiti, M., Milani, D., Zhu, W., Vicini, S., and Manev, H., 1995, Melatonin protects primary cultures of cerebellar granule neurons from kainate but not from *N*-methyl-D-aspartate excitotoxicity, *Exp. Neurol.* **131:**39–46.

Giusti, P., Francheschini, D., Petrone, M., Manev, H., and Floreanin, M., 1996, In vitro and in vivo protection against kainate-induced excitotoxicity by melatonin, *J. Pineal Res.* **20:**226–231.

Golstein, P., 1997, Controlling cell death, *Science* **275:**1081–1082.

Haendeler, J., Zeiher, A. M., and Dimmeler, S., 1996, Vitamin C and E prevent lipopolysaccharide-induced apoptosis in human endothelial cells by modulation of bcl-2 and bax, *Eur. J. Pharmacol.* **317:**407–411.

Halliwell, B., 1995, Antioxidant characterization: Methodology and mechanism, *Biochem. Pharmacol.* **49:**1341–1348.

Hamada, N., and Nagase, M., 1996, In vivo effect of OPC15161, a superoxide scavenger, on anti-Thy1 nephritis, *Eur. J. Pharmacol.* **317:**123–128.

Hardeland, R., Reiter, R. J., Poeggeler, B., and Tan, D.-X., 1993, The significance of the metabolism of the neurohormone melatonin: Antioxidative protection and formation of bioactive substances, *Neurosci. Biobehav. Rev.* **17:**347–357.

Harman, D., 1956, Aging: A theory based on free radical and radiation chemistry, *J. Gerontol.* **11:**298–300.

Harman, D., 1983, Free radical theory of aging: Consequences of mitochondrial aging, *Age* **6:**86–94.

Hirata, F., Hayaishi, O., Tokuyama, T., and Senoh, S., 1974, *In vitro* and *in vivo* formation of two new metabolites of melatonin, *J. Biol. Chem.* **249:**1311–1315.

Itoh, T., Kamisaki, K., Kitamura, Y., Omodani, H., and Hamada, T., 1992, Cystathionine, a physiological prodrug of cysteine, protects against acetaminophen-induced liver toxicity, in *Frontiers and New Horizons in Amino Acid Research* (K. Takai, ed.), pp. 451–455, Elsevier, Amsterdam.

Kane, D. J., Sarafian, T. A., Anton, R., Hahn, H., Gralla, E. B., Valentine, J. S., Ord, T., and Bredesen, D. E., 1993, Bcl-2 inhibition of neuronal death: Decreased generation of reactive oxygen species, *Science* **262:**1274–1277.

Kitamura, Y., Kamisaki, Y., and Itoh, T., 1989, Hepatoprotective effects of cystathionine against acetaminophen-induced necrosis, *J. Pharmacol. Exp. Ther.* **250**:667–671.

Kodama, H., Ikeda, H., Awat, S., and Nakayama, K., 1985, Cystathionine accumulation in various regions of brain after DL-propargylglycine-treated rats, *J. Neurochem.* **44**:1207–1209.

Kodama, H., Sakaki, K., Mizobuchi, N., and Kikuchi, R., 1988, Contents of cystathionine and taurine in various cerebellar regions of DL-propargylglycine-treated rats, *J. Neurochem.* **51**:1046–1048.

Kopin, I. J., Pare, C. M. B., Axelrod, J., and Weissbach, H., 1961, The fate of melatonin in animals, *J. Biol. Chem.* **236**:3072–3076.

Kothari, L. S., and Subramanian, S., 1992, A possible modulatory influence of melatonin on representative phase I and phase II drug metabolizing enzymes in 9,10-dimethyl- 1,2-benzanthracene induced rat mammary tumorigenesis, *Anti-Cancer Drugs* **3**:623–628.

Longoni, B., Salgo, M. G., Pryor, W. A., and Marchiafara, P. L., 1995, Melatonin inhibits oxygen radicals induced lipid peroxidation, *Soc. Neurosci. Abstr.* **21**:Abstract 196.15.

Machida, Y., Zhang, J. Y., Hashimoto, K., Wakiguchi, H., Kurashige, T., Masuoka, N., Ubuka, T., and Kodama, H., 1995, Identification of perhydro-1,4-thiazepine-3,5-dicarboxylic acid, cystathionine mono-ore acids, cystathionine ketimines, cystathionine sulfoxide and *N*-acetylcystathionine sulfoxide in the urine sample of D,L-propargylglycine-treated rats, *Physiol. Chem. Phys. Med. NMR* **27**:203–216.

Manchester, L. C., Poeggeler, B., Alvares, F. L., Odgen, G. B., and Reiter, R. J., 1995, Melatonin immunoreactivity in the photosynthetic prokaryote, *Rhodospirillum rubrum*: Implications for an ancient antioxidant system, *Cell. Mol. Biol. Res.* **41**(5):391–395.

Marshall, K.-A., Reiter, R. J., Poeggeler, B., Aruma, O. I., and Halliwell, B., 1996, Evaluation of the antioxidative activity of melatonin in vitro, *Free Radical Biol. Med.* **21**:307–315.

Meister, A., 1992, On the antioxidant effect of ascorbic acid and glutathione, *Biochem. Pharmacol.* **44**:1905–1915.

Melchiorri, M., Reiter, R. J., Attia, A. M., Hara, M., Burgess, A., and Nistico, G., 1994, Potent protective effect of melatonin on in vivo paraquat-induced oxidative damage, *Life Sci.* **56**:83–89.

Menendez-Peleaz, A., Poeggeler, B., Reiter, R. J., Barlow-Walden, L. R., Pablos, M. I., and Tan, D.-X., 1993, Nuclear localization of melatonin in different tissues: Immunocytochemical and radioimmunoassay evidence, *J. Cell. Biochem.* **53**:373–382.

Miller, J. W., Selhub, J., and Joseph, J. A., 1996, Oxidative damage caused by free radicals during catecholeamine autooxidation: Protective effects of O-methylation and melatonin, *Free Radical Biol. Med.* **21**:241–249.

Miquel, J., and Economos, A. C., 1979, Favorable effects of the antioxidant sodium and magnesium thiazolidine carboxylate on the vitality and life span of *Drosophila* and mice, *Exp. Gerontol.* **14**:279–285.

Nakano, Y., Kawaguchi, T., Sumitomo, J., Takizawa, T., Uetsuki, S., Sugawara, M., and Kido, M., 1991, Novel inhibitor of superoxide anion generation OPC-15160 and OPC-15161, taxonomy, fermentation, isolation, physicochemical properties, biological characteristics and structure determination, *J. Antibiot.* **44**:52–64.

Nath, K. A., and Salahudeen, A. K., 1993, Autooxidation of cysteine generates hydrogen peroxide: Cytotoxicity and attenuation of pyruvate, *Am. J. Physiol.* **264**:F306–F314.

Oakin-Bendahan, S., Anis, Y., Nir, I., and Zisapel, N., 1995, Effects of long-term administration of melatonin and a putative antagonist on the ageing rat, *NeuroReport* **6**:785–788.

Oeiru, S., and Vochitu, E., 1965, The effect of the administration of compounds containing sulfhydryl groups on the survival rates of mice, rats and guinea pigs, *J. Gerontol.* **20**:417–419.

Ohlenschläger, G., 1991, Reaktionen des G-SH als Scavenger (Radikalfänger) oder mit Scavengern, in *Das Glutathionsystem* (G. Ohlenschläger, ed.), pp. 78–96, Verlag für Medizin Dr. Ewald Fischer, Heidelberg.

Pieri, C., Marra, M., Moroni, F., Recchioni, R., and Marcheselli, F., 1994, Melatonin: A peroxyl radical scavenger more effective than vitamin E, *Life Sci.* **15**:PL271–PL276.

Pieri, C., Moroni, F., Marra, M., Marcheselli, F., and Recchioni, R., 1995, Melatonin is an efficient antioxidant, *Arch. Gerontol. Geriatr.* **20**:159–165.

Pierpaoli, W., and Regelson, W., 1994, Pineal control of aging: Effect of melatonin and pineal grafting on aging mice, *Proc. Natl. Acad. Sci. USA* **91**:781–791.

Pierrefieche, G., Topall, G., Courboin, G., Henriet, I., and Laborit, H., 1993, Antioxidant activity of melatonin in mice, *Res. Commun. Chem. Pathol. Pharmacol.* **80**:211–223.

Poeggeler, B., 1993, Melatonin and the light–dark zeitgeber in vertebrates, invertebrates and unicellular organisms. Introduction to the multi-author review on melatonin, *Experientia* **49**:611–613.

Poeggeler, B., and Hardeland, R., 1994, Detection and quantification of melatonin in a dinoflagellate, *Gonyaulax polyedra*: Solutions to the problem of methoxyindole destruction in non-vertebrate material, *J. Pineal Res.* **17**:1–10.

Poeggeler, B., Balzer, I., Hardeland, R., and Lerchl, A., 1991, Pineal hormone melatonin oscillates also in the dinoflagellate *Gonyaulax polyedra, Naturwissenschaften* **78**:268–269.

Poeggeler, B., Reiter, R. J., Tan, D.-X., Chen, L.-D., and Manchester, L. C., 1993, Melatonin, hydroxyl radical-mediated oxidative damage, and aging, *J. Pineal Res.* **14**:151–168.

Poeggeler, B., Saarela, S., Reiter, R. J., Tan, D.-X., Chen, L.-D., and Manchester, L. C., 1994, Melatonin—A highly potent endogenous radical scavenger and electron donor: New aspects of the oxidation chemistry of this indole assessed in vitro, *Ann. N.Y. Acad. Sci.* **738**:419–420.

Poeggeler, B., Reiter, R., Hardeland, R., Sewerynek, E., Melchiorri, D., and Barlow-Walden, L. R., 1995, Melatonin, a mediator of electron transfer and repair reactions, acts synergistically with the chain breaking antioxidants ascorbate, Trolox and glutathione, *Neuroendocrinol. Lett.* **195**:87–92.

Poeggeler, B., Reiter, R. J., Hardeland, R., Tan, D.-X., and Barlow-Walden, L. R., 1996, Melatonin and structurally related, endogenous indoles act as potent electron donors and radical scavengers in vitro, *Redox Rep.* **2**:179–184.

Pozo, D., Reiter, R. J., Calvo, J. P., and Guerrero, J. M., 1994, Physiological concentrations of melatonin inhibit nitric oxide synthetase in rat cerebellum, *Life Sci.* **55**:PL455–PL460.

Puka-Sundvall, M., Eriksson, P., Nilsson, M., Sandberg, M., and Lehmann, A., 1995, Neurotoxicity of cysteine: Interaction with glutamate, *Brain Res.* **705**:65–70.

Reed, D. J., 1995, Cystathionine, *Methods Enzymol.* **252**:92–102.

Reiter, R. J., 1980, The pineal and its hormones in the control of reproductive physiology in mammals, *Endocrine Rev.* **1**:109–131.

Reiter, R. J., Melchiorri, D., Sewerynek, E., Poeggeler, B., Barlow-Walden, L. R., Chuang, J., Ortiz, G. G., and Acuna-Castroviejo, D., 1995, A review of the evidence supporting melatonin's role as an antioxidant, *J. Pineal Res.* **18**:1–11.

Rice-Evans, C., and Miller, N. J., 1994, Total antioxidative status in plasma and body fluids, *Methods Enzymol.* **234**:279–293.

Rusakow, L. S., White, C. W., and Stabler, S. P., 1993, O_2-induced changes in lung and storage pool thiols in mice: Effects of superoxide dismutase, *J. Appl. Physiol.* **74**:989–999.

Sainz, R. M., Mayo, J. C., Uria, H., Kotler, M., Antolin, I., Rodriguez, C., and Menendez-Peleaz, A., 1995, The pineal hormone melatonin prevents in vivo and in vitro apoptosis in thymocytes, *J. Pineal Res.* **19**:178–188.

Sewerynek, E., Melchiorri, D., Ortiz, G. G., Poeggeler, B., and Reiter, R. J., 1995a, Melatonin reduces H_2O_2-induced lipid peroxidation in homogenates of different rat brain regions, *J. Pineal Res.* **19**:51–56.

Sewerynek, E., Melchiorri, D., Reiter, R. J., Ortiz, G., and Lewinski, A., 1995b, Lipopolysaccharide-induced hepatotoxicity is inhibited by the antioxidant melatonin, *Eur. J. Pharmacol. Environ. Toxicol. Pharmacol. Sect.* **293**:327–334.

Sewerynek, E., Poeggeler, B., Melchiorri, D., and Reiter, R. J., 1995c, H_2O_2-induced lipid peroxidation in rat brain homogenates is greatly reduced by melatonin, *Neurosci. Lett.* **195**:203–205.

Shigenaga, M. K., Hagen, T. M., and Ames, B. A., 1994, Oxidative damage and mitochondrial decay in aging, *Proc. Natl. Acad. Sci. USA* **91**:10771–10778.

Sohal, R. S., and Weinbruch, R., 1996, Oxidative stress, caloric restriction, and aging, *Science* **273**:59–63.

Tallan, H., Moore, S., and Stein, W. H., 1957, L-Cystathionine in human brain, *J. Biol. Chem.* **230**:707–716.

Tan, D.-X., Chen, L.-D., Poeggeler, B., Manchester, L. C., and Reiter, R. J., 1993a, Melatonin: A potent, endogenous hydroxyl radical scavenger, *Endocrine J.* **1**:57–60.

Tan, D.-X., Poeggeler, B., Reiter, R. J., Chen, L.-D., Chen, S., Manchester, L. C., and Barlow-Walden, L. R., 1993b, The pineal hormone melatonin inhibits DNA adduct formation induced by the chemical carcinogen safrole, *Cancer Lett.* **70**:65–71.

Tan, D.-X., Reiter, R. J., Chen, L.-D., Poeggeler, B., Manchester, L. C., and Barlow-Walden, L. R., 1994, Both physiological and pharmacological levels of melatonin reduce DNA adduct formation induced by the carcinogen safrole, *Carcinogenesis* **15**:215–218.

Thomas, J. N., and Smith-Sonneborn, J., 1996, Supplemental melatonin increases clonal lifespan in *Paramecium tetraaurelia, Gerontologist* **36**:204.

Tilden, A. R., Becker, M. A., Amma, L. L., Arciniega, J., and McGaw, A. K., 1997, Melatonin production in an aerobic photosynthetic bacteria: An evolutionary early association with darkness, *J. Pineal Res.* **22:**102–106.

Wada, K., Kamisaki, Y., Kitano, M., Nakamoto, K., and Itoh, T., 1995, Protective effect of cystathionine on acute gastric mucosal injury induced by ischemia–reperfusion in rats, *Eur. J. Pharmacol.* **294:**377–382.

Wada, K., Kamisaki, Y., Nakamoto, K., and Toh, T., 1996, Effect of cystathionine as a scavenger of superoxide generated from human leukocytes or derived from xanthine oxidase in vitro, *Eur. J. Pharmacol.* **296:**335–340.

Yu, B. P., 1994, Cellular defenses against damage from reactive oxygen species, *Physiol. Rev.* **74:**139–161.

Zhang, J.-R., Andrus, P. K., and Hall, E. D., 1993, Age-related regional changes in hydroxyl radical stress and antioxidants in gerbil brain, *J. Neurochem.* **61:**1640–1647.

Chapter 17

Ubiquinol

An Endogenous Lipid-Soluble Antioxidant in Animal Tissues

Patrik Andrée, Gustav Dallner, and Lars Ernster

1. INTRODUCTION

Ubiquinone (UQ) was first described by Morton and colleagues (Festenstein *et al.*, 1955) as a quinone with a ubiquitous occurrence in various tissues, hence its name (Figure 1). Two years later, Crane *et al.* (1957) identified a quinone that was proposed to be a component of the mitochondrial respiratory chain, functioning as a coenzyme for electron transfer from Complexes I and II to Complex III (Figure 2). As such, it was given the name *coenzyme Q*. Its structure was determined by Folkers and colleagues (Wolf *et al.*, 1958) and found to be identical to ubiquinone.

The biochemical function of ubiquinone as an electron carrier in the mitochondrial respiratory chain was not generally accepted until the late 1960s, when it was demonstrated that depletion of beef heart submitochondrial particles of ubiquinone by pentane extraction caused an inhibition of both the NADH and succinate oxidase activities, and that these could be restored by the reincorporation of the same amount of ubiquinone as was originally present in the particles (Ernster *et al.*, 1969; Norling *et al.*, 1974). A minor fraction of ubiquinone during respiration was found to occur as the semiquinone radical (Bäckström *et al.*, 1970). The finding that there is an approximately 10-fold molar excess

Patrik Andrée and Gustav Dallner Department of Biochemistry, Arrhenius Laboratories for Natural Sciences, Stockholm University, S-106 91 Stockholm, and Clinical Research Center, NOVUM, Karolinska Institute, S-141 86 Huddinge, Sweden. **Lars Ernster** Department of Biochemistry, Arrhenius Laboratories for Natural Sciences, Stockholm University, S-106 91 Stockholm, Sweden.

Reactive Oxygen Species in Biological Systems, edited by Gilbert and Colton. Kluwer Academic / Plenum Publishers, New York, 1999.

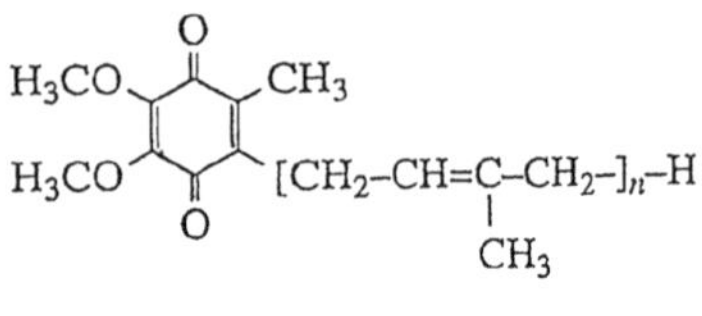

FIGURE 1. Structure of ubiquinone. The number of isoprene units, *n*, varies between 6 and 10 in different eukaryotes. It is predominantly 10 in human and 9 in rat.

of ubiquinone in the inner mitochondrial membrane as compared with other respiratory-chain carriers led Kröger and Klingenberg (1970, 1973a,b) to develop the concept of a "pool function" of ubiquinone in electron transport. These findings found their explanation through Mitchell's (1975, 1976) proposal of the protonmotive Q cycle (Figure 2) as the mechanism involved in energy conservation at coupling site 2 of the respiratory chain, which is now generally accepted (for review, see Trumpower, 1990).

The earliest observations of an antioxidant function for ubiquinol date back to the 1960s (for reviews, see Beyer and Ernster, 1990; Ernster and Dallner, 1995). In most studies, lipid peroxidation was the parameter demonstrating an antioxidant effect of ubiquinol, with isolated mitochondria or submitochondrial particles as the test objects. In recent years it was found that the agents used to initiate lipid peroxidation could also

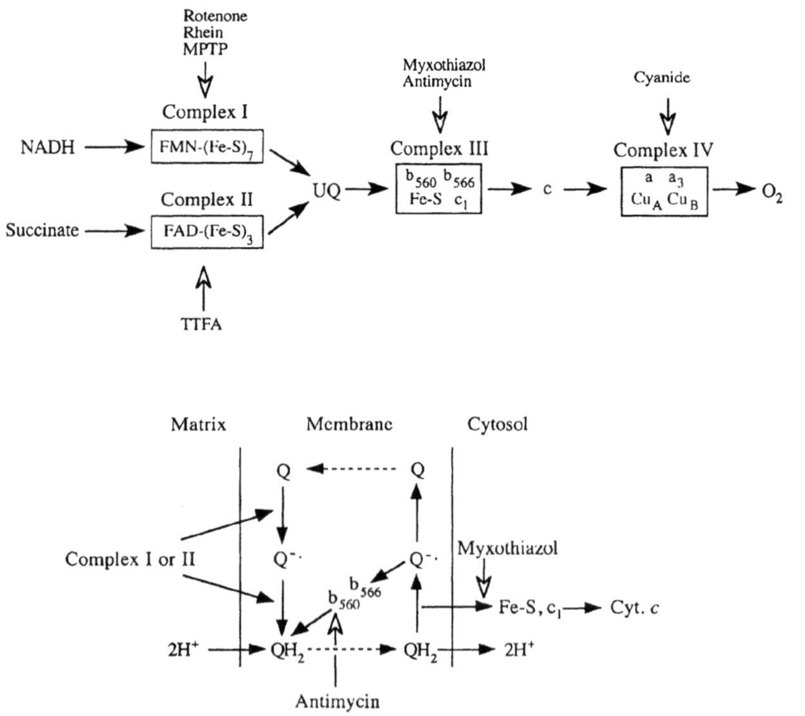

FIGURE 2. Schematic illustration of (top) the mitochondrial respiratory chain and (bottom) the protonmotive Q cycle. The open arrows indicate the sites of action of some inhibitors.

damage mitochondrial proteins and DNA, and that ubiquinol may prevent some of these effects. Moreover, it was shown that ubiquinone occurs also in cellular membranes other than mitochondria and in serum low-density lipoprotein (LDL), and that ubiquinol may protect these structures as well from oxidative damage. In general, ubiquinol appears to be the only known lipid-soluble antioxidant that animal cells can synthesize *de novo* and regenerate enzymically from its oxidized form in the course of its antioxidant function.

The present chapter is an overview of the above developments. It also summarizes available information concerning the role of ubiquinol as an antioxidant *in vivo* under various physiological and pathological conditions.

2. PROTECTIVE EFFECT OF UBIQUINOL AGAINST MITOCHONDRIAL LIPID PEROXIDATION, PROTEIN, AND DNA OXIDATION

2.1. Lipid Peroxidation: Effects of Ubiquinone and Vitamin E

Lipid peroxidation is one of the earliest recognized and most extensively studied manifestations of oxygen toxicity in biology (for review, see Ernster, 1993). Over the last three decades, peroxidation of polyunsaturated fatty acids induced by reactive oxygen species (ROS)—superoxide radical ($\cdot O_2^-$), hydrogen peroxide (H_2O_2), hydroxyl radical ($\cdot OH$), and singlet oxygen ($^1O_2$)—has been studied in great detail as it occurs in various cellular membranes, as well as in serum lipoproteins. In these studies, ROS were generated either nonenzymatically or, in the case of mitochondria and the endoplasmic reticulum, through enzymes present in the membranes of these organelles.

The peroxidation process can be divided into three separate phases: initiation, propagation, and termination.

Initiation takes place through a transition metal-induced (or radiation-induced) abstraction of a hydrogen atom from a methylene group of a fatty acid containing two or more separated double bonds, giving rise to a carbon-centered alkyl radical (·L), with a simultaneous rearrangement of the double bonds to become conjugated ("diene conjugation"). The ·L formed reacts with O_2 at a diffusion-controlled rate (Maillard *et al.*, 1983), giving rise to a peroxyl radical (·LOO).

Propagation involves the abstraction of a hydrogen atom from an adjacent unsaturated fatty acid by ·LOO, resulting in the formation of a lipid hydroperoxide (LOOH) and a new ·L radical. LOOH can react with Fe^{2+}, producing the alkoxyl radical (·LO). This radical, which is more reactive than ·LOO, can again reinitiate lipid peroxidation by hydrogen abstraction from an adjacent polyunsaturated fatty acid, with the formation of ·L and an alcohol (LOH) as the end product.

LOOH can also undergo degradation into hydrocarbons (ethane, pentane), alcohols, ethers, epoxides, and aldehydes. Among the latter, malondialdehyde (MDA) and 4-hydroxynonenal (4-HNE) are of special importance because they can cross-link phospholipids, proteins, and DNA (for review, see Esterbauer *et al.*, 1990).

Termination of the lipid peroxidation process is generally believed to take place by interaction between two radicals, to form a nonradical product.

Beginning in 1975, Takeshige and Minakami (1975, 1979) published a series of studies describing various parameters of lipid peroxidation in beef heart submitochondrial particles induced by NAD(P)H and ADP-Fe^{3+}, a system earlier employed to initiate lipid peroxidation in rat liver microsomes (Hochstein and Ernster, 1963; Hochstein *et al.*, 1964). Takeshige and collaborators (Takayanagi *et al.*, 1980; Takeshige *et al.*, 1980) reported that the lipid peroxidation in submitochondrial particles could be inhibited by the addition of succinate or high concentrations of NADH (but not NADPH). This protective effect was abolished on removal of the bulk of ubiquinone from the particles by lyophilization and pentane extraction, a procedure described by Ernster *et al.* (1969). Quantitative reincorporation of ubiquinone into the particles by the same procedure restored the ability of NADH or succinate to inhibit lipid peroxidation.

These results were confirmed by Glinn *et al.* (1991, 1997). They found, however, that the protective effect of high concentrations of NADH also occurs in the presence of rotenone and is therefore unlikely to be related to ubiquinol. Rhein, another inhibitor of Complex I, abolished the protection by NADH, and even stimulated lipid peroxidation in the pentane-extracted particles, indicating the occurrence of a lipid-soluble antioxidant component between the rhein- and rotenone-sensitive sites of Complex I (Glinn *et al.*, 1997).

The results reported by Takeshige and Minakami (1975, 1979) were also confirmed in our laboratory (Forsmark *et al.*, 1991; Ernster *et al.*, 1992) using ascorbate and ADP-Fe^{3+} to induce lipid peroxidation, and NADH or succinate in the presence of antimycin to reduce ubiquinone in the particles. It was found that the pentane extraction method used in these experiments also removed vitamin E (α-tocopherol) from the particles and, thus, that the restoration of the inhibition of lipid peroxidation on the reincorporation of ubiquinone did not require the presence of vitamin E. These findings eliminated a proposal by Kagan *et al.* (1990) that the antioxidant effect of ubiquinol is dependent on vitamin E.

Incorporation of vitamin E into the extracted particles resulted in an inhibition of lipid peroxidation in the absence of ubiquinone, the extent to which was enhanced by increasing ascorbate concentrations (Ernster *et al.*, 1992). Ubiquinone, when incorporated into the extracted particles together with vitamin E, promoted uptake of the latter, probably by increasing membrane fluidity (Katsikas and Quinn, 1983). After enzymatic reduction of ubiquinone, it also amplified the effect of vitamin E as observed at limiting ascorbate concentrations, apparently by regenerating the vitamin from the α-tocopheroxyl radical, in accordance with earlier observations (Maguire *et al.*, 1989, 1992; Kagan *et al.*, 1990; Mukai *et al.*, 1990, 1992).

In conclusion, these results demonstrated that ubiquinol can act as an antioxidant both directly, without mediation by vitamin E, and by promoting the antioxidant effect of the latter (Figure 3). Ubiquinol may exert its direct antioxidant effect by preventing initiation, propagation, or both, depending on the prevailing conditions.

2.2. Protein Oxidation and Its Prevention by Ubiquinol

Protein oxidation induced by ROS implies the oxidation of certain amino acid residues of a particular protein. It can take place through different mechanisms, depending on the way of generation of the ROS and the local environment of the protein. Stadtman (1993) has described a "site-specific" initiation based on the involvement of a transition metal, according to the following sequence of reactions: (1) a transition metal, usually

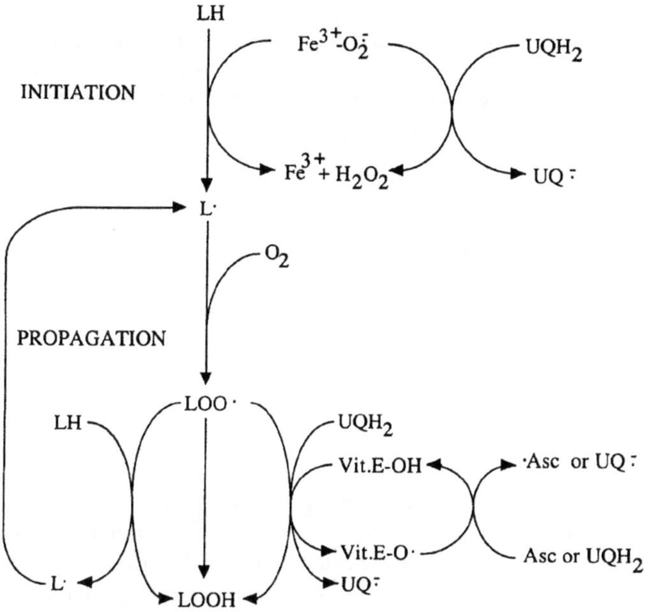

FIGURE 3. Possible sites of action of ubiquinone, vitamin E, and ascorbate on lipid peroxidation initiated by a perferryl radical. From Ernster *et al.* (1992). LH, polyunsaturated fatty acid; Fe^{3+}-·O_2^-, perferryl radical; UQH_2, ubiquinol; UQ^-·, ubisemiquinone; L·, carbon-centered radical; LOO·, lipid peroxyl radical; Vit. E-OH, vitamin E; Vit. E-O·, α-tocopheroxyl radical; Asc, ascorbate; Asc·, ascorbyl radical.

iron or copper, is bound to the protein; (2) the bound metal reacts with H_2O_2 to yield a hydroxyl radical, which oxidizes the amino acid residues at the metal-binding site.

Several authors have described this kind of inactivation of enzymes *in vitro*, and some of the modified amino acid residues have been identified. Davies and associates (Davies, 1987; Davies and Delsignore, 1987; Davies *et al.*, 1987a,b) have studied protein oxidation caused by pulse radiolysis. In this case, no bound transition metal is required. This type of oxidation has been demonstrated with both soluble and membrane-bound proteins (Davies, 1987; Davies and Delsignore, 1987; Davies *et al.*, 1987a,b; Zhang *et al.*, 1990; Grant *et al.*, 1993). In addition, it has been proposed that oxidation of membrane proteins can be mediated by lipid-derived radicals formed during lipid peroxidation (for review, see Wolff *et al.*, 1986).

Oxidative modification of proteins can also occur in a secondary manner. Aldehydes formed as degradation products of lipid peroxides, notably MDA and 4-HNE, can bind covalently to amino acid residues of proteins by Schiff's base or thioether linkage and cause intra- and intermolecular cross-linking (Esterbauer *et al.*, 1991).

Beef heart submitochondrial particles incubated with ascorbate and ADP-Fe^{3+} were found to give rise to protein carbonyl formation parallel to lipid peroxidation (Forsmark-Andrée *et al.*, 1995), the latter measured as formation of thiobarbituric acid reactive

Table I
Formation of TBARS and Protein Carbonyls in Native, UQ-Depleted, and UQ-Replenished Beef Heart Submitochondrial Particles after 30 min Incubation with ADP-Fe^{3+} and Ascorbate, in the Absence or Presence of Succinate and Antimycin[a]

SMP	Additions	TBARS	Carbonyls
Native	None	0.72	1.53
	ADP-Fe^{3+}, asc.	12.15	13.72
	ADP-Fe^{3+}, asc., succ. + antim.	0.63	1.32
UQ-depleted	ADP-Fe^{3+}, asc.	13.71	14.53
	ADP-Fe^{3+}, asc., succ. + antim.	12.22	13.76
UQ-replenished	ADP-Fe^{3+}, asc.	12.61	13.87
	ADP-Fe^{3+}, asc., succ. + antim.	1.20	2.15

[a]Data from Forsmark-Andrée *et al.* (1995).

substances (TBARS). Both processes could be prevented by the addition of succinate and antimycin, which caused a reduction of endogenous ubiquinol. This effect was abolished by the extraction, and restored by the reincorporation, of ubiquinone (Table I).

Attempts were made to estimate the extent to which the protein carbonyls originated from direct oxidation of amino acid residues and from secondary modification by way of Schiff's base or thioether formation with MDA or 4-HNE. From results reported by Forsmark-Andrée *et al.* (1995), it appears that the major part of the carbonyl derivatives under the conditions employed were products of direct protein oxidation. However, the direct oxidation of proteins may also be mediated by lipid-derived free radicals (·LOO, ·LO), in view of its close kinetic relationship to lipid peroxidation and the protective effect of ubiquinol on both processes. A schematic illustration of the interplay between these two processes is shown in Figure 4.

2.3. Identification of Oxidatively Modified Proteins

SDS-PAGE analysis revealed a complex pattern regarding the effect of ascorbate and ADP-Fe^{3+} on the individual protein components of the particles (Forsmark-Andrée *et al.*, 1995). The most striking change was a shift of a band corresponding to ~29 kDa toward a higher molecular mass position. This shift was prevented by succinate and antimycin, i.e., by the reduction of ubiquinone in the particles. The protein corresponding to this band was tentatively identified as the adenine nucleotide translocator (ANT), which is the most abundant protein component of the mitochondrial inner membrane. This protein had been shown to be particularly sensitive to lipid peroxidation (Zwizinski and Schmid, 1992), probably because of its close association with cardiolipin (Beyer and Klingenberg, 1985). This conclusion was subsequently substantiated with immunoblot analysis using a monoclonal antibody raised against beef heart ANT (Andrée, 1996).

As revealed by SDS-PAGE analysis, there were also a number of other proteins that were affected in the course of lipid peroxidation and protected by ubiquinol. One component, with a molecular mass of ~110 kDa, was subsequently identified by immunoblot analysis as the nicotinamide nucleotide transhydrogenase (NNT) (Forsmark-Andrée *et al.*, 1996).

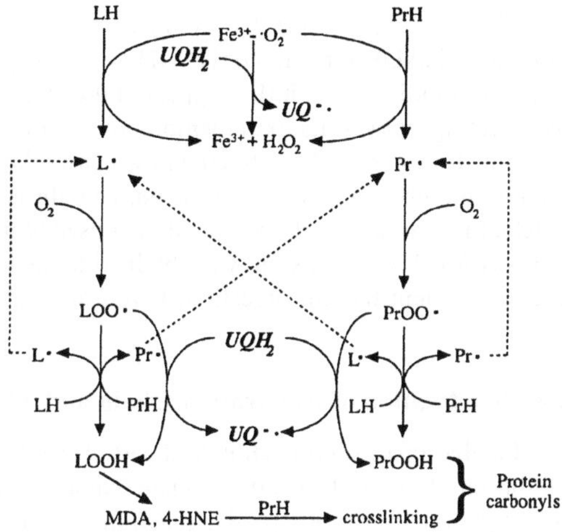

FIGURE 4. Schematic illustration of the sequence of reactions involved in lipid (LH) and protein (PrH) oxidation as initiated by a perferryl radical, and the conceivable sites of action of ubiquinol (UQH_2) as an antioxidant. From Forsmark-Andrée *et al.* (1995).

Mitochondrial proton-translocating NNT has been extensively studied (Hoek and Rydström, 1988; Hatefi and Yamaguchi, 1992) ever since the description of an "energy-dependent" NNT by Danielson and Ernster (1963). NNT is the largest protein component and the only known single-subunit proton-translocating enzyme in the mitochondrial inner membrane. Its structure consists of 14 transmembrane α-helices, connecting a 45-kDa N-terminal and a 25-kDa C-terminal domain, both located on the matrix side of the membrane and harboring the NAD(H) and NADP(H) binding sites, respectively. The enzyme catalyzes the reaction

$$NADH_m + NADP^+_m + H^+_c \leftrightarrow NAD^+_m + NADPH_m + H^+_m$$

where c and m denote the cytosol and the matrix, respectively. It utilizes a proton gradient to shift the equilibrium of the hydride transfer between the two nicotinamide nucleotides, and at the same time to enhance its rate in the forward (left to right) direction. According to a recent proposal by Hatefi and Yamaguchi (1996), the proton gradient acts by changing the binding energy of NADP(H) to the enzyme, in a manner similar to the "binding-change mechanism" proposed by Boyer (1993) for ATP synthase.

As recently shown by Forsmark-Andrée *et al.* (1996), treatment of submitochondrial particles with ascorbate and ADP-Fe^{3+} caused an inactivation of NNT that could be prevented by the addition of succinate and antimycin, i.e., by endogenous ubiquinol. In another series of experiments, the effect of peroxynitrite on the NNT reaction was investigated. It was found that treatment of the particles with peroxynitrite caused an inactivation of both the forward and reverse reactions catalyzed by NNT, and that these effects were not prevented by reducing the endogenous ubiquinone (Forsmark-Andrée *et al.*, 1996). Kinetic analysis revealed that treatment with ascorbate and ADP-Fe^{3+} increased the K_m for NADPH, but not for NADH, whereas peroxynitrite treatment increased both values.

These results suggest that lipid peroxidation is accompanied by an alteration of the binding of NADPH, but not of NADH, to the enzyme, whereas peroxynitrite affects both substrate-binding sites. It thus appeared that peroxynitrite treatment inhibits the enzyme by affecting both of its extramembranous domains and thereby its binding affinity for both substrates. In contrast, treatment with ascorbate and ADP-Fe^{3+} may act by perturbing the intramembranous region of the enzyme through lipid peroxidation, thereby causing a defect in proton translocation and a consequent decrease in NADPH-binding affinity. Support for this conclusion was obtained by analyzing the enzyme by immunoblotting and subsequent trypsin digestion, before and after exposure to different experimental conditions.

2.4. Inactivation of Respiratory Chain and ATP Synthase

Incubation of submitochondrial particles with ascorbate and ADP-Fe^{3+} resulted in an inhibition of the NADH and succinate oxidase activities (Forsmark-Andrée *et al.*, 1997). Investigation of various partial reactions of the respiratory chain revealed that those involving ubiquinone were primarily responsible for the inactivation of the NADH and succinate oxidases. Minor degrees of inhibition were also found with succinate dehydrogenase and cytochrome *c* oxidase, but these could not account for the overall inactivation of the respiratory chain. However, it was found that the content of ubiquinone decreased during lipid peroxidation, and that there was a close correlation between the inactivation of NADH and succinate oxidases and the decrease in ubiquinone content. Reduction of the endogenous ubiquinone, by succinate in the presence of cyanide, prevented both the inactivation of the oxidase activities and the decrease in ubiquinone content (Table II).

Attempts to reactivate the oxidatively damaged particles by incorporation of ubiquinone, using the pentane procedure (Ernster *et al.*, 1969), have so far been unsuccessful. Likewise, little information is available about the breakdown products of ubiquinone, although there are several earlier reports describing an oxidative breakdown of ubiquinone, and proposing various conceivable pathways (Morimoto *et al.*, 1969, 1970; Imada *et al.*, 1970). Preliminary experiments in our laboratory indicate that solanesol and an unidentified polar ring structure may be some of the breakdown products under our experimental conditions.

Table II
TBARS Formation, NADH and Succinate Oxidase Activities, and Ubiquinone Content after 30 min Incubation of Beef Heart Submitochondrial Particles with or without Lipid Peroxidation Induced by ADP-Fe^{3+} and Ascorbate in the Absence or Presence of Succinate and Cyanide[a]

Additions	TBARS formation (nmole/mg protein)	NADH oxidase (μatom O/min/mg protein)	Succinate oxidase (μatom O/min/mg protein)	Ubiquinone content (nmole/mg protein)
None	0.25	1.39	0.73	4.83
ADP-Fe^{3+}, asc.	14.25	0.15	0.21	1.48
ADP-Fe^{3+}, asc., succ. + cyanide	0.52	1.31	0.69	4.37

[a]Data from Forsmark-Andrée *et al.* (1997).

In a model system, using ubiquinone incorporated into phospholipid liposomes consisting of either saturated or polyunsaturated phospholipids, it was shown (Forsmark-Andrée *et al.*, 1997) that treatment with ascorbate and ADP-Fe^{3+} resulted in a breakdown of ubiquinone in the latter, but not in the former type of liposomes. These results provide strong evidence that the breakdown of ubiquinone found in the submitochondrial particles is mediated by lipid peroxidation and is nonenzymatic.

Experiments were also performed to assess the effect of lipid peroxidation on the ATP synthase (F_0F_1-ATPase) activity, in both beef heart submitochondrial particles (Andrée, 1996) and whole rat liver mitochondria (Forsmark-Andrée and Ernster, 1994).

In the case of particles, incubation with ascorbate and ADP-Fe^{3+} resulted in an inactivation of the ATPase that was not prevented by reduction of the endogenous ubiquinone. In intact mitochondria, the ATPase activity was also inhibited on incubation with ascorbate and ADP-Fe^{3+}, but in this case the inhibition was prevented by reduction of the ubiquinone with succinate and antimycin. These findings suggest that in the case of the particles, the radical attack occurs on both the intramembrane proton-translocating F_0 and the protruding catalytic F_1 moieties of the enzyme, the latter probably in a site-specific manner at the ADP-binding sites on the α and/or β subunits, which are inaccessible to ubiquinol. In the case of mitochondria, the oxidative damage is probably restricted to the F_0 moiety of the enzyme, which can be prevented by ubiquinol. In both mitochondria and submitochondrial particles, the ATPase activity was oligomycin-sensitive, indicating that the ATP hydrolysis was obligatorily coupled to proton translocation.

2.5. DNA Oxidation

ROS-dependent DNA oxidation is generally believed to proceed in a manner resembling the site-specific oxidation of proteins discussed above. It involves a reaction between H_2O_2 with a transition metal prebound to DNA (Halliwell and Auroma, 1991). The damage consists of base oxidation, which can be detected by analysis of the oxidized bases, e.g., 8-hydroxy-deoxyguanosine (8-OH-dG) (Floyd *et al.*, 1986). The measurement of strand breaks can also be used as an indicator of oxidative DNA damage (Whitaker *et al.*, 1991). In recent years, much attention has been directed to mitochondrial DNA, which is particularly susceptible to oxidative damage (Shigenaga *et al.*, 1994).

Incubation of rat liver mitochondria for 30 min in the presence of ascorbate and ADP-Fe^{3+} was found to result in an increased content of 8-OH-dG (Forsmark-Andrée and Ernster, 1994) and a marked induction of DNA strand breaks (Andrée, 1996) (Figure 5). Both effects were counteracted—the former completely, and the latter partially—by the addition of succinate and antimycin, i.e., by endogenous ubiquinol.

2.6. Anti- and Prooxidant Effects of Ubiquinone in Mitochondria

Electron leakage from the respiratory chain giving rise to $\cdot O_2^-$ and H_2O_2 has been shown by several investigators, using whole mitochondria, submitochondrial particles, or isolated respiratory-chain complexes (for review, see Chance *et al.*, 1979). The site and mechanism of this leakage have received much attention, and several components of the respiratory chain have been implicated. Among these, ubiquinone has been proposed to be the major site (Cadenas *et al.*, 1977; Turrens *et al.*, 1985), based on the following lines

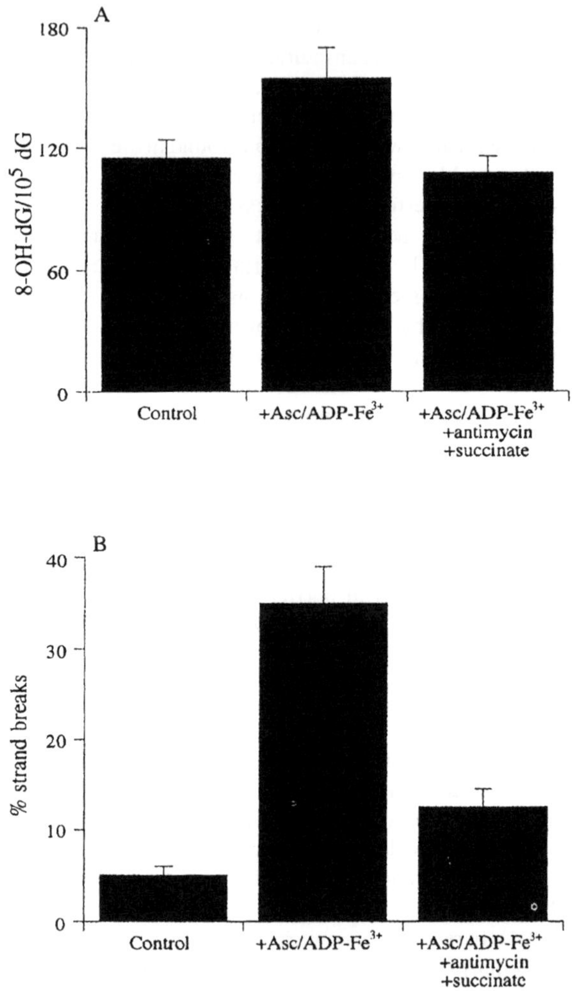

FIGURE 5. Formation of (A) 8-OH-dG and (B) strand breaks in rat liver mitochondria treated with ascorbate and ADP-Fe^{3+} for 30 min, in the absence or presence of succinate and antimycin. From Forsmark-Andrée and Ernster (1994) and Andrée (1996).

of evidence: (1) ubisemiquinone is known to occur during respiration (Bäckström *et al.*, 1970; Davies and Hochstein, 1982), and semiquinones in general react rapidly with oxygen, producing $\cdot O_2^-$; (2) NAD(P)H- and/or succinate-supported radical leakage is stimulated 10- to 15-fold by the presence of antimycin and an uncoupler, indicating that the electron leak occurs on the substrate side of the antimycin-sensitive site of the respiratory chain (Loschen and Flohé, 1971); and (3) extraction of ubiquinone from mitochondria inhibits, and reincorporation restores, H_2O_2 formation (Boveris and Chance, 1973).

However, the following information argues against the notion that ubisemiquinone is the principal source of electron leakage during mitochondrial respiration: (1) ubisemiquinone formed during mitochondrial respiration is maintained at a relatively high steady-state level, as revealed by EPR measurements (Bäckström *et al.*, 1970; Salerno and Ohnishi, 1980; Salerno *et al.*, 1990), and is bound to specific ubiquinone-binding proteins associated with Complexes I, II, and III (Yu and Yu, 1981; King, 1982) which stabilize it against autoxidation; agents that modify these complexes, such as TTFA in the case of Complex II (Ingledew and Ohnishi, 1997; Salerno and Ohnishi, 1980) or antimycin in the case of Complex III (Ohnishi and Trumpower, 1980; Bowyer and Trumpower, 1981), alter this stability of ubisemiquinone (Rich *et al.*, 1990), which may explain the stimulatory effect of TTFA on lipid peroxidation (Glinn *et al.*, 1991; Eto *et al.*, 1992) and of antimycin on H_2O_2 formation (Loschen and Flohé, 1971); (2) myxothiazol, another inhibitor of Complex III, suppresses H_2O_2 formation (Loschen *et al.*, 1973; von Jagow *et al.*, 1984; Nohl and Jordan, 1986), even though it does not prevent ubiquinone reduction; the difference between the effects of antimycin and myxothiazol on H_2O_2 formation is more likely to be related to their effects on the redox state of cytochrome b_{566}, the latter being reduced in the presence of antimycin but not in the presence of myxothiazol (see Figure 2); it thus appears likely that reduced cytochrome b_{566}, rather than ubisemiquinone, is responsible for the electron leak in Complex III; (3) electron leak during NADH oxidation has been demonstrated to occur in the presence of rotenone (Ramsay and Singer, 1992; Giulivi *et al.*, 1995), rhein (Floridi *et al.*, 1989; Glinn *et al.*, 1991, 1997), and MPTP (Hasegawa *et al.*, 1990; Ramsay and Singer, 1992; Murphy *et al.*, 1995), agents that inhibit Complex I and thus ubiquinone reduction.

In conclusion, the available evidence strongly indicates that electron leakage leading to $\cdot O_2^-$ and H_2O_2 formation occurs in Complexes I, II, and III of the respiratory chain, mainly through autoxidation of components other than ubisemiquinone. Like other antioxidants, ubiquinone may obviously also act as a prooxidant under special circumstances. However, this does not detract from its importance as a biological antioxidant, especially as this function of ubiquinol is not limited to mitochondria.

3. ANTIOXIDANT FUNCTION OF UBIQUINOL OUTSIDE MITOCHONDRIA

3.1. Intracellular Distribution of Ubiquinone

In view of its role in the respiratory chain, ubiquinone in eukaryotes was initially assumed to be located exclusively in the inner mitochondrial membrane. However, several investigations have revealed that ubiquinone is also present in various other cellular locations (Sastry, 1961; Jayaraman and Ramasarma, 1963; Lang *et al.*, 1986; Kalén *et al.*, 1987), including the Golgi vesicles, the lysosomes, the endoplasmic reticulum, the peroxisomes, the plasma membrane, and the outer mitochondrial membrane. Table III shows the distribution of ubiquinone in various fractions of rat liver homogenate. Notably, this distribution is markedly different from that of α-tocopherol, the ubiquinone/α-tocopherol ratio being highest in the inner and outer mitochondrial membranes and the plasma membrane, and lowest in the microsomes and the Golgi vesicles.

Table III
Levels of Ubiquinones and α-Tocopherol in Hepatic Subfractions Prepared from Rats[a]

Fraction	UQ	α-Tocopherol	UQ/α-tocopherol
	(pmole/mg protein)		
Homogenate	1139	165	6.9
Inner mitochondrial membranes	2654	77	34.5
Outer mitochondrial membranes	3800	142	26.8
Rough microsomes	128	102	1.3
Smooth microsomes	178	228	0.8
Golgi vesicles	897	1105	0.8
Lysosomes	1246	421	3.0
Plasma membranes	390	19	20.5
Nuclear fraction	215	95	2.3
Peroxisomes	134	49	2.7
Cytosol	117	47	2.5

[a]Data from Zhang *et al.* (1996).

It has also been shown that ubiquinone is discharged to a limited extent across the plasma membrane into the blood, where it is partially bound to LDL (Elmberger *et al.*, 1989). However, in contrast to cholesterol, ubiquinone does not seem to be distributed among different tissues via the circulation.

3.2. Tissue Distribution and Redox State

Analysis of its redox state in various tissues (Åberg *et al.*, 1992) (Figure 6) and in blood (Zhang *et al.*, 1995) has revealed that ubiquinone is present mainly in the reduced form, a fact that may be consistent with its function as an antioxidant. However, the mechanism by which ubiquinone is reduced in locations other than the inner mitochondrial membrane is unclear. Several enzymes have been considered to be involved in ubiquinone reduction, such as DT-diaphorase (Beyer *et al.*, 1996), microsomal NADH-cytochrome b_5 and NADPH-cytochrome P450 reductases (Cadenas *et al.*, 1992), and NADH dehydrogenases associated with the outer mitochondrial membrane (Sottocasa *et al.*, 1967) and the plasma membrane (Crane *et al.*, 1985). A cytosolic ubiquinone reductase different from DT-diaphorase has recently been described (Takahashi *et al.*, 1995). Reduction of ubiquinone in LDL has been proposed to take place by way of a quinone reductase located in the erythrocyte membrane, utilizing intracellular NADH as the electron donor (Stocker and Suarna, 1993). It has also been proposed that ubiquinone reduction may take place by temporary fusion between different membranes (Takeshige *et al.*, 1980).

In addition to its potential role as an antioxidant, the presence of ubiquinone in membranes other than mitochondria raises the question of possible other functions in these locations. Crane and Morré (1977) have proposed that ubiquinone located in the Golgi apparatus may be involved in vesicle migration associated with secretion. The same group has also shown an involvement of ubiquinone in cell growth via an NADH oxidoreductase-activated signal system (Sun *et al.*, 1992; Crane and Sun, 1993). Further-

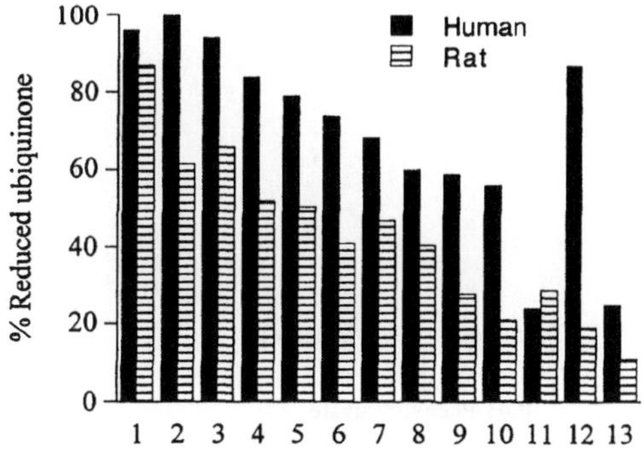

FIGURE 6. Extent of reduction of ubiquinone from various human and rat tissues. 1, liver; 2, pancreas; 3, intestine; 4, colon; 5, testis; 6, kidney; 7, thyroid; 8, muscle; 9, ventricle; 10, heart; 11, brain; 12, spleen; 13, lung. From Åberg *et al.* (1992).

more, the neutrophil redox system has been reported to contain ubiquinone (Crawford and Schneider, 1982).

3.3. Ubiquinone Biosynthesis and Its Regulation

The biosynthesis of ubiquinone involves four main processes: (1) synthesis of the benzene ring, (2) synthesis of the isoprenoid side chain, (3) condensation of the side chain with the ring, and (4) subsequent modifications of the ring structure, including hydroxylation, methylation, and decarboxylation steps (for reviews, see Olson and Rudney, 1983; Ericsson and Dallner, 1993; Grünler *et al.*, 1994). The reaction pathways involved in these processes are schematically illustrated in Figure 7.

The benzene ring is synthesized predominantly from tyrosine (in some instances from phenylalanine) and converted through a number of steps into 4-hydroxybenzoate. The polyisoprenoid side chain is synthesized from acetyl-CoA through a reaction sequence commonly referred to as the *mevalonate pathway*, leading to the formation of farnesyl-PP. This product constitutes a branching point, serving as a precursor for the side chain of ubiquinone, decaprenyl-PP, and for two other major products, cholesterol and dolichol. In addition, farnesyl-PP serves as the substrate for protein isoprenylation, either directly or through geranylgeranyl-PP. Decaprenyl-PP undergoes condensation with 4-hydroxybenzoate to form decaprenyl-4-hydroxybenzoate, which subsequently is converted through several steps to ubiquinone. However, there is still some uncertainity about the order in which these steps take place, and their intracellular localization.

The tissue levels of ubiquinone have been shown to vary according to the metabolic rate. Cold acclimation (Beyer *et al.*, 1962), thyroid hormone treatment (Pedersen *et al.*, 1963; Sterling *et al.*, 1977; Sterling, 1986; Mancini *et al.*, 1989), exercise (Beyer *et al.*, 1984), pancreatectomy (Boveris *et al.*, 1969), peroxisome proliferation (Kalén *et al.*, 1990; Åberg *et al.*, 1994), and myocardial reperfusion (Muscari *et al.*, 1995), all of which affect the overall rate of oxidative metabolism, and thus ROS production, have been shown to alter the tissue ubiquinone content in the same direction as the change in metabolic rate. Significantly, the increase in tissue ubiquinone content following thyroid hormone treatment occurs after the increase in metabolic rate, suggesting that it is an adaptation to, rather than a cause of, the increased oxidative activity (Pedersen *et al.*,

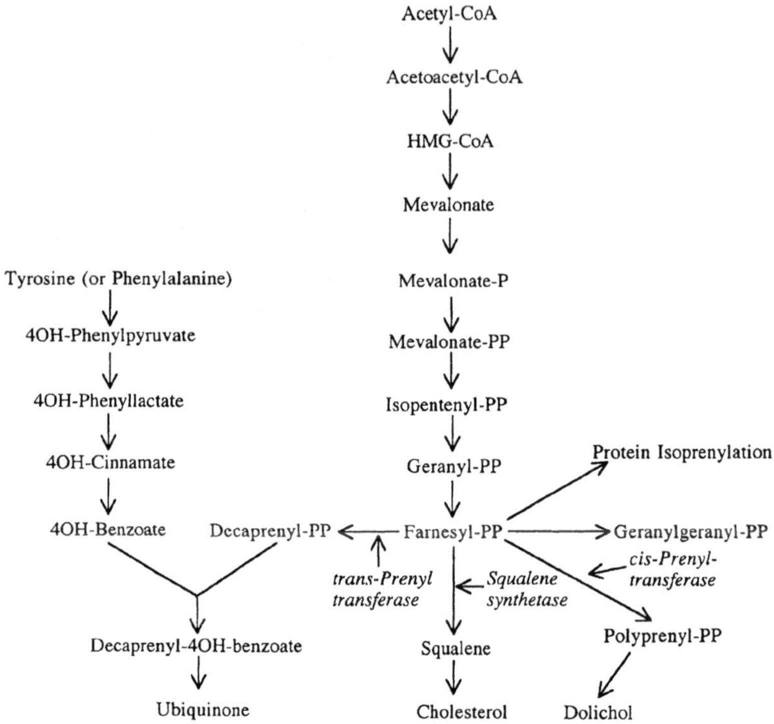

FIGURE 7. Reaction pathways of the biosynthesis of ubiquinone, cholesterol, and dolichol.

1963). In this context, it is of interest to note that no increase in the ubiquinone content of skeletal muscle was found in "Luft's disease" (Luft *et al.*, 1962), a state of severe hypermetabolism of nonthyroid origin, with a defect in mitochondrial respiratory control. In general, there is growing evidence for the existence of a coordinate regulation of pro- and antioxidant mechanisms according to the prevailing oxidative stress. In the case of peroxisome proliferator-treated rats, there is a positive correlation between the increase in β-oxidation of fatty acids and the ubiquinone content in liver (Figure 8A); and a negative correlation between the plasma cholesterol level and the liver ubiquinone content (Figure 8B) (Åberg *et al.*, 1996).

Ubiquinone levels have been reported to change with increasing age (Beyer *et al.*, 1985; Kalén *et al.*, 1989; Söderberg *et al.*, 1990; Matsura *et al.*, 1991; Edlund *et al.*, 1994), reaching a maximal level at age 20 in humans, followed by a decline (Kalén *et al.*, 1989; Söderberg *et al.*, 1990) (Table IV). This decrease in ubiquinone content on increasing age is consistent with Harman's "free radical theory of aging" (Harman, 1981, 1994), and may be a reflection of the organism's diminished capacity to maintain adequate ubiquinol levels in relation to the prevailing need for antioxidant defense. An inverse correlation between longevity and peroxide-producing potential in mammalian tissues has been reported by Cutler (1985).

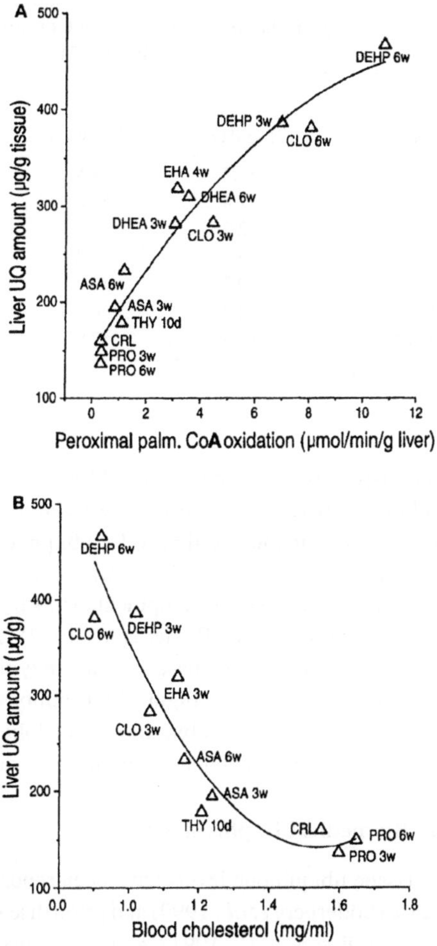

FIGURE 8. Effects of peroxisome proliferator treatment of rats on ubiquinone content and (A) peroxisomal palmitoyl-CoA oxidation in liver and (B) blood cholesterol level. From Åberg *et al.* (1996). CRL, control; PRO, probucol; ASA, acetylsalicylic acid; THY, thyroxine; CLO, clofibrate; DHEA, dehydroepiandrosterone; EHA, 2-ethylhexanoic acid; DEHP, di(2-ethylhexyl)phthalate; w refers to number of weeks of treatment, d to days.

The regulatory site in ubiquinone biosynthesis has been proposed to be in the mevalonate pathway, i.e., in the biosynthesis of the side chain (Aiyar and Olson, 1972). Agents that inhibit HMG-CoA reductase have been shown to cause a decrease of the tissue levels of ubiquinone in animals (Willis *et al.*, 1990; Appelkvist *et al.*, 1993) and the plasma levels in humans (Folkers *et al.*, 1990; Elmberger *et al.*, 1991; Watts *et al.*, 1993).

Much interest is currently focused on the effects of various HMG-CoA inhibitors—pravastatin, lovastatin (mevinolin), and related anticholestenemic drugs—on tissue ubiquinone levels *in vivo*. The picture that emerges is that these drugs may reduce ubiquinone levels to various extents depending on the test conditions used (Goldstein and Brown, 1990; Willis *et al.*, 1990; Löw *et al.*, 1992; Appelkvist *et al.*, 1993). An important

Table IV
Age-Related Changes in the Ubiquinone Content of Human Organs (in μg/g Wet Weight Tissue)[a]

	Age group				
Organ	1–3 days	0.7–2 years	19–21 years	39–41 years	77–81 years
Heart	36.7	78.5	110.0	75.0	47.2
Kidney	17.4	53.4	98.0	71.1	64.0
Liver	12.9	45.1	61.2	58.3	50.8
Pancreas	9.2	38.2	21.0	19.3	6.5
Spleen	20.7	30.2	32.8	28.6	13.1
Lung	2.2	6.4	6.0	6.5	3.1
Adrenal	17.5	57.9	16.1	12.2	8.5

[a]Data from Kalén *et al.* (1989).

future aspect of these studies would be to investigate the effects of these drugs under conditions where the organism is exposed to oxidative stress, e.g., by physical training, and thereby to an increased need for ubiquinone biosynthesis, or during aging, when the tissue ubiquinone levels, and thereby the antioxidant capacity, decrease.

An interesting new development in this field is the use of squalestatin 1 as an inhibitor of cholesterol synthesis (Baxter *et al.*, 1992). This fungal product is a potent and specific inhibitor of squalene synthase, i.e., an enzyme that is below the branching point of the mevalonate pathway (see Figure 7). It inhibits cholesterol synthesis selectively, without affecting the synthesis of ubiquinone and dolichol. Moreover, as recently found by Thelin *et al.* (1994), it even enhances ubiquinone synthesis by three- to fourfold (Table V).

3.4. Biomedical Implications

Tissue ubiquinone levels have been reported to increase in the brain in Alzheimer's disease (Söderberg *et al.*, 1990) and prion disease (Guan *et al.*, 1996), and in preneoplastic nodules (Olsson *et al.*, 1991), and to decrease in hepatocellular cancer (Eggens *et al.*, 1989), cardiomyopathy (Karlsson *et al.*, 1993), and muscle degeneration (Goda *et al.*, 1987; Yamamoto *et al.*, 1987).

Table V
Effect of Squalestatin 1 on the Relative Rates of Incorporation of [^{3}H]-Mevalonate into Cholesterol, Dolichol, and Ubiquinone in Cultures of Chinese Hamster Ovary Cells[a]

Squalestatin 1 (mM)	Cholesterol	Dolichol (% of control)	Ubiquinone
0.5	38	118	286
1.0	8	100	387
2.0	1.5	86	345

[a]Data from Thelin *et al.* (1994).

Over the last three decades, a large number of studies have been published describing beneficial effects of ubiquinone administration to humans and experimental animals in various diseased states. Among the clinical studies, the most striking effects have been reported on cardiovascular diseases (for review, see Mortensen, 1993) and certain mitochondrial disorders, including mitochondrial encephalomyopathy, lactic acidosis, and strokelike symptoms (MELAS) (Yamamoto *et al.*, 1987), Kearns–Sayre syndrome (Ogasahara *et al.*, 1986; Bet *et al.*, 1987), and Alper's disease (Fischer *et al.*, 1986). In animal models, an effect of ubiquinone in preventing reperfusion injury of the heart has been demonstrated (Atar *et al.*, 1993; Mortensen, 1993).

Initially, these effects were interpreted in terms of a ubiquinone deficiency and ensuing defect in mitochondrial ATP synthesis in the diseased tissues, which could be repaired by ubiquinone supplementation. An alternative explanation, considered in recent years, is based on the generally recognized antioxidant function of ubiquinol (for review, see Ernster, 1994), and on studies of the uptake of ubiquinone after oral administration (Mohr *et al.*, 1992; Zhang *et al.*, 1995). The latter studies have revealed an increase in ubiquinone level in blood, but not in heart, brain, kidney, or skeletal muscle; an increase was found in the liver ubiquinone content, but this was sequestered in the lysosomes. Decreased plasma levels of ubiquinone have been found in patients with mevalonate

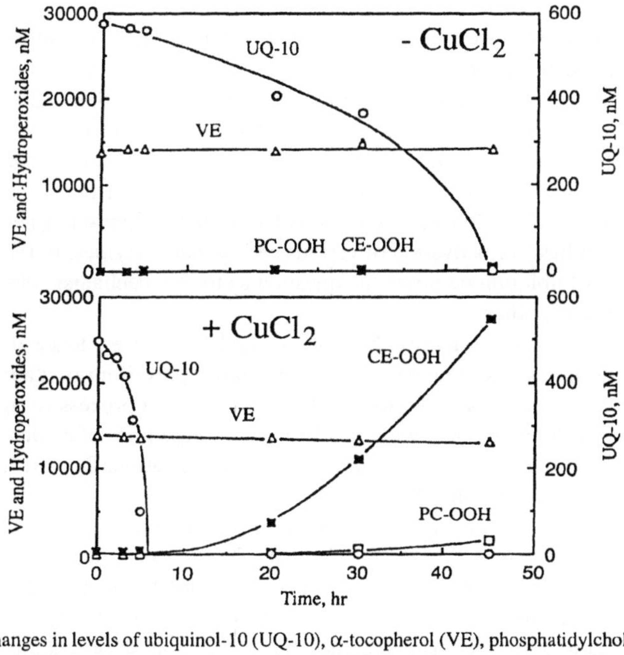

FIGURE 9. Changes in levels of ubiquinol-10 (UQ-10), α-tocopherol (VE), phosphatidylcholine hydroperoxide (PC-OOH), and cholesteryl ester hydroperoxide (CE-OOH) during the oxidation of human LDL in the absence (upper panel) or presence (lower panel) of 5 μM cupric chloride at 37°C under air. From Yamamoto *et al.* (1991).

kinase deficiency, suffering from psychomotor retardation, ataxia, and myopathy (Hübner *et al.*, 1993).

According to reports from two laboratories (Stocker *et al.*, 1991; Yamamoto *et al.*, 1991), ubiquinol-10 protects human LDL from lipid peroxidation more efficiently than does α-tocopherol (Figure 9). Dietary supplementation of humans with ubiquinone-10 has been shown to result in increased levels of ubiquinol-10 within circulating lipoproteins and increased resistance of LDL to the initiation of lipid peroxidation (Mohr *et al.*, 1992; Alleva *et al.*, 1995). Significantly increased LDL/ubiquinone ratios are found in patients suffering from ischemic heart disease (Hanaki *et al.*, 1991), which is not altered by treatment with the HMG-CoA reductase inhibitor pravastatin (Hanaki *et al.*, 1993).

On the basis of the available information, it has been proposed that ubiquinone as a dietary supplement may act primarily by elevating the ubiquinone level in blood (Ernster and Dallner, 1995), where it may serve several important functions. Among these are, in addition to an enhanced protection of LDL from oxidation, a prevention of ROS-induced damage caused by neutrophils in inflammatory diseases, and of oxidative injury by endothelial cells resulting from ischemia–reperfusion. These and possibly other protective functions against ROS-induced damage in the circulation may account for the majority of the reported beneficial effects of ubiquinone administration in experimental and clinical medicine.

4. UBIQUINONE AND REDOX SIGNALING: FUTURE PERSPECTIVES

The control of gene expression by oxidants, antioxidants, and other factors that influence the intracellular redox state has in recent years become a major field of research in molecular biology (Pahl and Baeuerle, 1994; Burdon, 1995; Sen and Packer, 1996). There is a growing literature describing the mechanisms involved in the adaptation of antioxidant defense to the prevailing oxidative stress, also referred to as *redox signaling*. A great deal of information is available about various transcription factors that are instrumental in this adaptation and are responsible for the regulation of gene expression leading to the synthesis or activation of various antioxidant enzymes. In this context, the antioxidant role of ubiquinone raises the question as to the mechanisms involved in the regulation of its biosynthesis.

As described in the foregoing, there are several lines of evidence indicating that ubiquinone biosynthesis is regulated in response to the prevailing oxidative challenge, and that several diseased states, as well as the physiological process of aging, may be consequences of a defect in this regulation. In view of these lines of evidence, it appears to be of interest to elucidate the mechanisms of this regulation. Does it involve redox signaling and, if so, at what stage?

The findings that mevalonate kinase deficiency leads to a decrease of ubiquinone content in plasma (Hübner *et al.*, 1993), and that inhibition of squalene synthase results in an enhanced ubiquinone biosynthesis (Thelin *et al.*, 1994), suggest that the rate-limiting step is between the former enzyme and the synthesis of farnesyl-PP. Any oxidant-induced stimulation of ubiquinone biosynthesis may thus require an upregulation of some enzyme(s) in this region of the mevalonate pathway. One intriguing possibility is that this regulation is mediated by ubiquinone catabolism. As already discussed, oxidative stress can cause a breakdown of ubiquinone (Forsmark-Andrée *et al.*, 1997). This may have a signaling function for enhanced ubiquinone biosynthesis, similar to the effect of the squalene synthase inhibitor, squalestatin 1 (Thelin *et al.*, 1994). Clearly, we need more information about the biosynthesis and catabolism of ubiquinone before the above questions can be answered.

In conclusion, the exploration of the possible role of ubiquinone in the regulation of antioxidant defense may open new perspectives for future research concerning biological redox signaling.

5. REFERENCES

Åberg, F., Appelkvist, E.-L., Dallner, G., and Ernster, L., 1992, Distribution and redox state of ubiquinones in rat and human tissues, *Arch. Biochem. Biophys.* **295**:230–234.

Åberg, F., Zhang, Y., Appelkvist, E.-L., and Dallner, G., 1994, Effects of clofibrate, phthalates and probucol on ubiquinone levels, *Chem. Biol. Interact.* **91**:1–14.

Åberg, F., Zhang, Y., Teclebrhan, H., Appelkvist, E.-L., and Dallner, G., 1996, Increases in tissue levels of ubiquinone in association with peroxisome proliferation, *Chem. Biol. Interact.* **99**:205–218.

Aiyar, A. S., and Olson, R. E., 1972, Enhancement of ubiquinone-9 biosynthesis in rat-liver slices by exogenous mevalonate, *Eur. J. Biochem.* **27**:60–64.

Alleva, R., Tomasetti, M., Battino, M., Curatola, G., Littarru, G. P., and Folkers, K., 1995, The roles of coenzyme Q_{10} and vitamin E on the peroxidation of human low density lipoprotein subfractions, *Proc. Natl. Acad. Sci. USA* **92**:9388–9391.

Andrée, P., 1996, Oxidative stress and mitochondrial function. Role of ubiquinol as antioxidant, Doctoral thesis, Karolinska Institute.

Appelkvist, E.-L., Edlund, C., Löw, P., Schedin, S., Kalén, A., and Dallner, G., 1993, Effects of inhibitors of hydroxymethylglutaryl coenzyme A reductase on coenzyme Q and dolichol biosynthesis, *Clin. Invest.* **71**:S97–S102.

Atar, D., Mortensen, S. A., Flachs, H., and Herzog, W. R., 1993, Coenzyme Q10 protects ischemic myocardium in an open-chest swine model, *Clin. Invest.* **71**:S103–S111.

Bäckström, D., Norling, B., Ehrenberg, A., and Ernster, L., 1970, Electron spin resonance measurement on ubiquinone-depleted and ubiquinone-replenished submitochondrial particles, *Biochim. Biophys. Acta* **197**:108–111.

Baxter, A., Fitzgerald, B. J., Hutson, J. L., McCarthy, A. D., Motteran, J. M., Ross, B. C., Sapra, M., Snowden, M. A., Watson, N. S., Williams, R. J., and Wright, C., 1992, Squalestatin 1, a potent inhibitor of squalene synthase, which lowers serum cholesterol *in vivo*, *J. Biol. Chem.* **267**:11705–11708.

Bet, L., Bresolin, N., Binda, A., Nador, F., and Ferrante, C., 1987, Cardiac improvement after coenzyme Q10 treatment with Kearns-Sayre syndrome, *Neurology* **37**:202–204.

Beyer, K., and Klingenberg, M., 1985, ADP/ATP carrier protein from beef heart mitochondria has high amounts of tightly bound cardiolipin, as revealed by ^{31}P nuclear magnetic resonance, *Biochemistry* **24**:3821–3826.

Beyer, R. E., and Ernster, L., 1990, The antioxidant role of coenzyme Q, in *Highlights in Ubiquinone Research* (G. Lenaz, O. Barnabei, A. Rabbi, and M. Battino, eds.), pp. 191–213, Taylor & Francis, London.

Beyer, R. E., Noble, W. M., and Hirschfield, T. J., 1962, Coenzyme Q (ubiquinone) levels of tissues of rats during acclimation to cold, *Can. J. Biochem. Physiol.* **40**:511–518.

Beyer, R. E., Morales-Corral, P. G., Ramp, B. J., Kreitman, K. R., Falzon, M. J., Rhee, S. Y., Kuhn, T. W., Stein, M., Rosenwasser, M. J., and Cartwright, K. J., 1984, Elevation of tissue coenzyme Q (ubiquinone) and cytochrome *c* concentrations by endurance exercise in the rat, *Arch. Biochem. Biophys.* **234**:323–329.

Beyer, R. E., Burnett, B. A., Cartwright, K. J., Edington, D. W., Falzon, M. J., Kreitman, K. R., Kuhn, T. W., Ramp, B. J., Rhee, S. Y., and Rosenwasser, M. J., 1985, Tissue coenzyme Q (ubiquinone) and protein concentrations over the life span of the laboratory rat, *Mech. Ageing Dev.* **32**:267–281.

Beyer, R. E., Segura Aguilar, J., DiBernardo, S., Cavazzoni, M., Fato, R., Fiorentini, D., Galli, M. C., Setti, M., Landi, L., and Lenaz, G., 1996, The role of DT-diaphorase in the maintenance of the reduced antioxidant form of coenzyme Q in membrane systems, *Proc. Natl. Acad. Sci. USA* **93**:2528–2532.

Boveris, A., and Chance, B., 1973, The mitochondrial generation of hydrogen peroxide. General properties and effect of hyperbaric oxygen, *Biochem. J.* **134**:707–716.

Boveris, A., Ramos, M. C. P., Stoppani, A. O. M., and Foglia, V. G., 1969, Phosphorylation, oxidation and ubiquinone content in diabetic mitochondria, *Proc. Soc. Exp. Biol. Med.* **132**:170–174.

Bowyer, J. R. ,and Trumpower, B. L., 1981, Rapid reduction of cytochrome c_1 in the presence of antimycin and its implication for the mechanism of electron transfer in the cytochrome bc_1 segment of the mitochondrial respiratory chain, *J. Biol. Chem.* **256**:2245–2251.

Boyer, P. D., 1993, The binding change mechanism for ATP synthase—Some probabilities and possibilities, *Biochim. Biophys. Acta* **1140**:215–250.

Burdon, R. H., 1995, Superoxide and hydrogen peroxide in relation to mammalian cell proliferation, *Free Radical Biol. Med.* **4**:775–794.

Cadenas, E., Boveris, A., Ragan, C. I., and Stoppani, A. O., 1977, Production of superoxide radicals and hydrogen peroxide by NADH-ubiquinone reductase and ubiquinol-cytochrome *c* reductase from beef-heart mitochondria, *Arch. Biochem. Biophys.* **180**:248–257.

Cadenas, E., Hochstein, P., and Ernster, L., 1992, Pro- and antioxidant functions of quinones and quinone reductases in mammalian cells, *Adv. Enzymol.* **65**:97–146.

Chance, B., Sies, H., and Boveris, A., 1979, Hydroperoxide metabolism in mammalian organs, *Physiol. Rev.* **59**:527–605.

Crane, F. L., and Morré, D. J., 1977, Evidence for coenzyme Q function in Golgi membranes, in *Biomedical and Clinical Aspects of Coenzyme Q*, Vol. 1 (K. Folkers and Y. Yamamura, eds.), pp. 3–14, Elsevier, Amsterdam.

Crane, F. L., and Sun, S. E. E., 1993, The essential functions of coenzyme Q, *Clin. Invest.* **71**:S55–S59.

Crane, F. L., Hatefi, Y., Lester, R. L., and Widmer, C., 1957, Isolation of a quinone from beef heart mitochondria, *Biochim. Biophys. Acta* **25**:220–221.

Crane, F. L., Sun, I. L., Clark, M. G., Grebing, C., and Löw, H., 1985, Transplasma-membrane redox systems in growth and development, *Biochim. Biophys. Acta* **811**:233–264.

Crawford, D. R., and Schneider, D. L., 1982, Identification of ubiquinone-10 in human neutrophils and its role in microbicidal events, *J. Biol. Chem.* **257**:6662–6668.

Cutler, R. G., 1985, Peroxide-producing potential of tissues: Inverse correlation with longevity of mammalian species, *Proc. Natl. Acad. Sci. USA* **82**:4798–4802.

Danielson, L., and Ernster, L., 1963, Demonstration of a mitochondrial energy-dependent pyridine nucleotide transhydrogenase reaction, *Biochem. Biophys. Res. Commun.* **10**:91–96.

Davies, K. J. A., 1987, Protein damage and degradation by oxygen radicals. I. General aspects, *J. Biol. Chem.* **262**:9895–9901.

Davies, K. J. A., and Delsignore, M. E., 1987, Protein damage and degradation by oxygen radicals. III. Modification of secondary and tertiary structure, *J. Biol. Chem.* **262**:9908–9913.

Davies, K. J. A., and Hochstein, P., 1982, Ubisemiquinone radicals in liver: Implications for a mitochondrial Q cycle *in vivo*, *Biochem. Biophys. Res. Commun.* **107**:1292–1299.

Davies, K. J. A., Delsignore, M. E., and Lin, S. W., 1987a, Protein damage and degradation by oxygen radicals. II. Modification of amino acids, *J. Biol. Chem.* **262**:9902–9907.

Davies, K. J. A., Lin, S. W., and Pacifici, R. E., 1987b, Protein damage and degradation by oxygen radicals. IV. Degradation of denatured protein, *J. Biol. Chem.* **262**:9914–9920.

Edlund, C., Söderberg, M., and Kristensson, K., 1994, Isoprenoids in aging and neurodegeneration, *Neurochem. Int.* **25**:35–38.

Eggens, I., Elmberger, P. G., and Löw, P., 1989, Polyisoprenoid, cholesterol and ubiquinone levels in human hepatocellular carcinomas, *Br. J. Exp. Pathol.* **70**:83–92.

Elmberger, P. G., Kalén, A., Brunk, U. T., and Dallner, G., 1989, Discharge of newly-synthesized dolichol and ubiquinone with lipoproteins to rat liver perfusate and to the bile, *Lipids* **24**:919–930.

Elmberger, P. G., Kalén, A., Lund, E., Reinhér, E., Eriksson, M., Berglund, L., Angelin, B., and Dallner, G., 1991, Effects of pravastatin and cholestyramine on products of the mevalonate pathway in familial hypercholesterolemia, *J. Lipid Res.* **32**:935–940.

Ericsson, J., and Dallner, G., 1993, Distribution, biosynthesis, and function of mevalonate pathway lipids, in *Subcellular Biochemistry, Vol. 21: Endoplasmic Reticulum* (N. Borgese and J. R. Harris, eds.), pp. 229–272, Plenum Press, New York.

Ernster, L., 1993, Lipid peroxidation in biological membranes: Mechanisms and implications, in *Active Oxygens, Lipid Peroxides, and Antioxidants* (K. Yagi, ed.), (Tokyo: Japan Sci. Soc. Press, and Boca Raton: CRC Press), pp. 1–38.

Ernster, L., 1994, Ubiquinol as a biological antioxidant: A review, in *Oxidative Processes and Antioxidants* (R. Paoletti, B. Samuelsson, A. L. Catapano, A. Poli, and M. Rinetti, eds.), pp. 185–198, Raven Press, New York.

Ernster, L., and Dallner, G., 1995, Biochemical, physiological and medical aspects of ubiquinone function, *Biochim. Biophys. Acta* **1271**:195–204.

Ernster, L., Lee, I. Y., Norling, B., and Persson, B., 1969, Studies with ubiquinone-depleted submitochondrial particles. Essentiality of ubiquinone for the interaction of succinate dehydrogenase, NADH dehydrogenase, and cytochrome *b*, *Eur. J. Biochem.* **9**:299–310.

Ernster, L., Forsmark, P., and Nordenbrand, K., 1992, The mode of action of lipid-soluble antioxidants in biological membranes: Relationship between the effects of ubiquinol and vitamin E as inhibitors of lipid peroxidation in submitochondrial particles, *BioFactors* **3**:241–248.

Esterbauer, H., Zollner, H., and Schaur, R. J., 1990, Aldehydes formed by lipid peroxidation: Mechanisms of formation, occurrence and determination, in *Membrane Lipid Oxidation*, Vol. I (C. D. Vigo-Pelfrey, ed.), pp. 239–283, CRC Press, Boca Raton.

Esterbauer, H., Schaur, R. F., and Zollner, H., 1991, Chemistry and biochemistry of 4-hydroxynonenal, malonaldehyde and related aldehydes, *Free Radical Biol. Med.* **11**:81–128.

Eto, Y., Kang, D., Hasegawa, E., Takeshige, K., and Minakami, S., 1992, Succinate-dependent lipid peroxidation and its prevention by reduced ubiquinone in beef heart submitochondrial particles, *Arch. Biochem. Biophys.* **295**:101–106.

Festenstein, G. N., Heaton, F. W., Lowe, J. S., and Morton, R. A., 1955, A constituent of the unsaponifiable portion of animal tissue lipids, *Biochem. J.* **59**:558–566.

Fischer, J. C., Ruitenbeck, W., Gabreels, F. J. M., Janssen, A. J. M., Renier, W. O., Sengers, R. C. A., Stadhouders, A. M., ter Lak, H. J., Trijbels, J. M. F., and Veerkamp, J. H., 1986, A mitochondrial encephalomyopathy: The first case with an established defect at the level of coenzyme Q10, *Eur. J. Pediatr.* **144**:441–447.

Floridi, A., Castiglione, S., and Bianchi, C., 1989, Sites of inhibition of mitochondrial electron transport by rhein, *Biochem. Pharmacol.* **38**:743–751.

Floyd, R. A., Watson, J. J., Wong, P. K., Altmiller, D. H., and Rickard, R. C., 1986, Hydroxyl free radical adduct of deoxyguanosine: Sensitive detection and mechanisms of formation, *Free Radical Res. Commun.* **1**:163–172.

Folkers, K., Langsjoen, P., Willis, R., Richardson, P., Xia, L. J., Ye, C. Q., and Tamagawa, H., 1990, Lovastatin decreases coenzyme Q levels in humans, *Proc. Natl. Acad. Sci. USA* **87**:8931–8934.

Forsmark, P., Åberg, F., Norling, B., Nordenbrand, K., Dallner, G., and Ernster, L., 1991, Inhibition of lipid peroxidation by ubiquinol in submitochondrial particles in the absence of vitamin E, *FEBS Lett.* **285**:39–45.

Forsmark-Andrée, P., and Ernster, L., 1994, Evidence for a protective effect of endogenous ubiquinol against oxidative damage to mitochondrial protein and DNA during lipid peroxidation, *Mol. Aspects Med.* **15**:S73–S81.

Forsmark-Andrée, P., Dallner, G., and Ernster, L., 1995, Endogenous ubiquinol prevents protein modification accompanying lipid peroxidation in beef heart submitochondrial particles, *Free Radical Biol. Med.* **19**:749–757.

Forsmark-Andrée, P., Persson, B., Radi, R., Dallner, G., and Ernster, L., 1996, Oxidative modification of mitochondrial nicotinamide nucleotide transhydrogenase in submitochondrial particles. Effect of endogenous ubiquinol, *Arch. Biochem. Biophys.* **336**:113–120.

Forsmark-Andrée, P., Lee, C. P., Dallner, G., and Ernster, L., 1997, Lipid peroxidation and changes in the ubiquinone content and the respiratory chain enzymes of submitochondrial particles, *Free Radical Biol. Med.* **22**:391–400.

Giulivi, C., Boveris, A., and Cadenas, E., 1995, Hydroxyl radical generation during mitochondrial electron transfer and the formation of 8-hydroxydeoxyguanosine in mitochondrial DNA, *Arch. Biochem. Biophys.* **316**:909–916.

Glinn, M., Ernster, L., and Lee, C. P., 1991, Initiation of lipid peroxidation in submitochondrial particles: Effects of respiratory inhibitors, *Arch. Biochem. Biophys.* **290**:57–65.

Glinn, M., Lee, C. P., and Ernster, L., 1997, Pro- and anti-oxidant activities of the mitochondrial respiratory chain: Factors influencing NAD(P)H-induced lipid peroxidation, *Biochim. Biophys. Acta* **1318:**246–254.

Goda, S., Hamada, T., Ishimoto, S., Kobayashi, T., Goto, I., and Kuroiwa, Y., 1987, Clinical improvement after administration of coenzyme Q10 in a patient with mitochondrial encephalomyopathy, *J. Neurol.* **234:**62–63.

Goldstein, J. L., and Brown, M. S., 1990, Regulation of the mevalonate pathway, *Nature* **343:**425–430.

Grant, A. J., Jessup, W., and Dean, R. T., 1993, Enhanced enzymatic degradation of radical damaged mitochondrial membrane components, *Free Radical Res. Commun.* **19:**125–134.

Grünler, J., Ericsson, J., and Dallner, G., 1994, Branch-point reactions in the biosynthesis of cholesterol, dolichol, ubiquinone and prenylated proteins, *Biochim. Biophys. Acta* **1212:**259–277.

Guan, Z., Söderberg, M., Sindelar, P., Prusiner, S. B., Kristensson, K., and Dallner, G., 1996, Lipid composition in scrapie-infected mouse brain: Prion infection increases the levels of dolichyl phosphate and ubiquinone, *J. Neurochem.* **66:**277–285.

Halliwell, B., and Auroma, O. I., 1991, DNA damage by oxygen-derived species. Its mechanism and measurement in mammalian systems, *FEBS Lett.* **281:**9–19.

Hanaki, Y., Sugiyama, S., Ozawa, T., and Ohno, M., 1991, Ratio of low-density lipoprotein cholesterol to ubiquinone as a coronary risk factor, *N. Engl. J. Med.* **325:**814–815.

Hanaki, Y., Sugiyama, S., Ozawa, T., and Ohno, M., 1993, Coenzyme Q10 and coronary artery disease, *Clin. Invest.* **71:**S112–S115.

Harman, D., 1981, The aging process, *Proc. Natl. Acad. Sci. USA* **78:**7124–7128.

Harman, D., 1994, Aging: Prospects for further increases in the functional life span, *Age* **17:**119–146.

Hasegawa, E., Takeshige, K., Oishi, T., Murai, Y., and Minakami, S., 1990, 1-Methyl-4-phenylpyridinium (MPP^+) induces NADH-dependent superoxide formation and enhances NADH-dependent lipid peroxidation in bovine heart submitochondrial particles, *Biochem. Biophys. Res. Commun.* **170:**1049–1055.

Hatefi, Y., and Yamaguchi, M., 1992, The energy-transducing nicotinamide nucleotide transhydrogenase, in *Molecular Mechanisms in Bioenergetics* (L. Ernster, ed.), pp. 265–281, Elsevier, Amsterdam.

Hatefi, Y., and Yamaguchi, M., 1996, Nicotinamide nucleotide transhydrogenase: A model for utilization of substrate binding energy for proton translocation, *FASEB J.* **10:**444–452.

Hochstein, P., and Ernster, L., 1963, ADP-activated lipid peroxidation coupled to the TPNH oxidase system of microsomes, *Biochem. Biophys. Res. Commun.* **12:**388–394.

Hochstein, P., Nordenbrand, K., and Ernster, L., 1964, Evidence for the involvement of iron in the ADP-activated peroxidation of lipids in microsomes and mitochondria, *Biochem. Biophys. Res. Commun.* **14:**323–328.

Hoek, J. B., and Rydström, J., 1988, Physiological roles of nicotinamide nucleotide transhydrogenase, *Biochem. J.* **254:**1–10.

Hübner, C., Hoffmann, G. F., Charpentier, C., Gibson, K. M., Finckh, B., Puhl, H., Lehr, H. A., and Kohlschutter, A., 1993, Decreased plasma ubiquinone-10 concentration in patients with mevalonate kinase deficiency, *Pediatr. Res.* **34:**129–133.

Imada, I., Watanabe, M., Matsumoto, N., and Morimoto, H., 1970, Metabolism of ubiquinone-7, *Biochemistry* **9:**2870–2878.

Ingledew, W. J., and Ohnishi, T., 1977, The probable site of action of thenoyltrifluoracetone on the respiratory chain, *Biochem. J.* **164:**617–620.

Jayaraman, J., and Ramasarma, T., 1963, Intracellular distribution of coenzyme Q in rat liver, *Arch. Biochem. Biophys.* **103:**258–266.

Kagan, V., Serbinova, E., and Packer, L., 1990, Antioxidant effects of ubiquinones in microsomes and mitochondria are mediated by tocopherol recycling, *Biochem. Biophys. Res. Commun.* **169:**851–857.

Kalén, A., Norling, B., Appelkvist, E.-L., and Dallner, G., 1987, Ubiquinone biosynthesis by the microsomal fraction from rat liver, *Biochim. Biophys. Acta* **926:**70–78.

Kalén, A , Appelkvist, E.-L., and Dallner, G., 1989, Age-related changes in the lipid compositions of rat and human tissues, *Lipids* **24:**579–584.

Kalén, A., Appelkvist, E.-L., and Dallner, G., 1990, The effects of inducers of the endoplasmic reticulum, peroxisomes and mitochondria on the amounts and synthesis of ubiquinone in rat liver subcellular membranes, *Chem. Biol. Interact.* **73:**221–234.

Karlsson, J., Liska, J., Gunnes, S., Koul, B., Semb, B., Åström, H., Diamant, B., and Folkers, K., 1993, Heart muscle ubiquinone and plasma antioxidants following cardiac transplantation, *Clin. Invest.* **71:**S76–S83.

Katsikas, H., and Quinn, P. J., 1983, Fluorescence probe studies of the distribution of ubiquinone homologues in bilayers of dipalmitoylglycerophosphocholine, *Eur. J. Biochem.* **131**:607–612.

King, T. E., 1982, Ubiquinone proteins in cardiac mitochondria, in *Functions of Quinones in Energy Conserving Systems* (B. L. Trumpower, ed.), pp. 3–25, Academic Press, New York.

Kröger, A., and Klingenberg, M., 1970, Quinones and nicotinamide nucleotides associated with electron transfer, *Vitam. Horm.* **28**:533–574.

Kröger, A., and Klingenberg, M., 1973a, The kinetics of the redox reactions of ubiquinone related to the electron-transport activity in the respiratory chain, *Eur. J. Biochem.* **34**:358–368.

Kröger, A., and Klingenberg, M., 1973b, Further evidence for the pool function of ubiquinone as derived from the inhibition of the electron transport by antimycin, *Eur. J. Biochem.* **39**:313–323.

Lang, J. K., Gohil, K., and Packer, L., 1986, Simultaneous determination of tocopherols, ubiquinols, and ubiquinones in blood, plasma, tissue homogenates, and subcellular fractions, *Anal. Biochem.* **157**:106–116.

Loschen, G., and Flohé, L., 1971, Respiratory chain linked H_2O_2 production in pigeon heart mitochondria, *FEBS Lett.* **18**:261–264.

Loschen, G., Azzi, A., and Flohé, L., 1973, Mitochondrial H_2O_2 formation: Relationship with energy conservation, *FEBS Lett.* **33**:84–88.

Löw, P., Andersson, M., Edlund, C., and Dallner, G., 1992, Effects of mevinolin treatment on tissue dolichol and ubiquinone levels in the rat, *Biochim. Biophys. Acta* **1128**:253–259.

Luft, R., Ikkos, D., Palmieri, G., Ernster, L., and Afzelius, B., 1962, A case of severe hypermetabolism of nonthyroid origin with a defect in the maintenance of mitochondrial respiratory control: A correlated clinical, biochemical and morphological study, *J. Clin. Invest.* **41**:1776–1804.

Maguire, J. J., Wilson, D. S., and Packer, L., 1989, Mitochondrial electron transport-linked tocopheroxyl radical reduction, *J. Biol. Chem.* **264**:21462–21465.

Maguire, J. J., Kagan, V., Ackrell, B. A., Serbinova, E., and Packer, L., 1992, Succinate-ubiquinone reductase linked recycling of alpha-tocopherol in reconstituted systems and mitochondria: Requirement for reduced ubiquinone, *Arch. Biochem. Biophys.* **292**:47–53.

Maillard, B., Ingold, K. U., and Scaniano, J. C., 1983, Rate constants for the reactions of free radicals with oxygen in solution, *J. Am. Chem. Soc.* **105**:5095–5099.

Mancini, A., De Marinis, L., Calabro, F., Sciuto, R., Oradei, A., Lippa, S., Sandric, S., Littarru, G. P., and Barbarino, A., 1989, Evaluation of metabolic status in amiodarone-induced thyroid disorders: Plasma coenzyme Q10 determination, *J. Endocrinol. Invest.* **12**:511–516.

Matsura, T., Yamada, K., and Kawasaki, T., 1991, Changes in the content and intracellular distribution of coenzyme Q homologs in rabbit liver during growth, *Biochim. Biophys. Acta* **1083**:277–282.

Mitchell, P., 1975, Protonmotive redox mechanism of the cytochrome bc_1 complex in the respiratory chain: Protonmotive ubiquinone cycle, *FEBS Lett.* **56**:1–6.

Mitchell, P., 1976, Possible molecular mechanisms of the protonmotive function of cytochrome systems, *J. Theor. Biol.* **62**:327–367.

Mohr, D., Bowry, V. W., and Stocker, R., 1992, Dietary supplementation with coenzyme Q10 results in increased levels of ubiquinol-10 within circulating lipoproteins and increased resistance of human low-density lipoprotein to the initiation of lipid peroxidation, *Biochim. Biophys. Acta* **1126**:247–254.

Morimoto, H., Imada, I., and Goto, G., 1969, Ubiquinone and related compounds. XV. Photochemical reaction of ubiquinone-7, *Liebigs Ann. Chem.* **729**:184–192.

Morimoto, H., Imada, I., and Goto, G., 1970, Ubiquinone and related compounds. XVI. Photo-oxidation of ubiquinone-7, *Liebigs Ann. Chem.* **735**:65–71.

Mortensen, S. A., 1993, Perspectives on therapy of cardiovascular diseases with coenzyme Q_{10} (ubiquinone), *Clin. Invest.* **71**:S116–S123.

Mukai, K., Kikuchi, S., and Urano, S., 1990, Stopped-flow kinetic study of the regeneration reaction of tocopheroxyl radical by reduced ubiquinone-10 in solution, *Biochim. Biophys. Acta* **1035**:77–82.

Mukai, K., Itoh, S., and Morimoto, H., 1992, Stopped-flow kinetic study of vitamin E regeneration reaction with biological hydroquinones (reduced forms of ubiquinone, vitamin K, and tocopherolquinone) in solution, *J. Biol. Chem.* **267**:22277–22281.

Murphy, M. P., Krueger, M. J., Sablin, S. O., Ramsay, R. R., and Singer, T. P., 1995, Inhibition of complex I by hydrophobic analogues of N-methyl-4-phenylpyridinium (MPP^+) and the use of an ion-selective electrode to measure their accumulation by mitochondria and electron-transport particles, *Biochem. J.* **306:**359–365.

Muscari, C., Biagetti, L., Stefanelli, C., Giordano, E., Guarnieri, C., and Caldarera, C. M., 1995, Adaptive changes in coenzyme Q biosynthesis to myocardial reperfusion in young and aged rats, *J. Mol. Cell Cardiol.* **27:**283–289.

Nohl, H., and Jordan, W., 1986, The mitochondrial site of superoxide production, *Biochem. Biophys. Res. Commun.* **138:**533–539.

Norling, B., Glazek, E., Nelson, B. D., and Ernster, L., 1974, Studies with ubiquinone-depleted submitochondrial particles. Quantitative incorporation of small amounts of ubiquinone and its effects on the NADH and succinate oxidase activities, *Eur. J. Biochem.* **47:**475–482.

Ogasahara, S., Nishikawa, Y., Yorifuji, S., Soga, F., Nakamura, Y., Takahashi, M., Hashimoto, S., Kono, N., and Tarui, S., 1986, Treatment of Kearns-Sayre syndrome with coenzyme Q10, *Neurology* **36:**45–53.

Ohnishi, T., and Trumpower, B. L., 1980, Differential effects of antimycin on ubisemiquinone bound in different environments in isolated succinate–cytochrome *c* reductase complex, *J. Biol. Chem.* **255:**3278–3284.

Olson, R. E., and Rudney, H., 1983, Biosynthesis of ubiquinone, *Vitam. Horm.* **40:**1–43.

Olsson, J. M., Eriksson, L. C., and Dallner, G., 1991, Lipid compositions of intracellular membranes isolated from rat liver nodules in Wistar rats, *Cancer Res.* **51:**3774–3780.

Pahl, H. L., and Baeuerle, P. A., 1994, Oxygen and the control of gene expression, *BioEssays* **16:**497–502.

Pedersen, S., Tata, J. R., and Ernster, L., 1963, Ubiquinone (coenzyme Q) and the regulation of basal metabolic rate by thyroid hormones, *Biochim. Biophys. Acta* **69:**407–409.

Ramsay, R. R., and Singer, T. P., 1992, Relation of superoxide generation and lipid peroxidation to the inhibition of NADH-Q oxidoreductase by rotenone, piericidin A, and MPP^+, *Biochem. Biophys. Res. Commun.* **189:**47–52.

Rich, P. R., Jeal, A. E., Madgwick, S. A., and Moody, A. J., 1990, Inhibitor effects on redox-linked protonations of the *b* haems of the mitochondrial bc_1 complex, *Biochim. Biophys. Acta* **1018:**29–40.

Salerno, J. C., and Ohnishi, T., 1980, Studies on the stabilized ubisemiquinone species in the succinate–cytochrome *c* reductase segment of the intact mitochondrial membrane system, *Biochem. J.* **192:**769–781.

Salerno, J. C., Osgood, M., Liu, Y. J., Taylor, H., and Scholes, C. P., 1990, Electron nuclear double resonance (ENDOR) of the $Qc^{-}\cdot$ ubisemiquinone radical in the mitochondrial electron transport chain, *Biochemistry* **29:**6987–6993.

Sastry, P. S., 1961, Distribution of coenzyme Q in rat liver cell fractions, *Nature* **189:**577.

Sen, C. K., and Packer, L., 1996, Antioxidant and redox regulation of gene transcription, *FASEB J.* **10:**709–720.

Shigenaga, M. K., Hagen, T., and Ames, B. N., 1994, Oxidative damage and mitochondrial decay in aging, *Proc. Natl. Acad. Sci. USA* **91:**10771–10778.

Söderberg, M., Edlund, C., Kristensson, K., and Dallner, G., 1990, Lipid compositions of different regions of the human brain during aging, *J. Neurochem.* **54:**415–423.

Sottocasa, G., Kuylenstierna, B., Ernster, L., and Bergstrand, A., 1967, An electron-transport system associated with the outer mitochondrial membrane of liver mitochondria: A biochemical and morphological study, *J. Cell Biol.* **32:**415–438.

Stadtman, E. R., 1993, Oxidation of free amino acids and amino acid residues in proteins by radiolysis and by metal-catalyzed reactions, *Annu. Rev. Biochem.* **62:**797–821.

Sterling, K., 1986, Direct thyroid hormone activation of mitochondria: The role of adenine nucleotide translocase, *Endocrinology* **119:**292–295.

Sterling, K., Milch, P. O., Brenner, M. A., and Lazarus, J. H., 1977, Thyroid hormone action: The mitochondrial pathway, *Science* **197:**996–999.

Stocker, R., and Suarna, C., 1993, Extracellular reduction of ubiquinone-1 and -10 by human Hep G2 and blood cells, *Biochim. Biophys. Acta* **1158:**15–22.

Stocker, R., Bowry, V. W., and Frei, B., 1991, Ubiquinol-10 protects human low density lipoprotein more efficiently against lipid peroxidation than does alpha-tocopherol, *Proc. Natl. Acad. Sci. USA* **88:**1646–1650.

Sun, I. L., Sun, E. E., and Crane, F. L., 1992, Stimulation of serum-free cell proliferation by coenzyme Q, *Biochem. Biophys. Res. Commun.* **189:**8–13.

Takahashi, T., Yamaguchi, T., Shitashige, M., Okamoto, T., and Kishi, T., 1995, Reduction of ubiquinone in membrane lipids by rat liver cytosol and its involvement in the cellular defence system against lipid peroxidation, *Biochem. J.* **309:**883–890.

Takayanagi, R., Takeshige, K., and Minakami, S., 1980, NADH- and NADPH-dependent lipid peroxidation in bovine heart submitochondrial particles. Dependence on the rate of electron flow in the respiratory chain and an antioxidant role of ubiquinol, *Biochem. J.* **192:**853–860.

Takeshige, K., and Minakami, S., 1975, Reduced nicotinamide adenine dinucleotide phosphate-dependent lipid peroxidation by beef heart submitochondrial particles, *J. Biochem.* **77:**1067–1073.

Takeshige, K., and Minakami, S., 1979, NADH- and NADPH-dependent formation of superoxide anions by bovine heart submitochondrial particles and NADH-ubiquinone reductase preparation, *Biochem. J.* **180:**129–135.

Takeshige, K., Takayanagi, K., and Minakami, S., 1980, Reduced coenzyme Q10 as an antioxidant of lipid peroxidation in bovine heart mitochondria, in *Biomedical and Clinical Aspects of Coenzyme Q,* Vol. 2 (Y. Yamamura, K. Folkers, and T. Ito, eds.), pp. 15–25, Elsevier, Amsterdam.

Thelin, A., Peterson, E., Hutson, J. L., McCarthy, A. D., Ericsson, J., and Dallner, G., 1994, Effect of squalestatin 1 on the biosynthesis of the mevalonate pathway lipids, *Biochim. Biophys. Acta* **1215:**245–249.

Trumpower, B. L., 1990, The protonmotive Q cycle. Energy transduction by coupling of proton translocation to electron transfer by the cytochrome bc_1 complex, *J. Biol. Chem.* **265:**11409–11412.

Turrens, J. F., Alexandre, A., and Lehninger, A. L., 1985, Ubisemiquinone is the electron donor for superoxide formation by complex III of heart mitochondria, *Arch. Biochem. Biophys.* **237:**408–414.

von Jagow, G., Ljungdahl, P. O., Ohnishi, T., and Trumpower, B. L., 1984, An inhibitor of mitochondrial respiration which binds to the cytochrome *b* and displaces quinone from the iron–sulfur protein of the cytochrome bc_1 complex, *J. Biol. Chem.* **259:**6318–6326.

Watts, G. F., Castelluccio, C., Rice-Evans, C. A., Taub, N. A., Baum, H., and Quinn, P. J., 1993, Plasma coenzyme Q (ubiquinone) concentrations in patients treated with simvastatin, *J. Clin. Pathol.* **46:**1055–1057.

Whitaker, S. J., Powell, S. N., and McMillan, T. J., 1991, Molecular assays of radiation-induced DNA damage, *Eur. J. Cancer* **27:**922–928.

Willis, R. A., Folkers, K., Tucker, J. L., Ye, C. Q., Xia, L. J., and Tamagawa, H., 1990, Lovastatin decreases coenzyme Q levels in rats, *Proc. Natl. Acad. Sci. USA* **87:**8928–8930.

Wolf, D. E., Hoffman, C. H., Trenner, N. R., Arison, B. H., Shunk, C. H., Linn, B. O., McPherson, J. F., and Folkers, K., 1958, Coenzyme Q. Structure studies on the coenzyme Q group, *J. Am. Chem. Soc.* **80:**4752.

Wolff, S. P., Garner, A., and Dean, R. T., 1986, Free radicals, lipids, and protein degradation, *Trends Biochem. Sci.* **11:**27–31.

Yamamoto, M., Sato, T., Anno, M., Ujike, H., and Takemoto, M., 1987, Mitochondrial myopathy, encephalopathy, lactic acidosis, and stroke-like episodes with recurrent abdominal symptoms and coenzyme Q10 administration, *J. Neurol. Neurosurg. Psychiatry* **50:**1475–1481.

Yamamoto, Y., Kawamura, M., Tatsuno, K., Yamashita, S., Niki, E., and Naito, C., 1991, Formation of lipid hydroperoxides in the cupric ion-induced oxidation of plasma and low density lipoprotein, in *Oxidative Damage and Repair: Chemical, Biological, and Medical Aspects* (K. J. A. Davies, ed.), pp. 287–291, Pergamon Press, New York.

Yu, C. A., and Yu, L., 1981, Ubiquinone-binding proteins, *Biochim. Biophys. Acta* **639:**99–128.

Zhang, Y., Marcillat, O., Giulivi, C., Ernster, L., and Davies, K. J. A., 1990, The oxidative inactivation of mitochondrial electron transport chain components and ATPase, *J. Biol. Chem.* **265:**16330–16336.

Zhang, Y., Aberg, F., Appelkvist, E.-L., Dallner, G., and Ernster, L., 1995, Uptake of dietary coenzyme Q supplement is limited in rats, *J. Nutr.* **125:**446–453.

Zhang, Y., Turunen, M., and Appelkvist, E.-L., 1996, Restricted uptake of dietary coenzyme Q is in contrast to the unrestricted uptake of α-tocopherol into rat organs and cells, *J. Nutr.* **126:**2089–2097.

Zwizinski, C. W., and Schmid, H. H. O., 1992, Peroxidative damage to cardiac mitochondria: Identification and purification of modified adenine nucleotide translocase, *Arch. Biochem. Biophys.* **294:**178–183.

Part VI

Specific Tissues

Chapter 18

Sources and Effects of Reactive Oxygen Species in Plants

C. Jacyn Baker and Elizabeth W. Orlandi

1. INTRODUCTION

Although basic oxygen metabolism is similar in plants and animals, there are several aspects unique to plants. Plants have photosynthetic electron transport in the chloroplast that produces oxygen and is accompanied by the production of reactive oxygen species (ROS). Other processes unique to plants that involve ROS are lignification of plant cell walls, which is important in normal growth as well as pathogenesis, and senescence, which is essential to fruit ripening, seed production, and overwintering. We have used a broad definition of the term ROS to include reactive species in which the active moiety involves oxygen (i.e., phenoxy radicals involved in lignin synthesis and acyl peroxides involved in membrane peroxidation).

In this review we will explore the many processes in plant physiology where ROS metabolism appears to be strongly associated. The study of ROS involvement in plant processes is being conducted in several laboratories and we are not able to address the findings of each group in the scope of this review. However, we will touch on a wide spectrum of topics currently of interest in plant systems. In addition, we have included a discussion of the use of transgenic plants and the unique insights they can provide in studies of oxidative stress.

C. Jacyn Baker U.S. Department of Agriculture, Agricultural Research Service, Molecular Plant Pathology Laboratory, Beltsville, Maryland 20705. **Elizabeth W. Orlandi** Department of Microbiology, University of Maryland, College Park, Maryland 20742-5815.

Reactive Oxygen Species in Biological Systems, edited by Gilbert and Colton. Kluwer Academic / Plenum Publishers, New York, 1999.

2. ROLE OF REACTIVE OXYGEN SPECIES IN NORMAL METABOLISM

2.1. Photosynthesis

The light phase of photosynthesis involves several steps: harvesting of the sunlight's energy captured in highly energized pigments, the release of molecular oxygen from water, and the flow of electrons through a series of electron donors and acceptors producing ATP and NADPH. A brief overview of the light phase of photosynthesis and the structure of the chloroplast will be helpful in understanding how ROS production is an integral part of this process and how the chloroplast protects itself from ROS.

2.1.1. Overview

The chloroplast has a double membrane, similar to the mitochondria, as well as a vast inner thylakoid membrane on which the light portion of photosynthesis takes place (Figure 1). The thylakoid contains the proteins and chlorophyll molecules associated with the two photosystems, PS I and PS II, as well as an ATP synthase. The two photosystems and intermediate electron carriers are chemically linked together in series to drive electrons from H_2O to $NADP^+$ (Figure 1). Absorption of light and excitation of PS II generates a strong oxidant that oxidizes water to oxygen. The reductant formed injects electrons into a series of electron transport carriers linking to PS I. Excitation of PS I generates a strong reductant that transfers electrons through a series of ferrodoxin (FD) carriers to $NADP^+$.

During the normal daily photoperiod, there will be periods of "photooxidative stress" when a plant is exposed to more light energy than it can handle. This results in a temporary period of overexcited pigments and an overload on the reduced electron carriers that can damage the photosynthetic apparatus if not controlled. The production of ROS during photooxidative stress is primarily associated with three processes: (1) singlet oxygen production by pigments, (2) superoxide production by electron carriers, and (3) acyl peroxide production in the thylakoid membrane, which will be discussed in more detail under senescence (Section 2.3). Fortunately, the chloroplast is able to scavenge most of the ROS produced under normal conditions as will be discussed below. However, under chronic stressed conditions, ROS accumulation may exceed the scavenging capacity in the chloroplast and severe photoinhibition can occur.

2.1.2. Singlet Oxygen Production by Pigments

The pigment molecules of the photosystems are viewed as part of an antenna system that absorbs light energy and transfers it to the different reaction centers containing the initial electron acceptors and donors. During periods of high light intensity, these antenna systems become overloaded and electrons cannot be passed on quickly enough to keep up with the excitation by light. During these periods, a fraction of the chlorophyll molecules remain excited for longer periods of time and may stand a greater chance of converting to the triplet state (Figure 2A). This is not desirable because they can now transfer energy to molecular oxygen, forming singlet oxygen (Figure 2B).

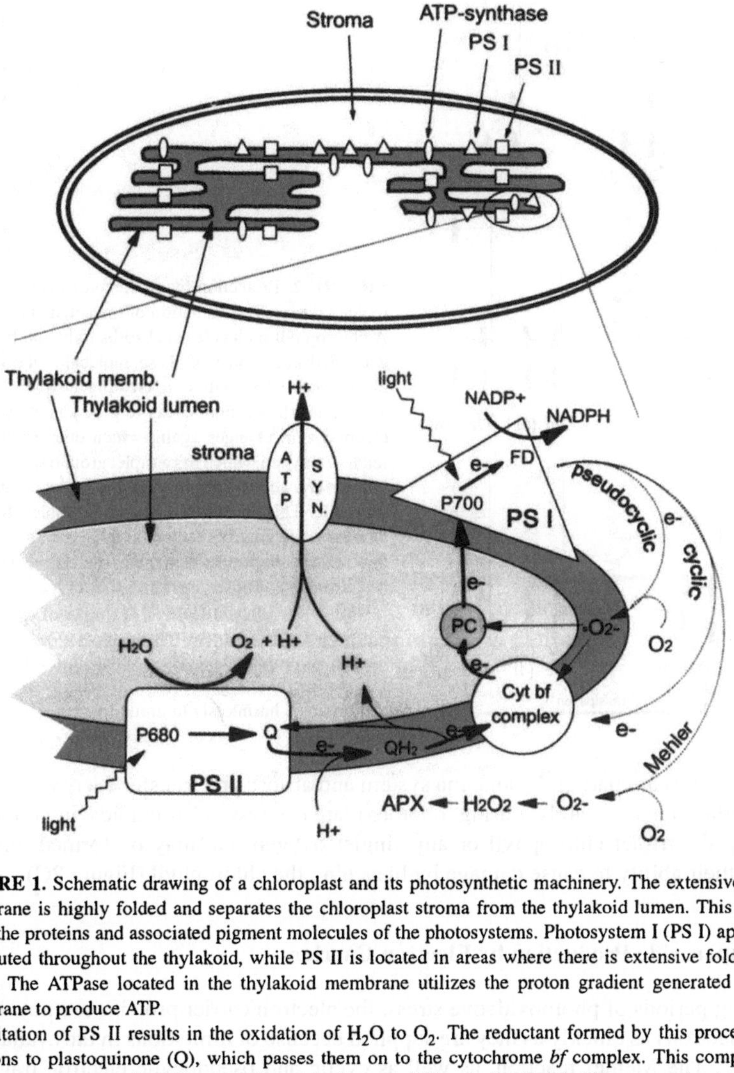

FIGURE 1. Schematic drawing of a chloroplast and its photosynthetic machinery. The extensive thylakoid membrane is highly folded and separates the chloroplast stroma from the thylakoid lumen. This membrane binds the proteins and associated pigment molecules of the photosystems. Photosystem I (PS I) appears to be distributed throughout the thylakoid, while PS II is located in areas where there is extensive folding, called *grana*. The ATPase located in the thylakoid membrane utilizes the proton gradient generated across the membrane to produce ATP.

Excitation of PS II results in the oxidation of H_2O to O_2. The reductant formed by this process donates electrons to plastoquinone (Q), which passes them on to the cytochrome *bf* complex. This complex passes electrons to plastocyanin (PC), a small protein that can diffuse in the thylakoid lumen and act as a mobile electron carrier to PS I. Excitation of PS I generates a strong reductant that transfers electrons through a series of ferrodoxin (FD) carriers to $NADP^+$. Protons that are translocated to the lumen in the process are used for photophosphorylation by the ATP synthase.

During periods of high light intensity the $NADPH/NADP^+$ ratio can increase to inhibitory levels. The Mehler reaction as well as cyclic and pseudocyclic electron transport provide alternate paths to relieve the highly reductive status. Cyclic transport donates electrons directly from FD to the cytochrome *bf* complex, while the Mehler reaction and pseudocyclic transport cycle utilize molecular oxygen to produce superoxide that can recycle electrons to the cytochrome *bf* complex or dismutate, forming hydrogen peroxide, which is scavenged by ascorbate peroxidase.

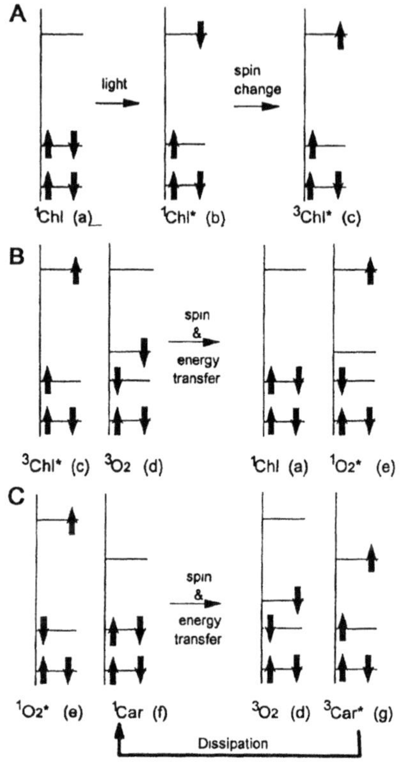

FIGURE 2. Protection by carotenoids against singlet oxygen during overexcitation of the antenna pigments. A chlorophyll molecule (a) absorbs light and becomes excited (b). In some of these molecules the excited electron will change its spin, creating a triplet state (c). The triplet state cannot decay to ground state until the electron spin changes again, which does not happen readily. Oxygen, which has a triplet ground state (d), can interact and accept the energy of the triplet chlorophyll molecule. This allows the chlorophyll molecule to return to its singlet ground state (a). However, the oxygen is now in its highly reactive singlet state (e), which is mobile and can cause damage in the chloroplast.

Carotene is singlet in ground state (f) and can interact with either triplet chlorophyll or singlet oxygen creating the triplet carotenoid (g). This excited molecule is harmless because its energy level is below singlet oxygen and can dissipate harmlessly to ground state.

Carotenoids are part of the antenna system and absorb and transfer energy. They can also dissipate energy safely during photooxidative stress. Carotenoids intervene by quenching the triplet chlorophyll or any singlet oxygen that may be formed, thereby reducing their ability to cause damage by bleaching the chlorophyll (Figure 2C).

2.1.3. Superoxide Production by Electron Carriers

During periods of photooxidative stress, the electron carrier proteins are not able to transfer electrons as quickly as they are supplied because of limitations in the availability of $NADP^+$. The Mehler reaction, as well as cyclic and pseudocylic electron transport (Figure 1), appear to be mechanisms by which molecular oxygen serves as an electron acceptor and damage to the photosynthetic components, or photoinhibition, can be avoided (Badger, 1985; Osmond and Grace, 1995).

Mehler (1951) and Good and Hill (1955) first described the photoreduction of oxygen by chloroplasts. The primary site for this photoreduction appears to be PS I (Figure 1) where the electrons can be passed directly to oxygen. Asada and Badger (1984) demonstrated a tight link between the superoxide formed by the Mehler reaction and its dismutation to hydrogen peroxide which was then scavenged by ascorbate peroxidase.

In cyclic transport, electrons are transferred from ferrodoxin back to the cytochrome *bf* complex and are recycled through the electron transport scheme (Figure 1). Takahashi and Asada (1988) demonstrated that a "pseudocyclic" transport also exists in which PS I donates electrons directly to molecular oxygen and the resulting superoxide is able to donate electrons to components of the cytochrome *bf* complex or plastocyanin carriers (Takahashi *et al.*, 1980).

Pseudocyclic electron transport and the Mehler reaction reduce the levels of reducing power by bypassing the terminal NADPH production, thereby increasing the $NADP^+$/NADPH ratio. Additionally, NADPH is expended in the ascorbate/glutathione cycle to scavenge the superoxide or hydrogen peroxide that is produced (see Section 2.1.4b and Figure 3). Thus, the production of ROS via these oxygen reduction pathways results in an increase in electron acceptors, helping to protect the photosynthetic process from photoinhibition (Sonike, 1996; Asada, 1994).

2.1.4. Reactive Oxygen Scavenging Mechanisms

As mentioned above, the carotenoids are important in regulation of singlet oxygen levels in the chloroplast. Scavenging of superoxide and hydrogen peroxide to lower their concentrations is essential to maintaining photosynthetic activity. The primary scavenging systems of plant cells will be discussed here, with greatest emphasis on those found in the chloroplast.

2.1.4a. Superoxide. Superoxide is scavenged via the disproportionation reaction catalyzed by superoxide dismutase that results in the production of hydrogen peroxide. There are three major types of SOD that differ mainly in their prosthetic metals: Cu/Zn, Mn, and Fe. Plants usually have a Cu/ZnSOD in the cytosol, a MnSOD in the mitochondria, and Cu/Zn and/or FeSOD in the chloroplast. SOD plays a very important role in reducing the superoxide levels in chloroplasts. Asada (1994) suggested that SOD reduces the level of superoxide from ~22 μM to ~2.4 nM in the stroma and lumen of chloroplasts, thus significantly reducing the potential for oxidative damage.

2.1.4b. Hydrogen Peroxide. Hydrogen peroxide for the most part is scavenged by either catalase [Eq. (1)] or peroxidase [Eq. (2)]:

$$H_2O_2 + H_2O_2 \rightarrow O_2 + 2\,H_2O \tag{1}$$

$$H_2O_2 + 2\,AH_2 \rightarrow 2\,H_2O + \cdot AH \tag{2}$$

Little catalase has been found in the chloroplast and because of its high K_m (generally > 10–100 mM) it would not function efficiently with the micromolar levels to which hydrogen peroxide must be reduced to avoid oxidative damage. The peroxidative mechanism generally requires a reductant (AH_2), which in chloroplasts and the cytoplasm is ascorbate. Levels of ascorbate can reach 10 mM in the chloroplast (Asada, 1994). Ascorbate peroxidase appears to be unique to plants. The K_m of this enzyme for hydrogen peroxide is much lower than that of catalase (5–100 μM), allowing it to effectively scavenge low levels of hydrogen peroxide.

Ascorbate peroxidase is part of the ascorbate–glutathione cycle, which involves successive enzymatic oxidations and reductions of ascorbate, glutathione, and NADPH, as shown in Figure 3. A portion of the ascorbate peroxidase appears to be membrane bound allowing it to effectively protect membranes and regenerate components of other membrane-associated antioxidant cycles such as the xanthophyll cycle components and tocopherol.

2.2. Lignification

Another metabolic process in which ROS may play an important role is lignification of plant cell walls. Lignification occurs during normal plant development as well as in response to wounding or stress through a series of steps involving the production of phenolic lignin radicals and the final polymerization of the radicals into lignin (Figure 4). The exact mechanism of the oxidation step resulting in phenolic radicals, however, is still unresolved. Freudenberg (1968) reported that a fungal polyphenol oxidase (often referred to as *laccase* because the first such enzyme was isolated from the Japanese lacquer tree, *Rhus vernicifera*) was able to form ligninlike polymers *in vitro* from precursors such as coniferyl alcohol. However, the involvement of laccase was discounted when the purified enzyme could not oxidize lignin precursors (Nakamura, 1967) and it

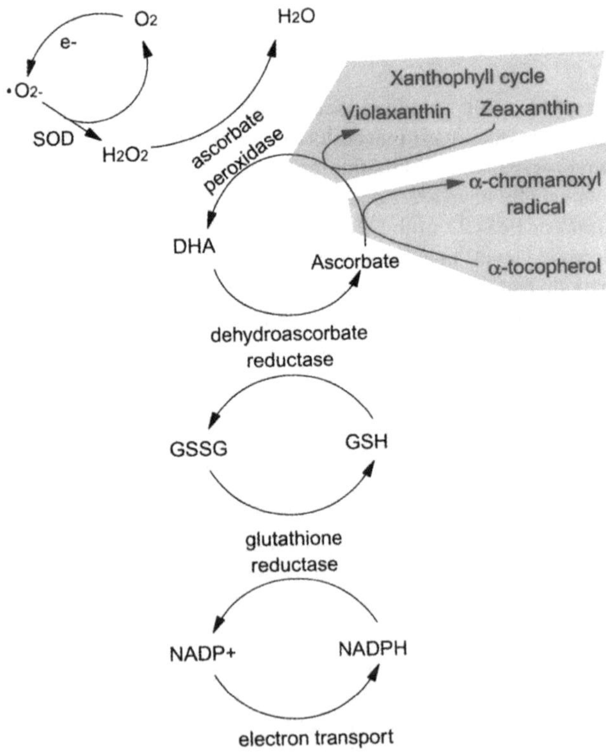

FIGURE 3. The ascorbate/glutathione cycle and associated reactions. DHA, dehydroascorbate; GSSG, oxidized glutathione; GSH, reduced glutathione; SOD, superoxide dismutase.

could not be found in lignifying xylem tissue (Harkin and Obst, 1973). It was then proposed that peroxidases and hydrogen peroxide were exclusively involved in the oxidative dehydrogenation of phenolics into lignin (Figure 4) (Gross *et al.*, 1977; Higuchi, 1985). Recent studies have also localized hydrogen peroxide in lignifying tissues (Czaninski *et al.*, 1993; Olson and Varner, 1993). However, the concept of plant laccases has been resurrected as a result of work with a purified laccase from suspension culture cells of *Acer pseudoplatanus* where lignin precursors were oxidized to form ligninlike polymers *in vitro* (Sterjiades *et al.*, 1992). This concept was further supported by Bao *et al.* (1993) who found laccase activity correlated with time and site of lignin deposition in differentiating xylem of loblolly pine.

As will be discussed in following sections, hydrogen peroxide appears to be produced under stress conditions. It therefore seems feasible that both laccase- and peroxidase-dependent mechanisms may operate under stress conditions and perhaps could play greater or lesser roles in different tissues during different periods of development.

2.3. Senescence

Senescence is an essential part of the life cycle of perennial plants. It allows fruit to ripen, seeds to develop, and leaves on deciduous plants to fall off so that the plant can overwinter. A key process of senescence is membrane deterioration that results in the loss of membrane integrity in cells and cellular organelles. Increasing evidence suggests that lipid peroxidation via reactive oxygen species contributes to the membrane deterioration (Thompson *et al.*, 1991). Studies using electron microscopy have revealed that peroxisomes, organelles that carry out highly oxidative metabolisms, increase in number during senescence, leading several authors to suggest that they may play a role in senescence through their production of reactive oxygen (Pastori and del Rio, 1994; Palma *et al.*, 1991).

Leaf peroxisomes are present in photosynthetic tissues that carry out the major reactions of photorespiration. The organelles contain glycollate-pathway enzymes that produce significant levels of hydrogen peroxide (Tolbert, 1981). Xanthine oxidase and urate oxidase, which produce superoxide as a result of nucleic acid breakdown, have been demonstrated in peroxisomes (Sandalio *et al.*, 1987, 1988; del Rio *et al.*, 1992). It would seem reasonable that peroxisomal metabolism could generate substantial levels of reac-

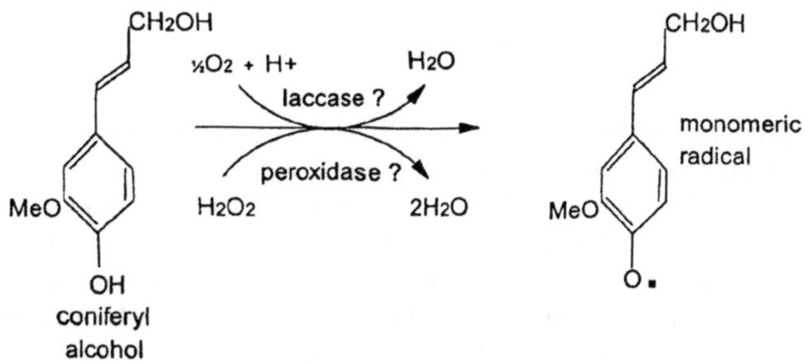

FIGURE 4. Two reported mechanisms for the production of phenoxy radical monomers used in lignin synthesis. Although coniferyl alcohol is shown, sinapyl and coumaryl alcohols are also commonly involved in this process.

tive oxygen during periods of senescence. In nonsenescent tissue, peroxisomes contain high concentrations of catalase, sometimes in a nearly crystalline state (Bilger and Björkman, 1991). During oxidative stress, antioxidant enzymes usually increase. However, Palma and del Rio (1994) found that during senescence, catalase activity decreased and MnSOD activity increased in leaf peroxisomes of pea (*Pisum sativum* L.). They hypothesized that these changes in antioxidant activities, coupled with increased oxidase activity, lead to increased H_2O_2 levels that could contribute to foliar senescence.

Lipid peroxidation and ROS production associated with senescence are partially attributed to increases in lipoxygenase activity (Lupu *et al.*, 1980). Lynch and Thompson (1984; Thompson *et al.*, 1991) found that as bean cotyledons age and eventually fall off, the lipoxygenase levels increase as do the levels of ROS. During senescence there is a decline in membrane phospholipid and the release of free unsaturated fatty acids that could be substrate for lipoxygenase. Lipoxygenase activity leads to the production of lipid radicals and, ultimately, lipid hydroperoxides (Chamulitrat and Mason, 1989). These lipoxyperoxides are, in effect, forms of ROS that can continue lipid peroxidation (Chamulitrat and Mason, 1989). In addition, lipoxygenase activity may give rise to superoxide and hydrogen peroxide that would continue lipid peroxidation through the formation of hydroxyl radicals (Thompson *et al.*, 1991).

3. ROLE OF REACTIVE OXYGEN SPECIES IN STRESSED METABOLISM

Plants are exposed to a number of environmental stresses in nature, such as excessive UV light, pollutants, pesticides, heavy metals, lack of nutrients, or attack by pathogens. With the exception of the application of herbicides, most natural abiotic stresses are generally chronic with exposures lasting or reoccurring over a long period of time. Although these stress conditions are often studied in isolation, it is important to realize that a plant is frequently subjected to multiple stresses at any given time. Exposure to one stressor often triggers defense mechanisms that will either protect the plant or alter its response to subsequent stressors. For example, plants that are prestressed by lack of nutrients or by disease were susceptible to lower levels of ozone than plants that were not previously stressed (Mehlhorn and Wellburn, 1994).

3.1. Abiotic Stresses

3.1.1. Ozone

Ozone is produced by the action of sunlight (UV) on air containing nitrogen oxides and volatile organic compounds. Other airborne pollutants, such as peroxyacetyl nitrate and hydroperoxides also appear to affect plant health. However, ozone has received the majority of attention. Early visible injury related to ozone toxicity can include wilting, watersoaking of tissue, and localized epidermal collapse. This is often followed by white or red necrotic spots on the foliage. When plants receive toxic levels of ozone (parts per billion range), they show signs of accelerated senescence.

Ozone is soluble in water. Therefore, once it enters through the stomata, it is thought to move into solution and generate a spectrum of ROS as well as organic radicals

(Mehlhorn *et al.*, 1990). Protection from ozone was obtained by pretreatment of the plants with antioxidants such as polyamines (Bors *et al.*, 1989), ascorbate (Freebairn and Taylor, 1960), and glutathione (Freebairn, 1957). One of the first responses to ozone exposure are changes in membrane permeability similar to lipid peroxidation associated with senescence. Ethylene production also increases with ozone exposure (Kangasjarvi *et al.*, 1994). Sensitivity to ozone has been correlated with rates of ethylene production, and resistant plants produce less ethylene. However, preexposure to ethylene appears to increase resistance to ozone by increasing antioxidant systems.

3.1.2. Drought

Water deficits vary in intensity from daily mild deficits related to high light and the resultant transpiration losses, to more acute drought conditions. Water deficits reduce the photosynthesis rate via stomatal closure and restriction of CO_2 fixation leading to excess excitation energy within the chloroplast. This excess excitation energy, much of which gets diverted into ROS, can be dissipated by the various scavenging systems of the chloroplast described above, including carotenes, cyclic and pseudocyclic electron transport, and other scavenging systems. Several studies have measured increases in ROS, decreases in antioxidant mechanisms, and signs of oxidative stress both in the chloroplast and in other parts of the cell (Smirnoff, 1993; Price *et al.*, 1989). In peas, water-stressed plants showed up to 78% reduction in photosynthesis; catalase and enzymes of the ascorbate–glutathione cycle were depressed 70–80%; lipid peroxidation increased and there was evidence of protein oxidation (Moran *et al.*, 1994).

However, as Smirnoff (1993) pointed out, although much evidence is consistent with the view that these symptoms may be related to drought, the alternative view exists that the symptoms of oxidative stress such as lipid peroxidation and loss of antioxidant capacity may result from senescence caused by excessive experimental treatments. He pointed out that under water deficit conditions normally encountered by plants, even if ROS formation is increased, the scavenging mechanisms either have sufficient capacity or can be induced to protect the plant. Often experimental treatments can push the plants beyond their normal limits and reveal processes that are active but not otherwise observed in nature. This same comment could apply to most environmental stress studies.

3.1.3. Herbicides

Many of the herbicides commonly used operate through excessive production of ROS in the chloroplast that overloads the scavenging capacity of the plant. Severe photooxidation and photoinhibition result and the death of the plant follows (Halliwell, 1991). One of the best-known herbicides is paraquat, a bipyridylium herbicide that is able to penetrate into chloroplasts. It can act as a redox-cycling agent by accepting electrons directly from PS I prior to ferrodoxin and then passing them on to oxygen (Figure 5). Therefore, electrons from photosynthesis are being diverted into producing active oxygen at the same time that the production of NADPH, which is needed for ROS scavenging by the ascorbate–glutathione cycle, is reduced. Some plants that are paraquat-resistant show

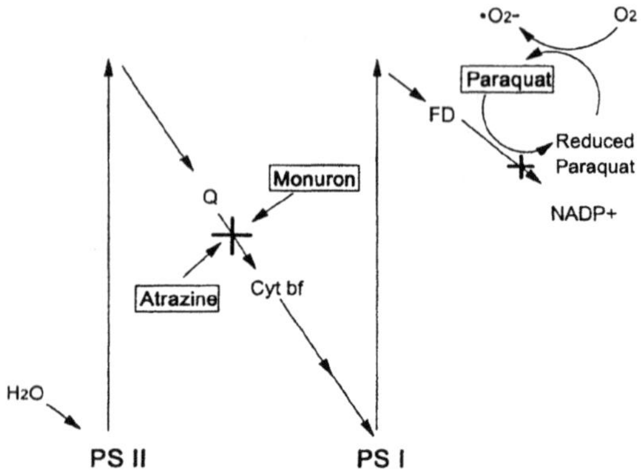

FIGURE 5. Interruption of photosynthetic electron flow by the herbicides atrazine, monuron, and paraquat.

increased activities of SOD and ascorbate–glutathione cycle enzymes (Shaaltiel *et al.* 1988).

Two other classes of herbicides, the triazines such as atrazine and the substituted ureas such as monuron, block electron transport in the chloroplast between Q and the cytochrome *bf* complex (Figure 5). This leads to photooxidation as described above in which singlet oxygen and superoxide would play a significant role in pigment bleaching, redox component destruction, and membrane deterioration (Dodge and Gillham, 1986).

3.2. Biotic Stresses

3.2.1. Introduction

Reactive oxygen is produced by plant cells on recognition of invading pathogens. Reactive oxygen production has been monitored in a variety of plant tissues including tubers, intact leaves, cell suspensions, and protoplasts. Superoxide and hydrogen peroxide are the primary ROS measured during pathogenesis, although there have been some reports of hydroxyl ion production. Plant cells have been shown to produce ROS during interactions with fungi, bacteria, and viruses and there are many similarities between the various microbial interactions.

3.2.1a. Fungi. Doke and colleagues were the first to extensively study the production of ROS during plant–pathogen interactions. They monitored the production of superoxide in potato tuber disks (Doke, 1983a) and protoplasts (Doke, 1983b) treated with zoospores or hyphal wall components of the fungus *Phytophthora infestans*. Production of superoxide in potato leaves inoculated with *P. infestans* occurred in two stages prior to and after fungal penetration of leaf tissues (Chai and Doke, 1987a). The second stage of superoxide production occurred only in incompatible interactions that resulted in the hypersensitive response (HR). The HR is characterized by rapid plant cell death in cells adjacent to pathogens and is an effective resistance response that results in localized necrosis but does not allow development of extensive disease symptoms in plant tissues.

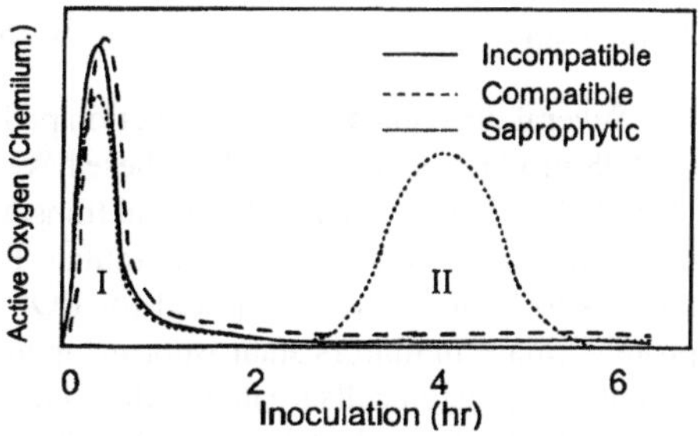

FIGURE 6. Active oxygen response of plant suspension cells during compatible (results in disease), incompatible (results in the hypersensitive response) and saprophytic (no symptoms) interactions with bacteria. I, phase I; II, phase II.

3.2.1b. Bacteria. ROS production during plant–bacteria interactions was first monitored in tobacco leaf disks (Adam *et al.*, 1989) and suspension cells (Keppler and Baker, 1989) inoculated with *Pseudomonas syringae* pv. *syringae*. Similar to fungus-infected potato leaves, the production of ROS in tobacco suspension cells inoculated with bacteria took place in two distinct phases (Figure 6) (Keppler *et al.*, 1989; Baker *et al.*, 1991). The first phase was an immediate response to compatible, incompatible, and saprophytic bacteria, whereas phase II occurred 1 1/2–3 hours after treatment with incompatible bacteria only. This two-phased response has also been monitored in suspension cells of other plant species (see review in Baker and Orlandi, 1995).

3.2.1c. Viruses. Fewer studies have measured ROS production during plant–virus interactions. Production of superoxide was monitored in tobacco leaves infected with tobacco mosaic virus (TMV) (Doke and Ohashi, 1988), and in cowpea leaves infected with tobacco ringspot virus and southern bean mosaic virus (El-Moshaty *et al.*, 1993). The production of ROS in TMV-infected leaf disks occurred in several cyclic phases over a 5-hr period (Doke and Ohashi, 1988).

3.2.1d. Elicitors. A variety of microbial components have been demonstrated to elicit ROS production in plant cells. Many of the elicitors prepared from pathogens have been crude extracts from fungal hyphae or intercellular fluids of infected plant tissues. The elicitor-active components of these preparations are often thought to be carbohydrates or glycoproteins. Studies with more highly purified microbial elicitors have demonstrated that a wide range of compounds have the potential to trigger ROS production during plant–pathogen interactions. Among these are a galactoglucomannan from *Colletotrichum lindemuthianum* (Anderson *et al.*, 1991), a proteoglucomannan from *Pyricularia oryzae* (Haga *et al.*, 1995), and harpin$_{Ea}$, a protein from *Erwinia amylovora* (Baker *et al.*, 1993b).

In addition to microbial elicitors, plant cells have been shown to produce ROS in response to their own cell wall fragments. Oligogalacturonides from plant cell walls were demonstrated to be efficient elicitors of ROS (Keppler *et al.*, 1988; Apostol *et al.*, 1989). These can be liberated from plant cell walls by pectin-degrading enzymes secreted by pathogens (Bateman and Bashman, 1976; Davis *et al.*, 1984). Plant cells are also sensitive to changes in medium osmolarity and mechanical perturbation of cells and both can result in the production of ROS (Glazener *et al.*, 1991; Qian *et al.*, 1993; Yahraus *et al.*, 1995).

3.2.2. Signal Transduction Events in ROS Production

Several different signal transduction pathways have been proposed for the activation of ROS production in plant cells. These have recently been reviewed by Low and Merida (1996). The components of the signal transduction pathway appear to vary depending on the pathogen or elicitor used to stimulate ROS production. However, one process that appears to be universally required for ROS elicitation is protein phosphorylation. The protein kinase inhibitors staurosporine and/or K-252a blocked the elicitation of ROS in various suspension cells treated with fungal elicitors (Schwacke and Hager, 1992; Viard *et al.*, 1994), oligogalacturonic acids (Chandra and Low, 1995), and harpin$_{Ea}$ (Baker *et al.*, 1993b). Similarly, both phases of ROS production induced by treatment of soybean and tobacco suspension cells with *Pseudomonas syringae* pv. *syringae* are blocked by K-252a (Orlandi and Baker, unpublished results). The majority of the above studies also demonstrated a requirement for Ca^{2+} efflux into the cell prior to ROS production.

GTP proteins are reportedly involved in the signaling events in potato tuber tissues treated with hyphal wall components from *Phytophthora infestans* (Kawakita and Doke, 1994), and both GTP proteins and phospholipase C activity are reportedly required for signal transduction after treatment of soybean suspension cells with oligogalacturonides (Legendre *et al.*, 1992). The stimulation of ROS production in soybean cells treated with oligogalacturonides and harpin$_{Ea}$ did not require phospholipase A, but a crude fungal elicitor from *Verticillium dahliae* showed an absolute requirement for this enzyme activity (Chandra *et al.*, 1996).

3.2.3. Source of ROS Production during Pathogenesis

Several enzymatic sources have been proposed as the source of ROS production during plant–pathogen interactions including NAD(P)H oxidases (Doke, 1985), peroxidases (Adam *et al.*, 1993; Peng and Kuc, 1992), xanthine oxidase (Montalbini, 1992), and lipoxygenase (Atkinson, 1993). Currently, the majority of studies on plant ROS production are focusing on the involvement of an NAD(P)H oxidase similar to that found in human neutrophils. Similar mechanisms of elicitation and signal transduction events are involved in human neutrophil and plant cell ROS production as reviewed by Low and Dwyer (1994). In addition, the ROS production in both systems is blocked by several of the same inhibitors (Levine *et al.*, 1994; Auh and Murphy, 1995). Recent studies have identified plant proteins immunoligically related to the p22 membrane-bound neutrophil oxidase component (Tenhaken *et al.*, 1995) and the cytosolic neutrophil oxidase components $p47^{phox}$ and $p67^{phox}$ (Dwyer *et al.*, 1996; Desikan *et al.*, 1996).

3.2.4. Roles of ROS Production in Pathogenesis

Because ROS can interact with and alter most of the biological building blocks of a cell, there are many potential roles for ROS in pathogenesis. ROS have been hypothesized to have direct effects on cellular components and processes and to act as a signal to elicit secondary reactions. We will briefly discuss the evidence for and against several potential roles for ROS during plant–pathogen interactions.

3.2.4a. Antimicrobial Activities. ROS are antimicrobial and it seems likely that they would play a role in limiting pathogen ingress. Exogenously added hydrogen peroxide or artificially generated superoxide have been shown to limit the growth of fungal (Peng and Kuc, 1992; Ouf *et al.*, 1993) and bacterial pathogens (Kiraly *et al.*, 1993). However, it is not clear that endogenously produced ROS plays an antimicrobial role during pathogenesis. Whereas Keppler *et al.* (1989) hypothesized that the viability of *P. syringae* was decreased by superoxide exposure during incompatible reactions, Minardi and Mazzucchi (1988) found no direct effect of superoxide on *P. syringae* in tobacco leaves. Likewise, Gönner and Schlösser (1993) found no antimicrobial role for ROS produced in oat leaf tissue but hypothesized that it actually facilitated colonization of plant tissue by the fungi *Drechslera* spp. Although ROS are clearly antimicrobial, it is possible that they do not accumulate to antimicrobial levels during some plant–pathogen interactions, perhaps because of antioxidant systems of plants and/or pathogens (Baker *et al.*, 1995, 1997).

3.2.4b. Antioxidant Increases. As discussed in previous sections of this review, antioxidant levels increase during stress responses of plant cells. Increased levels of antioxidant enzymes have also been demonstrated during incompatible plant–pathogen interactions. Recent studies have established a causal role for ROS production in increased antioxidant activity. *Arabidopsis thaliana* suspension cells showed measurable increases in glutathione synthesis in response to increased levels of endogenous hydrogen peroxide (May and Leaver, 1993). Similarly, Levine *et al.* (1994) demonstrated increases in glutathione *S*-transferase in soybean suspension cells during incompatible interactions between plant and bacterial cells as well as after exogenous addition of hydrogen peroxide.

3.2.4c. HR Induction. It has long been known that exogenously supplied ROS can induce plant cell death. However, production of ROS by plant cells is not consistently followed by cell death. As described in previous sections of this chapter, cells treated with various elicitors and bacteria produce an immediate burst of ROS production. However, such potent ROS elicitors as the saprophyte *Pseudomonas fluorescens* (Baker *et al.*, 1991), heat-killed *P. corrugata*, or $HgCl_2$ (Devlin and Gustine, 1992) do not cause hypersensitive cell death. Several studies have found a correlation between the later, more prolonged second phase of ROS production and the hypersensitive cell death that results from incompatible plant–pathogen interactions (Doke, 1983a; Adam *et al.*, 1989, 1990; Keppler *et al.*, 1989; Baker *et al.*, 1991, 1993a). This correlation seems to support the recent proposal that the ROS produced during the first few hours of plant–pathogen interactions is directly responsible for hypersensitive cell death (Levine *et al.*, 1994). However, a recent study with Tn*pho*A mutants of *P. s.* pv. *syringae* and *P. s.* pv. *fluorescens* demonstrated that hypersensitive cell death did not consistently follow this second phase of ROS production (Glazener *et al.*, 1996). In this study, tobacco cells treated with the mutant bacteria elicited a second phase ROS response comparable to that elicited by the wild-type bacteria but did not elicit hypersensitive cell death (Figure 7). Although several studies present corollary evidence supporting a role for phase II ROS production in the development of hypersensitive cell death in plants, it may not be a direct causal relationship.

3.2.4d. Cell Wall Strengthening. Hydrogen peroxide production during pathogenesis has also been proposed to play a role in cell wall strengthening. As described

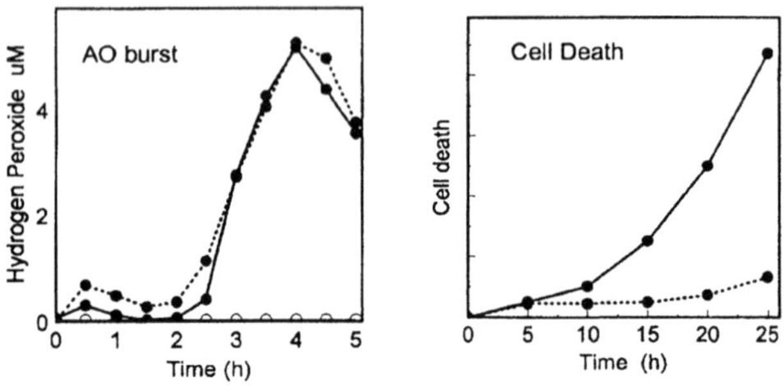

FIGURE 7. Tn*pho*A mutants (----) of *Pseudomonas syringae* pv. *syringae* caused a phase II burst of hydrogen peroxide nearly identical to the wild type (——). However, the mutants did not cause hypersensitive cell death, whereas the wild-type bacteria elicited cell death by approximately 10 h after inoculation.

above, hydrogen peroxide may be essential to the formation of lignin during normal and stress conditions. In addition, hydrogen peroxide produced during plant–bacteria interactions has been hypothesized to immobilize specific plant cell wall proteins through oxidative cross-linking (Bradley *et al.*, 1992; Wojtaszek *et al.*, 1995). Treatment with fungal elicitor induced the immobilization of specific wall proteins and almost completely inhibited the protoplastibility of soybean suspension cells (Brisson *et al.*, 1994). Therefore, the authors proposed that the oxidative cross-linking of cell wall proteins strengthened the cell wall and may play a role in protecting the cells against subsequent microbial attack.

3.2.4e. Phytoalexin Production. Phytoalexins are antimicrobial metabolites produced during many plant–pathogen interactions. Many of the earlier studies on ROS production by plant cells employed phytoalexin elicitors to induce ROS production (Doke, 1983a; Lindner *et al.*, 1988; Apostol *et al.*, 1989). The ability of known phytoalexin elicitors to induce ROS production led to speculation about the relationship between ROS production and the elicitation of phytoalexins. Although there have been many studies of this relationship, the results are conflicting and no clear consensus has been formed.

Exogenously added hydrogen peroxide or artificially generated superoxide or ·OH have been demonstrated to trigger phytoalexin production in soybean suspension cells, and hypocotyls (Montillet and Degousee, 1991; Degousee *et al.*, 1994; Gomez *et al.*, 1994). However, the addition of SOD and catalase did not inhibit the elicitation of the phytoalexin medicarpin in white clover suspension cells treated with *Pseudomonas corrugata* or with the abiotic elicitor $HgCl_2$ (Devlin and Gustine, 1992). Similarly, neither catalase, SOD, nor diphenyleneiodonium, an inhibitor of AO production, had an inhibitory effect on phytoalexin production by tobacco suspension cells treated with two fungal elicitins (Rusterucci *et al.*, 1996). Recently, Davis and colleagues were able to separate ROS-eliciting activity from phytoalexin-eliciting activity in a crude fungal elicitor preparation (Davis *et al.*, 1993), suggesting two separate elicitors for the two responses. In summary, although elicitation of ROS and phytoalexins is often correlated, and the exogenous addition of ROS can trigger the formation of phytoalexins, there is no clear causal relationship between the two defense responses during pathogenesis.

3.2.4f. Systemic Acquired Resistance. The pretreatment of plants with pathogens or pathogen-related agents can induce resistant reactions to subsequent challenges. This phenomenon of systemic acquired resistance has been hypothesized to involve the translocation of a highly mobile signal throughout the plant as a result of pretreatment. Several studies have investigated the involvement of ROS in the long-distance signaling leading to systemic acquired resistance. Chai and Doke (1987b) measured increased superoxide production in upper leaves of potato plants whose lower leaves had been treated with elicitors from *P. infestans*. These upper leaves showed enhanced superoxide production and resistance in response to subsequent challenge with *P. infestans* zoospores. These results led the authors to suggest that the signal resulting in systemic acquired resistance was also responsible for the systemic activation of superoxide production. Recently, it has been proposed that salicylic acid, a known inducer of systemic acquired resistance, binds to catalase and leads to increased levels of hydrogen peroxide in plant tissues (Chen *et al.*, 1993). The authors proposed that this increased hydrogen peroxide accumulation would lead to increased resistance to challenging pathogens. However, several studies have contradicted this hypothesis (Neuenschwander *et al.*, 1995; Rüffer *et al.*, 1995; Green and Fluhr, 1995) and a role for ROS in systemic acquired resistance is still under debate.

4. THE USE OF TRANSGENIC PLANTS TO STUDY OXIDATIVE STRESS

As described in previous sections, plants face numerous stresses caused by pathogens and various abiotic excesses or deficiencies. Despite the varied origins of these stressors, they have in common the triggering of oxidative damage in plant cells. Oxidative stress occurs when ROS accumulates to levels that the constitutive scavenging systems cannot regulate and is believed to be a major factor limiting plant productivity. In recent years, plant scientists have attempted to increase stress resistance by increasing cellular antioxidant levels. The use of transgenic plants that are endowed with genetic material from other plant species or that overexpress endogenous antioxidants has been an invaluable tool for studying the effects of antioxidant systems on stress resistance. Several reviews have been published on this topic in recent years (Rennenberg and Polle, 1994; Foyer *et al.*, 1994; Allen, 1995; Herouart *et al.*, 1993). We will touch on only a few recent studies relevant to this review.

4.1. Overexpression of Antioxidants

Transgenic plants have been made with enhanced levels of antioxidant enzymes, such as SOD, or of the enzymes that produce low molecular weight antioxidants, such as ascorbate. Many of these increases have been targeted to specific subcellular compartments. Overexpression of SOD in plants would be presumed to be a double-edged sword: It may provide increased scavenging of superoxide, although the resultant hydrogen peroxide may exceed other scavenging capacities and could actually lead to increased oxidative damage. Therefore, it is not surprising that the results of studies on transgenic plants overexpressing SOD have varied greatly (Allen, 1995). Many of these variations could be attributed to differences between plant species and the specific enzymes manipulated, as well as to the intensity of the challenge stress to which transgenic plants were subjected to measure changes in resistance.

In their study, Bowler *et al.* (1991) targeted overexpressed MnSOD to either the chloroplasts or mitochondria of tobacco plants and, likewise, targeted their challenge oxidative stresses to either of these subcellular locations. They found that large increases in SOD activity could preferentially protect the targeted organelles, whereas smaller increases could actually lead to increased damage. The type and location of the overexpressed enzyme as well as the existing scavenging mechanisms for the resultant hydrogen peroxide produced also impacted resistance. The importance of endogenous antioxidant systems in transgenic plants was recently illustrated by Sen Gupta *et al.* (1993). In this study, transgenic tobacco expressing a chloroplastic Cu/ZnSOD had enhanced tolerance of high light and cool temperatures. This enhanced stress resistance was apparently related to increased ascorbate peroxidase activity that accompanied the increased SOD activity.

4.2. Enhanced ROS Production

Transgenic plants have also been used to study the importance of hydrogen peroxide production as a mechanism of resistance to pathogens (Wu *et al.*, 1995). The authors genetically engineered potato plants to express fungal glucose oxidase in the apoplast and found increased hydrogen peroxidase production in leaf, tuber, and root tissues. The authors found increased resistance to the fungus *Phytophthora infestans* and the bacterial pathogen *Erwinia caratovora* sp. *caratovora*. The increased resistance that accompanied increased ROS production could be related to antimicrobial effects of the hydrogen peroxide or enhanced triggering of the defense responses discussed above such as phytoalexins.

5. CONCLUSIONS

Although ROS metabolism is similar in plants and animals, there are several unique processes in plants that involve ROS metabolism. In photosynthesis, ROS production plays a dual role. Under unstressed conditions, the chloroplast must be protected from damage by any singlet oxygen, superoxide, or hydrogen peroxide that may be produced as a result of excess light energy. However, during times of acute photooxidative stress, ROS production by pseudocyclic electron transport serves to protect the photosynthetic machinery from irreversible damage by providing an alternate acceptor for the electron flow. Lignification, which has been demonstrated to involve hydrogen peroxide and peroxidases to polymerize phenoxy radicals, fortifies the cell wall to provide structural support during normal growth as well as increased protection against pathogens.

Reactive oxygen production is believed to be involved in lipid peroxidation during senescence, a process that is vital to the survival and reproduction of plant species. In addition, ROS has been theorized to have a direct impact on disease resistance of plants and to act as a signal for other defense-related processes.

The presence of the cell wall, chloroplasts, and peroxisomes presents unique challenges in understanding the metabolism and roles of ROS in plant cells. One tool that may provide greater insights in this area is the transgenic plant. The genetic engineering of plants with modified ROS production or scavenging capabilities has greatly expanded our understanding of the intricate interactions of ROS metabolites and the enzymatic pathways involved. Because ROS production during plant stress is believed to be important in limiting plant growth and productivity, the development of plants with increased antioxidant mechanisms could have widespread implications for continued increases in worldwide food production.

6. REFERENCES

Adam, A., Farkas, T., Somlyai, G., Hevesi, M., and Kiraly, Z., 1989, Consequence of O_2 generation during a bacterially induced hypersensitive reaction in tobacco: Deterioration of membrane lipids, *Physiol. Mol. Plant Pathol.* **34:**13–26.

Adam, A., Ersek, T., Barna, B., and Kiraly, Z., 1990, Role of oxidative stress in plants on the development of necrosis induced by pathogens, *Radicals, Ions and Tissue Damage* **1990:**1–18.

Adam, A. L., Bestwick, C. S., Galal, A. A., Manninger, K., Barna, B., and Mansfield, J. W., 1993, What is the putative source of free radical generation during hypersensitive response in plants?, in *Oxygen Free Radicals and Scavengers in the Natural Sciences* (G. Mozsik, I. Emerit, J. Feher, B. Matkovics, and A. Vineze, eds.), pp. 35–43, Akademiai Kiado, Budapest.

Allen, R., 1995, Dissection of oxidative stress tolerance using transgenic plants, *Plant Physiol.* **107:**1049–1054.

Anderson, A., Rogers, K., Tepper, C., Blee, K., and Cardon, J., 1991, Timing of molecular events following elicitor treatment of plant cells, *Physiol. Mol. Plant Pathol.* **38:**1–13.

Apostol, I., Heinstein, P. F., and Low, P. S., 1989, Rapid stimulation of an oxidative burst during elicitation of cultured plant cells, *Plant Physiol.* **90:**109–116.

Asada, K., 1994, Production and action of active oxygen species in photosynthetic tissues, in *Causes of Photooxidative Stress and Amelioration of Defense Systems in Plants* (C. H. Foyer and P. M. Mullineaux eds.), pp. 77–105, CRC Press, Boca Raton.

Asada, K., and Badger, M. R., 1984, Photoreduction of $^{18}O_2$ and $H_2 18O_2$ with concomitant evolution of $^{16}O_2$ in intact spinach chloroplasts: Evidence for scavenging of hydrogen peroxide by peroxidase, *Plant Cell Physiol.* **25:**1169–1179.

Atkinson, M. M., 1993, Molecular mechanisms of pathogen recognition by plants, *Adv. Plant Pathol.* **10:**35–64.

Auh, C., and Murphy, T., 1995, Plasma membrane redox enzyme is involved in the synthesis of superoxide and hydrogen peroxide by Phytophthora elicitor-stimulated rose cells, *Plant Physiol.* **107:**1241–1247.

Badger, M. R., 1985, Photosynthetic oxygen exchange, *Annu. Rev. Plant Physiol.* **36:**27–53.

Baker, C. J., and Orlandi, E. W., 1997, Active oxygen in plant pathogenesis, *Annu. Rev. Phytopathol.* **33:**299–321.

Baker, C. J., O'Neill, N. R., Keppler, L. D., and Orlandi, E. W., 1991, Early Responses during plant–bacteria interactions in tobacco cell suspensions, *Phytopathology* **81:**1504–1507.

Baker, C. J., Mock, N. M., Glazener, J. A., and Orlandi, E. W., 1993a, Recognition responses in pathogen/non-host and race/cultivar interactions involving soybean (*Glycine max*) and *Pseudomonas syringae* pathovars, *Physiol. Mol. Plant Pathol.* **43:**81–94.

Baker, C. J., Orlandi, E. W., and Mock, N. M., 1993b, Harpin, an elicitor of the hypersensitive response in tobacco caused by *Erwinia amylovora*, elicits active oxygen production in suspension cells, *Plant Physiol.* **102:**1341–1344.

Baker, C. J., Harmon, G. L., Glazener, J. A., and Orlandi, E. W., 1995, A non-invasive technique for monitoring peroxidative and H_2O_2-scavenging activities during interactions between bacterial plant pathogens and suspension cells, *Plant Physiol.* **108:**353–359.

Baker C. J., Orlandi E. W., Anderson A. J., 1997, Oxygen metabolism in plant/bacteria interactions: Role of bacterial concentration and H_2O_2-scavenging in survival under biological and artificial oxidative stress, *Physiol. Mol. Plant Pathol.* **51:**401–415.

Bao, W., O'Malley, D. M., Whetten, R., and Sederoff, R. R., 1993, A laccase associated with lignification in Loblolly pine xylem, *Science* **260:**672–674.

Bateman, D. F., and Bashman, H. G., 1976, Degradation of plant cell walls and membranes by microbial enzymes, in *Encyclopedia of Plant Physiology New Series, Physiological Plant Pathology*, Volume 4 (R. Heitefuss and P.H. Williams, eds.), pp. 316–355, Springer-Verlag, Berlin.

Bilger, W., and Björkman, O., 1991, Temperature dependence of violaxanthin de-epoxidation and non-photochemical fluorescence quenching in intact leaves of *Gossypium hirsutum* L. and *Malva parviflora* L., *Planta* **184:**226.

Bors, W., Langebartels, C., Michel, C., and Sandermann, H., Jr., 1989, Polyamines as radical scavengers and protectants against ozone damage, *Phytochemistry* **28:**1589–1595.

Bowler, C., Slooten, L., Vandenbranden, S., DeRycker, R., Botteman, J., Sybesma, C., Van Montagu, M., and Inze, D., 1991, Manganese superoxide dismutase can reduce cellular damage mediated by oxygen radicals in transgenic plants, *EMBO J.* **10**(7):1723–1732.

Bradley, D. J., Kjellbom, P., and Lamb, C. J., 1992, Elicitor- and wound-induced oxidative cross-linking of a proline-rich plant cell wall protein: A novel, rapid defense response, *Cell* **1970:**21–30.

Brisson, L. F., Tenhaken, R., and Lamb, C. J., 1994, Function of oxidative cross-linking of cell wall structural proteins in plant disease resistance, *Plant Cell* **6**(12)**:**1703–1712.

Chai, H. B., and Doke, N., 1987a, Superoxide anion generation: A response of potato leaves to infection with *Phytophthora infestans*, *Physiol. Biochem.* **77**(5)**:**645–649.

Chai, H. B., and Doke, N., 1987b, Systemic activation of O_2 generating reaction, superoxide dismutase, and peroxidase in potato plants in relation to induction of systemic resistance to Phytophthora infestans, *Ann. Phytopathol. Soc. Japan* **53**(5)**:**585–590.

Chamulitrat, W., and Mason, R. P., 1989, Lipid peroxyl radical intermediates in the peroxidation of polyunsaturated fatty acids by lipoxygenase—Direct electron spin resonance investigations, *J. Biol. Chem.* **264:**20968–20973.

Chandra, S., and Low, P., 1995, Role of phosphorylation in elicitation of the oxidative burst in cultured soybean cells, *Proc. Natl. Acad. Sci. USA* **92:**4120–4123.

Chandra, S., Heinstein, P. F., and Low, P., 1996, Activation of phospholipase A by plant defense elicitors, *Plant Physiol.* **110:**979–986.

Chen, Z., Silva, H., and Klessig, D. F., 1993, Active oxygen species in the induction of plant systemic acquired resistance by salicylic acid, *Science* **262:**1883–1886.

Czaninski, Y., Sachot, R. M., and Catesson, A. M., 1993, Cytochemical localization of hydrogen peroxide in lignifying cell walls, *Ann. Bot.* **72**(6):547–550.

Davis, D., Merida, J., Legendre, L., Low, P. S., and Heinstein, P., 1993, Independent elicitation of the oxidative burst and phytoalexin formation in cultured plant cells, *Phytochemistry* **32**(3)**:**607–611.

Davis, K. R., Lyon, G. D., Darvill, A. G., and Albersheim, P., 1984, Host–pathogen interactions XXV. Endopolygalacturonic acid lyase from *Erwinia carotovora* elicits phytoalexin accumulation by releasing plant cell wall fragments, *Plant Physiol.* **74:**52–60.

Degousee, N., Triantaphylides, C., and Montillet, J.-L., 1994, Involvement of oxidative processes in the signaling mechanisms leading to the activation of glyceollin synthesis in soybean (*Glycine max*), *Plant Physiol.* **104:**945–952.

del Rio, L. A., Sandalio, L. M., Palma, J. M., Bueno, P., and Corpas, F. J., 1992, Metabolism of oxygen radicals in peroxisomes and cellular implications, *Free Radical Biol. Med.* **13:**557–580.

Desikan, R., Hancock, J., Coffey, M., and Neill, S., 1996, Generation of active oxygen in elicited cells of *Arabidopsis thaliana* is mediated by a NADPH oxidase-like enzyme, *FEBS Lett.* **382:**213–217.

Devlin, W. S., and Gustine, D. L., 1992, Involvement of the oxidative burst in phytoalexin accumulation and the hypersensitive reaction, *Plant Physiol.* **100:**1189–1195.

Dodge, A. D., and Gillham, D. J., 1986, Methods for assessing the action of light-activated herbicides. (i) The assessment of lipid peroxidation, *Aspects Appl. Biol.* **11:**97–104.

Doke, N., 1983a, Involvement of superoxide anion generation in the hypersensitive response of potato tuber tissues to infection with an incompatible race of *Phytophthora infestans* and to the hyphal wall components, *Physiol. Plant Pathol.* **23:**345–357.

Doke, N., 1983b, Generation of superoxide anion by potato tuber protoplasts during the hypersensitive response to hyphal wall components of *Phytophthora infestans* and specific inhibition of the reaction by suppressors of hypersensitivity, *Physiol. Plant Pathol.* **23:**359–367.

Doke, N., 1985, NADPH-dependent O_2 generation in membrane fractions isolated from wounded potato tubers inoculated with *Phytophthora infestans*, *Physiol. Plant Pathol.* **27:**311–322.

Doke, N., and Ohashi, Y., 1988, Involvement of an O_2 generating system in the induction of necrotic lesions on tobacco leaves infected with tobacco mosaic virus, *Physiol. Mol. Plant Pathol.* **32**:163–175.

Dwyer, S., Legendre, L., Heinstein, P., Low, P., and Leto, T., 1996, Plant and human neutrophil oxidative burst complexes contain immunologically related proteins, *Biochim. Biophys. Acta Gen. Subj.* **1289**(2):231–237.

El-Moshaty, F. I. B., Pike, S. M., Novacky, A. J., and Sehgal, O. P., 1993, Lipid peroxidation and superoxide production in cowpea (*Vigna unguiculata*) leaves infected with tobacco ringspot virus or southern bean mosaic virus, *Physiol. Mol. Plant Pathol.* **43**:109–119.

Foyer, C. H., Descourvieres, P., and Kunert, K. J., 1994, Protection against oxygen radicals: An important defense mechanism studied in transgenic plants, *Plant Cell Environ.* **17**:507–523.

Freebairn, H. T., 1957, Reversal of inhibitory effects of ozone on oxygen uptake of mitochondria, *Science* **126**:303–304.

Freebairn, H. T., and Taylor, O. C., 1960, Prevention of plant damage from air-borne oxidizing agents, *Proc. Am. Soc. Hortic. Sci.* **76**:693–699.

Freudenberg, K., 1968, The constitution and biosynthesis of lignin, in *Constitution and Biosynthesis of Lignin* (K. Freudenberg and A. C. Neish, eds.), pp. 45–122, Springer-Verlag, Berlin.

Glazener, J., Orlandi, E. W., Harmon, G. L., and Baker, C. J., 1991, An improved method for monitoring active oxygen in bacteria-treated suspension cells using luminol-dependent chemiluminescence, *Physiol. Mol. Plant Pathol.* **39**:123–133.

Glazener, J., Orlandi, E., and Baker, C., 1996, The active oxygen response of cell suspensions to incompatible bacteria is not sufficient to cause hypersensitive cell death, *Plant Physiol.* **110**:759–763.

Gomez, L., Braga, M., and Dietrich, S., 1994, Involvement of active oxygen species and peroxidases in phytoalexin production induced in soybean hypocotyls by an elicitor from a saprophytic fungus, *J. Brazil. Assoc. Adv. Sci.* **46**(3):153–156.

Gönner, M. V., and Schlösser, E., 1993, Oxidative stress in interactions between *Avena sativa* L. and *Drechslera* spp., *Physiol. Mol. Plant Pathol.* **42**:221–234.

Good, N., and Hill, R., 1955, Photochemical reduction of oxygen in chloroplast preparations. II. Mechanisms of the reaction with oxygen, *Arch. Biochem. Biophys.* **57**:355–366.

Green, R., and Fluhr, R., 1995, UV-B-induced PR-1 accumulation is mediated by active oxygen species, *Plant Cell* **7**:203–212.

Gross, G. G., Janse, C., and Elstner, E. F., 1977, Involvement of malate, monophenols, and the superoxide radical in H_2O_2 formation by isolated cell walls from horseradish (*Armoracia lapathifolia* Gilib), *Planta* **136**:271–276.

Haga, M., Kohno, Y., Iwata, M., and Sekizawa, Y., 1995, Superoxide anion generation in rice blade protoplasts with the blast fungus proteoglucomannan elicitor as determined by CLA-phenyl luminescence and its suppression by treating the elicitor with x-D-mannosidase, *Biosci. Biotechnol. Biochem.* **59**(6):969–973.

Halliwell, B., 1991, Oxygen radicals: Their formation in plant tissues and their role in herbicide damage, in *Herbicides* (N. R. Baker, and M. P. Percival, eds.), pp. 87–129, Elsevier, Amsterdam.

Harkin, J. M., and Obst, J. R., 1973, Lignification in trees: Indication of exclusive peroxidase participation, *Science* **180**:296–298.

Herouart, D., Bowler, C., Willekens, H., Van Camp, W., Slooten, L., Van Montagu, M., and Inze, D., 1993, Genetic engineering of oxidative stress resistance in higher plants, *Philos. Trans. R. Soc. London Ser. B* **342**: 235–240.

Higuchi, T., 1985, Biosynthesis of lignin, in *Biosynthesis and Biodegradation of Wood Components*, (T. Higuchi, ed.), pp. 141–160, Academic Press, Orlando.

Kangasjarvi, J., Talvinen, J., Utriainen, M., and Karjalainen, R., 1994, Plant defense systems induced by ozone, *Plant Cell Environ.* **17**:783–794.

Kawakita, K., and Doke, N., 1994, Involvement of a GTP-binding protein in signal transduction in potato tubers treated with the fungal elicitor from *Phytophthora infestans*, *Plant Sci.* **96**:81–86.

Keppler, L. D., and Baker, C. J., 1989, Superoxide-initiated lipid peroxidation in a bacteria-induced hypersensitive reaction in tobacco cell suspensions, *Phytopathology* **79**:555–562.

Keppler, L. D., Atkinson, M. M., and Baker, C J., 1988, Association of "active oxygen" with induction of increased extracellular pH in tobacco suspension cultures by pectic oligosaccharides, *Curr. Top Plant Biochem. Physiol.* **7**:230.

Keppler, L. D., Baker, C. J., and Atkinson, M. M., 1989, Active oxygen production during a bacteria-induced hypersensitive reaction in tobacco suspension cells, *Phytopathology* **79**:974–978.

Kiraly, Z., El-Zahaby, H., Galal, A., Abdou, S., Adam, A., Barna, B., and Klement, Z., 1993, Effect of oxy free radicals on plant pathogenic bacteria and fungi and on some plant diseases, in *Oxygen Free Radicals and Scavengers in the Natural Sciences* (G. Mozsik, I. Emerit, J. Feher, B. Matkovics, and A. Vincze, eds.), pp. 9–19, Akademiai Kiado, Budapest.

Legendre, L., Heinstein, P. F., and Low, P. S., 1992, Evidence for participation of GTP-binding proteins in elicitation of the rapid oxidative burst in cultured soybean cells, *J. Biol. Chem.* **267**:20140–20147.

Levine, A., Tenhaken, R., Dixon, R., and Lamb, C., 1994, H_2O_2 from the oxidative burst orchestrates the plant hypersensitive disease resistance response, *Cell* **79**:583–593.

Lindner, W. A., Hoffmann, C., and Grisebach, H., 1988, Rapid elicitor-induced chemiluminescence in soybean cell suspension cultures, *Phytochemistry* **27**(8):2501–2503.

Low, P. S., and Dwyer, S. C., 1994, Comparison of the oxidative burst signaling pathways of plants and human neutrophils, *Proc. 1994 Korean Bot. Soc.* **1994**:75–87.

Low, P. S., and Merida, J., 1996, The oxidative burst in plant defense: Function and signal transduction, *Physiol. Plant.* **96**:533–542.

Lupu, R., Grossman, S., and Cohen, Y., 1980, The involvement of lipoxygenase and antioxidants in pathogenesis of powdery mildew on tobacco plants, *Physiol. Plant Pathol.* **16**:241–248.

Lynch, D. V., and Thompson, J. E., 1984, Lipoxygenase-mediated production of superoxide anion in senescing plant tissue, *FEBS Lett.* **173**:251–254.

May, M.J., and Leaver, C. J., 1993, Oxidative stimulation of glutathione synthesis in Arabidopsis thaliana suspension cultures, *Plant Physiol.* **103**:621–627.

Mehler, A. H., 1951, Studies on reactions of illuminated chloroplasts. II. Stimulation and inhibition of the reaction with molecular oxygen, *Arch. Biochem. Biophys.* **34**:339–351.

Mehlhorn, H., and Wellburn, A. R., 1994, Man-induced causes of free radical damage to plants: O_3 and other gaseous pollutants, in *Causes of Photooxidative Stress and Amelioration of Defense Systems in Plants*, (C. H. Foyer and P. M. Mullineaux, eds.), pp. 155–175, CRC Press, Boca Raton.

Mehlhorn, H., Tabner, B. J., and Wellburn, A. R., 1990, Electron spin resonance evidence for the formation of free radicals in plants exposed to ozone, *Physiol. Plant.* **79**:377–383.

Minardi, P., and Mazzucchi, U., 1988, No evidence of direct superoxide anion effect in hypersensitive death of *Pseudomonas syringae* Van Hall in tobacco leaf tissue, *J. Phytopathol.* **122**:351–358.

Montalbini, P., 1992, Inhibition of hypersensitive response by allopurinol applied to the host in the incompatible relationship between *Phaseolus vulgaris* and *Uromyces phaseoli*, *J. Phytopathol.* **134**:218–228.

Montillet, J. L., and Degousee, N., 1991, Hydroperoxides induce glyceollin accumulation in soybean, *Plant Physiol. Biochem.* **29**(6):689–694.

Moran, J. F., Becana, M., Iturbe-Ormaetxe, I., Frechilla, S., Klucas, R.V., and Aparicio-Tejo, P., 1994, Drought induces oxidative stress in pea plants, *Planta* **194**(3):346–352.

Nakamura, W., 1967, Studies on the biosynthesis of lignin. I. Disproof against the catalytic activity of laccase in the oxidation of coniferyl alcohol, *J. Biochem.* **62**:54–61.

Neuenschwander, U., Vernooij, B., Friedrich, L., Uknes, S., Kessmann, H., and Ryals, J., 1995, Is hydrogen peroxide a second messenger of salicylic acid in systemic acquired resistance? *Plant J.* **8**(2):227–233.

Olson, P. D., and Varner, J. E., 1993, Hydrogen peroxide and lignification, *Plant J.* **4**(5):887–892.

Osmond, C. B., and Grace, S. C., 1995, Perspectives on photoinhibition and photorespiration in the field: Quintessential inefficiencies of the light and dark reactions of photosynthesis? *J. Exp. Bot.* **46**:1351–1362.

Ouf, M. F., Gazar, A. A., Shehata, Z. A., Abdou, E. S., Kiraly, Z., and Barna, B., 1993, The effect of superoxide anion on germination and infectivity of wheat stem rust (*Puccinia graminis* Pers. f. sp. *tritici* Eriks and Henn.) uredospores, *Cereal Res. Commun.* **21**(1):31–37.

Palma, J. M., Garrido, M., Rodriguez-Garcia, M. I., and del Rio, L.A., 1991, Peroxisome proliferation and oxidative stress mediated by activated oxygen species in plant peroxisomes, *Arch. Biochem. Biophys.* **287**:68–74.

Pastori, G. M., and del Rio, L. A., 1994, An activated-oxygen-mediated role for peroxisomes in the mechanism of senescence of *Pisum sativum* L. leaves, *Planta* **193**:385–391.

Peng, M., and Kuc, J., 1992, Peroxidase-generated hydrogen peroxide as a source of antifungal activity in vitro and on tobacco leaf disks, *Phytopathology* **82**:696–699.

Price, A. H., Atherton, N., and Hendry, G. A. F., 1989, Plants under drought stress generate active oxygen, *Free Radical Res. Commum.* **8:**61–66.

Qian, Y.-C., Nguyen, T., and Murphy, T. M., 1993, Effect of washing on the plasma membrane and on stress reactions of cultured rose cells, *Plant Cell Tissue Organ Culture* **35:**245–252.

Rennenberg, H., and Polle, A., 1994, Protection from oxidative stress in transgenic plants, *Biochem. Soc. Trans.* **22**(4):936–940.

Rüffer, M., Steipe, B., and Zenk, M., 1995, Evidence against specific binding of salicylic acid to plant catalase, *FEBS Lett.* **377:**175–180.

Rusterucci, C., Stallaert, V., Milat, M.-L., Pugin, A., Ricci, P., and Blein, J.-P., 1996, Relationship between active oxygen species, lipid peroxidation, necrosis, and phytoalexin production induced by elicitins in Nicotiana, *Plant Physiol.* **111:**885–891.

Sandalio, L. M., Palma, J. M., and del Rio, L. A., 1987, Localization of manganese superoxide dismutase in peroxidsomes isolated from *Pisum sativum* L., *Plant Sci.* **51:**1–8.

Sandalio, L. M., Fernandez, V. M., Ruperez, F. L., and del Rio, L. A., 1988, Superoxide free radicals are produced in glyoxisomes. *Plant Physiol.* **87:**1–4.

Schwacke, R., and Hager, A., 1992, Fungal elicitors induce a transient release of active oxygen species from cultured spruce cells that is dependent on Ca^{2+} and protein-kinase activity, *Planta* **187:**136–141.

Sen Gupta, A., Webb, R., Holaday, A., and Allen, R. D., 1993, Overexpression of superoxide dismutase protects plants from oxidative stress, *Plant Physiol.* **103:**1067–1073.

Shaaltiel, Y., Glazer, A., Bocion, P. F., and Gressel, J., 1988, Cross tolerance to herbicidal and environmental oxidants of plant biotypes tolerant to paraquat, sulfur dioxide, and ozone, *Pestic. Biochem. Physiol.* **31:**13–23.

Smirnoff, N., 1993, The role of active oxygen in the response of plants to water deficit and desiccation, *New Phytol.* **125:**27–58.

Sonike, K., 1996, Photoinhibition of photosystem I: Its physiological significance in the chilling sensitivity of plants, *Plant Cell Physiol.* **37:**239–247.

Sterjiades, R., Dean, J. F. D., and Eriksson, K. E. L., 1992, Laccase from sycamore maple (*Acer pseudoplatanus*) polymerizes monolignols, *Plant Physiol.* **99:**1162–1168.

Takahashi, M., and Asada, K., 1988, Superoxide production in aprotic interior of chloroplast thylakoids, *Arch. Biochem. Biophys.* **267:**714–722.

Takahashi, M., Kono, Y., and Asada, K., 1980, Reduction of plastocyanin with superoxide and superoxide dismutase-dependent oxidation of plastocyanin by hydrogen peroxide, *Plant Cell Physiol.* **21:**1431–1438.

Tenhaken, R., Levine, A., Brisson, L., Dixon, R., and Lamb, C., 1995, Function of the oxidative burst in hypersensitive disease resistance, *Proc. Natl. Acad. Sci. USA* **92:**4158–4163.

Thompson, J. E., Brown, J. H., Paliyath, G., Todd, J. F., and Yao, K., 1991., Membrane phospholipid catabolism primes the production of activated oxygen in senescing tissues, in *Active Oxygen/Oxidative Stress and Plant Metabolism.* (E. Pell and K. Steffen, eds.), pp. 57–66, American Society Of Plant Physiologists, Rockville.

Tolbert, N. E., 1981, Metabolic pathways in peroxisomes and glyoxysomes, *Annu. Rev. Biochem.* **50:**133–157.

Viard, M.-P., Martin, F., Pugin, A., Ricci, P., and Blein, J.-P., 1994, Protein phosphorylation is induced in tobacco cells by the elicitor cryptogein, *Plant Physiol.* **104:**1245–1249.

Wojtaszek, P., Trethowan, J., and Bolwell, G. P., 1995, Specificity in the immobilization of cell wall proteins in response to different elicitor molecules in suspension-cultured cells of French bean (*Phaseolus vulgaris* L.), *Plant Mol. Biol.* **28:**1075–1087.

Wu, G., Shortt, B., Lawrence, E., Levine, E. B., and Fitzsimmons, K. C., 1995, Disease resistance conferred by expression of a gene encoding H_2O_2-generating glucose oxidase in transgenic potato plants, *Plant Cell* **7:**1357–1368.

Yahraus, T., Chandra, S., Legendre, L., and Low, P. S., 1995, Evidence for a mechanically induced oxidative burst, *Plant Physiol.* **109:**1259–1266.

Chapter 19

The Production and Use of Reactive Oxidants by Phagocytes

Bernard M. Babior

One of the most noteworthy features of the professional phagocytes (neutrophils, eosinophils, and mononuclear phagocytes) is their ability to produce an extraordinary variety of highly reactive oxidizing agents in large quantities for use as antimicrobial and antiparasitic agents. As suggested by the purposes for which they are manufactured, these oxidants have the capacity to inflict great harm on biological systems. It is therefore a remarkable fact, and an illustration of the versatility of nature, that phagocytes have learned how to employ these very harmful and dangerous agents for beneficial aims.

In this chapter I will review some of what is known about these oxidants. The review will first survey the oxidants produced by the professional phagocytes, identifying them and describing some of their properties. The enzymes involved in their production will then be considered.

1. THE OXIDANTS

1.1. Superoxide and Hydrogen Peroxide

Apart from nitric oxide, which is not really an oxidant in the sense used here, all of the oxidizing agents produced by phagocytes are generated by the partial reduction of oxygen. The initial step in their generation is the reduction of oxygen by one electron to generate superoxide ($\cdot O_2^-$) (Babior *et al.*, 1973):

Bernard M. Babior Division of Biochemistry, Department of Molecular and Experimental Medicine, The Scripps Research Institute, La Jolla, California 92037.

Reactive Oxygen Species in Biological Systems, edited by Gilbert and Colton. Kluwer Academic / Plenum Publishers, New York, 1999.

$$O_2 + e^- \rightarrow \cdot O_2^-$$

In phagocytes, the electron comes from NADPH, so the overall reaction is

$$2\ O_2 + NADPH \rightarrow 2\ \cdot O_2^- + NADP^+ + H^+$$

$\cdot O_2^-$ is in equilibrium with its conjugate acid, the hydroperoxyl radical:

$$\cdot O_2^- + H^+ \rightleftharpoons \cdot HO_2$$

The *pK* for this reaction is 4.8, so at biological pH values, about 0.2–0.3% of the total is in the form of the acid (Fridovich, 1995), though in the vicinity of a biological membrane the acid fraction is larger because of the membrane's negative surface charge (Gilbert and Ehrenstein, 1984).

For a compound with an unpaired electron and a fairly low redox potential (–330 mV, comparable to that of a reduced pyridine nucleotide), $\cdot O_2^-$ is surprisingly unreactive. The major reaction in which it participates is its own dismutation,

$$\cdot O_2^- + \cdot HO_2 + H^+ \rightarrow O_2 + H_2O_2$$

which takes place at a near diffusion-limited rate when uncatalyzed, and at ambient tissue levels of $\cdot O_2^-$ is even more rapid when catalyzed by superoxide dismutase, the difference arising because the spontaneous reaction is second order in $\cdot O_2^-$ while the enzyme-catalyzed reaction is first order (Fridovich, 1986, 1995). The other biologically significant process in which $\cdot O_2^-$ participates is the liberation of Fe^{2+} from ferritin, the major iron storage protein of the body (Harris *et al.*, 1994), and from iron–sulfur proteins such as aconitase (Gardner *et al.*, 1995) and dihydroxy acid dehydratase (Brown *et al.*, 1995). Not only are the activities of iron–sulfur proteins affected by the loss of Fe^{2+} from their iron–sulfur clusters, but the iron released into the tissues from ferritin and the iron–sulfur clusters is potentially very harmful, for reasons that will be discussed below.

The H_2O_2 produced by phagocytes is generated by the dismutation of the $\cdot O_2^-$ manufactured by these cells in the course of their antimicrobial activities (Root *et al.*, 1975). H_2O_2 per se is relatively harmless; it is a rather weak oxidant, its high redox potential (1.77 V for the H_2O_2:H_2O couple) being offset by its low reactivity and by the fact that most organisms are well defended against H_2O_2 through their content of catalase, which dismutes H_2O_2 to oxygen and H_2O (Gaetani *et al.*, 1996):

$$2\ H_2O_2 \rightarrow O_2 + 2\ H_2O$$

and the glutathione peroxidase–glutathione reductase system, which reduces H_2O_2 to water at the expense of NADPH (Michiels *et al.*, 1994):

$$2\ GSH + H_2O_2 \xrightarrow{\text{(GSH peroxidase)}} GSSG + 2\ H_2O$$

$$NADPH + GSSG + H^+ \xrightarrow{\text{(GSSG reductase)}} 2\ GSH + NADP^+$$

$$\text{Sum: } NADPH + H^+ + H_2O_2 \rightarrow 2\ H_2O + NADP^+$$

The real danger in H_2O_2 is that it is the precursor of most of the truly harmful oxidants that are generated by phagocytes when they are engaging the enemy.

1.2. Oxidized Halogens

The principal microbicidal oxidants generated by phagocytes are the oxidized halogens. The first of these are produced when a halide anion is oxidized by H_2O_2, a reaction catalyzed by the enzymes myeloperoxidase (in neutrophils and some mononuclear phagocytes) and eosinophil peroxidase (in eosinophils):

$$X^- + H_2O_2 \rightarrow XO^- + H_2O$$

Myeloperoxidase is able to catalyze the oxidation of Cl^-, Br^-, and I^-, though of the three, Cl^- is the most important because of its concentration in serum (Thomas *et al.*, 1982, 1995; Klebanoff and Clark, 1975). The eosinophil peroxidase catalyzes the oxidations of Br^- and I^-, but not Cl^- (Thomas *et al.*, 1995; Turk *et al.*, 1983). The concentration of Br^- in normal serum, however, is surprisingly high, and its oxidation is an important feature because of its high reactivity (OBr^- is more powerful oxidant than OCl^-) and because of its role in the production of singlet oxygen (see below).

As highly reactive oxidants, the oxidized halogens act on a wide variety of biological molecules. They halogenate nucleotides, as demonstrated by the chlorination of NADH by the myeloperoxidase–H_2O_2–Cl^- system (Selvaraj *et al.*, 1980). They also oxidize thiol groups and thioethers (Johnson and Travis, 1979; Vissers and Winterbourn, 1995; Schraufstatter *et al.*, 1990), the former to disulfides and higher oxides, the latter to sulfoxides. Perhaps most important, they form halamines from any primary or secondary amine they encounter. Depending on their degree of lipid solubility, these halamines may be more or less toxic than the original hypohalite (Thomas *et al.*, 1982, 1983). For example, chloramine itself (NH_2Cl), a lipid-soluble halamine, is considerably more toxic than HOCl, whereas taurine chloramine (2-chloraminoethylsulfonic acid) is strikingly nonreactive (Green *et al.*, 1991), and its formation from taurine and HOCl is used by neutrophils as an intracellular detoxifying mechanism:

$$NH_2CH_2CH_2SO_3^{2-} + HOCl \rightarrow NHClCH_2CH_2SO_3^{2-} + H_2O$$

A number of studies have been carried out to identify the fatal lesion inflicted on bacteria by HOCl. A study in which *E. coli* were combined with HOCl in a rapid mixing apparatus showed that at the concentration of HOCl used in the study, the microorganism's ability to reproduce itself was destroyed over the course of a few milliseconds (Albrich and Hurst, 1982). As far as specific targets are concerned, the myeloperoxidase–H_2O_2–halide system was shown to inactivate succinoxidase (Rosen *et al.*, 1987) and the penicillin-binding protein, an enzyme necessary for cell wall synthesis (Rakita and Rosen, 1991), but the inactivation of these proteins was too slow to explain the death of the bacteria. A lesion that could account for bacterial killing, however, was the destruction of the membrane-associated protein complex to which the *E. coli* chromosome has to attach for replication to occur (Rosen *et al.*, 1990). The destruction of this complex and the death of the microorganisms occurred at the same rate. This is to date the only instance in which the molecular basis for the killing of a microorganism by a microbicidal oxidant has been elucidated.

1.3. Oxygen-Centered Radicals

Reactive oxygen-centered radicals arise through a number of different reactions. First is the metal-catalyzed Haber–Weiss reaction, which generates hydroxyl (·OH) and alkoxyl radicals (·RO) from H_2O_2 and alkyl hydroperoxides (ROOH), respectively (Halliwell and Gutteridge, 1986), the H_2O_2 arising chiefly by the dismutation of the $\cdot O_2^-$ produced by the leukocyte NADPH oxidase or the leakage of electrons from the respiratory chain of mitochondria (Giulivi *et al.*, 1995) and the ROOH arising in the course of lipid peroxidation. In the Haber–Weiss reaction, the peroxide accepts one electron from a reduced transition metal; in the case of H_2O_2:

$$H_2O_2 + M^{2+} \rightarrow \cdot OH + OH^- + M^{3+}$$

The oxidized transition metal is then rereduced in a subsequent step:

$$M^{3+} + RH \rightarrow M^{2+} + \cdot R$$

The net reaction is

$$H_2O_2 + RH \rightarrow \cdot OH + OH^- + \cdot R$$

The second step can be carried out by a number of biological reducing agents. Examples are ascorbate (Higson *et al.*, 1988), which is oxidized to the ascorbyl semiquinone, and $\cdot O_2^-$, which is converted to oxygen.

The transition metals that typically catalyze the Haber–Weiss reaction in biological systems are Fe and Cu. Normally these are carefully sequestered in proteins that block the participation of the metals in the Haber–Weiss reaction: Fe in ferritin, transferrin, lactoferrin, hemopexin (which binds heme iron) and various iron-containing enzymes, and Cu in cytochrome oxidase, ceruloplasmin, superoxide dismutase, and other copper-containing enzymes. Under some circumstances, however, these metals are released from their protein carriers—$\cdot O_2^-$, for example, can expel Fe from ferritin and from iron–sulfur proteins—and the liberated metals are able to catalyze the production of these very reactive and dangerous oxidizing radicals.

Other processes in phagocytes are capable of giving rise to ·OH. HOCl, produced from the myeloperoxidase-catalyzed oxidation of Cl^- by H_2O_2, can react with $\cdot O_2^-$ to form ·OH (Kettle and Winterbourn, 1994; Ramos *et al.*, 1992):

$$HOCl + \cdot O_2^- \rightarrow \cdot OH + Cl^- + O_2$$

·OH is also formed, but only in very small amounts, by the action of cyclooxygenase, which converts arachidonic acid to prostaglandin H_2 hydroperoxide (Rowe *et al.*, 1983).

·RO and particularly ·OH are extremely reactive, generally attacking the first molecule they encounter. Therefore, when produced at an arbitrary site within a cell, they would be likely to react with a nonessential molecule, with little effect on the biological system in which they were generated. Their actions, however, are amplified by two mechanisms. First is that when they react with another molecule, by abstracting a hydrogen atom, for example, or adding to a double bond,

$$\cdot OH + RH \longrightarrow H_2O + \cdot R$$

they produce another radical, which in a subsequent reaction will produce yet another radical. This sequence, known as a free radical chain reaction, will continue until two radicals react with each other to terminate the chain. In this way, a single initiating radical will cause many molecules to undergo chemical reactions, increasing the likelihood that a molecule essential for life (e.g., a DNA strand) will be damaged.

The second amplifying mechanism operates as a result of the tendency of transition metals to bind nonspecifically to macromolecules. H_2O_2, for example, may be reduced by a transition metal bound to a nucleosome. The effect of this event will be to release $\cdot OH$ in the neighborhood of the nucleosome, greatly increasing the likelihood of a reaction between the radical and the nucleosomal DNA.

Oxidizing radicals definitely participate in microbial killing by phagocytes. This is most clearly illustrated by the difference in the clinical pictures of patients with chronic granulomatous disease and subjects with myeloperoxidase deficiency. In the former case, the phagocytes are unable to produce reactive oxidants of any kind, and patients with chronic granulomatous disease are subject to recurrent infections with bacteria and fungi, including such unusual organisms as *Burkholderia cepacia* (formerly, *Pseudomonas cepacia*), a plant pathogen whose attentions are usually directed toward onions (Fischer *et al.*, 1993). In myeloperoxidase deficiency, however, the phagocytes are unable to produce oxidized halogens, but display if anything an increased ability to manufacture oxidizing radicals. Individuals with myeloperoxidase deficiency are almost always perfectly healthy, though there is a slight increase in their susceptibility to *Candida* infections (Nauseef, 1988). It is likely that the most important microbicidal oxidants produced by normal phagocytes are the oxidized halogens, but that when oxidized halogen production fails, reactive oxidizing radicals, serving as a backup system, are able to operate quite effectively as potent microbicidal oxidants.

1.4. Singlet Oxygen

In its ground state, oxygen exists as a diradical, possessing two unpaired electrons with parallel spins in separate orbitals. This state is called a *triplet state* because of the effect of a magnetic field on its emission spectrum. Under suitable conditions the spin on one of the unpaired electrons will flip, resulting in a pairing of the electrons in a single orbital.* Oxygen in this state is known as *singlet oxygen* ($^1O_2$) because a magnetic field no longer affects its emission spectrum. While the oxygen is in a

*There are actually two forms of singlet oxygen, one ($^1\Delta_g$) with the formerly unpaired electrons in a single orbital, and the other ($^1\Sigma_g^+$) with the electrons in their original orbitals but with their spins oriented in opposite directions. The $^1\Delta_g$ and $^1\Sigma_g^+$ forms are respectively 22.5 and 37.5 kcal/mole above ground state (Corey and Taylor, 1964). The $^1O_2$ produced by phagocytes is almost entirely the $^1\Delta_g$ form.

$$\uparrow\text{O-O}\uparrow \rightleftharpoons \downarrow\uparrow\text{O-O}$$

triplet singlet

singlet state it is highly reactive, but it spontaneously relaxes to the ground (triplet) state with a half time in the microsecond range.

Typical reactions of $^1O_2$ involve oxidations at double bonds. In biological systems, lipid peroxides are formed when $^1O_2$ attacks polyunsaturated fatty acids

$$\text{-CH=CH-CH}_2\text{-CH=CH-} + {}^1O_2 \rightarrow \text{-CH=CH-CH=CH-C-OOH}$$

or cholesterol (Korytowski *et al.*, 1992) (Figure 1):

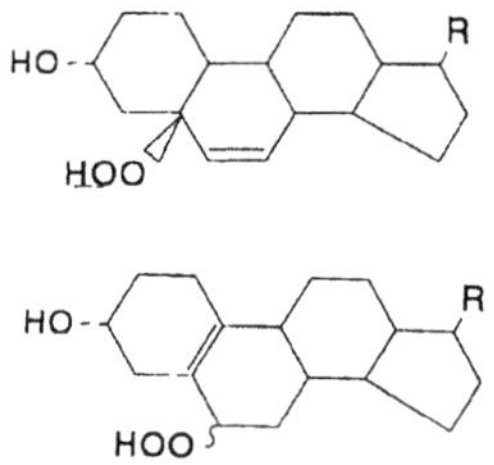

FIGURE 1. Products of oxidation of cholesterol by $^1O_2$. (Above) 3β-Hydroxy-5α-cholest-6-ene-5-hydroperoxide; (below) 3β-hydroxy-6α and β-cholest-5-ene-6-hydroperoxide. These products are specific for oxidations by $^1O_2$. Free radical oxidations occur at the 7-carbon of cholesterol.

In proteins, aromatic amino acids (particularly histidine) (Fornier de Violet *et al.*, 1984; Midi *et al.*, 1991; Hartman *et al.*, 1990), cysteine (Buettner and Hall, 1987), and methionine (Youngman *et al.*, 1985) are destroyed by $^1O_2$. Finally, $^1O_2$ attacks nucleotides and nucleic acids, chiefly by reacting with guanine residues (Kawanishi *et al.*, 1986; Epe, 1991). When DNA is the molecule under attack, mutations will arise if the DNA is not repaired before replication takes place.

In phagocytes, singlet oxygen is produced mainly by the reaction between oxidized halogens and H_2O_2 (Kanofsky *et al.*, 1984; Khan and Kasha, 1994; Weiss *et al.*, 1986):

$$OX^- + H_2O_2 \rightarrow {}^1O_2 + H_2O + X^-$$

Early studies looking for $^1O_2$ production by phagocytes used relatively nonspecific trapping reagents and gave ambiguous results. $^1O_2$ production by eosinophils, however, was unequivocally demonstrated by infrared emission spectroscopy (Kanofsky *et al.*, 1988), and its production by neutrophils and macrophages was later shown by specific assay methods (Kanofsky *et al.*, 1984; Steinbeck *et al.*, 1992, 1993). The oxidized halogen used for the production of $^1O_2$ by eosinophils was OBr^-, which reacts readily with H_2O_2 to produce $^1O_2$. Neutrophils and mononuclear phagocytes probably produce $^1O_2$ by the same route, because the production of $^1O_2$ from H_2O_2 and OCl^- is considerably slower than its production from OBr^- (Kanofsky *et al.*, 1984). At a surprising 2.5–5 mM (Jennings and Elia, 1996), the plasma Br^- concentration is probably high enough to support $^1O_2$ production by any of these cells.

1.5. Nitric Oxide and Peroxynitrite

Nitric oxide (·NO) is produced from L-arginine, oxygen, and NADPH by the enzyme nitric oxide synthase, which catalyzes the following two reactions:

$$R{-}NH{-}\underset{\displaystyle NH_2}{C}{=}NH_2^+ + 1\,1/2\,O_2 + 1/2\,NADPH \longrightarrow R{-}NH{-}\underset{\displaystyle NHOH}{C}{=}NH_2^+ + 1/2\,NADP + H_2O$$

N^G-hydroxyarginine

$$R{-}NH{-}\underset{\displaystyle NHOH}{C}{=}NH_2^+ + 1/2\,NADPH + O_2 \longrightarrow R{-}NH{-}\underset{\displaystyle NH_2}{C}{=}O + {\cdot}NO + 1/2\,NADP + H_2O$$

Citrulline

These reactions amount to a five-electron oxidation of arginine, one that consumes 6 e^- to reduce three oxygen atoms, and returns 1 e^- by oxidizing half an NADPH. In the presence of oxygen, ·NO can be oxidized to the highly toxic radical $\cdot NO_2$, but the accumulation of $\cdot NO_2$ seems to be largely forestalled in tissues by the rapid oxidation of ·NO to N_2O_3:

$$2\,\cdot NO + 1/2\,O_2 \rightarrow N_2O_3$$

Both $\cdot NO_2$ and N_2O_3 react with ambient sulfhydryl groups such as those found on proteins and especially glutathione to form nitrosothiols (Goldstein and Czapski, 1996; Kharitonov *et al.*, 1995):

$$\cdot NO_2 + RS^- \rightarrow \cdot RS + NO_2^-$$

$$\cdot RS + \cdot NO \rightarrow RSNO$$

and

$$N_2O_3 + RS^- \rightarrow RSNO + NO_2^-$$

The nitrosothiols are stable under normal biological conditions (pH 7.4, 37 °C), but they break down in the presence of trace amounts of transition metal ions with the liberation of ·NO and the formation of a disulfide compound (Singh *et al.*, 1996; Wink *et al.*, 1994). (Free transition metals are generally absent from biological systems, raising the question as to whether the release of ·NO from nitrosothiols might be catalyzed by an as yet undiscovered enzyme.) Finally, glutathione reacts with glutathione nitrosothiol to generate glutathione disulfide and nitrous oxide (N_2O), an inert gas of low toxicity that is excreted through the lungs (Hogg *et al.*, 1996).

·NO is manufactured by phagocytes, vascular endothelial cells, and certain central nervous system neurons. In blood vessels it acts as a vasodilator, and in the central nervous system as a neurotransmitter. It serves in phagocytes as a precursor for peroxynitrite. On a molecular level, its sole function is as an activator of guanylate cyclase. This enzyme contains a heme group, and when the sixth coordination position of the heme iron is occupied by ·NO, the enzyme is converted from a latent to a catalytically active form

(Stone and Marletta, 1996), producing cyclic guanosine monophosphate (cGMP) from GTP. The cGMP in turn activates a series of cGMP-dependent protein kinases whose actions are ultimately responsible for the physiological effects of ·NO (Francis and Corbin, 1994).

Peroxynitrite is produced by a reaction between ·NO and $\cdot O_2^-$ (Pryor and Squadrito, 1995):

$$\cdot NO + \cdot O_2^- \rightarrow ONOO^-$$

It will therefore be generated by cells that manufacture both ·NO and $\cdot O_2^-$, notably neutrophils (McCall *et al.*, 1989) and mononuclear phagocytes (Salvemini *et al.*, 1989). $ONOO^-$ is a highly reactive substance, inflicting damage directly on biological macromolecules and reacting with other small molecules to produce many other highly reactive oxidants. $ONOO^-$ appears to react with CO_2 to form a species with a significantly increased reactivity, perhaps to nitrocarbonate via the nitrosoperoxy-carbonate anion (Denicola *et al.*, 1996; Uppu *et al.*, 1996):

$$ONOO^- + CO_2 \rightarrow O{=}N\text{-}OOCO_2^- \rightarrow O_2N\text{–}O\text{–}CO_2^-$$

$ONOO^-$ reacts with radical scavengers including thiols and ascorbate to generate the corresponding thiyl and ascorbyl radicals, which in turn react with oxygen to form oxygen-centered radicals (Squadrito *et al.*, 1995; Karoui *et al.*, 1996; Vasquez-Vivar *et al.*, 1996; Shi *et al.*, 1994), and with methionine to produce methionine sulfoxide (Pryor *et al.*, 1994). $^1O_2$ is produced from $ONOO^-$ by the spontaneous decomposition of its conjugate acid (ONOOH, p*K* 6.8; Pryor and Squadrito, 1995) or from the reaction of $ONOO^-$ with H_2O_2 (Alvarez *et al.*, 1995; DiMascio *et al.*, 1994). The dissociation of ONOOH to $\cdot NO_2$ and ·OH has been postulated to occur from the observation that reactions involving $ONOO^-$ give rise to products similar to those generated by ·OH (Crow *et al.*, 1994; Pryor and Squadrito, 1995; van der Vliet *et al.*, 1994), but several direct attempts to find ·OH by spin-trapping (Pryor *et al.*, 1994; Pou *et al.*, 1995; Lemercier *et al.*, 1995) have met with failure, and the consensus seems to be that the postulated dissociation does not take place, but that the ·OH-like behavior of $ONOO^-$ is actually related to the high reactivity of the *trans* isomer of its conjugate acid, ONOOH.

As to its action on biological molecules, the best known reaction involving $ONOO^-$ is the nitration of tyrosine. Several routes of nitration have been proposed, all resulting in the nitration of tyrosine at the 3-position of the aromatic ring. One route of nitration involves the conjugate acid ONOOH, and appears to take place by a free radical mechanism in which $\cdot NO_2$ derived from ONOOH combines with a tyrosyl free radical (Crow and Beckman, 1995; Eiserich *et al.*, 1996). The conclusion that the nitration is a free radical reaction is based on the appearance of 3,3′-dityrosine in the course of the reaction, a product thought to arise from the coupling of two tyrosyl radicals. How the tyrosyl radicals arise and how $\cdot NO_2$ is generated from ONOOH without the production of a hydroxyl radical are not known. A second proposed mechanism involves nitrocarbonate ($O_2N\text{–}O\text{–}CO_2^-$) as the nitrating agent, but is otherwise generally similar to the mechanism proposed for nitration by unadorned $ONOO^-$ (Lymar *et al.*, 1996). In a third mechanism, the nitrating agent is nitryl chloride ($Cl\text{–}NO_2$), which is produced by the myeloperoxidase-catalyzed chlorination of NO_2^-, a product of oxidation of $\cdot NO_2$ (Eiserich *et al.*, 1996):

$$NO_2^- + Cl^+ \rightarrow Cl\text{–}NO_2$$

$$Cl\text{–}NO_2 \rightarrow \cdot Cl + \cdot NO_2$$

$$\cdot NO_2 + \cdot\phi(OH)\text{–}CH_2\text{–}CHNH\text{–}C{=}O \rightarrow O_2N\text{–}\phi(OH)\text{–}CH\text{–}CHNH\text{–}C{=}O$$

Other effects of $ONOO^-$ on proteins include the oxidation of tryptophan, cysteine, and methionine and the production of carbonyl groups that can be recognized by their reaction with 2,4-dinitrophenylhydrazine (Berlett *et al.*, 1996; Ischiropoulos and Al-Mehdi, 1995; Stadtman, 1995). Lipids also react with $ONOO^-$, its actions resulting in lipid peroxidation and nitration, the latter giving rise to a variety of nitrogen-containing lipids including nitrito-, nitro-, and nitrosoperoxo compounds (Rubbo *et al.*, 1994).

$ONOO^-$ also damages DNA, showing a predilection for guanine, which it oxidizes at the 8-position to 8-nitroguanine or 8-hydroxyguanine (Yermilov *et al.*, 1995; Inoue and Kawanishi, 1995). In whole cells this leads to breakage of the DNA, possibly as a consequence of the action of repair endonucleases, leading to the activation of poly-ADP ribose synthase and the consumption of large amounts of NAD^+ (Zingarelli *et al.*, 1996; Salgo *et al.*, 1995). With extensive DNA cleavage, intracellular NAD^+ can fall to levels too low to sustain the life of the cell (Schraufstatter *et al.*, 1985, 1986; Szabo *et al.*, 1996).

Not surprisingly, an attack on a protein by $ONOO^-$ can alter the protein's function significantly. Aconitase, and probably other iron–sulfur proteins, are inactivated by $ONOO^-$ (Hausladen *et al.*, 1994). After exposure to $ONOO^-$, low-density lipoprotein is recognized by the macrophage scavenger receptor[*] (Cassina and Radi, 1996). Mitochondrial function is impaired through the action of $ONOO^-$ on Complexes I–III and the ATP-synthesizing enzyme, but Complex IV (cytochrome oxidase) is spared (Cassina and Radi, 1996). In regulatory proteins, $ONOO^-$ may nitrate critical tyrosine residues whose phosphorylation or other modification is important for function. If this happens, phosphorylation may be blocked, and the regulatory behavior of the protein may be altered in other ways as well (Kong *et al.*, 1996; Berlett *et al.*, 1996).

$ONOO^-$ is an important antimicrobial oxidant of macrophages, acting against bacteria (Zhu *et al.*, 1992; Brunelli *et al*,, 1995), fungi (Vazquez-Torres *et al.*, 1996) and protozoa (Assreuy *et al.*,1994; Denicola *et al.*, 1993). $ONOO^-$ may also be important in pathogenesis, as indicated by the presence of nitrotyrosine in atherosclerotic lesions (especially in and around the foamy macrophages within the lesion) (Beckmann *et al.*, 1994) and in the lungs of patients who had died of the adult respiratory distress syndrome (Haddad *et al.*, 1994), an often fatal pulmonary disorder seen in severely ill patients,

*The scavenger receptor takes up LDL that has been chemically altered (e.g., acetylated or oxidized) (Krieger and Herz, 1994). Unlike the usual LDL receptor, which can only take up unmodified LDL, its concentration on the cell surface is not regulated by cholesterol, so it remains active regardless of the intracellular cholesterol level, and can cause the cell to accumulate very large quantities of cholesterol and cholesterol esters.

especially those on respirators. In these cases, nitration serves as an *in vivo* footprint, marking sites where phagocytes have inflicted damage on other cells and tissues.

Studies on nitric oxide production by phagocytes have been carried out principally with cells from mice or rats, and it is clear that these cells produce significant amounts of nitric oxide in response to appropriate stimuli. The production of nitric oxide by human phagocytes, however, is a highly controversial question. Investigators fall into one of two camps: those who claim that human phagocytes produce tiny amounts of nitric oxide (Dias-Da-Motta *et al.*. 1996; Larfars and Gyllenhammar, 1995; Paul-Eugene *et al.*, 1995; Catz *et al.*, 1995; Vouldoukis *et al.*, 1995) and those who claim that human phagocytes produce none at all (Condino-Neto *et al.*, 1996; Albina, 1995; Sakai and Milstein, 1993; Schneemann *et al.*, 1993; Bermudez, 1993; Colton *et al.*, 1996). Some investigators have a foot in both camps. The only conclusion that can be drawn at this point is that if human phagocytes make nitric oxide at all, they make very little, implying that nitric oxide production has only marginal significance for the *human* inflammatory response.

2. THE ENZYMES

All of the myriad oxidants described in the foregoing section are produced through the action of only three enzymes: the leukocyte NADPH oxidase, myeloperoxidase (in the eosinophil, eosinophil peroxidase), and nitric oxide synthase. In this section, the properties of these enzymes are briefly reviewed.

2.1. Leukocyte NADPH Oxidase

The leukocyte NADPH oxidase is the key enzyme in the oxygen-dependent microbicidal mechanisms of phagocytes. It is found in neutrophils, eosinophils, monocytes (but not macrophages except under special circumstances), and B lymphocytes (Nathan, 1987; Yazdanbakhsh *et al.*, 1987; Volkman *et al.*, 1984). In all of these cells, it catalyzes the production of $\cdot O_2^-$ from oxygen and NADPH:

$$2\,O_2 + NADPH \rightarrow 2\,\cdot O_2^- + NADP^+ + H^+$$

In the professional phagocytes, it participates in the destruction of invading microorganisms. Its purpose in B lymphocytes is not known, though it may be speculated that it assists in antigen presentation by rendering proteins more susceptible to degradation into peptides.

The leukocyte NADPH oxidase, a highly complex enzyme, is dormant in resting cells but comes to life when the cells are exposed to appropriate stimuli (Chanock *et al.*, 1994). In the resting state its components are distributed between the cytosol and the membranes of various organelles: secretory vesicles, specific granules, and the plasma membrane (Sengelov *et al.*, 1992; Borregaard *et al.*, 1983). The cytosolic components, which exist in the form of a complex (Park *et al.*, 1992, 1994), are the three subunits $p47^{phox}$ (Lomax *et al.*, 1989; Volpp *et al.*, 1989), $p67^{\mathrm{phox}}$ (Leto *et al.*, 1990) and $p40^{phox}$ (Wientjes *et al.*, 1993) (molecular masses 45, 60, and 39 kDa, respectively). The membrane-bound components are $gp91^{phox}$ (a glycoprotein) (Dinauer *et al.*, 1987) and $p22^{phox}$ (Parkos *et al.*, 1988), which together comprise a heterodimeric flavohemoprotein cytochrome b_{558}

(Parkos *et al.*, 1987; Rotrosen *et al.*, 1992; Segal *et al.*, 1992). In addition to these components, two guanine nucleotide-binding proteins are required for oxidase activity: Rac2, a cytosolic guanine nucleotide-binding protein of the Rho family (Knaus *et al.*, 1992), and Rap1A, a membrane-bound protein of the Ras family (Maly *et al.*, 1994). When the oxidase is activated, the cytosolic components, including both the complex and Rac2, migrate to the membrane, where they associate with cytochrome b_{558} to form the active oxidase (Clark *et al.*, 1990; El Benna *et al.*, 1994b).

Activation of the oxidase is a complex event that is as yet only poorly understood. It has been possible to activate the oxidase in a cell-free system with anionic detergents such as SDS (Bromberg and Pick, 1984; Pick *et al.*, 1987), and much has been accomplished using this system, including the cloning of the subunits of the oxidase and the demonstration that oxidase activity requires only four of these subunits—$p47^{phox}$, $p67^{phox}$, and the dimeric cytochrome b_{558}—plus Rac2 and GTP (Abo *et al.*, 1992). The reason for the presence of $p40^{phox}$ in the cytosolic complex remains unexplained, however. Furthermore, it is known that in whole cells, oxidase activation is associated with the phosphorylation of the C-terminal end of $p47^{phox}$ (El Benna *et al.*, 1994a, 1996), and there is good evidence that this phosphorylation is an essential element of activation in whole cells (Johnson *et al.*, 1996), but detergent-mediated activation in the cell-free system does not require the phosphorylation of any component of the oxidase. The relationship between the mechanism of activation of the oxidase by SDS and the mechanism of protein kinase-dependent activation in whole cells is obscure. Finally, little is known at a molecular level about the means by which Rac2 or Rap1A promote oxidase activation.

Oxidase activation is in principle a reversible process, the enzyme returning to its resting state when the stimulus is removed (Curnutte *et al.*, 1979; Badwey *et al.*, 1984). Deactivation of the enzyme is associated with the dephosphorylation of $p47^{phox}$, suggesting that a phosphatase may be responsible for the deactivation event (Hayakawa *et al.*, 1986). There is evidence that both activation and deactivation of the oxidase take place continuously, the level of oxidase activity being a function of the relative rates of these two reactions at the time of the measurement (Akard *et al.*, 1988).

Electron transport by the oxidase is somewhat better understood. NADPH initially associates with an NADPH binding site on $p67^{phox}$. From there it appears to be transferred to a weaker NADPH binding site on the $gp91^{phox}$ subunit of cytochrome b_{558}, the transfer presumably being facilitated by the proximity of the binding site on cytochrome b_{558} to the NADPH associated with $p67^{phox}$. (An advantage to this arrangement is that it greatly reduces the likelihood that the very dangerous $\cdot O_2^-$ radical will be accidentally produced by electron leakage from NADPH to oxygen through the resting enzyme, without compromising the ability of the activated enzyme to catalyze $\cdot O_2^-$ production from NADPH and oxygen.) The NADPH is subsequently oxidized by the flavin, which then transfers its electron to another oxidant, either directly to oxygen to form $\cdot O_2^-$, or to cytochrome b_{558}, which then passes the electron on to oxygen. Most workers believe that cytochrome b_{558} is the final electron carrier in a short redox chain between NADPH and oxygen, but there is considerable evidence against that conception (Cross *et al.* 1982, 1985; Gabig *et al.*, 1982; Foroozan *et al.*, 1992). Another area of uncertainty derives from the fact that a flavoprotein usually removes both electrons from a reduced pyridine nucleotide, storing one of them on the flavin that first accepts the electrons and the other one on a second redox carrier, e.g., an iron–sulfur center. Apart from the heme, however, no second redox carrier has been found on the oxidase, but a thorough search, especially for redox-active metals, has yet to be made.

A proton channel is also associated with the oxidase. This channel is necessary because the activity of the oxidase results in the rapid acidification of the cytosol, as $\cdot O_2^-$ is exported from the cell leaving a proton behind. If no mechanism were available to expel the protons responsible for this acidification, $\cdot O_2^-$ production would be quickly shut off. In the resting cell the proton channel is closed, but it opens when the oxidase is activated. There is some evidence that the proton channel may be located in cytochrome b_{558}, but this is controversial.

2.2. Myeloperoxidase

Myeloperoxidase is the enzyme responsible for the formation of the microbicidal oxidant HOCl, which is produced when the enzyme catalyzes the oxidation of Cl^- by H_2O_2 as described above. It is found in the azurophil granules of neutrophils, where it comprises 2–5% of the total neutrophil protein. It is also present in the granules of monocytes, but is partly or completely lost when monocytes evolve into macrophages, although macrophages lacking myeloperoxidase can use the enzyme by taking it up after it has been released into the extracellular environment by the degranulation or lysis of neutrophils.

Myeloperoxidase is composed of a pair of identical heterodimers, each consisting of two polypeptides of ~60 and ~15 kDa (Andrews and Krinsky, 1981; Zuurbier *et al.*, 1992; Zeng and Fenna, 1992). The heterodimer is formed by posttranslational processing of a single 88-kDa polypeptide, and its two subunits are held together by noncovalent forces. Crystallography shows that the prosthetic group, a tetrapyrrole, lies in a pocket of the heterodimer, its proximal side facing the large polypeptide and its distal side the small polypeptide, though R239, an important arginine from the large polypeptide, also lies near the distal side (Zeng and Fenna, 1992). Glycosylation at two positions on the heavy chain was also seen (Taylor *et al.*, 1992; Zeng and Fenna, 1992). In the full-size protein, the two heterodimers are held together by a single disulfide bond between the two heavy chains. Though comprising a single entity, the two heterodimers of the myeloperoxidase tetramer are functionally independent.

A feature of myeloperoxidase that has attracted a great deal of interest from investigators is its green color. This is an unusual color for a heme-containing protein, and led to numerous studies attempting to define the nature of the prosthetic group. The tetrapyrrole was proposed by some to be an iron chlorine (i.e., an iron tetrapyrrole with the β–β double bond of one of the pyrrole rings reduced) (Timcenko-Youssef *et al.*, 1985; Ikeda-Saito *et al.*, 1985; Babcock *et al.*, 1985) and by others to be a formyl-substituted porphyrin such as occurs in cytochrome oxidase (Harrison and Schultz, 1978; Sono *et al.*, 1991; Wever *et al.*, 1991). The question was finally answered by crystallography, which showed that the tetrapyrrole was a conventional heme residue covalently attached to the enzyme by three separate bonds: ester linkages involving methoxy groups on the A and C rings, and a sulfonium linkage involving a methionine and the ring A vinyl group (Fenna *et al.*, 1995).* Furthermore, the ring is not planar, but is bent toward the ester

*A sulfonium linkage between myeloperoxidase and its tetrapyrrole was identified by Taylor and colleagues before the crystallographic structure was available (Taylor *et al.*, 1992, 1995).

linkages (Figure 2, from Fenna *et al.*, 1995), a conformational feature that undoubtedly affects its spectral properties and redox potential.

Myeloperoxidase converts Cl^- to OCl^- by a two-electron oxidation that involves the following reactions. H_2O_2 removes two electrons from myeloperoxidase, which contains iron at the oxidation level of Fe^{III}, oxidizing it to Compound I, which now contains a ferryloxy group, with iron at a formal oxidation level of Fe^{IV}, and a π-cation radical created by the loss of one electron from the porphyrin ring (Harrison *et al.*, 1980). The formation of Compound I from myeloperoxidase and H_2O_2 is reversible (Marquez *et al.*, 1994), indicating that its redox potential is close to that of the very strongly oxidizing H_2O_2/H_2O couple (1.77 V). The Cl^- ion, held in the active site by an amino acid side chain [possibly the protonated imidazole of H95 (Zeng and Fenna, 1992; Lee *et al.*, 1991)], is stripped of two electrons by Compound I, producing OCl^- and regenerating the Fe^{III} form of myeloperoxidase. The whole sequence is

$$H_2O_2 + \text{MPO } (Fe^{III}) \rightarrow \text{Cpd I } (Fe^{IV}{=}O/\cdot R) + H_2O$$

$$Cl^- + \text{Cpd I } (Fe^{IV}{=}O/\cdot R) \rightarrow OCl^- + \text{MPO } (Fe^{III})$$

$\cdot R$ represents the porphyrin π-cation radical.

Myeloperoxidase, however, is also inactivated by H_2O_2. This occurs because H_2O_2 converts Compound I to the inactive Compound II (containing ferryloxy but no cation radical), giving up one electron to produce $\cdot O_2^-$:

$$\text{Cpd I } (Fe^{IV}{=}O/\cdot R) + H_2O_2 \rightarrow \text{Cpd II } (Fe^{IV}{=}O) + \cdot O_2^-$$

$\cdot O_2^-$ also gives an electron to Compound I, producing oxygen and more Compound II:

$$\text{Cpd I } (Fe^{IV}{=}O/\cdot R) + \cdot O_2^- \rightarrow \text{Cpd II } (Fe^{IV}{=}O) + O_2$$

Compound II is incapable of oxidizing Cl^-, so as far as OCl^- production is concerned, the enzyme is dead until it is able to extract another electron from some passing reductant, returning to its starting point as myeloperoxidase (Fe^{III}) (Marquez *et al.*, 1994). In fact, the presence in serum of substances that promote the H_2O_2-dependent conversion of MPO (Fe^{III}) to Compound II has led to the suggestion that the rate of production of HOCl by myeloperoxidase may be limited by the rate of conversion of Compound II back to MPO (Fe^{III}) (Kettle and Winterbourn, 1988). At the same time, myeloperoxidase is inhibited by Cl^-, which binds to the Fe^{III} form and interferes with its oxidation by H_2O_2. All of these processes are sensitive to pH, giving rise to very complicated kinetics, the net effect of which is to fix the pH optimum for myeloperoxidase in the vicinity of 5.0.

2.3. Nitric Oxide Synthase

$\cdot$NO is synthesized by nitric oxide synthase, a soluble enzyme that catalyzes the five-electron oxidation of arginine to $\cdot$NO and citrulline (see above). Nitric oxide synthase is more accurately referred to as the nitric oxide synthases, because there are two kinds: the constitutive synthases, which are found in endothelial cells and neurons and which produce $\cdot$NO at low levels; and the inducible synthases, which appear in phagocytes when appropriate stimuli are applied and which manufacture $\cdot$NO in large amounts. As de-

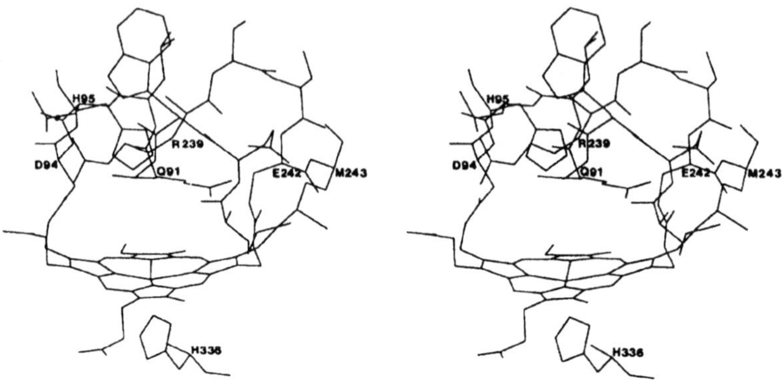

FIGURE 2. X-ray crystallography of the heme residue and nearby peptide sequences in myeloperoxidase. A stereo representation of the structure of myeloperoxidase in the vicinity of the heme residue, showing the bowing of the tetrapyrrole macrocycle, the three covalent bonds between the periphery of the tetrapyrrole and the myeloperoxidase polypeptide, and the histidine residues that comprise the proximal and distal ligands of the heme iron. Reproduced with permission from Fenna *et al.* (1995).

scribed above, ·NO produced by the inducible phagocyte nitric oxide synthases is used as a microbial killing agent after oxidation to $OONO^-$. The small amounts of ·NO produced by neurons and endothelial cells, however, act as regulators, activating guanylate cyclase by complexing with the iron of the heme residue attached to the enzyme (Stone and Marletta, 1996; Murad, 1994).

·NO was first shown to be a biological molecule by its identification as endothelium-derived relaxation factor (EDRF), an agent released from acetylcholine-treated endothelium that induced the relaxation of vascular smooth muscle (Palmer *et al.*, 1987). The first hint that it was produced by phagocytes was the finding that a macrophage cell line was able to generate nitrate and nitrite when activated with γ-interferon or lipopolysaccharide (Iyengar *et al.*, 1987). ·NO was soon shown to be the precursor of the nitrate and nitrite produced by these cells (Driever and Fishman, 1996), the guanidino nitrogen of arginine the source from which the ·NO was produced (Iyengar *et al.*, 1987; Pufahl *et al.*, 1992), and citrulline the other product (Iyengar *et al.*, 1987).

Nitric oxide synthase from phagocytes is a soluble enzyme with a molecular mass of ~260 kDa (Stuehr *et al.*, 1991; Hevel *et al.*, 1991). The active enzyme is a homodimer composed of two inactive ~130-kDa monomers. It is a highly unusual enzyme containing four redox cofactors: one FAD, one FMN (Hevel *et al.*, 1991; Stuehr *et al.*, 1991), one tetrahydrobiopterin (Tayeh and Marletta, 1989; Hevel and Marletta, 1992; Baek *et al.*, 1993) (a cofactor also found in aromatic hydroxylases such as phenylalanine hydroxylase; Kaufman, 1987) and one heme (White and Marletta, 1992). The same cofactors are found in the neuronal and endothelial nitric oxide synthases (Galli *et al.*, 1996; McMillan *et al.*, 1992; Klatt *et al.*, 1996; Marletta, 1993; Rodriguez-Crespo *et al.*, 1996). The heme forms the prosthetic group of the cytochrome P450 domain of the enzyme, the proximal iron ligand being the $-S^-$ of a cysteine residue (Masters *et al.*, 1996; McMillan and Masters, 1995).

The mechanism of action of nitric oxide synthase is something of a puzzle. The oxidation of arginine to N^G-hydroxyarginine is a straightforward cytochrome P450 reaction similar to reactions by which xenobiotic amines are oxidized (Karki and Dinnocenzo, 1995). On the other hand, the mechanism for the generation of ·NO and citrulline from N^G-hydroxyarginine is not so obvious. Clues to a possible mechanism, however, have been provided by two recent sets of observations. Xia and colleagues have shown that at low levels of arginine, nitric oxide synthase produces $\cdot O_2^-$ as well as ·NO (Xia *et al.*, 1996). In a study of the oxidation of *N*-hydroxyguanidines, Sennequier and colleagues showed that the C=NOH bonds of these compounds are cleaved by $\cdot O_2^-$ to generate ·NO (Sennequier *et al.*, 1995). It therefore seems reasonable to suppose that nitric oxide synthase catalyzes a cytochrome P450-like oxidation of arginine to N^G-hydroxyarginine using one of the flavins as reductant for the cytochrome P450 reaction, and that the N^G-hydroxyarginine is then passed on to another domain of the enzyme, where locally synthesized $\cdot O_2^-$ (possibly generated via the other flavin) is used to release ·NO from the hydroxyguanidine moiety of the N^G-hydroxyarginine. The role of biopterin in either the formation of hydroxyarginine or its conversion into ·NO and citrulline is obscure.

Calmodulin participates in the activation of all three forms of nitric oxide synthase. The constitutive forms of the enzyme are activated when the exposure of a cell to an appropriate agonist causes the intracellular Ca^{2+} concentration to rise. This results in the formation of a Ca^{2+}-calmodulin complex which then binds to the nitric oxide synthase, activating it by opening a channel through which electrons can pass from NADPH to the heme cofactor (Mateo *et al.*, 1996). The activity of the inducible nitric oxide synthase of phagocytes, however, is not affected by Ca^{2+} (Cho *et al.*, 1992; Stevens-Truss and Marletta, 1995). This is not because the inducible synthase is inert to calmodulin. On the contrary, calmodulin was shown to bind very tightly to the inducible nitric oxide synthase, remaining firmly attached to the synthase even at a Ca^{2+} concentration below that found in resting cells. The bound calmodulin was necessary for ·NO synthesis by the enzyme, but because the calmodulin–synthase complex was stable over the biological range of calcium concentrations, the activity of the enzyme was insensitive to calcium. Therefore, the activity of the constitutive nitric oxide synthases are regulated by both enzyme concentration (determined by the rate of biosynthesis and turnover of the enzyme) and intracellular Ca^{2+} levels, but the inducible nitric oxide synthase is regulated only by its concentration (Cho *et al.*, 1992; Stevens-Truss and Marletta, 1995; Wang and Marsden, 1995).

ACKNOWLEDGMENTS. The author's research reported herein was supported in part by USPHS Grants AI-24227, AI-28479, and RR-00833 and the Stein Endowment Fund.

3. REFERENCES

Abo, A., Boyhan, A., West, I., Thrasher, A. J., and Segal, A. W., 1992, Reconstitution of neutrophil NADPH oxidase activity in the cell-free system by four components: p67-phox, p47-phox, p21rac1, and cytochrome b-245, *J. Biol. Chem.* **267:**16767–16770.

Akard, L. P., English, D., and Gabig, T. G., 1988, Rapid deactivation of NADPH oxidase in neutrophils: Continuous replacement by newly activated enzyme sustains the respiratory burst, *Blood* **72:**322–327.

Albina, J. E., 1995, On the expression of nitric oxide synthase by human macrophages. Why no NO, *J. Leukocyte Biol.* **58:**643–649.

Albrich, J. M., and Hurst, J. K., 1982, Oxidative inactivation of Escherichia coli by hypochlorous acid. Rates and differentiation of respiratory from other reactive sites, *FEBS Lett.* **144:**157–161.

Alvarez, B., Denicola, A., and Radi, R., 1995, Reaction between peroxynitrite and hydrogen peroxide: Formation of oxygen and slowing of peroxynitrite decomposition, *Chem.Res.Toxicol.* **8:**859–864.

Andrews, P. C., and Krinsky, N. I., 1981, The reductive cleavage of myeloperoxidase in half, producing enzymically active hemi-myeloperoxidase, *J. Biol. Chem.* **256:**4211–4218.

Assreuy, J., Cunha, F. Q., Epperlein, M., Noronha-Dutra, A., O'Donnell, C. A., Liew, F. Y., and Moncada, S., 1994, Production of nitric oxide and superoxide by activated macrophages and killing of *Leishmania major*, *Eur. J. Immunol.* **24:**672–676.

Babcock, G. T., Ingle, R. T., Oertling, W. A., Davis, J. C., Averill, B. A., Hulse, C. L., Stufkens, D. J., Bolscher, B. G. J. M., and Wever, R., 1985, Raman characterization of human leukocyte myeloperoxidase and bovine spleen green haemoprotein. Insight into chromophore structure and evidence that the chromophores of myeloperoxidase are equivalent, *Biochim. Biophys. Acta* **828:**58–66.

Babior, B. M., Kipnes, R. S., and Curnutte, J.,T., 1973, Biological defense mechanisms: The production by leukocytes of superoxide, a potential bactericidal agent, *J. Clin. Invest.* **52:**741–744.

Badwey, J. A., Curnutte, J. T., Robinson, J. M., Berde, C. B., Karnovsky, M. J. and Karnovsky, M. L., 1984, Effects of free fatty acids on release of superoxide and on change of shape by human neutrophils: Reversibility by albumin, *J. Biol. Chem.* **259:**7870–7877.

Baek, K. J., Thiel, B. A., Lucas, S., and Stuehr, D. J., 1993, Macrophage nitric oxide synthase subunits. Purification, characterization, and role of prosthetic groups and substrate in regulating their association into a dimeric enzyme, *J. Biol. Chem.* **268:**21120–21129.

Beckmann, J. S., Ye, Y. Z., Anderson, P. G., Chen, J., Accavitti, M. A., Tarpey, M. M., and White, C. R., 1994, Extensive nitration of protein tyrosines in human atherosclerosis detected by immunohistochemistry, *Biol. Chem. Hoppe Seyler* **375:**81–88.

Berlett, B. S., Friguet, B., Yim, M. B., Chock, P. B., and Stadtman, E. R., 1996, Peroxynitrite-mediated nitration of tyrosine residues in Escherichia coli glutamine synthetase mimics adenylylation: Relevance to signal transduction, *Proc. Natl. Acad. Sci. USA* **93:**1776–1780.

Bermudez, L. E., 1993, Differential mechanisms of intracellular killing of *Mycobacterium avium* and *Listeria monocytogenes* by activated human and murine macrophages. The role of nitric oxide, *Clin. Exp. Immunol.* **91:**277–281.

Borregaard, N., Heiple, J. M., Simons, E. R., and Clark, R. A., 1983, Subcellular localization of the b-cytochrome component of the human neutrophil microbicidal oxidase: Translocation during activation, *J. Cell. Biol.* **97:**52–61.

Bromberg, Y., and Pick, E., 1984, Unsaturated fatty acids stimulate NADPH-dependent superoxide production by cell-free system derived from macrophages, *Cell. Immunol.* **88:**213–221.

Brown, O. R., Smyk-Randall, E., Draczynska-Lusiak, B., and Fee, J. A., 1995, Dihydroxy-acid dehydratase,a [4Fe-4S] cluster-containing enzyme in *Escherichia coli*: Effects of intracellular superoxide dismutase on its inactivation by oxidant stress, *Arch. Biochem. Biophys.* **319:**10–22.

Brunelli, L., Crow, J. P., and Beckman, J. S., 1995, The comparative toxicity of nitric oxide and peroxynitrite to Escherichia coli, *Arch. Biochem. Biophys.* **316:**327–334.

Buettner, G. R., and Hall, R. D., 1987, Superoxide hydrogen peroxide and singlet oxygen in hematoporphyrin derivative-cysteine, -NADH and -light systems, *Biochim. Biophys. Acta* **923:**501–507.

Cassina, A., and Radi, R., 1996, Differential inhibitory action of nitric oxide and peroxynitrite on mitochondrial electron transport, *Arch. Biochem. Biophys.* **328:**309–316.

Catz, S. D., Carreras, M. C., and Poderoso, J. J., 1995, Nitric oxide synthase inhibitors decrease human polymorphonuclear leukocyte luminol-dependent chemiluminescence, *Free Radical Biol. Med.* **19:**741–748.

Chanock, S. J., El Benna, J., Smith, R. M., and Babior, B. M., 1994, The respiratory burst oxidase, *J. Biol. Chem.* **269:**24519–24522.

Cho, H. J., Xie, Q., Calaycay, J., Mumford, R. A., Swiderek, K. M., Lee, T. D., and Nathan, C., 1992, Calmodulin is a subunit of nitric oxide synthase from macrophages, *J. Exp. Med.* **176:**599–604.

Clark, R. A., Volpp, B. D., Leidal, K. G., and Nauseef, W. M., 1990, Two cytosolic components of the human neutrophil respiratory burst oxidase translocate to the plasma membrane during cell activation, *J. Clin. Invest.* **85:**714-721.

Colton, C., Wilt, S., Gilbert, D., Chernyshev, O., Snell, J., and Dubois-Dalcq, M., 1996, Species differences in the generation of reactive oxygen species by microglia, *Mol. Chem. Neuropathol.* **28:**15–20.

Condino-Neto, A., Muscará, M. N., Bellinati-Pires, R., Carneiro-Sampaio, M. M. S., Brandao, A. C., Grumach, A. S., and De Nucci, G., 1996, Effect of therapy with recombinant human interferon-gamma on the release of nitric oxide by neutrophils and mononuclear cells from patients with chronic granulomatous disease, *J. Interferon Cytokine Res.* **16:**357–364.

Corey, E. J., and Taylor, W. C., 1964, A study of the peroxidation of organic compounds by externally generated singlet oxygen molecules, *J. Am. Chem. Soc.* **86:**3881–3882.

Cross, A. R., Higson, F. R., Jones, O. T. G., Harper, A. M., and Segal, A. W., 1982, The enzymic reduction and kinetics of oxidation of cytochrome b-245 of neutrophils, *Biochem. J.* **204:**479–485.

Cross, A. R., Parkinson, J. F., and Jones, O. T. G., 1985, Mechanism of the superoxide-producing oxidase of neutrophils. O_2 is necessary for the fast reduction of cytochrome b-245 by NADPH, *Biochem. J.* **226:**881–884.

Crow, J. P., and Beckman, J. S., 1995, The role of peroxynitrite in nitric oxide-mediated toxicity, *Curr. Top. Microbiol. Immunol.* **196:**57–73.

Crow, J. P., Spruell, C., Chen, J., Gunn, C., Ischiropoulos, H., Tsai, M., Smith, C. D., Radi, R., Koppenol, W. H., and Beckman, J. S., 1994, On the pH-dependent yield of hydroxyl radical products from peroxynitrite, *Free Radical Biol. Med.* **16:**331–338.

Curnutte, J. T., Babior, B. M., and Karnovsky, M. L., 1979, Fluoride-mediated activation of the respiratory burst in human neutrophils. A reversible process, *J. Clin. Invest.* **63:**637–647.

Denicola, A., Rubbo, H., Rodriguez, D., and Radi, R., 1993, Peroxynitrite-mediated cytotoxicity to Trypanosoma cruzi, *Arch. Biochem. Biophys.* **304:**279–286.

Denicola, A., Freeman, B. A., Trujillo, M., and Radi, R., 1996, Peroxynitrite reaction with carbon dioxide/bicarbonate: Kinetics and influence on peroxynitrite-mediated oxidations, *Arch. Biochem. Biophys.* **333:**49–58.

Dias-Da-Motta, P, Arruda, V. R., Muscara, M. N., Saad, S. T., De Nucci, G., and Costa, F. F., 1996, The release of nitric oxide and superoxide anion by neutrophils and mononuclear cells from patients with sickle cell anaemia, *Br. J. Haematol.* **93:**333–340.

DiMascio, P., Bechara, E. J., Medeiros, M. H., Briviba, K., and Sies, H., 1994, Singlet molecular oxygen production in the reaction of peroxynitrite with hydrogen peroxide, *FEBS Lett.* **355:**287–289.

Dinauer, M. C., Orkin, S. H., Brown, R., Jesaitis, A. J., and Parkos, C. A., 1987, The glycoprotein encoded by the X-linked chronic granulomatous disease locus is a component of the neutrophil cytochrome b complex, *Nature* **327:**717–720.

Driever, W., and Fishman, M. C., 1996, The zebrafish: Heritable disorders in transparent embryos, *J. Clin. Invest.* **97:**1788–1794.

Eiserich, J. P., Cross, C. E., Jones, A. D., Halliwell, B., and van der Vliet, A., 1996, Formation of nitrating and chlorinating species by reaction of nitrite with hypochlorous acid. A novel mechanism for nitric oxide-mediated protein modification, *J. Biol. Chem.* **271:**19199–19208.

El Benna, J., Faust, L. P., and Babior, B. M., 1994a, The phosphorylation of the respiratory burst oxidase component $p47^{phox}$ during neutrophil activation. Phosphorylation of sites recognized by protein kinase C and by proline-directed kinases, *J. Biol. Chem.* **269:**23431–23436.

El Benna, J., Ruedi, J. M., and Babior, B.M., 1994b, Cytosolic guanine nucleotide-binding protein Rac2 operates *in vivo* as a component of the neutrophil respiratory burst oxidase. Transfer of Rac2 and the cytosolic oxidase components p47*phox* and p67*phox* to the submembranous actin cytoskeleton during oxidase activation, *J. Biol. Chem.* **269:**6729–6734.

El Benna, J., Faust, L. P., Johnson, J. L., and Babior, B. M., 1996, Phosphorylation of the respiratory burst oxidase subunit p47phox as determined by 2-dimensional phosphopeptide mapping. Phosphorylation by protein kinase C, protein kinase A and a mitogen-activated protein kinase (MAP-kinase). *J. Biol. Chem.* **271:**6374–6378.

Epe, B., 1991, Genotoxicity of singlet oxygen, *Chem. Biol. Interact.* **80:**239–260.

Fenna, R., Zeng, J., and Davey, C., 1995, Structure of the green heme in myeloperoxidase, *Arch. Biochem. Biophys.* **316:**653–656.

Fischer, A., Segal, A. W., Seger, R., and Weening, R. S., 1993, The management of chronic granulomatous disease, *Eur. J. Pediatr.* **152:**896–899.

Fornier de Violet, P., Veyret, B., Vincendeau, P., and Caristan, A., 1984, Chemiluminescence induced by oxidation of tryptophan by singlet oxygen and by hypochlorous acid. Implications in the luminescence emitted in phagocytosis, *Photochem. Photobiol.* **39:**707–712.

Foroozan, R., Ruedi, J. M., and Babior, B. M., 1992, The reduction of cytochrome b_{558} and the activity of the respiratory burst oxidase from human neutrophils, *J. Biol. Chem.* **267:**24400–24407.

Francis, S. H., and Corbin, J. D., 1994, Structure and function of cyclic nucleotide-dependent protein kinases, *Annu. Rev. Physiol.* **56:**237–272.

Fridovich, I., 1986, Superoxide dismutases, *Adv. Enzymol.* **58:**61–97.

Fridovich, I., 1995, Superoxide radical and superoxide dismutases, *Annu. Rev. Biochem.* **64:**97–112.

Gabig, T. G., Schervish, E. W., and Santinga, J. T., 1982, Functional relationship of the cytochrome b to the superoxide-generating oxidase of human neutrophils, *J. Biol. Chem.* **257:**4114–4119.

Gaetani, G. F., Ferraris, A. M., Rolfo, M., Mangerini, R., Arena, S., and Kirkman, H. N., 1996, Predominant role of catalase in the disposal of hydrogen peroxide within human erythrocytes, *Blood* **87:**1595–1599.

Galli, C., MacArthur, R., Abu-Soud, H. M., Clark, P., Steuhr, D. J., and Brudvig, G. W., 1996, EPR spectroscopic characterization of neuronal NO synthase, *Biochemistry* **35:**2804–2810.

Gardner, P. R., Rainer, I., Epstein, L. B., and White, C. W., 1995, Superoxide radical and iron modulate aconitase activity in mammalian cells, *J. Biol. Chem.* **270:**13399–13405.

Gilbert, D. L., and Ehrenstein, G., 1984, Membrane surface charge, *Curr. Top. Membr. Transp.* **22:**407–421.

Giulivi, C., Boveris, A., and Cadenas, E., 1995, Hydroxyl radical generation during mitochondrial electron transfer and the formation of 8-hydroxydesoxyguanosine in mitochondrial DNA, *Arch. Biochem. Biophys.* **316:**909–916.

Goldstein, S., and Czapski, G., 1996, Mechanism of the nitrosation of thiols and amines by oxygenated ·NO solutions: The nature of the nitrosating intermediates, *J. Am. Chem. Soc.* **118:**3419–3425.

Green, T. R., Fellman, J. H., Eicher, A. L., and Pratt, K. L., 1991, Antioxidant role and subcellular location of hypotaurine and taurine in human neutrophils, *Biochim. Biophys. Acta Gen. Subj.* **1073:**91–97.

Haddad, I. Y., Pataki, G., Calliani, C., Beckman, J. S., and Matalon, S., 1994, Quantitation of nitrotyrosine levels in lung sections of patients and animals with acute lung injury, *J. Clin. Invest.* **94:**2407–2413.

Halliwell, B., and Gutteridge, J. M. C., 1986, Iron and free radical reactions: Two aspects of antioxidant protection, *Trends Biochem. Sci.* **11:**372–375.

Harris, L. R., Cake, M. H., and Macey, D. J., 1994, Iron release from ferritin and its sensitivity to superoxide ions differs among vertebrates, *Biochem. J.* **301:**385–389.

Harrison, J. E., and Schultz, J., 1978, Myeloperoxidase: Confirmation and nature of heme-binding inequivalence. Resolution of a carbonyl-substituted heme, *Biochim. Biophys. Acta* **536:**341–349.

Harrison, J. E., Araiso, T., Palic, M. M., and Dunford, H. B., 1980, Compound I of myeloperoxidase, *Biochem. Biophys. Res. Commun.* **94:**34–40.

Hartman, P. E, Hartman, Z., and Ault, K. T., 1990, Scavenging of singlet molecular oxygen by imidazole compounds: High and sustained activities of carboxy terminal histidine dipeptides and exceptional activity of imidazole-4-acetic acid, *Photochem. Photobiol.* **51:**59–66.

Hausladen, A., Fridovich, I., Castro, L., Rodriguez, M., and Radi, R., 1994, Superoxide and peroxynitrite inactivate aconitases, but nitric oxide does not. Aconitase is readily inactivated by peroxynitrite, but not by its precursor, nitric oxide, *J. Biol. Chem.* **269:**29405–29408, 29409–29415.

Hayakawa, T., Suzuki, K., Suzuki, S., Andrews, P. C., and Babior, B. M., 1986, A possible role for protein phosphorylation in the activation of the respiratory burst in human neutrophils, *J. Biol. Chem.* **261:**9109–9115.

Hevel, J. M., and Marletta, M. A., 1992, Macrophage nitric oxide synthase: Relationship between enzyme-bound tetrahydrobiopterin and synthase activity, *Biochemistry* **31:**7160–7165.

Hevel, J. M., White, K. A., and Marletta, M. A., 1991, Purification of the inducible murine macrophage nitric oxide synthase, *J. Biol. Chem.* **266:**22789–22791.

Higson, F. K., Kohen, R., and Chevion, M., 1988, Iron enhancement of ascorbate toxicity, *Free Radical Res. Commun.* **5:**107–115.

Hogg, N., Singh, R. J., and Kalyanaraman, B., 1996, The role of glutathione in the transport and catabolism of nitric oxide, *FEBS Lett.* **382:**223–228.

Ikeda-Saito, M., Argade, P. V., and Rousseau, D.I., 1985, Resonance Raman evidence of chloride binding to the heme iron in myeloperoxidase, *FEBS Lett.* **184:**52–55.

Inoue, S., and Kawanishi, S., 1995, Oxidative DNA damage induced by simultaneous generation of nitric oxide and superoxide, *FEBS Lett.* **371:**86–88.

Ischiropoulos, H., and Al-Mehdi, A. B., 1995, Peroxynitrite-mediated oxidative protein modifications, *FEBS Lett.* **364:**279–282.

Iyengar, R., Stuehr, D. J., and Marletta, M. A., 1987, Macrophage synthesis of nitrite, nitrate, and N-nitrosamines: Precursors and role of the respiratory burst, *Proc. Natl. Acad. Sci. USA* **84:**6369–6373.

Jennings, G., and Elia, M., 1996, Automated assay of plasma bromide after a single deproteinization step, *Clin. Chem.* **42:**1210–1213.

Johnson, D., and Travis, J., 1979, The oxidative activation of human α-1-proteinase inhibitor. More evidence for methionine at the reactive center, *J. Biol. Chem.* **254:**4022–4026.

Johnson, J. L., Park, J.- W., El Benna, J., Inanami, O., and Babior, B. M., 1996, Phosphorylation-dependent activation of the respiratory burst oxidase, *Blood* **88:**622a.

Kanofsky, J. R., Wright, J., Miles-Richardson, G. E., and Tauber, A. I., 1984, Biochemical requirements for singlet oxygen production by purified human myeloperoxidase, *J. Clin. Invest.* **74:**1489–1495.

Kanofsky, J. R., Hoogland, H., Wever, R., and Weiss, S. J., 1988, Singlet oxygen production by human eosinophils, *J. Biol. Chem.* **263:**9692–9696.

Karki, S. B., and Dinnocenzo, J. P., 1995, On the mechanism of amine oxidations by P450, *Xenobiotica* **35:**711–724.

Karoui, H., Hogg, N., Frejaville, C., Tordo, P., and Kalyanaraman, B., 1996, Characterization of sulfur-centered radical intermediates formed during the oxidation of thiols and sulfite by peroxynitrite. ESR-spin trapping and oxygen uptake studies, *J. Biol. Chem.* **271:**6000–6009.

Kaufman, S., 1987, Enzymology of the phenylalanine-hydroxylating system, *Enzyme* **38:**286–295.

Kawanishi, S., Inoue, S., Sano, S., and Aiba, H., 1986, Photodynamic guanine modification by hematoporphyrin is specific for single-stranded DNA with singlet oxygen as a mediator, *J. Biol. Chem.* **261:**6090–6095.

Kettle, A. J., and Winterbourn, C. C., 1988, The mechanism of myeloperoxidase-dependent chlorination of monochlorodimedon, *Biochim. Biophys. Acta* **957:**185–191.

Kettle, A. J., and Winterbourn, C. C., 1994, Superoxide-dependent hydroxylation by myeloperoxidase, *J. Biol. Chem.* **269:**17146–17151.

Khan, A. U., and Kasha, M., 1994, Singlet molecular oxygen evolution upon simple acidification of aqueous hypochlorite: Application to studies on the deleterious health effects of chlorinated drinking water, *Proc. Natl. Acad. Sci. USA* **91:**12362–12364.

Kharitonov, V. G., Sundquist, A. R., and Sharma, V. S., 1995, Kinetics of nitrosation of thiols by nitric oxide in the presence of oxygen, *J. Biol. Chem.* **270:**28158–28164.

Klatt, P., Pfeiffer, S., List, B. M., Lehner, D., Glatter, O., Bachinger, H. P., Werner, E. R., Schmidt, K., and Mayer, B., 1996, Characterization of heme-deficient neuronal nitric-oxide synthase reveals a role for heme in subunit dimerization and binding of the amino acid substrate and tetrahydrobiopterin, *J. Biol. Chem.* **271:**7336–7342.

Klebanoff, S. J., and Clark, R. A., 1975, Hemolysis and iodination of erythrocyte components by a myeloperoxidase-mediated system, *Blood* **45:**699–707.

Knaus, U. G., Heyworth, P. G., Kinsella, B. T., Curnutte, J. T., and Bokoch, G. M., 1992, Purification and characterization of Rac 2. A cytosolic GTP-binding protein that regulates human neutrophil NADPH oxidase, *J. Biol. Chem.* **267:**23575–23582.

Kong, S.- K., Yim, M. B., Stadtman, E. R., and Chock, P. B., 1996, Peroxynitrite disables the tyrosine phosphorylation regulatory mechanism: Lymphocyte-specific tyrosine kinase fails to phosphorylate nitrated cdc2(6–20)NH_2 peptide. *Proc. Natl. Acad. Sci. USA* **93:**3377–3382.

Korytowski, W., Bachowski, G. J., and Girotti, A. W., 1992, Photoperoxidation of cholesterol in homogeneous solution, isolated membranes, and cells: Comparison of the 5α- and 6β-hydroperoxides as indicators of singlet oxygen intermediacy, *Photochem. Photobiol.* **56:**1–8.

Krieger, M., and Herz, J., 1994, Structures and functions of multiligand lipoprotein receptors: Macrophage scavenger receptors and LDL receptor-related protein (LRP), *Annu. Rev. Biochem.* **63:**601–637.

Larfars, G., and Gyllenhammar, H., 1995, Measurement of methemoglobin formation from oxyhemoglobin. A real-time, continuous assay of nitric oxide release by human polymorphonuclear leukocytes, *J. Immunol. Methods* **184:**53–62.

Lee, H. C., Booth, K. S., Caughey, W. S., and Ikeda-Saito, M., 1991, Interaction of halides with the cyanide complex of myeloperoxidase: A model for substrate binding to compound I, *Biochim. Biophys. Acta* **1076:**317–320.

Lemercier, J.- N., Squadrito, G. L., and Pryor, W. A., 1995, Spin trap studies on the decomposition of peroxynitrite, *Arch. Biochem. Biophys.* **321:**31–39.

Leto, T. L., Lomax, K. J., Volpp, B. D., Nunoi, H., Sechler, J. M. G., Nauseef, W. M., Clark, R. A., Gallin, J. I., and Malech, H. L., 1990, Cloning of a 67-kD neutrophil oxidase factor with similarity to a noncatalytic region of p60^{c-}*src*, *Science* **248:**727–730.

Lomax, K. J., Leto, T. L., Nunoi, H., Gallin, J. I., and Malech, H. L., 1989, Recombinant 47kD cytosol factor restores NADPH oxidase in chronic granulomatous disease [published erratum appears in *Science* **246:**987 (1989)], *Science* **245:**409–412.

Lymar, S. V., Jiang, Q., and Hurst, J. K., 1996, Mechanism of carbon dioxide-catalyzed oxidation of tyrosine by peroxynitrite. *Biochemistry* **35:**7855–7861.

Maly, F. E., Quilliam, L. A., Dorseuil, O., Der, C. J., and Bokoch, G. M., 1994, Activated or dominant inhibitory mutants of Rap1A decrease the oxidative burst of Epstein–Barr virus-transformed human B lymphocytes, *J. Biol. Chem.* **269:**18743–18746.

Marletta, M. A., 1993, Nitric oxide synthase structure and mechanism. *J. Biol. Chem.* **268:**12231–12234.

Marquez, L. A., Huang, J. T., and Dunford, H. B., 1994, Spectral and kinetic studies on the formation of myeloperoxidase compounds I and II: Roles of hydrogen peroxide and superoxide, *Biochemistry* **33:**1447–1454.

Masters, B. S., McMillan, K., Sketa, E. A., Nishimura, J. S., Roman, L. J., and Martasek, P., 1996, Neuronal nitric oxide synthase, a modular enzyme formed by convergent evolution: Structure studies of a cysteine thiolate-liganded heme protein that hydroxylates L-arginine to produce NO as a cellular signal, *FASEB J.* **10:**552–558.

Mateo, R. B., Reichner, J. S., and Albina, J. E., 1996, NO is not sufficient to explain maximal cytotoxicity of tumoricidal macrophages against an NO-sensitive cell line, *J. Leukocyte Biol.* **60:**245–252.

McCall, T. B., Boughton-Smith, N. K., Palmer, R. M., Whittle, B. J., and Moncada, S., 1989, Synthesis of nitric oxide from L-arginine by neutrophils. Release and interaction with superoxide anion, *Biochem. J.* **261:**293–296.

McMillan, K., and Masters, B. S., 1995, Prokaryotic expression of the heme- and flavin-binding domains of rat neuronal nitric oxide synthase as distinct polypeptides: Identification of the heme-binding proximal thilate ligand as cysteine-415, *Biochemistry* **34:**3686–3693.

McMillan, K., Bredt, D. S., Hirsch, D. J., Snyder, S. H., Clark, J. E., and Masters, B. S., 1992, Cloned, expressed rat cerebellar nitric oxide synthase contains stoichiometric amounts of heme, which binds carbon monoxide, *Proc. Natl. Acad. Sci. USA* **89:**11141–11145.

Michiels, C., Raes, M., Toussaint, O., and Remacle, J., 1994, Importance of Se-glutathione peroxidase, catalase, and Cu/Zn-SOD for cell survival against oxidative stress, *Free Radical Biol. Med.* **17:**235–248.

Midi, T., Yu, L., and Yu, C. A., 1991, Hematoporphyrin-promoted photoinactivation of mitochondrial ubiquinol–cytochrome c reductase: Selective destruction of the histidine ligands of the iron–sulfur cluster and protective effect of ubiquinone, *Biochemistry* **30:**230–238.

Murad, F., 1994, Regulation of cytosolic guanylyl cyclase by nitric oxide: The NO–cyclic GMP signal transduction system, *Adv. Pharmacol.* **26:**19–33.

Nathan, C. F., 1987, Secretory products of macrophages, *J. Clin. Invest.* **79:**319–326.

Nauseef, W. M., 1988, Myeloperoxidase deficiency, in *Hematology/Oncology Clinics of North America* (J. T. Curnutte, ed.), pp. 135–158, Saunders, Philadelphia.

Palmer, R. M. J., Ferrige, A. G., and Moncada, S., 1987, Nitric oxide release accounts for the biological activity of endothelium-derived relaxing factor, *Nature* **327:**524–526.

Park, J.- W., Ma, M., Ruedi, J. M., Smith, R. M., and Babior, B. M., 1992, The cytosolic components of the respiratory burst oxidase exist as a Mr 240,000 complex that acquires a membrane-binding site during activation of the oxidase in a cell-free system, *J. Biol. Chem.* **267:**17327–17332.

Park, J.- W., El Benna, J., Scott, K. E., Christensen, B. L., Chanock, S. J., and Babior, B. M., 1994, Isolation of a complex of respiratory burst oxidase components from resting neutrophil cytosol, *Biochemistry* **33:**2907–2911.

Parkos, C. A., Allen, R. A., Cochrane, C. G., and Jesaitis, A. J., 1987, Purified cytochrome b from human granulocyte plasma membrane is comprised of two polypeptides with relative molecular weights of 91,000 and 22,000, *J. Clin. Invest.* **80:**732–742.

Parkos, C. A., Dinauer, M. C., Walker, L. E., Allen, R. A., Jesaitis, A. J., and Orkin, S. H., 1988, Primary structure and unique expression of the 22-kilodalton light chain of human neutrophil cytochrome b, *Proc. Natl. Acad. Sci. USA* **85:**3319–3323.

Paul-Eugene, N., Kolb, J. P., Sarfati, M., Arock, M., Ouaaz, F., Debre, P., Mossalayi, D. M., and Dugas, B., 1995, Ligation of CD23 activates soluble guanylate cyclase in human monocytes via an L-arginine-dependent mechanism, *J. Leukocyte Biol.* **57:**160–167.

Pick, E., Bromberg, Y., Sphungin, S., and Gadba, R., 1987, Activation of the superoxide forming NADPH oxidase in a cell-free system by sodium dodecyl sulfate. Characterization of the membrane-associated component, *J. Biol. Chem.* **262:**16476–16483.

Pou, S., Nguyen, S. Y., Gladwell, T., and Rosen, G. M., 1995, Does peroxynitrite generate hydroxyl radical? *Biochim. Biophys. Acta* **1244:**62–68.

Pryor, W. A., and Squadrito, G. L., 1995, The chemistry of peroxynitrite: A product from the reaction of nitric oxide with superoxide, *Am. J. Physiol. Lung Cell. Mol. Physiol.* **268:**L699–L722.

Pryor, W. A., Jin, X., and Squadrito, G. L., 1994, One- and two-electron oxidations of methionine by peroxynitrite, *Proc. Natl. Acad. Sci. USA* **91:**11173–11177.

Pufahl, R. A., Nanjappan, P. G., Woodard, R. W., and Marletta, M. A., 1992, Mechanistic probes of N-hydroxylation of L-arginine by the inducible nitric oxide synthase from murine macrophages, *Biochemistry* **31:**6822–6828.

Rakita, R. M., and Rosen, H., 1991, Penicillin-binding protein inactivation by human neutrophil myeloperoxidase, *J. Clin. Invest.* **88:**750–754.

Ramos, C. L., Pou, S., Britigan, B. E., Cohen, M. S., and Rosen, G. M., 1992, Spin trapping evidence for myeloperoxidase-dependent hydroxyl radical formation by human neutrophils and monocytes, *J. Biol. Chem.* **267:**8307–8312.

Rodriguez-Crespo, I., Gerber, N. C., and Ortiz de Montellano, P. R., 1996, Endothelial nitric-oxide synthase. Expression in Escherichia coli, spectroscopic characterization, and role of tetrahydrobiopterin in dimer formation, *J. Biol. Chem.* **271:**11462–11467.

Root, R. K., Metcalf, J., Oshino, N., and Chance, B., 1975, H_2O_2 release from human granulocytes during phagocytosis. I. Documentation, quantitation and some regulating factors, *J. Clin. Invest.* **55:**945–955.

Rosen, H., Rakita, R. M., Waltersdorph, A. M., and Klebanoff, S. J., 1987, Myeloperoxidase-mediated damage to the succinate oxidase system of Escherichia coli. Evidence for selective inactivation of the dehydrogenase component, *J. Biol. Chem.* **242:**15004–15010.

Rosen, H., Orman, J., Rakita, R. M., Michel, B. R., and VanDevanter, D. R., 1990, Loss of DNA–membrane interactions and cessation of DNA synthesis in myeloperoxidase-treated Escherichia coli, *Proc. Natl. Acad. Sci. USA* **87:**10048–10052.

Rotrosen, D., Yeung, C. L., Leto, T. L., Malech, H. L., and Kwong, C. H., 1992, Cytochrome b558: The flavin-binding component of the phagocyte NADPH oxidase, *Science* **256:**1459–1462.

Rowe, G. T., Manson, N. H., Caplan, M., and Hess, M. L., 1983, Hydrogen peroxide and hydroxyl radical mediation of activated leukocyte depression of cardiac sarcoplasmic reticulum. Participation of the cyclooxygenase pathway, *Circ. Res.* **53:**584–591.

Rubbo, H., Radi, R., Trujillo, M., Telleri, R., Kalyanaraman, B., Barnes, S., Kirk, M., and Freeman, B. A., 1994, Nitric oxide regulation of superoxide and peroxynitrite-dependent lipid peroxidation. Formation of novel nitrogen-containing oxidized lipid derivatives, *J. Biol. Chem.* **269:**26066–26075.

Sakai, N., and Milstein, S., 1993, Availability of tetrahydrobiopterin is not a factor in the inability to detect nitric oxide production by human macrophages, *Biochem. Biophys. Res. Commun.* **193:**378–383.

Salgo, M. G., Bermudez, E., Squadrito, G. L., and Pryor, W. A., 1995, Peroxinitrite causes DNA damage and oxidation of thiols in rat thymocytes [corrected] [published erratum appears in *Arch. Biochem. Biophys.* **324**(1)**:**200 (1995)], *Arch. Biochem. Biophys.* **322:**500–505.

Salvemini, D., Nucci, G., Gryglewski, R. J., and Vane, J. R., 1989, Human neutrophils and mononuclear cells inhibit platelet aggregation by releasing a nitric oxide-like factor, *Proc. Natl. Acad. Sci. USA* **86:**6328–6332.

Schneemann, M., Schoedon, G., Hofer, S., Blau, N., Guerrero, L., and Schaffner, A., 1993, Nitric oxide synthase is not a constituent of the antimicrobial armature of human mononuclear phagocytes, *J. Infect. Dis.* **167:**1358–1363.

Schraufstatter, I. U., Hinshaw, D. B., Hyslop, P. A., Spragg, R. G., and Cochrane, C. G., 1985, Glutathione cycle activity and pyridine nucleotide levels in oxidant-induced injury of cells, *J. Clin. Invest.* **76:**1131–1139.

Schraufstatter, I. U., Hyslop, P. A., Hinshaw, D. B., Spragg, R. G., Sklar, L. A., and Cochrane, C. G., 1986, Hydrogen peroxide-induced injury of cells and its prevention by inhibitors of poly(ADP-ribose) polymerase, *Proc. Natl. Acad. Sci. USA* **83:**4908–4912.

Schraufstatter, I. U., Browne, K., Harris, A., Hyslop, P. A., Jackson, J. H., Quehenberger, O., and Cochrane, C. G., 1990, Mechanisms of hypochlorite injury of target cells, *J. Clin. Invest.* **85:**554–562.

Segal, A. W., West, I., Wientjes, F., Nugent, J. H. A., Chavan, A. J., Haley, B., Garcia, R. C., Rosen, H., and Scrace, G., 1992, Cytochrome b245 is a flavocytochrome containing FAD and the NADPH-binding site of the microbicidal oxidase of phagocytes, *Biochem. J.* **284:**781–788.

Selvaraj, R. J., Zgliczynski, J. M., Paul, B. B., and Sbarra, A. J., 1980, Chlorination of reduced nicotinamide adenine dinucleotides by myeloperoxidase: A novel bactericidal mechanism, *J. Reticuloendothel. Soc.* **27:**31–38.

Sengelov, H., Nielsen, M. H., and Borregaard, N., 1992, Separation of human neutrophil plasma membrane from intracellular vesicles containing alkaline phosphatase and NADPH oxidase activity by free flow electrophoresis, *J. Biol. Chem.* **267:**14912–14917.

Sennequier, N., Boucher, J. L., Battioni, P., and Mansuy, D., 1995, Superoxide anion efficiently performs the oxidative cleavage of C=NOH bonds of amidoximes and N-hydroxyguanidines with formation of nitrogen oxides, *Tetrahedron Lett.* **36:**6059–6062.

Shi, X., Rojanasakul, Y., Gannett, P., Liu, K., Mao, Y., Daniel, L. N., Ahmed, N., and Saffiotti, U., 1994, Generation of thiyl and ascorbyl radicals in the reaction of peroxynitrite with thiols and ascorbate at physiological pH, *J. Inorg. Biochem.* **56:**77–86.

Singh, R. J., Hogg, N., Joseph, J., and Kalyanaraman, B., 1996, Mechanism of nitric oxide release from S-nitrosothiols, *J. Biol. Chem.* **271:**18596–18603.

Sono, M., Bracete, A. M., Huff, A. M., Ikeda-Saito, M., and Dawson, J. H., 1991, Evidence that a formyl-substituted iron porphyrin is the prosthetic group of myeloperoxidase: Magnetic circular dichroism similarity of the peroxidase to Spirographis heme-reconstituted myoglobin, *Proc. Natl. Acad. Sci. USA* **88:**11148–11152.

Squadrito, G. L., Jin, X., and Pryor, W. A., 1995, Stopped-flow kinetic study of the reaction of ascorbic acid with peroxynitrite, *Arch. Biochem. Biophys.* **322:**53–59.

Stadtman, E. R., 1995, Role of oxidized amino acids in protein breakdown and stability, *Methods Enzymol.* **258:**379–393.

Steinbeck, M. J., Khan, A. U., Karnovsky, M. J., and Hegg, G. G., 1992, Intracellular singlet oxygen generation by phagocytosing neutrophils in response to particles coated with a chemical trap, *J. Biol. Chem.* **267:**13425–13433.

Steinbeck, M. J., Khan, A. U., and Karnovsky, M. J., 1993, Extracellular production of singlet oxygen by stimulated macrophages quantified using 9,10-diphenylanthracene and perylene in a polystyrene film, *J. Biol. Chem.* **268:**15649–15654.

Stevens-Truss, R., and Marletta, M.A., 1995, Interaction of calmodulin with the inducible murine macrophage nitric oxide synthase, *Biochemistry* **34:**15638–15645.

Stone, J. R., and Marletta, M. A., 1996, Spectral and kinetic studies on the activation of soluble guanylate cyclase by nitric oxide, *Biochemistry* **35:**1093–1099.

Stuehr, D. J., Cho, H. J., Kwon, N. S., Weise, M. F., and Nathan, C. F., 1991, Purification and characterization of the cytokine-induced macrophage nitric oxide synthase: An FAD- and FMN-containing flavoprotein, *Proc. Natl. Acad. Sci. USA* **88:**7773–7777.

Szabo, C., Zingarelli, B., O'Connor, M., and Salzman, A. L., 1996, DNA strand breakage, activation of poly(ADP-ribose) synthetase, and cellular energy depletion are involved in the cytotoxicity in macrophages and smooth muscle cells exposed to peroxynitrite, *Proc. Natl. Acad. Sci. USA* **93:**1753–1758.

Tayeh, M. A., and Marletta, M. A., 1989, Macrophage oxidation of L-arginine to nitric oxide, nitrite, and nitrate, *J. Biol. Chem.* **264:**19654–19658.

Taylor, K. L., Pohl, J., and Kinkade, J. M., Jr., 1992, Unique autolytic cleavage of human myeloperoxidase. Implications for the involvement of active site MET409, *J. Biol. Chem.* **267:**25282–25288.

Taylor, K. L., Strobel, F., Yue, K. T., Ram, P., Pohl, J., Woods, A. S., and Kinkade, J. M., Jr., 1995, Isolation and identification of a protoheme IX derivative released during autolytic cleavage of human myeloperoxidase, *Arch. Biochem. Biophys.* **316:**635–642.

Thomas, E. L., Jefferson, M. M., and Grisham, M., 1982, Myeloperoxidase-catalyzed incorporation of amino acids into proteins: Role of hypochlorous acid and chloramines, *Biochemistry* **21:**6299–6308.

Thomas, E. L., Grisham, M. B., and Jefferson, M. M., 1983, Myeloperoxidase-dependent effect of amines on functions of isolated neutrophils, *J. Clin. Invest.* **72:**441–454.

Thomas, E. L., Bozeman, P. M., Jefferson, M. M., and King, C. C., 1995, Oxidation of bromide by the human leukocyte enzymes myeloperoxidase and eosinophil peroxidase. Formation of bromamines, *J. Biol. Chem.* **270:**2906–2913.

Timcenko-Youssef, L., Yamazaki, R. K., and Kimura, T., 1985, Subcellular localization of adrenal cortical glutathione peroxidase and protective role of the mitochondrial enzyme against lipid peroxidative damage, *J. Biol. Chem.* **260:**13355–13359.

Turk, J., Henderson, W. R., Klebanoff, S. J., and Hubbard, W. C., 1983, Iodination of arachidonic acid mediated by eosinophil peroxidase, myeloperoxidase, and lactoperoxidase: Identification and comparison of products, *Biochim. Biophys. Acta* **751:**189–200.

Uppu, R. M., Squadrito, G. L., and Pryor, W. A., 1996, Acceleration of peroxynitrite oxidations by carbon dioxide, *Arch. Biochem. Biophys.* **327:**335–343.

van der Vliet, A., O'Neill, C. A., Halliwell, B., Cross, C. E., and Kaur, H., 1994, Aromatic hydroxylation and nitration of phenylalanine and tyrosine by peroxynitrite. Evidence for hydroxyl radical production from peroxynitrite, *FEBS Lett.* **339:**89–92.

Vasquez-Vivar, J., Santos, A. M., Junqueira, V. B., and Augusto, O., 1996, Peroxynitrite-mediated formation of free radicals in human plasma: EPR detection of ascorbyl, albumin-thiyl and uric acid-derived radicals, *Biochem. J.* **314:**869–876.

Vazquez-Torres, A., Jones-Carson, J., and Balish, E., 1996, Peroxynitrite contributes to the candidacidal activity of nitric oxide-producing macrophages, *Infect. Immun.* **64:**3127–3133.

Vissers, M. C., and Winterbourn, C. C., 1995, Oxidation of intracellular glutathione after exposure of human red blood cells to hypochlorous acid, *Biochem. J.* **307:**57–62.

Volkman, D. J., Buescher, E. S., Gallin, J. I., and Fauci, A. S., 1984, B cell lines as models for inherited phagocytic diseases: Abnormal superoxide generation in chronic granulomatous disease and giant granules in Chediak-Higashi syndrome, *J. Immunol.* **133:**3006–3009.

Volpp, B. D., Nauseef, W. M., Donelson, J. E., Moser, D. R., and Clark, R. A., 1989, Cloning of the cDNA and functional expression of the 47-kilodalton cytosolic component of human neutrophil respiratory burst oxidase [published erratum appears in *Proc. Natl. Acad. Sci. USA* **86**(23)**:**9563 (1989)], *Proc. Natl. Acad. Sci. USA* **86:**7195–7199.

Vouldoukis, I., Riveros-Moreon, V., Dugas, B., Ouaaz, F., Becherel, P., Debre, P., Moncada, S., and Mossalayi, M. D., 1995, The killing of Leishmania major by human macrophages is mediated by nitric oxide induced after ligation of the Fc epsilon RII/CD23 surface antigen, *Proc. Natl. Acad. Sci. USA* **92:**7804–7808.

Wang, Y., and Marsden, P. A., 1995, Nitric oxide synthases: Gene structure and regulation, *Adv. Pharmacol.* **34:**71–90.

Weiss, S. J., Test, S. T., Eckmann, C. M., Roos, D., and Regiani, S., 1986, Brominating oxidants generated by human eosinophils, *Science* **234:**200–202.

Wever, R., Oertling, W. A., Hoogland, H., Bolscher, B. G., Kim, Y., and Babcock, G. T., 1991, Denaturation and renaturation of myeloperoxidase. Consequences for the nature of the prosthetic group, *J. Biol. Chem.* **266:**24308–24313.

White, K. A., and Marletta, M. A., 1992, Nitric oxide synthase is a cytochrome P-450 type hemoprotein, *Biochemistry* **31:**6627–6631.

Wientjes, F. B., Hsuan, J. J., Totty, N. F., and Segal, A. W., 1993, p40-phox, a third cytosolic component of the activation complex of the NADPH oxidase to contain src homology domains, *Biochem. J.* **296:**557–561.

Wink, D. A., Nims, R. W., Darbyshire, J. F., Christodoulou, D., Hanbauer, I., Cox, G. W., Laval, F., Laval, J., Cook, J. A., Krishna, M. E., and DeGraff, N. G., and Mitchell, J. B., 1994, Reaction kinetics for nitrosation of cysteine and glutathione in aerobic nitric oxide solutions at neutral pH. Insights into the fate and physiological effects of intermediates generated in the NO/O_2 reaction, *Chem. Res. Toxicol.* **7:**519–525.

Xia, Y., Dawson, V. L., Dawson, T. M., Snyder, S. H., and Zweier, J. L., 1996, Nitric oxide synthase generates superoxide and nitric oxide in arginine-depleted cells leading to peroxynitrite-mediated cellular injury, *Proc. Natl. Acad. Sci. USA* **93:**6770–6774.

Yazdanbakhsh, M., Tai, P.-C., Spry, C. J. F., Gleich, G. J., and Roos, D., 1987, Synergism between eosinophil cationic protein and oxygen metabolites in killing of schistosomula of Schistosoma mansoni, *J. Immunol.* **138:**3443–3447.

Yermilov, V., Rubio, J., Becchi, M., Friesen, M. D., Ignatelli, B., and Oshima, H., 1995, Formation of 8-nitroguanine by the reaction of guanine with peroxynitrite in vitro, *Carcinogenesis* **16:**2045–2050.

Youngman, R. J., Wagner, G. R., Kuhne, F. W., and Elstner, E. F., 1985, Biochemical oxygen activation as the basis for the physiological action of tetrachlorodecaoxide (TCDO), *Z. Naturforsch.* **40:**409–414.

Zeng, J., and Fenna, R. E., 1992, X-ray crystal structure of canine myeloperoxidase at 3 A resolution, *J. Mol. Biol.* **226:**185–207.

Zhu, L., Gunn, C., and Beckman, J. S., 1992, Bactericidal activity of peroxynitrite, *Arch. Biochem. Biophys.* **298:**452–457.

Zingarelli, B., O'Connor, M., Wong, H., Salzman, A. L., and Szabó, C., 1996, Peroxynitrite-mediated DNA strand breakage activates poly-adenosine diphosphate ribosyl synthetase and causes cellular energy depletion in macrophages stimulated with bacterial lipopolysaccharide, *J. Immunol.* **156:**350–358.

Zuurbier, K. W. M., van den Berg, J. D., Van Gelder, B. F., and Muijsers, A. O., 1992, Human hemi-myeloperoxidase. Initial chlorinating activity at neutral pH, compound II and III formation, and stability towards hypochlorous acid and high temperature, *Eur. J. Biochem.* **205:**737–742.

Chapter 20

Production and Effects of Reactive Oxygen Species by Spermatozoa

R. John Aitken

1. INTRODUCTION

The potential of human spermatozoa to generate reactive oxygen species (ROS) has been known since 1943, when the Scottish andrologist John MacLeod noted that catalase had the capacity to preserve the motility of human spermatozoa cultured *in vitro* (MacLeod, 1943). He concluded from this result that human spermatozoa must have an intrinsic capacity to generate hydrogen peroxide and yet were susceptible to the stress created by this oxidant. Definitive proof that spermatozoa have the capacity to generate hydrogen peroxide was obtained 3 years later by Tosic and Walton (1946). The biochemistry of ROS generation by mammalian spermatozoa and the physiological and pathological consequences of this activity were then ignored until a series of landmark papers by Thaddeus Mann and Roy Jones in the 1970s reawakened interest in the field (Jones and Mann, 1973, 1976, 1977a,b; Jones *et al.*, 1978, 1979). During the past 15 years, significant progress has been made in resolving the functional significance of ROS generated by mammalian spermatozoa and in determining the importance of oxidative stress in the etiology of male infertility. These developments are reviewed in this chapter.

2. ROS GENERATION BY MAMMALIAN SPERMATOZOA

2.1. Biochemical Mechanisms

The first demonstration that mammalian spermatozoa could generate ROS came from Tosic and Walton's (1946) pioneering studies on bovine spermatozoa. Using the benzidine

R. John Aitken MRC Reproductive Biology Unit, Centre for Reproductive Biology, Edinburgh EH3 9EW, Scotland.

Reactive Oxygen Species in Biological Systems, edited by Gilbert and Colton. Kluwer Academic / Plenum Publishers, New York, 1999.

peroxidase reaction, they were able to demonstrate the production of hydrogen peroxide by these cells and concluded that the excessive production of this oxidant suppressed respiration and sperm motility. The source of this activity was found to be an amino acid oxidase with an affinity for aromatic amino acids such as tryptophan, tyrosine, and phenylalanine, proceeding according to the general equation

$$RCH_2CH(NH_2)COOH + O_2 + H_2O = RCH_2COCOOH + H_2O_2 + NH_3$$

where R is a phenyl, *p*-hydroxyphenyl or indolyl group (Tosic and Walton, 1950).

Tosic and Walton (1950) were also able to demonstrate the generation of hydrogen peroxide by boar and ram spermatozoa and more recent studies have confirmed the production of ROS by the spermatozoa of all mammalian species examined, including mouse (Alvarez and Storey, 1984), rat (Kumar *et al.*, 1991), hamster (Bize *et al.*, 1991), and rabbit (Alvarez and Storey, 1982). In 1987 the production of ROS by human spermatozoa was reported independently by Aitken and Clarkson (1987) and Alvarez *et al.* (1987). The primary product of the free radical generating system in human spermatozoa appears to be superoxide anion, which then secondarily dismutates to hydrogen peroxide under the influence of superoxide dismutase (SOD) (Alvarez *et al.*, 1987).

There is currently no consensus as to the biochemical mechanisms responsible for the generation of ROS by mammalian spermatozoa. The amino acid oxidase discovered by Tosic and Walton (1950) in bovine spermatozoa has not been found in other species. In the rabbit there is evidence that at least some of the superoxide generated by the spermatozoa represents the leakage of electrons from the mitochondrial electron transport chain (Holland *et al.*, 1982). However, neither of these mechanisms contributes to ROS generation by human spermatozoa (Aitken and Clarkson, 1987). In the latter, there is some evidence to suggest that NADPH plays a pivotal role as the source of electrons for the generation of superoxide anion. Addition of high doses of NADPH to populations of intact, viable human spermatozoa has been found to stimulate superoxide anion production (Figure 1), in much the same way as the addition of this cofactor has been found to stimulate ROS production by other cell types, including glomerular mesangial cells or

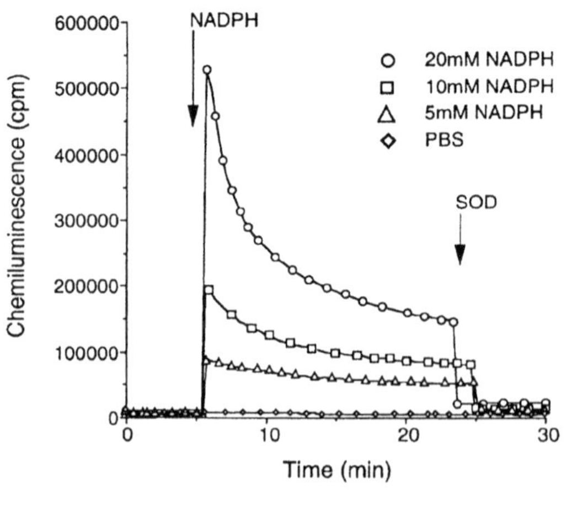

FIGURE 1. Mechanisms of ROS generation by human spermatozoa. High doses of NADPH induce ROS generation in populations of intact, functional spermatozoa (Adapted from Aitken *et al.*, 1996a).

fibroblasts (Meier *et al.*, 1991; Radeke *et al.*, 1991). Furthermore, strong correlations exist between the generation of ROS by human spermatozoa and the activity of an enzyme (glucose-6-phosphate dehydrogenase; G6PDH) that controls the intracellular generation of NADPH through the hexose monophosphate shunt (Aitken *et al.*, 1994b). The importance of glucose oxidation in the generation of ROS by human spermatozoa is also suggested by the ability of 2-deoxyglucose to completely suppress this activity. The enzymes responsible for effecting the transfer of electrons from NADPH to ground-state oxygen have not been characterized although they appear to be distinct from the NADPH oxidase complex of phagocytic leukocytes on the basis of their cross-reactivity with monoclonal antibodies directed against the α and β subunits of cytochrome b_{558} (unpublished observations).

2.2. ROS and Male Infertility

Clinical interest in the production of ROS by human spermatozoa stems from the observation that defective sperm function is frequently linked with oxidative stress. Jones *et al.* (1979) demonstrated that clinical conditions associated with the complete loss of sperm viability (necrozoospermia) were associated with increased peroxidative damage to the spermatozoa. Subsequent studies have demonstrated that less severe disruptions of sperm motility, as well as defects in the capacity of human spermatozoa to fuse with the vitelline membrane of the oocyte, were also correlated with high rates of lipid peroxidation (Aitken *et al.*, 1993a,b; Figure 2). In fact, a subpopulation of defective, peroxidized spermatozoa can be detected in every human ejaculate and can be recovered from the low-density region of discontinuous Percoll gradients (Aitken *et al.*, 1989a). The size of this defective population is inversely correlated with semen quality.

The origins of this peroxidative damage might involve the excessive generation of ROS by the spermatozoa themselves, exposure to cytotoxic oxygen metabolites generated by leukocytes migrating into the male reproductive tract, or defects in the antioxidant protection afforded to the spermatozoa by the genital tract secretions. Evidence for each of these sources of oxidative stress has been found.

2.2.1. Excessive Generation of ROS by Spermatozoa

One of the first pieces of evidence that male infertility might involve oxidative stress came from the observation that the ejaculates of infertile male patients generated significantly higher levels of ROS than samples recovered from fertile donors (Aitken and Clarkson, 1987; Iwasaki and Gagnon, 1992). Elevated ROS generation has been reported in around 25 to 40% of semen samples from unselected patients consulting for infertility (Iwasaki and Gagnon, 1992; Zini *et al.*, 1993) and significant negative correlations have been observed between ROS generation and various aspects of sperm function, including motility and the capacity for sperm–oocyte fusion (Aitken and Clarkson, 1987; Mazzilli *et al.*, 1994). For example, in an analysis of oligozoospermic patients, 55% of samples were characterized by defective sperm–oocyte fusion associated with elevated levels of ROS generation (Aitken *et al.*, 1989b; Figure 3). By contrast, a cohort of 70 fertile control donors displayed an overwhelming pattern of low levels of ROS generation coupled with a high competence for sperm–oocyte fusion (Aitken *et al.*, 1989b). Such negative

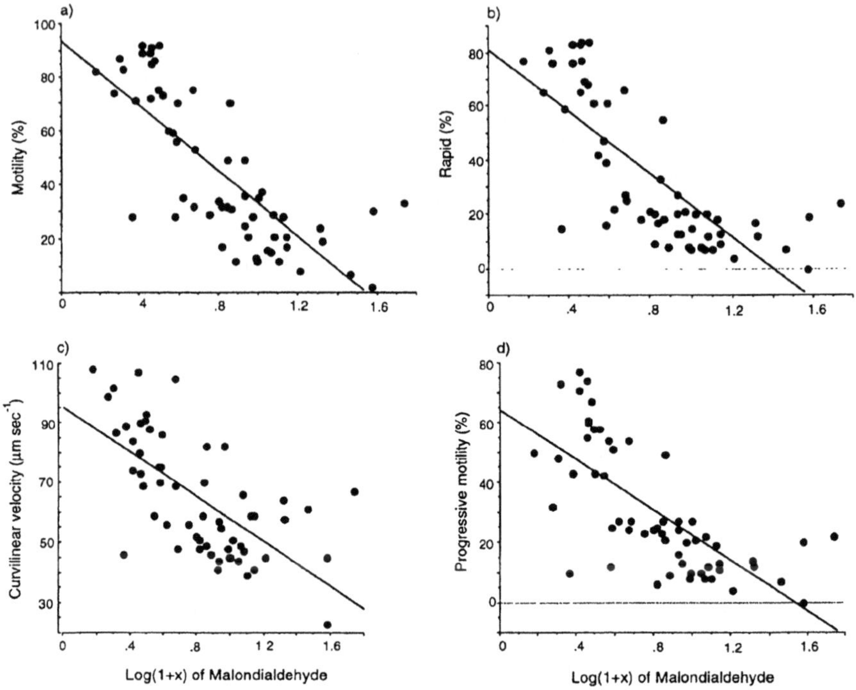

FIGURE 2. Negative relationships between a marker for peroxidative damage to the spermatozoa, malondialdehyde generation in the presence of a ferrous ion promoter, and different aspects of sperm motility (Adapted from Aitken *et al.*, 1993a).

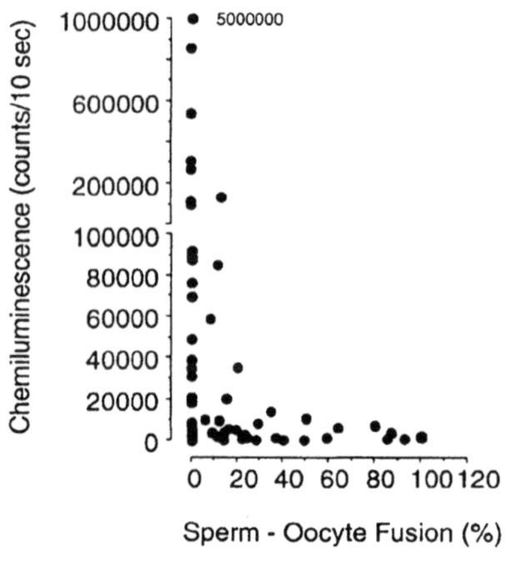

FIGURE 3. Analysis of the relationship between ROS generation and the competence of human spermatozoa to exhibit sperm-oocyte fusion in cases of oligozoospermia. Note the high proportion of samples exhibiting low levels of sperm–oocyte fusion in association with high levels of ROS production (Adapted from Aitken *et al.*, 1989b).

correlations between the competence of human sperm suspensions to generate ROS and their functional competence are not only of significance *in vitro*. In a long-term prospective study of the spontaneous conception rates recorded in couples characterized by a lack of detectable abnormalities in the female partner, ROS generation was shown to be significantly correlated with fertility after a follow-up period lasting a median of 17 months. The clinical significance of these data was emphasized by the fact that within the same data set, the conventional criteria of semen quality (sperm count, motility, and morphology) were of no diagnostic significance whatsoever (Aitken *et al.*, 1991).

Based on a lack of luminol-dependent chemiluminescent activity given by azoospermic samples (ejaculates lacking spermatozoa), it has been concluded that a significant source of ROS is the spermatozoa themselves (Iwasaki and Gagnon, 1992). However, it should also be recognized that every human semen sample is contaminated with leukocytes, and the predominant species is the neutrophil (Aitken *et al.*, 1994a). The latter are powerful generators of ROS and highly significant correlations exist between the luminol-dependent chemiluminescent signals given by unfractionated semen samples and the levels of leukocyte contamination (Aitken *et al.*, 1994a; Figure 4). As a consequence of the presence of such cells, it is impossible to determine whether studies reporting differences in ROS generation between fertile and infertile ejaculates, are referring to differences in the free radical generating capacity of the spermatozoa or simply variable levels of leukocyte contamination (Aitken *et al.*, 1989b; Weese *et al.*, 1993). To address this problem, careful studies have been conducted in which the levels of leukocyte contamination have been accurately quantified and strategies developed to effect the selective removal of these cells (Aitken and West, 1990; Aitken *et al.*, 1992, 1996b). As a result of such analyses, it is now clear that there are two major sources of ROS in the human ejaculate, namely, neutrophils and defective spermatozoa.

Thus, sperm populations recovered from the high- and low-density regions of Percoll gradients generate luminol-dependent chemiluminescent signals that vary over two log orders of magnitude in the absence of any significant leukocyte contamination (Aitken and West, 1990), suggesting that spermatozoa are a significant source of ROS. This conclusion was reinforced by a recent comparison of fertile donors and oligozoospermic patients (Aitken *et al.*, 1992) in which no significant differences were detected in the mean number of leukocytes contaminating the semen samples. However, the purified sperm

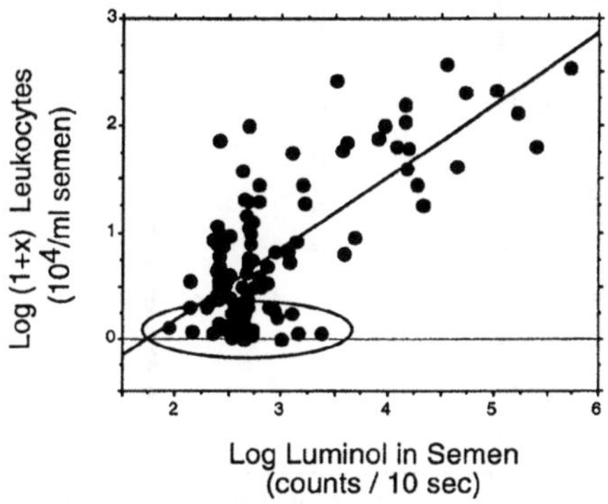

FIGURE 4. Relationship between the luminol-dependent chemiluminescent signals produced by human semen samples and the degree of leukocyte contamination. Although some samples are present (circled) in which variable ROS generation is evident despite negligible leukocyte contamination, in general, leukocyte infiltration and ROS production are highly correlated.

populations were found to have very different capacities for ROS generation; the oligozoospermic samples giving a median chemiluminescent signal that was 100-fold greater than the fertile specimens after stimulation with phorbol ester (Figure 5). This scattergram plot also emphasizes that whereas most cases of oligozoospermia were characterized by excessive ROS generation by the spermatozoa, this did not apply to every sample in the data set. In other words, *oligozoospermia* is a descriptive label encapsulating a variety of conditions with different underlying etiologies, and although oxidative stress appears to be a significant contributor to the loss of fertility in most cases, it is not the only pathogenic mechanism operating in this cohort of patients.

Additional studies involving the use of CD45-coated ferrofluids and magnetic beads to completely remove the leukocytes from human sperm suspensions have also confirmed the spermatozoon as a potential source of ROS (Aitken *et al.*, 1996b). The reasons behind the enhanced capacity for ROS generation exhibited by defective spermatozoa, frequently appear to involve defects during spermiogenesis leading to the retention of excess residual cytoplasm in the midpiece of the spermatozoa (Figure 6) . As spermatozoa complete their

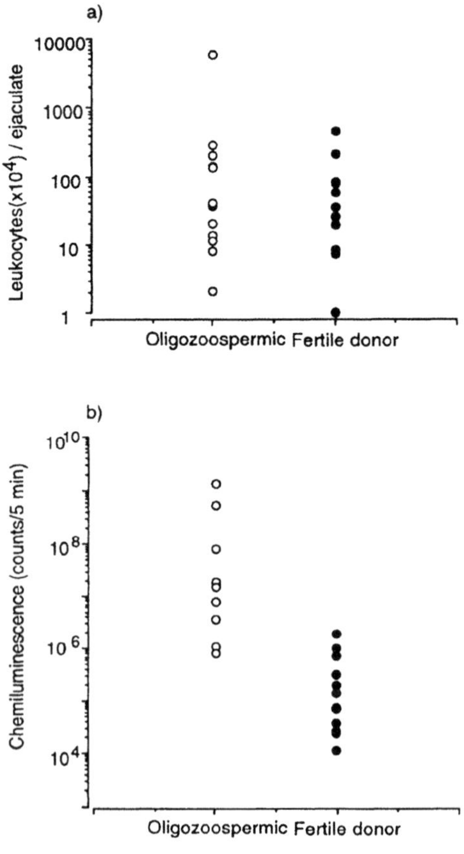

FIGURE 5. Sources of ROS in cases of oligozoosperma. (a) No differences are apparent between oligozoospermic patients and fertile donors in terms of the number of CD45-positive leukocytes in the ejaculate. (b) In contrast, the isolated spermatozoa exhibit considerable differences in ROS generation, measured by luminol-dependent chemiluminescence.

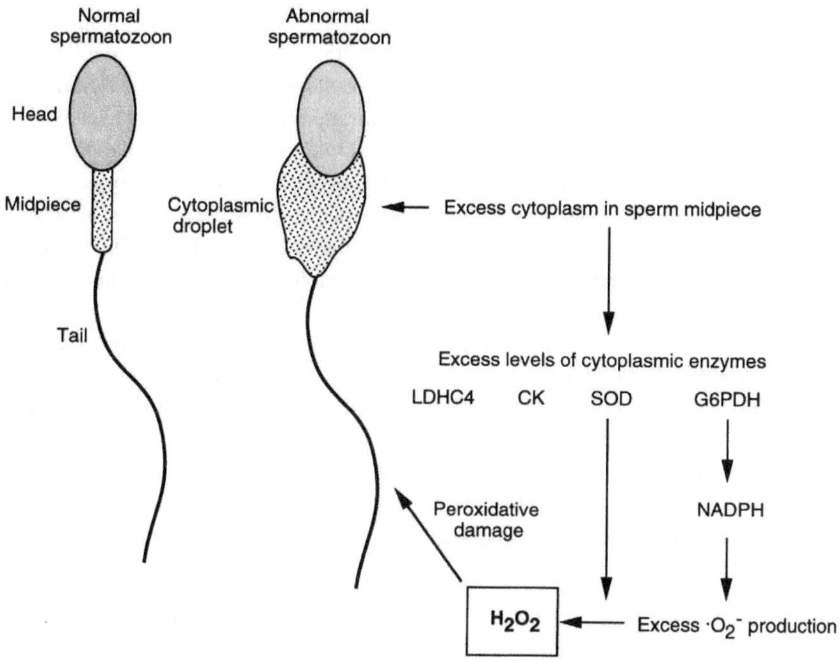

FIGURE 6. Proposed mechanism by which the retention of excess cytoplasm in the midpiece of the spermatozoon leads to the excessive generation of ROS and the stimulation of peroxidative damage. CK, creatine kinase; G6PDH, glucose-6-phosphate dehydrogenase; SOD, superoxide dismutase; LDHC4, testis-specific isoform of lactic acid dehydrogenase.

differentiation, they adopt the very unusual strategy of discarding most of their cytoplasm, just before they are released from the germinal epithelium. The small amount of residual cytoplasm that remains is concentrated in the midpiece of the cells in the vicinity of the mitochondria. Occasionally, this process is defective and excess cytoplasm is retained by the spermatozoa in the form of a "cytoplasmic droplet" (Figure 6).

That the retention of excess residual cytoplasm is detrimental to human spermatozoa was first suggested by data indicating that defective sperm are characterized by excessively high activities of certain cytoplasmic enzymes such as lactic acid dehydrogenase (LDHC4) and creatine kinase (Huszar *et al.*, 1988a,b; Casano *et al.*, 1991). The excessive activity of such fundamental enzymes could not, in itself, account for the loss of function seen in spermatozoa possessing additional residual cytoplasm. The first clue as to the mechanism by which this structural defect might impair the functional competence of human spermatozoa came from Rao *et al.* (1989) who found that defects in the midpiece of human spermatozoa were accompanied by a loss of motility associated with peroxidative damage.

The biochemical connection between the presence of excess cytoplasm and peroxidative damage was suggested by experiments indicating that creatine kinase activity was correlated with the activities of two enzymes involved in ROS metabolism, namely,

G6PDH and SOD (Aitken *et al.*, 1994b, 1996c; Figure 7). G6PDH is significant because this enzyme controls the availability of NADPH in the spermatozoon by regulating the rate of glucose flux through the hexose monophosphate shunt. Because NADPH is a substrate for superoxide anion production in human spermatozoa (Figure 1), an excellent correlation is seen between G6PDH activity and ROS generation by these cells (Aitken *et al.*, 1994b). The parallel increase in SOD is important because this enzyme converts an inert, membrane-impermeant reactive oxygen species (superoxide) to a membrane-permeant toxicant (hydrogen peroxide). It is for this reason that high SOD activity in populations of human spermatozoa is associated with defective sperm function involving high rates of peroxidative damage (Aitken *et al.*, 1996c).

This hypothesis of defective sperm function must somehow explain why the elevated levels of NADPH generated in the presence of excess cytoplasm are not used to detoxify the hydrogen peroxide generated by SOD, through the stimulation of glutathione peroxidase (GPx) activity (Alvarez and Storey, 1989). A possible reason for this is that although spermatozoa express GPx activity, it is strictly limited in terms of its capacity and activity (Alvarez and Storey, 1989), so that as hydrogen peroxide generation increases, its detoxifying capacity is easily overwhelmed.

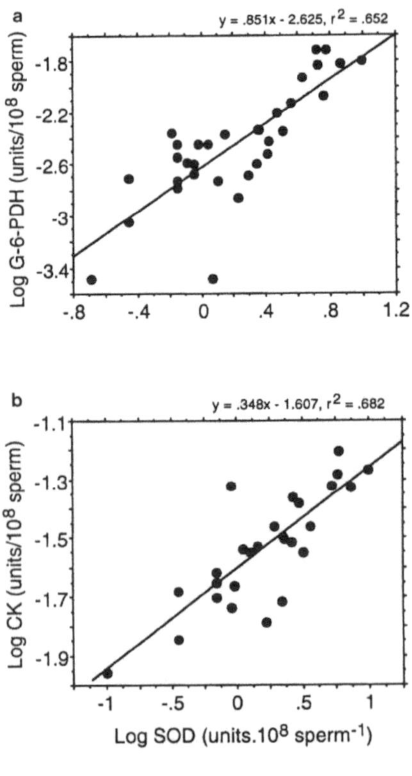

FIGURE 7. Correlation between different markers of the cytoplasmic space in human spermatozoa. (a) SOD and G6PDH; (b) SOD and CK. CK, creatine kinase; G-6-PDH, glucose-6-phosphate dehydrogenase; SOD, superoxide dismutase (Adapted from Aitken *et al.*, 1996c).

Further proof of this association between the retention of excess residual cytoplasm and oxidative stress has come from experiments in which image analysis techniques have been used to accurately quantify the cytoplasmic component of these cells (Gomez *et al.*, 1996). Such measurements were shown to be highly correlated with biochemical markers of the cytoplasmic space such as creatine kinase and G6PDH. Moreover, the presence of excess residual cytoplasm was positively correlated with defective sperm function and the generation of ROS. In light of these findings, we hypothesize that peroxidative damage to human spermatozoa is associated with errors of spermiogenesis leading to the retention of excess residual cytoplasm in the midpiece of the spermatozoa. This morphological defect is correlated with the excessive presence of cytoplasmic enzymes such as LDH, creatine kinase, SOD, and G6PDH. It is the latter two that are thought to fuel ROS generation in spermatozoa by increasing the availability of NADPH to a putative oxidase complex and ensuring the efficient conversion of the superoxide generated by this system to hydrogen peroxide (Figure 6).

2.2.2. Role of Extracellular Antioxidant Protection

Human spermatozoa are particularly vulnerable to oxidative stress because they contain high concentrations of unsaturated fatty acids (Jones *et al.*, 1979) and are characterized by a paucity of antioxidant defense enzymes. The high concentrations of unsaturated fatty acids are required to give the sperm plasma membrane the fluidity it needs to engage in the membrane fusion events (the acrosome reaction and fusion with the vitelline membrane of the oocyte) associated with fertilization. The paucity of antioxidant enzymes such as SOD, catalase, and GPx is simply a reflection of the fact that spermatozoa discard most of their cytoplasm just before they are discharged from the germinal epithelium. As a consequence, these cells are highly dependent on the presence of extracellular antioxidants for their protection against oxidative stress. Thus, as spermatozoa are passing through the epididymis, they are protected by antioxidant enzymes present in the epididymal lumen including unique secreted forms of SOD and GPx (Perry *et al.*, 1992, 1993). Following ejaculation, the spermatozoa are coated in a lactoferrinlike molecule derived from the seminal vesicles (Sato, 1995), the presumed function of which is to prevent iron gaining access to the lipid bilayer where it might promote the initiation and propagation of peroxidative damage. Seminal plasma contains a variety of other powerful antioxidants designed to protect the spermatozoa from oxidative stress, including a variety of antioxidant enzymes like catalase and SOD (Mennella and Jones, 1980; Zini *et al.*, 1993). Indeed, seminal plasma contains more extracellular SOD than any other fluid in biology, the world record being held by donkey semen (>5000 IU SOD/ml). The SOD activity in human seminal plasma does not appear to correlate with sperm motility or sperm number (Kobayashi *et al.*, 1991) although a recent report claimed to have detected a striking increase in seminal SOD activity in oligozoospermic patients (Sanocka *et al.*, 1996). The authors interpreted this rise as evidence for the hyperactivation of antioxidant systems in response to oxidative stress associated with the oligozoospermic condition (Sanocka *et al.*, 1996).

SOD is only one of the many antioxidant factors in seminal plasma, and low-molecular-weight scavengers such as vitamin C, uric acid, α-tocopherol, and albumin (Fraga *et al.*,

1991; Moilanen *et al.*, 1993; Thiele *et al.*, 1995) certainly contribute significantly to the total antioxidant activity of this fluid. Using a TRAP assay to monitor the total antioxidant activity in seminal plasma, striking differences in the protective properties of seminal plasma have been observed between infertile patients and fertile donors (Smith *et al.*, 1996). The TRAP values for the normal fertile controls were considerably higher than those recorded for groups of infertile patients in whom sperm counts were normal, but the spermatozoa were functionally or morphologically defective. Moreover, the TRAP values were inversely related to indices of peroxidative damage in the spermatozoa (Smith *et al.*, 1996). Such results emphasize just how dependent spermatozoa are on the antioxidant protection provided by the fluids of the male reproductive tract and how defects in this protective system can lead to reproductive pathology.

2.3. Physiological Function of ROS in Spermatozoa

Given the inherent susceptibility of human spermatozoa to ROS, it is reasonable to ask why spermatozoa should have developed a capacity for ROS generation. A number of studies have shown that scavenging the ROS generated by mammalian spermatozoa has a detrimental effect on sperm function. Thus, the addition of catalase to populations of hamster spermatozoa has been found to disrupt the ability of these cells to undergo the acrosome reaction, an exocytotic event that is an essential component of fertilization (Bize *et al.*, 1991). Catalase has also been found to suppress the fertilizing potential of human spermatozoa whether activated by the divalent cation ionophore, A23187, or through the extragenomic action of progesterone (Aitken *et al.*, 1995, 1996a). The implications of these findings are that hydrogen peroxide generated by the spermatozoa is an essential component of the mechanisms by which spermatozoa become activated during the early stages of fertilization. Specifically, hydrogen peroxide appears to be involved in a process known as *capacitation*. This term describes a maturational change that spermatozoa must undergo during their transport through the female reproductive tract, so that they are competent to respond to signals given out by the oocyte–cumulus complex at the site of fertilization (Griveau *et al.*, 1994; Aitken *et al.*, 1996a).

Biochemically, hydrogen peroxide production during capacitation appears to result in a dramatic increase in the levels of tyrosine phosphorylation exhibited by these cells (Aitken *et al.*, 1995). The induction of tyrosine phosphorylation can be achieved by the direct addition of hydrogen peroxide to populations of human spermatozoa or by the exposure of these cells to a hydrogen peroxide generating system (glucose oxidase). Conversely, tyrosine phosphorylation can be suppressed by the addition of catalase to the medium or by the addition of membrane-permeant thiols, such as 2-mercaptoethanol or dithiothreitol (Aitken *et al.*, 1995). In these experiments the capacitation status of the spermatozoa was found to change in concert with the levels of tyrosine phosphorylation, being stimulated or suppressed by oxidizing and reducing conditions, respectively (Aitken *et al.*, 1995, 1996a).

A major consequence of the peroxide-induced increase in tyrosine phosphorylation observed during capacitation is to render the spermatozoon exquisitely sensitive to calcium transients generated when this cell makes contact with agonists within the oocyte–cumulus complex, such as progesterone or the zona glycoprotein ZP3 (Aitken *et al.*, 1995, 1996a). The specific manner in which this sensitivity to calcium is effected is currently unknown. Hydrogen peroxide appears to induce tyrosine phosphorylation in multiple sperm proteins, none of which have been definitively identified.

Although hydrogen peroxide is an essential component of capacitation, an additional role for superoxide anion has also been proposed. Thus, Griveau *et al.* (1995) have suggested that following capacitation, spermatozoa develop the capacity to release a burst of superoxide anion in response to a calcium signal generated by A23187. If SOD was present in the medium when A23187 was added to the spermatozoa, then this superoxide burst could not be detected and the biological response of the spermatozoa to ionophore was disrupted (Griveau *et al.*, 1995). The ability of superoxide anion to promote the membrane fusion events associated with the acrosome reaction appears to involve the deesterification of membrane phospholipids. In the aprotic conditions pertaining in the interior of biological membranes, the deesterification induced by superoxide anion would lead to the generation of lysophospholipids and destabilization of the plasma membrane, as a consequence of which the membrane fusion events associated with the acrosome reaction would be facilitated.

Superoxide anion is also held to be involved in the induction of a specific form of motility expressed by mammalian spermatozoa once they have become capacitated, known as *hyperactivation* (de Lamirande and Gagnon, 1993a–c). This form of movement is characterized by the generation of large-amplitude, asymmetrical thrashings of the sperm tail, the function of which is to generate the propulsive forces necessary to drive the sperm head through the acellular, translucent shell (the zona pellucida) that surrounds the oocyte. Addition of SOD prevents the onset of this form of movement whereas the ability of biological fluids (seminal plasma, fetal cord serum, or follicular fluid) to stimulate hyperactivation is inversely related to their SOD content (de Lamirande and Gagnon, 1993a–c).

Finally, there is evidence that the promotion of low levels of lipid peroxidation in human spermatozoa, through the addition of a ferrous ion promoter, enhances the capacity of these cells to adhere to the zona pellucida (Aitken *et al.*, 1989a). Such observations are in keeping with a proposed role for ROS and lipid peroxidation in mediating cell adhesion events involving other cell types, such as neutrophils and platelets (Lafuze *et al.*, 1983; Bearpark *et al.*, 1988). Moreover, the peroxidation of liposomes containing unsaturated fatty acids has been shown to increase vesicle aggregation and fusion (Sevanian *et al.*, 1988).

3. CONCLUSIONS

In conclusion, mammalian spermatozoa are highly specialized cells that have evolved a capacity for generating ROS, including the superoxide anion and hydrogen peroxide. Both of these molecules appear to play important biological roles in the preparation of spermatozoa for fertilization (capacitation) and during the act of fertilization itself, in the adhesion of the spermatozoa to the zona surface and during the acrosome reaction. The mechanisms by which the ROS are generated and the precise mechanisms by which they exert their biological effects have not been resolved and may show species differences. In the case of human spermatozoa, there is evidence for the central involvement of NADPH in supplying electrons to a putative free radical generating enzyme complex, which appears to be distinct from the NADPH of phagocytic neutrophils. The mechanisms by which the ROS generated by human spermatozoa appear to induce their biological effects involve the stimulation of tyrosine phosphorylation and the deesterification of membrane phospholipids, although the mechanisms by which these activities are induced have still to be elucidated.

Despite the functional significance of ROS generation by human spermatozoa, it is also clear that perturbations of this system are involved in the etiology of male infertility. Specifically, defects in the terminal stages of spermiogenesis leading to the retention of excess residual cytoplasm by the spermatozoa are associated with high levels of ROS production. As a consequence of this excessive ROS generating activity, peroxidative damage is induced in the sperm plasma membrane, disrupting the motility of these cells and their capacity for sperm–oocyte fusion. Such findings are clinically important in terms of the diagnosis of male infertility and the design of appropriate methods of treatment. If peroxidative damage is involved in the development of male infertility, then antioxidants may be of value in addressing this condition. A variety of studies, involving the exposure of human spermatozoa to antioxidants *in vivo* and *in vitro*, have demonstrated the potential of this approach (Aitken and Clarkson, 1988; Lenzi *et al.*, 1993). If this promise is realized, then oxidative stress would be one of the first examples of a defined biochemical cause of male infertility, for which a rational form of therapy would be available.

4. REFERENCES

Aitken, R. J., and Clarkson, J. S., 1987, Cellular basis of defective sperm function and its association with the genesis of reactive oxygen species by human spermatozoa, *J. Reprod. Fertil.* **81**:459–469.

Aitken, R. J., and Clarkson, J. S., 1988, Significance of reactive oxygen species and antioxidants in defining the efficacy of sperm preparation techniques, *J. Androl.* **9**:367–376.

Aitken, R. J., and West, K. M., 1990, Analysis of the relationship between reactive oxygen species production and leucocyte infiltration in fractions of human semen separated on Percoll gradients, *Int. J. Androl.* **13**:433–451.

Aitken, R. J., Clarkson, J. S., and Fishel, S., 1989a, Generation of reactive oxygen species, lipid peroxidation and human sperm function, *Biol. Reprod.* **40**:183–197.

Aitken, R. J., Clarkson, J. S., Hargreave, T. B., Irvine, D. S., and Wu, F. C., 1989b, Analysis of relationship between defective sperm function and the generation of reactive oxygen species in cases of oligozoospermia, *J. Androl.* **10**:214–220.

Aitken, R. J., Irvine, D. S., and Wu, F. C., 1991, Prospective analysis of sperm–oocyte fusion and reactive oxygen species generation as criteria for the diagnosis of infertility, *Am. J. Obstet. Gynecol.* **164**:542–551.

Aitken, R. J., Buckingham, D., West, K., Wu, F. C., Zikopoulos, K., and Richardson, D. W., 1992, On the contribution of leucocytes and spermatozoa to the high levels of reactive oxygen species recorded in the ejaculates of oligozoospermic patients, *J. Reprod. Fertil.* **94**:451–462.

Aitken, R. J., Harkiss, D., and Buckingham, D., 1993a, Relationship between iron-catalysed lipid peroxidation potential and human sperm function, *J. Reprod. Fertil.* **98**:257–265.

Aitken, R. J., Harkiss, D., and Buckingham, D. W., 1993b, Analysis of lipid peroxidation mechanisms in human spermatozoa, *Mol. Reprod. Dev.* **35**:302–315.

Aitken, R. J., West, K., and Buckingham, D., 1994a, Leukocyte infiltration into the human ejaculate and its association with semen quality, oxidative stress and sperm function, *J. Androl.* **15**:343–352.

Aitken, R. J., Krausz, C., and Buckingham, D. W., 1994b, Relationship between the biochemical marker for human sperm quality, creatine kinase, and the mechanisms responsible for the generation of reactive oxygen species, *Mol. Reprod. Dev.* **39**:268–279.

Aitken, R. J., Paterson, M., Fisher, H., Buckingham, D. W., and van Duin, M., 1995, Redox regulation of tyrosine phosphorylation in human spermatozoa and its role in the control of human sperm function, *J. Cell Sci.* **108**:2017–2025.

Aitken, R. J., Buckingham, D. W., Harkiss, D., Paterson, M., Fisher, H., and Irvine, D. S., 1996a, The extragenomic action of progesterone on human spermatozoa is influenced by redox regulated changes in tyrosine phosphorylation during capacitation, *Mol. Cell. Endocrinol.* **117**:83–93.

Aitken, R. J., Buckingham, D. W., West, K., and Brindle, J., 1996b, On the use of paramagnetic beads and ferrofluids to assess and eliminate the leukocytic contribution to oxygen radical generation by human sperm suspensions, *Am. J. Reprod. Immunol.* **35**:541–551.

Aitken, R. J., Buckingham, D. W., Carreras, A., and Irvine, D. S., 1996c, Superoxide dismutase in human sperm suspensions: Relationships with cellular composition, oxidative stress and sperm function, *Free Radical Biol. Med.* **21**:495–504.

Alvarez, J. G., and Storey, B. T., 1982, Spontaneous lipid peroxidation in rabbit epididymal spermatozoa: Its effect on sperm motility, *Biol. Reprod.* **27**:1102–1108.

Alvarez, J. G., and Storey, B. T., 1984, Lipid peroxidation and the reactions of superoxide and hydrogen peroxide in mouse spermatozoa, *Biol. Reprod.* **30**:833–841.

Alvarez, J. G., and Storey, B. T., 1989, Role of glutathione peroxidase in protecting mammalian spermatozoa from loss of motility caused by spontaneous lipid peroxidation, *Gamete Res.* **23**:77–90.

Alvarez, J. G., Touchstone, J. C., Blasco, L., and Storey, B. T., 1987, Spontaneous lipid peroxidation and production of hydrogen peroxide and superoxide in human spermatozoa, *J. Androl.* **8**:338–348.

Bearpark, T., Salvemini, D., Sneddon, J. M., and Vane, J. R., 1988, Endothelium-derived relaxing factor (EDRF) and superoxide anions modulate platelet adhesion to endothelial cells, *J. Physiol (London)* **339**:12P.

Bize, I., Santander, G., Cabello, P., Driscoll, D., and Sharpe, C., 1991, Hydrogen peroxide is involved in hamster sperm capacitation in vitro, *Biol. Reprod.* **44**:398–403.

Casano, R., Orlando, C., Serio, M., and Forti, G., 1991, LDH and LDH-X activity in sperm from normospermic and oligozoospermic men, *Int. J. Androl.* **14**:257–263.

de Lamirande, E., and Gagnon, C., 1993a, Human sperm hyperactivation in whole semen and its association with low superoxide scavenging capacity in seminal plasma, *Fertil. Steril.* **59**:1291–1295.

de Lamirande, E., and Gagnon, G., 1993b, A positive role for the superoxide anion in triggering hyperactivation and capacitation of human spermatozoa, *Int. J. Androl.* **16**:21–25.

de Lamirande, E., and Gagnon, G., 1993c, Human sperm hyperactivation and capacitation as part of an oxidative process, *Free Radical Biol. Med.* **14**:157–166.

Fraga, C. G., Motchnik, P. A., Shigenaga, M. K., Helbock, H. J., Jacob, R. A., and Ames, B. N., 1991, Ascorbic acid protects against endogenous oxidative DNA damage in human sperm, *Proc. Natl. Acad. Sci. USA* **88**:11003–11006.

Gomez, E., Buckingham, D. W., Brindle, J., Lanzafame, F., Irvine, D. S., and Aitken, R. J., 1996, Development of an image analysis system to monitor the retention of residual cytoplasm by human spermatozoa: Correlation with biochemical markers of the cytoplasmic space, oxidative stress, and sperm function, *J. Androl.* **17**:276–287.

Griveau, J. F., Renard, P., and Le Lannou, D., 1994, An in vitro promoting role for hydrogen peroxide in human sperm capacitation, *Int. J. Androl.* **17**:300–307.

Griveau, J. F., Renard, P., and Le Lannou, D., 1995, Superoxide anion production by human spermatozoa as a part of the ionophore-induced acrosome reaction process, *Int. J. Androl.* **18**:67–74.

Holland, M. K., Alvarez, J. G., and Storey, B. T., 1982, Production of superoxide and activity of superoxide dismutase in rabbit epididymal spermatozoa, *Biol. Reprod.* **27**:1109–1118.

Huszar, G., Corrales, M., and Vigue, L., 1988a, Correlation between sperm creatine phosphokinase activity and sperm concentrations in normospermic and oligozoospermic men, *Gamete Res.* **19**:67–75.

Huszar, G., Vigue, L., and Corrales, M., 1988b, Sperm creatine phosphokinase quality in normospermic, variablespermic and oligospermic men, *Biol. Reprod.* **38**:1061–1066.

Iwasaki, A., and Gagnon, C., 1992, Formation of reactive oxygen species in spermatozoa of infertile patients, *Fertil. Steril.* **57**:409–416.

Jones, R., and Mann, T., 1973, Lipid peroxidation in spermatozoa, *Proc. R. Soc. London B Ser.* **184**:103–107.

Jones, R., and Mann, T., 1976, Lipid peroxides in spermatozoa: Formation, role of plasmalogen, and physiological significance, *Proc. R. Soc. London B Ser.* **193**:317–333.

Jones, R., and Mann, T., 1977a, Toxicity of exogenous fatty acid peroxides towards spermatozoa, *J. Reprod. Fertil.* **50**:255–260.

Jones, R., and Mann, T., 1977b, Damage to ram spermatozoa by peroxidation of endogenous fatty acids, *J. Reprod. Fertil.* **50**:261–268.

Jones, R., Mann, T., and Sherins, R. J., 1978, Adverse effects of peroxidized lipid on human spermatozoa, *Proc. R. Soc. London B Ser.* **201**:413–417.

Jones, R., Mann, T., and Sherins, R. J., 1979, Peroxidative breakdown of phospholipids in human spermatozoa: Spermicidal effects of fatty acid peroxides and protective action of seminal plasma, *Fertil. Steril.* **31**:531–537.

Kobayashi, T., Miyazaki, T., Natori, M., and Nozawa, S., 1991, Protective role of superoxide dismutase in human sperm motility: Superoxide dismutase activity and lipid peroxide in human seminal plasma and spermatozoa, *Hum. Reprod.* **6**:987–991.

Kumar, G. P., Laloraya, M., and Laloraya, M. M., 1991, Superoxide radical level and superoxide dismutase activity changes in maturing mammalian spermatozoa, *Andrologia* **23**:171–175.

Lafuze, J. E., Weisman, S. J., Ingraham, L. M., Butterick, C. J., Alpert, L. A., and Baehner, R. L., 1983, The effect of vitamin E on rabbit neutrophil activation, in *Biology of Vitamin E*, Ciba Foundation Symposium 101 (R. Porter and J. Whelan, eds.), pp. 130–140, Pitman Press, Bath.

Lenzi, A., Culasso, F., Gardini, L., Lombado, F., and Dondero, F., 1993, Placebo-controlled double-blind cross-over trial of glutathione therapy in male infertility, *Hum. Reprod.* **8**:1657–1662.

MacLeod, J., 1943, The role of oxygen in the metabolism and motility of human spermatozoa, *Am. J. Physiol.* **138**:512–518.

Mazzilli, F., Rossi, T., Marchesini, M., Ronconi, C., and Dondero, F., 1994, Superoxide anion in human semen related to seminal parameters and clinical aspects, *Fertil. Steril.* **62**:862–868.

Meier, B., Cross, A. R., Hancock, J. T., Kaup, F. J., and Jones, O. T. G., 1991, Identification of a superoxide-generating NADPH oxidase system in human fibroblasts, *Biochem. J.* **275**:241–245.

Mennella, M. R. F., and Jones, R., 1980, Properties of spermatozoal superoxide dismutase and lack of involvement of superoxides in metal-ion-catalysed lipid-peroxidation reactions in semen, *Biochem. J.* **191**:289–297.

Moilanen, J., Howatta, O., and Lindroth, L., 1993, Vitamin E levels in seminal plasma can be elevated by oral administration of vitamin E in infertile men, *Int. J. Androl.* **16**:165–166.

Perry, A. C. F., Jones, R., Niang, L. S. P., Jackson, R. M., and Hall, L., 1992, Genetic evidence for an androgen-regulated epididymal secretory glutathione peroxidase whose transcript does not contain a selenocysteine codon, *Biochem. J.* **285**:863–870.

Perry, A. C. F., Jones, R., and Hall, L., 1993, Isolation and characterization of a rat cDNA clone encoding a secreted superoxide dismutase reveals the epididymis to be a major site of its expression, *Biochem. J.* **293**:21–25.

Radeke, H. H., Cross, A. R., Hancock, J. T., Jones, O. T. G., Nakamura, M., Kaever, V., and Resch, K. 1991, Functional expression of NADPH oxidase components (α- and β-subunits of cytochrome b_{558} and 45-kDa flavoprotein) by intrinsic human glomerular mesangial cells, *J. Biol. Chem.* **266**:21025–21029.

Rao, B., Soufir, J. C., Martin, M., and David, G., 1989, Lipid peroxidation in human spermatozoa as related to midpiece abnormalities and motility, *Gamete Res.* **24**:127–134.

Sanocka, D., Miesel, R., Jedrzejczak, P., and Kurpisz, M. K., 1996, Oxidative stress and male infertility, *J. Androl.* **17**:449–454.

Sato, I., 1995, Characterization of the 84-kDa protein with ABH activity in human seminal plasma, *Jpn. J. Legal Med.* **49**:281–293.

Sevanian, A., Wratten, M. L., McLeod, L. L., and Kim, E., 1988, Lipid peroxidation and phopholipase A2 activity in liposomes composed of unsaturated phospholipids: A structural basis for enzyme activation, *Biochim. Biophys. Acta* **961**:316–327.

Smith, R., Vantman, D., Ponce, J., Escobar, J., and Lissi, E., 1996, Total antioxidant capacity of human seminal plasma, *Hum. Reprod.* **11**:1655–1660.

Thiele, J. J., Freisleben, H. J., Fuchs, J., and Ochensendorf, F. R., 1995, Ascorbic acid and urate in human seminal plasma: Determination and interrelationship with chemiluminescence in washed semen, *Hum. Reprod.* **10**:110–115.

Tosic, J., and Walton, A., 1946, Formation of hydrogen peroxide by spermatozoa and its inhibitory effect on respiration, *Nature* **158**:485.

Tosic, J., and Walton, A., 1950, Metabolism of spermatozoa. Formation of hydrogen peroxide by spermatozoa and its effects on motility and survival, *Biochem. J.* **47:**199–212.

Weese, D. L., Peaster, M. L., Himsl, K. K., Leach, G. E., Lad, P. M., and Zimmern, P. E., 1993, Stimulated reactive oxygen species generation in the spermatozoa of infertile men, *J. Urol.* **149:**64–67.

Zini, A., de Lamirande, E., and Gagnon, C., 1993, Reactive oxygen species in semen of infertile patients: Levels of superoxide dismutase- and catalase-like activities in seminal plasma and spermatozoa, *Int. J. Androl.* **16:**183–188.

Chapter 21

Respiratory Burst Oxidase of Fertilization Peroxidative Mechanisms in Sea Urchin Eggs and Human Phagocytes

Jay W. Heinecke

1. ACTIVATED SEA URCHIN EMBRYOS ASSEMBLE A FERTILIZATION ENVELOPE

Minutes after an echinoderm egg is fertilized, it develops an impenetrable coat that acts as a microincubator, creating and preserving a protective environment around the developing embryo (reviewed in Rothschild, 1959; Shapiro *et al.*, 1981; Shapiro, 1991). This protein coat, the fertilization envelope, provides the definitive block to polyspermy and shields the embryo from the mechanical, chemical, and bacterial damage that can occur in the inhospitable marine environment. It is so resilient that it can survive passage through the digestive tracts of predators, as demonstrated in experiments with tunicates (Shapiro *et al.*, 1981).

This chapter focuses on the oxidative reaction that cross-links the fertilization envelope. It also draws parallels between this reaction and oxidative reactions that may damage human tissues in atherosclerosis and other diseases.

The sea urchin fertilization envelope develops in several stages (reviewed in Shapiro *et al.*, 1981; Shapiro, 1991). It begins as a glycocalyx or vitelline layer that lies on the outer surface of the oocyte's plasma membrane. This layer is modified by the contents of thousands of vesicles—cortical granules—that initially lie quiescent beneath the plasma

---

Jay W. Heinecke Departments of Medicine and of Molecular Biology and Pharmacology, Washington University, St. Louis, Missouri 63110.

Reactive Oxygen Species in Biological Systems, edited by Gilbert and Colton. Kluwer Academic / Plenum Publishers, New York, 1999.

membrane. In response to a fertilization-induced rise in intracellular Ca^{2+}, the cortical granules fuse with the plasma membrane, expelling their contents by exocytosis. Among the discharged proteins is a serine protease, which separates the vitelline layer from the plasma membrane (Vacquier *et al.*, 1972), creating a perivitelline space and an elevated fertilization envelope. At this stage, the protein coat—the "soft" fertilization envelope—is noncovalently associated, and it disaggregates when exposed to chelators or reducing agents (Vernon *et al.*, 1977; Kay *et al.*, 1982). The deposition of proteins from the cortical granules restructures the fertilization envelope, changing its appearance under the scanning electron microscope. This morphological change fails to occur when inhibitors such as glycine ethyl ester or Zn^{2+} block transglutaminase activity (Vernon *et al.*, 1977; Battaglia and Shapiro, 1988).

An oxidative reaction then cross-links the protein coat polypeptides into one enormous macromolecule, the "hard" fertilization envelope. The hardening reaction is blocked by peroxidase inhibitors like azide (Foerder and Shapiro, 1977; Foerder *et al.*, 1978). It is catalyzed by ovoperoxidase, a 70-kDa enzyme that is spectroscopically similar to lactoperoxidase (Diets *et al.*, 1984). Secreted by cortical granules but initially inactive, ovoperoxidase cross-links tyrosine residues in the soft fertilization envelope (Foerder and Shapiro, 1977; Foerder *et al.*, 1978; Diets *et al.*, 1984). Cross-linking occurs because ovoperoxidase uses H_2O_2 to convert tyrosine residues to long-lived tyrosyl radicals (tyrosyl•). The radicals undergo phenolic coupling, generating *o,o′*-dityrosine bridges between adjacent polypeptides (Foerder and Shapiro, 1977; Foerder *et al.*, 1978).

$$H_2O_2 + \text{tyrosine} \rightarrow 2\ H_2O + \text{tyrosyl}^\bullet \tag{1}$$

$$2\ \text{tyrosyl}^\bullet \rightarrow o,o'\text{-dityrosine} \tag{2}$$

The peroxide used as oxidizing substrate by ovoperoxidase is produced when the egg consumes a large amount of oxygen shortly after fertilization. This "respiratory burst," first described 90 years ago (Warburg, 1908), is the archetype of cellular activation—and little affected by mitochondrial poisons and quantitatively accounting for H_2O_2 production (Figure 1), the burst is catalyzed by a Ca^{2+}-stimulated NADPH oxidase (Heinecke and Shapiro, 1989; Heinecke *et al.*, 1990).

Reactive intermediates such as H_2O_2 and tyrosyl radical could harm the nascent embryo if the hardening of the fertilization envelope was not carefully regulated. Sea urchins represent a particularly useful model for exploring this regulation because the female sheds vast quantities of quiescent eggs that are rapidly and synchronously activated at fertilization.

2. A Ca^{2+}-STIMULATED NADPH OXIDASE CATALYZES THE RESPIRATORY BURST

Because of its central role early in egg activation (Steinhardt and Epel, 1974; Vacquier, 1975), we reasoned that Ca^{2+} might be the intracellular signal that triggers the respiratory burst. We developed a glutathione peroxidase-dependent method able to measure peroxide in complex mixtures, and we used this method to quantify H_2O_2 production in homogenized sea urchin eggs (Heinecke and Shapiro, 1989). The Ca^{2+}-

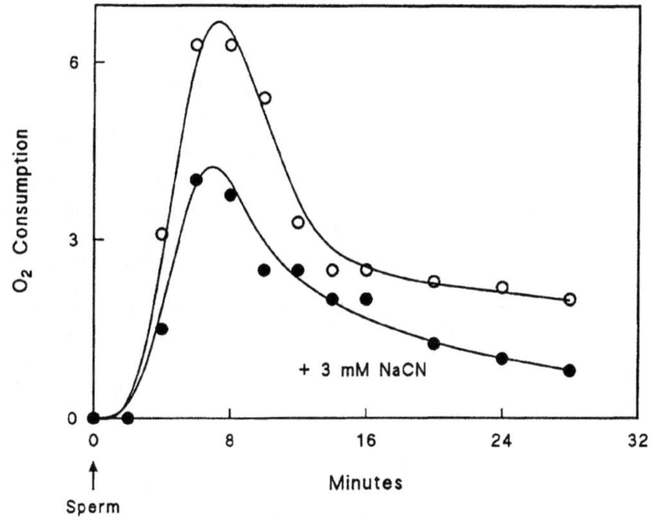

FIGURE 1. The respiratory burst of fertilization. Oxygen consumption is relatively cyanide-insensitive and is superimposed on accelerated mitochondrial respiration of the embryo.

stimulated oxidase identified by us in unfertilized eggs of the purple sea urchin, *Strongylocentrotus purpuratus*, was soluble following centrifugation of the egg homogenate at 10,000*g*, but it could be precipitated by adding ammonium sulfate to 20% saturation.

The oxidase fraction directly reduced molecular O_2 to H_2O_2 (Heinecke and Shapiro, 1989). It used NADPH, but not NADH, as a cofactor (K_ds 40 and > 400 μM, respectively):

$$NADPH + O_2 + H^+ \rightarrow H_2O_2 + NADP^+ \quad (3)$$

H_2O_2 synthesis by the oxidase fraction resembles that by intact eggs in that it is not inhibited by cyanide, azide, or aminotriazole.

The oxidase requires both Ca^{2+} and MgATP for activation, and Sr^{2+} can replace Ca^{2+} to yield approximately 40% of the activity. Mn^{2+}, Zn^{2+}, or Co^{2+} cannot substitute for Mg^{2+}, and nonhydrolyzable ATP analogues and other nucleotides cannot replace ATP.

When the oxidase reaction is initiated with Ca^{2+}, a lag phase of approximately 30 s precedes H_2O_2 production (Figure 2A) (Heinecke and Shapiro, 1989, 1990a). This lag also is observed when H_2O_2 production is monitored by O_2 consumption or by formation of horseradish peroxidase compound II, a specific H_2O_2–peroxidase complex. Moreover, it is present when the reaction is initiated with ATP (Figure 2B). It is absent, however, when NADPH initiates the reaction in a mixture that already contains ATP and Ca^{2+} (Figure 2C). These observations suggest that Ca^{2+} and MgATP are involved in activating the oxidase. Ca^{2+} also is necessary for the reaction to continue because addition of the calcium chelator EGTA rapidly inhibits H_2O_2 synthesis (Figure 2A).

We used inhibitors to determine whether protein kinases might be responsible for the ATP-dependent activation step. H-7 and staurosporine, potent but nonspecific protein

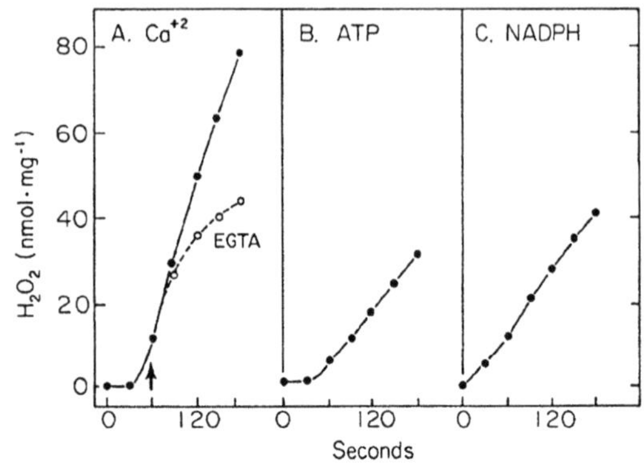

FIGURE 2. Progress curve of H_2O_2 generation by the oxidase fraction following initiation of the oxidase reaction with Ca^{2+}, ATP, or NADPH. (A) Reaction initiated with Ca^{2+}. The oxidase fraction was preincubated for 2 min with ATP and NADPH. At 0 time Ca^{2+} was added and H_2O_2 generation monitored using glutathione peroxidase. At the arrow, EGTA was added to lower the concentration of free Ca^{2+} to less than 0.01 μM. (B) Reaction initiated with ATP. During the preincubation period, ATP was replaced with Ca^{2+}, and the reaction was initiated by the addition of ATP. (C) Reaction initiated with NADPH. During the preincubation period, NADPH was replaced with Ca^{2+}, and the reaction was initiated by the addition of NADPH. When the oxidase fraction was preincubated with Ca^{2+} prior to initiation of the oxidase reaction with ATP or NADPH, the rate of H_2O_2 generation was lower than when the reaction was initiated with Ca^{2+}. This may reflect inactivation of the oxidase fraction by a Ca^{2+}-dependent process. Reproduced with permission from Heinecke and Shapiro, 1989, *Proc. Natl. Acad. Sci USA* **86:** 1259–1263.

kinase inhibitors (Schachtele *et al.*, 1988), block H_2O_2 synthesis by the oxidase fraction (Heinecke and Shapiro, 1989, 1990a). They also block cyanide-insensitive oxygen consumption by ionophore-stimulated eggs (Heinecke and Shapiro, 1989, 1990a). These results suggest that a protein kinase might regulate the NADPH oxidase and the respiratory burst of fertilization.

We used other inhibitors to test this idea further. The thiol-reactive reagent *N*-ethylmaleimide, but not iodoacetamide, inhibits both the oxidase reaction and the respiratory burst (Heinecke and Shapiro, 1989, 1990a). Similarly, the phenathiazines trifluooperazine and promethazine, which inhibit certain protein kinases and calmodulin-dependent enzymes (Mori *et al.*, 1980; Weiss *et al.*, 1980), suppress H_2O_2 synthesis both by the oxidase fraction and by eggs (Heinecke and Shapiro, 1989, 1990a). In both systems, trifluooperazine is more potent than promethazine. Procaine, a quaternary amine that affects phospholipid-dependent protein kinases (Mori *et al.*, 1980), blocks the oxidase activity *in vitro* and *in vivo*. Because procaine is a weak base, it should accumulate inside eggs at approximately tenfold its concentration in sea water, according to the pH gradient (Winkler and Grainger, 1978). Consistent with this observation, the half-maximal inhibitory concentration of procaine is ten-fold lower for eggs than for the oxidase fraction (Heinecke and Shapiro, 1989, 1990a).

The oxidase fraction is strongly pH-dependent, tripling its rate of H_2O_2 production as the pH rises from 6.4 to 7.0 (Heinecke and Shapiro, 1989). Thus, the enzyme would become more active after fertilization, which elevates the intracellular pH (pH_i) from 6.9 to 7.2 in *S. purpuratus* (Johnson *et al.*, 1976; Shen and Steinhardt, 1978).

Blocking this pH_i increase by replacing the Na^+ in artificial sea water with choline (Shen and Steinhardt, 1979; Johnson and Epel, 1981) inhibits O_2 consumption and H_2O_2 synthesis by ionophore-stimulated eggs (Heinecke and Shapiro, 1990a), and the respiratory burst can be partly restored by alkalinizing the cytosol of the eggs with NH_4^+ (Winkler and Grainger, 1978; Johnson and Epel, 1981). The oxidase activity also is blocked (Heinecke and Shapiro, 1990a) when the egg cytoplasm is weakly acidified with acetate (Johnson and Epel, 1981). These results show that a physiologic increase in pH both stimulates H_2O_2 synthesis by the NADPH oxidase and promotes cyanide-insensitive oxygen uptake by intact eggs. A Na^+-dependent increase in intracellular pH_i is known to be essential for the onset of late events in echinoderm egg activation such as protein and DNA synthesis (Johnson *et al.*, 1976; Johnson and Epel, 1981).

Therefore, four lines of evidence indicate that the sea urchin egg uses the Ca^{2+}-stimulated NADPH oxidase we have described to synthesize H_2O_2 (Heinecke and Shapiro, 1989, 1990a). First, a series of chemically unrelated reagents, including protein kinase inhibitors, suppress both the NADPH oxidase and the respiratory burst of fertilization. Second, changes in pH implicated in egg activation modulate NADPH oxidase activity and oxygen consumption by ionophore-activated eggs. Third, Epel has shown that the substrate for the oxidase reaction, NADPH, is produced from NAD^+ on fertilization (Epel, 1964). Fourth, the Ca^{2+} ionophore A23187 stimulates oxygen consumption by eggs, and the onset of this respiratory burst is associated in time with an increase in intracellular Ca^{2+} (Steinhardt and Epel, 1974; Vacquier, 1975; Foerder and Shapiro, 1977). Physiologically relevant concentrations of free Ca^{2+} are similarly required to initiate H_2O_2 synthesis by the NADPH oxidase.

3. PROTEIN KINASE C ACTIVATES THE RESPIRATORY BURST OXIDASE

Our observation that both Ca^{2+} and MgATP are required to activate the respiratory burst oxidase suggested the involvement of protein kinase C, a ubiquitous family of enzymes that play key roles in the control of cell behavior (reviewed in Nishizuka, 1988, 1995). To explore this possibility, we separated the NADPH oxidase into soluble and membrane-associated fractions (Heinecke *et al.*, 1990). Neither fraction had significant oxidase activity by itself. However, recombining the fractions restored H_2O_2 production.

We purified the soluble fraction by anion exchange chromatography. The resulting kinase fraction stimulated the membrane-associated fraction to generate H_2O_2 in a reaction that required both Ca^{2+} and MgATP. The reaction showed the characteristic lag phase between stimulation and H_2O_2 production (Heinecke *et al.*, 1990).

Several observations indicate that a protein kinase in oocytes is the only soluble protein necessary for activation of membrane-associated oxidase activity (Heinecke *et al.*, 1990). First, the activating kinase fraction exhibits histone kinase activity that is stimulated by a combination of Ca^{2+}, phospholipid, and diacylglycerol, a hallmark of protein kinase C (Nishizuka, 1988). Second, phorbol ester, a potent activator of protein kinase C (Nishizuka, 1988), lowers by a factor of 40 the concentration of free Ca^{2+} required to activate membrane-associated H_2O_2 synthesis. Third, apparently homogeneous protein kinase C from rat brain can also activate the oocyte oxidase, unlike calmodulin-dependent kinase II, the catalytic subunit of cAMP-dependent protein kinase, casein kinase II, or myosin light-chain kinase (Heinecke *et al.*, 1990). Finally, a synthetic peptide (residues 19–31 of the δ isozyme of rat brain protein kinase C) that is thought to specifically inhibit mammalian protein kinase C (House and Kemp, 1987) inhibits the kinase fraction's ability to activate the membrane-associated NADPH oxidase. Half-

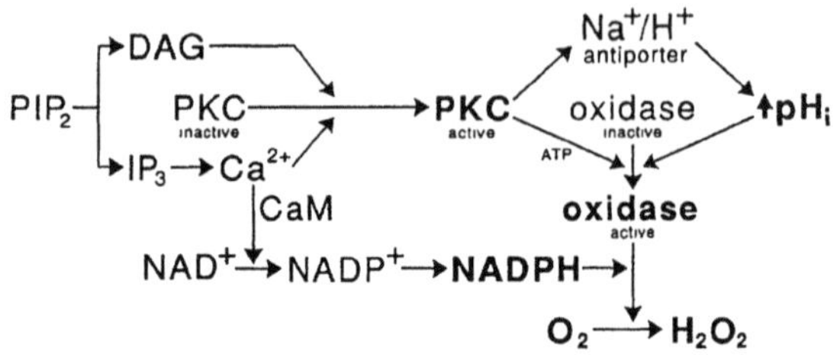

FIGURE 3. Proposed model for regulation of the respiratory burst oxidase of fertilization by protein kinase C (PKC). DAG, diacylglycerol; CaM, calmodulin; pH_i, intracellular pH. Reproduced with permission from Heinecke and Shapiro, 1992, *J. Biol. Chem.* **267:** 7959–7962.

maximal inhibition occurs at a peptide concentration of 420 nM, which is similar to the apparent K_i (360 nM) obtained when rat brain protein kinase C is substituted for the kinase fraction. In contrast, a synthetic peptide that inhibits cAMP-dependent kinase (Cheng *et al.*, 1986) fails to affect H_2O_2 synthesis when the membrane fraction is activated by either the kinase fraction or rat brain protein kinase C (Heinecke *et al.*, 1990).

Based on these results and on the signaling pathways known to be associated with fertilization, we proposed a model for the regulation of the early events that activate echinoderm eggs (Figure 3). Immediately following the union of egg and sperm, a phospholipase C is activated, liberating diacylglycerol and inositol trisphosphate from phosphatidylinositol bisphosphate (Turner *et al.*, 1984). Inositol trisphosphate triggers the release of Ca^{2+} from intracellular stores (Whitaker and Irvine, 1984), generating the necessary ionic signal for many of the early events in egg activation, including the calmodulin-dependent stimulation of NAD kinase (Epel *et al.*, 1981) and the exocytosis of cortical granules (Steinhardt and Epel, 1974; Vacquier, 1975). This kinase converts NAD^+ to $NADP^+$, while hexose monophosphate shunt activity accelerates, reducing $NADP^+$ to NADPH (Epel, 1964; Whitaker and Irvine, 1984). The latter is the reducing substrate for the respiratory burst oxidase (Heinecke and Shapiro, 1989; Heinecke *et al.*, 1990) as well as for ovothiol (1-methyl-4-thiohistidine) (Turner *et al.*, 1988), an amino acid that scavenges H_2O_2. The coordinated increase in Ca^{2+} and diacylglycerol stimulates protein kinase C, which in turn activates the membrane-associated NADPH oxidase, resulting in H_2O_2 synthesis (Heinecke and Shapiro, 1989; Heinecke *et al.*, 1990). The protein kinase C also may regulate the Na^+/H^+ antiporter that alkalinizes the egg cytoplasm because, in phorbol ester-treated eggs, pH_i increases in a Na^+-dependent manner (Swann and Whitaker, 1985; Shen and Burgart, 1986; Lau *et al.*, 1986). Such cytoplasmic alkalinization may modulate oxidase activity (Heinecke and Shapiro, 1989, 1990a).

4. FERTILIZED OOCYTES LIMIT OXIDATIVE STRESS

The fertilized oocyte is at the beginning of its developmental program and thus particularly susceptible to the toxic effects of H_2O_2, an uncharged species that should readily diffuse across the plasma membrane. Sea urchin embryos employ a number of mechanisms to limit their exposure to potentially lethal oxidative insults. For example, H_2O_2 synthesis is tightly controlled temporally, beginning shortly after the soft fertilization envelope forms and terminating immediately after the envelope is hardened by dityrosine cross-linking (Foerder and Shapiro, 1977; Foerder *et al.*, 1978).

Protein kinase C potentially represents an ideal mechanism for such regulation because its activity is controlled by intracellular Ca^{2+}, whose concentration increases and then decreases over the appropriate time frame (Turner *et al.*, 1986; Swann and Whitaker, 1986). The increase in pH_i that begins after cortical granule exocytosis (Johnson *et al.*, 1976; Shen and Steinhardt, 1978, 1979; Johnson and Epel, 1981) may further modulate the NADPH oxidase activity (Heinecke and Shapiro, 1989, 1990a).

Any H_2O_2 that does manage to diffuse into the embryo could oxidatively damage proteins, lipids, and nucleic acids (Fridovich, 1978). One pathway for initiating such reactions is the generation of highly reactive hydroxyl radicals (·OH) by metal ion (M^{n+})-catalyzed Fenton chemistry (Fridovich, 1978):

$$H_2O_2 + M^{n+} \rightarrow \cdot HO + HO^- + M^{n-1} \tag{4}$$

Sea urchins have evolved a powerful strategy to prevent H_2O_2-dependent damage, however. This involves ovothiol, which scavenges H_2O_2 nonenzymatically and is present at millimolar concentrations in egg cytoplasm (Turner *et al.*, 1987, 1988). First discovered in eggs, ovothiol (OSH) is an aromatic mercaptohistidine (Turner *et al.*, 1987, 1988; Palumbo *et al.*, 1982) that exhibits an unusually low thiol pK_a (Holler and Hopkins, 1988).

At physiological pH, ovothiol is present almost exclusively as the thiolate anion (Holler and Hopkins, 1988), making it an effective reducing agent for H_2O_2. The resulting oxidized ovothiol (OSSO) is reduced by glutathione (GSH) with the concomitant production of glutathione disulfide (GSSG), which itself is reduced by glutathione reductase at the expense of NADPH (Turner *et al.*, 1988).

$$2\ OSH + H_2O_2 \rightarrow OSSO + 2\ H_2O \tag{5}$$

$$OSSO + 2\ GSH \rightarrow 2\ OSH + GSSG \tag{6}$$

$$GSSG + NADPH \rightarrow 2\ GSH + NADP^+ \tag{7}$$

Because the reduction potential of ovothiol is 84 mV positive to glutathione, the equilibrium for the reaction between reduced GSH and oxidized OSSO [Eq. (6)] lies far to the right (Turner *et al.*, 1988; Holler and Hopkins, 1988). Sea urchin eggs lack glutathione peroxidase but have high levels of glutathione reductase, which maintains glutathione in the reduced form following fertilization (Turner *et al.*, 1988). Thus, ovothiol rapidly reduces H_2O_2 to H_2O, and NADPH ultimately serves both to generate H_2O_2 and to detoxify any oxidant that diffuses back into the egg.

The embryo also is threatened by the presence in sea water of significant concentrations of photochemically generated superoxide, $\cdot O_2^-$, and by high concentrations of redox-active transition metal ions like manganese and copper. $\cdot O_2^-$ spontaneously dismutates into H_2O_2, again providing the substrate for the production of hydroxyl radical, perhaps through Fenton chemistry (Fridovich, 1978).

We found that ovoperoxidase can catalyze the breakdown of $\cdot O_2^-$ in a reaction that requires H_2O_2 and is sensitive to inhibition by peroxidase inhibitors (Heinecke and Shapiro, 1990b). This reaction is likely to be quite different from that catalyzed by superoxide dismutase, whose active site uses $\cdot O_2^-$ to alternately reduce and oxidize the metal ion by one-electron chemistry (Fridovich, 1978). We proposed (Heinecke and Shapiro, 1990b) that the first step in the reaction mechanism is the oxidation of ferric ovoperoxidase by H_2O_2 to a ferryl-π-cation radical complex (compound I). Subsequently, compound I is reduced back to the native state by $\cdot O_2^-$ via sequential one-electron transfer reactions (Heinecke and Shapiro, 1990b). Each turn of this proposed cycle would consume 1 mole of H_2O_2 and 2 moles of $\cdot O_2^-$:

$$Fe(III) + H_2O_2 \rightarrow \text{compound I} + H_2O \tag{8}$$

$$\text{compound I} + \cdot O_2^- + H^+ \rightarrow \text{compound II} + O_2 \tag{9}$$

$$\text{compound II} + \cdot O_2^- + H^+ \rightarrow Fe(III) + H_2O + O_2 \tag{10}$$

This superoxide peroxidase activity of ovoperoxidase in the fertilization envelope may serve as a chemical shield that protects the developing embryo from photochemically generated oxidants.

5. MAMMALIAN FERTILIZATION AND THE RESPIRATORY BURST

Many of the mechanisms employed in sea urchin fertilization have been conserved by mammalian gametes during evolution (Shapiro *et al.*, 1981; Shapiro, 1991). Mouse eggs have cytoplasmic granules that contain a histochemically detectable peroxidase, and peroxidase inhibitors apparently slow the mouse zona hardening reaction (Schmell and Gulyas, 1980; Gulyas and Schmell, 1980). Perhaps peroxidative pathways play a role in the early developmental program of the mammalian embryo.

6. OXIDATIVE REACTIONS OF PHAGOCYTES

Activated mammalian phagocytes and fertilized echinoderm eggs exhibit remarkable metabolic similarities (Klebanoff *et al.*, 1979). These include chemiluminescence, acid production, and a surge in glucose-6-phosphate dehydrogenase activity, which reduces $NADP^+$ to NADPH.

Both types of cells also display a cyanide-insensitive respiratory burst. Moreover, there are striking parallels between the oxidases involved (reviewed in Segal and Abo, 1993, and Chapter 19). Both enzymes have membrane-associated and soluble components, and both use NADPH, but not NADH, as a source of reducing equivalents. However, the oocyte NADPH oxidase reduces oxygen directly to H_2O_2* in a reaction that

*We tested the possibility that the sea urchin egg also produces $\cdot O_2^-$ during the respiratory burst of fertilization (Heinecke, unpublished observations). $\cdot O_2^-$ reduces ferricytochrome *c* in a reaction that is inhibited by superoxide dismutase (Fridovich, 1978). Unfertilized eggs continuously reduce cytochrome *c*, but the rate remains constant following either activation or the addition of superoxide dismutase. Moreover, the respiratory burst of eggs that have been stripped of the fertilization envelope, a potential permeability barrier, is little affected by cytochrome *c*. If the oocyte oxidase generated $\cdot O_2^-$, it presumably would react with cytochrome *c*, and both oxygen consumption and H_2O_2 synthesis would be inhibited. Finally, the oocyte oxidase fails to generate $\cdot O_2^-$ as assessed by the superoxide dismutase-inhibitable reduction of cytochrome *c*, acetylated cytochrome *c*, or nitroblue tetrazolium. These results imply that the egg oxidase reduces O_2 directly to H_2O_2 by a two-electron transfer reaction.

involves the transfer of two electrons, whereas the neutrophil NADPH oxidase initially generates $\cdot O_2^-$, which spontaneously dismutates to form H_2O_2 and O_2.

The constituents of the electron transport chain of the oocyte NADPH oxidase have not been identified. Because apparently homogeneous mammalian protein kinase C can activate the membrane fraction to synthesize H_2O_2 (Heinecke *et al.*, 1990), all of the electron transport components of the sea urchin NADPH oxidase presumably are membrane-associated. Difference spectroscopy of dithionite-reduced versus native membranes reveals a cytochrome that cosediments with oxidase activity on sucrose gradients (Heinecke, unpublished observation). However, we were unable to demonstrate reduction of this cytochrome by NADPH (either aerobically or anaerobically) when the membrane fraction was stimulated with protein kinase C in the presence of cofactors required for H_2O_2 synthesis. Thus, the relation of this cytochrome to oxidase activity remains to be clarified.

Comparing activation of the two oxidases, we find that a protein kinase is the only cytosolic protein required to activate the egg oxidase (Heinecke *et al.*, 1990). The neutrophil oxidase is considerably more complex (reviewed in Segal and Abo, 1993, and Chapter 19) in that it contains at least three soluble components, p47-phox, p67-phox, and a GTP-binding protein. Phosphorylation may be involved, however, because p47-phox incorporates radiolabeled phosphate when neutrophils are activated with phorbol ester. Moreover, p47-phox fails to be phosphorylated in certain patients with chronic granulomatous disease, where neutrophil $\cdot O_2^-$ synthesis is impaired.

Because phorbol ester is a potent stimulator of H_2O_2 synthesis by human neutrophils, monocytes, and macrophages, protein kinase C may be one enzyme that regulates oxidant generation by phagocytes. However, other signaling pathways also trigger H_2O_2 synthesis in phagocytes. This complexity may make teleological sense because, unlike the oocyte, which uses H_2O_2 after fertilization specifically to harden the fertilization envelope, the phagocyte generates oxidants in response to a multitude of different stimuli.

7. PEROXIDATIVE MECHANISMS OF OOCYTES AND PHAGOCYTES

Another common feature is that phagocytes, like oocytes, exocytose a peroxidase that uses H_2O_2 from the respiratory burst as an oxidizing substrate. This enzyme is myeloperoxidase, which generates potent microbicidal products (Harrison and Schultz, 1976; Foote *et al.*, 1981). Like ovoperoxidase, myeloperoxidase uses H_2O_2 to convert tyrosine to tyrosyl radical (Heinecke *et al.*, 1993a,b; Savenkova *et al.*, 1994).

As in the oocyte, such reactive products have the capacity to be beneficial or harmful, for they can damage human tissues as well as kill pathogenic organisms. Our studies of the sea urchin fertilization envelope led us to suggest that phagocyte-generated tyrosyl radical might damage the artery wall, contributing to the initial events of atherosclerosis (Heinecke *et al.*, 1993a, 1993b; Savenkova *et al.*, 1994). Indeed, many lines of evidence indicate that low-density lipoprotein (LDL), the major carrier of blood cholesterol, must be oxidatively modified to be atherogenic (reviewed in Berliner and Heinecke, 1996).

We obtained evidence that myeloperoxidase is present in the damaged artery wall by demonstrating immunoreactivity between a monospecific polyclonal antibody and a protein in detergent extracts of surgically excised atherosclerotic lesions (Daugherty *et al.*, 1994). Immunostaining of lesions generated similar patterns to those we obtained

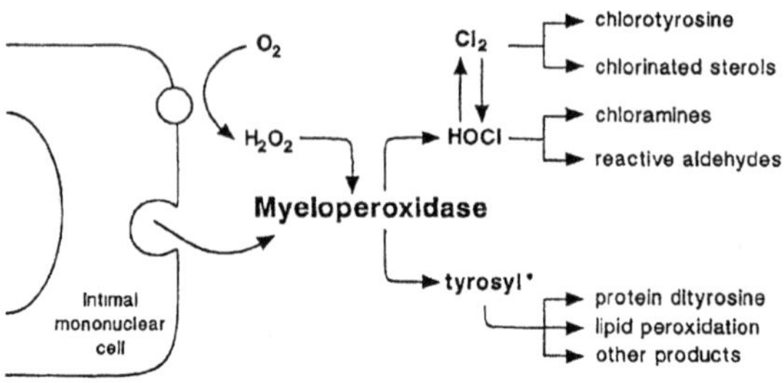

FIGURE 4. Potential pathways for LDL oxidation and tissue damage by myeloperoxidase secreted by activated phagocytes. The enzyme uses hydrogen peroxide (H_2O_2) to generate tyrosyl radical, which tyrosylates proteins and initiates LDL lipid peroxidation. In the presence of chloride ions, myeloperoxidase converts H_2O_2 to hypochlorous acid (HOCl), which is in equilibrium with molecular chlorine (Cl_2). HOCl and Cl_2 chlorinate reactive protein groups and cholesterol. By generating products that oxidatively modify LDL, myeloperoxidase may play a critical role early in atherogenesis.

with an antibody to macrophages. This finding contradicts the prevailing dogma that macrophages do not contain myeloperoxidase and suggests that the enzyme's gene is upregulated under certain pathological conditions.

In vitro experiments with myeloperoxidase have revealed that the phagocyte enzyme, like ovoperoxidase, uses H_2O_2 to oxidize L-tyrosine to tyrosyl radical, generating the stable cross-linked product *o,o'*-dityrosine (Heinecke *et al.*, 1993a, b). Activated human neutrophils also generate *o,o'*-dityrosine from L-tyrosine in a reaction catalyzed by myeloperoxidase. Moreover, the oxidation of L-tyrosine either by myeloperoxidase and H_2O_2 or by the myeloperoxidase system of activated human neutrophils leads, *in vitro*, to the oxidation of LDL (Savenkova *et al.*, 1994). Tyrosine oxidation also may occur *in vivo* because we detected a 100-fold higher concentration of *o,o'*-dityrosine in LDL extracted from human atherosclerotic lesions than in circulating LDL (Leeuwenburgh *et al.*, 1997). The concentration of this marker was 11-fold higher in fatty lesions and 6-fold higher in advanced lesions than in samples of normal artery wall.

In summary, we suggest that phagocytes, like oocytes, employ peroxidative mechanisms to generate tyrosyl radical, as well as other cytotoxic oxidants such as HOCl and chlorine gas (Figure 4). These reactive species promote survival by killing invading microorganisms, but they also may inadvertently damage proteins and lipids by cross-linking or chlorinating tyrosine (Heinecke *et al.*, 1993a; Hazen *et al.*, 1996a), generating reactive aldehydes from amino acids (Hazen *et al.*, 1996c), and converting cholesterol into epoxides and chlorohydrins (Heinecke *et al.*, 1994; Hazen *et al.*, 1996b). Such chemical modifications may contribute to diseases of aging, such as atherosclerosis. Perhaps they also may exacerbate acute conditions, such as stroke, in which phagocytes are implicated in tissue damage.

ACKNOWLEDGMENTS. I thank Dr. Linda Sage for critical comments on the manuscript. J.W.H. is the recipient of an Established Investigator Award from the American Heart Association. This research was supported by Grant RO1 HD29025 from the National Institutes of Health.

8. REFERENCES

Battaglia, D. E., and Shapiro, B. M., 1988, Hierarchies of protein cross-linking in the extracellular matrix, *J. Cell Biol.* **107**:2447–2452.

Berliner, J. A., and Heinecke, J. W., 1996, The role of oxidized lipoproteins in atherogenesis, *Free Radical Biol. Med.* **20**:707–727.

Cheng, H.-C., Kemp, B. E., Pearson, R. P., Smith, A. J., Misoni, L., VanPatten, S. M., and Walsh, D.A., 1986, A potent synthetic peptide inhibitor of cAMP-dependent protein kinase, *J. Biol. Chem.* **261**:989–992.

Daugherty, A., Dunn, J. L., Rateri, D. L., and Heinecke, J. W., 1994, Myeloperoxidase, a catalyst for lipoprotein oxidation, is expressed in human atherosclerotic lesions, *J. Clin. Invest.* **94**:437–444.

Diets, T., Farrance, M., Kay, E. S., Medill, L., Turner, E. E., Weidman, P. J., and Shapiro, B. M., 1984, Purification and properties of ovoperoxidase, the enzyme responsible for hardening the fertilization membrane of the sea urchin egg, *J. Biol. Chem.* **259**:13525–13533.

Epel, D., 1964, Simultaneous measurement of TPNH formation and respiration following fertilization of the sea urchin egg, *Biochem. Biophys. Res. Commun.* **17**:69–73.

Epel, D., Patton, C., Wallace, R. W., Cheung, W. Y., 1981, Calmodulin activates NAD kinase of sea urchin eggs: An early event of fertilization, *Cell* **23**:543–549.

Foerder, C. A., and Shapiro, B. M., 1977, Release of ovoperoxidase from sea urchin eggs hardens the fertilization membrane with tyrosine crosslinks, *Proc. Natl. Acad. Sci. USA* **74**:4214–4218.

Foerder, C. A., Klebanoff, S. J., and Shapiro, B. M., 1978, Hydrogen peroxide production, chemiluminescence, and the respiratory burst of fertilization: Interrelated events in early sea urchin development, *Proc. Natl. Acad. Sci. USA* **75**:3183–3187.

Foote, C. S., Goyne, T. E., and Lehrer, R. I., 1981, Assessment of chlorination by human neutrophils, *Nature* **301**:715–716.

Fridovich, I., 1978, The biology of oxygen radicals, *Science* **201**:875–880.

Gulyas, B. V., and Schmell, E. D., 1980, Ovoperoxidase activity in ionophore treated mouse eggs. I. Electon microscopic localization, *Gamete Res.* **3**:267–278.

Harrison, J. E., and Schultz, J., 1976, Studies on the chlorinating activity of myeloperoxidase, *J. Biol. Chem.* **251**:1371–1374.

Hazen, S. L., Hsu, F. F., Crowley, J. R., Mueller, D. M., Heinecke, J. W., 1996a, Human neutrophils employ chlorine gas as an oxidant during phagocytosis, *J. Clin. Invest.* **98**:1283–1289.

Hazen, S. L., Hsu, F. F., Duffin, K., Heinecke, J. W., 1996b, Molecular chlorine generated by the myeloperoxidase–hydrogen peroxide–chloride system of phagocytes converts low density lipoprotein cholesterol into a family of chlorinated sterols, *J. Biol. Chem.* **271**:23080–23088.

Hazen, S. L., Hsu, F. F., Heinecke, J. W., 1996c, p-Hydroxyphenylacetaldehyde is the major product of L-tyrosine oxidation by activated phagocytes: A chloride-dependent mechanism for the conversion of free amino acids into reactive aldehydes by myeloperoxidase, *J. Biol. Chem.* **271**:1861–1867.

Heinecke, J. W., and Shapiro, B. M., 1989, Respiratory burst oxidase of fertilization, *Proc. Natl. Acad. Sci. USA* **86**:1259–1263

Heinecke, J. W., and Shapiro, B. M., 1990a, Protein kinase C activates the respiratory burst of fertilization, but not cortical granule exocytosis, in ionophore stimulated sea urchin eggs, *Dev. Biol.* **142**:216–223.

Heinecke, J. W., and Shapiro, B. M., 1990b, Superoxide peroxidase activity of ovoperoxidase, the crosslinking enzyme of fertilization, *J. Biol. Chem.* **265**:9241–9246.

Heinecke, J. W., Meier, K. E., Lorenzen, J. A., and Shapiro, B. M., 1990, A specific requirement for protein kinase C in activation of the respiratory burst oxidase of fertilization, *J. Biol. Chem.* **265**:7717–7720.

Heinecke, J. W., Li, W., Daenke, H.L., III, and Goldstein, J. A., 1993a, Dityrosine, a specific marker of oxidation, is synthesized by the myeloperoxidase–hydrogen peroxide system of human neutrophils and macrophages, *J. Biol. Chem.* **268**:4069–4077

Heinecke, J. W., and Shapiro, B. M., 1992, The respiratory burst oxidase of fertilization, *J. Biol. Chem.* **267**:7959–7962.

Heinecke, J. W., Li, W., Francis, G. A., and Goldstein, J. A., 1993b, Tyrosyl radical generated by myeloperoxidase catalyzes the oxidative crosslinking of proteins, *J. Clin. Invest.* **91**:2866–2872.

Heinecke, J. W., Li, W., Mueller, D. M., Bohrer, A., Turk, J., 1994, Cholesterol chlorohydrin synthesis by the myeloperoxidase–hydrogen peroxide–chloride system: Potential markers for lipoproteins oxidatively damaged by phagocytes, *Biochemistry* **33**:10127–10136.

Holler, T. P., and Hopkins, P. B., 1988, Ovothiols as biological antioxidants, *J. Am. Chem. Soc.* **110:** 4837–4838.

House, C., and Kemp, B. E., 1987, Protein kinase C contains a pseudosubstrate prototope in its regulatory domain, *Science* **238:**1726–1728.

Johnson, C. H., and Epel, D., 1981, Intracellular pH of sea urchin eggs measured by the dimethyloxasolidinedione method, *J. Cell Biol.* **89:**284–291.

Johnson, J. D., Epel D., and Paul, M., 1976, Intracellular pH and activation of sea urchin eggs after fertilization, *Nature* **262:**661–664.

Kay, E., Eddy, E. M., and Shapiro, B. M., 1982, Assembly of the fertilization membrane of the sea urchin, *Cell* **29:**867–875.

Klebanoff, S. J., Foerder, C. A., Eddy, E. M., and Shapiro, B. M., 1979, Metabolic similarities between fertilization and phagocytosis, *Exp. Med.* **149:**938–953.

Lau, A. F., Ranyon, T. C., and Humphreys, T., 1986, Tumor promoters and diacylglycerol activate the Na^+/H^+ antiporter of sea urchin eggs, *Exp. Cell Res.* **166:**23–30.

Leeuwenburgh, C., Rasmussen, J. E., Hsu, F. F., Mueller, D. M., Pennathur, S., and Heinecke, J. W., 1997, Mass spectrometric quantification of markers for protein oxidation by tyrosyl radical, copper and hydroxyl radical in low density lipoprotein isolated from human atherosclerotic plaques, *J. Biol. Chem.* **272:**1433–1436.

Mori, T., Taki, R., Minakuchi, R., Yu, B., and Nishizuka, Y., 1980, Inhibitory action of chlorpromazine, dibucaine, and other phospholipid interacting drugs on calcium-activated phospholipid-dependent protein kinase, *J. Biol. Chem.* **255:**8378–8380.

Nishizuka, Y., 1988, The molecular heterogeneity of protein kinase C and its implications for cellular regulation, *Nature* **334:**661–665.

Nishizuka, Y., 1995, Protein kinase C and lipid signaling for sustained cellular responses, *FASEB J.* **9:**484–496.

Palumbo, A., Ol'Ischia, M., Misuraca, G., and Prota, G., 1982, Isolation and structure of a new sulphur-containing amino acid from sea urchin eggs, *Tetrahedron Lett.* **23:**3207–3208.

Rothschild, L., 1959, *Fertilization*, pp. 91–103, Methuen, London.

Savenkova, M. I., Mueller, D. M., and Heinecke, J. W., 1994, Tyrosyl radical generated by myeloperoxidase is a physiological catalyst for the initiation of lipid peroxidation in low density lipoprotein, *J. Biol. Chem.* **269:**20394–20400.

Schachtele, C., Seifert, R., and Osswald, H., 1988, Stimulus-dependent inhibition of platelet aggregation by the protein kinase C inhibitors polymyxin B, H-7, and staurosporine, *Biochem. Biophys. Res. Comm.* **151:**542–547.

Schmell, E. D., and Gulyas, B. J., 1980, Ovoperoxidase activity in ionophore treated mouse eggs. II. Evidence for the enzyme's role in hardening the zona pellucida, *Gamete Res.* **3:**279–290.

Segal, A. W., and Abo, A., 1993, The biochemical basis of the NADPH oxidase of phagocytes, *Trends Biochem. Sci.* **18:**43–47.

Shapiro, B. M., 1991, The control of oxidant stress at fertilization, *Science* **252:**533–536.

Shapiro, B. M., Shackmann, R. W., and Gabel, C. A., 1981, Molecular approaches to the study of fertilization, *Annu. Rev. Biochem.* **50:**815–843.

Shen, S. S., and Burgart, L. J., 1986, 1,2 - Diacylglycerols mimic phorbol 12-myristate 13-acetate activation of the sea urchin egg, *J. Cell. Physiol.* **127:**330–340.

Shen, S. S., and Steinhardt, R. A., 1978, Direct measurement of intracellular pH during metabolic derepression of the sea urchin egg, *Nature* **272:**253–254.

Shen, S. S., and Steinhardt, R. A., 1979, Intracellular pH and the sodium requirement at fertilization, *Nature* **282:**87–89.

Steinhardt, R. A., and Epel, D., 1974, Activation of sea urchin eggs by a calcium ionophore, *Proc. Natl. Acad. Sci. USA* **71:**1915–1919.

Swann, K., and Whitaker, M., 1985, Stimulation of the Na/H exchanger of sea urchin eggs by phorbol ester, *Nature* **314:**274–277.

Swann, K., and Whitaker, M., 1986, The part played by inositol trisphosphate and calcium in the propagation of the fertilization wave in sea urchin eggs, *J. Cell. Biol.* **103:**2333–2342.

Turner, E., Klevit, R., Hager, L. J., and Shapiro, B. M., 1987, Ovothiols, a family of redox-active mercaptohistidine compounds from marine invertebrate eggs, **26:**4028–4036.

Turner, E., Hager, L. J., and Shapiro, B. M., 1988, Ovothiol replaces glutathione peroxidase as a hydrogen peroxide scavenger in sea urchin eggs, *Science* **242:**939–941.

Turner, P. R., Sheetz, M. P., and Jaffe, L. A., 1984, Fertilization increases the polyphosphoinositide content of sea urchin eggs, *Nature* **310**:414–415.

Turner, P. R., Jaffe, L. A., and Fein, A., 1986, Regulation of cortical vesicle exocytosis in sea urchin eggs by inositol 1,4,5,-trisphosphate and GTP binding protein, *J. Cell Biol.* **102**:70–76.

Vacquier, V. D., 1975, The isolation of intact cortical granules from sea urchin eggs: Calcium ions trigger cortical granule discharge, *Devl. Biol.* **43**:62–74.

Vacquier, V. D., Epel, D., and Douglas, L. A., 1972, Sea urchin eggs release protease activity at fertilization, *Nature* **237**:34–36.

Vernon, M., Foerder, C., Eddy, E. M., and Shapiro, B. M., 1977, Sequential biochemical and morphological events during assembly of the fertilization membrane, *Cell* **10**:321–328.

Warburg, O., 1908, Boebachtungen über die Oxydationsprozesse im Seeegelei, *Physiol. Chem.* **57**:1–11.

Weiss, B., Prozialeck, W., Cimino, M., Barnette, M. S., and Wallace, T. L., 1980, Pharmacological regulation of calmodulin, *Ann. N.Y. Acad. Sci.* **536**:319–345.

Whitaker, M., and Irvine, R. F., 1984, Inositol 1,4,5-trisphosphate microinjection activates sea urchin eggs, *Nature* **312**:636–639.

Winkler, M. W., and Grainger, J. L., 1978, Mechanism of action of NH_4Cl and other weak bases in activation of sea urchin eggs, *Nature* **273**:536–538.

Chapter 22

Brain Chemiluminescence as an Indicator of Oxidative Stress

Alberto Boveris and Enrique Cadenas

1. INTRODUCTION

Brain chemiluminescence was detected (Jamieson *et al.*, 1986; Adamo *et al.*, 1989) as part of a research effort to characterize the photon emission from *in situ* organs, such as liver (Boveris *et al.*, 1980, 1983), lung (Turrens *et al.*, 1988; Giulivi *et al.*, 1995), skeletal muscle (Llesuy *et al.*, 1994), and small intestine (Roldán *et al.*, 1989). The spontaneous chemiluminescence of *in situ* organs provides a quantitative determination of the steady-state concentrations of molecules in the excited state, for these molecules give off photons in the process of returning to the ground state and the photons are quantitatively determined at the organ surface by the single photon count technique. The spectral analysis of organ chemiluminescence and the effect of traps and quenchers appear to indicate that the light emission from mammalian organs under physiological conditions is mainly related to singlet oxygen ($^1O_2$) and, in a minor proportion, to molecules with excited carbonyl groups (Cadenas and Sies, 1984; Cadenas *et al.*, 1994). Singlet oxygen was considered in the past as a reactive species generated in photosensitized oxidations in biological systems, but more recently, it was reported that other reactive oxygen species such as $\cdot O_2^-$, (Kanfer and Turro, 1981; Giulivi *et al.*, 1990) are involved in certain cases, for instance when rose bengal is the photosensitizer. Moreover, low-level chemiluminescence was proposed as an indicator of the physiological production of singlet oxygen by Cadenas and Sies (1984). The key step in the biological generation of $^1O_2$ seems to be

Alberto Boveris Department of Physical Chemistry, School of Pharmacy and Biochemistry, University of Buenos Aires, Buenos Aires, Argentina. **Enrique Cadenas** Department of Molecular Pharmacology and Toxicology, School of Pharmacy, University of Southern California, Los Angeles, California 90033.

Reactive Oxygen Species in Biological Systems, edited by Gilbert and Colton. Kluwer Academic / Plenum Publishers, New York, 1999.

the termination reaction, called the Russell reaction, in which two peroxyl radicals (·ROO) react and, through the formation of a transient tetroxide intermediate, lead to the production of either $^1O_2$ or ketones or aldehydes with excited carbonyl groups (Cadenas *et al.*, 1994). The electronically excited states of these ketones and aldehydes afford another chemical pathway for the production of $^1O_2$ through energy transfer, after molecular collision, from an excited carbonyl group to molecular oxygen (Cadenas *et al.*, 1994). Recent evidence by Giulivi *et al.* (1998) identified liver mitochondria as an effective source of nitric oxide (·NO) with rates of about 60 nmole ·NO/min per g liver, which are comparable to the organ rates of $\cdot O_2^-$ production, about 80 nmole $\cdot O_2^-$ /min per g liver (Chapter 3). In such conditions, the steady-state concentrations of $\cdot O_2^-$ (10^{-11} to 10^{-10} M) and of ·NO (10^{-8} to 10^{-7} M) in the tissue should sustain an effective peroxynitrite ($OONO^-$) formation through the Beckman reaction (Beckman, 1990). The recent report by Khan (1995) that $OONO^-$ yields $^1O_2$ on acidification provides an additional chemical pathway that has to be evaluated in terms of the physiological production of $^1O_2$ and of the occurrence of oxidative stress.

Oxidative stress is understood, from a biological viewpoint, as a situation in which the balance between oxidants and antioxidants in the biological system is shifted toward a prooxidative state (Sies, 1985). However, from a physicochemical viewpoint, oxidative stress is better understood as a situation in which the intracellular steady-state concentrations of the reactive oxygen species (namely, $\cdot O_2^-$, superoxide radical; H_2O_2, hydrogen peroxide; ·HO, hydroxyl radical; ·ROO, peroxyl radical; and $^1O_2$, singlet oxygen) are increased over their physiological values (González Flecha *et al.*, 1993). The term and concept of reactive oxygen species reveal that the reactivity of these chemical species toward biomolecules and cell constituents is far higher than the reactivity of molecular oxygen (O_2, dioxygen) toward the same targets because of the electronic spin restrictions of the latter molecule (Kanfer and Turro, 1981). An increased intracellular steady-state concentration of any of the reactive oxygen species unequivalently implies an increased rate of the free radical chain reaction in which reactive oxygen species are products and reactants.

The reactive oxygen species are produced sequentially in the free radical chain reaction occurring in cells and tissues in the order in which they are listed above, for kinetic and thermodynamic reasons. Thus, if one of these species is specifically overproduced, the oxidative stress situation implies an increased steady-state concentration of the specifically overproduced species and of all of the other species down the chain reaction of free radical and excited states. For instance, in a situation in which glucose oxidase is used to supplement a cell culture that includes glucose in the culture medium, the steady-state concentration of H_2O_2 will be primarily increased, and, secondarily, the steady state concentrations of ·HO, ·ROO, and $^1O_2$ will also be increased, whereas the $\cdot O_2^-$ steady-state concentration will be unaffected. Similarly, an ABAP-supplemented system will show increased levels of ·ROO (including the precursors ·R) and of $^1O_2$ but not of $\cdot O_2^-$ and H_2O_2. Thus, different biological situations of oxidative stress are to be found, even for the same type of cell or tissue, depending on the primarily overproduced reactive oxygen species. Moreover, if the situation of oxidative stress is extended over a period of time, the signaling system of intracellular defense against oxidative stress will be triggered. The regulatory response triggered either by the products of the univalent or bivalent biological reduction of oxygen ($\cdot O_2^-$ and H_2O_2), by other reactive oxygen species, or by the reaction products of reactive oxygen species and cell constituents will be superimposed on the original oxidative stress and both will define the biological situation.

2. CHEMILUMINESCENCE MEASUREMENTS

Rat brains were exposed, so as to place them in front of the photomultiplier, by three different surgical procedures depending on rat weight (see Figure 1): (1), neonatal rats, weighing 23 to 30 g and 2 to 3 weeks old, were anesthetized with 0.1 g urethane/kg ip, head skin was removed to expose the skull, and brain chemiluminescence was measured through the translucid parietal bones; (2) young and neonatal rats, weighing 30 to 80 g and about 3 weeks to 3 months old, were anesthetized with urethane as described above or with 50 mg sodium pentobarbital/kg ip, head skin was removed, the parietal bones were cut out with a curved scissor, meninges were removed, and chemiluminescence was measured from the exposed brain; and (3) adult rats, weighing 180 to 300 g and 8 months to 2 years old, were anesthetized with 50 mg sodium pentobarbital/kg ip, head skin was removed, and the adhered membranes were removed by rubbing with cotton balls. Two oval holes, about 1 × 0.5 cm, were carved with a dentist drill in the parietal bones; one in each parietal bone and avoiding damage to the sagittal suture that produces intense bleeding. Brain chemiluminescence was measured through the holes in the parietal bones. In all three cases, rats were covered with aluminum foil in which a window was cut to expose only the brain area to be measured. Chemiluminescence is expressed in counts per second (cps) per square centimeter of organ surface; this latter was determined from the window area in the aluminum foil.

Brain chemiluminescence was determined with a photon counter (Johnson Research Foundation, University of Pennsylvania, Philadelphia) specially adapted for organ photoemission; it consisted of a lighttight box thermostatized at 30°C with a shutter operated by remote control and an EMI 9658 red-sensitive phototube with an applied potential of −1.3 kV. Other experimental procedures were as described elsewhere (Boveris et *al.*, 1980, 1983).

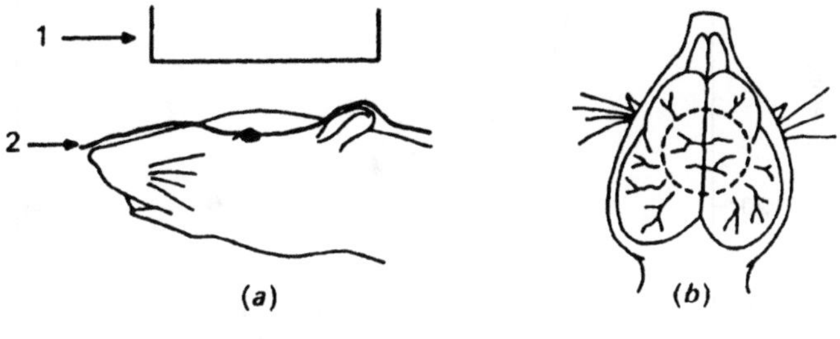

FIGURE 1. Scheme showing the positioning of the phototube and the surgical procedures to measure brain chemiluminescence. (a) Measurement through the translucid parietal bones: 1, Lucite light guide of the photon counter; 2, aluminun foil. (b) Measurement after removing the parietal bones. The circle indicates the area exposed through an aluminum foil window.

3. BRAIN CHEMILUMINESCENCE AND OXIDATIVE STRESS

3.1. Brain Spontaneous Chemiluminescence

The *in situ* rat brain was found to be a source of spontaneous photoemission in the range of 20 to 120 cps/cm^2. The high brain photoemission corresponded to neonatal rats (2 to 3 weeks old, weighing 25 to 30 g) and the low chemiluminescence to adult animals (20 to 30 months old, 200 to 300 g) (Figure 2). This phenomenon of higher values of organ chemiluminescence in young animals has also been found for the *in situ* liver and is understood to be related to the higher metabolic rate and to the higher oxygen uptake in younger animals. Brain spontaneous chemiluminescence was stable for a period of 10 to 30 min after which chemiluminescence started to decline, likely reflecting drying of the tissue and secondary vasoconstriction in the dried areas. Brain chemiluminescence decreased to zero photoemission 2 to 3 min after animal death. These two properties of brain photoemission were taken as an indication that bone fluorescence was not involved in brain chemiluminescence.

3.2. Hyperbaric Oxygen and Brain Chemiluminescence

Adult rats anesthetized with sodium pentobarbital and with the brain exposed were placed in a hyperbaric chamber (Bethlehem Steel, Pittsburgh, PA) and subjected to high oxygen pressures. A 2.5-cm-diameter Lucite light guide was placed inside the chamber

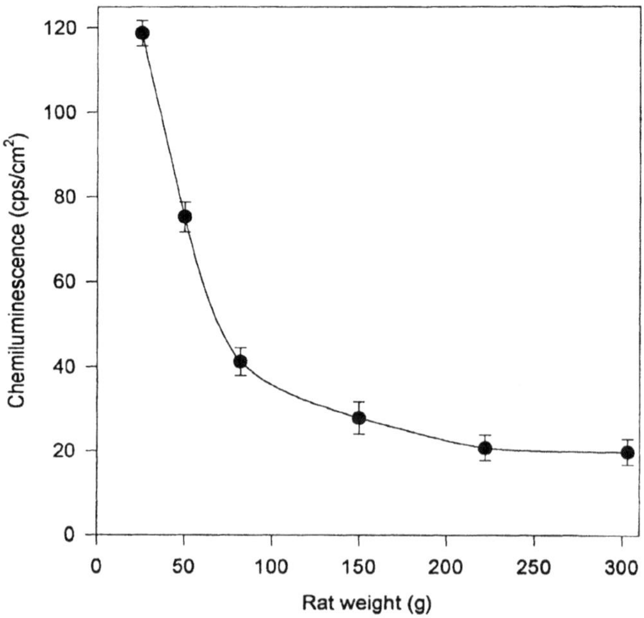

FIGURE 2. Rat brain chemiluminescence as a function of animal weight.

at about 2–3 mm from the brain surface, collected brain photoemission, and delivered it to the photon counter. The chamber was flushed with pure oxygen and then pressurized, in such a way that the rats were breathing hyperbaric oxygen pressures. The spontaneous brain chemiluminescence increased hyperbolically with increasing oxygen pressures, reaching a fourfold higher photoemission at 0.80 MPa (Figure 3) with a profile resembling blood pO_2 in animals breathing air and hyperbaric oxygen in the 0.02 to 0.80 MPa span. Similar data obtained for *in situ* liver chemiluminescence are included for purposes of comparison.

3.3. Hyperthyroidism and Brain Chemiluminescence

Neonatal hyperthyroidism causes marked changes in brain metabolism, such as an increase in oxygen uptake, a reduction in oligodendroglia number, and decreased myelinogenesis. The myelin deficit has been related to oxidative stress in the hyperthyroid brain, reflecting myelin susceptibility to oxidative attack (Adamo *et al.*, 1989). Newborn rats were made hyperthyroid by injecting them subcutaneously with triiodothyronine (25 μg at birth, 3 μg on the second day of life, and 0.5 and 1.0 μg on alternate days thereafter up to 18–20 days of life) as described before (Adamo *et al.*, 1989). Brain chemiluminescence was increased by 46 or 70% in experimental neonatal hyperthyroidism. The two values correspond to the two procedures utilized for the determination of brain photoemission: (a) from the exposed brain and (b) through the translucid parietal bones (Table I).

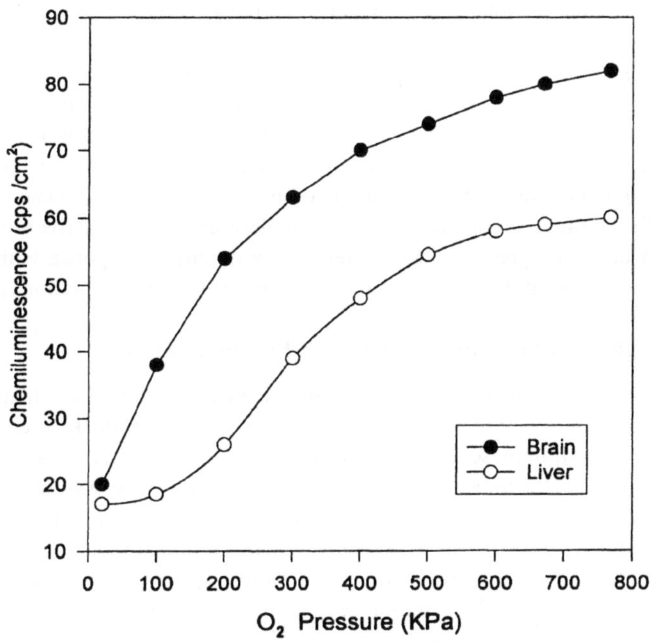

FIGURE 3. Effect of hyperbaric oxygen on brain chemiluminescence.

Table I
Brain Chemiluminescence, Oxygen Uptake, and Life Span in Hyperthyroid Neonatal Rats

	Control ($n = 6$)	Hyperthyroid ($n = 6$)
In situ brain chemiluminescence (cps/cm^2)		
(a) Exposed brain	130 + 13	190 + 30*
(b) Through the parietal bones	100 + 10	170 + 10*
Brain oxygen uptake (slices), (nmole O_2/min per g brain)	330 + 25	425 + 35*
Life span (weeks)	140 + 8	20 + 2*

*$p < 0.001$.

The latter procedure, which avoids surgical trauma, gave a 30% decreased brain chemiluminescence as a result of scattering of the emitted light by the parietal bones and a more marked effect of hyperthyroidism. Triiodothyronine-intoxicated neonatal rats showed a 30% increase in the oxygen uptake of brain slices and a dramatically decreased life span; from about 115 weeks to 14 weeks (Table I).

Hyperthyroid brains also showed increases of 18, 36, and 30% in the activities of Cu/Zn-superoxide dismutase (SOD), catalase, and Se-glutathione peroxidase, what can be understood as a regulatory cellular response to oxidative stress (Adamo *et al.*, 1989). On the other hand, MnSOD and non-Se-glutathione peroxidase activities were found unchanged in hyperthyroid brains (Adamo *et al.*, 1989).

Brain mitochondria isolated from hyperthyroid rats showed state 4 and state 3 rates of oxygen utilization (Adamo *et al.*, 1989). Accordingly, it seems that the increased oxygen uptake observed in the slices of hyperthyroid brains is related to an increase in the intracellular steady state of ADP that shifts respiration toward a more active (state 3) metabolic state. Mitochondria isolated from hyperthyroid brains showed a rate of production of H_2O_2 similar to that corresponding to normal animals (Adamo *et al.*, 1989). This is consistent with the unchanged activity of mitochondrial MnSOD and points to the cytosol as the cellular compartment that is undergoing the oxidative stress of the hyperthyroid brain in agreement with a regulatory defensive response with increased levels of cytosolic Cu/ZnSOD, peroxisomal catalase, and total Se-glutathione peroxidase.

3.4. Acute Ethanol Intoxication and Brain Chemiluminescence

Young rats, weighing 65 to 70 g, were intoxicated with 1 to 4 g ethanol/kg body weight, ip, as a 20% v/v ethanol solution in 145 mM NaCl. A marked increase, of about 175%, in brain chemiluminescence was recorded 3 hr after ethanol administration (Figure 4). Parallel experiments indicated that liver chemiluminescence was increased about 110%, 3 hr after alcohol intoxication with blood ethanol concentration in the range of 53 to 55 mM, stable in the 1- to 3-hr period after ethanol administration.

Rats pretreated with 4-methylpyrazole (1 mmole/kg, ip), an inhibitor of alcohol dehydrogenase, showed a 75% inhibition of the ethanol-induced increase in brain chemiluminescence. In contrast, pretreatment with disulfiram (2 mmole/kg, intragastrically), an inhibitor of acetaldehyde dehydrogenase, increased by 86% the ethanol-induced

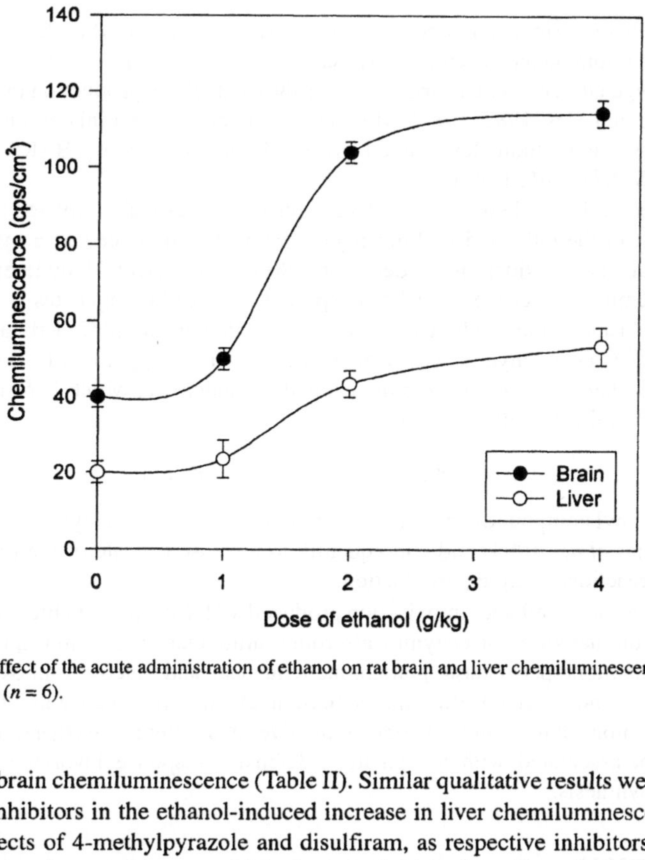

FIGURE 4. Effect of the acute administration of ethanol on rat brain and liver chemiluminescence. Values are means + SEM ($n = 6$).

increase in brain chemiluminescence (Table II). Similar qualitative results were observed with both inhibitors in the ethanol-induced increase in liver chemiluminescence (Table II). The effects of 4-methylpyrazole and disulfiram, as respective inhibitors of alcohol dehydrogenase that produces acetaldehyde and of aldehyde oxidase that utilizes acetaldehyde, appear to clearly indicate acetaldehyde as mainly responsible for the increased brain chemiluminescence. Acetaldehyde by itself does not seem able to sustain an increased rate of oxyradical formation in the organs. However, previous reports indicated

Table II
Effect of 4-Methylpyrazole and Disulfiram Treatment on Brain and Liver Alcohol-Induced Chemiluminescence

	Chemiluminescence (cps/cm^2)	
Treatment	Brain	Liver
None (control)	21 + 2	29 + 3
Ethanol	57 + 3*	42 + 3*
4-Methylpyrazole + ethanol	30 + 7**	33 + 4**
Disulfiram + ethanol	88 + 7**	71 + 3**

*$p < 0.01$ with respect to control rats; **$p < 0.01$ with respect to ethanol-treated rats.

that acetaldehyde promotes lipoperoxidation in isolated hepatocytes (Di Luzio, 1973), increases liver homogenate chemiluminescence (Stege, 1982), and enhances antioxidant-sensitive oxygen uptake (Videla and Villena, 1986) and alkane production in perfused rat liver (Muller and Sies, 1982, 1983). Moreover, acetaldehyde administration *in vivo* (0.3 g/kg) decreased glutathione levels and increased lipoperoxidation (TBARS production) in rat liver (Videla *et al.*, 1982).

Spontaneous brain chemiluminescence is understood as mainly related to $^1O_2$ dimol red emission in the 640- and 670-nm region, being $^1O_2$ produced as a by-product of physiological lipoperoxidation (Codenas and Sies, 1984). Acetaldehyde effects in increasing chemiluminescence have been reported in liver homogenates (Cadenas and Chance, 1981) and in the xanthine oxidase reaction (Puntarulo and Cederbaum, 1989). It is regarded that acetaldehyde reacts with an endogenous energy donor (A^*) produced in the lipoperoxidation process generating excited acetaldehyde, which is responsible for the enhanced brain chemiluminescence:

$$A^* + \text{acetaldehyde} \rightarrow A + \text{acetaldehyde}^*$$

Excited carbonyl compounds are able to abstract hydrogen from vinylic carbons (Turro, 1978; Cadenas *et al.*, 1994) and consequently to increase the rate of the physiological free radical reactions of lipid peroxidation.

In addition, acute ethanol intoxication produced a 31% decrease in brain glutathione; in contrast, the activities of enzymes affording antioxidant protection, namely, SOD, catalase, glutathione peroxidase, glutathione reductase, and glucose-6-phosphate dehydrogenase were not affected. Both the increase in chemiluminescence and the decrease in total glutathione content indicate brain oxidative stress after acute ethanol intoxication which is not associated with a regulatory defensive response involving antioxidant enzyme biosynthesis.

4. DISCUSSION

The concept of oxidative stress has been widely used during the last decade in the biological sciences with an intuitive understanding that it describes a prooxidant state in cells, tissues, and whole organisms. Sies (1985) reshaped the idea of an imbalance between oxidants and antioxidants into the concept of oxidative stress understood as an oxidative state that could lead to cell damage. The success of the concept was almost instantaneous and became extremely useful in the understanding of the complex biological phenomena related to increases in the rate of the chain reaction involving free radicals and excited states that constitutes the physiological process of lipid peroxidation.

The rates of H_2O_2 and hydroperoxide (ROOH) production (Chance *et al.*, 1979) as well as the "antioxidant-sensitive" oxygen uptake in rat liver (Videla *et al.*, 1984; Videla and Villena, 1986) suggest that about 2 to 3% of the total organ oxygen uptake is concerned with O_2-consuming reactions of the chain reaction of lipoperoxidation. The situation in other organs, including the brain, is thought to be quantitatively quite similar. Then, the question as to the existence of oxidative stress, that delicate imbalance between oxidants and antioxidants that is of great biological and medical interest, could be answered with a quantitative measurement of the rate of O_2 incorporation into the chain reaction of lipoperoxidation. Unfortunately, the chain reaction includes a complex series of partial reactions that generate a set of products and by-products. The determination of some of these products and by-products, such as the thiobarbituric acid-reactive substances (TBARS) that have been used for about 50 years, malonaldehyde, ethane, pentane, and other hydrocarbons, has been very useful in the past. As one step forward, what is needed today is methodology that could be used *in vivo* to establish (1) the existence of oxidative stress in *in situ* organs and (2) the pharmacological effect of antioxidants in *in situ* organs submitted to oxidative stress.

The chemical backbone of the physiological chain reaction of lipoperoxidation, with its initiation, propagation, and termination steps, is now well known. The reaction of the products of the univalent and bivalent reduction of O_2, that is, $\cdot O_{2s}^-$ and H_2O_2, where in addition the second is the product of the dismutation of the first, generates ·OH in a typical initiation reaction. The highly reactive ·OH reacts with unsaturated fatty acids producing the alkyl radicals (·R) that on collision with O_2 produce peroxyl radicals (ROO·), in turn generating other R· and ROO· in a typical set of propagation reactions. In this way, the peroxyl radicals of primary or secondary carbons, by a Russell termination reaction, produce either the excited state of oxygen ($^1O_2$) or a ketone or aldehyde with the =CO group in the excited state. The production of these excited species seems to afford the chemical basis for the assay of organ chemiluminescence as an indicator of oxidative stress.

Singlet molecular oxygen, the electronically excited state of normal atmospheric oxygen, is apparently produced as a by-product of physiological lipoperoxidation in cells, tissues, and organs, the spectral analysis of *in situ* organ chemiluminescence showing a mainly red emission in the 620- to 700-nm region that indicates $^1O_2$ as the main photon source. Considering a brain chemiluminescence of 50 cps/cm^2 (a mean value), with an efficiency of the photomultiplier in the red region of 10% and a 0.5 factor for the 180° geometry of the surface detection, the brain emission can be estimated as 1000 photons/s.cm^2. The active tissue depth from which photoemission can be detected may be estimated as 0.1 mm (Costa *et al.*, 1993), which takes photoemission to 1×10^5 photons/s·ml or 1×10^8 photons/liter·s or 1.6×10^{-15} einstein/liter·s (1 einstein = 6.02×10^{23} photons). Assuming that photoemission is related to the dimol emission of $^1O_2$ and that it is a diffusion-controlled process with $k_1 = 1 \times 10^{10}\ M^{-1}\ s^{-1}$

$$d(h\nu)/dt = k_1\ [^1O_2]\ [^1O_2]$$

$$1.6 \times 10^{-15}\ \text{einstein/s} = 1 \times 10^{10}\ M^{-1}\ s^{-1}\ [^1O_2]\ [^1O_2]$$

then the steady-state concentration of $^1O_2$ in the brain can be approximately estimated as: $[^1O_2] = 1.3 \times 10^{-13}$ M.

The rate of $^1O_2$ generation, which should be equal to the rate of deexcitation and return to the ground state, can be estimated as: $-d[^1O_2]/dt = k_2\ [^1O_2]$ from the steady-state concentration of $^1O_2$ and k_2, calculated as 0.2×10^6 s from a $t_{1/2}$ of 3×10^{-6} s:

$$d\ [^1O_2]/dt = 1.3 \times 10^{-13}\ M \times 0.2 \times 10^6\ s = 2.6 \times 10^{-8}\ M\ s^{-1}$$

This rate corresponds to about 1.5×10^{-6} M $^1O_2$/min or 1.5 nmole $^1O_2$/min per g brain, which can be regarded to be a high rate as it amounts to about 15% of the rate of mitochondrial $\cdot O_2^-$ generation (10 nmole $\cdot O_2^-$ /min per g brain) and to about 0.5% of the total rate of O_2 uptake in the brain (330 nmole O_2/min per g brain).

5. REFERENCES

Adamo, A. M., Llesuy, S. F., Pasquini, J. M., and Boveris, A., 1989, Chemiluminescence and oxidative stress in the brain of hyperthyroid rats, *Biochem. J.* **263**:273–277.

Beckman, J. S., 1990, Ischaemic injury mediator, *Nature* **345**:27–28.

Boveris, A., Fraga, C. G., Varsavsky, A. I., and Koch, O. R., 1983, Increased chemiluminescence and superoxide production in the liver of chronically-ethanol-treated rats, *Arch. Biochem. Biophys.* **227**:534–541

Boveris, A., Cadenas, E., Reiter, R., Filipkowski, M., Nakase, Y., and Chance, B., 1980, Organ chemiluminescence: Non-invasive assay for oxidative radical reactions, *Proc. Natl. Acad. Sci. USA* **77**:347–351

Cadenas, E., and Chance, B., 1981, Acetaldehyde-induced chemiluminescence of xanthine oxidase, *Biophys. J.* **33**:184a.

Cadenas, E., and Sies, H., 1984, Low level chemiluminescence as an indicator of singlet molecular oxygen in biological systems, *Methods Enzymol.* **105**:221–231.

Cadenas, E., Ursini, F., and Boveris, A., 1994, Electronically-excited state formation during lipid peroxidation, *Methods Toxicol.* **1**:384–399.

Chance, B., Sies, H., and Boveris, A., 1979, Hydroperoxide metabolism in mammalian organs, *Physiol. Rev.* **59**:527–605.

Costa, L. E., Llesuy, S., and Boveris, A., 1993, Active oxygen species in the liver of rats submitted to chronic hypobaric hypoxia, *Am. J. Physiol.* **264**:C1395–C1400.

Diluzio, N. R., 1973, Antioxidants, lipid peroxidation and chemical-induced liver injury, *Fed. Proc.* **32**:1875–1881.

Giulivi, C., Sarcansky, M., Rosenfeld, E., and Boveris, A., 1990, The photodynamic effect of rose bengal on proteins of the mitochondrial inner membrane, *Photochem. Photobiol.* **52**:745–751.

Giulivi, C., Lavagno, C., Lucesoli, F., Novoa Bermúdez, M. J., and Boveris, A., 1995, Lung damage in paraquat poisoning and hyperbaric oxygen exposure: Superoxide-mediated inhibition of phospholipase A_2, *Free Radical Biol. Med.* **18**:203–213.

Giulivi, C., Poderoso, J. J., and Boveris, A., 1998, Production of nitric oxide by liver mitochondria, *J. Biol. Chem.* **273**:11038–11043.

González Flecha, B., Cutrin, J. C., and Boveris, A., 1993, Time course and mechanism of oxidative stress and tissue damage in rat liver subjected to in vivo ischemia–reperfusion, *J. Clin. Invest.* **91**:456–464.

Jamieson, D., Chance, B., Cadenas, E., and Boveris, A., 1986, The relation of free radical production to hyperoxia, *Annu. Rev. Physiol.* **48**:703–719.

Kanfer, S., and Turro, N. J., 1981, Reactive forms of oxygen, in *Oxygen and Living Processes* (D. L. Gilbert, ed.), pp. 47–64, Springer-Verlag, Berlin.

Khan, A. U., 1995, Quantitative generation of singlet oxygen from acidified aqueous peroxynitrite produced by the reaction of nitric oxide and superoxide anion, *J. Biolumin. Chemilumin.* **10**:329–333.

Llesuy, S., Evelson, P., González Flecha, B., Peralta, J., Carreras, M. C., Poderoso, J. J., and Boveris, A., 1994, Oxidative stress in muscle and liver of rats with septic syndrome, *Free Radical Biol. Med.* **16**:445–451.

Muller, A., and Sies, H., 1982, Role of alcohol dehydrogenase activity and of acetaldehyde in the ethanol-induced ethane and pentane production by isolated perfused rat liver, *Biochem. J.* **206**:153–156.

Muller, A., and Sies, H., 1983, Ethane release during metabolism of aldehydes and monoamines in perfused rat liver, *Eur. J. Biochem.* **134**:599–602.

Puntarulo, S., and Cederbaum, A., 1989, Chemiluminescence from acetaldehyde oxidation by xanthine oxidase involves generation of and interactions with hydroxyl radicals, *Alcohol Clin. Exp. Res.* **13**:84–90.

Roldán, E. J. A., Turrens, J. F., Pinus, C., and Boveris, A., 1989, Chemiluminescence of ischemic and reperfused intestine *in vivo*, *Gut* **30**:184–187.

Sies, H., 1985, Oxidative stress: Introductory remarks, in *Oxidative Stress* (H. Sies, ed.), pp. 1–7, Academic Press, San Diego.

Stege, T. E., 1982, Acetaldehyde-induced lipid peroxidation in isolated hepatocytes, *Res. Commun. Chem. Pathol. Pharmacol.* **36**:287–292.

Turrens, J. F., Giulivi, C., Pinus, C., Lavagno, C., and Boveris, A., 1988, Spontaneous lung chemiluminescence upon paraquat administration, *Free Radical Biol. Med.* **5**:319–323.

Turro, N. J., 1978, Photoaddition and photosubstitution reactions, in *Modern Molecular Photochemistry: Photoaddition and Photosubstitution Reactions* (E. Cummings, ed.), pp. 362–413, Menlo Park Press, San Francisco.

Videla, L. A., and Villena, M. I., 1986, Effect of ethanol, acetaldehyde and acetate on the antioxidant-sensitive respiration in the perfused rat liver: Influence of fasting and diethylmaleate treatment, *Alcohol* **3**:163–167.

Videla, L. A., Fernandez, V., and de Marinis, A., 1982, Liver lipoperoxidative pressure and glutathione status following acetaldehyde and aliphatic alcohols pretreatment of rats, *Biochem. Biophys. Res. Commun.* **104**:965–970.

Videla, L. A., Villena, M. I., Donoso, G., Giulivi, C., and Boveris, A., 1984, Changes in oxygen consumption induced by t-butyl hydroperoxide in perfused rat liver, *Biochem. J.* **223**:879–883.

Chapter 23

Reactive Oxygen Species and Neuronal Function

Carol A. Colton and Daniel L. Gilbert

Dedicated to the memory of Joel S. Colton (1945–1997) who contributed significantly to the understanding of oxygen toxicity in the CNS.

1. THE BERT EFFECT

The first demonstration of the damaging effects on oxygen to neuronal tissue was made by Paul Bert. In a series of now classic experiments, Bert (Bert, 1872, 1878) placed animals in a hyperoxic environment and found that they exhibited tonic–clonic (grand-mal) seizures before death. He concluded that "the toxic action [of oxygen] produces its effect on the nervous centers, as do strychnine, phenol, and other poisons which cause convulsions" (Bert, 1878). Bert (1878) also wrote, "But the violent excitation, the constant convulsions [grand-mal seizures] which accompany this death are still unexplainable." The induction of seizures (Figure 1) and the alteration in neuronal function by exposure to high tensions of oxygen has been termed the "Bert effect" (Bean, 1945). Through the intervening years since these initial experiments, the interest in neuronal oxygen toxicity has waxed and waned but the Bert effect clearly established the exquisite sensitivity of nervous tissue to oxidative stress.

The enhanced susceptibility of the central nervous system (CNS) to oxygen stress has been confirmed by various investigators and numerous hypotheses have been put forward to explain this apparent tissue-specific effect. These include a high level of unsaturated lipids in neuronal membranes and hence increased ease of lipid peroxidation, high levels of transition metals such as iron and copper, and variable levels of antioxidant protection mechanisms (Choi and Yu, 1995; Gutteridge, 1994; Subbarao and Richardson,

Carol A. Colton Department of Physiology and Biophysics, Georgetown University Medical School, Washington, D.C. 20007. **Daniel L. Gilbert** Unit on Reactive Oxygen Species, BNP, NINDS, National Institutes of Health, Bethesda, Maryland 20892-4156.

Reactive Oxygen Species in Biological Systems, edited by Gilbert and Colton. Kluwer Academic / Plenum Publishers, New York, 1999.

FIGURE 1. Dog during the tonic convulsions of oxygen poisoning. "But let us come to the description of the convulsive attack itself. . . . When the animal is taken from the apparatus, it is generally in severe tonic convulsions, the four feet are still, the trunk is bent back or a little to one side. . . . attacks so violent that one can lift the animal by a single foot, stiff as a piece of wood" (Bert, 1943).

1990; Halliwell and Gutteridge, 1985; Cohen, 1994; Morriss *et al.*, 1983; Gotz *et al.*, 1994). The CNS also vigorously extracts oxygen from the blood for use in oxidative metabolism. The resultant high rate of ATP production can lead to an increased generation of reactive oxygen species as a by-product of the electron transport chain (Gotz *et al.*, 1994; Patole *et al.*, 1986).

2. THE REDOX ENVIRONMENT IN THE CNS

2.1. Normal Tissue Oxygen Levels

Like many other actively metabolizing tissues, the overall level of oxygen in the parenchyma of the brain is low. CNS oxygen tension has been measured using a variety of techniques including the placement of oxygen sensitive electrodes into various regions of the brain (Vanderkooi *et al.*, 1991; Silver, 1966). The values obtained depend, in part, on the area of the CNS and range from a low of 9 torr in the gray matter of the cortex to around 35 torr at the pial surface. This low oxygen tension is related to the high demand of the cells in the CNS for oxygen and can be changed by alterations in blood flow or by breathing a gas mixture containing a higher than normal oxygen level. Almost all of the oxygen is in the form of oxyhemoglobin and very little is in physical solution. For a standard man (70 kg, 170 cm, 1.8 m^2, 20 years old), the resting oxygen uptake is 251 ml/min or 15 liters/hr (Shephard, 1971). The brain takes up about 20% of this amount or about 3 liters/hr despite the fact that the mass of the brain is 1.4 kg or 2% of the total body mass (Sokoloff, 1960). Thus, the oxygen uptake per unit mass by the brain is 10 times greater than most tissues in the body. For children, this level is even more striking and the oxygen uptake can be as much as 50% of the total oxygen uptake of the body (Sokoloff, 1960). Most of the oxygen uptake occurs in the gray matter regions and accounts for the low oxygen tensions seen in those regions (Sokoloff, 1960).

2.2. Sources of Reactive Oxygen Species (ROS)

The redox environment of the CNS can be altered in favor of oxidation by increasing the production of ROS or by decreasing antioxidant protection mechanisms. The initial biological oxyradical produced by many cells is the superoxide anion, the single-electron reduction product of oxygen. Further sequential one-electron reduction steps generate hydrogen peroxide (H_2O_2), hydroxyl radical (·OH), and eventually water. Nitric oxide (·NO) is also considered an oxyradical because it has an unpaired electron associated with an oxygen atom and, like other ROS, has profound effects on neuronal function. For that reason ·NO is included in our discussion. Oxyradicals are not limited, however, to superoxide anion, H_2O_2, ·OH, or ·NO because these molecules are highly reactive and initiate chain reactions that result in the generation of a variety of carbon-based radicals or nitrosyls. The presence of transition metals such as Fe^{2+} or Cu^{2+} further promotes radical formation.

A number of excellent reviews have described the various sources of oxyradicals in tissues throughout the body, cellular antioxidant protection mechanisms, and the reactions in which ROS participate (Gutteridge, 1994; Cohen, 1994; Cohen and Werner, 1993; Halliwell, 1992; Halliwell and Gutteridge, 1985). The CNS is like other tissues and both enzymatic and nonenzymatic sources of oxyradicals are found. These range from intracellular enzymes such as xanthine oxidase located primarily in the capillary endothelial cells to the enzymes of the mitochondrial electron transport chain found in all cells. Table I provides a partial listing of these general sources for ROS and their putative cellular locations in the CNS. The level of oxyradical produced by each of these mechanisms is controversial and it is not clear when significant quantities of damaging species "leak" from the cell to injure either that cell or surrounding cells. Clearly the level of intracellular and extracellular antioxidant protection is a factor in the overall redox balance. Furthermore, cell injury is not a necessary endpoint of exposure to ROS. Activation of the innate immune system in the CNS, i.e., the microglia, however, may produce a burst of superoxide anion, which in the presence of transition metals may generate significant oxidative damage. Microglia in all animal species studied to date have been shown to produce a large transient increase in superoxide anion as a result of the activation of the "respiratory burst" NADPH oxidase pathway (Babior, 1984, this volume; Colton and Gilbert, 1987; Colton *et al.*, 1994; Giulian and Baker, 1986). Because of the widespread location of microglia throughout the CNS and their ability to respond to injury and infection, the microglia are prime candidates as a source of acute oxidative stress. Changes in other enzymatic pathways such as the ethanol-induced cytochrome P450 may serve as a more chronic source of oxidative stress (Hunt, 1993; Nordmann *et al.*, 1992).

Table I
CNS Sources of Reactive Oxygen Species

	Primary oxidant species generated	Primary location in brain	References
Enzymatic			
Xanthine oxidase/ hypoxanthine	$\cdot O_2^-$	Capillary endothelial cells	Betz (1985)
Lipoxygenase/ cyclooxygenase	$\cdot O_2^-$	Capillary endothelial cells, all cells	Kim *et al.* (1991), Naidu and Kulkami (1992)
Mitochondrial monoamine oxidase	H_2O_2	Neurons, astrocytes	Cohen (1994), Zoccarato *et al.* (1989), Gotz *et al.* (1990)
Microsomal cytochrome P450 oxidases	H_2O_2, $\cdot OH$	All cells	Cohen (1994), Halliwell and Gutteridge (1989)
Mitochondrial electron transport chain	$\cdot O_2^-$, H_2O_2, $\cdot OH$	All cells	See review by Gotz *et al.* (1990), Ramasarma (1982), Patole *et al.* (1986)
Nitric oxide synthase	$\cdot NO$, NO^-	Neurons, endothelial cells, glia	Dawson *et al.* (1994), Moncado and Higgs (1991)
NADPH oxidase (respiratory burst)	$\cdot O_2^-$	Microglia/macrophage	Babior *et al.* (1973), Colton and Gilbert (1987)
Nonenzymatic			
Neuromelanin	$\cdot OH$	Neurons in the substantia nigra	Gotz *et al.* (1990)
Catecholamine autoxidation	$\cdot O_2^-$, H_2O_2, $\cdot OH$	Neurons	Cohen and Werner (1993)
Other (redox cycling, metal-catalyzed oxidation, lipid peroxidation chain reactions)	$\cdot O_2^-$, H_2O_2, $\cdot OH$	All cells	

3. CONSEQUENCES OF OXIDATIVE STRESS

The generation of seizures in animals exposed to 100% oxygen at high pressure has been repeatedly demonstrated and clearly shows that synaptic function is altered by oxidative stress (Colton and Colton, 1982, 1985; O'Dell *et al.*, 1991; Harel and Lavy, 1871; Wood *et al.*, 1963). A number of electrophysiological studies have delineated the changes that occur during exposure to oxygen and ROS and have identified potential sites and mechanisms of action. These studies have included measurement of resting membrane properties, action potential characteristics, and both pre- and postsynaptic events during synaptic transmission. Current clamp, voltage clamp, and, more recently, patch clamp analyses have been done and indicate that multiple sites are affected although, in general, synaptic transmission is more readily altered by exposure to high levels of oxygen or ROS.

3.1. Changes in Resting Membrane Properties

Resting membrane properties of neurons and axons are largely unaffected by acute exposure to low oxidative stress. For example, isolated invertebrate axons or muscle fibers from either frog or lobster show no significant changes in resting membrane potential (V_m) or input resistance when treated with up to 5 mM H_2O_2 or with superoxide anion generated by the interaction of xanthine and xanthine oxidase (Davison *et al.*, 1984; Colton *et al.*, 1991). A similar phenomenon is found for mammalian cells, there being no change in either V_m or input resistance with 1 mM H_2O_2, H_2O_2 plus iron, or ionizing radiation in intracellular recordings from CA1 pyramidal neurons in the guinea pig hippocampus (Tolliver and Pellmar, 1987; Pellmar, 1986, 1987). However, alteration in resting properties can be seen under some circumstances. Exposure to 100% oxygen increased input resistance in the lobster muscle fiber with no change in V_m (Colton and Colton, 1982), whereas in CA1 pyramidal cells exposed to 100% oxygen, both an increase in input resistance and a slow depolarization were observed (Bingmann *et al.*, 1982). This change in input resistance accompanied by alteration in V_m suggests that channel function and ion flux were affected by the high oxygen tensions. Alternatively, the depolarization of V_m is consistent with the known inhibition of the Na^+/K^+-ATPase by oxidation (Kovachich *et al.*, 1981; Morriss *et al.*, 1983) and may result in a slow but unchecked rearrangement of ionic gradients across the neuronal membrane (Bingmann *et al.*, 1982; King and Parmentier, 1983).

3.2. Changes in Voltage-Dependent Channels

Deterioration of the axon membrane accompanied by loss of the action potential is seen within 1 hr of treatment with 20 mM 6-hydroxydopamine (6-OHDA) plus copper (Davison *et al.*, 1984) or with concentrations of H_2O_2 above 5 mM (Colton *et al.*, 1986). However, lower concentrations of oxidants are not associated with such catastrophic changes in axonal function. No change in action potential amplitude or rise time was seen in the squid giant axon exposed to 10 μM superoxide anion generated by the interaction of xanthine and xanthine oxidase or 1 mM H_2O_2 (Colton *et al.*, 1986, 1991). Similar levels of ROS generate small but finite changes in the frequency adaptation of CA1 pyramidal neurons in the guinea pig hippocampus (Pellmar, 1987; Pellmar *et al.*, 1994). The resultant lowered rate of action potential generation produced by 1 mM H_2O_2 was not related to alteration in the inward currents for Na^+ or Ca^{2+} or to alterations in the delayed rectifier, A-current, M-current, or Q-current carried by K^+.

Certain potassium currents are modulated by H_2O_2. For example, Vega-Saenz and Rudy (1992) have expressed alternatively spliced products of the *Drosophila* shaker gene in *Xenopus* oocytes. These voltage-gated K^+ channels (KShIIIC, KShIIID) are reminiscent of channels found in the carotid body (an oxygen sensor in mammals) and demonstrate a time-dependent, fast inactivation. Treatment with 100 to 200 μM H_2O_2 resulted in prolongation and an increase in the amplitude of these K^+ currents (Vega-Saenz and Rudy, 1992). Redox-sensitive cysteine residues have been found on the B1 subunit of specific K^+ channels (McCormack and McCormack, 1994). Furthermore, both the B1 and B2 subunits of this K^+ channel share homology with the aldo-keto reductase superfamily and may serve as functional oxoreductases, providing a link between cellular oxidative metabolism and K^+ channel function (McCormack and McCormack, 1994). Similar redox-sensitive K^+ channels have been found in neuroblastoma cells (NG108-15 hybrid cells) and in HEK-293 cells transfected with a rat Kv1.4 channel clone (Stephens *et al.*, 1996; Rouzaire-Dubois and Dubois, 1990). In the Kv1.4-transfected cells, oxidation with either chloramine-T or other sulfhydryl oxidizing agents decreased K^+ channel inactivation, which could be reversed by rereduction with dithiothreitol (DTT) (Stephens *et al.*, 1996). Rouzaire-Dubois and Dubois (1990) also demonstrated that the redox sensitivity found in the neuroblastoma cell was selective for a K^+ channel as whole-cell recording of barium currents demonstrated no change in Ca^{2+}-channel parameters.

The functional consequences of ROS-mediated changes in K^+-channel function are not clear but they may be involved in the response of neuronal tissue to high tensions of oxygen. For example, exposure of the CA1 neurons in guinea pig hippocampal slices to high oxygen tensions (600 mm Hg) generated bursting activity (Bingmann *et al.*, 1982). Blockade of K^+-channel inactivation is associated with spontaneous action potentials in a variety of axons. Similar changes were shown by Muller *et al.* (1993) who treated CA3 pyramidal neurons with the lymphokines α/β- or γ-interferon. Because the increase in neuronal excitability was blocked by treatment with superoxide dismutase (SOD) and catalase, the authors proposed that ROS were involved.

3.3. Changes in Synaptic Transmission

3.3.1. High Oxygen Tensions

Although changes are seen in resting membrane and axonal properties, the main effect of exposure to high oxygen tensions and to reactive oxygen intermediates such as superoxide anion, H_2O_2, and the hydroxyl radical is on synaptic transmission. As already mentioned, hippocampal neurons demonstrated increased excitability when exposed to high tensions of oxygen (Bingmann *et al.*, 1982; King and Parmentier, 1983). The resultant seizurelike activity in this preparation can also be explained by an increase in excitatory neurotransmission. Using an invertebrate model system, Colton and Colton (J. Colton and Colton, 1978; C. Colton and Colton, 1982) have shown that the frequency of excitatory miniature postsynaptic potentials (mEPSPs) and the amplitude of evoked excitatory postsynaptic potentials (eEPSPs) significantly increased during exposure to an atmosphere of 100% oxygen. The mechanism of the increase in spontaneous and evoked release of transmitter in high oxygen is not clear and may be related, in part, to the slow depolarization of membrane potential and K^+-channel changes described above. Alternatively, intraterminal Ca^{2+} levels may change. ROS are known to block the activity of ATPases including the Ca^{2+}-ATPase located on plasma and intracellular membranes (Kaneko *et al.*, 1989). Intraterminal Ca^{2+} levels would thus rise as a secondary consequence of pump inhibition and, in turn, would promote both spontaneous and evoked release (Stanley, 1986).

C. Colton and Colton (1986) have also demonstrated that evoked inhibitory synaptic potentials are depressed by high tensions of oxygen. Again using the crustacean neuromuscular junction as a model system, the amplitude of inhibitory synaptic potentials decreased rapidly during exposure to an atmosphere of 100% oxygen or to oxygen at high pressure. This effect is primarily presynaptic in nature as the postsynaptic response to the inhibitory neurotransmitter γ-aminobutyric acid (GABA) was unchanged from control values (C. Colton and Colton, 1986). The loss of inhibition is most likely related to reduction in the available pool of GABA in the nerve terminal. GABA-mediated transmission is well known to be inhibited by oxidation via inactivation of glutamic acid decarboxylase (GAD) (Wood and Watson, 1963; Wood *et al.*, 1963). GAD is a sulfhydryl-containing enzyme that catalyzes the production of GABA from glutamate in mammalian and invertebrate neurons.

3.3.2. Oxygen-Induced Seizures

As previously mentioned, seizures are one of the well-described effects of exposure of humans or other animals to high oxygen tensions. Loss of GABA from neurons has been correlated with oxygen toxicity and the seizure activity during exposure to 100% oxygen or to oxygen at high pressure is related, in part, to failure of inhibitory transmission (Wood *et al.*, 1969). It is clear, however, that changes in excitatory transmission and specifically glutamate-mediated transmission are also involved in the generation of seizures during exposure to high oxygen tensions. Pharmacological antagonists of the postsynaptic action of glutamate block oxygen-induced seizures in mice (C. Colton and Colton, 1985). These effects are readily reversed by removal of the high oxygen tension. The reversibility of this type of seizure, its clonic/tonic nature, and the characteristic changes in both GABA- and glutamate-mediated neurotransmission strongly suggest that oxygen-induced seizures may be a useful, but currently underutilized, model for the study of epilepsy in humans.

3.3.3. ROS

3.3.3a. Effects on Presynaptic Transmitter Release. Treatment of neurons with either superoxide anion or H_2O_2 also dramatically affects synaptic transmission but in a different fashion than described above for high oxygen tensions. The predominant actions, as summarized in Table II, are to decrease the amplitude of the evoked synaptic potential and to induce a small but significant increase in miniature endplate potential frequency.

Table II
Effect of Oxidation and Oxidizing Radicals on Synaptic Transmission[a]

Agent	Concentrations	Action	Preparation	Reference
Oxygen	760 mm Hg	Increased EPSP amplitude; decreased IPSP amplitude; increased frequency of mEPSPs	Crustacean neuromuscular junction	C. Colton and Colton (1982, 1986), J. Colton and Colton (1978)
	100%	Increased burst activity	Rat hippocampal slice	Bingmann *et al.* (1982)
	Hyperbaric	Decreased EPSP/IPSP amplitude; increased frequency of mEPSPs; seizures	Crustacean neuromuscular junction; rat, mouse, human cortex	C. Colton and Colton (1982), Harel and Lavy (1971), Wood *et al.* (1963)
Superoxide anion (X/XO)	10 μM	Decreased EPSP amplitude	Squid giant synapse	Colton *et al.* (1991)
	1 mM/0.05 U/ml	Blockade of NMDA-mediated increase in intracellular Ca^{2+}; decreased NMDA-mediated current	Rat cortical neurons	Aizenman *et al.* (1990)
	10 μM	Decreased NMDA-mediated current	NMDAR1-2A- or 2B-transfected HEK cells	Gbadegesin *et al.* (1997)
H_2O_2	1–100 μM	Decreased EPSP amplitude; increased frequency of mEPSPs	Crustacean neuromuscular junction	Colton *et al.* (1986)
	1–1000 μM	Decreased EPSP amplitude	Squid giant synapse; guinea pig hippocampal slice; rat hippocampal slice	Colton *et al.* (1989, 1991), Pellmar (1986, 1987)
	100 μM	Decreased NMDA-mediated current	NMDAR1-2A- or 2B-transfected HEK cells	Gbadegesin *et al.* (1997)
Dithio-threitol	2–4 mM	Increased NMDA-mediated current	NMDAR1-2A- or 2B-transfected HEK cells; rat cortical and cerebellar neurons; hippocampal slice	Gbadegesin *et al.* (1997), Tang and Aizenman (1993), Sucher *et al.* (1990), Tauck (1992)
	2 mM	Increased NMDA-mediated intracellular Ca^{2+}	Rat retinal ganglion cells	Sucher *et al.* (1990)
	2–4 mM	Induction of epileptiform activity	Guinea pig hippocampal slice	Tolliver and Pellmar (1987)
Dithio-bis(2-nitrobenzoic acid)	200 μM	Decreased NMDA-mediated current; decreased EPSC	Rat cortical neurons	Bernard *et al.* (1997)

[a]Abbreviations: X/XO, xanthine/xanthine oxidase; EPSP, excitatory postsynaptic potential; IPSP, inhibitory postsynaptic potential; EPSC, excitatory postsynaptic current.

The overall result is loss of neurotransmission. Depression of the excitatory postsynaptic potential has been seen in many preparations, including the squid giant synapse, the lobster neuromuscular junction, and the guinea pig and rat hippocampal slice (Colton *et al.*, 1986; Pellmar, 1986, 1987; Pellmar *et al.*, 1994). Inhibitory potentials in the CA1 region and the release of radioactive GABA from synaptosomes prepared from guinea pig brain are also depressed by hydroxyl radical (Pellmar *et al.*, 1994). Because excitatory and inhibitory transmission are affected by exposure to H_2O_2, a mechanism common to both may be altered. Evidence from several different synaptic preparations suggests that the influx of Ca^{2+} into the nerve terminal is depressed. For example, inward Ca^{2+} currents are blocked by H_2O_2 (Livengood and Miller, 1989) and Ca^{2+} influx into synaptosomes is decreased by ionizing radiation and ROS (Pellmar *et al.*, 1994).

The data on transmitter release and ROS are not always consistent and appear contradictory as a result, in part, of the technique employed in the study of release. Retinal cells cultured from chick demonstrate a slight but significant increase in Ca^{2+}-dependent and -independent radiolabeled GABA release when peroxidized with a mixture of ascorbic acid and Fe^{2+} (Oliveria *et al.*, 1994). Because the Ca^{2+}-independent effect predominates, Oliveria *et al.* (1994) concluded that peroxidation primarily promotes reversal of the Na^+/GABA cotransporter. This uptake mechanism is important in the neuronal recycling of GABA at synaptic junctions. Reduction in radiolabeled GABA uptake by neurons is further supported by a decrease in synaptosomal uptake of GABA during treatment with ADP-Fe ascorbate to initiate lipid peroxidation (Rafalowska *et al.*, 1989). Radioactive glutamate release from tissue slices has also been shown to increase in response to superoxide anion generated by the interaction of xanthine and xanthine oxidase (Pellegrini-Giampietro *et al.*, 1988). Again, both Ca^{2+}-dependent and -independent release were affected and reuptake mechanisms for glutamate are a likely site of action for the ROS. These studies are confounded by their inability to distinguish between neurons and glia, particularly astrocytes and microglia. Both of these latter cell types express glutamate transporters which are redox regulated (Volterra *et al.*, 1994). Trotti *et al.* (1996) demonstrated that all three types of glutamate transporters found in brain, GLT1, GLAST, and EAAC1, are inhibited by biological oxidants. The blockade of these transporters plus the inactivation of glutamine synthetase, the enzyme that catalyzes the conversion of glutamate to glutamine in astrocytes (Schor, 1988; Oliver *et al.*, 1990), promotes the accumulation of glutamate in the synaptic junction during various pathological states in which ROS-mediated damage is found. Synaptic transmission, however, is depressed.

3.3.3b. Postsynaptic Receptors. Karlin and Bartels (1966) demonstrated that reduction of the nicotinic receptor in electric fish (electroplax) by DTT decreased the postsynaptic response to acetylcholine. More recent work has demonstrated that other transmitter receptors can be modified by oxidation or reduction (Schwartz *et al.*, 1988; Colton *et al.*, 1989, 1991; Quesada *et al.*, 1996; Bernard *et al.*, 1997; Tauck, 1992) and it is now clear that transmitter receptors, as well as other important membrane proteins, have redox-sensitive amino acids such as cysteine located at sites critical to the function of that receptor. Currently, the most-studied receptor has been the *N*-methyl-D-aspartate (NMDA) receptor, a subtype of excitatory amino acid (EAA) receptors found in many regions of the CNS (McBain and Mayer, 1994). The NMDA receptor/channel is particularly interesting because of its putative involvement in a cellular form of learning and memory, i.e., long-term potentiation (LTP). Colton *et al.* (1989) were the first to demonstrate that exposure of the CA1 pyramidal neurons in the rat hippocampal slice to H_2O_2 blocked LTP, indicating that the NMDA receptor was redox sensitive. Subsequent studies have confirmed and extended this finding. As summarized in Table II, whole-cell currents in response to application of NMDA are increased by treatment of hippocampal or cortical

Table III
The Effect of ·NO and ·NO Donors on NMDA-Mediated Synaptic Transmissions

Agent	Concentration	Action	Preparation	Reference
NO gas	1–100 μM	Blockade of NMDA-mediated increase in intracellular Ca^{2+}	Rat forebrain neurons	Hoyt *et al.* (1992)
	5–10 nM; > 40 nM	Increased frequency of mEPSCs; loss of mEPSC and cell death	Rat hippocampal neurons	O'Dell *et al.* (1991)
Sodium nitroprusside (SNP)	100 μM, 1 mM	Blockade of NMDA-mediated increase in intracellular Ca^{2+}	Rat forebrain or cortical cultured neurons	Hoyt *et al.* (1992), Lei *et al.* (1992)
	1–100 μM	Inhibition of NMDA-mediated activation of phosphoinositide	Rat cerebellar granule cells	Yu and Chang (1996)
	100 μM	Decreased amplitude of EPSC and mEPSCs; decreased mEPSC frequency	*Xenopus* nerve–muscle	Wang *et al.* (1995)
S-Nitroso-*N*-acetyl-penicillamine (SNAP)	100–1000 μM	Blockade of NMDA-mediated increase in intracellular Ca^{2+}	Rat forebrain or cerebellar neurons	Hoyt *et al.* (1992)
	100 μM–1 mM	Decreased NMDA currents	Rat cerebellar neurons	Fagni *et al.* (1995)
	100 μM	Decreased amplitude of EPSC and mEPSCs; decreased mEPSC frequency	*Xenopus* nerve–muscle	Wang *et al.* (1995)
3-Morpholino-sydnonimine (SIN-1)	1 μM–10 mM	Blockade of NMDA-mediated increase in intracellular Ca^{2+}; decreased NMDA currents	Rat striatal or cerebellar neurons	Manzoni *et al.* (1992)
	100 μM–10 mM	Decreased NMDA currents, decreased P_o, decreased unitary conductance	Rat cerebellar neurons	Fagni *et al.* (1995)
Nitrosopyrrolidine	1 mM	Decreased NMDA currents	Rat cerebellar neurons	Manzoni *et al.* (1992)
Nitroglycerin	100–500 μM	Decreased NMDA currents; blockade of NMDA-mediated increase in intracellular Ca^{2+}	Rat cortical neurons	Lei *et al.* (1992)
	500–1000 μM	Decreased amplitude of EPSCs; decreased frequency of mEPSCs	Rat hippocampal neurons	Pan *et al.* (1996)
S-Nitroso-cysteine (SNOC)	100 μM	Blockade of NMDA-mediated increase in intracellular Ca^{2+};	Rat cortical neurons	Lei *et al.* (1992)

Table III (*Continued*)

Agent	Concentration	Action	Preparation	Reference
	250–500 μM	Decreased synaptic activity	Rat cortical neurons	Pan *et al.* (1996)
	100 μM	Decreased unitary conductance	Rat cerebellar neurons	Fagni *et al.* (1995)
Hemoglobin	100 μM	Increased NMDA-mediated increase in intracellular Ca^{2+}	Rat cerebellar neurons	Manzoni *et al.* (1992)
Nitrite/nitrate	10 μM–10 mM	Blockade of NMDA-mediated increase in intracellular Ca^{2+}	Rat forebrain cultured neurons	Hoyt *et al.* (1992)
Diethylamine–NO complex	30 μM–1 mM	Slight increase in EPSC amplitude, prolongation of EPSC; slight increase in mEPSC frequency	Rat hippocampal neurons	Pan *et al.* (1996)
PAPA–NO	100 μM	Increased NMDA-mediated current; increased P_o	NMDAR1-2A- or 2B-transfected HEK cells	Gbadegesin *et al.* (1997)
NOS inhibitors	100 μM–10 mM	Decreased NMDA-mediated glutamate release	Perfused guinea pig hippocampal slice; microdialysis of rat striatum	Bogdanov and Wurtman (1997), Nei *et al.* (1996)

Abbreviations: EPSCs, excitatory postsynaptic currents; mEPSCs, miniature (spontaneous) excitatory postsynaptic currents; P_o, probability of channel opening.

neurons with the sulfhydryl-reducing agent, DTT (Aizenman *et al.*, 1990; Gbadegesin *et al.*, 1997). This effect has been seen in patch-clamp studies of cultured neurons or of HEK cells transfected with recombinant receptor subunits. The increase in NMDA-induced peak current amplitude could be reversed to control levels by washing or, in some cases, by reoxidation of the receptor with 5,5′-dithiobis(2-nitrobenzoic acid) (DTNB). Treatment with oxidizing agents other than DTNB such as H_2O_2 plus iron or a mixture of xanthine and xanthine oxidase significantly decreased the NMDA response below untreated values. Single-channel patch recordings by Tang and Aizenman (1993) demonstrated no significant change in channel open time or single-channel conductance in the presence of DTT. Frequency of opening was increased, however.

Recombinantly expressed NMDA receptor subunits and site-specific mutagenesis have indicated that redox responsiveness can be found on more than one subunit of the NMDA receptor (Kohr *et al.*, 1994; Sullivan *et al.*, 1994). A functional NMDA receptor can be formed by the homomeric combination of the principal subunit, the NMDAR1, or by the heteromeric combination of the NMDAR1 subunit with one of four NR2 subunits (i.e., NR2A, NR2B, NR2C, or NR2D) (McBain and Mayer, 1994). DTT-induced sensitivity is found in all cases but the increased current found in the NMDAR1-NR2B, C, or D subunit combinations has a slow onset and is not readily reversible by washing. Once exposed to *N*-ethylmaleimide, a sulfhydryl alkylating agent, reoxidation of the response is eliminated. Mutation of two cysteines to alanine at Cys-744 and Cys-798 eliminated the response of the NMDAR1-NR2B, C, or D subunit combinations to DTT but not the response of the NMDAR1-NR2A heteromeric combination (Sullivan *et al.*, 1994). Thus, although these two cysteines are critical to redox modulation, the NR2A subunit may have distinct redox regulatory sites. A separate redox site on the NR2A component is also supported by the difference in kinetics seen during the response of the NMDAR1-NR2A combination. In this case, exposure to DTT results in a more rapid potentiation of the NMDA-mediated current which can be fully reversed by washing (Kohr *et al.*, 1994). Although the location of the cysteine residues involved in the redox regulation of this subunit has not been firmly established, recent work by Paoletti *et al.* (1997) indicates that the zinc binding site may be involved in the NR2A DTT sensitivity. Their data suggest that the rapid potentiation associated with the NR2A subunit in response to DTT is, in fact, related to chelation of zinc by DTT and, hence, alteration in zinc binding.

The changes in NMDA-mediated whole-cell currents in neurons are functionally significant because intracellular calcium fluxes are altered in a parallel fashion. Aizenman *et al.* (1990) demonstrated that DTT promoted calcium entry in single cortical neurons whereas oxidation with xanthine/xanthine oxidase decreased intracellular calcium. As NMDA-mediated calcium entry has been correlated with glutamate excitotoxicity, it is possible that redox regulation of the NMDA channel could play a role in modulating the level of excitotoxicity. Using rat cortical neurons at different days in culture, Sinor *et al.* (1997) have demonstrated that cultured neurons at 14 days *in vitro* are in a more oxidized, hence less responsive state than older (32 days *in vitro*) neuronal cultures. Furthermore, the younger cultures were more resistant to NMDA-mediated death. The potential implication is that expression of NR subunits is developmentally regulated and imparts a decreased sensitivity for fetal neurons to excess glutamate and oxidative damage. *In situ* analysis of the mRNA for each of the NMDA subunits has demonstrated that the NR2B and D appear prenatally whereas the NR2A and C appear after birth (Monyer *et al.*, 1994). Further evidence supporting the functional role for redox modulation of the NMDA receptor has been presented by Hirsch *et al.* (1996) and Quesada *et al.* (1996). In these studies, NMDA-dependent epileptiform discharges in hippocampal slices were controlled by treatment of the slice with oxidizing agents. *In vivo* sources of reducing and oxidizing agents are speculative but include the stimulated release of glutathione (GSH) and cysteine (Zangerle *et al.*, 1992), and a variety of potential oxidizing sources are available (Table I).

3.3.4. Reactive Nitrogen Intermediates: Effects on Pre- and Postsynaptic Events

The involvement of ·NO, another class of ROS, in neuronal function has recently become an area of active research interest. As in other tissues in the body, ·NO is generated in the CNS from the conversion of arginine to citrulline in a reaction catalyzed by nitric oxide synthase (NOS). Three well-described isoforms of NOS are found: immunological NOS (iNOS, NOSII) localized to glia and perhaps some neurons and two forms of constitutive NOS, neuronal NOS (nNOS, NOSI) localized to neurons and endothelial NOS (eNOS, NOSIII) localized to the endothelial cells of the cerebral blood vessels (Moncada and Higgs, 1991). An association between ·NO and synaptic transmission was first suggested by the rise of cyclic GMP in cerebellar granule cells exposed to NMDA

(Garthwaite and Garthwaite, 1988). This rise was essential for protection of the granule cells from excitotoxicity and could be mimicked by the putative ·NO donor sodium nitroprusside (SNP) and blocked by hemoglobin, a ·NO scavenger. Subsequently, a variety of studies have demonstrated that NMDA treatment of cortical, hippocampal, or cerebellar neurons results in the production of ·NO via entry of calcium and calcium-dependent induction of nNOS (Nicotera *et al.*, 1997; Samdani *et al.*, 1997; Schuman and Madison, 1994; Dawson *et al.*, 1994). Because activation of the NMDA receptor is an important step in one type of cellular learning and memory, the role of ·NO in this process was further investigated. ·NO is made during NMDA-mediated activation of the postsynaptic neuron and has been proposed as a retrograde messenger (Wang *et al.*, 1995; Dawson *et al.*, 1994; Samdani *et al.*, 1997). As the ·NO molecule is highly diffusible, it enters the extracellular space where it can interact with the presynaptic nerve terminal, thereby modulating nerve terminal function. The result is an enhanced release of glutamate from the nerve terminal and hence an enhancement of LTP and strengthening of the synaptic connection. This hypothesis is supported by four basic types of experiments: (1) the effect of direct application of ·NO on synaptic transmission, (2) measurement of glutamate release in response to NOS inhibitors or ·NO scavengers, (3) changes in LTP in response to NOS inhibitors or ·NO scavengers, and (4) changes in LTP in mice made genetically deficient in NOS (knockout mice). Each of these types of experiments has produced data supporting the role of ·NO as a retrograde messenger. For example, O'Dell *et al.* (1991) demonstrated that exposure of dissociated hippocampal neurons to NO gas increased the spontaneous release of transmitter. Furthermore, glutamate release was increased in response to high concentrations of NMDA (from 0.2 to 1 mM) and was decreased by NOS antagonists in microdialyzate from the striatum of freely moving rats or from supernatants of slices of dentate gyrus (Nei *et al.*, 1996; O'Dell *et al.*, 1991; Bogdanov and Wurtman, 1997). At least part of this enhanced glutamate release was Ca^{2+}-dependent and could be blocked by tetrodotoxin (TTX). NOS inhibitors such as *N*-monomethyl arginine (L-NMMA) or ·NO scavengers such as hemoglobin inhibit LTP in some cases (Schuman and Madison, 1994). Finally, double knockout mice for eNOS and nNOS demonstrate a significant loss of LTP in some regions of the hippocampus (Son *et al.*, 1996).

Despite this apparent wealth of data, the role of ·NO as a retrograde messenger has not been firmly established and a significant number of studies indicate that ·NO plays a very different role in LTP. Using many of the same types of experiments as described above, data have been provided that directly contradict the proposed enhancement of glutamate release and LTP by ·NO. For example, some putative ·NO donors increase spontaneous release of transmitter in neuronal cultures but block evoked release (Pan *et al.*, 1996). The overall effect is depression of transmission. In *Xenopus* nerve–muscle junction, ·NO donors reduced the frequency of spontaneous miniature currents and decreased evoked synaptic current (Wang *et al.*, 1995). Furthermore, the experiments on ·NO-mediated glutamate release do not firmly identify the site of ·NO action or of glutamate release. LTP is not always enhanced by ·NO donors (Schuman and Madison, 1994; Malen and Chapman, 1997). The most convincing evidence in favor of ·NO as a retrograde messenger is the loss of LTP in the striatum radiatum (CA1) region of the hippocampus in doubly mutant mice ($nNOS^-/eNOS^-$) lacking both neuronal and endothelial ·NOS. But even these experiments indicate that LTP in the CA3 region is

NOS-independent (Son *et al.*, 1996). These confusing results are compounded by experiments on NMDA receptor function. As summarized in Table III, most ·NO donors depress evoked synaptic currents or NMDA-mediated whole-cell currents. In addition, Fagni's group (Manzoni *et al.*, 1992) demonstrated that the putative ·NO donor, SIN-1, decreased single-channel conductance in a voltage-dependent fashion and decreased the probability of NMDA channel opening. These changes are functional because Ca^{2+} entry was also decreased (Hoyt *et al.*, 1992; Lei *et al.*, 1992; Fagni *et al.*, 1995). In addition, cGMP was not involved in this effect. Recent evidence from our laboratory, however, indicates that ·NO donors can increase NMDA-mediated whole-cell currents in HEK cells transfected with the NMDAR1-N2A subunits (Gbadegesin *et al.*, 1997). This effect is seen as an enhanced probability of opening of the NMDA channel with no change in single-channel conductance.

A variety of explanations for these contradictory data have been offered. Because ·NO is highly diffusible and has a long half-life, especially at low concentrations (Wink *et al.*, 1996b), it may act on multiple sites such as both pre- and postsynaptic NMDA receptors as well as other glutamate or other transmitter receptors. In addition, the amount and species of NO^x produced by the ·NO donor are highly variable. As a consequence, the effect of treatment with these donors may vary widely. Wink *et al.* (1995, 1996a) have compared the amount of ·NO generated by the most common ·NO donors. Of these, NONOates (a series of diethylene triamine adducts) generated significantly greater amounts than did *S*-nitrosoglutathione or *S*-nitroso-*N*-acetylpenicillamine (SNAP) (12 μM versus about 4 μM at 5 mins). SNP and 3-morpholinosydnonimine (SIN-1), agents that are widely used in the study of ·NO, failed to produce measurable amounts of ·NO in the same study and are also known to generate damaging by-products (Wink *et al.*, 1996b; Yu and Chang, 1996). The possibility also exists that N_2O_3 (NO^+) or NO^- are generated instead of ·NO with resulting differences in action (Lipton and Stamler, 1994). Alternatively, ·NO may not act as a retrograde transmitter especially if its main action on the postsynaptic receptor is to block channel function. Rather, ·NO may alter other functions of the cell by interacting with metal-centered proteins, thereby modulating their function, or by modulating apoptosis and cell survival (Wink *et al.*, 1996b).

3.4. Other Actions of ROS

3.4.1. ROS-Mediated Neuronal Death

Although induction of programmed cell death cannot be considered a normal function of individual neurons, neuronal death certainly plays an important role in the development of the normal, functioning nervous system (Oppenheim, 1991). The process of cell elimination is not unique to neurons, and cells throughout the body are lost during normal development of most, if not all, tissues. Neuronal death can also be found as a result of injury or inflammation in the CNS and both necrotic ("accidental") and apoptotic ("programmed") death are seen under these abnormal circumstances.

The role of ROS in neuronal cell death has been relatively well studied and is described in detail in several excellent studies and reviews (Bredesen, 1995; Oppenheim, 1991; Nicotera *et al.*, 1997; Ratan *et al.*, 1994; Rothstein *et al.*, 1994; Enokido and

Table IV
Criteria for ROS-Mediated Cell Death

High O_2 tension or high levels of $\cdot O_2^-$, H_2O_2, ·OH, or ·NO must be present
Evidence of oxidative "footprints" (e.g., lipid peroxidation) is found
Necrosis/apoptosis must be inhibited by antioxidants or NOS inhibitors
Decreased levels of antioxidants promote necrosis/apoptosis

Hatanaka, 1993). Essentially, as discussed by Bredesen (1995), ROS and oxidative stress fulfill all of the criteria used to commonly associate a specific cytotoxic agent with necrotic and/or apoptotic cell death (Table IV). ·NO has been included with high oxygen tensions, superoxide anion, H_2O_2, and the hydroxy radical as causative agents for both necrosis and apoptosis. In some cases, the interaction of ·NO and superoxide anion to generate peroxynitrite, another oxidant species, has been implicated (Bonfoco *et al.*, 1995; Nicotera *et al.*, 1997; Beckman, 1994; Cazevieille *et al.*, 1993). Because the biological chemistry of ·NO is complex and reaction pathways have not been completely identified, it is unclear what role, if any, nitrosation products or peroxynitrite might play in neuronal death (Wink *et al.*, 1996b). In addition, the apparent connection between ·NO

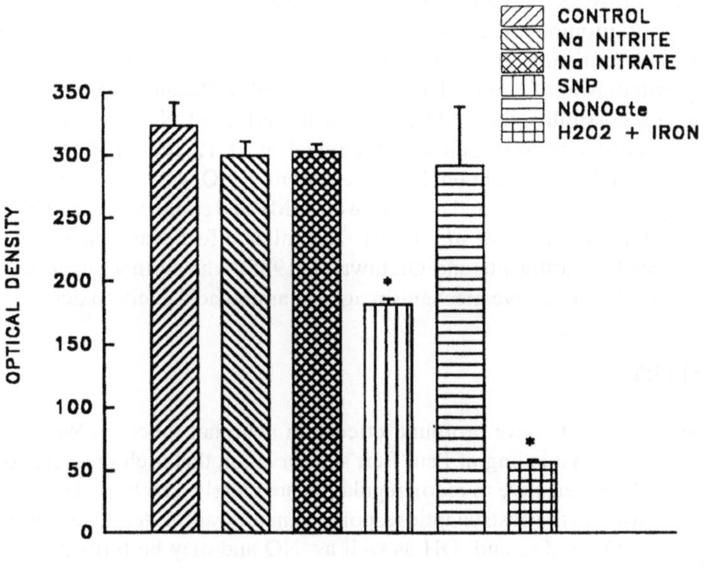

FIGURE 2. Comparison of ·NO donors on neuronal toxicity. Primary neuronal cultures were prepared from the cerebral cortices of neonatal hamster. After 7 days *in vitro*, the cultures were exposed to 50 μM Na nitrite, 50 μM Na nitrate, 50 μM sodium nitroprusside (SNP), or 50 μM DETA NONOate for 24 hr in serum-free media. Cell viability was measured using a standard MTT assay where optical density is directly related to the number of living cells. Data are presented as the average values of optical density (± SEM, $n = 8$ wells assayed). The combination of 10^{-5} M H_2O_2 and 5×10^{-5} M $FeSO_4$ was used as a positive control for cell toxicity mediated by oxidative stress. Asterisk indicates significance at $p < 0.001$.

and NMDA-mediated excitotoxicity has added an extra layer of complexity to the role of ·NO in neuronal death. The question as to whether ·NO is or is not cytotoxic remains open. We have shown that primary neurons cultured from neonatal hamster cerebral cortices or B103 cells, a mouse neuroblastoma cell line (data not shown), do not die when treated for 24 hr with DETA NONOate but do die when treated with SNP (Figure 2). As previously mentioned, SNP fails to generate significant quantities of ·NO and releases cytotoxic side products. However, DETA NONOate generates from 5 to 10 μM ·NO with a half-life of about 500 min (Wink *et al.*, 1996b). This effect is consistent with studies from other cells that demonstrate no ·NO toxicity and, in fact, a decreased toxicity to H_2O_2-mediated cell death (Wink *et al.*, 1995, 1996a) or to NMDA mediated cell death (Garthwaite and Garthwaite, 1988).

3.4.2. Inhibition of Proliferation

Proliferation and the response to nerve growth factor (NGF) in PC12 cells are regulated by treatment of the cells with ·NO donors. Peunova and Enikolopov (1995) demonstrated that ·NO produced by three different putative donors (SNP, SIN and *S*-nitrosocysteine) reversibly decreased radioactive thymidine uptake, indicating that ·NO inhibits DNA synthesis in these cells. Furthermore, ·NO donors have been shown to prevent PC12 cell death following NGF withdrawal and NGF has been shown to increase the expression of ·NOS (Farinelli *et al.*, 1996; Peunova and Enikolopov, 1995). NGF is widely known to promote differentiation in PC12 cells, stopping cell proliferation and initiating branch formation. ·NO also provides protection against apoptosis and cell death induced by withdrawal of serum (Farinelli *et al.*, 1996). Because NGF promotes NOS expression, the possibility that ·NO mediated the effect of NGF was explored. NOS inhibitors did not block NGF-dependent survival of PC12 cells, indicating that NGF action was not totally dependent on ·NO. The action of ·NO was, however, dependent on activation of guanylate cyclase and formation of cGMP and could be blocked by guanylate cyclase inhibitors (Farinelli *et al.*, 1996). A similar effect was demonstrated in rat cerebellar slices by Garthwaite and Garthwaite (1988) who further suggested that this action was part of a protective mechanism against amino acid excitotoxicity.

4. SUMMARY

ROS including ·NO have profound effects on neuronal function. We have come a long way in our understanding of Paul Bert's observation that high pressures of oxygen cause seizures. Sites sensitive to redox regulation are clearly a factor in both presynaptic transmitter release and the postsynaptic response to neurotransmitters. Transmitter release is decreased by $\cdot O_2^-$, H_2O_2, and ·OH as well as ·NO and may be further modulated by redox modification of K^+ channels in the neuronal membrane. The response of the NMDA receptor/channel is highly sensitive to oxidation and reduction and at least one of the redox sites is extracellular. Function of this important receptor may be fined-tuned by alterations in the local redox environment. As a result, overall neuronal excitability in various regions of the CNS may be affected. Unfortunately, the potential for interaction of the reactive molecules and the complexity of the biochemical pathways used by oxyradicals often make interpretation of experimental data difficult. Continued exploration with well-defined "donor" agents and inhibitors will help to resolve the issues of which ROS are neuroactive, how they regulate neuronal function, and when they become neurotoxic.

5. REFERENCES

Aizenman, E., Hartnett, K., and Reynolds, I., 1990, Oxygen free radicals regulate NMDA receptor function via a redox modulatory site, *Neuron* **5**:841–846.

Babior, B., 1984, Oxidants from phagocytes: Agents of defense and destruction, *Blood* **64**:959–966.

Babior, B., Kipnes, R., and Curnette, J., 1973, Biological defense mechanisms: The production by leukocytes of superoxide, a potential bactericidal agent, *J. Clin. Invest.* **52**:742–744.

Bean, J., 1945, Effects of oxygen at increased pressure, *Physiol. Rev.* **25**:1–147.

Beckman, J., 1994, Peroxynitrite versus hydroxyl radical: The role of nitric oxide in superoxide-dependent cerebral injury, *Ann. N.Y. Acad. Sci.* **738**:69–75.

Bernard, C., Hirsch, J., Khazipov, R., Ben-Ari, Y., and Gozlan, H., 1997, Redox modulation of synaptic responses and plasticity in rat CA1 hippocampal neurons, *Exp. Brain Res.* **113**:343–352.

Bert, P., 1872, Recherches experimentales sur l'influence que les changements dans la pression barometrique exercent sur les phenomenes de la vie. 3rd note, *C. R. Acad. Sci.* **73**:617–621.

Bert, P., 1878, *Barometric Pressure. Researches in Experimental Physiology* (M. A. Hitchcock and F. A. Hitchcock, Trans.), College Book Co., Columbus, Ohio, 1943.

Betz, A., 1985, Identification of hypoxanthine transport and xanthine oxidase activity in brain capillaries, *J. Neurochem.* **44**:574–579.

Bingmann, D., Kolde, G., and Speckman, E., 1982, Effects of elevated PO_2 values in the superfusate on the neural activity in hippocampal slices, in *Physiology and Pharmacology of Epileptogenic Phenomena* (M. Klee, ed.), pp. 97–104, Raven Press, New York.

Bogdanov, M., and Wurtman, R., 1997, Possible involvement of nitric oxide in NMDA-induced glutamate release in the rat striatum: An *in vivo* microdialysis study, *Neurosci. Lett.* **221**:197–201.

Bonfoco, E., Krainc, D., Ankarcrona, M., Nicotera, P., and Lipton, S., 1995, Apoptosis and necrosis: Two distinct events induced respectively by mild and intense insults with N-methyl-D-aspartate or nitric oxide/superoxide in cortical cell cultures, *Proc. Natl. Acad. Sci. USA* **92**:7162–7166.

Bredesen, D., 1995, Neural apoptosis, *Ann. Neurol.* **38**:839–851.

Cazevieille, C., Muller, A., Meynier, F., and Bonne, C., 1993, Superoxide and nitric oxide cooperation in hypoxia/reoxygenation-induced neuron injury, *Free Radical Biol. Med.* **14**:389–395.

Choi, J., and Yu, B., 1995, Brain synaptosomal aging: Free radicals and membrane fluidity, *Free Radical Biol. Med.* **18**:133–139.

Cohen, G., 1994, Enzymatic/non-enzymatic sources of oxyradicals and regulation of antioxidant defenses, *Ann. N.Y. Acad. Sci.* **738**:8–14.

Cohen, G., and Werner, P., 1993, Free radicals, oxidative stress and neurodegeneration, in *Neurodegenerative Disease* (D. Calne, ed.), pp. 139–161, Saunders, Philadelphia.

Colton, C., and Colton, J., 1982, An electrophysiological analysis of oxygen and pressure on synaptic transmission, *Brain Res.* **251**:221–227.

Colton, C., and Colton, J., 1985, Blockade of hyperbaric oxygen induced seizures by excitatory amino acid antagonists, *Can. J. Physiol. Pharmacol.* **63**:519–521.

Colton, C., and Colton, J., 1986, The action of oxygen and oxygen at high pressure on inhibitory transmission, *Brain Res.* **364**:151–158.

Colton, C., and Gilbert, D., 1987, Production of superoxide anions by a CNS macrophage, the microglia, *FEBS Lett.* **223**:284–288.

Colton, C., Colton, J., and Gilbert, D., 1986, Changes in synaptic transmission produced by hydrogen peroxide, *Free Radical Biol. Med.* **2**:141–148.

Colton, C., Fagni, L., and Gilbert, D., 1989, The action of hydrogen peroxide on paired pulse and long term potentiation in the hippocampus, *Free Radical Biol. Med.* **7**:3–8.

Colton, C., Yao, J., Grossman, Y., and Gilbert, D., 1991, The effect of xanthine/xanthine oxidase generated reactive oxygen species on synaptic transmission, *Free Radical Res. Commun.* **14:**385–393.

Colton, C., Cherynshev, O., Snell, J., and Gilbert, D., 1994, Induction of superoxide anion and nitric oxide production in cultured microglia, in *The Neurobiology of NO· and ·OH* (C. Cheiuh, D. Gilbert, and C. Colton, eds.), pp. 54–63, New York Academy of Sciences, New York.

Colton, J., and Colton, C., 1978, Effect of oxygen at high pressure on spontaneous transmitter release, *Am. J. Physiol.* **235:**C233–C237.

Davison, A., Wilson, B., and Belton, P., 1984, Deterioration of axonal membranes induced by phenolic pro-oxidants, *Biochem. Pharmacol.* **33:**3887–3891.

Dawson, T., Dawson, V., and Snyder, S., 1994, Molecular mechanisms of nitric oxide actions in the brain, *Ann. N.Y. Acad. Sci.* **738:**76–85.

Enokido, Y., and Hatanaka, H., 1993, Apoptotic cell death occurs in hippocampal neurons cultured in a high oxygen atmosphere, *Neuroscience* **57:**965–972.

Fagni, L., Olivier, M., Lafon-Cazal, M., and Bockaert, J., 1995, Involvement of divalent ions in the nitric oxide induced blockade of *N*-methyl-D-aspartate receptors in cerebellar granule cells, *Mol. Pharmacol.* **47:**1239–1247.

Farinelli, S., Park, D., and Greene, L., 1996, Nitric oxide delays the death of trophic factor-deprived PC12 cells and sympathetic neurons by a cGMP-mediated mechanism, *J. Neurosci.* **16:**2325–2334.

Garthwaite, G., and Garthwaite, J., 1988, Cyclic GMP and cell death in rat cerebellar slices, *Neuroscience* **26:**321–326.

Gbadegesin, M., Vicini, S., Wink, D., and Colton, C., 1997, Subunit dependent modulation of NMDA receptor function by reducing agents, oxyradicals and NO, *Soc. Neurosci. Abstr.* **23:**941.

Giulian, D., and Baker, T., 1986, Characterization of ameboid microglia isolated from developing mammalian brain, *J. Neurosci.* **6:**2163–2178.

Gotz, M., Kunig, G., Riederer, P., and Youdim, M., 1994, Oxidative stress: Free radical production in neural degeneration, *Pharmacol. Ther.* **63:**37–122.

Gutteridge, J., 1994, Hydroxyl radicals, iron, oxidative stress and neurodegeneration, *Ann. N.Y. Acad. Sci.* **738:**201–213.

Halliwell, B., 1992, Reactive oxygen species and the central nervous system, *J. Neurochem.* **59:**1609–1623.

Halliwell, B., and Gutteridge, J., 1985, Oxygen radicals and the nervous system, *Trends Neural Sci.* **8:**22–26.

Halliwell, B., and Gutteridge, J., 1989, *Free Radicals in Biology and Medicine*, 2nd ed., Oxford University Press, London.

Harel, D., and Lavy, S., 1971, Changes in electrical activity of the brain as a criterion of hyperbaric oxygen toxicity in central nervous system, *T. I. T. J. Life Sci.* **1:**111–114.

Hirsch, J., Quesada, O., Esclapez, M., and Gozlan, H., 1996, Enhanced NMDAR-dependent epileptiform activity is controlled by oxidizing agents in a chronic model of temporal lobe epilepsy, *J. Neurophysiol.* **76:**4185–4189.

Hoyt, K., Tang, L., Aizenman, E., and Reynolds, I., 1992, Nitric oxide modulates NMDA induced increases in intracellular Ca^{2+} in cultured rat forebrain neurons, *Brain Res.* **592:**310–316.

Hunt, W., 1993, Role of free radical reactions in ethanol-induced brain damage: An introduction, in *Alcohol-Induced Brain Damage* (W. Hunt and S. Nixon, eds.), pp. 327–338, National Institutes of Health, Bethesda.

Kaneko, M., Elimban, V., and Dhalia, N., 1989, Mechanism for depression of heart sarcolemmal Ca^{2+} pump by oxygen free radicals, *Am. J. Physiol.* **257:**H804–H811.

Karlin, A., and Bartels, E., 1966, Effects of blocking sulfhydryl groups and of reducing disulfide bonds on the acetylcholine-activated permeability system of the electroplax, *Biochim. Biophys. Acta* **126:**525–535.

Kim, J., Shih, J., Chen, K., Chen, L., Bao, S., Maren, S., Anagnostaras, S., Fanselow, M., De Maeyer, E., Seif, I., and Thompson, R., 1997, Selective enhancement of emotional but not motor learning in monoamine oxidase A-deficient mice, *Proc. Natl. Acad. Sci. USA* **94:**5929–5933.

King, G., and Parmentier, J., 1983, Oxygen toxicity of hippocampal tissue *in vitro*, *Brain Res.* **260:**139–142.

Kohr, G., Eckardt, S., Luddens, H., Monyer, H., and Seeburg, P., 1994, NMDA receptor channels: Subunit specific potentiation by reducing agents, *Neuron* **12:**1031–1040.

Kovachich, G., Mishra, O., and Clark, J., 1981, Depression of cortical Na^+, K^+ ATPase activity in rats exposed to hyperbaric oxygen, *Brain Res.* **206:**229–232.

Lei, S., Pan, Z., Aggarwal, S., Chen, H., Hartman, J., Sucher, N., and Lipton, S., 1992, Effect of nitric oxide production on the redox modulatory site of the NMDA receptor channel complex, *Neuron* **8:**1087–1099.

Lipton, S., and Stamler, J., 1994, Actions of redox related congeners of nitric oxide at the NMDA receptor, *Neuropharmacology* **33:**1229–1233.

Livengood, D., and Miller, D., 1989, Differential calcium current block by hydrogen peroxide in lobster cardiac ganglion neurons mediated by internal calcium, *Biophys. J.* **55:**309 (Abstract)

Malen, P., and Chapman, P., 1997, Nitric oxide facilitates long term potentiation but not long term depression, *J. Neurosci.* **17:**2645–2651.

Manzoni, O., Prezeau, L., Marin, P., Deshager, S., Bockaert, J., and Fagni, L., 1992, Nitric oxide-induced blockade of NMDA receptors, *Neuron* **8:**653–662.

McBain, C., and Mayer, M., 1994, *N*-Methyl-D-aspartate acid receptor structure and function, *Physiol. Rev.* **74:**723–760.

McCormack, T., and McCormack, K., 1994, Shaker K^+ channel b subunits belong to an NADPH-dependent oxidoreductase superfamily, *Cell* **79:**1133–1135.

Moncada, S., and Higgs, E., 1991, Endogenous nitric oxide: Physiology, pathology and clinical relevance, *Eur. J. Clin. Invest.* **21:**361–374.

Monyer, H., Burnashev, N., Laurie, D., Sakmann, B., and Seeburg, P., 1994, Developmental and regional expression in the rat brain and functional properties of four NMDA receptors, *Neuron* **12:**529–540.

Morriss, L., Colton, C., and Colton, J., 1983, Effect of hydrogen peroxide on sulfhydryl oxidation of brain ATPase, *Biophys. J.* **41:**410 (Abstract).

Muller, M., Fontana, A., Zbinden, G., and Gahwiler, B., 1993, Effects of interferons and hydrogen peroxide on CA3 pyramidal cells in rat hippocampal slice cultures, *Brain Res.* **619:**157–162.

Naidu, A., Naidu, A., and Kulkani, A., 1992, Dioxygenase and hydroperoxidase activities of rat brain cytosolic lipoxygenase, *Res. Commun. Chem. Pathol. Pharmacol.* **75:**347–356.

Nei, K., Matsuyama, S., Shuntoh, H., and Tanaka, C., 1996, NMDA receptor activation induces glutamate release through nitric oxide synthesis in guinea pig dentate gyrus, *Brain Res.* **728:**105–110.

Nicotera, P., Ankarcrona, M., Bonfoco, E., Orrenhius, S., and Lipton, S., 1997, Neuronal necrosis and apoptosis: Two distinct events induced by exposure to glutamate or oxidative stress, *Adv. Neurol.* **72:**95–101.

Nordmann, R., Ribiere, C., and Rouach, H., 1992, Implication of free radical mechanisms in ethanol-induced cellular injury, *Free Radical Biol. Med.* **12:**219–240.

O'Dell, T., Hawkins, R., Kandel, E., and Arancio, O., 1991, Tests of the roles of two diffusible substances in long-term potentiation: Evidence for nitric oxide as a possible early retrograde messenger, *Proc. Natl. Acad. Sci. USA* **88:**11285–11289.

Oliveira, C., Agostinho, P., Caseiro, P., Daurte, C., and Carvalho, A., 1994, Reactive oxygen species on GABA release, *Ann. N.Y. Acad. Sci.* **738:**130–140.

Oliver, C., Starke-Reed, P., Stadtman, E., Liku, G., Carney, J., and Floyd, R., 1990, Oxidative damage to brain proteins, loss of glutamine synthetase activity and production of free radicals during ischemia/reperfusion mediated injury to gerbils, *Proc. Natl. Acad. Sci. USA* **87:**5144–5147.

Oppenheim, R., 1991, Cell death during development of the nervous system, *Annu. Rev. Neurosci.* **14:**453–501.

Pan, Z., Segal, M., and Lipton, S., 1996, Nitric oxide related species inhibit evoked neurotransmission but enhance spontaneous miniature synaptic currents in central neuronal cultures, *Proc. Natl. Acad. Sci. USA* **93:**15423–15428.

Paoletti, P., Ascher, P., and Neyton, J., 1997, High-affinity zinc inhibition of NMDA NR1-NR2A receptors, *J. Neurosci.* **17:**5711–5725.

Patole, M., Swaroop, A., and Ramasarma, T., 1986, Generation of H_2O_2 in brain mitochondria, *J. Neurochem.* **47:**1–8.

Pellegrini-Giampietro, D., Cherici, G., Alesiani, M., Carla, V., and Moroni, F., 1988, Excitatory amino acid release from rat hippocampal slices as a consequence of free-radical formation, *J. Neurochem.* **51:**1960–1963.

Pellmar, T., 1986, Electrophysiological correlates of peroxide damage in guinea pig hippocampus *in vitro*, *Brain Res.* **364:**377–381.

Pellmar, T., 1987, Peroxide alters neuronal excitability in the CA1 region of guinea pig hippocampus *in vitro*, *Neuroscience* **23:**447–456.

Pellmar, T., Gilman, S., Keyser, D., Lee, K., Lepinski, D., Livengood, D., and Myers, L., 1994, Reactive oxygen species on neural transmission, *Ann. N.Y. Acad. Sci.* **738:**121–129.

Peunova, N., and Enikolopov, G., 1995, Nitric oxide triggers a switch to growth arrest during differentiation of neuronal cells, *Nature* **375**:68–72.

Quesada, O., Hirsch, J., Ben-Ari, Y., and Bernard, C., 1996, Redox sites of NMDA receptors can modulate epileptiform activity in hippocampal slices from kainic acid treated rats, *Neurosci. Lett.* **212**:171–174.

Rafalowska, U., Liu, G., and Floyd, R., 1989, Peroxidation induced changes in synaptosomal transport of dopamine and γ-aminobutyric acid, *Free Radical Biol. Med.* **6**:485–492.

Ramasarma, T., 1982, Generation of H_2O_2 in biomembranes, *Biochem. Biophys. Acta.* **694**:69–93.

Ratan, R., Murphy, T., and Baraban, J., 1994, Oxidative stress induces apoptosis in embryonic cortical neurons, *J. Neurochem.* **62**:376–379.

Rothstein, J., Bristol, L., Hosler, B., Brown, R., and Kuncl, R., 1994, Chronic inhibition of superoxide dismutase produces apoptotic death of spinal neurons, *Proc. Natl. Acad. Sci. USA* **91**:4155–4159.

Rouzaire-Dubois, B., and Dubois, J., 1990, Modification of electrophysiological and pharmacological properties of K channels in neuroblastoma cells induced by the oxidant chloramine-T, *Pflugers Arch.* **416**:393–397.

Samdani, A., Dawson, T., and Dawson, V., 1997, Nitric oxide synthase in models of focal ischemia, *Stroke* **28**:1283–1288.

Schor, N., 1988, Inactivation of mammalian brain glutamine synthetase by oxygen radicals, *Brain Res.* **456**:17–21.

Schuman, E., and Madison, D., 1994, Nitric oxide and synaptic function, *Annu. Rev. Neurosci.* **17**:153–183.

Schwartz, R., Skolnick, P., and Steven, P., 1988, Regulation of γ-aminobutyric acid/barbiturate receptor gated chloride ion flux in brain vesicles by phospholipase A2: Possible role of oxygen radicals, *J. Neurochem.* **50**:565–571.

Shephard, R., 1971, Part II: Oxygen consumption and ventilatory equivalent, in *Respiration and Circulation* (P. Altman and D. Dittmer, eds.), pp. 43–45, Fed. Am. Soc. Exp. Biol., Bethesda.

Silver, I., 1966, The measurement of oxygen tension in tissues, in *Oxygen Measurements in Blood and Tissues* (J. Payne and D. Hill, eds.), pp. 135–153, Little, Brown, Boston.

Sinor, J., Boeckman, F., and Aizenman, E., 1997, Intrinsic redox properties of *N*-methyl-D-aspartate receptor can determine the developmental expression of excitotoxicity in rat cortical neurons *in vitro*, *Brain Res.* **747**:297–303.

Sokoloff, L., 1960, Metabolism of the central nervous system *in vivo*, in *Handbook of Physiology: Neurophysiology* (J. Field, H. Magoun, and V. Hall, eds.), pp. 1843–1849, American Physiological Society, Bethesda.

Son, H., Hawkins, R., Martin, K., Kiebler, M., Huang, P., Fishman, M., and Kandel, E., 1996, Long term potentiation is reduced in mice that are doubly mutant in endothelial and neuronal nitric oxide synthase, *Cell* **87**:1015–1023.

Stanley, E. F., 1986, Decline in calcium cooperativity as the basis of facilitation at the squid giant synapse, *J. Neurosci.* **6**:782–789.

Stephens, G. D., Owen, D., and Robertson, B., 1996, Cysteine-modifying reagents alter the gating of the rat cloned potassium channel Kv1.4, *Pflugers Arch.* **431**:435–442.

Subbarao, K., and Richardson, J., 1990, Iron-dependent peroxidation rat brain: A regional study, *J. Neurosci. Res.* **26**:224–232.

Sucher, N., Wong, L., and Lipton, S., 1990, Redox modulation of NMDA receptor mediated Ca^{2+} flux in mammalian central neurons, *NeuroReport* **1**:29–32.

Sullivan, J., Traynelis, S., Chen, H., Escobar, W., Heinemann, S., and Lipton, S., 1994, Identification of two cysteine residues that are required for redox modulation of the NMDA subtype of glutamate receptor, *Neuron* **13**:929–936.

Tang, L. H., and Aizenman, E., 1993, Long lasting modification of the *N*-methyl-D-aspartate receptor channel by a voltage dependent sulfhydryl redox process, *Mol. Pharmacol.* **44**:473–478.

Tauck, D., 1992, Redox modulation of NMDA receptor mediated synaptic activity in the hippocampus, *NeuroReport* **3**:781–784.

Tolliver, J., and Pellmar, T., 1987, Ionizing radiation alters neuronal excitability in hippocampal slices of the guinea pig, *Radiat. Res.* **112**:555–563.

Trotti, D., Rossi, D., Gjesdal, O., Levy, L., Racagni, G., Danbolt, N., and Volterra, A., 1996, Peroxynitrite inhibits glutamate transporter subtypes, *J. Biol. Chem.* **271**:5976–5979.

Vanderkooi, J., Erecinska, M., and Silver, I., 1991, Oxygen in mammalian tissue: Methods of measurement and affinities of various reactions, *Am. J. Physiol.* **260**:C1131–C1150.

Vega-Saenz, E., and Rudy, B., 1992, Modulation of K^+ channels by hydrogen peroxide, *Biochem. Biophys. Res. Commun.* **186:**1681–1687.

Volterra, A., Trotti, D., Floridi, S., and Racagni, G., 1994, Reactive oxygen species inhibit high affinity glutamate uptake: Molecular mechanisms and neuropathological implications, *Ann. N.Y. Acad. Sci. USA* **738:**153–162.

Wang, T., Xie, Z., and Lu, B., 1995, Nitric oxide mediates activity dependent synaptic suppression at developing neuromuscular synapses, *Nature* **374:**262–266.

Wink, D., Cook, J., Krishna, M., Hanbauer, I., DeGraff, W., Gamson, J., and Mitchell, J., 1995, Nitric oxide protects against alkyl peroxide mediated cytotoxicity: Further insights into the role nitric oxide plays in oxidative stress, *Arch. Biochem. Biophys.* **319:**402–407.

Wink, D., Cook, J., Pacelli, R., DeGraff, W., Gamson, J., Liebmann, J., Krishna, M., and Mitchell, J., 1996a, The effect of various nitric oxide donor agents on hydrogen peroxide mediated toxicity: A direct correlation between nitric oxide formation and protection, *Arch. Biochem. Biophys.* **331:**241–248.

Wink, D., Hanbauer, I., Grisham, M., Laval, F., Cook, J., Pacelli, R., Liebmann, J., Krishna, M., Ford, P., and Mitchell, J., 1996b, Chemical biology of nitric oxide: Regulation and protective and toxic mechanisms, *Curr. Top. Cell. Regul.* **34:**159–187.

Wood, J., and Watson, W., 1963, Gamma aminobutyric acid levels in the brain of rats exposed to oxygen at high pressure, *Can. J. Biochem. Physiol.* **41:**1907–1913.

Wood, J. D., Watson, W. J., and Clydesdale, F., 1963, Gamma-aminobutyric acid and oxygen poisoning, *J. Neurochem.* **10:**625–633.

Wood, J., Watson, W., and Murray, G., 1969, Correlation between decreases in brain gamma aminobutyric acid levels and susceptibility to convulsions induced by hyperbaric oxygen, *J. Neurochem.* **16:**281–287.

Yu, O., and Chang, D., 1996, Inhibition of excitatory amino-acid-induced phosphoinositide hydrolysis as a possible mechanism of nitroprusside neurotoxicity, *J. Neurochem.* **66:**346–354.

Zangerle, L., Cuenod, M., Winterhalter, K., and Quang Do, K., 1992, Screening of thiol compounds: Depolarization induced release of glutathione and cysteine from rat brain slice, *J. Neurochem.* **59:**181–189.

Zoccarato, F., Deana, R., Cacallini, L., and Alexandre, A. 1989, Generation of hydrogen peroxide by cerebral cortex synaptosomes, *Eur. J. Biochem.* **180:**473–478.

Part VII

Pathological States and Aging

Chapter 24

Oxidative Stress and Parkinson's Disease

Gerald Cohen

1. CHARACTERISTICS OF PARKINSON'S DISEASE

Parkinson's disease (PD) is a progressive neurodegenerative disorder affecting primarily the dopamine (DA) neurons that arise in the midbrain (mesencephalon) and project to the putamen and caudate regions (the striatum) of the brain, areas concerned with the control of motor movements (Hornykiewicz and Kish, 1986). Unaffected (or minimally affected) by the disease are DA neurons that arise in the midbrain and project to cortical and limbic regions; overactivity of the latter neuronal circuits has been implicated in schizophrenia (e.g., Lozoncy *et al.*, 1987). Also unaffected are other monoaminergic neurons, specifically the norepinephrine (NE)-secreting and serotonin-secreting neurons of the brain.

The DA neurons that degenerate in PD arise in the substantia nigra, a local region of the midbrain. As its name implies, this region is normally heavily pigmented; the pigment is readily visible to the unaided eye as a brown-to-black region of the mesencephalon. The pigment is an insoluble polymer that is related to melanin of skin, and has been termed *neuromelanin*. However, unlike skin, the pigment is not derived by enzymatic synthesis from L-dopa catalyzed by tyrosinase, because tyrosinase is absent from the brain. Rather, neuromelanin appears to be formed via a slow, nonenzymatic process based on the autoxidation and spontaneous polymerization of the catecholamine DA (Graham, 1978). As with melanin of skin, quinoidal intermediates react with soluble tissue thiols, such as glutathione and cysteine, to incorporate sulfur residues into the matrix of the polymer (Carstam *et al.*, 1992).

Gerald Cohen Department of Neurology and Center for Neurobiology, Mount Sinai School of Medicine, New York, New York 10029.

Reactive Oxygen Species in Biological Systems, edited by Gilbert and Colton. Kluwer Academic / Plenum Publishers, New York, 1999.

A hallmark of PD is the disappearance of pigmentation from the substantia nigra. This is readily apparent visually to neuropathologists at autopsy and, because of this, absence of neuromelanin became an early marker for PD. It is now known that the marked diminution or absence of pigmentation does not represent a change in the chemical process that forms neuromelanin. Rather, loss of pigment reflects the loss of DA neurons via an underlying neurodegenerative process. The key findings that prove this point are: (1) a loss of the neurotransmitter DA and (2) loss of the biosynthetic enzyme tyrosine hydroxylase from autopsy specimens of brain. These two markers reflect loss of DA-secreting neurons. With the degeneration and loss of DA neurons, the neuromelanin associated with the cell bodies is phagocytized and disappears.

The reasons for the selective loss of nigrostriatal DA neurons in PD remain obscure. Current thoughts and research center around possible genetic predisposition, environmental factors, and an oxidative stress that may be derived directly from the natural utilization and turnover of the neurotransmitter DA. A recent conference stressed an interplay between all three factors (Gorrell *et al.*, 1996). At least one clinical trial has been directed toward attempting to slow progression driven by DA turnover or driven by exposure to putative environmental toxins (Parkinson Study Group, 1989a). Although intervention with the monoamine oxidase (MAO) B inhibitor deprenyl (selegiline) gave evidence for slowing of disease progression in subjects with early PD (Parkinson Study Group, 1989b), these results remain the subject of some debate. In clinical trials, disease progression is most frequently assessed from behavioral scores; however, to properly assess motor function, experimental drugs need to be first washed out, lest drug effects interfere with the assessment of motor scores. In the deprenyl trial (the so-called DATATOP study) a mild symptomatic effect was detected, which could have been responsible, in part, for the behavioral benefit of deprenyl. Considerable drug benefit was retained after washout, but disease progression was still evident (Parkinson Study Group, 1993; Olanow *et al.*, 1995). The costly and newly developed technique of PET scanning (positron emission tomography) can provide more direct assessment of changes to nigrostriatal neurons that are affected in PD.

Generally, the expression of symptomatology (tremor, akinesia) in PD is delayed until 80% or more of the nigrostriatal DA neurons have been lost. Progression with further loss of DA neurons occurs despite treatment, which is mainly directed at controlling symptomatology, i.e., to enhancing dopaminergic neurotransmission. Widely used treatments consist of L-dopa (to enhance brain synthesis of DA), deprenyl (to impede metabolism of DA), and DA agonists (to function in place of the missing DA). Considerable interest exists in identifying factors that contribute to disease progression because such information could lead to the development of new treatment approaches to curtail the progressive loss of DA neurons.

2. DOPAMINERGIC NEUROTOXINS

Concepts of oxidant stress in PD have been driven by experiences with two dopaminergic neurotoxins, namely, 6-hydroxydopamine (6-OHDA; 2,4,5-trihydroxyphenylethylamine) and 1-methyl-4-phenyl-1,2,3,6-tetrahydropyridine (MPTP). Both of these agents have been used experimentally to produce animal models of PD in which nigrostriatal

DA neurons are lesioned relatively selectively. Studies on the roles of superoxide and hydroxyl radicals in the mechanism of action of 6-OHDA played an important historical role in the early development of concepts of oxy-radical toxicity in biological systems (e.g., Cohen and Heikkila, 1974; Cohen *et al.*, 1976).

2.1. 6-Hydroxydopamine

6-OHDA was first discovered as a neurotoxic agent by Tranzer and Thoenen in the 1960s as an agent that disrupted brain slices that were prepared for electron microscopic examination of the vesicular binding properties of catecholamine analogues. It was subsequently determined that the agent produced relatively selective damage to peripheral sympathetic neurons (NE-secreting) when injected peripherally, but that it did not cross the blood–brain barrier (reviewed by Kostrzewa and Jacobowitz, 1974). However, injections into the cerebral ventricles or directly into target brain tissue (e.g., the substantia nigra or striatum) have been widely used to lesion targeted DA or NE neurons of the brain.

As a hydroxylated DA analogue, 6-OHDA is recognized by the catecholamine transporters present in the axonal membranes of catecholamine neurons and gains access to the neurons via this pathway (Kostrzewa and Jacobowitz, 1974). Axonal transport targets catecholamine neurons and explains the extraordinary specificity of neurotoxic action. Under the appropriate experimental conditions, only catecholamine neurons are detrimentally affected, while other neuronal types are spared. The use of selective inhibitors of either the DA transporter or the NE transporter can provide further experimental specificity to selectively spare either DA or NE neurons.

6-OHDA is highly unstable in aqueous solution at neutral pH and in the presence of oxygen. The reaction with molecular oxygen yields hydrogen peroxide (H_2O_2) and a red quinone [Eq. (1)]. The rapid formation of red quinoidal products over 10–20 s at pH 7.4 is readily apparent visually. This observation led many early workers to add ascorbic acid as a reducing agent to "protect" the 6-OHDA, i.e., to prevent discoloration of the solution containing 6-OHDA. However, the situation is more complicated and ascorbate plays a more prominent role as an agent that exacerbates the neurotoxic effects of 6-OHDA. Entry into catecholamine neurons requires that 6-OHDA be present in its reduced form, which is recognized by catecholamine transporters. Hence, ascorbate helps to maintain 6-OHDA in a transportable form. However, the same role is served by endogenous ascorbate (e.g., in blood plasma). Indeed, direct intravenous injection of 6-OHDA-quinones leads to the same destruction of peripheral sympathetic neurons as 6-OHDA (Heikkila *et al.*, 1973). Tissue ascorbate also promotes toxicity by recycling the quinones [Eq. (2)].

$$\text{6-OHDA} + O_2 \xrightarrow{\text{autoxidation}} o\text{- and } p\text{-quinones} + H_2O_2 \quad (1)$$

$$\text{ascorbate} + \text{quinone} \rightarrow \text{dehydroascorbate} + \text{6-OHDA} \quad (2)$$

$$Fe^{2+} + H_2O_2 \rightarrow Fe^{3+} + \cdot OH + OH^- \quad (3)$$

The H_2O_2 generated in Eq. (1) can lead to damage of DA neurons (Heikkila and Cohen, 1971); available evidence points to the formation of hydroxyl radicals [Eq. (3)] as neurotoxic intermediates (Cohen and Heikkila, 1974; Cohen *et al.*, 1976). The quinones can express toxicity via the formation of adducts with protein -SH groups (Graham *et al.*, 1978) leading to inactivation of enzymatic or structural properties of key proteins. Although both H_2O_2- and quinone-based mechanisms are probably operative, the available evidence favors the view that H_2O_2 plays the more dominant toxic role. For example, ascorbate added to tissue slices prevents the appearance of colored quinones, but potentiates neurotoxicity (Heikkila and Cohen, 1972). The reason for this effect is that ascorbate recycles the quinones by reduction [Eq. (2)] and, thereby, amplifies H_2O_2 production [Eq. (1)]. Indeed, in experiments carried out with tissue slices, formation of H_2O_2 by relatively small amounts of 6-OHDA was controlled by the level of ascorbate. In addition, analogues of 6-OHDA with methyl groups blocking the ring positions that would normally make adducts with -SH groups of proteins, still exhibit neurotoxicity (Graham *et al.*, 1978).

The generation of superoxide by 6-OHDA was detected in experiments with superoxide dismutase (Heikkila and Cohen, 1973). By scavenging superoxide radicals, superoxide dismutase suppressed both the formation of quinones and the accumulation of H_2O_2. Thus, although superoxide dismutase catalyzes a reaction that generates H_2O_2 (namely, the dismutation of superoxide to yield H_2O_2 and oxygen), its net effect is to suppress the overall formation of H_2O_2 and, subsequently, the derived hydroxyl radicals. The explanation for this apparently paradoxical situation can be found in the reaction mechanism for autoxidation of 6-OHDA, where Q stands for 6-OHDA-quinone, SQ· is the corresponding semiquinone, and $\cdot O_2^-$ is the superoxide radical anion:

$$\text{6-OHDA} + O_2 \rightarrow SQ\cdot + \cdot O_2^- + H^+ \qquad \text{(slow)} \tag{4}$$

$$\text{6-OHDA} + \cdot O_2^- + H^+ \rightarrow SQ\cdot + H_2O_2 \qquad \text{(fast)} \tag{5}$$

$$SQ\cdot + O_2 \rightarrow Q + \cdot O_2^- \qquad \text{(fast)} \tag{6}$$

The direct reaction of oxygen with 6-OHDA is relatively slow [Eq. (4)] whereas the oxidation of 6-OHDA by superoxide is much faster [Eq. (5)]. After initiation of the reaction sequence [Eqs. (4) and (5)], SQ· reacts rapidly with molecular oxygen to regenerate superoxide [Eq. (6)]. Equations (5) and (6) establish a radical chain reaction in which superoxide is consumed as it oxidizes 6-OHDA and then it is regenerated; the reaction chain bypasses the slow step [Eq. (4)] and promotes formation of H_2O_2 [Eq. (5)] and quinones [Eq. (6)]. The chain reaction can be intercepted by superoxide dismutase, or by catecholamines (such as DA or NE) which are also oxidized by superoxide (Sachs *et al.*, 1975; Cohen and Heikkila, 1977). In this sense, the catecholamines act as scavengers for superoxide, suppressing the overall rate of production of H_2O_2 and quinones by replacing the extraordinarily rapid autoxidation sequence for 6-OHDA with the much slower reactions for catecholamine autoxidation at neutral pH.

The neurotoxic analogue 6-aminodopamine behaves differently. Its autoxidation rate is not catalyzed by superoxide (Sachs *et al.*, 1975; Cohen and Heikkila, 1977). In experiments conducted *in vivo*, 6-OHDA and 6-aminodopamine were differentially affected by endogenous catecholamines, consistent with their relative dependence on (6-OHDA) or independence (6-aminodopamine) from superoxide as a catalyst: Prior depletion of NE (with α-methyl-*p*-tyrosine methyl ester) potentiated the neurotoxicity of 6-OHDA, but not 6-aminodopamine. Enhancing the level of NE protected against 6-OHDA, but not 6-aminodopamine. However, elevated tissue levels of octopamine (a phenolic monoamine that does not scavenge superoxide) failed to protect against either 6-OHDA or 6-aminodopamine neurotoxicity. These results are in keeping with the ability of NE in sympathetic neurons *in vivo* to suppress the catalytic formation of peroxide by 6-OHDA, but not 6-aminodopamine.

The hydroxyl radical (·OH) is generated from H_2O_2 by the Fenton reaction with iron in the ferrous form [Eq. (3)]. Hydroxyl radicals can lead to generalized cellular damage based on the oxidation of membrane lipids (lipid peroxidation), enzymes, structural proteins, and nucleic acids. Hydroxyl radicals were detected *in vitro* during the autoxidation of 6-OHDA and 6-aminodopamine (Cohen and Heikkila, 1974). Peripheral sympathetic nerves innervating the heart and iris were protected from neurodegeneration *in vivo* by scavengers of ·OH (Cohen *et al.*, 1976). Jonsson (1976) showed that ·OH scavengers did not inhibit the neuronal accumulation of $[^3H]$-6-OHDA, thereby confirming that protection was the result of interruption of the cytotoxic mechanism, and not an effect on axonal transport of 6-OHDA.

6-OH-dopa (2,4,5-trihydroxyphenylalanine), the amino acid analogue of 6-OHDA, is also neurotoxic, but exhibits some unusual properties. 6-OH-dopa readily crosses the blood–brain barrier and achieves access to central catecholamine neurons. Neurotoxicity requires prior conversion to 6-OHDA. The available evidence indicates that transformation to 6-OHDA takes place within catecholamine neurons and does not require the catecholamine transporter (Evans and Cohen, 1993). However, the unusual aspect is that 6-OH-dopa selectively targets NE-secreting neurons in the CNS, while sparing DA neurons (Kostrzewa and Jacobowitz, 1974). DA neurons are spared even though they accumulate quite high levels of 6-OHDA (Evans and Cohen, 1989) approaching those of endogenous DA (which are in the range of 50 mM). These observations indicate that central DA neurons are endowed with special protective mechanisms compared with NE neurons. The selectivity of 6-OH-dopa is opposite to that of MPTP (see below), which targets DA neurons but spares NE neurons.

2.2. MPTP

The MPTP story opened with a single case report in 1979 by a group of investigators at the National Institutes of Health (NIH). This important observation encountered difficulty in gaining access for publication, but was eventually published in the inaugural volume of *Psychiatry Research* (Davis *et al.*, 1979). Autopsy of a young adult who had overdosed on a narcotic drug showed extensive damage to the substantia nigra paralleling a motor defect resembling PD. Because PD is a disease of aging, loss of nigral DA neurons in a young adult was unusual. The subject had been in the habit of preparing his own "designer" drugs by organic synthesis and he kept a notebook with details. It was discovered that during the preparation of a meperidine analogue via a reverse synthesis pathway, he had used heat to speed a reaction and this had resulted in an unexpected chemical elimination of a side chain (dealkylation) with the resultant formation of a double bond in the 4–5 position of a piperidine ring. Other products were also formed, but the main product, MPTP, was considered to be the culprit for rapid onset of parkinsonism in a young adult. This tentative explanation was confirmed in 1983 in a publication in *Science* (Langston *et al.*, 1983). This second publication extended the observations to a broader series of young adults who had inadvertently been exposed to

MPTP as a contaminant in illicit drug preparations. The latter study strengthened considerably the link to MPTP, opening a new prospect for understanding PD. MPTP toxicity in human subjects is viewed as a chemically induced form of parkinsonism.

Unlike research on 6-OHDA, which was facilitated by the rapid response of rodents (rats, mice) to the injected neurotoxin, research into the mechanism of action of MPTP was severely hampered by an absence of animal models. Initial studies by the group at the NIH who made the first report drew a blank in studies with rats, mice, rabbits, and cats. It was not until monkeys were tested that symptoms of PD paralleling overt loss of central DA neurons appeared with relatively low doses of MPTP, commensurate with the doses to which human drug users had been exposed (Burns *et al.*, 1983). However, with only monkeys as an *in vivo* model, research in this important area was severely constrained until other investigators showed that mice could be used when the dose of MPTP was raised about 100-fold (Heikkila *et al.*, 1984; Hallman *et al.*, 1985). In addition, an important new tool, mesencephalic cell cultures (Mytilineou and Cohen, 1984), was added to the research armamentarium; DA neurons in cell cultures proved to be exquisitely sensitive to MPTP.

It has been established that the neurotoxicity of MPTP is directed relatively selectively at the dopaminergic nigrostriatal tract, while DA neurons in the limbic system, as well as central NE and serotonin neurons, are spared. Biochemical and histological studies verified the destruction of DA cell bodies in the substantia nigra and the loss of dopaminergic innervation in the caudate and putamen.

The neurotoxic mechanism involves the conversion of MPTP, which is a protoxin, to 1-methyl-4-phenylpyridinium (MPP^+), which is the ultimate toxin. Initial studies with mesencephalic cultures had implicated MAO, because deprenyl or pargyline (MAO-B inhibitors) protected the DA neurons (Mytilineou and Cohen, 1984). One possibility was that the MAO inhibitors protected by suppressing the metabolism of DA, which produces H_2O_2 as an end product. However, studies by Chiba *et al.* (1984) with isolated mitochondria established the link to MAO as transformation of MPTP to MPP^+ [Eq. (7)]. The form of MAO required to transform MPTP is MAO-B. *In vivo*, inhibitors of MAO-B, such as deprenyl (Cohen *et al.*, 1984) or pargyline (Langston *et al.*, 1984), prevent the destruction of nigrostriatal neurons in monkeys.

$$\text{MPTP} \xrightarrow{\text{MAO-B}} \text{MPDP}^+ \xrightarrow{\text{spontaneous}} \text{MPP}^+ \qquad (7)$$

There were several surprising aspects to this story. First, MAO, which normally deaminates substrates, was invoked for a dehydrogenation reaction to yield a double bond. Second, the transformation took place in astrocytes, which are rich in MAO-B, but astrocytes were not themselves damaged. Third, the product, the aromatic amine MPP^+, was recognized by the dopamine transporter, resulting in a marked accumulation of the toxin by DA neurons (Javitch *et al.*, 1985). In this latter regard, susceptibility to both 6-OHDA and MPP^+, and the targeting of vulnerable neuronal types, is based on accumulation of the neurotoxin mediated by the axonal membrane transporter systems for catecholamines.

The toxic effects of MPP^+ require an accumulation by mitochondria where electron transport is inhibited at the level of complex I (Nicklas *et al.*, 1985; Ramsay *et al.*, 1986). Poisoning of the electron transport chain is responsible for the eventual demise of nigrostriatal DA neurons. Spurred by the latter findings, yet another surprising fact was uncovered, namely, that PD appears to be characterized by a similar defect at the level of Complex I in the substantia nigra, but not in brain regions unaffected by the disease (e.g., Schapira *et al.*, 1990). This observation has given rise to concepts that PD may be caused by environmental mitochondrial toxins, perhaps related to MPTP (e.g., Mizuno *et al.*, 1995; Gorrell *et al.*, 1996).

Inhibition of mitochondrial electron transport (e.g., by rotenone or antimycin A) leads to increased release of reactive oxygen species, such as superoxide and H_2O_2, from mitochondria (Zoccarato *et al.*, 1988). In turn, these agents may spur cell death. Thus, MPTP toxicity may derive from an oxidative stress. In experiments with isolated mitochondria, Cleeter *et al.* (1992) observed that inactivation of Complex I by directly added MPP^+ required the presence of oxygen, in keeping with an oxidative mechanism. Moreover, protection was observed with added antioxidants (ascorbate, glutathione); catalase, a specific enzymatic scavenger of H_2O_2, was also protective. These observations point to an oxidative mechanism for inhibition of Complex I. Chiueh *et al.* (1992) used the salicylate trapping method and *in vivo* dialysis to detect hydroxyl radicals in the caudate nucleus during exposure to MPP^+. Because MPP^+ causes release of DA from neurons, the authors suggested that either DA autoxidation or its oxidative deamination by MAO may play a role in ·OH formation by MPTP/MPP^+ *in vivo*. It is of interest that transgenic mice with increased Cu/Zn-superoxide dismutase are resistant to the neurodegenerative effects of MPTP (Przedborski *et al.*, 1992). These observations, as a whole, indicate a prominent role for oxy-radicals and oxidative stress in parkinsonism induced by MPTP.

3. OXIDATIVE STRESS AND PARKINSON'S DISEASE

A state of oxidative stress exists when either oxidizing species (e.g., oxy-radicals, peroxides) are present in excess, or cellular antioxidant defenses are lowered. Oxidative stress drives cellular systems to an oxidized state. An increased metabolic demand by various cellular support or repair mechanisms can be detrimental to other biological needs. Many markers are available to detect oxidative changes in cellular systems, including lipid peroxidation, formation of protein carbonyls, loss of reducing substances (such as GSH), changes in antioxidant enzymes, and oxidative damage to DNA. Over the years considerable evidence has been amassed from autopsy studies and from experimental studies with animal models that point to the presence of an oxidant stress associated with PD (e.g., Jenner, 1991; Fahn and Cohen, 1992; Ames *et al.*, 1993). A number of authors have addressed the issue that oxidative stress and oxidative damage may play a critical role in neurodegenerative diseases in general, including PD (see Ames *et al.*, 1993; Coyle and Puttfarcken, 1993; Cohen and Werner, 1994; Beal, 1995).

3.1. The L-Dopa Question

From the outset it must be noted that many observations made on autopsy specimens of brain from parkinsonian subjects are partially compromised because most subjects were in treatment with L-dopa. L-Dopa autoxidizes slowly (Basma *et al.*, 1995) to form reactive quinones and H_2O_2. The process is similar to that described in Eqs. (4)–(6) and

Eq. (3), but at a much slower overall rate. In this process, oxy-radical intermediates, such as superoxide, hydroxyl, and semiquinone free radicals, are generated. Hence, the question arises: Does the appearance of markers of oxidative stress in autopsy specimens reflect the disease process itself or does it reflect a side effect of treatment with L-dopa?

The issue is complex and data are available to support both points of view. First, a limited number of observations have been made on patients not treated with L-dopa, which support the presence of an oxidative stress unrelated to treatment. Second, patients treated with L-dopa show better survival, and there is no direct evidence for toxicity of chronic L-dopa in animals models (Hefti *et al.*, 1980). On the other hand, L-dopa exhibits toxicity in cell culture experiments, directed relatively specifically at DA neurons (Olney *et al.*, 1990; Mytilineou *et al.*, 1993; Mena *et al.*, 1993). And, in two *in vivo* models, albeit under special experimental circumstances, evidence for detrimental effects of L-dopa on survival of DA neurons has been presented (Steece-Collier *et al.*, 1990; Blunt *et al.*, 1993). Adding fuel to the fire is the observation that chronic administration of L-dopa to rats produces a Complex I defect in dopaminergic regions of brain (Przedborski *et al.*, 1993). Supporting an opposite point of view is the observation that exposure of mesencephalic cultures to L-dopa induces a compensatory rise in GSH, which prevents a loss in cell viability during exposure to an organic hydroperoxide (Han *et al.*, 1996). Thus, the observed effects of L-dopa are double-edged, and it remains unclear whether L-dopa "therapy" carries a toxic potential, perhaps in selected subsets of PD patients.

3.2. Evidence for Oxidative Stress

A variety of studies have provided evidence for a condition of oxidative stress in the parkinsonian brain.

Lipid peroxidation is a radical chain reaction evoked by exposure to hydroxyl radicals or other oxidants. Once initiated, it can be transmitted to adjacent lipids in membranes. The biological quenching agent is vitamin E. Two major changes are evoked by lipid peroxidation: The conjugation of previously separated double bonds in unsaturated lipids (such as linoleic, arachidonic, and linolenic acid) makes rigid previously flexible segments of membranes. Rigidity is related to the fact that rotation is restricted in conjugated dienes because the double bonds must lie in the same plane. The second change is the accumulation of fatty acid hydroperoxides, capable of altering membrane structure and function. Potentially toxic aldehyde products, such as 4-hydroxynonenal, are also formed. Dexter *et al.* (1989) reported an increase in lipid peroxides in the parkinsonian brain. The increase was associated with a simultaneous decrease in polyunsaturated fatty acids, and it was localized to the substantia nigra. Confirmatory evidence was provided by HPLC and electron spin resonance spectroscopy (Dexter *et al.*, 1994). Increased levels of 4-hydroxynonenal bound as an adduct to proteins have also been reported in PD (Yoritaka *et al.*, 1996). Jenner (1991) also reported elevated lipid peroxides in the parkinsonian brain.

A catalyst for lipid peroxidation is iron. Iron, specifically iron in the ferrous state, is also required for the production of hydroxyl radicals from H_2O_2 via the Fenton reaction [Eq. (3)]. A number of reports have described elevated tissue levels of iron in the parkinsonian brain and this has engendered strong interest in the exact location of the iron and whether or not it can contribute to the progression of PD (e.g., Olanow *et al.*, 1992).

The iron that can be visualized by histochemical stain in normal brain is relatively high in the substantia nigra, but it is restricted to the pars reticularis, whereas the DA cell bodies are present in another subregion, the pars compacta. However, the iron that accumulates in PD is seen in the pars compacta (Sofic *et al.*, 1991). The stainable iron is distributed in astrocytes, glia, microglia, and non-DA neurons (Jellinger *et al.*, 1990). However, analyses conducted by laser microprobe mass analysis (Good *et al.*, 1992) and X-ray microanalysis (Jellinger *et al.*, 1992) have revealed an accumulation associated directly with neuromelanin. So, a portion of the iron is localized within the melanized DA neurons.

Glutathione (GSH) in conjunction with GSH peroxidase normally detoxifies H_2O_2 and lipid peroxides within cells [Eqs. (8) and (9)]. GSH also reacts with and removes quinones derived from L-dopa. There is some evidence for decreased levels of GSH in the parkinsonian brain, which could expose DA neurons to the oxidant attack of peroxides. Sian *et al.* (1994) reported a mean 40% decline in GSH in the substantia nigra in 16 PD subjects, extending an earlier observation (Sofic *et al.*, 1991) based on only 4 PD subjects. The diminished GSH was not a secondary effect of nigral pathology as similar changes were not seen in Huntington's disease or progressive supranuclear palsy. However, the fact that lower GSH levels, as well as a change in the ratio of oxidized to reduced forms of glutathione, were seen in both PD and multiple system atrophy, both treated with L-dopa, has raised the question of whether these changes reflect drug therapy (Cohen, 1994).

$$H_2O_2 + 2GSH \xrightarrow{\text{GSH peroxidase}} GSSG + 2H_2O \tag{8}$$

$$ROOH + 2GSH \xrightarrow{\text{GSH peroxidase}} GSSG + ROH + H_2O \tag{9}$$

Direct oxidant damage to DNA (hydroxylation) has also been observed in the PD brain (Sanchez-Ramos *et al.*, 1994). In cell culture, DA neurons from rat brain are also susceptible to damage by environmental conditions or endogenous enzymatic activities: The number of surviving neurons can be increased two- to fourfold by adding antioxidants and/or simply lowering the oxygen tension (Colton *et al.*, 1995).

4. THEORIES ABOUT PARKINSON'S DISEASE

Theories about the etiology of PD and disease progression have centered on the research experience with the two neurotoxins, 6-OHDA and MPTP. One point of view is that oxy-radicals and peroxides, generated by natural metabolic pathways, such as MAO activity, promote a background of oxidative stress that spurs loss of DA neurons. This concept suggests a mechanism for disease progression, but does not explain etiology. A second point of view is that exposure to environmental toxins, perhaps related to MPTP and producing the same mitochondrial defect at the level of complex I, underlies both the development and progression of PD. This point of view is supported by the observation that PD is associated with a defect in mitochondrial Complex I (e.g., Schapira *et al.*, 1990; Mizuno *et al.*, 1995). Either theory may be affected by a genetic component that predisposes DA neurons to either oxidative stress or mitochondrial defects. In addition, the accumulation of iron must be considered. At the present time, iron fits better as a catalyst for oxidative events (the MAO hypothesis) than it does for damage by environmental toxins (the MPTP hypothesis).

4.1. The MAO Hypothesis

The MAO hypothesis (Cohen, 1983, 1986) flows directly from the observations that H_2O_2 plays an essential role in the neurotoxic mechanism for 6-OHDA. If H_2O_2 is essential, a very prominent route for its production is via the oxidative deamination of DA by monoamine oxidase [Eq. (10)]. DA is a substrate for both MAO-A and MAO-B. The levels of DA in DA nerve terminals are quite high: An average concentration of 50 mM (Anden *et al.*, 1966) reflects a cytosolic pool in the range of 0.5–2.0 mM (e.g., Corrodi and Jonsson, 1967) and the larger amounts present in the vesicular storage pool.

$$DA + O_2 + H_2O \xrightarrow{\text{MAO}} \text{aldehyde} + H_2O_2 + NH_3 \tag{10}$$

The surviving nigrostriatal neurons in PD show increased turnover of DA (Hornykiewicz and Kish, 1986). Similar observations have been made experimentally in rats after partial lesioning of nigrostriatal neurons by 6-OHDA, particularly when the lesion encompassed greater than 80% of DA neurons (Hefti *et al.*, 1980; Altar *et al.*, 1987). These observations are in keeping with the operation of a feedback regulatory loop between the striatum and the substantia nigra: As dopaminergic neurotransmission falls off as a result of nigral pathology, the system becomes disinhibited and increased firing rates, associated with increased release of DA, characterize the surviving DA neurons. A portion of released DA that is recaptured by the DA nerve terminal (uptake via the transporter) is oxidized by MAO. Some of the released DA is metabolized postsynaptically by catechol-*O*-methyltransferase and MAO. For each mole of DA or 3-*O*-methyl-DA oxidized by MAO, 1 mole of H_2O_2 is formed [Eq. (10)]. Considerable H_2O_2 can be generated as the 50 mM DA present within DA terminals is turned over both pre- and postsynaptically in the immediate vicinity of DA neurons. Therefore, an oxidant stress is directed at surviving DA neurons by enhanced MAO activity with DA as substrate; this effect is a natural consequence of the partial loss of DA neurons. In this sense, DA acts as an endogenous neurotoxin in PD.

4.2. The Environmental Toxin (MPTP-like) Hypothesis

The MPTP hypothesis is more self-evident. MPTP, or an agent like it, can gain access to brain and become transformed metabolically into a mitochondrial poison. There has been considerable interest in the possibility that exposure to herbicides or pesticides may promote parkinsonism (Tanner and Langston, 1990; Gorrell *et al.*, 1996). MPTP, via its metabolite MPP^+, poisons Complex I of the respiratory chain. The as yet unexplained presence of a mitochondrial defect in Complex I in PD (Mizuno *et al.*, 1989; Schapira *et al.*, 1990) may reflect exposure to environmental MPTP-like agents.

4.3. The Link between the MAO and MPTP Hypotheses

MAO is a constituent of the outer mitochondrial membrane, while Complex I and the other electron transport enzymes are present in the inner mitochondrial membrane.

An intermembrane space intervenes. Nonetheless, a new study (Cohen *et al.*, 1997) has demonstrated a linkage between MAO activity in the outer membrane and damage to electron transport at the inner membrane.

Exposure of rat brain mitochondria to 500 μM DA (or other monoamines) suppressed mitochondrial electron flow. The effect was significantly greater (56% inhibition) when the metabolic substrate was pyruvate, which initiates electron flow at Complex I, than with succinate (28% inhibition), which initiates electron flow at Complex II. Mitochondria were completely protected by MAO inhibitors. Mitochondrial damage was also reversed during electron flow. A probable explanation is that MAO-generated H_2O_2 oxidizes glutathione to glutathione disulfide [GSSG, Eq. (8)], which undergoes thiol-disulfide interchange with protein thiols (Pr-SH) to form protein mixed-disulfides [Pr-SSG, Eq. (11)] thereby interfering reversibly with thiol-dependent enzymatic function. In agreement with this interpretation, MAO activity induces mitochondrial loss of GSH (Sandri *et al.*, 1990), elevation in GSSG (Werner and Cohen, 1993), and accumulation of protein mixed-disulfides (Cohen *et al.*, 1997); these effects are also blocked by inhibition of MAO. Reversal during electron flow may be mediated by reduction of GSSG and Pr-SSG by reducing equivalents (NADPH, via transhydrogenase) generated during metabolism of pyruvate or succinate.

$$\text{GSSG} + \text{Pr-SH} \rightarrow \text{Pr-SSG} + \text{GSH} \tag{11}$$

It follows, therefore, that defects in mitochondrial respiration associated with PD may reflect, in part, the established increase in DA turnover. This means that the MAO and MPTP (mitochondrial poisoning) hypotheses are linked. The potential for mitochondrial damage may be enhanced when DA turnover is further sustained during chronic treatment with L-dopa (e.g., Przedborski *et al.*, 1993).

5. NEW DIRECTIONS IN PARKINSON RESEARCH

Several new areas related to oxidative stress and cell death are emerging in Parkinson research.

One report (Hunot *et al.*, 1997) described an increase in the proportion of DA neurons in the PD brain at autopsy with NF-κB translocated to the nucleus. NF-κB is a transcription factor that activates the expression of certain genes, including those responsible for several antioxidant enzymes, such as the biosynthesis of GSH and the mitochondrial (manganese) form of superoxide dismutase. Translocation of NF-κB can be caused by an oxidative stress. The implication is that NF-κB labeling of DA nuclei confirms the presence of an oxidative stress in PD. Translocation of NF-κB is also part of a signaling mechanism that induces programmed cell death (apoptosis).

A gene defect has also been described in a Parkinson-prone kindred of Italian descent (Polymeropoulos *et al.*, 1997). Although this is only one subgroup in the Parkinson spectrum, it offers an opportunity to identify factors that may be part of a common thread leading to the loss of nigrostriatal neurons. The gene defect in PD has been localized to a presynaptic protein, α-synuclein, which was previously implicated in Alzheimer's disease. An interesting aspect is that the gene "defect" in amino acid sequence in human subjects is the "normal" form of rodent synuclein. Therefore, it is an enigma that these animals do not spontaneously show signs of parkinsonism during aging. Nonetheless, this genetic breakthrough opens opportunities for uncovering new details about the pathogenesis of PD.

A third new direction is really an old one. Does L-dopa carry a liability as well as a therapeutic benefit? This question still needs resolution. The therapeutic benefits of L-dopa are obvious and can be contrasted with experimental paradigms where damage is evoked, as well as the ever-present concern that indices of oxidant damage are directly affected by chronic treatment with L-dopa. To this turbulent mix, two new factors are added. One concerns the widespread belief that mitochondrial defects promote cellular demise, including apoptotic cell death. If true, then L-dopa exhibits a potential for promoting mitochondrial damage either directly (Przedborski *et al.*, 1993; Werner *et al.*, 1994) or indirectly via the metabolism of DA (Cohen *et al.*, 1997). The second factor concerns another widespread belief that peroxynitrite, formed in a reaction between superoxide and nitric oxide [Eq. (12)], can promote cellular damage and cell death:

$$\cdot O_2^- + \cdot NO \rightarrow ONOO^- \tag{12}$$

Treatment with L-dopa elevates tissue DA levels, and tissue DA can scavenge superoxide; therefore, peroxynitrite formation will be impeded. Are these events relevant in either a negative way (mitochondrial damage) or a positive way (suppression of peroxynitrite) to the progression of PD?

Basic research on PD has opened vistas into the pathways of oxidative stress and their basic role in neurodegenerative processes. One continues to hope that the time is either now or soon at hand when this new information will open possibilities for clinical application to block the relentless and debilitating loss of DA neurons that characterizes PD. The DATATOP study (Parkinson Study Group, 1989a,b, 1993) was the opening sally into the realm of "antioxidant" therapy, an attempt to block disease progression at its roots. Although clinical success in the trial with deprenyl was modest and temporary (Parkinson Study Group, 1993; Olanow *et al.*, 1995), it has clearly opened a window into an approach that will certainly see further development and improved clinical success.

6. REFERENCES

Altar, C. A., Marien, M. R., and Marshall, J. F., 1987, Time course of adaptations in dopamine biosynthesis, metabolism, and release following nigrostriatal lesions: Implications for behavioral recovery from brain injury, *J. Neurochem.* **48**:390–399.

Ames, B. N., Shigenaga, M. K., and Hagen, T. M., 1993, Oxidants, antioxidants, and the degenerative diseases of aging, *Proc. Natl. Acad. Sci. USA* **90**:7915–7922.

Anden, N. E., Fuxe, K., Hamberger, B., and Hokfelt, T., 1966, A quantitative study of the nigro-neostriatal dopamine neuron system in the rat, *Acta Physiol. Scand.* **67**:306–312.

Basma, A. N., Morris, E. J., Nicklas, W. J., and Geller, H. M., 1995, L-DOPA cytotoxicity to PC12 cells in culture is via its autoxidation, *J. Neurochem.* **64**:825–832.

Beal, M. F., 1995, Aging and oxidative stress in neurodegenerative disease, *Ann. Neurol.* **38**:357–366.

Blunt, S. B., Jenner, P., and Marsden, C. D., 1993, Suppressive effect of L-DOPA on dopamine cells remaining in the ventral tegmental area of rats previously exposed to the neurotoxin 6-hydroxydopamine, *Move. Disord.* **8**:129–133.

Burns, R. S., Chiueh, C. C., Markey, S. P., Ebert, M. P., Jacobowitz, D. M., and Kopin, I. J., 1983, A primate model of parkinsonism: Selective destruction of dopaminergic neurons in the pars compacta of the substantia nigra by N-methyl-4-phenyl-1,2,3,6-tetrahydropyridine, *Proc. Natl. Acad. Sci. USA* **80:**4546–4550.

Carstam, R., Brinck, C., Hindemith-Augustsson, H., Rorsman, H., and Rosengren, E., 1992, The neuromelanin of the human substantia nigra, *Biochim. Biophys. Acta* **1097:**152–160.

Chiba, K., Trevor, A., and Castagnoli, N., Jr., 1984, Metabolism of the neurotoxic tertiary amine, MPTP, by brain monoamine oxidase, *Biochem. Biophys. Res. Commun.* **120:**574–578.

Chiueh, C. C., Krishna, G., Tulsi, P., Obata, T., Lang, K., Huang, S. J., and Murphy, D. L., 1992, Intracranial microdialysis of salicylic acid to detect hydroxyl radical generation through dopamine autoxidation in the caudate nucleus: Effects of MPP^+, *Free Radical Biol. Med.* **13:**581–583.

Cleeter, M. W. J., Cooper, J. M., and Schapira, A. H. V., 1992, Irreversible inhibition of mitochondrial complex I by 1-methyl-4-phenylpyridinium: Evidence for free radical involvement, *J. Neurochem.* **58:**786–789.

Cohen, G., 1983, The pathobiology of Parkinson's disease: Biochemical aspects of dopamine neuron senescence, *J. Neural Transm. Suppl.* **19:**89–103.

Cohen, G., 1986, Monoamine oxidase, hydrogen peroxide, and Parkinson's disease, *Adv. Neurol.* **45:**119–125.

Cohen, G., 1994, Editorial: The brain on fire? *Ann. Neurol.* **36:**333–334.

Cohen, G., and Heikkila, R. E., 1974, The generation of hydrogen peroxide, superoxide radical, and hydroxyl radical by 6-hydroxydopamine, dialuric acid, and related cytotoxic agents, *J. Biol. Chem.* **249:**2447–2452.

Cohen, G., and Heikkila, R. E., 1977, In vivo scavenging of superoxide radicals by catecholamines, in *Superoxide and Superoxide Dismutases* (M. Michelson, J. M. McCord, and I. Fridovich, eds.), pp. 351–365, Academic Press, New York.

Cohen, G., and Werner, P., 1994, Free radicals, oxidative stress, and neurodegeneration, in *Neurodegenerative Disorders* (D. Calne, ed.), pp. 139–162, Academic Press, New York.

Cohen, G., Heikkila, R. E., Allis, B., Cabbat, F., Dembiec, D., MacNamee, D., Mytilineou, C., and Winston, B., 1976, Destruction of sympathetic nerve terminals by 6-hydroxydopamine: Protection by 1-phenyl-3-(2-thiazolyl)-2-thiourea, diethyldithiocarbamate, methimazole, cysteamine, ethanol and n-butanol, *J. Pharmacol. Exp. Ther.* **199:**336–352.

Cohen, G., Pasik, P., Cohen, B., Leist, A., Mytilineou, C., and Yahr, M. D., 1984, Pargyline and deprenyl prevent the neurotoxicity of 1-methyl-4-phenyl-1,2,3,6-tetrahydropyridine (MPTP) in monkeys, *Eur. J. Pharmacol.* **106:**209–210.

Cohen, G., Farooqui, R., and Kesler, N., 1997, Parkinson's disease: A new link between monoamine oxidase and mitochondrial electron flow, *Proc. Natl. Acad. Sci. USA* **94:**4890–4894.

Colton, C. A., Pagan, F., Snell, J., Colton, J. S., Cummins, A., and Gilbert, D. L., 1995, Protection from oxidation enhances the survival of cultured mesencephalic neurons, *Exp. Neurol.* **132:**54–61.

Corrodi, H., and Jonsson, G., 1967, The formaldehyde fluorescence method for the histochemical demonstration of biogenic amines. A review on the methodology, *J. Histochem. Cytochem.* **15:**65–78.

Coyle, J. T., and Puttfarcken, P., 1993, Oxidative stress, glutamate, and neurodegenerative disorders, *Science* **262:**689–695.

Davis, G. C., Williams, A. C., Markey, S. P., Ebert, M. H., Caine, E. D., Reichert, C. M., and Kopin, I. J., 1979, Chronic parkinsonism secondary to intravenous injection of meperidine analogues, *Psychiatry Res.* **1:**249–254.

Dexter, D. T., Carter, C. J., Wells, F. R., Javoy-Agid, F., Agid, Y., Lees, A., Jenner, P., and Marsden, C. D., 1989, Basal lipid peroxidation in substantia nigra is increased in Parkinson's disease, *J. Neurochem.* **52:**381–389.

Dexter, D. T., Holley, A. E., Flitter, W. D., Slater, T. F., Wells, F. R., Daniel, S. E., Lees, A. J., Jenner, P., and Marsden, C. D., 1994, Increased levels of lipid hydroperoxides in the parkinsonian substantia nigra: An HPLC and ESR study, *Move. Disord.* **9:**92–97.

Evans, J. M., and Cohen, G., 1989, Studies on the formation of 6-hydroxydopamine in mouse brain after administration of 6-hydroxydopa, *J. Neurochem.* **52:**1461–1467.

Evans, J. M., and Cohen, G., 1993, Catecholamine uptake inhibitors elevate 6-hydroxydopamine in brain after administration of 6-hydroxydopa, *Eur. J. Pharmacol.* **232:**241–245.

Fahn, S., and Cohen, G., 1992, The oxidant stress hypothesis in Parkinson's disease: Evidence supporting it, *Ann. Neurol.* **32:**804–812.

Good, P. F., Olanow, C. W., and Perl, D. P., 1992, Neuromelanin-containing neurons of the substantia nigra accumulate iron and aluminum in Parkinson's disease: A LAMMA study, *Brain Res.* **593:**343–346.

Gorrell, J. M., DiMonte, D., and Graham, D., 1996, The role of the environment in Parkinson's disease, *Environ. Health Perspect.* **104:**652–654.

Graham, D. G., 1978, Oxidative pathways for catecholamines in the genesis of neuromelanin and cytotoxic quinones, *Mol. Pharmacol.* **14:**633–643.

Graham, D. G., Tiffany, S. M., Bell, W. R., and Gutknecht, W. F., 1978, Autoxidation versus covalent binding of quinones as the mechanism of toxicity of dopamine, 6-hydroxydopamine, and related compounds toward C-1300 neuroblastoma cells in vitro, *Mol. Pharmacol.* **14:**644–653.

Hallman, H., Lange, J., Olson, L., Stromberg, I., and Jonsson, G., 1985, Neurochemical and histochemical characterization of the effects of 1-methyl-4-phenyl-1,2,3,6-tetrahydropyridine on brain catecholamine neurons in the mouse, *J. Neurochem.* **44:**117–127.

Han, S.-K., Mytilineou, C., and Cohen, G., 1996, L-DOPA up-regulates glutathione and protects mesencephalic cultures against oxidative stress, *J. Neurochem.* **66:**501–520.

Hefti, F., Melamed, E., and Wurtman, R. J., 1980, Partial lesions of the dopaminergic nigrostriatal system: Biochemical characterization, *Brain Res.* **195:**123–137.

Heikkila, R. E., and Cohen, G., 1971, Inhibition of biogenic amine uptake by hydrogen peroxide: A mechanism for toxic effects of 6-hydroxydopamine, *Science* **172:**1257–1258.

Heikkila, R. E., and Cohen, G., 1972, Further studies on the generation of hydrogen peroxide by 6-hydroxydopamine: Potentiation by ascorbic acid, *Mol. Pharmacol.* **8:**241–248.

Heikkila, R. E., and Cohen, G., 1973, 6-Hydroxydopamine: Evidence for superoxide radical as an oxidative intermediate, *Science* **181:**456–457.

Heikkila, R. E., Mytilineou, C., Cote, L. J., and Cohen, G., 1973, Evidence for degeneration of sympathetic nerve terminals caused by the ortho and para-quinones of 6-hydroxydopamine, *J. Neurochem.* **20:**1345–1350.

Heikkila, R. E., Hess, A., and Duvoisin, R., 1984, Dopaminergic neurotoxicity of 1-methyl-4-phenyl-1,2,3,6-tetrahydropyridine in mice, *Science* **224:**1451–1453.

Hornykiewicz, O., and Kish, S. J., 1986, Biochemical pathophysiology of Parkinson's disease, *Adv. Neurol.* **45:**19–34.

Hunot, S., Brugg, B., Ricard, D., Michel, P. P., Muriel, M.-P., Ruerg, M., Faucheux, B. A., Agid, Y., and Hirsch, E. C., 1997, Nuclear translocation of NF-kappaB is increased in dopaminergic neurons of patients with Parkinson's disease, *Proc. Natl. Acad. Sci. USA* **94:**7531–7533.

Javitch, J. A., D'Amato, R. J., Strittmatter, S. M., and Snyder, S. H., 1985, Parkinsonism-inducing neurotoxin, N-methyl-4-phenyl-1,2,3,6-tetrahydropyridine: Uptake of the metabolite N-methyl-4-phenylpyridine by dopamine neurons explains selective toxicity, *Proc. Natl. Acad. Sci. USA* **82:**2173–2177.

Jellinger, K., Paulus, W., Grundke-Iqbal, I., Riederer, P., and Youdim, M. B. H., 1990, Brain iron and ferritin in Parkinson's and Alzheimer's diseases, *J. Neural Transm. Park. Dis. Dement. Sect.* **2:**327–340.

Jellinger, K., Kienzl, E., Rumpelmair, G., Riederer, P., Ben-Shachar, D., and Youdim, M. B. H., 1992, Iron–melanin complex in substantia nigra of parkinsonian brains: An X-ray microanalysis, *J. Neurochem.* **59:**1168–1171.

Jenner, P., 1991, Oxidative stress as a cause of Parkinson's disease, *Acta Neurol. Scand.* **84**(Suppl. 136)**:**6–15.

Jonsson, G., 1976, Studies on the mechanisms of 6-hydroxydopamine cytotoxicity, *Med. Biol.* **54:**406–420.

Kostrzewa, R. M., and Jacobowitz, D. M., 1974, Pharmacological actions of 6-hydroxydopamine, *Pharmacol. Rev.* **26:**199–288.

Langston, J. W., Ballard, P., Tetrud, J. W., and Irwin, I., 1983, Chronic parkinsonism in humans due to a product of meperidine-analog synthesis, *Science* **219:**979–980.

Langston, J. W., Irwin, I., Langston, E. B., and Forno, L., 1984, Pargyline prevents MPTP-induced parkinsonism in primates, *Science* **225:**1480–1482.

Lozoncy, M. F., Davidson, M., and Davis, K. F., 1987, The dopamine hypothesis of schizophrenia, in *Psychopharmacology: A Second Generation of Progress* (H. Y. Meltzer, ed.), pp. 715–726, Raven Press, New York.

Mena, M. A., Pardo, B., Paino, C. L., and de Yebenes, J. G., 1993, Levodopa toxicity in foetal rat midbrain neurones in culture: Modulation by ascorbic acid, *NeuroReport* **4:**438–440.

Mizuno, Y., Ohta, S., Tanaka, M., Takamiya, S., Suzuki, K., Sato, T., Oya, H., Ozawa, T., and Kagawa, Y., 1989, Deficiencies in complex I subunits of the respiratory chain in Parkinson's disease, *Biochem. Biophys. Res. Commun.* **163:**1450–1455.

Mizuno, Y., Ikebi, S. I., Hattori, N., Nakagawa-Hattori, Y., Mochizuki, H., Tanaka, M., and Ozawa, T., 1995, Role of mitochondria in the etiology and pathogenesis of Parkinson's disease, *Biochim. Biophys. Acta* **1271**:265–274.

Mytilineou, C., and Cohen, G., 1984, 1-Methyl-4-phenyl-1,2,3,6-tetrahydropyridine destroys dopamine neurons in explants of rat embryo mesencephalon, *Science* **225**:529–531.

Mytilineou, C., Han, S.-K., and Cohen, G., 1993, Toxic and protective effects of L-dopa on mesencephalic cell cultures, *J. Neurochem.* **61**:1470–1478.

Nicklas, W. J., Vyas, I., and Heikkila, R. E., 1985, Inhibition of NADH-linked oxidation in brain mitochondria by MPP^+, a metabolite of the neurotoxin, MPTP, *Life Sci.* **36**:2503–2508.

Olanow, C. W., Cohen, G., Perl, D. P., and Marsden, C. D., (eds.), 1992, Role of Iron and Oxidant Stress in the Normal and Parkinsonian Brain, *Ann. Neurol.* **32**(Suppl.).

Olanow, C. W., Hauser, R. A., Gauger, L., Malapira, T., Koller, W., Hubble, J., Bushenbark, K., Lilienfeld, D., and Esterlitz, J., 1995, The effect of deprenyl and levodopa on the progression of Parkinson's disease, *Ann. Neurol.* **38**:771–777.

Olney, J. W., Zorumski, C. F., Stewart, G. R., Price, M. T., Wang, G., and Labruyere, J., 1990, Excitotoxicity of L-DOPA and 6-OH-DOPA: Implications for Parkinson's and Huntington's diseases, *Exp. Neurol.* **108**:269–272.

Parkinson Study Group, 1989a, DATATOP: A multicenter controlled clinical trial in early Parkinson's disease, *Arch. Neurol.* **46**:1052–1060.

Parkinson Study Group, 1989b, Effect of deprenyl on the progression of disability in early Parkinson's disease, *N. Engl. J. Med.* **321**:1364–1371.

Parkinson Study Group, 1993, Effects of tocopherol and deprenyl on the progression of disability in early Parkinson's disease, *N. Engl. J. Med.* **328**:183.

Polymeropoulos, M. H., Levadan, C., Leroy, E., Ide, S. E., Dehejia, A., Dutra, A., Pike, B., Root, H., Rubenstein, J., Boyer, R., Stenroos, E. S., Chandrasekhaprappa, S., Athanassiadou, A., Papetropoulos, T., Johnson, W. G., Lazzarini, A. M., Duvoisin, R. C., Di Iorio, G., Golbe, L. I., and Nussbaum, R. L., 1997, Mutation in the alpha-synuclein gene identified in families with Parkinson's disease, *Science* **276**:2045–2047.

Przedborski, S., Kostic, V., Jackson-Lewis, V., Naini, A. B., Simonetti, S., Fahn, S., Carlson, E., Epstein, C. J., and Cadet, J. L., 1992, Transgenic mice with increased Cu/Zn-superoxide dismutase activity are resistant to 1-methyl-4-phenyl-1,2,3,6-tetrahydropyridine-induced neurotoxicity, *J. Neurosci.* **12**:1658–1661.

Przedborski, S., Jackson-Lewis, V., Muthane, U., Jiang, H., Ferreira, M., Naini, A. B., and Fahn, S., 1993, Chronic levodopa administration alters cerebral mitochondrial respiratory chain activity, *Ann. Neurol.* **34**:715–723.

Ramsay, R. R., Salach, J. I., Dadgar, J., and Singer, T. P., 1986, Inhibition of mitochondrial NADH dehydrogenase by pyridine derivatives and its possible relationship to experimental and idiopathic Parkinsonism, *Biochem. Biophys. Res. Commun.* **135**:259–275.

Sachs, C., Jonsson, G., Heikkila, R. E., and Cohen, G., 1975, Control of the neurotoxicity of 6-hydroxydopamine by intraneuronal noradrenaline in rat iris, *Acta Physiol. Scand.* **93**:345–351.

Sanchez-Ramos, J. R., Overvik, E., and Ames, B. N., 1994, A marker of oxyradical-mediated DNA damage (8-hydroxy-2′deoxyguanosine) is increased in nigro-striatum of parkinson's disease brain, *Neurodegeneration* **3**:197–204.

Sandri, G., Panfili, E., and Ernster, L., 1990, Hydrogen peroxide production by monoamine oxidase in isolated rat-brain mitochondria: Its effect on glutathione levels and Ca^{2+} efflux, *Biochim. Biophys. Acta Gen. Subj.* **1035**:300–305.

Schapira, A. H. V., Cooper, J. M., Dexter, D., Clark, J. B., Jenner, P., and Marsden, C. D., 1990, Mitochondrial complex I deficiency in Parkinson's disease, *J. Neurochem.* **54**:823–827.

Sian, J., Dexter, D. T., Lees, A. J., Daniel, S., Agid, Y., Javoy-Agid, F., Jenner, P., and Marsden, C. D., 1994, Alterations in glutathione levels in Parkinson's disease and other neurodegenerative disorders affecting the basal ganglia, *Ann. Neurol.* **36**:348–355.

Sofic, E., Paulus, W., Jellinger, K., Riederer, P., and Youdim, M. B. H., 1991, Selective increase of iron in substantia nigra zona compacta of parkinsonian brains, *J. Neurochem.* **56**:978–982.

Steece-Collier, K., Collier, T. J., Sladek, C. D., and Sladek, J. R., Jr., 1990, Chronic levodopa impairs morphological development of grafted embryonic dopamine neurons, *Exp. Neurol.* **110**:201–208.

Tanner, C., and Langston, J. W., 1990, Do environmental toxins cause Parkinson's disease? A critical review, *Neurology* **40**(Suppl. 3):17–30.

Werner, P., and Cohen, G., 1993, Glutathione disulfide (GSSG) as a marker of oxidative injury to brain mitochondria, *Ann. N.Y. Acad. Sci.* **679:**364–369.

Werner, P., Mytilineou, C., Cohen, G., and Yahr, M. D., 1994, Impaired oxidation of pyruvate in human embryonic fibroblasts after exposure to L-dopa, *Eur. J. Pharmacol.* **263:**157–162.

Yoritaka, A., Hattori, N., Uchida, K., Tanaka, M., Stadtman, E. R., and Mizuno, Y., 1996, Immunohistochemical detection of 4-hydroxynonenal protein adducts in Parkinson disease, *Proc. Natl. Acad. Sci. USA* **93:**2696–2701.

Zoccarato, F., Cavallini, L., Deana, R., and Alexandre, A., 1988, Pathways of hydrogen peroxide generation in guinea pig cortex mitochondria, *Biochem. Biophys. Res. Commun.* **154:**727–734.

Chapter 25

Alzheimer's β-Amyloid Peptide and Free Radical Oxidative Stress

D. Allan Butterfield

1. INTRODUCTION

Given the increasing mean age of the populace in the United States, especially as the Baby Boomer generation enters the next century, aging and age-related neurodegenerative disorders, including Alzheimer's disease (AD), will pose significant public health questions for Americans. Currently, there are an estimated 5 million Americans with AD, and that number is expected to rise to nearly 14 million in the next century. AD is a disorder affecting the entire family, not just the patient. The human suffering imposed by a disease that steals the essence of what makes us human—the ability to think, reason, remember—is motivation enough for researchers to know more of AD. But there are also financial considerations that impact Americans: Because health care costs associated with AD are already extremely high, the situation will only worsen in the future. Consequently, it is imperative that more is known of the molecular basis of AD, and with that knowledge, that new therapeutic approaches to AD be developed.

In the past few years, a plethora of new studies relevant to AD have been reported (some reviews include Markesbery, 1997; Butterfield, 1997, 1996; Butterfield and Stadtman, 1997; Butterfield *et al.*, 1996a; Hensley *et al.*, 1996a; Selkoe, 1996, 1994; Katzman and Saitoh, 1991), and a consensus is emerging that oxidative stress may play an important role in the pathogenesis of AD. As is outlined in more detail below, β-amyloid (Aβ), the central constituent of one of the hallmark pathological lesions of AD

D. Allan Butterfield Department of Chemistry, Center of Membrane Sciences, and Sanders-Brown Center on Aging, University of Kentucky, Lexington, Kentucky 40506-0055.

Reactive Oxygen Species in Biological Systems, edited by Gilbert and Colton. Kluwer Academic / Plenum Publishers, New York, 1999.

[senile plaques (SP)], was shown to be associated with oxygen-dependent free radicals, and these free radicals are a significant source of oxidative stress to neuronal and glial membrane systems. A model for neuronal cell loss in AD brain based on Aβ-associated free radical damage to membrane systems was developed (Butterfield, 1997; Butterfield *et al.*, 1996a,b, 1994a; Hensley *et al.*, 1996a,1994b), and it is chiefly this research on Aβ-related free radical oxidative stress and its relevance to AD that is reviewed here. Because of constraints of space, studies of all possible relevance to oxidative stress may not be examined, but many of those involving Aβ will be.

Other sources of oxidative stress of relevance to AD include: advanced glycation end products (AGE) (M. A. Smith *et al.*, 1995; Yan *et al.*, 1994); these AGEs are formed in complex reactions thought to include free radical reactions and are associated primarily with the skeletal protein tau, a major constituent of a second hallmark pathological lesion in AD [neurofibrillary tangles (NFT)], but also associated with SP (M. A. Smith *et al.*, 1995, 1994); decreased levels of antioxidant enzyme systems (Pappolla *et al.*, 1992); activated microglia (Colton *et al.*, 1994); mitochondrial metabolic impairment with consequent electron leakiness (Nutisya *et al.*, 1994; Parker *et al.*, 1994; Beal, 1992); activation of NMDA receptors by the excitatory neurotransmitter, glutamate, leading to accumulation of superoxide radical anion, hydrogen peroxide, and peroxynitrite (Estevez *et al.*, 1995; Mattson *et al.*, 1995a; Lafon-Cazal *et al.*, 1993); redox-active metal ions, such as iron and copper, that can catalyze formation of hydroxyl radicals (Multhaup *et al.*, 1996; Hensley *et al.*, 1994a; Markesbery *et al.*, 1994); nitric oxide, formed by nitric oxide synthetase and capable of reaction with $\cdot O_2^-$ to form highly reactive peroxynitrite (Estevez *et al.*, 1995); and possibly 4-hydroxy-2-nonenal (4-HNE) a lipid peroxidation product that can modify protein structure and function (Mark *et al.*, 1997; Subramaniam *et al.*, 1997; Esterbauer *et al.*, 1991). There is reason to suspect that Aβ may have influence on or a role to play in the oxidative stress on neuronal and glial systems generated by these species, and as such, the Aβ orientation of this review may have even more importance relative to free radical oxidative stress in AD.

2. HOW DO FREE RADICALS REACT AND LEAD TO MEMBRANE DYSFUNCTION?

2.1. Lipid Bilayer-Resident Free Radicals or Their Breakdown Products Can Bind to or Cross-Link Membrane-Bound Proteins

Free radicals are molecular species, usually transient and usually highly reactive, with one or more unpaired electrons (Butterfield, 1982). This definition precludes transition metal atoms or ions, many of which have unpaired electrons. Molecules with unpaired electrons, free radicals, especially oxygen free radicals, are especially reactive. Oxyradicals, particularly hydroxyl or peroxyl free radicals, are easily able to abstract a hydrogen atom from carbon atoms on carbon–carbon double bonds, leading to carbon-centered free radicals. Such radicals then nearly instantaneously can add molecular oxygen, which contains two unpaired electrons, to form peroxyl radicals, and the cycle can be repeated. In biological membranes, oxygen, being nonpolar, is soluble in the hydrocarbon core of lipid bilayers. Hence, if free radicals are formed by some moiety in the lipid bilayer of neuronal or glial membranes, lipid radicalization can quickly ensue. Such lipid radicals can lead to lipid peroxidation products, such as 4-hydroxy-*trans*-2-nonenal (4-HNE) (Esterbauer *et al.*, 1991), or lead to the formation of free fatty acids (Koppal *et al.*, 1998; Prasad *et al.*, 1994). In either case, the composition and function of neuronal lipid bilayers can be compromised by free radical attack. In addition, lipid free radicals can abstract hydrogen atoms from adjacent proteins, leading to free radicals on proteins thereby potentially modifying the structure, function, and interactions of membrane proteins. Further, 4-HNE is capable of binding to membrane proteins via Michael addition, leading to modified proteins, and can, in appropriate cases, further react via its terminal aldehydic functionality with lysine NH_2 groups to form Schiff bases, i.e., two proteins can be cross-linked to 4-HNE (recently reviewed in Butterfield and Stadtman, 1997).

2.2. Some Methods for Assessing Free Radical Oxidation-Induced Alterations in the Physical State of Neuronal and Glial Membranes

There are several ways to monitor alterations in the physical and biochemical states of neuronal membranes caused by free radicals. Stadtman (1992) has reviewed the formation of protein carbonyl functionalities when proteins are oxidized by free radical oxidative stress. Although the precise mechanisms for carbonyl formation are still unclear, the increased levels of protein carbonyls are invariably linked to oxidative stress (Stadtman, 1992), and in aging and neurodegenerative disorders, protein carbonyl levels are reportedly increased (Aksenova *et al.*, 1998; Butterfield *et al.*, 1997a; Hensley *et al.*, 1995b; Carney *et al.*, 1991; C. D. Smith *et al.*, 1991). Protein carbonyls are reacted with amines, such as 2,4-dinitrophenylhydrazine (DNPH), to form Schiff bases, and these adducts are quantified by UV spectroscopy (Butterfield and Stadtman, 1997; Stadtman, 1992). Carbonyl levels also can be monitored by a new, highly sensitive histofluorescence method in which a derivative of biotin hydrazide is reacted with protein carbonyls to form the Schiff base, which is subsequently reacted with fluorescein isothiocyanate-linked streptavidin (Harris *et al.*, 1996; 1995a). The resulting fluorescence, and hence the protein carbonyl content, is quantified by confocal fluorescence microscopy.

The highly sensitive techniques of electron paramagnetic resonance (EPR) spin labeling in conjunction with lipid- and protein-specific spin labels, are excellent means of detecting structural alterations in brain membranes following free radical oxidative stress (Butterfield *et al.*, 1996a,b, 1994a; Howard *et al.*, 1996; Hall *et al.*, 1995 a–c; Hensley *et al.*, 1995a,b, 1994a; Butterfield, 1986, 1982). Spin labeling is an EPR technique in which a paramagnetic spin label is added to the system of interest (Butterfield, 1982). In most biological systems, the spin label is the only source of paramagnetism, and therefore, the only source of the EPR spectrum. Spin labels are generally of the nitroxide type, i.e., three-electron, two-atom centers in molecules containing various functionality groups that direct the spin label to specific sites in the membrane. The most commonly used protein-specific spin label is MAL-6 (2,2,6,6-tetramethyl-4-maleimidopiperidin-1-oxyl), which covalently binds to SH groups on membrane proteins (Butterfield, 1982). We have extensively characterized the EPR spectrum of MAL-6 in erythrocyte and synaptosomal membranes (Hensley *et al.*, 1994a; Umhauer *et al.*, 1992; Butterfield, 1982). Based on antibody and selective protein isolation studies, MAL-6 binds nearly exclusively to cytoskeletal proteins in erythrocyte membranes, particularly spectrin; in synaptosomal membranes, cytoskeletal proteins and some transmembrane protein bind MAL-6 (Umhauer *et al.*, 1992). This protein-specific spin label yields an EPR spectrum in both membrane systems that reflects at least two kinds of SH-binding sites: (1) those in which the spin label is relatively free to rotate, and, therefore, gives a narrower resonance line (designated W sites) and (2) those in which the spin label is motionally restricted (e.g., in a narrow cleft), and, therefore, gives a broad line (designated S sites) (Figure 1). The analysis of the spectrum is performed by measurement of the W/S ratio (Figure 1), which is highly sensitive to conformation of membrane proteins (reviewed in Butterfield, 1990, 1986, 1982).

Structural changes in membranes after interaction with oxidizing systems, i.e., differences in protein–protein interactions, can be assessed by MAL-6 EPR spectra. Changes in the W/S ratio are known to be strong indicators of perturbations in the normal interactions of cytoskeletal proteins (Butterfield, 1990, 1986, 1982). An increase or decrease in the W/S ratio is conceptualized as arising from protein structural changes that decrease or increase, respectively, the steric hindrance to segmental motion in the region of protein to which the spin label is attached (Figure 1). For example, covalent or electrostatic cross-linking of cytoskeletal proteins leads to a decrease in the W/S ratio, whereas proteolysis or physical disruption of the membrane increases the W/S ratio (Hensley *et al.*, 1993; Butterfield, 1990).

Much data has been accumulated to suggest that oxidation of synaptosomal and erythrocyte membranes produces protein structural changes that hinder spin label motion and decrease the W/S ratio. Oxidative stress [caused *in vitro* by addition of free-radical-generating compounds (Bellary *et al.*, 1995; Hensley *et al.*, 1994a; Trad and Butterfield, 1994) or *in vivo* by hyperoxia (Howard *et al.*, 1996; Hensley *et al.*, 1995a), glutathione depletion (Hall *et al.*, 1997), or natural aging (Butterfield *et al.*, 1997a; Hensley *et al.*, 1995a)] leads to a marked decrease in the W/S ratio of MAL-6-labeled synaptosomes, indicating that this EPR parameter of this protein-specific spin label is a sensitive biomarker of synaptosomal membrane protein oxidative damage. In diet-restricted, aged animals, less oxidative stress occurs (Aksenova *et al.*, 1998; Gabbita *et al.*, 1998, 1997).

In contrast to the isotropic motion exhibited by MAL-6, lipid bilayer-specific stearic acid spin labels (Figure 2) exhibit anisotropic motion related to their preferential alignment in the lipid bilayer with the carboxylic acid portion of the spin label near the polar head groups of the lipids and the acyl chain directed to the hydrophobic portion of the bilayer (Butterfield, 1986, 1982). These lipid-specific spin probes have been employed to investigate the structure and function of biological membranes and interactions of pharmacological agents with these membranes (Butterfield, 1995, 1982; Butterfield and Rangachari, 1991). The paramagnetic nitroxide groups can be placed anywhere along the carbon chain of stearic acid so as to probe different depths of the lipid bilayer (Butterfield, 1982). 5-Nitroxide stearate (5-NS), used to investigate the lipid bilayer near the lipid–water interface, and 12-nitroxide stearate (12-NS), employed to study the region of the lipid bilayer near the most common sites of lipid unsaturation deep within the bilayer, are two common lipid-specific spin labels (Figure 2). Free radical reduction of the paramagnetism of the nitroxide groups of 5-NS and 12-NS, located at different depths of the lipid bilayer, can be used to determine the regional vulnerability of the lipid bilayer to free radical oxidative stress. By monitoring the EPR signal peak heights as a function of time, one can directly determine the kinetics of interaction of free radicals with the spin labels. Only by interaction with the free radicals will the spin label's paramagnetic center be

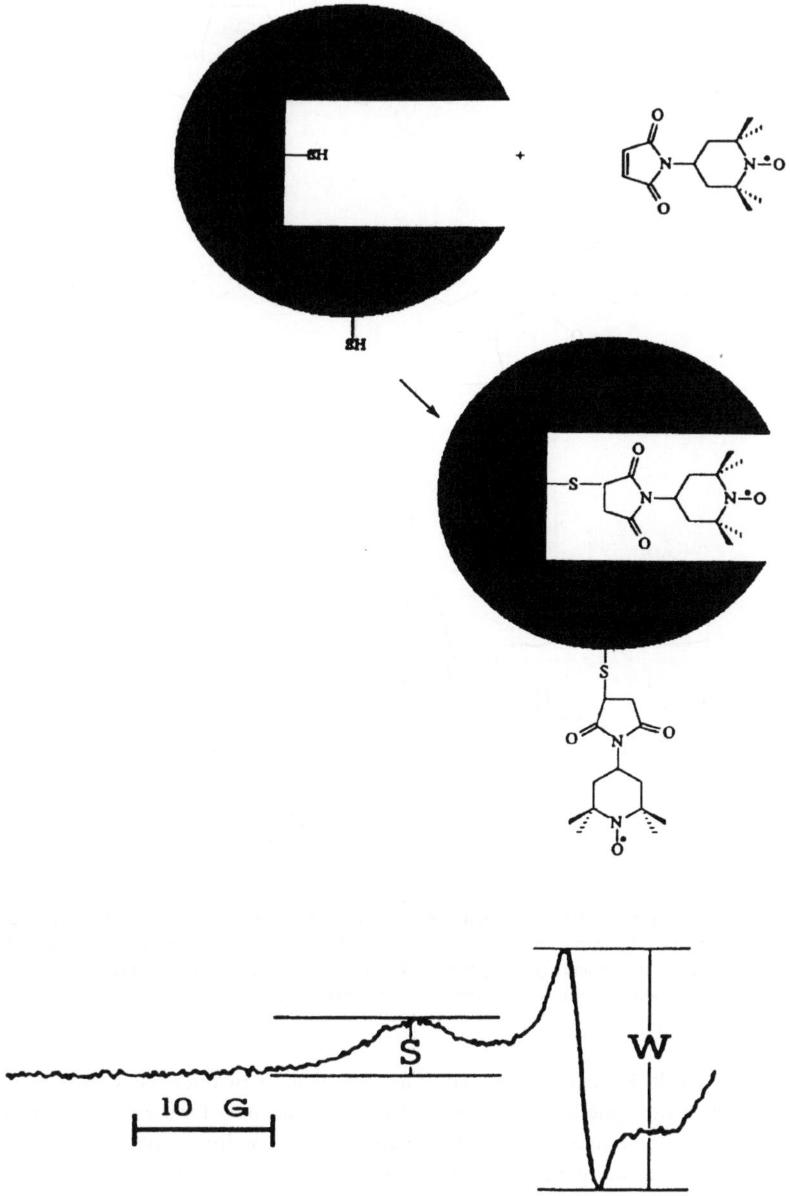

FIGURE 1. (Top) Reaction of MAL-6 with membrane proteins. Strongly (S) and weakly (W) immobilized reaction sites of MAL-6 are represented as being in a cleft that restricts the motion of the spin label and on the surface of the protein, respectively. (Bottom) A typical EPR spectrum of the low-field $M_I = +1$ resonance lines of MAL-6 covalently bound to cortical synaptosomal membrane proteins. W and S EPR signal heights are indicated.

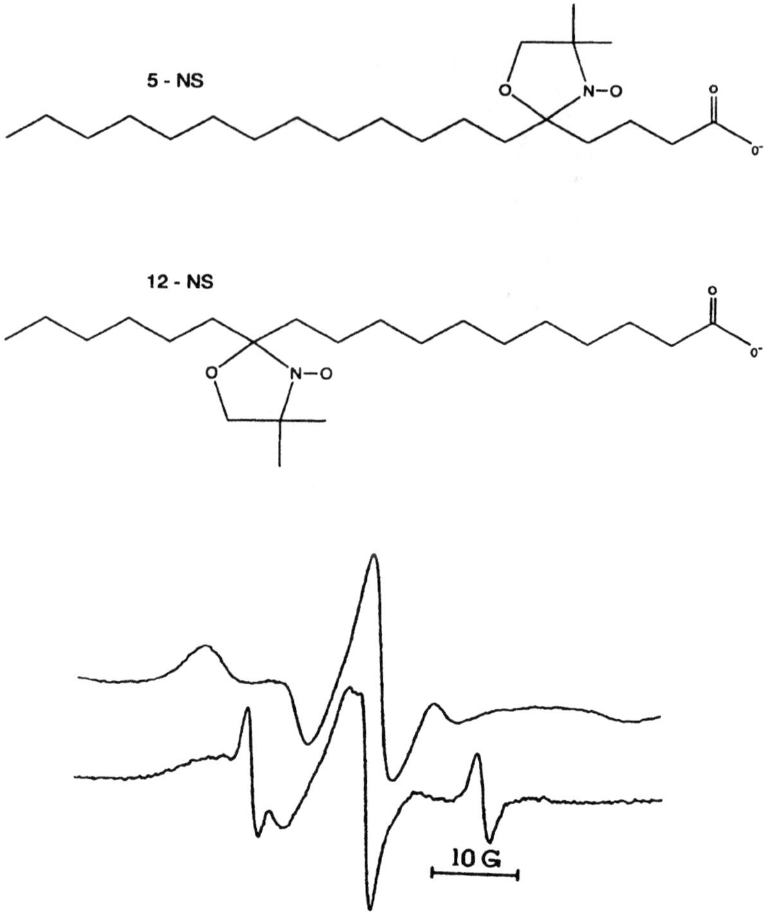

FIGURE 2. (Top) Structure of the lipid-specific spin labels, 5- and 12-nitroxyl stearates (NS). The region of the lipid bilayer near the lipid/water interface is probed by 5-NS, while a region deep within the lipid bilayer, near sites of lipid unsaturation and high oxygen solubility, is probed by 12-NS. (Bottom) EPR spectra of 5-NS (upper) and 12-NS (lower) in cortical synaptosomal membranes.

reduced and thus its intensity decreased. If no change in linewidth occurs [thereby ruling out dipolar or paramagnetic relaxation mechanisms to account for the decrease in signal intensity of the membrane-bound EPR spectrum on interaction of synaptosomal membranes with free radical generators (Butterfield, 1982)], peak heights serve well as an index of signal intensity.

As Figure 2 shows, the spectrum of 5-NS incorporated into synaptosomal membranes consists purely of an anisotropic component; the $M_I = +1$, 0, and -1 lines have no overlapping peaks or shoulders reflective of unbound 5-NS because the spin label is completely transferred to the membrane during the labeling procedure. However, as we have observed with other membrane systems (Cader *et al.*, 1995; Butterfield, 1982), 12-NS consistently partitions between the lipid (membrane) and aqueous (extrasynaptosomal) phases, and one can resolve in the 12-NS spectrum both a "bound" and a "free" population (Figure 2). The bound population predominates in untreated 12-NS-labeled synaptosomes, with the three sharp lines of the free population providing only a minor contribution to the total integrated spectral envelope.

As an example of the use of EPR spin labeling in free radical modification, hydroxyl free radical formation via Fenton chemistry was shown to lead to compromised structural integrity of cortical synaptosomal membranes (Hensley *et al.*, 1994a). In addition, the structure of both the lipid bilayer and membrane proteins of cortical synaptosomal membranes is altered in ischemia/reperfusion-associated free radical damage (Hall *et al.*, 1997, 1995a–c), and this damage can be prevented by preinjection of the brain-accessible free radical scavenger and spin trap, α-phenyl-*tert*-butylnitrone (PBN) (Hall *et al.*, 1995b). Hyperoxia has been considered a good model of aging (Stadtman, 1992), and membrane changes in aging are hypothesized to be a consequence of free radical damage (Harman, 1994; Stadtman, 1992). Both aging and hyperoxia led to an altered physical state of membrane proteins in cortical synaptosomes (Butterfield *et al.*, 1997a; Howard *et al.*, 1996; Hensley *et al.*, 1995a).

3. Aβ-ASSOCIATED FREE RADICAL OXIDATIVE STRESS: A MODEL FOR NEUROTOXICITY IN AD BRAIN

With the research from our laboratory and those of others detailed below, a model for neuronal death in AD brain based on Aβ-associated free radical oxidative stress has been developed (Butterfield, 1997a; Butterfield, *et al.*, 1996a,b, 1994a; Hensley *et al.*, 1996a, 1994b).

In the model (Figure 3), Aβ, produced on the external side of neuronal membranes from the transmembrane amyloid precursor protein (APP) in ways not completely understood but thought to involve a thiol protease (β-secretase) or other proteases, inserts into the plasma membrane of neuronal or glial cells, forming oxygen-dependent, reactive free radicals. These free radicals induce lipid peroxidation, forming reactive lipid radicals and lipid by-products, e.g., 4-HNE, both of which in turn lead to protein oxidation and/or modification, and subsequent disruption of membrane function, including inhibition of ion-motive ATPases, loss of Ca^{2+} homeostasis, inhibition of glial cell Na^{+}-dependent glutamate uptake system with consequences on neuronal excitatory NMDA receptors, and disruption of signaling pathways. Ultimately, these membrane dysfunctions lead to death of the neuron. One significant advancement of this Aβ-associated free radical model of AD neurotoxicity is its unification of the AD literature into a self-consistent theoretical framework. This model is consistent with the myriad of reports citing altered membrane enzymes, transport proteins, structural and cytoskeletal proteins, lipids, and so forth in AD (reviewed in Selkoe, 1994, 1991; Corain *et al.*, 1993; Katzman and Saitoh, 1991; Butterfield, 1986). A "shower" of Aβ free radicals (shrapnel) interacting with proximal membrane moieties would account for many of these membrane alterations. In addition, this Aβ free radical shrapnel model is congruous with the age dependence of AD: Younger persons have greater antioxidant capacity (C. D. Smith *et al.*, 1992, 1991; Carney *et al.*,

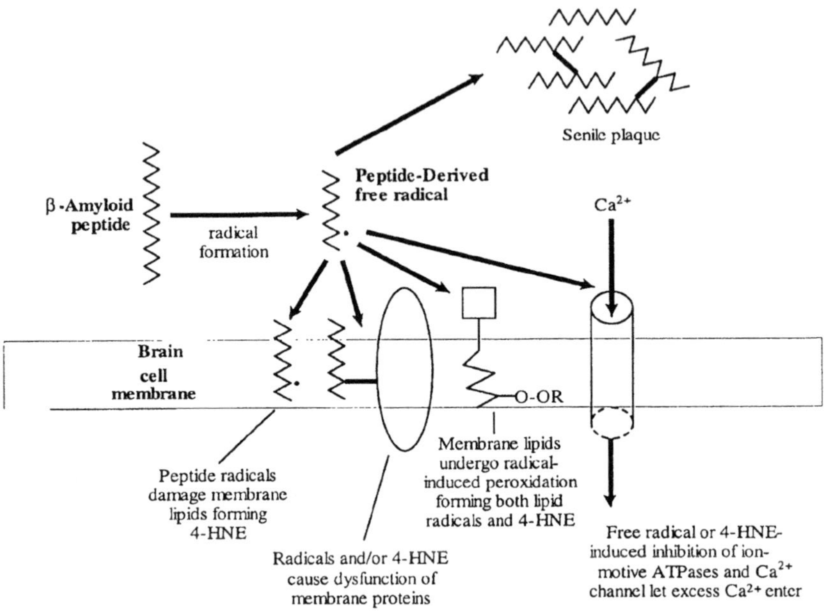

FIGURE 3. Aβ-associated free radical model of Alzheimer's disease. Aβ forms free radicals that can interact with various moieties in the membrane. Aβ-associated free radicals can directly cause membrane protein dysfunction. Or 4-hydroxyl-2-nonenal (4-HNE), formed subsequent to Aβ-induced lipid peroxidation, can lead to altered membrane protein function. Note that this model is consistent with reports of multiple membrane alterations of enzymes, proteins, transport proteins, and so forth, as well as lipid and protein oxidation and calcium ion accumulation in neurons and subsequent cell death in AD. See text.

1991; Starke-Reed and Oliver, 1989) and thus can withstand amyloid-induced free radical oxidative stress; aging, possibly combined with environmental insults that diminish antioxidant capacity, could partially account for membrane alterations reported in AD. Genetic factors contributing either to decreased antioxidant status or to altered binding to chaperon proteins, such as apoE4 (Soto *et al.*, 1996; Corder *et al.*, 1993), might also predispose AD persons to these processes. Moreover, this model for Aβ free radical-based aggregation and neurotoxicity in AD offers a molecular rationale for possible therapeutic strategies in this dementing disorder involving appropriate brain-accessible free radical scavengers. In our laboratory, this possibility is under active investigation, and has been encouraged by our recent success in preventing free radical damage to brain synaptosomal membranes in stroke, hyperoxia, and accelerated aging (Butterfield *et al.*, 1997a; Howard *et al*, 1996; Hall *et al.*, 1995b). Further, high-dose vitamin E treatment is reported to be beneficial in AD (Sano *et al.*, 1997). In addition, our model may provide perspicacity into prion protein disorders, also known to involve amyloidlike plaques and neurodegeneration (Prusiner, 1992): It is conceivable that mechanisms similar to those described in our AD research may be applicable to prion disorders as well.

The remainder of this review provides evidence of support for the Aβ-associated free radical model of AD neurotoxicity and demonstration that the predictions of the model have been realized.

4. ARE FREE RADICALS ASSOCIATED WITH Aβ?

4.1. Spin Trapping

The most direct way of detecting transient, reactive free radicals is by the magnetic resonance technique of EPR spin trapping. In EPR spin trapping studies, a nonparamagnetic species [usually a nitrone (the trap)] reacts with a transient free radical (the spin) to produce a stable paramagnetic nitroxide (the spin adduct) (Figure 4) (Butterfield, 1982; Janzen, 1980). In the case of PBN as the trap, the reaction of an oxygen- or carbon-centered free radical adduct normally produces a six-line EPR spectrum (Figure 4). The magnitude of the doublet splitting (A_H) depends on several factors including the size of

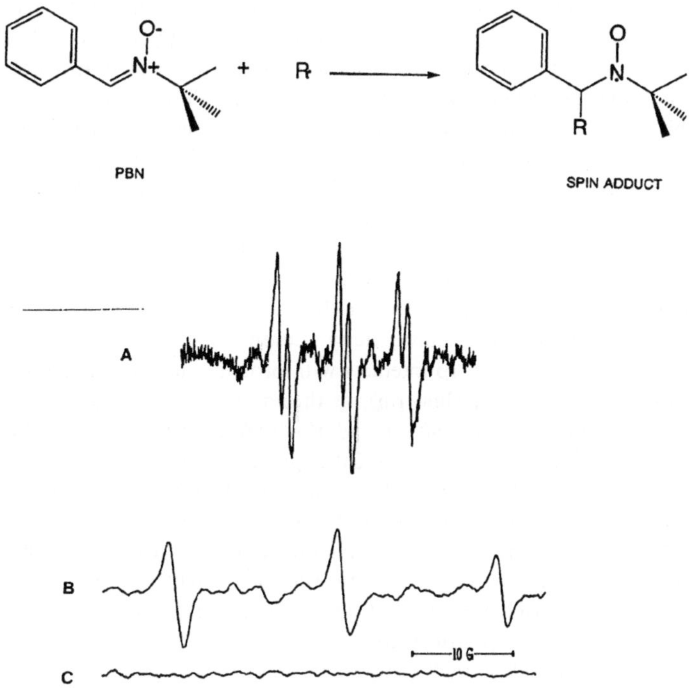

FIGURE 4. (Top) Reaction between PBN and a transient free radical R· to form a nitroxide spin adduct. (A) EPR spectrum of the hydroxyl free radical adduct with PBN. Note thc six lines of this spectrum. (B) EPR spectrum of Aβ(1–40) reacted with PBN. Note that only three lines are present in this spectrum. (C) EPR spectrum of PBN alone. Note that no signal is observed, indicating that the EPR signal in B originates from the peptide.

the adduct (Butterfield, 1982; Janzen, 1980). If on binding to PBN a radical is able to cause decomposition of the trap, then three-line (a nitroxide) or four-line (a hydronitroxide) EPR spectra can result. After ascertaining that the PBN itself is unable to generate a free radical under the conditions of the experiment, an EPR spectrum is prima facie evidence for the presence of a free radical. Researchers have used EPR spin trapping to investigate free radical involvement in metabolism, toxicology, and pharmacology (Clerici *et al.*, 1996; Harris *et al.*, 1995a,b; Hensley *et al.*, 1995c–e, 1994a,b; Chan *et al.*, 1994; Butterfield, 1982; Janzen, 1980). EPR is an extremely sensitive technique, rivaling the detection limits of fluorescence, and EPR spectra can be easily quantified by double integration techniques (Butterfield, 1982; Janzen, 1980). The power of EPR over optical methods such as fluorescence involves its extreme sensitivity, the fact that opaque samples can be used (i.e., no light scattering effects), the insight into the polarity and motion of the local microenvironment near the paramagnetic center of the free radical that can be gained, and, generally, the biological system is EPR silent, i.e., only the free radical trapped by PBN gives rise to an EPR spectrum.

4.2. Amyloid

Amyloid, hydrophobic peptides of 39–43 amino acids in length and the main component of SP, are derived from the transmembrane glycoprotein APP. Genetic considerations implicate Aβ in the pathogenesis of AD. Mutations in APP correlate with some familial forms of AD (Selkoe, 1996, 1994), and APP-overexpressing, transgenic mice exhibit some brain pathology reminiscent of AD (Games *et al.*, 1995). Mutations in presenilin-1 and presenilin-2, thought to be involved with APP processing, lead to early-onset AD, and person with Down's syndrome invariably develop AD after sufficient time (Selkoe, 1996). In contrast to fresh, nontoxic Aβ, Aβ aged over a period of hours or days spontaneously aggregates to form large, neurotoxic, fibrillar entities that fail to dissociate in the presence of denaturants (Pike *et al.*, 1993; Burdick *et al.*, 1992). These fibrillar aggregates are toxic to neurons, and this has led to the claim that aggregated Aβ is an *absolute* requirement for neurotoxicity (Lorenzo and Yankner, 1994). However, recent studies have shown that Aβ interacts with proteins such as glutamine synthetase (GS), apolipoprotein J (apo J, clusterin), or thrombin, to yield *increased Aβ-induced hippocampal neurotoxicity in the absence of fibrils* (Aksenov *et al.*, 1996, 1995; Oda *et al.*, 1995; Smith-Swintosky *et al.*, 1995). Other proteins can inhibit fibril formation, but not affect the toxicity of Aβ(1–40) (Aksenova *et al.*, 1996). These results, though not inconsistent with the neurotoxic properties of fibrillar Aβ, are inconsistent with the claim of an *absolute* requirement for fibril formation before Aβ toxicity can be displayed (Lorenzo and Yankner, 1994). These results are consistent with the notion that SPs in AD brain form in a protein-rich environment, and different proteins may interact with and modulate the properties of Aβ differently.

In contrast to the usual "aging" period for full-length Aβ, certain peptide sequences within Aβ, especially an 11-amino acid subset, Aβ(25–35), exhibit rapid aggregation and high neurotoxicity soon after dissolution (Pike *et al.*, 1993; Burdick *et al.*, 1992; Yankner *et al.*, 1990). Kinetics studies of aggregation suggest a nucleation event dependent on the hydrophobic carboxyl-terminal residues (Jarrett and Lansbury, 1993; Jarrett *et al.*, 1992; Tomski and Murphy, 1992), and in hippocampal neuronal cultures, Aβ tends to accumulate near the membrane surface, producing local microenvironments of high Aβ concentration (Pike *et al.*, 1993; Mattson *et al.*, 1993). Aβ can disrupt Ca^{2+} homeostasis in a variety of ways and is known to result in high intracellular Ca^{2+} accumulation in neuronal cultures (Mattson *et al.*, 1993). Free radical antioxidants can moderate many of the Aβ-induced cell-damaging effects (Bruce *et al.*, 1996; Goodman and Mattson, 1996, 1994a,b; Goodman *et al.*, 1996, 1994; Harris *et al.*, 1996, 1995a,b; Kelly *et al.*, 1996; Puttfarcken *et al.*, 1996; Tomiyama *et al.*, 1996, 1994; Manelli and Puttfarcken, 1995; Schorderet, 1995; Smith-Swintosky *et al.*, 1995; Behl *et al.*, 1994; Kumar *et al.*, 1994). Consequently, research has tended to focus on the role of Aβ peptides as mediators and initiators of oxyradical stress-induced cell damage (Bush *et al.*, 1998; Daniels *et al.*, 1998; Gridley *et al.*, 1998; Koppal *et al.*, 1998; Subramaniam *et al.*, 1998, 1995; Butterfield, 1997, 1996; Butterfield *et al.*, 1996a,b, 1994a; Harris *et al.*, 1996, 1995a,b; Hensley *et al.*, 1996, 1995b–e, 1994b; Tomiyama *et al.*, 1996; Mark *et al.*, 1995; Mattson *et al.*, 1995a).

4.3. Spin Trapping Studies of Aβ Peptides

The hypothesis that Aβ could be a source of free radical damage in cultured neuronal systems, and by extension, in AD brain was investigated (Butterfield, 1997, 1996; Butterfield *et al.*, 1996a–c, 1994a; Harris *et al.*, 1996, 1995a,b; Hensley *et al.*, 1996, 1995b–e, 1994b; Tomiyama *et al.*, 1996; Mark *et al.*, 1995; Mattson *et al.*, 1995a; Subramaniam *et al.*, 1995). The EPR technique of spin trapping was used to detect transient Aβ-derived free radicals (Harris *et al.*, 1996, 1995a,b; Tomiyama *et al.*, 1996; Hensley *et al.*, 1995c–e, 1994b; Subramaniam *et al.*, 1995). Other methods were used to assess Aβ-derived free radicals including: nitroxide spin labels, whose paramagnetism, and, hence, EPR signal intensity is lost on reaction with a free radical (Koppal *et al.*, 1998; Butterfield *et al.*, 1996b, 1994a); salicylate, which participates in electrophilic addition reactions with oxyradicals to produce 2,5- and 2,3-dihydroxybenzoic acids (Hensley *et al.*, 1994b); sensitive fluorescence or colorimetric probes that indicate protein oxidation or the presence of ROS (Harris *et al.*, 1996, 1995a); ROS-specific enzymes, namely, superoxide dismutase (SOD) or catalase, to help identify the radical type (Harris *et al.*, 1995b); and brain-resident oxidatively sensitive biomolecules, such as GS or creatine kinase (CK), as markers of ROS presence (Hensley *et al.*, 1995b, 1994b).

Synthetic Aβ(1–42), Aβ(1–40), its highly reactive and rapidly toxic 11-mer, Aβ(25–35), or the C-terminus fragment Aβ(32–35), on reaction with the spin trap PBN in metal-chelated, oxygenated buffers, lead to EPR-detectable nitroxide spin adducts that are not observed in the absence of either peptide or oxygen (Butterfield *et al.*, 1998; Harris *et al.*, 1995a; Hensley *et al.*, 1995c–e, 1994b; Subramaniam *et al.*, 1995) (Figure 4), showing that the peptide is likely the source of the radicals generated. Although all components in this mixture were pure as judged by HPLC and NMR, it is possible that trace amounts of non-chelated metals and/or PBN impurities contribute to these EPR spectra. The appearance of an oxygen-dependent three-line EPR spectrum in each case (Harris *et al.*, 1995a,b; Subramaniam *et al.*, 1995; Hensley *et al.*, 1994b) coincides with the reported temporal requirements for neurotoxicity of these two species, i.e., hours to days for Aβ(1–40) and minutes to hours for Aβ(25–35). Sometimes, four-line EPR spectra of Aβ–PBN spin adducts were observed (see below). Our spin trapping results were recently confirmed in another laboratory (Tomiyama *et al.*, 1996).

The normally expected six-line EPR spectra of PBN spin adducts of Aβ (Butterfield, 1982; Janzen, 1980) were not observed, suggesting that Aβ-peptide-derived free radicals were either highly reactive or unusual. Alkylnitroxides or alkoxylnitroxides yield three-line EPR spectra whereas hydronitroxides yield four-line EPR spectra. Should highly reactive Aβ-derived free radicals cause decomposition of the PBN spin trap leading to such products, three- or four-line EPR spectra would be expected. To test this hypothesis, [^{13}C]-PBN, with the labeled carbon in the α-position, was used (Hensley *et al.*, 1995d). If Aβ-derived free radicals did decompose the PBN spin adduct, there would be no ^{13}C–N hyperfine coupling, i.e., there would still remain three- or four-line EPR spectra. This is exactly what was observed (Hensley *et al.*, 1995d), indicating that the Aβ-derived free radicals caused decomposition of the spin traps used. The four-line generating Aβ(1–40) or Aβ(25–35) peptide species were invariably less toxic than the respective three-line-generating species (Hensley *et al.*, 1995c). The four-line-generating species was confirmed to be a hydronitroxide based on use of D_2O, with subsequent deuterium–hydrogen exchange resulting in the predicted nine-line EPR spectrum (Hensley *et al.*, 1995d). The hyperfine coupling constants also matched those of known hydronitroxide, further confirming the identity of the four-line-generating species. That different EPR spectra resulted from variably toxic Aβ samples is consistent with the known neurotoxic variability of different lots of commercial amyloid (Simmons *et al.*, 1994; May *et al.*, 1992). Others have shown recently than Aβ can interact with apolipoprotein E differently, depending on the solution conformation adopted by Aβ (Soto *et al.*, 1996). Circular dichroism studies suggested that the structure of Aβ peptides in oxygenated buffers [required for free radical production (Hensley *et al.*, 1994b)] was different than Aβ peptide structure in deoxygenated buffers (which prevent free radical formation) (Harris *et al.*, 1995b).

Evidence to support the notion that decomposition of the Aβ–PBN spin adduct of oxygen-dependent, Aβ-derived free radicals as discussed above is consistent with the formation of peroxyl-type free radicals was obtained (Butterfield *et al.*, 1996b; Harris *et al.*, 1995b). SOD did not affect the Aβ–PBN EPR spectrum, whereas the EPR signal was completely abolished in the presence of catalase (Harris *et al.*, 1995b). Peroxyl-type free radicals derived from Aβ were confirmed by use of a sensitive colorimetric assay specific for peroxyl species (Butterfield *et al.*, 1996b). Other laboratories have confirmed subsequently that catalase will mitigate against Aβ-induced free radical damage to cells (Puttfarcken *et al.*, 1996; Manelli and Puttfarcken, 1995), consistent with our observations (Butterfield *et al.*, 1996b; Harris *et al.*, 1995b).

Bush and co-workers (1998) showed radicals associated with the peptide using other methods. These researchers showed that Aβ(1–42) could reduce Cu^{2+} to Cu^{+}. That is, the electron originated from the peptide to reduce the metal ion, confirming by a different method what we and others have shown by EPR spin trapping (Butterfield, 1997; Tomiyama *et al.*, 1996). The reduced Cu^{+} was shown to lead to H_2O_2 production, i.e., oxidative stress. These authors pointed out that such reactions are not operable with the shorter peptide, Aβ(25–35). In that case, the C-terminal methionine residue is important in oxidative stress associated with this peptide (see below). Methionine may also be important in the full-length Aβ, as shown by high-field solution-phase NMR studies in which Met-35 of the Aβ(1–42) monomer residues over Phe-19 (Zagorski *et al.*, 1998), providing potential stabilization for sulfur-based radicals.

4.4. Other Biomarkers of Aβ Free Radical Generation in Solution

Other biomarkers of free radicals to confirm that Aβ generated free radical oxidative stress were employed. For example, Aβ-derived free radical ROS were trapped by salicylate to yield dihydroxybenzoic acids (Hensley *et al.*, 1994b), consistent with the EPR results. Rodent brain GS and CK, both oxidatively sensitive enzymes (Smith *et al.*, 1992; Oliver *et al.*, 1987), whose activity is decreased in AD brain (Hensley *et al.*, 1995b), were inactivated by Aβ incubation (Harris *et al.*, 1995a; Hensley *et al.*, 1994b). Aβ(1–40) interaction with purified GS inhibited the enzyme activity and introduced carbonyls into GS in a cell-free system in a process that could be blocked by a free radical scavenger (Aksenov *et al.*, 1997), consisting with the notion that Aβ is a prooxidant. Using EPR and a SH-specific spin label (MTS), real-time kinetics of covalent incorporation of MTS into GS were investigated (Butterfield *et al.*, 1997b). Upon oxidation of GS by Fe^{2+}/H_2O_2, a three-fold decreased rate of uptake of the spin label by GS was observed. Aβ(1–40) addition to purified GS from sheep brain showed a nearly 90% reduction in MTS incorporation kinetics. Further, GS isolated and purified from human control and AD brain showed a similar 2–3-fold decreased MTS uptake rate in AD GS compared to that of control GS, suggesting that this enzyme is oxidized in AD brain (Butterfield *et al.*, 1997b) and consistent with the lower activity of GS in AD brain (Hensley *et al.*, 1995b) (see below). ROS were also detected in neurons and astrocytes by oxidatively sensitive fluorescent dyes (Harris *et al.*, 1996, 1995a).

4.5. Importance of Methionine in Free Radical Production by Aβ

The precise chemical mechanism(s) involved in Aβ-derived free radical ROS in oxygenated buffers is not yet known and is under active investigation in our laboratory. Based on amino acid analysis of neurotoxic, free-radical-generating Aβ incubates (Hensley *et al.*, 1995c; Subramaniam *et al.*, 1995), the C-terminal methionine (residue 35) is converted to a species that coelutes on HPLC with methionine sulfoxide (Figure 5) (Hensley *et al.*, 1995c; Subramaniam *et al.*, 1995). This conversion occurs in the absence of redox-active metals and is completely prevented by millimolar levels of the free radical scavenger PBN (Hensley *et al.*, 1995c). Methionine, like other dialkyl sulfides, is known to participate in unusual free radical reaction chemistry (Schoneich *et al.*, 1994). In addition, oxidation of methionine residues in model peptides is known to significantly alter secondary structure (Dado and Gellman, 1994); namely, Met oxidation to the sulfoxide leads to predominately β-sheet conformation, which is the conformation adopted by toxic Aβ (Selkoe, 1994). Consistent with the importance of Met-35 in the ability of Aβ peptides to generate reactive free radicals, we found in preliminary experiments that Aβ(25–34), which lacks C-terminal methionine, yielded no EPR spectrum when incubated with PBN. Initial attempts to model the structure of Aβ(1–40), using the CHARM molecular modeling program, suggested that Met-35 was exposed to the outside of the peptide, where it could be readily available for reaction or alignment with adjacent Aβ peptides. In preliminary experiments, Aβ(25–35)-Met-sulfoxide, in which the methionine residue was already oxidized, gave no EPR spectrum with PBN, nor was it toxic to GS (*vide infra*).

The research discussed above suggests that Aβ free radicals and neurotoxicity appear to be related and Met-35 may be important in this relationship. In agreement with this

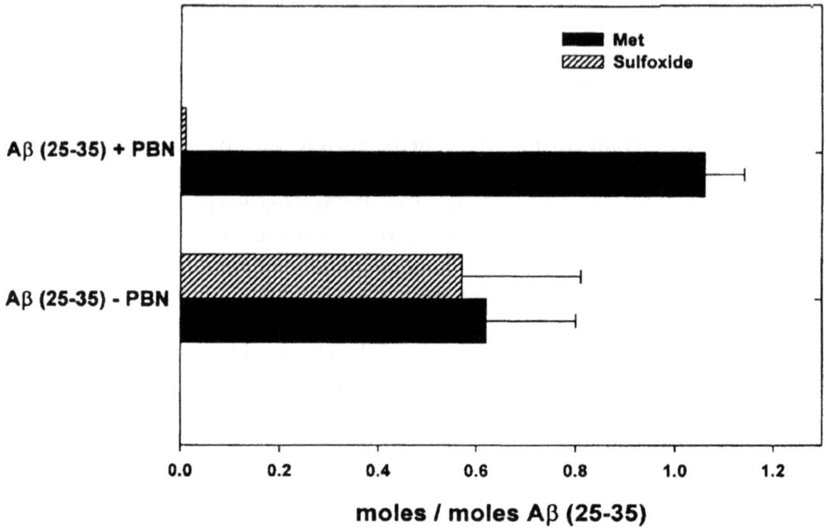

FIGURE 5. Formation of a species that coelutes with Met-sulfoxide on HPLC traces following amino acid hydrolysis. The free radical scavenger and spin trap PBN prevents formation of this species.

notion, Pike *et al.* (1995) reported that the C-terminal region of Aβ(25–35) was critical in its neurotoxicological properties and that modifications of the 33–35 region of the amyloid peptide led to a loss of peptide aggregation. Also in agreement with the potential importance of Met in Aβ chemistry and pathology, Snyder *et al.* (1995) reported that synthetic Aβ(1–40) containing Met-sulfoxide in residue 35 formed fibrils at twice the rate of unmodified Aβ(1–40) (Snyder *et al.*, 1995), and Naslund *et al.* (1994) found that SP Aβ(1–40) in AD brain is rich in Met-sulfoxide. Substitution of methionine by norleucine in Aβ(1–42) is reported to modulate the toxicity of the peptide, suggesting that methionine was important to the neurotoxic properties of Aβ (Manelli and Puttfarcken, 1995).

Based on these findings, including those of Pike *et al.* (1995), who, as noted above, showed that the C-terminal end of Aβ(25–35) was important in the neurotoxic properties of this amyloid peptide, we wondered what was the shortest C-terminal fragment of Aβ(25–35) that was capable of generating free radicals. No free radical PBN spin adduct was observable with Aβ(33–35) nor was this tripeptide toxic to GS. However, Aβ(32–35) (sequence Ile-Gly-Leu-Met) was found to form free radicals as judged by EPR studies with PBN spin trapping (Subramaniam *et al.*, 1995). Aβ(32–35), after 24 h incubation with PBN, yielded a three-line EPR spectrum, similar to those of Aβ(1–40) and Aβ(25–35), but requiring a longer time period for detection (Subramaniam *et al.*, 1995). The iron ion chelator, deferoxamine, was not effective in preventing this EPR signal, analogous to other Aβ peptides. Also similar to Aβ(25–35), catalase prevented the PBN spin adduct, consistent with a peroxyl radical (Subramaniam *et al.*, 1995). Neurotoxicity toward hippocampal neurons was demonstrated by Aβ(32–35), like Aβ(1–40) and Aβ(25–35), and this tetrapeptide inhibited GS activity, which was partially protected by the free radical antioxidant, vitamin E (Subramaniam *et al.*, 1995). HPLC amino acid analysis showed that Aβ(32–35), like Aβ(25–35), yielded a peak that coelutes with methionine sulfoxide and loss of the methionine peak. Aβ(32–35) with methionine-35 substituted by norleucine (same length, same hydrophobicity, but no S atom) showed no EPR signal with PBN and no toxicity to GS or hippocampal neurons (*vide infra*). Periodically, some commercial samples of Aβ(25–35) did not produce three-line EPR spectra with PBN; such samples were invariably nontoxic to GS, similar to the variable toxicity of other Aβ peptides reported previously (Hensley *et al.*, 1995c; Simmons *et al.*, 1994; May *et al.*, 1992).

We conclude that Aβ-derived free radical chemistry is oxygen-dependent and can be modulated by free radical scavengers, consistent with the shrapnel model noted above, and that oxyradical chemistry may be significant in the role played by Aβ in the AD brain. Based on the oxygen dependency of Aβ-peptide free radical generation, the apparent peroxyl nature of these free radicals, and putative involvement of methionine in this process, we formulated the following tentative mechanism by which these findings might be synthesized (Figure 6). This paradigm, which must be emphasized is tentative and subject to later revision as more is learned, follows the known chemistry of methionine-containing model peptide systems (Schoneich *et al.*, 1994), a reactivity that might be encouraged by adjacent aligned hydrophobic peptides in the lipid bilayer of neuronal and glial membranes. Such an alignment is consistent with the molecular modeling studies cited above and supported by the results of Mason *et al.* (1996), who reported from

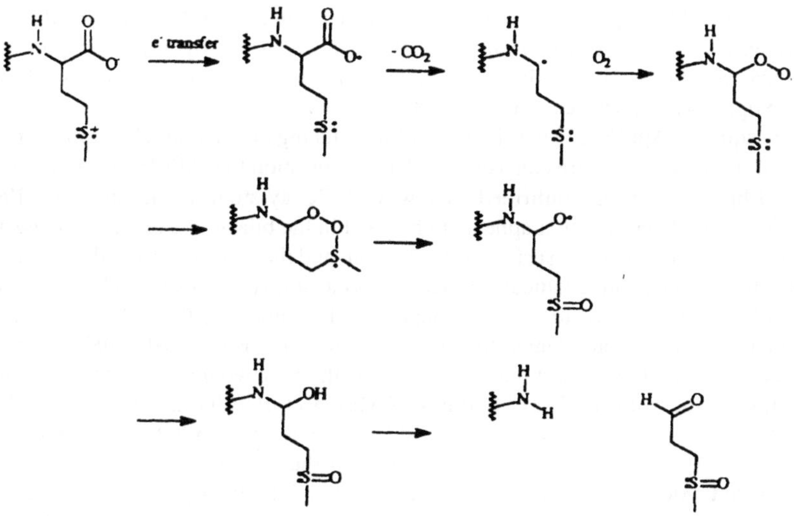

FIGURE 6. Tentative mechanism, based on known methionine free radical chemistry, to account for the involvement of methionine in the generation of an Aβ-derived free radical and to account for both a peroxyl radical and formation of a sulfoxide. See text.

low-angle X-ray findings that Aβ(25–35) inserts into lipid bilayers with the methionine terminus deep in the bilayer, near common sites of lipid unsaturation, and consequent free radical reactivity. Following known methionine chemistry, various electron transfer reactions and rearrangements of Aβ peptides can lead to both a sulfoxide and a peroxyl radical, characteristics of the Aβ free radical reactions discussed above (Figure 6). Explicit tests of this tentative mechanism are in progress.

5. PREDICTIONS OF AND EVIDENCE FOR THE Aβ FREE RADICAL SHRAPNEL MODEL OF AD

5.1. Aβ-Induced Lipid Oxidation

A key prediction of the Aβ-associated free radical model for neurotoxicity in AD is that Aβ induces lipid peroxidation. To test this prediction, Aβ(25–35) was added to gerbil neocortical synaptosomal membranes that had been previously spin labeled with the lipid-specific spin label derivatives of stearic acid, namely, 5-NS or 12-NS (Butterfield *et al.*, 1994a), which differ only in the location of the paramagnetic center of the spin probe (Butterfield, 1986, 1982). As an Aβ-associated reactive free radical, or the lipid radical that is formed from it, encounters the paramagnetic center of the lipid-specific spin label, the paramagnetism of the spin label is lost and the EPR signal intensity decreases. Hence, by following the kinetics of the loss of the EPR signal intensity, one can determine if free radical oxidation of lipids has occurred. Aβ(25–35), consistent with the model, led to a free radical-mediated, 60% reduction of the 12-NS signal (Butterfield *et al.*, 1994a). The paramagnetic nitroxide moiety of 12-NS is located deep in the lipid bilayer, near the most common sites of unsaturation, i.e., near the sites of lipid radical formation. In addition, nonpolar oxygen, which was required for Aβ free radical generation (Hensley *et al.*, 1994b), is highly soluble deep in the hydrophobic portion of biological membranes. Interestingly, the paramagnetic nitroxide moiety of 5-NS, located near the lipid–water interface, was essentially unaffected by Aβ(25–35), and the nontoxic reverse sequence Aβ(35–25) was ineffective in reducing either spin label (Butterfield *et al.*, 1994a). These results are consistent with the suggestion that Aβ(25–35) is soluble in the lipid bilayer, a result confirmed by low-angle X-ray studies (Mason *et al.*, 1996). Aβ-induced lipid peroxidation appears to be focal in the bilayer, i.e., more damage deep within the bilayer than at the surface of the bilayer, and the results support the notion that the structure of the peptide radical is related to its capability to cause membrane damage, i.e., Aβ(35–25) is nontoxic to cortical synaptosomal membranes. These Aβ-induced lipid peroxidation data are consistent both with the prediction of Kang *et al.* (1987), who used molecular modeling to suggest that C-terminal methionine would reside deep within the lipid bilayer, and with the X-ray findings of Mason *et al.* (1996), who confirmed this suggestion. CD spectra, discussed above, showing a correlation between structure and reactivity of Aβ peptides also are consistent with the lipid peroxidation studies. The lipophilic antioxidant, vitamin E, was able to greatly inhibit the Aβ-induced reduction of the 12-NS signal, consistent with a free radical process (Koppal *et al.*, 1998). The gene product for the protooncogene Bcl-2 is thought to have antioxidant properties (Hockenbery *et al.*, 1993). In PC-12 cells overexpressing Bcl-2, plasma membrane and mitochondrial membrane lipid peroxidation was prevented following Aβ(25–35) addition as assessed by the EPR stearic acid spin label reduction studies and the thiobarbituric acid-reactive substance (TBARS) assay (Bruce-Keller *et al.*, 1998), consistent with the above results. In addition to EPR studies, free fatty acid release from cortical synaptosomal membranes, another marker of lipid peroxidation (Prasad *et al.*, 1994), was found following Aβ(25–35) addition, and this Aβ-induced process could be blocked by the free radical antioxidant, vitamin E (Koppal *et al.*, 1998).

Another measure of cortical synaptosomal membrane lipid peroxidation caused by Aβ(25–35) is an increase over control levels of a lipid peroxidation product, conjugated dienes. Consistent with Aβ(25–35)-induced lipid peroxidation monitored by 12-NS spin labeling studies (Butterfield *et al.*, 1994a), an approximately 70% elevation in conjugated dienes over controls was found ($p < 0.02$, $N = 3$). Aβ-induced lipid free radicals and lipid oxidation products, such as 4-HNE, are some means by which protein modification with subsequent dysfunction of enzymes, transport proteins, and so forth can occur (Figure 2). Consistent with this idea, 4-HNE significantly altered the physical state of cortical synaptosomal membrane proteins and lipids in similar ways as Aβ (Subramaniam *et al.*, 1998, 1997), and like Aβ (Mark *et al.*, 1995), 4-HNE alters ion-motive ATPases and induces other synaptosomal membrane abnormalities (Mark *et al.*, 1997), which likely result from 4-HNE-induced modification of protein conformation (Subramaniam, *et al.*, 1997).

5.2. Aβ-Induced Free Radical Oxidation of Brain Membrane Proteins

A prediction of the Aβ-derived free radical model for neuronal death in AD is that protein oxidation in brain membranes should occur, either directly or secondarily to lipid radical formation. Oxidation of hippocampal cultured neuronal cell membrane proteins on addition of Aβ(1–40) was demonstrated using a histofluorescence method described above (Harris *et al.*, 1995a). Oxidized proteins have increased levels of protein carbonyls (Butterfield and Stadtman, 1997; C. D. Smith *et al.*, 1992, 1991; Stadtman, 1992; Carney *et al.*, 1991). These carbonyls were reacted with an amino-derivatized biotin moiety, and the imine formed was reacted with fluorescein isothiocyanate-conjugated streptavidin. We used confocal fluorescence microscopy and computerized image analysis to quantify the resulting fluorescence (Harris *et al.*, 1995a). Membrane proteins in rat cultured hippocampal neurons were highly oxidized by Aβ (Figure 7). That this process involved free radial oxidative stress was supported by the abrogation of the Aβ-induced increased fluorescence if cultures were first treated with the free radical scavenger propyl gallate (Figure 7). The maximum fluorescence intensity induced by Aβ(1–40) was obtained in a time frame that gave the maximum EPR signal intensity in PBN spin trapping experiments (Harris *et al.*, 1995a), suggesting that Aβ-induced free radicals and neuronal membrane protein oxidation occurred by a common process. The nontoxic reverse sequence, Aβ(40–1), did not cause protein oxidation (Harris *et al.*, 1995a). Membrane proteins in cultured astrocytes were also oxidized by Aβ(1–40) as determined by this histofluorescence method (Harris *et al.*, 1996). As with neuronal cultures, astrocytic membrane protein oxidation could be inhibited by free radical scavengers (Harris *et al.*, 1996).

Other fluorescence methods to demonstrate free radical ROS production following exposure of cells to Aβ were used (Harris *et al.*, 1996, 1995a). The redox-sensitive dye,

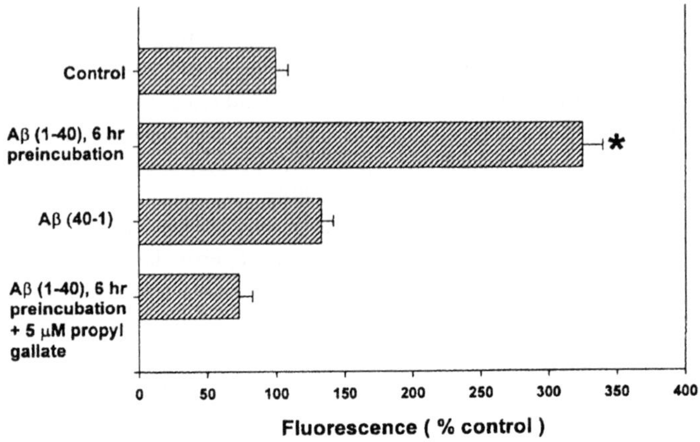

FIGURE 7. Histofluorescence imaging of protein carbonyl formation following Aβ addition to cultured neurons. Fluorescence intensity of FITC-labeled streptavidin following binding to biotin hydrazide, which was previously reacted with carbonyls of oxidized proteins of cultured hippocampal neurons. Aβ(1–40), preincubated 6 hr to give a strong free radical signal when incubated separately with PBN, resulted in a highly significant, approximately three-fold increase in the fluorescence intensity compared with controls [no peptide; nontoxic Aβ(40–1)]. The free radical scavenger, propyl gallate, inhibited increased fluorescence over controls, consistent with a free radical process. See text.

2′,7′-dichlorofluorescein diacetate, once transported into hippocampal neuronal or astrocytic cultures, is converted by esterases to 2′,7′-dichlorofluorescein (DCF), which following reaction with peroxyl radicals leads to the appearance of fluorescence. In the time frame in which PBN spin trapping studies yielded maximal EPR signal intensity, Aβ led to fluorescence in neuronal (Harris *et al.*, 1995a) and astrocytic (Harris *et al.*, 1996) cell cultures, indicating that Aβ-induced ROS production had occurred. In both cell types, fluorescence was inhibited by free radical scavengers (Harris *et al.*, 1996, 1995a). Behl *et al.* (1994), in agreement with our results on neurons (Harris *et al.*, 1995a), reported similar findings on immortalized PC12 cells using the redox-sensitive dye 3-(4,5-dimethylthiazol-2-yl)-2,5-diphenyltetrazolium bromide (MTT). Apparent mitochondrial reductive processes convert MTT to a colored formazan product. Treatment of cells with H_2O_2 or Aβ diminished this conversion, consistent with a more oxidizing intracellular environment (Behl *et al.*, 1994). Many antioxidants have been reported to inhibit the effect of Aβ peptides (Koppal *et al.*, 1998; Subramaniam *et al.*, 1998; Bruce *et al.*, 1996; Goodman and Mattson, 1996, 1994a,b; Goodman *et al.*, 1996, 1994; Harris *et al.*, 1996, 1995a,b; Kelly *et al.*, 1996; Puttfarcken *et al.*, 1996; Tomiyama *et al.*, 1996, 1994; Manelli and Puttfarcken, 1995; Schorderet, 1995; Smith-Swintosky *et al.*, 1995; Subramaniam *et al.*, 1995; Behl *et al.*, 1994; Kumar *et al.*, 1994), in concordance with our model of a free radical process associated with Aβ.

In addition to fluorescence methods, Aβ-induced synaptosomal membrane protein oxidation also was detected by EPR protein-specific spin labeling with MAL-6 (Butterfield *et al.*, 1996b), which had been demonstrated previously to be sensitive to protein oxidation (Butterfield *et al.*, 1997a; Howard *et al.*, 1996; Bellary *et al.*, 1995; Hall *et al.*,

1995a–c; Hensley *et al.*, 1995a, 1994a; Trad and Butterfield, 1994). Based on these results, the effect of Aβ peptides on brain synaptosomal membranes as monitored by the W/S ratio of MAL-6 was investigated. Aβ(25–35) significantly decreased the W/S ratio of subsequently added MAL-6, suggesting that membrane proteins are also a target of Aβ(25–35) free radical attack, consistent with the model proposed and consistent with the fluorescence studies noted above (Subramaniam *et al.*, 1998; Butterfield, 1997). Vitamin E prevented this effect induced by Aβ but was ineffective against 4-HNE, which is a downstream product of a free radical reaction (Subramaniam *et al.*, 1998).

Protein oxidation in neuronal and astrocytic cultures by Aβ and its inhibition by free radical scavengers (Subramaniam *et al.*, 1998; Harris *et al.*, 1996, 1995a) occurs in a complex milieu. To investigate whether protein oxidation would occur in a cell-free system, Aβ was added to GS, whose activity is decreased in AD brain (Hensley *et al.*, 1995b), and to GS solutions (Aksenov *et al.*, 1996, 1995; Hensley *et al.*, 1994b), and protein carbonyl content of GS determined by an immunoassay procedure. Highly toxic Aβ gave pronounced increased carbonyl content, a measure of protein oxidation (Butterfield and Stadtman, 1997; Stadtman, 1992), whereas commercial lots of nontoxic Aβ produced no excess carbonyl content of GS (Aksenov *et al.*, 1997). The ability of toxic forms of Aβ to produce increased carbonyl content of GS was inhibited by the free radical scavenger and spin trap, sulfonated PBN, consistent with the concept that Aβ can directly oxidize proteins as well as induce protein oxidation via Aβ-generated, lipid-derived free radicals or lipid peroxidation by-products.

5.3. Multiple Transmembrane Protein Alterations: Increase of Intracellular Ca^{2+}, Alterations in Ion-Motive ATPases, and Inhibition of Glutamate Uptake

The free radical shrapnel model for AD predicts that Aβ, once formed outside the neuron, inserts into the lipid bilayer of neuronal and glial membranes, where oxygen-dependent free radicals are generated and cause lipid peroxidation. Lipid-derived reactive free radicals or their breakdown products such as 4-HNE then could bind to multiple adjacent transmembrane proteins (enzymes, transporters, G-proteins, other signaling pathways, and so forth) and alter their function. That the membrane is a target for Aβ damage is supported by lipid peroxidation monitored by EPR (Butterfield *et al.*, 1994a), small-angle X-ray studies showing the insertion of Aβ(25–35) into the lipid domain of brain membranes (Mason *et al.*, 1996), and by electron microscopic immunolocalization of Aβ to the neuronal plasma membrane of cultured cells (Mattson *et al.*, 1993). The model is consistent with Aβ-induced, large increases in intracellular Ca^{2+} levels in hippocampal neuronal and astrocytic cultures (Harris *et al.*, 1996, 1995a; Mattson *et al.*, 1993). Such alterations may be a consequence of, or contribute to, altered activity of several ion-motive ATPases in cultured cells exposed to amyloid (Harris *et al.*, 1996; Mark *et al.*, 1995). Neurons exposed to micromolar levels of Aβ(25–35) or Aβ(1–40) lose Na^+/K^+-ATPase and Ca^{2+}-ATPase activities; loss of Mg^{2+}-ATPase required longer times, while the Na^+/Ca^{2+}-exchanger was unaffected by Aβ peptides (Mark *et al.*, 1995). Based on studies with ouabain, which affects K^+ flux, tetrodotoxin, a specific inhibitor of voltage-dependent Na^+ channels, or Na^+-deficient media, we suggested that increased intracellular Ca^{2+} levels were secondary to increased Na^+ influx. Alterations in ion homeostasis, particularly Ca^{2+}, following Aβ free radical oxidative damage could have serious consequences on cell function, ranging from disruption of various signaling pathways and second messenger levels, to alterations in membrane cytoskeletal proteins following Ca^{2+}-activated proteolysis, to compromised mitochondrial function and loss of ATP, and/or to activation of DNA endonucleases. The free radical scavengers, vitamin E, propyl gallate, or PBN, inhibited the Aβ-induced impairment of Na^{+}/K^{+}-ATPase, Ca^{2+} influx activities, and cell death (Mark *et al.*, 1995), in agreement with predictions of the Aβ-derived free radical shrapnel model for AD. In addition to neuronal cultures, human autopsy Na^{+}/K^{+}-ATPase and Ca^{2+}-ATPase activities were also impaired by synthetic amyloid: 70 and 30% loss of activity, respectively, on exposure to 50 μM Aβ(25–35) for 1 hr (Mark *et al.*, 1995). As noted above, 4-HNE addition to synaptosomal membranes was able to elicit the same effects as Aβ (Mark *et al.*, 1997), consistent with the Aβ-associated free radical model for neuronal death in AD brain. Also in support of the model predicting that Aβ-associated free radicals will produce dysfunctioning transmembrane proteins, recent studies have shown that Aβ disrupts carbachol-induced muscarinic signal transduction in cortical neurons and that this effect is abrogated by free radical scavengers (Kelly *et al.*, 1996).

Glutamate exhibits the properties of both an excitatory neurotransmitter and a toxin, and the term *excitotoxin* is often used to describe this brain-resident ion (Rothman and Olney, 1995). Mattson *et al.* (1993) have shown that glutamate excitotoxicity is exacerbated by low-dose exposure of mixed astrocytes and neurons to amyloid. Normally, excitotoxic glutamate is sequestered from neurons by the astrocyte-resident, Na^{+}-dependent glutamate transport system. In astrocytes, glutamate reacts with ammonia in a reaction catalyzed by the oxidatively sensitive enzyme, GS, to form glutamine. Aβ peptides inhibit both the uptake of glutamate by astrocytes (Harris *et al.*, 1996, 1995b) and the activity of GS (Harris *et al.*, 1995a; Hensley *et al.*, 1994b). Hydroxyl radicals were shown by others to inhibit both systems (Volterra *et al.*, 1994; Stadtman, 1992). As noted above, our laboratory (Harris *et al.*, 1996, 1995b) reported that Aβ addition to astrocytes led to a marked elevation of intracellular ROS (based on DCF fluorescence) and a large increase in protein carbonyls, a measure of protein oxidation (Stadtman, 1992), and that these alterations were prevented by the free radical scavenger, Trolox (Harris *et al.*, 1996, 1995b), also consistent with the Aβ-associated free radical shrapnel model of AD (Butterfield, 1997; Butterfield *et al.*, 1994a). The confluence of three oxidative modalities may exist in the glutamate transporter and GS systems: Aβ inhibits GS activity (Aksenov *et al.*, 1997, 1996; Harris *et al.*, 1995a; Hensley *et al.*, 1994b) and the astrocytic glutamate uptake system (Harris *et al.*, 1996, 1995b), but, as mentioned previously, the neurotoxicity of Aβ is increased in the presence of GS (Aksenov *et al.*, 1996, 1995). Relevant to these studies, a 42-kDa ATP-binding protein in AD CSF, not present in controls and later identified as GS, was proposed as a diagnostic marker for AD (Gunnersen and Haley, 1992). Noting that GS is normally resident in astrocytic cytosol, finding AD GS in CSF may reflect the finding of free radical oxidative damage to astrocytes, GS, and glutamate uptake enumerated above (Aksenov *et al.*, 1997, 1996, 1995; Harris *et al.*, 1996, 1995b; Hensley *et al.*, 1994b).

The notion of a synergy between GS and Aβ-associated free radical neurotoxicity was further supported by spin labeling studies (Butterfield *et al.*, 1997). We hypothesized that GS, an oxidatively sensitive enzyme whose activity is diminished by Aβ as described above (Harris *et al.*, 1995a; Hensley *et al.*, 1994b) and in AD brain (Hensley *et al.*, 1995b), when oxidized, would react differently with a SH-specific protein spin label directed at

Table I
Second-Order Kinetic Rate Constants of the Reaction of MTS with Glutamine Synthetase[a]

System	Rate constant ($M^{-1}s^{-1}$)
Unoxidized sheep brain GS	1.55×10^{-3}
Oxidized sheep brain GS	0.36×10^{-3}
Human control brain GS	1.94×10^{-3}
Human AD brain GS	0.84×10^{-3}

[a](GS).

the 88 sulfhydryl groups per GS molecule. MTS [(1-oxyl-2,2,5,5-tetramethylpyrroline-3-methly)-methanethiosulfonate], a spin label specific for protein SH groups (Bhardwaj *et al.*, 1996; Trad *et al.*, 1995; Butterfield and Lee, 1994; Butterfield *et al.*, 1994b; Hensley *et al.*, 1994a), was used to determine the kinetics of binding to ovine brain oxidized and control GS (Butterfield *et al.*, 1997b). Oxidized GS reacted with MTS by second-order kinetics three times slower than unoxidized GS (Table I). Similar studies were performed with human GS highly purified from AD and control brain, and the rate of MTS reaction with AD GS was analogous to that of oxidized GS (Table I). The lower activity of GS that had been exposed to Aβ (Harris *et al.*, 1995a; Hensley *et al.*, 1994b) and the diminished activity of GS in AD brain (Hensley *et al.*, 1995b) conceivably could be a consequence of an altered structure (and hence reactivity with MTS) resulting from free radical oxidation. These results also suggest that free radical oxidation can significantly affect the properties of both Aβ and one of its targets.

5.4. Aβ-Induced Damage to Neurons and Glial Cells Is Modulated by Free Radical Scavengers

If Aβ induces brain cell membrane damage in AD by free radical mechanisms, then a prediction of the shrapnel model for this disorder is that appropriate free radical scavengers may provide protection against this membrane damage. Throughout this review, many references were cited that are consistent with this concept (Koppal *et al.*, 1998; Subramaniam *et al.*, 1998, 1995; Bruce *et al.*, 1996; Butterfield *et al.*, 1996a; Goodman and Mattson, 1996, 1994a,b; Goodman *et al.*, 1996, 1994; Harris *et al.*, 1996, 1995a,b; Hensley *et al.*, 1996, 1995c; Kelly *et al.*, 1996; Zhou *et al.*, 1996; Mark *et al.*, 1997, 1995; Puttfarcken *et al.*, 1996; Tomiyama *et al.*, 1996, 1994; Manelli and Puttfarcken, 1995; Schorderet, 1995; Smith-Swintosky *et al.*, 1995; Behl *et al.*, 1994; Kumar *et al.*, 1994; Yatin *et al.*, 1999). For example, in their study using EPR spin trapping to confirm our finding that Aβ generated free radicals, Tomiyama *et al.* (1996) showed that rifampicin was effective in preventing the PBN EPR spin adduct, i.e., its semiquinone component was thought to trap the free radicals produced. Cocktails containing catalase are reported to protect neurons from Aβ damage (Puttfarcken *et al.*, 1996; Manelli and Puttfarcken, 1995), consistent with our results (Harris *et al.*, 1995b). Other antioxidants ranging from new experimental antioxidants such as EUK-8 (Bruce *et al.*, 1996), U-83836E (Zhou *et al.*, 1996), and U-78517F (Kumar *et al.*, 1994) to the spin trapping antioxidant compound PBN (Behl *et al.*, 1994) are reported to prevent Aβ-induced neurotoxicity. Other compounds with antioxidant capability, such as nordihydroguaiaretic acid (Goodman *et al.*, 1994) and estrogens (Goodman *et al.*, 1996), are reported to offer protection to neurons from Aβ toxicity. In disagreement with these many positive effects of antioxidants against Aβ-induced membrane dysfunctions and neuronal toxicity, two reports failed to find protective effects of antioxidants against Aβ (Pike *et al.*, 1997; Lockhart *et al.*, 1994). However, these two studies, at odds with a huge literature showing that antioxidants are protective against Aβ neurotoxicity (see above), may have methodoligcal differences with many such investigations. For example, Pike *et al.* (1997) used 1–3-day-old, still developing cells to test their antioxidants used, in contrast to the 9–11-day-old cultures used in our and many other studies. Cells early in development such as those used by Pike *et al.* (1997) have not fully expressed receptors, transports proteins (i.e., glutamate receptors), etc. Further, in modeling effects of age-dependent neurodegenerative disorders, old cultures such as ours, in which all receptors and transport proteins have been expressed, seems more germane. Another difference in the Pike *et al.* (1997) study compared to others is the high level of vitamin E used (100–1000 mM) compared to others, including our laboratory, who used 50 μM vitamin E. Such high concentrations of vitamin E might have their own membrane-altering effects that affect the interaction of Aβ. The sum total of all studies overwhelmingly supports the notion of efficacious effects of antioxidants against Aβ-induced ROS and cell damage, strongly support the free radical shrapnel model of AD (Butterfield, 1997; Butterfield *et al.*, 1996, 1994a; Hensley *et al.*, 1996a, 1994b). These considerations support the view that appropriate brain-accessible free radical scavengers may be a promising therapeutic intervention in AD (Butterfield, 1997; Butterfield *et al.*, 1996a,b, 1994a; Hensley *et al.*, 1996, 1 994b).

5.5. Protein Oxidation in AD Brain Is Correlated with Regions of Aβ-Rich, High Senile Plaque Density

Thus far, this chapter has largely reviewed the strong evidence of Aβ-associated free radical formation and resultant toxicity to neuronal and glial cultures. If these same processes occur in AD brain, then, according to the Aβ-derived free radical shrapnel model of AD, focal brain regions of high SP density, containing large amounts of aggregated amyloid, e.g., the hippocampus and inferior parietal lobule regions, should exhibit protein oxidation, while in the cerebellum, poor in SP, membrane protein oxidation should be relatively absent. This prediction of the model was confirmed (Hensley *et al.*, 1995b). Control and AD brain, obtained from populations of similar demographic characteristics, were examined immediately after rapid autopsy (Hensley *et al.*, 1995b). Four biomarkers of neuronal protein oxidation (W/S ratio of MAL-6-spin-labeled synaptosomes, protein carbonyl content, and GS and CK activities) in three brain regions of AD-demented and age-matched control subjects were evaluated. All four biomarkers indicated protein oxidation in AD hippocampal and inferior parietal regions, rich in SP, but no evidence of protein oxidation in AD cerebellum, in which SP were absent. In contrast, control brain showed no increased oxidation in any of the three regions, i.e., no difference in hippocampal and inferior parietal regions relative to cerebellum (Hensley *et al.*, 1995b). Evidence for lipid peroxidation in AD brain recently was reported and included increased TBARS (Lovell *et al.*, 1995) and increased levels of 4-HNE in ventricular fluid (Lovell *et al.*, 1997). All these results are in complete agreement with the predictions of the Aβ free radical shrapnel model for neuronal death in AD brain and strongly support the concept of a principal role of Aβ in AD neurodegeneration, a role that is highly likely to involve free radical oxidative stress.

6. Aβ-ASSOCIATED FREE RADICAL OXIDATION IN AD BRAIN: WHERE DO WE GO FROM HERE?

Based on the research described above, much of which has been confirmed by and/or is consistent with findings of other researchers, we have formulated an Aβ-associated free radical unifying model for AD that accounts for three major observations for which any theory of AD would have to be in agreement. First, multiple membrane alterations of lipids, enzymes, transmembrane transport proteins, and so forth, reported in the literature, can be rationalized as discrete manifestations of Aβ-associated free radical interaction with moieties in neuronal and glial cell membranes. The age dependence of AD is the second observation that the model addresses. This age dependence of AD credibly could be explained by this model based on the known loss of antioxidant capacity with age. Lastly, the genetic mutations of APP and the presenilins that lead to familial AD can be explained by the model; more Aβ is produced in these mutations, leading to more free radical neurodegeneration and subsequent development of AD. Predictions of the Aβ-derived free radical shrapnel model of AD, including lipid oxidation, protein oxidation, neurotoxicity, Ca^{2+} accumulation, inhibition of multiple transport proteins, abrogation of many of these effects by free radical scavengers, and correlation of protein oxidation in AD brain regions rich in SP, have all been borne out by experiment, and most have been confirmed by others. Consistent with our oxidative stress model for neuronal death in AD brain, APP overexpressing mice have increased deposition of Aβ and increased oxidative stress (Smith *et al.*, 1998).

The molecular mechanisms by which Aβ forms free radicals are under inquiry in our laboratory and remain to be elucidated. However, it now seems highly probable that Aβ-associated free radical damage is a fundamental process in AD. The model provides a rational strategy for therapeutic intervention in AD, and studies of oxidative stress models are under active investigation.

NOTE ADDED IN PROOF. Consistent with the hypothesis that Met-35 is important in Aβ-associated free radical oxidative stress, we recently showed that *C. elegans* transgenic animals expressing human, full-length Aβ(1-42) showed strong evidence of protein oxidation (Yatin *et al.*, 1998), a key marker of free radical oxidative stress. Mutation of Met-35 to Cys-35 resulted in healthy animals that showed no evidence of protein oxidation.

ACKNOWLEDGMENTS. The excellent research of Drs. Kenneth Hensley, Michael Aksenov, Marina Aksenova, Marni Harris, Nathan Hall, Ram Subramaniam, Sridhar Varadarajan, and Robert Mark, and Pamela Cole, M.S., Beverly Howard, M.S., and graduate students Servet Yatin, Michael Lafontaine, Tanuja Koppal, and Marsha Cole is gratefully acknowledged. Collaborations with faculty colleagues John Carney, Mark Mattson, and William Markesbery are also gratefully acknowledged. I acknowledge Professor Christian Schoneich for helpful discussions about this research. This work was supported in part by grants from the National Institutes of Health (AG-10836, AG-05119). I gratefully acknowledge Centaur Pharmaceuticals for their generous gift of [^{13}C]-PBN and Professor Boyd Haley for his gift of highly purified GS from AD and control brains.

7. REFERENCES

Aksenov, M. Y., Aksenova, M. V., Harris, M. E., Hensley, K., Butterfield, D. A., and Carney, J. M., 1995, Enhancement of Aβ(1–40) neurotoxicity by glutamine synthetase, *J. Neurochem.* **65**:1899–1902.

Aksenov, M. Y., Aksenova, M. V., Butterfield, D. A., Hensley, K., Vigo-Pelfrey, C., and Carney, J. M., 1996, Glutamine synthetase-induced enhancement of β-amyloid peptide Aβ(1–40) neurotoxicity accompanied by abrogation of fibril formation and Aβ fragmentation, *J. Neurochem.* **66**:2050–2056.

Aksenov, M. Y., Aksenova, M. V., Carney, J. M., and Butterfield, D. A., 1997, Oxidative modification of glutamine synthetase by amyloid beta peptide, *Free Rad. Res.* **27**:267–281.

Aksenova, M. V., Aksenov, M. Y., Butterfield, D. A., and Carney, J. M., 1996, Alpha 1-antichymotrysin interaction with Aβ(1–40) inhibits fibril formation but does not affect the peptide toxicity, *Neurosci. Lett.* **211**:45–47.

Aksenova, M. V., Aksenov, M. Y., Carney, J. M., and Butterfield, D. A., 1998, Protein oxidation and enzyme activity decline in old brown Norway rats are reduced by dietary restriction, *Mech. Aging Dev.* **100**:157–168.

Beal, M. F., 1992, Does impairment of energy metabolism result in excitotoxic neuronal death in neurodegenerative illnesses? *Ann. Neurol.* **31**:119–130.

Behl, C., Davis, J. B., Lesley, R., and Shubert, D., 1994, Hydrogen peroxide medicates amyloid β protein toxicity, *Cell* **77**:817–827.

Bellary, S. S., Anderson, K. W., Arden, W. A., and Butterfield, D. A., 1995, Effect of lipopolysaccharide on the physical conformation of the erythrocyte cytoskeletal proteins, *Life Sci.* **56**:91–98.

Bhardwaj, A., Lee, J., Glauner. K., Ganapathi, S., Bhattacharyya, D., and Butterfield, D. A., 1996, Biofunctional membranes: An EPR study of active site structure and stability of papain non-covalently immobilized on the surface of modified poly(ether)sulfone membranes through the avidin–biotin linkage, *J. Membr. Sci.* **119**:241–252.

Bruce, A. J., Malfroy, B., and Baudry, M., 1996, β-Amyloid toxicity in organotypic hippocampal cultures: Protection by EUK-8, a synthetic catalytic free radical scavenger, *Proc. Natl. Acad. Sci. USA* **93**:2312–2316.

Burdick, D., Soreghan, B., Kwon, M., Kosmoski, J., Knauer, M., Henshen, A., Yates, J., Cotman, C., and Glabe, C., 1992, Assembly and aggregation properties of synthetic Alzheimer's A4/B-amyloid peptide analogs, *J. Biol. Chem.* **267**:546–554.

Bush, A. I., Huang, X., Atwood, C. S., Cherny, R. A., Moir, R. D., Goldstein, L. E., O'Malley, C. M., Saunders, A. J., Multhap, G., Beyreuther, K., Masters, C. L., and Tanzi, R. E., 1998, Interactions with ionic zinc, copper, and iron, govern Aβ redox activity and accumulation in Alzheimer's disease, *Neurobiol. Aging* **19**:S40–41.

Butterfield, D. A., 1982, Spin labeling in disease, *Biol. Magn. Reson.* **4**:1–78.

Butterfield, D. A., 1986, Spectroscopic methods in degenerative neurological diseases, *Crit. Rev. Neurobiol.* **2**:169–240.

Butterfield, D. A., 1990, Principles of biological membrane structure: Alteration in the physical state of one side of the membrane by modulation of the physical state of the other side, *J. Membr. Sci.* **53**:3–17.

Butterfield, D. A., 1995, Membrane cytoskeletal protein alterations in Alzheimer's disease, in *Non-Neuronal Cells in Alzheimer's Disease* (P. Zatta and M. Nicolini, eds.), pp. 78–94, World Scientific Publishers, London.

Butterfield, D. A., 1996, Alzheimer's disease: A disorder of oxidative stress, *Alzheimer's Disease Rev.* **1**:68–70.

Butterfield, D. A., 1997, β-Amyloid-associated free radical oxidative stress and neurotoxicity: Implications for Alzheimer's disease, *Chem. Res. Toxicol.* **10**:495–506.

Butterfield, D. A., and Lee, J., 1994, Active site structure and stability of the thiol protease, papain, studied by electron paramagnetic resonance employing a methanethiosulfonate spin label, *Arch. Biochem. Biophys.* **310:**167–171.

Butterfield, D. A., and Rangachari, A., 1991, Membrane-altering effects of velnacrine and N-methylacridinium: Relevance to THA (tacrine) and Alzheimer's disease, *Biochem. Biophys. Res. Commun.* **185:**596–603.

Butterfield, D. A., and Stadtman, E. R., 1997, Protein oxidation processes in aging brain, in *The Aging Brain* (M. P. Mattson and J. W. Geddes, eds.), pp. 161–191, JAI Press, Greenwich, CT.

Butterfield, D. A., Hensley, K., Harris, M., Mattson, M. P., and Carney, J. M., 1994a, β-Amyloid peptide free radical fragments initiate synaptosomal lipoperoxidation in a sequence-specific fashion: Implications to Alzheimer's disease, *Biochem. Biophys. Res. Commun.* **200:**710–715.

Butterfield, D. A., Lee, J., Ganapathi, S., and Bhattacharyya, D., 1994b, Biofunctional membranes IV. Active site structure and stability of an immobilized enzyme, papain, on modified polysulfone membranes studied by electron paramagnetic resonance and kinetics, *J. Membr. Sci.* **91:**47–64.

Butterfield, D. A., Hensley, K., Hall, N., Subramaniam, R., Howard, B. J., Cole, P., Yatin, S., Lafontaine, M., Harris, M. E., Aksenova, M., Aksenov, M., and Carney, J., 1996a, β-Amyloid-derived free radical oxidation: A fundamental process in Alzheimer's disease, in *Molecular Models of Dementia* (R. Tanzi, and W. Wasco, eds.), pp. 145–167, Humana Press, Totowa, NJ.

Butterfield, D. A., Martin, L., Carney, J. M., and Hensley, K., 1996b, Aβ(25–35) peptide displays H_2O_2-like reactivity towards aqueous Fe^{2+}, nitroxide spin probes, and synaptosomal membrane proteins, *Life Sci.* **58:**217–228.

Butterfield, D. A., Hensley, K., Cole, P., Subramaniam, R., Aksenov, M., Aksenova, M., Bummer, P. M., Haley, B. E., and Carney, J. M., 1997b, Oxidatively induced structural alteration of glutamine synthetase assessed by analysis of spin-labeled incorporation kinetics: Relevance to Alzheimer's disease, *J. Neurochem.* **68**:2451–2457.

Butterfield, D. A., Howard, B. J., Yatin, S., Allen, K. L., and Carney, J. M., 1997a, Free radical oxidation of brain proteins in accelerated senescence and its modulation by N-tert-butyl-α-phenylnitrone, *Proc. Natl. Acad. Sci. USA* **94**:674–678.

Butterfield, D. A., Subramaniam, R., Cole, P., Yatin, S., Hensley, K., Hall, N. C., and Carney, J. M., 1997c, Electron paramagnetic resonance studies of amyloid β-peptide- and ischemia/reperfusion-associated oxygen free radicals and the membrane damage they cause: Relevance to Alzheimer's disease and stroke, in *Oxygen Free Radicals and the Disease Process* (C. E. Thomas, and B. Kalyanaraman, eds.), pp. 41–64, Harwood Academic Publishers, Reading, United Kingdom.

Butterfield, D. A., Varadarajan, S., LaFontaine, M., Subramaniam, R., Koppal, T., Yatin, S., Hensley, K., Aksenova, M., and Aksenov, M., 1998, Alzheimer's amyloid β-peptide-associated free radical oxidative stress and neurotoxicity, *Neurobiol. Aging* **19**:S262.

Cader, A. A., Butterfield, D. A., Watkins, B. A., Chung, B. H., and Hennig, B., 1995, Lipid-induced alterations in membrane fluidity contribute to endothelial barrier dysfunction, *Int. J. Biochem. Mol. Biol.* **27:**665–673.

Carney, J. M., Starke-Reed, P. E., Oliver, C. N., Landum, R. W., Cheng, M., Wu, J. F., and Floyd, R. A., 1991, Reversal of age-related increase in brain protein oxidation, decrease in enzyme activity, and loss of temporal and spatial memory by chronic administration of the spin-trapping compound N-*tert*-butyl-α-phenylnitrone, *Proc. Natl. Acad. Sci. USA* **88**:3633–3636.

Chan, W. K. M., Decker, E. A., Lee, J., and Butterfield, D. A., 1994, EPR spin trapping studies of the hydroxyl radical scavenging activity of carnosine and related dipeptides, *J. Agric. Food Chem.* **42:**1407–1410.

Clerici, W. J., Hensley, K., DiMartino, D. L., and Butterfield, D. A., 1996, Direct detection of ototoxicant-induced reactive oxygen species generation in cochlear explants, *Hearing Res.* **98:**116–124.

Colton, C. A., Snell, J., Chernyshev, O., and Gilbert, D. L., 1994, Induction of superoxide anion and nitric oxide production in cultured microglia, *Ann NY Acad. Sci.* **738:**54–63.

Corain, B., Iqbal, K., Nicolini, M., Winblad, B., Wisniewski, H., and Zatta, P. (eds.), 1993, *Alzheimer's Disease: Advances in Clinical and Basic Research*, Wiley, New York.

Corder, E. H., Saunders, A. M., Strittmatter, W. J., Schmechel, D. E., Gaskell, P. C., Small, G. W., Roses, A. D., and Pericak-Vance, M. A., 1993, Gene dose of apolipoprotein E type 4 allele and the risk of Alzheimer's disease in late onset families, *Science* **261:**921–923.

Dado, G. P., and Gellman, S. H., 1994, Redox control of secondary structure in a designed peptide, *J. Am. Chem. Soc.* **115:**12609–12610.

Daniels, W. M., van Rensbury, S. J., van Zyl, J. M., and Taljaard, J. J., 1998, Melatonin prevents beta-amyloid-induced lipid peroxidation, *J. Pineal Res.* **24**:78–82.

Esterbauer, H., Schaur, R. J., and Zollner, H., 1991, Chemistry and biochemistry of 4-hydroxynonenal, malonaldehyde, and related aldehydes, *Free Radical Biol. Med.* **11:**81–128.

Estevez, A. G., Radi, R., Barbeito, L., Shin, J. T., Thompson, J. A., and Beckman, J. S., 1995, Peroxynitrite-induced cytotoxicity in PC12 cells: Evidence for an apoptotic mechanism differentially modulated by neurotrophic factors, *J. Neurochem.* **65:**1543–1550.

Gabbita, S. P., Hensley, K., Butterfield, D. A., and Carney, J. M., 1997, Aging and caloric restriction affect mitochondrial respiration and lipid membrane status: An electron paramagnetic resonance investigation, *Free Rad. Biol. Med.* **23**:191–201.

Gabbita, S. P., Subramaniam, R., Allouch, F., Carney, J. M., and Butterfield, D. A., 1998, Effects of mitochondrial respiratory stimulation on membrane lipids and proteins: An electron paramagnetic resonance investigation, *Biochem. Biophys. Acta.* **1372**:163–173.

Games, D., Adams, D., Alessandrini, R., Barbour, R., Berthelette, P., Blackwell, C., Carr, T., Clemens, J., Donaldson, T., Gillespie, F., Guido, T., Hagoplan, S., Johnson-Wood, K., Khan, K., Lee, M., Leibowitz, P., Lieberburg, I., Little, S., Masliah, E., McConlogue, L., Montoya-Zavala, M., Mucke, L., Paganini, L., Penniman, E., Power, M., Schenk, D., Seubert, P., Snyder, B., Soriano, F., Tan, F., Vital, J., Wadsworth, S., Wolozin, B., and Zhao, J., 1995, Alzheimer-type neuropathology in transgenic mice overexpressing V717F β-amyloid precursor protein, *Nature* **373:**523–527.

Goodman, Y., and Mattson, M. P., 1994a, Staurosporine and K-252 compounds protect hippocampal neurons against amyloid β-peptide toxicity and oxidative injury, *Brain Res.* **650:**170–174.

Goodman, Y., and Mattson, M. P., 1994b, Secreted forms of β-amyloid precursor protein protect hippocampal neurons against amyloid β-peptide-induced oxidative injury, *Exp. Neurol.* **128:**1–12.

Goodman, Y., and Mattson, M. P., 1996, K^+ channel openers protect hippocampal neurons against oxidative injury and amyloid-β toxicity, *Brain Res.* **706:**328–332.

Goodman, Y., Steiner, M. R., Steiner, S. M., and Mattson, M. P., 1994, Nordihydroguaiaretic acid protects hippocampal neurons against amyloid β-peptide toxicity, and attenuates free radical and calcium accumulation, *Brain Res.* **654:**171–176.

Goodman, Y., Bruce, A. J., Cheng, B., and Mattson, M. P., 1996, Estrogens attenuate and corticosterone exacerbates excitotoxicity, oxidative injury, and amyloid-β toxicity in hippocampal neurons, *J. Neurochem.* **66:**1836–1844.

Gridley, K. E., Green, P. S., and Simpkins, J. W., 1997, Low concentrations of estradiol reduce beta amyloid (25–35)-induced toxicity, lipid peroxidation, and glucose utilization in human SK-N-SH cells, *Brain Res.* **778**:158–165.

Gunnersen, D., and Haley, B., 1992, Detection of glutamine synthetase in the cerebrospinal fluid of Alzheimer diseased patients: A potential diagnostic marker, *Proc. Natl. Acad. Sci. USA* **89:**11949–11953.

Hall, N. C., Carney, J. M., Cheng, M. S., and Butterfield, D. A., 1995a, Ischemia/reperfusion induced changes in membrane proteins and lipids of gerbil cortical synaptosomes, *Neuroscience* **64:**81–89.

Hall, N. C., Carney, J. M., Cheng, M., and Butterfield, D. A., 1995b, Prevention of ischemia/reperfusion-induced alterations in synaptosomal membrane-associated proteins and lipids by N-tert-butyl-α-phenylnitrone and difluormethylornithine, *Neuroscience* **69:**591–600.

Hall, N. C., Dempsey, R. J., Carney, J. M., Donaldson, D. L., and Butterfield, D. A., 1995c, Structural alterations in synaptosomal membrane-associated proteins and lipids by transient middle cerebral artery occlusion in the cat, *Neurochem. Res.* **20:**1161–1169.

Hall, N. C., Carney, J. M., Plante, O. J., Cheng, M., and Butterfield, D. A., 1997, Effect of 2-cyclohexene-1-one-induced glutathione diminution on ischemia/reperfusion-induced alterations in the physical state of brain synaptosomal membrane proteins and lipids, *Neuroscience* **77**:283–290.

Harman, D., 1994, Free radical theory of aging: Increasing the functional life span. *Ann. N. Y. Acad. Sci.* **717:**1–15.

Harris, M. E., Hensley, K., Butterfield, D. A., Leedle, R. E., and Carney, J. M., 1995a, Direct evidence of oxidative injury by the Alzheimer's amyloid β peptide in cultured hippocampal neurons, *Exp. Neurol.* **131:**193–202.

Harris, M. E., Carney, J. M., Cole, P., Hensley, K., Howard, B. J., Martin, L., Bummer, P., Wang, Y., Pedigo, N., and Butterfield, D. A., 1995b, β-Amyloid peptide-derived, oxygen-dependent free radicals inhibit glutamate uptake in cultured astrocytes: Implications to Alzheimer's disease, *NeuroReport* **6:**1875–1879.

Harris, M. E., Wang, Y., Pedigo, N. W. Jr., Hensley, K., Butterfield, D. A., and Carney, J. M., 1996, Amyloid β peptide (25–35) inhibits Na^+-dependent glutamate uptake in rat hippocampal astrocyte cultures, *J. Neurochem.* **67**:277–286.

Hensley, K., Postlewaite, J. J., Dobbs, P., and Butterfield, D. A., 1993, Alteration of the physical state of erythrocyte membranes by selective enzymatic degradation of ankyrin (band 2.1): Subcellular surgery characterized by EPR spectroscopy, *Biochim. Biophys. Acta* **1145**:205–211.

Hensley, K., Hall, N., Shaw, W., Carney, J. M., and Butterfield, D. A., 1994a, Electron paramagnetic resonance investigation of free radical induced alterations in neocortical synaptosomal membrane protein infrastructure, *Free Radical Bio. Med.* **17**:321–331.

Hensley, K., Carney, J. M., Mattson, M. P., Aksenova, M., Harris, M., Wu, J. F., Floyd, R. A., and Butterfield, D. A., 1994b, A model for β-amyloid aggregation and neurotoxicity based on free radical generation by the peptide: Relevance to Alzheimer disease, *Proc. Natl. Acad. Sci. USA* **91**:3270–3274.

Hensley, K., Howard, B. J., Carney, J. M., and Butterfield, D. A., 1995a, Membrane protein alterations in rodent erythrocytes and synaptosomes due to aging and hyperoxia, *Biochim. Biophys. Acta* **1270**:203–206.

Hensley, K., Hall, N., Subramaniam, R., Cole, P., Harris, M., Aksenov, M., Aksenova, M., Gabbita, P., Wu, J. F., Carney, J. M., Lovell, M., Markesbery, W. R., and Butterfield, D. A., 1995b, Brain regional correspondence between Alzheimer's disease histopathology and biomarkers of protein oxidation, *J. Neurochem.* **65**:2146–2156.

Hensley, K., Aksenova, M., Carney, J. M., and Butterfield, D. A., 1995c, Amyloid β-peptide spin trapping I: Enzyme toxicity is related to free radical spin trap reactivity, *NeuroReport* **6**:489–493.

Hensley, K., Aksenova, M., Carney, J. M., and Butterfield, D. A., 1995d, Amyloid β-peptide spin trapping II: Evidence for decomposition of the PBN spin adduct, *NeuroReport* **6**:493–496.

Hensley, K., Butterfield, D. A., Aksenova, M., Harris, M., Wu, J., Floyd, R., Mattson, M., and Carney, J. M., 1995e, A model for β-amyloid aggregation and neurotoxicity based on the free radical generating capacity of the peptide: Implications of peptide-derived free radicals to Alzheimer's disease, *Proc. Western Pharmacol. Soc.* **38**:113–120.

Hensley, K., Butterfield, D. A., Hall, N., Cole, P., Subramaniam, R., Mark, R., Mattson, M. P., Markesbery, W. R., Harris, M. E., Aksenov, M., Aksenova, M., Wu, J. F., and Carney, J. M., 1996a, Reactive oxygen species as causal agents in the neurotoxicity of the Alzheimer's disease-associated amyloid beta peptide, *Ann. N. Y. Acad. Sci.*, **786**:120–134.

Hockenbery, D. M., Oltvai, Z. N., Xiao-Ming, Y., Milliman, C. L., and Korsmeyer, S. J., 1993, Bcl-2 functions in an antioxidant pathway to prevent apoptosis, *Cell* **75**:241–251.

Howard, B. J., Yatin, S., Hensley, K., Allen, K. L., Kelly, J. P., Carney, J. M., and Butterfield, D. A., 1996a, Prevention of hyperoxia-induced alterations in synaptosomal membrane-associated proteins by N-tert-butyl-α-phenylnitrone (PBN) and 4-hydroxy-2,2,6,6-tetramethylpiperidine-1-oxyl (Tempol), *J. Neurochem.* **67**:2045–2050.

Janzen, E. G., 1980, A critical review of spin trapping in biological systems, in *Free Radicals in Biology*, Volume 4 (W. A. Pryor, ed.), pp. 115–154, Academic Press, New York.

Jarrett, J. T., and Lansbury, P. T., 1993, One-dimensional crystallization of amyloid: A pathogenic mechanism in Alzheimer's disease? *Cell* **73**:1053–1058.

Jarrett, J. T., Berber, E. P., and Lansbury, P. T., 1992, The carboxy terminus of the β-amyloid protein is critical in the seeding of amyloid formation: Implications for the pathogenesis of Alzheimer's disease, *Biochemistry* **32**:4693–4697.

Kang, J., Lemair, H. G., Unterbeck, A., Salbaum, J. M., Masters, C. L., Grzeshik, K. H., Multhaup, G., Beyreuther, K., and Müller-Hill, B., 1987, The precursor of Alzheimer' disease A4 protein resembles a cell-surface receptor, *Nature* **325**:733–736.

Katzman, R., and Saitoh, T., 1991, Advances in Alzheimer's disease, *FASEB J.* **4**:278–286.

Kelly, J., Furukawa, K., Barger, S. W., Mark, R. J., Rengen, M. R., Roth, G., and Mattson, M. P., 1996, Amyloid β-peptide disrupts carbachol-induced muscarinic cholinergic signal transduction in cortical neurons: Involvement of free radicals, *Proc. Natl. Acad. Sci. USA* **93**:6753–6758.

Koppal, T., Subramaniam, R., Drake, J., Prasad, M. R., Dhillon, H., and Butterfield, D. A., 1998, Vitamin E protects against Alzheimer's amyloid peptide (25–35)-induced changes in neocortical synaptosomal membrane lipid structure and composition, *Brain Res.* **786**:270–273.

Kumar, U., Dunlop, D. M., and Richardson, J. S., 1994, The acute neurotoxic effect of β-amyloid on mature cultures of rat hippocampal neurons is attenuated by the antioxidant U-78517F, *Int. J. Neurosci.* **79:**185–190.

Lafon-Cazal, M., Pietri, S., Culcasi, M., and Bockaert, J., 1993, NMDA-dependent superoxide production and neurotoxicity, *Nature* **364:**535–537.

Lockhart, B. P., Benicourt, C., Junien, J.-L., and Privat, A., 1994, Inhibitors of free radical formation fail to attenuate direct β-amyloid 25–35 peptide mediated neurotoxicity in rat hippocampal cultures, *J. Neurosci. Res.* **39**:494–505.

Lorenzo, A., and Yankner, B. A., 1994, β-Amyloid neurotoxicity requires fibril formation and is inhibited by Congo red, *Proc. Natl. Acad. Sci. USA* **91:**12243–12247.

Lovell, M. A., Ehmann, W. D., Butler, S. M., and Markesbery, W. R., 1995, Elevated thiobarbituric acid-reactive substances and antioxidant activity in the brain in Alzheimer's disease, *Neurology* **45**:1594–1601.

Lovell, M. A., Ehmann, W. D., Mattson, M. P., and Markesbery, W. R., 1997, Elevated 4-hydroxynonenal in ventricular fluid in Alzheimer's disease, *Neurobiol. Aging* **18**:457–461.

Manelli, A. M., and Puttfarcken, P. S., 1995, β-Amyloid-induced toxicity in rat hippocampal cells: In vitro evidence for the involvement of free radicals, *Brain Res. Bull.* **38:**569–576.

Mark, R. J., Hensley, K., Butterfield, D. A., and Mattson, M. P., 1995, Amyloid β-peptide impairs ion-motive ATPase activities: Evidence for a role in loss of neuronal Ca^{2+} homeostasis and cell death, *J. Neurosci.* **15:**6239–6249.

Mark, R. J., Lovell, M. A., Markesbery, W. R., Uchida, K., and Mattson, M. P., 1997, A role for 4-hydroxynonenal, an aldehydic product of lipid peroxidation, in disruption of ion homeostasis and neuronal death induced by amyloid β-peptide, *J. Neurochem.* **68**:255–264.

Markesbery, W. R., 1997, Oxidative stress hypothesis in Alzheimer's disease, *Free Rad. Biol. Med.* **23**:134–147.

Markesbery, W. R., Lovell, M. A., and Ehmann, W. D., 1994, Brain trace metals in Alzheimer disease, in *Alzheimer Disease* (R. D. Terry, R. Katzman, and K. L. Bick, eds.), pp. 353–367, Raven Press, New York.

Mason, R. P., Estermyer, J. D., Kelly, J. F., and Mason, P. E., 1996, Alzheimer's disease amyloid β peptide 25–35 is localized in the membrane hydrocarbon core: X-ray diffraction analysis, *Biochem. Biophys. Res. Commun.* **222:**78–82.

Mattson, M. P., Barger, S. W., Cheng, B., Lieberburg, I., Smith-Swintosky, V. L., and Ryel, R. E., 1993, β-Amyloid precursor protein metabolites and loss of neuronal Ca^{2+} homeostasis in Alzheimer's disease, *Trends Neurosci.* **16:**409–414.

Mattson, M. P., Carney, J. M., and Butterfield, D. A., 1995a, Glycation: A tombstone in AD? *Nature* **373:**481.

Mattson, M. P., Lovell, M. A., Furukawa, K., and Markesbery, W. R., 1995b, Neurotrophic factors attenuate glutamate-induced accumulation of peroxides, elevation of $[Ca^{2+}]_i$ and neurotoxicity, and increase in antioxidant enzyme activities in hippocampal neurons, *J. Neurochem.* **65:**1740–1751.

May, P. C., Gitter, B. D., Waters, D. C., Simmons, L. K., Becker, G. W., Small, J. S., and Robinson, P. M., 1992, β-Amyloid peptide in vitro toxicity: Lot-to-lot variability, *Neurobiol. Aging* **13:**605–607.

Multhaup, G., Schlicksupp, A., Hesse, L., Beher, D., Ruppert, T., Masters, C. L., and Beyreuther, K., 1996, The amyloid precursor protein of Alzheimer's disease in the reduction of copper (II) to copper (I), *Science,* **271:**1406-1409.

Naslund, J., Schierhorn, A., Hellman, U., Lanfelt, L., Roses, A. D., Tjernberg, L. O., Siberring, J., Gandy, S. E., Winblad, B., Greengard, P., Nordstedt, C., and Terenius, L., 1994, Relative abundance of Alzheimer Aβ amyloid peptide variants in Alzheimer disease and normal aging, *Proc. Natl. Acad. Sci. USA* **91:**8378–8382.

Nutisya, E. M., Bowling, A. C., and Beal, M. F., 1994, Cortical cytochrome oxidase activity is reduced in Alzheimer's disease, *J. Neurochem.* **63:**2179–2184.

Oda, T., Wals, P., Osterbung, H., Johnson, S., Pasinetti, G., Morgan, T., Rozovosky, I., Stein, W. B., Synder, S., Holzman, T., Krafft, G., and Finch, C., 1995, Clusterin (apo J) alters the aggregation of amyloid β peptide (Aβ 1–42) and forms slowly sedimenting Aβ complexes that cause oxidative stress, *Exp. Neurol.* **136:**22–31.

Oliver, C. N., Ahn, B. W., Moerman, E. J., Boldstein, S., and Stadtman, E. R., 1987, Age-related changes in oxidized proteins, *J. Biol. Chem.* **262:**5488–5491.

Pappolla, M. A., Omar, R. A., Kim, K. S., and Robakis, N. K., 1992, Immunohistochemical evidence of antioxidant stress in Alzheimer's disease, *Am. J. Pathol.* **140:**621–628.

Parker, W. D., Jr., Parks, J., Filley, C. M., and Kleinschmidt-Demasters, B. K., 1994, Electron transport chain defects in Alzheimer's disease brain, *Neurology* **44:**1090–1096.

Pike, C. J., Burdick, D., Walencewicz, A. J., Glabe, C. G., and Cotman, C. W., 1993, Neurodegeneration induced by β-amyloid peptides in vitro: Role of peptide assembly state, *J. Neurosci.* **13:**1676–1687.

Pike, C. J., Walencewicz-Wasserman, A. J., Kosmoski, J., Cribbs, D. H., Glabe, C. G., and Cotman, C. W., 1995, Structure–activity analyses of β-amyloid peptides: Contributions of the β25–35 region to aggregation and neurotoxicity, *J. Neurochem.* **64:**253–265.

Pike, C. J., Ramezan-Arab, N., and Cotman, C. W., 1997, β-amyloid neurotoxicity in vitro: Evidence of oxidative stress but not protection by antioxidants, *J. Neurochem.* **69**:1601–1611.

Prasad, M. R., Dhillon, H. S., Carbary, T., Dempsey, R. J., and Scheff, S. W., 1994, Enhanced phosphodiestric breakdown of phosphatidylinositol bisphosphate after experimental brain injury, *J. Neurochem.* **63:**773–776.

Prusiner, S. B., 1992, Biology and genetics of prion diseases, *Biochemistry* **31:**12277–12288.

Puttfarcken, P. S., Manelli, A. M., Neilly, J., and Frail, D. E., 1996, Inhibition of age-induced β-amyloid neurotoxicity in rat hippocampal cells, *Exp. Neurol.* **138:**73–81.

Rothman, S. M., and Olney, J. W., 1995, Excitotoxicity and the NMDA receptor—still lethal after eight years, *Trends Neurosci.* **18**:57–58.

Sano, M., Ernesto, C., Thomas, R. G., Klauber, M. R., Schafer, K., Grundman, M., Woodbury, P., Growdon, J., Cotman, C. W., Pfeiffer, E., Scheider, L. S., and Thal, L. J., 1997, A controlled trial of selegiline alpha-tocopherol, or both as a treatment for Alzheimer's disease: The Alzheimer's disease cooperative study, *New Eng. J. Med.* **336**:1216–1222.

Schoneich, C., Zhao, F., Madden, K. P., and Bobrowski, K., 1994, Side chain fragmentation of N-terminal threonine or serine residue induced through intramolecular protein transfer to hydroxyl sulfuranyl radical formed at neighboring methionine in dipeptides, *J. Am. Chem. Soc.* **116:**4641–4652.

Schorderet, M., 1995, Alzheimer's disease: Fundamental and therapeutic aspects, *Experientia* **51:**99–105.

Selkoe, D. J., 1991, The molecular pathology of Alzheimer's disease, *Neuron* **6:**487–498.

Selkoe, D. J., 1994, Alzheimer's disease: A central role of amyloid, *J. Neuropathol. Exp. Neurol.* **53**:438–447.

Selkoe, D. J., 1996, Amyloid β-protein and the genetics of Alzheimer's disease, *J. Biol. Chem.* **271:**18295–18298.

Simmons, L. K., May, P. C., Tomaselli, K. J., Ryel, R. E., Fuson, K. S., Brigham, E. F., Wright, S., Lieberburg, I., Becker, G. W., Brems, D. N., and Li, W. Y., 1994, Secondary structure of amyloid β peptide correlates with neurotoxic activity in vitro, *Mol. Pharmacol.* **45:**373–379.

Smith, C. D., Carney, J. M., Starke-Reed, P. E., Oliver, C. N., Stadtman, E. R., Floyd, R. A., and Markesbery, W. R., 1991, Excess brain protein oxidation and enzyme dysfunction in normal aging and Alzheimer disease, *Proc. Natl. Acad. Sci. USA* **88:**10540–10543.

Smith, C. D., Carney, J. M., Tatsumo, T., Stadtman, E. R., Floyd, R. A., and Markesbery, W. R., 1992, Protein oxidation in aging brain, *Ann. N.Y. Acad. Sci.* **663:**110–119.

Smith, M. A., Taneda, S., Richey, P. L., Mikyata, S., Yan, S., Stern, D., Sayer, L., Monnier, V. M., and Perry, G., 1994, Advanced maillard reaction end products are associated with Alzheimer disease pathology, *Proc. Natl. Acad. Sci. USA* **91:**5710–5714.

Smith, M. A., Sayre, L. M., Monnier, V. M., and Perry, G., 1995, Radical AGEing in Alzheimer's disease, *Trends Neurosci.* **18:**172–176.

Smith, M. A., Hirai, K., Hsaio, K., Pappolla, M. A., Harris, P. L. R., Siedlak, S. L., Tabaton, M., and Perry, G., 1998, Amyloid-β deposition in Alzheimer transgenic mice is associated with oxidative stress, *J. Neurochem.* **70**:2212–2215.

Smith-Swintosky, V. L., Zimmer, S., Fenton, J. W., and Mattson, M. P., 1995, Opposing actions of thrombin and protease nexin-1 on amyloid β-peptide toxicity and on accumulation of peroxides and calcium in hippocampal neurons, *J. Neurochem.* **65:**1415–1418.

Snyder, S. W., Ladror, U. S., Wade, W. S., Wang, G. T., Barrett, L. W., Matayoshi, E. D., Huffaker, J. H., Krafft, G. A., and Holzman, T. F., 1995, Amyloid-β aggregation: Selective inhibition of aggregation in mixtures of amyloid with different chain lengths, *Biophys. J.* **67:**1216–1228.

Soto, C., Golabek, A., Wisniewski, T., and Castano, E. M., 1996, Alzheimer's β-amyloid peptide is conformationally modified by apolipoprotein E in vitro, *NeuroReport* **7:**721–725.

Stadtman, E. R., 1992, Protein oxidation and aging, *Science* **257:**1220–1224.

Starke-Reed, P. E., and Oliver, C. N., 1989, Protein oxidation and proteolysis during aging and oxidative stress, *Arch. Biochem. Biophys.* **275**:559–567.

Subramaniam, R., Howard, B. J., Hensley, K., Aksenova, M., Carney, J. M., and Butterfield, D. A., 1995, β-Amyloid (32–35) generates reactive free radicals that are toxic to biomolecules: Implications to Alzheimer's disease, *Alzheimer's Research*, **1**:141–144.

Subramaniam, R., Roediger, F., Jordan, B., Mattson, M. P., Keller, J. N., Waeg, G., and Butterfield, D. A., 1997, The lipid peroxidation product, 4-hydroxy-2-trans-nonenal, alters the conformation of cortical synaptosomal membrane proteins, *J. Neuorchem.* **69**:1161–1169.

Subramaniam, R., Koppal, T., Green, M., Yatin, S., Jordan, B., Drake, J., and Butterfield, D. A., 1998, The free radical antioxidant vitamin E protects cortical synaptosomal membranes from amyloid beta-peptide (25–35) toxicity but not from hydroxynonenal toxicity: Relevance to the free radical hypothesis of Alzheimer's disease, *Neurochem. Res.*, in press.

Tomiyama, T., Asano, S., Suwa, Y., Moria, T., Kataoka, K., Mori, H., and Endo, N., 1994, Rifampicin prevents the aggregation and neurotoxicity of amyloid β protein in vitro, *Biochem. Biophys. Res. Commun.* **204**:76–83.

Tomiyama, T., Shoji, A., Kataoka, K.-I., Suwa, Y., Asano, S., Kaneko, H., and Endo, N., 1996, Inhibition of amyloid β protein aggregation and neurotoxicity by rifampicin, *J. Biol. Chem.* **271**:6839–6844.

Tomski, S., and Murphy, R. M., 1992, Kinetics of aggregation of synthetic β-amyloid peptides, *Arch. Biochem. Biophys.* **294**:630–638.

Trad, C. H., and Butterfield, D. A., 1994, Menadione induced cytotoxicity effects of human erythrocyte membranes studies by electron paramagnetic resonance, *Toxicol. Lett.* **73**:145–155.

Trad, C. H., James, W., Bhardwaj, A., and Butterfield, D. A., 1995, Selective labeling of membrane protein sulfhydryl groups with methanethisulfonate spin label, *J. Biochem. Biophys. Methods* **30**:287–299.

Umhauer, S. A., Isbell, B. I., and Butterfield, D. A., 1992, Spin labeling of membrane proteins in mammalian brain synaptic plasma membranes: Partial characterization, *Analy. Lett.* **25**:1201–1215.

Volterra, A., Trotti, D., Tromba, C., Floridi, S., and Racagni, G., 1994, Glutamate uptake inhibition by oxygen free radicals in rat cortical astrocytes, *J. Neurosci.* **14**:2924–2932.

Yan, S. D., Chen, X., Schmidt, A.-M., Brett, J., Godman, G., Zou, Y.-S., Scott, C. W., Caputo, C., Frappier, T., Smith, M. A., Perry, M. A., Yen, S. H., and Stern, D., 1994, Glycated tau protein in Alzheimer disease: A mechanism for induction of oxidant stress, *Proc. Natl. Acad. Sci. USA* **91**:7787–7791.

Yankner, B. A., Duffy, L. K., and Kirschnier, D. A., 1990, Neurotrophic and neurotoxic effects of amyloid β-protein: Reversal by tachykinin neuropeptides, *Science* **250**:279–282.

Yatin, S. M., Link, C. D., and Butterfield, D. A., 1998, *In-vitro* and *in-vivo* oxidative stress associated with Alzheimer's amyloid β-peptide (1-42), Submitted for publication.

Yatin, S. M., Aksenov, M., and Butterfield, D. A., 1999, The antioxidant vitamin E modulates amyloid β-peptide-induced creatine kinase inhibition and increased protein oxidation: Implications for the free radical hypothesis of Alzheimer's disease, *Neurochem. Res.*, in press.

Zagorski, M. G., Shao, H., Ma, K., Yang, J., and Papolla, M., 1998, The solution structures of Amyloid β-(1–40) and β-(1–42) peptides and the utility of NMR in the design of β-amyloid inhibitors, *Neurobiol. Aging* **19**:S44.

Zhou, Y., Gopalakrishnan, V., and Richardson, J. S., 1996, Actions of neurotoxic β-amyloid on calcium homeostasis and viability of PC12 cells are blocked by antioxidants but not by calcium channel antagonists, *J. Neurochem.* **67**:1419–1425.

Chapter 26

Oxidative Pathology in Amyotrophic Lateral Sclerosis

Robert H. Brown, Jr.

1. CLINICAL INTRODUCTION

Amyotrophic lateral sclerosis (ALS) is a devastating neurological disease caused by death of motor neurons in the brain and spinal cord. Its essential clinical aspect is unrelenting paralysis that usually begins focally and then spreads. Most cases of ALS reveal evidence of concurrent corticospinal (upper) and spinal (lower) motor neuron death. The mean age of onset in ALS is 55 years; the duration is typically 3–5 years. About 10% of cases are inherited as an autosomal dominant trait (Mulder *et al.*, 1986); the familial and sporadic cases are clinically indistinguishable. The primary pathological feature of ALS is motor neuron death (Brownell *et al.*, 1970). Affected neurons often reveal cytoskeletal pathology such as ubiquitin-positive, cytoskeletal inclusions in the cytoplasm and proximal axons (Leigh and Swash, 1991; Leigh *et al.*, 1988).

2. GENETIC ANALYSIS IN FAMILIAL ALS

Several factors are hypothesized to trigger motor neuron cell death in ALS. A partial list encompasses infections (by either typical or atypical viral agents); exogenous toxins such as heavy metals; autoimmune reactions to calcium channels on motor neurons (Appel *et al.*, 1994); abnormal DNA repair; and enhanced sensitivity to excitotoxins such as glutamate (Coyle and Puttfarcken, 1993). Whatever the speculated causes of ALS, it

Robert H. Brown, Jr. Day Neuromuscular Research Laboratory, Massachusetts General Hospital, Charlestown, Massachusetts 02129.

Reactive Oxygen Species in Biological Systems, edited by Gilbert and Colton. Kluwer Academic / Plenum Publishers, New York, 1999.

is clear that some cases are familial (FALS). There are, for example, two types of juvenile ALS. The type that is recessively inherited begins in the first decade and progresses slowly over decades (Ben Hamida *et al.*, 1990); this type is genetically associated with a locus on human chromosome 2q33 (Hentati *et al.*, 1994). A second form is dominantly inherited and is genetically linked to chromosome 9. Perhaps most importantly, the major subset of FALS is an adult-onset, rapidly progressive disorder that is clinically indistinguishable from sporadic ALS. In 1993, a large, collaborative group reported that affected individuals in some families with this type of ALS have mutations in the gene encoding the protein copper–zinc superoxide dismutase (SOD1) (Rosen *et al.*, 1993). More than 50 such mutations have now been identified selectively in familial ALS.

3. SUPEROXIDE DISMUTASE AND FAMILIAL ALS—THE FREE RADICAL HYPOTHESIS

The SOD1 gene and protein have been extensively studied. Highly conserved in evolution, the Cu/Zn SOD gene probably evolved before the divergence of bacteria and eukaryotes (Imlay and Imlay, 1996). The SOD1 gene is constitutively expressed in all cells in eukaryotes (Fridovich, 1986a,b, 1997). It is one of a family of at least three different SOD molecules including MnSOD (SOD2, prominently expressed in mitochondria) and a glycosylated form of Cu/ZnSOD (SOD3) that binds to extracellular matrix and thus is potentially an important extracellular antioxidant. The SOD1 protein is composed of 153 amino acids. Each molecule of functional SOD1 has one atom each of copper and zinc. The protein functions as a homodimer. As illustrated in Figure 1, SOD1 catalyzes the dismutation of the superoxide anion $\cdot O_2^-$ to hydrogen peroxide (H_2O_2). H_2O_2 is then converted to H_2O by either glutathione peroxidase or catalase. Because each superoxide anion has an unpaired electron, $\cdot O_2^-$ is a free radical. This suggested that the mutations in SOD1 might injure motor neurons by disturbing free radical homeostasis or, more generally, that the biochemical processes killing motor neurons in SOD1-associated ALS, and possibly all ALS, might entail oxidative injury, perhaps to subcellular organelles and/or specific types of molecules within motor neurons. This "free radical hypothesis" in ALS is also predicated on the observation that $\cdot O_2^-$ interacts extremely rapidly with nitric oxide to form peroxynitrite ($ONOO^-$). The latter is reactive by itself and reacts with metals to generate highly reactive hydroxyl radicals.

Several experiments have studied the free radical hypothesis in FALS and to characterize the properties of mutant SOD1. Why more than 50 different mutations in the SOD1 molecule all cause motor neuron cell death remains problematic, but a few salient points have emerged from studies of mutant SOD1. The first is that the mutations do not uniformly diminish the normal dismutation activity of SOD1. When assessed in red cells and in brain tissues, total SOD1 activity of ALS patients with SOD1 mutations is variably reduced (Bowling *et al.*, 1993; Robberecht *et al.*, 1993), in part because the mutant SOD1 protein is reduced in quantity (Bowling *et al.*, 1995). When expressed in *cos* cells, some SOD1 mutants (e.g., G37R) have normal specific activity, while the activity of others (e.g., G85R) is reduced. However, the half lives of all of the mutants are significantly reduced (Borchelt *et al.*, 1994). It does not appear that the mutant SOD1 protein exerts a

Normal

(1) Dismutation

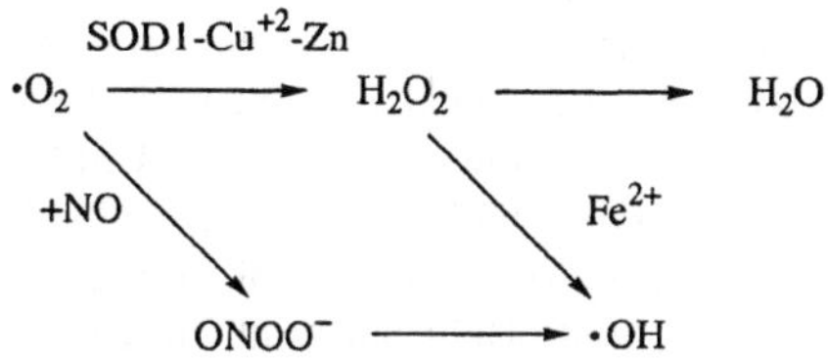

$$
\begin{array}{lrcl}
\text{a.} & \cdot O_2^- + SOD\text{–}Cu^{2+} & \rightarrow & O_2 + SOD\text{–}Cu^{1+} \\
\text{b.} & \cdot O_2^- + SOD\text{–}Cu^{1+} + H^+ & \rightarrow & HO_2^- + SOD\text{–}Cu^{2+} \\
\text{c.} & HO_2^- + H^+ & \rightarrow & H_2O_2 \\
\hline
 & 2\cdot O_2^- + 2H^+ & \rightarrow & H_2O_2
\end{array}
$$

Mutant

(2) Tyrosine Nitration

$$
\begin{array}{lrcl}
\text{a.} & SOD\text{–}Cu^{2+} + ONOO^- & \rightarrow & SOD\text{–}CuO\text{–}NO_2^+ \\
\text{b.} & SOD\text{–}CuO\text{–}NO_2^+ + H\text{–tyr–}R & \rightarrow & SOD\text{–}Cu^{2+} + OH^- \\
\hline
 & ONOO^- + H\text{–tyr–}R & \rightarrow & NO_2\text{–tyr–}R + OH^-
\end{array}
$$

(3) Peroxidation

$$
\begin{array}{lrcl}
\text{a.} & H_2O_2 + SOD\text{–}Cu^{2+} & \rightarrow & SOD\text{–}Cu^{1+} + O_2^- + 2H^+ \\
\text{b.} & H_2O_2 + SOD\text{–}Cu^{1+} & \rightarrow & SOD\text{–}Cu^{2+}(\cdot OH) + OH^- \\
\text{c.} & SOD\text{–}Cu^{2+}(\cdot OH) + DMPO & \rightarrow & DMPO(\cdot OH) + SOD\text{–}Cu^{2+} \\
\hline
 & 2H_2O_2 + DMPO & \rightarrow & DMPO(\cdot OH) + O_2^- + OH^- + 2H^+
\end{array}
$$

FIGURE 1. Selected reactions of superoxide dismutase. SOD1 normally dismutates the superoxide anion to form hydrogen peroxide. An alternate reaction to the normal dismutation activity of SOD1 is formation of peroxynitrite. Two alternate pathways for the formation of hydroxyl radicals are illustrated. The reaction schemes illustrate the component reactions that comprise normal dismutation [Reaction (1), parts a, b and c] and either tyrosine nitration [Reaction (2)] or peroxidation [Reaction (3)]. $ONOO^-$, peroxynitrite; ·OH, hydroxyl radical; GSHpx, glutathione peroxidase; tyr, tyrosine. DMPO is the hydroxyl radical trapping agent 5,5-dimethy-1-pyrroline N oxide.

dominant negative inhibitory influence on the wild-type molecule in heterodimeric SOD1 proteins (Borchelt *et al.*, 1995).

The argument that the mutant SOD1 protein does not impair motor neurons through loss of dismutation activity is supported by the finding that transgenic mice that express high levels of mutant SOD1 proteins develop an adult-onset motor neuron disease (Gurney, 1994). The mutants studied in transgenic mice include G93A (Gurney, 1994), G95R (Ripps *et al.*, 1995), G37R (Wong *et al.*, 1995), and A4V (J. Rothstein, personal communication). The motor neuron paralysis in these animals occurs in the face of above normal levels of SOD1 activity (Gurney, 1994); control mice expressing equivalent levels of wild-type SOD1 do not develop motor neuron pathology. To a remarkable extent, these mouse models show many aspects of human ALS. The mouse motor neuron disease begins in adult mice, typically starting focally, and spreading to be fatal within weeks. In addition, the murine ALS is specific to spinal motor neurons. Like human ALS, this mouse ALS is minimally but reproducibly slowed by riluzole. Unlike human ALS, the motor neuron disease in mice expressing high levels of the G93A and G37R mutants is accompanied by vacuolar degeneration in the neuropil, probably starting in mitochondrial membranes (Mourelatos *et al.*, 1996; Dal Canto and Gurney, 1995; Wong *et al.* , 1995). This feature may be a consequence of extremely high levels of the mutant SOD1 molecule; there is not an obvious counterpart in human ALS pathology. Dramatic cystoskeletal pathology arises in the G85R mice (Bruijn *et al.*, 1997).

More recently, the argument that loss of superoxide anion dismutation does not cause ALS has been strengthened by the finding that mice with targeted inactivation of the SOD1 gene, and no tissue SOD1 activity, do not develop an ALS phenotype. Early in life they show very subtle evidence of motor neuron dysfunction and enhanced motor neuron death following axonal injury but do not spontaneously become paralyzed with fulminant motor neuron disease (Reaume *et al.*, 1996). At about 1 year of life they develop a slowly progressive, nonlethal motor neuropathy; strikingly, this is not accompanied by any pathology in spinal motor neurons; electrophysiological studies indicate that aged motor neurons devoid of SOD1 activity develop an axonopathy (Flood *et al.*, 1996).

It is therefore probable that the mutant SOD1 protein is injurious to motor neurons through some aberrant property conferred by the mutations. The nature of this acquired, cytotoxic property has not been definitively identified. Many hypotheses share the premise that the mutant protein is less stable in the mutant SOD1 molecule as indicated by the shortened half-life of the mutant protein (Borchelt *et al.* , 1994) and its diminished affinity for copper and zinc (Carri *et al.*, 1994; Nishida *et al.*, 1994).

An important predication of the presumed loss of stability of the mutant protein is that it may be promiscuous, interacting with substrates that would normally be sterically excluded from the active channel. For example, it has been demonstrated that wild-type SOD1 interacts with peroxynitrite to form nitronium ions that donate nitrate groups to tyrosines (Ischiropoulos *et al.*, 1992). It is suggested that the conformation of the mutant active channel is less rigidly defined than wild-type and more open to peroxynitrite as a potential substrate (Beckman *et al.*, 1993). This is predicted to enhance tyrosine nitration. If the nitrated tyrosines are functionally important, for example by blocking assembly of neurofilaments (Beckman *et al.*, 1994) or phosphorylation of tyrosine kinase receptors (Kong *et al.*, 1996), the increased nitration may be toxic to motor neurons. [See Reaction (2) of Figure 1).

It has also been postulated that mutant SOD1 has an enhanced ability to interact with hydrogen peroxide, normally the product of the dismutation reaction [Figure 1, Reaction (3)]. It has long been recognized that wild-type SOD1 can function as a peroxidase, reacting with hydrogen peroxide to form hydroxyl radicals (Hodgson and Fridovich, 1975). Within the last 2 years, two groups have employed hydroxyl radical trapping methods to demonstrate that this peroxidation reaction is facilitated in the SOD1 proteins with FALS-associated mutations such as A4V (Wiedau-Pazos *et al.*, 1996) and G93A

(Yim *et al.*, 1996, 1997). Although the capacities of the mutant and wild-type SOD1 proteins for dismutation (V_{max}, k_{cat}) are identical, the mutant proteins have increased affinities for H_2O_2 (reduced k_m). Indeed, there is a loose inverse correlation between k_m and the severity of the resulting disease; thus, the most dramatic reduction in k_m is seen with the A4V mutant (13 mM versus 44 mM for wild type) that typically is correlated with survival of only about 1 year (Cudkowicz *et al.*, 1997). A critical component of this peroxidation reaction is reduction of the valence of copper from +2 to +1 [Figure 1, Reactions (3a) and (3b)]. The copper ion is apparently more available for reduction by ascorbate in the mutant as compared with the wild-type SOD1 protein (Lyons *et al.*, 1996); this may be a consequence of altered diminished affinity of the mutant protein for zinc. If, indeed, mutant SOD1 molecules have enhanced peroxidase activity *in vivo*, this may be injurious to neurons. Direct reactions of the highly reactive hydroxyl radical can be damaging. Moreover, enhanced peroxidation may be indirectly damaging through formation of reactive intermediates with longer half lives than the hydroxyl radical.

A third mechanism to explain the toxicity of mutant SOD1 to motor neurons is that the mutant protein somehow activates programmed cell death pathways. This view is broadly consistent with several recent investigations implicating SOD1 in events that signal events upstream of activation of caspase activity essential to apoptotic death. There is now abundant evidence that reductions in levels of wild-type SOD1 predispose neurons to apoptosis *in vivo* (Greenlund *et al.*, 1995; Rothstein *et al.*, 1994; Troy and Shelanski, 1994). Particularly striking is evidence that mutant SOD1 protein is also "pro-apoptotic" in a transformed neuronal cell line, even without a reduction in levels of total SOD1 activity (Rabizadeh *et al.*, 1995). Although the mechanism of this pro-apoptotic influence is not known, it apparently entails generation of one or more reactive oxygen species and requires copper (Wiedau-Pazos *et al.*, 1996). Forced expression of the mutant SOD1 protein induces apoptosis in primary neuronal cultures, whether the mutant cDNA is injected into the neuronal nucleus or administered to the cytoplasm (e.g., via transfection or infection using viral vectors) (Durham *et al.*, 1997; Roos *et al.*, 1997). This effect is blocked by inhibition of cell death enzymes of the caspase family such as interleukin-1β converting enzyme (ICE) (Roos *et al.*, 1997). Presumably, reactive oxygen species signal events that activate one or more apoptotic pathways.

Two recent studies of apoptotic cell death pathways in FALS mice support the proposal that the mutant SOD1 molecule is pro-apoptotic. In one study, FALS mice were bred to coexpress the mutant SOD1 molecule and a dominant negative inhibitor of ICE in which the active-site cysteine is replaced by glycine. The caspase inhibitor did not alter the timing of disease onset but did prolong survival in the FALS mice by approximately 2 weeks (27.0 versus 11.7 days, $p < 0.0002$) (Friedlander *et al.*, 1997). A second study analyzed mice doubly transgenic for mutant SOD1 and bcl-2, the human proto-oncogene that inhibits programmed cell death. FALS mice with the bcl-2 transgene showed onset at 203 ± 6 days, whereas in control FALS mice without the bcl-2 transgene, onset was at 170 ± 3 days ($p < 0.001$) (Kostic *et al.*, 1997). Once triggered, the disease advanced at a similar pace in both groups. Although the bcl-2 transgene did not affect the numbers of motor neurons present at birth, it reduced spinal motor neuron death as gauged by cell counts and two markers of cell injury, c-jun and ubiquitin (Kostic *et al.*, 1997).

Given these observations in the murine ALS model, it is intriguing that apoptotic cells have been detected in postmortem ALS spinal cord (Troost *et al.*, 1995). Apoptosis was evident in all of 12 ALS cords as measured by *in situ* end labeling of fragmented DNA. It was not confined to motor systems. The authors state that "none of the non-neurologic controls showed as much apoptosis as any of the ALS cases." Whether other investigators will validate these findings remains to be seen. Another postmortem study used *in situ* hybridization to estimate levels of bcl-2 and the pro-apoptotic cell death gene, bax (Mu *et al.*, 1996). Hybridization to bcl-2 mRNA was prominent in spinal cord motor neurons and evident, but less prominent, in Clarke's column nuclei and dorsal horn sensory neurons. Levels in ALS patients were diminished relative to controls. Bax mRNA was evident in many cell types but was increased in ALS motor neurons. Although these observations also await independent confirmation, they are notably consistent with the findings *in vitro* and in ALS mice pointing to a role for cell death gene activation as one component of the death process in SOD1-induced motor neuron death.

Another explanation of the neurotoxicity of the mutant SOD1 molecule is that the mutant enzyme has a heightened affinity for some cellular constituent and that binding of the mutant SOD1 protein to the abnormal target molecules or structures disrupts cellular function. Yeast two-hybrid methods have demonstrated that mutant but not wild-type SOD1 binds to Lys tRNA synthetase and to a membrane protein in the endoplasmic reticulum (Kunst *et al.*, 1997).

It is also possible that the mutant SOD1 protein is cytotoxic by virtue of its diminished buffering of copper or zinc, or its precipitation to form protein aggregates. SOD1-positive aggregates have been observed within the neuronal cytoplasm in both sporadic and familial ALS. They are colocalized with brain NOS, neurofilament subunits, calmodulin and cyclic GMP (Chou *et al.*, 1996a,b). Moreover, these aggregates are positive for nitrotyrosine, suggesting an intimate relationship between NOS, nitrotyrosine, and aggregated SOD1 (see above). It is conceivable that these protein aggregates are toxic, either indirectly by promoting adverse reactions such as nitration chemistry, or perhaps directly through disruption of one or more systems for protein degradation and reutilization.

4. EVIDENCE FOR OXIDATIVE TOXICITY IN ALS

Whether there is one predominant mechanism for the lethality of the SOD1 defects is entirely unclear at present. It is certainly conceivable that the mutant SOD1 protein may have more than one toxic property. The importance of different adverse properties may depend on the stage of the death process in any given cell. Despite the precise mechanisms at play, several of these proposed mechanisms entail generation of free radical species. Accordingly, an important prediction of the free radical hypothesis in ALS is that there may be chemical evidence of oxidative injury in ALS brain tissues or that measures that reduce oxidative stress may ameliorate some aspect of disease progression in ALS. To the extent the pathogenesis of sporadic and familial ALS overlap, sporadic cases may also entail oxidative pathology. Indeed, Bredesen and associates recently proposed that some cases of sporadic ALS may be caused by age-dependent biochemical modifications of the wild-type SOD1 molecule that mimic the toxic chemical reactions generated by mutations in SOD1 in FALS (Bredesen *et al.*, 1997). Although an initial investigation, prior to knowledge of SOD1 mutations in FALS, failed to find evidence of oxidative pathology in ALS (Mitchell *et al.*, 1987), several subsequent reports have begun to delineate a pattern of findings to the contrary.

Proteins undergo a variety of reactions with reactive oxygen. The peptide backbone may be attacked to form carbon-based radicals that enter into reactions such as cross-linking or cleavage. The amino acids may be attacked to form derivatives such as methionine sulfoxide or nitrated tyrosines. Through several pathways, carbonyl derivatives may be formed (Berlett and Stadtman, 1997). With the exception of reductases to salvage methionine from methionine sulfoxides, biological systems do not have mechanisms to reverse oxidative protein alteration (Berlett and Stadtman, 1997). For this reason, quantitative measures of protein modifications provide an index of the level of oxidative injury in a tissue. Accordingly, protein carbonyl groups have been used to gauge oxidative stress in cells and tissues in aging and after injury (Smith *et al.*, 1991). To date, there have been relatively few analyses of oxidative protein modification in ALS. A study in 1993 reported that carbonyl proteins were elevated in frontal cortex from brains of sporadic but not familial ALS patients (Bowling *et al.*, 1993). Analogously, carbonyl groups were found to be increased in the spinal cords of sporadic ALS patients in a second study (Shaw *et al.*, 1995). More recently, carbonyl proteins were found to be elevated by about 50% in motor cortex but not parietal cortex or cerebellum from 16 patients with sporadic ALS as compared with 16 controls. No increases in the same region were detected in 7 patients with familial ALS (including 4 with the A4V mutation) (Ferrante *et al.*, 1997a).

Like proteins, DNA is also subject to diverse oxidation reactions that modify its bases or sugars, cross-link DNA to itself or to protein, and cause strand breaks and other defects (Henle and Linn, 1997; Dizdaroglu, 1993). One useful measure of oxidative DNA injury is the level of the guanine adduct 8-hydroxy-2-deoxyguanine, as quantitated by HPLC. This adduct was elevated in samples of total DNA from spinal cord tissues of two cases of sporadic ALS as compared with two controls (Fitzmaurice *et al.*, 1996). A larger study of 16 ALS cases documented a 40% increase in this substance in nuclear DNA from motor but not parietal or cerebellar cortex of sporadic ALS patients. Levels were not increased in FALS (Ferrante *et al.*, 1997a).

It is also possible to quantify measures of oxidative alteration to lipid molecules by quantifying levels of derivatives such as thiobarbituric acid-reactive substances. In a single report, levels of modified lipids were not elevated in sporadic or familial ALS when measured quantitatively using HPLC; however, there was a tendency toward increased immunostaining of large neurons and endothelial cells in both sporadic and familial ALS (Ferrante *et al.*, 1997a).

As described above, another chemical pathway deriving from an oxidative intermediate (superoxide anion) is the formation of peroxynitrite. The presence of peroxynitrite is usually assessed indirectly through analysis of levels of its products, such as nitrotyrosine. An important prediction of the nitration hypothesis presented above is that nitrotyrosine levels may be elevated in ALS tissues. This possibility has now been examined both in ALS mice and in human ALS. In the G37R ALS mice, levels of free 3-nitrotyrosine in spinal cord are double or triple those in wild-type littermates or in mice that express high levels of wild-type SOD1. The levels were elevated throughout the life span of the mice. In the G93A mice, 3-nitrotyrosine levels were also increased in both the spinal cord and the cerebral cortex but not in other regions (Ferrante *et al.*, 1997b). As assessed by salicylate trapping, hydroxyl radical production was not enhanced in either the G37R or the G93A animals (Bruijn *et al.*, 1997a; Ferrante *et al.* , 1997b). Levels of malondialdehyde, a marker for lipid peroxidation, were normal in the G37R animals in spinal cord (Bruijn *et al.*, 1997) but increased in frontal cortex in the G93A animals.

Analogous studies of nitrotyrosine formation have been described in a study of human ALS. In 16 ALS patients, lumbosacral tissue levels of 3-nitrotyrosine were nearly double control levels; the increases were evident in both sporadic and familial ALS cords. Elevated to a greater degree were levels of 3-nitro-4-hydroxyphenacetic acid, formed at least in part from the decomposition of 3-nitrotyrosine (Beal *et al.*, 1997). By contrast with the 3-nitrotyrosine results, but consistent with the findings in the G37R mice, levels of salicylate-trapped hydroxyls were not increased in spinal cord tissues from sporadic or SOD1-associated ALS patients (Bruijn *et al.* , 1997a).

In considering the role of free radicals in ALS pathogenesis in ALS, it is instructive to examine factors that may predispose to oxidative cytotoxicity. One such factor may be the tissue levels of transition metals such as iron, manganese, and copper. Several investigations have concluded that levels of brain tissue iron are elevated in ALS. This has been described in analyses of bulk iron levels in cervical cord of ALS/Parkinson's disease patients (Mitzumoto *et al.*, 1980) and in patients with sporadic ALS (Kurlander and Patten, 1979). It has also been reported in analyses of subcellular metal concentration and location. In a study of ventral cervical cord from 5 patients with sporadic ALS, levels of iron determined with laser microprobe mass spectrometry were increased nearly twice those of controls (Karsarskis *et al.*, 1995); the increase was prominent in neuronal cytoplasm and the nucleus but not in capillaries. As assessed by neutron activation analysis, mean spinal cord iron levels were increased in 38 ALS cases as compared with 22 controls (Ince *et al.*, 1994). It has also been reported that iron levels are elevated in ALS kidney (see discussion in Karsarskis *et al.*, 1995). These biochemical findings are supported by MRI analyses indicating an increased deposition of iron in ALS brains (Imon *et al.*, 1995; Oba *et al.*, 1993). Levels of iron in the cerebrospinal fluid (CSF) have not been described extensively; measurements of iron in CSF of 25 patients with neurological diseases (including 4 with ALS) revealed elevations in 2 patients (1 with ALS and 1 with Parkinson's disease) (Kjellin, 1967). In the context of the free radical hypothesis in ALS, it is tempting to speculate that increased tissue and perhaps CSF levels of iron may initiate toxic oxidative reactions. That is, it is at least conceivable that iron might play a causal or catalytic role in the pathogenesis of neural cell death. On the other hand, it remains plausible that the iron reflects a secondary consequence of neural injury mediated, for example, by injury-provoked increases in iron binding proteins such as transferrin (Graeber *et al.*, 1989). One such iron-binding protein documented to be elevated within glial cells in ALS cortex is lactotransferrin, a 80-kDa glycoprotein that binds not only iron but also aluminum, manganese, magnesium, zinc, and copper (Leveugle *et al.*, 1994).

Studies of other transition metals have been less conclusive. In an analysis of manganese levels in 24 brain areas, a single ALS case showed no abnormalities (Larsen *et al.*, 1981). An evaluation of spinal cord from 5 cases of sporadic ALS revealed no abnormalities of total manganese levels using inductively coupled plasma spectroscopy; however, these investigators described an alteration in the distribution of this metal, with relatively higher concentrations in the anterior cords in ALS patient samples (Kihira *et al.*, 1990). By contrast, in 2 ALS cases studied with neutron activation analysis, manganese levels were increased anteriorly and laterally (Miyata *et al.*, 1980). No consistent abnormalities of copper levels have been identified. A recent study failed to find abnormalities of total or free copper or zinc levels in blood, plasma, and red blood cells of 12 FALS patients with seven different SOD1 mutations (Radunovic *et al.*, 1997). In reviewing these data, as in evaluating the studies of iron, one must again consider the possibility that potential alterations in metal levels may arise secondarily to altered levels of metal-binding proteins. In one study of 18 control and 24 ALS autopsies, kidney tissues from the ALS patients had heightened levels of metallothioneins, a group of cysteine-rich proteins that bind multiple metals including copper, cadmium, and zinc (Sillevis-Smit *et al.*, 1992).

Another approach to judging the likelihood that free radical toxicity is important in ALS is to examine the adequacy of both small molecule and protein antioxidant defense systems in ALS patients. Unfortunately, there are currently almost no data on this point. A single group has reported that levels of both α-tocopherol and α-tocopherol quinone (α-TQ) are reduced in CSF of sporadic ALS patients (Tohgi *et al.*, 1996); particularly dramatic is the observation that all 14 controls were above and all 13 ALS patients were below a cutoff α-TQ level of 3.0 nM. This is consonant with the observation that the G93A ALS mice fail to show the expected, age-dependent increase in brain vitamin E levels, a finding indicating there may be ongoing oxidative pathology in these animals (Gurney *et al.*, 1996). More specifically, because vitamin E is lipophilic, the relative deficiency of vitamin E may denote oxidation of membrane lipids.

Data concerning activities of the protein antioxidant glutathione peroxidase are conflicting. In an analysis of 9 ALS brains, glutathione peroxidase activity was reduced in the precentral gyrus but not cerebellar cortex in the ALS tissues (Przedborski *et al.*, 1992). Two analogous studies documented unchanged (Shaw *et al.* , 1995) or increased (Ince *et al.*, 1994) activity of glutathione peroxidase in ALS spinal cord. Whole blood glutathione peroxidase levels have been reported to be decreased in ALS (Mitchell *et al.*, 1993). Levels of glutathione have not been examined in CSF or tissues of ALS patients. However, a single report of increased binding sites for glutathione in 5 ALS brains was interpreted to indicate a deficiency of glutathione in these tissues (Lanius *et al.*, 1993). Activity levels of Cu/ZnSOD (SOD1), manganese SOD (SOD2), and catalase are normal in sporadic ALS brains (Przedborski *et al.*, 1996).

Still another approach to assessing whether free radical chemistry is important in neuronal death in ALS is to test the efficacy of antioxidant compounds on disease onset and progression. At least two antioxidant compounds have a mildly palliative effect in the mouse ALS model. Vitamin E significantly slows the onset of weakness in the ALS mice but does not slow the rate of progression of the illness once it begins (Gurney *et al.*, 1996). A novel antioxidant, buckminsterfullerene, also slightly prolongs survival in the same ALS mice (Dugan *et al.*, 1996). As described above, overexpression of human bcl-2 ameliorated the disease course in these mice. Although this substance acts to inhibit programmed cell death, it also has antioxidant properties (Hockenbery *et al.*, 1993). It is therefore possible that the primary action of bcl-2 in the mice is as an antioxidant, independently of its anti-apoptotic properties (although it is not yet known whether these properties of bcl-2 are fully dissociable).

In human studies, a trend toward improved survival was detected with the antioxidant *N*-acetylcysteine in 35 ALS patients (Louwerse *et al.*, 1995); this did not achieve a level of significance (*p* 0.06). Although there are scattered anecdotal reports in the literature to

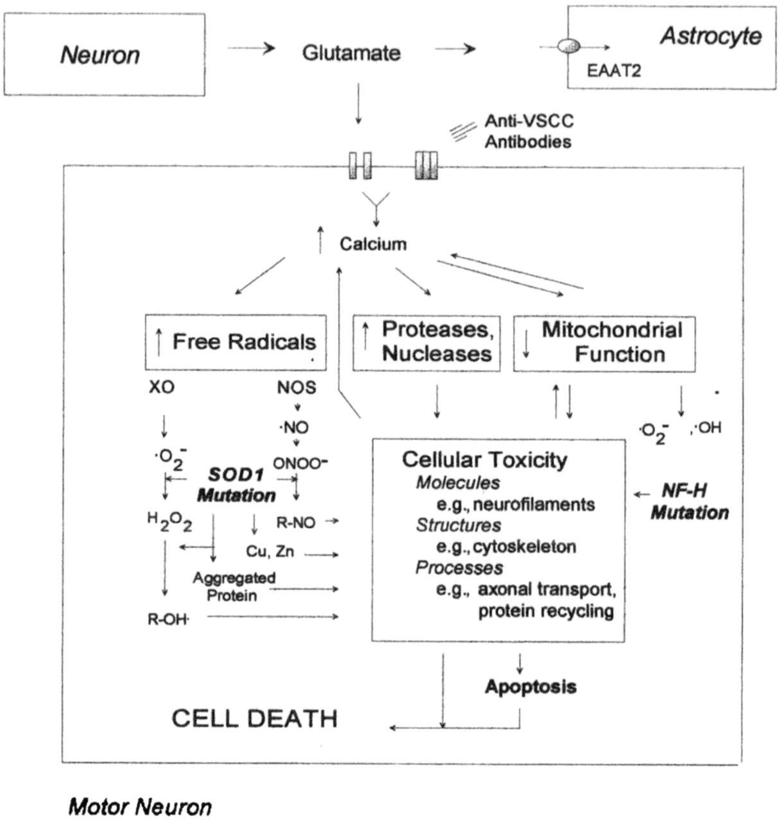

FIGURE 2. Motor neuron cell death in amyotrophic lateral sclerosis. This illustrates events that may underlie motor neuron death in ALS (see text). Glutamate triggers or sustains this cascade of events, via elevated levels of cytosolic calcium. Heightened levels of free calcium invoke the chemistry of normal and mutant SOD1 by increasing generation of both superoxide anion and nitric oxide. Through one or more pathways, mutant SOD1 mediates toxicity to subcellular organelles and molecules. Ultimately, this compromises neuronal viability leading to death, at least in part via activation of one or more apoptotic pathways.

Increased levels of cytosolic calcium may also be injurious through activation of proteases and nucleases and impairment of mitochondrial function. Two important consequences of mitochondrial failure include release of reactive intermediates from mitochondria (superoxide anion, hydrogen peroxide) and diminished ATP production. The latter may exaggerate sensitivity to glutamate by causing rundown of cellular ion pumps and slow depolarization. This process may be self-sustaining: as the cellular gradients run down, depolarization advances, further enhancing the action of glutamate. Moreover, as proteases are activated and cell membranes become leaky, calcium fluxes into the cell are likely to increase.

At least two reactive intermediates (hydrogen peroxide and nitric oxide) are lipophilic and thus can diffuse across the cell membrane and into the extracellular compartment to affect neighboring cells. It is conceivable that they might diffuse into adjacent neurons and instigate reactions analogous to those depicted here. It is also possible that a gradient of reactive substances emanating from an affected motor neuron could inactivate glutamate transporters (EAAT2) on adjacent astroctyes, diminishing astrocytic glutamate uptake, and enhancing the excitotoxic stimulus.

Shown in bold are two proteins whose gene defects have been implicated as possible causes of ALS: SOD1 (Cu/Zn cytosolic superoxide dismutase) and NF-H (neurofilament heavy subunit).

the contrary, vitamin E has never proven effective in patients with ALS (Reider and Paulson, 1997). A controlled trial of vitamin E with deprenyl in sporadic ALS also did not show benefit (D. Lange, personal communication).

5. OVERVIEW OF CELL DEATH IN THE INHERITED MOTOR NEURON DISEASES

These considerations define a series of pathways that may interact to initiate motor neuron cell death in ALS (Figure 2). Presumably one or more relatively upstream factors initiate the death process; these might include elevated levels of extracellular glutamate or antibodies directed against voltage sensitive calcium channels (VSCC). For example, glutamate levels may be elevated because of enhanced release from presynaptic terminals or diminished uptake activity of the major astrocytic glutamate transporter, EAAT2. Either stimulus will potentially elevate levels of free, cytosolic calcium (Choi, 1988). This, in turn, is envisioned to trigger parallel events including activation of enzymes that generate free radicals such as nitric oxide synthase (NOS) and xanthine oxidase (XA), activation of proteases and nucleases, and increased flux of calcium into mitochondria. Calcium-mediated stimulation of NOS and XA will directly augment levels of nitric oxide and superoxide anion and indirectly increase levels of peroxynitrite and possibly hydroxyl radical. Experimental data (described above) suggest that formation of hydroxyl radical, peroxynitrite, and nitrated tyrosines may be accelerated in the presence of mutant SOD1. Mitochondrial calcium loading will enhance release of hydroxyl radicals and superoxide anion and impair ATP generation (Dykens, 1993). These events may become self-sustaining. For example, as ATP generation fails, neuronal sensitivity to glutamate is enhanced (Beal, 1990, 1995). Protease activation and the newly generated reactive species may damage the cellular membrane (e.g., via lipid peroxidation), further increasing calcium entry and sustaining these cytotoxic pathways. Moreover, if there are alterations in the redox status of neighboring cells, this, too, may intensify the glutamate toxicity. For example, function of the astrocytic glutamate transporter EAAT2 is reduced by intracellular oxidation of this protein (Volterra *et al.*, 1994).

The basis for the selective loss of motor neurons in ALS is difficult to comprehend, as SOD1 is constitutively expressed in all cell types. The motor selectivity may be a consequence of several factors: the formidably large size of the motor neuron and a corresponding high level of metabolic activity and cytoskeletal complexity; poor calcium buffering in the motor neuron (Alexianu *et al.*, 1994); and a high density of glutamate receptors on the motor neuron surface.

6. CONCLUSIONS

Within the last 5 years, genetic and biochemical lines of investigation have begun to converge to favor the view that one or more categories of oxidative chemistry occur during motor neuron cell death in ALS. The former demonstrate a causal relationship between defects in the SOD1 gene and motor nerve death but by no means prove that this death process is fundamentally oxidative. The latter provide increasingly credible evidence that oxidative modification of proteins and DNA occurs in ALS but do not yet establish that such modifications either instigate or sustain motor neuron death. At least four types of information will help determine whether specific oxidative reactions are required (and sufficient) to provoke motor neuron death in ALS. First, it will be important to demonstrate a direct relationship between the putative causal reactions (or levels of resulting toxic substances) and the disease course. Manipulations that augment or inhibit the offending reactions should enhance or ameliorate motor neuron death. Such measures might include inhibition of critical enzymes or administration of appropriate antioxidants. Second, measures that block events downstream of the critical oxidative activity should also ameliorate the disease course. For example, genetic inactivation of molecules important in downstream reactions should retard motor neuronal death. Third, as a corollary, one would anticipate identifying FALS pedigrees whose primary genetic defect impairs proteins acting in these downstream reactions. Fourth, if a particular family of molecules is essential as a trigger of neuron death, then entirely independent methods of generating and delivering such toxic molecules should result in the motor neuron death phenotype encountered in ALS.

Although studies of oxidative cytopathology in ALS have multiplied in the last few years, our present knowledge in this area is still quite limited. Several types of basic information should prove immediately useful as one applies the above criteria to assess the free radical hypothesis in ALS. (1) Ideally, one would like, for example, to assay brain and spinal cord directly for different reactive oxygen intermediates. Because this is difficult, it will be instructive to extend studies of biochemical markers of these intermediates along the lines enumerated above (carbonyl proteins, DNA adducts, lipid peroxides, nitrotyrosines), possibly including techniques that trap and quantity short-lived reactive intermediates. (2) An accurate description of the status of antioxidant defenses (protein and small molecule) will almost certainly help refine hypotheses about how free radical pathology arises. (3) It will be imperative to examine these broad profiles of oxidative markers and intermediates in related and very different diseases, in effect testing the hypothesis that a specific set of reactants is involved in any neurodegenerative disease that, like ALS, targets such a specific set of neurons. (4) Equally important are studies of variations in the patterns of oxidative markers and antioxidant defenses over the course of the illness and in different cell types. Such studies are essential in humans with ALS and particularly important in the ALS mice in which presymptomatic analyses are possible. The temporal and anatomic patterns of marker expression may illuminate how the disease begins (Does it start focally and spread or is there widespread oxidative pathology at the outset? Does a deficiency in a particular category of antioxidant correlate with onset?) (5) In the long term, it will be helpful to identify markers of the disease in readily available tissues and fluids (blood cells, CSF, urine, perhaps fibroblasts). This may provide surrogate markers with which one can not only easily gauge disease progression but also test possible subclinical efficacy of therapeutic agents.

7. REFERENCES

Alexianu, M., Ho, B., Mohamed, A., Bella, V. L., Smith, R., and Appel, S., 1994, The role of calcium-binding proteins in selective motoneuron vulnerability in amyotrophic lateral sclerosis, *Ann. Neurol.* **36**:846–858.

Appel, S., Smith, R., Engelhardt, J., and Stefani, E., 1994, Evidence of autoimmunity in amyotrophic lateral sclerosis [review], *J. Neurol. Sci.* **124** Suppl:14–19.

Beal, M. F., 1990, Does impairment of energy metabolism result in excitotoxic neuronal death in neurodegenerative illnesses? *Ann. Neurol.* **31:**119–130.

Beal, M., 1995, Aging, energy, and oxidative stress in neurodegenerative diseases [review], *Ann. Neurol.* **38:**357–366.

Beal, M., Ferrante, R., Browne, S., Matthews, R., Kowall, N., and Brown, R., Jr., 1997, Increased 3-nitrotyrosine in both sporadic and familial amyotrophic lateral sclerosis, *Ann. Neurol.* in press.

Beckman, J. S., Carson, M., Smith, C. D., and Kuppenol, W. H., 1993, ALS, SOD, and peroxynitrite, *Nature* **364:**584.

Beckman, J. S., Ye, Y. Z., Anderson, P. G., Chen, J., Accavitti, M. A., Tarpey, M. M., and White, C. R., 1994, Extensive nitration of protein tyrosines in human atherosclerosis detected by immunohistochemistry, *Biol. Chem. Hoppe-Seyler* **375:**81–88.

Ben Hamida, M., Hentati, F., and Ben Hamida, C., 1990, Hereditary motor system diseases (chronic juvenile amyotrophic lateral sclerosis). Conditions combining a bilateral pyramidal syndrome with limb and bulbar atrophy, *Brain* **113:**347–363.

Berlett, B., and Stadtman, E., 1997, Protein oxidation in aging, disease, and oxidative stress, *J. Biol. Chem.* **272:**20313–20316.

Borchelt, D. R., Lee, M. K., Slunt, H. S., Guarnieri, M., Xu, Z. S., Wong, P.-C., Brown, R. H., Jr., Price, D. L., Sisodia, S. S., and Cleveland, D. W., 1994, Superoxide dismutase 1 with mutations linked to familial amyotrophic lateral sclerosis possesses significant activity, *Proc. Natl. Acad. Sci. USA* **91:**8292–8296.

Borchelt, D. R., Guarnieri, M., Wong, P.-C., Lee, M. K., Slunt, H. S., Xu, Z.-S., Sisodia, S. S., Price, D. L., and Cleveland, D. W., 1995, Superoxide dismutase 1 subunits with mutations linked to familial amyotrophic lateral sclerosis do not affect wild-type subunit function, *J. Biol. Chem.* **270** (6):1–5.

Bowling, A. C., Schulz, J. B., Brown, R. H., Jr., and Beal, M. F., 1993, Superoxide dismutase activity, oxidative damage and mitochondrial energy metabolism in familial and sporadic amyotrophic lateral sclerosis, *J. Neurochem.* **61:**2322–2325.

Bowling, A., Barkowski, E., McKenna-Yasek, D., Sapp, P., Horvitz, H., Beal, M., and Brown, R. H. Jr., 1995, Superoxide dismutase concentration and activity in familial amyotrophic lateral sclerosis, *J. Neurochem.* **64:**2366–2369.

Bredesen, D., Ellerby, L., Hart, P., Wiedau-Pazos, M., and Selverstone, J., 1997, Do post-translational modifications of CuZnSOD lead to sporadic amyotrophic lateral sclerosis? *Ann. Neurol.* **42:**135–137.

Brownell, B., Oppenheimer, D. R., and Hughes, J. T., 1970, The central nervous system in motor neurone disease, *J. Neurol. Neurosurg. Psychiatry* **33:**338–357.

Bruijn, L., Beal, M., Becher, M., Schulz, J., Wong, P., Price, D., and Cleveland, D., 1997a, Elevated free nitrotyrosine levels but not protein-bound nitrotyrosine or hydroxyl radicals, throughout amyotrophic lateral sclerosis (ALS)-like disease implicate tyrosine nitration as an aberrant in vivo property of one familial ALS-liked superoxide dismutase 1 mutant, *Proc. Natl. Acad. Sci. USA* **94:**7606–7611.

Bruijn, L., Becher, M., Lee, M., Anderson, K., Jenkins, N., Copeland, N., Sisodia, S., Rothstein, J., Borchelt, D., Price, D., and Cleveland, D., 1997b, ALS-linked SOD1 mutant G85R mediates damage to astrocytes and promotes rapidly progressive disease with SOD1-containing inclusion, *Neuron* **18:**327–338.

Carri, M., Battistoni, A., Polizio, F., Desideri, A., and Rotillo, G., 1994, Impaired copper binding by H46R mutant of human Cu,Zn superoxide dismutase, involved in amyotrophic lateral sclerosis, *FEBS Lett.* **356:**314–316.

Choi, D. W., 1988, Glutamate toxicity and diseases of the nervous system, *Neuron* **1:**623–634.

Chou, S., Wang, H., and Komai, K., 1996a, Colocalization of NOS and SOD1 in neurofilament accumulations within motor neurons of amyotrophic lateral sclerosis: An immunohistochemical study, *J. Chem. Neuroanat.* **10:**249–258.

Chou, S., Wang, H., and Taniguchi, A., 1996b, Role of SOD-1 and nitric oxide/cyclic GMP cascade in neurofilament aggregation in ALS/MND, *J. Neurol. Sci.* **139:**16–26.

Coyle, J. T., and Puttfarcken, P., 1993, Oxidative stress, glutamate and neurodegenerative disorders, *Science* **262:**689–695.

Cudkowicz, M., McKenna-Yasek, D., Sapp, P., Chin, W., Geller, B., Hayden, D., Horvtiz, H., and Brown, R. H., Jr., 1997, Epidemiology of SOD1 mutations in amyotrophic lateral sclerosis, *Ann. Neurol.* **41:**210–212.

Dal Canto, M., and Gurney, M., 1995, Neuropathological changes in two lines of mice carrying a transgene for mutant human Cu,Zn SOD, and in mice overexpressing wild type human SOD: A model of familial amyotrophic lateral sclerosis, *Brain Res.* **676**:25–40.

Dizdaroglu, M., 1993, Chemistry of free radical damage to DNA and nucleoprotein, in *DNA and Free Radicals* (B. Halliwell and O. Aruoma, eds.), pp. 19–39, Ellis Horwood Chichester.

Dugan, L., Turetsky, D., Du, C., Lin, T., Lobner, D., Almli, R., Wheeler, M., and Choi, C., 1996, Carboxy-buckminsterfullerenes: Novel anti-oxidants with neuroprotective efficacy in vitro and in a mouse model of ALS, *Soc. Neurosci. Abstr.* **22**:2142.

Durham, H., Roy, J., Dong, L., and Figlewicz, D., 1997, Aggregation of mutant Cu/Zn superoxide dismutase proteins in a culture model of ALS, *J. Neuropathol. Exp. Neurol.* **56**:523–530.

Dykens, J. A., 1993, Isolated cerebral and cerebellar mitochondria produce free radicals when exposed to elevated calcium and sodium: Implications for neurodegeneration, *J. Neurochem.* **63**:584–591.

Ferrante, R., Browne, S., Shinobu, L., Bowling, A., Baik, M., MacGarvey, U., Kowall, N., Brown, R. H., Jr, and Beal, M., 1997a, Evidence of increased oxidative damage in both sporadic and familial amyotrophic lateral sclerosis, *J. Neurochem.* in press.

Ferrante, R., Shinobu, L., Schultz, J., Matthews, R., Thomas, C., Kowall, N., Gurney, M., and Beal, M., 1997b, Increased 3-nitrotyrosine and oxidative damage in mice with human copper zinc superoxide dismutase mutations, *Ann. Neurol.* in press.

Fitzmaurice, P., Shaw, I., HE, K., Miller, R., Monks, T., Lau, S., Mitchell, J., and Lynch, P., 1996, Evidence for DNA damage in amyotrophic lateral sclerosis, *Muscle Nerve* **19**:797–798.

Flood, D., Reaume, A., Lin, Y., Hoffman, E., Gruner, J., Kowall, N., Brown, R. H., Jr, and Scott, R., 1996, SOD1 deficient mice exhibit a slowly progressive neuromuscular deficit, *Abstracts 7th International Symposium on ALS/MND, 11–13 November, Chicago,* p. 163.

Fridovich, I., 1986a, Biological effects of the superoxide radical, *Arch. Biochem. Biophys.* **247**:1–11.

Fridovich, I., 1986b, Superoxide dismutases, *Adv. Enzymol.* **58**:61–97.

Fridovich, I., 1997, Superoxide anion radical ($\cdot O_2^-$), superoxide dismutases and related matters, *J. Biol. Chem.* **272**:18515–18517.

Friedlander, R., Brown, R. H., Jr., Gagliardini, V., Wang, J., and Wang, J., 1997, Inhibition of ICE slows ALS in mice, *Nature* **388**:31.

Graeber, M., Raivich, G., and Kreutzberg, G., 1989, Increase of transferrin receptors and iron uptake in regenerating motor neurons, *J. Neurosci. Res.* **23**:342–345.

Greenlund, L., Deckwerth, T., and Johnson, E., 1995, Superoxide dismutase delays neuronal apoptosis: A role for reactive oxygen species in programmed neuronal death, *Neuron* **14**:303–315.

Gurney, M., 1994, Mutant mice, Cu,Zn superoxide dismutase, and motor neuron degeneration: Response, *Science* **266**:1586.

Gurney, M., Cutting, F., Zhai, P., Doble, A., Taylor, C., Andrus, P., and Hall, E., 1996, Benefit of vitamin E, riluzole and gabapentin in a transgenic model of familial ALS, *Ann. Neurol.* **39**:147–158.

Henle, E., and Linn, S., 1997, Formation, prevention, repair of DNA damage by iron/hydrogen peroxide, *J. Biol. Chem.* **272**:19095–19098.

Hentati, A., Bejaoui, K., Pericak-Vance, M. A., Hentati, F., Speer, M. C., Hung, W.-Y., Figlewicz, D. A., Haines, J., Rimmler, J., Ben Hamida, C., Ben Hamida, M., Brown, R. H., Jr., and Siddique, T., 1994, Linkage of recessive familial amyotrophic lateral sclerosis to chromosome 2q33–35, *Nature Gene.* **7**:425–428.

Hockenbery, D. M., Oltavi, Z. N., Yin, X. M., Milliman, C. L., and Korsmeyer, S. J., 1993, Bcl-2 functions in an antioxidant pathway to prevent apoptosis, *Cell* **75**:241–251.

Hodgson, E., and Fridovich, I., 1975, The interaction of bovine erythrocyte superoxide dismutase with hydrogen peroxide: Inactivation of the enzyme, *Biochemistry* **14**:5294–5299.

Imlay, K., and Imlay, J., 1996, Cloning and analysis of sodC encoding the copper, zinc superoxide dismutase of Escherichia coli, *J. Bacteriol.* **178**:2564–2571.

Imon, Y., Yamaguchi, S., Yamamura, Y., Tsuji, S., Kajima, T., Ito, K., and Nakamura, S., 1995, Low intensity areas observed on T2-weighted magnetic resonance imaging of the cerebral cortex in various neurological diseases, *J. Neurol. Sci.* **134**:27–32.

Ince, P., Shaw, P., Candy, J., Mantle, D., Tandon, L., Ehmann, W., and Markesbery, W., 1994, Iron, selenium and glutathione peroxidase activity are elevated in sporadic motor neuron disease, *Neurosci. Lett.* **182**:87–90.

Ischiropoulos, H., Zhu, L., Chen, J., Tsai, M., Martin, J. C., Smith, C. D., and Beckman, J. S., 1992, Peroxynitrite-mediated tyrosine nitration catalyzed by superoxide dismutase, *Arch. Biochem. Biophys.* **298:**431–437.

Karsarskis, E., Tandon, L., Lovell, M., and Ehmann, W., 1995, Aluminum, calcium, and iron in the spinal cord of patients with sporadic amyotrophic lateral sclerosis using laser microprobe mass spectroscopy: A preliminary study, *J. Neurol. Sci.* **130:**203–208.

Kihira, T., Mukoyama, M., Ando, K., Yase, Y., and Yasui, M., 1990, Determination of manganese concentrations in the spinal cords from amyotrophic lateral sclerosis patients by inductively coupled plasma emission spectroscopy, *J. Neurol. Sci.* **98:**251–258.

Kjellin, K., 1967, The CSF iron in patients with neurological diseases, *Acta Neurol Scand.* **43:**299–313.

Kong, S., Yim, M., Stadtman, E., and Chock, P., 1996, Peroxynitrite disables to tyrosine phosphorylation regulatory mechanism, *Proc. Natl. Acad. Sci. USA* **93:**3377–3382.

Kostic, V., Jackson-Lewis, V., Bilbao, F. d., Dubois-Dauphin, M., and Przedborski, S., 1997, Bcl-2: Prolonging life in a transgenic mouse model of familial amyotrophic lateral sclerosis, *Science* **277:**559–562.

Kunst, C., Mezey, E., Brownstein, M., and Patterson, D., 1997, Mutations in SOD1 associated with familial ALS cause novel protein interactions, *Nature Gene.* **15:**91–94.

Kurlander, H., and Patten, B., 1979, Metals in spinal cord tissue of patients dying of motor neuron disease, *Ann. Neurol.* **6:**21–24.

Lanius, R., Krieger, C., Wagey, R., and Shaw, C., 1993, Increased 35-S-glutathione binding sites in spinal cords from patients with sporadic amyotrophic lateral sclerosis, *Neurosci Lett.* **163:**89–92.

Larsen, N., Pakkenberg, H., Damsgaard, E., Heydorn, K., and Wold, S., 1981, Distribution of arsenic, manganese and selenium in the human brain in chronic renal insufficiency, Parkinson's disease, and amyotrophic lateral sclerosis, *J. Neurol. Sci.* **51:**437–446.

Leigh, P. N., and Swash, M., 1991, Cytoskeletal pathology in motor neuron diseases, in *Amyotrophic Lateral Sclerosis and Other Motor Neuron Diseases* (L. P. Rowland, ed.), pp. 3–23, Raven Press, New York.

Leigh, P. N., Anderton, B. H., Dodson, A., Gallo, J. M., Swash, M., and Power, D. M., 1988, Ubiquitin deposits in anterior horn cells in motor neurone disease, *Neurosci. Lett.* **93:**197–203.

Leveugle, B., Spik, G., Perl, D., Bouras, C., Fillit, H., and Hof, P., 1994, The iron-binding protein lactotransferrin is present in pathologic lesions in a variety of neurodegenerative disorders: A comparative immunohistochemical analysis, *Brain Res.* **650:**20–31.

Louwerse, E., Weverling, G., Bossuyt, P., Meyjes, F., and Jong, J. V. d., 1995, Randomized, double-blind, controlled trial of acetylcysteine in amyotrophic lateral sclerosis, *Arch. Neurol.* **52:**559–564.

Lyons, T., Liu, H., Goto, J., Nerissian, A., Roe, J., Graden, J., Cafe, C., Ellenby, L., Bredesen, D., Gralla, E., and Valentine, J., 1996, Mutations in copper zinc superoxide dismutase that cause amyotrophic lateral sclerosis alter the zinc binding site and the redox behavior of the protein, *Proc. Natl. Acad. Sci., USA* **93:**12240–12244.

Mitchell, J. D., Jackson, M. J., and Pentland, B., 1987, Indices of free radical activity in the cerebrospinal fluid of motor neuron disease, *J. Neurol. Neurosurg. Psychiatry* **50:**919–912.

Mitchell, J., Gatt, J., Phillips, T., Houghton, E., Roston, G., and Wignall, C., 1993, Cu/Zn superoxide dismutase, free radicals, and motoneurone disease, *Lancet* **342:**1051–1052.

Mitzumoto, Y., Iwata, S., Sasajima, K., Yase, Y., and Yoshida, S., 1980, Alpha particle excited X-ray fluorescence analysis for trace elements in cervical cord of amyotrophic lateral sclerosis, *Radioisotopes* **29:**385–389.

Miyata, S., Nakamura, S., Toyoshima, M., Hirata, Y., Saito, M., Kameyama, M., Matsushita, R., and Koyama, M., 1980, Determination of manganese in tissues by neutron activation analysis using an antimony pentoxide column, *Clin Chim. Acta* **106:**235–242.

Mourelatos, Z., Gonatas, N., Stieber, A., Gurney, M., and Canto, M. D., 1996, The Golgi apparatus of spinal cord motor neurons in transgenic mice expressing mutant Cu,Zn superoxide dismutase becomes fragmented in early, preclinical stages of the disease, *Proc. Natl. Acad. Sci. USA* **93:**5472–5477.

Mu, X., He, J., Anderson, D., Trojanowski, J., and Springer, J., 1996, Altered expression of bcl-2 and bax mRNA in amyotrophic lateral sclerosis spinal cord motor neurons, *Ann. Neurol.* **40:**379–386.

Mulder, D., Kurland, L., Offord, K., and Beard, C., 1986, Familial adult motor neuron disease: Amyotrophic lateral sclerosis, *Neurology* **36:**511–517.

Nishida, C., Gralla, E., and Valentine, J., 1994, Characterization of three yeast copper-zinc superoxide dismutase mutants analogous to those coded for in familial amyotrophic lateral sclerosis, *Proc. Natl. Acad. Sci. USA* **91:**9906–9910.

Oba, H., Araki, T., Ohtomo, K., Monzawa, S., Uchiyama, G., Koizumi, K., Nigata, Y., Kachi, K., Shiozawa, Z., and Kobayashi, M., 1993, Amyotrophic lateral sclerosis: T2 shortening in motor cortex at MR imaging, *Radiology* **189:**843–846.

Przedborski, S., Jackson-Lewis, V., Kostic, V., Carlson, E., Epstein, C. J., and Cadet, J. L., 1992, Superoxide dismutase, catalase and glutathione peroxidase activities in Cu/Zn superoxide dismutase transgenic mice, *J. Neurochem.* **58** (5):1760–1767.

Przedborski, S., Donaldson, D., Murphy, P., Lange, D., Latov, N., McKenna-Yasek, D., and Brown, R. H. Jr., 1996, Erythrocyte superoxide dismutase, catalase, and glutathione peroxidase activities in familial and sporadic amyotrophic lateral sclerosis, *Neurodegeneration* **5:**57–64.

Rabizadeh, S., Gralla, E., Borchelt, D., Gwinn, R., Valentine, J., Sisodia, S., Wong, P., Lee, M., Hahn, H., and Bredesen, D., 1995, Mutations associated with amyotrophic lateral sclerosis convert superoxide dismutase from an antiapoptotic gene to a proapoptotic gene: studies in yeast and neural cells, *Proc. Natl. Acad. Sci. USA* **92:**3024–3028.

Radunovic, A., Delves, H., Robberecht, W., Tilkin, P., Enayat, Z., Shaw, C., Stevic, Z., Apostolski, S., Powell, J., and Leigh, P., 1997, Copper and zinc levels in familial amyotrophic lateral sclerosis patients with Cu/Zn gene mutations, *Ann. Neurol.* **42:**130–131.

Reaume, A., Elliott, J., Hoffman, E., Kowall, N., Ferrante, R., Siwek, D., Wilcox, H., Flood, D., Beal, M., Brown, R., Scott, R., and Snyder, W., 1996, Motor neurons in Cu/Zn superoxide dismutase-deficient mice develop normally but exhibit enhanced cell death after axonal injury, *Nature Genet.* **13:**43–47.

Reider, C., and Paulson, G., 1997, Lou Gehrig and amyotrophic lateral sclerosis. Is vitamin E to be revisited? *Arch. Neurol.* **54:**527–528.

Ripps, M. E., Huntley, G. W., Hof, P. R., Morrison, J. H., and Gordon, J. W., 1995, Transgenic mice expressing an altered murine superoxide dismutase gene provide an animal model of amyotrophic lateral sclerosis, *Proc. Natl. Acad. Sci. USA* **92:**689–693.

Robberecht, W., Sapp, P., Viaene, M. K., Rosen, D., McKenna-Yasek, D., Haines, J., Horvitz, R., Theys, P., and Brown, R. H., Jr., 1993, Cu/Zn superoxide dismutase activity in familial and sporadic amyotrophic lateral sclerosis, *J. Neurochem.* **62:**384–387.

Roos, R., Lee, J., Bindokas, V., Jordan, J., Miller, R., Ma, L., Weihl, C., Habib, A., and Ghadge, G., 1997, Gene delivery by replication-deficient recombinant adenoviruses (AdVs) in the study of Cu,Zn superoxide dismutase type 1 (SOD-1)-linked familial amyotrophic lateral sclerosis (FALS), *Neurology* **48:**A150.

Rosen, D. R., Siddique, T., Patterson, D., Figlewicz, D. A., Sapp, P., Hentati, A., Donaldson, D., Goto, J., O'Regan, J. P., Deng, H. X., Rahmani, Z., Krizus, A., McKenna-Yasek, D., Cayabyab, A., Gaston, S. M., Berger, R., Tanzi, R. E., Halperin, J. J., Herzfeldt, B., Van den Bergh, R., Hung, W. Y., Bird, T., Deng, G., Mulder, D. W., Smyth, C., Laing, N. G., Soriano, E., Pericak-Vance, M., Haines, J., Rouleau, G. A., Gusella, G. S., Horvitz, H. R., and Brown, R. H., Jr., 1993, Mutations in Cu/Zn superoxide dismutase gene are associated with familial amyotrophic lateral sclerosis, *Nature* **362:**59–62.

Rothstein, J. D., Bristol, L. A., Hosler, B. A., Brown, R. H., Jr., and Kuncl, R. W., 1994, Chronic inhibition of superoxide dismutase produces apoptotic death of spinal neurons, *Proc. Natl. Acad. Sci. USA* **91:**4155–4159.

Shaw, P., Ince, P., Falkous, G., and Mantle, D., 1995, Oxidative damage to protein in sporadic motor neurone disease spinal cord, *Ann. Neurol.* **38:**691–695.

Sillevis-Smit, P., Beek, H. v., Baars, A.-J., Troost, D., Louwerse, E., Krops-Hermus, A., Wolff, F. d., and Jong, J. V. d., 1992, Increased metallothionein in the liver and kidney of patients with amyotrophic lateral sclerosis, *Arch. Neurol.* **49:**721–724.

Smith, C. D., Carney, J. M., Starke-Reed, P. E., Oliver, C. N., Stadtman, E. R., Floyd, R. A., and Markesbery, W. R., 1991, Excess brain oxidation and enzyme dysfunction in normal aging and in Alzheimer's disease, *Proc. Natl. Acad. Sci. USA* **8:**10540–10543.

Tohgi, H., Abe, T., Saheki, M., Yamazaki, K., and Takahasi, S., 1996, α-Tocopherol quinone level is remarkably low in the cerebrospinal fluid of patients with sporadic amyotrophic lateral sclerosis, *Neurosci. Lett.* **207:**5–8.

Troost, D., Aten, J., Morsink, F., and Jong, J. d., 1995, Apoptosis in not restricted to motoneurons: Bcl-2 expression is increased in post-central cortex, adjacent to affected motor cortex, *J. Neurol. Sci.* **129** (Suppl):79–80.

Troy, C. M., and Shelanski, M., 1994, Down-regulation of copper/zinc superoxide dismutase causes apoptotic death in PC12 neuronal cells, *Proc. Natl. Acad. Sci. USA* **91:**6384–6387.

Volterra, A., Trotti, D., Tromba, C., Floridi, S., and Racagni, G., 1994, Glutamate uptake inhibition by oxygen free radicals in rat cortical astrocytes, *J. Neurosci.* **14**:2924–2932.

Wiedau-Pazos, M., Goto, J., Rabizadeh, S., Gralla, E., Roe, J., Lee, M., Valentine, J., and Bredesen, D., 1996, Altered reactivity of superoxide dismutase in familial amyotrophic lateral sclerosis, *Science* **271**:515–518.

Wong, P., Pardo, C., Borchelt, D., Lee, M., Copeland, N., Jenkins, N., Sisodia, S., Cleveland, D., and Price, D., 1995, An adverse property of a familial ALS-linked SOD1 mutation causes motor neuron disease characterized by vacuolar degeneration of mitochondria, *Neuron* **14**:1105–1116.

Yim, H.-S., Kang, J.-H., Chock, P. B., Stadtman, E. R., and Yim, M., 1997, A familial amyotrophic lateral sclerosis-associated A4V Cu,Zn dismutase mutant has a lower K_m for hydrogen peroxide, *J. Biol. Chem.* **272**:8861–8863.

Yim, M., Kang, J., Yim, H., Kwak, H., Chock, P., and Stadtman, E., 1996, A gain-of-function mutation of an amyotrophic lateral sclerosis-associated Cu,Zn-superoxide dismutase mutant: An enhancement of free radical formation due to a decrease in Km for hydrogen peroxide, *Proc. Natl. Acad. Sci. USA* **93**:5709–5714.

Chapter 27

Reactive Oxygen-Mediated Protein Oxidation in Aging and Disease

Earl R. Stadtman and Barbara S. Berlett

1. INTRODUCTION

Our knowledge concerning the oxidation of proteins by oxygen free radicals comes largely from the results of pioneering studies of Garrison (1987; Garrison *et al.*, 1962), Swallow (1960), and Schuessler and Schilling (1984). Working under conditions where the radiolysis of water leads almost exclusively to net formation of either $\cdot OH$ or $\cdot O_2^-$ or a mixture of $\cdot OH$ and $\cdot O_2^-$, they were able to define the roles of $\cdot OH$ and $\cdot O_2^-$ in protein oxidation. Results of these studies show that the oxidative modification of proteins is initiated by the $\cdot OH$-mediated abstraction of a hydrogen atom either from the α-CH group of the polypeptide backbone or from carbon atoms in amino acid side chains. This leads ultimately to intra- or interprotein cross-linked derivatives, peptide bond cleavage reactions, and the formation of hydroxyl or carbonyl derivatives of amino acid side chains.

2. OXIDATION OF THE POLYPEPTIDE BACKBONE

Based on the results obtained from the radiation experiments, a general mechanism for the oxidation of the polypeptide chain has been elucidated (Garrison, 1987). As is illustrated in Figure 1, oxidation is initiated by the $\cdot OH$-mediated abstraction of hydrogen from the α-carbon atom of an amino acid residue, leading to the formation of a carbon-centered radical (reaction *a*, Figure 1). In the absence of oxygen, two such

Earl R. Stadtman and Barbara S. Berlett Laboratory of Biochemistry, National Heart, Lung, and Blood Institute, National Institutes of Health, Bethesda, Maryland 20892-0342.

Reactive Oxygen Species in Biological Systems, edited by Gilbert and Colton. Kluwer Academic / Plenum Publishers, New York, 1999.

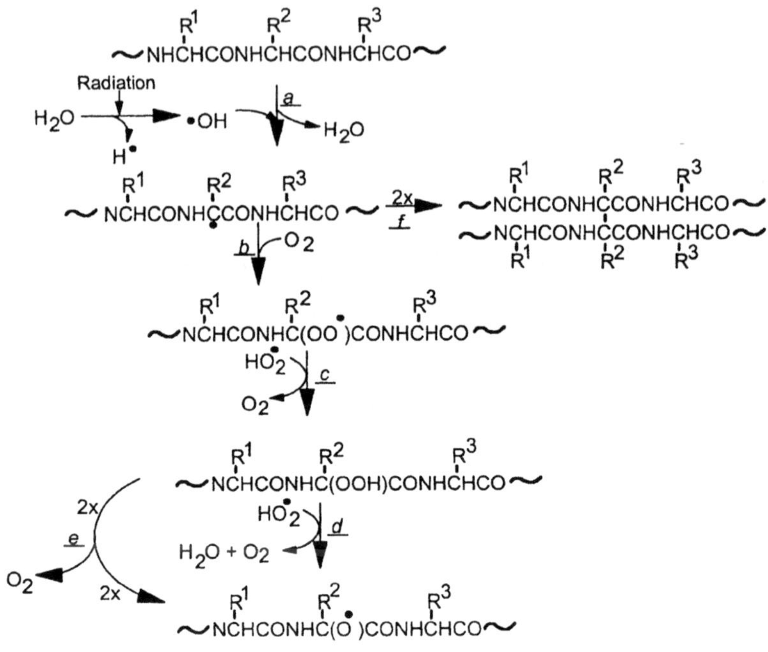

FIGURE 1. Mechanisms of protein oxidation by hydroxyl radical-generating systems.

carbon-centered radicals may combine to form intra- or interprotein cross-linked derivatives (reaction *f*). But oxygen, when present, will add to the carbon-centered radical to form an alkylperoxyl radical (reaction *b*). Reaction of the alkylperoxyl radical with a hydroperoxyl radical leads to the formation of a protein alkylperoxide derivative (reaction *c*), which can react further with the hydroperoxyl radical (reaction *d*) or undergo dismutation (reaction *e*) to form an alkoxyl radical. Significantly, Fe(II) and H^+ can replace the hydroperoxyl radical ($\cdot HO_2$) in reactions *c* and *d* in Figure 1, in which case Fe(III) is also formed.

3. PEPTIDE BOND CLEAVAGE

Formation of the alkoxyl radical (Figure 1) sets the stage for peptide bond cleavage by either the α-amidation (reactions *a* and *b,* Figure 2) or the diamide pathway (reactions *f* and *g,* Figure 2). Characteristically, the primary fragment derived from the N-terminal portion of the oxidized protein obtained in the diamide pathway exists as a diamide derivative (Fragment III, Figure 2), and the product derived from the C-terminal portion of the protein exists as an isocyanate derivative (substance IV, Figure 2). For diagnostic purposes, hydrolysis of these products yields a carboxylic acid, NH_3, and CO_2 (reactions *c* and *d,* Figure 2). Alternatively, as shown in Figure 2, cleavage of the protein-alkoxyl radical derivative by the α-amidation pathway leads to one fragment (Fragment I) in which the C-terminal amino acid is present as an amide derivative, and to Fragment II in

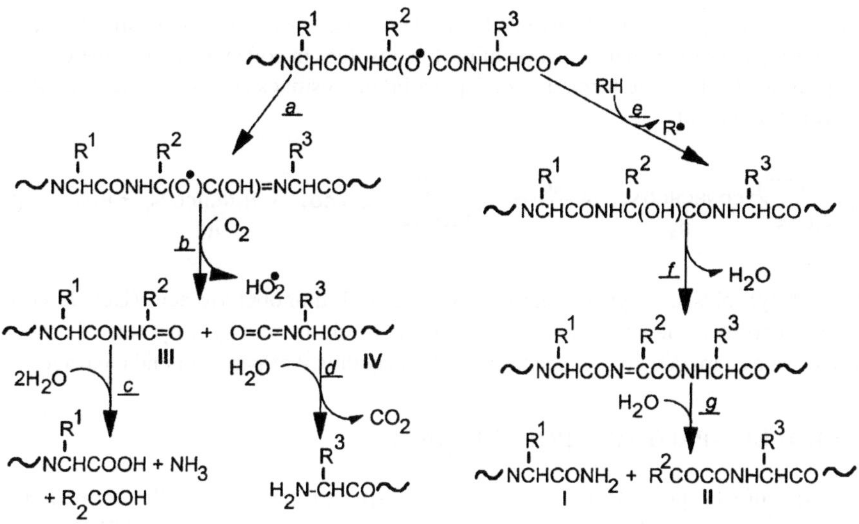

FIGURE 2. Peptide bond cleavage by the α-amidation and diamide pathways.

which the N-terminal amino acid residue is present as an α-ketoacyl derivative. On acid hydrolysis, Fragments I and II would yield NH_3 and an α-keto acid, respectively.

In addition to these more general mechanisms, peptide bond cleavage can occur also by free radical attack of the side chains of glutamyl (or aspartyl) or prolyl residues, according to the overall reactions (1) and (2). A plausible mechanism for peptide bond cleavage by the glutamyl pathway as described by Garrison (1987) involves the ·OH-dependent abstraction of a hydrogen atom from the γ-carbon of the glutamyl side chain, followed by reactions with O_2 and $HO_2\cdot$ (analogous to reactions *b* and *c* in Figure 2) forming alkylperoxyl and alkoxyl radical intermediates, and leading eventually to cleavage of the polypeptide chain by a mechanism in which oxalic acid is formed and the N-terminal amino acid of the peptide fragment derived from the C-terminal portion of the original protein will exist as an *N*-pyruvyl derivative:

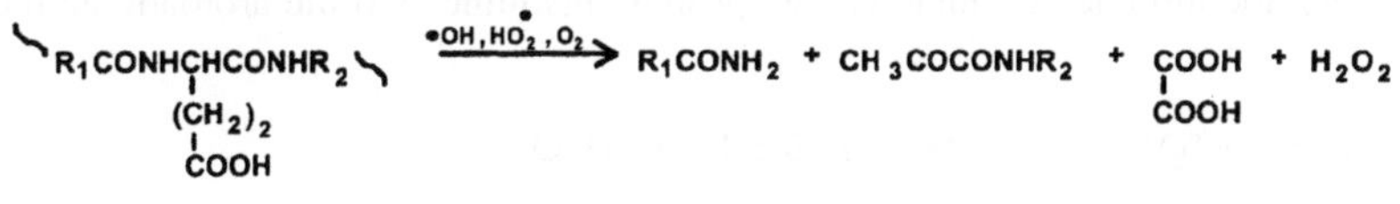

(1)

Aspartyl residues of proteins may undergo an analogous series of reactions leading to peptide bond cleavage (Garrison, 1987).

Schuessler and Schilling (1984) observed that the number of peptide fragments formed by radiolysis of bovine serum (BSA) is approximately equal to the number of prolyl residues in the protein. Based on this and the fact that tertiary bonds are more susceptible to oxidation than secondary amide bonds, they proposed that proline residues are preferred sites of radical-mediated peptide bond cleavage. Subsequently, Uchida *et al.* (1990) have shown that oxidation of proline residues in collagen and proline-containing peptides leads to the formation of 2-pyrrolidone residues and concomitant cleavage of the peptide bond:

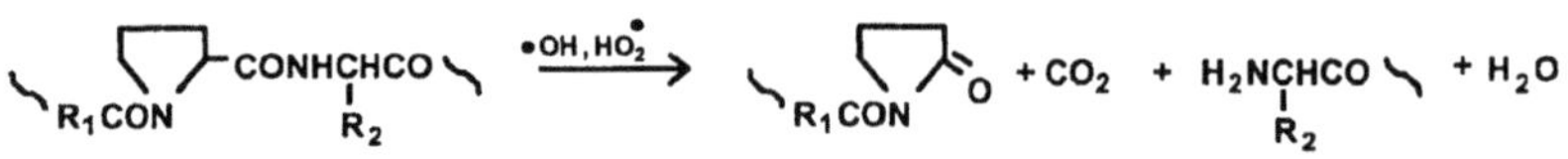
(2)

On acid hydrolysis, 2-pyrrolidone is converted to 4-aminobutyric acid (Uchida *et al.*, 1990). Therefore, the presence of 4-aminobutyric acid in acid hydrolysates of proteins may serve as a marker for the occurrence of this pathway of peptide bond cleavage.

4. PROTEIN–PROTEIN CROSS-LINKAGE

Exposure of proteins to reactive oxygen species can lead to the formation of protein–protein cross-linkages by any one of a number of mechanisms, illustrated in Figure 3: (1) the formation of disulfide bridges via oxidation of the sulfhydryl groups of cysteine residues in two different molecules; (2) the interaction of the N^{ε}-amino group of a lysine residue in one protein molecule with the carbonyl group of another protein molecule to form a Schiff base cross-link (various mechanisms for the introduction of carbonyl groups into proteins are described below); (3) the interaction of lysine amino groups in two different protein molecules with the aldehyde groups of malondialdehyde, which is produced in the ROS-mediated oxidation of polyunsaturated fatty acids; and (4) the formation of carbon-carbon covalent linkages by the interaction of carbon-centered radicals in two different protein molecules as shown in Figure 1, reaction *f*.

5. SIDE CHAIN MODIFICATIONS

All amino acid residues of proteins are potential targets for attack by reactive oxygen species (ROS) produced in the radiolysis of water; however, in only a few cases have the oxidation products been fully characterized. Moreover, under most physiological conditions, cysteine, methionine, arginine, lysine, proline, histidine, and the aromatic amino

$$2PSH + \text{"O"} \longrightarrow P\text{-}S\text{-}S\text{-}P + H_2O$$

$$PNH_2 + PCHO \longrightarrow PN{=}CHP + H_2O$$

$$2PNH_2 + CH_2(CHO)_2 \longrightarrow PN{=}CHCH_2CH{=}NP + 2H_2O$$

$$2P\dot{C}HR \longrightarrow \begin{array}{c} PCHR \\ | \\ PCHR \end{array}$$

FIGURE 3. Formation of protein–protein cross-linkages.

acids are primary targets for ROS-mediated oxidation. Some of the products that are formed in the oxidation of these amino acids are listed in Table I.

5.1. Sulfur-Containing Amino Acid Residues

Cysteine and methionine residues are particularly sensitive to oxidation by almost all forms of ROS. Cysteine residues are converted mainly to disulfide derivatives and methionine residues are converted to methionine sulfoxide (MeSOX) derivatives. In contrast to the oxidation of other amino acids, oxidative modification of cysteine and methionine residues can be reversed by enzymes that catalyze reduction of disulfides and MeSOX back to their original forms (Brot and Weissbach, 1983; Vogt, 1995; Moskovitz *et al.*, 1995). Indeed, based on results of recent studies showing that surface-exposed methionine residues of some proteins are preferred targets for oxidation by ROS, it was proposed that the cyclic oxidation and reduction of methionine residues in proteins may serve an antioxidant function to protect proteins from more extensive oxidative damage and loss of biological activity (Levine *et al.*, 1997b).

5.2. Oxidation of Histidine Residues

Histidine residues are readily converted to asparagine and/or aspartic acid residues (Farber and Levine, 1986; Berlett *et al.*, 1996b) and 2-oxohistidine (Uchida and Kawak-

Table I
ROS-Mediated Oxidation of Amino Acid Residues

Amino acid	Products (references)
Histidine	Aspartic acid, asparagine, 2-oxohistidine (Uchida and Kawakishi, 1993; Berlett *et al.*, 1996b; Farber and Levine, 1986)
Arginine	Glutamic semialdehyde (Amici *et al.*, 1989)
Cysteine	Cy S-S Cy; CySOH, CySOOH, $CySO_2H$, CySSG (Garrison, 1987; Swallow, 1960; Zhou and Gafni, 1991; Brodie and Reed, 1990; Takahashi and Goto, 1990)
Leucine	3-, 4-, and 5-hydroxyleucine (Garrison, 1987; Kopoldova and Liebster, 1963)
Methionine	Methionine sulfoxide, methionine sulfone (Garrison, 1987; Garrison *et al.*, 1962; Berlett *et al.*, 1996a,b; Vogt, 1995; Pryor *et al.*, 1994; Berlett and Stadtman, 1996)
Phenylalanine	2,3-Dihydroxyphenylalanine, 2-, 3-, and 4-hydroxyphenylalanine (Maskos *et al.*, 1992a; Solar, 1985; Beckman *et al.*, 1992; Fletcher and Okada, 1961; Gieseg *et al.*, 1993; Davies *et al.*, 1987)
Tyrosine	3,4-Dihydroxyphenylalanine, tyrosine–tyrosine cross-links, 3-nitrotyrosine (Berlett *et al.*, 1996a; Berlett and Stadtman, 1996; Maskos *et al.*, 1992a; Beckman *et al.*, 1992; Huggins *et al.*, 1993; van der Vliet *et al.*, 1995; Ischiropoulos and Al-Medi, 1995; Giulivi and Davies, 1993; Heinecke *et al.*, 1993; Guptasarma *et al.*, 1992)
Tryptophan	2-, 4-, 5-, 6-, and 7-hydroxytryptophan; formyl kynurenine, 3-hydroxy-kynurenine, nitrotryptophan (Armstrong and Swallow, 1969; Winchester and Lynn, 1970; Kikugawa *et al.*, 1994; Maskos *et al.* 1992b)
Threonine	2-Amino-3-ketobutyric acid (Taborsky, 1973)
Proline	Glutamylsemialdehyde, 2-pyrrolidone, 4- and 5-hydroxyproline; pyroglutamic acid (Amici *et al.*, 1989; Creeth *et al.*, 1983; Poston, 1988; Uchida *et al.*, 1990)
Glutamic acid	Oxalic acid, pyruvic acid (Garrison, 1987)
Lysine	α-Aminodipylsemialdehyde[a]

[a]B. S. Berlett, D. G. Miller, L. Szweda, and E. R. Stadtman (unpublished data).

ishi, 1993) by various ROS. Both the susceptibility of histidine residues to oxidation and the nature of the products formed are dependent on the primary, secondary, and quaternary structures of the protein. Thus, although glutamine synthetase (GS) and BSA contain nearly identical amounts of histidine residues per subunit (50–60 kDa), the rate of histidine destruction in BSA by ozone is more than two times greater than in GS. Moreover, whereas the rate of histidine oxidation in tripeptides containing two alanine and one histidine residues is independent of the amino acid sequence, the yields of aspartate and/or asparagine are 32, 75, and 100% for peptides in which the histidine residue occupies the N-terminal, C-terminal, and central positions, respectively (Berlett *et al.*, 1996b). In contrast to the more or less random attack of histidine residues in proteins by hydroxyl radical or by ozone, the modification of histidine residues in proteins by metal-catalyzed oxidation (MCO) systems is a highly site specific event; amino acid residues at metal binding sites on the protein are preferential targets (Climent *et al.*, 1989; Stadtman *et al.*, 1991) (see below).

5.3. Oxidation of Phenylalanine Residues

Phenylalanine residues are oxidized to 2-, 3-, and 4-hydroxy and 2,3-dihydroxy derivatives (Huggins *et al.*, 1993; Maskos *et al.*, 1992a; Wells-Knecht *et al.*, 1993; Gieseg *et al.*, 1993; Dean *et al.*, 1993). Because *o*-tyrosine is the major phenylalanine product formed when proteins are exposed to an MCO-radical generating system [H_2O_2/Cu(II)] or to radicals produced by high energy radiation, it was proposed that *o*-tyrosine in protein hydrolysates may be used as a marker of radical-mediated damage to proteins (Huggins *et al.*, 1993).

5.4. Oxidation of Tyrosine Residues

The oxidation of tyrosine residues by MCO systems or high-energy radiation has been shown to yield 3,4-dihydroxyphenylalanine (Dopa) and tyrosine–tyrosine cross-linked (dityrosine) derivatives (Giulivi and Davies, 1993; Huggins *et al.*, 1993; Davies *et al.*, 1987; Heinecke *et al.*, 1993; Wells-Knecht *et al.*, 1993). Because the dityrosine derivatives are highly fluorescent, analytical procedures based on fluorescence measurements have been widely used as a measure of dityrosine formation in proteins exposed to various conditions of oxidative stress (Dean *et al.*, 1993; Davies *et al.*, 1987; Davies, 1988). In the absence of other supporting evidence, estimates of dityrosine by these procedures are not reliable because several other modifications of proteins also yield derivatives having similar fluorescence characteristics (Friguet *et al.*, 1994; Szweda, 1994; Guptasarma *et al.*, 1992). Compared with other kinds of protein modification, the oxidation of tyrosine residues under physiological conditions is quantitatively a minor event; nevertheless, it may prove to be a reliable marker of oxidative protein damage under some conditions of oxidative stress (Huggins *et al.*, 1993; Giulivi and Davies, 1993).

5.5. Nitration of Tyrosine Residues

Proteins are very susceptible to modification on exposure to peroxynitrite, which is produced endogenously by reaction of nitric oxide with superoxide anion:

$$\cdot NO + \cdot O_2^- \rightarrow ONOO^- \qquad (3)$$

In earlier studies, it was demonstrated that treatment of proteins with peroxynitrite leads to nitration of tyrosine residues (Beckman *et al.*, 1992; Ischiropolous and Al-Medi, 1995; Berlett *et al.*, 1996a) and that peroxynitrite is in equilibrium with an activated form of unknown structure ($ONOO^{-}*$), which reacts readily with methionine residues of proteins to form MeSOX residues (Pryor and Squadrito, 1995; Pryor *et al.*, 1994). In addition, Lymar and Hurst (1995) reported that peroxynitrite reacts almost instantly with carbon dioxide to form a derivative, possibly nitrosoperoxycarbonate ($O = NOOCO_2^-$) or nitrocarbonate ($O_2NOCO_2^-$), that is able to nitrate aromatic residues. It was subsequently demonstrated in several laboratories that the carbon dioxide derivative reacts with tyrosine residues of proteins to form the 3-nitrotyrosine derivatives (Lymar *et al.*, 1996; Lymar and Hurst, 1996; Uppu *et al.*, 1996; Denicola *et al.*, 1996; Berlett *et al.*, 1996a; Gow *et al.*, 1996a). Any one of these peroxynitrite-mediated modifications may lead to loss of biological function of some enzymes; however, the nitration of tyrosine residues deserves special consideration because nitration is an irreversible process and precludes the ability of tyrosine residues of regulatory proteins to undergo interconversion between unmodified and nucleotidylated forms, as occurs in regulation of the bacterial GS cascade (Stadtman *et al.*, 1981), or between unphosphorylated and phosphorylated forms, as occurs in a myriad of signal transduction pathways in mammals (reviewed by Hunter, 1995). The singular importance of nitration in these instances is underscored by the demonstration that when exposed to peroxynitrite in the presence of CO_2, *Escherichia coli* GS is converted to a form that possesses regulatory properties almost identical to those obtained *in vivo* by adenylylation of a single tyrosyl residue in each subunit of the enzyme (Berlett *et al.*, 1996b); moreover, peroxynitrite-mediated nitration of the tyrosine residues in model peptide substrates prevents the phosphorylation of the tyrosine residues in these peptides by protein tyrosine kinases (Kong *et al.*, 1996; Gow *et al.*, 1996b).

It is evident from these studies that the peroxynitrite-mediated oxidation of methionine and tyrosine residues of proteins are competitive processes in which the availability of CO_2 plays a pivotal role as illustrated below:

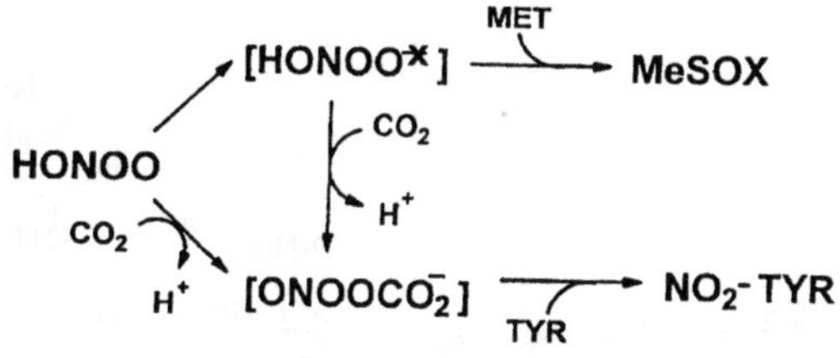

Consistent with the above scheme, the oxidation of methionine residues of GS by peroxynitrite in the complete absence of CO_2, and the nitration of tyrosine residues as occurs at pH 7.4 in the presence of 5% CO_2, are mutually exclusive processes; yet both processes convert GS to a form that is comparable to that obtained by adenylylation of the enzyme (Berlett and Stadtman, 1996).

6. FORMATION OF CARBONYL DERIVATIVES

Of considerable interest is the fact that ROS-mediated modification of proteins leads in some cases to the formation of carbonyl derivatives, which can be generated by at least five different mechanisms: (1) by direct oxidation of side chains of lysine, arginine, proline, and threonine residues to aldehyde or ketone derivatives as is noted in Figure 3; (2) by oxidative cleavage of polypeptide bonds by the α-amidation pathway (Figure 2) or cleavage by oxidation of glutamate (or aspartate) residue side chains [Reaction 1], (3) by Michael addition of lysine amino groups, cysteine sulfhydryl groups, or histidine imidazole groups to α,β-unsaturated aldehydes, as illustrated in Figure 4 for reactions with the lipid peroxidation product 4-hydroxy-2-nonenal (Schuenstein and Esterbauer, 1979; Uchida and Stadtman, 1993; Friguet *et al.*, 1994; Nadkarni and Sayre, 1995; Sayre *et al.*, 1993; Bruenner *et al.*, 1994, 1995); (4) by reaction of the lipid peroxidation product malondialdehyde with amino groups of lysine residues to form Schiff base derivatives (Burcham and Kuhan, 1996); (5) by reaction of reducing sugars or their oxidation products with amino groups of lysine residues in a process known as glycation (Figure 5) (Monnier *et al.*, 1995; Mullarkey *et al.*, 1990; Wolf and Dean, 1987; Kristal and Yu, 1992). Glycation involves reaction of the carbonyl groups of sugars with the N-terminal amino group of proteins or with the N^{ε}-amino groups of lysine residues to form Schiff base adducts that undergo Amadori rearrangements to form keto-amines (Figure 5). In the presence of transition metals, the Amadori products undergo oxidation (glycoxidation) leading eventually to *N*-carboxymethyl lysine residues (Reddy *et al.*, 1995) or to intermediates capable of reacting with arginine residues to form pentosidine protein cross-linked adducts (Figure 5) (Sell and Monnier, 1989; Monnier *et al.*, 1995). The Amadori products may also undergo dehydration and fragmentation to yield deoxyosones that give rise to fluorescent derivatives of ill-defined structure. These are collectively referred to as Maillard products, or advanced glycosylation end products (AGEs), that accumulate during aging and the development of various diseases, including, diabetes, atherogenesis, Alzheimer's disease, and some eye disorders (review: Monnier, 1990; Cerami *et al.*, 1987; Hunt, 1996; Reddy *et al.*, 1995; Baynes, 1991; Smith *et al.*, 1994;

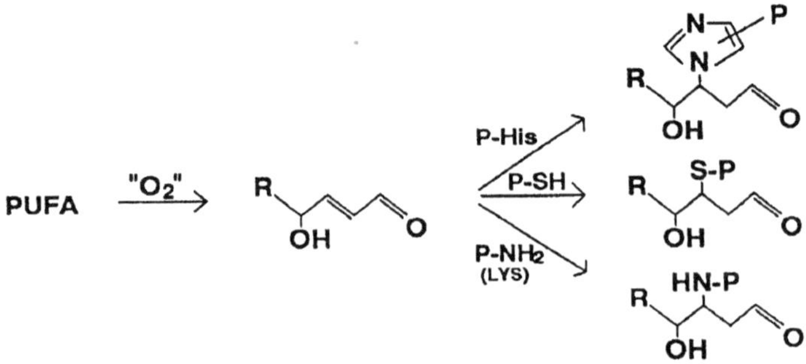

FIGURE 4. Michael addition of histidine (P-HIS), lysine (P-NH_2), and cysteine (P-SH) residues of proteins with 4-hydroxy-2-nonenal (HNE), a product of polyunsaturated fatty acid (PUFA) oxidation.

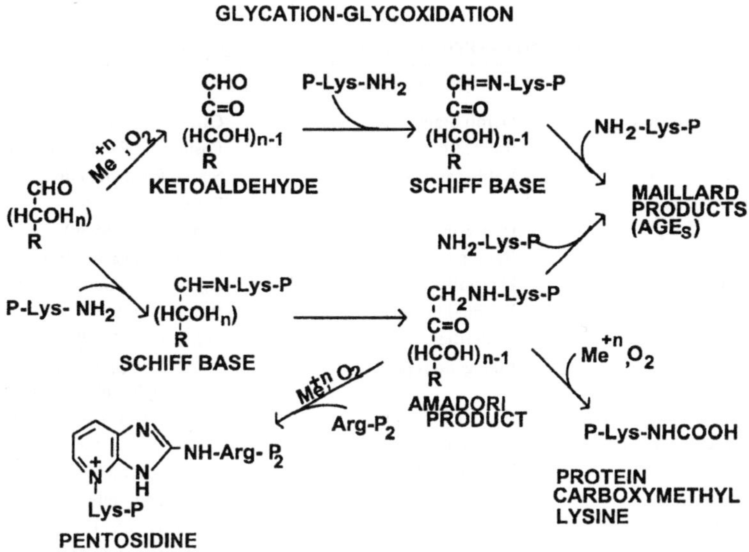

FIGURE 5. Reaction of sugars and sugar oxidation products with protein lysine amino groups (P-Lys-NH_2).

Harrington and Colaco, 1994). It is also reported that some Maillard products can give rise to free radicals by metal ion-independent mechanisms (Mullarkey *et al.*, 1990).

7. PROTEIN CARBONYLS SERVE AS MARKERS OF OXIDATIVE STRESS

Because ROS-mediated modification of proteins leads in some cases to the generation of carbonyl derivatives, and because several highly sensitive techniques are available for quantitation of the protein carbonyl content, the presence of carbonyl groups in proteins is a widely used marker of oxidative stress-induced cellular damage. Validity of this marker is supported by results of studies showing that exposure of animals and/or cultured cells to various conditions of oxidative stress leads to an increase in protein carbonyl content. Thus, as summarized in Table II, exposure to either hyperoxia, high-energy radiation, enforced exercise, cigarette smoke, ischemia–reperfusion, magnesium deficiency, activated neutrophils, paraquat toxicity, ozone, or MCO systems can lead to an increase in protein carbonyl content of cellular protein. Moreover, a role of ROS-mediated tissue damage in aging and in the etiology and/or progression of a number of diseases is suggested by the demonstration that these physiological processes are accompanied by an increase in the carbonyl content of tissue proteins. Thus, elevated levels of protein carbonyls are associated with Alzheimer's disease (Carney *et al.*, 1994; Harris *et al.*, 1995; Chauhan *et al.*, 1991; M. A. Smith *et al.*, 1994, 1995, 1996), amyotrophic lateral sclerosis (Bowling *et al.*, 1993), rheumatoid arthritis (Chapman *et al.*, 1989), muscular dystrophy (Murphy and Kherer, 1989), cataractogenesis (Garland *et al.*, 1988), respiratory distress syndrome (Gladstone and Levine, 1994), iron-induced renal carcinogenesis (Toyokuni *et*

Table II
Oxidative Stress-Provoked Protein Carbonyl Formation

Oxidative stress	Tissue affected	Reference
Ischemia–reperfusion	Gerbil brain	Oliver *et al.* (1990)
	Rat heart	Poston and Parentcau (1992)
Hyperoxia	Rat hepatocytes	Starke-Reed and Oliver (1989)
		Starke *et al.* (1987)
	Houseflies	Sohal *et al.* (1993b)
	Rat lung	Winter and Liehr (1991)
X-irradiation	Houseflies	Agarwal and Sohal (1993)
Exercise	Rat hind leg muscle	Witt *et al.* (1992)
	Houseflies	Sohal *et al.* (1993b)
Neutrophil activation	Endogenous protein	Oliver *et al.* (1987), Karsek-Staples and Webster (1993)
	Endothelial cells	Karsek-Staples and Webster (1993)
Rapid correction of hyponatremia	Rat brain	Mickel *et al.* (1990)
Estrogen administration	Hamster kidney	Winter and Liehr (1991)
Paraquat toxicity	Rat lung	Winter and Liehr (1991)
Magnesium deficiency	Rat kidney	Stafford *et al.* (1993)
	Rat brain	Stafford *et al.* (1993)
Cigarette smoke	Human plasma	Reznick *et al.* (1992)
Xanthine oxidase/xanthine	Endothelial cells	Karsek-Staples and Webster (1993)

al., 1994; Uchida *et al.*, 1995), cardiovascular disease (Kelley and Birch, 1993; Uchida *et al.*, 1994) and probably Parkinson's disease (Yoritaka *et al.*, 1996).

8. METAL-CATALYZED SITE-SPECIFIC MODIFICATION OF PROTEINS

It is generally accepted that hydroxyl radical is the most damaging radical species formed under most physiological conditions. As noted above, most of our knowledge of the basic mechanisms involved in the modification of proteins and other biological molecules comes from detailed studies in which hydroxyl radicals were generated by radiolysis of water; however, under most physiological conditions $\cdot OH$ is derived mainly by the metal-catalyzed heterolytic cleavage of hydrogen peroxide, i.e., by the Fenton reaction:

$$H_2O_2 + Fe(II) \text{ or } Cu(I) \rightarrow \cdot OH + OH^- + Fe(III) \text{ or } Cu(II) \quad (4)$$

Hydrogen peroxide needed to fuel this reaction is derived mainly from the dismutation of $\cdot O_2^-$ [Reaction (5)], which is produced as an unavoidable by-product of electron transport through the respiratory chain to cytochrome *c*. Superoxide is also a source of reducing equivalents needed to regenerate catalytic levels of Fe(II) and Cu(I) consumed in Reaction (2) [Reaction (6)].

$$2 \cdot O_2^- + 2H^+ \rightarrow H_2O_2 + O_2 \quad (5)$$

$$\cdot O_2^- + Fe(III) \rightarrow Fe(II) + O_2 \tag{6}$$

However, the production of H_2O_2 and the reduction of Fe(III) are also catalyzed by a number of enzymatic and nonenzymatic mixed-function oxidation (MFO) systems by both superoxide-dependent and -independent pathways (Stadtman, 1988, 1986; Stadtman and Oliver, 1991). Enzymes that have been shown to participate in MFO systems include NADH and NADPH oxidases (Fucci *et al.*, 1983), xanthine and nicotinic acid oxidases (Stadtman and Wittenberger, 1985), and various cytochrome P450 oxidases (Fucci *et al.*, 1983; Oliver *et al.*, 1982), all of which, in the presence of oxygen and their oxidizable substrates, are able to generate H_2O_2 and reduce Fe(III) and Cu(II) (Oliver *et al.*, 1982; Nakamura *et al.*, 1985).

Among the nonenzymatic systems capable of generating H_2O_2 and reducing transition metals are systems comprised of O_2 and Fe(III) or Cu(II) together with either ascorbate (Levine, 1983; Chevion, 1988) or sulfhydryl compounds (Kim *et al.*, 1985). In earlier reports, these diverse systems were all correctly referred to as MFO systems because, in addition to oxygen, they required oxidizable cosubstrates [namely, NAD(P)H, ascorbate, sulfhydryl compounds, hypoxanthine, and others] that serve as electron donors for the reduction of oxygen. But to avoid confusion between reactions catalyzed between these systems and those catalyzed by mixed-function *oxidases*, these systems are often referred to as metal-catalyzed oxidation (MCO) systems (Climent *et al.*, 1989; Stadtman, 1990). All of these systems have been shown to modify enzymes (Fucci *et al.*, 1983; Levine, 1983; Oliver *et al.*, 1987b; Stadtman, 1990).

In theory, the source of ·OH production should not affect the pattern of protein oxidation. In fact, the modification of proteins by ·OH generated in the presence of high, nonphysiological concentrations of Fe(II) and H_2O_2 is similar to that obtained by ionizing radiation; i.e., almost all amino acid residues are targets, peptide bond cleavage occurs, and protein–protein cross-links are formed (Huggins *et al.*, 1993; Neuzil *et al.*, 1993). However, in the presence of low concentrations of Fe(II) and H_2O_2, as might occur under physiological conditions, the modification of proteins is limited to residues at metal-binding sites on the protein (Farber and Levine, 1986; Levine, 1989; Rivett and Levine, 1990). This and the fact that protein modification under these conditions is resistant to inhibition by ·OH scavengers, such as mannitol or dimethyl sulfoxide (Levine, 1983; Oliver *et al.*, 1982; Fucci *et al.*, 1983; Friguet *et al.*, 1994), supports the hypothesis that under physiological conditions the metal-catalyzed modification of protein is a site-specific process in which the Fe(II) and H_2O_2 react at metal-binding sites to generate ·OH, which preferentially attacks amino acid residues in close proximity to the metal-binding sites (Oliver *et al.*, 1982; Stadtman, 1990). This hypothesis is confirmed by the demonstration that the oxidative modification of GS is restricted to the oxidation of amino acid residues located at one or another, or both, of two metal-binding sites on the enzyme (Climent and Levine, 1991; Sahakian *et al.*, 1991).

9. PROTEIN OXIDATION IN AGING

During aging, there in an accumulation of catalytically inactive or less active forms of many enzymes (Dreyfus *et al.*, 1978; Rothstein, 1977). Various lines of evidence summarized below indicate that ROS-mediated protein modifications are responsible for some of these age-related changes.

1. In several animal models, the intracellular level of oxidized protein (carbonyl content) has been shown to increase exponentially as a function of animal age. Thus: (i) The carbonyl content of protein in cultured human fibroblasts increases as a function of the fibroblast donor age and is independent of the culture passage number (Oliver *et al.*, 1987a). (ii) There is an age-dependent increase in the protein carbonyl content of human brain (C. Smith *et al.*, 1991), human eye lens (Garland *et al.*, 1988), rat hepatocytes (Starke-Reed and Oliver, 1989), whole body protein of houseflies (Sohal *et al.*, 1995), mouse brain and kidneys (Sohal *et al.*, 1994), gerbil brain (Carney *et al.*, 1991; Sohal *et al.*, 1995), and human erythrocytes (Oliver *et al.*, 1987a). Based on reasonable assumptions, it has been estimated that the oxidized forms of proteins may account for 20–50% of the total protein in an 80-year-old human (Starke-Reed and Oliver, 1989; Stadtman, 1992).

2. Old animals are more susceptible than young animals to oxidative stress-induced protein oxidation. For example, when treated with X rays, there is a greater increase in the protein carbonyl content and a greater loss of glucose-6-phosphate dehydrogenase in old houseflies as compared with young flies (Agarwal and Sohal, 1993). Likewise, protein in tissue homogenate of old Mongolian gerbils is more susceptible to oxidation by X rays than is protein in homogenate from young gerbils (Sohal *et al.*, 1995).

3. Circumstances that extend the life span of animals lead to a decrease in the level of oxidized proteins and vice versa. Thus, when examined at the same chronological age, short-lived houseflies contain a higher level of protein carbonyls than do their long-lived cohorts (Sohal *et al.*, 1993b). The life span of mice exposed to hyperbaric oxygen decreases as a function of increasing oxygen pressure over the range of 1 to 9 atm (Gerschman *et al.*, 1958). Caloric restriction of rats leads to an increase in life span and to an increase in resistance to X-irradiation and to lower levels of oxidized protein (Youngman et al., 1992). Caloric restriction of mice leads to extension of life span, prolongation of mortality rate doubling time, and decrease in the level of oxidized protein in the brain, heart, and kidney, relative to that observed in ad libitum-fed animals (Sohal *et al.*, 1994). The protein carbonyl content of the cortical synaptosomal membranes from a strain of senescence-prone mice is higher than that in a strain of senescence accelerated-resistant mice (Butterfield *et al.*, 1997).

4. The age-related changes in activities and stabilities of enzymes can be mimicked by exposure of young animals to oxidative stress. For example, exposure of young rats to 100% oxygen leads to an increase in the carbonyl content of hepatocyte protein and to losses of GS, glucose-6-phosphate dehydrogenase, and the multicatalytic proteinase activities, i.e., changes comparable to those observed during aging (Starke-Reed and Oliver, 1989; Starke *et al.*, 1987). Also, *in vitro* exposure of purified enzymes to MCO systems leads to changes in activity and heat stability and to the generation of carbonyl groups similar to that observed with these enzymes during aging (Fucci *et al.*, 1983; Chevion, 1988; Taborsky, 1973; Gordillo *et al.*, 1988; Szweda and Stadtman, 1992; Takahashi and Goto, 1990; Zhou and Gafni, 1991; Oliver *et al.*, 1987b; Cook and Gafni, 1988; De La Cruz *et al.*, 1996; Musci *et al.*, 1993; Mordente *et al.*, 1988).

10. WHY DO OXIDIZED FORMS OF PROTEIN ACCUMULATE?

The oxidation and degradation of proteins is a dynamic process. The steady-state level of oxidized protein under any physiological condition is determined by the relative rates of protein oxidation and of oxidized protein degradation. The overall rate of oxidation is a complex function of numerous biological and environmental factors that govern the formation of ROS, the availability of antioxidants, and the susceptibility of proteins to oxidation. These include: the availability of iron, copper, and oxygen; the intracellular levels of mixed-function oxidation systems; the redox state (i.e., the NADPH/NADP, NADH/NAD, GSH/GSSG ratios); the concentrations of lipid and glycoxidation products; aberrations in the flow of electrons to cytochrome *c*; exposure to high-energy radiation; occurrence of inflammatory processes induced by "oxygen-burst" activation of neutrophils and macrophages; the level of oxidase and nitric oxide synthetase activities; and the concentration of pollutants in the atmosphere. Oxidation is also a function of the concentrations of antioxidant enzymes (SOD, glutathione peroxidase, thiol peroxidase, methionine sulfoxide reductase, catalase, ceruloplasmin) as well as various nonenzymatic antioxidants [glutathione, ascorbate, vitamine E, β-carotene, metal chelators, bilirubin, and metal-binding proteins (ferritin, transferrin)] as well as factors that regulate their concentrations and activities. The level of oxidized protein is also determined by the concentration of factors that govern the susceptibility of proteins to oxidative modification; e.g., substrates and allosteric effectors of enzymes, and the divalent cations, Mn(II), Mg(II), and Zn(II).

Though not as well studied, the degradation of oxidized proteins is governed: by genetic and regulatory factors that control the intracellular concentration of the proteases that preferentially degrade oxidized proteins; by the concentrations of inhibitors and activators that control the activities of the proteases; by oxidative modifications that render proteins resistant to proteolytic degradation (Rivett, 1986; Friguet *et al.*, 1994; Grune *et al.*, 1995; Grant *et al.*, 1993). It is the balance between all of these many factors that specify the steady-state level of oxidized protein during aging and in some pathological disorders. There is, in fact, evidence that during aging the level of prooxidant activities is accelerated (Sohal and Dubey, 1994) and that the levels and activities of some protease activities decline (Starke-Reed and Oliver, 1989; Carney *et al.*, 1991). This may account for the observed progressive, age-dependent increase in the level of oxidized protein. In any case, it is noteworthy that each of the numerous parameters that affect this balance is in the last analysis subject to genetic control. The accumulation of oxidized protein during aging and some diseases could therefore reflect the accumulation of chromosomal damage that leads one by one, in a random manner, to deficiencies in one or more of the many factors that favor protein oxidation or protein degradation. However, the steady-state level of oxidized protein is determined by the collective contributions of a very large number of different enzymes, regulatory proteins, and metabolites. Therefore, except for a few rare diseases (namely, progeria, Werner's syndrome), it may not be possible to attribute age-related disorders to a particular genetic aberration. For example, a loss of protease activity in one individual might be phenotypically indistinguishable from mutations that lead to a loss of antioxidant defenses or to an increase in the rate of ROS generation in another individual.

11. REFERENCES

Agarwal, S., and Sohal, R. S., 1993, Relationship between aging and susceptibility to protein oxidative damage, *Biochem. Biophys. Res. Commun.* **194:**1203–1206.

Amici, A., Levine, R. L., Tsai, L., and Stadtman, E. R., 1989, Conversion of amino acid residues in proteins and amino acid homopolymers to carbonyl derivatives by metal-catalyzed reactions, *J. Biol. Chem.* **264:**3341–3346.

Armstrong, R. C., and Swallow, A. J., 1969, Pulse- and gamma-radiolysis of aqueous solutions of tryptophan, *Radiat. Res.* **41:**563–579.

Baynes, J. W., 1991, Perspectives in diabetes. Role of oxidative stress in development of complications in diabetes, *Diabetes* **40:**405–411.

Beckman, J. S., Ischiropoulos, H., Zhu, L., van der Woerd, M., Smith, C., Chen, J., Harrison, J., Martin, J. C., and Tsai, M., 1992, Kinetics of superoxide dismutase- and iron-catalyzed nitration of phenolics by peroxynitrite, *Arch. Biochem. Biophys.* **298:**438–445.

Berlett, B. S., and Stadtman, E. R., 1996, Carbon dioxide stimulates nitration of tyrosine residues and inhibits oxidation of methionine residues in *Escherichia coli* glutamine synthetase by peroxynitrite, *FASEB J.* **10,** Abstract #585.

Berlett, B. S., Friguet, B., Yim, M. B., Chock, P. B., and Stadtman, E. R., 1996a, Peroxynitrite-mediated nitration of tyrosine residues of *Escherichia coli* glutamine synthetase mimic adenylylation: Relevance to signal transduction, *Proc. Natl. Acad. Sci. USA* **93:**1776–1780.

Berlett, B. S., Levine, R. L., and Stadtman, E. R., 1996b, A comparison of the effects of ozone on the modification of amino acid residues in glutamine synthetase and bovine serum albumin, *J. Biol. Chem.* **271:**4177–4182.

Bowling, A. C., Schultz, J. B., Brown, Jr., R. H., and Beal, M. F., 1993, Superoxide dismutase activity, oxidative damage, and mitochondrial energy metabolism I familial and sporadic amyotrophic lateral sclerosis, *J. Neurochem.* **61**:2322–2325.

Brodie, E., and Reed, D. J., 1990, Cellular recovery of glyceraldehyde-3-phosphate dehydrogenase activity and thiol status after exposure to hydroperoxide, *Arch. Biochem. Biophys.* **277**:228–233.

Brot, N., and Weissbach, H., 1983, Biochemistry and physiological role of methionine sulfoxide reductase in proteins, *Arch. Biochem. Biophys.* **233:**271–288.

Bruenner, B. A., Jones, A. D., and German, J. B., 1994, Maximum entropy deconvolution of heterogeneity in protein modification: Protein adducts of 4-hydroxy-2-nonenal, *Rapid Communs, Mass Spectrom.* **8:**509–512.

Bruenner, B. A., Jones, A. D., and German, J. B., 1995, Direct characterization of protein adducts of the lipid peroxidation product 4-hydroxy-2-nonenal using electrospray mass spectrometry, *Chem. Res. Toxicol.* **8:**552–559.

Burcham, P. C., and Kuhan, Y. T., 1996, Introduction of carbonyl groups into proteins of the lipid peroxidation product, malondialdehyde, *Biochem. Biophys. Res. Commun.* **220:**996–1001.

Butterfield, D. A., Howard, B. J., Yatin, S., Allen, K. L., and Carney, J. M., 1997, Free radical oxidation of brain proteins in accelerated senescence and its modulation by N-*tert*-butyl-α-phenylnitrone, *Proc. Natl. Acad. Sci. USA* **94:**674–678.

Carney, J. M., Starke-Reed, P. E., Oliver, C. N., Landum, R. W., Cheng, M. S., Wu, J. F., and Floyd, R. A., 1991, Reversal of age-related increase in brain protein oxidation, decrease in enzyme activity, and loss of temporal and spatial memory by chronic administration of the spin-trapping compound N-*tert*-butyl-α-phenylnitrone, *Proc. Natl. Acad. Sci. USA* **88:**3633–3636.

Carney, J. M., Smith, C. D., Carney, A. M., and Butterfield, D. A., 1994, Aging- and oxygen-induced modifications in brain biochemistry and behavior, in *Aging and Cellular Defense Mechanisms*, (C. Franceschi, G., Crepaldi, V.J. Cristofalo, and J. Vijg, eds.), pp. 110–119, New York Academy of Sciences, New York.

Cerami, A., Vlassara, H., and Brownlee, M., 1987, Glucose and aging, *Sci. Am.* **256:**90–96.

Chapman, M. L., Rubin, B. R., and Gracy, R. W., 1989, Increased carbonyl content of proteins in synovial fluid of patients with rheumatoid arthritis, *J. Rheumatol.* **16:**15–18.

Chauhan, A., Chauhan, V. P. S., Brockerhoff, H., and Wisniewski, H. M., 1991, Action of amyloid β-protein on protein kinase C activity, *Life Sci.* **49:**1555–1556.

Chevion, M., 1988, A site-specific mechanism for free radical induced biological damage: The essential role of redox-active transition metals, *Free Radical Biol. Med.* **5**:27–37.

Climent, I., and Levine, R. L., 1991, Oxidation of the active site of glutamine synthetase: Conversion of arginine-344 to γ-glutamyl semialdehyde, *Arch. Biochem. Biophys.* **189**:371–375.

Climent, I., Tsai, L., and Levine, R. L., 1989, Derivatization of γ-glutamyl semialdehyde residues in oxidized proteins by fluoresceinamine, *Anal. Biochem.* **182**:226–232.

Cook, L. L., and Gafni, A., 1988, Protection of phosphoglycerate kinase against *in vitro* aging by selective cysteine methylation. *J. Biol. Chem.* **263**:13991–13993.

Creeth, J. M., Cooper, B., Donald, A. S. R., and Clamp, J. R., 1983, Studies of the limited degradation of mucous glycoproteins, *Biochem. J.* **211**:323–332.

Davies, K. J. A., 1988, Protein oxidation, protein cross-linking, and proteolysis in the formation of lipofuscin: Rationale and methods for the measurement of protein degradation, In *Lipofuscin – 1987: State of the Art*, (I. Zs-Nagy, ed.), pp. 109–133, Elsevier, Amsterdam.

Davies, K. J. A., Delsignore, M. E., and Lin, S. W., 1987. Protein damage by oxygen radicals. II. Modification of amino acids, *J. Biol. Chem.* **262**:9902–9907.

Dean, R. T., Gieseg, S., and Davies, M. J., 1993, Reactive species and their accumulation on radical-damaged proteins, *Trends Biochem. Sci.* **18**:437–441.

De La Cruz, C. P., Revilla, E., Venero, J. L., Ayala, A., Cano, J., and Machado, A., 1996, Oxidative inactivation of tyrosine hydroxylase in *substantia nigra* of aged rat, *Free Radical Biol. Med.* **20**:53–61.

Denicola, A., Freeman, B. A., Trujillo, M., and Radi, R., 1996, Peroxynitrite reaction with carbon dioxide/bicarbonate: Kinetics and influence on peroxynitrite-mediated oxidations, *Arch. Biochem. Biophys.* **333**:49–58.

Dreyfus, J. C., Kahn, A., and Schapira, F., 1978, Post translational modifications of enzymes, *Curr. Top. Cell. Regul.* **14**:243–297.

Farber, J. M., and Levine, R. L., 1986, Sequence of a peptide susceptible to mixed-function oxidation: Probable cation binding site in glutamine synthetase, *J. Biol. Chem.* **261**:4574–4578.

Fletcher, G. L., and Okada, S., 1961, Radiation-induced formation of dihydroxy phenylalanine from typrosine-containing peptides in aqueous solution, *Radiat. Res.* **15**:349–351.

Friguet, B., Szweda, L., and Stadtman, E. R., 1994, Susceptibility of glucose-6-phosphate dehydrogenase modified by 4-hydroxy-2-nonenal and metal-catalyzed oxidation to proteolysis by the multicatalytic protease, *Arch. Biochem. Biophys.* **311**:168–173.

Fucci, L., Oliver, C. N., Coon, M. J., and Stadtman, E. R., 1983, Inactivation of key metabolic enzymes by mixed-function oxidation reactions: Possible implications in protein turnover and aging, *Proc. Natl. Acad. Sci. USA* **80**:1521–1525.

Garland, D., Russell, P., and Zigler, J. S., 1988, The oxidative modification of lens proteins. In *Oxygen Radicals in Biology and Medicine* (M. G. Simic, K. S. Taylor, J. F. Ward, and V. von Sontag, eds.), pp 347–353, Plenum Press, New York.

Garrison, W. M., 1987, Reaction mechanisms in the radiolysis of peptide, polypeptides, and proteins, *Chem. Rev.* **87**:381–398.

Garrison, W. M., Jayko, M. E., and Bennett, W., 1962, Radiation-induced oxidation of proteins in aqueous solution, *Radiat. Res.* **16**:487–502.

Gerschman, R., Gilbert, D.I., and Caccamise, D., 1958, Effect of various substances on survival times of mice exposed to different high oxygen tension, *Am. J. Physiol.* **192**:563–571.

Gieseg, S. P., Simpson, J. A., Charlton, T. S., Duncan, M. W., and Dean, R. T., 1993, Protein-bound 3,4-dihydroxyphenylalanine is a major product formed during hydroxyl radical damage to proteins, *Biochemistry* **32**:4780–4786.

Giulivi, C., and Davies, K. J. A., 1993, Dityrosine and tyrosine oxidation products are endogeneous markers for the selective proteolysis of oxidatively modified red blood cell hemoglobin by the 19S proteosome, *J. Biol. Chem.* **268**:8752–8759.

Gladstone, I. M., and Levine, R. L., 1994, Oxidation of proteins in neonatal lungs, *Pediatrics* **93**:764–768.

Gordillo, E., Ayala, A., F-Lobato, M., Bautista, J., and Machada, A., 1988, Possible involvement of histidine residues in loss of enzymatic activity of rat liver malic enzyme during aging, *J. Biol. Chem.* **263**:8053–8057.

Gow, A.J., Duran, D., Malcolm, S., and Ischiropoulos, H., 1996a, Effects of peroxynitrite-induced protein modification on tyrosine phosphorylation and degradation, *FEBS Lett.* **385**:63–66.

Gow, A., Duran, D., Thom, S. R., and Ischiropoulos, H., 1996b, Carbon dioxide enhancement of peroxynitrite-mediated protein tyrosine nitration, *Arch. Biochem. Biophys.* **333**:42–48.

Grant, A. J., Jessup, W., and Dean, R. J., 1993, Inefficient degradation of oxidized regions of protein molecules, *Free Radical Res. Commun.* **18**:259–267.

Grune, T., Reinheckel, T., Joshi, M., and Davies, K. J. A., 1995, Proteolysis in cultured liver epithelial cells during oxidative stress, *J. Biol. Chem.* **270**:2344–2351.

Guptasarma, P., Balasubramanian, D., Matsugo, S., and Saito, I., 1992, Hydroxyl radical mediated damage to proteins, with special reference to the crystallins, *Biochemistry* **31**:4296–4302.

Harrington, C. R., and Colaco, C. A. L. S., 1994, A glycation connection, *Nature* **370**:247–248.

Harris, M., Hensley, K., Butterfield, D. A., Leedle, R. A., and Carney, J. M., 1995, Direct evidence of oxidative injury produced by Alzheimer's β-amyloid peptide (1–40) in cultured hippocampal neurons, *Exp. Neurol.* **131**:193–202.

Heinecke, J. W., Li, W., Daehnke, H. L., III, and Goldstein, J. A., 1993, Dityrosine, a specific marker of oxidation, is synthesized by the myeloperoxidase–hydrogen peroxide system of human neutrophils and macrophages. *J. Biol. Chem.* **268**:4069–4077.

Huggins, T. G., Wells-Knecht, M C., Detorie, N. A., Baynes, J. W., and Thorpe, S. R., 1993, Formation of O-tyrosine and dityrosine in proteins during radiolytic and metal-catalyzed oxidation, *J. Biol. Chem.* **268**:12341–12347.

Hunt, J. V., 1996, The assessment of protein glycation in human atherosclerotic plaques by affinity chromatography, *Redox Rep.* **2**:5–8.

Hunter, T., 1995, Protein kinases and phosphatases: The Yin and Yang of protein phosphorylation and signaling, *Cell* **80**:225–236.

Ischiropoulos, H., and Al-Medi, A. B., 1995, Peroxynitrite-mediated oxidative protein modifications, *FEBS Lett.* **364**:279–282.

Karsek-Staples, J. A., and Webster, R. O., 1993, Caeruplasmin inhibits carbonyl formation in endogeneous cell proteins, *Free Radical Biol. Med.* **14**:115–125.

Kelley, F. J., and Birch, S., 1993, Ozone exposure inhibits cardiac protein synthesis in the mouse, *Free Radical Biol. Med.* **14**:443–446.

Kikugawa, K., Kato, T., and Okamoto, Y., 1994, Damage of amino acids and proteins induced by nitrogen dioxide, a free radical toxin in air, *Free Radical Biol. Med.* **16**:373–382.

Kim, K., Rhee, S. G., and Stadtman, E. R., 1985, Nonenzymatic cleavage of proteins by reactive oxygen species generated by dithiothreitol and iron, *J. Biol. Chem.* **260**:15394–15397.

Kong, S.-K., Yim, M. B., Stadtman, E. R., and Chock, P. B., 1996, Peroxynitrite disables the tyrosine phosphorylation regulatory mechanism: Lymphocyte-specific tyrosine kinase fails to phosphorylate nitrated cdc2(6–20)NH_2 peptide, *Proc. Natl. Acad. Sci. USA* **93**:3377–3382.

Kopoldova, J., and Liebster, J., 1963, The mechanism of radiation chemical degradation of amino acids V, *Int. J. Appl. Radiat. Isotopes* **14**:493–498.

Kristal, B. S., and Yu, B. P., 1992, An emerging hypothesis: Synergistic induction of aging by free radicals and Maillard reactions, *J. Gerontol.* **47**:B104–B107.

Levine, R. L., 1983, Oxidation of glutamine synthetase. II. Characterization of the ascorbate model system, *J. Biol. Chem.* **258**:11828–11833.

Levine, R. L., 1989, Proteolysis induced by metal-catalyzed oxidation, *Cell. Biol. Rev.* **21**:347–360.

Levine, R. L., Mosoni, L., Berlett, B. S., and Stadtman, E. R., 1996, Methionine residues as endogenous antioxidants in proteins, *Proc. Natl. Acad. Sci. USA* **93**:15036–15040.

Lymar, S. V., and Hurst, J. K., 1995, Rapid reaction between peroxynitrite ion and carbon dioxide: Implications for biological activity, *J. Am. Chem. Soc.* **117**:8867–8868.

Lymar, S. V., and Hurst, J. K., 1996, Carbon dioxide: Physiological catalyst for peroxynitrite-mediated cellular damage or cellular protectant? *Chem. Res. Toxicol.* **9**:845–850.

Lymar, S. V., Jiang, Q., and Hurst, J. K., 1996, Mechanisms of carbon dioxide-catalyzed oxidation of tyrosine by peroxynitrite, *Biochemistry* **35**:7855–7861.

Maskos, Z., Rush, J. D., and Koppenol, W. H., 1992a, The hydroxylation of phenylalanine and tyrosine: A comparison with salicylate and tryptophan, *Arch. Biochem. Biophys* **296**:521–529.

Maskos, Z., Rush, J. D., and Koppenol, W. H., 1992b, The hydroxylation of tryptophan, *Arch. Biochem. Biophys.* **296**:514–520.

Mickel, H. S., Oliver, C. N., and Starke-Reed, P. E., 1990, Protein oxidation and myelinolysis occur in brain following rapid correction of hyponatremia, *Biochem. Biophys. Res. Commun.* **172:**92–97.

Monnier, V., 1990, Nonenzymatic glycosylation, the Maillard reaction and the aging process, *J. Gerontol.* **45:**B106–B111.

Monnier, V., Gerhardinger, C., Marion, M. S., and Taneda, S., 1995, Novel approaches toward inhibition of the Maillard reaction *in vivo*: Search, isolation, and characterization of prokaryotic enzymes which degrade glycated substrates, in *Oxidative Stress and Aging*, (R. G. Cutler, L. Packer, J. Bertram, and A. Mori, eds.), pp 141–149, Birkhauser Verlag, Basel.

Mordente, A., Martorana, G. E., Miggiano, G. A. D., Meucci, E., Santini, S. A., and Castelli, A., 1988, Mixed function oxidation and enzymes: Kinetic and structural properties of oxidatively modified alkaline phosphatase, *Arch. Biochem. Biophys.* **264:**502–509.

Moskovitz, J., Rahman, M. A., Strassman, J., Yancey, S. O., Kushner, S. R., Brot, N., and Weissbach, H., 1995, *Escherichia coli* peptide methionine sulfoxide reductase gene: Regulation of expression and role in protecting against oxidative damage, *J. Bacteriol.* **177:**502–507.

Mullarkey, C. J., Edelstein, D., and Brownlee, M., 1990, Free radical generation by early glycation products. A mechanism for accelerated atherogenesis in diabetes, *Biochem. Biophys. Res. Commun.* **173:**932–939.

Murphy, M. E., and Kherer, J. P., 1989, Oxidation state of tissue thiol groups and content of protein carbonyl groups in chickens with inherited muscular dystrophy, *Biochem. J.* **260:**359–364.

Musci, G., Bonaccorsi di Patti, M. C., Fagiolo, U., and Calabrese, L., 1993, Age-related changes in human ceruloplasmin, *J. Biol. Chem.* **268:**13388–13395.

Nadkarni, D. V., and Saayre, L. M., 1995, Structural definition of early lysine and histidine adduction chemistry of 4-hydroxynonenal, *Chem. Res. Toxicol.* **8:**284–291.

Nakamura, K., Oliver, C., and Stadtman, E. R., 1985, Inactivation of glutamine synthetase by a purified rabbit liver microsomal cytochrome P-450 system, *Arch. Biochem. Biophys.* **240:**319–329.

Neuzil, J., Gebiki, J. M., and Stocker, R., 1993, Radical-induced chain oxidation of proteins and its inhibition by chain-breaking antioxidants, *Biochem. J.* **293:**601–606.

Oliver, C. N., Fucci, L., Levine, R. L., Wittenberger, M. E., and Stadtman, E. R., 1982, Inactivation of key metabolic enzymes by P450 linked mixed function oxidation systems, in *Cytochrome P-450, Biochemistry, Biophysics, and Environmental Implications* (E. Heitanen, M. Laitinen, and O. Hanninen, eds.), pp 531–539, Elsevier, Amsterdam.

Oliver, C. N., Ahn, B.-W., Moerman, E. J., Goldstein, S., and Stadtman, E. R., 1987a, Age-related changes in oxidized proteins, *J. Biol. Chem.* **262:**5488–5491.

Oliver, C. N., Levine, R. L., and Stadtman, E. R., 1987b, A role of mixed-function oxidation reactions in the accumulation of altered enzyme forms during aging, *J. Geriatric Bioscience* **35:**947–956.

Oliver, C. N., Starke-Reed, P. E., Stadtman, E. R., Liu, G. J., Carney, J. M., and Floyd, R. A., 1990,. Oxidative damage to brain proteins, loss of glutamine synthetase activity, and production of free radicals during ischemia-reperfusion-induced injury to gerbil brain, *Proc. Natl. Acad. Sci. USA* **87:**5144–5147.

Poston, J. M., 1988, Detection of oxidized amino acid residues using *P*-aminobenzoic acid adducts, *Fed. Proc.* **46:**1979 (Abstract).

Poston, J. M., and Parenteau, G. L., 1992, Biochemical effects of ischemia on isolated perfused rat heart tissues, *Arch. Biochem. Biophys.* **295:**35–41.

Pryor, W. A., and Squadrito, G. L., 1995. The chemistry of peroxynitrite: A product from the reaction of nitric oxide with superoxide, *Am. J. Physiol.* **268:**L699–L722.

Pryor, W. A., Jin, X., and Squadrito, G. L., 1994, One- and two-electron oxidations of methionine by peroxynitrite, *Proc. Natl. Acad. Sci. USA* **91:**11173–11177.

Reddy, S., Bichler, J., Wells-Knecht, K. J., Thorpe, S. R., and Baynes, J. W., 1995, Nε-(carboxymethyl)lysine is a dominant advanced glycation end product (AGE) antigen in tissue proteins, *Biochemistry* **34:**10872–10878.

Reznick, A. Z., Cross, C. E., Hu, M.-L., Suzuki, Y. J., Khwaja, S., Safadi, A., Motchnik, P. A., Packer, L., and Halliwell, B., 1992, Modification of plasma proteins by cigarette smoke as measured by protein carbonyl formation, *Biochem. J.* **286:**607–611.

Rivett, A. J., 1986, Regulation of intracellular protein turnover: Covalent modification as a mechanism of marking proteins for degradation, Curr. Top. Cell. Regul. **28:**291–337.

Rivett, A. J., and Levine, R. L., 1990, Metal-catalyzed oxidation of *Escherichia coli* glutamine synthetase: Correlation of structural and functional changes, *Arch. Biochem. Biophys.* **278:**26–34.

Rothstein, M., 1977, Recent developments in age-related alteration of enzymes, *Mech. Aging and Dev.* **6**:241–257.

Sahakian, J. A., Shames, B. D., and Levine, R. L., 1991, Metal-catalyzed oxidation of glutamine synthetase. Susceptibility to proteolysis requires oxidation of histidine residue at both metal binding sites, *FASEB J.* **5**:A1177 (Abstract).

Sayre, L. M., Arora, P. K., Iyer, R. S., and Salomon, R. G., 1993, Pyrrole formation from 4-hydroxynonenal and primary amines, *Chem. Res. Toxicol.* **6**:19–22.

Schuenstein, E., and Esterbauer, H., 1979, Formation and properties of reactive aldehydes, in *Submolecular Biology of Cancer*, pp 225–244, CIBA Foundation Series 67, Excerpta Medica, Elsevier, Amsterdam.

Schuessler, H., and Schilling, K., 1984, Oxygen effect in radiolysis of proteins. Part 2. Bovine serum albumin, *Int. J. Radiat. Biol.* **45**:267–281.

Sell, D. R., and Monnier, V. M., 1989, Structure elucidation of a senescence cross-link from human extracellular matrix. Implication of pentoses in the aging process, *J. Biol. Chem.* **264**:21597–21602.

Smith, C. D., Carney, J. M., Starke-Reed, P. E., Oliver, C. N., Stadtman, E. R., and Floyd, R. A., 1991, Excess brain protein oxidation and enzyme dysfunction in normal and Alzheimer's disease, *Proc. Natl. Acad. Sci. USA* **88**:10540–10543.

Smith, C. D., Carney, J. M., Tatsuno, T., Stadtman, E. R., Floyd, R. A., and Marksbery, W. R., 1994, Protein oxidation in aging brain, in *Aging and Cellular Defense Mechanisms* (C. Franceschi, G. Cerpaldi, V. G. Cristofalo, and J. Vijg, eds.), pp 110–119, New York Academy of Sciences, New York.

Smith, M. A., Taneda, S., Richey, P. L., Miyata, S., Yan, S.-D., Stern, D., Sayre, L. M., Monnier, V., and Perry, G., 1994, Advanced Maillard reaction end products are associated with Alzheimer's disease pathology, *Proc. Natl. Acad. Sci. USA* **91**:5710–5714.

Smith, M. A., Rudnicka-Nawrot, M., Richey, P. L., Praprotnik, D., Mulvihill, P., Miller, C. A., Sayre, L. M., and Perry, G., 1995, Carbonyl-related posttranslational modification of neurofilament protein in neurofibrillary pathology of Alzheimer's disease, *J. Neurochem.* **64**:2660–2666.

Smith, M. A., Perry, P. L., Sayre, L. M., Anderson, V. E., Beal, M.F., and Kowall, N., 1996, Oxidative damage in Alzheimer's disease, *Nature* **382**:120–121.

Sohal, R. S., Agarwal, S., Dubey, A., and Orr, W. C., 1993b, Protein oxidative damage is associated with life expectancy of houseflies, *Proc. Natl. Acad. Sci. USA* **90**:7255–7259.

Sohal, R. S., Su, H.-H., Agarwal, S., Forster, M. J., and Lal, H., 1994, Oxidative damage, mitochondrial oxidant generation, and antioxidant defenses during aging and response to food restriction, *Mech. Aging and Dev.* **74**:121–133.

Sohal, R. S., Ku, H.-H., and Agarwal, S., 1993, Biochemical correlates of longevity in two closely related rodent species, *Biochem. Biophys. Res. Commun.* **196**:7–11.

Sohal, R. S., Agarwal, S., and Sohal, B. H., 1995, Oxidative stress and aging in the Mongolian gerbil (*Meriones unguiculatus*), *Mech. Aging Dev.* **81**:15–25.

Sohal, R. S., and Dubey, A., 1994, Mitochondrial oxidative damage, hydrogen peroxide release, and aging, *Free Rad. Biol. Med.* **16**:621–626.

Solar, S., 1985, Reactions of ·OH with phenylalanine in neutral solutions, *Radiat. Phys. Chem.* **26**:103–108.

Stadtman, E. R., 1986, Oxidation of proteins by mixed-function oxidation systems: Implication in protein turnover, aging, and neutrophil function, *Trends Biochem. Sci.* **11**:11–12.

Stadtman, E. R., 1988, Oxygen radical-mediated inactivation of enzymes: Biological implications, in *Medical, Biochemical, and Chemical Aspects of Free Radicals* (O. Hayaishi, E. Niki, Kondo, and T. Yoshikawa, eds.), pp. 11–19, Elsevier, Amsterdam.

Stadtman, E. R., 1990, Metal ion-catalyzed oxidation of proteins: Biochemical mechanism and biological consequences, *Free Radical Biol. Med.* **9**:315–325.

Stadtman, E. R., 1992, Protein oxidation and aging, *Science* **257**:1220–1224.

Stadtman, E. R., and Oliver, C. N., 1991, Metal catalyzed oxidation of proteins: Physiological consequences, *J. Biol. Chem.* **266**:2005–2008.

Stadtman, E. R., and Wittenberger, M.E., 1985, Inactivation of *Escherichia coli* glutamine synthetase by xanthine oxidase, nicotinate hydroxylase, horseradish peroxidase, or glucose oxidase: Effects of ferredoxin, putidaredoxin, and menadione, *Arch. Biochem. Biophys.* **239**:379–387.

Stadtman, E. R., Chock, P. B., and Rhee, S. G., 1981, Interconvertible enzyme cycles in cellular regulation, *Curr. Top. Cell. Regul.* **18**:79–83.

Stafford, R. E., Mak, T. M., Kramer, J. H., and Weglicki, W. B., 1993, Protein oxidation in magnesium deficient rat brains and kidneys, *Biochem. Biophys. Res. Commun.* **196:**596–600.

Starke, P. E., Oliver, C. N., and Stadtman, E. R., 1987, Modification of hepatic proteins in rats exposed to high oxygen concentration, *FASEB J.* **1:**36–39.

Starke-Reed, P. E., and Oliver, C. N., 1989, Protein oxidation and proteolysis during aging and oxidative stress, *Arch. Biochem. Biophys.* **275:**559–567.

Swallow, A. J., 1960, Effect of ionizing radiation on proteins. RCO groups, peptide bond cleavage, inactivation, -SH oxidation, in *Radiation Chemistry of Organic Compounds* (A. J. Swallow, ed.), pp. 211–224, Pergamon Press, New York.

Szweda, L. I., 1994, Age-related increase in liver retinyl palmitate: Relationship to lipofuscin, *J. Biol. Chem.* **269:**8712–8715.

Szweda, L. I., and Stadtman, E. R., 1992, Iron-catalyzed oxidative modification of glucose-6-phosphate dehydrogenase from *Leuconostoc mesenteroides*, *J. Biol. Chem.* **267:**3096–3100.

Taborsky, G., 1973, Oxidative modification of proteins in the presence of ferrous iron and air. Effect of ionic constituents of the reaction medium on the nature of the oxidation products, *Biochemistry* **12:**1341–1348.

Takahashi, R., and Goto, S., 1990, Alteration of aminoacyl-tRNA synthetase with age: Heat labilization of the enzyme by oxidative damage, *Arch. Biochem. Biophys.* **277:**228–233.

Toyokuni, S. Uchida, K., Okamoto, K., Hattori-Nakakuki, Y., Hiai, H., and Stadtman, E. R., 1994, Formation of 4-hydroxy-2-nonenal-modified proteins in the renal proximal tubules of rats treated with a renal carcinogen ferric nitrilotriacetate, *Proc. Natl. Acad. Sci. USA* **91:**2616–2620.

Uchida, K., and Kawakishi, S., 1993, 2-oxohistidine as a novel biological marker for oxidatively modified proteins, *FEBS Lett.* **332:**208–210.

Uchida, K., and Stadtman, E. R., 1993, Covalent modification of 4-hydroxynonenal to glyceraldehyde-3-phosphate, *J. Biol. Chem.* **268:**6388–6393.

Uchida, K., Kato, Y., and Kawakishi, S., 1990, A novel mechanism for oxidative damage of prolyl peptides induced by hydroxyl radicals, *Biochem. Biophys. Res. Commun.* **169:**265–271.

Uchida, K., Toyokuni, S., Nishikawa, K., Kawakishi, S., Oda, H., Hiai, H., and Stadtman, E. R., 1994, Michael addition-type 4-hydroxy-2-nonenal adducts in modified low density lipoproteins: Markers for atherosclerosis, *Biochemistry* **33:**12487–12494.

Uchida, K., Fukuda, A., Kawakishi, S., Hiai, H., and Toyokuni, S., 1995, A renal carcinogen ferric nitriloacetate mediates a temporary accumulation of aldehyde-modified proteins within cytosolic compartments of rat kidney, *Arch. Biochem. Biophys.* **317:**405–411.

Uppu, R. M., Squadrito, G. L., and Pryor, W. A., 1996, Acceleration of peroxynitrite oxidation by carbon dioxide, *Arch. Biochem. Biophys.* **327:**335–343.

van der Vliet, A., Eiserich, J. P., O'Neill, C. A., Halliwell, B., and Cross, C. E., 1995, Tyrosine modification by reactive nitrogen species. A closer look, *Arch. Biochem. Biophys.* **319:**341–349.

Vogt, W., 1995, Oxidation of methionine residues in proteins: Tools, targets, and reversal, *Free Rad. Biol. Med.* **18:**93–105.

Wells-Knecht, M. C., Huggins, T. G., Dyer, G., Thorpe, S. R., and Baynes, J. W., 1993, Oxidized amino acids in lens protein with age, *J. Biol. Chem.* **268:**12348–12352.

Winchester, R. V., and Lynn, K. R., 1970, X- and γ-radiolysis of some tryptophan dipeptides, *Int. J. Radiat. Biol.* **17:**541–549.

Winter, M. L., and Liehr, J. G., 1991, Free radical-induced carbonyl content in protein of estrogen-treated hamsters assayed by sodium boro[^{3}H]hydride reduction, *J. Biol. Chem.* **266:**14446–14450.

Witt, E. H., Reznick, A. Z., Viguie, C. A., Starke-Reed, P. E., and Packer, L., 1992, Exercise, oxidative damage, and effects of antioxidant manipulation, *J. Nutr.* **122:**766–773.

Wolf, S. P., and Dean, R. T., 1987, Glucose autoxidation and protein modification, *Biochem. J.* **245:**243–250.

Yoritaka, A. Hattori, N., Uchida, A., Tanaka, M., and Stadtman, E. R., 1996, Immuochemical detection of 4-hydroxy-nonenal protein adducts in Parkinson's disease, *Proc. Natl. Acad. Sci. USA* **93:**2696–2701.

Youngman, L. D., Park, J.-Y. K., and Ames, B., 1992, Protein oxidation associated with aging is reduced by dietary restriction of protein calories, *Proc. Natl. Acad. Sci. USA* **89:**9112–9116.

Zhou, J. Q., and Gafni, A., 1991, Exposure of rat muscle phosphoglycerate kinase to a nonenzymatic MFO system generates the old form of enzyme, *J. Gerontol.* **46:**B217–B221.

Part VIII

Conclusion

Chapter 28

An Overview of Reactive Oxygen Species

Daniel L. Gilbert and Carol A. Colton

1. INTRODUCTION

It is commonly believed that the Earth's atmosphere at one time had no free oxygen (Gilbert, 1996). As the atmospheric oxygen concentration increased via photosynthetic processes, various types of antioxidant defenses were induced in the anaerobic cell. These anaerobic cells are the facultative anaerobes and can exist in the absence or presence of oxygen. Aerobic life evolved as additional antioxidant defenses were developed to combat the increased atmospheric oxygen. For those organisms that can utilize oxygen, there are special advantages. All multicellular organisms require oxygen (Gilbert, 1996), and humans from the beginnings of recorded history have recognized that the "breath of life" (oxygen) is necessary to maintain life (Chapter 1). Oxygen is a sluggish oxidizing agent despite the fact that it possesses a high thermodynamic potential, the Gibbs free energy of formation of water being 474 kJ per mole of dioxygen (Hoare, 1985). The oxidation of sugar to carbon dioxide and water is 480 kJ per mole of dioxygen. What prevents the free oxygen from oxidizing the essential cellular ingredients?

The answer is antioxidant defenses (Tanswell and Freeman, 1995). The energy that supports aerobic life also provides the stress that has to be endured during the lifetime of the organism. Transition metals catalyze the release of this energy (Götz *et al.*, 1994; Chapter 4). Ionizing radiation is also a catalyst for the release of this energy. Fifty percent of all humans can be killed by 5 grays (Gy) of radiation, which increases the temperature

Daniel L. Gilbert Unit on Reactive Oxygen Species, BNP, NINDS, National Institutes of Health, Bethesda, Maryland 20892-4156. **Carol A. Colton** Department of Physiology and Biophysics, Georgetown University Medical School, Washington, D.C. 20007.

Reactive Oxygen Species in Biological Systems, edited by Gilbert and Colton. Kluwer Academic / Plenum Publishers, New York, 1999.

a negligible 0.0012°C (Gilbert, 1972). Oxygen stress is unique, it is different from other types of stress, such as heat stress (Perdrizet, 1995) or cold stress, although other types of stress might invoke oxygen stress. Essentially, the energy is derived from the splitting of water into oxygen and hydrogen; the oxygen is released into the atmosphere and the hydrogen is taken up by carbon to form carbohydrates (Gilbert, 1996).

In this chapter, the fundamental aspects of reactive oxygen species (ROS) are summarized so as to highlight their essential nature.

2. SPECIFICITY OF ROS

The ROS that are generated during respiration are not always deleterious. The biosphere has learned to use oxygen not only for energy but also for synthesis. It is not that surprising to learn that the less reactive oxygen species can also be used in cell signaling (Chapter 8). Figure 1 gives an oversimplified scheme of this signaling aspect. The superoxide radical anion and hydrogen peroxide can act as signaling molecules (Chapters 5–7), inducing the production of the antioxidant enzymes superoxide dismutase and catalase. Activated macrophages in the extracellular environment (Chapters 19 and 23) and mitochondria in the intracellular environment (Chapter 3) both release ROS. Nitric oxide is one such species that is released by the intracellular nitric oxide synthase (Chapter 9). It is significant that nitroxides can act as superoxide dismutases (Chapter 11). ROS such as the superoxide radical anion, hydrogen peroxide, and nitric oxide are all relatively inactive and can diffuse some distance from the site where they are generated to their sites of action. Iron and copper, in trace amounts, can catalyze superoxide and hydrogen peroxide by the Fenton reaction to produce the highly toxic hydroxyl radical (Chapters 2 and 4). Nitric oxide can diffuse across cellular membranes to activate soluble guanylate cyclase, a heme protein (Chapter 10). Nitric oxide inhibits xanthine dehydrogenase/xanthine oxidase, an iron–sulfur enzyme containing sulfhydryl groups (Hassoun *et al.*, 1995). Oxygen can diffuse across almost all cellular membranes, with the exception of the swimbladder of fish (Gilbert, 1996). For example, oxygen can combine with hemoglobin to form oxyhemoglobin. Lander (1997) has pointed out that ROS can act as signal transducers. However, the most reactive oxygen species, the hydroxyl radical, is so reactive that as soon as it contacts anything, it reacts with it (Halliwell and Gutteridge, 1989). Thus, the hydroxyl radical is always nonspecific. In the presence of a transition metal, hydrogen peroxide can generate hydroxyl radicals (Chapter 2), damaging cellular constituents (Figure 1). ROS, with the exception of the hydroxyl radical, do have chemical properties enabling reaction with sulfhydryl groups, iron–sulfur proteins, and heme (iron) proteins. Elevated oxygen concentrations can attack sulfhydryl groups on enzymes, inactivating them (Stadie *et al.*, 1944; Haugaard, 1946). Hence, it is not surprising that sulfhydryl groups are sensitive to oxygen and under redox control. There is evidence that hydrogen peroxide and the superoxide radical anion can activate specific sites and become second messengers as well (Suzuki *et al.*, 1997). Thioredoxin, a small sulfhydryl protein, is capable of "redox regulation," and is induced by many types of stress (Nakamura *et al.*, 1997). It regulates the cytoplasmic NF-κB (Hayashi *et al.*, 1993). Thioredoxin is found in bacteria, plants, and animals.

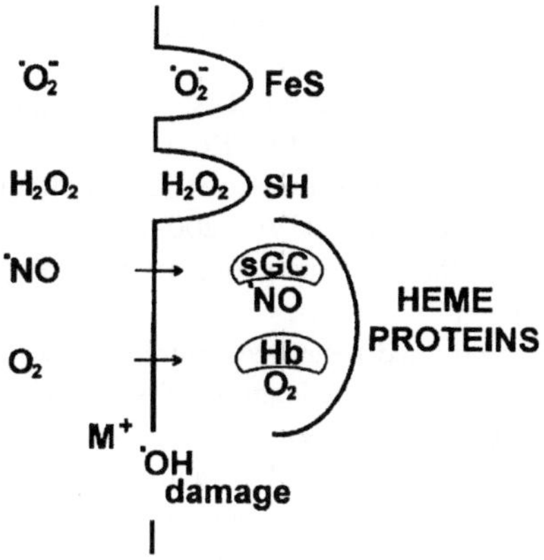

FIGURE 1. Site-specific action of ROS. Some examples are given, but it is possible that other modes exist. The receptor for the superoxide radical anion is the transcription activator iron–sulfur protein, whereas the receptor for hydrogen peroxide is composed of cysteine sulfhydryl groups. Nitric oxide binds to the soluble guanylate cyclase (sGC) and molecular oxygen binds to hemoglobin (Hb); both of these are heme proteins. In the presence of transition metals (M) such as iron or copper ions, hydroxyl radicals are formed. The hydroxyl radical is so reactive that it attacks all chemical groups without much discrimination; it is radically different from the other common ROS.

3. ROS RESPONSE TO DECREASED OXYGEN

If the mammalian brain does not receive adequate oxygen, then the brain cells will die. The carotid body acts to monitor the level of oxygen; it uses a cytochrome P450 family to sense the oxygen tension (Hatton and Peers, 1996). The carotid body is located at the bifurcation of the carotid artery. When the arterial oxygen is below the normal range of 85 to 95 torr, the heme protein is possibly activated in the chemoreceptor cells, i.e., the glomus cells (type I carotid body cells), of the carotid body (Bunn and Poyton, 1996). There are other chemoreceptors in the pulmonary smooth muscle (López-Barneo, 1996). It seems that potassium channels are of at least two groups; one is calcium activated and inhibited by charybdotoxin, and the other one is calcium independent. The charybdotoxin-sensitive ones contribute to the resting potential. When the cells are subjected to hypoxia, the charybdotoxin channels cause depolarization (Wyatt *et al.*, 1995), which leads to opening of voltage-gated calcium channels. The calcium channels are mainly L-type and some N-type (Peers *et al.*, 1996). Low oxygen closes the potassium channels (López-Barneo, 1994). Similar events occur in the lung, resulting in pulmonary arterial vasoconstriction (HPV) (Ureña *et al.*, 1996). The superoxide anion produced by an NADH oxidoreductase may also be involved (Mohazzab-H *et al.*, 1995).

Bunn and Poyton (1996) have theorized that during hypoxia, the oxygen binds to a heme protein giving rise to the superoxide radical anion, which can then dismutate and form hydrogen peroxide. The transcription factor, hypoxia-inducible factor-1, in the reduced sulfhydryl form (SH) is active and in the oxidized disulfide form (SS) is inactive.

4. TISSUES NORMALLY SUBJECTED TO HIGH OXYGEN TENSIONS

The skin and cornea are exposed to the highest concentrations of oxygen. At sea level, the ambient oxygen concentration is 158 torr. Both the skin and the cornea are subjected to this high oxygen tension, as well as ultraviolet (UV) radiation. Classifications of UV radiation according to wavelength are: UVC (200–280 nm), UVB (280–320 nm), and UVA (320–400 nm). UVC is the most energetic, but is blocked by the ozone layer in the stratosphere (Chapter 12).

UVA is the least energetic and penetrates more deeply into human skin than does UVB (Emerit, 1992). Emerit classified aging of skin into chronological aging and photoaging. ROS is involved in both types of aging. Melanins are free radical pigments, which are formed after exposure of skin to UV radiation. Tyrosine oxidation to 3,4-dihydroxyphenylalanine (Dopa) results in eumelanin formation. Pheomelanin is formed in the presence of sulfhydryl compounds, such as cysteine and reduced glutathione. Pheomelanin is more easily photooxidized than eumelanin. Red-haired individuals, who not only have less total melanin but also a higher percentage of pheomelanin than dark-haired persons, have more skin cancer. Superoxide radical anion and hydrogen peroxide are also formed during the photooxidation. Photodynamic therapy has been successfully used in treating skin cancers. Photosensitizers such as 5-aminolevulinic acid have demonstrated their efficacy in the treatment of actinic keratoses (Jeffes *et al.*, 1997; Moore *et al.*, 1997). When epidermal Langerhans cells were exposed to UVB, superoxide dismutase protected these cells from being killed, indicating that superoxide anions were causing their death (Horio and Okamoto, 1987). Packer *et al.* (1997) found that skin homogenates exposed to UV generated the ascorbyl radical.

The cornea can metabolize arachidonate to prostaglandins and hydroxy arachidonate derivatives (HETEs), which function in the inflammatory response (Hurst *et al.*, 1989). Dimethylthiourea, an antioxidant, improved inflammation induced by sodium hydroxide (Alio *et al.*, 1995), Nitric oxide, an air pollutant, produces nitrite in aqueous solution (Snell *et al.*, 1996). Experiments show that nitrite in the presence of UVA oxidized the corneal thiols and reduced glutathione, and that this outcome was prevented by ascorbate (Varma *et al.*, 1997). Also, nuclear ferritin induced in chick cornea cells might be acting like an antioxidant by binding iron (Cai *et al.*, 1997).

The risk of cataract formation, i.e., opacity in the aging lens, is decreased by antioxidants (Zigler and Hess, 1985; Trevithick and Dzialoszynski, 1997). Trevithick and Dzialoszynski found that rat lens and retina homogenates have higher endogenous antioxidant activity than vitreous humor homogenates. The retina, being part of the brain, is highly oxygenated, but also is subject to concentrated irradiation (Hunt *et al.*, 1996a). Normally, iron, a prooxidant, is sequestered by apotransferrin (apo-Tf) and by apohemopexin (apo-HPX) (Hunt *et al.*, 1996b). In addition, hemoglobin, which promotes lipid peroxidation by serving as an iron source and by the heme promoting peroxidation (Halliwell and Gutteridge, 1989), is bound to haptoglobin (Hunt *et al.*, 1996b), which inhibits the peroxidation. Apo-Tf and apo-HPX are synthesized in the retina (Hunt *et al.*, 1996b); the heme and apo-HPX form the complex, heme–HPX, which dissociates in the liver, releasing bilirubin, an antioxidant, under certain conditions (Halliwell and Gutteridge, 1989). Aging increases oxidative damage to the retina (Ohia *et al.*, 1994; Trevithick and Dzialoszynski, 1997). ROS increase the vascular endothelial growth factor in the retina, which was prevented by antioxidants in reperfusion of ischemic retina (Kuroki *et al.*, 1996). Aged human retina contains fluorescent material in lipofuscin granules of the retinal pigment epithelium (Gaillard *et al.*, 1995; Rózanowska *et al.*, 1995). When illuminated with blue light, lipofuscin acts as a photosensitizer and generates ROS including singlet oxygen. Vitamins C and E prevent the retina of diabetic rats from undergoing a decrease in antioxidant defenses (Kowluru *et al.*, 1997). Retinopathy of prematurity is a retinal disease that produces blindness in many premature infants (Flower and Patz, 1981) and is caused by high oxygen levels. Using a rat model of this disease, Niesman *et al.* (1997) found that when rats were exposed to 0.8 bar of oxygen, the level of superoxide dismutase (SOD) in the retina was decreased. The intraperitoneal administration of SOD in liposomes did increase the level of retinal SOD.

Although the lung is exposed to approximately 60% of the oxygen that is in the ambient air, its oxygen concentration is still greater than that of the other internal organs. When rats breath 1 atm of oxygen, their survival time is only 111.3 ± 1.4 hr (Gerschman *et al.*, 1958). The mitochondrial manganese superoxide dismutase (MnSOD) activity is decreased when rats are exposed to oxygen concentrations exceeding 95%. Transgenic mice overexpressing MnSOD can resist the toxicity of the increased oxygen (Camhi *et al.*, 1995). Diseases such as asthma demonstrate that there is an increase in ROS (Smith *et al.*, 1997).

The brain is not exposed to high oxygen concentrations, although the oxygen demand is high (Chapter 23). In addition, the brain contains high amounts of ascorbate (Rebec and Pierce, 1984; Katsuki, 1996) and iron (Conner *et al.*, 1994). The ascorbate in the presence of iron ions is potentially a prooxidant mixture.

5. SOURCES OF ROS IN THE MAMMALIAN ORGANISM

There are many major sources of ROS in the body. If nitric oxide is included, then other cells, such as astrocytes and neurons, should also be included (Minc-Golumb and Schwartz, 1994). Cellular organelles, such as mitochondria and peroxisomes, also contribute to ROS production. In plants, the chloroplasts manufacture the oxygen that animals require for respiration. The specific enzymes responsible for ROS production include the cytochrome P450 enzymes, the membrane NADPH enzymes, and the monoamine oxidase enzymes. Polyunsaturated fatty acids, such as the 20-carbon arachidonic acid, are also responsible for ROS generation.

Neutrophils in the blood can release superoxide anions into vacuoles produced by phagocytosis. In addition, myeloperoxidase is released into the vacuoles: In the presence of hydrogen peroxide and chloride anions, hypochlorous acid (HOCl) is formed (Halliwell and Gutteridge, 1989).

Macrophages reside in many parts of the body. In the brain, the resident macrophages are the microglia (Streit and Kincaid-Colton, 1995). Activated microglia produce superoxide anion when stimulated with various agents (Giulian and Baker, 1986; Colton and Gilbert, 1987, 1989, 1993; Colton *et al.*, 1990, 1992; Yao *et al.*, 1990; Zielasek *et al.*, 1993; Colton, 1994, 1995; Colton and Chernyshev, 1996; Zielasek and Hartung, 1996; Zielasek *et al.*, 1996). Rat and mouse microglia also produce nitric oxide (Colton *et al.*, 1994). In contrast, microglia from hamsters and humans produce significantly less superoxide than do rat and mouse microglia; in addition, there is very little or no nitric oxide output by human and hamster microglia (Colton *et al.*, 1996). However, Díng *et al.* (1997) have reported 10 to 20 μM nitric oxide in stimulated human microglia. Microglia contain a very high level of iron, principally as ferritin (Conner *et al.*, 1994).

The lung contains two types of macrophages, namely, alveolar and interstitial (Franke-Ullmann *et al.*, 1996). *In vitro* experiments have shown that the activated alveolar macrophage produces more superoxide anions than does the interstitial macrophage. Asbestos fibers that are too long for alveolar macrophages to phagocytize stimulate the macrophages to release ROS (Quinlan *et al.*, 1994). Collagenlike polypeptides activate alveolar macrophages to produce ROS (Laskin *et al.*, 1994); Kupffer cells are the resident macrophages in the liver and do not release much ROS (Kausalya *et al.*, 1996). Macrophages resident in the spleen, bone marrow, and liver contain substantial quantities of iron, chiefly as ferritin and hemosiderin (Senozan and Christiano, 1997). Activated peritoneal macrophages produce ROS, the level of which is decreased when cod liver oil is present (Joe and Lokesh, 1994). When neurons are grown under hypoxic conditions, they release superoxide anions (Daval *et al.*, 1995).

5.1. Cellular Organelles

Mitochondria are cellular organelles that convert the energy released by the reduction of oxygen to water into ATP production (Papa, 1996). The oxygen is reduced in univalent steps, first producing the superoxide anion. Several studies have found that about 2 to 5% of the oxygen uptake results in the formation of the superoxide anion. The rat liver mitochondrion produces about 3×10^7 radicals daily (Richter *et al.*, 1995). This superoxide production takes place in complexes I and III in the inner mitochondrial membrane (Götz *et al.*, 1994). However, MnSOD, the mitochondrial enzyme, dismutates the superoxide anion into dioxygen and hydrogen peroxide. The mitochondrial DNA is continuously being replaced even in postmitotic nerve cells (Papa, 1996) and is subject to oxidative damage. *N*-Methyl-D-aspartate produced ROS in cortical neurons; when inhibitors of mitochondrial transport were used in this system, the level of ROS was reduced (Dugan *et al.*, 1995). Cytokines such as tumor necrosis factor generate a sphingolipid (ceramide) that leads to ROS production in mitochondria (García-Ruiz *et al.*, 1997).

When lipid peroxidation occurs, singlet oxygen can be produced. During lipid peroxidation, peroxyl radicals (·ROO) are formed. Singlet oxygen can arise from the following Russell-type mechanism (Halliwell and Gutteridge, 1989; Sharov *et al.*, 1996):

$$RO_2\cdot + RO_2\cdot \rightarrow RO^*(RO) + {}^1O_2 \tag{1}$$

According to Boveris and Cadenas (Chapter 22), about 15% of the superoxide radicals generated by the mitochondria in the brain give rise to singlet oxygen.

Chloroplasts in higher plants are the sites where photosynthesis occurs (Mauzerall and Piccioni, 1981; Chapter 18). Photosystem II splits water into oxygen and hydrogen; the hydrogen, actually an electron, then is carried to Photosystem I where ferrodoxin is formed, and finally NADPH is generated (Chapter 18).

Like mitochondria, peroxisomes are cellular organelles that produce hydrogen peroxide and catalase, and oxidize very-long-chain fatty acids (Naidu and Moser, 1994). Unlike mitochondria, peroxisomes have only a single leaky membrane and no DNA; they reproduce by fission, just like mitochondria and chloroplasts. They function in lipid metabolism. The diameter of these organelles can be as small as 0.1 μm in the brain or as large as 0.5 μm in the liver and kidney. During the first 2 weeks of life, cerebral and cerebellar neurons contain many peroxisomes, which rapidly disappear (Naidu and Moser, 1990). Oligodendrocytes synthesize myelin (Gogate *et al.*, 1994); peroxisomes in the oligodendrocytes take part in the synthesis of myelin sheaths (Naidu and Moser, 1990).

Much of the cytochrome P450 family of enzymes is located in the peroxisomes (Halliwell and Gutteridge, 1989).

5.2. Cellular Enzymes

The superoxide anion is generated by membrane NADPH oxidases and nitric oxide is generated by both inducible and constitutive nitric oxide synthases. Cytochrome P450 refers to a broad family of different enzymes, which catalyze reactions involving ROS (Bernhardt, 1995). The P450 level in the brain is higher in mitochondria than in microsomes (Bhagwat *et al.*, 1995a), but is less than in liver (Bhagwat *et al.*, 1995b); P450 is also present in both rat and human spinal cord (Bhagwat *et al.*, 1995c). Other investigators have found P450 in the rat brain (Sasame *et al.*, 1997).

P450 monooxygenases have the capacity for xenobiotic metabolism (Ravindranath *et al.*, 1989). Bhagwat's group (Bhagwat *et al.*, 1996a) has reported the presence of a flavin-containing monooxygenase in human brain that is not the same as P450 (Bhagwat *et al.*, 1996b).

Monoamine oxidase (MAO), localized in the mitochondria, oxidatively deaminates neurotransmitters, resulting in the production of hydrogen peroxide. There are two forms of the enzyme: MAO-A and MAO-B. The two forms can be distinguished from each other by the use of metabolic inhibitors, such as clorgyline, which acts on MAO-A, and deprenyl, which acts on MAO-B. Much of MAO resides in the astrocytes (Carlo *et al.*, 1996a,b). MAO-A metabolizes norepinephrine and serotonin in the frontal cortex, hippocampus, and cerebellum (Kim *et al.*, 1997). Cohen and his collaborators (Cohen *et al.*, 1997; Chapter 24) theorize that MAO is involved in Parkinson's disease, a brain mitochondrial disease for which there is evidence that hydrogen peroxide is generated.

Xanthine oxidase/dehydrogenase can generate the superoxide anion (Chapter 15).

5.3. Endogenous Chemicals

Arachidonic acid is the precursor for the cyclooxygenase and lipogenase pathways. The cyclooxygenase pathway produces both prostaglandins and thromboxanes whereas the lipogenase pathway produces the unstable monohydroperoxy-eicosatetraenoic acids (HPETEs), which quickly reduce to monohydro-eicosatetraenoic acids (HETEs) (Halliwell and Gutteridge, 1989).

Studying brain injury in cats, Wei *et al.* (1981) noted that prostaglandin synthesis resulted in ROS production. This ROS production was decreased when indomethacin, a blocker of prostaglandin synthesis, was used. These experimenters (Kontos, 1985, 1987, 1989; Kontos and Wei, 1986) showed that arachidonate metabolism results in the formation of ROS. Prostaglandin synthase in the presence of NADH or NADPH (cytochrome P450) results in superoxide production during experimental brain injury. During ischemia, phospholipase A_2 releases arachidonic acid and platelet-activating factor (PAF), which subsequently induces AP-1; during reperfusion, arachidonic acid is converted to prostaglandins and leukotrienes (Bazan *et al.*, 1995). PAF synthesis occurs in the brain macrophage, the microglia (Bazan, 1989), which promotes the release of arachidonic acid (Bazan and Allan, 1996).

Catecholamines and hemoglobin can also be sources of ROS (Halliwell and Gutteridge, 1989).

6. REPERFUSION INJURY

Reperfusion injury occurs in many organs, such as heart, brain, small intestine, kidney, lung, pancreas, stomach, and skeletal muscle (Chapter 15). Although the mechanism of this injury has not been clearly elucidated, it is evident that ROS are involved. Because allopurinol, a known inhibitor of xanthine oxidase, attenuated the damaging effects of reperfusion injury to the small intestine, xanthine oxidase is implicated. In addition, there is evidence that neutrophils accumulate in the mucosa, which can be inhibited by allupurinol or SOD.

Coronary artery atherosclerosis is a condition marked by proneness to heart attacks. The risk for the disease is reduced by oral supplementation of at least 100 IU vitamin E daily (Hodis *et al.*, 1995). Lipid peroxidation occurs as a result of reperfusion (Kramer *et al.*, 1994), partly a consequence of NADH production of superoxide anion during posthypoxic reoxygenation (Mohazzab-H *et al.*, 1996). Calcium ion channel blockers and β-adrenoceptor antagonists also have been found to possess antioxidant properties. Nicardipine and propranolol are good antioxidants, but not nearly as effective as vitamin E. Carvedilol, a new antihypertensive drug, is a β blocker that possesses antioxidant properties as effective as vitamin E. BM-918228 (Boehringer-Mannheim GmbH, Mannheim, Germany), a hydroxylated analogue of carvedilol, was found to be much more potent than vitamin E as an antioxidant (Yue *et al.*, 1994; Kramer and Weglicki, 1996).

Although the heart and brain are the organs that are most easily affected by ischemia–reperfusion events, almost all of the other organs of the body can also be affected. In the brain, disruption of the blood flow leads to cell damage and extracellular potassium increase, which stimulates the NMDA and AMPA kainate glutamate receptors; this in turn allows entry of more calcium into the neurons. The calcium buildup causes excitotoxicity, which is mediated by ROS (Giroux and Scatton, 1996). In addition, ROS induce the formation of the cerebrospinal fluid SOD, which takes at least 1 day to increase, reaching a peak after 1 to 2 weeks and then returning to normal (Gruener *et al.*, 1994).

7. MAINTAINING A PROPER BALANCE

For many years, mothers have instructed their children to eat a balanced diet. Diets rich in fruits and vegetables contain many antioxidants, especially vitamins C and E, carotenoids, and flavonoids (Chapter 14). Fish oil fatty acids are generally antioxidative (Chandrasekar and Fernandes, 1994; Crosby *et al.*, 1996), but when given in excess they can become prooxidative (Saito and Nakatsugawa, 1994). Moderation is essential.

Parkinsonian dementia on Guam (Garruto, 1991) has been linked to high aluminum and manganese in the water supply (Yasui *et al.*, 1993). One of the effects of acid rain is an increase in the concentration of aluminum (Garruto *et al.*, 1997). Aluminum can increase the rate of lipid peroxidation (Halliwell and Gutteridge, 1989). Thus, food and water should be obtained from many sources. It is important to note that the blood pH is relatively close to neutral. Any change in acidity or alkalinity is not tolerated.

Antioxidant defenses can be grouped into (1) antioxidant defense enzymes that can be induced by transcription factors regulating gene function and (2) antioxidant defense substances that the body is unable to synthesize, such as vitamins. Moderate exercise increases ROS, which in turn induces the activation of the genes controlling the antioxidant enzymes, such as glutathione peroxidase (Giuliani and Cestaro, 1997). Without substrates for the enzyme reactions, the enzymes will not be effective. Intake of antioxidant defense substances, such as vitamins, should also be increased to take advantage of the increased antioxidant enzymes (Fielding and Meydani, 1997). On the other hand, strenuous exercise leads to increased ROS, which are not fully compensated by the increased antioxidant defenses (Giuliani and Cestaro, 1997).

It is often difficult to tell when a prooxidant becomes an antioxidant and vice versa (Gilbert, 1963). An example is xanthine oxidoreductase. This molybdenum enzyme is mainly present as xanthine dehydrogenase (XDH), which converts to xanthine oxidase (XO) during ischemia–reperfusion. XO is both a prooxidant, when it oxidizes purines resulting in superoxide anion formation, and an antioxidant, when it produces uric acid. Uric acid is a water-soluble antioxidant, which is the main antioxidant in blood plasma (Chapter 15). Packer *et al.* (1997) have demonstrated that vitamin E, a lipid-soluble antioxidant, quenches lipid peroxidation in membranes, resulting in the formation of the vitamin E radical. The latter is regenerated to vitamin E by ascorbate (vitamin C) in the aqueous phase and in the process becomes dehydroascorbate. It takes two reduced glutathiones (GSH) to react with the dehydroascorbate to yield ascorbate and one oxidized glutathione (GSSG) (Packer *et al.*, 1997). In the presence of dihydrolipoate, GSSG reacts to form two molecules of GSH and α-lipoate. Finally, NAD(P)H regenerates dihydrolipoate by reacting with α-lipoate to produce $NAD(P)^+$. Both reduced glutathione and dihydrolipoate possess sulfhydryl groups. Dihydrolipoate can regenerate the dehydroascorbate to form ascorbate directly (Packer *et al.*, 1997). Ubiquinol is a lipid-soluble antioxidant that is present in animal membranes and can act either independently of vitamin E or in conjunction with vitamin E (Packer, 1994; Chapter 17). In addition, melatonin, an indoleamine neurohormone synthesized in the pineal gland, has been shown to possess antioxidant properties (Chapter 16). L-Cystathionine, a thioether, is a precursor to L-cysteine, glutathione, and taurine; however, it possesses antioxidant properties independent of these amino acids (Chapter 16).

Ionizing radiation has been used successfully in cancer treatment (Bergman and Harris, 1997). However, radiation is also toxic to normal tissue (Harris *et al.*, 1997); in fact, it has been shown to be a carcinogenic agent (Hall *et al.*, 1982). The principal effects are known to be related to free radicals. Photodynamic therapy has also been used successfully to treat cancer. This therapy is performed with a photosensitizer and a light source such as a laser. The light source is more efficient when the wavelength is greater than 600 nm. A common photosensitizer is a hematoporphyrin derivative. The photodynamic effect seems to be related mainly to singlet oxygen, which has a half-life in the range of 3 to 25 μs. Thus, little diffusion occurs and the damaging effect on the tumor depends principally on the photosensitizer distribution ratio of tumor to normal tissue. No matter how large this ratio is, there is always some damage to normal tissue (Moore *et al.*, 1997).

The porphyria diseases are attributed to malfunctions of heme synthesis (Ellefson and Ford, 1996; Billi de Catabbi *et al.*, 1997; Crimlisk, 1997; Schreiber, 1997) and might be related to ROS (Thunell *et al.*, 1995). Red hemoglobin and green chlorophyll are both porphyrins; over 40 metals can comprise complex porphyrins (Hendry and Jones, 1980). Porphyrins have a long evolutionary history (Hodgson, 1972; Simionescu *et al.*, 1978). The first step in heme biosynthesis is the production of 5-aminolevulinic acid (ALA), which is followed by the production of porphobilinogen (PBG). As previosly mentioned, ALA is used as a photosensitizer in the treatment of cutaneous abnormalities. Overabundance of ALA and/or PBG as a result of enzyme deficiencies in tissues is characteristic of neuropathic porphyrias. Steps 3 to 7 lead to heme synthesis. Neuropathic porphyrias are also produced by abnormalities in steps 5 and 6. Enzyme deficiencies in these last five steps result in cutaneopathic porphyrias, which are caused by the porphyrins acting as photosensitizers (Ellefson and Ford, 1996; Crimlisk, 1997). Porphyrinogenic drugs are ones that intensify the porphyrias and include antibiotics such as erythromycin, sedatives such as barbiturates, and antiepileptics such as phenytoin and carbamazepine (Crimlisk, 1997).

Spermatozoa are sensitive to the damaging influence of ROS, but without ROS, they cannot function (Chapter 20). Sea urchin eggs produce hydrogen peroxide (Chapter 21). Thus, at the beginning of life, ROS are involved. As pointed out in Chapter 22, the production of singlet oxygen in the brain can be used as a measure of oxidative stress. ROS probably have roles in diseases occurring in older individuals, such as Parkinson's disease (Chapter 24), Alzheimer's disease (Chapter 25; Smith, 1998) and amyotrophic lateral sclerosis or Lou Gehrig's disease (Chapter 26). ROS are active in the aging process (Chapter 27).

Ozone in the stratosphere filters out ultraviolet radiation from reaching the ground (Chapter 12), whereas ozone in the troposphere, i.e., ground level, is damaging to animals (Chapter 13) and plants (Bae *et al.*, 1996; Sharma and Davis, 1997). In other words, ozone must be kept in its proper place. Similarly, the substances that make up ROS must be kept in their proper place within the organism (Gilbert, 1994). There exists the danger of depleting stratospheric ozone, but ground-level ozone is one of the leading causes of atmospheric pollution. Thus, by being outdoors one is subject to the harmful effects of UV radiation plus breathing in the damaging ozone.

The chloroplasts in plants pollute the atmosphere with molecular oxygen (Chapter 18). Animals use oxygen to produce ATP.

Oxidative stress can lead to programmed cell death (apoptosis). Scrapie, a prion disease, is a consequence of apoptosis (Kretzschmar *et al.*, 1997), preceded by microglial activation (Giese *et al.*, 1998). When the stress is not large, glutathione is increased, probably via compensatory mechanisms. If the stress is continued, then the level of glutathione decreases. The mitochondrial membrane decreases before there is a rise in the intracellular level of calcium. Cells that have high amounts of the Bcl-2 protein are protected against ROS (Backway *et al.*, 1997).

In summary, to continue to exist, the biosphere is like a tightrope walker who must maintain good balance; too little oxygen and the biosphere dies, too much oxygen and the biosphere also dies. The sword of Damocles hangs over the biosphere. Gerschman paraphrased the sixteenth century figure, Paracelsus: *Everything is poison, it only depends upon the dose* (Gerschman, 1981).

8. SUMMARY

The biosphere is in a constant battle. Every organism is struggling against a variety of stresses, whether from other organisms or from the natural environment. Humans have adjusted to a larger range of environmental variables than other organisms and can be found in temperatures ranging from –60 to 55°C, from 0 to 160 torr (0.02 MPa) oxygen pressure, from 0 to 100% humidity, from 0 to 110 MPa, and in absolute darkness to bright light. The oldest human on record was a woman who was born on February 21, 1875, and died on August 4, 1997 (Young, 1998), a life span of 122 years and 164 days. Cutler (1982) presumed that the maximum life span potential (MLP) of a human was approximately 110 years; Spirduso (1995) gave the MLP of a human as 115 years, but pointed out that females live longer than males, and for a female, the MLP is 120 years. Cutler (1978) speculated that the MLP can increase to 200 years. Other species survive in extreme environments, but no other species can survive in such a range of environments.

By considering the broad range of ROS interactions with life processes, we can more fully understand how they interact in a vital way in the biosphere's struggle for existence.

ACKNOWLEDGMENT. Acknowledgment is given to Dr. Claire Gilbert for her helpful discussions, criticisms, and editing.

9. REFERENCES

Alio, J. L., Ayala, M. J., Mulet, M. E., Artola, A., Ruiz, J. M., and Bellot, J., 1995, Antioxidant therapy in the treatment of experimental acute corneal inflammation, *Ophthalmic Res.* **27**:136–143.

Backway, K. L., McCulloch, E. A., Chow, S., and Hedley, D. W., 1997, Relationships between the mitochondrial permeability transition and oxidative stress during ara-C toxicity, *Cancer Res.* **57**:2446–2451.

Bae, G. Y., Nakajima, N., Ishizuka, K., and Kondo, N., 1996, The role in ozone phytotoxicity of the evolution of ethylene upon induction of 1-aminocyclopropane-1-carboxylic acid synthase by ozone fumigation in tomato plants, *Plant Cell Physiol.* **37**:129–134.

Bazan, N., 1989, Arachidonic acid in the modulation of excitable membrane function and at the onset of brain damage, *Ann. N.Y. Acad. Sci.* **559**:1–15.

Bazan, N. G., and Allan, G., 1996, Platelet-activating factor is both a modulator of synaptic function and a mediator of cerebral injury and inflammation, *Adv. Neurol.* **71**:475–484.

Bazan, N. G., Marcheselli, V. L., and Mukherjee, P. K., 1995, Inducible prostaglandin synthase in cell injury, in *Advances in Prostaglandin, Thromboxase, and Leukotriene Research, Volume 23* (B. Samuelsson, ed.), pp. 317–323, Raven Press, New York.

Bergman, P. J., and Harris, D., 1997, Radioresistance, chemoresistance, and apoptosis resistance. The past, present, and future, *Vet. Clin. North Am. Small Anim. Pract.* **27**:47–57.

Bernhardt, R., 1995, Cytochrome P450: structure, function, and generation of reactive oxygen species, *Rev. Physiol. Biochem. Pharmacol.* **127**:137–221.

Bhagwat, S. V., Boyd, M. R., and Ravindranath, V., 1995a, Rat brain cytochrome P450. Reassessment of monooxygenase activities and cytochrome P450 levels, *Drug Metabol. Dispos.* **23**:651–654.

Bhagwat, S. V., Boyd, M. R., and Ravindranath, V., 1995b, Brain mitochondrial cytochomes P450: Xenobiotic metabolism, presence of multiple forms and their selective inducibility, *Arch. Biochem. Biophys.* **320:**73–83.

Bhagwat, S. V., Leelavathi, B. C., Shanker, S. K., Boyd, M. R., and Ravindranath, V., 1995c, Cytochrome P450 and associated monooxygenase activities in the rat and human spinal cord: Induction, immunological characterization and immunocytochemical localization, *Neuroscience* **68:**593–601.

Bhagwat, S. V., Bhamre, S., Boyd, M. R., and Ravindranath, V., 1996a, Cerebral metabolism of imipramine and a purified flavin-containing monooxygenase from human brain, *Neuropsychopharmacology* **15:**133–142.

Bhagwat, S. V., Bhamre, S., Boyd, M. R., and Ravindranath, V., 1996b, Further characterization of rat brain flavin-containing monooxygenase. Metabolism of imipramine to its N-oxide, *Biochem. Pharmacol.* **51:**1469–1475.

Billi de Catabbi, S., Sterin-Speziale, N., Fernandez, M. C., Minutolo, C., Aldonatti, C., and San Martin de Viale, L., 1997, Time course of hexachlorobenzene-induced alterations of lipid metabolism and their relation to porphyria, *Int. J. Biochem. Cell Biol.* **29:**335–344.

Bunn, H. F., and Poyton, R. O., 1996, Oxygen sensing and molecular adaptation to hypoxia, *Physiol. Rev.* **76:**839–885.

Cai, C. X., Birk, D. E., and Linsenmayer, T. F., 1997, Ferritin is a developmentally regulated nuclear protein of avian corneal epithelial cells, *J. Biol. Chem.* **272:**12831–12839.

Camhi, S. L., Lee, P., and Choi, A. M. K., 1995, The oxidative stress response, *New Horiz.* **3:**170–182.

Carlo, P., Del Rio, M., Violani, E., Sciaba, L., and Picotti, G. B., 1996a, Influence of culture conditions on monoamine oxidase A and B activity in rat astrocytes, *Cell Biochem. Funct.* **14.:**19–25.

Carlo, P., Violani, E., Del Rio, M., Olasmaa, M., Santagati, S., Maggi, A., and Picotti, G. B., 1996b, Monoamine oxidase B expression is selectively regulated by dexamethasone in cultured rat astrocytes, *Brain Res.* **711:**175–183.

Chandrasekar, B., and Fernandes, G., 1994, Decreased pro-inflammatory cytokines and increased antioxidant enzyme gene expression by omega-3 lipids in murine lupus nephritis, *Biochem. Biophys. Res. Commun.* **200:**893–898.

Cohen, G., Farooqui, R., and Kesler, N., 1997, Parkinson disease: A new link between monoamine oxidase and mitochondrial electron flow, *Proc. Natl. Acad. Sci. USA* **94:**4890–4894.

Colton, C. A., 1994, Microglial oxyradical production: Causes and consequences, *Neuropathol. Appl. Neurobiol.* **29:**208–209.

Colton, C. A., 1995, Induction of nitric oxide in cultured microglia: Evidence for a cytoprotective role, *Adv. Neuroimmunol.* **5:**491–503.

Colton, C. A., and Chernyshev, O. N., 1996, Inhibition of microglial superoxide anion production by isoproterenol and dexamethasone, *Neurochem. Int.* **29:**43–53.

Colton, C. A., and Gilbert, D. L., 1987, Production of superoxide anions by a CNS macrophage, the microglia, *FEBS Lett.* **223:**284–288.

Colton, C. A., and Gilbert, D. L., 1989, An endogenous source of the superoxide anion in the central nervous system, in *Oxygen Radicals in Biology and Medicine* (M. G. Simic, K. A. Taylor, J. F. Ward, and C. von Sonntag, eds.), pp. 1005–1010, Plenum Press, New York.

Colton, C. A., and Gilbert, D. L., 1993, Microglia, an *in vivo* source of reactive oxygen species in the brain, in *Advances in Neurology, Volume 59* (F. J. Seil, ed.), pp. 321–326, Raven Press, New York.

Colton, C. A., Yao, J., Gilbert, D., and Oster-Granite, M. L., 1990, Enhanced production of superoxide anion by microglia from trisomy 16 mice, *Brain Res.* **519:**236–242.

Colton, C. A., Yao, J., Keri, J. E., and Gilbert, D., 1992, Regulation of microglial function by interferons, *J. Neuroimmunol.* **30:**89–98.

Colton, C. A., Chernyshev, O., Snell, J., and Gilbert, D., 1994, Induction of superoxide anion and nitric oxide production in cultured microglia, in *Neurobiology of NO· and ·OH* (C. C. Chiueh, D. L. Gilbert, and C. Colton, eds.), pp. 54–63, New York, Academy of Sciences, New York.

Colton, C., Wilt, S., Gilbert, D., Chernyshev, O., Snell, J., and Dubois-Dalcq, M., 1996, Species differences in the generation of reactive oxygen species by microglia, *Mol. Chem. Neuropathol.* **28:**15–20.

Conner, J. R., Boeshore, K. L., Benkovic, S. A., and Menzies, S. L., 1994, Isoforms of ferritin have a specific cellular distribution in the brain, *J. Neurosci. Res.* **37:**461–465.

Crimlisk, H. L., 1997, The little imitator—porphyria: A neuropsychiatric disorder, *J. Neurol. Neurosurg. Psychiatry* **62:**319–328.

Crosby, A. J., Wahle, K. W., and Duthie, G. G., 1996, Modulation of glutathione peroxidase activity in human vascular endothelial cells by fatty acids and the cytokine interleukin-1 beta, *Biochim. Biophys. Acta* **1303**:187–192.

Cutler, R. G., 1978, Evolutionary biology of senescence, in *The Biology of Aging* (J. A. Behnke, C. E. Finch, and G. B. Moment, eds.), pp. 311–360, Plenum Press, New York.

Cutler, R. G., 1982, Longevity is determined by specific genes: Testing the hypothesis, in *Testing the Theories of Aging* (R. C. Adelman and G. S. Roth, eds.), pp. 25–114, CRC Press, Boca Raton.

Daniell, W. E., Stockbridge, H. L., Labbe, R. F., Woods, J. S., Anderson, K. E., Bissell, D. M., Bloomer, J. R., Ellefson, R. D., Moore, M. R., Pierach, C. A., Schreiber, W. E., Tefferi, A., and Franklin, G. M., 1997, Environmental chemical exposures and disturbances of heme synthesis, *Environ. Health Perspect.* **105**(Suppl. 1):37–53.

Daval, J. L., Ghersi-Egea, J. F., Oillet, J., and Koziel, V., 1995, A simple method for evaluation of superoxide radical production in neural cells under various culture conditions: Application to hypoxia, *J. Cereb. Blood Flow Metab.* **15**:71–77.

Ding, M., St. Pierre, B. A., Parkinson, J. F., Medberry, P., Wong, J. L., Rogers, N. E., Ignarro, L. J., and Merrill, J. E., 1997, Inducible nitric-oxide synthase and nitric oxide production in human fetal astrocytes and microglia, *J. Biol. Chem.* **272**:11327–11335.

Dugan, L. L., Sensi, S. L., Canzoniero, L. M. T., Handran, S. D., Rothman, S. M., Lin, T. S., Goldberg, M. P., and Choi, D. W., 1995, Mitochondrial production of reactive oxygen species in cortical neurons following exposure to *N*-methyl-D-aspartate, *J. Neurosci.* **15**:6377–6388.

Ellefson, R. D., and Ford, R. E., 1996, The polyphyrias: Characteristics and laboratory tests, *Regul. Toxicol. Pharmacol.* **24**:S119–S125.

Emerit, I., 1992, Free radicals and aging of the skin, in *Free Radicals and Aging* (I. Emerit and B. Chance, eds.), pp. 328–341, Birkhäuser Verlag, Basel.

Fielding, R. A., and Meydani, M., 1997, Exercise: free radical generation, and aging, *Aging Clin. Exp. Res.* **9**:12–18.

Flower, R. W., and Patz, A., 1981, Retinopathy of prematurity and the role of oxygen, in *Oxygen and Living Processes: An Interdisciplinary Approach* (D. L. Gilbert, ed.), pp. 368–375, Springer-Verlag, Berlin.

Franke-Ullmann, G., Pfortner, C., Walter, P., Steinmuller, C., Lohmann-Matthes, M. L., and Kobzik, L., 1996, Characterization of murine lung interstitial macrophages in comparison with alveolar macrophages in vitro, *J. Immunol.* **157**:3097–3104.

Gaillard, E. R., Atherton, S. J., Eldred, G., and Dillon, J., 1995, Photophysical studies on human retinal lipofuscin, *Photochem. Photobiol.* **61**:448–453.

García-Ruiz, C., Colell, A., Marí, M., Morales, A., and Fernández-Checa, J. C., 1997, Direct effect of ceramide on the mitochondrial electron transport chain leads to generation of reactive oxygen species. Role of mitochondrial glutathione, *J. Biol. Chem.* **272**:11369–11377.

Garruto, R. M., 1991, Pacific paradigms of environmentally-induced neurological disorders: Clinical, epidemiological and molecular perspectives, *Neurotoxicology* **12**:347–378.

Garruto, R. M., Flaten, T. P., and Wakayama, I., 1997, Effects of acid rain: A natural model of aluminum toxicity in fish, in *Mineral and Metal Neurotoxicology* (M. Yasui, ed.), pp. 27–34, CRC Press, Boca Raton.

Gerschman, R., 1981, Historical introduction to the "free radical theory" of oxygen toxicity, in *Oxygen and Living Processes: An Interdisciplinary Approach* (D. L. Gilbert, ed.), pp. 44–46, Springer-Verlag, Berlin.

Gerschman, R., Gilbert, D. L., and Caccamise, D., 1958, Effect of various substances on survival times of mice exposed to different high oxygen tensions, *Am. J. Physiol.* **192**:563–571.

Giese, A., Brown, D. R., Groschup, M. H., Feldmann, C., Haist, I., and Kretzchmar, H. A., 1998, Role of microglia in neuronal cell death in prion disease, *Brain. Pathol.* **8**:449–457.

Gilbert, D. L., 1963, The role of pro-oxidants and antioxidants in oxygen toxicity, *Radiat. Res. Suppl.* **3**:44–53.

Gilbert, D. L., 1972, Introduction: Oxygen and life, *Anesthesiology* **37**:100–111.

Gilbert, D. L., 1994, Keeping reactive oxygen species (ROS) in their proper place, in *Neurobiology of NO· and ·OH* (C. C. Chiueh, D. L. Gilbert, and C. A. Colton, eds.), pp. 1–7, Academy of Sciences, New York.

Gilbert, D. L., 1996, Evolutionary aspects of atmospheric oxygen and organisms, in *Handbook of Physiology: Section 4. Adaptation to the Environment: Volume II* (M. J. Fregly and C. M. Blatteis, eds.), pp. 1059–1094, Oxford University Press, London.

Giroux, C., and Scatton, B., 1996, Ischemic stroke: Treatment on the horizon, *Eur. Neurol.* **36**:61–64.

Giulian, D., and Baker, T. J., 1986, Characterization of Ameboid microglia isolated from developing mammalian brain, *J. Neurosci.* **6**:2163–2178.

Giuliani, A., and Cestaro, B., 1997, Exercise, free radical generation and vitamins, *Eur. J. Cancer Prev.* **6** (Suppl. 1):S55–S67.

Gogate, N., Verma, L., Zhou, J. M., Milward, E., Rusten, R., O'Connor, M., Kufta, C., Kim, J., Hudson, L., and Dubois-Dalcq, M., 1994, Plasticity in the adult human oligodendrocyte lineage, *J. Neurosci.* **14**:4571–4587.

Götz, M. E., Küng, G., Riederer, P., and Youdim, M. B. H., 1994, Oxidative stress: Free radical production in neural degeneration, *Pharmacol. Ther.* **63**:37–122.

Gruener, N., Gross, B., Gozlan, O., and Barak, M., 1994, Increase in superoxide dismutase after cerebrovascular accident, *Life Sci.* **11**:711–713.

Hall, E. J., Miller, R. C., Osmak, R., and Zimmerman, M., 1982, Comparison of the incidence of oncogenic transformation produced by X rays, misonidazole, and chemotherapy agents, *Radiology* **145**:521–523.

Halliwell, B., and Gutteridge, J. M. C., 1989, *Free Radicals in Biology and Medicine*, 2nd ed., Oxford University Press, London.

Harris, D., King, G. K., and Bergman, P. J., 1997, Radiation therapy toxicities, *Vet. Clin. North Am. Small Anim. Pract.* **27**:37–46.

Hassoun, P. M., Yu, F. S., Zulueta, J. J., White, A. C., and Lanzillo, J. J., 1995, Effect of nitric oxide and cell redox status on the regulation of endothelial cell xanthine dehydrogenase, *Am. J. Physiol.* **268**:L809–L817.

Hatton, C. J., and Peers, C., 1996, Effects of cytochrome *P*-450 inhibitors on ionic currents in isolated rat type I carotid body cells, *Am. J. Physiol.* **271**:C85–C92.

Haugaard, N., 1946, Oxygen poisoning, XI. The relation between inactivation of enzymes by oxygen and essential sulfhydryl groups, *J. Biol. Chem.* **164**:265–270.

Hayashi, T., Ueno, Y., and Okamoto, T., 1993, Oxidoreductive regulation of nuclear factor B. Involvement of a cellular reducing catalyst thioredoxin, *J. Biol. Chem.* **268**:11380–11388.

Hendry, G. A. F., and Jones, O. T. G., 1980, Haems and chlorophylls: Comparison of function and formation. *J. Med. Genet.* **17**:1–14.

Hoare, J. P., 1985, Oxygen, in *Standard Potentials in Aqueous Solution* (A. J. Bard, R. Parsons, and J. Jordan, eds.), pp. 49–66, Dekker, New York.

Hodgson, G. W., 1972, Cosmochemical evolution of large organic molecules: Illustrative laboratory simulations for porphyrins, *Ann. N. Y. Acad. Sci.* **194**:86–97.

Hodis, H. N., Mack, W. J., LaBree, L., Cashin-Hemphill, L., Sevanian, A., Johnson, R., and Azen, S. P., 1995, Serial coronary angiographic evidence that antioxidant vitamin intake reduces progression of coronary artery atherosclerosis, *J. Am. Med. Assoc.* **273**:1849–1854.

Horio, T., and Okamoto, H., 1987, Oxygen intermediates are involved in ultraviolet radiation-induced damage of Langerhans cells., *J. Invest. Dermatol.* **88**:699–702.

Hunt, R. C., Handy, I., and Smith, A., 1996a, Heme-mediated reactive oxygen species toxicity to retinal pigment epithelial cells is reduced by hemopexin, *J. Cell. Physiol.* **168**:81–86.

Hunt, R. C., Hunt, M., Gaur, N., and Smith, A., 1996b, Hemopexin in the human retina: Protection of the retina against heme-mediated toxicity, *J. Cell. Physiol.* **168**:71–80.

Hurst, J. S., Paterson, C. A., and Short, C. S., 1989, Oxidant and anti-oxidant effects on arachidonate metabolism by rabbit ocular tissues, *J. Ocul. Pharmacol.* **5**:51–64.

Jeffes, E. W., McCullough, J. L., Weinstein, G. D., Fergin, P. E., Nelson, J. S., Shull, T. F., Simpson, K. R., Bukaty, L. M., Hoffman, W. L., and Fong, N. L., 1997, Photodynamic therapy of actinic keratosis with topical 5-aminolevulinic acid. A pilot dose-ranging study, *Arch. Dermatol.* **133**:727–732.

Joe, B., and Lokesh, B. R., 1994, Role of capsaicin, curcumin and dietary n-3 fatty acids in lowering the generation of reactive oxygen species in rat peritoneal macrophages, *Biochim. Biophys. Acta* **1224**:255–263.

Katsuki, H., 1996, Vitamin C and nervous tissue, *In vivo* and *in vitro* aspects, in *Ascorbic Acid: Biochemistry and Biomedical Cell Biology* (J. R. Harris, ed.), pp. 293–311, Plenum Press, New York.

Kausalya, S., Kaur, S., Malla, N., Ganguly, N. K., and Mahajan, R. C., 1996, Microbicidal mechanisms of liver macrophages in experimental visceral leishmaniasis, *APMIS* **104**:171–175.

Kim, J. J., Shih, J. C., Chen, K., Chen, L., Bao, S., Maren, S., Anagnostaras, S. G., Fanselow, M. S., De Maeyer, E., Seif, I., and Thompson, R. F., 1997, Selective enhancement of emotional, but not motor, learning in monoamine oxidase A-deficient mice, *Proc. Natl. Acad. Sci. USA* **94:**5929–5933.

Kontos, H. A., 1985, George E. Brown Memorial Lecture. Oxygen radicals in cerebral vascular injury, *Circ. Res.* **57:**508–516.

Kontos, H. A., 1987, Oxygen radicals from arachidonate metabolism in abnormal vascular responses, *Am. Rev. Respir. Dis.* **136:**474–477.

Kontos, H. A., 1989, Oxygen radicals in CNS damage, *Chem. Biol. Interact.* **72:**229–255.

Kontos, H. A., and Wei, E. P., 1986, Superoxide production in experimental brain injury, *J. Neurosurg.* **64:**803–807.

Kowluru, R. A., Kern, T. S., and Engerman, R. L., 1997, Abnormalities of retinal metabolism in diabetes or experimental galactosemia. IV. Antioxidant defense system, *Free Radical Biol. Med.* **22:**587–592.

Kramer, J. H., and Weglicki, W. B., 1996, A hydroxylated analog of the β-adrenoceptor antagonist, carvedilol, affords exceptional antioxidant protection to postischemic rat hearts, *Free Radical Biol. Med.* **21:**813–825.

Kramer, J. H., Misík, V., and Weglicki, W. B., 1994, Lipid peroxidation-derived free radical production and postischemic myocardial reperfusion injury, *Ann. N. Y. Acad. Sci.* **723:**180–196.

Kretzschmar, H. A., Giese, A., Brown, D. R., Herms, J., Keller, B., Schmidt, B., and Groschup, M., 1997, Cell death in prion disease, *J. Neural Transm. Suppl.* **50.:**191–210.

Kuroki, M., Voest, E. E., Amano, S., Beerepoot, L. V., Takashima, S., Tolentino, M., Kim, R. Y., Rohan, R. M., Colby, K. A., Yeo, K.T., and Adamis, A. P., 1996, Reactive oxygen intermediates increase vascular endothelial growth factor expression in vitro and in vivo, *J. Clin. Invest.* **98:**1667–1675.

Lander, H. M., 1997, An essential role for free radicals and derived species in signal transduction, *FASEB J.* **11:**118–124.

Laskin, D. L., Soltys, R. A., Berg, R. A., and Riley, D. J., 1994, Activation of alveolar macrophages by native and synthetic collagen-like polypeptides, *Am. J. Respir. Cell. Mol. Biol.* **10:**58–64.

López-Barneo, J., 1994, Oxygen-sensitive ion channels: How ubiquitous are they? *Trends Neurosci.* **17:**133–135.

López-Barneo, J., 1996, Oxygen-sensing by ion channels and the regulation of cellular functions, *Trends Neurosci.* **19:**435–440.

Mauzerall, D. C., and Piccioni, R. G., 1981, Photosynthetic oxygen production, in *Oxygen and Living Processes. An Interdisciplinary Approach* (D. L. Gilbert, ed.), pp. 102–123, Springer-Verlag, Berlin.

Minc-Golumb, D., and Schwartz, J. P., 1994, Expression of both constitutive and inducible nitric oxide synthases in neuronal and astrocyte cultures, in *Neurobiology of NO· and ·OH* (C. C. Chiueh, D. L. Gilbert, and C. A. Colton, eds.), pp. 462–467, New York Academy of Sciences, New York.

Mohazzab-H, K. M., Fayngersh, R. P., Kaminski, P. M., and Wolin, M. S., 1995, Potential rold of NADH oxidoreductase-derived reactive O_2 species in calf pulmonary arterial Po_2-elicited responses, *Am. J. Physiol.* **269:**L637–L644.

Mohazzab-H, K. M., Kaminski, P. M., Fayngersh, R. P., and Wolin, M. S., 1996, Oxygen-elicited responses in calf coronary arteries: Role of H_2O_2 production via NADH-derived superoxide, *Am. J. Physiol.* **270:**H1044–H1053.

Moore, J. V., West, C. M., and Whitehurst, C., 1997, The biology of photodynamic therapy, *Phys. Med. Biol.* **42:**913–935.

Naidu, S., and Moser, H. W., 1990, Peroxisomal disorders, *Neurol. Clin.* **8:**507–528.

Naidu, S., and Moser, H. W., 1994, Peroxisomal disorders, *Neurol. Clin.* **12:**727–739.

Nakamura, H., Nakamura, K., and Yodoi, J., 1997, Redox regulation of cellular activation, *Annu. Rev. Immunol.* **15:**351–369.

Niesman, M. R., Johnson, K. A., and Penn, J. S., 1997, Therapeutic effect of liposomal superoxide dismutase in an animal model of retinopathy of prematurity, *Neurochem. Res.* **22:**597–605.

Ohia, S. E., Bagchi, M., and Stohs, S. J., 1994, Age-related oxidative damage in Long-Evans rat retina, *Res. Commun. Mol. Pathol. Pharmacol.* **85:**21–31.

Packer, L., 1994, Antioxidant properties of lipoic acid and its therapeutic effects in prevention of diabetes complications and cataracts, in *Neurobiology of NO· and ·OH* (C. C. Chiueh, D. L. Gilbert, and C. A. Colton, eds.), pp. 257–264, New York Academy of Sciences, New York.

Packer, L. P., Kitazawa, M., Thiele, J., Saliou, C., Witt, E., and Traber, M. G., 1997, Vitamin E and the metabolic antioxidant network, in *Molecular Mechanisms of Signalling and Membrane Transport* (K. W. A. Wirtz, ed.), pp. 281–304, Springer, Berlin.
Papa, S., 1996, Mitochondrial oxidative phosphorylation changes in the life span. Molecular aspects and physiopathological implications, *Biochim. Biophys. Acta* **1276**:87–105.
Peers, C., Carpenter, E., Hatton, C. H., Wyatt, C. B., and Bee, D., 1996, Ca^{2+} channel currents in type I carotid body cells of normoxic and chronically hypoxic neonatal rats, *Brain Res.* **739**:251–271.
Perdrizet, G. A., 1995, Heat shock and tissue protection, *New Horiz.* **3**:312–320.
Quinlan, T., Spivack, S., and Mossman, B. T., 1994, Regulation of antioxidant enzymes in lung after oxidant injury, *Environ. Health Perspect.* **102** (Suppl. 2):79–87.
Ravindranath, V., Anandatheerhavarada, H. K., and Shankar, S. K., 1989, Xenobiotic metabolism in human brain—Presence of cytochrome *P*-450 and associated mono-oxygenases, *Brain Res.* **496**:331–335.
Rebec, G. V., and Pierce, R. C., 1994, A vitamin as neuromodulator: Ascorbate release into the extracellular fluid of the brain regulates dopaminergic and glutamatergic transmission, *Prog. Neurobiol.* **43**:537–565.
Richter, C., Gogvadze, V., Laffranchi, R., Schlapbach, R., Schweizer, M., Suter, M., Walter, P., and Yaffee, M., 1995, Oxidants in mitochondria: From physiology to diseases, *Biochim. Biophys. Acta* **1271**:67–74.
Rózanowska, M., Jarvis-Evans, J., Korytowski, W., Boulton, M. E., Burke, J. M., and Sarna, T., 1995, Blue light-induced reactivity of retinal age pigment. In vitro generation of oxygen-reactive species, *J. Biol. Chem.* **270**:18825–18830.
Saito, M., and Nakatsugawa, K., 1994, Increased susceptibility of liver to lipid peroxidation after ingestion of a high fish oil diet, *Int. J. Vitam. Nutr. Res.* **64**:144–151.
Sasame, H. A., Ames, M. M., and Nelson, S. D., 1977, Cytochrome P-450 and NADPH cytochrome *c* reductase in rat brain; Formation of catechols and reactive catechol metabolites, *Biochem. Biophys. Res. Commun.* **78**:919–926.
Schreiber, W. E., 1997, A molecular view of the neurologic porphyrias, *Clin. Lab. Med.* **17**:73–83.
Senozan, N. M., and Christiano, M. P., 1997, Iron as nutrient and poison, *J. Chem. Educ.* **74**:1060–1063.
Sharma, Y. K., and Davis, K. R., 1997, The effects of ozone on antioxidant responses in plants, *Free Radical Biol. Med.* **23**:480–488.
Sharov, V. S., Briviba, K., and Sies, H., 1996, Assessment of the C-525 laser dye as a chemiluminescence sensitizer for lipid peroxidation in biological membranes: A comparison with chlorophyll-α, *Free Radical Biol. Med.* **21**:833–843.
Simionescu, C. I., Simionescu, B. C., Mora, R., and Leanca, M., 1978, Porphyrin-like compounds genesis under simulated abiotic conditions, *Orig. Life* **9**:103–114.
Smith, L. J., Shamsuddin, M., Sporn, P. H., Denenberg, M., and Anderson, J., 1997, Reduced superoxide dismutase in lung cells of patients with asthma, *Free Radical Biol. Med.* **22**:1301–1397.
Smith, M. A., 1998, Alzheimer disease, *Internat. Rev. Neurobiol.* **42**:1–54.
Snell, J., Colton, C. A., Chernyshev, O. N., and Gilbert, D. L., 1996, Location dependent artifact for NO measurement using multiwell plates, *Free Radical Biol. Med.* **20**:361–363.
Spirduso, W. W., 1995, *Physical Dimensions of Aging*, Human Kinetics, Champaign, IL.
Stadie, W. C., Riggs, B. C., and Haugaard, N., 1944, Oxygen poisoning, *Am J. Med. Sci.* **207**:84–114.
Streit, W. J., and Kincaid-Colton, C. A., 1995, The brain's immune system, *Sci. Am.* **273** (**5**):54–55, 58–61.
Suzuki, Y. J., Forman, H. J., and Sevanian, A., 1997, Oxidants as stimulators of signal transduction, *Free Radical Biol. Med.* **22**:269–285.
Tanswell, A. K., and Freeman, B. A., 1995, Antioxidant therapy in critical care medicine, *New Horiz.* **3**:330–341.
Thunell, S., Andersson, C., Carlmark, B., Floderus, Y., Grönqvist, S. O., Harper, P., Henrichson, A., and Lindh, U., 1995, Markers for vulnerability in acute porphyria. A hypothesis paper, *Eur. J. Clin. Chem. Clin. Biochem.* **33**:179–194.
Trevithick, J. R., and Dzialoszynski, T., 1997, Endogenous superoxide-like species and antioxidant activity in ocular tissues detected by luminol luminescence, *Biochem. Mol. Biol. Interact.* **41**:695–705.
Ureña, J., Franco-Obregón, A., and López-Barneo, J., 1996, Contrasting effects of hypoxia on cytosolic Ca^{2+} spikes in conduit and resistance myocytes of the rabbit pulmonary artery, *J. Physiol. (London)* **496**:103–109.
Varma, S. D., Ali, A. H., Devamanoharan, P. S., and Morris, S. M., 1997, Nitrite-induced photo-oxidation of thiol and its implications in smog toxicity to the eye: Prevention by ascorbate, *J. Ocul. Pharmacol. Ther.* **13**:179–187.

Wei, E. P., Kontos, H. A., Dietrich, W. D., Povlishock, J. T., and Ellis, E. F., 1981, Inhibition by free radical scavengers and by cyclooxygenase inhibitors of pial arteriolar abnormalities from concussive brain injury in cats, *Circ. Res.* **48:**95–103.

Wyatt, C. N., Wright, C., Bee, D., and Peers, C., 1995, O_2-sensitive K^+ currents in carotid body chemoreceptor cells from normoxic and chronically hypoxic rats and their roles in hypoxic chemotransduction, *Proc. Natl. Acad. Sci. USA* **92:**295–299.

Yao, J., Harvath, L., Gilbert, D. L., and Colton, C. A., 1990, Chemotaxis by a CNS macrophage, the microglia, *J. Neurosci. Res.* **27:**36–42.

Yasui, M., Ota, K., and Garruto, R. M., 1993, Concentrations of zinc and iron in the brains of Guamanian patients with amyotrophic lateral sclerosis and parkinsonism-dementia, *Neurotoxicology* **14:**445–450.

Young, M. A. (Ed.), 1998, *The Guiness Book of World Records, 1998 Edition*, p. 360, Bantam, New York.

Yue, T. L., Lysko, P. G., Barone, F. C., Gu, J. L., Ruffolo, R. R., Jr., and Feuerstein, G. Z., 1994, Carvedilol, a new antihypertensive drug with unique antioxidant activity: Potential role in cerebroprotection, in *Neurobiology of NO· and ·OH* (C. C. Chiueh, D. L. Gilbert, and C. A. Colton, eds.), pp. 230–242, Academy of Sciences, New York.

Zielasek, J., and Hartung, H. P., 1996, Molecular mechanisms of microglial activation, *Adv. Neuroimmunol.* **6:**191–222.

Zielasek, J., Archelos, J. J., Toyka, K. V., and Hartung, H. P., 1993, Expression of intercellular adhesion molecule-1 on rat microglial cells, *Neurosci. Lett.* **153:**136–139.

Zielasek, J., Müller, B., and Hartung, H. P., 1996, Inhibition of cytokine-inducible nitric oxide synthase in rat microglia and murine macrophages by methyl-2,5-dihydroxycinnamate, *Neurochem. Int.* **29:**83–87.

Zigler, J. S., Jr., and Hess, H. H., 1985, Cataracts in the Royal College of Surgeons rat: Evidence for initiation peroxidation products, *Exp. Eye Res.* **41:**67–76.

Index

Page numbers in **bold type** indicate illustrations and page numbers in *italic type* indicate tables.

MIX
Papier aus verantwortungsvollen Quellen
Paper from responsible sources
FSC® C105338

If you have any concerns about our products,
you can contact us on
ProductSafety@springernature.com

In case Publisher is established outside the EU,
the EU authorized representative is:
Springer Nature Customer Service Center GmbH
Europaplatz 3, 69115 Heidelberg, Germany

Printed by Libri Plureos GmbH
in Hamburg, Germany